BASIC CIRCUIT THEORY

Series in Computer Applications in Electrical Engineering

Franklin F. Kuo, editor

PRENTICE-HALL INTERNATIONAL, INC., *London*
PRENTICE-HALL OF AUSTRALIA. PTY. LTD., *Sydney*
PRENTICE-HALL OF CANADA, LTD., *Toronto*
PRENTICE-HALL OF INDIA PRIVATE LTD., *New Delhi*
PRENTICE-HALL OF JAPAN, INC., *Tokyo*

BASIC CIRCUIT THEORY WITH DIGITAL COMPUTATIONS

LAWRENCE P. HUELSMAN

Department of Electrical Engineering
University of Arizona

PRENTICE-HALL, INC., *Englewood Cliffs, N. J.*

10 9 8 7 6 5 4 3 2 1

ISBN: 0-13-057430-9

Library of Congress Catalog Card Number: 78-39007
Printed in the United States of America

To **Jo** and **David**

Preface

This book was written to provide an introduction to the basic concepts of modern circuit theory. The level at which the developments are made and the quantity of material which is included make this book suitable for a one-year course beginning either in the Sophomore or Junior year. Such a course is usually referred to as a "traditional" circuit theory course. In addition to presenting the basic methods of circuit anylysis, such a course usually has as its goal the development of fundamental concepts and techniques which are needed in subsequent courses in electronic circuits, automatic control, and communication systems.

It is almost trite to say that there are many texts available which cover the same general area of subject matter that this one does. It seems probable that, if all the introductory circuit theory books that have been written in the last decade were gathered on a single shelf, the shelf would break. This being the case, before undertaking to write still another circuit theory book, the author should have strong personal convictions that his efforts will result in a work that, in some way, is an improvement over the existing ones. If his convictions are to have reasonable validity, they must not be based on trite subjective concepts such as "better organization of topics" or on vague attributes such as "improved clarity of exposition," since, if such criteria are not successfully satisfied, the fledgling text can never hope to replace the old favorites in their well-established positions on the list of best sellers. A far more significant rationale for the writing of a new text is that the author recognizes some new development in his field, or some major change in the techniques or methods of his subject, important enough to require exposition and treatment not presently available at introductory levels. In circuit theory, as well as in all other engineering subjects, such a major change in technique is currently upon us. It may be more dramatically referred to as a revolution rather than a change. The cause of the revolution is the modern high-speed digital computer, a device which has the capabilities of performing, in seconds, computations which otherwise might easily take years. One of the author's primary purposes in the writing of this book was to recognize this revolution, i.e., to show how the digital computer may be treated as a powerful and useful tool which has a most significant role in any treatment of modern circuit theory. There are several other goals which also motivated the writing of this text. One of these was the change in the state-of-the-art of circuit theory brought about by the ever increasing use of active circuit elements. One example of this is the field of active *RC* circuits, in which arbitrary network functions are realized by inductanceless networks. Such developments seriously weaken the foundations of the traditional circuit theory barrier which for years has separated the treatment of networks containing passive elements from those containing active ones. To thoroughly demolish such a barrier, it was felt that there was a need for a text in which the properties of both passive and active circuit elements were developed simultaneously. Another goal motivating the author in the writing of this book was to provide recognition of the fact that in the real world all networks are composed to some extent of time-varying and non-linear elements. Thus, it was felt that a book which was to be a significant improvement over existing ones in this area should include introductory material to illustrate the problems produced by such real-world phenomena, and to treat some of the basic means of solving circuits which include these effects.

As indicated in the opening paragraphs of this preface, this book is aimed at the sophomore or junior undergraduate engineering student. In making the many decisions involved in selecting and eliminating matrial to be included in the text it was assumed that such a student had taken the usual lower division courses in physics and mathematics. Specifically, these courses were assumed to include:

1. A basic physic course including a treatment of the electrical variables of voltage and current, the circuit elements of resistance, capacitance and inductance, and the use of Kirchhoff's laws to describe simple networks.
2. The usual introductory engineering mathematics courses including differential and integral calculus.
3. A course in Fortran IV programming covering the basic operations and treating the use of functions and subroutines.

In the writing of this text, several departures were made from the procedure followed in most of the existing basic circuit theory texts. One such departure is the extended use which is made in the text of matrix methods for writing and solving network equations. The decision to include this type of material was based on two factors: First, the convenience of such methods and the resulting simplicity of the mathematical operations, and, second, the very real need to develop in the student, early in his undergraduate career, a facility for the use of such methods. This is most important, since such techniques become increasingly useful in later studies. In general, it has been found that most students entering courses at this level have already encountered enough applications of matrices to be familiar with the matrix operations of addition and multiplication and the mechanics of writing a set of simultaneous equations in matrix form. Thus, this material has been assumed to be part of the student's background. For the individuals who do not have such a background, an introductory treatment of matrix algebra, suitable for either study or review, is given in an appendix. The more difficult and less familiar topics of the operations connected with the solution of a matrix equation, i.e., the process of matrix inversion, is covered in detail in the text at the point the problem arises.

The subject of complex numbers is treated in a manner similar in philosophy to that chosen for the matrix algebra material. Thus, it is assumed that most of the students at this level will have a working knowledge of the use of complex numbers. For this reason it was not felt desirable to break the flow of text material by inserting a treatment of this subject in the body of the book. An appendix describing this topic, however, is included, and this may be assigned by the instructor if it is required.

The topics covered in this book have been ordered so as to emphasize a logical or algorithmic approach to the procedures of network analysis. Thus, the first three chapters emphasize the mechanics of formulating network equations. The resistance network is used as a vehicle to illustrate the techniques. In this group of chapters, Chapter 1 serves as an introduction to the book and also provides a review of some of the basic concepts such as Kirchhoff's law. Also in this chapter, the all-important concept of reference directions is heavily emphasized since experience has shown this to be a stumbling block for many beginning students. In Chapter 2 sources and resistors are introduced and simple interconnections of these elements are discussed. In the third chapter, mesh and node equations are presented. Topics pertaining to controlled sources and other multi-terminal devices such as operational amplifiers, gyrators and NIC's are included. Digital computational methods for finding the solutions for the network variables in arbitrary resistance networks are presented.

The next group of chapters extends the formulation methods developed in the first three chapters to the time-domain treatment of circuits which include capacitors and inductors. Chapter 4 presents the characteristics of these elements and discusses the constraints they impose on various time-domain waveforms. Chapter 5 treats the solution of first-order circuits, and Chapter 6 covers the properties of higher-ordered ones. In these chapters, Secs. 5-8 (on convolution), 6-8 (on dual and analog circuits), and 6-9 (on active *RC* circuits) may be omitted without loss of continuity. The solutions of the differential equations describing these circuits are presented by ordinary mathematical techniques as well as by numerical ones implemented on the digital computer. It is shown how the latter may be applied to networks which include time-varying and non-linear elements. The final group of chapters is concerned with the behavior of networks in the frequency domain.

Chapter 7 covers the sinusoidal steady-state case. In this chapter, Secs. 7-12 (on three-phase circuits) and 7-13 (on active *RC* circuits) may be omitted without loss of continuity. Chapter 8 introduces the Laplace transformation and its applications. Chapter 9 discusses two-port parameters and presents a treatment of the Fourier series and the Fourier transformation. In all these chapters the basic procedures are presented both from the viewpoint of hand computation and also as topics which may be implemented on the digital computer.

The majority of the material contained in this book may be covered in a two-semester course of three one-hour periods per week. In note form it has been used in this manner for the past two years at the University of Arizona in a course on basic circuit theory presented in the Junior year. The pace may be slowed if desired by eliminating some or all of the sections referred to above which treat more advanced topices. For the instructor's convenience, these sections are shown with an asterisk in the Table of Contents. Other variations, such as eliminating the chapter on two-port parameters or some of the material on the Fourier transformation are also possible. Some rearrangement of the order of the chapters to meet individual preferences or satisfy particular curriculum requirements may also be made. To make such a rearrangement easier, an effort has been made to keep the separate treatments of time-domain and frequency-domain topics as self-contained as possible, although complete independence of these topics is obviously not practical or desirable.

The implementation of the basic solution algorithms by digital computational techniques has been found to be of considerable value to the students. One of the reasons for this is that to understand the computer implementation the student is forced to thoroughly understand all the ramifications of the related circuit theory material. Thus, far from detracting from the emphasis on basic circuit theory topics, the inclusion of the digital computational material has been found to reinforce and improve the student's understanding of such topics. The basic software package consisting of the programs described in this book has been found to provide the student with a quite complete set of numerical tools which have considerable application in other undergraduate courses and also in many graduate courses. To make such application as simple as possible, a decision was made to implement the various digital computational techniques as subroutines and let the student supply his own main program for the input and output of data. This has proven to be an effective procedure both from the viewpoint of permitting easy cascading of subroutines to achieve complex computational goals, and also from the viewpoint of minimizing the errors made by the student in trying to input his own data to match a set of format specifications prepared by someone else. It should be noted that, in the development of these subroutines, every effort has been made to make them as simple as possible. The rationale for doing this is to avoid excess use of computer time (an important factor when hundreds of students are simultaneously running the same program), and also to provide maximum encouragement to the student to learn for himself how the program actually works and to experiment with his own modifications and improvements of the programs. In summary, the techniques presented in this text not only have been found to teach the student the principles of network theory but also to provide him with an up-to-date arsenal of computing power which may be successfully applied to a wide range of engineering problems.

The author would like to express his thanks to the many persons who assisted him in the preparation of this book, especially: Dr. Roy H. Mattson, Head of the Department of Electrical Engineering of the University of Arizona whose contagious enthusiasm and continuing support made the project possible; Dr. James Melsa of Southern Methodist University and Dr. Frank Kuo of the University of Hawaii who provided extremely thorough and detailed reviews of the manuscript; Dr. L. Schooley, who gave many helpful comments while teaching from the original classroom notes; and Messrs. Richard Miles, Daniel Nuñez, and Jeff Quintez who collectively spent many man-months punching cards, carrying decks to the computer center, typing dictation, redrawing figures, checking examples, working problems, and, in general, providing a host of invaluable services. Finally, the author would like to acknowledge his debt of gratitude to the many students who suffered through the original versions of the classroom notes and who enthusiastically accepted his offer of "one point for each error you find." The final form of the book owes much to their collective efforts.

L. P. HUELSMAN

Contents

CHAPTER 1

Introduction

An electrical network may be considered as a particular example of a broad class of entities called systems. There are many different kinds of systems, such as hydraulic systems, mechanical systems, air traffic control systems, nuclear reactors, computers, etc. In order to be able to discuss the behavior of a system, it is necessary to specify a set of variables which can be used to describe the conditions which prevail in that particular system. Thus, for a mechanical system we might be interested in variables such as the position and velocity of a mass. In an air traffic control system, the density of airplanes would be an important variable. In a nuclear reactor, the fission rate would be a most critical variable, etc. In this chapter we shall consider some of the variables which are most useful in characterizing those electrical systems referred to as circuits or networks.[1] In addition, we shall see that certain constraining relations called Kirchhoff's Laws exist relating the variables of a given network.

1-1 NETWORK VARIABLES

In this section we present a brief description of the basic electrical variables that we shall use throughout the text. There are several systems of units which may be used to describe electrical network variables. The one which is most commonly used by electrical engineers is the MKS system, which is based on the use of the meter, the kilogram, and the second as basic units of length, mass, and time, respec-

[1]In modern practice the words "network" and "circuit" are used interchangeably. We will follow such practice throughout this text.

tively.[2] In actual physical situations, the ranges of these variables may vary from very small numbers to quite large numbers. In such cases, it is customary to avoid the use of large numbers of zeros by using an appropriate multiplier of ten raised to the proper positive or negative power. Thus, 0.0000035 second (s) is more rapidly visualized as 3.5×10^{-6} s. In addition, it is sometimes convenient to replace the multiplying factor of ten by redefining the units. For example, 3.5×10^{-6} s may also be expressed as 3.5 microseconds (μs). A table of the commonly accepted multiplying factors, the associated prefix for the units, and the symbol used are given in Table 1-1.1. The names in this table apply to all the network variables that we shall deal with in this text.

Now let us consider in some detail the variables that are used in network studies.

TABLE 1-1.1 Common Multiplying Factors for Network Theory Variables*

Multiplying Factor	Name of Prefix	Symbol
10^{12}	tera	T
10^{9}	giga	G
10^{6}	mega	M
10^{3}	kilo	k
10^{-3}	milli	m
10^{-6}	micro	μ
10^{-9}	nano	n
10^{-12}	pico	p

*A more complete tabulation of multipliers, as well as a listing of current practice for units to be used in scientific and technical work, may be found in "IEEE Recommended Practice for Units in Published Scientific and Technical Work," *IEEE Spectrum*, Vol. 3, No. 3, pp. 169–173, March, 1966.

CHARGE

This electrical quantity is a property of the atomic particles of which all matter consists. We know that the atom may be pictured as composed of a nucleus with a positive charge surrounded by negatively charged particles called electrons. When there is a balance between positive and negative charges in a given quantity of matter, we say that it is uncharged or neutral. If electrons are removed from a neutral quantity of matter, the result is to leave it positively charged. Similarly, if

[2]Actually, there are two MKS systems, the "rationalized" and the "unrationalized" MKS system. The units used for electrical networks are the same no matter which of these systems is used.

electrons are added to a quantity of matter, the result is to leave it negatively charged. The MKS unit of charge is the coulomb (abbreviated C). It is the charge contained in 6.24×10^{18} electrons. When referring to the charge that a given quantity of matter possesses, we shall use the literal symbol $q(t)$. It is a basic physical postulate that charge cannot be created or destroyed, only transferred. This is referred to as the *conservation of charge*.

CURRENT

When charged particles are transferred through a given two-dimensional surface (such as the cross section of a conductor), the net time-rate of transference of charge is referred to as the flow of current. Frequently, especially in solid state phenomena, charge may be transferred by the movement of both positively charged particles and negatively charged ones. Thus it is important to keep in mind the fact that current refers to the *net* transference of charge. The unit of current in the MKS system is the ampere (abbreviated A). It is the transference of one coulomb of charge per second. We will use the literal symbol $i(t)$ for current. From the above definition, we see that

$$i(t) = \frac{dq}{dt} \tag{1-1}$$

Similarly,

$$q(t) = \int_{-\infty}^{t} i(\tau)\, d\tau \tag{1-2}$$

ENERGY

It is a basic physical postulate that energy cannot be created, only transformed. This is referred to as the *conservation of energy*. Electrical energy may be produced from many other types of energy such as chemical energy (as in a battery), mechanical energy (as in a hydroelectric generator), atomic energy (as in a nuclear reactor), etc. The basic MKS unit of energy (or work) is the joule (abbreviated J). The literal symbol that we shall use for energy is $w(t)$.

VOLTAGE

If energy is expended (as work) on a quantity of charge, the ratio of work to charge is given the name voltage. For example, a battery uses chemical processes to do work on charged particles, thus a voltage appears across its terminals. The unit of voltage in the MKS system is the volt (abbreviated V). It is equal to an energy of one joule given to a charge of one coulomb.[3] We shall use the literal symbol $v(t)$ for voltage.

POWER

The time rate of performing work is defined as power. The MKS unit of power is the watt (abbreviated W). It is equal to one joule of work per second. We shall use the literal symbol $p(t)$ for power. Thus, in terms of energy, we see that

$$p(t) = \frac{dw}{dt} \tag{1-3}$$

Similarly,

$$w(t) = \int_{-\infty}^{t} p(\tau)\, d\tau \tag{1-4}$$

Since the specification of conditions at $t = -\infty$ (as required by the lower limit of the integral in the above relation) is not always convenient, an alternative integral relation between power and energy that is frequently seen is

$$w(t) = w(t_0) + \int_{t_0}^{t} p(\tau)\, d\tau \tag{1-5}$$

Similar alternative expressions may be written for any of the other integral relations used in this section. Since current is the time-rate of transference of charge (dq/dt), and voltage is a quantity giving the ratio of work to charge (dw/dq), we see that the product of voltage and current is dw/dt, i.e., power. Thus we may write

$$p(t) = v(t)i(t) \tag{1-6}$$

This is a relation for which we shall have frequent use in this text.

FLUX LINKAGES

A current flowing in a conductor will produce magnetic lines of flux which link the conductor as shown in Fig. 1-1.1. The MKS unit for such flux linkages is the weber-turn. This is the product of the flux in webers and the number of turns of the circuit which are linked by such a flux. Thus, a single loop of wire with a current $i(t)$ flowing in it as shown in Fig. 1-1.2(a) will produce the same number of flux linkages as the two-turn loop with a current $i(t)/2$ flowing in it as shown in Fig. 1-1.2(b). When the flux linkages of an electric circuit are changed, either by mov-

[3]In other words, if a total energy of one joule is required to move a group of charged particles with a total charge of one coulomb from one point to another in a given circuit, then a potential difference of one volt is produced between the two points.

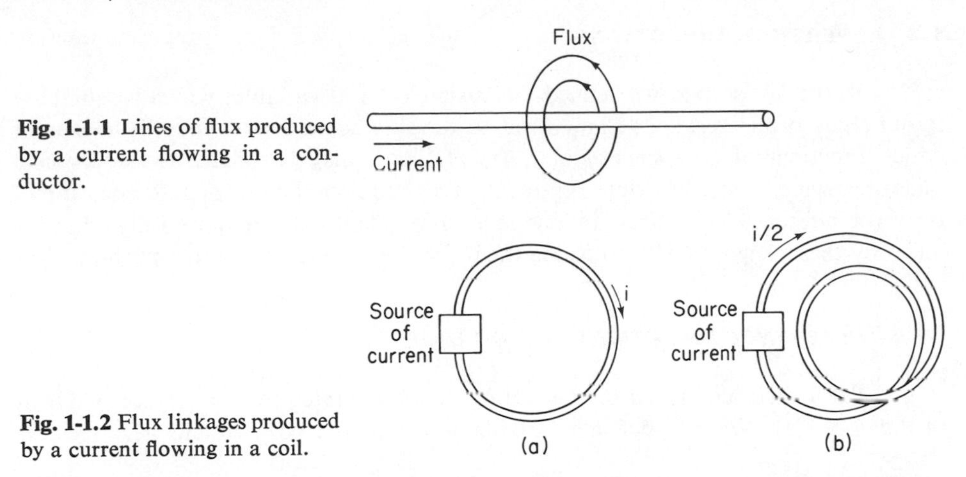

Fig. 1-1.1 Lines of flux produced by a current flowing in a conductor.

Fig. 1-1.2 Flux linkages produced by a current flowing in a coil.

ing the conductors of the circuit so that fewer lines of flux are enclosed, or by changing the flux, then a voltage is induced in the circuit. Using the literal symbol $\phi(t)$ for flux, the relation between flux linkages and voltage is

$$v(t) = \frac{d\phi}{dt} \tag{1-7}$$

The corresponding integral relation is

$$\phi(t) = \int_{-\infty}^{t} v(\tau)\, d\tau \tag{1-8}$$

The above expressions are frequently referred to as Faraday's law.

The basic electrical variables with which we shall be concerned in this text are summarized in Table 1-1.2.

TABLE 1-1.2 The Basic Electrical Variables

Variable	Unit	Abbreviation	Symbol	Relation
charge	coulomb	C	$q(t)$	$q(t) = \int i(t)\, dt$
current	ampere	A	$i(t)$	$i(t) = \frac{dq}{dt}$
energy	joule	J	$w(t)$	$w(t) = \int p(t)\, dt$
voltage	volt	V	$v(t)$	$v(t) = \frac{d\phi}{dt}$
power	watt	W	$p(t)$	$p(t) = v(t)i(t) = \frac{dw}{dt}$
flux linkages	weber-turn	Wb-turn	$\phi(t)$	$\phi(t) = \int v(t)\, dt$

1-2 REFERENCE DIRECTIONS

In the last section we defined the basic electrical variables which we shall use in our study of networks. We now need to develop some conventions by means of which functions of time such as $q(t)$, $i(t)$, $v(t)$, etc., may be related to the physical quantities which these functions represent. This requires the use of reference directions (or reference polarities). In this section we shall indicate more fully what we mean by these "signposts," and define them for the various electrical variables.

REFERENCE DIRECTIONS FOR CHARGE

Let us consider a pair of conducting plates separated by an air space as shown in Fig. 1–2.1(a). We may define a variable $q(t)$ with the positive and negative ref-

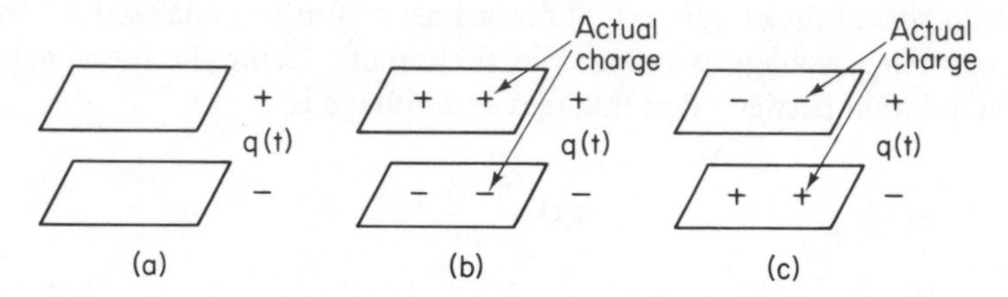

Fig. 1-2.1 (a) Reference directions for charge $q(t)$ on a pair of conducting plates, (b) $q(t) > 0$, (c) $q(t) < 0$.

erence markings shown in the figure. The reference polarities have nothing to do with the actual presence (or absence) of charge on the plates. They simply provide a convention by means of which we can correlate the parity (or sign) of the *function* $q(t)$ at a given instant of time with the *actual* charge polarity. Thus, as shown in Fig. 1–2.1(b), if the actual charge distribution at a given instant of time is such that the upper plate is positively charged while the bottom plate is negatively charged, since the actual charge distribution agrees with the reference markings, we say that at the given instant of time $q(t)$ is positive. Similarly, if at some other given instant of time the top plate is charged negatively while the bottom plate is charged positively, as shown in Fig. 1–2.1(c), then, since the charge distribution is the opposite of that indicated by the reference polarity markings, we say that at the given instant of time $q(t)$ is negative. The opposite viewpoint may also be taken, namely, if for a certain value of t, the function $q(t)$ as given by some expression is positive, then we say that the actual charge distribution must be as shown in Fig. 1–2.1(b), etc.

REFERENCE DIRECTIONS FOR CURRENT

Let us consider a section of some conductor. We may define a variable $i(t)$ and place a reference direction arrow adjacent to the conductor, as shown in Fig. 1–2.2(a). If, at a given instant of time, the actual net current flow in the conductor

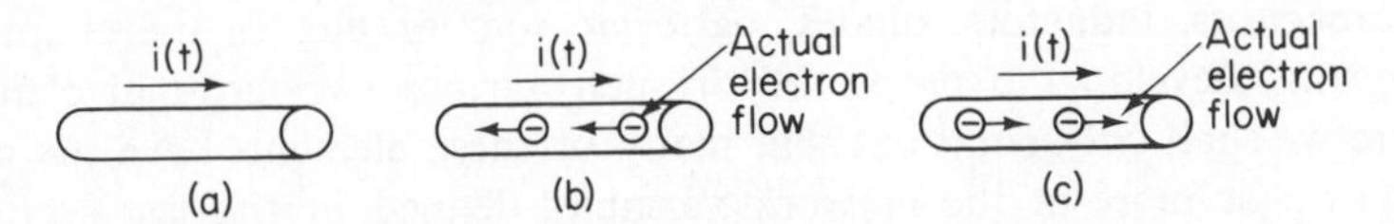

Fig. 1-2.2 (a) Reference directions for current $i(t)$ flowing in a conductor, (b) $i(t) > 0$, (c) $i(t) < 0$.

is from left to right (as indicated by the movement of *negatively* charged particles from *right to left*), then, as shown in Fig. 1–2.2(b), we say that the variable $i(t)$ is positive at that instant of time. Similarly, if at some other given instant of time the actual net current flow is from right to left, as shown in Fig. 1–2.2(c), we say that $i(t)$ is negative at that instant of time. We may also take the opposite viewpoint, and say that if the function $i(t)$ is positive at a given instant of time, then the net current flow must be in the direction indicated by the reference arrow, as shown in Fig. 1–2.2(b), etc.

REFERENCE POLARITIES FOR VOLTAGE

Let us consider a pair of terminals connected to some network as shown in Fig. 1–2.3(a). We may define a variable $v(t)$ by marking the terminals with positive

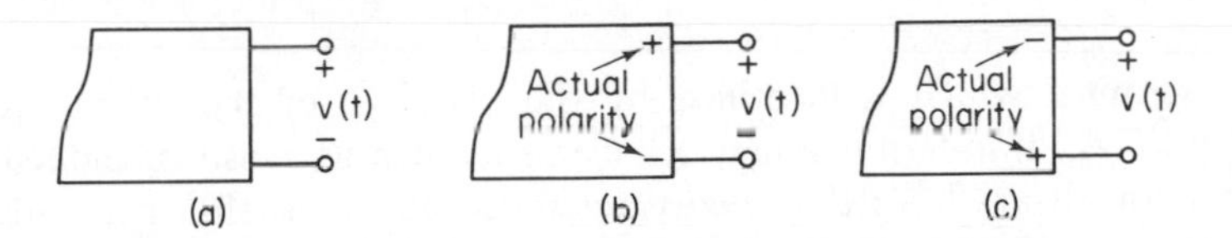

Fig. 1-2.3 (a) Reference directions for the voltage $v(t)$ at a pair of terminals, (b) $v(t) > 0$, (c) $v(t) < 0$.

and negative reference markings as shown in the figure. When the actual potential difference between the two terminals at a given instant of time is such that the upper terminal is at a higher potential than the lower terminal, as shown in Fig. 1–2.3(b), then we say that the variable $v(t)$ is positive. Similarly, if, at a given instant of time, the potential difference between the two terminals is such that the lower terminal is at the higher potential, as shown in Fig. 1–2.3(c), we then say that the voltage $v(t)$ is negative. The same logic holds true if we consider the function $v(t)$ itself. At any instant of time when this function is positive, then the relative polarities of the two terminals must be as indicated in Fig. 1–2.3(b). If the function is negative, then the relative polarities must be as shown in Fig. 1–2.3(c).

REFERENCE CONVENTIONS FOR TWO-TERMINAL ELEMENTS

The first type of network element that we shall study in this text is the two-terminal network element. Some examples of two-terminal network elements are resistors, capacitors, inductors, diodes, batteries, sources, etc. In the chapters that follow, we shall develop the properties of such various two-terminal elements in detail. Here we shall only point out that many of these elements have the property of relating two or more of the network variables defined in the last section. For example, the resistor provides a relation between voltage and current specified by Ohm's law. Thus, we may write

$$v(t) = Ri(t) \tag{1-9}$$

Any such element must have reference polarities specified for each of the variables that is involved in its defining relation. Furthermore, the *relative* directions of these reference polarities with respect to each other must be chosen in such a way that the relation between the variables is satisfied. For two-terminal elements, such as a resistor, in which a relation is specified between the variables of voltage and current, the relative reference directions of voltage and current are taken to be as shown in Fig. 1-2.4. As we see from this figure, the reference arrow for the current variable

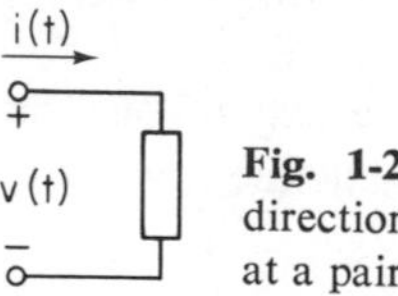

Fig. 1-2.4 Associated reference directions for voltage and current at a pair of terminals.

enters the terminal which is identified by the *plus* sign of the voltage reference polarity markings. Two-terminal network elements such as resistors, inductors, and capacitors must always have the *relative* reference directions shown in this figure. They are usually referred to as *associated* reference directions. We shall see, however, that this convention need not be followed if there is no defining expression relating the two variables and identifying the two-terminal element. We shall show that such a situation occurs for the case of ideal sources.

REFERENCE POLARITIES FOR POWER

One of the relations that we used in the preceding section to define power was

$$p(t) = v(t)i(t) \tag{1-10}$$

Since power is a function of two variables, namely, voltage and current, the relative reference directions of both variables must be taken into account in determining a reference direction for power. Let us consider the situation where two networks are connected together at a pair of terminals with the reference polarities for $v(t)$ and $i(t)$ defined as shown in Fig. 1-2.5(a). If, at a given instant of time, $v(t)$ and $i(t)$ have polarities such that $p(t)$ is positive, then there will be a flow of energy from the

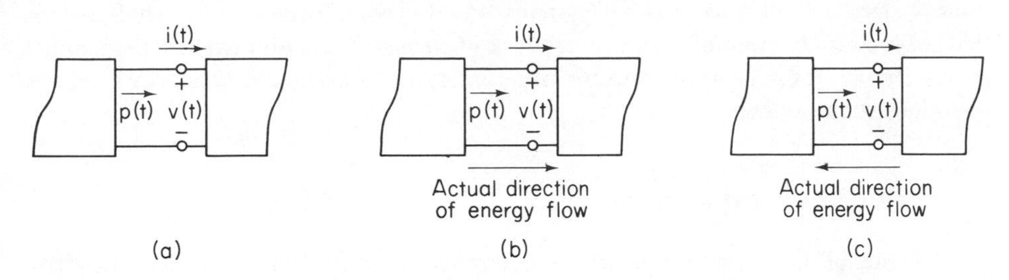

Fig. 1-2.5 (a) Reference directions for power $p(t)$, (b) $p(t) > 0$, (c) $p(t) < 0$.

network on the left to the network on the right at that instant of time. Thus we may use a reference arrow for the positive direction of energy flow as shown in Fig. 1–2.5(b). Similarly, if the actual flow of energy is from right to left as indicated in Fig. 1–2.5(c), then $p(t)$ will be negative. If the product $v(t)i(t)$ is zero at a given instant of time, then there will be no transfer of energy between the two networks. Note that this will occur when either $v(t)$ or $i(t)$ is zero. Note also that if the network shown on the right in Fig. 1–2.5(a) is a simple two-terminal element, the relative reference directions for $v(t)$ and $i(t)$ used to define the reference direction for $p(t)$ are the same as those specified for a two-terminal element as shown in Fig. 1–2.4. Thus we see that, for a two-terminal element, a positive value of $p(t)$ at a given instant of time indicates that energy is flowing into the two-terminal element. There are two things that can happen to this energy, depending on the type of two-terminal element. If the element is a resistor, the energy that flows into it is dissipated as heat. Thus, the power associated with a resistor is always positive.[4] On the other hand, two-terminal elements such as an inductor, a capacitor, a battery, or a source, are able to store this energy and deliver it back to the network at some later time. Thus, the power associated with such elements may be positive or negative. We shall explore such considerations in more detail when we discuss the properties of these elements.

1-3 CLASSIFICATION OF NETWORK ELEMENTS

Before we develop tools to use in the analysis of networks, we have to specify the types of networks which are to be analyzed. For example, we may logically assume that the techniques used to analyze linear networks are much simpler than those which must be used to analyze non-linear ones. In this section, therefore, we shall define some basic characterizations which may be used to divide networks into classes. It should be noted that the classifications which we shall develop here will be directly applied to the *individual elements* of networks and systems in later chapters. They may also, however, be used to define various similar classifications for entire networks. For example, a linear network may be considered as one comprised of linear elements. There are, however, some pitfalls that may occur when we attempt to use classifica-

[4]We shall see that it is possible to define a negative-valued resistor for which the power is always negative.

tions of elements to define classes of systems. At this point, therefore, we shall assume that these classifications refer only to network *elements*. We shall illustrate the manner in which they may be applied to the different types of complete networks when we introduce such networks.

LINEAR VS. NON-LINEAR

One of the most frequently encountered classifications used to categorize network elements is based on whether or not the network elements exhibit *linear* behavior. Actually, in the physical or "real-life" world, there is no such thing as a linear element or a linear system. Apply enough voltage to any electrical component and it will be ruined. Place too much stress on a beam and it will buckle. Allow too many airplanes to be under the control of an air traffic control system, and a collision will result. There are many other examples. Such examples might be termed "destructive" non-linearities. If we avoid these extreme cases, then the elements of most systems fall into two classes: (1) those which over a useful range of their defining variables exhibit a behavior which is so close to linear that it can be treated (within this range) as linear, and (2) those which over even the smallest useful range of their variables display characteristics which can only be described as non-linear. Thus, we see that in a practical sense linearity implies a consideration of how an element is used as well as a consideration of the characteristics of the element itself. In treating the situation specified in (1) above, we say that we may represent an actual physical element, i.e., one which is non-linear, by an *idealized model* which is *defined* as being linear for all ranges of its variables. This modeling concept enables us to use linear analysis techniques for a large portion of our system studies.

The property of linearity may be defined in terms of the following test. Consider a network element as shown symbolically in Fig. 1-3.1(a), where an input $i_1(t)$

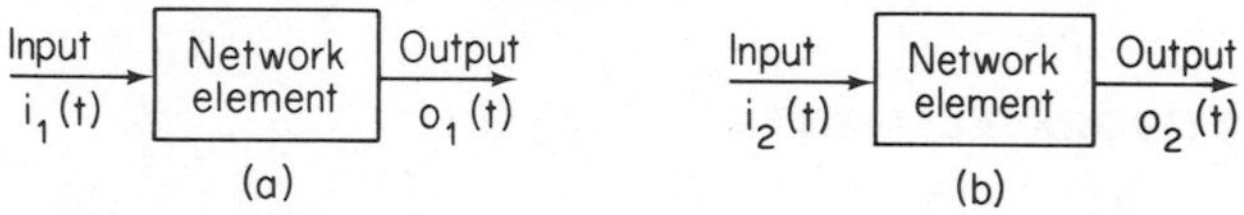

Input $i_1(t)$ → Network element → Output $o_1(t)$
(a)

Input $i_2(t)$ → Network element → Output $o_2(t)$
(b)

Fig. 1-3.1 Two tests of a network element.

is applied, and an output $o_1(t)$ is produced as a result. Similarly, assume that a second (different) input $i_2(t)$ produces an output $o_2(t)$ as shown in Fig. 1-3.1(b). Now let us apply an input $i_1(t) + i_2(t)$ to the element. If it is a linear element, then the output must have the form $o_1(t) + o_2(t)$.[5] This must hold true for any arbitrary inputs $i_1(t)$ and $i_2(t)$. In describing this test, we may simply say that the property of *superposition* holds true for linear elements.[6] The assumption of linearity will apply to most of the networks to be studied in this text. The resulting implications, such as super-

[5]We assume that there are no initial conditions present in the element.

[6]If such a test is made for the case where $i_2(t)$ is some multiple of $i_1(t)$, i.e., where $i_2(t) = ki_1(t)$, and if the output $o_2(t) = ko_1(t)$, the system is sometimes said to satisfy the property of *homogeneity*.

position, provide one of the most important bases on which the theory of network analysis is founded. Thus, we shall have much more to say about linearity in the following chapters.

An element which does not satisfy the above test is said to be *non-linear*, i.e., it is one in which superposition does not hold true. Most of our attention in this text will be on the treatment of linear elements. The analysis of systems containing such elements is considerably simpler than that of a system containing non-linear ones, since the response of a linear element to a complicated series of excitations is simply the sum (or superposition) of the responses to each of the excitations considered individually. We shall, however, introduce some digital computational techniques which are directly applicable to non-linear elements. Such applications will be covered in later chapters.

TIME-INVARIANT VS. TIME-VARYING

A second characteristic which we may use to classify network elements is the characteristic of time-invariance. If an element has parameters whose values do not change with time, then it may be said to be a time-invariant element. Actually, in the physical or "real-life" world, there is no such thing as a time-invariant element, in the same sense (as we pointed out in the paragraphs above) that there is no such thing as a linear element. Over a sufficiently long span of time, the parameters of all physical elements must change. For purposes of analysis, however, such elements may be modeled by time-invariant elements by assuming that their parameters remain substantially constant over the time range of interest. Thus, we consider such elements as *time-invariant*. On the other hand, we may define elements in which the change in their parameters is so rapid as to affect our analysis during the time span of interest as *time-varying*.[7] The major emphasis of the treatment in this text will be given to time-invariant elements. We shall, however, show that the same digital computation techniques which may be applied to non-linear cases are also directly applicable to the time-varying situation.

LUMPED VS. DISTRIBUTED

The propagation of signal information, whether it be an electrical signal in a conductor, a shock wave in a metal beam, or a flow of fluid down a pipe, is never accomplished instantaneously. Thus, in the physical or real-life world the dimensions of the elements of a system must be taken into consideration. For example, a resistor with a total resistance of 1000 ohms (Ω) and a length of 1 in would be expected to have quite different properties than a wire with the same resistance but with a length of 1 mi. When the physical dimensions of an element of a system become significant

[7]Note that the classification of time-varying or time-invariant refers to the *parameters* of the element, not the *variables* associated with the element. The latter, being functions of time, will normally vary in any but the simplest systems.

with respect to the propagation of signal information in the element, then we refer to such an element as a *distributed* element. When the physical dimensions are of little significance, and may be safely ignored without invalidating the analysis of the system, then such a component is referred to as a *lumped* element. The mathematics that must be used to treat distributed elements is, in general, considerably more complicated than the mathematics which is necessary to describe lumped elements. In brief, *partial* differential equations must be used to describe distributed elements, while *ordinary* differential equations can be used to treat lumped ones. Thus, to simplify our treatment, and to prevent ourselves from becoming enmeshed in a welter of mathematics which may actually tend to obscure the fundamental concepts we will be developing in this text, we shall assume that the network elements which we treat may be considered as being represented by lumped models. Such a representation is a valid one over a great portion of the frequency spectrum for a large number of network elements.

ACTIVE VS. PASSIVE

The final classification that may be applied to the network elements we shall study in this text is the classification of active vs. passive. The defining criterion is most easily stated in terms of energy. Such a definition, of course, must be based on the reference conventions for power and energy which were developed in Sec. 1–2. If the total energy supplied to a given network element is always non-negative regardless of the type of circuit to which the element is connected, i.e., if

$$w(t) = \int_{-\infty}^{t} p(\tau)\, d\tau \geq 0 \tag{1-11}$$

then we shall say that such an element is passive.

The majority of the material presented in this text will be concerned with the study of the properties of networks comprised of linear, time-invariant, lumped, and passive network elements. We shall, however, also include some discussions of a few cases of networks containing time-varying and non-linear elements, as well as some which include active elements.

1-4 KIRCHHOFF'S LAWS

The network variables which were introduced in Sec. 1–1 may have many interrelationships among themselves. Some of these relationships are due to the nature of the variables. For example, there is a relation between current and charge, namely $i(t) = dq/dt$, which is the result of the definition of these variables. A different class of relationship occurs because of the restriction that some specific type of network element places on the variables. For example, for a two-terminal resistor the variables of voltage and current are related by the equation $v(t) = Ri(t)$, which is usually called Ohm's law. Still another class of relationship is one between several variables of the same type which occurs as the result of the network configuration, i.e., the manner in which the various elements of the network are interconnected. Such a relation is said to be based on the *topology* of the network. In this section we shall introduce some of the most basic concepts of network topology. We shall also define two of the most well known topological relationships, namely, Kirchhoff's voltage law and Kirchhoff's current law.

The first topological relationship among network variables that we shall introduce is Kirchhoff's current law, which we shall refer to as KCL. Before presenting this relationship, let us introduce the concept of a node and a branch. A *node* is a point in a circuit where two or more network elements are connected together. A *branch* is any two-terminal network element.[8]

As an example of the concepts of branches and nodes, consider Fig. 1-4.1.

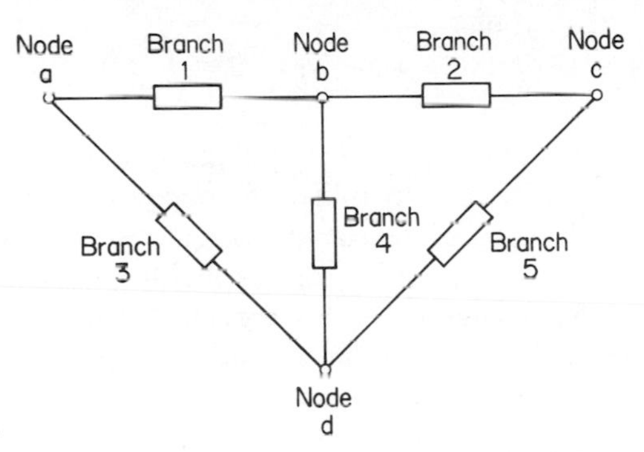

Fig. 1-4.1 Nodes and branches.

Node a is readily seen to be the point where branches 1 and 3 are connected. Similarly, node b is the point where branches 1, 2, and 4 are connected, etc. Now let us see how KCL applies to branches and nodes. If we define a set of reference currents in each of the branches connected to a node, then KCL tells us that there is a relation among these currents which may be summarized as follows:

SUMMARY 1-4.1

Kirchhoff's Current Law (KCL): The algebraic sum of the branch currents at a node is zero at every instant of time.

Thus, we see that a linear constraint exists among the branch currents at a given node.

As an example of an application of KCL, consider the three branches con-

[8]Another term which will be used interchangeably in this text for node is *terminal*. Similarly, we will frequently refer to a branch as an *element*. In the literature, the terms *vertex* and *junction* are also used to refer to a node, and the term *edge* is used to refer to a branch.

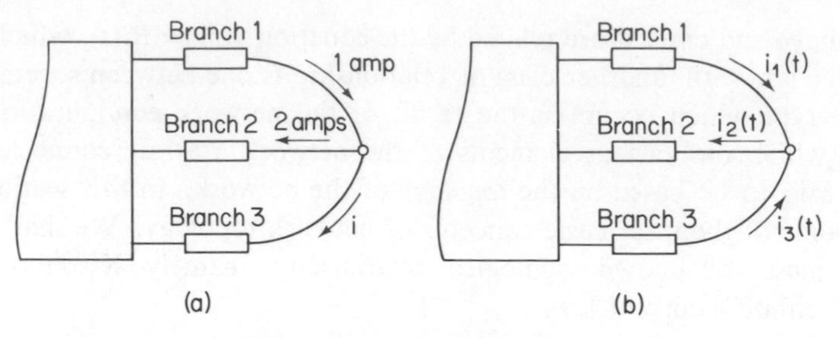

Fig. 1-4.2 Examples of KCL.

nected to the node shown in Fig. 1–4.2(a) (we are not interested in the remainder of the network so this is merely shown as a box). If, at a given instant of time, the currents in branches 1 and 2 have the values and directions indicated on the figure, then we may find the current i by applying KCL. We obtain

$$1 - 2 - i = 0. \tag{1–12}$$

From this we see that the current i at the specified instant of time equals -1 A, i.e., a current of 1 A flows in a direction opposite to that of the reference arrow for i in branch 3. The preceding example considered constant currents, however, as indicated in Summary 1–4.1, KCL also applies to the case where the currents are functions of time. As an example of this, consider the set of branches and the node shown in Fig. 1–4.2(b). At every instant of time, KCL says that the three currents shown in the figure must obey the relationship

$$i_1(t) - i_2(t) + i_3(t) = 0 \tag{1–13}$$

For example, if $i_1(t)$ and $i_2(t)$ were specified by the relations

$$i_1(t) = 1 - 3 \sin t \text{ A} \quad \text{and} \quad i_2(t) = 4 \sin 3t \text{ A}$$

then from (1–13) we see that

$$i_3(t) = 4 \sin 3t - 1 + 3 \sin t \text{ A}$$

In addition to having constant values, or being specified by mathematical expressions, the currents in a network may also be specified by graphical plots. For example, for the network branches shown in Fig. 1–4.2(b), if the currents $i_2(t)$ and $i_3(t)$ are specified by the plots shown in Fig. 1–4.3(a), then the plot for $i_1(t)$ must be as shown in Fig. 1–4.3(b).

Another equivalent way in which KCL is frequently stated is summarized as follows:

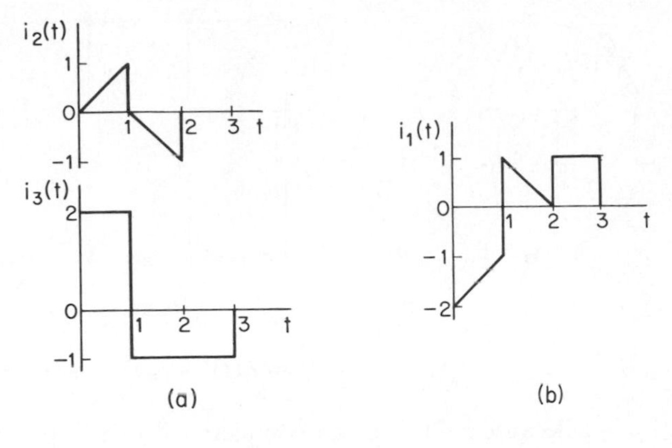

Fig. 1-4.3 A set of currents for the network shown in Fig. 1-4.2.

SUMMARY 1-4.2

Kirchhoff's Current Law (alternative form): The sum of the branch currents entering a node at a given instant of time is equal to the sum of the currents leaving the node at that instant of time.

As an example of the application of the above, (1–13) may be written in the form

$$i_1(t) + i_3(t) = i_2(t) \tag{1–14}$$

An interesting corollary which results from KCL may be obtained by integrating both sides of the above equation. Thus, we obtain

$$\int i_1(t)\,dt + \int i_3(t)\,dt = q_1(t) + q_3(t) \tag{1–15}$$

$$= \int i_2(t)\,dt = q_2(t) \tag{1–16}$$

From Eqs. (1–15) and (1–16) we see that KCL basically implies the conservation of charge at a node, i.e., the charge delivered to the node equals the charge taken from the node.

It should be noted that, in drawing schematics of networks, nodes can be defined clearly by connecting all branches directly to a given node, as shown in Fig. 1–4.4(a). The use of the curved connecting lines shown in this figure, however, is impractical in any but the simplest of networks. Thus we will usually use the more conventional drawing form shown in Fig. 1–4.4(b). Note that in this figure, the entire horizontal connecting line at the top of the drawing constitutes a "node" to which the four branches are connected. Thus, for either Fig. 1–4.4(a) or Fig. 1–4.4(b) we may write

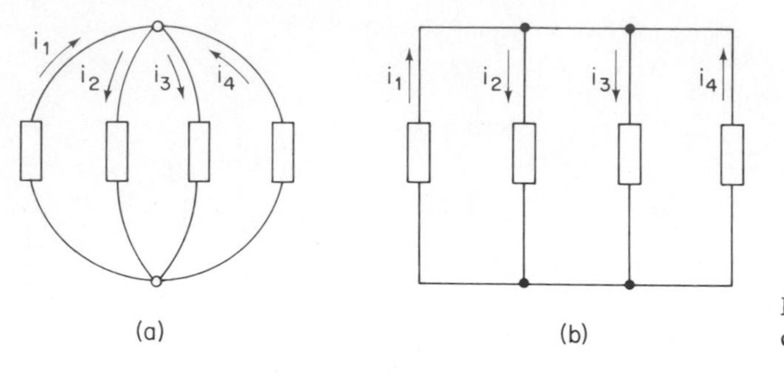

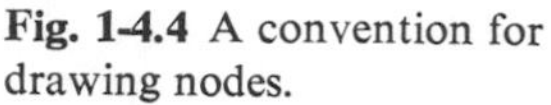

Fig. 1-4.4 A convention for drawing nodes.

$$i_1(t) - i_2(t) - i_3(t) + i_4(t) = 0$$

Because KCL is based completely on the topology of the network, and not on the properties of the network elements, it applies to all networks with lumped elements, independent of whether the elements are linear or non-linear, time-varying or time-invariant. By implication, KCL tells us that the current into one terminal of a two-terminal element equals the current out of the other terminal at every instant of time. Such a conclusion, however, does not apply to distributed elements. For example, consider a microwave antenna placed on the roof of an automobile for use in mobile communication. When the antenna is being used for transmission, the current at the base of the antenna must certainly be non-zero. However, at the other end of the antenna, the current must always be zero. If we consider the two ends of the antenna as comprising a two-terminal element, we see that because of the distributed nature of the antenna (at the frequency at which it is used), it will not satisfy KCL. We conclude that networks containing distributed elements may not satisfy KCL.

A generalization of KCL is frequently useful in many network situations. This requires the definition of the topological term called a cutset. To see what such a term implies, let us construct the *graph* of a given network by replacing every two-terminal network with a simple line as shown in Figs. 1–4.5(a) and (b). If we now

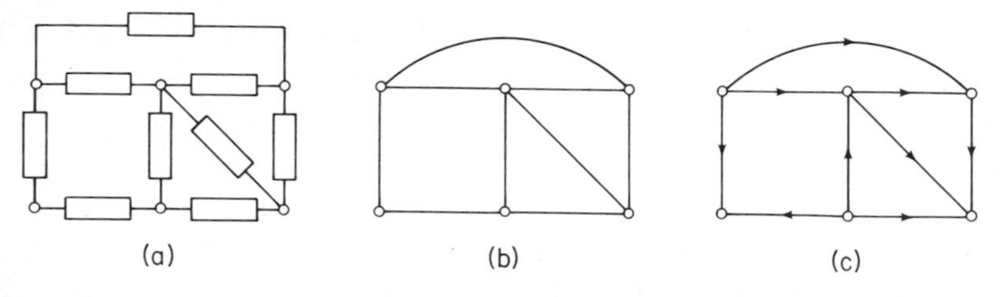

Fig. 1-4.5 A network and its graph.

assign a reference direction to each of the branches in the graph, then we may refer to the result as an *oriented graph.* A convenient way of doing this is by drawing an arrowhead on each branch. Such arrowheads can be thought of as establishing the reference directions for the currents in the branches. Such directions may be established arbitrarily; however, once they have been chosen, then the associated voltage reference polarities for the branches are fixed, as was pointed out in Sec. 1–2. An oriented graph for the network shown in Fig. 1–4.5(a) is drawn in Fig. 1–4.5(c). Now let us look for collections (a mathematical term for a collection is a *set*) of branches such that if they are removed from the network, the branches and nodes that remain form two separate parts, i.e., the network is *cut* into two parts. Such a collection of branches shall be called a *cutset*. It must have the property that if any branch is excluded from the cutset, i.e., if any branch of the set is replaced in the network, then the network is no longer separated. Thus we may define a cutset by the following statement:

SUMMARY 1-4.3

> *Definition of a Cutset*: A cutset is a set of n branches with the property that if all n branches are removed from the network graph, it is separated into two parts, but if any $n - 1$ branches of the set are removed, the graph remains connected.

In the network graph shown in Fig. 1–4.6, for example, we have indicated the various cutsets by drawing dashed lines through the graph intersecting the branches of the different cutsets. Some examples of these various cutsets are: cutset A, branches 1-2; cutset B, branches 2-3-4; cutset D, branches 4-5; cutset F, branches 1-3-5; etc. From a study of these cutsets, two facts should be apparent concerning the cutsets that exist for a given network graph. The first of these is that the number of branches in the various cutsets will not be the same. Cutset A for example contains two branches, while cutset B contains three branches. The second fact which should be noticed concerning cutsets is that a single node is considered as a part of the network graph; thus, a cutset, such as cutset A, which separates this node from the rest of the network is simply the set of branches which are directly connected to that node.

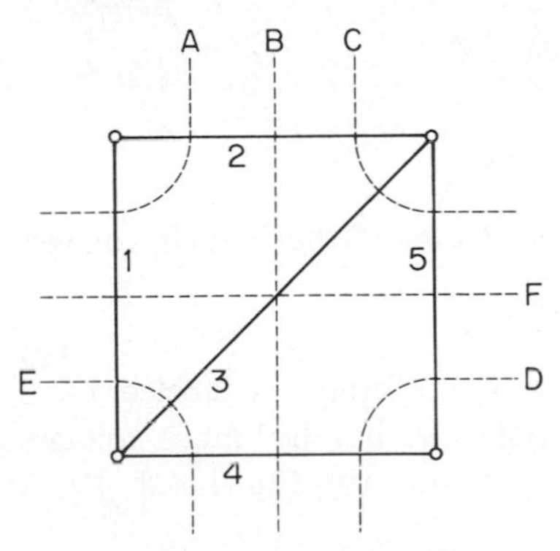

Fig. 1-4.6 Examples of cutsets.

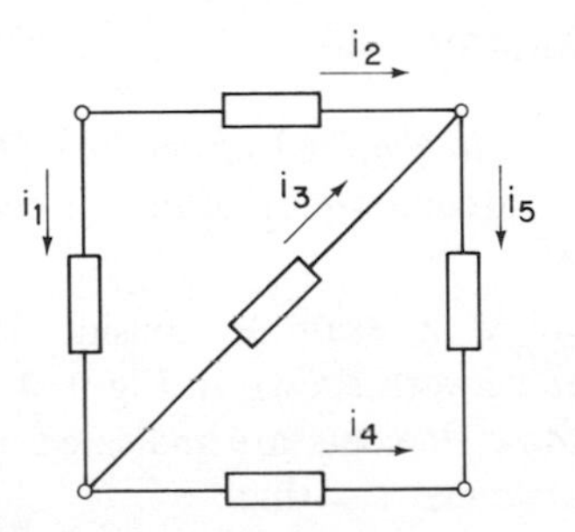

Fig. 1-4.7 Branch current reference directions for the network graph shown in Fig. 1-4.6.

One of the reasons for learning about cutsets is that KCL may be applied to cutsets as well as to simple nodes. The reasoning for this may be developed intuitively. A cutset separates the network into two parts; therefore the *net* current flow between these two parts must be zero, since a movement of charge in one direction "through" the cutset along a branch must be matched by a movement of charge in an opposite direction through the cutset along some other branch or branches. Thus we may state the following version of KCL:

SUMMARY 1-4.4

Kirchhoff's Current Law (*generalized to cutsets*): The algebraic sum of the branch currents of a cutset is zero at every instant of time.

When the cutset is such as to separate a single node from the rest of the network, then KCL as applied to cutsets simply becomes the standard version of KCL as given in the previous definition.

As an example of the application of KCL to cutsets, in Table 1–4.1 we have listed the cutsets for the network graph shown in Fig. 1–4.6 with the reference current directions for the branches shown in Fig. 1–4.7. Application of KCL to these cutsets yields the equations listed in the right column of the table.

The second of the two topological relationships among network variables which we will present in this section is Kirchhoff's voltage law. We shall refer to this as KVL. Before introducing this relationship, let us introduce the concept of a loop. A *loop* is a set of branches which forms a connected path in the network, and which has the property that exactly two branches of the set are connected to every node which is encountered in the path. This latter restriction prevents the possibility of multiple paths being included in the definition of a loop. For example, for the network graph shown in Fig. 1–4.6 there are three possible loops, namely a loop formed by branches 1-2-3, a loop formed by branches 3-4-5, and a loop formed by branches 1-2-4-5. We may now state KVL as follows:

SUMMARY 1-4.5

Kirchhoff's Voltage Law (*KVL*): The algebraic sum of the branch voltages around a loop is zero at every instant of time.

As an example, consider the loop consisting of the branches labeled 1-2-3 of the network shown in Fig. 1–4.6. The reference polarities and the branch voltages for these elements are indicated as $v_1(t)$, $v_2(t)$, $v_3(t)$ as shown in Fig. 1–4.8. From KVL we may say that

$$v_1(t) - v_2(t) + v_3(t) = 0 \tag{1-17}$$

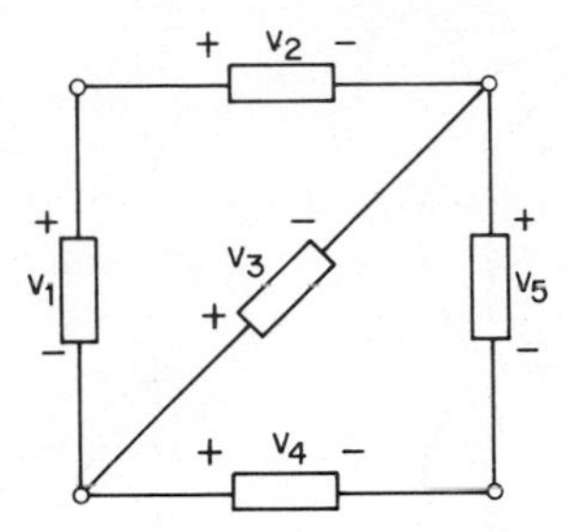

Fig. 1-4.8 An example of KVL.

for all values of t. For example, if $v_1(t) = t_3 + 1$ V, and $v(t) = \sin t + 3$ V, then $v_3(t) = \sin t - t^2 + 2$ V. Similarly, for this loop, if $v_1(t)$ and $v_3(t)$ are defined by the graphical plots indicated in Fig. 1–4.9(a), then $v_2(t)$ will have the waveshape indicated in Fig. 1–4.9(b). Another equivalent way of stating KVL is as follows:

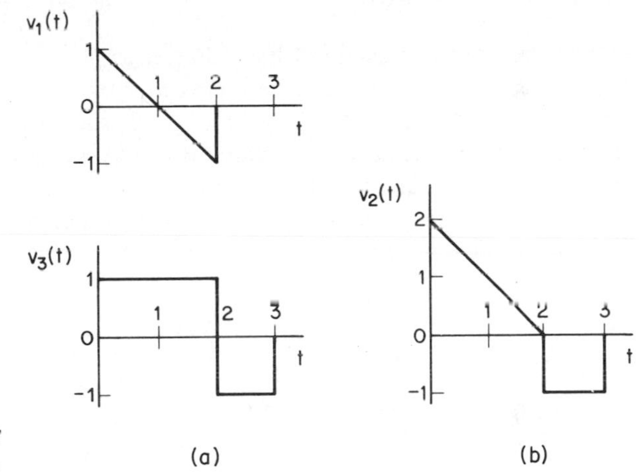

Fig. 1-4.9 A set of voltages for the network shown in Fig. 1-4.8.

SUMMARY 1-4.6

Kirchhoff's Voltage Law (*alternative form*): The sum of the voltage rises around a loop is equal to the sum of the voltage drops around the loop at every instant of time.

It should be noted that KVL, like KCL, does not apply to circuits containing distributed elements. It does, however, apply irrespective of whether the network is linear or non-linear, time-varying or time-invariant, active or passive.

There is one final classification of network graphs which is frequently of interest. This is the distinction between planar and non-planar graphs. A *planar graph* is one which has the property (or can be redrawn so that it has the property) that no lines cross. Correspondingly, a *non-planar graph* is one that cannot be redrawn so

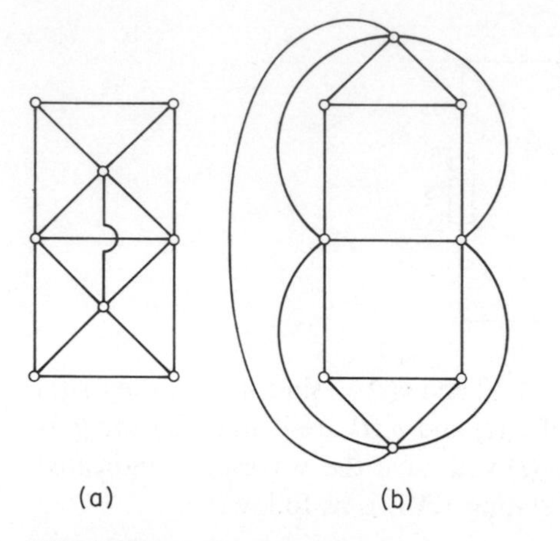

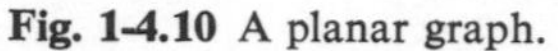

Fig. 1-4.10 A planar graph.

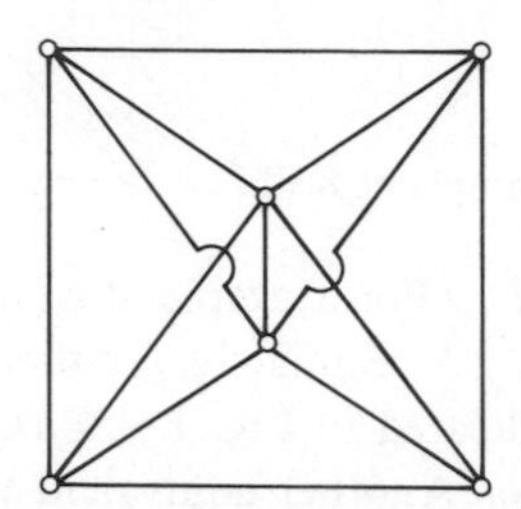

Fig. 1-4.11 A non-planar graph.

as to eliminate the crossing of lines. For example, the graph shown in Fig. 1–4.10(a) is a planar graph, since it can be redrawn as shown in Fig. 1–4.10(b) without any crossing lines. On the other hand, the graph shown in Fig. 1–4.11 is a non-planar graph, since there is no way in which it can be redrawn so that no lines cross. We shall see in a later chapter that, in some cases, the analysis techniques which we use for networks with planar graphs must be modified when networks with non-planar graphs are encountered.

TABLE 1-4.1

Cutset	*Branches*	*KCL Applied to Cutset*
A	1-2	$i_1(t) + i_2(t) = 0$
B	2-3-4	$i_2(t) + i_3(t) + i_4(t) = 0$
C	2-3-5	$i_2(t) + i_3(t) - i_5(t) = 0$
D	4-5	$i_4(t) + i_5(t) = 0$
E	1-3-4	$i_1(t) - i_3(t) - i_4(t) = 0$
F	1-3-5	$i_1(t) - i_3(t) + i_5(t) = 0$

1-5 CONCLUSION

The major result of this chapter has been the presentation of Kirchhoff's laws. The importance of these laws cannot be overemphasized. They are based only on the topological form of the network, i.e., on the manner in which the various branches of the network are interconnected. Thus, they are completely independent of the nature of the network elements, requiring only that they be lumped. In the following chapters, we shall see that Kirchhoff's laws form the basis of all our network analysis techniques.

Problems

Problem 1-1 (*Sec. 1-1*)

The charge $q(t)$ in coulombs present in a two-terminal network element is defined by the waveshape shown in Fig. P1-1.

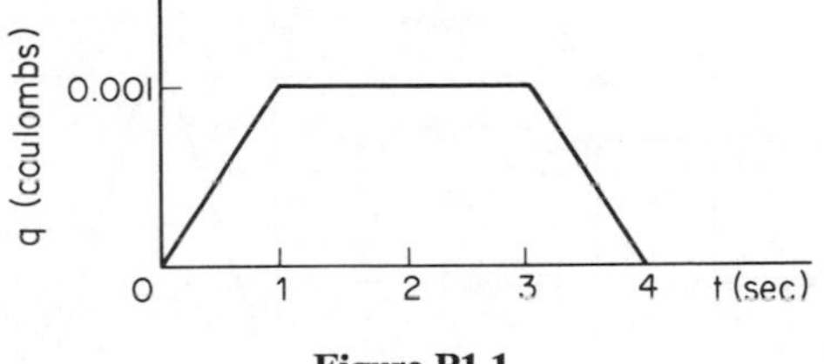

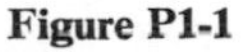

Figure P1-1

(a) Find a waveshape of current (in mA) which must be applied to the element to produce the waveshape of $q(t)$.
(b) Find another different waveshape of charge which could be produced by the current waveshape which was found in part (a).

Problem 1-2 (*Sec. 1-1*)

Find an expression for the total charge $q(t)$ in coulombs which has passed a point in a branch of a circuit, if the current in the branch is equal to $i(t) = 4e^{-4t}$ for $t \geq 0$.

Problem 1-3 (*Sec. 1-1*)

The voltage and current variables at a pair of network terminals are defined with the relative reference polarities shown in Fig. P1-3. It is assumed that the network

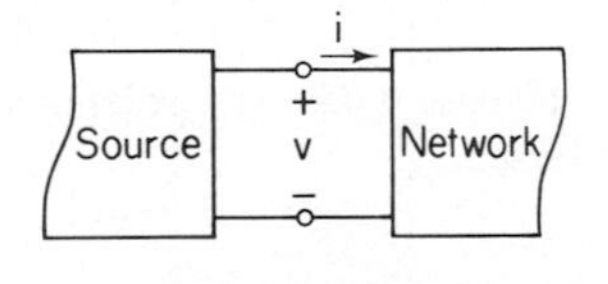

Figure P1-3

has no initial energy stored in it. If for the period $0 \leq t \leq 3$ ms, the voltage $v(t) = 10 - 200t$ V and the current $i(t) = 10t^2$ mA find: (a) the power $p(t)$ transferred from the source to the network; (b) the energy $w(t)$ supplied to the network; (c) the total energy stored in the network at $t = 3$ ms.

Problem 1-4 (Sec. 1-1)

Waveshapes of voltage and current defined as shown in Fig. P1-4 are measured at the terminals of the network shown in Fig. P1-3. Draw waveshapes for the power $p(t)$ and the energy $w(t)$. Identify the units used on the ordinates of these plots.

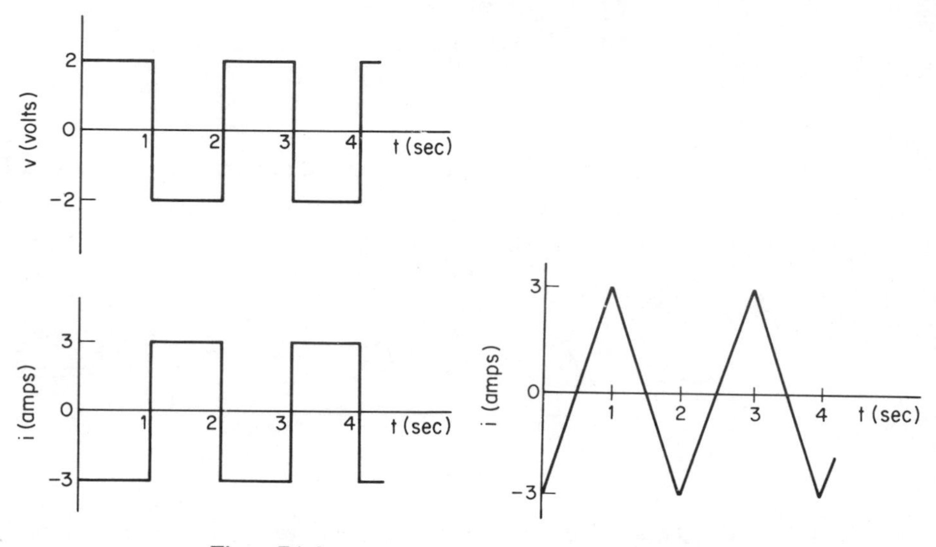

Figure P1-4

Figure P1-5

Problem 1-5 (Sec. 1-1)

Repeat Problem 1-4 using the waveshape of current shown in Fig. P1-5.

Problem 1-6 (Sec. 1-2)

A voltage $v(t) = \sin 10t$ V and a current $i(t) = 5 \sin 10t$ A are the variables (with associated reference directions) of a two-terminal element.

(a) Find an expression for the power dissipated in the two-terminal element.

(b) Find the total energy supplied to the element over the time range $0 \le t \le 0.2$ s.

(c) Repeat parts (a) and (b) for the case where the current $i(t) = 5 \cos 10t$ A.

Problem 1-7 (Sec. 1-2)

Repeat Problem 1-3 for the case where the reference direction for the current $i(t)$ shown in Fig. P1-3 is reversed.

Problem 1-8 (Sec. 1-3)

The source shown in Fig. P1-3 applies a series of test voltages as identified below to the network shown in that figure. The current that results in each of these tests is also tabulated. On the basis of this table, comment on whether the network is linear and time-invariant. It is assumed that there is no energy stored in the network prior to the start of each test.

TABLE P1-8

Test	t (s)	$v(t)$ (V)	$i(t)$ (A)
1	$0 \leq t \leq 7$	3	$2 + e^{-t} \sin t$
2	$10 \leq t \leq 17$	6	$4 + 2e^{-t} \sin t$

Problem 1-9 (Sec. 1-3)

Repeat Problem 1-8 for the data given below.

TABLE P1-9

Test	t (s)	$v(t)$ (V)	$i(t)$ (A)
1	$t > 0$	1	$3 + 4e^{-t}$
2	$t > 0$	t	$2 + t$
3	$t > 0$	$1 + t$	$5 + t + 4e^{-t}$

Problem 1-10 (Sec. 1-4)

Write the equations determined by applying KCL at nodes a, b, c, and d of the network shown in Fig. P1-10.

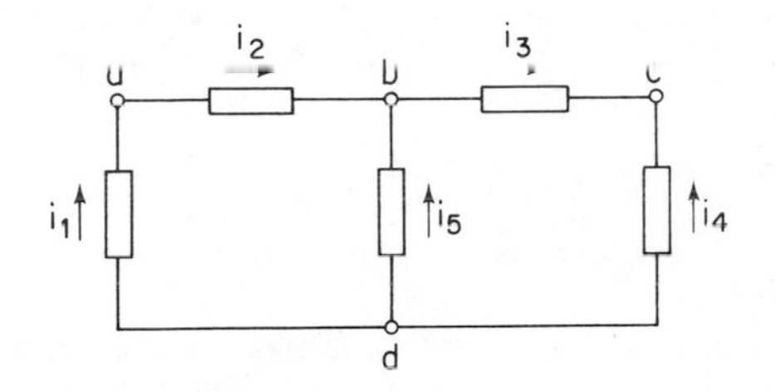

Figure P1-10

Problem 1-11 (Sec. 1-4)

Add the four equations found in Problem 1-10. What conclusion can you make from the result?

Problem 1-12 (Sec. 1-4)

In the circuit shown in Fig. P1-10, $i_2(t) = 3 + 2t$, and $i_3(t) = -3 + \sin t$ A. Find the currents $i_1(t)$, $i_4(t)$, and $i_5(t)$.

Problem 1-13 (Sec. 1-4)

In the circuit shown in Fig. P1-13 the waveshapes for $i_1(t)$, $i_2(t)$, and $i_3(t)$ are shown. Make a sketch of the current $i_4(t)$.

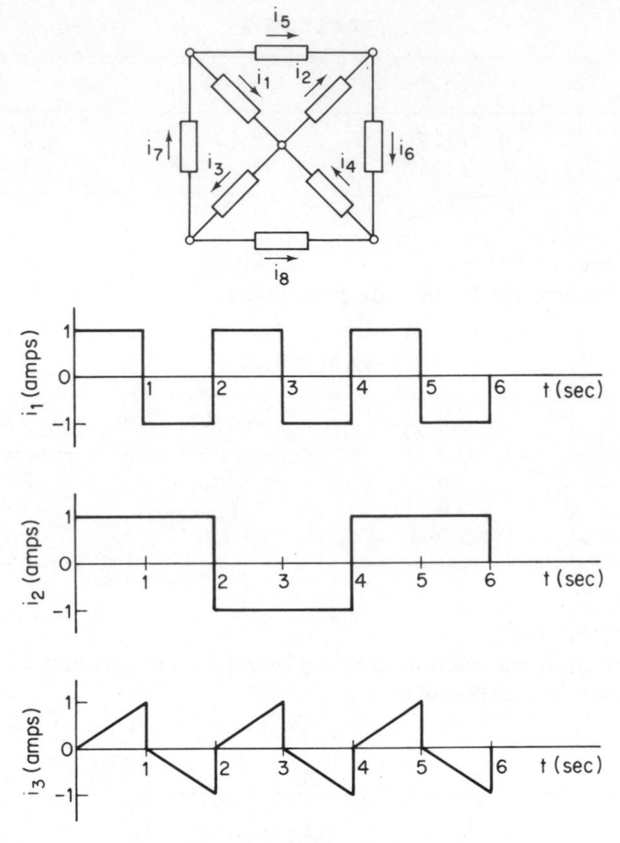

Figure P1-13

Problem 1-14 (Sec. 1-4)

Apply KVL to obtain three different equations for the circuit shown in Fig. P1-14.

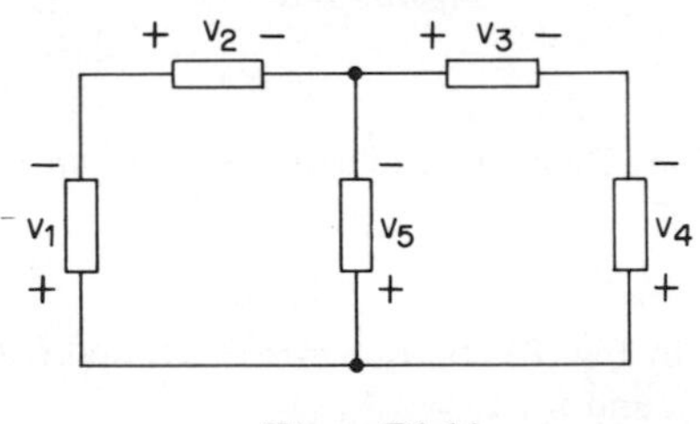

Figure P1-14

Problem 1-15 (Sec. 1-4)

Add the three KVL equations determined in Problem 1-14. What conclusion can you make from the result?

Problem 1-16 (Sec. 1-4)

In the circuit shown in Fig. P1-14, $v_1(t) = 2 + 7t$, $v_4(t) = -4 + \sin t$, and $v_5(t) = 3 + e^{-t}$ V. Find $v_2(t)$ and $v_3(t)$.

Problem 1-17 (Sec. 1-4)

Draw an oriented graph for the circuit shown in Fig. P1-10.

Problem 1-18 (Sec. 1-4)

Draw an oriented graph for the circuit shown in Fig. P1-13.

Problem 1-19 (Sec. 1-4)

Determine all possible cutsets for the circuit shown in Fig. P1-10 and write the KCL equations for each cutset.

Problem 1-20 (Sec. 1-4)

Determine all possible cutsets for the circuit shown in Fig. P1-13 and write the KCL equations for each cutset.

Problem 1-21 (Sec. 1-4)

In the circuit shown in Fig. P1-13, $i_1(t) = 2 + 3t$, $i_6(t) = -1 + 2 \sin t$, and $i_7(t) = -3t + t^2$ A. Find $i_2(t)$.

Problem 1-22 (Sec. 1-4)

In the network graph shown in Fig. P1-22, four of the branch currents have the indicated values. Find the value of the current i.

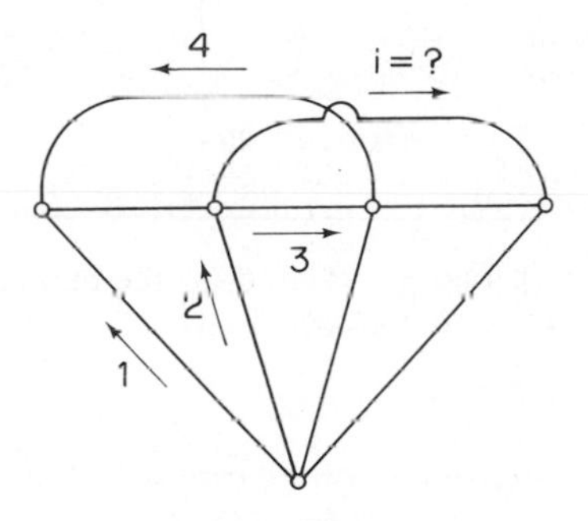

Figure P1-22

Problem 1-23 (Sec. 1-4)

Find all the cutsets for the network graph shown in Fig. P1-23 and write the KCL equations for each.

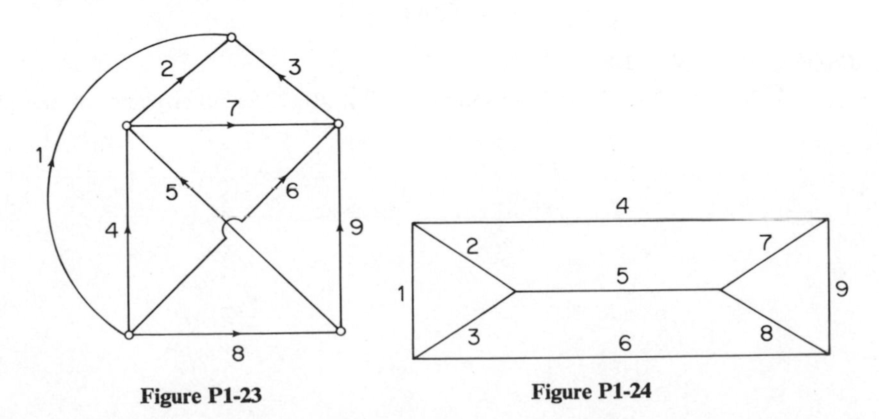

Figure P1-23 **Figure P1-24**

Problem 1-24 (Sec. 1-4)

Identify all possible cutsets in the network graph shown in Fig. P1-24 by listing the numbers of the branches in each such cutset.

Problem 1-25 (Sec. 1-4)

In the network graph shown in Fig. P1-25, the currents $i_1(t)$ and $i_2(t)$ have the waveshapes illustrated. Sketch the waveshape for $i_3(t)$. Give all the necessary values on the ordinate of the sketch.

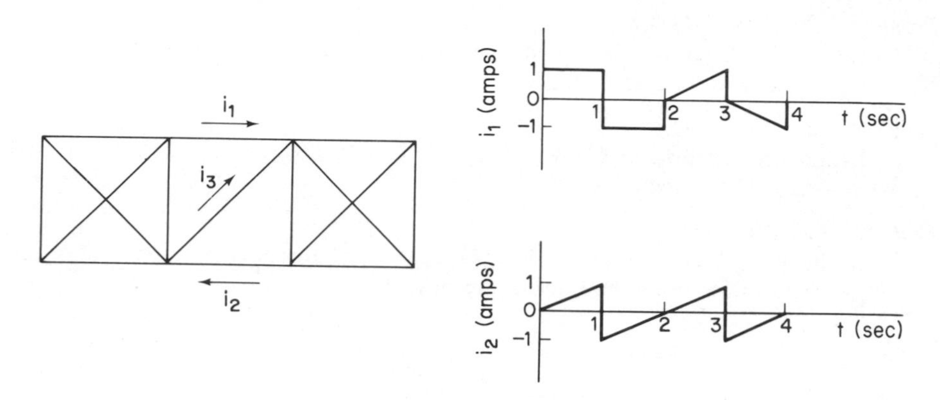

Figure P1-25

Problem 1-26 (Sec. 1-4)

For the network shown in Fig. P1-26, find (a) the number of nodes, (b) the number of branches, and (c) the number of cutsets.

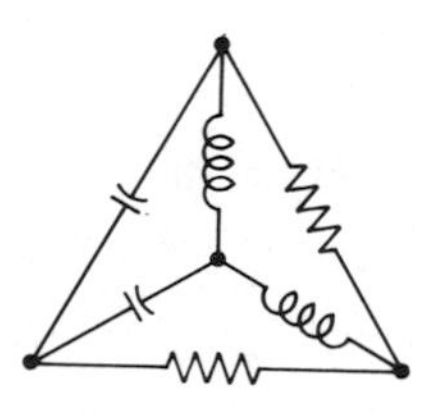

Figure P1-26

Problem 1-27 (Sec. 1-4)

Classify each of the networks shown in Fig. P1-27 as either planar or non-planar.

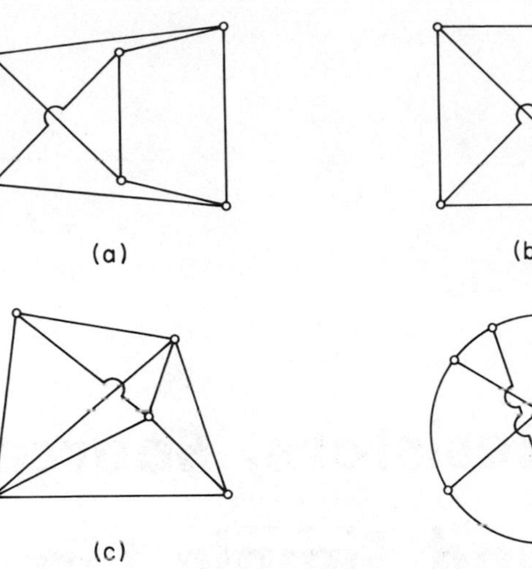

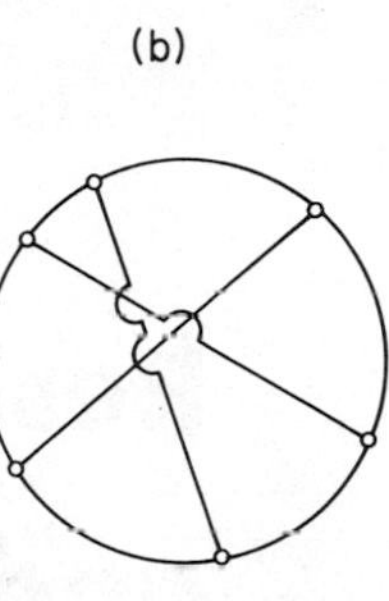

Figure P1-27

CHAPTER 2

Resistors, Sources, and Simple Circuits

In order to determine the properties of any specific type of system, we must start by determining the answers to two general questions. The first of these is: What are the basic elements or building blocks of which the system is constituted? The second is: How are these basic elements interconnected? Once the answers to these questions have been found, then we may turn our attention to determining the properties of the system, i.e., finding how it operates under various conditions of input. In this and the following chapter we shall apply this method of approach to determine the properties of resistance networks. We will begin by examining the basic elements of such networks, namely resistors and sources.

2-1 THE RESISTOR

A *resistor* is a two-terminal element which has the property that its branch voltage and current variables are related by the expression

$$v(t) = Ri(t) \tag{2-1}$$

This expression is frequently referred to as *Ohm's law*, and the unit used for specifying the value of the *resistance* R is the *ohm* (abbreviated Ω), so named in honor of the German physicist Georg Simon Ohm, who originally discovered the relationship.

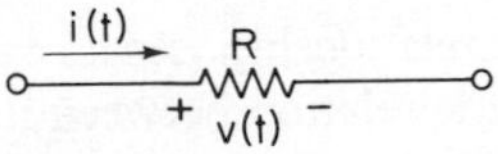

Fig. 2-1.1 A resistor and its associated reference directions.

The circuit symbol and the reference polarities for a resistor are shown in Fig. 2–1.1. Note that the reference polarities are associated, i.e., they follow the convention for two-terminal elements which was set up in Sec. 1–2. A similar relation which is frequently used to define a resistor is

$$i(t) = Gv(t) \tag{2–2}$$

where $G (= 1/R)$ gives the value of the *conductance* in *mhos*, i.e., reciprocal ohms. We will use both of these relationships in the developments that follow.

We may illustrate the properties of a resistor more fully by plotting the locus of Eq. (2–1) on an *i-v* plane [this is a two-dimensional plot in which the current variable $i(t)$ is plotted along the "x" axis and the voltage variable $v(t)$ is plotted along the "y" axis] as shown in Fig. 2–1.2(a). A *linear* resistor is characterized by a

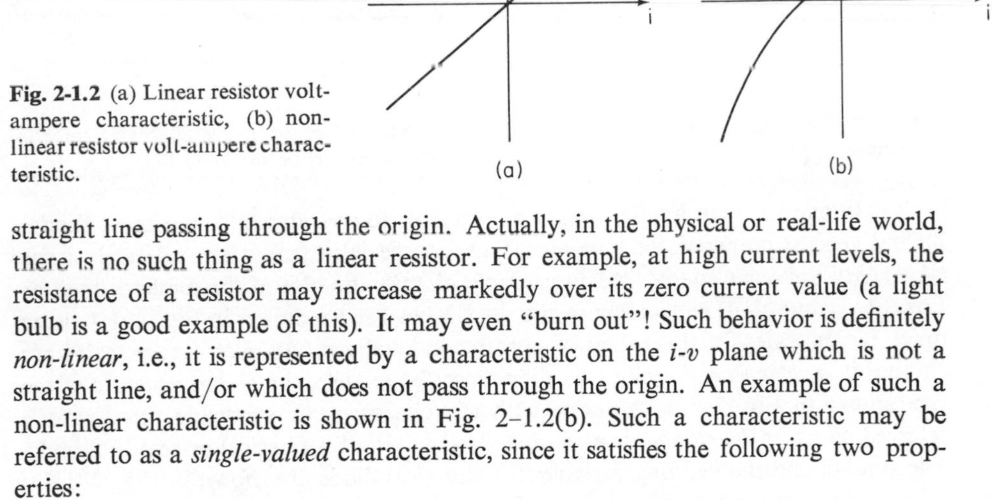

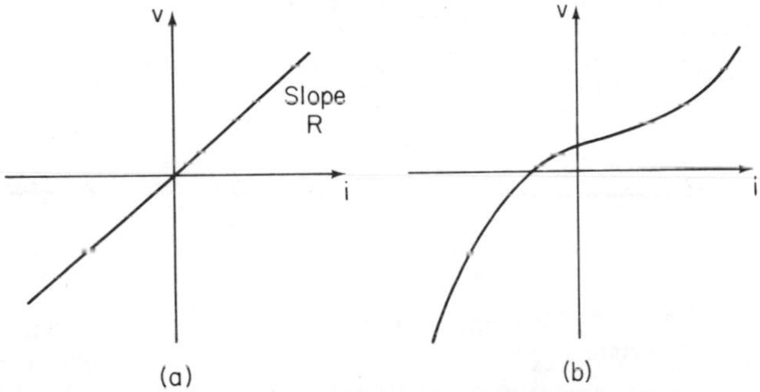

Fig. 2-1.2 (a) Linear resistor volt-ampere characteristic, (b) non-linear resistor volt-ampere characteristic.

straight line passing through the origin. Actually, in the physical or real-life world, there is no such thing as a linear resistor. For example, at high current levels, the resistance of a resistor may increase markedly over its zero current value (a light bulb is a good example of this). It may even "burn out"! Such behavior is definitely *non-linear*, i.e., it is represented by a characteristic on the *i-v* plane which is not a straight line, and/or which does not pass through the origin. An example of such a non-linear characteristic is shown in Fig. 2–1.2(b). Such a characteristic may be referred to as a *single-valued* characteristic, since it satisfies the following two properties:

1. For any value of current, a single unique value of voltage is defined by the characteristic;
2. For each value of voltage, a single unique value of current is defined by the characteristic.

In such a case we may write the defining relation either as $v = f(i)$, or as $i = h(v)$, where $f(i)$ and $h(v)$ specify the relationship. As an example of such a device, consider a solid-state diode. It has a static relation between its terminal variables of voltage v and current i that may be closely approximated by the expression[1]

$$i = I_o(e^{v/V_o} - 1) \tag{2-3}$$

where I_o and V_o are constants whose values depend on the given diode. Equation (2-3) is clearly an expression of the form $i = h(v)$. The opposite relationship in the form $v = f(i)$ is easily derived from Eq. (2-3). It is

$$v = V_o \ln\left(\frac{i}{I_o} + 1\right) \tag{2-4}$$

A plot of the relations given in Eqs. (2-3) and (2-4) is shown Fig. 2-1.3.

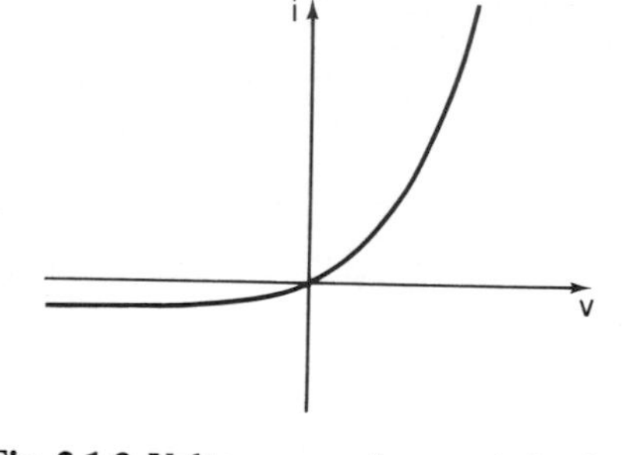

Fig. 2-1.3 Volt-ampere characteristic for a solid-state diode.

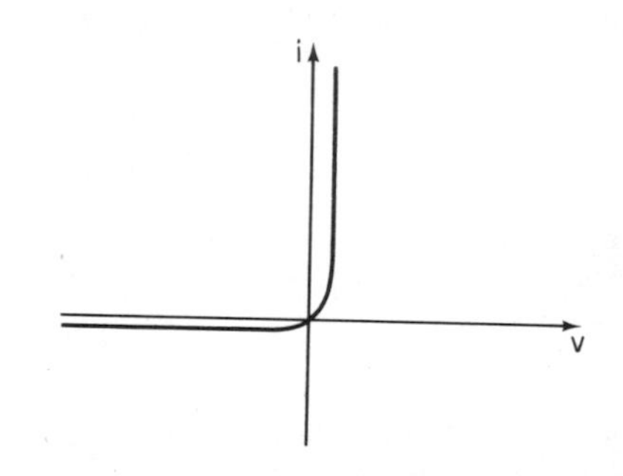

Fig. 2-1.4 Effect of increased scales on the diode characteristic shown in Fig. 2-1.3.

In many circuit applications it is desirable to consider only the most fundamental general properties of a device. This is especially true when relatively large signals are to be considered. As an example of this, the plot of the diode defined by Eqs. (2-3) and (2-4) and shown in Fig. 2-1.3 has been redrawn in Fig. 2-1.4 using much larger scales for the values of current and voltage. As even larger scales are chosen, it is clear that the plot of a diode may be modeled by a characteristic of the type shown in Fig. 2-1.5. A device with the characteristic shown in this figure is referred to as an *ideal diode*, i.e., an *idealized model* of a physical or real-life diode. The symbol and the defining variables for the ideal diode are shown in Fig. 2-1.6.

There are many other types of non-linear characteristics. As an example, consider the non-linear resistance curve shown on the v-i plane in Fig. 2-1.7. This is the static characteristic curve for a two-terminal *gas diode*. We note that for any given specified value of current i, there is a single unique value of the voltage v. Thus

[1]By "static" here we refer to the case where the voltage and current variables have constant values, i.e., they are not functions of time.

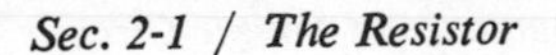

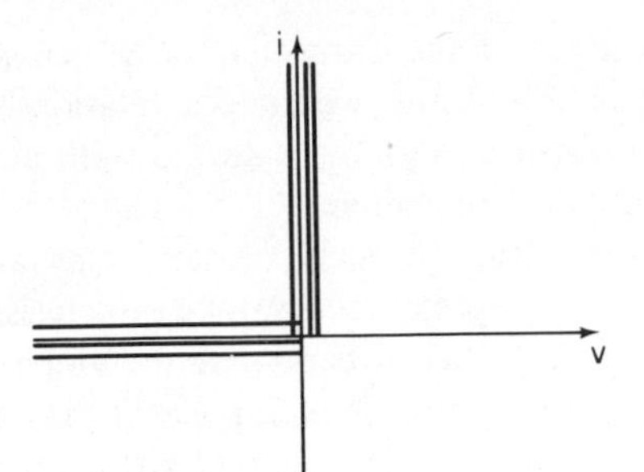

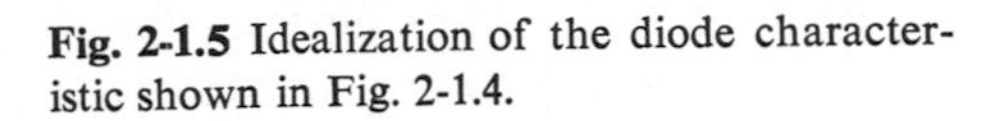
Fig. 2-1.5 Idealization of the diode characteristic shown in Fig. 2-1.4.

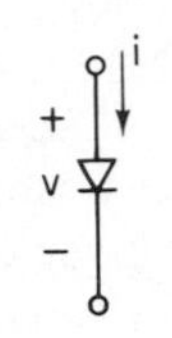

Fig. 2-1.6 Circuit symbol and reference directions for an ideal diode.

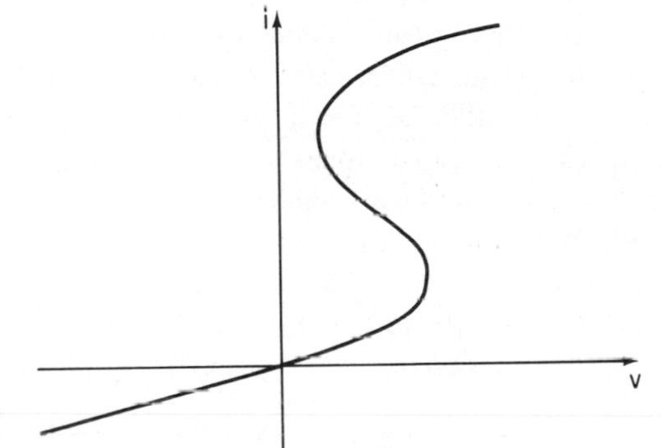

Fig. 2-1.7 Volt-ampere characteristic for a gas diode.

we may write that $v = f(i)$. For such a characteristic, however, it is not possible to define a unique relationship $i = h(v)$, since for some ranges of v, the device has a characteristic which is multivalued in the variable i. This type of a characteristic is referred to as a *current-controlled* characteristic, since, for a given value of current, a unique value of voltage exists, but the opposite is not true for all values of voltage.

A second type of non-linear, multivalued characteristic is shown in Fig. 2–1.8. This is a static characteristic curve for a two-terminal solid-state device called a *tunnel diode*. For such a characteristic, a specific value of current exists for every

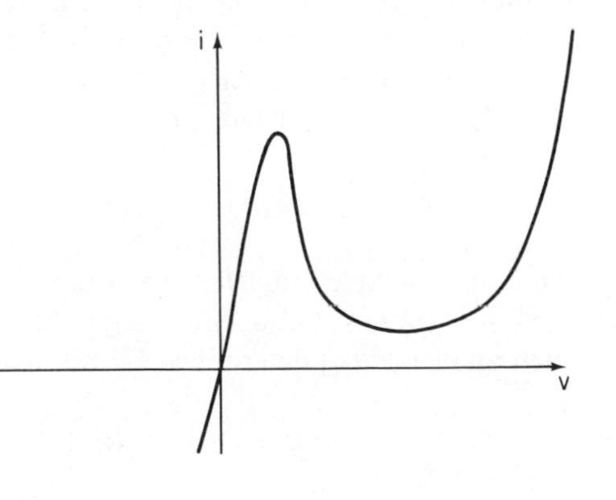

Fig. 2-1.8 Volt-ampere characteristic for a tunnel diode.

value of the voltage variable v, but there are ranges of the current i for which the characteristic is multivalued in the voltage variable v. Thus, we have a relationship of the type $i = h(v)$, but no relationship of the form $v = f(i)$. A device with such a characteristic is said to be a *voltage-controlled* non-linear device.

In addition to the possibility of being non-linear, physical resistors may also be *time-varying*. A *linear*, time-varying resistor is represented by a characteristic which is exactly the same as that shown in Fig. 2–1.2(a), except that the slope of the straight-line characteristic varies with time. In this case, R of Eq. (2–1) may be written as $R(t)$. Correspondingly, a *non-linear*, time-varying characteristic might appear at a given instant of time, as shown in Fig. 2–1.2(b), and be different for all other values of time. For many applications, the terminal behavior of physical resistors so closely approximates that of a linear or ideal resistor that little is gained by trying to include the non-linear and/or time-varying behavior. Indeed, much may be lost, since the mathematical treatment required to characterize the physical element may become so complicated that it completely conceals the overall properties of the circuit in which the element appears. To avoid this, we will assume that physical resistors are "modeled" by ideal elements obeying the relations given in Eqs. (2–1) and (2–2). Thus, from this point on, when we use the word resistor, we imply a *lumped, linear, time-invariant* model of some physical resistor, unless otherwise specified.

The instantaneous power $p(t)$ which is supplied to a resistor is easily determined from the relation $p(t) = v(t)i(t)$. Substituting for $v(t)$ from Eq. (2–1), we obtain

$$p(t) = i^2(t)R \tag{2–5}$$

Similarly, using (2–2), we see that an alternative expression for power is

$$p(t) = v^2(t)G \tag{2–6}$$

Since the expressions for power given in Eqs. (2–5) and (2–6) are always non-negative for positive values of R (and G), the energy supplied to a resistor must also always be non-negative. Thus, from the discussion given in Sec. 1–3, we conclude that a resistor must be a passive element. In future developments of this text we shall see that it is possible to realize two-terminal network elements which have the property that they may be modeled as negative-valued resistors (over certain ranges of their variables). Such an element has the terminal relation

$$v(t) = -|R|i(t) \tag{2–7}$$

where, for emphasis, we have indicated that the value of the resistor is less than zero by using magnitude symbols and a minus sign. The instantaneous power which is supplied to such an element is determined by the relation

$$p(t) = -|R|i^2(t) \tag{2–8}$$

Thus we see that the power being supplied to such a negative-valued resistor is negative. From the conventions established for the reference polarity of power in Sec. 1–2, we may say that such a resistor *supplies* energy to the circuit to which it is connected. Thus, it may be considered to be an active element. We shall discuss the means by which negative-valued resistors may be realized in later portions of this text.

The passive resistor is sometimes referred to as a *dissipative* element in that the energy $w(t)$ (where $w = \int p\,dt$) which is absorbed by it is turned into heat, and cannot be reclaimed. We shall see that two other types of basic network elements, namely, the inductor and the capacitor, have a quite different property in that they can store energy, which can then be used to excite a network at some future time.

Since real-world resistors are called upon to dissipate energy, they must be rated not only with respect to their resistance, i.e., their nominal value in ohms, but also with respect to the maximum amount of power that they can dissipate without reaching a temperature which is so high that their characteristics may drastically change, or that they may even "burn out." For radio, television, and communication applications, wattage ratings are usually in fractions of a watt, such as $\frac{1}{4}$ W, $\frac{1}{2}$ W, etc. For industrial control and power applications, the rating of a resistor may be many watts. Obviously, the power rating greatly influences the size and physical construction of a given resistor.

The values of resistors used in most practical circuit applications range from a few ohms to many megohms (MΩ). The practical difficulties of making numerical calculations involving large numbers, however, makes it undesirable to use resistors with large values in our examples. Thus, in general, we shall study circuits which contain resistors restricted to values of a few ohms. Such a restriction still gives us complete generality in the analysis of circuits, however, since we shall see in Sec. 8–8 that a process called "normalization" may be applied to transform such circuits to ones having resistors of any desired range of values. Thus, the use of small numbers gives us the advantage of numerical simplicity without sacrificing practicality.

2-2 SOURCES

In this section we shall introduce a basic network element which may be used to provide inputs to the resistance network. It is called a *source*. The type of source which we shall consider here is referred to as an *independent* source, since its output characteristics are *not dependent* on any other network variable such as a current or a voltage. Their characteristics, however, may be time-varying. There are two types of independent sources. The first of these is called a *voltage* source. It is a two-terminal element which has the property that a specified voltage appears across its terminals, and that the value of this voltage at any instant of time is *independent of the value or direction of the current that flows through it*. Since there is no relation between the voltage and the current, it is not necessary for us to observe any relative reference directions between these variables. A circuit symbol for a voltage source

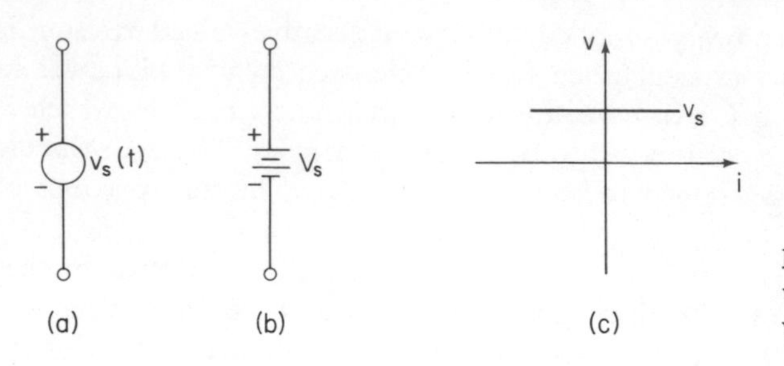

Fig. 2-2.1 Circuit symbols and volt-ampere characteristic for a voltage source.

is shown in Fig. 2–2.1(a). It should be noted that the terminal voltage $v_s(t)$ may be a constant, or it may be some specified function of time. Thus, we may refer to a *constant* (dc) source or a *time-varying* source. The constant-valued (dc) voltage source is sometimes referred to as a model for an ideal battery, which is indicated by the circuit symbol shown in Fig. 2–2.1(b). We may illustrate the properties of the voltage source more fully by drawing its characteristic on the *i-v* plane. As shown in Fig. 2–2.1(c), this is simply a horizontal line. If the source is time-varying, the position of the horizontal-line characteristic will vary for different values of time. When the terminal voltage is zero, the characteristic coincides with the abscissa. In this case it is exactly the same as the characteristic of a resistor which has zero resistance, i.e., a *short circuit*. Thus, we may say that the *internal resistance* of a voltage source, i.e., the resistance measured between its terminals when its output voltage is zero, is zero. This, of course, is implied in the statement that the terminal voltage of such a source is independent of the current flowing through it.

A second kind of independent source is the *current* source. A current source is a two-terminal element which has the property that a specified current flows through it, and that the value and direction of this current at any instant of time is *independent of the value or direction of the voltage that appears across the terminals of the source*. Just as in the case of the voltage source, there is no required relation between the reference polarities of voltage and current for such an element. A circuit symbol for the current source is shown in Fig. 2–2.2(a). The output current $i_s(t)$ may be constant, or it may be a function of time, thus we may refer to a

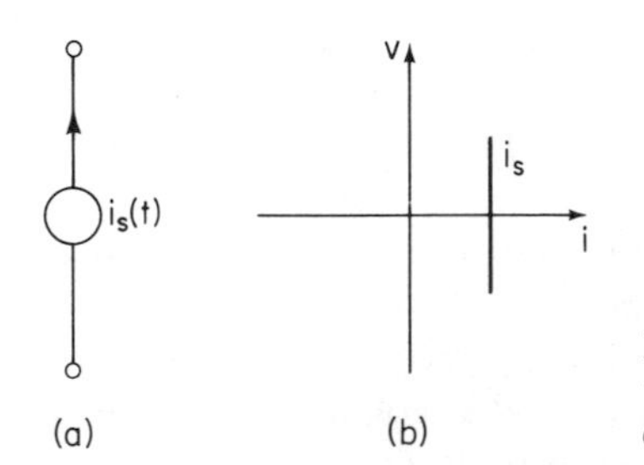

Fig. 2-2.2 Circuit symbol and volt-ampere characteristic for a current source.

constant (dc) or a *time-varying* current source. On the *i-v* plane, the characteristics of the current source are indicated by a vertical line as shown in Fig. 2–2.2(b).

For the case of a time-varying source, the position of this vertical line will vary with time. When the source current is zero, then the characteristic coincides with the ordinate. In this case, the source has properties that are identical with those of a resistor of infinite resistance, i.e., an *open circuit*. Thus we may say that the internal resistance of a current source, i.e., the resistance measured between its terminals when its output current is zero, is infinite. This, of course, is implied in the statement that the current flowing through such a source is independent of the value of the voltage which is impressed across it.

Sources are generally considered to be active elements. For example, consider the connection of a constant voltage source and a resistor as shown in Fig. 2–2.3.

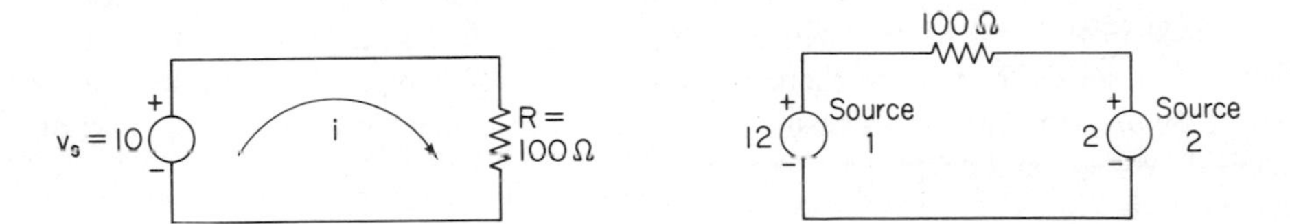

Fig. 2-2.3 A simple circuit with a voltage source. **Fig. 2-2.4** A circuit with two voltage sources.

For the indicated variables, application of KVL around the loop tells us that $v_s = iR$. Thus, for a 10-V dc source, 0.1 A flows through the 100-Ω resistor. The power dissipated in the resistor is found from the expression $p = i^2R$. It is equal to 1 W. In calculating the power for the two-terminal voltage source, however, we see that, following the associated reference polarity conventions for voltage and current specified in the definition of power in Sec. 1–2, the power for the source is given by the expression $p = -iv_s$, and thus the source power is -1 W. From our definition, power *into* a two-terminal element is positive; therefore we see that 1 W of power flows *out of* the source in this case. Note that it is possible for a source to receive power from the circuit. For example, for the network shown in Fig. 2–2.4, application of KVL shows that power associated with source 1 is -1.2 W, and that the power associated with source 2 is 0.2 W. Thus, source 2 is, in effect, being "charged," i.e., it is *receiving* energy from source 1. In such a case, unlike a resistor, the source does not dissipate this energy. Instead it converts it into some other form of energy. This might be chemical energy (as in the case of a battery), or mechanical energy (as in the case of a motor-driven generator), etc.

Similar conclusions hold true for the case of a current source. For example, let us analyze the network shown in Fig. 2–2.5 using KCL. We see that $i_s = vG$. For the 0.1-A source, a voltage of 10 V appears across the 0.01-mho resistor. The power dissipated in the resistor is found from the expression $p = v^2G$, and is equal to 1 W. It is readily shown that the power for the current source is -1 W, thus the current source supplies power to the circuit.

In general we will refer to voltage and current sources as *active* elements, since they are capable of supplying energy to a network. We should remember, however, that in a given situation, such sources may actually receive energy from the

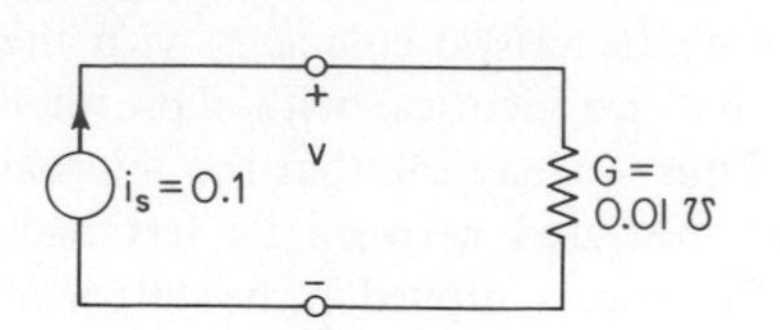

Fig. 2-2.5 A simple circuit with a current source.

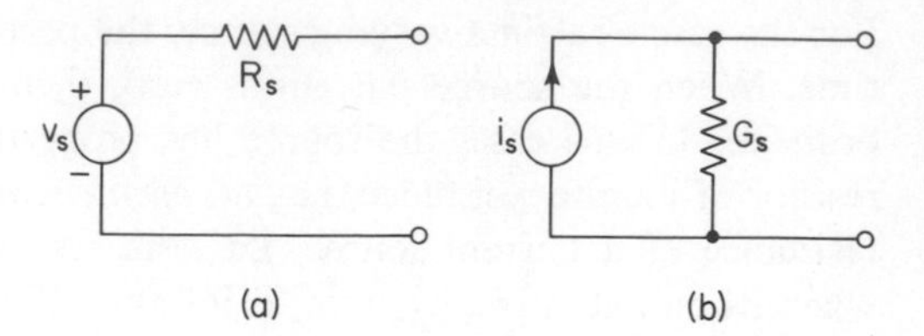

Fig. 2-3.1 Non-ideal voltage and current sources.

remainder of the circuit. Since sources are not defined by a relation between two variables, but simply by a specification of a single variable, the question of linear vs. non-linear behavior does not arise. Specifically, a test for superposition as outlined in Sec. 1–3 is not defined, since there is only one variable associated with the source. Thus, we may say that the classification of linear vs. non-linear behavior does not apply to voltage and current sources.

2-3 NON-IDEAL SOURCES

The voltage and current sources which were discussed in the preceding section should actually be considered as *ideal* sources in the sense that they have no dissipation associated with them. We will now define *non-ideal* voltage and current sources as ones which include some internal dissipation. The non-ideal voltage source may be shown schematically by a resistor in series with an ideal voltage source, as shown in Fig. 2–3.1(a), and the non-ideal current source by a resistor in parallel (or shunt) with an ideal current source, as shown schematically in Fig. 2–3.1(b).[2] Non-ideal sources come considerably closer to modeling the properties of physical energy-supplying devices than do the ideal sources described in the last section. For example, an automobile battery may be modeled (approximately) as an ideal voltage source of about 12 V, in series with a resistor of about 0.05 Ω. Thus, the non-ideal voltage source representation shown in Fig. 2–3.1(a) is appropriate for use in modeling such a physical device. Similarly, a sinusoidal signal generator as used in testing electronic circuits might be modeled as an ideal voltage source which has a terminal voltage $v(t) = V_0 \sin(\omega_0 t + \phi)$ where V_0 is the magnitude of the sinusoidal output (this may usually be varied by an "amplitude" or "gain" control), and a series resistance (if the resistor value is 600 Ω, the signal generator is said to have an internal resistance of 600 Ω).

[2]It should be noted that a non-ideal source consisting of an ideal voltage source with a resistor in *parallel* with it is indistinguishable from an ideal voltage source alone, since the specified voltage will always be present at the terminals and it will be independent of the current (the voltage source, however, will have to supply the extra current required by the resistor). Similarly, a non-ideal source consisting of an ideal current source in *series* with a resistor is indistinguishable from an ideal current source by itself, since it will still supply the specified current to any externally connected circuit independent of any voltage that appears across it (there will, however, be a voltage drop across the resistor).

Actually, the two non-ideal source models shown in Fig. 2–3.1 may be used interchangeably in network studies. To see this, consider the effect of connecting each of them to a pair of terminals of some arbitrary network. Let the network variables be $v_0(t)$ and $i_0(t)$ as shown in Figs. 2–3.2 and 2–3.3. For the non-ideal voltage

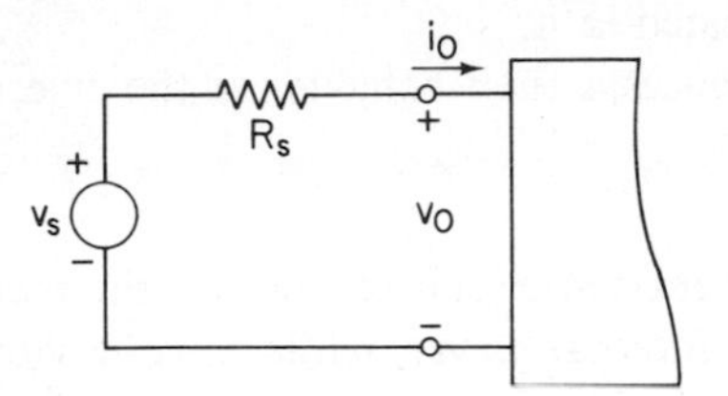

Fig. 2-3.2 A non-ideal voltage source connected to a network.

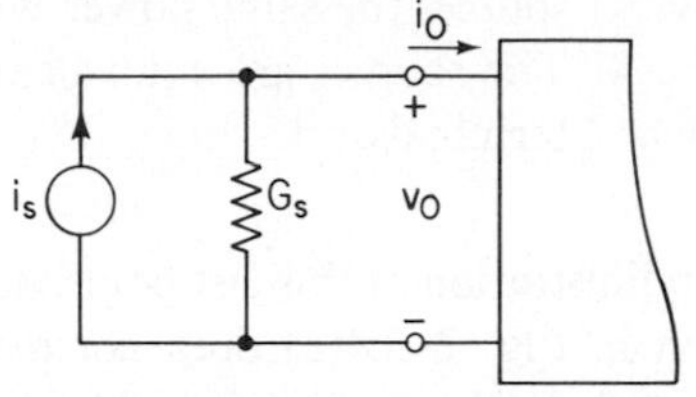

Fig. 2-3.3 A non-ideal current source connected to a network.

source, we may apply KVL to obtain the relation

$$v_0(t) = v_s(t) - i_0(t)R_s \tag{2–9}$$

For the non-ideal current source, we may apply KCL to obtain

$$v_0(t) = \frac{i_s(t) - i_0(t)}{G_s} \tag{2–10}$$

If we compare the two relations given above, we see that they will be the same for all values of $v_0(t)$ and $i_0(t)$, providing that

$$v_s(t) = \frac{i_s(t)}{G_s} \quad \text{and} \quad G_s = \frac{1}{R_s} \tag{2–11}$$

Thus, *at a given pair of terminals*, either of the non-ideal sources shown in Fig. 2–3.1 may be replaced by the other without affecting the behavior of the rest of the network. As an example of such an equivalence, the automobile battery referred to above may also be represented by an ideal current source of $i_s = v_sG_s = (12)(20) = 240$ A in parallel with a conductance of $G_s = 1/R_s = 1/0.05 = 20$ mhos. The two equivalent models are shown in Fig. 2–3.4. To emphasize this equivalence note the following:

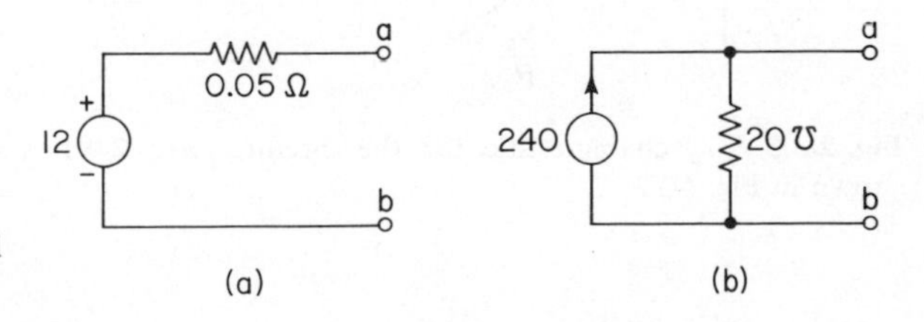

Fig. 2-3.4 Examples of non-ideal sources.

1. The open-circuit voltage that appears at the terminals *a-b* for the two sources is the same (12 V).
2. The short-circuit current that results when the terminals *a-b* of either source are shorted together is the same (240 A).
3. If an arbitrary resistor is connected across the output terminals of either source, the same power will be dissipated in it.
4. The sources are equivalent only as concerns their behavior at the external terminals.

As an illustration of this last point, note that under open-circuit conditions, the model shown in Fig. 2–3.4(a) does not dissipate any internal power, while the one shown in Fig. 2–3.4(b) dissipates 2880 W! Under short-circuit conditions the situation is reversed. Thus, even though the two models have identical properties in terms of their behavior at their external terminals, their internal behavior is considerably different.

Let us consider that a source has a constant open-circuit voltage of $v_s(t) = V_s$ V and an internal resistance of R_s Ω. If we connect such a source to a pair of terminals of some arbitrary network, and define the variables at the pair of terminals as $v_0(t)$ and $i_0(t)$ as shown in Fig. 2–3.2, then, applying KVL to the loop formed by the non-ideal source and the pair of terminals, we see that

$$v_0(t) = i_0(t)(-R_s) + V_s \tag{2-12}$$

This equation defines a straight line on an i_0-v_0 plane, i.e., a plane formed by plotting $v_0(t)$ along the ordinate and $i_0(t)$ along the abscissa. The plot is shown in Fig. 2–3.5. This plot may be used to determine the voltage appearing at the terminals of the non-ideal source for any given value of output current $i_0(t)$. We shall see an example of such an application when we disuss the "biasing" of a non-linear resistive element later in this chapter.

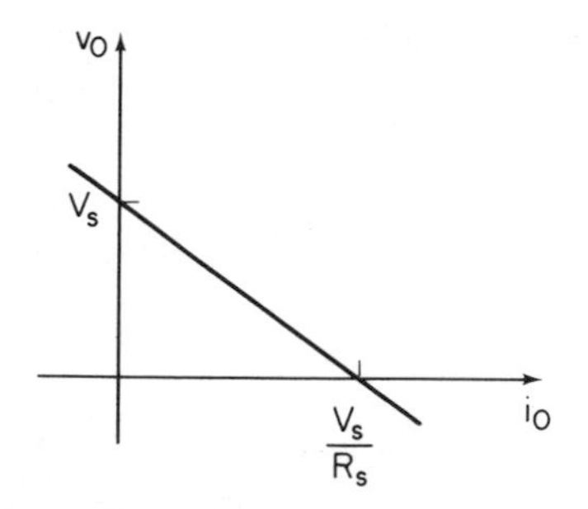

Fig. 2-3.5 I_0-v_0 characteristic for the circuit shown in Fig. 2-3.2.

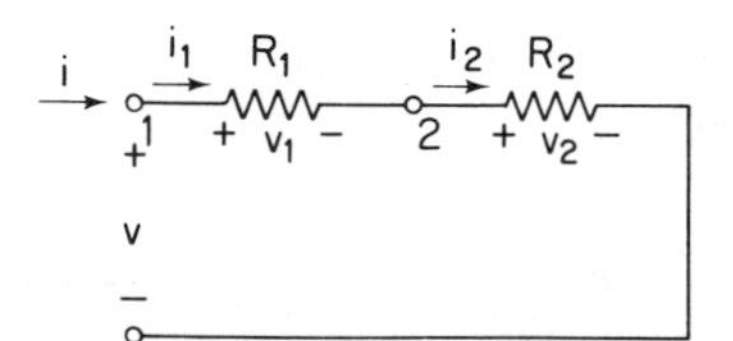

Fig. 2-4.1 A series connection of two resistors.

2-4 SERIES AND PARALLEL CONNECTIONS OF RESISTORS

In this section we shall discuss some simple circuits formed by the interconnection of resistors. There are two basic connections to be considered, namely, series connections and parallel connections. In Fig. 2-4.1 we consider the series connection of two resistors R_1 and R_2. The two resistive branches form an overall two-terminal branch with variables $v(t)$ and $i(t)$ as shown. If we apply KCL at node 1, we see that the branch current variables $i(t)$ and $i_1(t)$ are equal. Similarly, at node 2 we see that $i_1(t)$ equals $i_2(t)$. Thus, we may write

$$i(t) = i_1(t) = i_2(t) \tag{2-13}$$

Now let us apply KVL to the "loop" formed by the voltages $v(t)$, $v_1(t)$, and $v_2(t)$. We obtain

$$v(t) = v_1(t) + v_2(t) \tag{2-14}$$

There are two other relations that may be written for the elements of the circuit shown in Fig. 2-4.1. These are the defining relations for the voltage and current variables for each of the resistive branches as given by Ohm's law. These have the form

$$v_1(t) = R_1 i_1(t) \tag{2-15a}$$
$$v_2(t) = R_2 i_2(t) \tag{2-15b}$$

If we combine the relations of (2-13), (2-14), and (2-15), we obtain

$$v(t) = (R_1 + R_2)i(t) \tag{2-16}$$

We conclude that the overall two-terminal network can be considered as being comprised of an equivalent resistance R_{eq}, which relates the terminal variables $v(t)$ and $i(t)$ according to the relation

$$v(t) = R_{eq}i(t) \tag{2-17}$$

where R_{eq} is simply defined by the relation

$$R_{eq} = R_1 + R_2 \tag{2-18}$$

The above discussion is easily generalized to the case where n resistors R_i ($i = 1, 2, \ldots, n$) are connected in series. Such a connection is shown in Fig. 2-4.2. Applying KCL at all the nodes, we conclude that

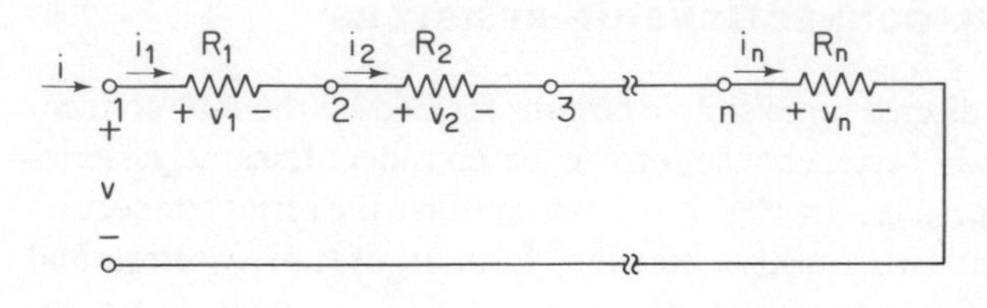

Fig. 2-4.2 A series connection of n resistors.

$$i(t) = i_1(t) = i_2(t) = \cdots = i_n(t) \tag{2–19}$$

Now let us apply KVL to a "loop" formed by $v(t)$ and by the voltages $v_i(t)$ ($i = 1, 2, \ldots, n$). We obtain the relation

$$v(t) = \sum_{i=1}^{n} v_i(t) \tag{2–20}$$

Each resistor shown in the figure may be thought of as comprising a two-terminal element or branch of the network. In terms of the indicated branch variables (which follow the convention for two-terminal elements introduced in Sec. 1–2), we see that we have n equations which may be written in the form

$$v_i(t) = R_i i_i(t) \qquad i = 1, 2, \ldots, n \tag{2–21}$$

From the relations (2–19), (2–20), and (2–21), we see that we may write

$$v(t) = \sum_{i=1}^{n} R_i i_i(t) = i(t) \sum_{i=1}^{n} R_i \tag{2–22}$$

We conclude that if we define an equivalent resistor R_{eq}, where

$$R_{eq} = \sum_{i=1}^{n} R_i \tag{2–23}$$

then the variables $v(t)$ and $i(t)$ are related by the equation

$$v(t) = R_{eq} i(t) \tag{2–24}$$

Thus, we may say that a series connection of resistors R_i may be replaced by a single equivalent resistor R_{eq} with no change in the terminal variables of the overall two-terminal network.

A similar analysis holds for the case where two resistors are connected in parallel as shown in Fig. 2–4.3. Such a connection may be treated as a single overall two-terminal element. If we apply KVL around the "loop" consisting of the voltage $v(t)$ on the one side and the branch voltage $v_1(t)$ on the other side, we see that $v(t)$ equals $v_1(t)$. Similarly, we see that $v_1(t)$ equals $v_2(t)$. Thus, we may write

$$v(t) = v_1(t) = v_2(t) \tag{2–25}$$

Now let us sum the currents at the upper node of the network. Applying KCL, we see that

$$i(t) = i_1(t) + i_2(t) \tag{2-26}$$

The defining relations for the two resistive branches are conveniently written using units of conductance rather than resistance; thus, we obtain

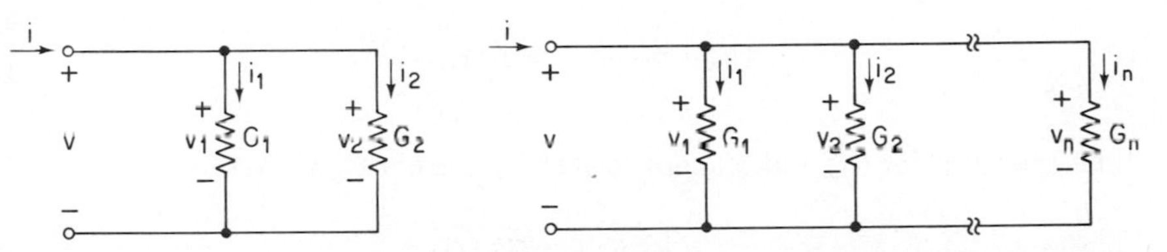

Fig. 2-4.3 A parallel connection of two resistors. **Fig. 2-4.4** A parallel connection of n resistors.

$$\begin{aligned} i_1(t) &= G_1 v_1(t) \\ i_2(t) &= G_2 v_2(t) \end{aligned} \tag{2-27}$$

If we combine the relations (2–25), (2–26), and (2–27), we obtain

$$i(t) = (G_1 + G_2)v(t) \tag{2-28}$$

We conclude that the overall two-terminal network shown in Fig. 2–4.3 can be considered as being comprised of a single equivalent resistance with a value of G_{eq} mhos, where

$$G_{eq} = G_1 + G_2 \tag{2-29}$$

The above discussion is easily generalized to the case where n resistors of value G_i $(i = 1, 2, \ldots, n)$ are connected in parallel as shown in Fig. 2–4.4. If we apply KVL around the first loop indicated in the network, we see that $v(t)$ equals $v_1(t)$. Similarly, applying KVL to all the other loops, we see that

$$v(t) = v_1(t) = v_2(t) = \cdots = v_n(t) \tag{2-30}$$

Now let us apply KCL to the upper node. We obtain

$$i(t) = \sum_{i=1}^{n} i_i(t) \tag{2-31}$$

The branch relations for the resistive elements yield a set of equations which have the form

$$i_i(t) = G_i v_i(t) \qquad i = 1, 2, \ldots, n \tag{2-32}$$

From the above three sets of relations we see that we may write

$$i(t) = \sum_{i=1}^{n} G_i v_i(t) = v(t) \sum_{i=1}^{n} G_i \tag{2-33}$$

We conclude that if we define a constant G_{eq}, where

$$G_{eq} = \sum_{i=1}^{n} G_i \tag{2-34}$$

then the variables $i(t)$ and $v(t)$ are related by the equation

$$i(t) = G_{eq} v(t) \tag{2-35}$$

Thus we may say that a parallel connection of a set of resistors G_i may be replaced by a single equivalent resistor G_{eq} with no change in the overall network variables.

Sometimes it is desirable to deal with resistance values in ohms while computing the equivalent resistance of a set of paralleled resistors. In such a case, Eq. (2–34) may be written as

$$\frac{1}{R_{eq}} = \sum_{i=1}^{n} \frac{1}{R_i} \tag{2-36}$$

For the frequently encountered case of two resistors R_1 and R_2 in parallel, Eq. (2–36) becomes

$$R_{eq} = \frac{R_1 R_2}{R_1 + R_2} \tag{2-37}$$

It should be noted that the rules given above for the replacing of a given set of series or parallel resistors by a single equivalent resistor apply to time-varying resistors as well as to time-invariant ones. For time-varying resistors, Eq. (2–23) may be written in the form

$$R_{eq}(t) = \sum_{i=1}^{n} R_i(t) \tag{2-38}$$

In the above paragraphs we have concluded that if resistors are connected in series, their resistances add, while if resistors are connected in parallel, their conductances add. Thus, in some analyses it is convenient to treat resistance values in ohms, while in other analyses the computations may be simplified by utilizing values in mhos. This is only the first of many illustrations that we shall make showing how calculations may be simplified by having available reciprocal units to describe a given network element. We shall see more examples of the convenience of such different units in the later chapters of this text. Here, however, we insert a word of caution, in that to take advantage of this flexibility, network elements must be labeled not only with their numerical values, but also with an indication such as the word "ohms" or the word "mhos" to indicate which type of units are being used. The symbol Ω is also used for ohms and the symbol ℧ for mhos. The only case where such a distinction will not be required is for the case of unity-valued elements.

Networks which consist of various configurations of resistors in series and parallel may sometimes be reduced to a single equivalent resistance with respect to a specific pair of nodes by successive application of the series and parallel rules given above. For example, consider the network shown in Fig. 2-4.5(a). Working from

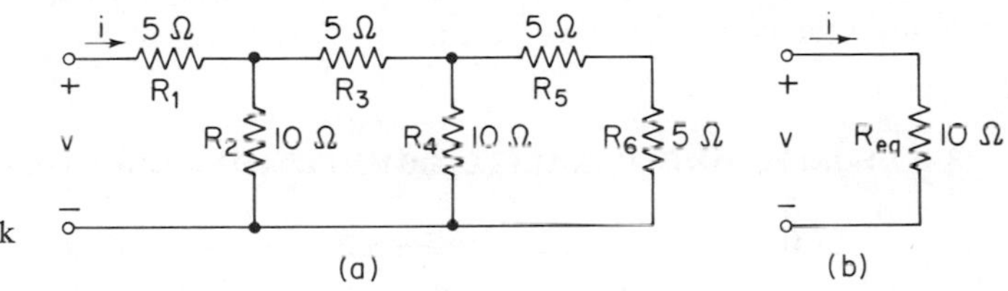

Fig. 2-4.5 A ladder network of resistors.

right to left we see that the series resistance of R_5 and R_6 is 10 Ω. Similarly, the equivalent resistance of this 10 Ω in parallel with R_4 is 5 Ω. Continuing, the equivalent resistance of R_3 in series with the 5 Ω is 10 Ω, and the parallel combination of the 10 Ω with R_2 gives an equivalent 5 Ω. Finally, the series combination of R_1 and the 5 Ω gives 10 Ω. We conclude that an equivalent single resistance of 10 Ω may be placed between nodes *a-b* as shown in Fig. 2-4.5(b), and that this single resistance will have the same effect as the original resistance network in relating the variables $v(t)$ and $i(t)$ defined for this node-pair.

The type of network illustrated in Fig. 2-4.5(a) is called a *ladder* network. The process of finding the equivalent resistance of a ladder network described above is easily generalized. Consider the network shown in Fig. 2-4.6. Starting with the resistor R_1, and proceeding through the network from left to right, we see that we may write an expression for the input resistance R_{in} in the form

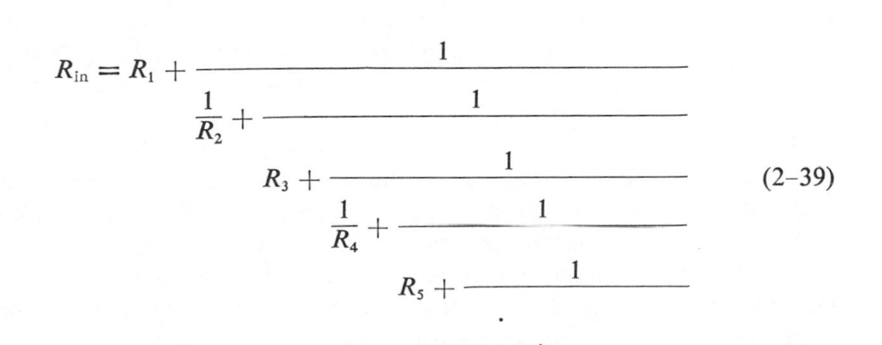

$$R_{\text{in}} = R_1 + \cfrac{1}{\cfrac{1}{R_2} + \cfrac{1}{R_3 + \cfrac{1}{\cfrac{1}{R_4} + \cfrac{1}{R_5 + \cfrac{1}{\ddots}}}}} \qquad (2\text{–}39)$$

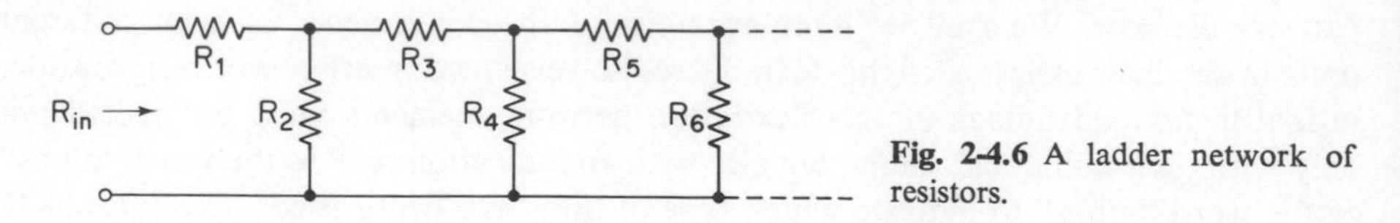

Fig. 2-4.6 A ladder network of resistors.

Such an expression is called a *continued fraction.* We shall see other examples of continued fractions in later portions of this text.

The procedure described above may always be applied to find the equivalent resistance of a ladder network. This equivalent resistance determines the ratio between the voltage and current variables at the indicated nodes; as such it is sometimes referred to as the *input resistance* of the network (as seen at the specified nodes). In a following section, we shall investigate more general techniques for determining the input resistance of a network at a pair of nodes.

2-5 SERIES AND PARALLEL CONNECTIONS OF SOURCES

In the last section we considered interconnections of resistors. In this section we will consider the effects of interconnecting various kinds of sources. In Fig. 2–5.1

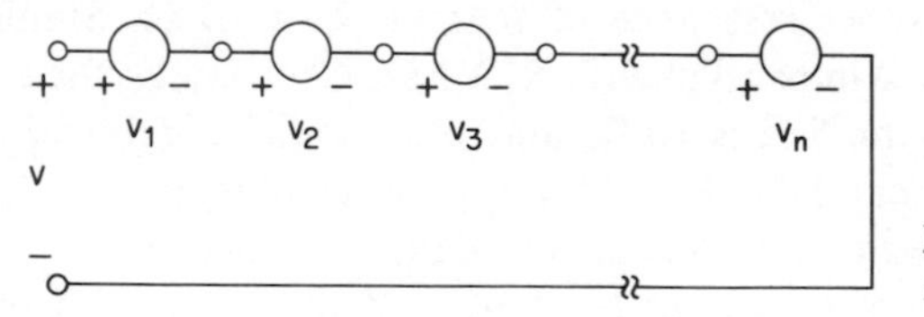

Fig. 2-5.1 A series connection of n ideal voltage sources.

we show a series connection of n voltage sources. The overall voltage $v(t)$ may be found in terms of the voltages of the individual sources from KVL. Thus, we obtain

$$v(t) = \sum_{i=1}^{n} v_i(t) \tag{2–40}$$

We conclude that for such a series connection the overall voltage is the sum of the voltages from the individual sources. The voltage resulting from a parallel connection of voltage sources is not defined, except for the case where all the sources have exactly the same voltage for all values of time. To see why this is so, consider what would happen if a 1-V source and a 2-V source were connected in parallel. Applying KVL around the loop formed by the two sources, there would be a net voltage of 1 V. However, the total resistance of the loop, i.e., the sum of the internal resistances of the two sources, is zero. Thus, the loop current that would flow is $i = v/R = 1/0 = \infty$, which is clearly not possible.

Now let us consider the current source. A parallel connection of n current sources is shown in Fig. 2–5.2. The overall current may be found in terms of the currents of the individual sources from KCL. Thus, we obtain

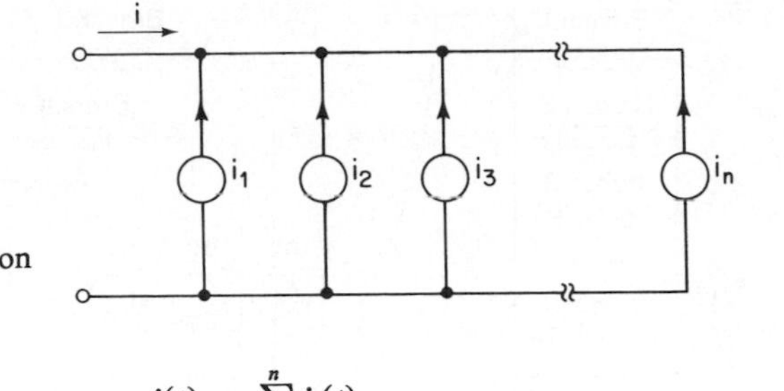

Fig. 2-5.2 A parallel connection of n ideal current sources.

$$i(t) = \sum_{i=1}^{n} i_i(t) \tag{2-41}$$

We conclude that the overall current is the sum of the currents from the individual sources. The current resulting from a series connection of current sources is not defined except for the case where all the sources have exactly the same output current for all values of time, since otherwise such a connection would violate KCL at any node connecting two sources.

In discussing both of the above connections, namely, the series connection of voltage sources and the parallel connection of current sources, it has been assumed that the reference directions of all the individual sources are the same. The case where some of the reference directions are opposite is easily accommodated by subtracting the values of any sources whose reference directions do not agree with the reference directions of the overall variables.

In the above paragraphs we have discussed the replacement of several sources by a single source. Frequently in our studies of network analysis it will be convenient to follow the opposite procedure, namely to replace a single source with an equivalent connection of more than one source. For the case of a voltage source, such an equivalence is shown in Figs. 2-5.3(a) and (b). From these figures we see that we may replace a single voltage source with output voltage $v(t)$ as shown in Fig. 2-5.3(a) with a parallel connection of an arbitrary number of voltage sources which have the same output voltage $v(t)$ as shown in Fig. 2-5.3(b). If some branches of a network are connected to the single source in the manner shown in Fig. 2-5.4(a), then an equivalent connection is that shown in Fig. 2-5.4(b). This is readily verified by application of Kirchhoff's laws, since the potential difference between nodes a and b in Fig. 2-5.4(b) must be zero; therefore, if a short circuit is placed between these two points, no current will flow through it. Thus, it is immaterial to the operation

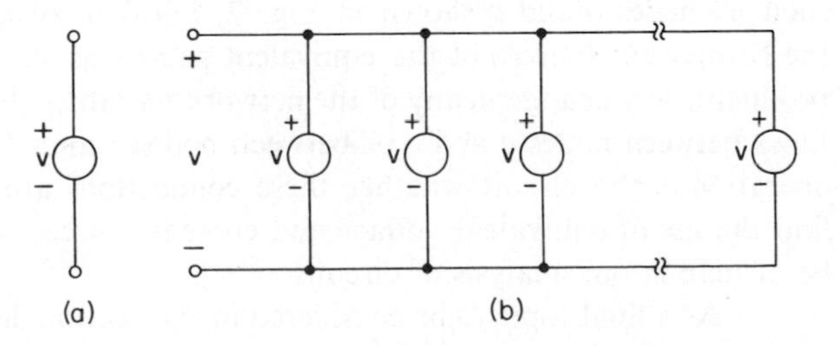

Fig. 2-5.3 A parallel connection of ideal voltage sources.

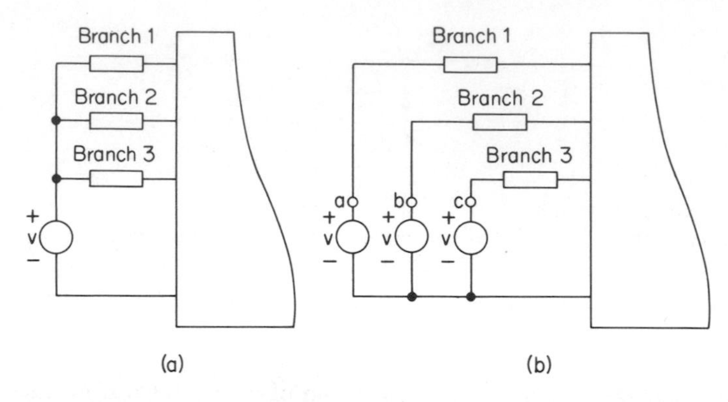

Fig. 2-5.4 Replacing a single ideal voltage source with several ideal voltage sources.

of the circuit whether such a conductor is present or absent. The same conclusion holds true for nodes *b* and *c*.

A similar equivalence holds true for the case of current sources. The single current source shown in Fig. 2–5.5(a) is equivalent to the series connection of current sources which have the same output current as shown in Fig. 2–5.5(b). At each in-

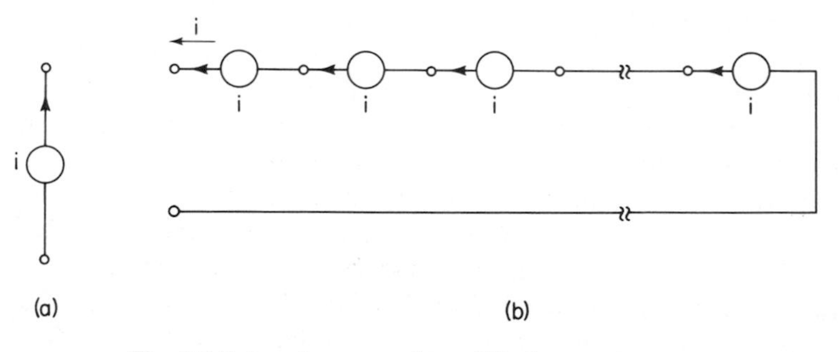

Fig. 2-5.5 A series connection of ideal current sources.

ternal node of the equivalent source, KCL is completely satisfied by the source currents. Since the voltage between the sources is not related to the currents, any arbitrary network nodes which are not originally connected to the current sources such as nodes *a* and *b* shown in Fig. 2–5.6(a), may be connected between two of the component sources of the equivalent source as shown in Fig. 2–5.6(b) without producing any change in any of the network variables. Note especially that no current flows between nodes *a* and *c* or between nodes *b* and *d*. It is thus immaterial to the operation of the circuit whether these connections are present or absent. We shall find the use of equivalent voltage and current sources of the type described above to be of help in our analysis of circuits.

As a final topic to be considered in this section, let us investigate what happens when a voltage source and a current source are connected in parallel as shown in Fig.

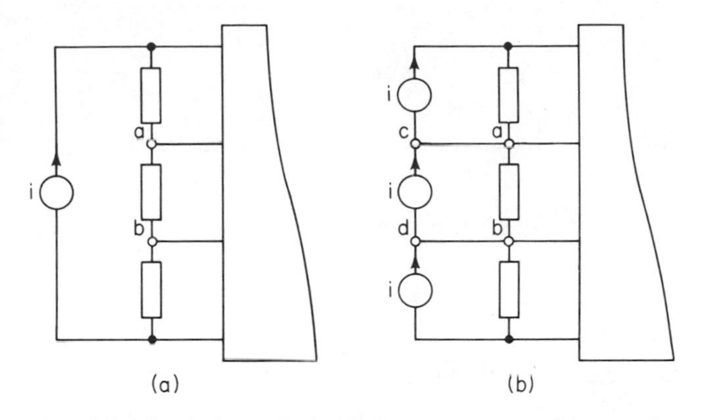

Fig. 2-5.6 Replacing a single ideal current source with several ideal current sources.

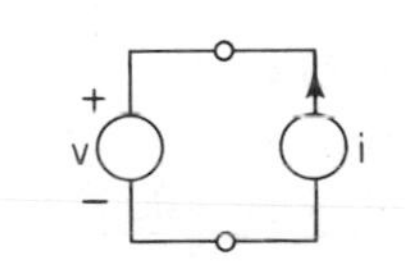

Fig. 2-5.7 A voltage source and a current source in parallel.

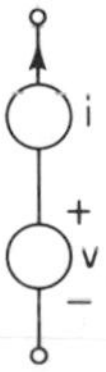

Fig. 2-5.8 A voltage source and a current source in series.

2–5.7. In such a configuration, since the internal resistance of the voltage source is zero, the output current from the current source flows through the voltage source without any effect on its output voltage. Thus the combination has exactly the same properties as a voltage source alone. Similarly, we may examine the behavior of a series-connected voltage source and current source as shown in Fig. 2–5.8. In this case, since the voltage across the current source is undefined, the addition of the output voltage of the voltage source still leaves a quantity which is undefined, and the combination has the same properties as a current source alone.

2-6 SOME SIMPLE CIRCUITS

In the preceding sections of this chapter, we have discussed the properties of various combinations of resistors, and various combinations of sources. Now let us investigate some simple circuits in which both resistors *and* sources are present. The first of these is shown in Fig. 2–6.1(a). It may be referred to as either a *single-loop* circuit or as a *single node-pair* circuit, since topologically speaking it consists both of a pair of nodes and also of a single loop. Application of KVL to this circuit yields the result that

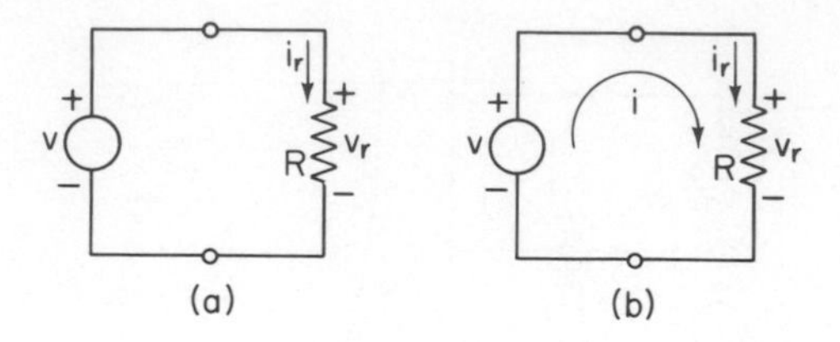

Fig. 2-6.1 A simple circuit with a voltage source.

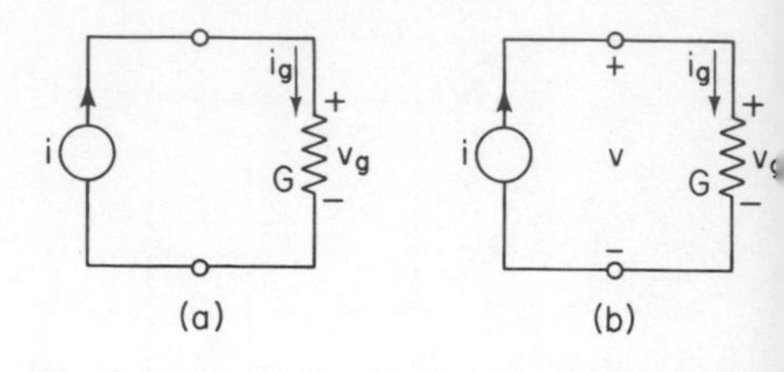

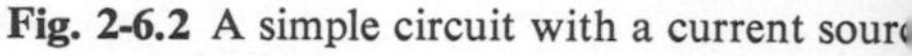

Fig. 2-6.2 A simple circuit with a current source

$$v(t) = v_r(t) \tag{2-42}$$

The resistor has the branch variable relationship

$$v_r(t) = Ri_r(t) \tag{2-43}$$

Thus, we may write

$$v(t) = Ri_r(t) \tag{2-44}$$

Now let us define a loop current $i(t)$ as shown in Fig. 2–6.1(b). It is easily seen that $i_r(t)$ is equal to the loop current $i(t)$. The relation of (2–44) now becomes

$$v(t) = Ri(t) \tag{2-45}$$

where $v(t)$ is the source voltage and $i(t)$ is the loop current. Note that we have eliminated the branch variables $v_r(t)$ and $i_r(t)$, using instead the source voltage $v(t)$ and the loop current $i(t)$. This is a simple example of a method of writing network equations that may be called *mesh* or *loop analysis.* It will be developed more fully in the following chapter.

As a second example of a simple circuit, consider the one-loop or single node-pair circuit consisting of a current source and a resistor as shown in Fig. 2–6.2(a). Application of KCL to this circuit yields the result that

$$i(t) = i_g(t) \tag{2-46}$$

Since the resistor has the branch relationship

$$i_g(t) = Gv_g(t) \tag{2-47}$$

we may write

$$i(t) = Gv_g(t) \tag{2-48}$$

Now let us define the node voltage $v(t)$ as shown in Fig. 2–6.2(b).[3] By KVL, $v_g(t)$ is equal to the node voltage $v(t)$. Thus, the relation of (2–48) becomes

$$i(t) = Gv(t) \tag{2–49}$$

where $i(t)$ is the source current and $v(t)$ is the node voltage. Note that just as in the previous example, the branch variables $v_g(t)$ and $i_g(t)$ have been eliminated to develop an equation in terms of some overall circuit variables, namely the source current $i(t)$ and the node voltage $v(t)$. This is a simple example of a method of writing network equations that we shall call *node analysis.* Although both of the examples discussed above may appear to have produced a quite similar result, we shall see in the next chapter that they are representative of two quite different approaches to the analysis of networks.

Another example of a simple resistive network is the *voltage divider.* Such a network is shown in Fig. 2–6.3. The current $i(t)$ that flows from the source $v_1(t)$ is determined by the relation

$$i(t) = \frac{v_1(t)}{R_1 + R_2} \tag{2–50}$$

The voltage $v_2(t)$ (sometimes referred to as the *output* of the voltage divider) is determined by the expression

$$v_2(t) = i(t)R_2 \tag{2–51}$$

We may combine the two expressions given above to determine the relationship between the input voltage $v_1(t)$ and the output voltage $v_2(t)$. We obtain

$$\frac{v_2(t)}{v_1(t)} = \frac{R_2}{R_1 + R_2} \tag{2–52}$$

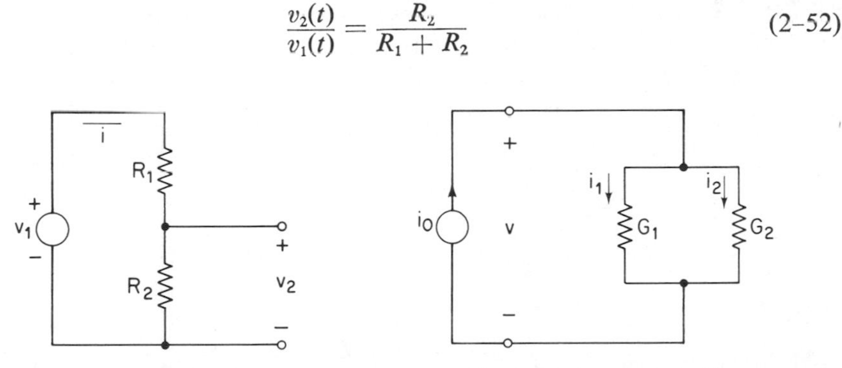

Fig. 2-6.3 A voltage divider.

Fig. 2-6.4 A current divider.

[3]This voltage is actually a node-pair voltage, i.e., the potential difference between a pair of nodes. It is convenient, however, to treat the lower node as a reference (or ground) node. Thus, we will refer to the *node voltage* implying the voltage of the upper node with respect to the lower node.

It should be noted that this expression is only valid if the current $i(t)$ flows through both of the resistors, i.e., if no current flows at the output terminals at which $v_2(t)$ is measured.

Another frequently encountered basic circuit is the *current divider*. Such a circuit is shown in Fig. 2–6.4. The voltage $v(t)$ that appears across the current source $i_0(t)$ is determined by the relation

$$v(t) = \frac{i_0(t)}{G_1 + G_2} \tag{2–53}$$

The current $i_1(t)$ is determined by the relation

$$i_1(t) = G_1 v(t) \tag{2–54}$$

Combining the above two expressions, we obtain a relation for the ratio between the input current $i_0(t)$ and the current $i_1(t)$, namely,

$$\frac{i_1(t)}{i_0(t)} = \frac{G_1}{G_1 + G_2} \tag{2–55}$$

Similarly, the expression for the ratio between the input current $i_0(t)$ and the current $i_2(t)$ is

$$\frac{i_2(t)}{i_0(t)} = \frac{G_2}{G_1 + G_2} \tag{2–56}$$

As a final example of a simple circuit, consider the network shown in Fig. 2–6.5(a), consisting of a voltage source, a linear resistor R_1, and a non-linear resistor R_2. Let R_2 have the non-linear characteristic defined in terms of its branch variables $v_2(t)$ and $i_2(t)$ by the curved locus shown in Fig. 2–6.5(b). If we apply KVL to this circuit, we obtain

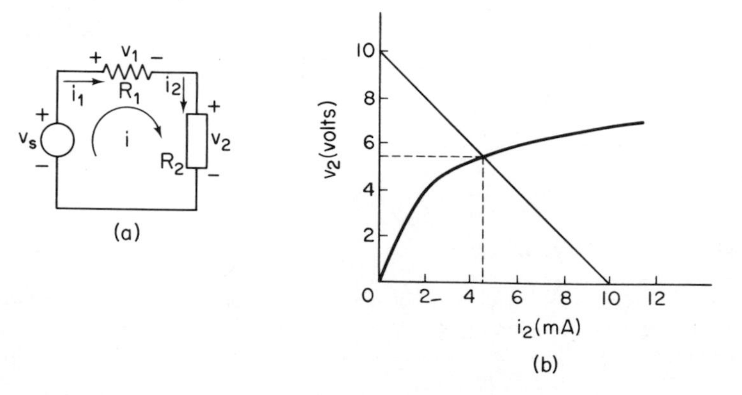

Fig. 2-6.5 Analysis of a non-linear circuit.

$$v_s = v_1 + v_2(i_2) \tag{2-57}$$

where v_s is the source voltage and v_1 and $v_2(i_2)$ are the branch voltages for the two resistors. Let us now define a loop current i as shown in Fig. 2-6.5(a). From an examination of the figure we see that $i = i_1 = i_2$, therefore we may now write (2-57) in the form

$$v_s - iR_1 = v_2(i) \tag{2-58}$$

The above equation is a non-linear one since v_2 is a non-linear function of the loop current i. We may, however, solve this equation graphically by drawing two plots, one of the left-hand member of the equation, as was done in Sec. 2-3, and the other of the right-hand member, putting both plots on the same graph. The intersection of the two loci indicates the values of v_2 and i which will satisfy the equality of (2-58). This has been done for the values $v_s = 10$ V and $R_1 = 1$ kΩ, in Fig. 2-6.5(b). The resulting solution is $v_2 = 5.5$ V, $i = 4.6$ mA. Such a solution is frequently referred to as the *operating point* for the non-linear characteristic of R_2. Note that if a specific operating point is desired, then a straight line may be drawn through the operating point to find the values of the voltage source v_s and the resistor R_1 which will determine that operating point. Specifically, the intercept with the v_2 axis gives the value of v_s and the intercept with the i axis gives the value v_s/R_1. This is referred to as a solution to the *biasing problem* of a non-linear resistive element. Since the straight line may be drawn with any slope, there are an infinite number of solutions to the problem unless either the value of the source voltage v_s or the value of R_1 is specified. Such a graphical technique is suitable for circuits containing a single non-linear resistive element. If the network contains more than a single non-linearity, however, we must use more sophisticated techniques to solve the biasing problem.

Once an operating point has been found, then, for small variations in the values of the variables of the non-linear element around such an operating point, we can treat the non-linear device as a linear one. This may be justified mathematically by expressing the non-linearity in the form of a Taylor series around the operating point, and retaining only the first two terms. As an example of this, consider the non-linear tunnel diode characteristic shown in Fig. 2-6.6. This is a voltage-controlled characteristic, since, for any given value of voltage appearing across the terminals of the diode, there exists a unique value of current through the device as specified by the characteristic on the v-i plane. The characteristic may be indicated functionally as $i = g(v)$, where g is the function specified by the plot shown in Fig. 2-6.6. In the vicinity of some operating point (v_0, i_0), we may write (the first two terms of a Taylor series)

$$i(t) \approx g(v_0) + g'(v_0)[v(t) - v_0] \tag{2-59}$$

where $g'(v_0)$ is the derivative of $g[v(t)]$ evaluated for $v(t) = v_0$. Thus, $g'(v_0)$ is simply the slope of the tangent to the characteristic curve at the operating point. Now let us connect such a two-terminal device into a circuit as shown in Fig. 2-6.7. The

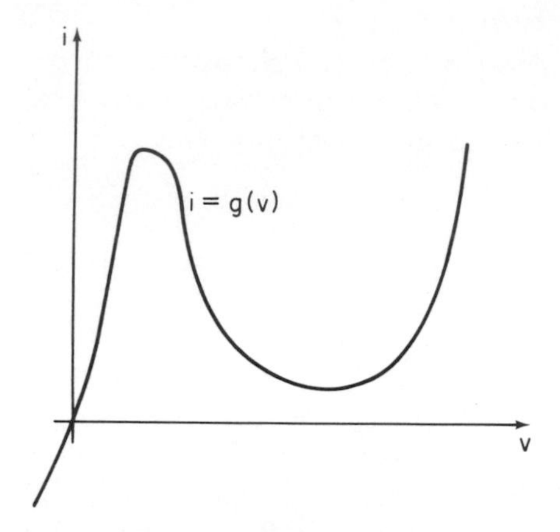

Fig. 2-6.6 A non-linear tunnel diode characteristic.

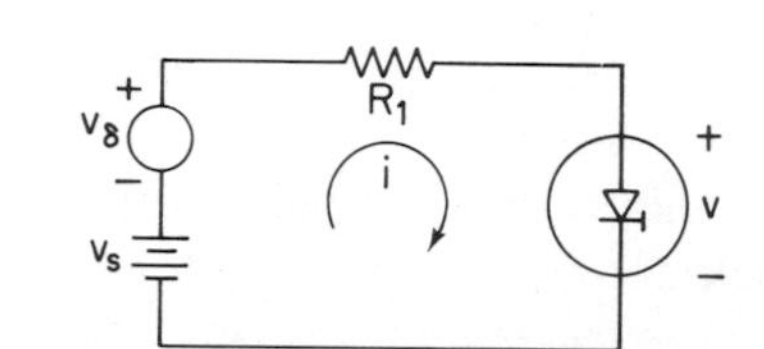

Fig. 2-6.7 A simple circuit with a tunnel diode.

symbol shown in this figure is the one customarily used for a tunnel diode. If we let the source $v_\delta(t) = 0$, then the operating point (v_0, i_0) will be determined by the values of v_s and R_1. Applying KVL in this case, we see that

$$v_0 = v_s - i_0 R_1 \tag{2-60}$$

The operating point may be determined by graphical means as shown in Fig. 2-6.8. Now let us apply an input $v_\delta(t)$ to the circuit. Let its magnitude be small enough so that the first-order approximation given in Eq. (2-59) is valid. The result will be to produce a small signal current $i_\delta(t)$, representing a deviation from the steady-state dc bias current i_0. Thus, we may write

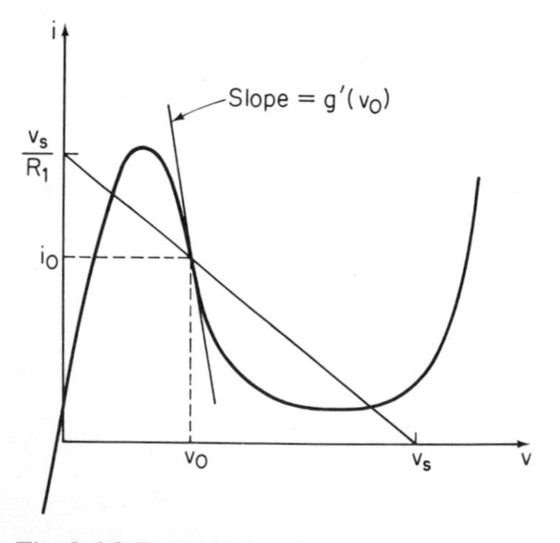

Fig. 2-6.8 Determining the dynamic resistance of a tunnel diode.

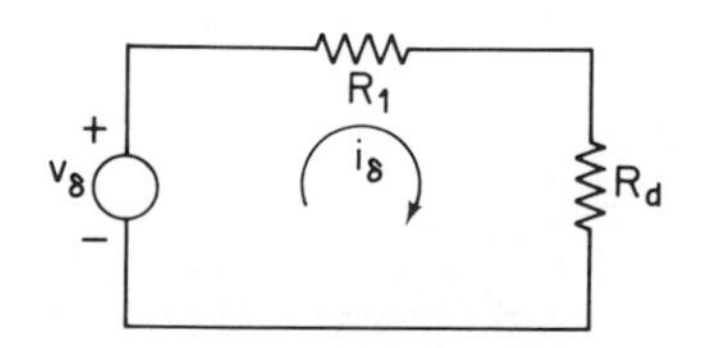

Fig. 2-6.9 A small-signal-equivalent circuit.

$$i_\delta(t) = i(t) - i_0 = i(t) - g(v_0) \approx g'(v_0)[v(t) - v_0] \tag{2-61}$$

From the circuit shown in Fig. 2-6.7, by applying KVL we may write

$$v(t) = v_s + v_\delta(t) - [i_0 + i_\delta(t)]R_1 \tag{2-62}$$

Substituting in this relation from (2-60) for $-i_0 R_1$ we obtain

$$v(t) - v_0 = v_\delta(t) - i_\delta(t)R_1 \tag{2-63}$$

If this result is substituted for the term in brackets in (2-61), we see that

$$i_\delta(t) = g'(v_0)[v_\delta(t) - i_\delta(t)R_1] \tag{2-64}$$

This may be rewritten in the form

$$i_\delta(t) = \frac{v_\delta(t)}{R_d + R_1} \tag{2-65}$$

where $R_d = 1/g'(v_0)$ represents the *dynamic resistance* of the tunnel diode at the operating point (v_0, i_0) for small magnitude signals. The above information is frequently summarized by constructing a *small-signal equivalent circuit* as shown in Fig. 2-6.9, eliminating the biasing voltage source v_s, and the resultant component i_0 of the current due to this source. What remains is then simply a linear circuit with variables $i_\delta(t)$ and $v_\delta(t)$, relating the small-signal or variational components of the actual physical variables $v(t)$ and $i(t)$. The use of such small-signal equivalent circuits is of considerable importance in analyzing electronic circuits containing non-linear solid-state devices.

Note that if the operating point has been set on a portion of the non-linear characteristic where the slope is negative, as shown in Fig. 2-6.8, then the value of R_d will be negative. Thus, we may use a tunnel diode to produce a negative-valued (small-signal) resistor. We shall see other methods of producing negative-valued resistors in the chapters which follow.

2-7 CONCLUSION

In this chapter, we have introduced some simple basic network elements, namely, resistors and sources. For a resistor, we have seen that the terminal variables are related by Ohm's law. This relation was applied to simple circuits in which we showed that if a voltage or current source is connected to the terminals of a resistor and thus used to determine one of the terminal variables, the value of the other variable is found by applying Ohm's law. Also in this chapter, we applied Kirchhoff's laws to various series and parallel combinations of resistors and sources to determine their properties. The results obtained in this chapter from our study of network elements will be greatly extended in the following chapter, where we will develop methods (including some which will require the use of a digital computer) for analyzing networks containing an arbitrary number of resistors and sources.

Problems

Problem 2-1 (Sec. 2-1)

A voltage 10 sin $2\pi t$ V is applied across a 10-kΩ resistor.

(a) Find an expression for the current through the resistor.

(b) Find an expression for the power dissipated in the resistor.

(c) Draw plots of the waveforms of the three quantities.

Problem 2-2 (Sec. 2-1)

A voltage 10 sin $2\pi t$ V is applied across a diode which has the characteristic shown in Fig. P2-2.

(a) Draw a plot of the waveform of the current which flows through the diode.

(b) Draw a plot of the power dissipated in the diode.

Note: Although the voltage has positive and negative polarity the current only has one polarity. In this case, we say that *rectification* has taken place.

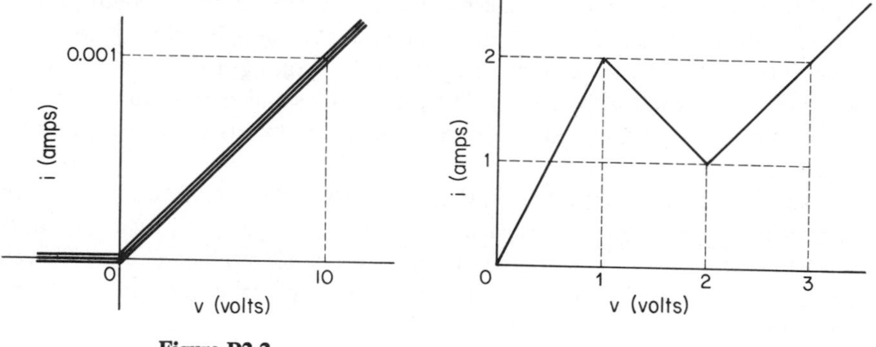

Figure P2-2

Figure P2-3

Problem 2-3 (Sec. 2-1)

To illustrate one of the differences between linear and non-linear elements, assume that two non-linear, multivalued, voltage-controlled resistors with the characteristic shown in Fig. P2-3 are connected in series. Draw the characteristics of the resulting two-terminal, non-linear element.

Problem 2-4 (Sec. 2-2)

Two sources and a single resistor are shown in Fig. P2-4. Determine the sign and the value of the power associated with each of the sources.

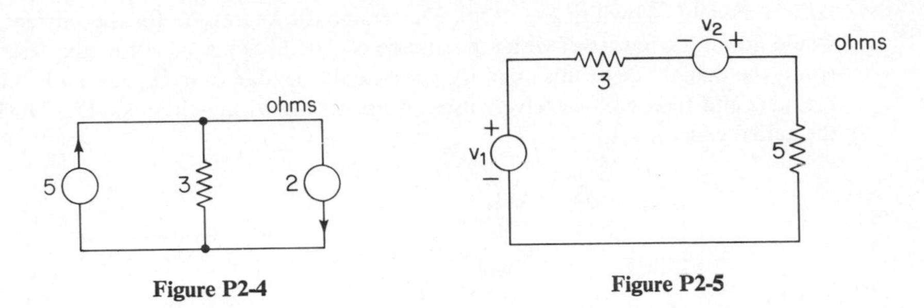

Figure P2-4

Figure P2-5

Problem 2-5 (Sec. 2-2)

The circuit shown in Fig. P2-5 consists of two sources and two resistors. Determine the polarity and the value of the power associated with each of the elements for the case where $v_1 = 10$, $v_2 = -5$ V.

Problem 2-6 (Sec. 2-2)

Repeat Problem 2-5 for the case where v_1 and v_2 have the waveshapes shown in Fig. P2-6.

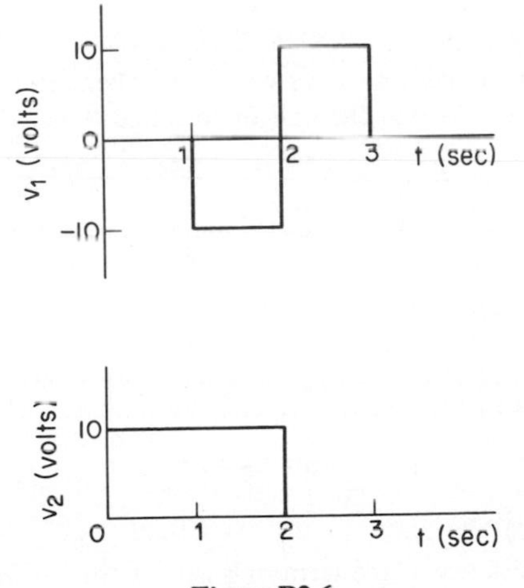

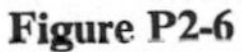

Figure P2-6

Problem 2-7 (Sec. 2-3)

A dry-cell battery has an open-circuit voltage of 1.5 V. When a short circuit is applied across the terminals of the battery, a current of 5 A flows through the short.

(a) Derive two circuit models for the battery.

(b) Which of the models is the "better" one from the viewpoint of matching the internal dissipation in the battery?

Problem 2-8 (Sec. 2-3)

The ac power available at the terminals of a standard household wall outlet may be modeled by an ideal sinusoidal voltage source in series with a very small resistance.

Such a model is shown in Fig. P2-8. Determine the expression for the current that would flow if a screwdriver with a resistance of 0.01 Ω is placed across the terminals (from the magnitude of this quantity you should be able to understand why circuit breakers and fuses are extensively used in household wiring circuits). DO NOT try the actual experiment.

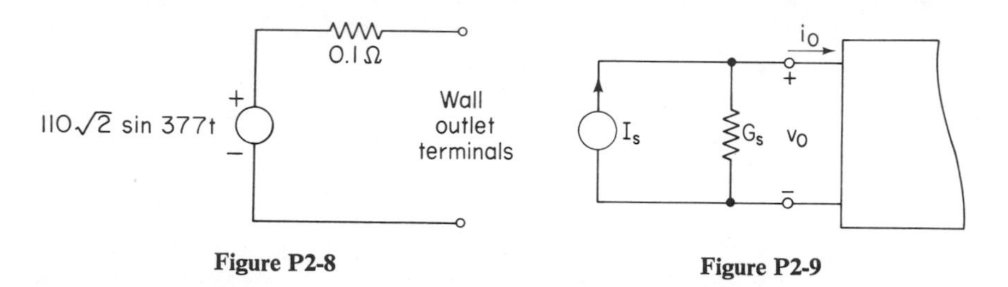

Figure P2-8 **Figure P2-9**

Problem 2-9 (Sec. 2-3)

Find the locus of operating points on a v_0-i_0 plane of the terminal characteristics of the network shown in Fig. P2-9 in which a current source in parallel with a resistor is connected to some arbitrary network.

Problem 2-10 (Sec. 2-4)

Find an equivalent single resistance which may be used to replace the resistance network shown in Fig. P2-10 at the pair of terminals labeled *a-b*.

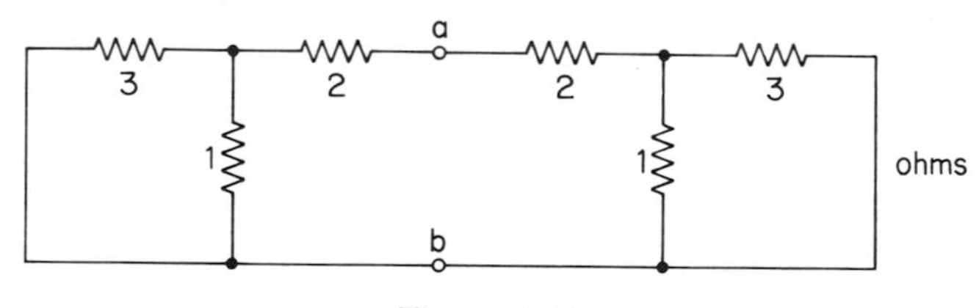

Figure P2-10

Problem 2-11 (Sec. 2-4)

Find the input resistance at the terminals *a-b* of the network of resistors shown in Fig. P2-11.

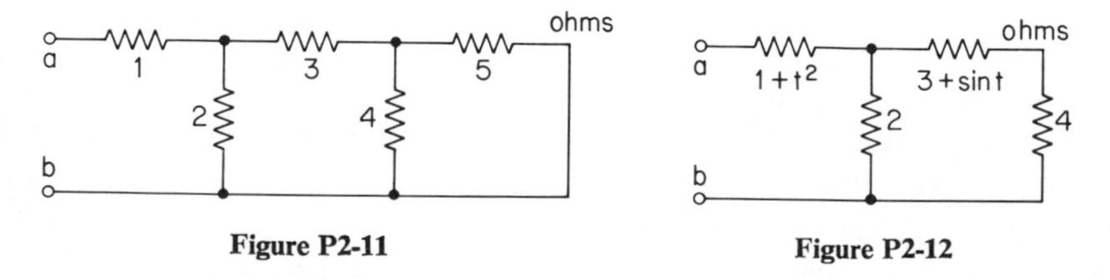

Figure P2-11 **Figure P2-12**

Problem 2-12 (Sec. 2-4)

Find the input resistance at the terminals *a-b* of the network containing two time-varying resistors shown in Fig. P2-12.

Problem 2-13 (*Sec. 2-4*)

For the network shown in Fig. P2-13, find the value of the resistor R_0 such that the input resistance at the pair of terminals labeled a-b is 2 Ω.

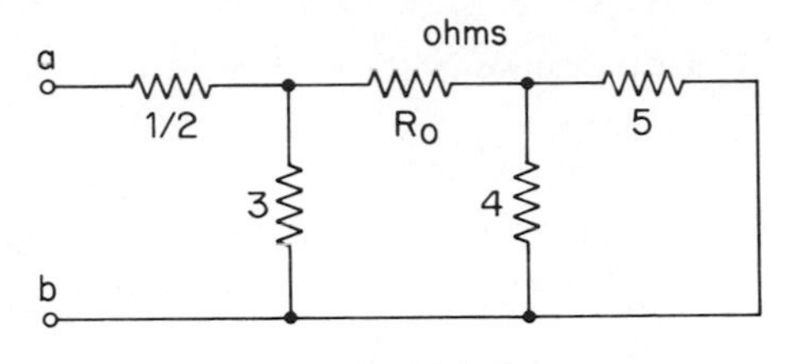

Figure P2-13

Problem 2-14 (*Sec. 2-5*)

Replace the network of sources shown in each part of Fig. P2-14 with a single equivalent source.

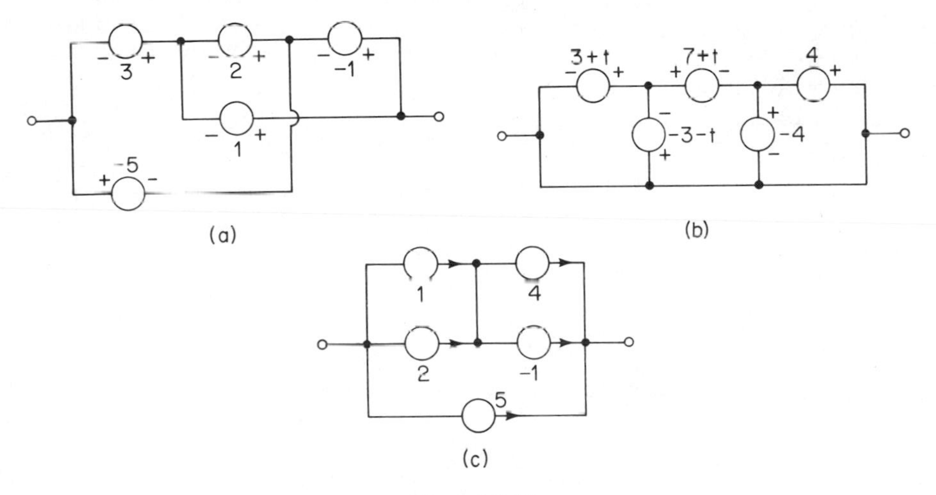

Figure P2-14

Problem 2-15 (*Sec. 2-5*)

Will the networks shown in Figs. P2-15(a) and (b) produce the same current in the

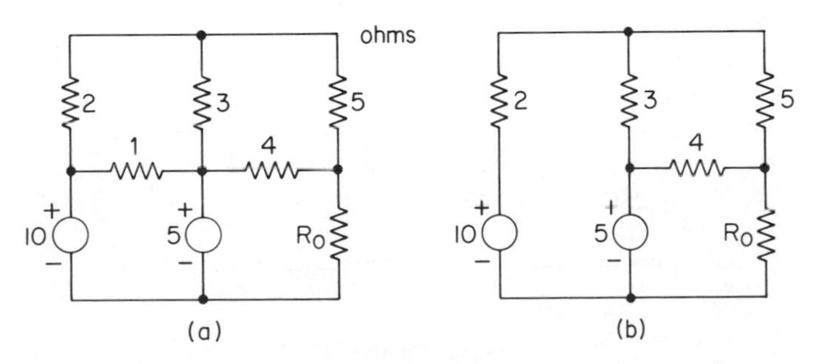

Figure P2-15

resistance R_0? If not, modify the network shown in part (b) of the figure by making any necessary changes, so that the currents will be the same. Do *not* attempt to solve the network.

Problem 2-16 (*Sec. 2-5*)

Repeat Problem 2-15 for the networks shown in Figs. P2-16(a) and (b).

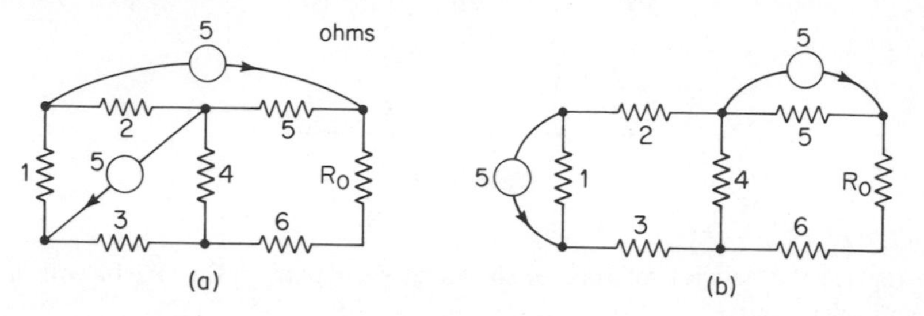

Figure P2-16

Problem 2-17 (*Sec. 2-5*)

The network shown in Fig. P2-17(a) may be reduced to the form shown in Fig. P2-17(b) by appropriate source transformations. Find the value of R_e and v_e.

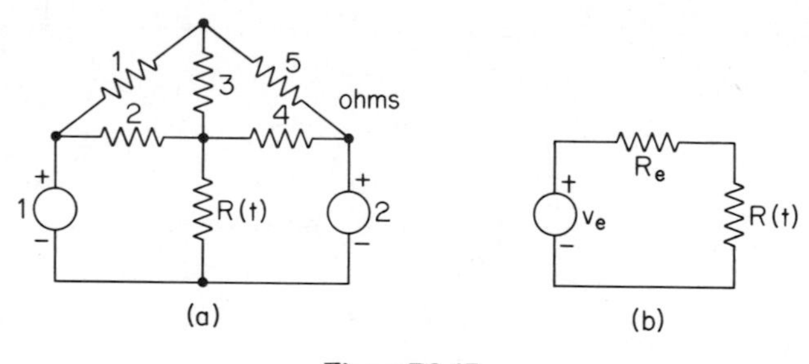

Figure P2-17

Problem 2-18 (*Sec. 2-5*)

The network shown in Fig. P2-18(a) may be reduced to the form shown in Fig. P2-18(b) by appropriate source transformations. Find the values of G_e and i_e.

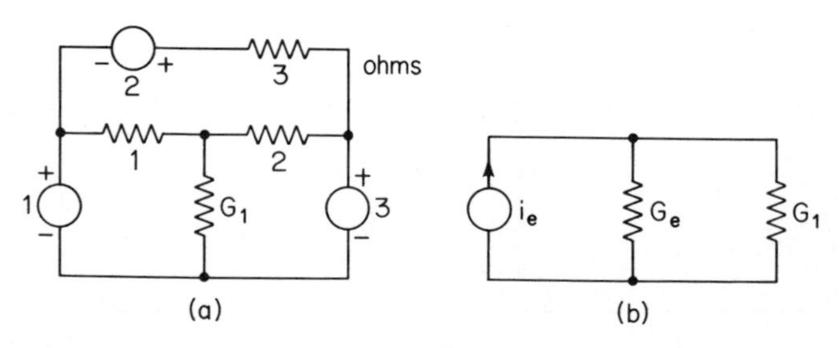

Figure P2-18

Problem 2-19 (*Sec. 2-6*)

A current generator of value $i_s(t) = 3 \cos 2t$ A is connected to a resistance network as shown in Fig. P2-19. Find the expression for the current $i_0(t)$.

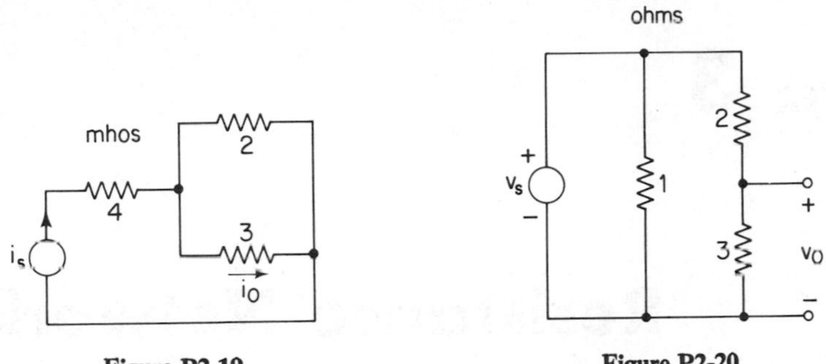

Figure P2-19

Figure P2-20

Problem 2-20 (*Sec. 2-6*)

An ideal voltage source with an output $v_s(t) = 7 + 3e^{-t}$ V is connected to a resistance network as shown in Fig. P2-20. Find an expression for the voltage $v_0(t)$.

Problem 2-21 (*Sec. 2-6*)

A tunnel diode characteristic is shown in Fig. P2-21(a). The diode is connected in series with a source and resistor as shown in Fig. P2-21(b).

(a) Determine the three possible operating points if $v_s = 0.5$ V and $R_1 = 625\ \Omega$.

(b) Compute the dynamic resistance of the tunnel diode at each of these operating points.

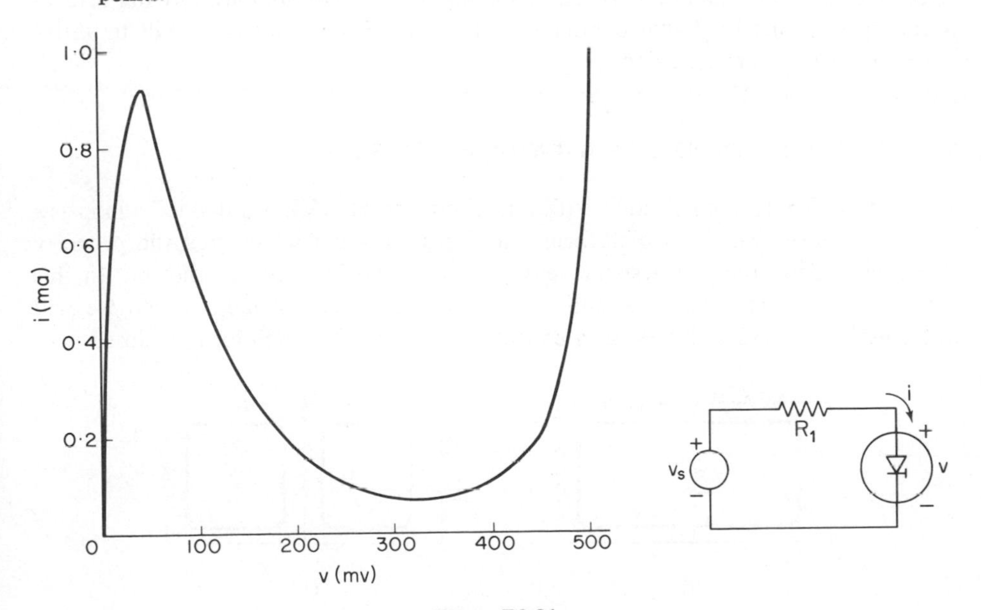

Figure P2-21

CHAPTER 3

Resistance Networks

In the preceding chapter, we defined the resistor, and showed how circuits consisting of a single loop (or a single node-pair) could be formulated by connecting such an element to a voltage or a current source. In this chapter, we shall extend our ideas to the case of a general resistance network. We shall see that the computational mechanics which are involved in solving any but the simplest circuits lead us to consider the use of digital computer techniques. Such techniques will be introduced as part of our discussion.

3-1 MESH EQUATIONS FOR A TWO-MESH CIRCUIT

In Chap. 1, in connection with our discussion of KVL, we defined a loop as a connected set of branches which formed a closed path and which had the property that each node in the path was incident to exactly two branches of the set. In this chapter, we shall restrict our attention to a particular type of loop, which we will call a *mesh*. A mesh is defined as a collection of branches which forms a closed path

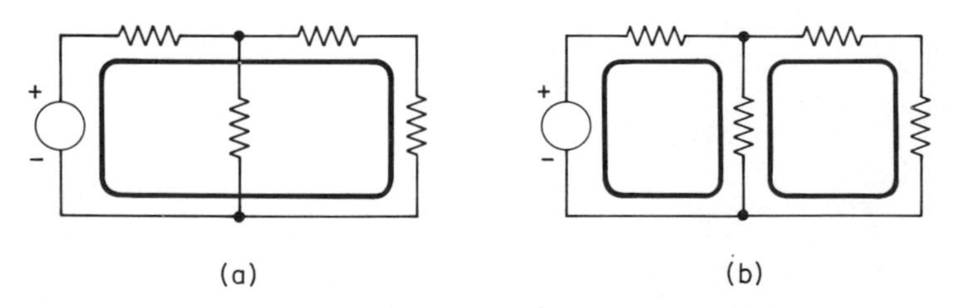

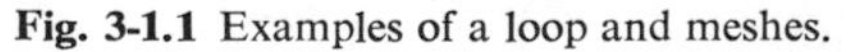
Fig. 3-1.1 Examples of a loop and meshes.

and which has the properties: (1) every node in the path is incident to exactly two branches; and (2) no other branches are enclosed by the collection. For example, in Fig. 3–1.1(a) the closed path indicated by the heavy line forms a loop, while in Fig. 3–1.1(b), the closed paths indicated by the heavy lines form meshes. Since a mesh

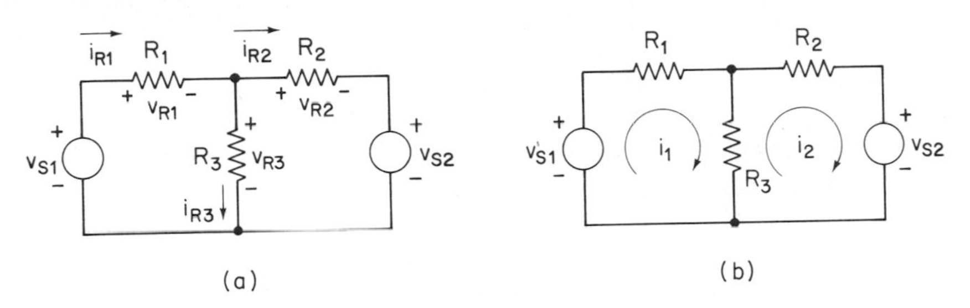

Fig. 3-1.2 A two-mesh resistance network.

is simply a loop which satisfies the additional condition given in (2) of the above definition, it should be clear that KVL is satisfied for any mesh.[1]

Now let us consider a resistance network containing two meshes, with a voltage source in each mesh as shown in Fig. 3–1.2(a). For each mesh we may define a mesh current as shown in Fig. 3 1.2(b). Note that the positive reference direction for both meshes is chosen the same, namely, clockwise. Also note that for the resistors R_1 and R_2, the mesh currents $i_1(t)$ and $i_2(t)$ are exactly the same as the defined branch currents $i_{R1}(t)$ and $i_{R2}(t)$. For the resistor R_3, however, the branch current $i_{R3}(t)$ defined in the figure is actually equal to the difference of the mesh currents, i.e., it is equal to $i_1(t) - i_2(t)$. We may now apply KVL around each mesh. In terms of the branch voltages and the source voltages we obtain

$$v_{S1}(t) = v_{R1}(t) + v_{R3}(t) \tag{3-1a}$$

$$-v_{S2}(t) = v_{R2}(t) - v_{R3}(t) \tag{3-1b}$$

Note that in both the equations given above we have chosen to put the source voltages on one side of the equation and the branch voltages on the other. In addition, we have set up the equation so that in traveling around the mesh in the direction indicated by the mesh reference current, a source whose reference polarity represents a voltage rise is treated as positive. This is a practice we shall find convenient to follow in all of our future discussions. The branch relations for the three resistors are

$$v_{R1}(t) = R_1 i_{R1}(t) \tag{3-2a}$$

[1]One of the reasons for the use of meshes to determine the unknown current variables is that the resulting equations describing the network will always be independent. If, on the other hand, an arbitrary set of loops is chosen (without taking some special precautions), the resulting equations may not be independent and thus they will not have a solution.

$$v_{R2}(t) = R_2 i_{R2}(t) \tag{3–2b}$$
$$v_{R3}(t) = R_3 i_{R3}(t) \tag{3–2c}$$

Substituting these branch relations in Eqs. (3–1a) and (3–1b), we obtain

$$v_{S1}(t) = R_1 i_{R1}(t) + R_3 i_{R3}(t) \tag{3–3a}$$
$$-v_{S2}(t) = R_2 i_{R2}(t) - R_3 i_{R3}(t) \tag{3–3b}$$

Finally, we may put Eqs. (3–3a) and (3–3b) into a still more useful form by substituting the relations between the branch currents and the mesh currents, namely,

$$i_{R1}(t) = i_1(t) \tag{3–4a}$$
$$i_{R2}(t) = i_2(t) \tag{3–4b}$$
$$i_{R3}(t) = i_1(t) - i_2(t) \tag{3–4c}$$

Substituting the above relations in Eqs. (3–3a) and (3–3b) and rearranging terms, we obtain

$$v_{S1}(t) = (R_1 + R_3)i_1(t) - R_3 i_2(t) \tag{3–5a}$$
$$-v_{S2}(t) = -R_3 i_1(t) + (R_2 + R_3)i_2(t) \tag{3–5b}$$

The relations given in Eqs. (3–5a) and (3–5b) are usually referred to as a set of *simultaneous equations*. We may now make some observations about these relations. In the first equation (which is the one expressing KVL for mesh 1), note the following points:

1. The reference polarity of the voltage source is such as to produce a reference voltage *rise* when traveling around the mesh in the direction of the mesh reference current $i_1(t)$. Thus, this source voltage appears as a positive quantity in the left member of the equation.
2. The term $(R_1 + R_3)$ multiplying the quantity $i_1(t)$ is simply the sum of the resistances which are present in mesh 1.
3. The term $-R_3$ which multiplies the quantity $i_2(t)$ is simply the resistance which is common to both mesh 1 and mesh 2 (multiplied by a negative sign). Note that the presence of the negative sign in this term is due to the fact that the mesh 2 reference current $i_2(t)$ passes through the branch containing R_3 in the *opposite* direction to that of the mesh 1 reference current $i_1(t)$, i.e., that there is a negative sign in (3–4c). This, of course, is the result of choosing the same positive reference direction for both meshes.

A similar set of observations may be made for the KVL equation for mesh 2 as given by (3–5b). We note that:

1. The left member, i.e., the term $-v_{S2}(t)$, represents the effect of a voltage source with its reference polarity so oriented that it produces a reference voltage *drop* when traveling around the mesh in the direction of the mesh reference current $i_2(t)$.
2. The term $(R_2 + R_3)$ multiplying the quantity $i_2(t)$ is simply the sum of the resistances which are present in the mesh.
3. The term $-R_3$ which multiplies the quantity $i_1(t)$ is simply the resistance which is common to both mesh 1 and mesh 2 (multiplied by a negative sign). As before, the negative sign in this term is due to the fact that the mesh reference currents $i_1(t)$ and $i_2(t)$ pass through the branch in opposite directions.

Now let us write Eqs. (3–5a) and (3–5b) in matrix format.[2] We obtain

$$\begin{bmatrix} v_{S1}(t) \\ -v_{S2}(t) \end{bmatrix} = \begin{bmatrix} R_1 + R_3 & -R_3 \\ -R_3 & R_2 + R_3 \end{bmatrix} \begin{bmatrix} i_1(t) \\ i_2(t) \end{bmatrix} \tag{3–6}$$

This is a specific example of a general matrix expression for the equations describing a two-mesh resistance network. Such an expression will always have the form

$$\begin{bmatrix} v_1(t) \\ v_2(t) \end{bmatrix} = \begin{bmatrix} r_{11} & r_{12} \\ r_{21} & r_{22} \end{bmatrix} \begin{bmatrix} i_1(t) \\ i_2(t) \end{bmatrix} \tag{3–7}$$

where $v_1(t)$ and $v_2(t)$ may be considered as the elements of a column matrix $\mathbf{v}(t)$, $i_1(t)$ and $i_2(t)$ may be considered as the elements of a column matrix $\mathbf{i}(t)$, and the quantities r_{ij} may be considered as elements of a square matrix $\mathbf{R}$, with the subscript i being used to represent the row of $\mathbf{R}$ in which the element occurs and the subscript j being used to represent the column of $\mathbf{R}$ in which the element occurs. Thus, we may rewrite the matrix equation (3–7) using the matrix symbols $\mathbf{v}(t)$, $\mathbf{i}(t)$, and $\mathbf{R}$ as

$$\mathbf{v}(t) = \mathbf{R}\mathbf{i}(t) \tag{3–8}$$

We may directly compare Eq. (3–6) with Eq. (3–7) by noting that the element $v_1(t)$ represents the source in mesh 1, the element $v_2(t)$ represents the source (the orientation of the reference polarity must be taken into consideration) in mesh 2, the element r_{11} represents the sum of the values of the resistors in mesh 1, the element r_{22} represents the sum of the values of the resistors in mesh 2, and the elements r_{12} and

[2]An introductory treatment of matrices is given in Appendix A.

r_{21} (which are equal) represent the value of the resistor which is common to meshes 1 and 2 (multiplied by a negative sign).

In the above paragraphs we have seen that for a two-mesh network, we may develop a matrix equation giving the relations between the voltage sources (or excitations) that are found in each of the meshes (these may be referred to as the "known" variables), and the mesh currents (the "unknown" variables). We would now like to solve these equations. By the word "solve" we mean that we desire to find the values of the mesh currents for some specified values of the voltage sources and the resistors. In effect, we desire to find the inverse set of relations to those given in Eq. (3–7), i.e., we desire to find a set of relations which has the form

$$\begin{bmatrix} i_1(t) \\ i_2(t) \end{bmatrix} = \begin{bmatrix} a_{11} & a_{12} \\ a_{21} & a_{22} \end{bmatrix} \begin{bmatrix} v_1(t) \\ v_2(t) \end{bmatrix} \tag{3–9}$$

Using the symbols $\mathbf{i}(t)$ and $\mathbf{v}(t)$ as previously defined, and letting $\mathbf{A}$ be the matrix with elements a_{ij} we may write Eq. (3–9) in the form

$$\mathbf{i}(t) = \mathbf{A}\mathbf{v}(t) \tag{3–10}$$

The matrix $\mathbf{A}$ of (3–10) is referred to as the *inverse* of the matrix $\mathbf{R}$. This may be written as

$$\mathbf{A} = \mathbf{R}^{-1} \tag{3–11}$$

The elements a_{ij} of the matrix $\mathbf{A}$ are easily found in terms of the elements r_{ij} of the matrix $\mathbf{R}$ by solving the set of simultaneous equations given in Eq. (3–7). For the simple set of two equations given here this may be done by simply substituting one equation into the other. To see this, consider the second equation of the set given in Eq. (3–7). This may be written as

$$v_2(t) = r_{21}i_1(t) + r_{22}i_2(t) \tag{3–12}$$

Solving for $i_2(t)$, we obtain

$$i_2(t) = \frac{v_2(t) - r_{21}i_1(t)}{r_{22}} \tag{3–13}$$

If we substitute this result into the first equation of (3–7) and rearrange terms we obtain

$$i_1(t) = \frac{1}{r_{11}r_{22} - r_{12}r_{21}}[r_{22}v_1(t) - r_{12}v_2(t)] \tag{3–14}$$

Similarly, solving the first equation of (3–7) for $i_1(t)$ and substituting the result into the second equation, we obtain

$$i_2(t) = \frac{1}{r_{11}r_{22} - r_{12}r_{21}}[r_{11}v_2(t) - r_{21}v_1(t)] \tag{3-15}$$

The two equations given above may be expressed in matrix form as

$$\begin{bmatrix} i_1(t) \\ i_2(t) \end{bmatrix} = \frac{1}{r_{11}r_{22} - r_{12}r_{21}} \begin{bmatrix} r_{22} & -r_{12} \\ -r_{21} & r_{11} \end{bmatrix} \begin{bmatrix} v_1(t) \\ v_2(t) \end{bmatrix} \tag{3-16}$$

If we compare (3–9) and (3–16), we see that the elements of the matrix **A** (where $\mathbf{A} = \mathbf{R}^{-1}$) are defined in terms of the elements of **R** by the relation

$$\begin{bmatrix} a_{11} & a_{12} \\ a_{21} & a_{22} \end{bmatrix} = \frac{1}{r_{11}r_{22} - r_{12}r_{21}} \begin{bmatrix} r_{22} & -r_{12} \\ -r_{21} & r_{11} \end{bmatrix} \tag{3-17}$$

Note that for this second-order case the inverse is easily found by interchanging the "11" and "22" elements of the original matrix **R**, multiplying the "12" and "21" elements by a minus sign, and dividing all the elements by the quantity $(r_{11}r_{22} - r_{12}r_{21})$. This latter quantity is simply the *determinant* of the array of elements specified by the matrix **R**. It is a theorem of linear algebra that the necessary and sufficient condition for a solution to exist, i.e., for the inverse of a matrix **R** to exist, is that the determinant of **R** be non-zero. Thus the relations of Eq. (3–16) provide a general solution for the two-mesh resistance network. An example follows.

EXAMPLE 3-1.1. A two-mesh resistance network is shown in Fig. 3-1.3. The voltage source in mesh 1 represents a voltage rise of 8 V, and the voltage source in mesh 2 represents a

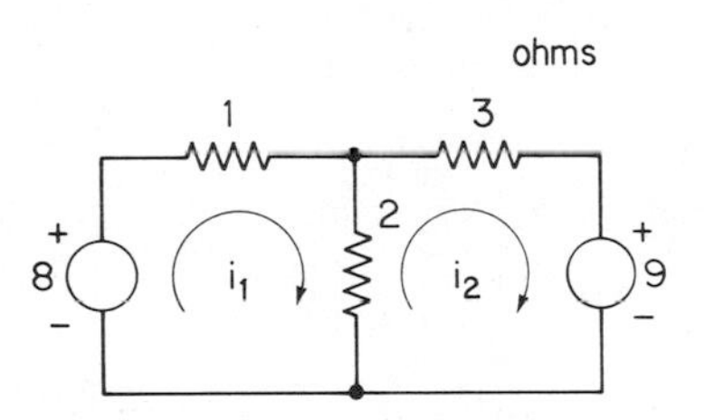

Fig. 3-1.3 A two-mesh resistance network.

voltage drop of 9 V. The total resistance in mesh 1 is 3 Ω. The total resistance in mesh 2 is 5 Ω. There is a 2-Ω resistor common to both mesh 1 and mesh 2. From the discussion given above, we see that we may write the KVL equations for this resistance network as

$$\begin{bmatrix} 8 \\ -9 \end{bmatrix} = \begin{bmatrix} 3 & -2 \\ -2 & 5 \end{bmatrix} \begin{bmatrix} i_1 \\ i_2 \end{bmatrix}$$

The solution is found by computing the inverse matrix. Following the relations given in (3-16), we obtain

$$\begin{bmatrix} i_1 \\ i_2 \end{bmatrix} = \frac{1}{11}\begin{bmatrix} 5 & 2 \\ 2 & 3 \end{bmatrix}\begin{bmatrix} 8 \\ -9 \end{bmatrix} = \begin{bmatrix} 2 \\ -1 \end{bmatrix}$$

Thus, the current in mesh 1 is 2 A flowing in the indicated reference direction, and the current in mesh 2 is -1 A, i.e., 1 A flowing in a direction opposite to the reference direction.

The method of analysis described above for the two-mesh resistance network is easily extended to cases where there are additional voltage sources in the meshes. The process is best illustrated by an example.

EXAMPLE 3-1.2. A two-mesh resistance network is shown in Fig. 3-1.4. The net effect of the two voltage sources that are encountered in mesh 1 is such as to produce a voltage rise

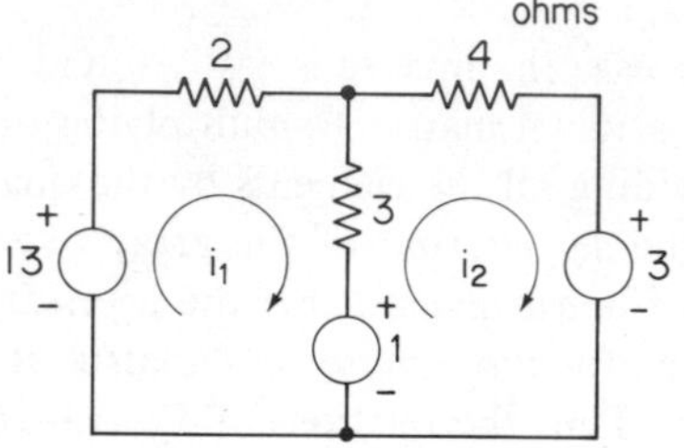

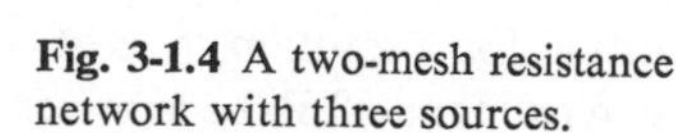
Fig. 3-1.4 A two-mesh resistance network with three sources.

of 12 V. The net effect of the two voltage sources that are encountered in mesh 2 is such as to produce a voltage drop of 2 V. Evaluating the values of the total resistance in both meshes and the resistance common to the two meshes, we may write the KVL equations for the network as

$$\begin{bmatrix} 12 \\ -2 \end{bmatrix} = \begin{bmatrix} 5 & -3 \\ -3 & 7 \end{bmatrix}\begin{bmatrix} i_1 \\ i_2 \end{bmatrix}$$

The inverse set of relations is easily found to be

$$\begin{bmatrix} i_1 \\ i_2 \end{bmatrix} = \frac{1}{26}\begin{bmatrix} 7 & 3 \\ 3 & 5 \end{bmatrix}\begin{bmatrix} 12 \\ -2 \end{bmatrix} = \begin{bmatrix} 3 \\ 1 \end{bmatrix}$$

Thus the current in mesh 1 is 3 A flowing in the indicated reference direction, and the current in mesh 2 is 1 A flowing in the indicated reference direction.

A resistance network such as the two-mesh ones described in the preceding examples is an illustration of a *linear* network. Such a network must have the property that superposition (as defined in Sec. 1-3) applies for any combination of inputs and for any variable considered as an output. For example, if we use the inverse matrix found in solving for the mesh currents of the network in Example 3-1.1, and leave the values of the two voltage sources unspecified, we may write

$$\begin{bmatrix} i_1(t) \\ i_2(t) \end{bmatrix} = \frac{1}{11}\begin{bmatrix} 5 & 2 \\ 2 & 3 \end{bmatrix}\begin{bmatrix} v_1(t) \\ v_2(t) \end{bmatrix} \tag{3-18}$$

If we now apply an excitation to the network only in mesh 1, by letting $v_1 = 8$ V, and setting v_2 to zero, then from Eq. (3–18) we see that $i_1 = \frac{40}{11}$ A and $i_2 = \frac{16}{11}$ A. Now let us change our inputs and apply an excitation to the network only in mesh 2, setting $v_2 = -9$ V, and setting v_1 to zero. In this case, from Eq. (3–18), we find that $i_1 = -\frac{18}{11}$ A and $i_2 = -\frac{27}{11}$ A. The currents for the case where excitations are applied in both of the meshes simultaneously, i.e., where $v_1 = 8$ and $v_2 = -9$ V, are simply found by the superposition of the results of the above two cases, namely $i_1 = (40 - 18)/11 = 2$, and $i_2 = (16 - 27)/11 = -1$ A. These are the results which were obtained in the example.

We could also have made use of the technique of superposition directly in the physical network, without using the inverse matrix relationships given in Eq. (3–18). As an example of this, in Fig. 3–1.5(a), we have redrawn the network used

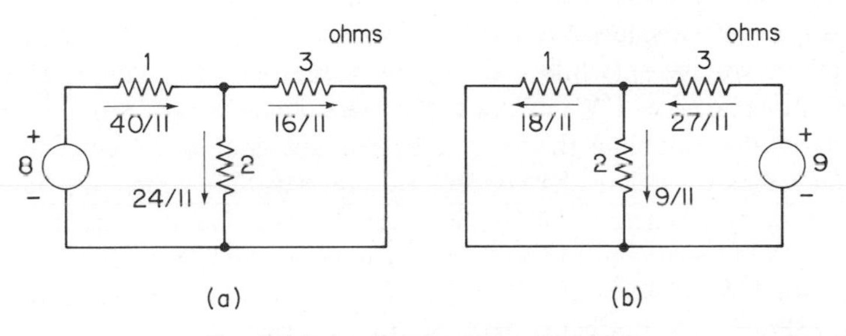

Fig. 3-1.5 Application of superposition.

in Example 3–1.1 for the case where $v_1 = 8$ V and $v_2 = 0$. Since the characteristic of an ideal voltage source with an output voltage of zero is exactly that of a short circuit, we have replaced the source in mesh 2 by a short circuit. Such a replacement is always implied when we state that a voltage source is set to zero. The resistance seen by the 8 V source is $\frac{11}{5}$ Ω, thus the current from this source is $\frac{40}{11}$ A (which is i_1). When this current reaches the junction of the 2- and 3-Ω resistors, these resistors will act as a current divider. The current through the 3-Ω resistor will thus be $(\frac{2}{5})(\frac{40}{11}) = \frac{16}{11}$ A (which is i_2). These results, of course, agree with the results found from the inverse matrix relation given in Eq. (3–18). A similar analysis may be made to find the contributions to the currents i_1 and i_2 when the source $v_2 = -9$ V and $v_1 = 0$. The physical situation is shown in Fig. 3–1.5(b). The addition of the results from the two separate cases yields the same results as were given in Example 3–1.1.

For a system with multiple inputs such as is discussed above, there is a right way and a wrong way to apply the tests for superposition. As an example, let us consider the two-mesh resistance network shown in Fig. 3–1.6. The mesh equations for this network have the form

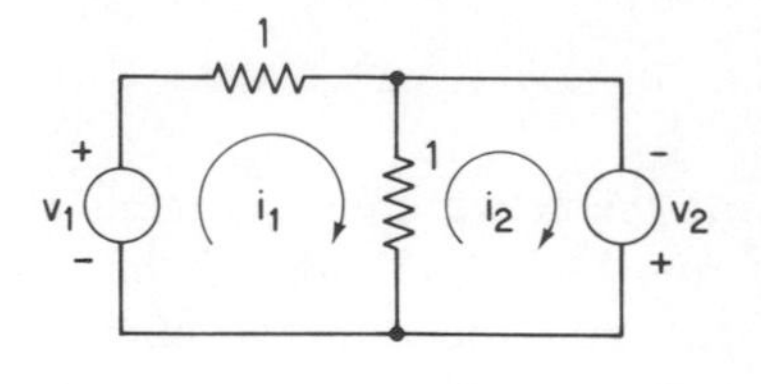

Fig. 3-1.6 A two-mesh resistance network.

$$\begin{bmatrix} v_1(t) \\ v_2(t) \end{bmatrix} = \begin{bmatrix} 2 & -1 \\ -1 & 1 \end{bmatrix} \begin{bmatrix} i_1(t) \\ i_2(t) \end{bmatrix} \tag{3-19}$$

The inverse set of equations is easily found to be

$$\begin{bmatrix} i_1(t) \\ i_2(t) \end{bmatrix} = \begin{bmatrix} 1 & 1 \\ 1 & 2 \end{bmatrix} \begin{bmatrix} v_1(t) \\ v_2(t) \end{bmatrix} \tag{3-20}$$

Now let us consider the case where $v_1 = 1$ V and $v_2 = 1$ V. We see that the currents i_1 and i_2 have the values 2 A and 3 A respectively. Suppose that we now double the value of the voltage v_1 (while keeping the value of v_2 the same). Inserting the values $v_1 = 2$ V and $v_2 = 1$ V in Eq. (3–20), we obtain the results that $i_1 = 3$ A and $i_2 = 4$ A. Note that doubling the voltage v_1 did not double the currents i_1 and i_2. This at first seems to violate the concept of superposition which we have used to define the property of linearity. Now, however, let us make a different test. Let us set $v_1 = 1$ V, while setting v_2 to zero. In this case we find from Eq. (3–20) that i_1 and i_2 are both 1 A. If we double the value of v_1 (while keeping v_2 at zero) we see that i_1 and i_2 both have the value of 2 A. In this case, doubling the input to the network has indeed doubled the outputs, i.e., the currents i_1 and i_2. It is easily shown that the same result holds true for the other source, namely, if v_2 is multiplied by some factor while v_1 is held equal to zero, then the currents i_1 and i_2 will also be multiplied by the same factor.

Considering the above results, we see that a linear resistance network may be characterized as one comprised of linear resistors and sources. It has the following properties:

1. The value of any variable may be found as the sum of the values of that variable produced by each of the excitation sources acting separately.
2. If *all the sources except one are set to zero*, then, multiplying the value of the source by a constant multiplies the values of all the network variables by the same constant.

This definition applies to the case of resistance networks excited by either voltage or current sources. The latter case will be discussed in a later section.

3-2 MESH EQUATIONS FOR THE GENERAL CIRCUIT

In this section we shall extend the results obtained for the two-mesh resistance circuit to the general case. First let us consider a network containing three meshes as shown in Fig. 3-2.1. As before, we have assigned the same (clockwise) positive

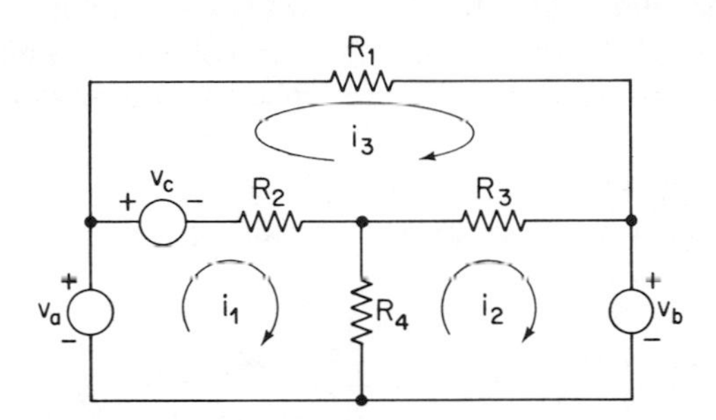

Fig. 3-2.1 A three-mesh resistance network.

reference directions for all the mesh currents. To determine the network equations for this circuit, we may apply the same procedure as was used for the two-mesh example, namely:

1. Write the KVL equations for each loop in terms of the branch voltage variables (the voltages across the resistive branches) and the voltage sources.
2. Substitute the branch relationships (Ohm's law for each branch) so that the KVL equations are expressed in terms of the branch currents.
3. Substitute the relationships defining the branch currents in terms of the mesh currents $i_1(t)$, $i_2(t)$, and $i_3(t)$ defined in the figure.

The details are left as an exercise for the reader. The result is the following set of equations:

$$\begin{bmatrix} v_a(t) - v_c(t) \\ -v_b(t) \\ v_c(t) \end{bmatrix} = \begin{bmatrix} R_2 + R_4 & -R_4 & -R_2 \\ -R_4 & R_3 + R_4 & -R_3 \\ -R_2 & -R_3 & R_1 + R_2 + R_3 \end{bmatrix} \begin{bmatrix} i_1(t) \\ i_2(t) \\ i_3(t) \end{bmatrix} \tag{3-21}$$

The set of equations given above is an example of a set of matrix equations of the form $\mathbf{v}(t) = \mathbf{R}\mathbf{i}(t)$, which may be written as

$$\begin{bmatrix} v_1(t) \\ v_2(t) \\ v_3(t) \end{bmatrix} = \begin{bmatrix} r_{11} & r_{12} & r_{13} \\ r_{21} & r_{22} & r_{23} \\ r_{31} & r_{32} & r_{33} \end{bmatrix} \begin{bmatrix} i_1(t) \\ i_2(t) \\ i_3(t) \end{bmatrix} \tag{3-22}$$

If we compare the elements of (3–22) with those of Eq. (3–21), we may make a set of conclusions similar to those made for the two-mesh network, namely, we see that the elements $v_i(t)$ of the matrix $\mathbf{v}(t)$ each represent the sum (with appropriate note taken of the signs) of the voltage sources that appear in the ith mesh of the network, the elements r_{ii} on the main diagonal of the matrix $\mathbf{R}$ represent the total resistance of the resistors in the ith mesh, and the elements $r_{ij} (i \neq j)$ represent the negative of the values of the resistors common to both the ith and the jth meshes. An example of determining the elements of such an equation follows.

EXAMPLE 3-2.1. A three-mesh resistance network is shown in Fig. 3-2.2. If we apply the rules specified above for determining the algebraic summation of the voltage sources in

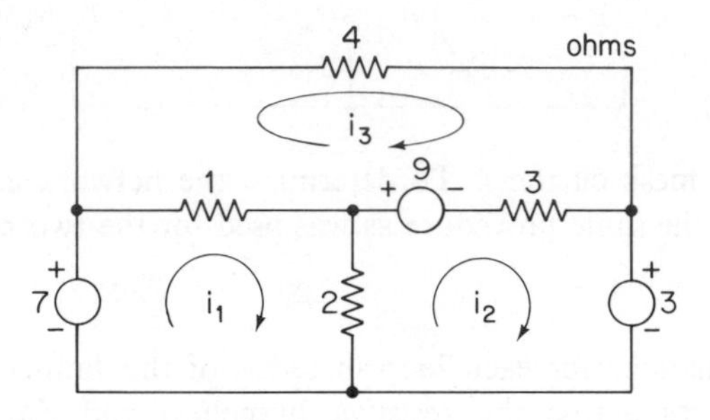

Fig. 3-2.2 A three-mesh resistance network.

each mesh, we see that $v_1(t)$ of (3-22) is 7 V, $v_2(t) = -9 - 3 = -12$ V, and $v_3(t) = 9$ V. The total resistance in the first mesh is 3 Ω, and this mesh has a 2-Ω resistor which is also in mesh 2, and a 1-Ω resistor which is also in mesh 3. Continuing in this manner, we obtain a set of simultaneous equations for the network which may be written in matrix form as

$$\begin{bmatrix} 7 \\ -12 \\ 9 \end{bmatrix} = \begin{bmatrix} 3 & -2 & -1 \\ -2 & 5 & -3 \\ -1 & -3 & 8 \end{bmatrix} \begin{bmatrix} i_1 \\ i_2 \\ i_3 \end{bmatrix}$$

We shall defer until the following section a discussion of the methods by means of which such a set of equations may be solved, i.e., the techniques by means of which the values of the currents i_1, i_2, and i_3 may be found.

The development made above may be easily extended to the case where there are an arbitrary number of meshes in the network. Consider a network containing n meshes and an arbitrary number of sources. In addition, let all the mesh currents have the same reference direction (we will assume that this is clockwise). An extension of the development made above shows that we may write the network equations for this network in the form

$$\begin{bmatrix} v_1(t) \\ v_2(t) \\ \vdots \\ v_n(t) \end{bmatrix} = \begin{bmatrix} r_{11} & r_{12} & \cdots & r_{1n} \\ r_{21} & r_{22} & \cdots & r_{2n} \\ \vdots & \vdots & & \vdots \\ r_{n1} & r_{n2} & \cdots & r_{nn} \end{bmatrix} \begin{bmatrix} i_1(t) \\ i_2(t) \\ \vdots \\ i_n(t) \end{bmatrix} \tag{3–23}$$

The procedure for determining the elements of the matrices of Eq. (3–23) may be summarized as follows:

SUMMARY 3-2.1

Writing the Mesh Equations for a Resistance Network: A set of mesh equations having the form given in Eq. (3–23) may be written for a planar[3] resistance network by defining a set of mesh currents all of which have the same reference direction, and determining the elements of the component matrices as follows:

$v_i(t)$ the summation of all the voltage sources in the ith mesh, with sources whose reference polarity constitutes a voltage rise (in the positive reference direction of the ith mesh current) treated as positive, and with sources whose reference polarity constitutes a voltage drop treated as negative.

$r_{ij}(i = j)$ the summation of the values of resistors that occur in the ith mesh.

$r_{ij}(i \neq j)$ the negative of the sum of the values of any resistors which are common to the ith and the jth mesh.[4]

$i_i(t)$ the unknown mesh currents which are the unknown variables for the network.

It is left to the reader as an exercise to determine that these rules lead to a set of equations which are exactly the same as those which will result if KVL is applied around each mesh of the network in terms of the branch voltage variables, and if the relations for the mesh currents are substituted into these relations. These rules may be applied to determine the mesh equations for any planar network. Techniques which may be applied to non-planar networks will be developed in Secs. 3–6 and 3–7. An example of the method outlined in the above summary follows.

EXAMPLE 3-2.2. As an example of the application of the rules given above to directly determine the matrix relations for a multi-mesh resistance network, consider the circuit shown

[3]A planar network is one with a planar graph. The planar graph is defined in Sec. 1–4.

[4]The quantity $r_{ij}(i = j)$ is sometimes called the *self-resistance* of the ith mesh, while the quantity $r_{ij}(i \neq j)$ is called the *mutual resistance* between the ith and jth mesh.

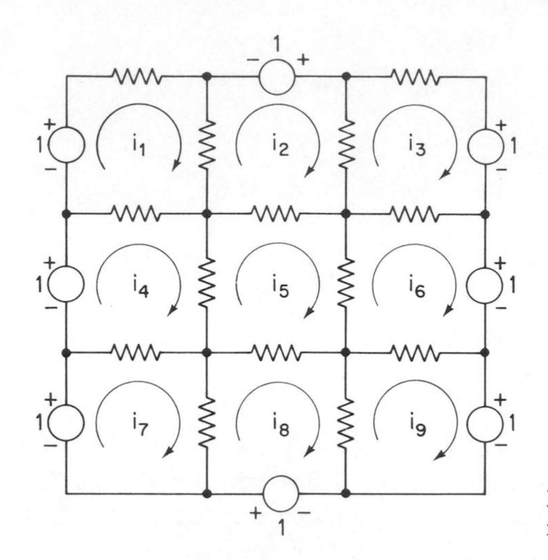

Fig. 3-2.3 A nine-mesh resistance network.

in Fig. 3-2.3. Assume that all the sources have a value of 1 V (with the indicated polarities) and that all resistors have unity value. The matrix relations are easily determined to be

$$\begin{bmatrix} 1 \\ 1 \\ -1 \\ 1 \\ 0 \\ -1 \\ 1 \\ 1 \\ -1 \end{bmatrix} = \begin{bmatrix} 3 & -1 & 0 & -1 & 0 & 0 & 0 & 0 & 0 \\ -1 & 3 & -1 & 0 & -1 & 0 & 0 & 0 & 0 \\ 0 & -1 & 3 & 0 & 0 & -1 & 0 & 0 & 0 \\ -1 & 0 & 0 & 3 & -1 & 0 & -1 & 0 & 0 \\ 0 & -1 & 0 & -1 & 4 & -1 & 0 & -1 & 0 \\ 0 & 0 & -1 & 0 & -1 & 3 & 0 & 0 & -1 \\ 0 & 0 & 0 & -1 & 0 & 0 & 2 & -1 & 0 \\ 0 & 0 & 0 & 0 & -1 & 0 & -1 & 3 & -1 \\ 0 & 0 & 0 & 0 & 0 & -1 & 0 & -1 & 2 \end{bmatrix} \begin{bmatrix} i_1 \\ i_2 \\ i_3 \\ i_4 \\ i_5 \\ i_6 \\ i_7 \\ i_8 \\ i_9 \end{bmatrix}$$

Note that for a general network such as the one analyzed in the above example, the mesh currents as defined by the unknown variables $i_i(t)$ may actually not be directly, physically measurable in the circuit. For example, the mesh current $i_5(t)$ shown in Fig. 3–2.3 always appears in connection with some other mesh current. Therefore, there is no way in which a single current-measuring instrument such as an ammeter may be used to physically observe the value of this mesh current. This will always be the case when a mesh is completely surrounded by other meshes.

The methods for writing the mesh equations of a network as presented in this section apply irrespective of whether the network resistors are time-in variant or time-varying. An example of the latter case follows.

EXAMPLE 3-2.3. In the two-mesh resistance network shown in Fig. 3-2.4, the values of the resistive elements are defined by expressions which are functions of time. The network equations may be written directly by applying the method defined above. Thus, the element

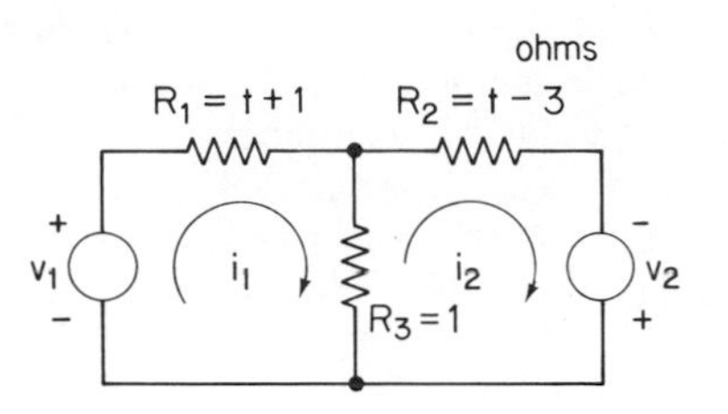

Fig. 3-2.4 A time-varying resistance network.

r_{11} of the resistance matrix **R** may be written as $r_{11}(t)$, and it is found by adding the expressions for the values of the two resistors R_1 and R_3 which are in the first mesh. Thus, $r_{11}(t) = t + 2$. Continuing in this manner, we may obtain the following matrix expression relating the voltage and current variables:

$$\begin{bmatrix} v_1(t) \\ v_2(t) \end{bmatrix} = \begin{bmatrix} t+2 & -1 \\ -1 & t-2 \end{bmatrix} \begin{bmatrix} i_1(t) \\ i_2(t) \end{bmatrix}$$

The inverse set of equations defining the mesh currents $i_1(t)$ and $i_2(t)$ in terms of the excitation voltages $v_1(t)$ and $v_2(t)$ may be found by forming the inverse matrix using the relations given in Eq. (3-17). We find that

$$\begin{bmatrix} i_1(t) \\ i_2(t) \end{bmatrix} = \frac{1}{t^2 - 5} \begin{bmatrix} t-2 & +1 \\ +1 & t+2 \end{bmatrix} \begin{bmatrix} v_1(t) \\ v_2(t) \end{bmatrix}$$

```
C     PORTION OF PROGRAM TO COMPUTE ELEMENTS OF RESISTANCE
C     MATRIX R FROM THE VALUES OF THE RESISTORS AND THE
C     NUMBERS OF THE MESHES IN WHICH THESE RESISTORS ARE
C     LOCATED.  EACH OF THE MESHES MUST HAVE THE SAME
C     REFERENCE DIRECTION, AND NO RESISTOR CAN BE PRESENT
C     IN MORE THAN TWO MESHES
C                N - NUMBER OF MESHES
C               NR - NUMBER OF RESISTORS
C              RES - ARRAY OF RESISTOR VALUES
C            MESH1 - FIRST ARRAY OF MESH NUMBERS
C            MESH2 - SECOND ARRAY OF MESH NUMBERS (IF RESISTOR
C                        OCCURS ONLY IN A SINGLE MESH, THE
C                        CORRESPONDING ELEMENT IN MESH2 IS ZERO)
C                R - OUTPUT RESISTANCE MATRIX
C
      DO 11 I = 1, NR
      J = MESH1(I)
      K = MESH2(I)
C
C     TEST TO DETERMINE WHETHER THE RESISTOR OCCURS
C     ONLY IN A SINGLE MESH
      IF (K) 11, 11, 8
    8 R(K,K) = R(K,K) + RES(I)
      R(J,K) = R(J,K) - RES(I)
      R(K,J) = R(J,K)
   11 R(J,J) = R(J,J) + RES(I)
```

Fig. 3-2.5 A set of statements for forming the array R.

It should be noted that the determinant of the set of equations given in the above example is equal to $t^2 - 5$. Thus, no solution exists for the set of equations when $t = \pm\sqrt{5}$. This fact is easily verified physically by determining the input resistance of the network at the terminals to which the sources are connected for these values of time. For example, let R_{in} be the resistance at the terminals to which source 1 is connected (assuming source 2 is set to zero). We see that

$$R_{\text{in}}(t)\Big|_{t=\sqrt{5}} = (t+1) + \frac{t-3}{t-2}\Big|_{t=\sqrt{5}} = 0$$

Thus, the ideal voltage source is effectively short-circuited, and an indeterminate (infinite) current flows. Similarly, it may be shown that source 2 has zero resistance across its terminals at $t = \sqrt{5}$. Thus, there is no (finite) solution at this instant of time.

The operations involved in forming the resistance matrix from the values of time-invariant resistors and the identifying numbers of the meshes in which these resistors are located may easily be implemented for the digital computer. If the values of the resistors are stored in a one-dimensional array RES, and the corresponding values of the mesh numbers in which the resistors are located are stored in one-dimensional arrays MESH1 and MESH2, then if there are NR resistors, a portion of

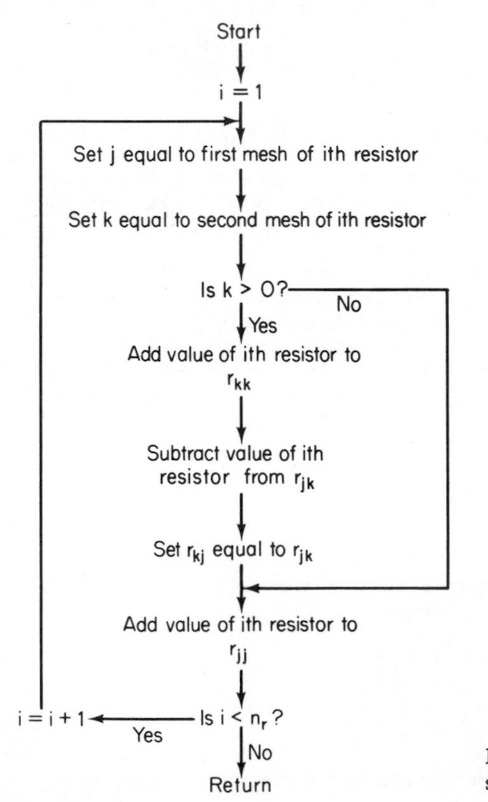

Fig. 3-2.6 Flow chart for the statements given in Fig. 3-2.5.

a program for computing the effect of these resistors in determining the elements of a two-dimensional array R will have the form shown in Fig. 3–2.5. A flow chart for the logic of this program segment is shown in Fig. 3–2.6. In programming the algorithm, it has been assumed that a resistor which occurs only in a single mesh will be identified by having the corresponding element of the MESH1 array set equal to the mesh number, and the element of the MESH2 array set to zero.

The program segment shown in Fig. 3–2.5 may be more conveniently used by including it as part of a subroutine. If we use the variable N for the number of meshes, then the argument listing for the subroutine may be put in the form (N, NR, RES, MESH1, MESH2, R). If we use the name RMESH for this subroutine, its identifying statement will have the form

SUBROUTINE RMESH (N, NR, RES, MESH1, MESH2, R)

A listing of the statements of the subroutine RMESH is given in Fig. 3–2.7. In addition to including the program segment shown in Fig. 3–2.5, a DO loop has been

```
      SUBROUTINE RMESH (N, NR, RES, MESH1, MESH2, R)
C     SUBROUTINE TO COMPUTE ELEMENTS OF RESISTANCE MATRIX R
C     FROM THE VALUES OF THE RESISTORS AND THE NUMBERS OF
C     THE MESHES IN WHICH THESE RESISTORS ARE LOCATED.
C     EACH OF THE MESHES MUST HAVE THE SAME REFERENCE
C     DIRECTION, AND NO RESISTOR CAN BE PRESENT IN MORE
C     THAN TWO MESHES
C            N - NUMBER OF MESHES
C           NR - NUMBER OF RESISTORS
C          RES - ARRAY OF RESISTOR VALUES
C        MESH1 - FIRST ARRAY OF MESH NUMBERS
C        MESH2 - SECOND ARRAY OF MESH NUMBERS (IF RESISTOR
C                    OCCURS ONLY IN A SINGLE MESH, THE
C                    CORRESPONDING ELEMENT IN MESH2 IS ZERO)
C            R - OUTPUT RESISTANCE MATRIX
      DIMENSION RES(20), MESH1(20), MESH2(20), R(10,10)
C
C     SET THE ELEMENTS OF THE RESISTANCE MATRIX R TO ZERO
      DO 3  I = 1, N
      DO 3  J = 1, N
    3 R(I,J)  = 0.
C
C     MODIFY THE ELEMENTS OF THE RESISTANCE MATRIX R FOR
C     EACH OF THE RESISTORS
      DO 11 I = 1, NR
      J = MESH1(I)
      K = MESH2(I)
C
C     TEST TO DETERMINE WHETHER THE RESISTOR OCCURS
C     ONLY IN A SINGLE MESH
      IF (K) 11, 11, 8
    8 R(K,K) = R(K,K) + RES(I)
      R(J,K) = R(J,K) - RES(I)
      R(K,J) = R(J,K)
   11 R(J,J) = R(J,J) + RES(I)
      RETURN
      END
```

Fig. 3-2.7 Listing of the subroutine RMESH.

TABLE 3-2.1 Summary of the Characteristics of the Subroutine RMESH

Identifying Statement: SUBROUTINE RMESH (N, NR, RES, MESH1, MESH2, R)

Purpose: To compute the elements of the matrix **R** which describes a resistance network.

Additional Subprograms Required: None.

Input Arguments:

- N The number of meshes in the network.
- NR The number of resistors in the network.
- RES The one-dimensional array of variables RES(I) in which are stored the values of the resistors.
- MESH1 The one-dimensional array of integer variables MESH1(I) in which are stored the values of the first mesh in which the Ith resistor is connected.
- MESH2 The one-dimensional array of integer variables MESH2(I) in which are stored the values of the second mesh in which the Ith resistor is connected. If the Ith resistor occurs only in a single mesh, the variable MESH2(I) must be set equal to zero.

Output Argument:

- R The two-dimensional array of variables R(I,J) in which are stored the elements r_{ij} of the resistance matrix **R**.

Note: The variables of the subroutine are dimensioned as follows: RES(20), MESH1(20), MESH2(20), R(10,10).

```
C     MAIN PROGRAM FOR EXAMPLE 3.2-4
C            N - NUMBER OF MESHES
C           NR - NUMBER OF RESISTORS
C          RES - ARRAY OF RESISTOR VALUES
C        MESH1 - FIRST ARRAY OF MESH NUMBERS
C        MESH2 - SECOND ARRAY OF MESH NUMBERS
C            R - RESISTANCE MATRIX
      DIMENSION R(10,10), RES(20), MESH1(20), MESH2(20)
C
C     READ INPUT DATA
      READ 2, N, NR
    2 FORMAT (2I2)
      READ 4, (MESH1(I), MESH2(I), RES(I), I = 1, NR)
    4 FORMAT (2I1, 8X, E10.0)
C
C     CALL THE SUBROUTINE RMESH TO CONSTRUCT
C     THE RESISTANCE MATRIX R
      CALL RMESH (N, NR, RES, MESH1, MESH2, R)
C
C     PRINT A LISTING OF THE RESISTANCE MATRIX R
      PRINT 7
    7 FORMAT (10X,35H* * *  RESISTANCE MATRIX  * * * */)
      DO 11  I = 1, N
      PRINT 10, (R(I,J), J = 1, N)
   10 FORMAT (10F6.1)
   11 CONTINUE
      STOP
      END
```

Fig. 3-2.8 Main program for determining a resistance matrix.

added to the subroutine to set the elements of the R array to zero, before adding the effects of the resistors to the appropriate elements of this array. A summary of the properties of the subroutine RMESH is given in Table 3–2.1. An example of its use follows.

EXAMPLE 3-2.4. As an example of the use of the subroutine RMESH, we will use it to corroborate the values found for the resistance matrix in Example 3-2.2. To do this, a main program is needed to read values for the number of meshes, the number of resistors, and the mesh numbers and the value of resistance for each resistor. The main program then simply calls the subroutine RMESH, and prints the resulting values of the elements of the resistance matrix. A listing of such a main program is given in Fig. 3-2.8. A listing of the input data and the resulting output is given in Fig. 3-2.9. A comparison of the output data with the resistance matrix given in Example 3-2.2 readily verifies the accuracy of the results.

```
INPUT DATA FOR EXAMPLE 3.2-4

 914
10          1.
12          1.
23          1.
30          1.
14          1.
25          1.
36          1.
45          1.
56          1.
47          1.
58          1.
69          1.
78          1.
89          1.
```

```
OUTPUT DATA FOR EXAMPLE 3.2-4

              * * * *  RESISTANCE MATRIX  * * * *

  3.0  -1.0   0.0  -1.0   0.0   0.0   0.0   0.0   0.0
 -1.0   3.0  -1.0   0.0  -1.0   0.0   0.0   0.0   0.0
  0.0  -1.0   3.0   0.0   0.0  -1.0   0.0   0.0   0.0
 -1.0   0.0   0.0   3.0  -1.0   0.0  -1.0   0.0   0.0
  0.0  -1.0   0.0  -1.0   4.0  -1.0   0.0  -1.0   0.0
  0.0   0.0  -1.0   0.0  -1.0   3.0   0.0   0.0  -1.0
  0.0   0.0   0.0  -1.0   0.0   0.0   2.0  -1.0   0.0
  0.0   0.0   0.0   0.0  -1.0   0.0  -1.0   3.0  -1.0
  0.0   0.0   0.0   0.0   0.0  -1.0   0.0  -1.0   2.0
```

Fig. 3-2.9 Input and output data for Example 3-2.4.

3-3 SOLVING THE MESH EQUATIONS

In the preceding section we showed how a resistance network comprised of voltage sources and resistors could be described by a set of simultaneous equations in which the mesh currents were the unknowns. Solving this set of equations, i.e., finding the values of the mesh currents for a given set of values of voltage sources, may be said to give us a *solution for the network*, since, if the mesh currents are known, we may easily find any individual branch current and/or voltage, and thus we may determine the value of any variable in the network.

There are many techniques which may be applied to solve a set of simultaneous equations. In this section we shall describe the use of determinants to perform such a solution. A different technique, which is more easily adapted to digital computer solution, will be covered in the following section.

If we consider a set of simultaneous equations written in matrix form as

$$\mathbf{v}(t) = \mathbf{Ri}(t) \tag{3-24}$$

where the various matrices are defined in (3–23), then the *determinant* of the set of equations has the symbolic representation

$$\det \mathbf{R} = \begin{vmatrix} r_{11} & r_{12} & \cdots & r_{1n} \\ r_{21} & r_{22} & \cdots & r_{2n} \\ \cdot & \cdot & \cdots & \cdot \\ \cdot & \cdot & \cdots & \cdot \\ \cdot & \cdot & \cdots & \cdot \\ r_{n1} & r_{n2} & \cdots & r_{nn} \end{vmatrix} \tag{3-25}$$

This is frequently referred to as a *determinant of order n.* Such a determinant has a definite (scalar) value which is determined by the elements r_{ij}. The rules for finding the value of a determinant are summarized in Appendix A.

A second scalar quantity which may be defined for a square matrix is the *cofactor*. For a matrix **R** with elements r_{ij} defined as in Eq. (3–23) it is customary to use the symbol R_{ij} to designate the cofactors. Specifically, the cofactor R_{ij} is the determinant formed by deleting the ith row and jth column from the array of elements, taking the determinant of the remaining array, and multiplying by $(-1)^{i+j}$. A more complete description of cofactors is given in Appendix A.

One of the most important uses of determinants and cofactors is in finding the solutions of sets of simultaneous equations such as arise when the mesh equations are written for a resistance network. An example of such a solution was given for a two-mesh network in Sec. 3–1. Let us now consider again a set of equations describing a three-mesh network. These will always have the form

$$\begin{bmatrix} v_1(t) \\ v_2(t) \\ v_3(t) \end{bmatrix} = \begin{bmatrix} r_{11} & r_{12} & r_{13} \\ r_{21} & r_{22} & r_{23} \\ r_{31} & r_{32} & r_{33} \end{bmatrix} \begin{bmatrix} i_1(t) \\ i_2(t) \\ i_3(t) \end{bmatrix} \tag{3-26}$$

The solution to such a set of simultaneous equations may be expressed in the form

$$\begin{bmatrix} i_1(t) \\ i_2(t) \\ i_3(t) \end{bmatrix} = \frac{1}{\det \mathbf{R}} \begin{bmatrix} R_{11} & R_{21} & R_{31} \\ R_{12} & R_{22} & R_{32} \\ R_{13} & R_{23} & R_{33} \end{bmatrix} \begin{bmatrix} v_1(t) \\ v_2(t) \\ v_3(t) \end{bmatrix} \tag{3-27}$$

where the elements R_{ij} are cofactors. The square matrix of cofactors is sometimes referred to as the *adjoint* matrix. Note that the indices of the cofactors are reversed, i.e., the element in the ith row and jth column of the square adjoint matrix of Eq. (3–27) is the cofactor formed by deleting the jth row and the ith column from the original array of the elements of **R**. As an example of the use of Eq. (3–27) to determine the unknown mesh currents, consider the case where it is desired to find only $i_1(t)$ in a three-mesh network specified by the relations of Eq. (3–26). From Eq. (3–27), we obtain

$$i_1(t) = \frac{v_1(t)\begin{vmatrix} r_{22} & r_{23} \\ r_{32} & r_{33} \end{vmatrix} - v_2(t)\begin{vmatrix} r_{12} & r_{13} \\ r_{32} & r_{33} \end{vmatrix} + v_3(t)\begin{vmatrix} r_{12} & r_{13} \\ r_{22} & r_{23} \end{vmatrix}}{\det \mathbf{R}} \tag{3-28}$$

A more convenient form for expressing the above relation is

$$i_1(t) = \frac{\begin{vmatrix} v_1(t) & r_{12} & r_{13} \\ v_2(t) & r_{22} & r_{23} \\ v_3(t) & r_{32} & r_{33} \end{vmatrix}}{\det \mathbf{R}} \tag{3-29}$$

Thus, we may say that the value of the mesh current $i_1(t)$ may be found by replacing the first column of the matrix **R** by the elements of the matrix **v**, taking the determinant of the result, and dividing this by the determinant of the matrix **R**. A necessary and sufficient condition for a solution to exist is that the determinant of **R** be non-zero. In this case, the matrix **R** is said to be non-singular. An example follows.

EXAMPLE 3-3.1. We may apply the procedure described above to obtain a solution for the mesh currents of the three-mesh resistance network analyzed in Example 3-2.1. The **R** matrix is

$$\mathbf{R} = \begin{bmatrix} 3 & -2 & -1 \\ -2 & 5 & -3 \\ -1 & -3 & 8 \end{bmatrix}$$

The value of the determinant of the resistance matrix for this network is computed in Appendix A and has a value of 44. Inserting the values of the elements of the matrix into Eq. (3-28), we obtain

$$i_1(t) = \frac{v_1(t)\begin{vmatrix} 5 & -3 \\ -3 & 8 \end{vmatrix} - v_2(t)\begin{vmatrix} -2 & -1 \\ -3 & 8 \end{vmatrix} + v_3(t)\begin{vmatrix} -2 & -1 \\ 5 & -3 \end{vmatrix}}{44}$$

Evaluating the second-order determinants, we obtain the result

$$i_1(t) = (\tfrac{31}{44})v_1(t) + (\tfrac{19}{44})v_2(t) + (\tfrac{11}{44})v_3(t)$$

If we now insert the known values of the source voltages specified in Example 3-2.1, namely, $v_1 = 7$ V, $v_2 = -12$ V, and $v_3 = 9$ V, we obtain the result that $i_1 = 2$ A. Following a similar procedure, we see that

$$i_2 = \frac{-7\begin{vmatrix} -2 & -3 \\ -1 & 8 \end{vmatrix} - 12\begin{vmatrix} 3 & -1 \\ -1 & 8 \end{vmatrix} - 9\begin{vmatrix} 3 & -1 \\ -2 & -3 \end{vmatrix}}{44} = -1\text{A}$$

$$i_3 = \frac{7\begin{vmatrix} -2 & 5 \\ -1 & -3 \end{vmatrix} + 12\begin{vmatrix} 3 & -2 \\ -1 & -3 \end{vmatrix} + 9\begin{vmatrix} 3 & -2 \\ -2 & 5 \end{vmatrix}}{44} = 1\text{A}$$

Thus we have obtained the solution for the mesh currents of a three-mesh resistance network.

The above procedure is frequently referred to as *Cramer's rule*. We may easily generalize this to the case of solving an nth order array produced by writing the mesh equations for an n-mesh resistance network. In such a case, the equations will have the form

$$\begin{bmatrix} v_1(t) \\ v_2(t) \\ v_3(t) \\ \cdot \\ \cdot \\ \cdot \\ v_n(t) \end{bmatrix} = \begin{bmatrix} r_{11} & r_{12} & r_{13} & \cdots & r_{1n} \\ r_{21} & r_{22} & r_{23} & \cdots & r_{2n} \\ r_{31} & r_{32} & r_{33} & \cdots & r_{3n} \\ \cdot & \cdot & \cdot & \cdots & \cdot \\ \cdot & \cdot & \cdot & \cdots & \cdot \\ \cdot & \cdot & \cdot & \cdots & \cdot \\ r_{n1} & r_{n2} & r_{n3} & \cdots & r_{nn} \end{bmatrix} \begin{bmatrix} i_1(t) \\ i_2(t) \\ i_3(t) \\ \cdot \\ \cdot \\ \cdot \\ i_n(t) \end{bmatrix} \tag{3-30}$$

If we use the usual symbols for the matrices defined above, we may write

$$\mathbf{v}(t) = \mathbf{R}\mathbf{i}(t) \tag{3-31}$$

The solution to the above set of equations will then have the form

$$\mathbf{i}(t) = \mathbf{A}\mathbf{v}(t) \tag{3-32}$$

where the matrix **A** is referred to as the inverse of the matrix **R**, namely

$$\mathbf{A} = \mathbf{R}^{-1} \tag{3-33}$$

and where the elements of **A** have the dimensions of reciprocal resistance, namely conductance. The matrix equation of (3–32) may be written in the form

$$\begin{bmatrix} i_1(t) \\ i_2(t) \\ i_3(t) \\ \cdot \\ \cdot \\ \cdot \\ i_n(t) \end{bmatrix} = \begin{bmatrix} a_{11} & a_{12} & a_{13} & \cdots & a_{1n} \\ a_{21} & a_{22} & a_{23} & \cdots & a_{2n} \\ a_{31} & a_{32} & a_{33} & \cdots & a_{3n} \\ \cdot & \cdot & \cdot & \cdots & \cdot \\ \cdot & \cdot & \cdot & \cdots & \cdot \\ \cdot & \cdot & \cdot & \cdots & \cdot \\ a_{n1} & a_{n2} & a_{n3} & \cdots & a_{nn} \end{bmatrix} \begin{bmatrix} v_1(t) \\ v_2(t) \\ v_3(t) \\ \cdot \\ \cdot \\ \cdot \\ v_n(t) \end{bmatrix} \tag{3–34}$$

From Cramer's rule, the elements r_{ij} of the matrix **R** of Eq. (3–30) are related to the elements a_{ij} of the matrix **A** of Eq. (3–34) (which is the inverse of **R**) by the equation

$$a_{ij} = \frac{R_{ji}}{\det \mathbf{R}} \tag{3–35}$$

where R_{ji} is the cofactor formed by taking the determinant of the array **R** after the jth row and ith column have been deleted, and multiplying its value by $(-1)^{j+i}$. If we compare Eq. (3–35) with the relations for the inverse of the third-order matrix given in Eq. (3–27) it is readily observed that these relations follow the general rule given above. It may be shown that the necessary and sufficient condition under which a solution for Eq. (3–30) exists, i.e., for which **R** has an inverse, is that det $\mathbf{R} \neq 0$. Note that, in verification of this, if det **R** should be equal to zero, the elements of **A** as defined by Eq. (3–35) are indeterminate.

The use of Cramer's rule to determine the elements of the adjoint matrix, and thus to provide a solution to a set of simultaneous equations, provides a general means for solving for the mesh currents of a resistance network. The major application of this rule is found in solving two- and three-mesh networks. To solve for the mesh currents of more complicated networks, techniques which are easily implemented on the digital computer are required. Such a technique is described in the following section.

3-4 DIGITAL COMPUTER TECHNIQUES FOR SOLVING THE MESH EQUATIONS

In the preceding section we presented a general method for solving the mesh equations of a resistance network. The method involved the use of determinants. Although, in theory, such an approach may be applied to quite large networks, i.e., to sets of equations for resistance networks with many meshes, serious practical difficulties arise when this is attempted. The logical solution is to turn the details of the computation of the determinants over to a computer. Here again, however, difficulties arise in that the use of determinants is a relatively inefficient way of solving simultaneous equations on a computer. We are led, therefore, to seek for other methods of finding the solution to a set of simultaneous equations, which are more amenable to digital computation. In this section we shall discuss such a method. As an illustration of the relative efficiencies of the two methods, let us consider the number of multiplications required in determining the solution of a set of n equations, i.e., of an nth order system. For large values of n, the method to be described here requires approximately $n^3/2$ multiplications.[5] For the use of determinants, however, $(n-1)(n+1)!$ are required. As an example of the size of this latter number, for $n = 10$, it has the value 359,251,200.[6] Such a number of operations requires over a day of computation time even on the modern, fast computers currently available.

The method of solving the mesh equations of a resistance network to be presented here is called the *Gauss-Jordan reduction method.* To show how this method is used, we will rework the problem given as Example 3–1.1 for the resistance network shown in Fig. 3–1.3. The matrix equations for a two-mesh network have the general form

$$\mathbf{v} = \mathbf{R}\mathbf{i} \tag{3–36}$$

For the network shown in the figure, this equation becomes

$$\begin{bmatrix} 8 \\ -9 \end{bmatrix} = \begin{bmatrix} 3 & -2 \\ -2 & 5 \end{bmatrix} \begin{bmatrix} i_1 \\ i_2 \end{bmatrix} \tag{3–37}$$

As a first step in applying the Gauss-Jordan method to the solution of this set of simultaneous equations, we shall define an *augmented* matrix $\mathbf{R}_a$ consisting of the elements of the matrix $\mathbf{R}$ with the elements of the matrix $\mathbf{v}$ added as a third column. Thus we see that for this example

$$\mathbf{R}_a = \begin{bmatrix} 3 & -2 & 8 \\ -2 & 5 & -9 \end{bmatrix} \tag{3–38}$$

Now let us multiply the first row of $\mathbf{R}_a$ by some constant so as to make the first element in the row unity. In this case the constant is clearly $\frac{1}{3}$. Performing this operation, leaving the second row invariant, and calling the result $\mathbf{R}'_a$, we obtain

$$\mathbf{R}'_a = \begin{bmatrix} 1 & -\frac{2}{3} & \frac{8}{3} \\ -2 & 5 & -9 \end{bmatrix} \tag{3–39}$$

[5]Actually, there are even more efficient algorithms available. For example, the Gauss elimination technique requires approximately $n^3/3$ multiplications, although it is slightly more complicated to program than the method given here. See A. Ralston, *A First Course in Numerical Analysis*, Chap. 9, McGraw-Hill Book Company, New York, 1965.

[6]G. E. Forsythe, *Modern Mathematics for the Engineer*, edited by E. F. Beckenbach, p. 436, McGraw-Hill Book Company, New York, 1956.

As a second step in this process, let us add some multiple of the first row of $\mathbf{R}'_a$ in Eq. (3–39) to the second row in such a manner as to make the first element in the second row zero. The multiple in this case is 2. Performing this operation, leaving the first row invariant, and calling the result $\mathbf{R}''_a$, we obtain

$$\mathbf{R}''_a = \begin{bmatrix} 1 & -\dfrac{2}{3} & \dfrac{8}{3} \\ -2+2(1) & 5+2\left(-\dfrac{2}{3}\right) & -9+2\left(\dfrac{8}{3}\right) \end{bmatrix}$$

$$= \begin{bmatrix} 1 & -\dfrac{2}{3} & \dfrac{8}{3} \\ 0 & \dfrac{11}{3} & -\dfrac{11}{3} \end{bmatrix}$$

As a third step in this process, let us multiply the second row of $\mathbf{R}''_a$ by a constant so as to make the first non-zero element in the row unity. The constant is $\frac{3}{11}$. Leaving the first row invariant and calling the result $\mathbf{R}'''_a$, we obtain

$$\mathbf{R}'''_a = \begin{bmatrix} 1 & -\dfrac{2}{3} & \dfrac{8}{3} \\ 0 & 1 & -1 \end{bmatrix}$$

As our final step, let us add a multiple of the second row of $\mathbf{R}'''_a$ to the first row in such a manner as to make the second element in the first row zero. The multiple in this case is $\frac{2}{3}$. Performing this operation, leaving the second row invariant, and calling the result $\mathbf{R}''''_a$, we obtain

$$\mathbf{R}''''_a = \begin{bmatrix} 1 & 0 & \left(\dfrac{8}{3}\right)-1\left(\dfrac{2}{3}\right) \\ 0 & 1 & -1 \end{bmatrix} = \begin{bmatrix} 1 & 0 & 2 \\ 0 & 1 & -1 \end{bmatrix} \tag{3–40}$$

It should be noted that the result of the above manipulations has been to convert the original $\mathbf{R}$ matrix (the left two columns of the augmented matrix $\mathbf{R}_a$) to an identity matrix, i.e., one with ones on the main diagonal and zeros elsewhere. As a result of these operations, the *right column* of the augmented matrix $\mathbf{R}_a$ has been converted to a matrix *giving the solution to the set of simultaneous equations*, i.e., this column gives the values of the elements of the matrix $\mathbf{i}$ of Eq. (3–36) which satisfy the simultaneous equations. Thus we see from Eq. (3–40) that $i_1 = 2$ A and $i_2 = -1$ A. The validity of these values is easily checked with the solution given in Example 3–1.1.

The above set of operations is easily applied to more complex sets of equations. As an illustration, consider the equations for the three-mesh network given in Example 3–2.1. For this case, the augmented matrix $\mathbf{R}_a$ is

$$\left[\begin{array}{rrr|r} 3 & -2 & -1 & 7 \\ -2 & 5 & -3 & -12 \\ -1 & -3 & 8 & 9 \end{array}\right] \tag{3-41}$$

Setting the leading element of the first row to unity, we obtain

$$\left[\begin{array}{rrr|r} 1 & -\frac{2}{3} & -\frac{1}{3} & \frac{7}{3} \\ -2 & 5 & -3 & -12 \\ -1 & -3 & 8 & 9 \end{array}\right]$$

Multiplying the first row by 2 and adding it to the second row, and also adding the first row to the third row, we obtain

$$\left[\begin{array}{rrr|r} 1 & -\frac{2}{3} & -\frac{1}{3} & \frac{7}{3} \\ 0 & \frac{11}{3} & -\frac{11}{3} & -\frac{22}{3} \\ 0 & -\frac{11}{3} & \frac{23}{3} & \frac{34}{3} \end{array}\right]$$

Multiplying the second row by the factor $\frac{3}{11}$ so that the leading non-zero element is unity, and adding appropriate multiples of the second row to the first and third rows so that the second element of each is set to zero, we obtain

$$\left[\begin{array}{rrr|r} 1 & 0 & -1 & 1 \\ 0 & 1 & -1 & -2 \\ 0 & 0 & 4 & 4 \end{array}\right]$$

Multiplying the last row by the factor $\frac{1}{4}$ so that the leading non-zero element is unity, and adding this row to the first and the second rows so that the third element of each is set to zero, we obtain

$$\left[\begin{array}{rrr|r} 1 & 0 & 0 & 2 \\ 0 & 1 & 0 & -1 \\ 0 & 0 & 1 & 1 \end{array}\right] \tag{3-42}$$

Following the reasoning used for the two-mesh case, we conclude that the solutions for the mesh currents are given by the elements of the right column, namely $i_1 = 2$ A, $i_2 = -1$ A, $i_3 = 1$ A. These are the values found by the use of determinants in Example 3–3.1.

The above sequence of operations is easily extended to solve any set of simultaneous equations regardless of the number of equations (we assume that a solution does exist). Before making this extension, let us examine why such a procedure provides a solution. If, for a three-mesh resistance network, we write out the simultaneous equations in general form, we obtain

$$\begin{aligned} i_1 r_{11} + i_2 r_{12} + i_3 r_{13} &= v_1 \\ i_1 r_{21} + i_2 r_{22} + i_3 r_{23} &= v_2 \\ i_1 r_{31} + i_2 r_{32} + i_3 r_{33} &= v_3 \end{aligned} \qquad (3\text{–}43)$$

These three equations may also be written using the following notation

$$i_1 r_{i1} + i_2 r_{i2} + i_3 r_{i3} = v_i \qquad i - 1, 2, 3 \qquad (3\text{–}44)$$

If a certain set of values of the currents i_1, i_2, and i_3 satisfy all of the equations given above, they must also satisfy an equation formed by multiplying both sides of any of the equations by an arbitrary constant, i.e., they are also solutions of the equations

$$ki_1 r_{i1} + ki_2 r_{i2} + ki_3 r_{i3} = kv_i \qquad i = 1, 2, 3 \qquad (3\text{–}45)$$

Thus, we conclude that we may multiply any of the simultaneous equations of (3–43) by a non-zero constant without changing the values of the solutions for the set of equations. This is the first of the two operations that we have used in the solution process illustrated above. Similarly, if a set of currents i_1, i_2, and i_3 satisfies the first equation of (3–43), and also satisfies some multiple of the second equation of (3–43), then it may be shown that it also satisfies an equation formed by adding the two equations together, i.e., that it satisfies the equation

$$i_1(r_{11} + kr_{21}) + i_2(r_{12} + kr_{22}) + i_3(r_{13} + kr_{23}) = v_1 + kv_2 \qquad (3\text{–}46)$$

This result is easily extendable to equations formed by other combinations of Eqs. (3–43). We conclude that multiplying any equation by a constant and adding it to any other equation does not change the values of the solutions to the set of equations. Finally, if, by the above operations, the equations have been put into the form

$$\begin{aligned} 1i_1 + 0i_2 + 0i_3 &= c_1 \\ 0i_1 + 1i_2 + 0i_3 &= c_2 \\ 0i_1 + 0i_2 + 1i_3 &= c_3 \end{aligned} \qquad (3\text{–}47)$$

then the solutions are obviously $i_1 = c_1$, $i_2 = c_2$, and $i_3 = c_3$. The operations on the augmented matrix $\mathbf{R}_a$ are exactly the operations justified above; thus, the elements of the right column of the augmented matrix, after the other columns have been transformed to an identity matrix, are exactly the elements c_1, c_2, and c_3 of Eq. (3–47), i.e., they are the solutions for the mesh currents i_1, i_2, and i_3.

The procedure described above is easily generalized and implemented on the digital computer. Consider the resistance matrix $\mathbf{R}$ which results for an n-mesh resistance network. It will have the form

$$\mathbf{R} = \begin{bmatrix} r_{11} & r_{12} & \cdots & r_{1n} \\ r_{21} & r_{22} & \cdots & r_{2n} \\ \cdot & \cdot & \cdots & \cdot \\ \cdot & \cdot & \cdots & \cdot \\ \cdot & \cdot & \cdots & \cdot \\ r_{n1} & r_{n2} & \cdots & r_{nn} \end{bmatrix} \tag{3–48}$$

The augmented matrix $\mathbf{R}_a$ may be written in the form

$$\mathbf{R}_a = \begin{bmatrix} r_{11} & r_{12} & \cdots & r_{1n} & r_{1,n+1} \\ r_{21} & r_{22} & \cdots & r_{2n} & r_{2,n+1} \\ \cdot & \cdot & \cdots & \cdot & \cdot \\ \cdot & \cdot & \cdots & \cdot & \cdot \\ \cdot & \cdot & \cdots & \cdot & \cdot \\ r_{n1} & r_{n2} & \cdots & r_{nn} & r_{n,n+1} \end{bmatrix} \tag{3–49}$$

where the elements of the right column are defined by the relations

$$r_{i,n+1} = v_i \qquad i = 1, 2, \ldots, n \tag{3–50}$$

i.e., they are the elements of the matrix $\mathbf{v}$ of Eq. (3–36).

In terms of the matrix of Eq. (3–49), we may implement the solution procedure by some explicit relations. First of all, consider the normalization of a given row, which we shall refer to as the ith row. This involves setting the diagonal element of the row to unity (it is assumed that the elements to the left of the diagonal element have already been set to zero by the operations on the preceding rows). Each element in the row may be replaced by a normalized element by defining a constant $\alpha = r_{ii}$ and by making the operations

$$r_{ij} \longrightarrow \frac{r_{ij}}{\alpha} \qquad j = i, i+1, \ldots, n+1 \tag{3–51}$$

The arrow in the above relation should be read "is replaced by."[7] Note that the

[7]If the Gauss-Jordan technique described in this section is applied to find a solution for an arbitrary set of simultaneous equations, i.e., a set which is *not* derived from the mesh equations for a resistance network, the numerical accuracy of the results will be improved if the equations are arranged so that the diagonal elements of the matrix are as large as possible. This is frequently referred to as choosing the largest "pivot" elements. Such a feature is usually included as a part of sophisticated programs for solving simultaneous equations. A set of mesh equations automatically possesses this property, thus such a feature is not necessary in our operations in this chapter.

original value of r_{ij} is destroyed by the first such replacement operation $(j = i)$; therefore, we use the constant α to retain the original value of r_{ii} for use in normalizing the other elements of the row. After the normalization described above has been completed, we may use the new elements of the ith row to compute a new set of elements for every other row in the array. These new rows will have the property that the ith element of each has been set to zero. For example, for the kth row, we may define a constant $\beta = r_{ki}$, and then make the replacement operations

$$r_{kj} \longrightarrow r_{kj} - \beta r_{ij} \qquad j = i, i + 1, \ldots, n + 1 \qquad (3\text{-}52)$$

Such a replacement is made for every row except the ith row, i.e., for all values of k from 1 to n, except $k = i$.

It is interesting to observe what happens in the sequence of operations described above in the case of a singular matrix. As an example of this situation, consider the extended matrix

$$\begin{bmatrix} 2 & 1 & 1 \\ 4 & 2 & 0 \end{bmatrix}$$

The matrix formed by the two left columns above is obviously singular since its determinant is zero. Applying the first step of the Gauss-Jordan procedure, we obtain

$$\begin{bmatrix} 1 & \frac{1}{2} & \frac{1}{2} \\ 4 & 2 & 0 \end{bmatrix}$$

Applying the second step, we obtain

$$\begin{bmatrix} 1 & \frac{1}{2} & \frac{1}{2} \\ 0 & 0 & -2 \end{bmatrix}$$

If we now apply Eq. (3–51) to normalize the elements of the second row, we find that we must divide by zero. This will bring any computer to a grinding halt! The only results that are obtained will be a diagnostic message. In general, this is what always happens when one attempts to take the inverse of a singular matrix.

The operations described above are easily programmed for a digital computer using the FORTRAN language and a nest of DO loops. An outer loop is used to successively apply the replacement operations of Eqs. (3–51) and (3–52) to all rows from the first to the nth. An inner DO loop is then used to normalize the elements of the ith row, and a second inner DO loop is used to change the elements of every row except the ith so that the ith elements become zero. The operations are summarized by the flow chart shown in Fig. 3–4.1. A listing of a set of FORTRAN

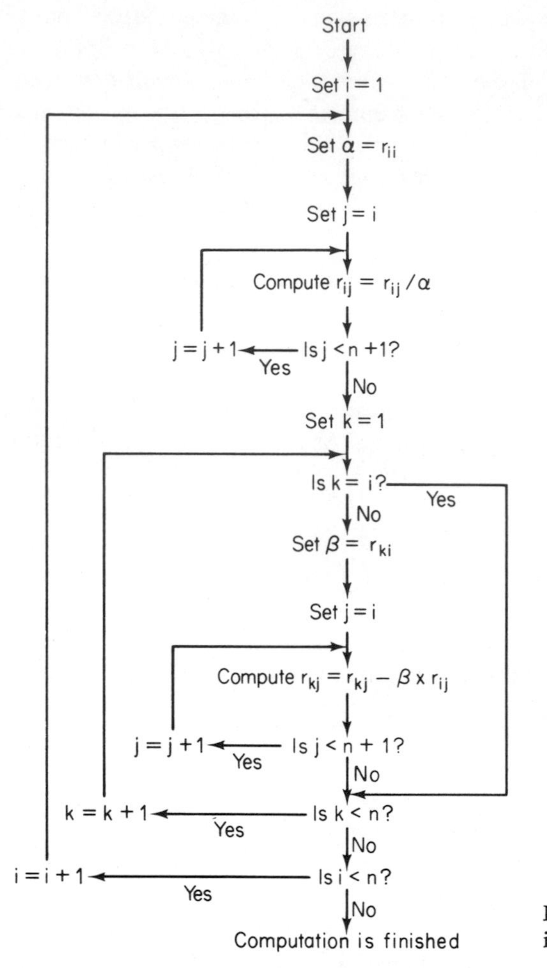

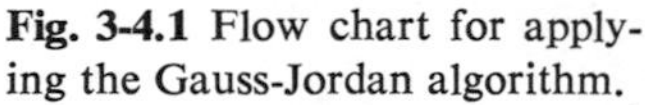
Fig. 3-4.1 Flow chart for applying the Gauss-Jordan algorithm.

statements for performing these operations is given in Fig. 3–4.2. These statements may be adapted to find a solution for the mesh currents of a resistance network. In using the algorithm defined in Fig. 3–4.2, considerably greater flexibility is achieved if the statements listed in the figure are written as a subroutine. Such a subroutine can then be used in many different types of programs, without requiring rewriting, and without the danger of duplicating variable names or statement numbers. Since the fundamental problem considered in this section is the solution of the matrix equation $\mathbf{v} = \mathbf{Ri}$, we may use the FORTRAN variables V(I) for the mesh voltages, the variables AMPS(I) for the mesh currents, and the variables R(I,J) for the elements of the resistance matrix **R**. It should be noted that the algorithm defined in Fig. 3–4.2 destroys the original values of the elements of the matrix **R**, since it converts **R** to an identity matrix. This may be undesirable in many applications. Therefore, in our subroutine we will use an internal two-dimensional array of variables

```
C     PORTION OF PROGRAM TO REDUCE AUGMENTED MATRIX RA
C     TO SOLVE A SET OF N SIMULTANEOUS EQUATIONS.
C     NP IS NO. OF COLUMNS IN AUGMENTED MATRIX,
C     ALFA AND BETA ARE TEMPORARY STORAGE
        NP = N + 1
        DO 12  I = 1, N
C
C     SET MAIN DIAGONAL ELEMENTS TO UNITY.
        ALFA = RA(I,I)
        DO  5  J = I, NP
      5 RA(I,J) = RA(I,J) / ALFA
C
C     SET ELEMENTS OF ITH COLUMN TO ZERO
        DO 11  K = 1, N
        IF (K - I) 8, 11, 8
      8 BETA = RA(K,I)
        DO 10  J = I, NP
     10 RA(K,J) = RA(K,J) - BETA * RA(I,J)
     11 CONTINUE
     12 CONTINUE
```

Fig. 3-4.2 Statements for applying the Gauss-Jordan algorithm.

RA(I,J) (for R Augmented matrix) on which all our reduction operations will take place, and preserve the values of the **R** matrix in a second two-dimensional array of variables R(I,J). The subroutine will perform the following operations:

1. Store the values of the elements of the R array in the RA array.
2. Store the values of the variables V(I) as the $(n + 1)$st column of the RA array.
3. Apply the algorithm shown in Fig. 3–4.2 to reduce the RA array.
4. Output the values of the $(n + 1)$st column of the reduced RA array as the variables AMPS(I).

Thus, the subroutine must be supplied with input information giving the values of the variables V(I), R(I, J), and N (the number of simultaneous equations). As output it will supply the n values of the variables AMPS(I). If we use the name GJSEQ (for Gauss-Jordan Simultaneous EQuation solution) for the subroutine, it must be identified by a statement of the form

SUBROUTINE GJSEQ (V, R, AMPS, N)

A listing of such a subroutine is given in Fig. 3–4.3. Note that the statements from Fig. 3–4.2 appear without change in the listing. A summary of the characteristics of the subroutine GJSEQ is given in Table 3–4.1.

The subroutine listed in Fig. 3–4.3 is easily incorporated as part of a general program for solving for the mesh currents of a resistance network. An example of such an application follows.

```
      SUBROUTINE GJSEQ (V, R, AMPS, N)
C     SUBROUTINE TO SOLVE A SET OF SIMULTANEOUS EQUATIONS V=R*AMPS
C     USING GAUSS-JORDAN REDUCTION. THE R MATRIX IS PRESERVED
C            V - COLUMN MATRIX OF KNOWN VARIABLES
C            R - SQUARE NON-SINGULAR MATRIX
C         AMPS - OUTPUT COLUMN MATRIX
C            N - NUMBER OF EQUATIONS
      DIMENSION V(10), R(10,10), AMPS(10), RA(10,11)
C
C     ENTER R MATRIX IN RA ARRAY AND ENTER V AS
C     N+1TH COLUMN OF RA ARRAY
      DO 23  I = 1, N
      DO 22  J = 1, N
   22 RA(I,J) = R(I,J)
   23 RA(I,N+1) = V(I)
C
C     PORTION OF PROGRAM TO REDUCE AUGMENTED
C     MATRIX RA TO SOLVE A SET OF N SIMULTANEOUS EQUATIONS.
C     NP IS NO. OF COLUMNS IN AUGMENTED MATRIX,
C     ALFA AND BETA ARE TEMPORARY STORAGE
      NP = N + 1
      DO 12  I = 1, N
C
C     SET MAIN DIAGONAL ELEMENTS TO UNITY.
      ALFA = RA(I,I)
      DO  5  J = I, NP
    5 RA(I,J) = RA(I,J) / ALFA
C
C     SET ELEMENTS OF ITH COLUMN TO ZERO
      DO 11  K = 1, N
      IF (K - I) 8, 11, 8
    8 BETA = RA(K,I)
      DO 10  J = I, NP
   10 RA(K,J) = RA(K,J) - BETA * RA(I,J)
   11 CONTINUE
   12 CONTINUE
C
C     SET OUTPUT MATRIX AMPS EQUAL TO LAST COLUMN OF
C     AUGMENTED MATRIX RA
      DO 31  I = 1, N
   31 AMPS(I) = RA(I,NP)
   32 RETURN
      END
```

Fig. 3-4.3 Listing of the subroutine GJSEQ.

EXAMPLE 3-4.1. A five-mesh resistance network is shown in Fig. 3-4.4. We may use the subroutine GJSEQ given in Fig. 3-4.3 to solve this network, i.e., to find the values of the mesh currents for the specified values of the voltage sources. To do this requires a FORTRAN main program which will perform the following functions:

1. Read as data the values of the resistances and the identifying numbers of the meshes in which these resistances are located.
2. Call the subroutine RMESH to operate on the data read in part (1) above, and compute the elements of the resistance matrix **R**, storing the results as the variables R(I, J).

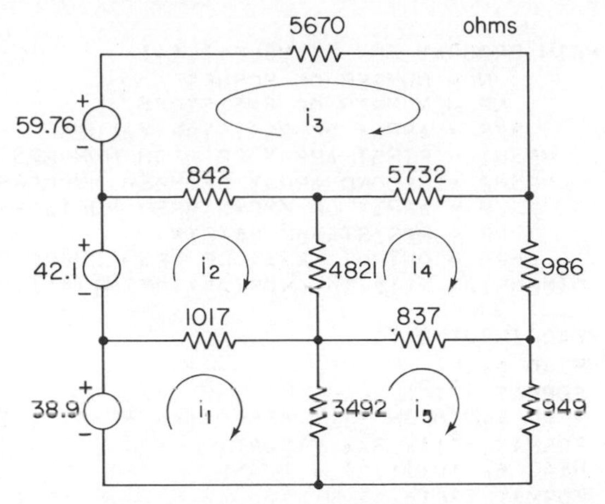

Fig. 3-4.4 A five-mesh resistance network.

3. Read as data the values of the equivalent mesh voltage sources and store these as the variables V(I).
4. Call the subroutine GJSEQ to find the solution of the simultaneous equations and store these as the variables AMPS(I).
5. Print the resulting values of the mesh currents.

A listing of a FORTRAN program for accomplishing the above is shown in Fig. 3-4.5. The separate sections of the program corresponding with the functions outlined above

TABLE 3-4.1 Summary of the Characteristics of the Subroutine GJSEQ

Identifying Statement: SUBROUTINE GJSEQ (V, R, AMPS, N)

Purpose: To solve a set of simultaneous equations having the form

$$\mathbf{v} = \mathbf{Ri}$$

i.e., given the elements of the matrices **v** and **R**, to find the elements of **i**.

Additional Subprograms Required: None.

Input Arguments:

- V The one-dimensional array of variables V(I) in which are stored the values of the known quantities v_i, the elements of the matrix **v**.
- R The two-dimensional array of variables R(I,J) in which are stored the values of the coefficients r_{ij} of the square matrix **R**.
- N The order of the matrix **R**, i.e., the number of simultaneous equations to be solved.

Output Argument:

- AMPS The one-dimensional array of variables AMPS(I) in which are stored the values of the unknown variables i_i of the matrix **i**.

Notes:

1. The dimensions of the variables used in this subroutine are V(10), R(10,10), and AMPS(10).
2. The subroutine uses a Gauss-Jordan reduction algorithm. The original **R** matrix is preserved.

```
C        MAIN PROGRAM FOR EXAMPLE 3.4-1
C                N - NUMBER OF MESHES
C               NR - NUMBER OF RESISTORS
C              RES - ARRAY OF RESISTOR VALUES
C            MESH1 - FIRST ARRAY OF MESH NUMBERS
C            MESH2 - SECOND ARRAY OF MESH NUMBERS
C                V - ARRAY OF KNOWN MESH VOLTAGES
C                R - RESISTANCE MATRIX
C             AMPS - OUTPUT ARRAY OF MESH CURRENTS
      DIMENSION R(10,10),RES(20),MESH1(20),MESH2(20),V(10),AMPS(10)
C
C        READ INPUT DATA
         READ 2, N, NR
       2 FORMAT (2I2)
         READ 4, (MESH1(I), MESH2(I), RES(I), I = 1, NR)
       4 FORMAT (2I1, 8X, E10.0)
         READ 6, (V(I), I = 1, N)
       6 FORMAT (8E10.0)
C
C        CALL THE SUBROUTINE RMESH TO CONSTRUCT
C        THE RESISTANCE MATRIX R
         CALL RMESH(N,NR,RES,MESH1,MESH2,R)
C
C        PRINT A LISTING OF THE VALUES OF THE
C        ELEMENTS OF THE RESISTANCE MATRIX R
         PRINT 9
       9 FORMAT (///, 22X, 30H* * * *   R   MATRIX   * * * *       /)
         DO 13 J = 1, N
         PRINT 12, (R(J,I), I = 1, N)
      12 FORMAT (5E13.5)
      13 CONTINUE
C
C        CALL THE SUBROUTINE GJSEQ TO SOLVE
C        THE SET OF SIMULTANEOUS EQUATIONS
         CALL GJSEQ (V, R, AMPS, N)
C
C        PRINT AN OUTPUT LISTING OF THE VALUES
C        OF THE MESH CURRENTS FOR THE NETWORK.
         PRINT 16
      16 FORMAT (///, 19X, 35H* * * *   MESH   CURRENTS   * * * *       /)
         PRINT 18, (I, AMPS(I), I = 1, N)
      18 FORMAT (1X, 2HI(, I1, 2H)=, E14.7)
         STOP
         END
```

Fig. 3-4.5 Main program for analyzing a resistance network.

are identified by appropriate "comment" statements. A listing of the input data and the output from the program is given in Fig. 3-4.6. The output lists the elements of the resistance matrix **R** (in ohms) as well as the values of the mesh currents (in amperes) calculated by the program. To check the values of the mesh currents computed by the program, a separate program was written to insert these values into the original matrix equation, and to compute the values of the resulting equivalent mesh voltage sources. The results are listed in Fig. 3-4.6. They show good agreement with the actual values of the voltage sources originally entered as data for the program. When using numerical techniques, it is always desirable to perform such a test to insure the validity of the results.

```
INPUT DATA FOR EXAMPLE 3.4-1

0509
12          1017.
23          842.
30          5670.
34          5732.
24          4821.
40          986.
45          837.
15          3492.
50          949.
38.9        42.1        59.76


OUTPUT DATA FOR EXAMPLE 3.4-1

                    * * * *   R   MATRIX   * * * *

  4.50900E+03 -1.01700E+03  0.            0.            -3.49200E+03
 -1.01700E+03  6.68000E+03 -8.42000E+02 -4.82100E+03  0.
  0.          -8.42000E+02  1.22440E+04 -5.73200E+03  0.
  0.          -4.82100E+03 -5.73200E+03  1.23760E+04 -8.37000E+02
 -3.49200E+03  0.           0.          -8.37000E+02  5.27800E+03


                    * * * *   MESH   CURRENTS   * * * *

 I(1)= 3.6338300E-02
 I(2)= 2.8965759E-02
 I(3)= 1.6625657E-02
 I(4)= 2.0833109E-02
 I(5)= 2.7345710E-02


TEST OF RESULTS OF EXAMPLE 3.4-1

              * * * *   COMPUTED   MESH   VOLTAGES   * * * *

 V(1)= 3.8900000E+01
 V(2)= 4.2100000E+01
 V(3)= 5.9760000E+01
 V(4)=-1.2505552E-12
 V(5)= 1.8189894E-12
```

Fig. 3-4.6 Input and output data for Example 3-4.1.

The maximum order of the matrix **R**, i.e., the maximum number of simultaneous equations, to which the Gauss-Jordan solution method may be applied depends on the number of significant figures (the word length) used by the particular computer in performing the calculations. For a computer with an accuracy of 8–10 decimal digits, the method can usually be applied successfully to matrices with an order of up to 10–15. Other numerical techniques are available which may be successfully applied to the solution of sets of simultaneous equations with over a thousand independent variables. A treatment of such techniques, however, is beyond the scope of this text.

3-5 A DIGITAL COMPUTER ALGORITHM FOR FINDING THE INVERSE RESISTANCE MATRIX

The algorithm presented in the last section for finding the solutions of a set of simultaneous equations may be easily extended to determine the inverse of a given matrix **R**. To show how this may be done, let us consider again the mesh equations for the two-mesh resistance network used in Example 3–1.1 and shown in Fig. 3–1.3. For this network, the resistance matrix **R** was shown to be

$$\mathbf{R} = \begin{bmatrix} 3 & -2 \\ -2 & 5 \end{bmatrix} \tag{3-53}$$

Let us now define a 4 × 2 *extended resistance matrix* $\mathbf{R}_e$ whose first two columns contain the elements of the resistance matrix **R** and whose last two columns contain the elements of a 2 × 2 identity matrix, i.e., a matrix whose main diagonal elements are unity and whose other elements are zero. Thus we see that

$$\mathbf{R}_e = \begin{bmatrix} 3 & -2 & 1 & 0 \\ -2 & 5 & 0 & 1 \end{bmatrix} \tag{3-54}$$

We will now proceed to apply the same row operations which were used in the last section to reduce the first two columns of $\mathbf{R}_e$ to the form of an identity matrix. We first multiply the first row of $\mathbf{R}_e$ by $\frac{1}{3}$ so as to make the leading element of the row unity. Thus, we obtain

$$\begin{bmatrix} 1 & -\dfrac{2}{3} & \dfrac{1}{3} & 0 \\ -2 & 5 & 0 & 1 \end{bmatrix}$$

We next modify the second row so that its leading element is zero. This is done by multiplying the first row by 2 and adding it to the second row. The first row is left invariant. The result is

$$\begin{bmatrix} 1 & -\dfrac{2}{3} & \dfrac{1}{3} & 0 \\ 0 & \dfrac{11}{3} & \dfrac{2}{3} & 1 \end{bmatrix}$$

If we now multiply the second row by $\frac{3}{11}$ so as to make the leading non-zero element unity, we obtain

$$\begin{bmatrix} 1 & -\frac{2}{3} & \frac{1}{3} & 0 \\ 0 & 1 & \frac{2}{11} & \frac{3}{11} \end{bmatrix}$$

Finally, we multiply the second row by $\frac{2}{3}$ and add it to the first row. The second row is left invariant. Thus, we obtain

$$\begin{bmatrix} 1 & 0 & \frac{5}{11} & \frac{2}{11} \\ 0 & 1 & \frac{2}{11} & \frac{3}{11} \end{bmatrix} \tag{3-55}$$

The last two columns of the matrix given in (3–55) above are the same as the elements of $\mathbf{R}^{-1}$, i.e., they define the inverse of the **R** matrix given in (3–53). Thus, we may write

$$\mathbf{R}^{-1} = \frac{1}{11}\begin{bmatrix} 5 & 2 \\ 2 & 3 \end{bmatrix} \tag{3-56}$$

This inverse may be verified by comparing it with the results given in Example 3-1.1.

The above set of operations is easily extended to find the inverse matrix of a more complex set of equations. As an illustration of this, consider the equations for the three-mesh network given in Example 3–2.1. For this example, the extended resistance matrix is

$$\begin{bmatrix} 3 & -2 & -1 & 1 & 0 & 0 \\ -2 & 5 & -3 & 0 & 1 & 0 \\ -1 & -3 & 8 & 0 & 0 & 1 \end{bmatrix} \tag{3-57}$$

Setting the leading element of the first row to unity, we obtain

$$\begin{bmatrix} 1 & -\frac{2}{3} & -\frac{1}{3} & \frac{1}{3} & 0 & 0 \\ -2 & 5 & -3 & 0 & 1 & 0 \\ -1 & -3 & 8 & 0 & 0 & 1 \end{bmatrix}$$

Multiplying the first row by 2 and adding it to the second row, and also multiplying the second row by 1 and adding it to the third row, we obtain

$$\begin{bmatrix} 1 & -\frac{2}{3} & -\frac{1}{3} & \frac{1}{3} & 0 & 0 \\ 0 & \frac{11}{3} & -\frac{11}{3} & \frac{2}{3} & 1 & 0 \\ 0 & -\frac{11}{3} & \frac{23}{3} & \frac{1}{3} & 0 & 1 \end{bmatrix}$$

Multiplying the second row by $\frac{3}{11}$ so that the leading non-zero element is set to unity, and adding appropriate multiples of the second row to the first and third rows so that the second element of each of these is set to zero, we obtain

$$\begin{bmatrix} 1 & 0 & -1 & \frac{5}{11} & \frac{3}{11} & 0 \\ 0 & 1 & -1 & \frac{2}{11} & \frac{3}{11} & 0 \\ 0 & 0 & 4 & 1 & 1 & 1 \end{bmatrix}$$

Finally, multiplying the last row by the factor $\frac{1}{4}$, so that the leading non-zero element is set to unity, and adding this equation to the first and second rows so that the third element of each is set to zero, we obtain

$$\begin{bmatrix} 1 & 0 & 0 & \frac{31}{44} & \frac{19}{44} & \frac{11}{44} \\ 0 & 1 & 0 & \frac{19}{44} & \frac{23}{44} & \frac{11}{44} \\ 0 & 0 & 1 & \frac{11}{44} & \frac{11}{44} & \frac{11}{44} \end{bmatrix} \tag{3-58}$$

We conclude that the inverse matrix $\mathbf{R}^{-1}$ is

$$\mathbf{R}^{-1} = \frac{1}{44}\begin{bmatrix} 31 & 19 & 11 \\ 19 & 23 & 11 \\ 11 & 11 & 11 \end{bmatrix} \tag{3-59}$$

As a check on the accuracy of our calculations, we may employ the well-known relation that a matrix multiplied by its inverse is equal to the identity matrix. Thus, we see that

$$\mathbf{RR}^{-1} = \begin{bmatrix} 3 & -2 & -1 \\ -2 & 5 & -3 \\ -1 & -3 & 8 \end{bmatrix} \frac{1}{44}\begin{bmatrix} 31 & 19 & 11 \\ 19 & 23 & 11 \\ 11 & 11 & 11 \end{bmatrix} = \begin{bmatrix} 1 & 0 & 0 \\ 0 & 1 & 0 \\ 0 & 0 & 1 \end{bmatrix} = \mathbf{I} \tag{3-60}$$

which verifies the result.

The above sequence of operations is easily extended to find the inverse of a matrix of any order. In addition, the computational algorithm developed in the last section is easily extended to make this determination. Before making this extension, let us show why such a procedure actually provides the inverse matrix. We will show this for a 3×3 matrix case by first writing out the mesh equations which involve the resistance matrix in the form

$$\begin{aligned} i_1 r_{11} + i_2 r_{12} + i_3 r_{13} &= v_1 a_{11} + v_2 a_{12} + v_3 a_{13} \\ i_1 r_{21} + i_2 r_{22} + i_3 r_{23} &= v_1 a_{21} + v_2 a_{22} + v_3 a_{23} \\ i_1 r_{31} + i_2 r_{32} + i_3 r_{33} &= v_1 a_{31} + v_2 a_{32} + v_3 a_{33} \end{aligned} \tag{3-61}$$

where all of the coefficients a_{ij} are zero except a_{11}, a_{22}, and a_{33}, which are equal to unity. Thus the coefficients a_{ij} determine a matrix $\mathbf{A}$, where $\mathbf{A}$ is equal to the identity matrix. The same operations which were introduced in the preceding section, namely, multiplying any of the equations by a constant, or adding some multiple of one of the equations to another equation, may be applied to the equations of (3–61) without changing the values of the solutions. Thus, these operations may be applied to convert the set of equations given above to the form

$$\begin{aligned} 1i_1 + 0i_2 + 0i_3 &= v_1 k_{11} + v_2 k_{12} + v_3 k_{13} \\ 0i_1 + 1i_2 + 0i_3 &= v_1 k_{21} + v_2 k_{22} + v_3 k_{23} \\ 0i_1 + 0i_2 + 1i_3 &= v_1 k_{31} + v_2 k_{32} + v_3 k_{33} \end{aligned} \tag{3-62}$$

An examination of the equations given in (3–62) shows that each equation gives a relation for one of the mesh current (unknown) variables i_1, i_2, or i_3 as a linear function of the mesh voltages v_1, v_2, and v_3. Thus, these equations represent a general solution for the mesh currents for any set of applied voltages. This is exactly the role represented by the inverse matrix, therefore we conclude that the coefficients k_{ij} represent the elements of the inverse matrix.

An algorithm for computing the inverse matrix for an arbitrary resistance matrix is easily defined. It is quite similar to the algorithm for finding the solution to a set of simultaneous equations developed in the preceding section. Let $\mathbf{R}$ be the resistance matrix which results for an n-mesh resistance network. It will have the form

$$\mathbf{R} = \begin{bmatrix} r_{11} & r_{12} & \cdots & r_{1n} \\ r_{21} & r_{22} & \cdots & r_{2n} \\ \cdot & \cdot & \cdots & \cdot \\ \cdot & \cdot & \cdots & \cdot \\ \cdot & \cdot & \cdots & \cdot \\ r_{n1} & r_{n2} & \cdots & r_{nn} \end{bmatrix} \tag{3-63}$$

We may now define an $n \times 2n$ extended resistance matrix $\mathbf{R}_e$ which has the form

$$\mathbf{R}_e = \begin{bmatrix} r_{11} & \cdots & r_{1n} & r_{1,n+1} & \cdots & r_{1,2n} \\ r_{21} & \cdots & r_{2n} & r_{2,n+1} & \cdots & r_{2,2n} \\ \cdot & \cdots & \cdot & \cdot & \cdots & \cdot \\ \cdot & \cdots & \cdot & \cdot & \cdots & \cdot \\ \cdot & \cdots & \cdot & \cdot & \cdots & \cdot \\ r_{n1} & \cdots & r_{nn} & r_{n,n+1} & \cdots & r_{n,2n} \end{bmatrix} \tag{3-64}$$

where the elements of the last n columns on the right are defined by the relations

$$r_{ij} = \begin{Bmatrix} 1 & i = j - n \\ 0 & i \neq j - n \end{Bmatrix} \qquad \begin{matrix} i = 1, 2, \ldots, n \\ j = n + 1, n + 2, \ldots, 2n \end{matrix}$$

The "replacement" operations defined in Eqs. (3–51) and (3–52) for finding the solution of a set of simultaneous equations are easily modified to determine the ele-

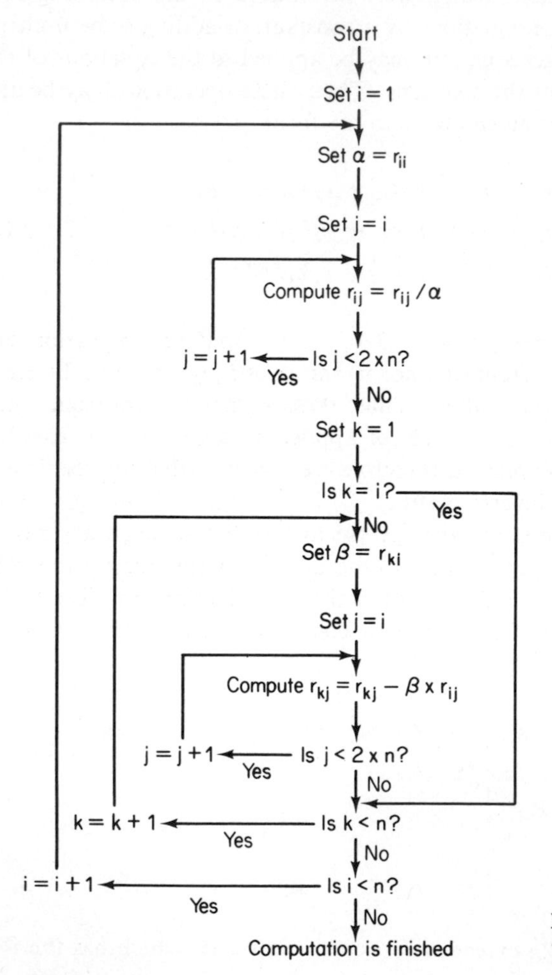

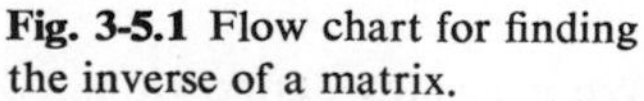

Fig. 3-5.1 Flow chart for finding the inverse of a matrix.

ments of the inverse matrix by changing the upper limit of the range of the indices j from $n + 1$ to $2n$. If this is done, the flow chart for determining the inverse matrix has the form shown in Fig. 3–5.1. The similarity between this flow chart and the one shown in Fig. 3–4.1 should be noted. A FORTRAN program to implement this flow chart is readily written. To provide maximum flexibility in its use, it has been prepared as a subroutine named MXINV (for MatriX INVersion). Since our task is to find the inverse of a matrix **R**, we shall define the input variables as R(I,J) and the output variables as RI(I,J). To avoid destroying the values of the **R** matrix, we shall make all the reduction operations indicated in the flow chart shown in Fig. 3–5.1 on a two-dimensional array named RA defined internally in the subroutine. The identifying statement for the subroutine can be written as

SUBROUTINE MXINV (R, N, RI)

where R is the input array, RI is the output array, and N is the order of the matrix to be inverted. A listing of a subroutine MXINV for performing matrix inversion is given in Fig. 3–5.2. The reader should note the similarity between this subroutine and the one for the solution of a set of simultaneous equations given in Fig. 3–4.3. A summary of the characteristics of the subroutine MXINV is given in Table 3–5.1.

The subroutine defined in Fig. 3–5.2 may easily be incorporated as part of a general program for finding the inverse of a matrix. An example follows.

EXAMPLE 3-5.1. The subroutine MXINV described above may be applied to find the inverse of the resistance matrix for the five-mesh resistance network presented in Example 3-4.1 and shown in Fig. 3-4.4. To do this, a FORTRAN main program is required which will perform the following functions:

1. Read as data the values of the elements of the resistance matrix **R** and store these as the variables R(I,J).

TABLE 3-5.1 Summary of the Characteristics of the Subroutine MXINV

Identifying Statement: SUBROUTINE MXINV (R, N, RI)

Purpose: To determine the inverse of a non-singular square matrix **R** which has real elements.

Additional Subprograms Required: None.

Input Arguments:

R The two-dimensional array of the real variables R(I,J) in which are stored the elements r_{ij} of the matrix **R**.

N The order of the matrix **R**.

Output Argument:

RI The two-dimensional array of real variables RI(I,J) in which are stored the values of the elements of the inverse of the matrix **R**.

Notes:

1. The dimensioning of the variables is R(10,10), RI(10,10).
2. The **R** matrix is preserved invariant.

```
      SUBROUTINE MXINV (R, N, RI)
C     SUBROUTINE TO FIND THE INVERSE OF A GIVEN MATRIX R USING
C     GAUSS-JORDAN REDUCTION.  R MATRIX IS PRESERVED
C            R - MATRIX WHICH IS TO BE INVERTED
C            N - ORDER OF MATRIX
C           RI - INVERSE MATRIX (OUTPUT)
      DIMENSION R(10,10), RA(10,20), RI(10,10)
C
C     ENTER R ARRAY INTO RA ARRAY AND SET
C     LAST N COLUMNS OF RA ARRAY TO IDENTITY MATRIX
      DO 26  I = 1, N
      DO 24  J = 1, N
      RA(I,J) = R(I,J)
      NJ = N + J
   24 RA(I,NJ) = 0.
      NI = N + I
   26 RA(I,NI) = 1.
C
C     REDUCE MATRIX RA SO THAT FIRST N COLUMNS
C     ARE SET EQUAL TO THE IDENTITY MATRIX
      NP = 2 * N
      DO 12  I = 1, N
C
C     SET MAIN DIAGONAL ELEMENT TO UNITY
      ALFA = RA(I,I)
      DO  5  J = I, NP
    5 RA(I,J) = RA(I,J) / ALFA
C
C     SET ELEMENTS OF ITH COLUMN TO ZERO
      DO 11  K = 1, N
      IF (K - I) 8, 11, 8
    8 BETA = RA(K,I)
      DO 10  J = I, NP
   10 RA(K,J) = RA(K,J) - BETA * RA(I,J)
   11 CONTINUE
   12 CONTINUE
C
C     SET INVERSE MATRIX RI EQUAL TO LAST
C     N COLUMNS OF RA ARRAY
      DO 33  J = 1, N
      JN = J + N
      DO 33  I = 1, N
   33 RI(I,J) = RA(I,JN)
      RETURN
      END
```

Fig. 3-5.2 Listing of the subroutine MXINV.

2. Call the subroutine MXINV to find the elements of the inverse matrix and store these as the variables RI(I,J).
3. Print the resulting values of the elements of the inverse matrix.

A listing of a FORTRAN program for accomplishing the functions listed above is given in Fig. 3-5.3. A listing of the input data and the output produced by this program is given in Fig. 3-5.4. As a check on the accuracy of the results of this computation, a separate program was used to mutiply the original matrix by its inverse. The results of this

```
C     MAIN PROGRAM FOR EXAMPLE 3.5-1
C            R - MATRIX TO BE INVERTED
C            N - ORDER OF MATRIX
C           RI - INVERSE MATRIX
      DIMENSION R(10,10), RI(10,10)
      N = 5
C
C     READ THE ELEMENTS OF THE MATRIX TO BE INVERTED
      READ    3, ((R(I,J), J = 1, N), I = 1, N)
    3 FORMAT (5E13.5)
C
C     CALL THE SUBROUTINE MXINV TO COMPUTE THE INVERSE
      CALL MXINV (R, N, RI)
C
C     PRINT AN OUTPUT LISTING OF THE INVERSE MATRIX
      PRINT 6
    6 FORMAT (///,17X, 35H* * * *  INVERSE  MATRIX  * * * *      /)
      PRINT   3, ((RI(I,J), J = 1, N), I = 1, N)
      STOP
      END
```

Fig. 3-5.3 Main program for finding an inverse matrix.

```
INPUT DATA FOR EXAMPLE 3.5-1

  4.50900E+03 -1.01700E+03  0.            0.           -3.49200E+03
 -1.01700E+03  6.68000E+03 -8.42000E+02 -4.82100E+03  0.
  0.          -8.42000E+02  1.22440E+04 -5.73200E+03  0.
  0.          -4.82100E+03 -5.73200E+03  1.23760E+04 -8.37000E+02
 -3.49200E+03  0.           0.          -8.37000E+02  5.27800E+03


OUTPUT DATA FOR EXAMPLE 3.5-1

                 * * * *  INVERSE  MATRIX  * * * *

  5.86207E-04  2.04726E-04  8.22604E-05  1.45641E-04  4.10939E-04
  2.04726E-04  3.36255E-04  1.14551E-04  1.95296E-04  1.66421E-04
  8.22604E-05  1.14551E-04  1.43961E-04  1.16226E-04  7.28562E-05
  1.45641E-04  1.95296E-04  1.16226E-04  2.19581E-04  1.31180E-04
  4.10939E-04  1.66421E-04  7.28562E-05  1.31180E-04  4.82152E-04


TEST OF RESULTS OF EXAMPLE 3.5-1

         * * * *  PRODUCT OF MATRIX TIMES INVERSE  * * * *

  1.00000E+00 -3.55271E-15 -1.77636E-14  8.88178E-15  2.84217E-14
 -1.06581E-14  1.00000E+00 -2.13163E-14  2.57572E-14  7.10543E-15
 -5.32907E-15 -3.55271E-15  1.00000E+00  9.32587E-15  1.77636E-15
 -1.42109E-14 -7.10543E-15 -7.10543E-15  1.00000E+00  3.55271E-15
 -2.13163E-14  0.          -7.10543E-15 -1.77636E-15  1.00000E+00
```

Fig. 3-5.4 Input and output data for Example 3-5.1.

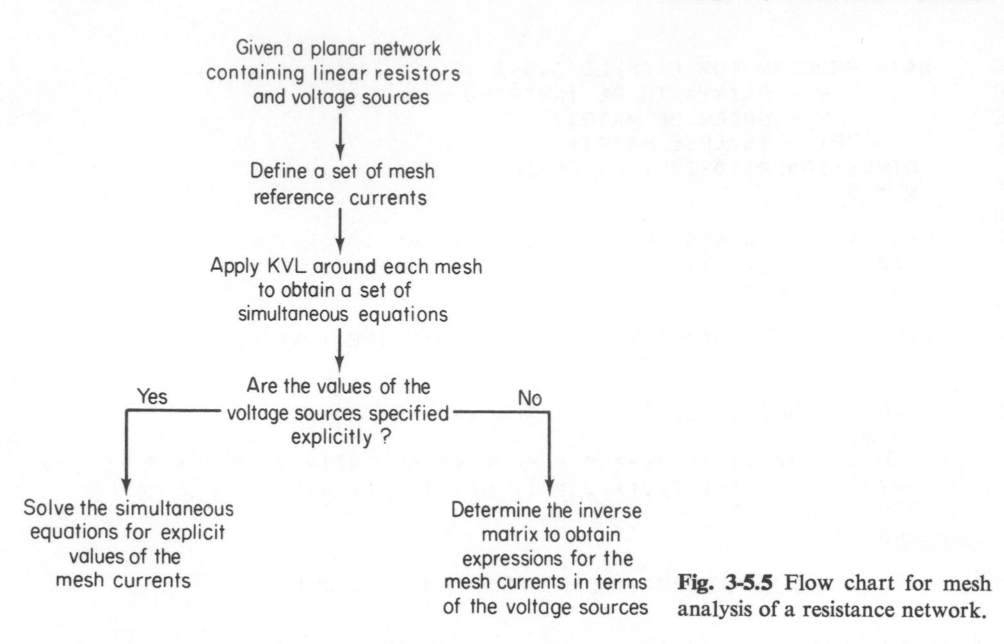

Fig. 3-5.5 Flow chart for mesh analysis of a resistance network.

check are also shown in the figure. It is readily apparent that the product of the two matrices yields a matrix which closely approximates the identity matrix.

The developments given to this point in this chapter have provided us with a general method of formulating the equations for a resistance network, and two methods of solving for the unknown values of the mesh currents. The latter two methods are given to cover the cases (1) when the values of the voltage sources are given explicitly (in which case we need merely solve the simultaneous equations) and (2) when the values of the sources are unspecified (in which case the simplest approach is to determine the inverse matrix). The general procedure is summarized in the flow chart shown in Fig. 3–5.5.

3-6 NODE EQUATIONS FOR A TWO-NODE CIRCUIT

In the early sections of this chapter we described methods by means of which a resistance network could be described using mesh currents as the unknown variables. In this section we shall introduce a second method of describing resistance networks, using node voltages as the unknown variables. To begin, let us consider a resistance network with two node voltages (both the node voltages are defined with respect to the same reference or ground node).[8] If we connect a current source from the reference node to each of the nodes defining the node voltages as shown in Fig. 3–6.1, and apply KCL at node 1 and node 2, we obtain the equations

[8]The reference node is also sometimes referred to as a *datum* node; thus, the node voltages are sometimes called node-to-datum voltages.

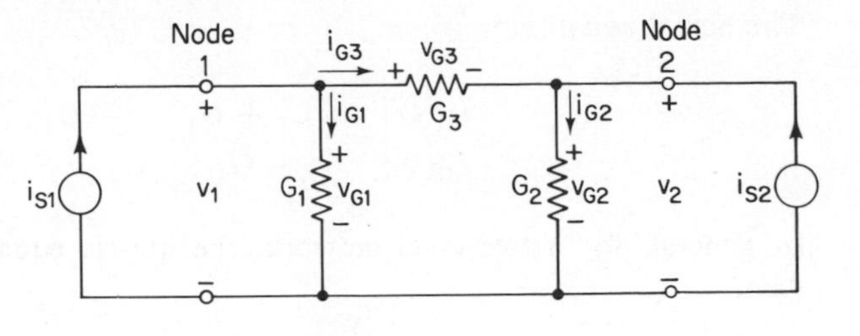

Fig. 3-6.1 A two-node resistance network.

$$i_{S1}(t) = i_{G1}(t) + i_{G3}(t)$$
$$i_{S2}(t) = i_{G2}(t) - i_{G3}(t) \qquad (3\text{–}65)$$

In the above equations we have elected to put the source current on one side of the equation and the branch currents in the resistive elements on the other. Now let us write the branch relations. These are most conveniently written using units of conductance (mhos). Thus, we obtain

$$i_{G1}(t) = G_1 v_{G1}(t)$$
$$i_{G2}(t) = G_2 v_{G2}(t) \qquad (3\text{–}66)$$
$$i_{G3}(t) = G_3 v_{G3}(t)$$

Substituting these branch relations in the KCL equations of (3–65), we obtain

$$\begin{aligned} i_{S1}(t) &= G_1 v_{G1}(t) + G_3 v_{G3}(t) \\ i_{S2}(t) &= G_2 v_{G2}(t) - G_3 v_{G3}(t) \end{aligned} \qquad (3\text{–}67)$$

Finally, we may put the above equations in a still more useful form by substituting the relations between the node voltages $v_1(t)$ and $v_2(t)$ and the branch voltages. These are easily seen to be

$$v_{G1}(t) = v_1(t)$$
$$v_{G2}(t) = v_2(t) \qquad (3\text{–}68)$$
$$v_{G3}(t) = v_1(t) - v_2(t)$$

Substituting the above relations in (3–67) and rearranging terms, we obtain

$$\begin{aligned} i_{S1}(t) &= (G_1 + G_3)v_1(t) - G_3 v_2(t) \\ i_{S2}(t) &= -G_3 v_1(t) + (G_2 + G_3)v_2(t) \end{aligned} \qquad (3\text{–}69)$$

These equations may be written as a matrix equation of the form

$$\mathbf{i}(t) = \mathbf{G}\mathbf{v}(t) \qquad (3\text{–}70)$$

The actual equation is

$$\begin{bmatrix} i_{S1}(t) \\ i_{S2}(t) \end{bmatrix} = \begin{bmatrix} G_1 + G_3 & -G_3 \\ -G_3 & G_2 + G_3 \end{bmatrix} \begin{bmatrix} v_1(t) \\ v_2(t) \end{bmatrix} \tag{3-71}$$

In general, for a two-node network, the matrix equation of (3-70) will have the form

$$\begin{bmatrix} i_1(t) \\ i_2(t) \end{bmatrix} = \begin{bmatrix} g_{11} & g_{12} \\ g_{21} & g_{22} \end{bmatrix} \begin{bmatrix} v_1(t) \\ v_2(t) \end{bmatrix} \tag{3-72}$$

For this case we may make the following observations about the elements of the matrices $\mathbf{i}(t)$, $\mathbf{G}$, and $\mathbf{v}(t)$:

1. The element $i_1(t)$ represents the sum of the effects of any current sources connected to node 1. The contribution of an individual source is positive if its reference current flows towards the node and negative if its reference current flows away from the node. The element $i_2(t)$ represents the sum of the effects of any current sources connected to node 2.
2. The element g_{11} of the matrix $\mathbf{G}$ represents the sum of all the conductances which are connected to the first node. The elements g_{12} and g_{21} (which are equal) represent the negative of the value (in mhos) of any resistor which is connected between nodes 1 and 2. The element g_{22} represents the sum of all the conductances which are connected to node 2.
3. The negative sign that appears in the term $-G_3$ is the result of having chosen the negative reference polarity of both the node voltages $v_1(t)$ and $v_2(t)$ at the same node.

To solve a resistance network of the type given in Fig. 3-6.1, i.e., to find the values of the node voltages for a given set of source currents, we must put the matrix equations in the form

$$\mathbf{v}(t) = \mathbf{G}^{-1}\mathbf{i}(t) \tag{3-73}$$

i.e., we must find the inverse of the conductance matrix $\mathbf{G}$. Since the relations for the inverse of any 2×2 matrix are the same, regardless of what physical quantities the matrix represents, we may use the relations developed in Sec. 3-1 to find the inverse. Thus, we obtain

$$\mathbf{G}^{-1} = \frac{1}{g_{11}g_{22} - g_{12}g_{21}} \begin{bmatrix} g_{22} & -g_{12} \\ -g_{21} & g_{11} \end{bmatrix} \tag{3-74}$$

The necessary and sufficient condition for a solution to exist, i.e., for the inverse of the $\mathbf{G}$ matrix to exist, is that the expression $g_{11}g_{22} - g_{12}g_{21}$ (the determinant of the matrix **G**) be non-zero. An example of the solution of a resistance matrix described by nodal voltage variables follows.

EXAMPLE 3-6.1. A two-node resistance network is shown in Fig. 3-6.2. The current source connected between the reference node and the first node produces a current of 8 A

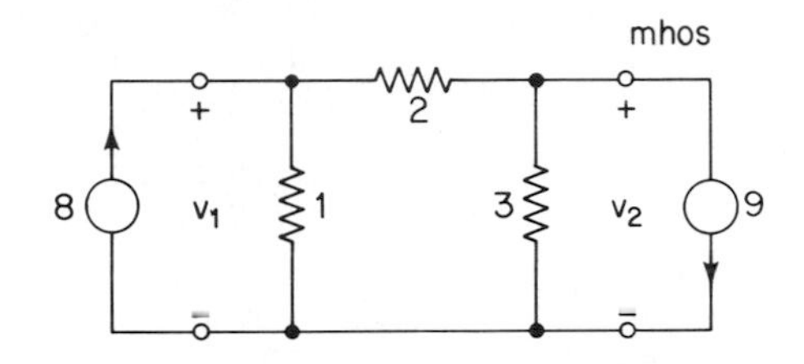

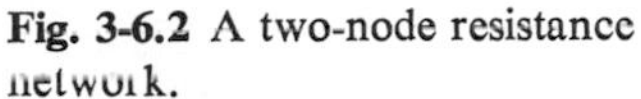
Fig. 3-6.2 A two-node resistance network.

flowing into the node. The current source connected between the reference node and the second node represents a current of -9 A flowing into the second node. The total conductance connected to the first node is 3 mhos, and the total of the conductance connected to the second node is 5 mhos. There is a 2-mho conductance connected between the first and the second nodes. From the discussion given above, we see that we may write the KCL equations for this network as

$$\begin{bmatrix} 8 \\ -9 \end{bmatrix} = \begin{bmatrix} 3 & -2 \\ -2 & 5 \end{bmatrix} \begin{bmatrix} v_1 \\ v_2 \end{bmatrix}$$

The solution is given by computing the inverse matrix. From the relations given in (3-74), we obtain

$$\begin{bmatrix} v_1 \\ v_2 \end{bmatrix} = \frac{1}{11} \begin{bmatrix} 5 & 2 \\ 2 & 3 \end{bmatrix} \begin{bmatrix} 8 \\ -9 \end{bmatrix} = \begin{bmatrix} 2 \\ -1 \end{bmatrix}$$

Thus the voltage at the first node is 2 V, while the voltage at the second node is -1 V (both measured with respect to the reference node).

The numerical computations for the example given above exactly parallel the computations for the development given in Example 3-1.1 in which we obtained a solution for the mesh currents of the two-mesh resistance network shown in Fig. 3-1.3. When a pair of circuits, such as the ones shown in Figs. 3-1.3 and 3-6.2, is characterized by a set of equations in which the numerical coefficients have the same values (although the physical significance of the variables may be different), the circuits are said to be *dual* to each other. We will treat dual circuits in more detail in Sec. 6-8.

The method of analysis described above is easily extended to the case where there are additional current sources in the network. An example follows.

EXAMPLE 3-6.2. A two-node resistance network is shown in Fig. 3-6.3. The net effect of the two current sources that are connected to node 1 is to provide a positive input cur-

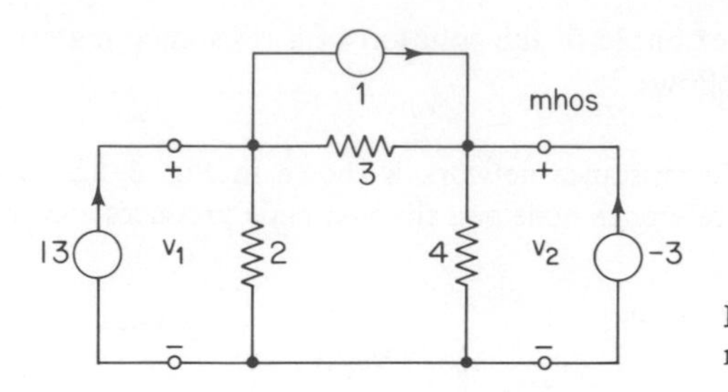

Fig. 3-6.3 A two-node resistance network with three sources.

rent to the node of 12 A. Similarly, the effect of the two sources connected to node 2 produces an input current to the node of −2 A. Evaluating the values of the total conductance connected to each node, and the conductance connected between the nodes, we may write the KCL equations for the two nodes in the form

$$\begin{bmatrix} 12 \\ -2 \end{bmatrix} = \begin{bmatrix} 5 & -3 \\ -3 & 7 \end{bmatrix} \begin{bmatrix} v_1 \\ v_2 \end{bmatrix}$$

The inverse set of relations is easily found to be

$$\begin{bmatrix} v_1 \\ v_2 \end{bmatrix} = \frac{1}{26} \begin{bmatrix} 7 & 3 \\ 3 & 5 \end{bmatrix} \begin{bmatrix} 12 \\ -2 \end{bmatrix} = \begin{bmatrix} 3 \\ 1 \end{bmatrix}$$

Thus, the voltage at the first node is 3 V, while that at the second node is 1 V.

The same conclusions that were made with respect to the linearity of networks described on a mesh basis in Sec. 3–1 apply to networks described on a nodal basis. The only difference is that when a current source is set to zero, it in effect becomes an open circuit, whereas a voltage source set to zero effectively becomes a short circuit. Thus, the definition for a linear resistance network given in Sec. 3–1 applies irrespective of whether the network is described on a mesh basis or on a node basis.

3-7 NODE EQUATIONS FOR THE GENERAL CIRCUIT

In this section we shall extend the discussion given in the preceding section for a two-node resistance network to the general case. First, let us consider a network containing three nodes as shown in Fig. 3–7.1. To determine the node equations for this network, we may follow a procedure similar to that used for the two-node case, namely: (1) write the KCL equations at each node in terms of the branch current variables; (2) substitute the branch relationships for each branch so that the KCL equations are expressed in terms of the branch voltage variables; and (3) substitute the relationships defining the branch voltages in terms of the node voltages $v_1(t)$, $v_2(t)$, and $v_3(t)$ defined in the figure. The result is the following set of equations:

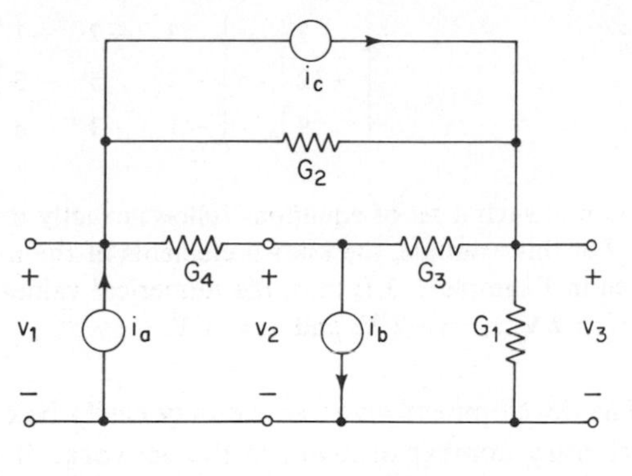

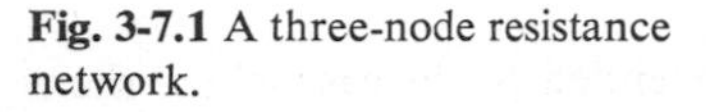
Fig. 3-7.1 A three-node resistance network.

$$\begin{bmatrix} i_a(t) - i_c(t) \\ -i_b(t) \\ i_c(t) \end{bmatrix} = \begin{bmatrix} G_2 + G_4 & -G_4 & G_2 \\ -G_4 & G_3 + G_4 & -G_3 \\ -G_2 & -G_3 & G_1 + G_2 + G_3 \end{bmatrix} \begin{bmatrix} v_1(t) \\ v_2(t) \\ v_3(t) \end{bmatrix} \tag{3–75}$$

This is an example of a set of matrix equations of the general form

$$\begin{bmatrix} i_1(t) \\ i_2(t) \\ i_3(t) \end{bmatrix} = \begin{bmatrix} g_{11} & g_{12} & g_{13} \\ g_{21} & g_{22} & g_{23} \\ g_{31} & g_{32} & g_{33} \end{bmatrix} \begin{bmatrix} v_1(t) \\ v_2(t) \\ v_3(t) \end{bmatrix} \tag{3–76}$$

If we compare the elements of Eq. (3–76) with those of Eq. (3–75), we may make a set of conclusions similar to those which were made for the two-node network. That is, we see that the elements $i_i(t)$ of Eq. (3–76) each represent the sum (with appropriate note taken of the reference polarity) of the currents incident to node i, the elements g_{ii} on the main diagonal of the conductance matrix **G** represent the sum of values of the conductances connected to the ith node, and the elements $g_{ij} (i \neq j)$ represent the negative of the value of any conductance connected between the ith and jth nodes. An example of such a determination follows.

EXAMPLE 3-7.1. A three-node resistance network is shown in Fig. 3-7.2. If we apply the rules specified above for determining the matrix elements, we see that the KCL equations for the network may be written in the form

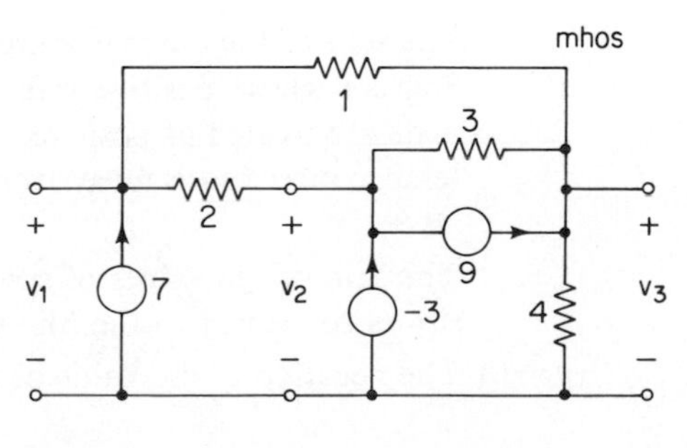

Fig. 3-7.2 A three-node resistance network.

$$\begin{bmatrix} 7 \\ -12 \\ 9 \end{bmatrix} = \begin{bmatrix} 3 & -2 & -1 \\ -2 & 5 & -3 \\ -1 & -3 & 8 \end{bmatrix} \begin{bmatrix} v_1 \\ v_2 \\ v_3 \end{bmatrix}$$

The solution to such a set of equations follows exactly the development given in Secs. 3-3 and 3-4. For this example, the known elements of the matrix equation are identical with those given in Example 3-3.1; thus, the numerical values of the solution are also identical, namely, $v_1 = 2$ V, $v_2 = -1$ V, and $v_3 = 1$ V.

The development given above may easily be extended to the case where there are an arbitrary number of nodes in the network. If there are n nodes (plus a reference node), then we may define n node voltage variables, all defined with respect to the same reference node [the $(n + 1)$st node]. An extension of the development made above readily shows that the KCL equations for the various nodes will produce a set of equations of the form

$$\begin{bmatrix} i_1(t) \\ i_2(t) \\ \cdot \\ \cdot \\ \cdot \\ i_n(t) \end{bmatrix} = \begin{bmatrix} g_{11} & g_{12} & \cdots & g_{1n} \\ g_{21} & g_{22} & \cdots & g_{2n} \\ \cdot & \cdot & \cdots & \cdot \\ \cdot & \cdot & \cdots & \cdot \\ \cdot & \cdot & \cdots & \cdot \\ g_{n1} & g_{n2} & \cdots & g_{nn} \end{bmatrix} \begin{bmatrix} v_1(t) \\ v_2(t) \\ \cdot \\ \cdot \\ \cdot \\ v_n(t) \end{bmatrix} \tag{3-77}$$

The procedure for determining the elements of the matrices of (3–77) may be summarized as follows:

SUMMARY 3-7.1

Writing the Node Equations for a Resistance Network: A set of node equations of the form given in (3–77) may be written for a given resistance network by defining a set of node voltages with their negative reference polarities all at the same common reference node, and determining the elements of the component matrices as follows:

$i_i(t)$ The sum of the current sources connected to the ith node. Sources whose positive reference direction is towards the node are treated as positive, and sources whose positive reference direction is away from the node are treated as negative.

g_{ij} $(i = j)$ The sum of the values of conductance (in mhos) of all resistors connected to the ith node.

g_{ij} $(i \neq j)$ The negative of the value of the conductance (in mhos) of any resistor which is directly connected between the ith and jth nodes.[9]

$v_i(t)$ The unknown node voltage variables which are to be found by solving the matrix equation.

It should be noted that, unlike the procedure for finding the mesh equations given in Summary 3-2.1, the procedure given above may be applied to any resistance network, irrespective of whether it is planar or non-planar. As an example of the use of this procedure for finding the KCL equations for a network, consider the following.

EXAMPLE 3-7.2. A resistance network with nine node voltage variables is shown in Fig. 3-7.3. We may apply the rules given above to directly write the KCL equations for the node voltages. For simplicity, we shall assume that all the resistors have unity value. We obtain

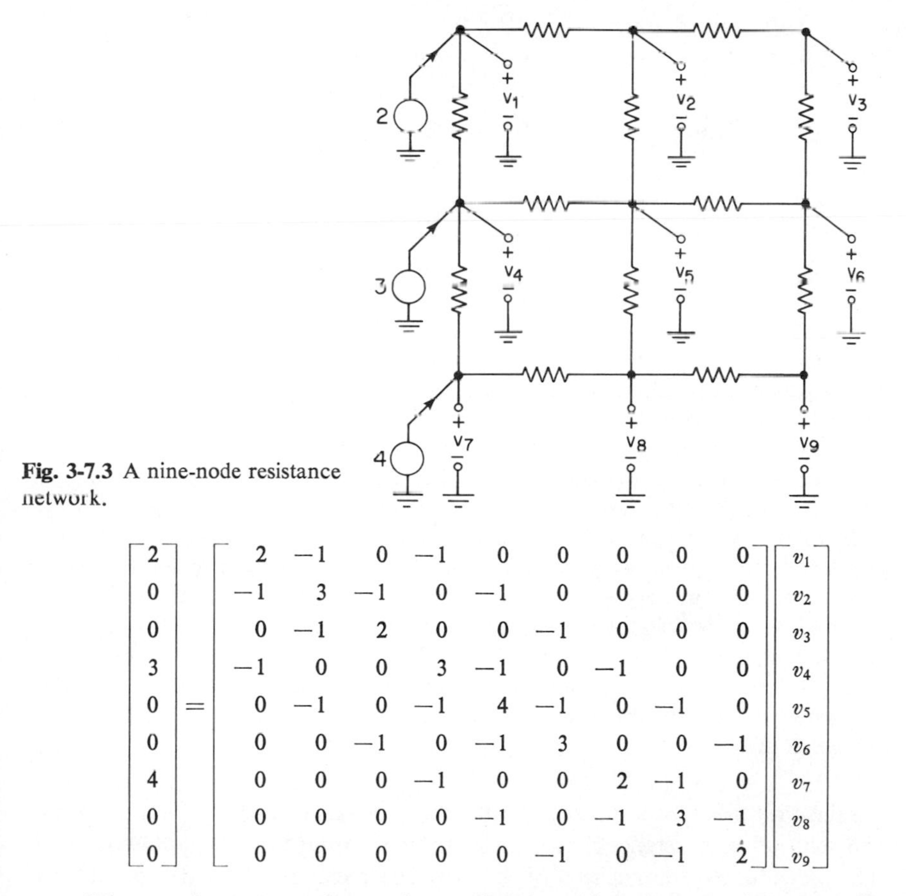

Fig. 3-7.3 A nine-node resistance network.

$$\begin{bmatrix} 2 \\ 0 \\ 0 \\ 3 \\ 0 \\ 0 \\ 4 \\ 0 \\ 0 \end{bmatrix} = \begin{bmatrix} 2 & -1 & 0 & -1 & 0 & 0 & 0 & 0 & 0 \\ -1 & 3 & -1 & 0 & -1 & 0 & 0 & 0 & 0 \\ 0 & -1 & 2 & 0 & 0 & -1 & 0 & 0 & 0 \\ -1 & 0 & 0 & 3 & -1 & 0 & -1 & 0 & 0 \\ 0 & -1 & 0 & -1 & 4 & -1 & 0 & -1 & 0 \\ 0 & 0 & -1 & 0 & -1 & 3 & 0 & 0 & -1 \\ 0 & 0 & 0 & -1 & 0 & 0 & 2 & -1 & 0 \\ 0 & 0 & 0 & 0 & -1 & 0 & -1 & 3 & -1 \\ 0 & 0 & 0 & 0 & 0 & -1 & 0 & -1 & 2 \end{bmatrix} \begin{bmatrix} v_1 \\ v_2 \\ v_3 \\ v_4 \\ v_5 \\ v_6 \\ v_7 \\ v_8 \\ v_9 \end{bmatrix}$$

[9]The quantity g_{ij} $(i = j)$ is sometimes called the *self conductance* of the ith node, while the quantity g_{ij} $(i \neq j)$ is called the *mutual conductance* between the ith and jth nodes.

The numerical techniques which may be applied to find the node voltages for a network defined on a node basis are exactly the same as those used to solve for the mesh currents of a resistance network which is described on a mesh basis. Thus, we may apply the identical digital computer program developed in Sec. 3–4 to find a solution for any resistance network which is described on a node basis. An example follows.

EXAMPLE 3-7.3. We may solve for the five unknown node voltages of the resistance network shown in Fig. 3-7.4 by using the digital computer program shown in Fig. 3-4.5

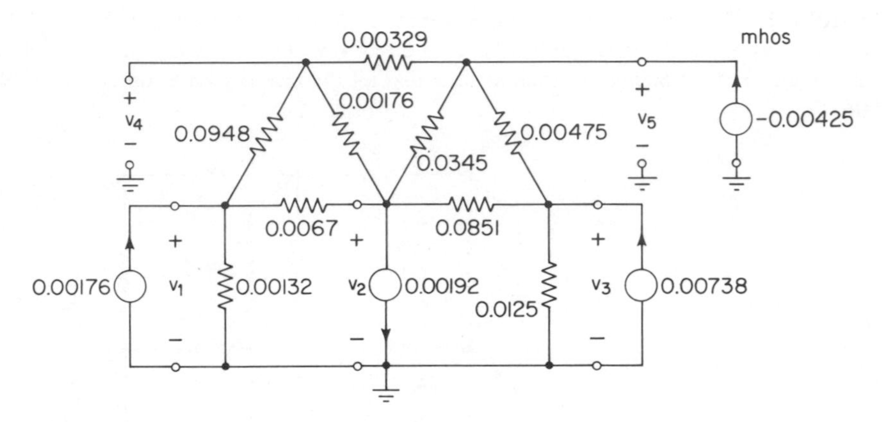

Fig. 3-7.4 A five-node resistance network.

(which was used to solve for the mesh currents of the resistance network of Example 3-4.1). The input data will consist of the values of the resistances (in mhos) and the identifying numbers for the nodes to which the resistances are connected. The values of the current sources connected to the various nodes must also be read as input data. The output will simply consist of the values of the node voltages, i.e., the solution to the set of simultaneous equations. A listing of the output from a program to accomplish this is given in Fig. 3-7.5. The program is identical with the one shown in Fig. 3-4.5, except for the fact that the Hollerith statements identifying the output data have been changed to reflect the fact that the matrix is a conductance matrix **G** rather than a resistance matrix **R**, and that the output variables are node voltages rather than mesh currents. To check the values of the node voltages computed by the program, these values have been inserted in the original set of equations to determine the values of the resultant input currents to the nodes. The results are shown in Fig. 3-7.5. It is easily seen that they demonstrate good agreement with the original data.

In the preceding two sections we have discussed techniques for the formulation and solution of resistance networks using node voltages as the unknown variables. The procedure is summarized in the flow chart shown in Fig. 3–7.6. In order to appreciate the similarity between the various mathematical steps used in node analysis and those used in mesh analysis, the reader should compare this flow chart with the one given in Fig. 3–5.5.

```
INPUT DATA FOR EXAMPLE 3.7-3

0509
10          .00132
12          .0067
14          .0948
24          .00176
25          .0345
23          .0851
30          .0125
45          .00329
35          .00475
.00176     -.00192     .00738                 -.00425


OUTPUT DATA FOR EXAMPLE 3.7-3

                    * * * *   G   MATRIX   * * * *

 1.02820E-01 -6.70000E-03  0.                -9.48000E-02  0.
-6.70000E-03  1.28060E-01 -8.51000E-02 -1.76000E-03 -3.45000E-02
 0.          -8.51000E-02  1.02350E-01  0.           -4.75000E-03
-9.48000E-02 -1.76000E-03  0.           9.90500E-02 -3.29000E-03
 0.          -3.45000E-02 -4.75000E-03 -3.29000E-03  4.25400E-02


                    * * * *   NODE   VOLTAGES   * * * *

V(1)= 2.6110246E-01
V(2)= 1.6173246E-01
V(3)= 2.1002758E-01
V(4)= 2.5319565E-01
V(5)= 7.4292771E-02


TEST OF RESULTS OF EXAMPLE 3.7-3

                * * * *   COMPUTED   NODE   CURRENTS   * * * *

I(1)= 1.7600000E-03
I(2)=-1.9200000E-03
I(3)= 7.3800000E-03
I(4)= 1.3357371E-16
I(5)=-4.2500000E-03
```

Fig. 3-7.5 Input and output data for Example 3-7.3.

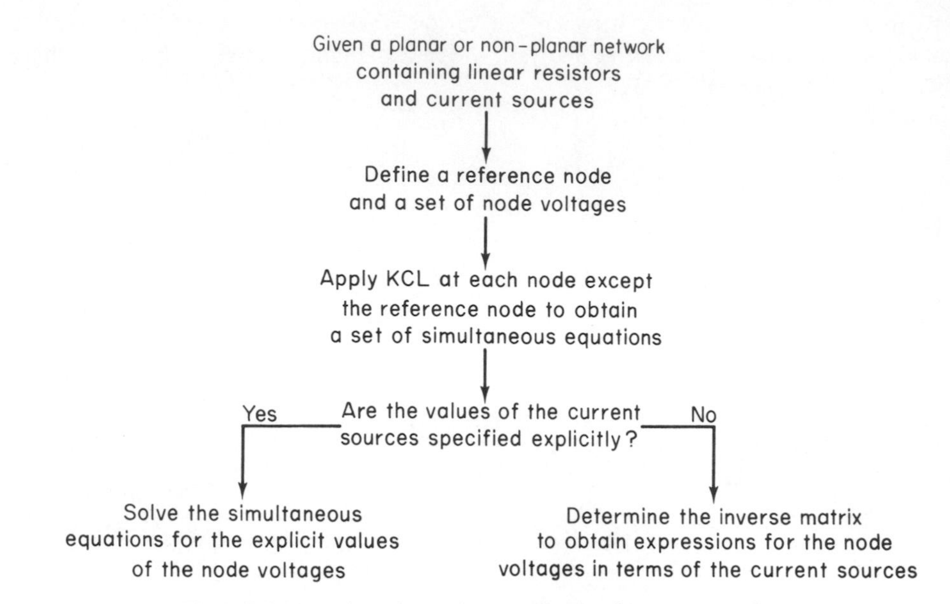

Fig. 3-7.6 Flow chart for node analysis of resistance networks.

3-8 NETWORKS CONTAINING BOTH VOLTAGE AND CURRENT SOURCES

In the preceding sections of this chapter we have discussed methods for writing and solving sets of simultaneous equations which describe networks composed of resistors and sources. These methods have covered two important cases, namely: (1) the case where all the sources are voltage sources (and the equations are written on a mesh basis); and (2) the case where all the sources are current sources (and the equations are written on a node basis). In this section we shall discuss the extension of these methods to the case where voltage *and* current sources are present in a given resistance network.

Let us assume that we desire to write a set of equations describing the resistance network shown in Fig. 3–8.1(a). Note that this network contains two voltage sources and one current source. Note also that the current source has a resistor in parallel with it. If it is desired to write the equations describing such a network on a mesh basis, we may use the concept of equivalent sources demonstrated in Sec. 2–3 to replace the 1–A current source and the 2–Ω shunt resistor R_0 shown in Fig. 3–8.1(a) by an equivalent 2–V voltage source and a series 2–Ω resistor R_0' as shown in Fig. 3–8.1(b). The mesh equations for the circuit of Fig. 3–8.1(b) may be written following the method described in Sec. 3–1. Thus, with the mesh currents defined as shown in Fig. 3–8.1(b), we obtain

$$\begin{bmatrix} 6 \\ -8 \end{bmatrix} = \begin{bmatrix} 4 & -2 \\ -2 & 6 \end{bmatrix} \begin{bmatrix} i_1 \\ i_2 \end{bmatrix} \tag{3-78}$$

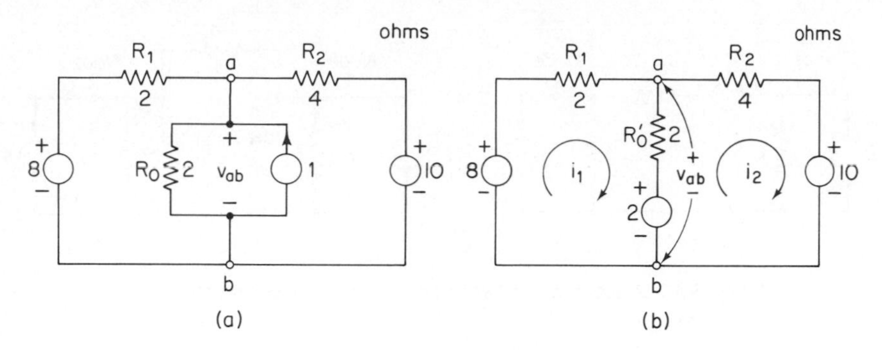

Fig. 3-8.1 A resistance network containing voltage and current sources.

The inverse set of equations is

$$\begin{bmatrix} i_1 \\ i_2 \end{bmatrix} = \frac{1}{20}\begin{bmatrix} 6 & 2 \\ 2 & 4 \end{bmatrix}\begin{bmatrix} 6 \\ -8 \end{bmatrix} = \begin{bmatrix} 1 \\ -1 \end{bmatrix} \tag{3-79}$$

Thus we see that $i_1 = 1$ A and $i_2 = -1$ A. The current in the resistor R_0' in Fig. 3–8.1(b) is thus 2 A (flowing downward). The voltage v_{ab} from node a to node b is 6 V. Since the voltage v_{ab} must be identical for both circuits, the voltage across the resistor R_0 in Fig. 3–8.1(a) must be 6 V. Thus, the current through this resistor is 3 A (flowing downward). From this discussion, we see that the *internal* conditions in the two equivalent (non-ideal) sources are not the same (as was originally pointed out in Sec. 2–3). Note that, since the internal conditions in the sources are not the same, the total power dissipated in the two networks is not the same. For example, in Fig. 3–8.1(a), 2 W is dissipated in R_1, 4 W is dissipated in R_2, and 18 W is dissipated in R_0, making a total of 24 W dissipated in the network. In the circuit shown in Fig. 3–8.1(b), however, the power dissipated in R_1 and R_2 is 2 and 4 W, as before, but the power dissipated in R_0' is 8 W, making a total of 14 W. This is a conclusion which might have been expected, since equivalent non-ideal sources are only equivalent with respect to their *terminal* behavior, not with respect to their internal conditions. Since the terminal variables of the two sources are identical, however, all variables in the resistance network external to the equivalent sources have the same values in both circuits.

As another example of analyzing resistance networks with both voltage and current sources, consider the resistance network shown in Fig. 3–8.2(a). This network contains two current sources and one voltage source. It may be modified so that it can be described by equations written on a node basis. To do this, the voltage source and the series resistance connected between nodes 1 and 2 may be replaced by an equivalent current source with a shunt resistance, as shown in Fig. 3–8.2(b). The equations describing this circuit are readily found using the methods outlined in Sec. 3–6. We obtain

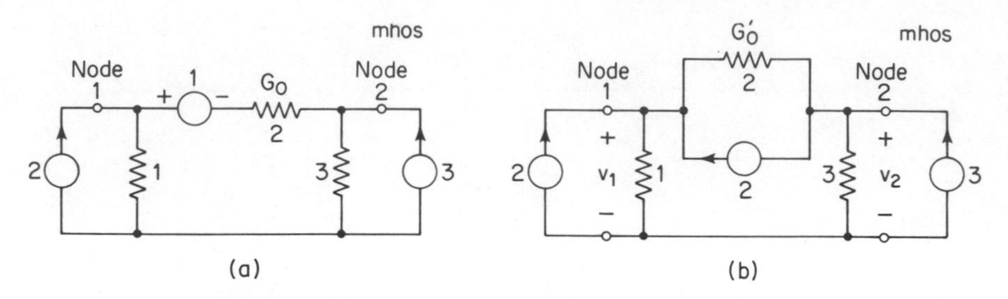

Fig. 3-8.2 A resistance network containing voltage and current sources.

$$\begin{bmatrix} 4 \\ 1 \end{bmatrix} = \begin{bmatrix} 3 & -2 \\ -2 & 5 \end{bmatrix} \begin{bmatrix} v_1 \\ v_2 \end{bmatrix} \tag{3-80}$$

The inverse set of equations is

$$\begin{bmatrix} v_1 \\ v_2 \end{bmatrix} = \frac{1}{11} \begin{bmatrix} 5 & 2 \\ 2 & 3 \end{bmatrix} \begin{bmatrix} 4 \\ 1 \end{bmatrix} = \begin{bmatrix} 2 \\ 1 \end{bmatrix} \tag{3-81}$$

We see that the nodal voltages in Fig. 3–8.2 are $v_1 = 2$ V and $v_2 = 1$ V. Thus, there is a potential difference of 1 V across the 2–mho resistor G_0' in Fig. 3–8.2(b), and we see that a current of 2 A flows through this resistor (from node 1 to node 2). In Fig. 3–8.2(a), however, a potential of 1 V must also exist between nodes 1 and 2; therefore, the voltage across the 2–mho resistor G_0 must be zero. Thus, the current through the branch is zero. The overall current summation at nodes 1 and 2, however, is the same as it was in the circuit shown in Fig. 3–8.2(b), even though the internal conditions in the equivalent sources connected between nodes 1 and 2 in the two figures are quite different. For example, it is easily shown that a different total amount of power is dissipated in the two networks.

The techniques described in the above paragraphs are obviously applicable to the case where an arbitrary number of current and voltage sources are present in a network as long as every voltage source has a resistor connected in series with it or every current source has a resistor in parallel with it. Such networks may be described by a set of equations written on a node basis or a mesh basis, respectively. We may summarize this by the following statement:

SUMMARY 3-8.1

Analysis of Resistance Networks Containing Voltage and Current Sources: Networks containing both voltage and current sources in which all voltage sources have a series resistor may be described on a node basis by appropriate substitution of equivalent sources. Similarly, networks in which all current sources have a shunt resistor may be described on a mesh basis.

Note that the development given above is also applicable to the case of a resistance network containing only voltage sources (with series resistors) for which it is desired to write node equations, or to the case of a network containing only current sources (with shunt resistors) for which it is desired to write mesh equations.

Since an arbitrary resistance network satisfying the characteristics given above may obviously be described either by mesh equations or by node equations, it is desirable to have some sort of a guide as to which formulation will provide the least number of equations, and thus the easier method of solution. To do this, we may proceed as follows.

1. Replace all sources by their internal resistance (a short circuit for a voltage source, and an open circuit for a current source).
2. In the resulting network, count the total number of nodes and the total number of meshes.

The number of nodes (minus one for the reference node) gives the number of node voltage variables, and thus the number of equations which will be required to describe the network on a node basis. Correspondingly, the number of meshes determines the number of equations which will be required to describe the network on a mesh basis.

Resistance networks which do not fall into the class defined above, i.e., those in which there are voltage sources which do not have a series resistor or current sources without a resistor in parallel, may easily be analyzed through the use of the techniques described in Sec. 2–5 for adding parallel voltage sources and/or series current sources without changing the network variables. As an example of the application of such techniques, consider the network shown in Fig. 3–8.3(a). This network contains a voltage source without a series resistor and also a current source without any shunt resistor. Suppose that it is desired to find the voltage v_{ab} across the 2–Ω resistor. We might proceed in our analysis by the following steps:

1. Replace the 3–V voltage source (which does not have a series resistor) with two 3–V sources connected in parallel as shown in Fig. 3–8.3(b).
2. Since the voltage sources are equal, the connection between them may be broken, leaving each source with a series resistance as shown in Fig. 3–8.3(c).
3. Replace each of the voltage source and series resistor combinations by an equivalent current source and shunt resistor as shown in Fig. 3–8.3(d).

The resulting circuit is now simply a two-node circuit in which the voltage v_{ab} across the resistor is simply v_1. The node equations may now be written as

$$\begin{bmatrix} 3 \\ 3 \end{bmatrix} = \begin{bmatrix} \frac{7}{4} & -\frac{1}{4} \\ -\frac{1}{4} & \frac{7}{12} \end{bmatrix} \begin{bmatrix} v_1 \\ v_2 \end{bmatrix} \tag{3-82}$$

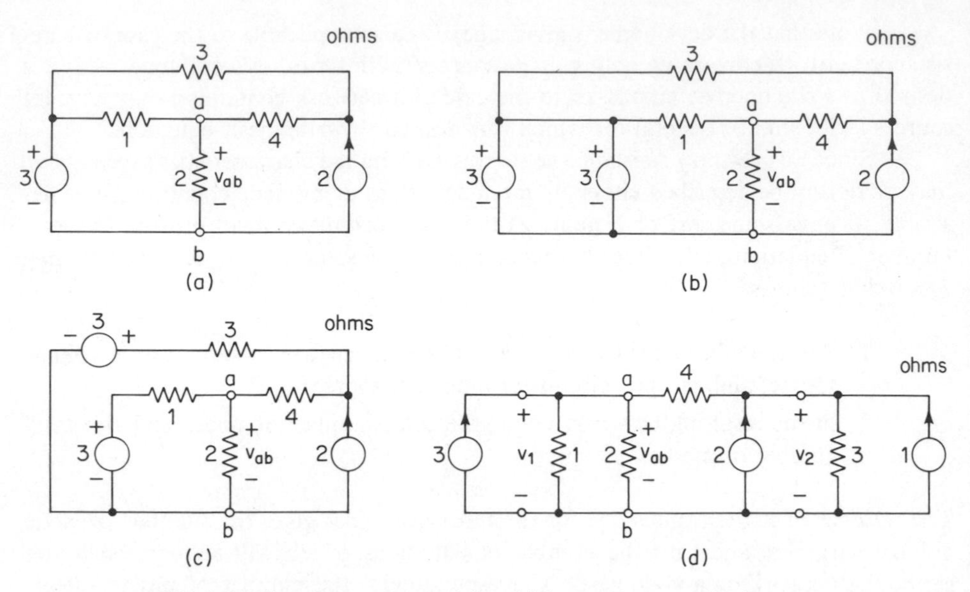

Fig. 3-8.3 A resistance network in which source conversion is not directly applicable.

The solution for v_1 using the standard techniques for the solution of simultaneous equations developed in Sec. 3–1 provides the desired answer for the value of v_{ab}.

As a second example of the application of the technique of adding additional sources to solve a resistance network, suppose that in the circuit shown in Fig. 3–8.4(a) it is desired to solve for the current i_{ab} in the 1–Ω resistor. The analysis might proceed as follows:

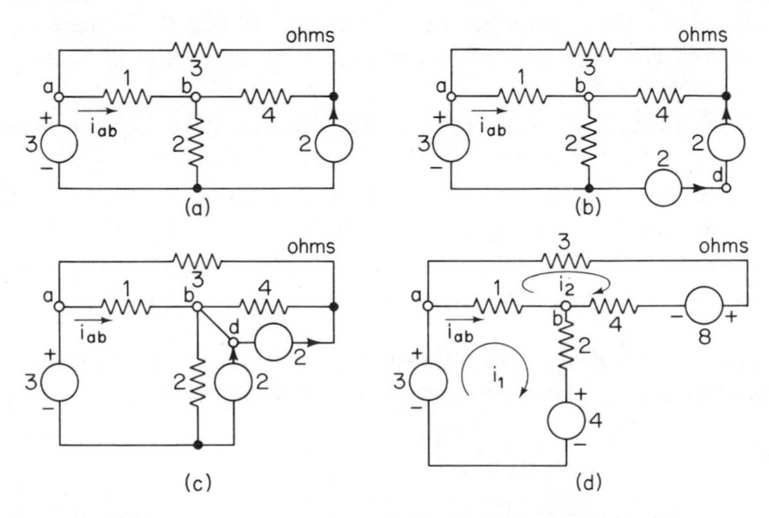

Fig. 3-8.4 A resistance network in which source conversion is not directly applicable.

1. Replace the 2–A current source by a series connection of two 2–A current sources as shown in Fig. 3–8.4(b).
2. Since KCL is satisfied by the two sources at the node *d* between them, and since the voltage at this node is unspecified, we may connect it to the node *b* at the junction of the 2–, 1–, and 4–Ω resistors as shown in Fig. 3–8.4(c) without changing any of the network variables.
3. Each of the current sources and its parallel resistor may now by replaced by a voltage source and a series resistor as shown in Fig. 3–8.4(d).

The remaining network is a two-mesh network. The simultaneous equations are easily written in terms of the mesh currents i_1 and i_2 defined in the figure. They are

$$\begin{bmatrix} -1 \\ -8 \end{bmatrix} = \begin{bmatrix} 3 & -1 \\ -1 & 8 \end{bmatrix} \begin{bmatrix} i_1 \\ i_2 \end{bmatrix} \tag{3–83}$$

The solution for i_1 and i_2 using standard techniques for the solution of simultaneous equations provides the desired solution for the current i_{ab} using the relation $i_{ab} = i_1 - i_2$. A summary of the various possibilities that may be used in analyzing a resistance network is given in the flow chart shown in Fig. 3–8.5.

To conclude this section, we shall introduce a final technique which is frequently of use as an aid in solving resistance networks containing both current and voltage sources. First, let us give a name to a circuit configuration of three nodes and three resistors, which has the property that each pair of nodes is connected by a single resistor. Such a configuration, consisting of the nodes numbered 1, 2, and 3, and the resistors labeled R_a, R_b, and R_c, is shown in Fig. 3–8.6(a). This circuit configuration is called a *delta network*, since it has the geometrical appearance of an upside-down Greek letter Δ. We will show that such a network may be replaced by a *wye connection* (which resembles the geometrical appearance of the letter **Y**) of three resistors R_1, R_2, and R_3 as shown in Fig. 3–8.6(b), which has identical properties at nodes 1, 2, and 3. To find the values of R_1, R_2, and R_3, we need only equate the input resistance seen between the various pairs of terminals in the two networks. For example, equating the resistance measured between terminals 1 and 3 in the two networks, we obtain

$$R_1 + R_3 = \frac{R_b(R_a + R_c)}{R_a + R_b + R_c} \tag{3–84}$$

Similarly, for the input resistance between nodes 1 and 2 to be equal in the two circuits, we must have

$$R_1 + R_2 = \frac{R_c(R_a + R_b)}{R_a + R_b + R_c} \tag{3–85}$$

Finally, for the input resistance between nodes 2 and 3 to be equal, we require

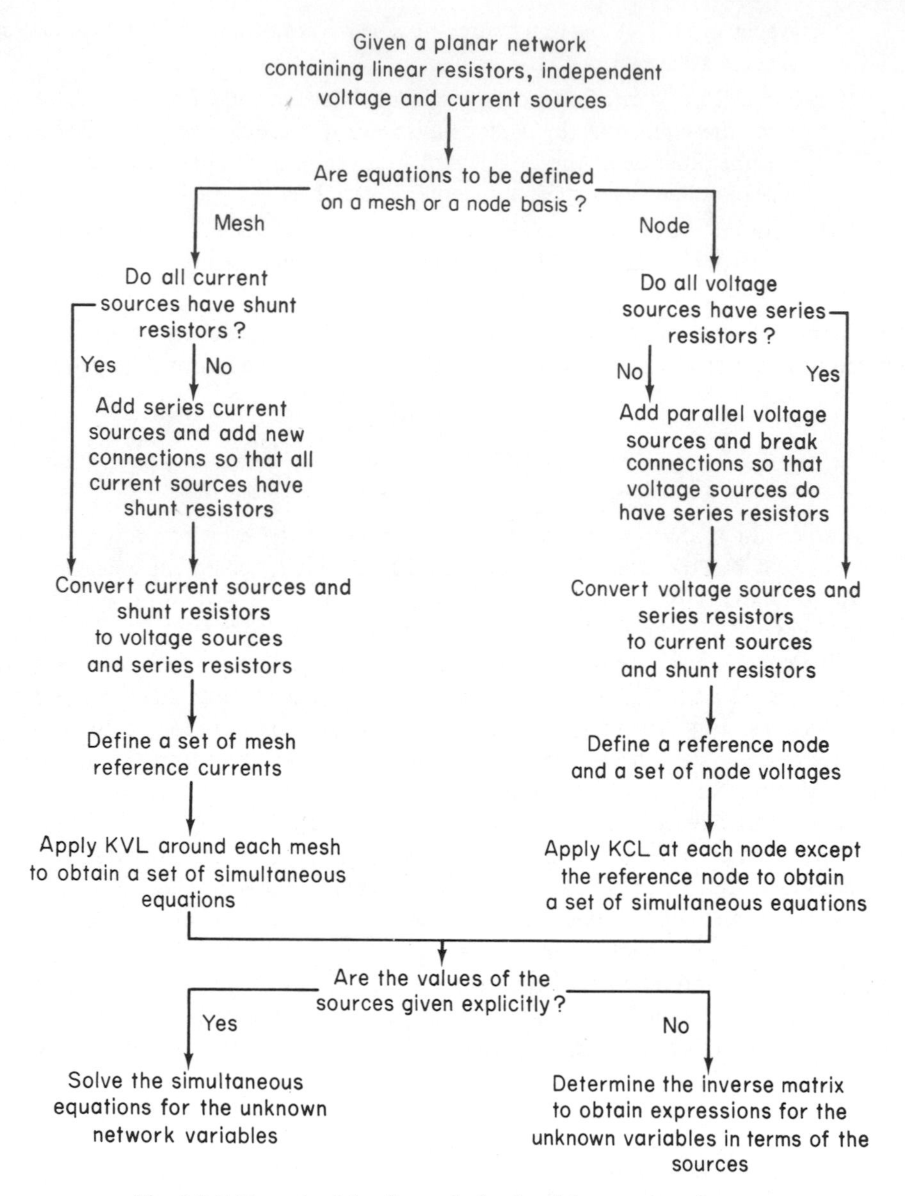

Fig. 3-8.5 Flow-chart for the analysis of resistance networks containing voltage and current sources.

$$R_2 + R_3 = \frac{R_a(R_b + R_c)}{R_a + R_b + R_c} \tag{3-86}$$

To solve the last three equations to obtain an expression for R_1, we may first multiply Eq. (3-86) by -1, and then add it to Eq. (3-84). Doing this, we obtain

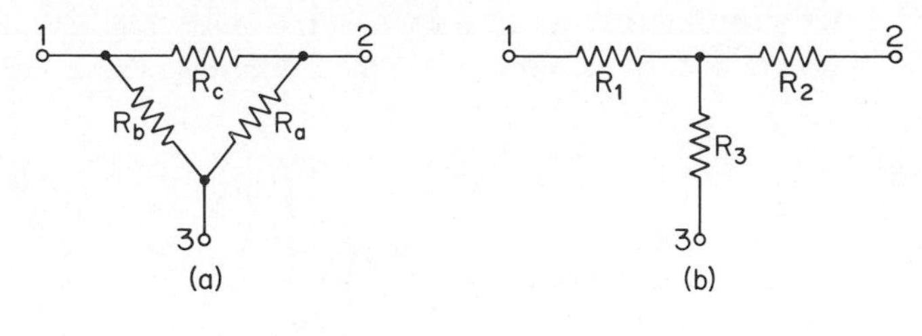

Fig. 3-8.6 Delta and wye resistance networks.

$$R_1 - R_2 = \frac{R_c(R_b - R_a)}{R_a + R_b + R_c} \tag{3-87}$$

If we now add Eqs. (3–85) and (3–87), we obtain

$$R_1 = \frac{R_c R_b}{R_a + R_b + R_c} = \frac{G_a}{G_aG_b + G_aG_c + G_bG_c} \tag{3-88a}$$

where, in the second relation, the values of the resistors labeled R_a, R_b, and R_c in Fig. 3 8.6(a) are given in units of conductance. By similar procedures, we may obtain the relations

$$R_2 = \frac{R_a R_c}{R_a + R_b + R_c} = \frac{G_b}{G_aG_b + G_aG_c + G_bG_c} \tag{3-88b}$$

$$R_3 = \frac{R_a R_b}{R_a + R_b + R_c} = \frac{G_c}{G_aG_b + G_aG_c + G_bG_c} \tag{3-88c}$$

Thus, we have found expressions for determining the elements of a wye of resistors in terms of the elements of a delta of resistors in such a manner that the terminal properties of the two resistance networks are identical. Such a process is referred to as a *delta-wye transformation.* By a similar process, we may define the inverse relations, i.e., the ones that determine a *wye-delta transformation.* The relations are

$$\begin{aligned} G_a &= \frac{G_2 G_3}{G_1 + G_2 + G_3} = \frac{R_1}{R_1R_2 + R_1R_3 + R_2R_3} \\ G_b &= \frac{G_1 G_3}{G_1 + G_2 + G_3} = \frac{R_2}{R_1R_2 + R_1R_3 + R_2R_3} \\ G_c &= \frac{G_1 G_2}{G_1 + G_2 + G_3} = \frac{R_3}{R_1R_2 + R_1R_3 + R_2R_3} \end{aligned} \tag{3-89}$$

where G_1, G_2, and G_3 are the values of conductance of the resistors labeled R_1, R_2, and R_3 in Fig. 3–8.6(b).

The relations of Eq. (3–88) and Eq. (3–89) may frequently be used to simplify the analysis of resistance networks. Note that the relations involve addition, multiplication, and division, but not subtraction. Thus, if the resistors in one of the circuits shown in Fig. 3–8.6 are positive-valued, then the resistors in the equivalent circuit will also be positive-valued. An example of the use of these relations follows.

EXAMPLE 3-8.1. As an example of the use of these relations, let us assume that it is desired to find the voltage v_1 across the 5-A current source shown in Fig. 3-8.7(a). The $\frac{3}{11}$-,

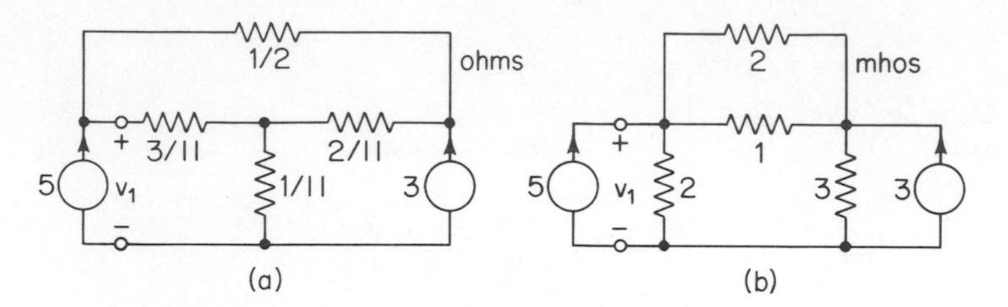

Fig. 3-8.7 Using a delta-wye transformation to analyze a resistance network.

$\frac{2}{11}$-, and $\frac{1}{11}$-Ω resistors form a wye; thus using (3-89), we find that the elements of the equivalent delta are $G_a = 3$, $G_b = 2$, $G_c = 1$ mhos. Thus, the circuit may be redrawn as shown in Fig. 3-8.7(b). The circuit now has two nodes rather than the original three. Routine node analysis may be applied to the circuit to show that $v_1 = \frac{13}{7}$ V.

3-9 THEVENIN AND NORTON'S THEOREMS FOR RESISTANCE NETWORKS

Frequently in our analysis of a resistance network we are not interested in the solution for all of the unknown network variables, but only in one specific variable. Thus, we may wish to calculate a single branch current or a single branch voltage. In addition, many problems call for the calculation of the variables in a given branch for several different values of the resistance of the branch. In such a case, it is desirable to be able to separate the branch from the rest of the network, and apply some type of solution to the rest of the network that will not have to be repeated for each different value of the resistance of the branch. In this section we shall introduce some methods for solving such a problem.

As an example of this type of a problem, consider the network shown in Fig. 3–9.1(a) in which a resistor R_0 is connected to nodes a and b. Let us assume that we desire to compute the current through the resistor for any specified value of resistance. Let us first attack this problem by conventional mesh analysis techniques. Defining two mesh currents $i_1(t)$ and $i_2(t)$ as shown in Fig. 3–9.1(b), we obtain the relations

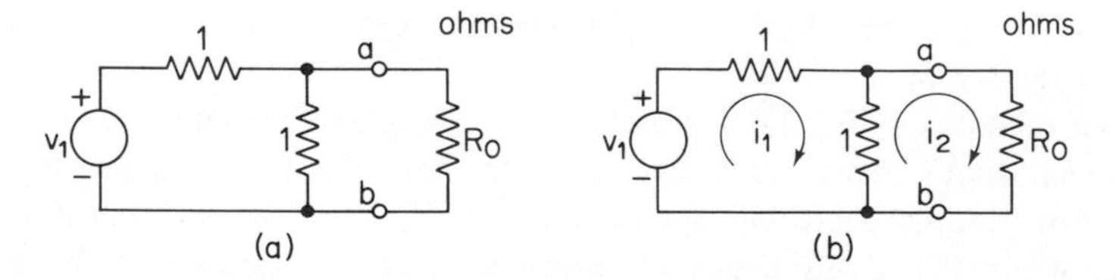

Fig. 3-9.1 A two-mesh resistance network.

$$\begin{bmatrix} v_1(t) \\ 0 \end{bmatrix} = \begin{bmatrix} 2 & -1 \\ -1 & 1 + R_0 \end{bmatrix} \begin{bmatrix} i_1(t) \\ i_2(t) \end{bmatrix} \tag{3-90}$$

The inverse set of relations is

$$\begin{bmatrix} i_1(t) \\ i_2(t) \end{bmatrix} = \frac{1}{1 + 2R_0} \begin{bmatrix} 1 + R_0 & 1 \\ 1 & 2 \end{bmatrix} \begin{bmatrix} v_1(t) \\ 0 \end{bmatrix} \tag{3-91}$$

The current through the resistor R_0 is $i_2(t)$. From the above set of equations, we see that

$$i_2(t) = \frac{v_1(t)}{1 + 2R_0} \tag{3-92}$$

This equation obviously gives us a solution for the current through the resistor R_0 for any value of resistance. As an introduction to a second means for solving this problem, let us connect a resistor R_0 to nodes c and d of a circuit consisting of a single voltage source of value $v_1(t)/2$ and a resistor of value $\frac{1}{2}\ \Omega$ as shown in Fig. 3-9.2. Applying KVL to this circuit, we may write

$$i_0(t) = \frac{v_1(t)/2}{\frac{1}{2} + R_0} = \frac{v_1(t)}{1 + 2R_0} \tag{3-93}$$

Comparing the last two equations, we see that, for a given expression for $v_1(t)$, the same current will flow in the resistor R_0 whether it is connected to nodes a and b of the circuit shown in Fig. 3-9.1 [in this case the current is $i_2(t)$], or to nodes c and d of the circuit in Fig. 3-9.2 [in this case the current is $i_0(t)$]. Furthermore, this statement is true for any value of R_0. Thus we may say that the two circuits (not including R_0) shown in the two figures are equivalent with respect to their behavior at the labeled pair of nodes. The circuit shown in Fig. 3-9.2, consisting of a single voltage source in series with a single resistor, is called the *Thevenin equivalent* of the resistance network (not including R_0) shown in Fig. 3-9.1. Such an equivalent circuit may be derived for any network consisting of linear resistors and sources at a specific pair of nodes. To see this, consider the properties that such an arbitrary network must have if all of its internal sources are set to zero, i.e., if all current sources are replaced by open circuits and all voltage sources are replaced by short circuits. Since the resulting (sourceless) network contains only resistors, it must be able to be replaced at a

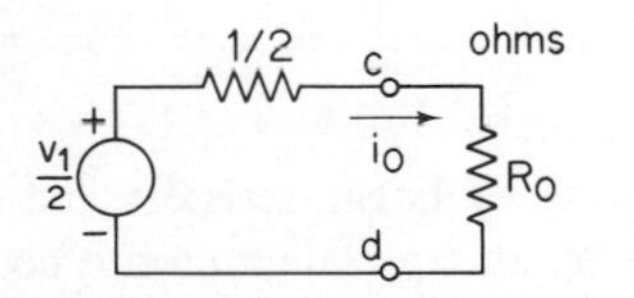

Fig. 3-9.2 A Thevenin equivalent circuit for the network shown in Fig. 3-9.1.

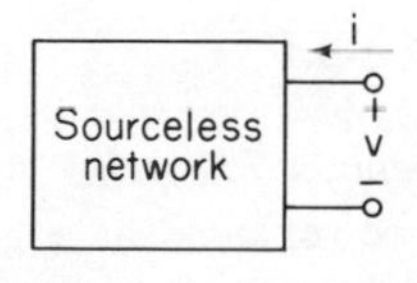

Fig. 3-9.3 Voltage and current variables at the terminals of a sourceless network.

given pair of nodes by a single resistor, the value of which is the input resistance of the network seen at that pair of nodes. Thus, if we define variables $v(t)$ and $i(t)$ at the terminals of such a sourceless network, as shown in Fig. 3–9.3, the input characteristics must be represented by a straight line passing through the origin of an i–v plane as shown in Fig. 3–9.4. The slope of this line gives the value of the input resistance. Note that if an external current is applied to the terminals of this network, a voltage appears across the terminals. The relation of the voltage to the current is determined by the slope of the characteristic.

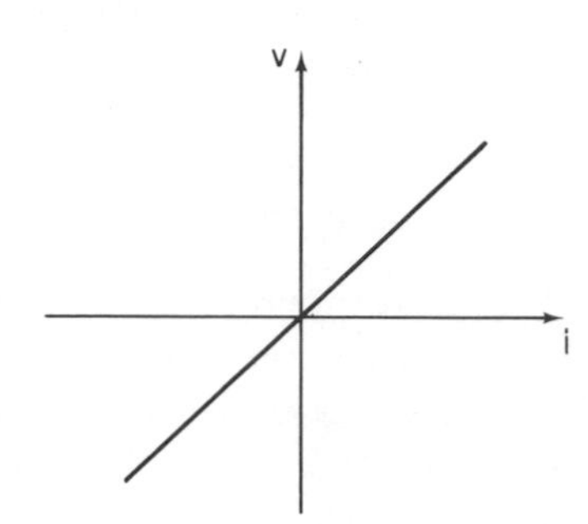

Fig. 3-9.4 Volt-ampere characteristic for the network shown in Fig. 3-9.3.

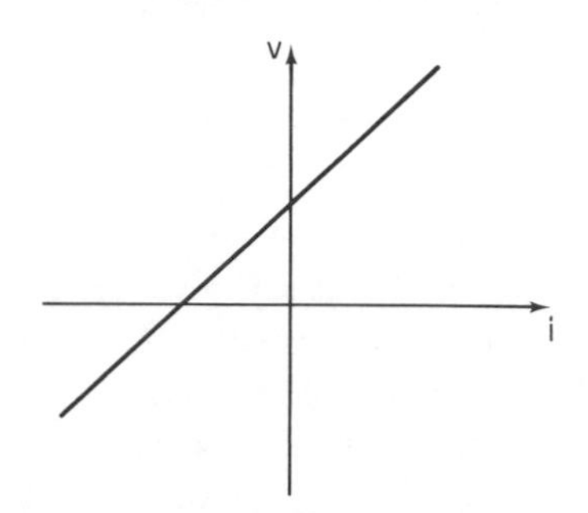

Fig. 3-9.5 Volt-ampere characteristic for a resistance network with sources.

Now let us consider again the original network including its sources. In general, a voltage will be present across the terminals, due to the effects of the sources, even though no current is flowing at the terminals. If, in addition, we apply a current to the network terminals (from an external current source), then, due to the linearity of the resistance network, this source must produce the same effect as it did when it was used to excite the sourceless network. In this case, however, such an effect is superimposed upon the effect of the network's internal sources. The additional effect is determined by the value of the equivalent resistance of the network, i.e., the slope of the i–v characteristic. Thus, we see that the original network must have an overall characteristic consisting of a straight line on the i–v plane *which does not pass* through the origin as shown in Fig. 3–9.5. Such a representation was shown in Sec. 2–3 to be equivalent to that of an ideal voltage source in series with a linear resistor. Thus we conclude that our original network may be replaced by an equivalent circuit consisting of a voltage source in series with a single resistor. We may summarize our results as follows:

SUMMARY 3-9.1

Thevenin's Theorem: Any network consisting of linear resistors and sources may be replaced at a given pair of nodes by an equivalent circuit consisting of a single voltage source and a series resistor. Such a circuit is called a

Thevenin equivalent circuit. The value of the resistor is the input resistance seen at the pair of nodes when all the sources of the original network are set to zero. The value of the voltage source is that seen at the open-circuited terminals.

As an example of the use of a Thevenin equivalent circuit in analyzing a resistance network, consider the following example.

EXAMPLE 3-9.1. It is desired to find the current i_R in the resistor R in the circuit shown in Fig. 3-9.6(a) for any value of the resistor. We may use Thevenin's theorem to replace

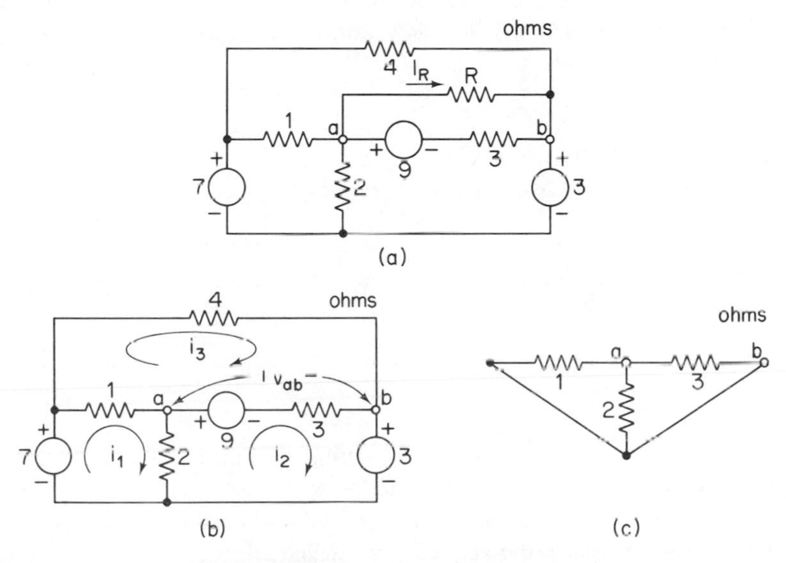

Fig. 3-9.6 Analysis of a resistance network.

the resistance network (excluding the resistor R) by a single source and a resistor to simplify the computations. First let us find the (open-circuit) voltage v_{ab} across the nodes a-b when the resistor R is removed as shown in Fig. 3-9.6(b). The network equations are the same as those given in Example 3-3.1, and the solution for the mesh currents i_2 and i_3 was shown in that example to be -1 A and 1 A, respectively. Thus, there is a net current of 2 A flowing through the 3-Ω resistor from right to left and the voltage across it is 6 V with a polarity opposite to that of the 9-V voltage source. Adding the two voltages, we see that the voltage v_{ab} as defined in Fig. 3-9.6(b) is 3 V. This is also the voltage of the voltage source in the Thevenin equivalent circuit. Now let us find the input resistance of the sourceless network. Replacing the voltage sources by short circuits and redrawing the network, as shown in Fig. 3-9.6(c), we see that the equivalent resistance of the network as seen at nodes a-b is $\frac{6}{11}$ Ω. This is the resistance of the Thevenin equivalent circuit. Combining the above results, the Thevenin equivalent circuit for the resistance network is shown in Fig. 3-9.7. Adding the resistance R, as shown in the figure, we see that the current i_R is given by the relation

$$i_R = \frac{3}{R + \frac{6}{11}}$$

The Thevenin equivalent circuit is not the only simple circuit that may be used to replace a given resistance network. To see this, consider again the equivalent terminal characteristic for a linear resistance network and its sources that is shown on the i–v plane in Fig. 3–9.5. This representation may also be considered as defining a single current source with a shunt resistor. The current source has a value which is given by the current flowing through a short circuit connected across the terminals of the original resistance network. The resistance is again the input resistance of the network when the sources are set to zero. The equivalent circuit is named a *Norton equivalent circuit.* We may summarize its description as follows.

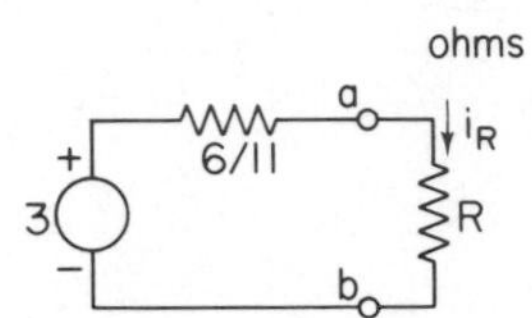

Fig. 3-9.7 Thevenin equivalent circuit for the resistance network shown in Fig. 3-9.6(a).

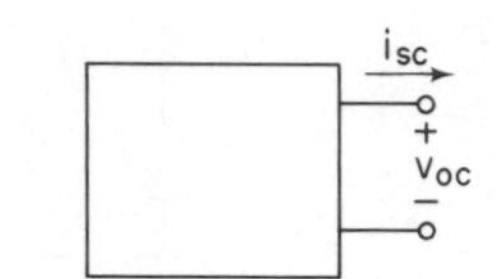

Fig. 3-9.8 Reference polarities for equation (3-94).

SUMMARY 3-9.2

Norton's Theorem: Any network consisting of linear resistors and sources may be replaced at a given pair of nodes by an equivalent circuit consisting of a single current source and a shunt resistor. The value of the resistor is the input resistance seen at the pair of nodes when all sources of the original network are set to zero. The value of the current source is the current which flows in a short circuit connecting the nodes.

It should be noted that the Norton equivalent circuit is easily derived directly from the Thevenin equivalent circuit or vice versa using the non-ideal source transformations developed in Sec. 2–3. It should also be noted that, if we let v_{oc} be the voltage measured at a pair of open-circuited terminals and i_{sc} be the current measured when the terminals are shorted (with the relative reference polarities shown in Fig. 3–9.8), then the value of the resistance R_s that is required in either a Thevenin or a Norton equivalent circuit is given by the relation

$$R_s = \frac{v_{oc}}{i_{sc}} \tag{3–94}$$

An example of the direct determination of a Norton equivalent circuit follows.

EXAMPLE 3-9.2. We may use basic network analysis techniques to derive a Norton equivalent circuit for the network shown in Fig. 3-9.9(a) at the nodes *a-b* shown in the figure. This is the same network that was used in Example 3-9.1. Thus, the input resistance at the terminals *a-b* found when all the internal sources are set to zero will be $\frac{6}{11}\ \Omega$, just as it was in the preceding example. If we now short-circuit the terminals *a-b* as shown in Fig. 3-9.9(b), we find that the short completely isolates the network branch consisting of the 9-V source and the series 3-Ω resistor from the rest of the network. Thus, this branch will produce a contribution to the current i_{ab} of 3 A, independent of the effects of the rest of the network. To determine the effects of the rest of the network, we may redraw the circuit without this branch, as shown in Fig. 3-9.9(c), where i'_{ab} is defined as $i_{ab} - 3$. By using the concept of equivalent, paralleled voltage sources developed in Sec. 2-5, the network may be redrawn as shown in Fig. 3-9.9(d). Obviously, the branch consisting of the right 3-V

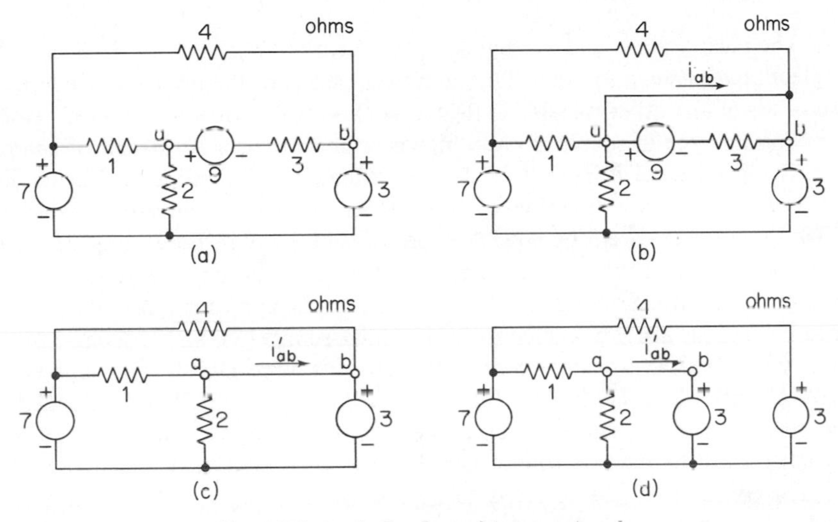

Fig. 3-9.9 Analysis of a resistance network.

source and the 4-Ω resistor makes no contribution to the current i_{ab}. We may solve for the contribution to the current i_{ab} from the remainder of the circuit. Routine mesh analysis applied to this two-mesh circuit shows that this contribution to i_{ab} is $\frac{5}{2}$ A. The superposition of the above effects gives a total short-circuit current i_{ab} of $\frac{11}{2}$ A. Thus the Norton equivalent network for the resistance network as viewed from the terminals *a-b* is a current source of $\frac{11}{2}$ A and a shunt resistance of $\frac{6}{11}\ \Omega$ as shown in Fig. 3-9.10. Comparison of this Norton equivalent circuit with the Thevenin equivalent circuit shown in Fig. 3-9.7 (without the resistor *R*) readily verifies the equivalence of the two circuits.

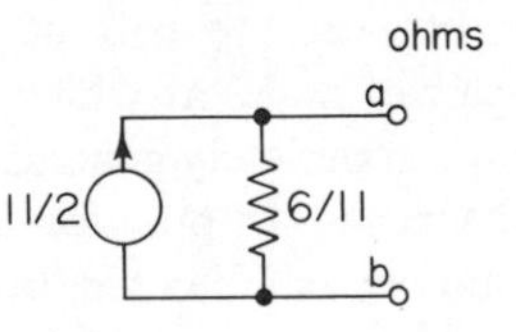

Fig. 3-9.10 Norton equivalent circuit for the resistance network shown in Fig. 3-9.9(a).

In conclusion, it should be noted that Thevenin and Norton equivalent circuits may be constructed for any linear network. Thus, they are directly applicable to cases where the sources are not constant but have outputs which are functions of time. They are also applicable to situations in which the resistors of the resistance network are time-varying. It should be emphasized that these equivalent circuits are only equivalent with respect to their terminal behavior, i.e., with respect to the values of the voltage and current variables at the network terminals. Internal conditions in the original network and in the equivalent circuit will, in general, be completely different.

3-10 DEPENDENT SOURCES

The ideal voltage and current sources which were introduced in Sec. 2-2 are examples of *independent* sources. They are so named since their output is independent of the values of any other variable in the network in which they are located. Another very important type of source is one which is referred to as a *dependent* or *controlled* source. Such a source is an active device defined by the property that its output voltage (or current) is a function of some other voltage or current variable in the network. Thus, there are four possible types of such sources, depending on whether the controlling variable is a voltage or a current, and depending on whether the source itself is of the voltage or current type. The names of these four types of sources and their abbreviations are: voltage-controlled voltage source (VCVS); voltage-controlled current source (VCIS); current-controlled voltage source (ICVS); and current-controlled current source (ICIS). In this text we shall be primarily concerned with *linear, time-invariant* dependent sources, i.e., ones in which the output of the source is linearly related to the controlling variable by some positive or negative real constant. The constant will be referred to as the *gain* of the source. A set of circuit symbols which may be used to represent the four different types of controlled sources is shown in Fig. 3-10.1. In the schematic representation used for these sources, a

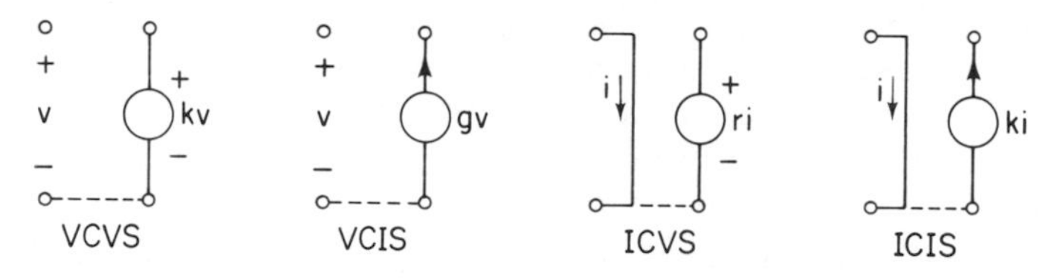

Fig. 3-10.1 Types of controlled sources.

complete separation is shown between the pair of nodes at which the controlling parameter is defined and the pair of nodes at which the output is presented. Such a separation is useful in developing completely general analysis techniques for networks which include such controlled sources. In practice, however, one node from each of these node-pairs is always common, as indicated by the dashed lines in the figure. Thus, these controlled sources are usually considered as three-terminal network elements. Note that, as indicated in the figure, the gain constant has different dimensions for the various types of sources. Thus, for the VCIS it has the dimensions of conductance (represented by the symbol g), for the ICVS it has the dimensions of resistance (represented by the symbol r), and for the VCVS and the ICIS it is dimensionless (represented by the symbol k). The constants associated with the VCIS and the ICVS are sometimes referred to as a *transfer conductance* (or *transconductance*) and a *transfer resistance*, respectively, where the word "transfer" is used to indicate that a transfer of signal is made from one portion of the circuit to another. It should be noted that the type of source, i.e., current or voltage, determines the output characteristic of the device. For example, both the VCVS and the ICVS have their output taken from a voltage source; thus, these network elements will provide the specified output voltage (as controlled by their input variables of voltage and current, respectively) independent of any current which is flowing through the source itself. Similarly, the ICIS and VCIS will provide a specified output current (as controlled by their input variables of current and voltage, respectively) independent of any voltage which may be present across the source itself. This property of controlled sources, is, of course, identical with that of independent sources.

Controlled sources provide our first contact with network elements which have more than two terminals. They play a very important role in the study of networks because they provide a means of modeling many physical active devices. Such modeling cannot be accomplished by two-terminal elements alone. For example, a transistor operating in the "active" region of its characteristic may be modeled (crudely, to be sure) by the circuit shown in Fig. 3-10.2(a), consisting of an ICIS of

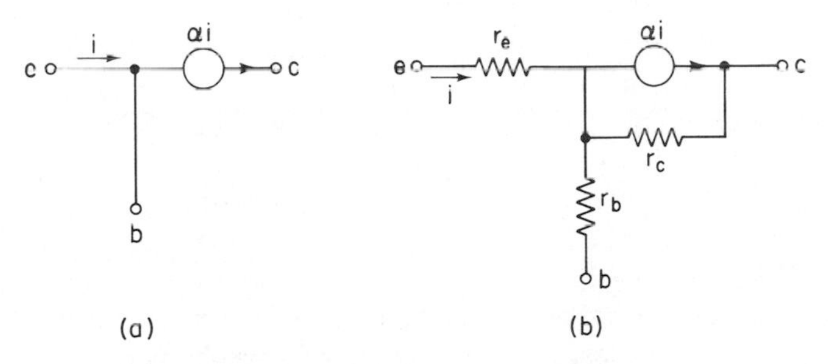

Fig. 3-10.2 Model of a transistor.

gain α, where α has a value slightly less than unity, and where the terminals labeled e, b, and c are those corresponding to the emitter, base, and collector of an actual transistor. A somewhat more sophisticated model for a transistor is shown in Fig. 3-10.2(b). In this figure, the resistors labeled r_e, r_b, and r_c, representing the internal emitter resistance, base resistance, and collector resistance of the transistor, have been added to the basic circuit. Typical values of these components are $\alpha = 0.99$, $r_e = 25\ \Omega$, $r_b = 100\ \Omega$, and $r_c = 10^6\ \Omega$. Another example of the use of controlled sources in modeling physical devices is shown in Fig. 3-10.3(a). The VCIS shown in this schematic represents a crude but useful model of the small-signal behavior of a pen-

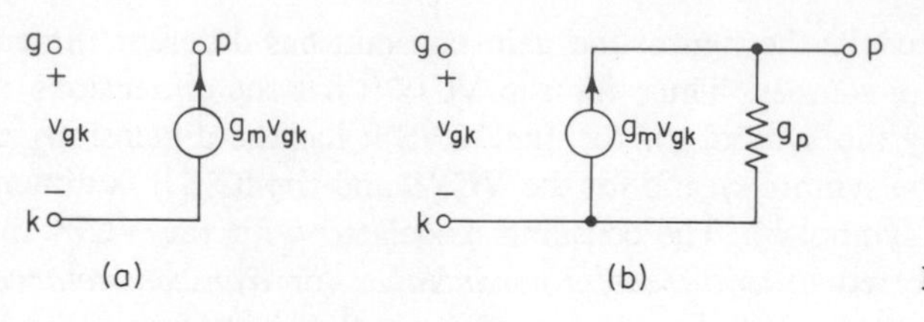

Fig. 3-10.3 Model of a pentode.

tode vacuum tube. The terminals labeled, g, k, and p represent the grid, cathode, and plate connections, respectively. A slightly more sophisticated model for such a tube might include a resistor g_p representing the dynamic plate resistance of the tube as shown in Fig. 3-10.3(b). Typical values of the components are $g_m = 10^{-3}$ mho and $g_p = 10^{-6}$ mho. In this section we shall consider the ways in which controlled sources affect the techniques and conclusions which we have developed for resistance networks in the preceding sections.

Let us begin by considering the two-mesh network shown in Fig. 3-10.4.

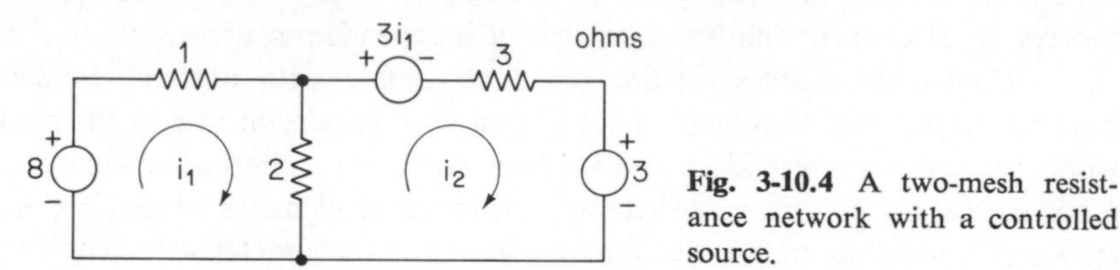

Fig. 3-10.4 A two-mesh resistance network with a controlled source.

This network contains an ICVS, which has an output voltage $3i_1$, i.e., a voltage numerically equal to three times the value of the mesh reference current i_1. We may use the techniques of Sec. 3-1 to write the mesh equations for this network in matrix form, adding the contribution from the two voltage sources in mesh 2, and putting their sum as an element of the voltage matrix **v**. Thus, we obtain

$$\begin{bmatrix} 8 \\ -3 - 3i_1 \end{bmatrix} = \begin{bmatrix} 3 & -2 \\ -2 & 5 \end{bmatrix} \begin{bmatrix} i_1 \\ i_2 \end{bmatrix} \tag{3-95}$$

In the above equation, a term which is a function of the mesh current i_1 appears in the left member. We may easily move this term to the right member of the equation. The resulting set of equations then has the form

$$\begin{bmatrix} 8 \\ -3 \end{bmatrix} = \begin{bmatrix} 3 & -2 \\ 1 & 5 \end{bmatrix} \begin{bmatrix} i_1 \\ i_2 \end{bmatrix} \tag{3-96}$$

The techniques given in Sec. 3-2 are readily applied to find the inverse set of equations, and thus the solution for the mesh currents. The inverse relations are

$$\begin{bmatrix} i_1 \\ i_2 \end{bmatrix} = \frac{1}{17} \begin{bmatrix} 5 & 2 \\ -1 & 3 \end{bmatrix} \begin{bmatrix} 8 \\ -3 \end{bmatrix} = \begin{bmatrix} 2 \\ -1 \end{bmatrix} \tag{3-97}$$

Thus, we have solved the network for the mesh currents, and we find that $i_1 = 2$ A and $i_2 = -1$ A.

The procedure illustrated above is easily generalized to apply to any arbitrary resistance network which is to be defined by mesh equations, and which contains controlled sources of the ICVS type.

The steps are summarized as follows:

SUMMARY 3-10.1

Solution of Resistance Networks Containing Voltage Sources and ICVS's: A set of simultaneous equations of the form $\mathbf{v}(t) = \mathbf{Ri}(t)$ may be written for a resistance network containing independent voltage sources and dependent ICVS's by the following steps:

1. Define a set of mesh reference currents.
2. Write the KVL equations for each mesh using the procedure given in Sec. 3-2 and putting all terms representing ICVS's in the appropriate elements of $\mathbf{v}(t)$.
3. Rewrite the simultaneous equations putting all terms representing the contributions from ICVS's into the appropriate elements of the matrix $\mathbf{R}$.
4. Solve the resulting set of simultaneous equations.

An example of this procedure follows.

EXAMPLE 3-10.1. A three-mesh resistance network is shown in Fig. 3-10.5. In mesh 2, there is an ICVS controlled by the current in mesh 1. In addition, in mesh 3 there is an

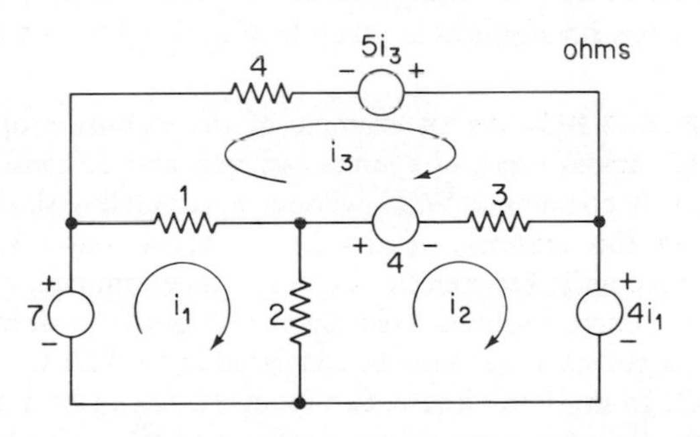

Fig. 3-10.5 A three-mesh resistance network with two controlled sources.

ICVS controlled by the current in that mesh. Applying the techniques described in Sec. 3-2, we may apply KVL around each of the three meshes to obtain the relations

$$\begin{bmatrix} 7 \\ -4 - 4i_1 \\ 4 + 5i_3 \end{bmatrix} = \begin{bmatrix} 3 & -2 & -1 \\ -2 & 5 & -3 \\ -1 & -3 & 8 \end{bmatrix} \begin{bmatrix} i_1 \\ i_2 \\ i_3 \end{bmatrix}$$

The terms in the left-hand member of this equation which are functions of i_1 and i_3 may be moved into the right-hand member of the equation. If this is done, we obtain the relations

$$\begin{bmatrix} 7 \\ -4 \\ 4 \end{bmatrix} = \begin{bmatrix} 3 & -2 & -1 \\ 2 & 5 & -3 \\ -1 & -3 & 3 \end{bmatrix} \begin{bmatrix} i_1 \\ i_2 \\ i_3 \end{bmatrix}$$

The set of simultaneous equations given above may be solved by any of the techniques discussed in Secs. 3-3 and 3-4. The resulting mesh currents are $i_1 = 2$ A, $i_2 = -1$ A, and $i_3 = 1$ A for this example.

It should be noted, from the example given above, that if the mesh equations describing a resistance network which contains controlled sources are put in the form $\mathbf{v}(t) = \mathbf{Ri}(t)$, then the matrix $\mathbf{R}$, in general, will not be symmetric.

Several extensions of the method described above should be readily apparent to the reader. For example, since any branch current can be expressed as a linear combination of the mesh currents, any ICVS which is controlled by a branch current that is not a mesh current can have its output voltage expressed as a function of mesh currents; thus, the development given above applies to this case. As another extension, consider the case where ICIS's are present in the network. If such sources have resistors in parallel with them, they may be converted to ICVS's with series resistors using the source conversion methods described in Sec. 2-3. If such shunt resistors are not present, then the techniques given in Sec. 3-8 for adding additional series current sources may be applied to provide the ICIS's with shunt resistors. Similarly, to treat VCVS's, it is usually possible to express the controlling voltage in terms of some branch current, which, in turn, can be expressed in terms of the unknown variables, the mesh currents. A flow chart of a general procedure to include these various possibilities is given in Fig. 3-10.6. An example follows.

EXAMPLE 3-10.2. As an example of the extension of the method for including the effects of various types of sources outlined above, consider the network shown in Fig. 3-10.7(a). It contains a VCIS without a paralleling shunt resistor. As a first step in the solution of this network, we may add a second series VCIS as shown in Fig. 3-10.7(b). Since the potential between the VCIS's is undetermined, we may connect this point so that each of the current sources has a shunt resistor as shown in Fig. 3-10.7(c). The VCIS's and their shunt resistors may now be converted to VCVS's with series resistors as shown in Fig. 3-10.7(d). In this latter figure, two mesh reference currents i_1 and i_2 have been defined. In terms of these mesh currents, we see that $v_0 = 3(i_1 - i_2)$. Applying KVL to the two meshes, we obtain the relations

$$\begin{bmatrix} 5 - 2v_0 \\ 6v_0 \end{bmatrix} = \begin{bmatrix} 5 - 6(i_1 - i_2) \\ 18(i_1 - i_2) \end{bmatrix} = \begin{bmatrix} 4 & -4 \\ -4 & 10 \end{bmatrix} \begin{bmatrix} i_1 \\ i_2 \end{bmatrix}$$

These equations may be rewritten in the form

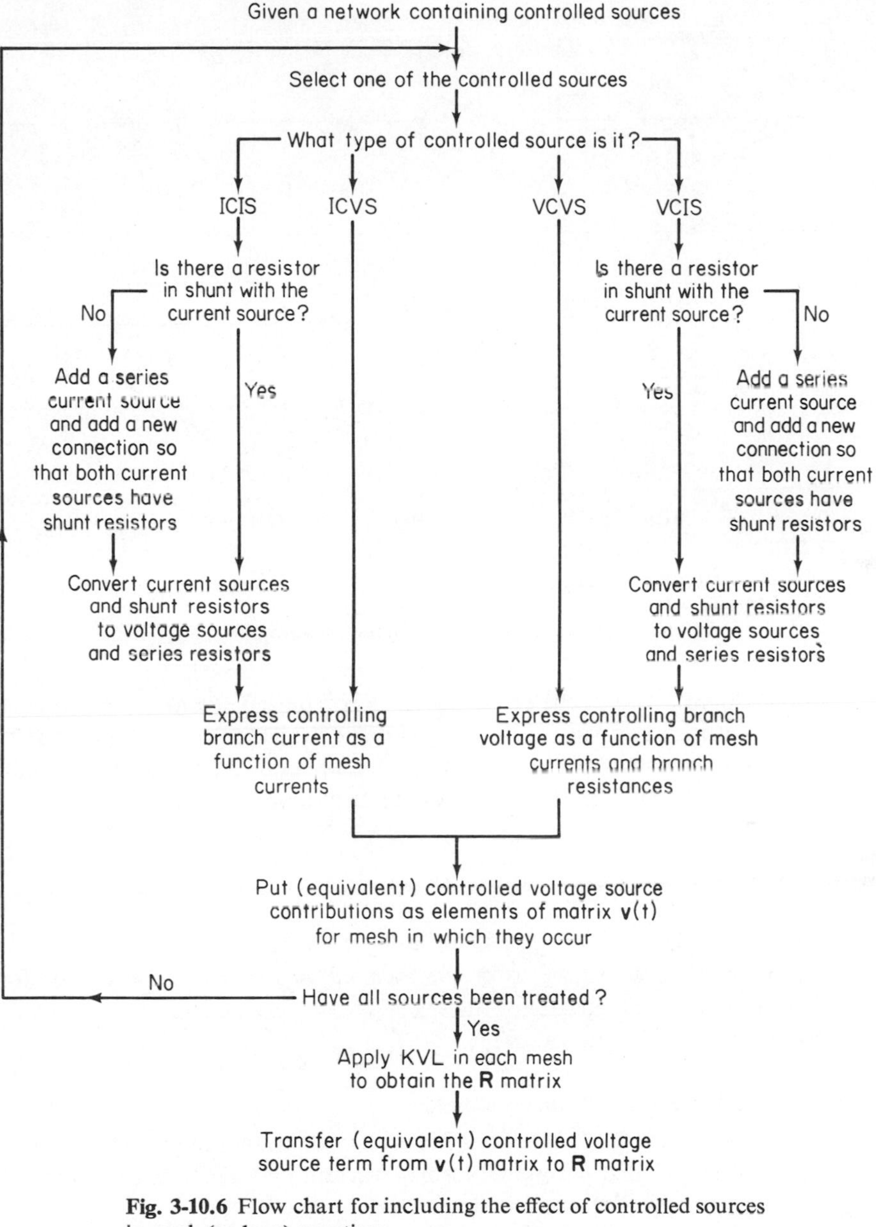

Fig. 3-10.6 Flow chart for including the effect of controlled sources in mesh (or loop) equations.

$$\begin{bmatrix} 5 \\ 0 \end{bmatrix} = \begin{bmatrix} 10 & -10 \\ -22 & 28 \end{bmatrix} \begin{bmatrix} i_1 \\ i_2 \end{bmatrix}$$

The solution for the mesh currents is easily accomplished, and is left to the reader as an exercise.

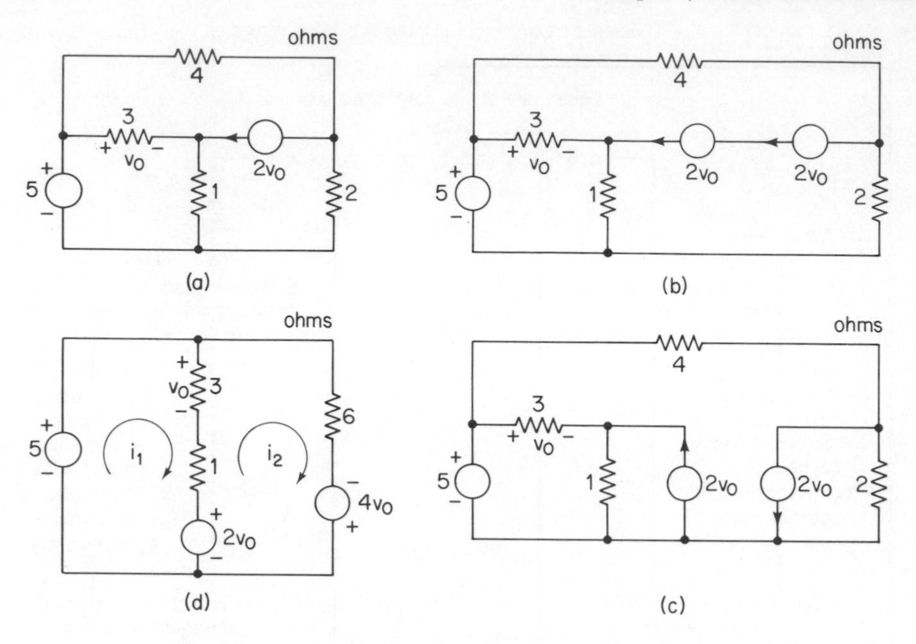

Fig. 3-10.7 A resistance network with a controlled source.

If one desires to write the equations for a resistance network containing controlled sources using node voltages as the unknown variables, he may use a method similar to that given above. If the controlled sources are VCIS's, the network equations may be written directly. An outline of the procedure follows:

SUMMARY 3-10.2

Solution of Resistance Networks Containing Current Sources and VCIS's: A set of simultaneous equations of the form $\mathbf{i}(t) = \mathbf{G}\mathbf{v}(t)$ may be written for a resistance network containing independent current sources and dependent VCIS's by the following steps:

1. Define a set of node voltages.
2. Write the KCL equations for each node using the procedure given in Sec. 3–7 and putting all terms representing VCIS's in the appropriate elements of the matrix $\mathbf{i}(t)$.
3. Rewrite the matrix equation putting all terms representing contributions from the VCIS's into the appropriate elements of the matrix $\mathbf{G}$.
4. Solve the resulting set of simultaneous equations.

The procedure given above is easily modified to take account of various other types of controlled sources. The modifications are similar to those used in writing the mesh equations. A flow chart of the procedure to be followed in writing node equations for networks containing various types of controlled sources is given in Fig. 3–10.8. The reader should note the similarity between this flow chart and the one given for the mesh analysis case in Fig. 3–10.6. An example follows.

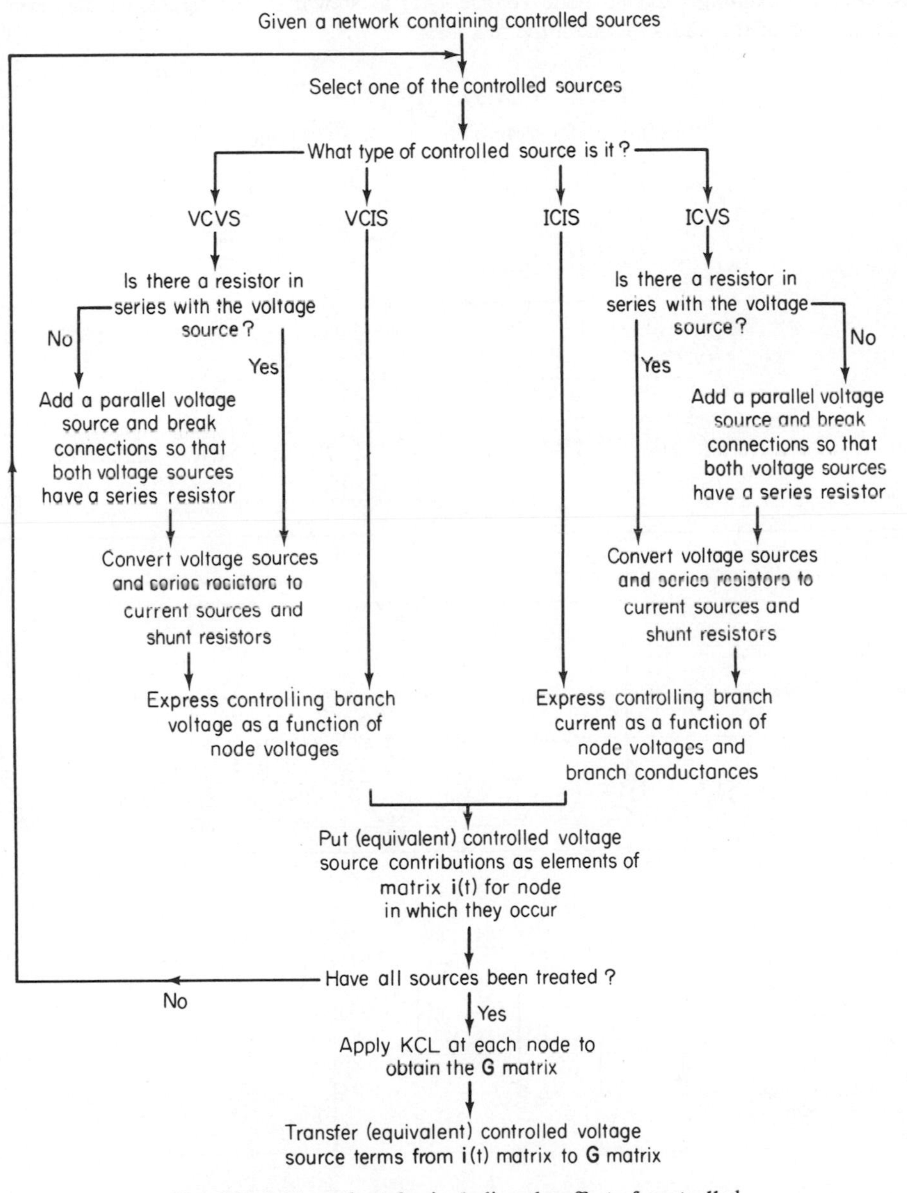

Fig. 3-10.8 Flow chart for including the effect of controlled sources in node equations.

EXAMPLE 3-10.3. The resistance network shown in Fig. 3-10.9(a) contains two controlled sources, a VCIS and a VCVS. To analyze this network on a node basis, we may begin by noting that the VCVS does not have a series resistor, therefore we will parallel this controlled source by a second and then break the connection between them as shown in Fig. 3-10.9(b). The two VCVS's and their series resistors may now be replaced by ICIS's and shunt resistors as shown in Fig. 3-10.9(c). Using the voltage $v_1(t)$ as one of our node voltages, and defining a second node voltage $v_2(t)$ as shown in the figure, we may apply KCL at each of the nodes to obtain the relations

$$\begin{bmatrix} i_1(t) + 2v_1(t) - 3v_1(t) \\ i_2(t) + 3v_1(t) + 4v_1(t) \end{bmatrix} = \begin{bmatrix} 7 & -2 \\ -2 & 7 \end{bmatrix} \begin{bmatrix} v_1(t) \\ v_2(t) \end{bmatrix}$$

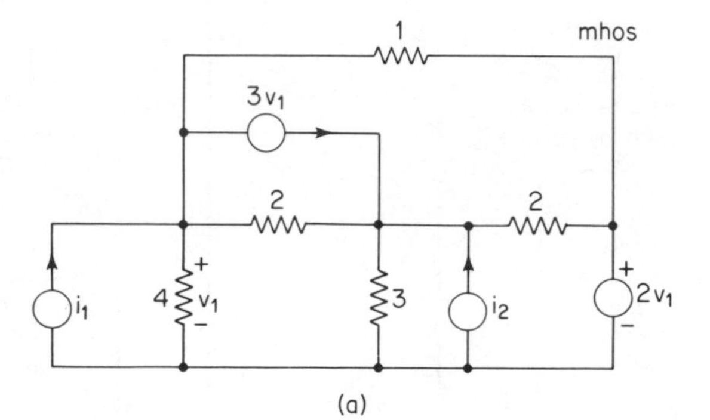

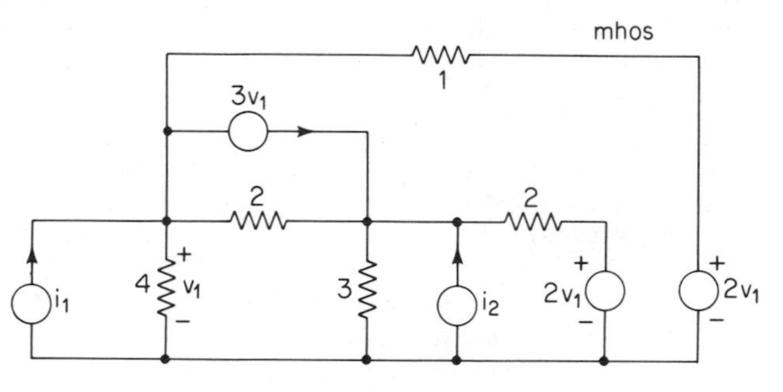

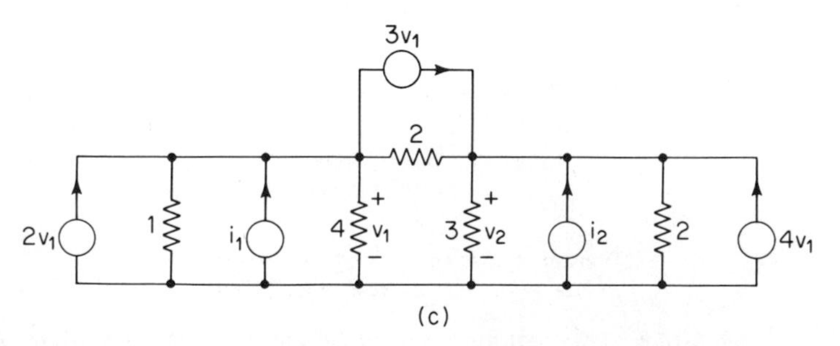

Fig. 3-10.9 A resistance network with a pair of controlled sources.

Rearranging the terms of this equation, we obtain a set of equations relating the node voltages $v_1(t)$ and $v_2(t)$ and the input currents from the independent current sources, namely, $i_1(t)$ and $i_2(t)$. These relations have the form

$$\begin{bmatrix} i_1(t) \\ i_2(t) \end{bmatrix} = \begin{bmatrix} 8 & -2 \\ -9 & 7 \end{bmatrix} \begin{bmatrix} v_1(t) \\ v_2(t) \end{bmatrix}$$

The inverse set of relations which gives the values of the node voltages as functions of the input currents is easily obtained to yield a solution for the network for any arbitrary set of input currents.

Another topic concerning controlled sources is of interest for future application. This concerns circuits formed by an interconnection of controlled sources and resistors which do not contain any independent sources. Any such circuit may be replaced at a given pair of terminals by a single equivalent resistor. To see how the value of such a resistor is determined, consider the following example.

EXAMPLE 3-10.4. The circuit shown in Fig. 3-10.10(a) consists of a single controlled source and three resistors. The source is an ICVS controlled by the current $i_0(t)$ in resistor

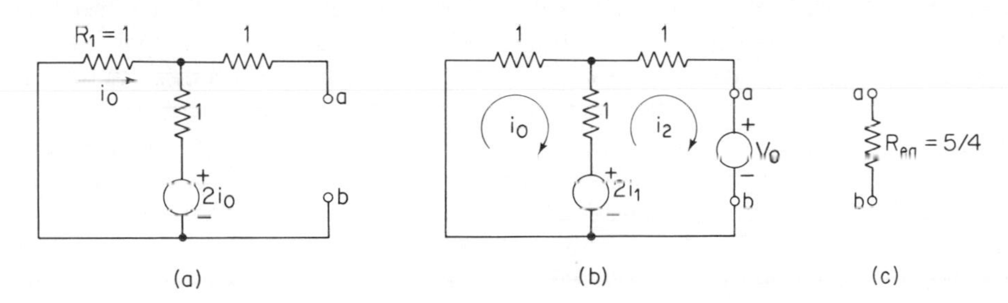

Fig. 3-10.10 An equivalent resistance for a resistance network with a controlled source.

R_1. It is desired to find a single equivalent resistor to replace the network, so that the relation between the variables at terminals *a-b* will be unchanged. To find the value of such an equivalent resistor, we may apply a voltage source with a constant output voltage V_0 to the terminals *a-b*. As indicated in Fig. 3-10.10(b), this in effect defines a two-mesh resistance network with mesh current variables $i_0(t)$ and $i_2(t)$. Applying KVL to the two meshes, we obtain the relations

$$\begin{bmatrix} 0 \\ -V_0 \end{bmatrix} = \begin{bmatrix} 4 & -1 \\ -3 & 2 \end{bmatrix} \begin{bmatrix} i_0(t) \\ i_2(t) \end{bmatrix}$$

Solving for $i_2(t)$, we find that $i_2(t) = -4V_0/5$. The relation between the input voltage V_0 provided by the voltage source and the current flowing out of the positive terminal of this source [which is $-i_2(t)$] provides the value of the equivalent resistance R_{eq}. Thus, we see that

$$R_{eq} = \frac{V_0}{4V_0/5} = \frac{5}{4}$$

We conclude that the entire circuit shown in Fig. 3-10.10(a) may be replaced at terminals a-b by an equivalent resistor of value $\frac{5}{4}$ Ω, as shown in Fig. 3-10.10(c), without any change in the relation between the terminal variables as observed at those terminals.

The procedure described above is easily generalized. It should be especially noted that the value of the source voltage applied cancels when the value of R_{eq} is determined. Thus, any numerical value of voltage may be applied in calculating the value of R_{eq} (unity is usually a convenient value). An alternative method of computation would be to apply a current source to the terminals a–b and determine the voltage appearing across it. The ratio of voltage to current will also be $\frac{5}{4}$ in this case. It should be noted that some connections of controlled sources and resistors may produce values of an equivalent resistance which are negative. Such a circuit will supply energy to any network to which it may be connected.

Now let us consider the case where, in addition to having resistors and controlled (i.e., dependent) sources present, as was discussed in the above development, we also have *independent* sources present. We may modify the development given above to obtain a Thevenin or Norton equivalent circuit for the entire network, including both types of sources. The value of the resistance to be used in such an equivalent circuit is found by using the method given above. When applying this method, all independent voltage sources must be replaced by short circuits, and all independent current sources by open circuits. In addition, to construct a Thevenin equivalent circuit, it is merely necessary to find the open-circuit voltage appearing at the specified terminals of the original network with all the independent sources replaced. The Thevenin equivalent circuit then consists of an independent voltage source, whose value is equal to the open-circuit voltage, in series with the equivalent resistance found when all independent sources are set to zero. An example of this procedure follows.

EXAMPLE 3-10.5. The network used in Example 3-10.4 and shown in Fig. 3-10.10(a) has been modified by the addition of a voltage source to provide the network shown in Fig. 3-10.11(a). To find a Thevenin equivalent circuit for this network at terminals a-b, we first set the (independent) voltage source to zero and find the equivalent resistance seen

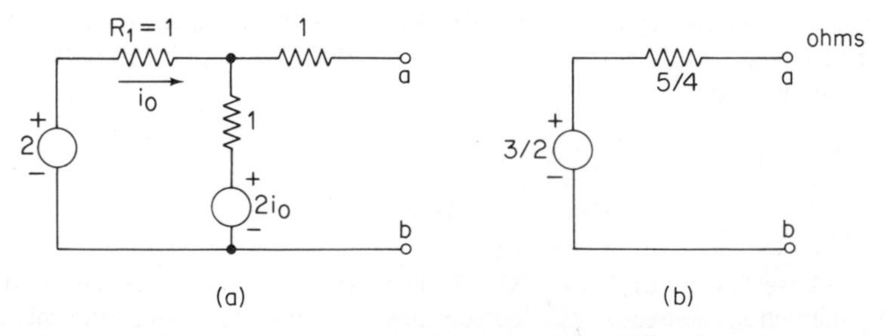

Fig. 3-10.11 Thevenin equivalent circuit for a resistance network with dependent and independent sources.

at terminals a-b. From Example 3-10.4, we know that this is $\frac{5}{4}\ \Omega$. We next find the open-circuit voltage appearing at the terminals a-b with the independent voltage source restored to the network. It is easily seen from Fig. 3-10.11(a) that this voltage is $\frac{3}{2}$ V (positive at terminal a). Thus, the Thevenin equivalent circuit which includes the effects of both the dependent and the independent sources from the original network consists of the independent voltage source and resistor shown in Fig. 3-10.11(b).

In view of the above development, we may modify the statement of Thevenin's theorem given in Sec. 3-9 to include the effect of controlled sources. We obtain the following:

SUMMARY 3-10.3

Thevenin's Theorem (for resistance networks containing dependent sources): Any network consisting of linear resistors, independent sources, and linear dependent sources may be replaced at a given pair of nodes by an equivalent circuit consisting of a single voltage source and a series resistor. The value of the resistor is the input resistance seen at the pair of nodes when all the independent sources of the network are set to zero. The value of the voltage source is the voltage seen at the open-circuited terminals.

A parallel development may be made to arrive at a more general statement of Norton's theorem. Thus, we obtain the following:

SUMMARY 3-10.4

Norton's Theorem (for resistance networks containing dependent sources): Any circuit consisting of linear resistors, independent sources, and linear dependent sources may be replaced at a given pair of nodes by an equivalent circuit consisting of a single current source and a shunt resistor. The value of the resistor is the input resistance seen at the pair of nodes when all the independent sources of the network are set to zero. The value of the current source is the current flowing through a short circuit placed across the terminals.

3-11 MULTI-TERMINAL DEVICES

The dependent or controlled sources introduced in the last section may be collectively referred to as multi-terminal devices, where "multi" is interpreted here to mean "three or more." In this respect, controlled sources differ considerably from the two-terminal elements such as resistors and independent sources which were treated in Chap. 2. In this section, we shall discuss some other linear, time-invariant, multi-terminal devices which are frequently encountered in modern network theory. These devices are newer in concept and implementation than the more traditional controlled sources; however, they provide the key to many new and exciting concepts in network theory.

As an introduction to the first multi-terminal device that we shall discuss in this section, consider a VCVS which has two input terminals rather than one, and in which the output voltage is proportional to the difference between the voltages applied to the input terminals. Such a device is shown in Fig. 3–11.1. The gain con-

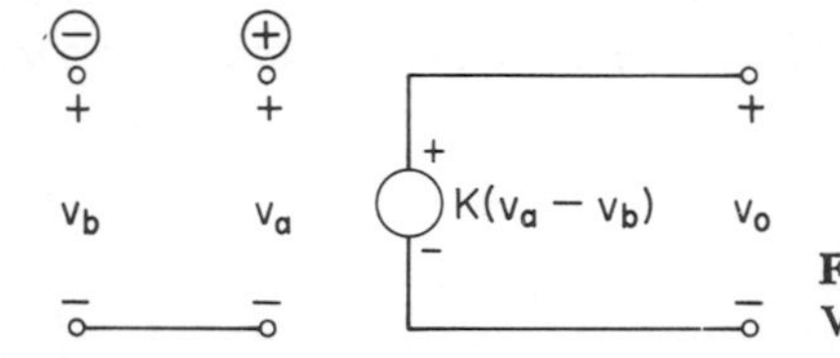

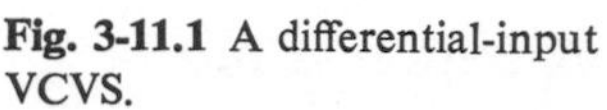

Fig. 3-11.1 A differential-input VCVS.

stant K is assumed to be positive, thus the input terminal labeled ⊕ may be considered as a non-inverting input terminal, and the one labeled ⊖ as an inverting one. Note that the negative reference polarity terminals for the input and output voltages are the same. Such a device is called a *differential-input VCVS.* If K approaches infinity, then the resulting multi-terminal device is referred to as a *differential-input operational amplifier*. This is, of course, an active device since it consists of a VCVS. A model of such a device is shown in Fig. 3–11.2(a) and a circuit symbol which is frequently used to represent it is shown in Fig. 3–11.2(b). It is readily noted from

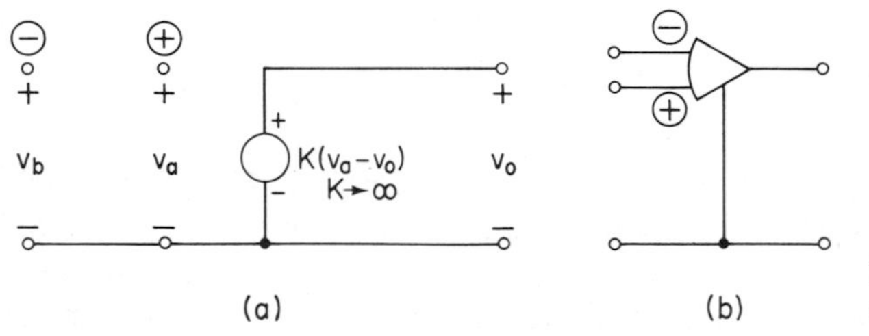

Fig. 3-11.2 A differential-input operational amplifier.

the model that no currents may flow into either of the input terminals of the operational amplifier. Thus, we may refer to the device as having infinite input resistance. In addition, as is always the case for voltage sources, the output voltage will be independent of the output current, Thus, the device may be said to have zero output resistance. In many applications, the ⊕ terminal of the differential-input operational amplifier is grounded. A model and a circuit symbol for this case are shown in Fig. 3–11.3. The result is referred to simply as an *operational amplifier*. Note that such a device has a gain of *minus* infinity.

In practical applications of operational amplifiers, the infinite gain property is usually not used directly. Instead, a connection from the output terminal of the VCVS to one or the other of the input terminals through some circuit element is usually made. Such a connection is called a *feedback path.* As an example of such a connection, consider the circuit using a differential-input operational amplifier and

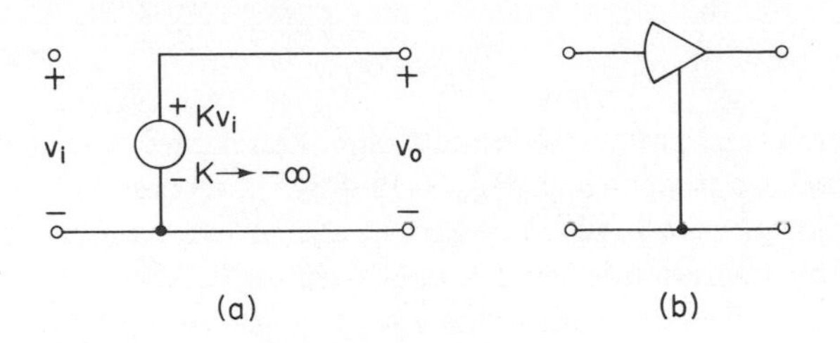

Fig. 3-11.3 A single-input operational amplifier.

two resistors shown in Fig. 3–11.4(a). Using the model for the amplifier shown in Fig. 3–11.2(a), we see that

$$v_o(t) = K[v_a(t) - v_b(t)] \qquad K \rightarrow \infty \tag{3–98}$$

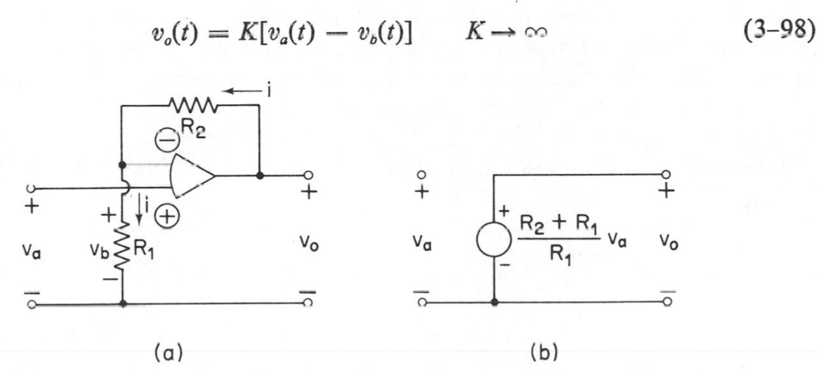

Fig. 3-11.4 A noninverting VCVS.

If, as shown in the figure, we define the current in the resistor R_2 as $i(t)$, then, since no current flows into the ⊖ input terminal of the differential-input operational amplifier, we see that the current $i(t)$ also flows as indicated in the resistor R_1. Thus, applying KVL, the output voltage $v_o(t)$ is

$$v_o(t) = (R_2 + R_1)i(t) \tag{3–99}$$

Equating the above two equations and dividing by K, we obtain

$$v_a(t) - v_b(t) = \frac{R_2 + R_1}{K} i(t) \qquad K \rightarrow \infty \tag{3–100}$$

In the limit as K approaches infinity, the right member of (3–100) approaches zero. Thus, the left member also goes to zero, which implies

$$v_a(t) = v_b(t) \tag{3–101}$$

Since from Ohm's law $v_b(t) = i(t)R_1$, it must also be true that $v_a(t) = i(t)R_1$. Thus, using (3–99), the ratio of the input and output voltages for the circuit shown in Fig. 3–11.4(a) is

$$\frac{v_o(t)}{v_a(t)} = \frac{R_1 + R_2}{R_1} \tag{3-102}$$

Such a circuit, with infinite input resistance and zero output resistance, may be modeled as shown in Fig. 3-11.4(b). It represents an ideal VCVS of positive (non-infinite) gain in which the input and output terminals have a common ground. Such a configuration is called a *non-inverting* VCVS.

The conclusion that $v_a(t) = v_b(t)$, which was shown to be true for the circuit of Fig. 3-11.4(a), always applies to a circuit in which a differential-input operational amplifier has a feedback path from the output to either of the input terminals. This conclusion together with the fact that no currents flow into the input terminals of the operational amplifier provides a convenient basis for analyzing circuits containing differential-input operational amplifiers. As an example of the use of these conclusions, consider the circuit shown in Fig. 3-11.5(a). In this circuit, the active element

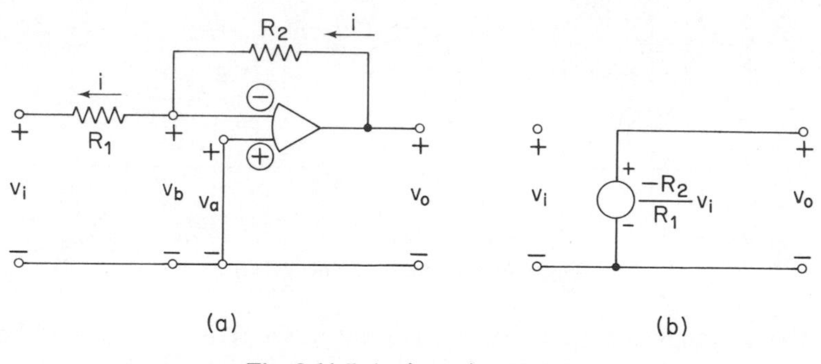

Fig. 3-11.5 An inverting VCVS.

is an ordinary operational amplifier without differential-input (since the ⊕ terminal is grounded). Thus, the voltage $v_a(t)$ is 0. From the conclusion reached in (3-101), namely that $v_a(t) = v_b(t)$, we therefore see that the voltage $v_b(t)$ must also be 0. Thus the ⊖ terminal is effectively grounded. However, we also know that no current flows into the ⊖ terminal of a differential-input operational amplifier (because of the high input resistance). This situation, in which no voltage is present, and no current flows to ground, is referred to as a *virtual ground*. The analysis of the circuit is now easily performed by defining the current $i(t)$ in the resistor R_2 as shown in the figure. As indicated, this current must also flow in the resistor R_1. Thus, the equations defining the input and output voltages $v_i(t)$ and $v_o(t)$ are

$$v_i(t) = -i(t)R_1 \qquad v_o(t) = i(t)R_2 \tag{3-103}$$

The ratio of these two equations is

$$\frac{v_o(t)}{v_i(t)} = -\frac{R_2}{R_1} \tag{3-104}$$

If the resistor R_1 is large in value relative to the resistance of any circuit connected to the input terminals of this device so that the magnitude of the current $i(t)$ is small, this circuit provides a very accurate realization of a VCVS of negative (non-infinite) gain, as shown in Fig. 3–11.5(b). In practice, such a circuit is referred to as an *inverting VCVS.*

One of the advantages of using operational amplifiers in circuit applications is that these devices are readily available in a wide range of prices, qualities, and sizes, ranging from small, completely integrated units which sell for a few dollars to precision, hybrid units which may have a price of $100 or more. In addition, operational amplifiers are readily used to produce other kinds of multi-terminal devices. Two examples of such uses have already been given above, namely, the inverting and non-inverting VCVS's. Some other types of multi-terminal elements which may be produced by operational amplifiers are discussed in the following paragraphs.

As a first example of the use of operational amplifiers to produce other multi-terminal network elements, consider the circuit shown in Fig. 3–11.6(a). Two currents

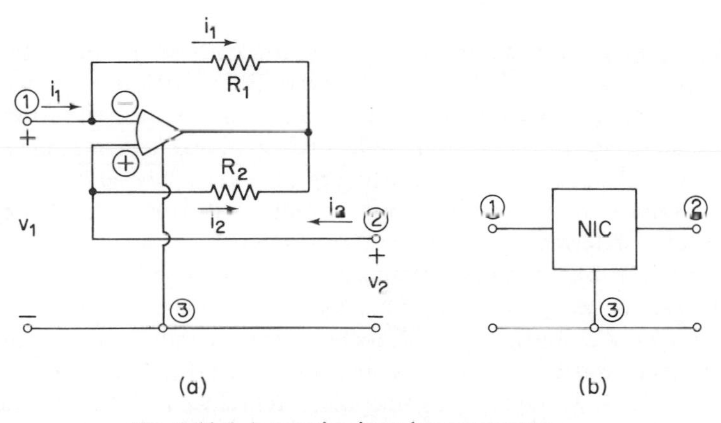

Fig. 3-11.6 A negative impedance converter.

are labeled in this figure, namely, $i_1(t)$ and $i_2(t)$. Since no current flows into the ⊖ terminal of the operational amplifier, the entire input current $i_1(t)$ must flow through the resistor R_1. Similarly, all of $i_2(t)$ must flow through the resistor R_2. Now consider the input voltages $v_1(t)$ and $v_2(t)$. From the discussion of the preceding paragraphs, we see that these voltages are equal, namely, $v_1(t) = v_2(t)$. This implies that the voltage drop across R_1 must be the same as that across R_2. Therefore, the relation $R_1 i_1(t) = R_2 i_2(t)$ must hold. Thus, this circuit defines a three-terminal device in which the terminal variables of voltage and current are related by the following set of equations.

$$\begin{aligned} v_1(t) &= v_2(t) \\ R_1 i_1(t) &= R_2 i_2(t) \end{aligned} \tag{3–105}$$

In order to determine the properties of such a device, let us represent it by the symbol shown in Fig. 3–11.6(b). The meaning of the letters NIC used in this figure will become apparent later in this discussion. Now let us connect a resistor R_L from terminal 2 to terminal 3 as shown in Fig. 3–11.7. The variables $v_2(t)$ and $i_2(t)$ must now satisfy the relation

$$v_2(t) = -R_L i_2(t) \tag{3–106}$$

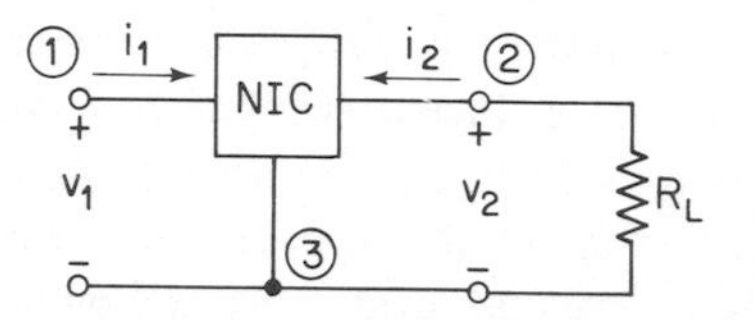

Fig. 3-11.7 A stable NIC network application.

Substituting for $v_2(t)$ and $i_2(t)$ from (3–105), this equation may be put in the form

$$\frac{v_1(t)}{i_1(t)} = -\frac{R_1}{R_2} R_L \tag{3–107}$$

The ratio $v_1(t)/i_1(t)$, however, defines the input resistance seen between the terminals 1 and 3 of the circuit shown in Fig. 3–11.7. From Eq. (3–107), we see that this input resistance is a negative-valued constant, namely, $-R_1/R_2$ times the resistance R_L. Thus, the circuit shown in Fig. 3–11.7 effectively produces a two-terminal, *negative-valued* resistor. This means that the portion of the circuit labeled NIC has *converted* the positive-valued resistor R_L to a negative-valued resistor. Thus, the circuit may be appropriately referred to as a *negative-resistance converter*. In Chap. 7 we shall show that the term *impedance* is a more general term which includes resistance as a special case. Anticipating this result, we may call the device shown in Fig. 3–11.6 a *negative-impedance converter* or NIC for short. The magnitude of the multiplicative constant R_1/R_2 is usually referred to as the *gain* of the NIC. Since this quantity need not be unity-valued, we see that we can obtain a resistance multiplication at the same time as the negative conversion is obtained. A development similar to that given above may be used to show that if a resistor R_0 is connected between terminals 1 and 3 of the NIC shown in Fig. 3–11.6(b), the resistance measured between terminals 2 and 3 is $(-R_2/R_1)R_0$. Thus, a negative-conversion operation also takes place from left to right through the NIC, although the magnitude of the gain is the reciprocal of that from right to left. Since an NIC produces a negative-valued resistance, it obviously must function as an active device. In actual circuit applications, such devices must always be used carefully to avoid the possibility of unstable operation. As an example of this, for the circuit shown in Fig. 3–11.6(a), stable operation will result only if the resistance connected between terminals 1 and 3 is greater than that connected between terminals 2 and 3. Thus, the configuration shown in Fig. 3–11.7 represents a stable situation since the open circuit present between terminals 1 and 3 clearly has a higher value of resistance (infinity) than any R_L (finite). To emphasize these properties, terminals 1 and 3 are frequently referred to as an OCS (open-circuit stable) pair of terminals and terminals 2 and 3 are referred to as an SCS (short-circuit stable) pair of terminals.

The concept of negative resistance has many applications in practical circuits. For example, consider a telephone line consisting of a pair of wires linking two stations as shown schematically in Fig. 3–11.8(a). The individual wires of such a line

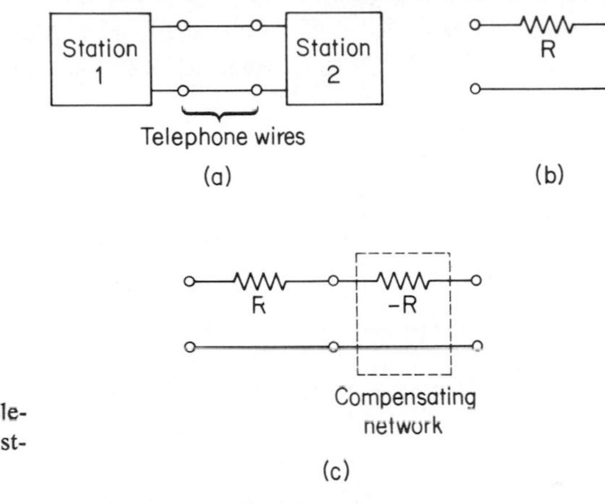

Fig. 3-11.8 Compensating a telephone line with a negative resistance.

have resistance which tends to attenuate any electrical signal traveling along the pair of wires. This resistance may be effectively modeled by a single resistor of value R as shown in Fig. 3–11.8(b). If at one end of the telephone line we now add a compensating network consisting of a series resistor of value $-R$ as shown in Fig. 3–11.8(c), then the combination of the telephone line plus the compensating network gives us a network which has zero series resistance. Thus, there will be no attenuation of the electrical signals passing through such a combination. This is the basic principle of the *negative-resistance amplifier* which is used in many modern telephone circuit applications. The $-R$ used in the compensating network of Fig. 3–11.8(c) can obviously be produced by an NIC and a positive-valued resistor.

Another multi-terminal device which is frequently encountered in network studies is the *gyrator*. In terms of the voltage and current variables shown in Fig. 3–11.9(a), it is defined by the equations

$$
\begin{aligned}
v_1(t) &= -Ri_2(t) \\
v_2(t) &= Ri_1(t)
\end{aligned}
\tag{3–108}
$$

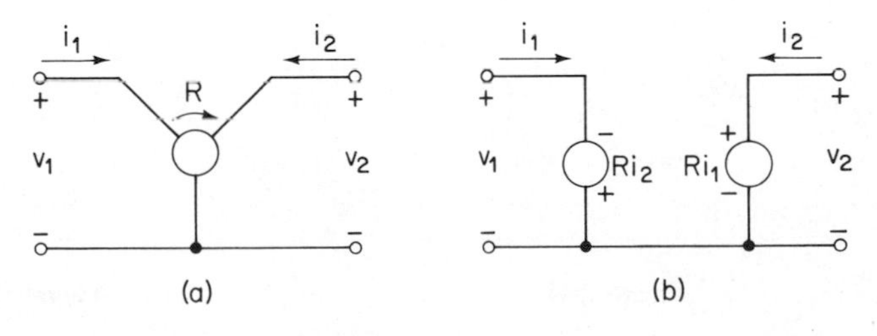

Fig. 3-11.9 A gyrator.

where R is called the *gyration resistance* of the device. The name "gyrator" comes about since this device may be shown to be the electrical analog of a familiar mechanical device, the gyroscope. The relations of Eq. (3–108) can be modeled using two ICVS's. Such a model is shown in Fig. 3–11.9(b). A circuit which realizes the relations of Eq. (3–108) is shown in Fig. 3–11.10. As illustrated, it requires the use of two NIC's. It is assumed that both of these have unity gain, i.e., in the circuit shown in Fig. 3–11.6(a), $R_1 = R_2$. The details of analyzing this circuit are left as an exercise for the reader. In Chap. 7, we shall describe some applications of the gyrator.

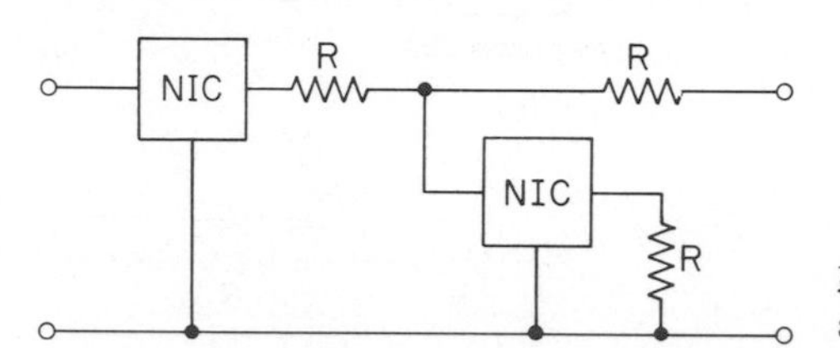

Fig. 3-11.10 Using NICs to realize a gyrator.

Both the NIC and the gyrator can also be realized by direct use of solid state devices such as transistors. Examples of such realizations may be found in problems at the end of this chapter. Additional information on these devices and their use may be found in the references given in the bibliography at the end of this book.

3-12 CONCLUSION

This chapter has presented our first definitive study of the analysis of networks. Although the techniques presented in it have been developed only for resistance networks, we shall see in later chapters that the same techniques apply with only slight modifications to a considerably broader class of circuits, namely, those which include inductors and capacitors, the network elements which are described in the following chapter.

Problems

Problem 3–1 (Sec. 3–1)

Find the mesh currents i_1 and i_2 in the resistance network shown in Fig. P3-1.

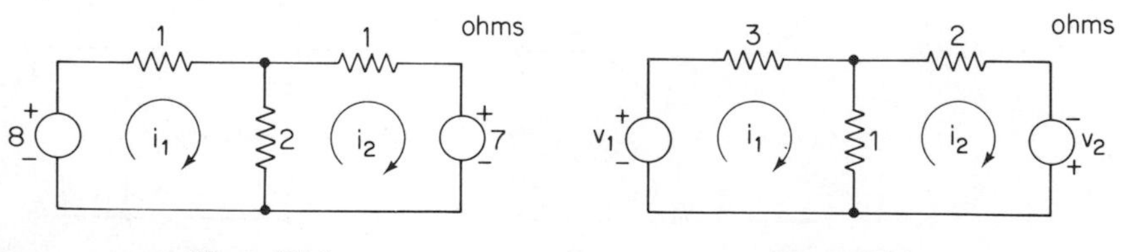

Figure P3-1 **Figure P3-2**

Problem 3–2 (*Sec. 3–1*)

In the circuit shown in Fig. P3-2, it is desired to have $i_1 = 2$ A and $i_2 = -3$ A. Find the voltages v_1 and v_2 which will achieve this.

Problem 3–3 (*Sec. 3–1*)

For the circuit shown in Fig. P3-3, find values of R_1 and R_2 such that $i_1 = 1$ A and $i_2 = 2$ A.

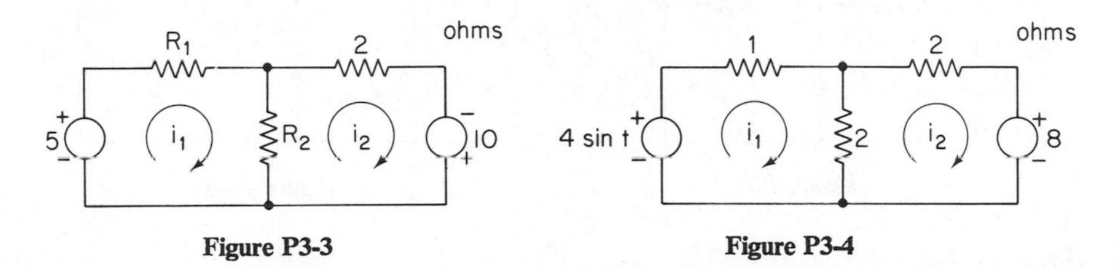

Figure P3-3

Figure P3-4

Problem 3–4 (*Sec. 3–1*)

For the circuit shown in Fig. P3-4, find the mesh currents $i_1(t)$ and $i_2(t)$.

Problem 3–5 (*Sec. 3–1*)

(a) Find the value of v_0 in the network shown in Fig. P3-5.

(b) Add 1 V to the value of all three sources and repeat the problem.

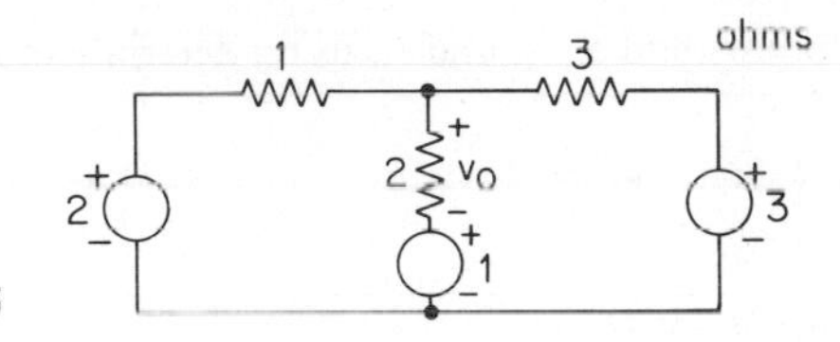

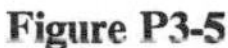

Figure P3-5

Problem 3–6 (*Sec. 3–2*)

Write (but do not solve) the matrix equations for the resistance network shown in Fig. P3-6.

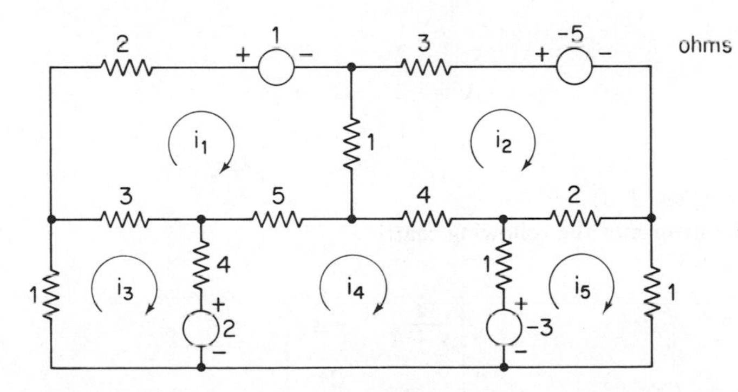

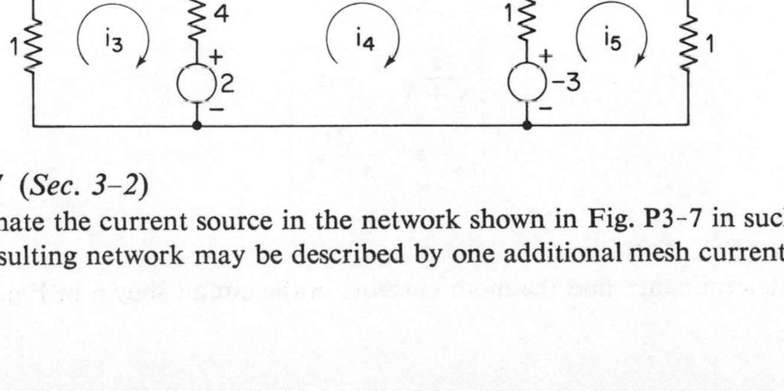

Figure P3-6

Problem 3–7 (*Sec. 3–2*)

Eliminate the current source in the network shown in Fig. P3-7 in such a way that the resulting network may be described by one additional mesh current (in addition to the two already shown). Write the mesh equations in matrix form for the resulting network. Do not solve the equations.

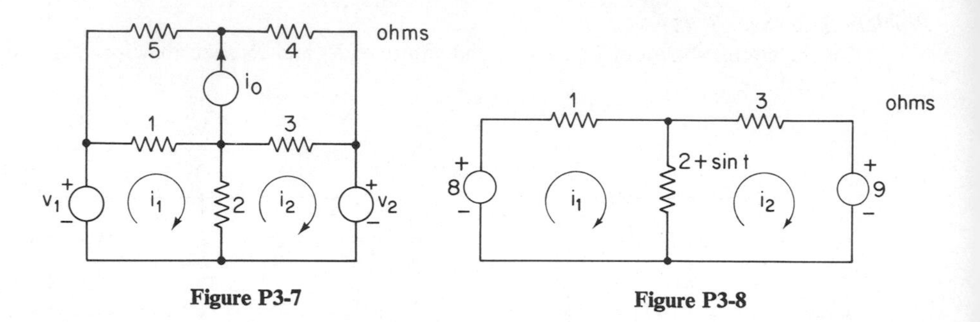

Figure P3-7

Figure P3-8

Problem 3–8 (Sec. 3–2)

Find the mesh currents $i_1(t)$ and $i_2(t)$ for the circuit shown in Fig. P3-8 in which one of the resistors is time-varying.

Problem 3–9 (Sec. 3–2)

In Problem 3-8, does superposition apply? Prove your answer by separately computing the response for each of the sources (when the other is set to zero).

Problem 3–10 (Sec. 3–3)

In the set of equations which follow, find i_1, i_2, and i_3, using determinants.

$$\begin{aligned} 2i_1 + 3i_2 - 4i_3 &= 1 \\ i_1 + 2i_2 - i_3 &= 12 \\ 4i_1 + i_2 - 3i_3 &= 7 \end{aligned}$$

Problem 3–11 (Sec. 3–3)

Find the inverse of the matrix **A** using determinants.

$$\mathbf{A} = \begin{bmatrix} 3 & -1 & 0 \\ 2 & 7 & 1 \\ 1 & -1 & 1 \end{bmatrix}$$

Problem 3–12 (Sec. 3–3)

Find the inverse of the following matrix.

$$\begin{bmatrix} \dfrac{s}{s+1} & -s \\ -s & \dfrac{2s}{s+1} \end{bmatrix}$$

Problem 3–13 (Sec. 3–3)

Using determinants, find the mesh currents in the circuit shown in Fig. P3-13.

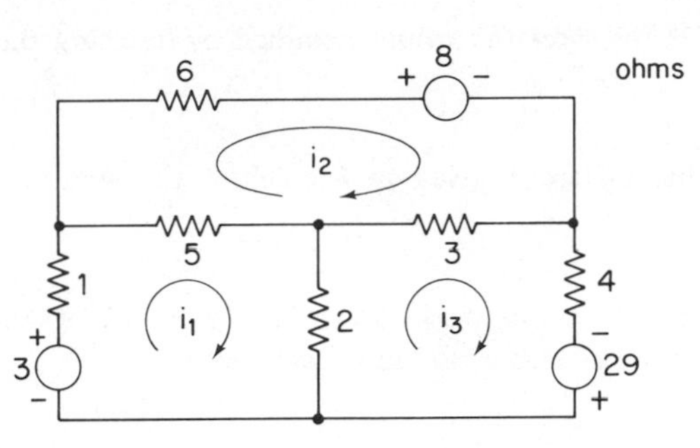

Figure P3-13

Problem 3–14 (Sec. 3–4)

Repeat Problem 3-1 using the Gauss-Jordan method for solving the simultaneous equations.

Problem 3–15 (Sec. 3–4)

Repeat Problem 3-4 using the Gauss-Jordan method for solving the simultaneous equations.

Problem 3–16 (Sec. 3–4)

Repeat Problem 3-5 using the Gauss-Jordan method for solving the simultaneous equations.

Problem 3–17 (Sec. 3–4)

Find the mesh currents of Problem 3-13 using the Gauss-Jordan method for solving the simultaneous equations.

Problem 3–18 (Sec. 3–4)*[10]

Verify the last section of data given in Fig. 3-4.6 by adding the necessary statements to the FORTRAN program given in Fig. 3-4.5.

Problem 3–19 (Sec. 3–4)*

Use the subroutines RMESH and GJSEQ to solve for the mesh currents in the network shown in Fig. P3-6. Test the values of the mesh currents obtained by inserting them in the network equations.

Problem 3–20 (Sec. 3–4)*

Use the subroutines RMESH and GJSEQ to solve for the node voltages in the net-

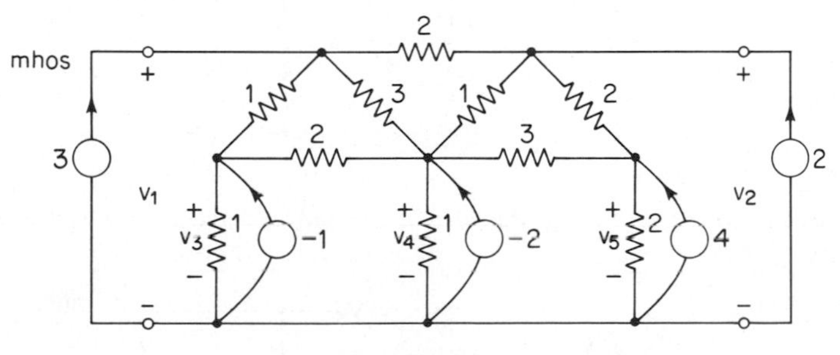

Figure P3-20

[10]Problems which require the use of digital computational techniques are identified by an asterisk.

work shown in Fig. P3-20. Test the values obtained by inserting them in the network equations.

Problem 3-21 (Sec. 3-5)

Find the inverse of the matrix **A** given in Problem 3-11 using Gauss-Jordan reduction.

Problem 3-22 (Sec. 3-5)

Find the mesh currents for the network shown in Fig. P3-13 by using the Gauss-Jordan method to find the inverse of the resistance matrix.

Problem 3-23 (Sec. 3-5)

Find the elements $a, b, c, d,$ and e in the matrix shown at the right below such that it is the inverse of the matrix shown at the left. Note that the constant $\frac{1}{44}$ multiplies all the elements of the right array.

$$\begin{bmatrix} 3 & -2 & -1 \\ -2 & 5 & -3 \\ -1 & -3 & 8 \end{bmatrix} \qquad \frac{1}{44}\begin{bmatrix} a & d & e \\ b & 23 & 11 \\ c & 11 & 11 \end{bmatrix}$$

Problem 3-24 (Sec. 3-5)*

Find the inverse of the following matrix by using the digital computer subroutine MXINV. Verify the result by multiplying the inverse by the original matrix.

$$\begin{bmatrix} 4 & 1 & 1 & 0 \\ 1 & 7 & 1 & 1 \\ 2 & 6 & 4 & 1 \\ -5 & 2 & -1 & 5 \end{bmatrix}$$

Problem 3-25 (Sec. 3-5)*

Verify the data given in the last section of Fig. 3-5.4 by adding the necessary statements to the FORTRAN program given in Fig. 3-5.3.

Problem 3-26 (Sec. 3-6)

Find the node voltages for the circuit shown in Fig. P3-26.

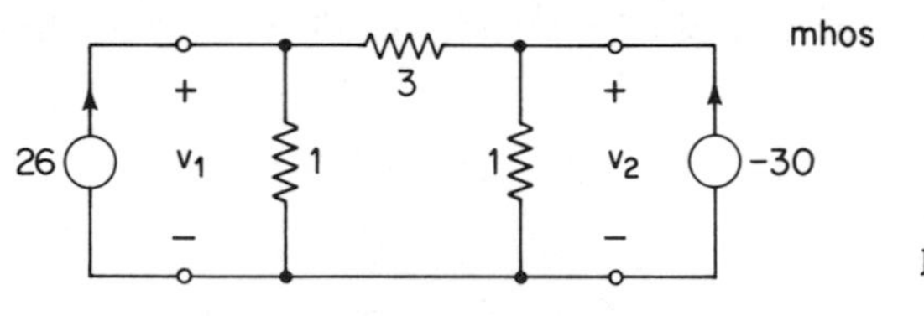

Figure P3-26

Problem 3-27 (Sec. 3-6)

In the circuit shown in Fig. P3-27 it is desired to have $v_1 = 3$ and $v_2 = 1$ V. Find i_1 and i_2.

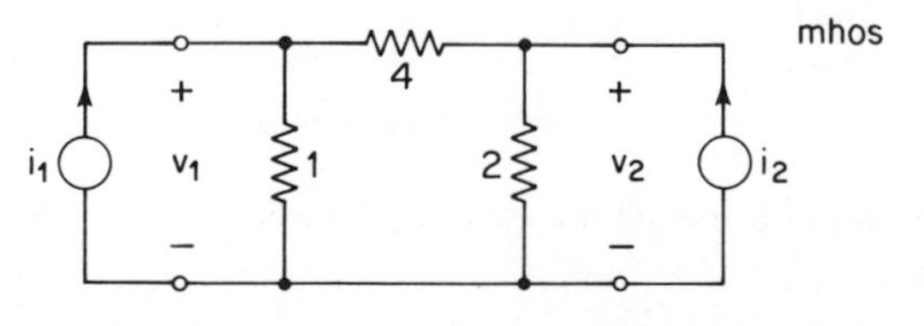

Figure P3-27

Problem 3–28 (*Sec. 3–6*)

In the circuit shown in Fig. P3-28, find G_1 and G_2 such that $v_1 = 1$ and $v_2 = 2$ V.

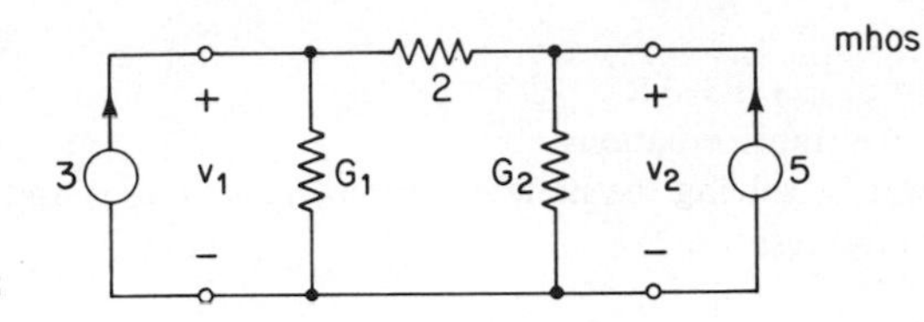

Figure P3-28

Problem 3–29 (*Sec. 3–6*)

For the circuit shown in Fig. P3-29, find the values of the node voltages $v_1(t)$ and $v_2(t)$.

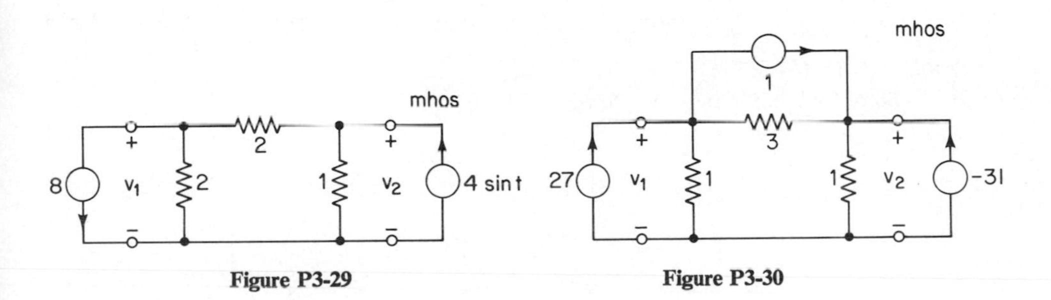

Figure P3-29 **Figure P3-30**

Problem 3–30 (*Sec. 3–6*)

Find the node voltages in the network shown in Fig. P3-30. Compare the results with those obtained in Problem 3-26.

Problem 3–31 (*Sec. 3–7*)

Using determinants, find the node voltages for the circuit shown in Fig. P3-31.

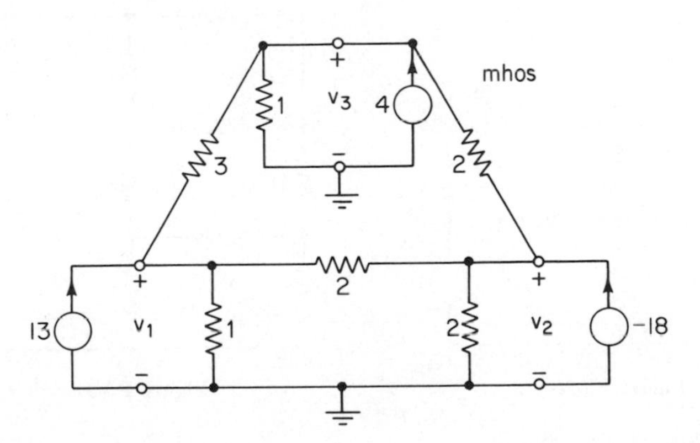

Figure P3-31

Problem 3–32 (*Sec. 3–7*)

Find the node voltages for the circuit shown in Fig. P3-31 using Gauss-Jordan reduction.

Problem 3–33 (Sec. 3–7)

Find the inverse of the resistance matrix for the circuit shown in Fig. P3-31 using Gauss-Jordan reduction.

Problem 3–34 (Sec. 3–8)

For the circuit shown in Fig. P3-34:

(a) Find i_a using source transformations.

(b) Verify your answer by solving the node equations for v_1 and v_2 and finding the current in the 2-mho resistor.

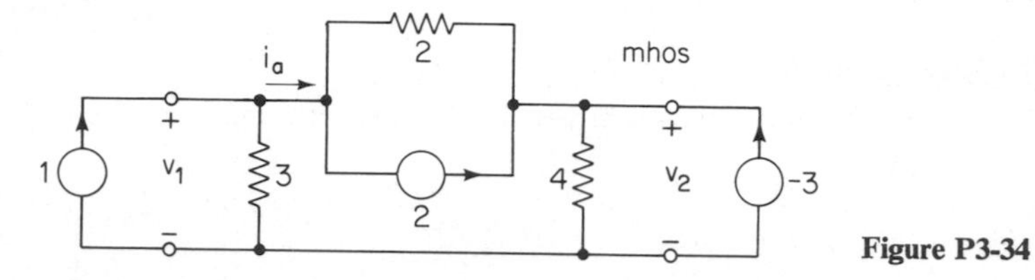

Figure P3-34

Problem 3–35 (Sec. 3–8)

Solve for i in the network shown in Fig. P3-35.

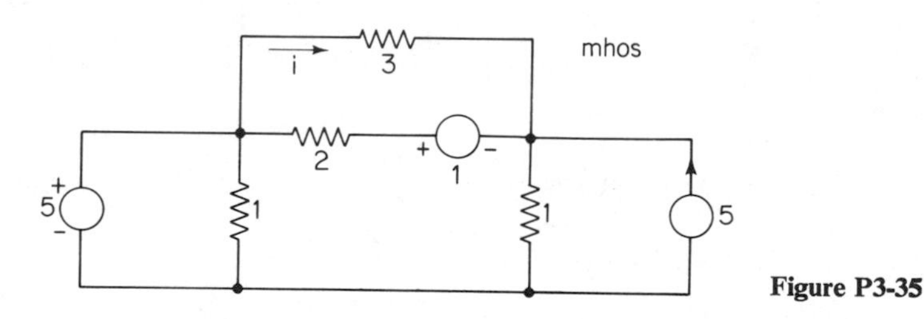

Figure P3-35

Problem 3–36 (Sec. 3–8)

Find i in the circuit shown in Fig. P3-36.

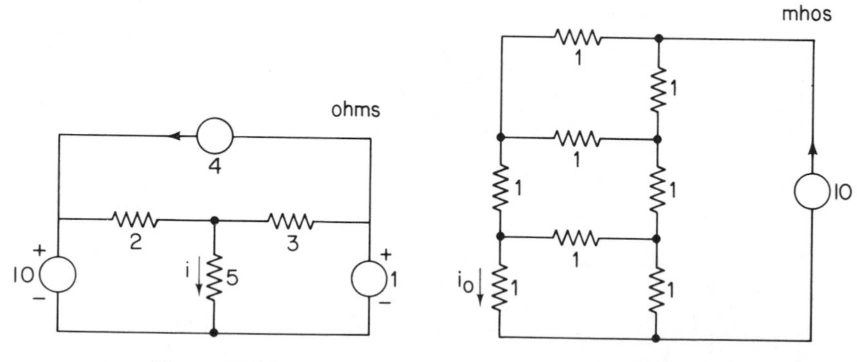

Figure P3-36

Figure P3-37

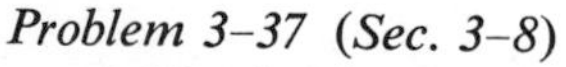

Problem 3–37 (Sec. 3–8)

For the circuit shown in Fig. P3-37, find i_0. (*Hint:* Use source transformations.)

Problem 3–38 (Sec. 3–8)

For the circuit shown in Fig. P3-38, find v_0. (*Hint:* Use source transformations.)

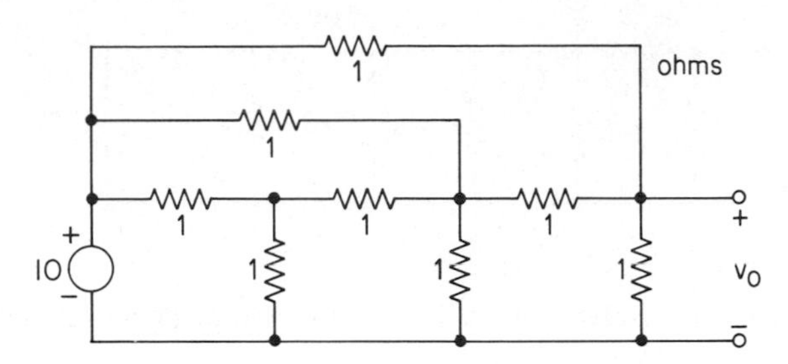

Figure P3-38

Problem 3–39 (Sec. 3–8)

For the circuit shown in Fig. P3-39:

(a) Find v_{ab} using source transformations.

(b) Verify the results of (a) using mesh equations to find i_1 and i_2 and thus find the voltage drop across the 2-Ω resistor.

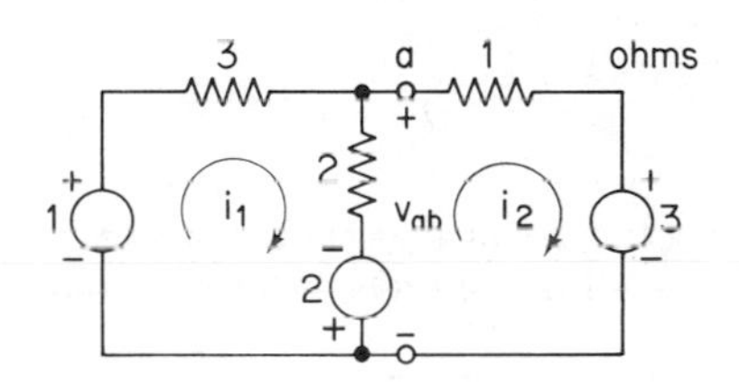

Figure P3-39

Problem 3–40 (Sec. 3–8)

In the circuit shown in Fig. P3-40, replace the five sources with a single equivalent source such that the currents and the voltages across resistors R_1 through R_6 remain unchanged.

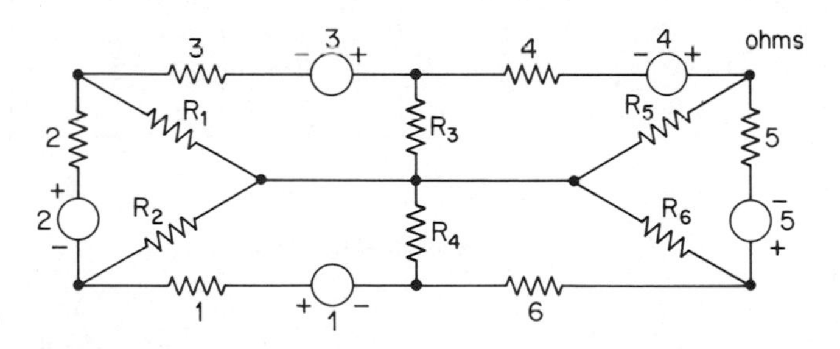

Figure P3-40

Problem 3–41 (Sec. 3–8)

For the circuit shown in Fig. P3-41, replace the three sources with a single source in such a way that the currents and the voltages across resistors R_1 through R_4 remain the same.

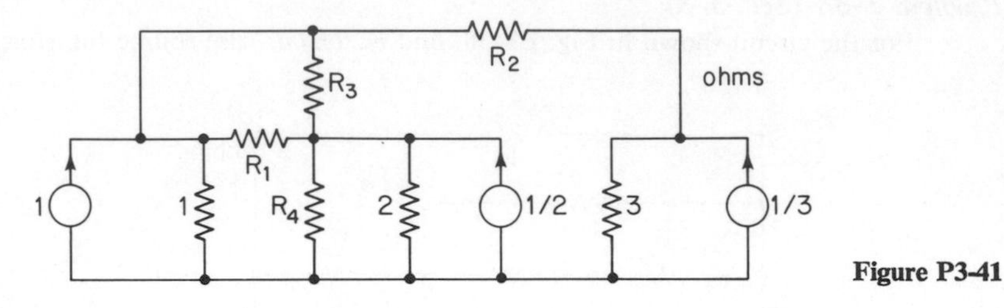

Figure P3-41

Problem 3-42 (Sec. 3-8)

Find delta and wye equivalent circuits for the wye and delta networks shown in Fig. P3-42.

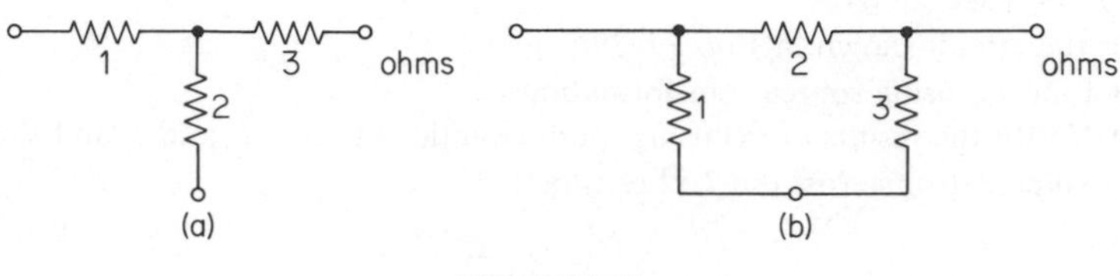

Figure P3-42

Problem 3-43 (Sec. 3-8)

Simplify the expressions given in (3-88) and (3-89) for the case where all the resistors of the wye network (and thus all the resistors of the delta network) have the same value.

Problem 3-44 (Sec. 3-9)

For the circuits shown in Fig. P3-44:

(a) Find the Thevenin equivalent circuit at terminals *a-b*.

(b) Find the Norton equivalent circuit at terminals *a-b*.

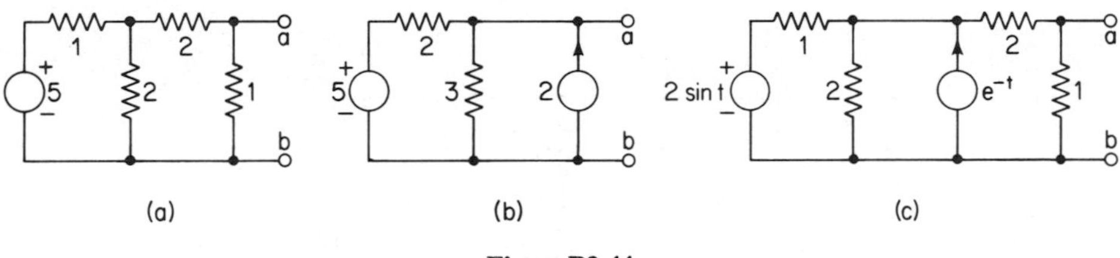

Figure P3-44

Problem 3-45 (Sec. 3-9)

For the circuits found in Problem 3-44, find the current that will flow under each of the following conditions:

(a) In a short circuit connected between terminals *a-b*.

(b) In a 1-Ω resistor connected between terminals *a-b*.

Problem 3-46 (Sec. 3-9)

For the circuit shown in Fig. P3-44(a), find the power dissipated in the original

circuit, the power dissipated in the Thevenin equivalent, and the power dissipated in the Norton equivalent for the following conditions:
(a) The terminals a-b are open-circuited.
(b) The terminals a-b are short-circuited.

Problem 3–47 (Sec. 3–9)

Find Thevenin and Norton equivalent circuits for the network shown in Fig. P3-47 at the terminals a-b.

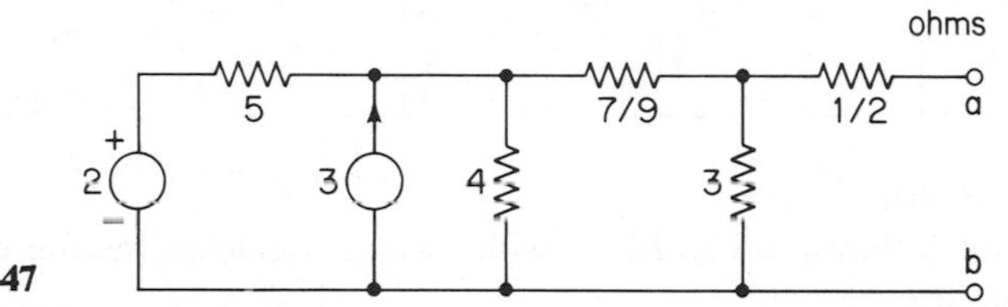

Figure P3-47

Problem 3–48 (Sec. 3–10)

Find the mesh currents i_1 and i_2 for the network shown in Fig. P3-48.

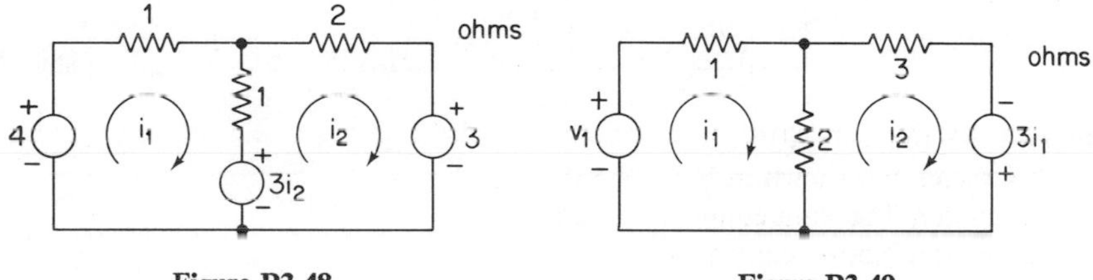

Figure P3-48 **Figure P3-49**

Problem 3–49 (Sec. 3–10)

Write and solve the matrix equations for the circuit shown in Fig. P3-49.

Problem 3–50 (Sec. 3–10)

Find the node voltages for the network shown in Fig. P3-50 for the case where $k = -2$.

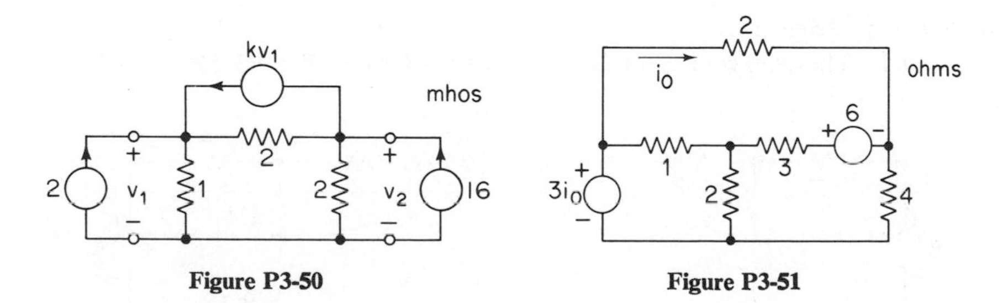

Figure P3-50 **Figure P3-51**

Problem 3–51 (Sec. 3–10)

For the circuit shown in Fig. P3-51 find i_0.

Problem 3–52 (Sec. 3–10)

The three-loop network shown in Fig. P3-52 includes dependent, current-controlled voltage sources. Find the loop currents i_1, i_2, and i_3.

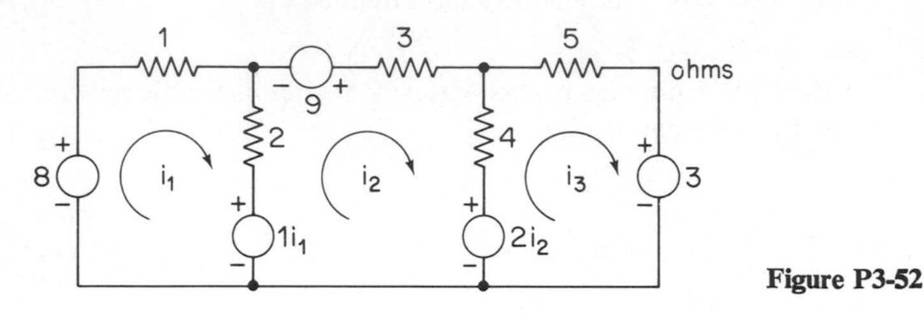

Figure P3-52

Problem 3–53 (Sec. 3–10)

Replace the circuit shown in Fig. P3-53 with a single equivalent resistor connected across terminals *a-b*.

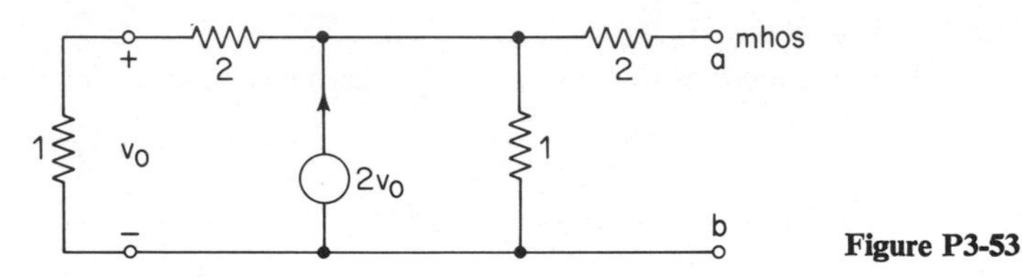

Figure P3-53

Problem 3–54 (Sec. 3–10)

For the circuit shown in Fig. P3-54:

(a) Find a Thevenin equivalent circuit.

(b) Find a Norton equivalent circuit.

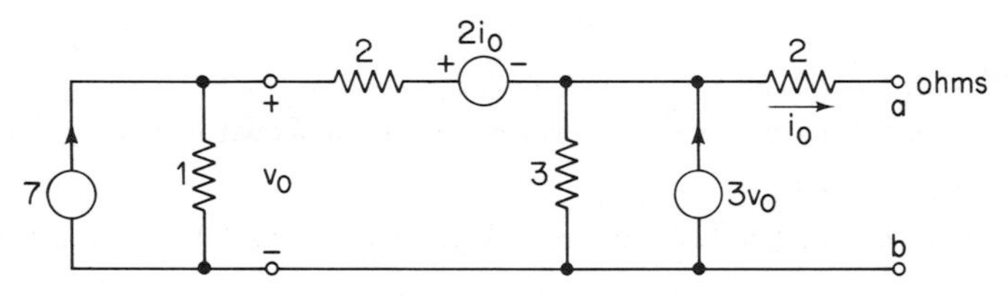

Figure P3-54

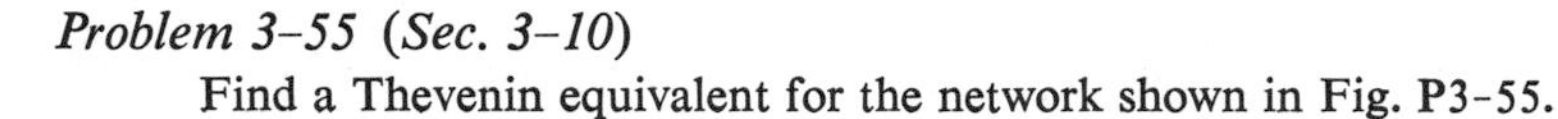

Problem 3–55 (Sec. 3–10)

Find a Thevenin equivalent for the network shown in Fig. P3-55.

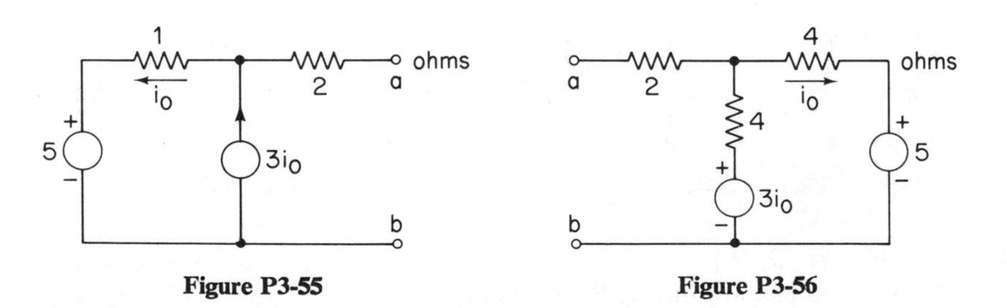

Figure P3-55 **Figure P3-56**

Problem 3–56 (*Sec. 3–10*)

Find a Thevenin equivalent for the circuit shown in Fig. P3–56.

Problem 3–57 (*Sec. 3–10*)

Prove that the ratio of the open-circuit voltage to the short-circuit current at a pair of network terminals gives the value of the resistance in a Thevenin equivalent circuit for the network (assuming that the network is not sourceless).

*Problem 3–58** (*Sec. 3–10*)

Add the necessary statements to the program given in Fig. 3-4.5 to include the effects of the ICVS's shown in the network in Fig. P3–58 and solve for the mesh currents of this network. Verify your answers by inserting the mesh current values in the actual network equations.

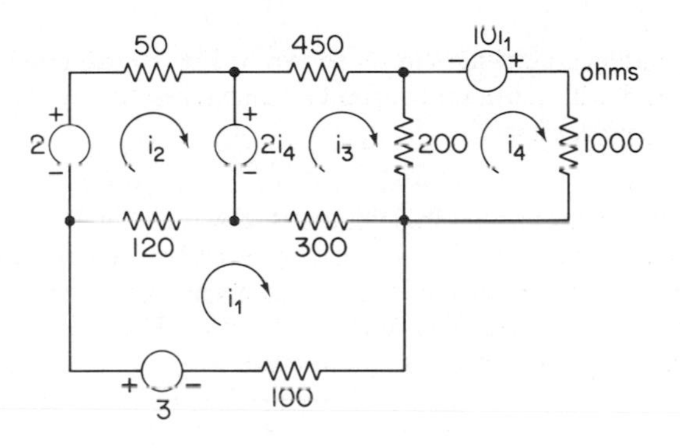

Figure P3-58

*Problem 3–59** (*Sec. 3–10*)

For the resistance network shown in Fig. P3–59, use digital computer methods to find the input resistance at the terminals *a–b*.

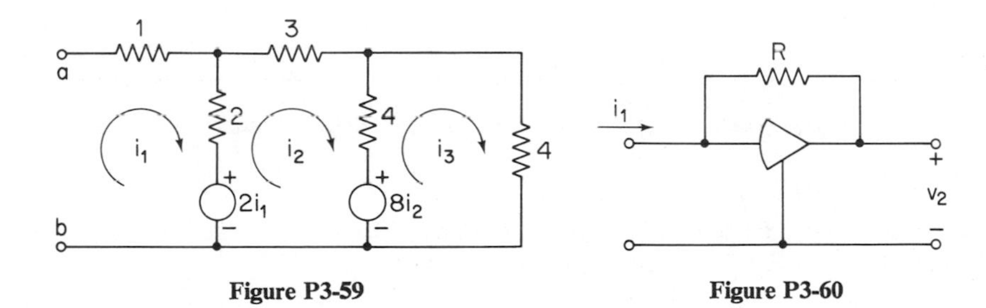

Figure P3-59 **Figure P3-60**

Problem 3–60 (*Sec. 3–11*)

Prove that the operational amplifier circuit shown in Fig. P3–60 provides a realization for an ideal inverting ICVS. What is the value of the gain constant?

Problem 3–61 (*Sec. 3–11*)

Prove that the operational amplifier circuit shown in Fig. P3–61 provides a realization for an ideal VCIS with the output current i_2 being taken through the load. What is the value of the gain constant?

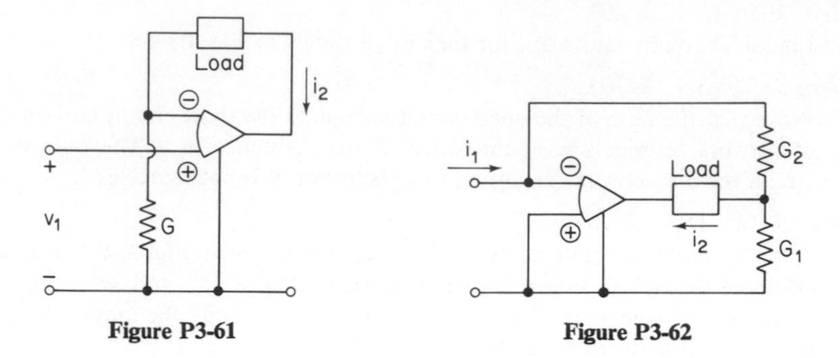

Figure P3-61 **Figure P3-62**

Problem 3-62 (Sec. 3-11)

Prove that the operational amplifier circuit shown in Fig. P3-62 provides a realization for an ideal ICIS with the output current i_2 being taken through the load. What is the value of the gain constant?

Problem 3-63 (Sec. 3-11)

The transistor circuit (minus its biasing elements) shown in Fig. P3-63 may be analyzed by using the transistor model given in Fig. 3-10.2(a). Using this model, find the input resistance, the output resistance, and the open-circuit voltage gain for the circuit as the alpha of the transistors approaches unity. For what ideal active element does this transistor circuit provide a realization?

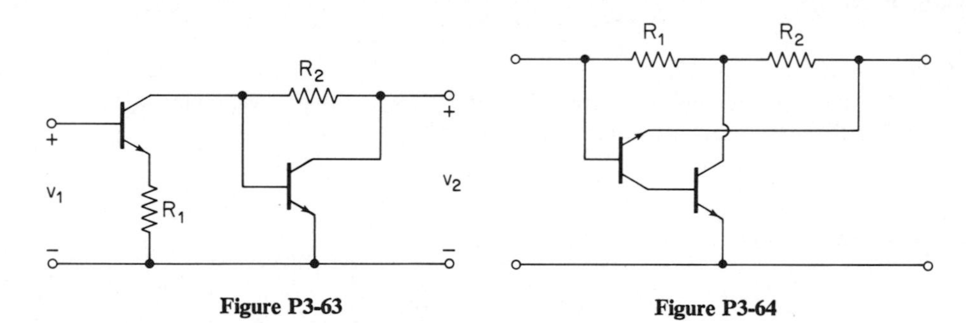

Figure P3-63 **Figure P3-64**

Problem 3-64 (Sec. 3-11)

Using the model for a transistor given in Fig. 3-10.2(a), prove that the transistor circuit shown in Fig. P3-64 provides a realization for an NIC.

Problem 3-65 (Sec. 3-11)

Prove that the circuit shown in Fig. P3-65 containing an ICVS and an ICIS may be used to realize a gyrator.

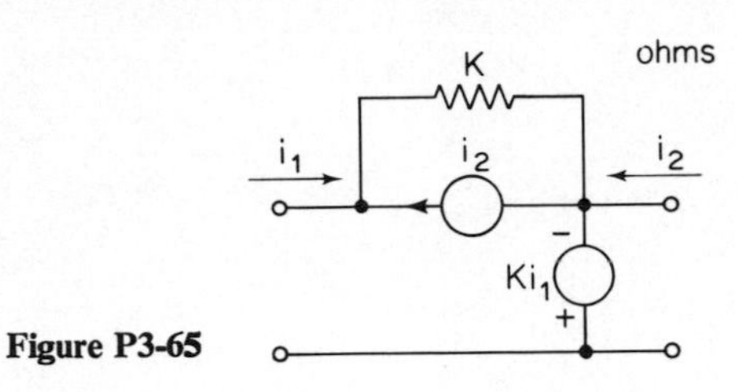

Figure P3-65

Problem 3–66 (Sec. 3–11)

Prove that the circuit shown in Fig. 3-11.10 provides a realization for a gyrator.

CHAPTER 4

Capacitors and Inductors

In Chap. 2 we introduced some of the basic elements of networks, namely, resistors and sources. We showed that the resistor was a two-terminal element in which the terminal variables of voltage and current were related by a linear algebraic expression called Ohm's law. In Chap. 3 we introduced some methods for analyzing circuits comprised of resistors and sources. We found that such networks were characterized by sets of simultaneous linear *algebraic* equations, and we discussed methods for solving such sets of equations.

In this chapter we shall introduce some two-terminal elements which have properties which are quite different than those of the resistor. These elements are the capacitor and the inductor. The terminal variables of voltage and current for these elements are related by integral and differential equations rather than algebraic equations. The equations describing circuits in which these elements are placed are *integro-differential* equations. The solution of such equations is a task of quite different complexity than the solution of sets of simultaneous algebraic equations covered in the preceding section. The rewards for attacking and solving this more complex problem, however, are great. We shall find that the properties of networks which contain capacitors and inductors are so interesting that resistance networks will appear drab and dull by comparison. Surprisingly, however, many of the techniques used to analyze resistance networks may be applied to these more interesting networks with relatively little modification. Such applications will be the subject of several of the following chapters.

4-1 THE CAPACITOR

A capacitor is a two-terminal element which has the property that its branch voltage and current variables are related by the expression

$$v(t) = \frac{1}{C}\int_{-\infty}^{t} i(\tau)\, d\tau \tag{4-1}$$

where the constant C is used to specify the value of the capacitor in farads (abbreviated F). The circuit symbol and the reference polarities for a capacitor are shown in Fig. 4-1.1. From Eq. (4-1), we see that the value of the voltage which is present

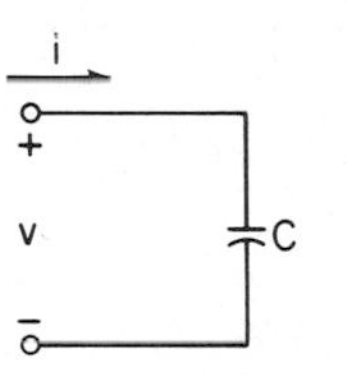

Fig. 4-1.1 Circuit symbol and reference polarities for a capacitor.

at the terminals of a capacitor depends on the values of current starting at $t = -\infty$. Since it is difficult to maintain a history of a variable such as current over such a long time range, we may use the following alternative defining relation for the branch variables of a capacitor:

$$v(t) = \frac{1}{C}\int_{0}^{t} i(\tau)\, d\tau + v(0) \tag{4-2}$$

Either of the equations given above may be referred to as the *integral* relation for the terminal variables of a capacitor. By differentiating both sides of either equation, we may easily derive the *differential* relation for the terminal variables of a capacitor. Thus, we obtain

$$i(t) = C\frac{dv}{dt} \tag{4-3}$$

It is frequently convenient to be able to use reciprocal units to express the value of a capacitor. In such a case we shall use units of *darafs* (abbreviated D), which is simply "farad" spelled backwards. We shall refer to such a unit as a unit of *elastance*, and use the symbol S when we desire to refer to capacitors defined on a reciprocal basis. Such usage exactly parallels the use of R (for resistors whose values are given in ohms), and G (for resistors whose values are given in mhos), as introduced in Sec. 2-1.

We may illustrate the properties of a capacitor more fully by substituting the relation between charge and current, namely, $i(t) = dq/dt$, in Eq. (4-1). We obtain

$$q(t) = Cv(t) \tag{4-4}$$

Thus, we see that the value of the capacitance determines the relationship between the value of the charge stored in the capacitor and the value of the voltage appearing across its teminals. Thus, to represent a capacitor we may define a v-q plane, i.e., a plane in which the voltage variable $v(t)$ is plotted along the abscissa or "x" axis and the charge variable $q(t)$ is plotted along the ordinate or "y" axis. Such a representation

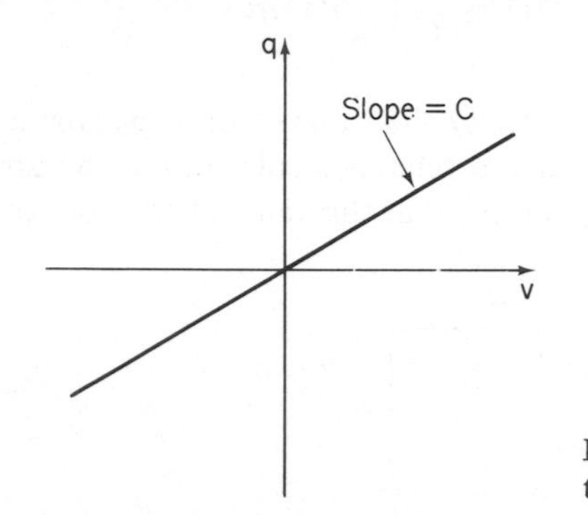

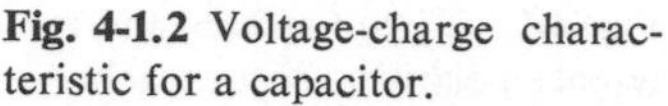
Fig. 4-1.2 Voltage-charge characteristic for a capacitor.

follows closely that used for a resistor in Sec. 2–1. We may now define a *linear* capacitor as one characterized by a straight line passing through the origin of the v-q plane, as shown in Fig. 4–1.2. Note that the slope of the characteristic specifies the value of the capacitor. A non-linear capacitor is thus one whose characteristic is not a straight line and/or does not pass through the origin. As an example of such a characteristic, a solid-state diode, when reverse biased (i.e., with the diode not conducting) exhibits a "space-charge" capacitance which is of the general form

$$C = Kv^{-1/3} \tag{4–5}$$

For such a diode, we see that Eq. (4–4) must be written in the form $q(t) = C[v(t)]$, i.e., that the relation between charge and voltage is a non-linear one. In addition to the possibility of being nonlinear, physical capacitors may also be time-varying. A linear time-varying capacitor would have a characteristic identical with the one shown in Fig. 4–1.2, except that the slope would vary as a function of time. For a linear time-varying capacitor, the defining relation of Eq. (4–4) becomes

$$q(t) = C(t)v(t) \tag{4–6}$$

For such a capacitor, we must modify the terminal relations given at the beginning of this section. For example, differentiating both sides of (4–6), we obtain

$$\frac{dq}{dt} = i(t) = C(t)\frac{dv}{dt} + v(t)\frac{dC}{dt} \tag{4–7}$$

This relationship actually includes both the time-varying and the time-invariant cases, since for a time-invariant capacitor $dC/dt = 0$ and Eq. (4–7) simplifies to Eq.

(4–3). Similarly, for a linear time-varying capacitor, the integral relationship for the terminal variables has the form

$$v(t) = \frac{1}{C(t)} \int_0^t i(\tau)\, d\tau + \frac{q(0)}{C(t)} \tag{4–8}$$

Actually, in the real-life or physical world, there is no such thing as a linear time-invariant capacitor. For many applications, however, the terminal properties of physical capacitors so closely approximate the v-q characteristic shown in Fig. 4–1.2 that little is gained by trying to include the non-linear and/or time-varying behavior. Indeed, much may be lost, since the more complicated mathematical treatment required to describe such a capacitor may tend to obscure the fundamental properties of the circuit in which the element is found. To avoid this, we will assume that all physical capacitors are "modeled" by ideal elements obeying the relations given in Eqs. (4–1) through (4–4). Thus, from this point on, when we use the word "capacitor" we imply a *lumped, linear, time-invariant model of some physical capacitor*, unless otherwise specified.

The instantaneous power $p(t)$ which is supplied to a capacitor is easily determined from the relation $p(t) = v(t)i(t)$. Substituting for $i(t)$ from Eq. (4–3), we obtain

$$p(t) = Cv(t)\frac{dv}{dt} \tag{4–9}$$

From this expression we note that, for a positive-valued capacitor, $p(t)$ may be positive or negative, depending on the sign of the quantity $v(t)\, dv/dt$. Thus we see that it is possible for a capacitor to receive energy from the circuit to which it is connected [in which case $p(t)$ will be positive], or to provide energy to the circuit [in which case $p(t)$ will be negative]. These conclusions are based on the associated reference directions developed in Sec. 1–2 for the transferring of energy.

We may readily compute the total energy which has been supplied to a capacitor using the relationship

$$w(t) = \int_{-\infty}^{t} p(\tau)\, d\tau$$

Using the relation for $p(t)$ derived above, we obtain

$$w(t) = \int_{-\infty}^{t} Cv(\tau)\frac{dv}{d\tau}\, d\tau = \int_{v(-\infty)}^{v(t)} Cv(t)\, dv = \frac{Cv^2(t)}{2} - \frac{Cv^2(-\infty)}{2} \tag{4–10}$$

In this relation, the quantity $v(-\infty)$, i.e., the voltage across the capacitor at $t = -\infty$, must be zero, since at $t = -\infty$ the integral of current supplied to the capacitor must be zero. Therefore, we conclude that the energy $w(t)$ stored in a capacitor is given by the relation

$$w(t) = \frac{Cv^2(t)}{2} \tag{4-11}$$

From the above relation we see that even though the *instantaneous* power $p(t)$ may be positive or negative, the total energy stored in the capacitor as given by (4–11) is always positive (or zero). Thus, we see that a capacitor is a passive network element.[1] It should be noted that the capacitor is a non-dissipative element, i.e., all the energy supplied to a capacitor is stored in the capacitor (actually, more properly, it is stored in the electric field associated with the capacitor). As such, it may be left in storage or it may be reclaimed in the circuit, but it is not dissipated, i.e., it is not converted to heat. Thus, we may expect that circuits containing capacitors will have markedly different properties than circuits which contain only dissipative elements such as resistors. The study of such properties will be one of our major concerns in the remainder of this text.

Just as it was possible to define a negative-valued resistor (see Sec. 2–1), so we may define a *negative-valued capacitor*. Such an element is defined by the relation

$$i(t) = -|C|\frac{dv}{dt} \tag{4-12}$$

where we have emphasized that the value of the capacitor is less than zero by using magnitude symbols and a minus sign. Such a capacitor becomes "charged" by supplying energy to the circuit. As such, it is an *active* device and the total energy $w(t)$ defined in (4–11) is always negative (or zero).

Practical capacitors, such as those used in radios and television sets, may range in value from a few picofarads (10^{-12} F) to hundreds of microfarads (10^{-6} F). As was the case for resistors, however, in our analyses we shall in general use values which are conveniently normalized to lie near unity. In addition to being rated as to their capacitance, a maximum voltage rating, which from Eq. (4–11) determines the maximum energy storage, is also given. Some capacitors, due to the type of materials used in their construction, can only be used in circuit applications in which the polarity of the applied voltage does not change. All real-world capacitors also have some leakage resistance associated with their capacitance. Such a parasitic effect may be modeled by using a shunt resistor in parallel with an ideal capacitor, if the accuracy of the circuit analysis which is being made requires it. In most applications, however, the value of such a resistor is so large as to be negligible.

[1]This, of course, only applies to a linear, time-invariant, positive-valued capacitor. It is interesting to observe that a linear, positive-valued, *time-varying* capacitor may be active! See C. A. Desoer and E. S. Kuh, *Basic Circuit Theory*, Chap. 19, McGraw-Hill Book Company, New York, 1969.

4-2 SOME FREQUENTLY ENCOUNTERED TIME FUNCTIONS

In our studies of resistance networks in Chaps. 2 and 3 we found that the voltage and current variables could have any desired time dependence since they are always related by *algebraic* equations. A quite different situation exists for networks containing capacitors, however, since from Eqs. (4–2) and (4–3) we see that *derivative and integral* relations for the terminal variables replace the algebraic relations which characterize resistors. Thus, we must concern ourselves with the derivatives and integrals of various time functions. In this section we shall define some commonly encountered time functions, and discuss their derivatives and integrals. In the following section we shall investigate what happens when these basic waveshapes are applied as excitations to capacitors.

One of the most frequently encountered time functions used in studies of circuits and systems is the unit step function. It has the waveform shown in Fig. 4–2.1. As shown in the figure, this function has a constant positive value of unity

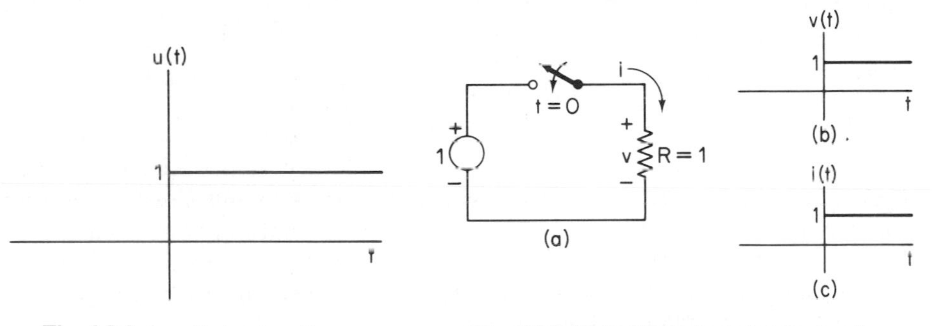

Fig. 4-2.1 A unit step function.

Fig. 4-2.2 Waveshapes in a simple circuit.

for all time greater than zero, and a value of zero for all time less than zero. Such a function is easily generated physically by a switch. As an example of this, consider the circuit shown in Fig. 4–2.2(a). If the switch is closed at the instant of time when t equals 0, then the waveshape of the voltage across the 1-Ω resistor R has the form shown in Fig. 4–2.2(b), i.e., it is a step function. Similarly, the waveshape of the current in the resistor, as shown in Fig. 4–2.2(c), is also a step function. Although the step function is easily generated physically, it is difficult to describe mathematically. This is the result of the fact that the function does not have a definite value at $t = 0$, even though it does have well-defined values for $t > 0$, and for $t < 0$. Such a function is referred to as a discontinuous function. We may use the symbol $u(t)$ to represent a *unit step function* occurring at $t = 0$. Thus, we may define

$$u(t) = \begin{cases} 0, & t < 0 \\ 1, & t > 0 \end{cases} \tag{4-13}$$

Its value at $t = 0$ may be taken as 0, $\frac{1}{2}$, or 1 without affecting the results that follow. Such a definition is easily generalized to provide a representation for step functions occurring at times other than zero. Thus, for a unit step occurring at time t_0 as shown in Fig. 4–2.3, we may define the function $u(t - t_0)$ by the relations

$$u(t - t_0) = \begin{Bmatrix} 0, & t - t_0 < 0 \\ 1, & t - t_0 > 0 \end{Bmatrix} = \begin{cases} 0, & t < t_0 \\ 1, & t > t_0 \end{cases} \tag{4–14}$$

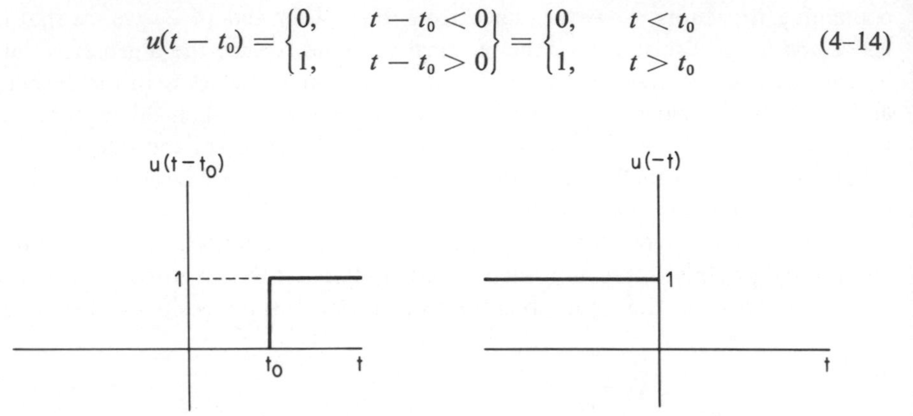

Fig. 4-2.3 A delayed unit step function.

Fig. 4-2.4 The function $u(-t)$.

Note that the significant feature of the definitions given in (4–13) and (4–14) above is the argument of $u(\cdot)$, i.e., the quantity in parentheses following u.[2] Both definitions state the same fact, namely, that when the *quantity or expression* listed as an argument is positive, the function $u(\cdot)$ is unity, and when the quantity is negative, $u(\cdot)$ is zero. Such an interpretation opens the way to extending the definition of the unit step function to many other useful situations. For example, the waveshape shown in Fig. 4–2.4 is simply defined by the expression $u(-t)$.

It is easy to add several unit step functions together to synthesize quite complicated discontinuous waveforms. Some examples follow.

EXAMPLE 4-2.1. The pulse $f(t)$ shown in Fig. 4-2.5(a) can be constructed from two step functions, a unit step applied at $t = 0$, as shown in Fig. 4-2.5(b), and a step function with a value of -1 applied at $t = 2$, as shown in Fig. 4-2.5(c). Thus, we may write

$$f(t) = u(t) - u(t - 2)$$

Note that for all values of $t > 2$, the sum of the unit value of $u(t)$ and the -1 value of $-u(t - 2)$ equals zero, as specified for the given function.

EXAMPLE 4-2.2. The notch waveshape $g(t)$ shown in Fig. 4-2.6(a) can be constructed from two step functions, one an inverted step of value 2 which stops at $t = 1$, as shown in Fig. 4-2.6(b), the other a step function with a value of 2 which is applied at $t = 4$, as shown in Fig. 4-2.6(c). Thus, we may write

[2]The symbolism $u(\cdot)$ is used when it is desired to represent the complete functional relationship, whereas $u(t)$ implies the *value* of the function as determined at some particular value t of its argument.

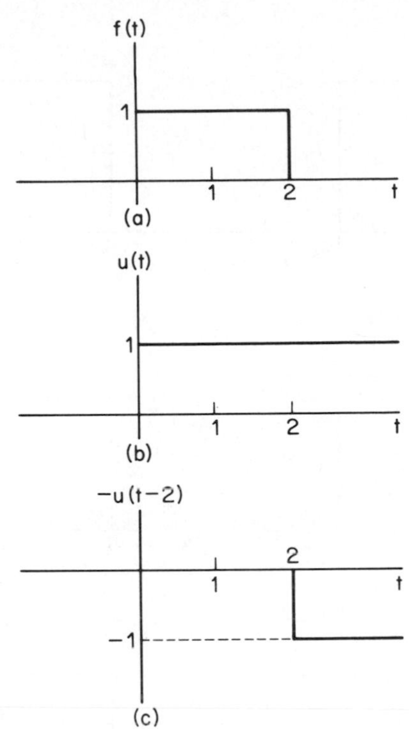

Fig. 4-2.5 A rectangular pulse and its components.

$$g(t) = 2u(-t + 1) + 2u(t - 4)$$

EXAMPLE 4-2.3. The "staircase" function $h(t)$ shown in Fig. 4 2.7 can be constructed from a series of unit step functions. Thus, we may write

$$h(t) = u(t) + u(t - 2) + u(t - 4) + \dots$$

Now let us consider a second type of function, the ramp function. A *unit ramp function* $r(t)$ starting at $t = 0$ is shown in Fig. 4–2.8. Such a function is characterized by the properties that it equals t for $t > 0$ and it equals 0 for $t < 0$. This function is easily defined by using the unit step function $u(t)$ to "turn the function off," i.e., set it to zero, for $t < 0$. Thus, we may describe the unit ramp function $r(t)$ shown in Fig. 4–2.8 by the relation

$$r(t) = tu(t) \tag{4–15}$$

More generally, a ramp of slope k which begins at $t = t_0$ as shown in Fig. 4–2.9 may be defined by the expression

$$(t - t_0)ku(t - t_0) \tag{4–16}$$

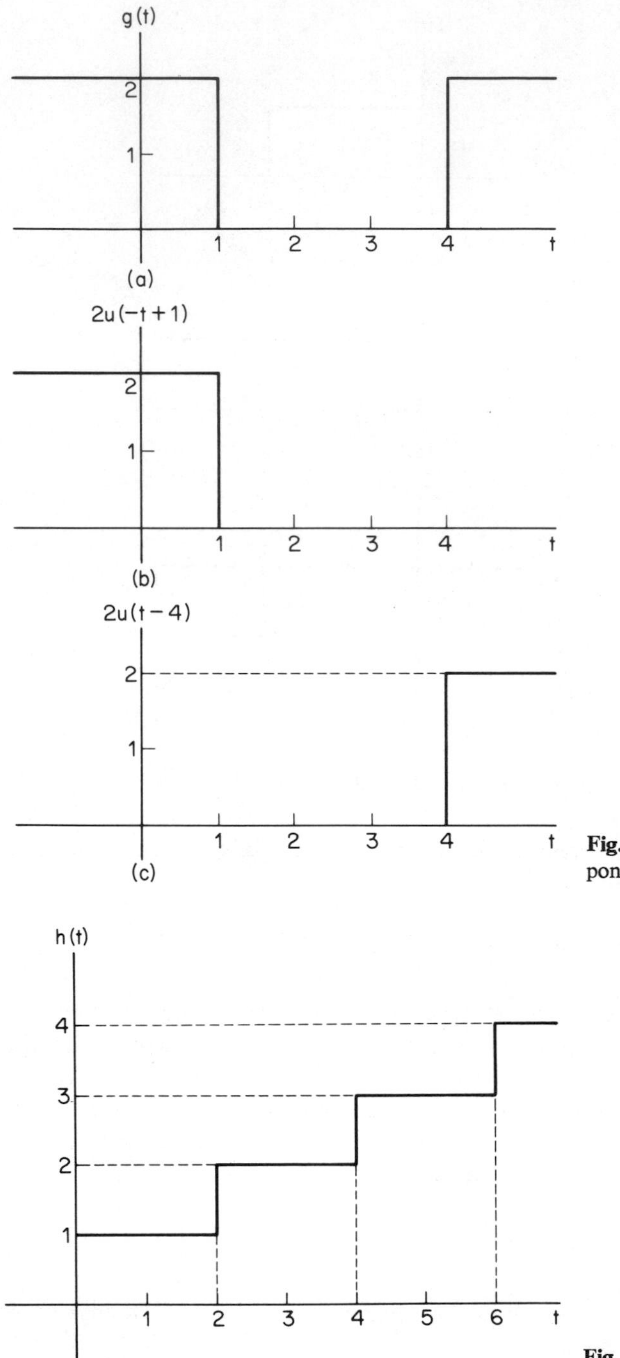

Fig. 4-2.6 A notch and its components.

Fig. 4-2.7 A staircase function.

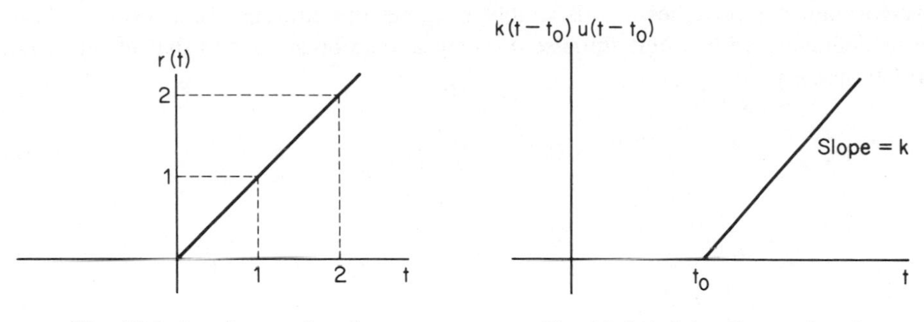

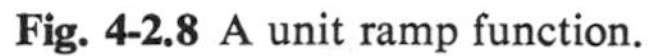

Fig. 4-2.8 A unit ramp function.

Fig. 4-2.9 A delayed ramp function.

It is easy to show that the unit ramp function $r(t)$ is equal to the integral of the unit step function. Thus we may write

$$r(t) = \int_{-\infty}^{t} u(\tau)\, d\tau \tag{4–17}$$

Since a ramp function may be considered as the *integral* of a step function, what sort of function is the *derivative* of a step function? Let us give the symbol $\delta(t)$ to such a function. It must have the property that its integral yields the step function, i.e., that the following relationship holds.

$$u(t) = \int_{-\infty}^{t} \delta(\tau)\, d\tau \tag{4–18}$$

This relation defines the *unit impulse function* $\delta(t)$. The opposite relationship may be found by differentiating both sides of (4–18). Thus, we obtain

$$\delta(t) = \frac{du(t)}{dt} \tag{4–19}$$

The "slope" of $u(t)$, i.e., the value of its derivative, is zero at every value of t except $t = 0$. At this value of its argument, it is undefined. However, in a non-rigorous sense, we may refer to it as being *infinite*. In such a case, we may define the unit impulse $\delta(t)$ as having a value of zero for all values of its argument t except $t = 0$, and having a value of infinity at $t = 0$. In addition, it must satisfy the constraint

$$\int_{-\epsilon}^{\epsilon} \delta(t)\, dt = 1 \tag{4–20}$$

where ϵ is some arbitrary positive constant. The definition of the unit impulse function $\delta(t)$ given above is more easily verified on a limit basis. To do this, let us define a modified unit step function $u_\epsilon(t)$ as shown in Fig. 4–2.10(a). In the limit as the positive constant ϵ approaches 0, this simply becomes the unit step function $u(t)$. Now let us define a modified unit impulse $\delta_\epsilon(t)$ by a relation similar to that given in Eq. (4–19), namely,

$$\delta_\epsilon(t) = \frac{d}{dt} u_\epsilon(t) \tag{4–21}$$

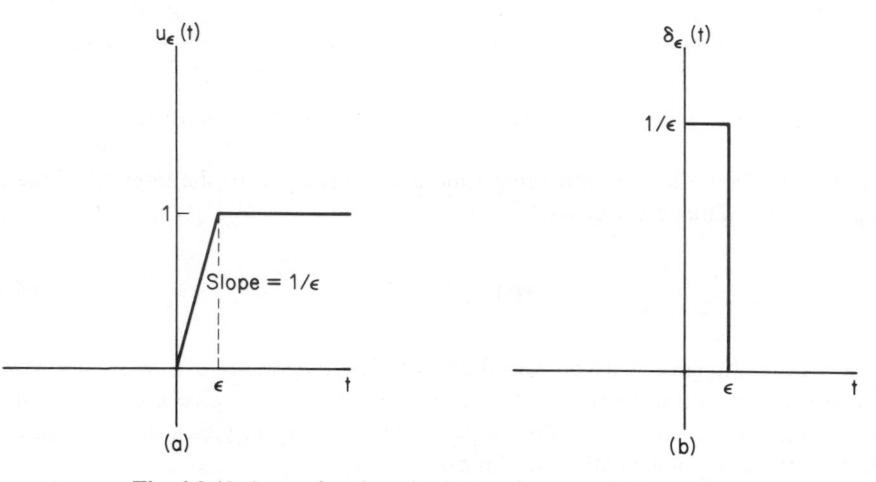

Fig. 4-2.10 Approximations for (a) a unit step, and (b) a unit impulse.

The form of $\delta_\epsilon(t)$ is shown in Fig. 4–2.10(b). Note that this function satisfies the relation

$$\int_{-\epsilon}^{\epsilon} \delta_\epsilon(t)\, dt = 1 \tag{4–22}$$

As ϵ is made smaller, the function $\delta_\epsilon(t)$ approaches a pulse with very small width, very large height, and an area of unity. Thus, in the limit as $\epsilon \to 0$, we may say that the modified function $\delta_\epsilon(t)$ becomes an impulse function of "zero" width and "infinite" magnitude. Such an interpretation will prove adequate for our applications in this text. The reader who is interested in a more rigorous definition of the properties of the impulse function is referred to the advanced texts listed in the references in the Bibliography.

We may treat impulses occurring at values of time other than zero in a manner similar to that used for unit step functions occurring at other values of time, e.g., the function $\delta(t - t_0)$ represents a unit impulse occurring at $t = t_0$. Similarly, impulses of non-unity value may be thought of as representing the derivative of step functions of non-unity value. Thus, in general we may write

$$k\delta(t - t_0) = k\frac{d}{dt}u(t - t_0) \tag{4-23}$$

A symbol which is frequently used to represent the impulse function is shown in Fig. 4-2.11. The number written alongside the arrowhead refers to the area or "strength" of the impulse, i.e., the value of k in Eq. (4-23).

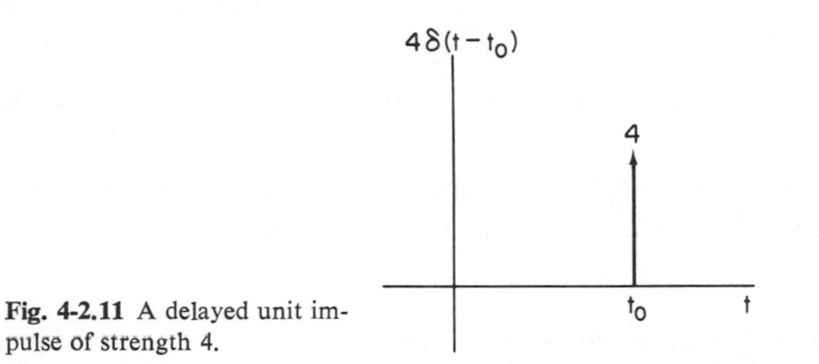

Fig. 4-2.11 A delayed unit impulse of strength 4.

In some network situations, we may desire to consider the derivative of an impulse. Such a function is called a *doublet*. Similarly, the derivative of a doublet is referred to as a *triplet*, etc. The treatment of such functions in any detail is beyond the scope of this text.

The impulse function, unlike the step function, is quite difficult to generate physically. Its characterization of near infinite magnitude (even though we say that this occurs for almost zero time) makes it an undesirable waveshape for any variable that might be encountered in a physical circuit. This is so because exceptionally large values of voltages and currents have a disconcerting habit of destroying circuit components long before they reach anything even close to "infinite" magnitude. In many theoretical situations, however, the mathematical use of impulse functions provides considerable insight into the behavior of networks; thus, even though the impulse is physically unattainable, we shall make considerable application of it in the chapters that follow. The reader should be alert to maintain the correct perspective in such discussions to differentiate between the "mathematical" world and the physical or "real-life" world.

4-3 WAVESHAPES FOR A CAPACITOR

In this section we shall consider the effect of exciting a capacitor by different waveshapes of voltage and current. Such a discussion will provide a clearer picture of the operation of a capacitor in a circuit, and will serve as preparation for a study of the more complicated situations which will be covered in the sections which follow. First of all, let us consider the situation that results when an ideal current source is connected to a capacitor as shown in Fig. 4-3.1. The voltage that appears across the terminals of the capacitor for $t > 0$ is determined by the relation given in

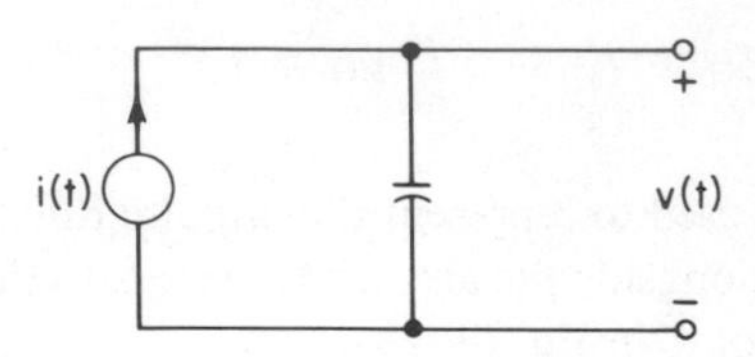

Fig. 4-3.1 A capacitor excited by a current source.

Sec. 4–1, namely,

$$v(t) = \frac{1}{C}\int_0^t i(\tau)\,d\tau + v(0) \tag{4–24}$$

Let us now consider the effects of applying various waveforms of current $i(t)$ to the circuit. For example, let $i(t) = u(t)$, a unit step. From the relation given above, we see that in this case

$$v(t) = \frac{1}{C}\int_0^t d\tau + v(0) = \frac{t}{C} + v(0) \qquad t > 0 \tag{4–25}$$

Thus, the voltage that appears across the terminals of the capacitor is a ramp of slope $1/C$ (starting at $t = 0$) plus the constant $v(0)$ representing the value of voltage that was present on the terminals of the capacitor before the unit step of current was applied. The value of a variable present before the application of some excitation is referred to as an *initial condition.* Clearly, an initial condition for the voltage across the capacitor must be known before the voltage across the capacitor can be fully defined. If we assume that the initial condition on the voltage is zero, then the current and voltage waveforms for the circuit shown in Fig. 4–3.1 for the case where a unit step of current is applied are shown in Fig. 4–3.2(a).

The presence or absence of an initial voltage on a capacitor has a direct bearing on linearity considerations for circuits which include capacitors. For example, suppose that a 1-F capacitor with an initial condition $v(0)$ of 1 V is excited by a unit

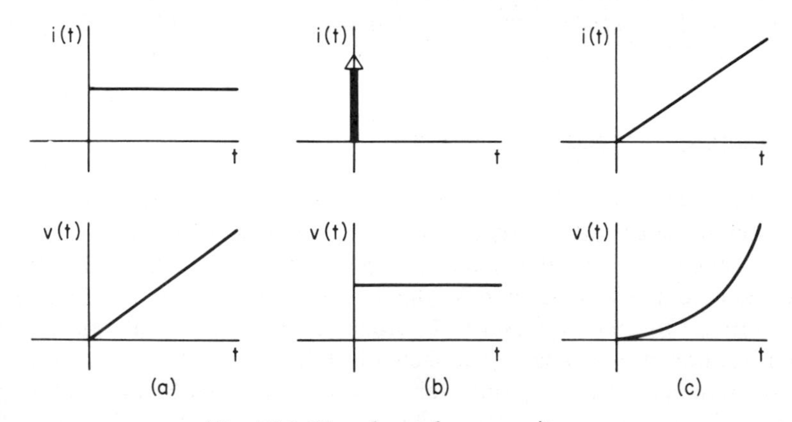

Fig. 4-3.2 Waveshapes for a capacitor.

step of current $u(t)$ A. The resulting capacitor voltage $v(t)$ is easily seen to be $t + 1$ V ($t > 0$). If we double the input excitation by applying an input current of value $2u(t)$ A, the resulting voltage $v(t)$ is $2t + 1$ V ($t > 0$). Note that by doubling our input we have *not* doubled the resulting voltage, even though the circuit element is a *linear* capacitor. Our conclusion is that superposition with respect to input currents applies *only if the capacitor is uncharged*, i.e., its initial condition is zero. Such considerations will become important when we consider circuits comprised of capacitors and other elements in Chap. 5.

In a similar manner, we may determine the waveshapes of voltage that result for the application of an impulse of current and a ramp of current to a capacitor. These are shown in Figs. 4–3.2(b) and 4–3.2(c), respectively. The waveshapes shown in Fig. 4–3.2(b), representing the situation where an impulse of current is applied to a capacitor, show that the resulting waveshape of voltage is a step function, i.e., a discontinuous waveshape. This result could also have been predicted from the differential relationship for the voltage and current variables of a capacitor. This has been shown to have the form

$$i(t) = C\frac{dv}{dt} \tag{4–26}$$

Thus, we see that if the voltage waveshape is discontinuous (in which case its derivative must have an infinite magnitude), then the resulting current must be an impulse. This conclusion provides a most important condition on the waveform of the voltage that appears across the terminals of a capacitor. It is known as the *continuity condition*, and it may be summarized as follows:

SUMMARY 4-3.1

Continuity Condition for Capacitors: The voltage appearing across the terminals of a linear time-invariant capacitor must always be a continuous function. This assumes that the current through the capacitor is bounded, i.e., that impulses of current are not permitted.

We may also express the continuity condition by a mathematical relationship. To do this, let us first define the quantities $0-$ and $0+$ as values of the independent variable t at which some function $f(t)$ is to be evaluated by the relations

$$f(0-) = \lim_{\substack{t\to 0\\ t<0}} f(t)$$

$$f(0+) = \lim_{\substack{t\to 0\\ t>0}} f(t)$$

Thus $f(0-)$ may be seen to be the limiting value of $f(t)$ as t approaches the value of 0 from "below," i.e., from values of t less than 0. Similarly, $f(0+)$ is the limiting value of $f(t)$ as t approaches 0 from "above." More generally, if we let t_0 be any value of t, then we may define

$$f(t_0-) = \lim_{\substack{t\to t_0 \\ t<t_0}} f(t)$$

$$f(t_0+) = \lim_{\substack{t\to t_0 \\ t>t_0}} f(t)$$

The continuity condition may now be expressed in the form

$$v_C(t_0-) = v_C(t_0) = v_C(t_0+) \tag{4-27}$$

where t_0 is any value of t, and where $v_C(t)$ is the voltage appearing across the terminals of a capacitor. The implication of the above equation is that the voltage $v_C(t)$ is a continuous function of time.

Now let us consider again the waveshapes shown for the application of different waveshapes of current to the terminals of a capacitor as illustrated in Fig. 4–3.2. We note that, for the complete specification of the voltage waveshape across a given capacitor, two items of information are necessary. These are: (1) the explicit nature of the waveshape of current applied to the capacitor; and (2) the initial *state* that the capacitor is in, i.e., the initial condition on the value of the voltage appearing across the capacitor. It should be noted that the use of the word "initial" is relative. If we know the value of the capacitor voltage at any given time t_0, then t_0 becomes our starting point for values of the independent variable t, and the "initial" condition is evaluated at t_0. For this case, Eq. (4–24) becomes

$$v(t) = \frac{1}{C}\int_{t_0}^{t} i(\tau)\, d\tau + v(t_0) \tag{4-28}$$

In most of the discussion which follows we shall assume, for simplicity of notation, that our interest in a given network situation begins at $t = 0$, and our initial condition will be specified at that point.

To further illustrate the manner in which voltage and current waveshapes are related in a capacitor, consider the network configuration shown in Fig. 4–3.3 in which a voltage produced by a voltage source is applied across the terminals of a capacitor. The resulting current that flows into the upper terminal of the capacitor is determined by the differential equation given in Eq. (4–26). Some examples of various waveshapes are shown in Fig. 4–3.4. For example, in Fig. 4–3.4(a) we show the effect of applying a parabolic waveshape of voltage $v(t) = t^2u(t)$. The result is a

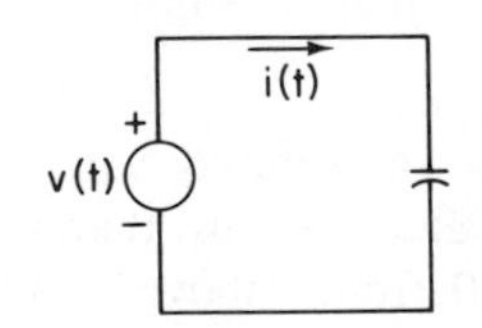

Fig. 4-3.3 A capacitor excited by a voltage source.

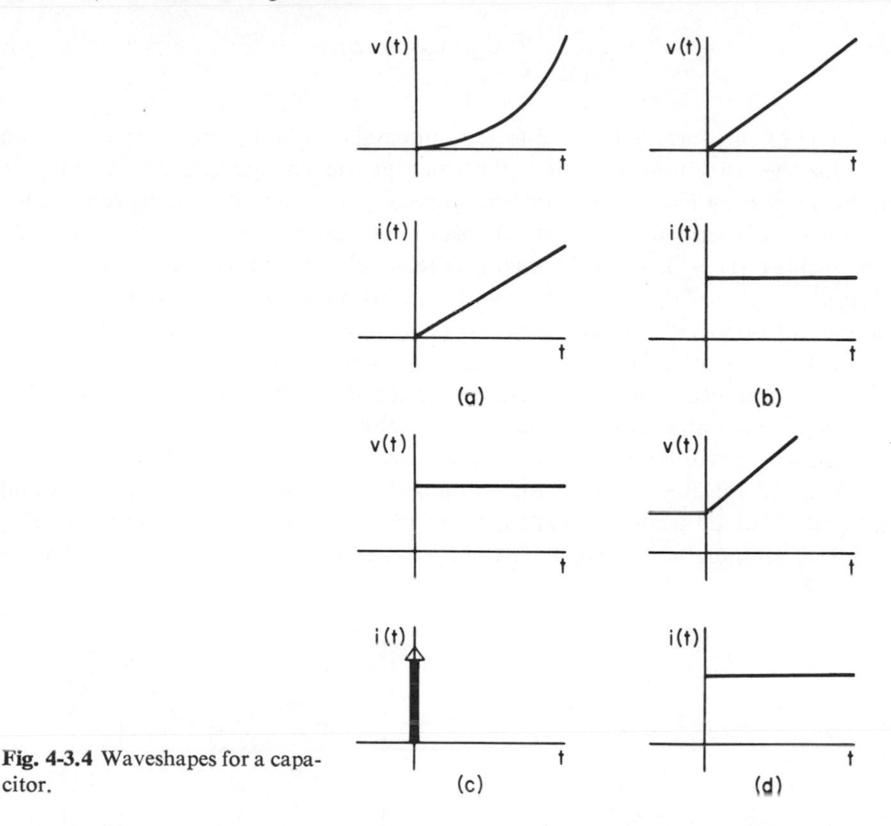

Fig. 4-3.4 Waveshapes for a capacitor.

ramp function of current $i(t)$. The actual value of the slope, of course, depends on the value of the capacitance. Similarly, in Fig. 4–3.4(b), a ramp voltage $v(t) = tu(t)$ is applied. The result is a step function of current. Finally, in Fig. 4–3.4(c), a step function of voltage is applied. The result, as was discussed above, is an impulse of current. This latter situation, of course, is not possible in physical or real-life situations. It should be noted that all the waveshapes of current shown in Fig. 4–3.4 may apply to many different waveshapes of voltage, all of which represent different initial conditions on the capacitor. As an example of this, in Fig. 4–3.4(d), the effect of applying a ramp of voltage at $t = 0$ in addition to a constant voltage which has been applied at $t = -\infty$ is shown. The waveshape of current (for the same value of capacitance) is indistinguishable from that resulting from a ramp alone as shown in Fig. 4–3.4(b). This, of course, is the result of the fact that differentiation of the constant term in the expression for the voltage applied to the capacitor gives zero. Therefore, such a term has no effect on the value of the current.

4-4 NUMERICAL INTEGRATION

The relation between the terminal variables of a capacitor has been shown to have the form

$$v(t) = \frac{1}{C}\int_0^t i(\tau)\,d\tau + v(0) \tag{4-29}$$

An equation of this form is referred to as an integral equation, since, if an expression is given for the current variable $i(t)$, the value of the voltage variable $v(t)$ may be found by integrating $i(t)$. Such a process proceeds very smoothly for expressions for $i(t)$ which have a simple mathematical form. For example, if $i(t)$ is specified by a relation such as $i(t) = 3t + \sin 2t$, then it is relatively easy to determine that $v(t) = (1/C)[(\frac{3}{2})t^2 - (\frac{1}{2})\cos 2t] + v(0)$. In practice, however, many situations arise in which the function describing $i(t)$ is not so easily integrable. As a matter of fact, the form of $i(t)$ may not even be mathematically describable, since it may consist of some waveshape determined from experimental measurements. In situations such as these, the process of integration may be accomplished more easily by the use of numerical techniques, which will be the subject of our discussion in this section.

As a general illustration of the techniques, suppose that it is desired to find the integral of some function $y(t)$ as shown in Fig. 4-4.1(a) between the limits of t_a and t_b. Such an integral is equal to the shaded area shown in the figure. As an ap-

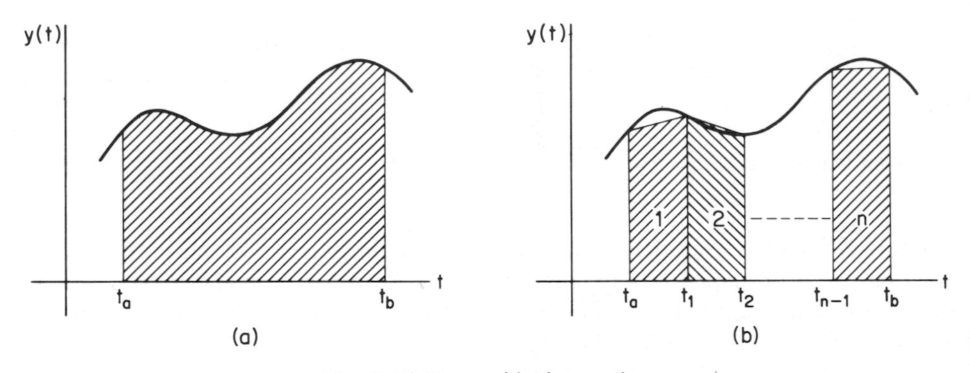

Fig. 4-4.1 Trapezoidal integration.

proximation to this area, we may divide the interval from t_a to t_b into n evenly spaced intervals and construct a set of n trapezoids as shown in Fig. 4-4.1(b). As we make the value of n large, the total area of the trapezoids approaches the area under the curve of $y(t)$, i.e., it approaches the value of the integral of $y(t)$ between the specified limits. We may express these ideas in mathematical notation by first defining a quantity Δt by the relation

$$\Delta t = \frac{t_b - t_a}{n} \tag{4-30}$$

Next, let us represent the $n - 1$ evenly-spaced intermediate values of t between t_a and t_b as t_i ($i = 1, 2, \ldots, n - 1$), as shown in Fig. 4-4.1(b). These values may be defined in terms of the quantity Δt as

$$t_i = t_a + i\,\Delta t \quad i = 1, 2, \ldots, n-1 \tag{4-31}$$

The value of the function $y(t)$ at the first intermediate value of time t_1 is thus simply $y(t_1)$, and the area A_1 of the first trapezoid is given by the expression

$$A_1 = \frac{y(t_a) + y(t_1)}{2}\,\Delta t \tag{4-32}$$

Similarly, the areas A_2 through A_{n-1} of the corresponding trapezoids are found by the relations

$$A_i = \frac{y(t_{i-1}) + y(t_i)}{2}\,\Delta t \qquad i = 2, 3, \ldots, n-1 \tag{4-33}$$

The area of the final trapezoid (the nth one) is found by the expression

$$A_n = \frac{y(t_{n-1}) + y(t_b)}{2}\,\Delta t \tag{4-34}$$

The total area of the n trapezoids is found by summing the above three expressions. If we call this area $A(n)$, we see that

$$A(n) = \left[\frac{y(t_a)}{2} + \frac{y(t_b)}{2} + \sum_{i=1}^{n-1} y(t_i)\right]\Delta t \tag{4-35}$$

In the limit, as the number of trapezoids is made large, the value of $A(n)$ approaches the value of the integral of the function $y(t)$. Thus, we may write

$$\lim_{n\to\infty} A(n) = \int_{t_a}^{t_b} y(t)\,dt \tag{4-36}$$

The procedure described above is usually referred to as *trapezoidal integration*. It is one of the simplest schemes of numerical integration, and is relatively easy to implement on the digital computer.

There are two types of errors which occur in implementing such a numerical process on the digital computer. The first of these is the error that results from using only a finite number of trapezoidal sections in approximating the area. This is frequently referred to as a *truncation error*. For a given function, the truncation error per trapezoidal section may be shown to be proportional to the cube of the quantity Δt.[3] Thus, halving Δt, in effect, should reduce the error by one-fourth (the error per section is one-eighth as large but there are twice as many sections). This conclusion only holds true if the effects of computer round-off error are not con-

[3]See M. L. James, G. M. Smith and J. C. Wolford, *Applied Numerical Methods for Digital Computation*, Chap. 6, International Textbook Co., Scranton, Pa., 1967.

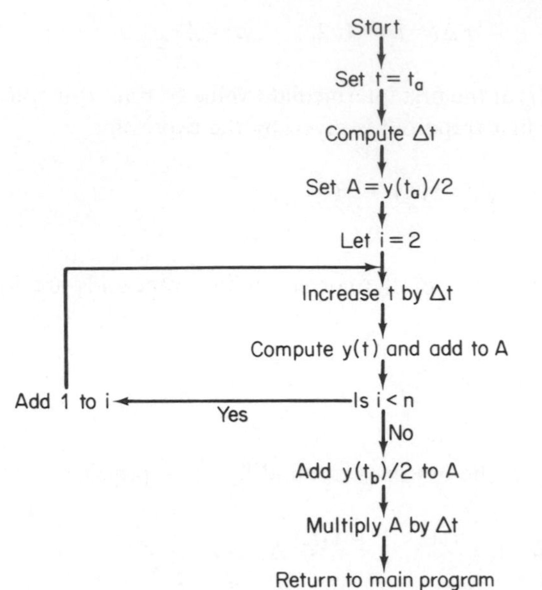

Fig. 4-4.2 Flow chart for trapezoidal integration.

sidered. Computer *round-off error*, the second type of error, occurs because the computer only has a limited accuracy in performing calculations. In other words, only a certain number of significant figures are retained in specifying the value of a given number. The details of this round-off process vary with different computers, as does the number of significant figures used. Later in this section we shall give an illustration of the effect of these two types of errors in performing an example integration.

```
C             TA - INITIAL VALUE OF INDEPENDENT VARIABLE T
C             TB - FINAL VALUE OF INDEPENDENT VARIABLE T
C              N - NUMBER OF TRAPEZOIDAL SECTIONS USED
C              A - VALUE OF INTEGRAL FROM TA TO TB
C       N MUST BE 2 OR GREATER
C
C       INITIALIZE THE VALUE OF TIME T TO THE VALUE TA
C       AND CALCULATE DT, THE INCREMENT IN TIME
          T = TA
          TN = N
          DT = (TB - TA) / TN
C
C       INITIALIZE THE VALUE OF THE INTEGRAL
          A = Y(TA) / 2.
C
C       COMPUTE THE N-1 TERMS IN THE SUMMATION
          DO  8  I = 2, N
          T = T + DT
          A = A + Y(T)
        8 CONTINUE
C
C       FINISH THE COMPUTATION OF THE INTEGRAL
          A = (A + Y(TB) / 2.) * DT
```

Fig. 4-4.3 Statements for implementing the flow chart shown in Fig. 4-4.2.

A flow chart for performing trapezoidal integration on a digital computer is shown in Fig. 4-4.2. This flow chart represents an implementation of Eq. (4-35). The flow chart is readily implemented by a set of FORTRAN statements. A listing of such a set of statements is given in Fig. 4-4.3. In the listing, note that the fourth statement contains the term Y(TA). Such a term tells the compiler to compute the value of some FORTRAN function Y under the conditions that the input argument to the function has the value TA. Thus, the value of $y(t_a)$ is computed. A subprogram which defines the function Y must, of course, be supplied by the user.

To provide maximum flexibility in the use of trapezoidal integration, it is convenient to incorporate the statements defining the trapezoidal integration operations as a FORTRAN subroutine. This can be accomplished by preceding the statements in Fig. 4-4.3 by a subroutine identification statement and following them by the usual RETURN and END statements. If we choose the name ITRPZ (for Integration by TRaPeZoidal method) for the subroutine, and we note that values for TA, TB, and N must be supplied to it and the value of A is provided as an output, then the identification statement will have the form

SUBROUTINE ITRPZ (TA, TB, N, A)

A complete listing of the trapezoidal integration subroutine ITRPZ is shown in Fig. 4-4.4. A summary of its characteristics is given in Table 4-4.1 (on page 185).

```
      SUBROUTINE ITRPZ (TA, TB, N, A)
C     TRAPEZOIDAL INTEGRATION SUBROUTINE
C           TA - INITIAL VALUE OF INDEPENDENT VARIABLE T
C           TB - FINAL VALUE OF INDEPENDENT VARIABLE T
C            N - NUMBER OF TRAPEZOIDAL SECTIONS USED
C            A - VALUE OF INTEGRAL FROM TA TO TB
C     A FUNCTION Y(T) MUST BE USED TO DEFINE THE INTEGRAND.
C     N MUST BE 2 OR GREATER
C
C     INITIALIZE THE VALUE OF TIME T TO THE VALUE TA
C     AND CALCULATE DT, THE INCREMENT IN TIME
      T = TA
      TN = N
      DT = (TB - TA) / TN
C
C     INITIALIZE THE VALUE OF THE INTEGRAL
      A = Y(TA) / 2.
C
C     COMPUTE THE N-1 TERMS IN THE SUMMATION
      DO 8 I = 2, N
      T = T + DT
      A = A + Y(T)
    8 CONTINUE
C
C     FINISH THE COMPUTATION OF THE INTEGRAL
      A = (A + Y(TB) / 2.) * DT
      RETURN
      END
```

Fig. 4-4.4 Listing of the subroutine ITRPZ.

An example of the use of the subroutine ITRPZ to determine the integral of a function which is not easily integrable directly follows.

EXAMPLE 4-4.1. A current $i(t) = e^{-t} \sin 6t$ A is applied to an uncharged 1-F capacitor during the period $0 \le t \le 3$ s. The resulting voltage v appearing across the capacitor at t equals 3 is given by the expression

$$v = \int_0^3 e^{-t} \sin 6t \, dt$$

We may easily write a main program to perform such an integration. First of all, we must define a function $y(t)$ to serve as the integrand. The defining statements are given in Fig. 4-4.5. The remainder of the listing shown in this figure constitutes a main program to provide the integral of the function $y(t)$ over the specified limits.[4] The value of the resulting integral, as given by the output from the program, is printed as the bottom of the listing.

```
      FUNCTION Y(T)
C     INTEGRAND FOR EXAMPLE 4.4-1
      Y = EXP (-T) * ABSF (SINF(6. * T))
      RETURN
      END
C     MAIN PROGRAM FOR EXAMPLE 4.4-1
C          TA - INITIAL VALUE OF TIME
C          TB - FINAL VALUE OF TIME
C           N - NUMBER OF TRAPEZOIDAL SECTIONS
C           A - VALUE OF INTEGRAL
C
C     INITIALIZE THE VARIABLES
      TA = 0.
      TB = 3.
      N = 100
C
C     CALL THE SUBROUTINE ITRPZ TO INTEGRATE G(T)
C     BETWEEN THE LIMITS TA AND TB
      CALL ITRPZ (TA, TB, N, A)
C
C     PRINT THE VALUE OF THE INTEGRAL
      PRINT 6, A
    6 FORMAT (1X, 6HAREA =, E15.8)
      STOP
      END


OUTPUT DATA FOR EXAMPLE 4.4-1

 AREA = 2.01048790E-01
```

Fig. 4-4.5 Main program and output data for Example 4-4.1.

[4]Some of the function names specified in Fig. 4-4.5 may have slightly different forms in different compilers. For example, the exponential to the base e specified as EXP in the listing is named EXPF in some compilers.

Note that in the above example a value of 100 was chosen for the FORTRAN variable N. Thus, 100 trapezoidal sections were used to approximate the area defining the integral. There is no simple general rule for determining the "best" number of sections to use in trapezoidal integration. To see this, let us examine some of the factors which must be considered in selecting the number of trapezoidal sections to approximate the integral of a given function. Not the least of these factors is the nature of the function which it is desired to integrate. For example, if too few sections are used, the approximation will obviously be a poor one. Surprisingly, however, *the use of too many sections may also produce poor results* due to the accumulation of round-off errors that occur in the numerical process. This, in turn, depends upon the word-length and round-off process used by the computer in performing the computations. As an illustration of some of these factors, consider the following example.

EXAMPLE 4-4.2. To see how the accuracy of the trapezoidal integration process will vary depending on the number of sections used and on the computer word length, let us consider the following integral.

$$\int_0^4 (3 + 6x + 6x^2 + x^3)\,dx$$

The integrand of this function has been picked so that the value of the integral may be easily calculated directly. The value is 252. Now let us consider the values obtained by using the trapezoidal integration approach developed in this section to calculate the value of the integral. In Fig. 4-4.6(a) we have listed the results of using various numbers of trapezoidal sections on a computer with a word length (or accuracy) of eight digits.[5] From this figure we see that for the specified integrand and interval of integration, the smallest error occurred when 300 sections were used. Thus, the actual number of sections which would produce the least possible error must lie somewhere between 200 and 400. The use of fewer than 200 sections obviously produces a larger error in determining the value of the integral. Surprisingly, the use of a larger number of sections than 400 also produces poorer results, due to the accumulation of round-off errors in the computer. Thus, we see that it is not correct to blindly assume that better results are produced by larger numbers of sections in the trapezoidal integration process. Now let us repeat the same problem on a computer with a word length of about 15 digits.[6] The results are shown in Fig. 4-4.6(b). Note that, in this case, the error continues to decrease even when the number of sections is increased to the absurdly extravagant value of 7000. The rate at which the error decreases is, however, very slow. Another surprising fact lurks in the data from these two computer runs. For the trial which used 200 sections, the error which resulted from the use of the machine which used eight digits was less than the error from the machine which used fifteen digits! The same is true at 300, 400, and 500 sections. Thus, we see another

[5]This portion of the problem was run on an IBM 7072 computer, a decimal machine with a word length of eight digits.

[6]This portion of the problem was run on a CDC 6400 computer with a word length of 60 bits, i.e., 60 binary numbers. This is approximately equivalent to a word length of 15 (decimal) digits.

SECTIONS	INTEGRAL	FRROR
10	253.28000	1.28000
20	252.32000	0.32000
30	252.14208	0.14208
40	252.08000	0.08000
50	252.05113	0.05113
60	252.03514	0.03514
70	252.02570	0.02570
80	252.01995	0.01995
90	252.01537	0.01537
100	252.01271	0.01271
200	252.00298	0.00298
300	252.00007	0.00007
400	252.00023	0.00023
500	251.99977	0.00023
600	251.99617	0.00383
700	251.99446	0.00554

SECTIONS	INTEGRAL	ERROR
10	253.279999999993	1.279999999993
20	252.319999999985	.319999999985
30	252.142222222206	.142222222206
40	252.079999999972	.079999999972
50	252.051199999979	.051199999979
60	252.035555555524	.035555555524
70	252.026122448945	.026122448945
80	252.019999999958	.019999999958
90	252.015802469095	.015802469095
100	252.012799999939	.012799999939
200	252.003199999896	.003199999896
300	252.001422221949	.001422221949
400	252.000799999731	.000799999731
500	252.000511999722	.000511999722
600	252.000355555122	.000355555122
700	252.000261224010	.000261224010
800	252.000199999541	.000199999541
900	252.000158023752	.000158023752
1000	252.000127999318	.000127999318
2000	252.000031998854	.000031998854
3000	252.000014219402	.000014219402
4000	252.000007997211	.000007997211
5000	252.000005118159	.000005118159
6000	252.000003551077	.000003551077
7000	252.000002607308	.000002607308

Fig. 4-4.6 The effect of word length on the accuracy of trapezoidal integration.

anomaly of numerical processes, namely, that although the use of increased numbers of digits will usually lead to better results, situations may occur where this is not the case.

From the above example it is possible to draw only general conclusions, but ones that are nevertheless very important. The results discussed in the example are caused by the basic nature of numerical processes. Such anomalies occur in the matrix inversion process discussed in Chap. 3, the differential equation solving techniques to be discussed in Chap. 5, and integration techniques of the type discussed in this section. As such, they require that the solution of a given problem take into consideration the numerical technique being used, the computer on which the program is to be run, and the nature of the problem itself in order to correctly determine the number of steps in numerical processes such as trapezoidal integration. The treatment of such considerations is, in general, beyond the scope of an introductory text such as this. When complex demanding problems are to be solved, however, the considerations which have been introduced in this section become of prime importance. A practical approach to determining the proper number of sections in a given situation is to make a run of the problem with some trial number of sections, then repeat the problem using a larger number. If both results show good agreement, this consititutes good evidence that the number of sections is correctly matched to the problem, the numerical technique being used, the computer on which the problem is being run, and the desired accuracy of the final answer. Such an approach gives almost as much information as a detailed error analysis and is considerably simpler to implement.

In many engineering applications, waveforms which are determined by experimental data and which are thus not defined by an explicit mathematical relationship must be treated. For such waveforms, a modification of the trapezoidal integration method presented in this section may be used to determine the integral of the waveform over some specified range of the independent variable. As an example of this method, suppose that the waveform $i(t)$ shown in Fig. 4-4.7(a) has been reproduced from an oscilloscope. To find the integral of this waveform over the range of time from zero to approximately 12 ms we may choose a series of points from the curve as shown in Fig. 4-4.7(b). That is, we choose closely spaced points in the regions of the curve where the radius of curvature is small, and more widely spaced points for those portions of the curve where the radius of curvature is large. Let us assume that there are n data points with abscissa values t_i and ordinate values h_i [where $h_i = i(t_i)$]. Joining the data points with straight lines as shown in the figure, we see that $n - 1$ trapezoids are formed. If we let A_i be the area of the ith trapezoid, then we see that

$$A_i = \frac{h_{i+1} + h_i}{2}(t_{i+1} - t_i) \qquad i = 1, 2, \ldots, n - 1 \tag{4-37}$$

Let us define $A(n)$ as the sum of the $n - 1$ relations of (4-37). If we make such a sum and rearrange the terms, we may write

$$A(n) = \frac{h_1(t_2 - t_1) + \sum_{i=2}^{n-1} h_i(t_{i+1} - t_{i-1}) + h_n(t_n - t_{n-1})}{2} \tag{4-38}$$

Thus, if we know the values of the quantities h_i and t_i which define a waveform in

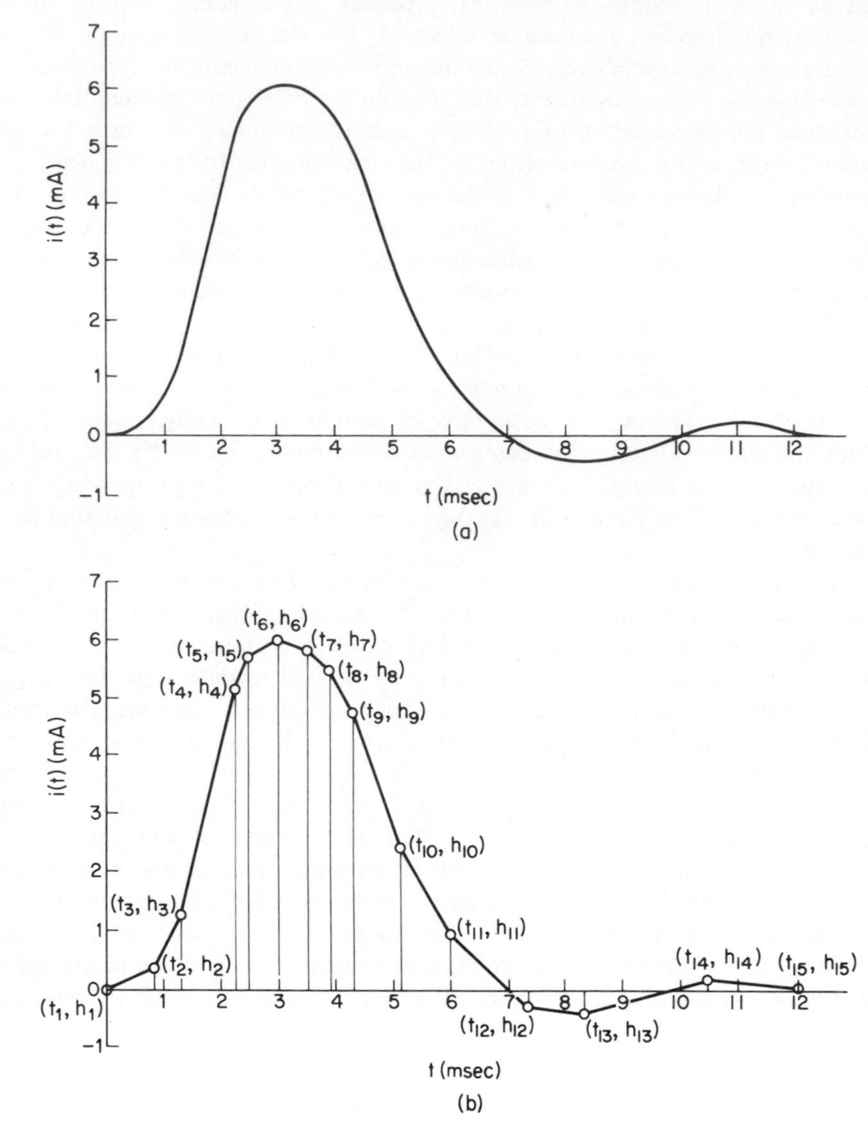

Fig. 4-4.7 A waveshape defined by a set of data points.

the manner shown in Fig. 4–4.7(b), the integral of the waveform may be calculated by using Eq. (4–38).

The algorithm for finding the integral of an arbitrary waveform as defined by (4–38) is readily programmed for the digital computer. Let us assume that two one-dimensional arrays H and T are defined, in which are stored the values of the

```
      SUBROUTINE DINTG (H, T, N, A)
C     SUBROUTINE FOR INTEGRATING A FUNCTION DEFINED BY A
C     SET OF N VALUES H(T) WITH RESPECT TO T
C            H - ARRAY OF VALUES OF DEPENDENT VARIABLE
C            T - ARRAY OF VALUES OF INDEPENDENT VARIABLE
C            N - NUMBER OF DATA POINTS
C            A - OUTPUT VALUE OF INTEGRAL
C     N MUST BE GREATER THAN 2.
      DIMENSION H(60), T(60)
      A = H(1) * (T(2) - T(1))
      NM = N - 1
      DO 5   I = 2, NM
      A = A + H(I) * (T(I+1) - T(I-1))
    5 CONTINUE
      A = A + H(N) * (T(N) - T(N-1))
      A = A / 2.
      RETURN
      END
```

Fig. 4-4.8 Listing of the subroutine DINTG.

quantities h_i and t_i, respectively. A listing of a set of FORTRAN statements which perform the operations defined in Eq. (4–38) is shown in Fig. 4–4.8. For convenience, these statements have been written as part of a subroutine named DINTG (for Data INTeGration). In addition to the input arrays H and T, an input variable N is used to indicate the number of data points which are furnished. The output variable giving the area is A. A summary of the characteristics of the subroutine DINTG is given in Table 4–4.2. An example of the use of this subroutine follows.

EXAMPLE 4-4.3. Assume that the waveshape shown in Fig. 4-4.7 is the waveshape of a current $i(t)$ applied to a 1-μF capacitor which is initially uncharged. The voltage across the capacitor that results from this excitation can be found by using the DINTG subroutine defined above. A set of 15 points is chosen as indicated in part (b) of the figure to represent the waveshape. The values of these points are scaled directly from the figure and punched on cards to serve as data. A FORTRAN main program may be written to perform the following functions:

1. Read the data from the punched cards, storing the values of current as the variables AMPS(I) and the values of time as the variables T(I).
2. Call the subroutine DINTG to integrate the curve defined by the data points.
3. Divide the area found by the subroutine DINTG by the factor 10^{-6} (the value of the capacitor).
4. Print the resulting quantity as the variable VOLTS.

A listing of a FORTRAN program to provide the functions defined above is given in Fig. 4-4.9. A listing of the data cards and the value of the output voltage is also shown in this figure.

```
C     MAIN PROGRAM FOR EXAMPLE 4.4-3
C             N - NUMBER OF DATA POINTS
C          AMPS - ARRAY OF VALUES OF I(T)
C             T - ARRAY OF VALUES OF TIME
C             A - VALUE OF INTEGRAL
      DIMENSION AMPS(20), T(20)
      N = 15
C
C     READ THE INPUT DATA DEFINING THE
C     CURVE OF I(T) VERSUS T
      READ 3, (AMPS(I), T(I), I = 1, N)
    3 FORMAT (2E10.0)
C
C     CALL THE SUBROUTINE DINTG TO INTEGRATE
C     THE CURVE DEFINING I(T), DIVIDE THE
C     RESULT BY C, AND PRINT ITS VALUE
      CALL DINTG (AMPS, T, N, A)
      VOLTS = A / 1.E-6
      PRINT 7, VOLTS
    7 FORMAT (1X, 16HOUTPUT VOLTAGE =, E11.4)
      STOP
      END
```

```
INPUT DATA FOR EXAMPLE 4.4-3

     .00E-03    .00E-03
     .40E-03    .80E-03
    1.30E-03   1.30E-03
    5.25E-03   2.25E-03
    5.80E-03   2.45E-03
    6.10E-03   3.00E-03
    5.95E-03   3.50E-03
    5.60E-03   3.90E-03
    4.85E-03   4.30E-03
    2.50E-03   5.15E-03
    1.00E-03   6.00E-03
   -0.25E-03   7.35E-03
   -0.40E-03   8.35E-03
     .20E-03  10.50E-03
     .05E-03  12.10E-03
```

```
OUTPUT DATA FOR EXAMPLE 4.4-3

 OUTPUT VOLTAGE = 2.0264E+01
```

Fig. 4-4.9 Listing of the program and data for Example 4-4.3.

TABLE 4-4.1 Summary of the Characteristics of the Subroutine ITRPZ

Identifying Statement: SUBROUTINE ITRPZ (TA, TB, N, A)
Purpose: To find the definite integral of a specified function $y(t)$, that is, to find A, where

$$A = \int_{t_a}^{t_b} y(t)\,dt$$

by using trapezoidal integration.
Additional Subprograms Required: This subroutine calls the function identified by the statement

FUNCTION Y(T)

This function must be used to define $y(t)$, i.e., the integrand.
Input Arguments:
- TA The lower limit t_a of the variable of integration t.
- TB The upper limit t_b of the variable of integration t.
- N The number n of trapezoidal segments used to approximate the area, i.e., the number of iterations used in performing the trapezoidal integration.

Output Argument:
- A The value of the integral.

TABLE 4-4.2 Summary of the Characteristics of the Subroutine DINTG

Identifying Statement: SUBROUTINE DINTG (H, T, N, A)
Purpose: To find the definite integral of a waveform $h(t)$ defined by a set of values h_i which specify the waveform at corresponding values of time t_i.
Additional Subroutines Required: None.
Input Arguments:
- H The one-dimensional array of variables H(I) in which are stored the values h_i which define the function.
- T The one-dimensional array of variables T(I) which specifies the values of time t_i at which the values of the function are known.
- N The number of data points used to define the function.

Output Argument:
- A The value of the integral.

Note: The variables of the subroutine are dimensioned as follows: H(60), T(60).

4-5 NUMERICAL SOLUTION OF INDEFINITE INTEGRALS—PLOTTING

In the previous section we showed how to use numerical techniques to determine the value of the voltage across a capacitor at a certain instant of time as the result of applying a specified current to the capacitor. Thus, we solved an integral equation of the form

$$v(t_b) = \int_{t_a}^{t_b} i(t)\,dt + v(t_a) \tag{4-39}$$

where the specification for $i(t)$ was made either by an explicit function (in which case, we used the subroutine ITRPZ) or by a set of data points (in which case, we used the subroutine DINTG). In mathematical terms, such a procedure is referred to as finding the value of a *definite integral*. By the word "definite" we imply that the upper limit of the variable of integration is fixed.

Frequently in engineering studies, we are more interested in determining the values of the voltage across a capacitor for a range of values of time, rather then just at a single instant. Thus, we are interested in solving an integral equation of the form

$$v(t) = \int_{t_a}^{t} i(t)\,dt + v(t_a) \qquad t > t_a \tag{4-40}$$

Such a procedure is referred to as finding the *indefinite integral* of $i(t)$. By the word "indefinite" we imply that a solution is desired for a continuous range of values of the independent variable t. In general, numerical processes are not suited to providing an explicit closed-form solution for such a problem.

Since closed-form results are, in general, not possible with numerical methods, we turn to graphical techniques for the display of such results. The data to be displayed graphically can be obtained by treating an indefinite integral as a series of definite integrals. For example, suppose that it is desired to plot the function $v(t)$, where

$$v(t) = \int_{0}^{t} i(t)\,dt + v(0) \tag{4-41}$$

over a range of 0–0.5 s. The data for such a plot can be prepared by choosing 50 evenly spaced points for the independent variable t and computing a corresponding 50 values for $v(t)$. The first point is $v(0.01)$. This value may be found by making a (definite) integration of $i(t)$ from 0–0.01 s and adding this value to $v(0)$. The next point is $v(0.02)$. This may be found by integrating $i(t)$ from 0.01–0.02 s, and adding the result to $v(0.01)$. The process can obviously be continued to generate 50 values of $v(t)$ for a range of t from 0.01 to 0.5 s. A plot of these values of $v(t)$ vs. t plus a 51st value specifying $v(0)$ then defines the indefinite integral.

There are many techniques currently available to the computer user which may be used to make a plot from a set of data generated by the procedure outlined above. These include sophisticated methods such as cathode ray tube displays at consoles of time-shared remote computer terminals as well as the more common off-line plotters which use input data prepared by a separate computer program, and which draw scaled and labeled inked plots on pre-printed graph paper. Considerably less sophisticated are the on-line techniques which use the computer system printer to prepare a plot at the same time that the program listing and output data are printed. This latter type of graphical display is more common than the other types described above. It has the advantage of requiring little extra programming, being low in operating cost, and minimizing time lost by having to wait in a plotter queue as well as in a computer queue. Most such plotting programs are written as subroutines and provide for the plotting of one or more functions simultaneously, using evenly spaced increments of the independent variable. In one common approach, the values of the variables to be plotted are stored in a two-dimensional array, and the plotting is accomplished by "calling" the plotting subroutine after the appropriate variables have been computed and stored. For example, let the increments of time at which a plot is desired be $k\,\Delta t$ ($k = 0, 1, \ldots, n$). If the functions which are to be plotted are $y_1(t)$, $y_2(t), \ldots$, and if the two-dimensional array in which the variables are to be stored is Y, we might store $y_1(0)$ in Y(1, 1), $y_1(\Delta t)$ in Y(1, 2), $y_1(2\,\Delta t)$ in Y(1, 3), ..., $y_2(0)$ in Y(2, 1), $y_2(\Delta t)$ in Y(2, 2), ..., etc. A simple plotting subroutine which will plot the values of data stored in this manner is described in Appendix C. This subroutine will be used for all the plotting of variables that is done in this text. The reader who desires a more elaborate plot may readily modify the program included here, or may consult his computer center for information on more sophisticated plotting subroutines which are available. The plotting subroutine used here has the name PLOT5. It is called by the statement

```
CALL PLOT5 (Y, M, N, MAX)
```

The Y input argument is a two-dimensional array dimensioned (5, 101) in which is stored up to 101 values of each of the variables (from 1 to 5) that it is desired to plot. Quantities stored in locations Y(1, I) (I = 1, 2, 3, ...) are plotted with the character A, quantities stored in locations Y(2, I) (I = 1, 2, 3, ...) are plotted using the character B, etc. The input argument M specifies the number of variables (from 1 to 5) that it is desired to plot and the input argument N specifies the number of values of each variable (from 1 to 101) that it is desired to plot. The input argument MAX specifies the maximum value of the ordinate (the minimum value is set 100 units lower).

Frequently in the computation and plotting of functions, the problem of correctly scaling the data before plotting it becomes a very critical one. To aid the user in this matter, an automatic scaling option has been included in the PLOT5 subroutine. This option is called by setting the final argument MAX of the subroutine to the value 999. If this is done, PLOT5 will automatically scale each of the functions being plotted so that it fills the entire ordinate range of the plot. In addition, the actual minimum and maximum values and the range of each of the functions before scaling is printed at the edge of the plot. A summary of the characteristics of the subroutine PLOT5 may be found in Table 4-5.1 (on page 193). An example illustrating the use of definite integration techniques to determine the plot of the solution of an integral equation follows.

EXAMPLE 4 5.1. A current defined as $i(t) = 20(1 - \cos 2\pi t)$ A is applied to the terminals of a 1-F capacitor. The voltage $v(t)$ across the capacitor is 0 at $t = 0$. We may use the integration techniques described above to obtain a plot of the values of $v(t)$ for 51 successive values of t spaced 0.1 s apart, and the subroutine PLOT5 to display the results. A flow chart describing the process is shown in Fig. 4-5.1. A listing of a FORTRAN main program for implementing this flow chart is shown in Fig. 4-5.2. Note that this program calls the subroutine ITRPZ defined in Sec. 4-4 and the subroutine PLOT5 discussed in this section. Five iterations in ITRPZ have been used for each point. The plot that results is shown in Fig. 4-5.3. It shows both the output voltage and the input current. Note that the values of the resulting voltage are printed along the upper margin of the figure. For convenience in using the plot, the values of the points printed by the computer have been linked by a shaded line, and information on the scales has been added along the border of the plot. For this simple example, it is easily verified that the values of voltage $v(t)$ displayed by the plot correspond with those found by directly integrating the defined expression for the current. In this case, we obtain $20[t - (1/2\pi) \sin 2\pi t]$.

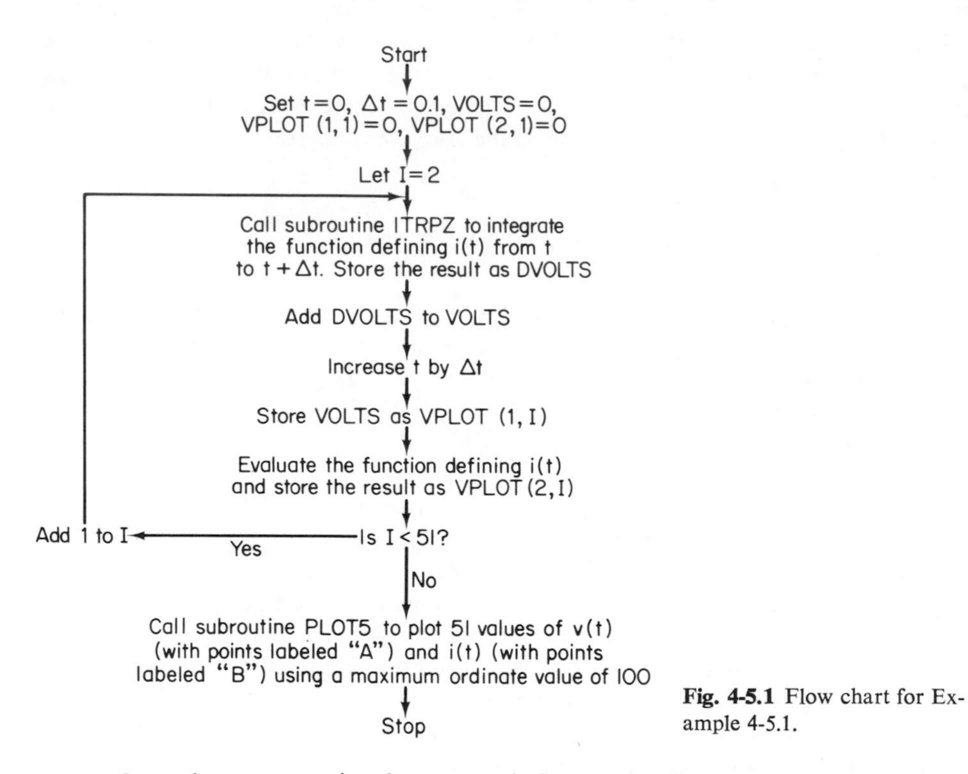

Fig. 4-5.1 Flow chart for Example 4-5.1.

One other computational accessory is frequently of use when large quantities of data are prepared for plotting. This is a subroutine which prints the values of the data stored in the basic two-dimensional array which is dimensioned (5, 101). This subroutine is named PRINT5, and it is called by the statement

CALL PRINT5 (Y, M, N)

where Y is the (5, 101) array, M is the number of functions whose values are to be printed, and N is the number of values of each function which is to be printed. A listing of the subroutine PRINT5 is given in Fig. 4-5.4. A summary of its characteristics is given in Table 4-5.2.

```
      FUNCTION Y(T)
C     THIS FUNCTION DEFINES I(T) IN EXAMPLE 4.5-1
      Y = 20. * (1. - COS (6.28318 * T) )
      RETURN
      END
C     MAIN PROGRAM, EXAMPLE 4.5-1
C           T - TIME
C          DT - INCREMENT FOR INCREASING T
C       VOLTS - VALUE OF VOLTAGE V(T)
C      DVOLTS - INCREMENTAL CHANGE IN VOLTAGE
C       VPLOT - ARRAY FOR STORING VALUES OF V(T) AND I(T)
C                      FOR PLOTTING
      DIMENSION VPLOT(5,101)
C
C     INITIALIZE PROGRAM VARIABLES
      T = 0.
      DT = .1
      VOLTS = 0.
      VPLOT(1,1) = VOLTS
      VPLOT(2,1) = 0.
C
C     USE ITRPZ TO COMPUTE 50 VALUES OF VOLTAGE
      DO 10  I = 2, 51
      CALL ITRPZ (T, T+DT, 5, DVOLTS)
      VOLTS = VOLTS + DVOLTS
C
C     INCREMENT THE VALUE OF TIME
      T = T + DT
C
C     STORE VALUES OF VOLTAGE AND CURRENT FOR PLOTTING
      VPLOT(1,I) = VOLTS
      VPLOT(2,I) = Y(T)
   10 CONTINUE
C
C     CALL PLOTTING SUBROUTINE TO PLOT VOLTAGE AND CURRENT
      PRINT 12
   12 FORMAT (1H1)
      CALL PLOTS (VPLOT, 2, 51, 100)
      STOP
      END
```

Fig. 4-5.2 Listing of the program implementing the flow chart of Fig. 4-5.1.

In this and the preceding section we have discussed the use of numerical techniques to perform integration. Numerical techniques are also readily implemented to perform differentiation. To illustrate this, let us consider the (time) derivative of a function $g(t)$. Let the value of the function at t_i be $g(t_i)$. Now let us use a Taylor series expansion to express the value t_{i+1}, where $t_{i+1} = t_i + \Delta t$. Thus, we may write

$$g(t_{i+1}) = g(t_i + \Delta t) = g(t_i) + \left.\frac{dg}{dt}\right|_{t=t_i} \Delta t + \left.\frac{d^2 g}{dt^2}\right|_{t=t_i} \frac{(\Delta t)^2}{2} + \cdots \tag{4-42}$$

Similarly, if we let $t_{i-1} = t_i - \Delta t$, then

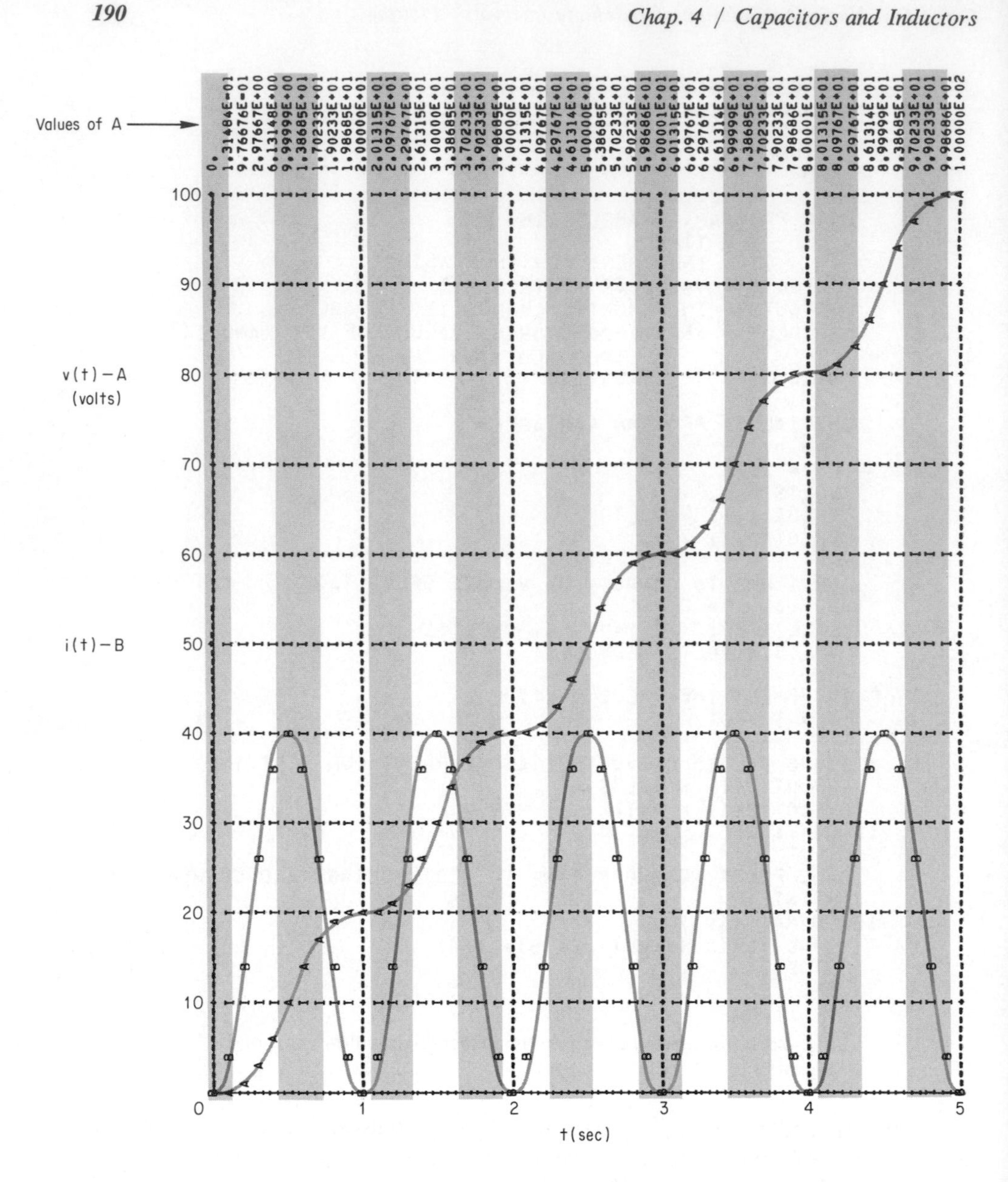

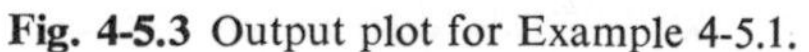
Fig. 4-5.3 Output plot for Example 4-5.1.

```
      SUBROUTINE PRINT5 (A, M, N)
C     SUBROUTINE FOR PRINTING THE VALUES OF VARIABLES STORED
C     IN A 5 X 101 ARRAY
C           A - THE TWO-DIMENSIONAL ARRAY USED TO STORE THE VARIABLES
C           M - THE NUMBER OF VARIABLES STORED IN THE ARRAY
C           N - THE NUMBER OF VALUES OF EACH VARIABLE
      DIMENSION A(5,101), L(5)
      DATA (L(I),I=1,5) / 1HA,1HB,1HC,1HD,1HE/
C
C     PRINT THE CAPTIONS FOR THE COLUMNS OF DATA
      PRINT 2, (L(I), I = 1, M)
    2 FORMAT(9X,5(5X,A1,9H - VALUES,5X))
C
C     PRINT THE OUTPUT DATA IN COLUMNS
      DO 7   I = 1, N
      IM = I - 1
      PRINT 6, IM, (A(J,I), J = 1, M)
    6 FORMAT(1X,I5,5E20.8)
    7 CONTINUE
      RETURN
      END
```

Fig. 4-5.4 Listing of the subroutine PRINT5.

$$g(t_{i-1}) = g(t_i - \Delta t) = g(t_i) - \left.\frac{dg}{dt}\right|_{t=t_i} \Delta t + \left.\frac{d^2 g}{dt^2}\right|_{t=t_i} \frac{(\Delta t)^2}{2} - \cdots \tag{4-43}$$

If we ignore terms of degree $(\Delta t)^3$ and higher, we may subtract the last two equations and rearrange terms to obtain the expression

$$g'(t_i) = \left.\frac{dg}{dt}\right|_{t=t_i} = \frac{g(t_{i+1}) - g(t_{i-1})}{2(\Delta t)} \tag{4-44}$$

Thus, we have obtained an expression for the derivative of $g(t)$ which is directly amenable to digital computation.[7] Obviously, a sequence of values of $g'(t)$ can be computed and plotted to display a graph of the function vs. time for any desired range of time. In general, numerical differentiation is considerably more error prone than numerical integration. Much of the reason for this is that the integration process tends to smooth out irregularities in data, whereas the differentiation process tends to increase such irregularities. At any rate, differentiation is encountered considerably less frequently than integration in digital computational techniques, and we shall not consider it further in this text.

If it is desired to determine the indefinite integral for a function defined by a set of data points, i.e., to make a plot of the integral of such a function, the subroutine DINTG discussed in Sec. 4-4 is not suitable, since, in general, the values of t which determine the set of points used by this subroutine will not be evenly spaced as is required for plotting. To treat this situation, we now introduce a function which uses piecewise linear interpolation to compute the values of a function

[7]The expression in (4-44) is called a first-order *central difference* approximation for $g'(t_i)$.

for values of its independent variable that lie between the specific values for which the function is specified. Thus, if we let $f(t)$ be a function defined at a set of values $t = t_i$ and also let $f_i = f(t_i)$ be the values of the function, then, assuming that the t_i are arranged in ascending order, for a value of t between t_{i-1} and t_i, we may write the value of $f(t)$ as

$$f(t) = f_{i-1} + \frac{f_i - f_{i-1}}{t_i - t_{i-1}}(t - t_{i-1}) \tag{4-45}$$

This relation is readily implemented as a FORTRAN function subprogram. It must be supplied with the values t_i at which the value of the function is known. We shall assume that these are stored in ascending order in a one-dimensional array named TDATA. The corresponding values of the function will be assumed to be stored in a one-dimensional array FDATA. If there are NDATA such points, then an appropriate identifying statement for a function PWL (for Piece Wise Linear determination of the value of a function) will be

FUNCTION PWL (T, NDATA, FDATA, TDATA)

where T is the value of t at which the function is to be evaluated and PWL is the resulting value of the function. A flow chart for the logic which may be used in implementing such a function is given in Fig. 4–5.5. A listing of the subprogram which follows this logic is shown in Fig. 4–5.6. A summary of the characteristics of the subprogram is given in Table 4–5.3. An example of its use follows.

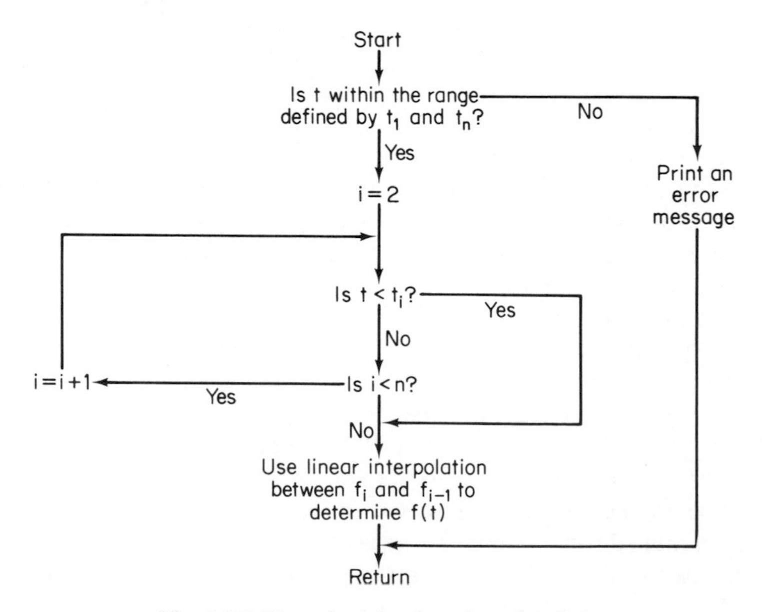

Fig. 4-5.5 Flow chart for the subroutine PWL.

TABLE 4-5.1 Summary of the Characteristics of the Subroutine PLOT5

Identifying Statement: SUBROUTINE PLOT5 (Y, M, N, MAX)
Purpose: To provide a simultaneous plot of several functions $y_i(t)$ for discrete, evenly spaced values of the independent variable t.
Additional Subroutines Required: None.
Input Arguments:

- Y The two-dimensional array of variables Y(I, J) giving the values of the functions $y_i(t)$ at the jth value of the independent variable t.
- M The number of functions $y_i(t)$ which are to be plotted.
- N The maximum number of values of t to be used in plotting the functions.
- MAX The upper value of the ordinate scale used to plot the functions. If the value of this argument is set to 999, the subroutine will automatically scale the plot of each function so that it includes the entire ordinate range.

Output: A printer-constructed plot is provided of the functions. The positive direction of the independent variable is taken as downward on the printed page. The letter A is used to indicate the points for y_1, the letter B for y_2, etc. The numerical values of the function y_1 are also printed along the edge of the sheet as a convenience. If the value of a function exceeds the ordinate range of the plot, the symbol $ is printed along the left or right edge of the plot, depending on whether the function was less than or greater than the permitted range.
Notes:

1. The subroutine is dimensioned Y(5, 101). Thus, it permits plotting up to five functions for up to 101 values of the independent variable.
2. An ordinate scale of 100 points is provided. Thus, the lower bound of the scale will be MAX − 100. The scale is printed automatically on the ordinate.

TABLE 4-5.2 Summary of the Characteristics of the Subroutine PRINT5

Identifying Statement: SUBROUTINE PRINT5 (Y, M, N)
Purpose: To provide an output listing of the values of a double-subscripted variable, i.e., to list the values of several functions $y_i(t)$ for discrete, evenly spaced values of the variable t.
Additional Subroutines Required: None.
Input Arguments:

- Y The two-dimensional array of variables Y(I, J) giving the values of the functions $y_i(t)$ at the jth value of the independent variable t.
- M The range of the subscript I, that is, the number of functions $y_i(t)$ which are to be listed.
- N The range of the subscript J, that is, the number of values of the independent variable t for which listings are to be made.

Output: The subroutine provides a listing of the values of the M quantities Y(1, J), Y(2, J), etc., on a given line. It continues this for N lines, i.e., for the N values of J.
Note: The subroutine is dimensioned Y(5, 101). Thus it permits up to five functions to be listed for up to 101 values of the independent variable t.

```
      FUNCTION PWL(T, NDATA, FDATA, TDATA)
C     SUBPROGRAM FOR FINDING VALUES OF A FUNCTION
C     SPECIFIED BY A SERIES OF DATA POINTS
C          PWL - VALUE OF FUNCTION
C            T - VALUE OF INDEPENDENT VARIABLE
C        NDATA - NUMBER OF DATA POINTS
C        FDATA - VALUES OF F AT DATA POINTS
C        TDATA - VALUES OF T AT DATA POINTS. THESE
C                MUST BE IN ASCENDING ORDER
      DIMENSION FDATA(60), TDATA(60)
C
C     TEST FOR T WITHIN SPECIFIED RANGE
      IF(T.LT.TDATA(1)) GO TO 13
      IF(T.GT.TDATA(NDATA)) GO TO 13
C
C     FIND FIRST VALUE OF TDATA GREATER THAN T
    7 DO 10  I = 2, NDATA
      K = I
      IF (TDATA(I).GT.T) GO TO 11
   10 CONTINUE
C
C     EXTRAPOLATE TO FIND VALUE OF FUNCTION
   11 KM = K - 1
      PWL = FDATA(KM) + (FDATA(K) - FDATA(KM)) *
     1(T - TDATA(KM)) / (TDATA(K) - TDATA(KM))
      RETURN
C
C     PRINT ERROR MESSAGE
   13 PRINT 14, T
   14 FORMAT(4H0T =,E15.8,26HNOT WITHIN SPECIFIED RANGE)
      RETURN
      END
```

Fig. 4-5.6 Listing of the subroutine PWL.

TABLE 4-5.3 Summary of the Characteristics of the Function PWL

Identifying Statement: FUNCTION PWL (T, NDATA, FDATA, TDATA)

Purpose: To construct a piecewise linear approximation for a function $f(t)$ from a set of n points (f_i, t_i) and to use this approximation to give a numerical value for the function $f(t)$ at a given value of t.

Additional Subprograms Required: None.

Input Arguments:

- T The value of the independent variable t for which a piecewise linear approximation to the value of $f(t)$ is to be found.
- NDATA The number of points of data.
- FDATA The one-dimensional array of variables FDATA(I) in which are stored the values f_i of the function.
- TDATA The one-dimensional array of variables TDATA(I) in which are stored (in ascending order) the values of the quantities t_i at which the value of the function $f(t)$ is known.

Output Argument:

- PWL The approximation to the value of the function $f(t)$.

Note: The variables of this function are dimensioned as follows:
FDATA (60), TDATA (60).

EXAMPLE 4-5.2. As an example of the use of the function PWL, consider the waveshape of current shown in Fig. 4-5.7. This current is applied at the terminals of a unity-

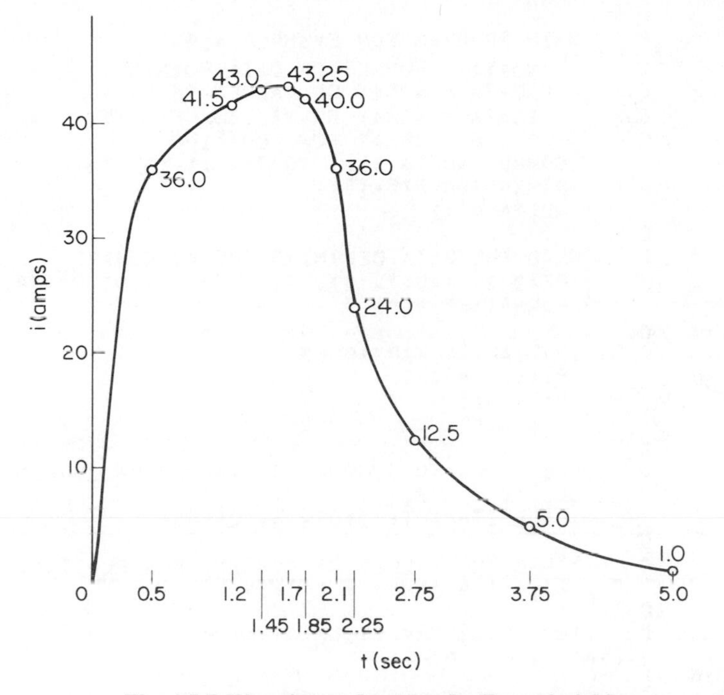

Fig. 4-5.7 Waveshape of current for Example 4-5.2.

valued capacitor (with a zero initial condition) and it is desired to find the resulting voltage across the capacitor. To do this we will use the trapezoidal integration subroutine ITRPZ. In connection with this subroutine, we must, of course, have a function defined as Y(T) which specifies the integrand. For this example, since the integrand is specified by the data points shown in Fig. 4-5.7, the function Y(T) is simply used to call the function PWL and to store the resulting value of the function as Y. In implementing such an approach to the solution of this problem, one difficulty arises. This is the fact that the function Y(T) must be supplied with values of the quantities NDATA, ADATA, and TDATA, so that these variables can be used as arguments when the function PWL is called. The argument list for the function Y(T), however, cannot be changed (without also changing the statements in the subroutine ITRPZ which calls the function). Thus, we must find some means of entering the quantities NDATA, ADATA, and TDATA into the function Y without using its argument list. A simple method of accomplishing this is by the use of a COMMON statement in the function Y(T) and also in the main program. Such an approach is illustrated in the listing of the main program and the function given in Fig. 4-5.8. The resulting output plot giving the waveshape of the voltage across the capacitor is shown in Fig. 4-5.9.

```
      FUNCTION Y(T)
C     FUNCTION FOR EXAMPLE 4.5-2
      COMMON ADATA(11),TDATA(11),NDATA
      Y = PWL(T, NDATA, ADATA, TDATA)
      RETURN
      END

C     MAIN PROGRAM FOR EXAMPLE 4.5-2
C        NDATA - NUMBER OF DATA POINTS
C        ADATA - ARRAY OF VALUES OF V(T)
C        TDATA - ARRAY OF VALUES OF TIME
C            P - ARRAY FOR PLOTTING
      COMMON ADATA(11), TDATA(11), NDATA
      DIMENSION P(5,101)
      NDATA = 11
C
C     READ THE DATA DEFINING THE WAVESHAPE
      READ 3, (ADATA(I), TDATA(I), I =1,NDATA)
    3 FORMAT(2F10.0)
C
C     INITIALIZE VARIABLES
      P(1,1) = 0.0
      DT = 0.1
      T = 0.0
C
C     USE ITRPZ TO COMPUTE 50 VALUES OF CURRENT
      DO 10 I = 2, 51
      CALL ITRPZ(T, T+DT, 5, DAMPS)
C
C     STORE THE VALUES OF CURRENT FOR PLOTTING
      P(1,I) = DAMPS + P(1,I-1)
C
C     INCREMENT THE VALUE OF TIME
   10 T = T + DT
C
C     CALL THE PLOTTING SUBROUTINE
      PRINT 12
   12 FORMAT(1H1)
      CALL PLOT5(P, 1, 51, 100)
      STOP
      END

INPUT DATA
 0.0       0.0
36.0       0.5
41.5       1.2
43.0       1.45
43.25      1.7
40.0       1.85
36.0       2.1
24.0       2.25
12.5       2.75
 5.0       3.75
 1.0       5.0
```

Fig. 4-5.8 Listing of the program for Example 4-5.2.

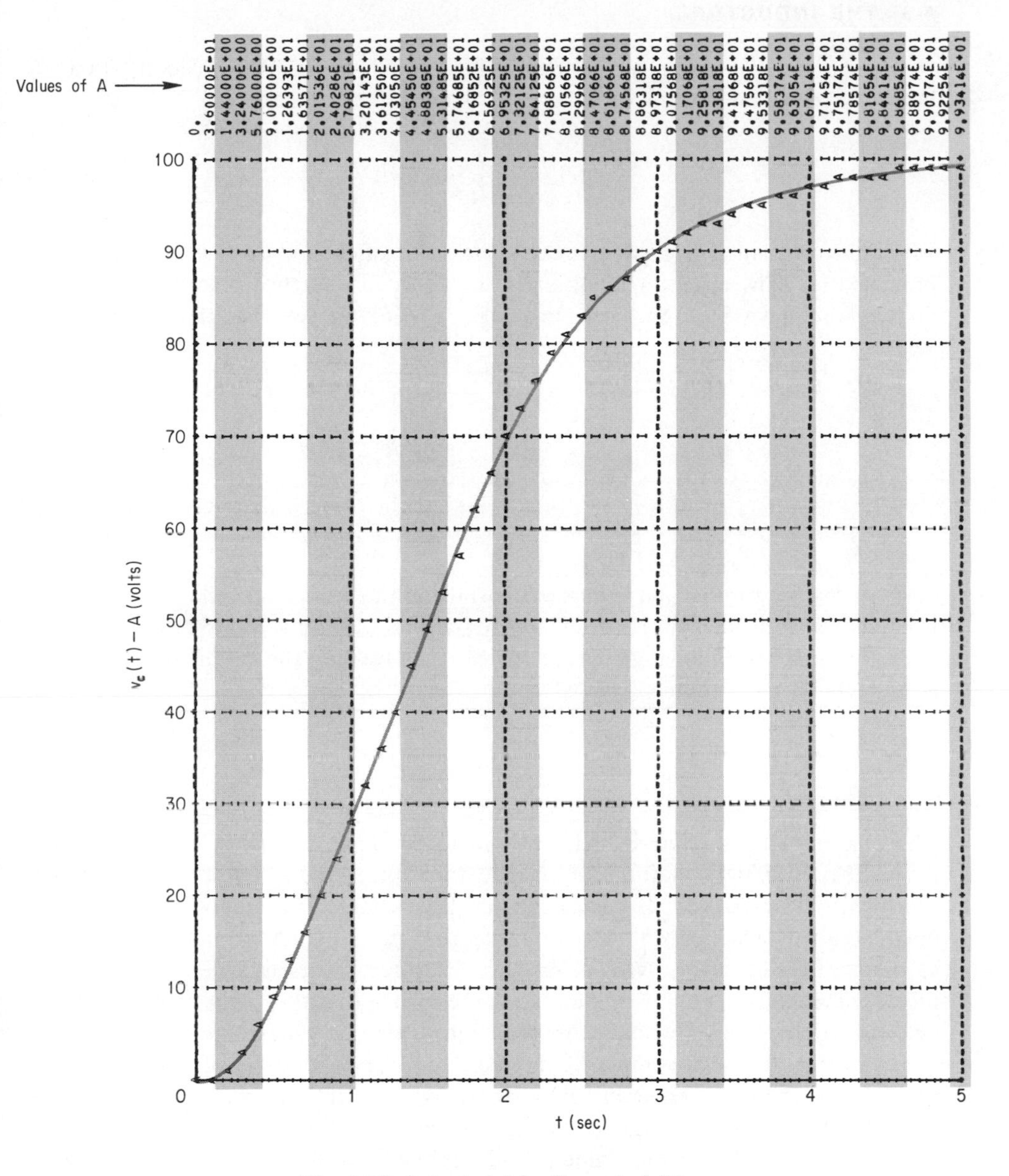

Fig. 4-5.9 Output plot for Example 4-5.2.

4-6 THE INDUCTOR

An inductor is a two-terminal element which has the property that its branch voltage and current variables are related by the expression

$$i(t) = \frac{1}{L}\int_{-\infty}^{t} v(\tau)\, d\tau \tag{4-46}$$

where the constant L is used to specify the value of the inductance in henries (abbreviated H). The circuit symbol and the associated reference polarities for an inductor are shown in Fig. 4-6.1. From Eq. (4-46) we see that the value of the

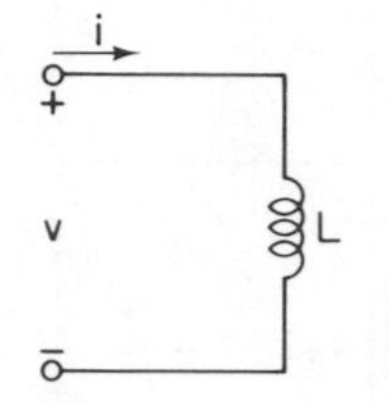

Fig. 4-6.1 Circuit symbol and reference polarities for an inductor.

current flowing through an inductor depends on the values of voltage which have been applied to it for values of time starting at $-\infty$. Since it is difficult to maintain such a history, the following alternative definition relating the terminal variables of an inductor is more frequently used

$$i(t) = \frac{1}{L}\int_{0}^{t} v(\tau)\, d\tau + i(0) \tag{4-47}$$

The quantity $i(0)$ is referred to as the *initial condition.* It should be noted by the reader that an initial condition for an inductor is preserved in quite a different way than that for a capacitor. If a capacitor has been charged to some voltage, then the open-circuited capacitor will retain this charge. As a matter of fact, high quality capacitors found in many types of modern electronic apparatus such as radar equipment, television sets, etc., will retain a charge so long that they may actually electrocute the careless worker who touches their terminals. This may happen even though the equipment has been completely disconnected from all sources of power for some time! In an inductor, however. an initial condition or "charge" is preserved by *shorting* the terminals of the inductor. The current (if the inductor is ideal) then simply keeps merrily flowing along. Although such a statement may at first seem strange due to the concept of the continuing "motion" of the charged particles, inductors maintained at cryogenic temperatures (very close to absolute zero) so as to make their resistance exceedingly small have actually been observed to maintain a current for years without any application of external power sources. The integral equations given in (4-46) and (4-47) are quite similar to those defining a capacitor which were presented in Sec. 4-1. As a matter of fact, if we interchange the variables of voltage and current, and substitute the constant C for the constant L, the relations are identical. Network elements with terminal relations whose form is identical if the variables of voltage and current are interchanged are said to be dual to each other. We shall see some interesting applications of the concept of duality in Sec. 6–8.

The relations given above are referred to as the integral relations for the terminal variables of an inductor. By differentiating both sides of either equation, we obtain the *differential* relation for the terminal variables of an inductor. This has the form

$$v(t) = L\frac{di}{dt} \tag{4-48}$$

It is frequently convenient to be able to use reciprocal-valued units when dealing with inductors in a given circuit. In such a case the Greek letter Γ (capital gamma) is usually used. Unlike the resistive and capacitive cases where compact names of "mhos" and "darafs" have been developed for reciprocal-valued elements, the units for Γ are referred to by the clumsy title of "reciprocal henries." If you're looking for an easy chance to become famous, you might try to invent a better name. Obviously, "yrnehs" is not a good choice.

When we described the characteristics of a capacitor, we noted that its properties were based on the relation that this element imposes between the charge stored in the capacitor and the voltage appearing across its terminals. In a similar manner, the properties of an inductor may be defined by the relation that exists between the flux linkages $\phi(t)$ (in weber-turns) and the current flowing in the element. The relation is

$$\phi(t) = Li(t) \tag{4-49}$$

The properties of an inductor are actually a direct result of Faraday's law, introduced in Sec. 1–1, which states that the induced voltage in a circuit is proportional to the time rate of change of flux linkages, namely,

$$v(t) = \frac{d\phi}{dt} \tag{4-50}$$

Combining the two expressions given above yields the differential relation given in Eq. (4–48).

We may illustrate the properties of an inductor more fully by defining an i–ϕ plane, this is, a plane in which the current variable $i(t)$ is plotted along the abscissa or "x" axis, and the flux-linkage variable $\phi(t)$ is plotted along the ordinate or "y" axis. Such a representation follows closely that used for the resistor in Sec. 2–1 and the capacitor in Sec 4–1. We may now define a *linear* inductor as one characterized by a straight line passsing through the origin of the i–ϕ plane. The slope of the line is the value of the inductance. An example of such a characteristic is shown in Fig. 4–6.2. Thus, a *time-varying* linear inductor is readily seen to be one with a straight-line characteristic passing through the origin but having a slope which varies as a

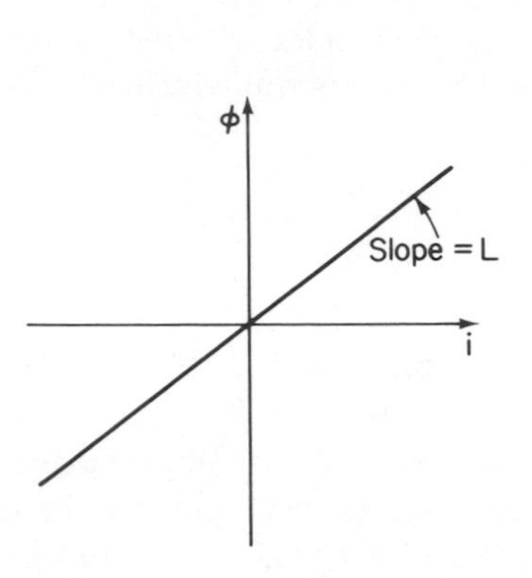

Fig. 4-6.2 Current flux-linkage characteristic for an ideal inductor.

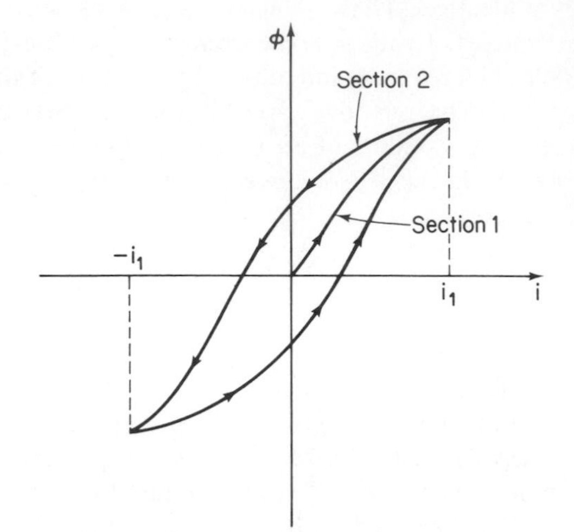

Fig. 4-6.3 Current flux-linkage characteristic for a nonlinear inductor with hysterisis.

function of time. For this case, the relation of (4–49) may be written in the form

$$\phi(t) = L(t)i(t) \tag{4–51}$$

If we differentiate both sides of the above equation, then, using (4–50), we see that

$$v(t) = L(t)\frac{di}{dt} + i(t)\frac{dL}{dt} \tag{4–52}$$

If the inductor is not time-varying, dL/dt equals 0, and the above equation takes the form shown in (4–48).

A *non-linear* inductor is characterized by a locus on the i–ϕ plane which is not a straight line and/or does not pass through the origin. A type of non-linear behavior which is encountered frequently is that shown in the i–ϕ plane of Fig. 4–6.3. This locus was generated by connecting a current generator to a previously unexcited non-linear inductor, and by increasing the current from 0 to some value i_1. The characteristic generated on the i–ϕ plane is labeled as Section 1 of the curve shown in Fig. 4–6.3. If the current is now returned to 0, then as shown in Section 2 of the curve, the flux linkages do not go to zero! If this at first seems surprising, the reader should note that this is exactly the type of behavior exhibited by a "permanent" magnet, i.e., it exhibits a flux even though there is no current flow. Additional variations of current between the limits of $-i_1$ and $+i_1$ generate the remaining portions of the characteristic curve in the order indicated by the arrows. This phenomenon is referred to as *hysteresis*. It is of considerable significance in determining the properties of magnetic devices such as solenoids, magnetic cores, magnetic tapes, transformers, and motors.

In the real-life world, all inductors demonstrate non-linear, time-varying behavior. In some applications, such as storage of information in computers, it is the non-linearity itself which provides the mechanism by means of which the desired operation is accomplished. Our primary interest in this text, however, will be with networks whose operation is based on the linear properties of inductors. Thus, we are lead to consider an "ideal" element which obeys the relations given in Eqs. (4–46) through (4–50). From this point on, when we use the word "inductor," we imply a lumped, linear, time-invariant model of some physical inductor, unless otherwise specified.

The power and energy considerations for an inductor are quite similar to those developed for a capacitor. For the power, we obtain

$$p(t) = i(t)v(t) = i(t)L\frac{di}{dt} \tag{4–53}$$

From this relation, we see that the instantaneous power supplied to an inductor may be positive or negative. Thus, at a given instant of time, an inductor may be receiving energy from the circuit to which it is connected, or supplying energy to that circuit. The total energy stored in an inductor is found from the relation

$$\begin{aligned} w(t) &= \int_{-\infty}^{t} p(\tau)\,d\tau = \int_{-\infty}^{t} Li(\tau)\frac{di}{d\tau}\,d\tau = \int_{i(-\infty)}^{i(t)} Li(t)\,di \\ &= \frac{Li^2(t)}{2} - \frac{Li^2(-\infty)}{2} \end{aligned} \tag{4–54}$$

The current though the inductor at $t = -\infty$, however, must be zero. Therefore, we may write $i(-\infty) = 0$. Thus, the development given above simplifies to

$$w(t) = \frac{Li^2(t)}{2} \tag{4–55}$$

From this relation, we see that the energy stored in a positive-valued inductor is always non-negative. Thus, such an inductor is a passive network element.[8] It is possible to define a negative-valued or active inductor just as we did for the resistive and capacitive cases. Unlike these latter cases, however, little practical usage has been found for the negative-valued inductor, and it will not be discussed further in this text. Like the capacitor, the inductor is a non-dissipative element, i.e., all the energy supplied to an inductor may be reclaimed, and none of it is converted to heat.

As a result of the duality which exists for the inductor and the capacitor, we may readily extend the conclusions covering the application of various simple waveshapes for a capacitor to the case of an inductor. For example, if a step input of voltage is applied to an inductor, a ramp of current results. The circuit configuration

[8]As was pointed out for the capacitor, a positive-valued, linear, *time-varying* inductor, however, may be active.

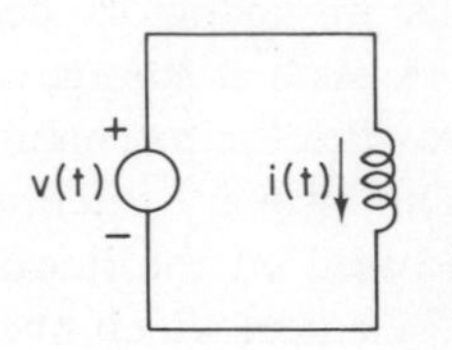

Fig. 4-6.4 An inductor excited by a voltage source.

for such an experiment is shown in Fig. 4–6.4. The waveshapes of voltage and current (assuming that the initial condition is zero) are shown in Fig. 4–6.5(a). Note that these are similar to the waveshapes shown in Fig. 4–3.2(a) for the case where a step input of current was applied to an unexcited capacitor, except for the fact that the voltage and current waveshapes are interchanged. Similarly, the waveshapes that result for the situations where an impulse of voltage and a ramp of voltage, respectively, are applied to an inductor are shown in Figs. 4–6.5(b) and 4–6.5(c). Note

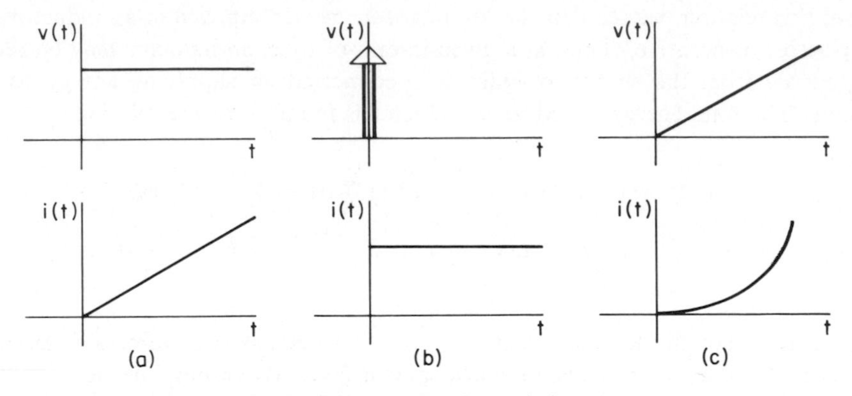

Fig. 4-6.5 Waveshapes for an inductor.

that an impulse of voltage creates a discontinuous function of current. This is readily verified from the derivative relationship given in (4–48). In the real-life world, impulses of voltage cannot exist; thus, we may state a *continuity condition* for inductors exactly paralleling the continuity condition developed for capacitors in Sec. 4–3. The condition may be summarized as follows:

SUMMARY 4-6.1

Continuity Condition for Inductors: The current which flows through a linear time-invariant inductor must always be a continuous function. This assumes that the voltage across the inductor is bounded, i.e., that impulses of voltage are not permitted.

We may also express the continuity condition by the relationship

$$i_L(t_0-) = i_L(t_0) = i_L(t_0+) \tag{4-56}$$

where $i_L(t)$ is the current flowing through an inductor and t_0 is any specific value of t. The notation follows that used for Eq. (4-27). Let us see what happens when we try to violate the continuity condition using the circuit shown in Fig. 4-6.6. We will

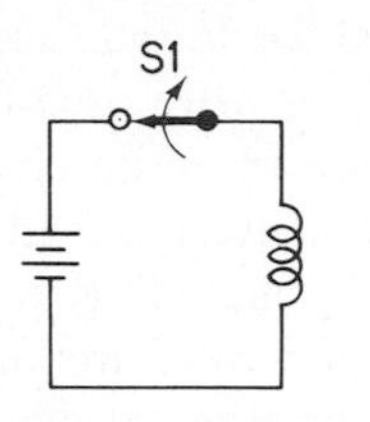

Fig. 4-6.6 Violating the continuity condition.

assume that the battery (a non-ideal source) is producing a constant current flow through the inductor (the details of how such a current flow originates will be treated in the next chapter). Now assume that the switch $S1$ shown in the figure is opened. From the circuit diagram it appears as if this action would instantaneously set the current to zero, thus producing an impulse of voltage across the terminals of the inductor. If such an experiment is actually attempted, however, an arc will be observed across the switch contacts as they are opened. The following is a simplified explanation of the phenomenon: As the switch begins to open and the current begins to change, the voltage induced in the inductor appears across the switch contacts. This voltage is large enough to provide an ionized path which permits the current to continue to flow until the gap between the switch contacts becomes so large that the arc is extinguished. The whole process takes place very quickly, but the net effect is that the current flow is stopped gradually (here "gradually" means in a small fraction of a second); thus the continuity condition is not violated. In effect, the process described above happens every time any switch is opened, since there is some inductance present even in straight wires.

Another application of the dual nature of the equations relating the terminal variables of inductors and capacitors may be found from the fact that the identical numerical techniques used to solve definite and indefinite integral or derivative problems involving a single capacitor may be directly applied to solve similar problems with a single inductor. An example follows.

EXAMPLE 4-6.1. A voltage defined as $v(t) = 20(1 - \cos 2\pi t)$ V is applied to the terminals of a 1-H inductor. The current through the inductor at $t = 0$ is 0. The numerical techniques developed in the preceding section may be applied to find the current at 50 successive values of time from 0.1 to 5 s. The program for performing this is identical with the one shown in Fig. 4-5.2, which was used for Example 4-5.1. The resulting plot is identical with the one shown in Fig. 4-5.3 if the identifiers for the "A" locus and the "B" locus are interchanged. Since the voltage $v(t)$ is expressed as a relatively simple function, we may also solve this problem by direct integration using (4-47). Thus, we see that $i(t) = 20[t - (\frac{1}{2}\pi) \sin 2\pi t]$.

The values of inductance which are frequently encountered in most circuit analyses range from the self inductance of a short length of straight wire (this is sometimes called a *lead inductance*), which may be a few microhenries, to the values found in power transformers and large filtering chokes, which may be several henries. As was the case for resistors and capacitors, however, in our analyses we shall, in general, use values which are conveniently normalized to lie near unity. Just as was the case for real-world capacitors, real-world inductors always have some dissipation. This is the result of the resistance of the wire used for winding the coil of the inductor. Also, when an inductor is used in an ac circuit, there is additional dissipation caused by core losses in the magnetic material on which the wire is wound. These effects are sometimes modeled by including resistors in series or parallel with the (ideal) inductor. In general, inductors have proportionately more dissipation than do capacitors, thus, a greater error may be introduced in an analysis due to the neglect of these effects. Successively higher levels of accuracy in a given analysis may be obtained by first including models for the dissipation in the inductors and then including models for the dissipation in capacitors.

4-7 SERIES AND PARALLEL COMBINATIONS OF CAPACITORS AND INDUCTORS

In this section we shall discuss the situations that occur when capacitors or inductors are connected in various series and parallel combinations. There are several interesting situations that arise when such connections are made. First of all, let us consider the case defined by the interconnection of capacitors in which all the initial conditions are zero.

In Fig. 4-7.1 a *parallel* connection of two capacitors of value C_1 and C_2 farads is shown. In terms of the variables defined in the figure, we may write the relations

$$i_1(t) = C_1 \frac{dv}{dt} \qquad i_2(t) = C_2 \frac{dv}{dt} \tag{4-57}$$

Application of KCL to the current variables shown in the figure shows that $i(t) = i_1(t) + i_2(t)$. Therefore, using the relations of Eq. (4-57), we may write

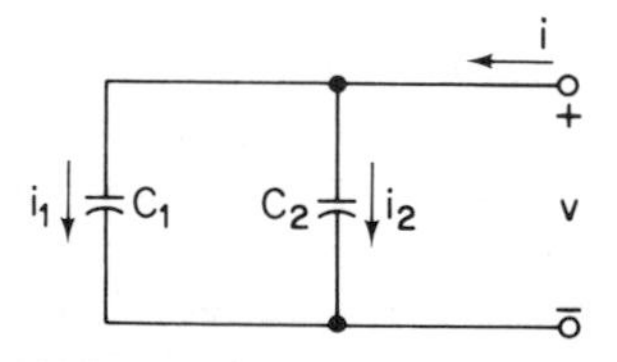

Fig. 4-7.1 A parallel connection of two capacitors.

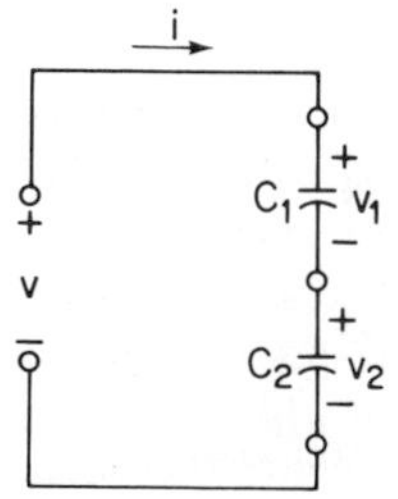

Fig. 4-7.2 A series connection of two capacitors.

$$i(t) = (C_1 + C_2)\frac{dv}{dt} \tag{4-58}$$

From the above relation, we may conclude that if two uncharged capacitors C_1 and C_2 are connected in parallel, they may be replaced by a single (uncharged) capacitor of value $C_1 + C_2$, without any noticeable difference in the relation that exists between the overall terminal variables $v(t)$ and $i(t)$. This conclusion is easily extended to the case where an arbitrary number of capacitors are connected in parallel. Thus, we may conclude that if n uncharged capacitors of value C_i farads ($i = 1, 2, \ldots, n$) are connected in parallel, they may be replaced by a single equivalent capacitor of value C_{eq}, where

$$C_{eq} = \sum_{i=1}^{n} C_i \tag{4-59}$$

Now let us consider the *series* connection of two uncharged capacitors as shown in Fig. 4-7.2. In terms of the variables defined in the figure, we may write

$$v_1(t) = \frac{1}{C_1}\int_0^t i(\tau)\,d\tau \qquad v_2(t) = \frac{1}{C_2}\int_0^t i(\tau)\,d\tau \tag{4-60}$$

Applying KVL to the circuit, we see that $v(t) = v_1(t) + v_2(t)$. Therefore, using the relations of (4-60) we may write

$$v(t) = \left(\frac{1}{C_1} + \frac{1}{C_2}\right)\int_0^t i(\tau)\,d\tau = \frac{C_1 + C_2}{C_1 C_2}\int_0^t i(\tau)\,d\tau \tag{4-61}$$

Thus, we conclude that two uncharged capacitors connected in series may be replaced by a single capacitor of value $C_1C_2/(C_1 + C_2)$. This relation is more easily expressed by using values of reciprocal capacitance, i.e., elastance (the units are darafs). If the first capacitor is of value S_1 darafs and the second is of value S_2 darafs, then the resulting equivalent capacitor has an elastance of value $S_1 + S_2$ darafs. This conclusion is easily extended to the case where an arbitrary number of capacitors are connected in series. Namely, if n uncharged capacitors of value S_i darafs ($i = 1, 2, \ldots, n$) are connected in series, they may be replaced by a single equivalent capacitor of value S_{eq}, where

$$S_{eq} = \sum_{i=1}^{n} S_i \tag{4-62}$$

The corresponding relation for the case where the values of the n capacitors are expressed as C_i farads ($i = 1, 2, \ldots, n$) is

$$\frac{1}{C_{eq}} = \sum_{i=1}^{n} \frac{1}{C_i} \tag{4-63}$$

The relations given above are easily combined to determine the value of simple series-parallel combinations of uncharged capacitors. An example follows:

EXAMPLE 4-7.1. The value C_{eq} of an equivalent capacitor which may be used to replace the four uncharged capacitors shown in Fig. 4-7.3 is easily found by making appropriate replacements of series and parallel capacitors. Thus, the parallel combination of C_1 and C_2 may be replaced by a single equivalent capacitor of 3 F. The series combination of this capacitor and C_3 is equivalent to a single capacitor of $\frac{6}{5}$ F, and the parallel combination of the latter with C_4 yields an equivalent capacitance for the entire network of $\frac{11}{5}$ F.

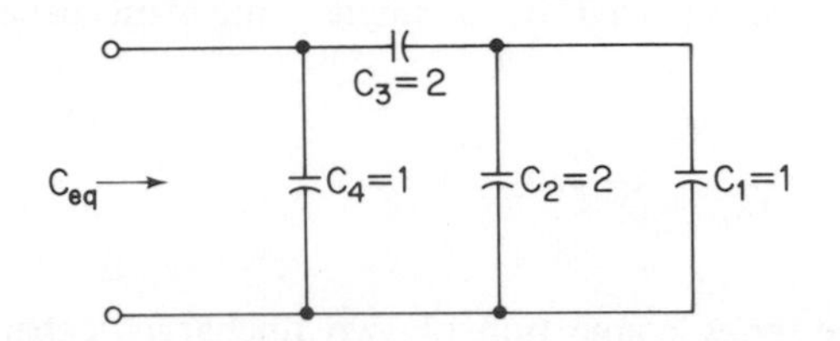

Fig. 4-7.3 A network of capacitors.

The method described in the above example is always applicable to capacitor networks which have the form of a ladder. Some techniques which may be applied to find the equivalent capacitance of more complicated networks may be found in the problems at the end of this section.

The discussion given in the above paragraphs is restricted to the case of uncharged capacitors, i.e., ones in which the initial condition is zero. We will now make some remarks about the interconnection of charged capacitors. First let us consider the parallel connection of two capacitors, one charged, the other uncharged, as shown in Fig. 4-7.4(a). The connection is made by means of the switch $S1$ shown

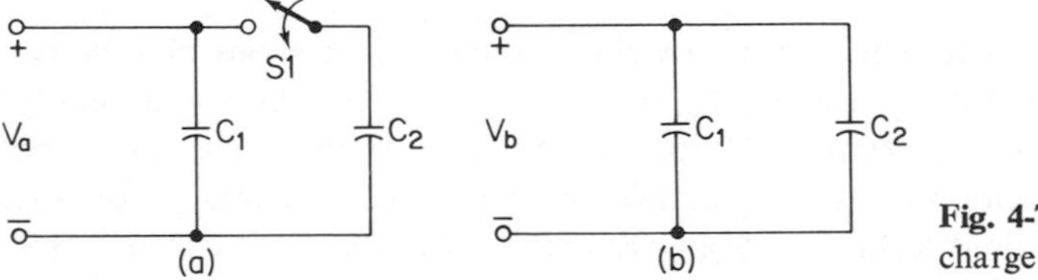

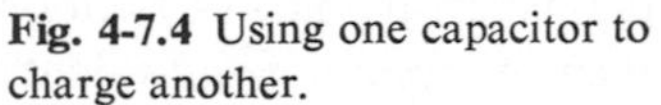
Fig. 4-7.4 Using one capacitor to charge another.

in the figure. Before the switch is closed, let us assume that the capacitor labeled C_1 has been charged to V_a volts. Thus, it has $Q_0 = C_1V_a$ coulombs of charge stored on its plates. When the switch is closed, since the two capacitors are effectively paralleled, the voltage across both must be the same. Let this voltage be V_b as shown in Fig. 4-7.4(b). The total charge that is to be found on the plates of both capacitors must still be Q_0 since charge is always conserved (see Sec. 1-1). Therefore, we may calculate the resulting voltage V_b by the relation

$$V_b = \frac{Q_0}{C_1 + C_2} = \frac{C_1V_a}{C_1 + C_2} \tag{4-64}$$

It should be noted that the closing of the switch presents a discontinuous waveshape of voltage on capacitor C_2. Thus, from our discussion in Sec. 4-3, an impulse of current must flow into the capacitor. We conclude that the readjustment of voltage takes place instantaneously if we are considering ideal capacitors. In the case of physical capacitors, there will always be some resistance present, and the transfer of energy will take place in a non-zero (but still very small) interval of time. Some other interesting facts about the values of the variables that result during this operation may be found in Problem 4–38 at the end of this chapter.

Now let us consider the behavior of inductors when these elements are connected in various series and parallel combinations. We shall see that the results are very similar to those previously discussed for the capacitor. First we will treat the case where the inductors are "uncharged," i.e., the initial condition of current through them is zero. In Fig. 4–7.5, a *series* connection of two inductors of value L_1 and L_2 henries is shown. In terms of the variables defined in the figure, we may write the relations

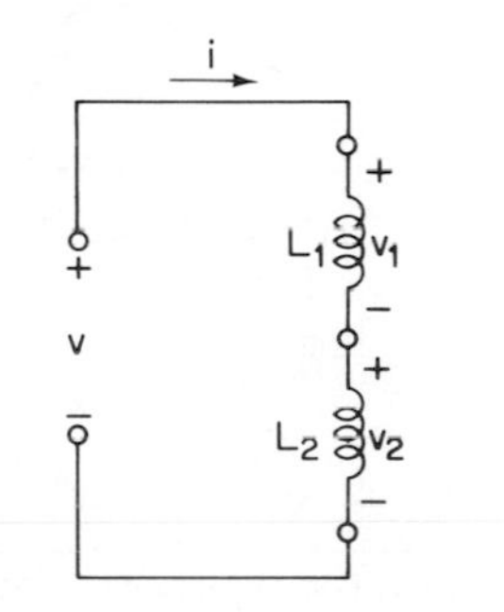

Fig. 4-7.5 A series connection of two inductors.

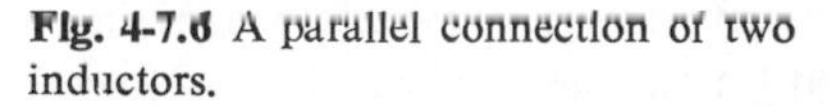

Fig. 4-7.6 A parallel connection of two inductors.

$$v_1(t) = L_1 \frac{di}{dt} \qquad v_2(t) = L_2 \frac{di}{dt} \tag{4–65}$$

From KVL, we see that $v_1(t) = v_1(t) + v_2(t)$. Therefore, using the relations of (4–65), we may write

$$v(t) = (L_1 + L_2) \frac{di}{dt} \tag{4–66}$$

From (4–66) we may conclude that two inductors connected in series and each having a zero initial condition may be replaced by a single equivalent inductor of value $L_1 + L_2$. Extending this result, we conclude that if n uncharged inductors of value L_i henries ($i = 1, 2, \ldots, n$) are connected in series, they may be replaced by a single equivalent inductor of value L_{eq}, where

$$L_{eq} = \sum_{i=1}^{n} L_i \tag{4–67}$$

Now let us consider the *parallel* connection of two uncharged inductors as shown in Fig. 4–7.6. In terms of the variables defined in the figure, we may write

$$i_1(t) = \frac{1}{L_1}\int_0^t v(\tau)\,d\tau \qquad i_2(t) = \frac{1}{L_2}\int_0^t v(\tau)\,d\tau \tag{4-68}$$

From KCL we see that $i(t) = i_1(t) + i_2(t)$. Therefore, using the relations of (4–68), we may write

$$i(t) = \left(\frac{1}{L_1} + \frac{1}{L_2}\right)\int_0^t v(\tau)\,d\tau = \frac{L_1 + L_2}{L_1 L_2}\int_0^t v(\tau)\,d\tau \tag{4-69}$$

Thus, we conclude that when uncharged inductors are connected in parallel, their reciprocal values of inductance add. In general, we see that if n uncharged inductors of value Γ_i reciprocal henries ($i = 1, 2, \ldots, n$) are connected in parallel, they may be replaced by a single equivalent inductor of value Γ_{eq} reciprocal henries, where

$$\Gamma_{eq} = \sum_{i=1}^{n} \Gamma_i \tag{4-70}$$

The corresponding relation for the case where the values of the inductors is expressed in henries is

$$\frac{1}{L_{eq}} = \sum_{i=1}^{n} \frac{1}{L_i} \tag{4-71}$$

The relations given above may be readily applied to find the equivalent inductance of many simple combinations of uncharged inductors in a manner similar to that used for the case of capacitors. Note that the relations for combining inductors are dual to those for combining capacitors, i.e., inductors *in series* add their individual values (in henries), while capacitors in *parallel* add their individual values (in farads). A summary of the properties of series and parallel connections of resistors, capacitors, and inductors is given in Table 4–7.1.

TABLE 4-7.1 Series and Parallel Connections of Resistors, Capacitors, and Inductors

Type of Element	Value of Elements	Units Used for Value	Total Value for a Series Connection	Total Value for a Parallel Connection
resistor	R_i G_i	ohms mhos	$R_{total} = \sum R_i$	$G_{total} = \sum G_i$
capacitor	C_i S_i	farads darafs	$C_{total} = \sum C_i$	$S_{total} = \sum S_i$
inductor	L_i Γ_i	henries reciprocal henries	$L_{total} = \sum L_i$	$\Gamma_{total} = \sum \Gamma_i$

In treating the case where inductors which have non-zero initial conditions are interconnected, we recall first that an inductor with an initial condition must have its terminals shorted, since there must be a closed path for the non-zero current to follow. The voltage appearing across such a short circuit will, of course, be zero. Now consider a circuit consisting of two inductors and a switch $S1$ which is initially closed, as shown in Fig. 4-7.7(a). Let us assume that the inductor of value L_1

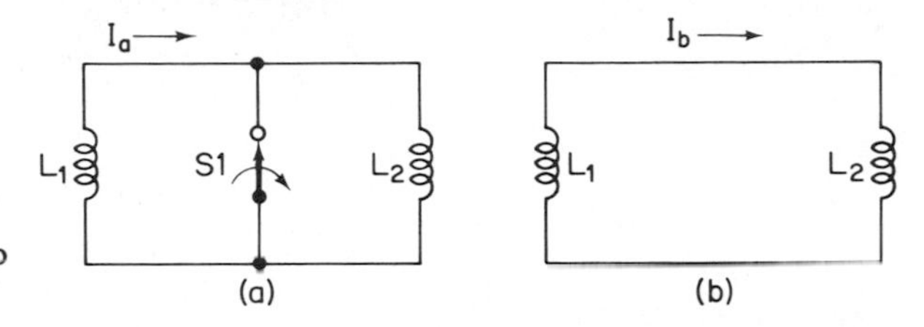

Fig. 4-7.7 Using one inductor to charge another.

henries has a non-zero initial current of value I_a flowing through it. In addition, let us assume that the inductor of value L_2 henries is uncharged. Thus, the current flowing through the (closed) switch $S1$ has the value I_a. From (4-51) we see that there must be $\phi_0 = L_1 I_a$ flux linkages with the coil of the charged inductor. When the switch $S1$ is opened, as shown in Fig. 4-7.7(b), the two inductors are connected in series. Thus, the current through both of them must be the same. Let this current be I_b as shown in the figure. The total flux linkages that exist for both of the inductors must still be ϕ_0, since flux linkages are conserved in a magnetic circuit, just as charge is conserved in a capacitive circuit (see Sec. 1-1). Therefore, we may determine the resulting current I_b by the relation

$$I_b = \frac{\phi_0}{L_1 + L_2} = \frac{L_1 I_a}{L_1 + L_2} \tag{4-72}$$

In a manner similar to that explained for the capacitor, the opening of the switch will produce an impulse of voltage across the terminals of the two inductors (since the current through the inductors is discontinuous). Some other interesting properties of this circuit may be found in Problem 4-39 at the end of this chapter.

4-8 MUTUAL INDUCTANCE—THE TRANSFORMER

In Sec. 3-11 we introduced three multi-terminal network elements: the differential-input operational amplifier, the NIC, and the gyrator. In addition to being characterized as having more than two terminals, these devices all have the property that their voltage and current variables are related by *algebraic* equations. In this section, we shall discuss a quite different multi-terminal device in which the voltage and current variables are related by *differential* equations. Such a device is called a set of coupled coils or, more generally, a *transformer*. Its principles of operation are a logical extension of the discussions pertaining to the inductor given in Sec. 4-6.

To begin our study of a pair of coupled coils, let us consider first the single coil shown in Fig. 4–8.1. The symbol ϕ_1 has been used for the flux linkages of this coil. The positive reference direction for this quantity follows the usual right-hand rule relating current and flux.[9] If we assume this coil is linear and time-invariant, the relation between the current $i_1(t)$ and the flux linkages $\phi_1(t)$ is

$$\phi_1(t) = L_1 i_1(t) \tag{4–73}$$

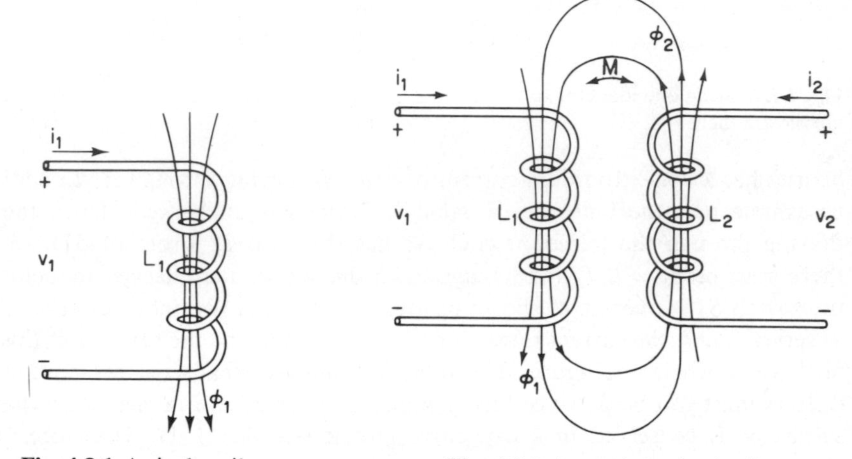

Fig. 4-8.1 A single coil.

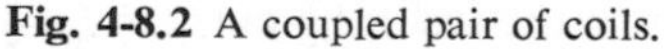
Fig. 4-8.2 A coupled pair of coils.

where the quantity L_1 is called the *self inductance* of the coil. Now consider the effect of adding a second coil in the vicinity of the first as shown in Fig. 4–8.2. Some of the lines of flux produced by the current in this now coil will also link the first coil. Thus, the flux linkages $\phi_1(t)$ of the first coil are now determined by the current $i_2(t)$ as well as by the current $i_1(t)$. If we again assume that the relations between flux and current are linear and time-invariant, we may write an expression for $\phi_1(t)$ for this circuit as follows:

$$\phi_1(t) = L_1 i_1(t) + M_{12} i_2(t) \tag{4–74}$$

where L_1 is the self inductance of coil 1 and the quantity M_{12} is called the *mutual inductance* of coil 1 with respect to coil 2. The unit for both the self inductance and the mutual inductance is the henry. The voltage $v_1(t)$ appearing across the terminals of coil 1 is easily found by applying Faraday's law. Thus, we have

[9]The right-hand rule, as formulated in basic physics courses, specifies that if the fingers of the right hand are closed around a coil in a direction which follows the positive reference direction of current in the winding, the thumb will point in the positive reference direction for the fiux.

$$v_1(t) = \frac{d\phi_1}{dt} = L_1 \frac{di_1}{dt} + M_{12} \frac{di_2}{dt} \tag{4–75}$$

Now let us consider the second coil shown in Fig 4–8.2. The flux $\phi_2(t)$ linking this coil is produced by the current $i_1(t)$ as well as the current $i_2(t)$. Thus, we may write

$$\phi_2(t) = M_{21}i_1(t) + L_2 i_2(t) \tag{4–76}$$

where L_2 is the self inductance of coil 2 and M_{21} is the mutual inductance of coil 2 with respect to coil 1. The voltage $v_2(t)$ appearing across the terminals of coil 2 is given by the expression

$$v_2(t) = \frac{d\phi_2}{dt} = M_{21} \frac{di_1}{dt} + L_2 \frac{di_2}{dt} \tag{4–77}$$

The equations of (4–75) and (4–77) determine the relations between the voltage and current variables for the multi-terminal network element comprised of the pair of coupled coils shown in Fig. 4–8.2. We shall examine these relations in more detail in the following paragraphs.

As a first investigation of the properties of the pair of coupled coils shown in Fig. 4–8.2, we shall use energy considerations to prove that the mutual inductance terms M_{12} and M_{21} are equal. To show this, let us connect a current source to the terminals of coil 1 and raise the current $i_1(t)$ from 0 to I_1 over the period $0 \le t < t_1$ as shown in Fig. 4–8.3(a). It is assumed that the change of current is made without producing any discontinuities in $i_1(t)$. The reason for this is to avoid any resultant impulses in $v_1(t)$. If we assume, as shown in Fig. 4–8.3(b), that the current $i_2(t)$ is

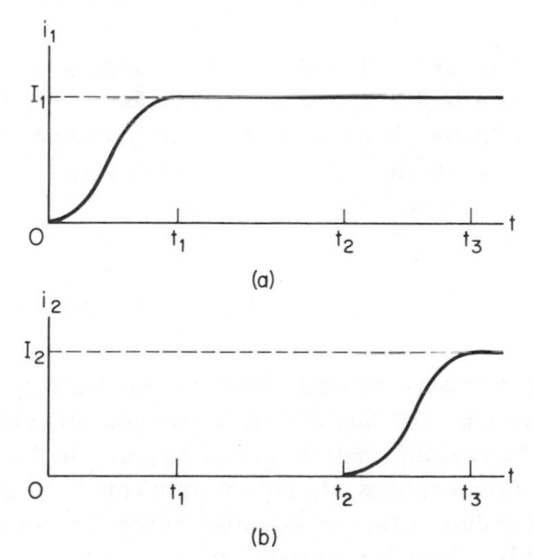

Fig. 4-8.3 Waveshapes in a pair of coupled coils.

zero over this time interval, we may compute the total energy W_a stored in coil 1 at time t_1 as follows.

$$W_a = \int_0^{t_1} v_1(t) i_1(t)\, dt = \int_0^{I_1} L_1 i_1\, di_1 = \tfrac{1}{2} L_1 I_1^2 \tag{4–78}$$

Now consider the time interval $t_1 \leq t \leq t_2$. During this period, both $v_1(t)$ and $v_2(t)$ are zero, so no energy is supplied to the coils. Now, leaving $i_1(t)$ constant at the value I_1, let us raise $i_2(t)$ from 0 to some value I_2 over a time period $t_2 \leq t \leq t_3$, as shown in Fig. 4–8.3(b). During this interval, since $i_2(t)$ is changing and $i_1(t)$ is constant, the voltages $v_1(t)$ and $v_2(t)$ are found from (4–75) and (4–77) to be

$$v_1(t) = M_{12} \frac{di_2}{dt} \qquad v_2(t) = L_2 \frac{di_2}{dt} \tag{4–79}$$

Thus, the energy W_b from both current sources supplied to the coils during this period is

$$\begin{aligned} W_b &= \int_{t_2}^{t_3} v_1(t) i_1(t)\, dt + \int_{t_2}^{t_3} v_2(t) i_2(t)\, dt \\ &= \int_0^{I_2} I_1 M_{12}\, di_2 + \int_0^{I_2} L_2 i_2\, di_2 \\ &= M_{12} I_1 I_2 + \tfrac{1}{2} L_2 I_2^2 \end{aligned} \tag{4–80}$$

The total energy W supplied to the pair of coupled coils over the time range $0 \leq t \leq t_3$ is simply the sum of (4–78) and (4–80). Thus, we may write

$$W = W_a + W_b = \tfrac{1}{2} L_1 I_1^2 + M_{12} I_1 I_2 + \tfrac{1}{2} L_2 I_2^2 \tag{4–81}$$

Now let us conduct an experiment similar to the one described above, again starting from a zero energy state. This time we will first raise the current $i_2(t)$ to a value I_2 and then raise the current $i_1(t)$ to the value I_1. Typical waveforms for such a process are shown in Fig. 4–8.4. An energy analysis directly parallel to that given above shows that in this case the total energy W supplied over the time period $0 \leq t \leq t_3$ is

$$W = \tfrac{1}{2} L_1 I_1^2 + M_{21} I_1 I_2 + \tfrac{1}{2} L_2 I_2^2 \tag{4–82}$$

In both of the cases described above, the final energy supplied to the network must be the same since the final values of $i_1(t)$ and $i_2(t)$ are the same, namely, I_1 and I_2. Thus, the expressions given in Eqs. (4–81) and (4–82) must be equal. This can only be true if the quantities M_{12} and M_{21} are equal. Although we have arrived at this conclusion from a particular example, it may be shown to be a general property of coupled coils. Therefore, from this point in the text we will drop the subscript nota-

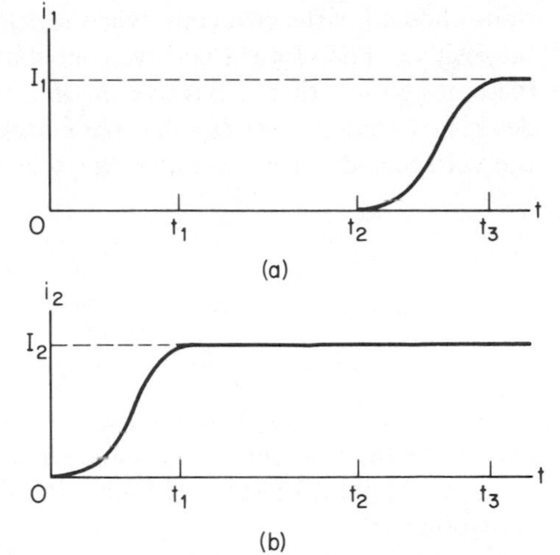

Fig. 4-8.4 Waveforms in a pair of coupled coils.

tion and simply refer to the quantity M as the mutual inductance of a pair of coupled coils with respect to each other. Thus, an expression for the energy stored in such a pair of coils may be written in the form

$$w(t) = \tfrac{1}{2} L_1 i_1^2(t) + M i_1(t) i_2(t) + \tfrac{1}{2} L_2 i_2^2(t) \tag{4-83}$$

The mutual inductance M discussed above may be positive or negative. To see this, consider the circuit shown in Fig. 4–8.5(a). To emphasize the path followed

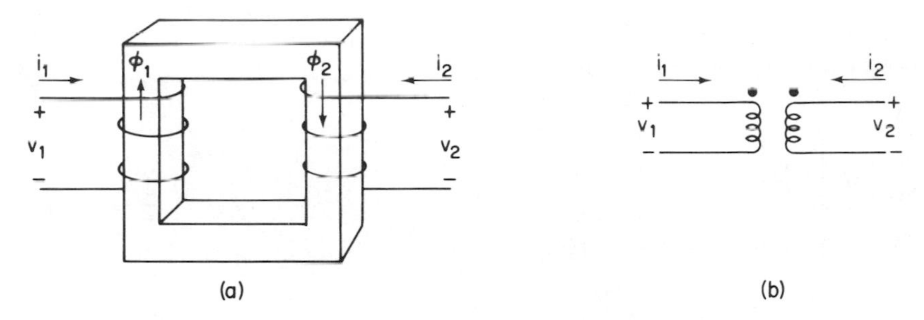

Fig. 4-8.5 A pair of coupled coils with positive mutual inductance.

by the lines of magnetic flux in this circuit, we have shown the coils as being wound on a core of some high-permeability magnetic material. The usual associated reference directions have been used for the voltage and current variables of each winding, and the reference direction for the fluxes follow the right-hand rule. First observe that in coil 1, if di_1/dt is positive, $d\phi_1/dt$ will also be positive. In addition, of course, $v_1(t)$ will be positive because the self inductance L_1 of coil 1 is always positive. Now consider the effect of the quantity di_2/dt on $d\phi_1/dt$. Because of the winding directions chosen for the two coils, when di_2/dt is positive, $d\phi_1/dt$ as well as $d\phi_2/dt$ will be positive. Therefore, a positive contribution to the voltage $v_1(t)$ will result. Similar reasoning shows that a positive di_1/dt or di_2/dt will produce a positive $v_2(t)$. To describe such a case we say that *the mutual inductance is positive.* The relations for the voltage and current variables for this case are

$$\begin{aligned} v_1(t) &= L_1 \frac{di_1}{dt} + |M| \frac{di_2}{dt} \\ v_2(t) &= |M| \frac{di_1}{dt} + L_2 \frac{di_2}{dt} \end{aligned} \tag{4-84}$$

where we have emphasized the positive nature of the mutual inductance M by using magnitude signs. It should be noted that, in order for the mutual inductance to be positive, the positive reference direction for $\phi_1(t)$ and $\phi_2(t)$ must be such as to aid each other.

Frequently, coupled coils are completely encapsulated or are constructed in such a way as to make it very difficult to actually observe the winding directions of the coils. In such a case, if we apply a positive increasing current $i_1(t)$, i.e., if we make di_1/dt positive and if we then find that $v_2(t)$ is also positive (assuming di_2/dt is zero), we can conclude that the mutual inductance M is positive. To indicate this, it is customary to use the symbol shown in Fig. 4–8.5(b) in which dots are used to indicate the relative coil winding directions. The significance of these dots is that a positive increasing current into the dot of one coil produces a voltage at the terminals of the other coil such that the dotted terminal is positive.

To illustrate the effect of a negative mutual inductance, consider the circuit shown in Fig. 4–8.6(a). In this circuit, if di_1/dt is positive, a positive $d\phi_1/dt$ and,

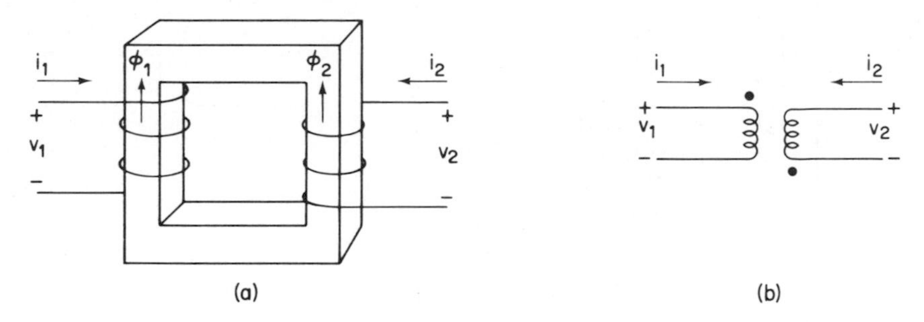

Fig. 4-8.6 A pair of coupled coils with negative mutual inductance.

thus, a positive $v_1(t)$ will be produced as a result. However, applying the right-hand rule, we see that a positive di_2/dt will produce a negative $d\phi_1/dt$ and thus a negative contribution to the voltage $v_1(t)$. Similarly, a positive di_1/dt will produce a negative contribution to $v_2(t)$. To describe this situation, we say that the mutual inductance is negative. The equations relating the voltage and current variables now have the form

$$v_1(t) = L_1 \frac{di_1}{dt} - |M| \frac{di_2}{dt}$$
$$v_2(t) = -|M| \frac{di_1}{dt} + L_2 \frac{di_2}{dt} \qquad (4\text{–}85)$$

It should be noted that for negative mutual inductance, the positive reference directions for the two fluxes $\phi_1(t)$ and $\phi_2(t)$ oppose each other in the magnetic path. The circuit symbol for the case of negative mutual inductance is shown in Fig. 4–8.6(b). The dot convention used here has the identical significance as it had in Fig. 4–8.5(a). In general, it should be noted that *the dot positions are determined by the actual winding direction of the coil* and thus they are invariant for a given set of coupled coils. The sign of the mutual inductance, however, *is determined by the relative reference directions chosen for the voltage and current variables at the terminals of the two coils.* Thus, we may reverse the sign of the mutual inductance by reversing the voltage and current reference directions at the terminals of either coil. As an example of this, note that the coupled coils shown in Fig. 4–8.7(a) have the same

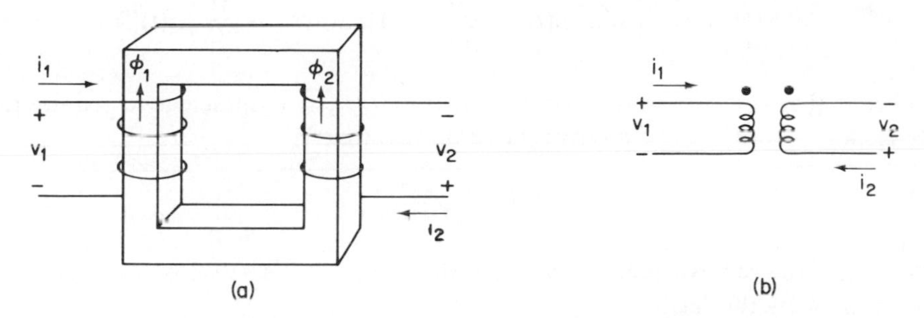

Fig. 4-8.7 A pair of coupled coils with negative mutual inductance.

winding configuration as those shown in Fig. 4–8.5(a). However, the reference directions for the variables $v_2(t)$ and $i_2(t)$ have been reversed. In this case, the positive reference direction for the two fluxes are such that the two fluxes oppose each other; thus, the mutual inductance is negative. The dot positions, however, are still the same and the coil may, therefore, be represented by the circuit symbol in Fig. 4–8.7(b), which is the same as that used in Fig. 4–8.5(b). The results given above may be summarized as follows:

SUMMARY 4-8.1

Coupled Coils: The properties of a pair of coupled coils as shown in Fig. 4–8.2 is defined by specifying the values of the self-inductance of each coil and the mutual inductance which couples them. The relations between the voltage and current variables are

$$v_1(t) = L_1 \frac{di_1}{dt} + M \frac{di_2}{dt}$$

$$v_2(t) = M \frac{di_1}{dt} + L_2 \frac{di_2}{dt}$$

The mutual inductance M may be positive or negative. A pair of dots is used to identify the relative winding directions. If the positive reference current directions both flow into (or both out of) the dotted terminals of the two coils, the mutual inductance is positive; otherwise it is negative.

The magnitude of the value of the mutual inductance found for a given set of coils has a definite upper bound which is determined by the values of the self inductances L_1 and L_2. This is a result of the fact that the total energy stored in a set of coupled coils can never be negative. To investigate this upper bound, let us write the expression for the energy $w(t)$ given in Eq. (4-83) in a different form by completing the square in the variable $i_1(t)$. We obtain

$$w(t) = \frac{1}{2}\left\{\left[\sqrt{L_1}\, i_1(t) + \frac{M}{\sqrt{L_1}} i_2(t)\right]^2 + \left(L_2 - \frac{M^2}{L_1}\right) i_2^2(t)\right\} \tag{4-86}$$

To insure that $w(t)$ be positive for all values of $i_1(t)$, it is sufficient to require that the term $L_2 - (M^2/L_1)$ be positive (or zero), i.e., that[10]

$$|M| \leq \sqrt{L_1 L_2} \tag{4-87}$$

To indicate the relative amount by which this inequality is satisfied, we may define a constant k by the relation

$$k = \frac{|M|}{\sqrt{L_1 L_2}} \tag{4-88}$$

The constant k is called the *coefficient of coupling*. It is a measure of the amount of flux generated by a current flowing in one coil which links the turns of the other coil. If this amount is small, k is small and we say that the coils are *loosely coupled*. On the other hand, if *all* the flux generated by one coil links the turns of the other, we say that the coils are *perfectly coupled*. In this case, the coefficient of coupling k is unity and the mutual inductance M is equal to the geometric mean of the self-inductances L_1 and L_2. This case is covered in more detail in the following paragraphs.

To determine the properties of a perfectly coupled set of coils, consider a set of coils in which $\phi(t)$ is the flux that would link a one-turn coil located anywhere in the flux path. Let us also assume that the mutual inductance is positive. If we let n_1

[10]Actually this condition is necessary as well as sufficient, as may be seen by putting (4-86) in the standard form of a parabola.

be the number of turns in coil 1 and n_2 be the number of turns in coil 2, then $\phi_1(t)$ and $\phi_2(t)$, the flux linkages with coil 1 and coil 2, respectively, are given as

$$\phi_1(t) = n_1\phi(t) \qquad \phi_2(t) = n_2\phi(t) \tag{4-89}$$

The voltages $v_1(t)$ and $v_2(t)$ are determined by Faraday's law and have the form

$$v_1(t) = \frac{d\phi_1}{dt} = n_1\frac{d\phi}{dt} \qquad v_2(t) = \frac{d\phi_2}{dt} = n_2\frac{d\phi}{dt} \tag{4-90}$$

Taking the ratio of these two latter equations, we obtain

$$\frac{v_2(t)}{v_1(t)} = \frac{n_2}{n_1} \tag{4-91}$$

Thus, we see that for a perfectly coupled set of coils, and for a positive mutual inductance, the ratio of the voltages at the terminals of the two coils is directly proportional to the ratio of the turns in the windings of the coils. Now let us assume that the magnetic permeability of the core on which the two coils are wound is infinite. This, of course, implies that the self inductances of the two coils as defined in Sec. 4-6 are also infinite. Since the magnetic reluctance $\mathscr{R}$ is inversely proportional to the permeability, it will be zero. This means that the "Ohm's law" for a magnetic circuit becomes

$$\text{mmf} = \phi\mathscr{R} = 0 \tag{4-92}$$

The magnetomotive force or mmf, however, is simply the sum (if the mutual inductance was negative, it would be the difference) of the mmf's which are applied to the two coils. This has the form

$$\text{mmf} = n_1 i_1(t) + n_2 i_2(t) = 0 \tag{4-93}$$

Thus, for the infinite permeability case, we may write the ratio of the two coil currents as

$$\frac{i_2(t)}{i_1(t)} = -\frac{n_1}{n_2} \tag{4-94}$$

Thus, the ratio of the magnitudes of the coil currents is inversely proportional to the magnitude of the turns ratio. The expressions given in Eqs. (4-91) and (4-94) determine the relations between the terminal variables of voltage and current for a perfectly coupled set of coils with infinite magnetic permeability and positive mutual inductance. Such a multi-terminal device is called an *ideal transformer*. If the mutual inductance is negative, similar relations apply. However, a minus sign is introduced into each of the equations. The results given above are summarized as follows:

SUMMARY 4-8.2

The Ideal Transformer: An ideal transformer may be considered as a pair of coupled coils in which the coefficient of coupling is unity and in which the self and mutual inductances are infinite. As a circuit device, it may be treated as a linear, time-invariant, multi-terminal element in which the relations between the terminal variables of voltage and current are defined by

$$\frac{v_2(t)}{v_1(t)} = \frac{n_2}{n_1} \qquad \frac{i_2(t)}{i_1(t)} = -\frac{n_1}{n_2}$$

where n_1 and n_2 are the turns of the windings numbered 1 and 2.

The ideal transformer is a very important multi-terminal network element since it models the actual performance of many physical transformers quite closely. Such a model can be made even more accurate by adding parasitic elements to represent some of the additional physical phenomena which are associated with an actual transformer.

Frequently it is desirable, for analysis purposes, to replace a pair of coupled coils by an equivalent circuit. As an example of such a circuit, consider the "T" of inductors shown in Fig. 4–8.8(a). In terms of the indicated variables, applying KVL we may write

$$\begin{aligned} v_1(t) &= (L_1 - M)\frac{di_1}{dt} + M\frac{d}{dt}[i_1(t) + i_2(t)] \\ v_2(t) &= (L_2 - M)\frac{di_2}{dt} + M\frac{d}{dt}[i_2(t) + i_1(t)] \end{aligned} \tag{4–95}$$

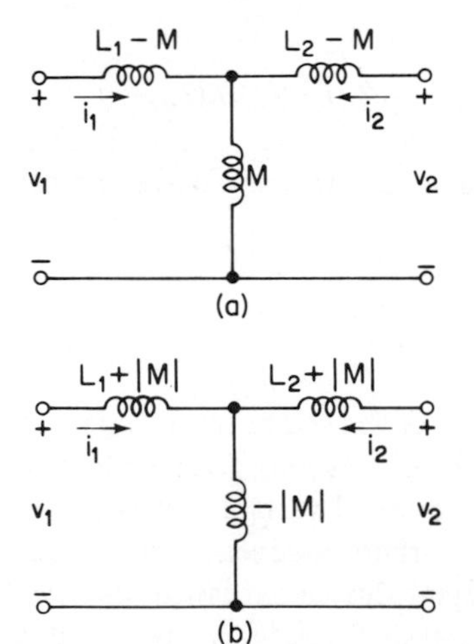

Fig. 4-8.8 A T-equivalent circuit for a pair of coupled coils.

Rearranging terms we obtain

$$v_1(t) = L_1 \frac{di_1}{dt} + M \frac{di_2}{dt}$$
$$v_2(t) = M \frac{di_1}{dt} + L_2 \frac{di_2}{dt} \qquad (4\text{–}96)$$

If M is positive, then these equations are the same as those given in (4–84), thus we conclude that this T of inductors may be used to replace a set of coupled coils. Since the mutual inductance M may be greater than L_1 or L_2 (but not greater than both), it is possible for one of the inductors in such an equivalent circuit to be negative, however this in no way affects the validity of the method. A similar T of inductors which may be used to model a set of coupled coils in which the mutual inductance is negative is shown in Fig. 4–8.8(b). In the figure we have used magnitude signs to emphasize the parity of the various terms. Analysis of this circuit yields a set of equations identical to those given in Eq. (4–85). It should be noted that the T-equivalent circuits described above fail to model the behavior of a set of coupled coils in one important respect, i.e., they do not provide the physical isolation which the coupled coils do. Such an isolation may, however, be added by using an ideal 1 : 1 turns ratio transformer at either end of the network. Some other examples of circuits which provide isolation may be found in Problem 4–43.

4-9 CONCLUSION

The material contained in this section opens a whole new series of challenges in our study of circuit theory. The integral and differential equations that relate the terminal variables of capacitors and inductors require the development of an entirely new set of techniques for the analysis of circuits containing such elements. These new techniques, however, will not replace the analysis methods that we presented for resistance networks in Chap. 3. Rather, they will supplement and generalize them. Thus, in the chapters that follow, we shall make frequent applications of Kirchhoff's laws to enable us to write network equations. The results of such applications, however, will be simultaneous sets of integro-differential equations rather than the simultaneous sets of algebraic equations which were encountered for resistance networks. In general, some entirely new techniques will have to be developed to enable us to solve such sets of integro-differential equations. For certain special situations, however, such as the sinusoidal steady-state one discussed in Chap. 7, we shall see that solution methods using determinants and Gauss-Jordan reduction, which are almost identical to those originally presented in Chap. 3, may be applied to analyze networks containing capacitors and inductors.

Problems

Problem 4–1 (Sec. 4–1)

The voltage across a capacitor of value C farads is described by the waveform shown in Fig. P4-1. Sketch the current and charge as functions of time.

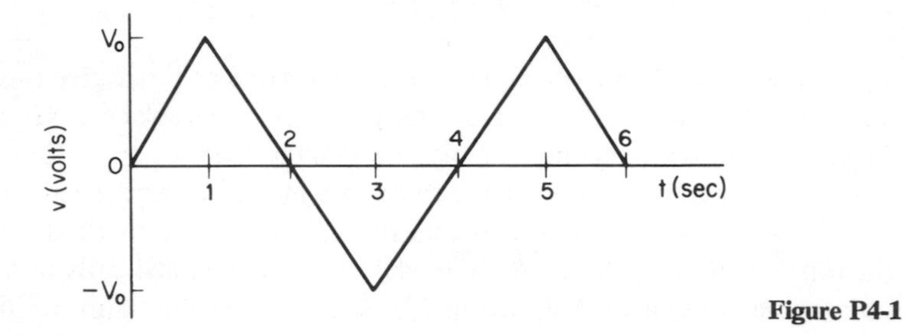

Figure P4-1

Problem 4–2 (Sec. 4–1)

If the voltage across a 5-F capacitor has the waveform drawn in Fig. P4-2, find the current through the capacitor.

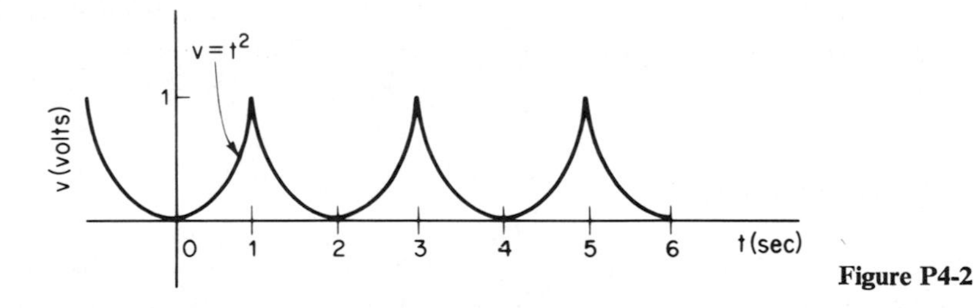

Figure P4-2

Problem 4–3 (Sec. 4–1)

If the current through a $\frac{1}{4}$-F capacitor is described by the waveform shown in Fig. P4-3, find the voltage across the capacitor.

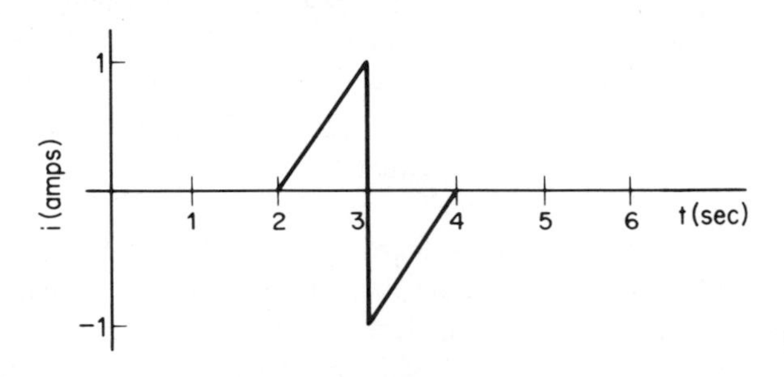

Figure P4-3

Problem 4–4 (Sec. 4–1)

Prove that the energy stored in a capacitor is equal to $w(t) = \frac{1}{2}[q^2(t)/C]$ by using the relation $w(t) = \int_{-\infty}^{t} v(t)i(t)\,dt$.

Problem 4–5 (Sec. 4–1)

Let the voltage across a particular capacitor be equal to $v(t) = \frac{1}{4}q^4(t)$, where $q(t)$ is the charge. Find an expression for $w(q)$, the energy stored.

Problem 4–6 (Sec. 4–1)

The circuit shown in Fig. P4-6 has reached steady-state conditions (the current to the capacitor is zero) with the switch in position 1. If the switch is thrown to position 2, and the circuit again reaches a steady-state condition, find the value of the total energy dissipated in all the intervening time in the circuit at the right (consisting of the 500 kΩ resistor and the 5-V source).

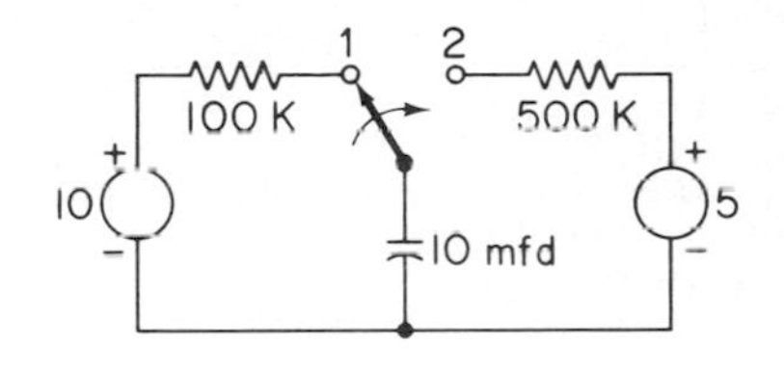

Figure P4-6

Problem 4–7 (Sec. 4–1)

A time-varying capacitor of value $C(t) = 3 + t$ F is excited by a voltage source having the value $v(t) = \sin 2t$ V. Find an expression for the input current $i(t)$ to the capacitor.

Problem 4–8 (Sec. 4–1)

For the network shown in Fig. P4-8, (a) apply KVL and write a set of mesh current equations, and (b) apply KCL and write a set of node equations.

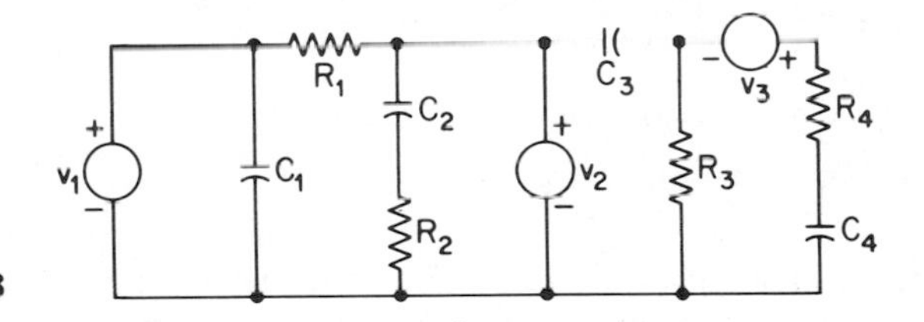

Figure P4-8

Problem 4–9 (Sec. 4–1)

Repeat Problem 4-1 for the case where the capacitor has a negative value $-|C|$ farads.

Problem 4–10 (Sec. 4–2)

Write a mathematical expression describing each of the functions shown in Fig. P4-10.

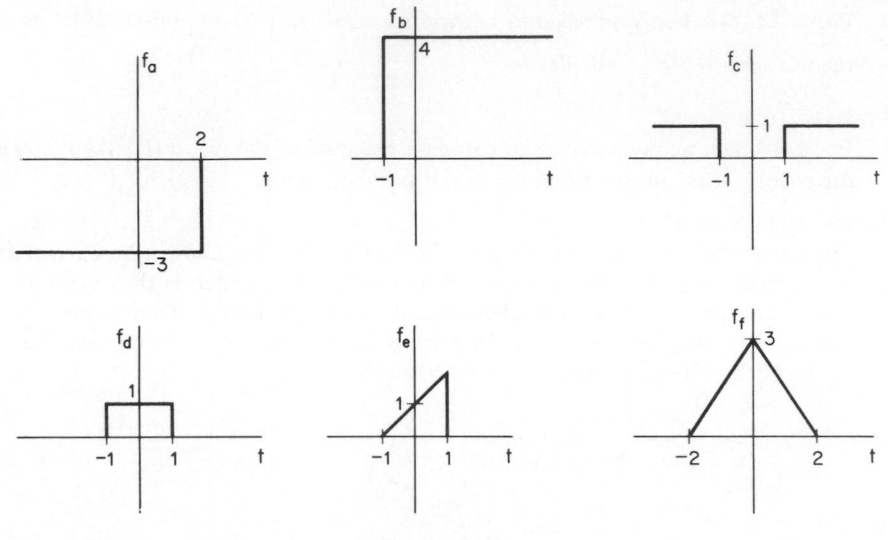

Figure P4-10

Problem 4–11 (Sec. 4–2)

Write mathematical expressions which describe the derivatives of the functions shown in Fig. P4-10.

Problem 4–12 (Sec. 4–2)

Write mathematical expressions which describe the integrals of the functions shown in Fig. P4-10, parts (b), (d), (e), and (f).

Problem 4–13 (Sec. 4–2)

Construct a sketch of each of the following functions.

(a) $f_a(t) = 2u(t + 3)$.

(b) $f_b(t) = 3u(3 - t)$.

(c) $f_c(t) = u(t - 2) + 2(t - 1)u(t - 1)$.

(d) $f_d(t) = u(t^2 - 1)$.

Problem 4–14 (Sec. 4–3)

If the waveshapes shown in Fig. P4-14 are applied as input currents to an uncharged 1-F capacitor, plot the resulting voltage $v(t)$ that appears across the terminals of the capacitor for each of the cases.

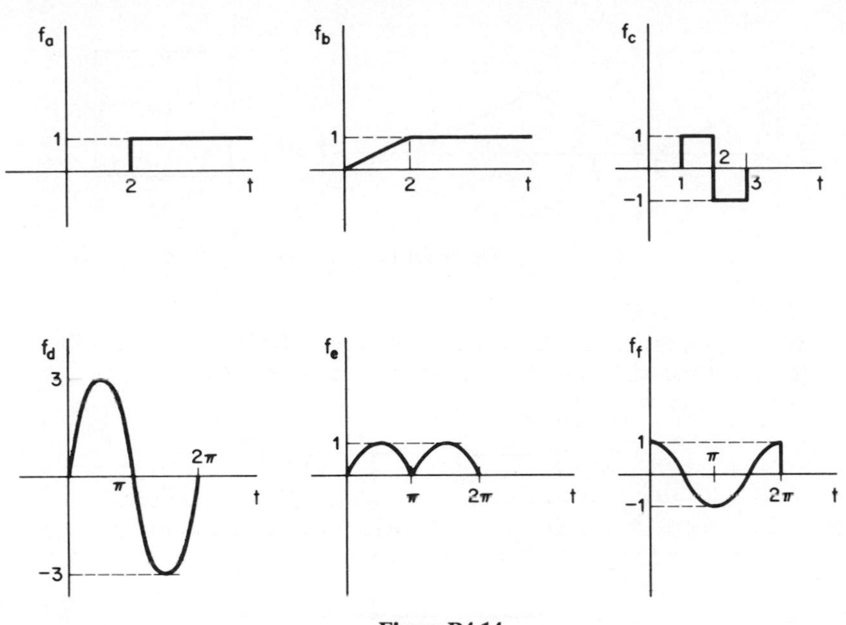

Figure P4-14

Problem 4–15 (Sec. 4–3)

If the waveshapes shown in Fig. P4-15 are applied as input voltages to the terminals of a 2-F capacitor, plot the resulting current $i(t)$ that flows into the capacitor in each case.

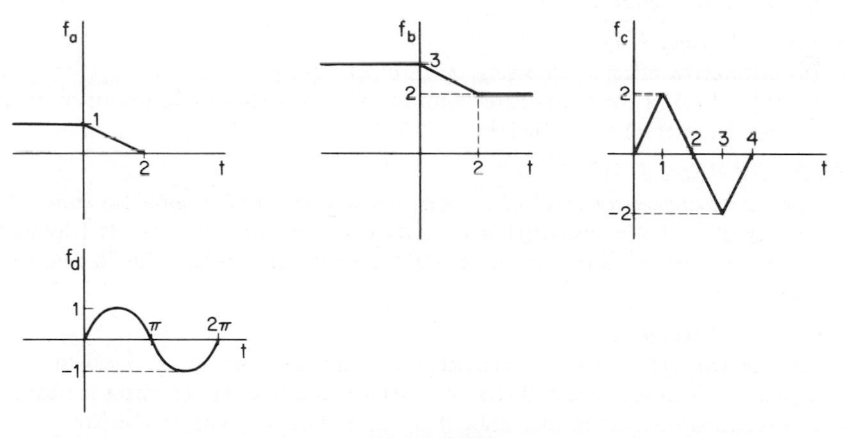

Figure P4-15

Problem 4–16 (Sec. 4–3)

The capacitor in the circuit shown in Fig. P4-16 varies with time as shown, but the voltage across the capacitor remains constant at 1 V. Sketch $i(t)$.

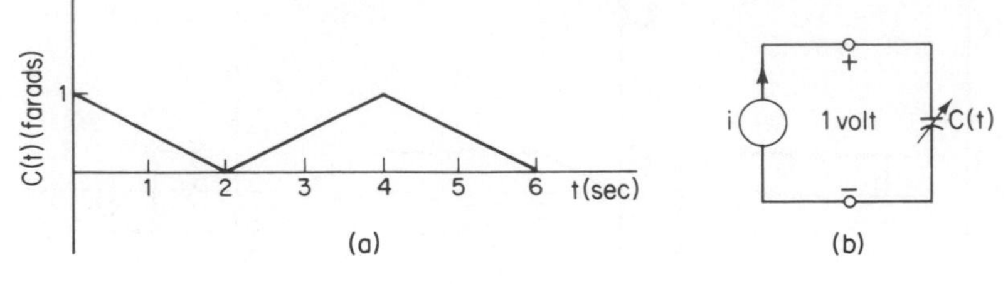

Figure P4-16

Problem 4–17 (Sec. 4–3)

The switch in the circuit shown in Fig. P4-17 is closed at $t = 0$. It is found that $v_c(0+) = 0$ and $(dv_c/dt)(0+) = 10$. What is the value of C?

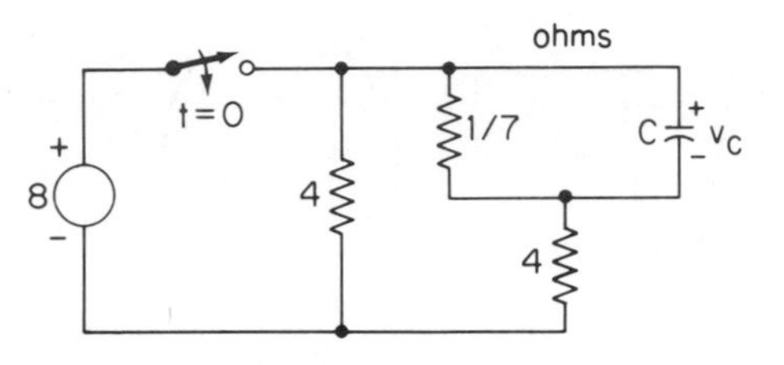

Figure P4-17

Problem 4–18 (Sec. 4–4)*[11]

Investigate the effect of the number of iterations used in trapezoidal integration on the accuracy of the resulting answer by integrating the function $f(x) = x^{3/2} - x^{5/2}$ over the period $0 \leq x \leq 1$ for the following numbers of iterations: 10, 20, 50, 100, and 200. (Compute the actual value of the integral by using analytical, i.e., non-numerical techniques.)

Problem 4–19 (Sec. 4–4)*

Investigate the effect of choosing every other data point for the function used in Example 4-4.3 (8 points compared with the 15 points used in the example) in determining the integral of the function.

Problem 4–20 (Sec. 4–4)*

Plot the absolute error involved in using various numbers of iterations from 2 to 40 in integrating the function $f(t) = x^2 \sin \pi x$ over the period $0 \leq t \leq 1$. (Compute a reference value of the integral using 200 iterations and use this value in determining the error.)

Problem 4–21 (Sec. 4–5)*

The current $i(t) = 0.09(2 - t) \cos \pi t$ mA is applied starting at $t = 0$ to a 1-μF capacitor which has no initial charge. Plot the resulting voltage across the capacitor as a function of time for the period $0 \leq t \leq 2$. Use 51 points in the plot.

Problem 4–22 (Sec. 4–5)*

A non-linear resistor has a value of resistance which is determined by the current flowing through it. Thus, $R = R(i)$. The relation between R and i is specified by

[11]Problems which require the use of digital computational techniques are identified by an asterisk.

the data points listed in Fig. P4-22. If a current $i(t) = 0.1t$ A is passed through such a resistor, plot the energy $w(t)$ dissipated in the resistor over a range of time from 0 to 0.5 s. Use 51 points in the plot. Scale the output by 60 for plotting.

I (AMPS)	R(I) (OHMS)
0.00000000E 00	0.00000000E 00
2.00000000E-03	2.00480000E 03
4.00000000E-03	2.01920000E 03
6.00000000E-03	2.04320000E 03
8.00000000E-03	2.07680000E 03
1.00000000E-02	2.12000000E 03
1.20000000E-02	2.17280000E 03
1.40000000E-02	2.23520000E 03
1.60000000E-02	2.30720000E 03
1.80000000E-02	2.38880000E 03
2.00000000E-02	2.48000000E 03
2.20000000E-02	2.58080000E 03
2.40000000E-02	2.69120000E 03
2.60000000E-02	2.81120000E 03
2.80000000E-02	2.94080000E 03
3.00000000E-02	3.08000000E 03
3.20000000E-02	3.22880000E 03
3.40000000E-02	3.38720000E 03
3.60000000E-02	3.55520000E 03
3.80000000E-02	3.73280000E 03
4.00000000E-02	3.92000000E 03
4.20000000E-02	4.11680000E 03
4.40000000E-02	4.32320000E 03
4.60000000E-02	4.53920000E 03
4.80000000E-02	4.76480000E 03
5.00000000E-02	5.00000000E 03

Figure P4-22

*Problem 4–23** (*Sec. 4–5*)

Use appropriate digital computer subprograms to construct a plot of the waveshape of the voltage in Example 4-4.3. Use 61 points to cover a time range from 0 to 12 ms. Scale the output data by 4.0.

*Problem 4–24** (*Sec. 4–5*)

Repeat Example 4-5.2 using 21 data points rather than the 11 given. Estimate values for the intermediate points from the plot given in Fig. 4-5.7. Compare your results with those given in the example.

Problem 4–25 (*Sec. 4–6*)

The waveshapes shown in Fig. P4-14 are applied as input voltages to a 1-H inductor. Draw the waveshapes of the resulting current. It is assumed that the inductor is initially uncharged.

Problem 4–26 (*Sec. 4–6*)

The waveshapes shown in Fig. P4-15 are applied as input currents to a 2-H inductor. Draw the resulting waveshapes of voltage that will appear at the terminals of the inductor for each of the cases shown.

Problem 4–27 (Sec. 4–6)

The current $i(t)$ in the circuit shown in Fig. P4-27 has the variation indicated. Find (a) $q(10)$, (b) $\phi(10)$, and (c) $v(10)$.

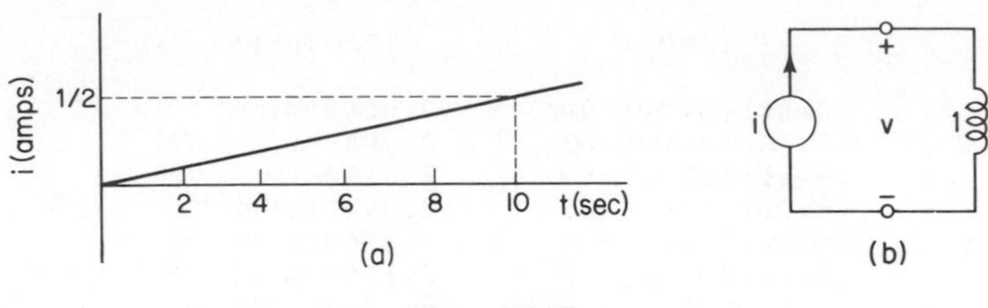

Figure P4-27

Problem 4–28 (Sec. 4–6)

The current through an inductor has the form $i(t) = 10 \sin [t(\pi/2)]$ A. Sketch $v(t)$ if $L = \frac{1}{2}$. Does the voltage sinusoid lead or lag the current?

Problem 4–29 (Sec. 4–6)

Let the current through a $\frac{1}{2}$-H inductor, as provided by a current source, be equal to 2A. Find the energy stored in the inductor. If the terminals of the inductor are suddenly short-circuited, what happens to the stored energy?

Problem 4–30 (Sec. 4–6)

Prove that the energy stored in an inductor is equal to $w(t) = \frac{1}{2}[\phi^2(t)/L]$ by using the relation $w(t) = \int_{-\infty}^{t} v(t)i(t)\,dt$.

Problem 4–31 (Sec. 4–6)

Switch $S1$ in the circuit shown in Fig. P4-31 closes at $t = 0$, and switch $S2$ simultaneously opens. Find $i(0+)$ and $(di/dt)(0+)$.

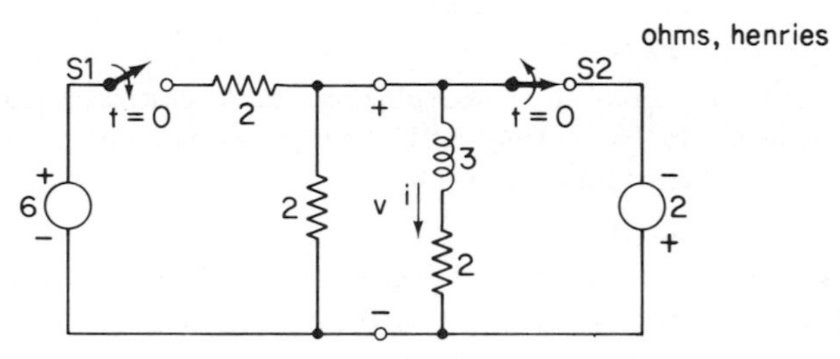

Figure P4-31

Problem 4–32 (Sec. 4–6)

For the network shown in Fig. P4-32, (a) apply KVL and write a set of mesh current equations, and (b) apply KCL and write a set of node equations.

Problem 4–33 (Sec. 4–6)

For the network shown in Fig. P4-33, (a) apply KVL and write a set of mesh current equations, and (b) apply KCL and write a set of node equations.

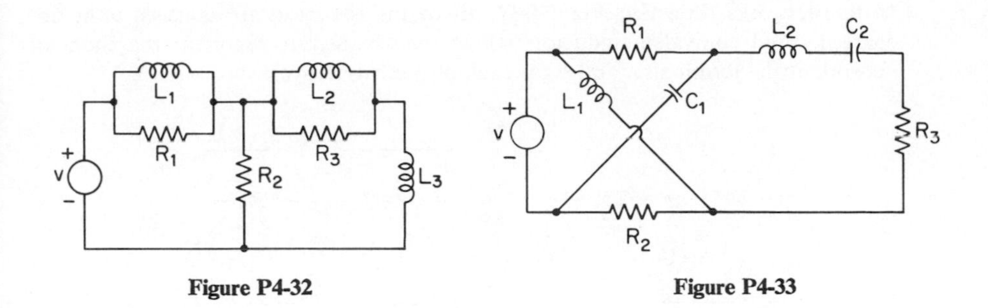

Figure P4-32

Figure P4-33

*Problem 4–34** (*Sec. 4–6*)

A voltage $v(t) = 3 + 3t$ volts is applied to a unity-valued inductor over the period of time $0 \leq t \leq 2$ s. Plot the voltage $v(t)$ as *C*, the current through the inductor $i(t)$ as *B*, and the power $p(t)$ supplied to the inductor as *A*. Assume the initial condition is zero. Use 51 points in the plot and scale the current, the voltage, and the power by 5, 10, and $\frac{1}{2}$, respectively. Verify the results at $t = 2$ s by actual conventional integration.

*Problem 4–35** (*Sec. 4–6*)

The subroutines ITRPZ and DINTG can be used in conjunction with each other to achieve multiple integration. To illustrate this, consider the case where a voltage $v(t) = 3 + 3t$ volts is applied to a unit-valued inductor and it is desired to find the resulting energy $w(t)$ stored in the inductor at $t = 2$ s. To do this we may use the subroutine ITRPZ to find the current $i(t)$ at intervals of 0.04 s over the range $0 \leq t \leq 2$ s. The subroutine DINTG may simultaneously be used to determine the total energy stored in the inductor by integrating a function defined by the computed values of power at these same times. Use this procedure to find $w(2)$, assuming that the initial current in the inductor $i(0)$ is zero. Verify the resulting value of stored energy by hand computation.

Problem 4–36 (*Sec. 4–7*)

In the networks shown in Fig. P4-36, all of the capacitors are assumed to be uncharged. Find equivalent capacitors which may be used to represent the capacitor networks at the terminals *a* and *b* for each of the two networks.

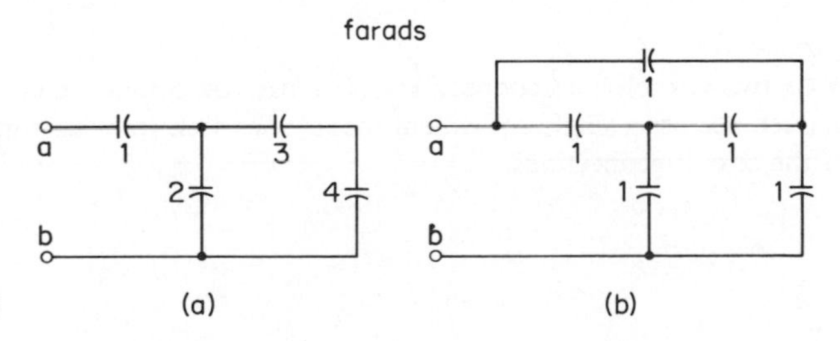

Figure P4-36

Problem 4–37 (*Sec. 4–7*)

In the networks shown in Fig. P4-37, all of the inductors are assumed to be uncharged. Find equivalent inductors which may be used to represent the inductor network at the terminals a and b for each of the two networks.

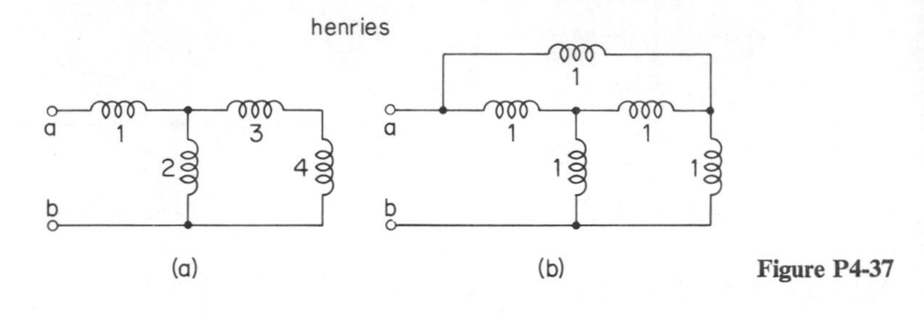

Figure P4-37

Problem 4–38 (*Sec. 4–7*)

In the circuit shown in Fig. P4-38, the capacitor C_1 is initially charged to 10 V and the capacitor C_2 is uncharged. The switch is closed at $t = 0$. Find (a) the voltage in the circuit after the switch is closed; (b) the energy stored in C_1 before the switch is closed; and (c) the total energy stored in C_1 and C_2 after the switch is closed. Explain any difference in the values found in parts (b) and (c).

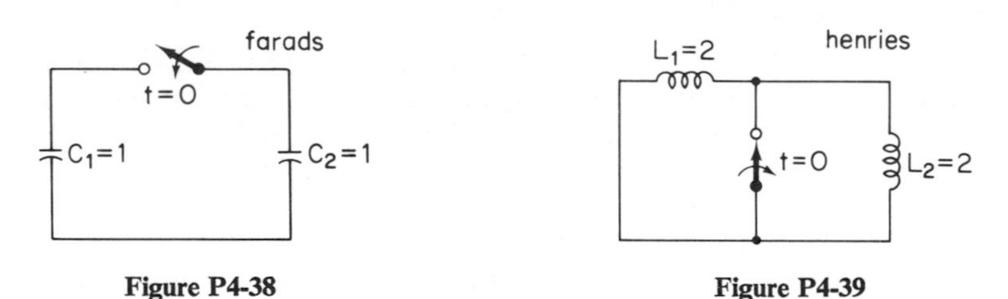

Figure P4-38 **Figure P4-39**

Problem 4–39 (*Sec. 4–7*)

The inductor L_1 shown in Fig. P4-39 is "charged" in the sense that a current of 1 A is flowing in the loop formed by it and the closed switch for values of $t < 0$. No current is flowing through L_2. The switch is opened at $t = 0$. Find (a) the current in the circuit after the switch is opened; (b) the energy stored in the inductor L_1 before the switch is opened; and (c) the total energy stored in both L_1 and L_2 after the switch is opened. Explain any difference in the values computed in parts (b) and (c).

Problem 4–40 (*Sec. 4–8*)

In Fig. P4-40, two possible interconnections of a pair of coupled coils are shown. Find the inductance for a single equivalent inductor as seen from terminals a and b for each of the circuit connections.

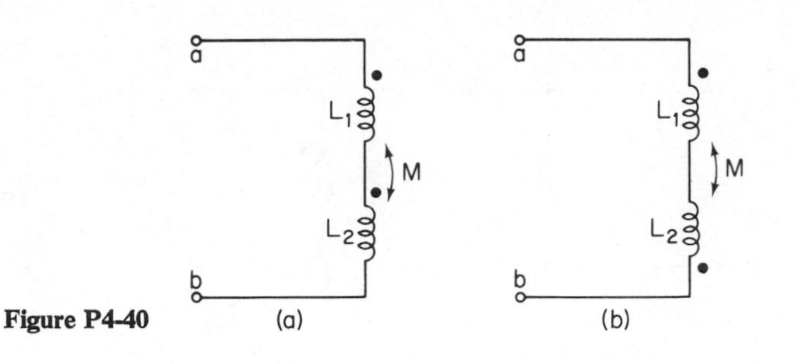

Figure P4-40

Problem 4-41 (*Sec. 4-8*)

For each of the circuits shown in Fig. P4-41, do the following:

(a) Write the equations indicating the sign of the mutual inductance between the various coils by using plus and minus signs and magnitude symbols.

(b) Identify the winding sense between coils 1 and 2 by dots, between coils 2 and 3 by squares, and between coils 1 and 3 by triangles.

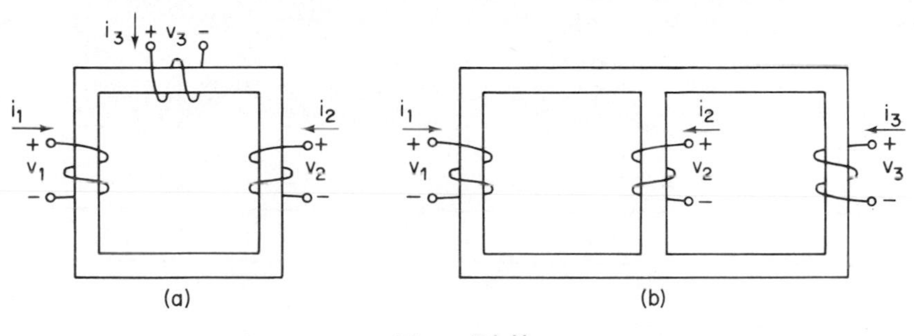

Figure P4-41

Problem 4-42 (*Sec. 4-8*)

Find an equivalent circuit representation for an ideal transformer which uses two controlled sources, one a VCVS, the other an ICIS.

Problem 4-43 (*Sec. 4-8*)

It is possible to replace a set of coupled coils with a "T" of inductors and an ideal transformer in such a way that the individual inductors are always positive-valued. Two such configurations are shown in Fig. P4-43. The circuits are drawn assuming a positive value of M. For a negative value of M it is only necessary to change the position of the dot on one of the windings of the ideal transformer. Verify the validity of the circuits.

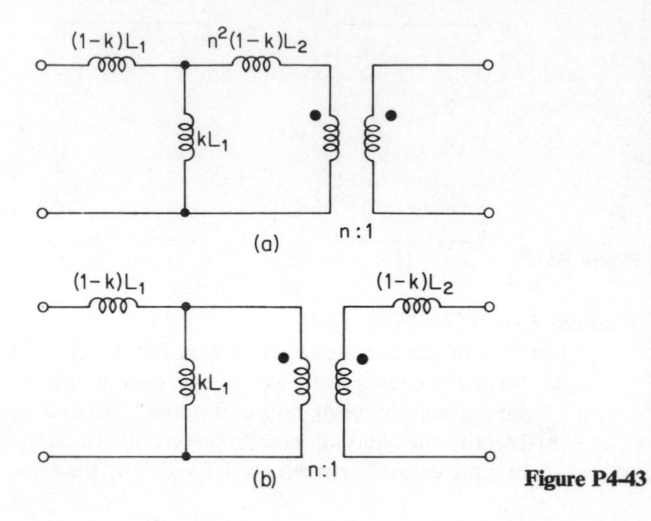

Figure P4-43

CHAPTER 5

First-Order Circuits

In the last chapter we showed that capacitors and inductors both have the property that they are capable of storing energy. In this chapter we shall introduce the study of circuits characterized by a single energy-storage element. Thus, the type of circuit to be discussed in this chapter is one which contains a single capacitor or a single inductor. In addition, the circuit may contain any number of resistors and sources. It will be shown that the equations describing such a circuit may be put in a form involving an unknown variable and its first derivative. Such an equation is referred to as a *first-order* differential equation, thus we shall refer to circuits which contain only a single energy-storage element as *first-order circuits*. The techniques which we shall discuss in this chapter will be ones that may readily be adapted to the study of higher-order circuits, i.e., circuits containing more than one energy-storage element. This will be done in succeeding chapters.

5-1 EXCITATION BY INITIAL CONDITIONS

If we consider a circuit in which all the variables are identically zero, we may say that such a circuit is *unexcited*. Conversely, circuits in which at least some of the variables are non-zero for some range of time are referred to as circuits which have been *excited*. In this chapter we shall consider the ways in which circuits can be excited and the results of such excitation. There are two basic ways of exciting a circuit. The first is by initial conditions, the second is by sources. In this section we shall study the effects of using initial conditions to excite a first-order circuit. Later in this chapter we shall study the effects of using sources.

As an example of a circuit which is excited only by initial conditions, consider the circuit consisting of a capacitor, a resistor, and a battery (a dc voltage source) shown in Fig. 5-1.1(a). If switch $S1$ is closed for all values of $t < 0$, then an initial

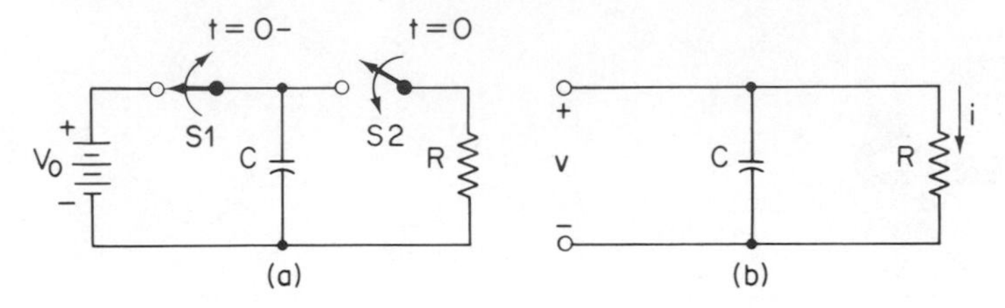

Fig. 5-1.1 A first-order *RC* circuit excited by an initial condition.

voltage V_0 is established on the capacitor by the battery. During this same period, it is assumed that switch $S2$ is open. If we open switch $S1$ just before $t = 0$, and close switch $S2$ at $t = 0$, then for all $t \geq 0$, we may ignore the battery shown in Fig. 5-1.1(a). Thus, for $t \geq 0$ we need only consider the circuit comprised of a resistor and a capacitor shown in Fig. 5-1.1(b). The initial condition on the capacitor voltage $v(t)$ for this circuit is $v(0) = V_0$. For the range of time $t \geq 0$, such a circuit may be considered as excited only by the initial condition which is present on the capacitor at $t = 0$. Applying KCL to this circuit, we obtain the differential equation

$$C\frac{dv}{dt} + \frac{1}{R}v(t) = 0 \qquad t \geq 0 \tag{5-1}$$

We may use the following words and phrases to describe this equation:

Ordinary This term applies since the differentiation is made with respect to only one independent variable, namely t. *Note:* A differential equation which has derivatives taken with respect to more than one independent variable is known as a *partial* differential equation rather than an ordinary differential equation.

First-order This term applies since the highest order derivative found in the equation is the first.

Linear This descriptive term tells us that the independent variable and its derivatives appear only as terms of the first degree, i.e., that there are no terms of the type $v^2(t)$ or $v(t) \times dv/dt$ (these are both second-degree terms).

Constant coefficients This term applies since the coefficients C and $1/R$ are constants, i.e., they are *not* functions of the variable t. *Note:* When the coefficients are functions of the variable t, they are referred to as *time-varying* coefficients.

Homogeneous This term applies since there is no term in the equation which does not include $v(t)$ or its derivative dv/dt. *Note:* An equation in

$v(t)$ and its derivatives which includes a term such as $g(t)$, where $g(t)$ may be a constant or some known function of t, is said to be *non-homogeneous*.

The complete description of Eq. (5–1) is thus: *an ordinary, first-order, linear, homogeneous, differential equation with constant coefficients*. This is too much for an author to write more than once (or for a reader to read more than once), so we shall assume that all the differential equations to be discussed in this text are ordinary, linear, and have constant coefficients unless otherwise stated. Thus we shall refer to Eq. (5–1) as a *first-order, homogeneous, differential equation.* Such an equation will always result when KVL or KCL is applied to a first-order circuit excited only by initial conditions. The first-order nature of the resulting differential equation comes from the fact that a single-storage element introduces only a single differential relation between the various branch variables of the circuit. The homogeneous nature of the resulting differential equation comes from the fact that, without sources, there can be no term which does not include some network variable or its derivative. Thus, if KCL and/or KVL is applied to a first-order circuit excited only by initial conditions, the equation that must be solved to determine any variable is a first-order homogeneous, differential equation.

The solution of a first-order, homogeneous, differential equation of the type given in Eq. (5–1) is easily found. First, for convenience, let us divide Eq. (5–1) by the quantity C. We obtain

$$\frac{dv}{dt} + \frac{1}{RC}v(t) = 0 \qquad t > 0 \tag{5–2}$$

We may employ a mathematical technique called *separation of variables* to solve the above equation. To do this, we first write Eq. (5–2) in the form

$$\frac{dv}{v} = \frac{-1}{RC}\,dt \qquad t \geq 0 \tag{5–3}$$

Integrating both sides of (5–3) we obtain

$$\ln v(t) = \frac{-1}{RC}t + K \qquad t \geq 0 \tag{5–4}$$

where ln is the natural logarithm. Now let us use both sides of (5–4) as an exponent for the quantity e, where e is the base of the natural logarithm. We obtain[1]

$$e^{\ln v(t)} = v(t) = e^{(K - t/RC)} = K'e^{-t/RC} \qquad t \geq 0 \tag{5–5}$$

where K' is a new constant equal, by definition, to e^K. This constant is easily given

[1]In this equation, we use the relation $e^{\ln x} = \ln(e^x) = x$.

a physical significance by noting that when $t = 0$, $v(0) = K'$. Thus, we may write (5–5) in the form

$$v(t) = v(0)e^{-t/RC} \qquad t \geq 0 \tag{5–6}$$

This last equation provides the solution for the first-order, homogeneous, differential equation given in Eq. (5–1). The fact that it is a solution is easily verified by substituting Eq. (5–6) into (5–1). This is left as an exercise for the reader.

The steps given above to find the solution of (5–1) are exactly the same for the solution of any other first-order, homogeneous, differential equation. Thus, we may generalize as follows:

SUMMARY 5-1.1

Solution of a First-Order Circuit Excited Only by Initial Conditions: A first-order circuit excited only by initial conditions is characterized by a first-order, homogeneous, differential equation having the form

$$\frac{df}{dt} + af(t) = 0 \tag{5–7}$$

The solution to this equation is

$$f(t) = f(0)e^{-at} \qquad t \geq 0$$

This result allows us to readily solve any first-order, homogeneous, differential equation which may occur in our studies of circuits. Two examples follow.

EXAMPLE 5-1.1. Assume that the circuit components shown in Fig. 5-1.1 have the following values: $C = 1\ \mu\text{F}$, $R = 10\ \text{k}\Omega$; and that $V_0 = v(0) = 10$ V. From (5-6) we see that the solution for the voltage $v(t)$ across the capacitor for $t \geq 0$ is $v(t) = 10e^{-100t}$ V (where t

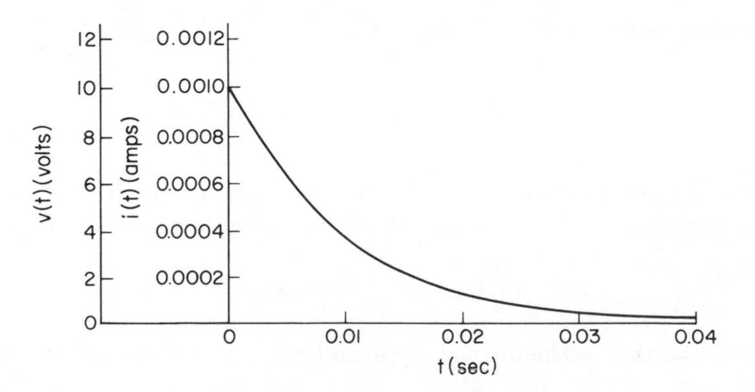

Fig. 5-1.2 Waveforms for Example 5-1.1.

is in seconds). The expression for the current $i(t)$ through the resistor is easily found using Ohm's law. Thus we see that $i(t) = 0.001e^{-100t}$ A. Plots of the waveshapes for $v(t)$ and $i(t)$ are given in Fig. 5-1.2. The scales for these plots have been chosen so as to make them collinear.

EXAMPLE 5-1.2. The circuit shown in Fig. 5-1.3 consists of an inductor with a resistor connected across its terminals. A switch $S1$ is also connected directly across the two terminals. It is assumed that the switch is closed for all $t < 0$. Thus, for this range of time the node-pair voltage $v(t)$ between the terminals must be zero, and consequently, no current will flow through the resistor. (From Ohm's law $v(t) = Ri_R(t)$, therefore if $v(t) = 0$ and $R \neq 0$, $i_R(t)$ must equal 0.) Let us assume that the inductor current, however, is non-zero.[2] Let this current be I_0. If we open the switch $S1$ at $t = 0$, then, for $t \geq 0$, we may apply KVL around the mesh consisting of the inductor and the resistor to obtain the first-order, homogeneous, differential equation. After dividing through by L, we obtain

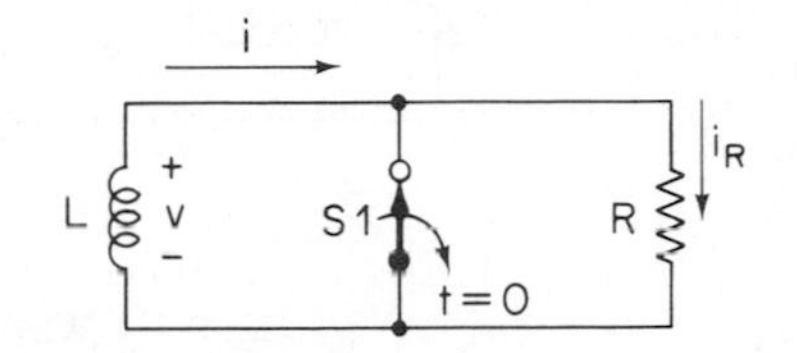

Fig. 5-1.3 A first-order *RL* circuit excited by an initial condition.

$$\frac{di}{dt} + \frac{R}{L}i(t) = 0 \qquad t \geq 0$$

The continuity condition for an inductor tells us that the current in the inductor at the instant the switch opens must be the same as it was just before the switch opened; thus, we see that $i(0) = I_0$. From the preceding summary, we find that the solution of the differential equation is

$$i(t) = I_0 e^{-Rt/L} \qquad t \geq 0$$

To make this example more concrete, suppose that $L = 1$ mH, $R = 10$ kΩ, and $I_0 = 0.001$ A. The solution for the inductor current is $i(t) = 0.001e^{-10^7 t}$ A (where t is in seconds). The voltage across the resistor can be found (from Ohm's law) to be $v(t) = 10e^{-10^7 t}$ V. This is also the voltage which appears across the inductor. An alternative means of finding the voltage across the inductor is to use the relation $v(t) = -L\,di/dt$. (The minus sign is required because the variables $i(t)$ and $v(t)$, shown in Fig. 5-1.3, have relative reference directions which are opposite to those used for defining the terminal relations for the inductor.) The waveshapes for $v(t)$ and $i(t)$ for this example are shown in Fig. 5-1.4. The scales have been chosen so that the two plots are collinear.

An interesting and most important fact may be observed in the above examples. In Example 5-1.1, the waveshapes for both the voltage and the current variables have the same exponential form, that is, they both have the form Ke^{-100t} (where the

[2]Remember that an initial current in an ideal inductor will flow indefinitely through a short circuit such as is provided by $S1$. If this is not clear, the student should review the properties of an inductor presented in Sec. 4-6.

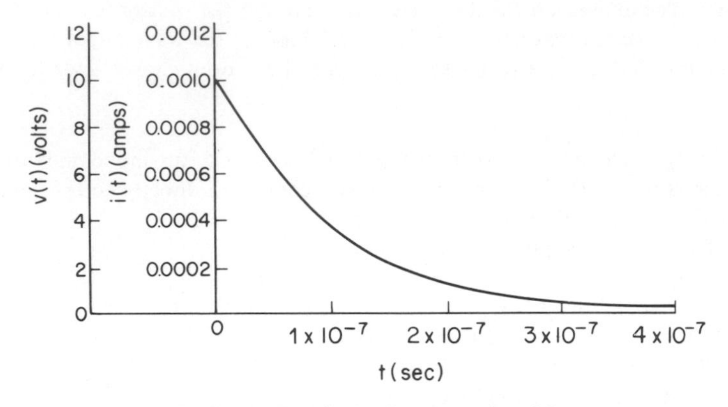

Fig. 5-1.4 Waveforms for Example 5-1.2.

value of K is different for the two variables). It may be easily verified that any other variable, such as the charge on the capacitor or the current through the capacitor, also has the same form. Similarly, in Example 5–1.2 the current and the voltage variable both have the form $Ke^{-10^7 t}$, as do the other variables in the circuit. Thus, we conclude that in each of these two circuits *all variables have an exponential behavior characterized by the same exponential term*, and they differ only in the magnitude (and sign) of the constant specifying the initial value of the variable. We may easily see that this must be true in general. Recall that all circuit variables are related by either of two types of relations: (1) algebraic relations such as those characterizing the resistor; and (2) differential or integral relations such as those characterizing the inductor and the capacitor. The derivative or integral of an exponential function Ke^{-at}, however, is still an exponential function $K'e^{-at}$, where the multiplicative constant K' is equal to $-aK$ or $-K/a$. Therefore, once we have found the exponential waveshape for any variable (in a first-order circuit excited only by initial conditions), we know the waveshape for all other variables. We need merely find the initial values of the other variables to completely characterize them. To emphasize this point, we state the following:

SUMMARY 5-1.2

Exponential Form of the Variables in a First-Order Circuit Excited Only by Initial Conditions: In a circuit characterized by a first-order, homogeneous, differential equation, all variables have an exponential behavior characterized by the same exponential term. The initial magnitude and polarity of each variable will, in general, be different.[3]

[3]When we refer to "all variables" here, we exclude quantities such as power and energy which are found as a product of basic variables such as voltage and current.

Since the variables of first-order circuits excited only by initial conditions are all exponential in form, it is worth our time to note some useful properties of the exponential waveshape. These properties apply to all variables in first-order circuits. First consider the circuit shown in Fig. 5–1.1. The solution for $v(t)$ is

$$v(t) = v(0)e^{-t/RC}$$

When t has the value RC, this becomes

$$v(RC) = v(0)e^{-1} = 0.368v(0) \tag{5–8}$$

The quantity $(-t/RC)$, which is the exponent of e, must be dimensionless; therefore, the quantity RC must have the dimensions of time. This is readily verified by direct dimensional analysis. Such a quantity is referred to as the *time constant* of the given circuit. From (5–8) we see that we may define the time constant of a given circuit as the time required for *any* variable to decay to 36.8 per cent of its initial value when the circuit is excited only by initial conditions. It is equal to the reciprocal of the quantity a in Eq. (5–7).

The time constant may be expressed either in literal form or in numerical form, whichever is appropriate. Thus, for the numerical values given in Example 5–1.1 the time constant for the circuit is 0.01 s. For the circuit shown in Example 5–1.2, the time constant is L/R, and, for the numerical values used in the example, the time constant is 10^{-7} s. Note that, for a given circuit, all variables have the same time constant. Since the time constant is a property of the circuit itself, it is convenient to refer to it as a *natural* property of the circuit and its variables (as opposed to a property *forced* on the circuit by the excitation from some source). Thus, we may refer to the *natural time constant* of a first-order circuit. Further, since frequency and time are reciprocal quantities, we may say that the circuit shown in Fig. 5–1.1 has a *natural frequency* of value $1/RC$, while the one shown in Fig. 5 1.3 has a natural frequency of value R/L. We shall see considerable application of the concept of natural frequencies in a later chapter. Note that the use of the word "frequency" here does not imply any connection with the usual "sinusoidal frequency."

In the last paragraph it was shown that an exponential function decays to about 36.8 per cent of its initial value in one time constant. Similar calculations show that in two time constants the exponential function decays to about 13.5 per cent of its initial value; in three time constants to about 5 per cent, in four time constants to about 1.8 per cent, and in five time constants to about 0.67 per cent. Thus, we see that, from a practical viewpoint, for values of time greater than four or five time constants, the values of the variables in a first-order circuit may be thought of as being zero, although in a mathematical sense, zero value is reached only when the variable t goes to infinity.

The techniques described in this section may also be applied to the case where the application of KCL or KVL produces an integral equation rather than a differential equation. The following example will serve to illustrate the process.

EXAMPLE 5-1.3. We may solve directly for the voltage $v(t)$ across the resistor for the network shown in Fig. 5-1.3 and discussed in Example 5-1.2. For values of $t \geq 0$, we note that $v(t)$ is also the voltage across the inductor. Thus, we may apply KCL to either of the nodes connecting the resistor and the inductor to obtain

$$\frac{1}{R}v(t) + \frac{1}{L}\int_0^t v(\tau)\,d\tau + i_L(0) = 0 \qquad t \geq 0$$

Differentiating the above equation and multiplying both sides by R, we obtain

$$\frac{dv}{dt} + \frac{R}{L}v(t) = 0 \qquad t \geq 0$$

This equation should be compared with the first equation given in Example 5-1.2. The coefficients are the same, but the variable is different. The solution is directly seen to be

$$v(t) = v(0)e^{-Rt/L} \qquad t \geq 0$$

The constant $v(0)$ is easily evaluated by noting that when the switch $S1$ is opened at $t=0$, all of the current I_0 formerly flowing through the switch must now flow through the resistor. Thus we see that $v(0) = I_0 R$. The numerical values for the circuit parameters used in the circuit shown in Example 5-1.2 are $I_0 = 0.001$ A, $R = 10$ kΩ, and $L = 1$ mH. Substituting these values in the relation given above, we obtain $v(t) = 10e^{-10^7 t}$, which is the same result that was found Example 5-1.2.

The techniques given in this section are easily extended to the case where more than one resistor is present in the circuit. This situation is best illustrated by an example.

EXAMPLE 5-1.4. A first-order circuit consisting of a capacitor and three resistors is shown in Fig. 5-1.5(a). An initial condition $i(0) = 4$ A is specified for the $\frac{1}{2}$-Ω resistor. It is de-

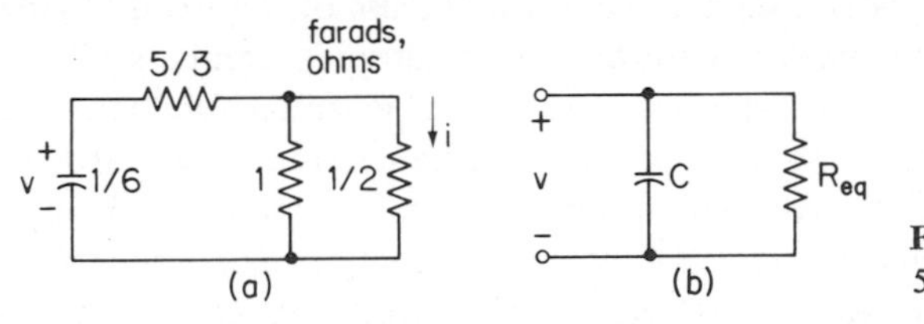

Fig. 5-1.5 Circuit for Example 5-1.4.

sired to find an expression for the current through the $\frac{1}{2}$-Ω resistor for all $t \geq 0$. This may be done by first replacing the network of three resistors connected to the capacitor by a single equivalent resistor, R_{eq}. Such a resistor has a value of 2 Ω. The circuit is now as shown in Fig. 5-1.5(b). From KCL we may write for this circuit,

$$C\frac{dv}{dt} + \frac{1}{R_{eq}}v(t) = 0 \qquad t \geq 0$$

Substituting the values $C = \frac{1}{6}$ F, $R_{eq} = 2\ \Omega$, we find that the solution for the voltage across the capacitor is $v(t) = v(0)e^{-3t}$. The exponential behavior of all variables in the circuit must be the same. Therefore, we conclude that the expression for the current in the $\frac{1}{2}$-Ω resistor is $i(t) = 4e^{-3t}$.

We conclude this section by briefly considering the linearity aspects of first-order circuits. For our purposes in this section, a linear first-order circuit may be considered as one comprised of linear two-terminal elements, namely, resistors, and a single energy-storage element. From the preceding discussion in this section, it should be clear that the response of such a circuit is linearly related to the initial condition used as an excitation, i.e., if the initial condition is multiplied by some constant, the response is multiplied by the same constant. As an example of this result, consider again the circuit analyzed in Example 5-1.1. Since an initial voltage $v(0) = 10$ V resulted in a current $i(t) = 0.001e^{-100t}$ A, and since the circuit is linear, we may directly conclude that an initial voltage $v(0) = 20$ V will produce a current $i(t) = 0.002e^{-100t}$ A. We shall explore some other considerations of the property of linearity when we discuss the excitation of a first-order circuit by sources in the following section.

In this section we have presented techniques for finding the solution of a first-order circuit excited only by initial conditions. The overall process is summarized by the flow chart shown in Fig. 5-1.6. In this flow chart the starting point is the

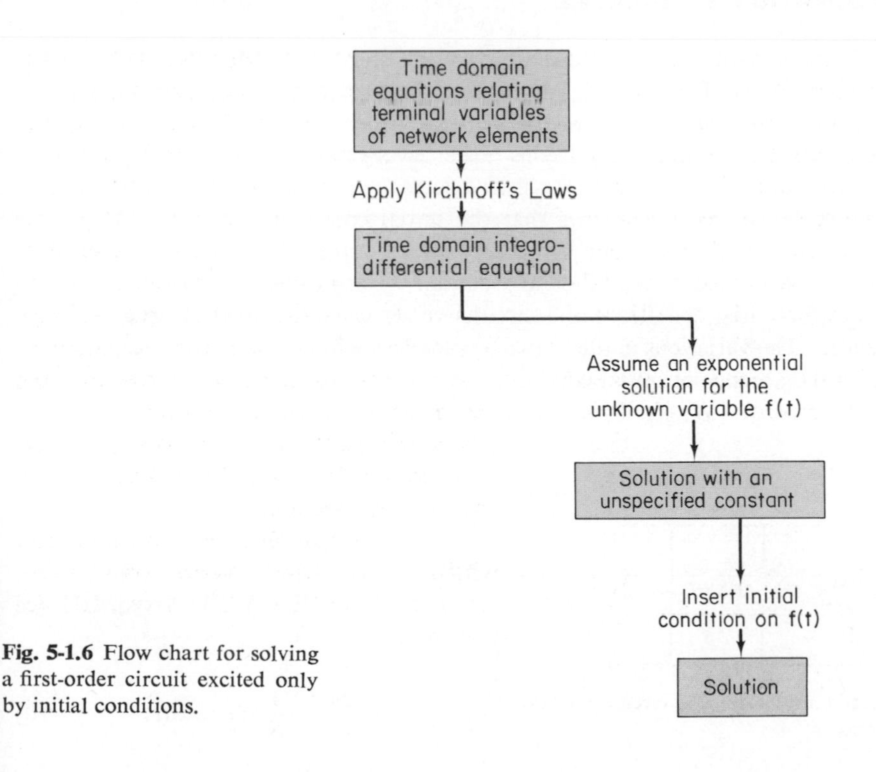

Fig. 5-1.6 Flow chart for solving a first-order circuit excited only by initial conditions.

equations relating the terminal variables of the independent network elements, i.e., the equations

$$v(t) = Ri(t) \qquad i(t) = C\frac{dv}{dt} \qquad v(t) = L\frac{di}{dt} \tag{5-9}$$

As indicated in the flow chart, if we apply Kirchhoff's laws to circuits containing elements defined by these relations, we obtain an integro-differential equation describing the network. This equation may be put in a form where it has as its unknown quantities some network variable $f(t)$ and its derivative. We then assume an exponential form of solution for $f(t)$ and we insert some specific initial condition to obtain the solution. It should be noted that the flow chart shown in Fig. 5-1.6 has its blocks placed in a somewhat irregular arrangement. This has been done in order to anticipate a merger of this flow chart with others which will be used to illustrate material which will be covered later in this chapter. To obtain the maximum benefit from this flow chart, the student should restudy the examples given in this section and he should identify each step in the solution process with the corresponding step in the flow chart. Such a study process is excellent for review and it is also a fine means of preparing for an examination.

5-2 EXCITATION BY SOURCES

In the last section we discussed the excitation of first-order circuits by means of initial conditions. The resulting variation of the circuit variables is sometimes referred to as the *zero-input response* of the network, since the variables respond to the excitation provided by initial conditions rather than responding to some input excitation from a source. In this section we shall discuss the excitation of a first-order circuit by a source (or sources), assuming that the initial conditions are zero. All of the sources referred to in this section are *independent* sources. The effect of dependent sources will be treated in the following section. A term which is frequently used to imply that the initial conditions of a circuit are zero is to say that the circuit is in the "zero state." The variations in the network variables which result from excitation by sources in the absence of initial conditions may correspondingly be referred to as the *zero-state response.* We shall make additional usage of this terminology when we discuss the excitation of a network by initial conditions *and* sources. This will be done in the following section.

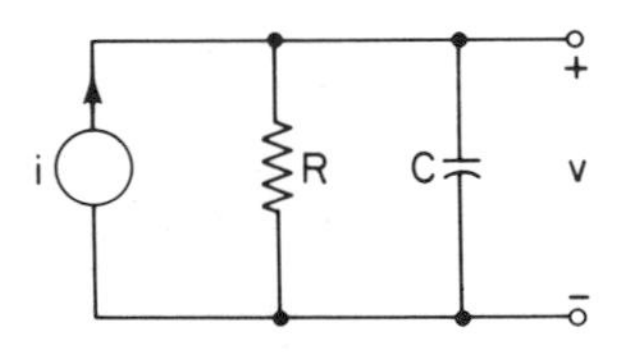

Fig. 5-2.1 A first-order *RC* circuit excited by a source.

As an example of a first-order circuit which is excited by a source, consider the circuit shown in Fig. 5-2.1. From KCL we may write

$$C\frac{dv}{dt} + \frac{1}{R}v(t) = i(t) \tag{5-10}$$

If we refer to the descriptive phrases given for differential equations in the last section, we see that Eq. (5–10) may be characterized as an ordinary first-order, linear, non-homogeneous, differential equation with constant coefficients. Thus, the difference between this equation and the ones treated in the last section is that this equation is *non-homogeneous*, i.e., a term $i(t)$ appears which is not a function of the dependent variable $v(t)$ or its derivative. In general, we shall refer to an equation such as Eq. (5–10) more compactly as a *first-order, non-homogeneous, differential equation.* The solution to such an equation consists of two parts. The first of these parts is the solution to the related homogeneous equation, i.e., the solution to Eq. (5–10) when $i(t)$ is set to zero. We will refer to this part of the solution as $v_h(t)$ (the h stands for homogeneous). This component of the solution is given the mathematical appellation of *complementary solution* or complementary function. To determine it, we simply apply the methods of the last section. Thus, we see that

$$v_h(t) = Ke^{-t/RC} \tag{5–11}$$

Note that in the above equation we have written the multiplicative constant as K rather than writing it as $v(0)$. The reason for this, of course, is that we must take both parts of the solution of (5–10) into account when evaluating the constant K so as to determine the required value of $v(0)$. We may generalize our discussion to this point by considering a differential equation of the form

$$\frac{df}{dt} + af(t) = g(t) \tag{5–12}$$

The complementary solution for this differential equation is defined as follows:

SUMMARY 5-2.1

Complementary Solution of a First-Order Differential Equation: The complementary solution is one of the parts of the overall solution of a non-homogeneous, first-order, differential equation. It is the solution to the related homogeneous equation. If the differential equation has the form given in Eq. (5–12), then the complementary solution $f_h(t)$ has the form Ke^{-at}, where K is a constant which must be evaluated in the complete solution.

The complementary solution given in Eq. (5–11) for the non-homogeneous differential equation of Eq. (5–10) is independent of the particular form of the excitation function $i(t)$. However, the second part of the overall solution, which we will now discuss, has a form which is quite similar to the particular form of the excitation function being used. Thus, it is logically called the *particular solution.* For example, suppose that $i(t)$ is a constant I_0 applied at $t = 0$. If we use $v_p(t)$ to designate the particular solution for $v(t)$, then $v_p(t)$ will also be a constant. Let us call the constant A. Thus, we may write $v_p(t) = A$. To show that this is indeed a solution, and also to determine the value of A in terms of the coefficients of the differential equation and the constant I_0, we need merely substitute A for $v(t)$ in Eq. (5–10). The derivative of A is, of course, zero. Thus, the first term in (5–10) disappears, and we obtain

$$\frac{1}{R}A = I_0 \qquad (5\text{–}13)$$

The solution for A is thus RI_0, and we see that the particular solution is given as

$$v_p(t) = RI_0 \qquad t \geq 0 \qquad (5\text{–}14)$$

As another example, suppose that $i(t) = 2 + t$ for all values of $t \geq 0$. In this case, $v_p(t)$ will have the form $A + Bt$, where A and B are constants. To show this and to find the values of A and B, we need merely substitute this expression for $v(t)$ in Eq. (5–10). Thus, we obtain

$$CB + \frac{1}{R}(A + Bt) = 2 + t \qquad (5\text{–}15)$$

For this equation to hold for all values of t, the coefficients of corresponding powers of t on both sides of the equation must be the same. This requires that

$$CB + \frac{A}{R} = 2 \qquad \frac{B}{R} = 1 \qquad (5\text{–}16)$$

Solving the two equations, we obtain

$$B = R \qquad A = R(2 - RC) \qquad (5\text{–}17)$$

Thus, the particular solution is

$$v_p(t) = R(2 - RC) + Rt \qquad t \geq 0 \qquad (5\text{–}18)$$

The method of determining the value of $v_p(t)$ given in the preceding examples is called the method of *undetermined coefficients*. It is easily extended to a large range of excitation functions. The discussion given above may be summarized as follows:

SUMMARY 5-2.2

> *Particular Solution of a First-Order Differential Equation:* If a non-homogeneous differential equation has the form $(df/dt) + af(t) = g(t)$, given in Eq. (5–12), the particular solution $f_p(t)$ may be found by substituting an assumed form for $f(t)$ and solving for any unknown constants. A listing of the assumed forms to use for $f(t)$ for various types of excitation functions $g(t)$ is given in Table 5–2.1.

Once the complementary solution and the particular solution have been found for a given differential equation, then the overall or *complete solution* is easily found as the sum of the two solutions. For the circuit shown in Fig. 5–2.1, for the case where $i(t)$ is a constant current I_0, the solutions for $v_h(t)$ and $v_p(t)$ are given in (5–11) and (5–14), respectively. Thus we see that the complete solution for $v(t)$ is

TABLE 5-2.1 Forms for the Particular Solution for a Non-Homogeneous Differential Equation $(df/dt) + af(t) = g(t)$

Form of the Excitation Function $g(t)$	Form of the Particular Solution $f_p(t)$
K_0	A
$K_0 t$	$A + Bt$
$K_0 + K_1 t$	$A + Bt$
$K_0 + K_1 t + K_2 t^2$	$A + Bt + Ct^2$
$K_0 e^{-bt}$ $(b \neq a)$	$K_0 e^{-bt}$
$K_0 e^{-at}$	$K_0 t e^{-at}$
$K_0 \sin bt$	$A \sin bt + B \cos bt$
$K_0 \cos bt$	$A \sin bt + B \cos bt$

$$v(t) = v_h(t) + v_p(t) = Ke^{-t/RC} + RI_0 \qquad t \geq 0 \tag{5–19}$$

The constant K may now be evaluated by inserting the initial condition on $v(t)$. For this case, let us assume that $v(0) = 0$, i.e., that there is no initial condition. From (5–19) we see that

$$v(0) = K + RI_0 = 0 \tag{5–20}$$

Thus, $K = -RI_0$, and the complete solution for $v(t)$ is

$$v(t) = RI_0(1 - e^{-t/RC}) \qquad t \geq 0 \tag{5–21}$$

A plot of this waveshape is shown in Fig. 5–2.2.

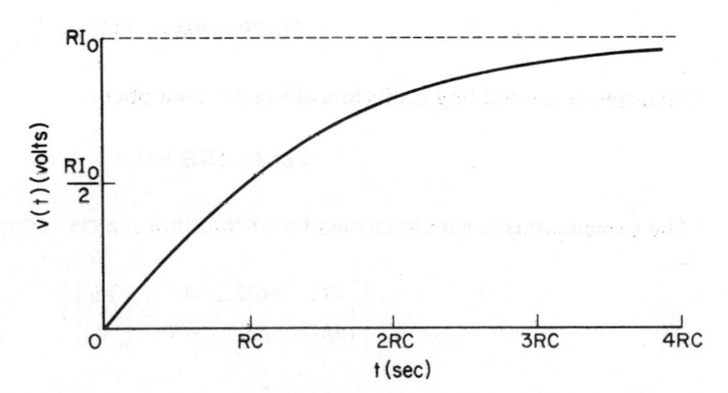

g. 5-2.2 Waveshapes for the cuit shown in Fig. 5-2.1 for) = i_0.

The above steps are easily applied to solve any first-order non-homogeneous differential equation. The procedure is summarized as follows:

SUMMARY 5-2.3

Solution of a First-Order Circuit Excited by Sources: A first-order circuit which is excited by sources is characterized by a first-order, non-homogeneous, differential equation. The complete solution of such an equation is found by adding the complementary solution and the particular solution. The constant (which is part of the complementary solution) is evaluated in the complete solution to match the specified initial condition.

An example follows:

EXAMPLE 5-2.1. A first-order *RL* circuit is excited by a voltage source as shown in Fig. 5-2.3. If we assume that $v(t)$, the output of the voltage source, is zero for $t < 0$, and is equal to $V_0 \sin \omega t$ for $t \geq 0$, then, applying KVL, we obtain the first-order, non-homogeneous, differential equation

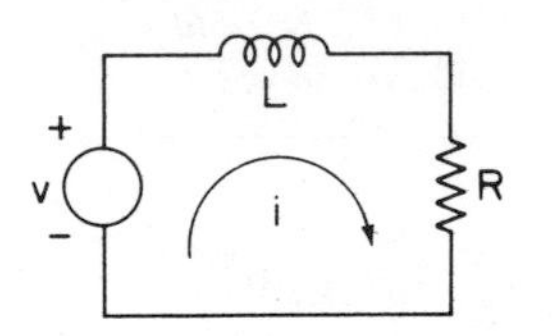

Fig. 5-2.3 A first-order *RL* circuit excited by a source.

$$L\frac{di}{dt} + Ri(t) = V_0 \sin \omega t \qquad t \geq 0$$

From Table 5-2.1, we see that the correct assumed form for the particular solution $i_p(t)$ is

$$i_p(t) = A \sin \omega t + B \cos \omega t$$

Substituting this expression into the non-homogeneous differential equation describing the circuit, we obtain

$$L\omega[A \cos \omega t - B \sin \omega t] + R[A \sin \omega t + B \cos \omega t] = V_0 \sin \omega t$$

Equating corresponding coefficients of $\sin \omega t$ on both sides of the above equation, we see that

$$-L\omega B + RA = V_0$$

Equating corresponding coefficients of $\cos \omega t$, we obtain

$$L\omega A + RB = 0$$

The two equations given above may be written in a matrix form as follows:

$$\begin{bmatrix} R & -\omega L \\ \omega L & R \end{bmatrix} \begin{bmatrix} A \\ B \end{bmatrix} = \begin{bmatrix} V_0 \\ 0 \end{bmatrix}$$

This is simply a set of simultaneous equations which may be solved to determine the constants A and B by means of the methods given in Sec. 3-1. Thus, we obtain the inverse set of relations

$$\begin{bmatrix} A \\ B \end{bmatrix} = \frac{1}{R^2 + (\omega L)^2} \begin{bmatrix} R & \omega L \\ \omega L & R \end{bmatrix} \begin{bmatrix} V_0 \\ 0 \end{bmatrix}$$

The solutions for A and B are readily seen to be

$$A = \frac{V_0 R}{R^2 + (\omega L)^2}$$

$$B = \frac{-\omega L V_0}{R^2 + (\omega L)^2}$$

Let us assume that the circuit parameters have the following values: $R = 2\,\Omega$, $L = 1$ H, $\omega = 1$ rad/s, $V_0 = 10$ V. Then, the constants have the values $A = 4$, $B = -2$. Thus, the particular solution $i_p(t)$ becomes

$$i_p(t) = 4 \sin t - 2 \cos t \qquad t \geq 0$$

The complementary solution $i_h(t)$ is easily found from the homogeneous equation. It is

$$i_h(t) = Ke^{-Rt/L} = Ke^{-2t} \qquad t \geq 0$$

The complete solution for $i(t)$ is simply written as the sum of the complementary solution and the particular solution. Thus, we obtain

$$i(t) = Ke^{-2t} + 4 \sin t - 2 \cos t \qquad t \geq 0$$

To determine the value of the constant K, let us assume that there is no initial condition present, i.e., that $i(0) = 0$. From the preceding equation, we see that $i(0) = K - 2 = 0$; therefore, $K = 2$. The complete solution for $i(t)$ is thus found to be[4]

$$i(t) = 2e^{-2t} + 4 \sin t - 2 \cos t \qquad t \geq 0$$

To verify the entries in Table 5-2.1 (or to determine particular solutions for functions not listed in the table), we may note that multiplying both sides of (5-12) by an appropriate "integrating factor" makes it possible to readily integrate both members of the equation. Specifically, multiplying both sides of (5-12) by e^{at}, we obtain

$$e^{at}\frac{df}{dt} + af(t)e^{at} = g(t)e^{at} \tag{5-22}$$

[4]Any expression involving both a sine term and a cosine term (with the same argument) can be rewritten using a single term of the form $A \sin(\omega t + \phi)$ or $B \cos(\omega t + \theta)$. A method for doing this will be given in Sec. 7-1.

We now observe that the integrating factor was chosen so as to make the left member of (5–22) expressible as the derivative of the product of two functions. Thus, we may write

$$\frac{d}{dt}[f(t)e^{at}] = e^{at}\frac{df}{dt} + af(t)e^{at} = g(t)e^{at} \tag{5–23}$$

Integrating both sides of the above equation and multiplying by the factor e^{-at}, we obtain

$$f(t) = e^{-at}\int g(t)e^{at}\,dt + Ke^{-at} \tag{5–24}$$

The integral term in the right member of this result is called the particular integral. Evaluating the integral and multiplying by e^{-at} gives us the particular solution. The second term in the right member of (5–24) is the complementary solution. The constant K is determined so as to match the specified initial condition. Thus, this is the general form of the complete solution to the first-order, homogeneous, differential equation given in (5–12).

The techniques defined in this section are easily applied to a circuit containing several voltage and current sources and/or several resistors. In such a case, we need merely reduce the portion of the circuit which contains the sources and the resistors to a single source and a single resistor, that is, we need simply use the techniques of Sec. 3–9 to find a Thevenin or Norton equivalent for the portion of the network consisting of the sources and resistors. This equivalent network may then be used to solve for the network variable associated with the energy-storage element. KCL and KVL may then be applied to find any other network variable. The procedure is illustrated by the following example.

EXAMPLE 5-2.2. The first-order circuit shown in Fig. 5-2.4(a) contains a voltage source, a current source, two resistors, and a capacitor. It is desired to find the current $i(t)$ flowing

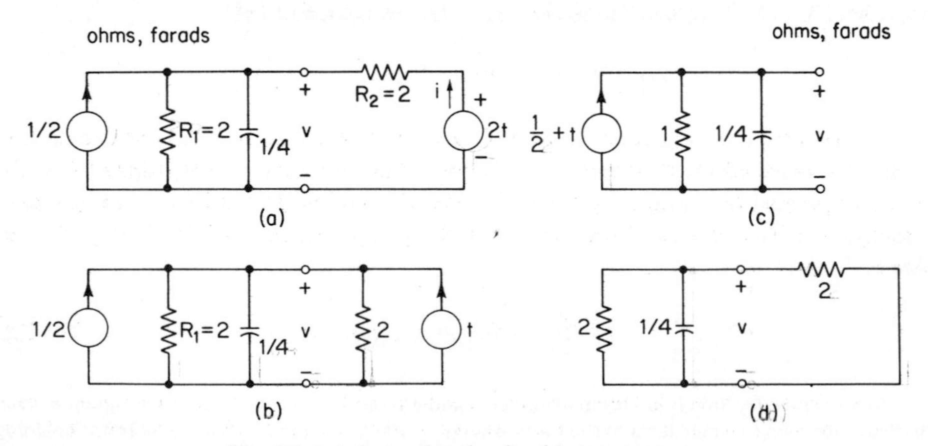

Fig. 5-2.4 A first-order circuit with two sources.

from the voltage source for $t \geq 0$. It is assumed that the circuit is initially in the zero state, i.e., that the initial voltage $v(0)$ across the capacitor is zero. To solve this problem, it is convenient to first solve for the voltage $v(t)$ rather than the current $i(t)$. To do this we must first reduce the circuit to the form where a single source and a single resistor are used to replace the two sources and the two resistors. First let us replace the voltage source and the resistor in series with it by a Norton equivalent source. Thus we obtain the circuit shown in Fig. 5-2.4(b). The two sources and the two resistors may now be combined to yield the circuit shown in Fig. 5-2.4(c). For this circuit, using KCL, we may write

$$\frac{1}{4}\frac{dv}{dt} + v(t) = \frac{1}{2} + t \qquad t \geq 0$$

From Table 5-2.1, the assumed form for the particular solution $v_p(t)$ is

$$v_p(t) = A + Bt$$

Substituting this expression in the above differential equation and equating like coefficients of t, we readily obtain the values $A = \frac{1}{4}$, $B = 1$. Thus the particular solution is known. The complementary solution $v_h(t)$ is easily seen to be Ke^{-4t}. The complete solution, with the value of K determined such that $v(0) = 0$, is thus

$$v(t) = \tfrac{1}{4}(1 - e^{-4t}) + t \qquad t \geq 0$$

We may now solve for the current $i(t)$ by noting that this variable may be expressed as a function of the capacitor voltage $v(t)$ and the source voltage $v_s(t)$. Applying KVL around the right hand mesh of the circuit shown in Fig. 5-2.4(a) we obtain

$$i(t) = \frac{v_s(t) - v(t)}{R_2}$$

Substituting the result for $v(t)$ determined above, the specified expression for $v_s(t)$, and the value of R_2 in the above equation, we obtain the result

$$i(t) = \tfrac{1}{2}t - \tfrac{1}{8}(1 - e^{-4t}) \qquad t \geq 0$$

It has been pointed out in the discussion of this section and in the examples that the complementary portion of the complete solution to a non-homogeneous differential equation is found by solving the related homogeneous equation. If the differential equation is actually known and has the form given in (5-12), this is readily accomplished by setting the right member $g(t)$ to zero. It is also possible, however, to find the homogeneous equation directly from the physical network and in many cases this provides a simpler procedure. Since the complementary part of the complete solution does not include the effect of sources, to find the homogeneous equation we need only set all the sources to zero, replace all the remaining resistors with a single equivalent resistance, and apply KVL or KCL. It should be recalled from Sec. 2-2 that when a voltage source is set to zero, it may be replaced by a short circuit.

Similarly, when a current source is set to zero, it may be replaced by an open circuit. Thus, if we set the sources in the circuit shown in Fig. 5–2.4(a) to zero, we obtain the circuit shown in Fig. 5–2.4(d). Replacing the two resistors shown in this circuit with a single equivalent resistor, we see that the homogeneous differential equation has the form $\frac{1}{4}\, dv/dt + v(t) = 0$. This is the same expression as was found for this circuit in the discussion given in Example 5–2.2, although in the example we reduced the entire network, including the sources, to find the equation. To emphasize this alternative method of finding the homogeneous equation for a circuit we present the following summary:

SUMMARY 5-2.4

Method for Finding the Complementary Solution: The homogeneous equation whose solution yields the complementary portion of the complete solution for a variable in a first-order circuit may be found by setting all independent sources in the network to zero, i.e., replacing voltage sources by short circuits and current sources by open circuits, and writing the differential equation for the remaining circuit.

As was pointed out at the beginning of this section, it should be noted in the method given above that only the *independent* sources are set to zero. The effect of dependent sources will be treated in Sec. 5–3.

The linearity aspects of circuits excited only by sources are similar to those discussed in the preceding section for circuits excited only by initial conditions. For example, consider the problem discussed in Example 5–2.1. If the source excitation $v(t)$ is changed to $20 \sin \omega t$ (twice the value given in the example), then it is readily verified that the resulting response for $i(t)$ is $4e^{-2t} + 8 \sin t - 4 \cos t$, namely, twice the value found in the example. More generally, for a first-order circuit with a zero initial condition, it may be shown that if a single source is used to excite the circuit, and if an excitation $e_1(t)$ produces a response $r_1(t)$ and an excitation $e_2(t)$ produces a response $r_2(t)$, then an excitation $e_1(t) + e_2(t)$ will produce a response $r_1(t) + r_2(t)$. If there is more than one (independent) source in the network, then considerations similar to those developed in Sec. 3–1 for the resistance network with multiple inputs apply. That is, the relation between an excitation provided by any one of the sources and the response of any network variable to it, when all the other sources are set to zero, will be a linear relation. By a linear relation, we mean that the properties of superposition and homogeneity defined in Sec. 1–3 apply.

As an example of this result, consider again the circuit discussed in Example 5–2.2. If we let $i_i(t)$ be the complete solution for $i(t)$ when only the current source is used as an excitation, i.e., when the voltage source is set to zero, we obtain $i_i(t) = (e^{-4t} - 1)/4$ A. Similarly, if we let $i_v(t)$ be the complete solution for $i(t)$ when only the voltage source is used as an excitation, we find that $i_v(t) = (t/2) + (1 - e^{-4t})/8$ A. The sum of $i_i(t)$ and $i_v(t)$ obviously gives the answer for $i(t)$ found in the example. If we now consider a new problem where the current source excitation is $\frac{3}{2}$ A (three times the value used in the example), the voltage source output is $4t$ V (two times the value used in the example), and both sources are applied simultaneously, then, since the network is linear, the resulting current $i(t) = 3i_i(t) + 2i_v(t)$. This result is readily verified by direct solution of the network.

In this section we have presented techniques for finding the solution of a first-order circuit which is excited only by sources. The overall process is summarized by the flow chart shown in Fig. 5–2.5. In this flow chart, both solid and shaded lines

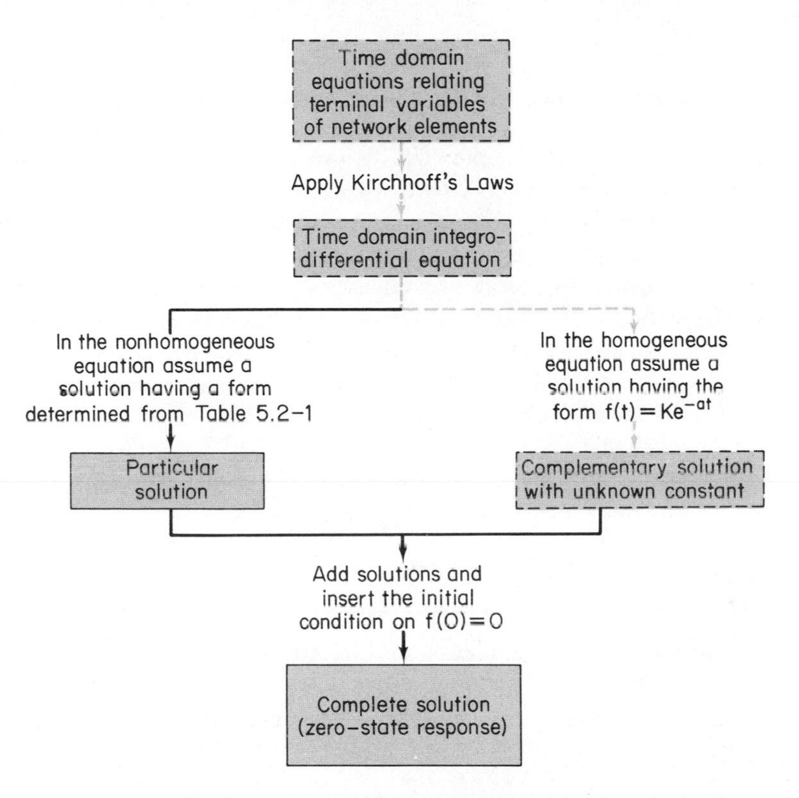

Fig. 5-2.5 Flow chart for solving a first-order circuit excited only by sources.

are used to outline the blocks and to indicate the paths. The parts of the flow chart which are presented with shaded lines are the same as those which are used in the flow chart for the solution of a first-order circuit excited only by initial conditions (Fig. 5–1.6). The positions of the blocks and lines which were used in that flow chart have been preserved, although some of the wording used has been modified to include the terminology presented in this section. The parts of the flow chart which use solid lines are the ones which have been added to cover the methods presented in this section. Thus we note that the integro-differential equation describing the network may be directly solved to find a particular solution and that the same equation may be converted to homogeneous form to find the complementary solution. Finally, we see that the sum of the two component solutions gives the complete solution, i.e., the zero-state response. It is strongly recommended that the student study the flow chart given in Fig. 5–2.5 using two approaches. First, he should compare it with the flow chart given in Fig. 5–1.6, carefully noting any changes and additions. Second, he should restudy the examples given in this section and he should identify each step of the solution procedure with the appropriate step shown in the flow chart. The use of both these approaches is an excellent preparation for an examination on this material.

5-3 EXCITATION BY INITIAL CONDITIONS AND SOURCES

In the first section of this chapter we discussed the procedure to be used in solving for the network variables of a first-order circuit which was excited *only by initial conditions*. One of the major steps in this procedure was the solution of a first-order, homogeneous, differential equation. In the following section we showed how to solve for the variables of a network excited *only by sources*. In this case, it was necessary to solve a first-order, non-homogeneous, differential equation. The next topic to be discussed follows very logically, namely, what happens when sources *and* initial conditions are present. This will be the subject of our discussion in this section. In addition, we shall present some further properties of first-order circuits which follow from our discussion.

In the last section we showed that the solution for a non-homogeneous, first-order, differential equation could be found by the following steps:

1. Find the complementary solution (with an unspecified constant).
2. Find the particular solution (based on the form of the excitation function).
3. Add the complementary solution and the particular solution to obtain the complete solution.
4. Evaluate the unspecified constant so as to satisfy some known initial condition.

These same steps, which were used in the last section for problems in which the initial condition was zero, may be directly applied to circuits in which the initial condition is non-zero. The only change in the form of the solution will be that the value of the constant in the complementary portion of the complete solution will be different. An example follows.

EXAMPLE 5-3.1. The first-order circuit shown in Fig. 5-2.3 and discussed in Example 5-2.1 was shown to have a complete solution for the current $i(t)$ which was of the form

$$i(t) = Ke^{-2t} + 4 \sin t - 2 \cos t \qquad t \geq 0$$

The example stipulated that the initial condition on $i(t)$ was zero. It was found that $K = 2$ satisfied this condition. Suppose that we now specify a non-zero initial condition; for example, let $i(0) = I_0$. We may evaluate the equation given above at $t = 0$ to find the new value of K. We obtain $i(0) = K - 2 = I_0$. Thus we find that in this case $K = 2 + I_0$. The complete solution for the current $i(t)$, including the effect of the initial condition, now becomes

$$i(t) = (2 + I_0)e^{-2t} + 4 \sin t - 2 \cos t \qquad t \geq 0$$

From the example given above, we see that the procedure for finding the complete solution for a first-order circuit which is excited both by initial conditions and by a source (or sources) is identical with that used for finding the complete solution for the case where the network is excited only by sources. To emphasize this, we repeat Summary 5-2.3 in the following form:

SUMMARY 5-3.1

Solution of a First-Order Circuit Excited by Initial Conditions and Sources: The complete solution for a first-order circuit excited by both initial conditions and sources may be found by solving the non-homogeneous differential equation describing the circuit. The complete solution is the sum of the complementary solution and the particular solution. The constant (which is part of the complementary solution) is evaluated in the complete solution to match the specified initial condition.

The expression for $i(t)$ given in the above example may be broken into separate parts to illustrate some of the properties of first-order circuits. For example, we may write

$$i(t) = \underbrace{I_0 e^{-2t}}_{\substack{\text{Zero-input response} \\ \text{(term due to the} \\ \text{excitation of the} \\ \text{network by a non-} \\ \text{zero initial condition)}}} + \underbrace{2e^{-2t} + 4 \sin t - 2 \cos t}_{\substack{\text{Zero-state response} \\ \text{(terms due to the excitation} \\ \text{of the network by a source)}}} \qquad t \geq 0 \tag{5-25}$$

As indicated by the above decomposition, we note that the complete response is the sum of the response produced by the initial condition and the response produced by the source. It should be noted, however, that the quantity characterizing the *natural response* of the network, namely, the exponential e^{-2t}, appears both in the term giving the response from the initial condition (the zero-input response) and in the term giving the response from the source (the zero-state response). Another way of expressing this fact is to say that the network's natural frequency is excited both by the initial condition and by the source.

Now let us consider a different decomposition of the complete solution for $i(t)$. This is

$$i(t) = \underbrace{I_0 e^{-2t} + 2e^{-2t}}_{\substack{\text{Transient response} \\ \text{(term due to the natural} \\ \text{response of the circuit} \\ \text{which disappears for} \\ \text{large values of time)}}} + \underbrace{4 \sin t - 2 \cos t}_{\substack{\text{Forced response} \\ \text{(term which is independent} \\ \text{of the natural response} \\ \text{of the circuit and which} \\ \text{depends only on the} \\ \text{form of the excitation} \\ \text{from the source)}}} \tag{5–26}$$

As indicated by Eq. (5–26) we note that a second way of expressing the complete response is as the sum of the transient response and the forced response. If the excitation is a constant or a sinusoid, then, after the transient response has disappeared, the remaining portion of the complete solution has the value of a constant (if the excitation is a constant) or a sinusoid of constant magnitude (if the excitation is such a sinusoid). In either of these cases, we say that the circuit has reached "steady-state" conditions, and thus, the forced response term on the right in Eq. (5–26) may also be referred to as the *steady-state response*. This term is only used for the case of excitations which are either constant or sinusoidal.

The conclusions reached above for Example 5–3.1 may easily be shown to be true for first-order circuits in general. Let us consider a first-order circuit described by a non-homogeneous differential equation of the form

$$\frac{df}{dt} + af(t) = g(t) \qquad t \geq 0 \tag{5–27}$$

In the following summary, we describe the complete solution of such a differential equation in terms of its component parts based on the manner in which the circuit is excited [see (5–25) as an example]:

SUMMARY 5-3.2

Solution of a First-Order Circuit (*zero-input response and zero-state response*): The complete solution of a non-homogeneous differential equation of the form given in (5–27) describing a first-order circuit will consist of a term resulting from the initial condition (the zero-input response) which has the form $K_0 e^{-at}$, and a term produced by the excitation function (the zero-state fesponse) which has the form $K_1 e^{-at} + g_0(t)$, where $g_0(t)$ is a function having the same general form as the excitation function $g(t)$, and may be determined from Table 5–2.1.

Similarly, in the following summary, we describe the complete solution of a first-order circuit in terms of its component parts based on the natural and forced behavior of the network [see (5–26) as an example]:

SUMMARY 5-3.3

Complete Solution of a First-Order Circuit (transient response and forced response): The complete solution of a non-homogeneous differential equation of the form given in (5–27) describing a first-order circuit will consist of a transient response term related to the natural behavior of the network and having the form Ke^{-at}, where the value of K is the result of the excitation from both the initial condition and the application of some excitation function $g(t)$. A second term in the complete response will also be produced by the excitation function $g(t)$. It will have a form similar to $g(t)$ and may be determined from Table 5–2.1.

A most important conclusion to be reached from the descriptions of the complete solution given above is that if an excitation is applied to a network from some source, a term of the form Ke^{-at} is always produced in the complete response.[5] This is true no matter what the waveshape of the source excitation is. In other words, *any type of excitation* will produce a transient response term, that is, it will excite the natural frequency of the network.

The techniques described above for finding the complete solution of a first-order circuit containing resistors and independent sources are easily extended to include the case where controlled sources are present. In such a situation, we need merely apply the techniques developed in Sec. 3–10 to reduce the portion of the network which contains the two types of sources and the resistors to an equivalent network containing a single resistor and a single independent source. Next, a differential equation describing this network may be written and solved to find an expression for some network variable. Finally, KVL and KCL may be applied to find the expressions for any other variables in the network. The process is illustrated by the following example.

EXAMPLE 5-3.2. The first-order circuit shown in Fig. 5-3.1(a) contains both a dependent source and an independent source. It is desired to find the current $i_1(t)$ flowing out of the independent source for all values of $t \geq 0$. It is assumed that the initial condition $v(0)$ on the capacitor is zero. The first step in solving this problem is to replace the portion of the network containing the resistors and the sources by a Thevenin equivalent circuit, following the procedure outlined in Summary 3-10.3. This portion of the network is the same as the one used in Example 3-10.5. Using the results of that example, we obtain the equivalent circuit shown in Fig. 5-3.1(b). Using KVL, for this circuit, we may write

$$\tfrac{3}{2} = \tfrac{5}{4}\, i(t) + v(t)$$

[5]In some cases, this term may be suppressed. As an illustration of this, note that in Example 5-3.1, if we choose the value of -2 for the constant I_0, then K will be zero and the exponential term will not be present. For a detailed discussion of how specific natural frequencies may be excited or suppressed, see L. P. Huelsman, *Circuits, Matrices, and Linear Vector Spaces,* Chap. 6, McGraw-Hill Book Company, New York, 1963.

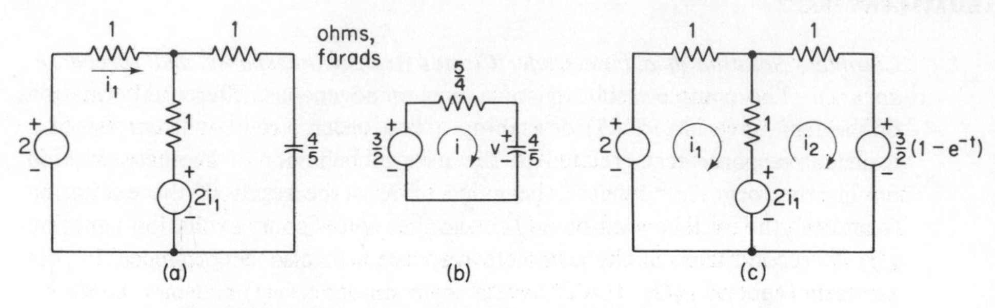

Fig. 5-3.1 A first-order circuit with a dependent and an independent source.

Substituting the relation $i(t) = \frac{4}{5}\,dv/dt$ for the terminal variables of the capacitor, the above relation may be written

$$\frac{dv}{dt} + v(t) = \frac{3}{2} \qquad t \geq 0$$

where $v(t)$ is the voltage across the capacitor. The solution for $v(t)$ is

$$v(t) = \tfrac{3}{2}(1 - e^{-t}) \qquad t \geq 0$$

Since the voltage across the capacitor is fixed by this expression, we may replace the capacitor by a voltage source of value $\frac{3}{2}(1 - e^{-t})$ V in the original circuit. Thus we obtain the circuit shown in Fig. 5-3.1(c). Using the techniques for solving resistance networks with controlled sources given in Sec. 3-10, we see that the equations describing this circuit may be put in the form

$$\begin{bmatrix} 2 \\ -\frac{3}{2}(1 - e^{-t}) \end{bmatrix} = \begin{bmatrix} 4 & -1 \\ -3 & 2 \end{bmatrix} \begin{bmatrix} i_1(t) \\ i_2(t) \end{bmatrix}$$

The solution for this set of equations is

$$\begin{bmatrix} i_1(t) \\ i_2(t) \end{bmatrix} = \frac{1}{5} \begin{bmatrix} 2 & 1 \\ 3 & 4 \end{bmatrix} \begin{bmatrix} 2 \\ -\frac{3}{2}(1 - e^{-t}) \end{bmatrix}$$

Thus, we find that

$$i_1(t) = \tfrac{1}{2} + \tfrac{3}{10}e^{-t} \qquad t \geq 0$$

This is the solution for the current flowing out of the independent source.

The linearity considerations given in Sec. 5-1 for the case of a first-order circuit excited only by initial conditions (the zero-input case) and in Sec. 5-2 for a network excited only by sources (the zero-state case) are easily extended to the case where such a circuit is excited by both initial conditions and sources. As an illustration of the principles involved, consider the following example.

EXAMPLE 5-3.3. The linear first-order circuit shown in Fig. 5-3.2 is excited by three separate effects, namely, the initial condition $v(0) = 1$ V on the capacitor, the voltage source with output voltage $v_s(t) = t$ V, and the current source with output current $i_S(t) = 1$ A (both applied at $t = 0$). Let us consider the following cases of reduced excitation:

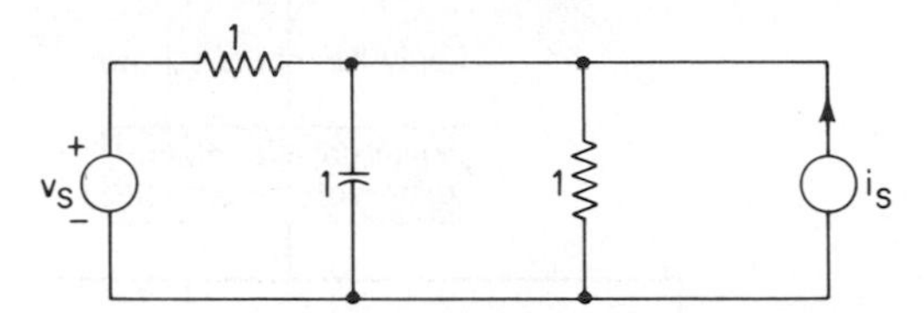

Fig. 5-3.2 A first-order circuit.

Case 1 (example of zero-input response): If both sources are set to zero, then the solution for the voltage $v(t)$ across the capacitor yields $v(t) = 1e^{-2t}$. Now suppose the initial condition on the capacitor is doubled, so that $v(0) = 2$. We find in this case that $v(t) = 2e^{-2t}$. We conclude that the variable $v(t)$ is linearly related to the excitation provided by the initial condition. This is the same conclusion given in Summary 5-1.1.

Case 2 (example of zero-state response): If the initial condition and the voltage source are both set to zero but $i_S(t) = 1$ A (applied at $t = 0$), then solving for $v(t)$, we find that $v(t) = -\frac{1}{2}e^{-2t} + \frac{1}{2}$ V. If we double the excitation by changing $i_S(t)$ from 1 to 2 A, then we find than $v(t) = -1e^{-2t} + 1$. We conclude that for a single source, the variable $v(t)$ is linearly related to the excitation if the initial condition and the other source are set to zero.

Case 3 (another example of zero-state response): If the initial condition and the current source are set to zero but $v_S(t) = t$ V (starting at $t = 0$), then, solving for $v(t)$, we find that $v(t) = \frac{1}{4}e^{-2t} - \frac{1}{4} + \frac{1}{2}t$ V. If the excitation from the voltage source is changed from t V to $2t$ V, we find that the response is also doubled, namely, $v(t) = \frac{1}{2}{}^{-2t} - \frac{1}{2} + t$. Just as in case 2, we conclude that the response $v(t)$ is linearly related to the excitation. The results for these two cases are the same as given in Summary 5-2.3.

Case 4 (example of general response): If the initial condition and both sources are used to excite the network, then we find that $v(t) = \frac{3}{4}e^{-2t} + \frac{1}{4} + \frac{1}{2}t$. Note that superposition applies, since this response is the sum of the three responses obtained for cases 1, 2, and 3 (for the original excitations). If we now double any of the excitations, however, we find that our response has not doubled. For example, if the initial condition is changed from 1 to 2 V, we find that $v(t) = \frac{7}{4}e^{-2t} + \frac{1}{4} + \frac{1}{2}t$. Of course, if all excitations are doubled, then the resulting expression for $v(t)$ will also be doubled.

Generalizing the results given in the above example, we see that the response of any variable in a first-order circuit consisting of linear elements is linearly related to the following excitations: (1) the excitation from an initial condition when all sources are set to zero; and (2) the excitation from any source when the initial condition and all other sources are set to zero. In general, the response is linearly related to any single excitation, and is not linearly related to a given excitation if other excitations are present.

In this section we have presented techniques for finding the solution of a first-order circuit which is excited by sources and initial conditions. The overall procedure is illustrated by the flow chart shown in Fig. 5–3.3. This flow chart is an

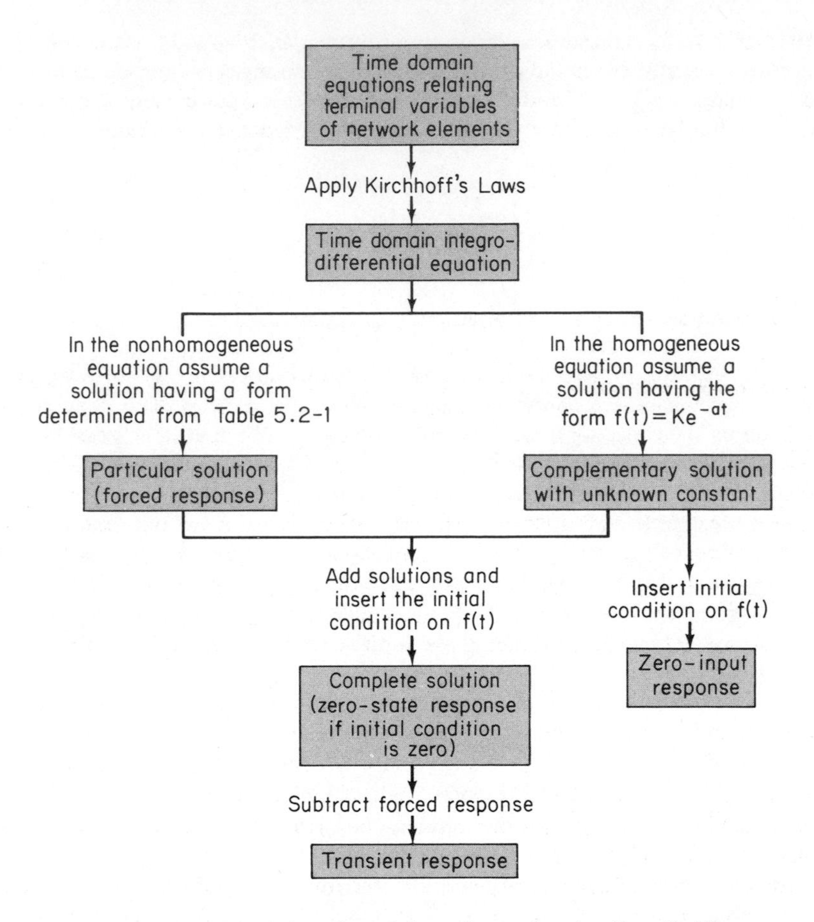

Fig. 5-3.3 Flow chart for solving a first-order circuit excited by sources and initial conditions.

extension of the one previously given for a first-order circuit excited only by sources (Fig. 5–2.5). In addition, some modifications of the earlier flow chart have been made to reflect the decomposition of the complete response into forced response and transient response, and also its decomposition into zero-input response and zero-state response as presented in this section. It is strongly recommended that the student make his own detailed comparison of this flow chart and the one shown in Fig. 5–2.5. It will also be very helpful to the student if, as a review, he restudy the examples given in this section and identify each step in their solution with a corresponding step in the flow chart.

5-4 RESPONSE TO SOURCES WITH CONSTANT EXCITATION

A simplified method may be applied to solve first-order circuit problems in which all sources have a constant output. For such a case, let us assume that the first-order differential equation describing the circuit has been put in the form

$$\frac{df}{dt} + af(t) = G \qquad t \geq 0 \tag{5–28}$$

where $a > 0$ and G is a constant. From the preceding discussion, we know that the solution must have the form

$$f(t) = Ke^{-at} + G_0 \tag{5–29}$$

where G_0 is also a constant. If we let the initial and final values of $f(t)$ be $f(0)$ and $f(\infty)$, respectively, then, from (5–29), the following relations must hold.

$$f(0) = K + G_0 \qquad f(\infty) = G_0 \tag{5–30}$$

From these two equations, we may express the constant K of (5–29) as

$$K = f(0) - f(\infty) \tag{5–31}$$

Substituting the values of K and G_0 in (5–29), we obtain the following general solution for (5–28):

$$f(t) = [f(0) - f(\infty)]e^{-at} + f(\infty) \tag{5–32}$$

To emphasize this method of solution we may state it in the following form:

SUMMARY 5-4.1

Solution of a First-Order Circuit with Constant Excitation: If a first-order circuit is excited by initial conditions and constant-valued sources, and if the complementary solution for the circuit has the form Ke^{-at} $(a > 0)$, then the complete solution for $f(t)$ for $t > 0$ is given by

$$f(t) = [f(0) - f(\infty)]e^{-at} + f(\infty)$$

where $f(t)$ is any variable in the circuit.

The utility of this method lies in the fact that it merely requires that we know the initial and final values of any circuit variable (and the natural exponential behavior of the circuit) to completely determine the value of that variable. Thus, we conclude that the circuit variables all change exponentially from their value when the network is in some initial state to their value when it is in some final state.

This method is easily implemented directly from the circuit diagram for a given network. We have already seen how to find the form of the complementary solution by setting the independent sources to zero. Now let us see how to determine the initial and final values of the network variables. First we will consider the initial values. The continuity conditions developed in Chap. 4 tell us that capacitor voltages and inductor currents cannot be discontinuous. Therefore, for a network with zero initial conditions at $t = 0$, we need simply replace a capacitor by a voltage source of zero voltage (this is simply a short circuit) or an inductor by a current source of zero output current (this is an open circuit), and solve for the initial values of any network variable in which we are interested. Note that if some initial condition happens to be present on either of these elements, then the replacement source will simply have the appropriate value of the initial voltage or current. Two examples follow.

EXAMPLE 5-4.1. The method described above is easily applied to find the initial value of the current through the capacitor in the circuit used in Example 5-2.2 and shown in Fig. 5-2.4(a). Since the initial voltage across the capacitor is zero, at $t = 0$ the capacitor may be replaced by a voltage source of zero voltage, namely, a short circuit. The current source provides $\frac{1}{2}$ A through this short, and the independent voltage source, since it has zero output voltage at $t = 0$, provides zero output voltage. Thus, we see that the initial value of current through the capacitor is $\frac{1}{2}$ A.

EXAMPLE 5-4.2. It is assumed that the initial condition at $t = 0$ for the inductor current $i_L(t)$ in the circuit shown in Fig. 5-4.1(a) is -5 A. It is desired to find the initial voltage

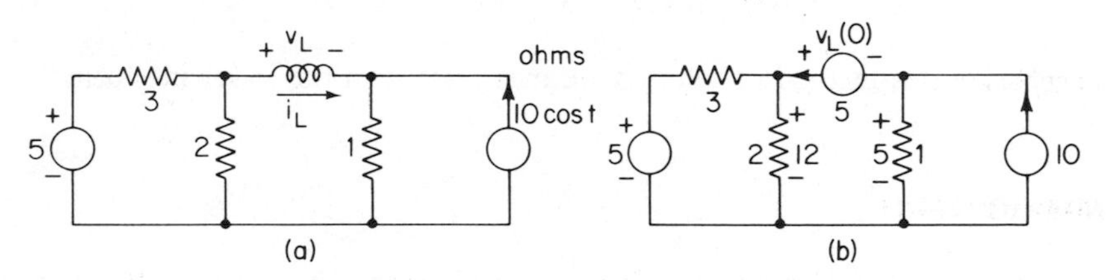

Fig. 5-4.1 Circuit for Example 5-4.2.

$v_L(0)$ across the inductor. Since the initial current in the inductor is -5 A, at $t = 0$ the inductor may be replaced by a current source of value 5 A. Thus, at $t = 0$, the circuit appears as shown in Fig. 5-4.1(b). From the figure we see that the voltage across the 2 Ω resistor is 12 V, and the voltage across the 1 Ω resistor is 5 V. From KVL, we find that the voltage across the current source is 7 V. This is also the initial voltage $v_L(0)$ across the inductor.

The method described above for finding the initial values of network variables is quite general. It may be applied to a circuit which is excited by sources having any time dependence, not just constant-valued sources. In addition, we will see in later chapters that this method is applicable to circuits of higher than first-order, i.e., circuits containing more than one energy-storage element. To emphasize this, we present the following:

SUMMARY 5-4.2

Initial Values of Circuit Variables: The initial values of any variables in a circuit may be found by replacing the capacitors with voltage sources of value equal to the initial voltage across them, replacing the inductors with current sources of value equal to the initial current through them, and solving the resistance network. If the circuit was unexcited for $t \leq 0$, then all the initial conditions on the energy-storage elements are zero. In this case, capacitors may be replaced by short circuits, and inductors by open circuits.

Now let us consider finding the *final* values of circuit variables. If the excitations provided by the independent sources are all constant in value, then as t approaches infinity, all the circuit variables must reach some constant value. This means that all the derivatives on the circuit variables must go to zero. Since the voltage across an inductor depends on the derivative of the current, any such voltage must also go to zero. The current through the inductor, however, may have any value. We conclude that as t approaches infinity, every inductor will be characterized as having zero voltage across it independent of the current flowing through it. This is exactly the property possessed by a short circuit. Therefore, if we desire to evaluate the circuit variables at $t = \infty$, we may replace all inductors by short circuits. Similar reasoning tells us that at $t = \infty$ we may replace capacitors by open circuits. Having done this, the values of any of the network variables are easily found. This technique is applicable to circuits containing any number of inductors and capacitors, as long as the sources exciting the circuit are all constant-valued. To emphasize this, we state the following:

SUMMARY 5-4.3

Final Values of Circuit Variables: If all the independent sources in a circuit are constant-valued, then the final values of any network variables may be found by replacing inductors with short circuits and capacitors with open circuits.[6]

Note that the replacement operations in the two cases described above are opposite, i.e., to determine *initial* values we replace inductors by open circuits and capacitors by short circuits (assuming that the initial conditions are zero), but to

[6]It is also necessary to require that the magnitude of the complementary solution decays exponentially. The reason for this stipulation will become clear when circuits containing inductors and capacitors are studied in the following chapter.

determine *final* values we replace inductors by short circuits and capacitors by open circuits. An example of finding the final value of a network variable follows.

EXAMPLE 5-4.3. The final value of the voltage $v_C(t)$ across the capacitor in the circuit shown in Fig. 5-4.2(a) is easily found by the method described above. To simplify our

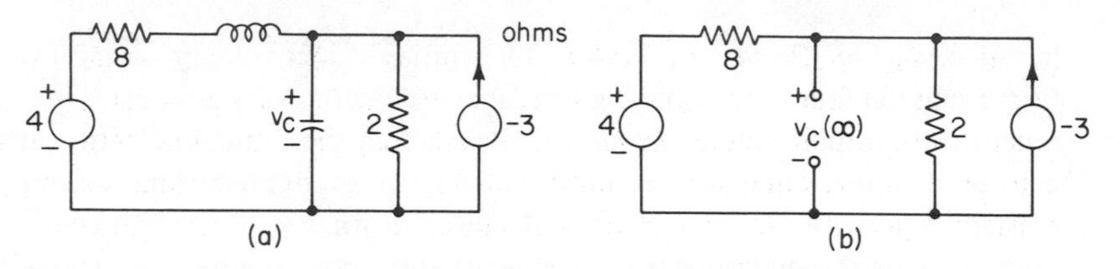

Fig. 5-4.2 Circuit for Example 5-4.3.

visualization of the process, the circuit has been redrawn in Fig. 5-4.2(b), with the inductor replaced by a short circuit and the capacitor replaced by an open circuit. This schematic represents the conditions that are found in the circuit as t approaches infinity. By routine analysis of the resulting resistance network, we see that $v_C(\infty) = -4$ V.

By the methods described above, we may easily find the initial and final values of any variable in a first-order circuit excited by constant-valued sources. Thus, we may apply the techniques specified in Summary 5-4.1 to determine the time variation of any variable directly. An example follows.

EXAMPLE 5-4.4. The first-order circuit shown in Fig. 5-4.3(a) is unexcited for all values of $t < 0$. At $t = 0$, excitations of 7 V and 3 A are applied from the sources as indicated. It is desired to find a solution for the current $i(t)$ flowing out of the voltage source for all values of $t \geq 0$. To do this, we first find the complementary solution by setting the sources to zero, using the method given in Summary 5-2.4. The resulting network is shown in Fig.

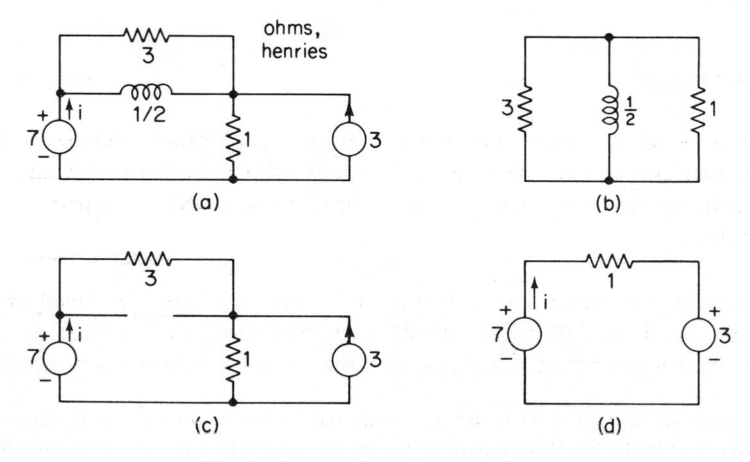

Fig. 5-4.3 Circuit for Example 5-4.4.

5-4.3(b). The first-order differential equation for this circuit is easily written in terms of the variable $i_L(t)$ by replacing the two resistors by an equivalent resistor and applying KVL. Thus, we obtain

$$\frac{1}{2}\frac{di_L}{dt} + \frac{3}{4}i_L(t) = 0$$

We conclude that the complementary solution for any variable in the network must have the form $Ke^{-3t/2}$. Now we may find the initial value of the current $i(t)$. This is done by drawing the circuit as it appears at $t = 0$ as shown in Fig. 5-4.3(c). In this figure, the inductor has been replaced by an open circuit (using the method given in Summary 5-4.2). Routine circuit analysis shows that $i(0) = 1$ A. To find the final value of the source current, we simply redraw the circuit as it appears at $t = \infty$ as shown in Fig. 5-4.3(d) (using the method given in Summary 5-4.3). We see that $i(\infty) = 4$ A. The complete solution for the current $i(t)$ flowing in the voltage source is thus found from Summary 5-4.1 to be

$$i(t) = 4 - 3e^{-3t/2} \qquad t \geq 0$$

A plot of the waveshape for $i(t)$ is shown in Fig. 5-4.4. In this figure, the exponential nature of the change in the value of $i(t)$ from its initial value to its final value is readily apparent. Note that the time constant of the circuit shown in Fig. 5-4.3 is $\frac{2}{3}$ s. It is readily verified from Fig. 5-4.4 that, when $t = \frac{2}{3}$ s, only 36.8 per cent of the total change in the value of $i(t)$ remains to be accomplished.

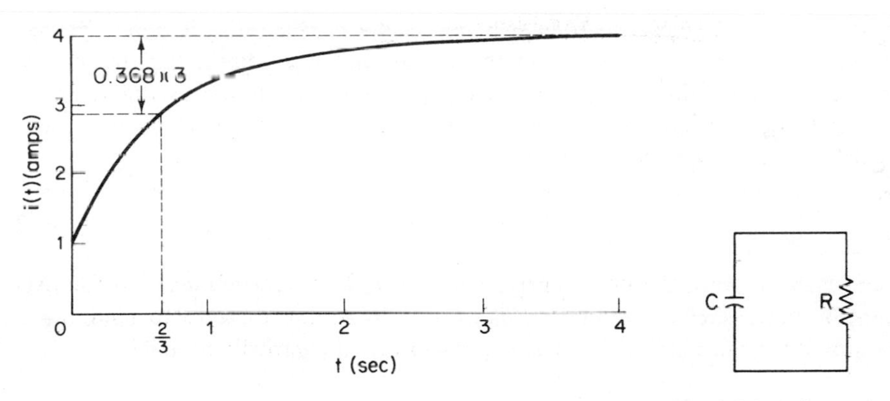

Fig. 5-4.4 Output waveform for Example 5-4.4.

Fig. 5-5.1 A first-order RC circuit.

5-5 INITIAL CONDITIONS APPLIED AT $t \neq 0$

In all our discussions of first-order circuits to this point, we have used initial conditions which are specified at $t = 0$. In this section we shall extend our treatment to cover the case of "initial" conditions which are specified for values of t not equal to zero. As a first example of such a case, consider the circuit shown in Fig. 5-5.1. Let us assume that this circuit is excited only by an initial condition on the capacitor. In Sec. 5-1 we pointed out that the time constant for this circuit, i.e., the time required for a variable to decay to 36.8 per cent of its initial value, is equal to *RC*. Since the circuit is time-invariant, this fact must hold true no matter at what instant we start our count of time. Therefore, if the voltage $v(t)$ across the capacitor is equal to V_0 at some time t_0, it will be equal to $0.368V_0$ at time $t_0 + RC$. In addition, all following values of $v(t)$ will exhibit a similar exponential behavior. For example, $v(t)$ will be equal to $0.018V_0$ at a value of t equal to four time constants after t_0, i.e., when $t = t_0 + 4RC$. To illustrate this, the waveshape of $v(t)$ for an initial condition V_0 applied at $t = t_0$ has been sketched in Fig. 5-5.2 using a solid line. For convenience,

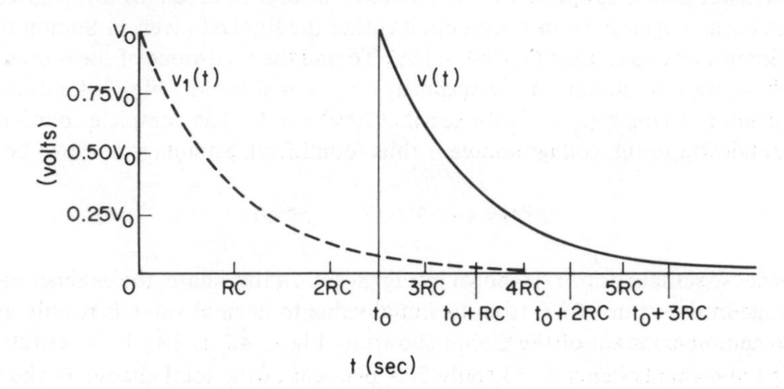

Fig. 5-5.2 Effect of an initial condition applied at $t \neq 0$.

the curve $V_0 e^{-t/RC} u(t)$ has also been sketched using a dashed line. It should be apparent, both from the sketch and from the discussion given above, that the curves are identical in shape, although they are displaced along the abscissa by the quantity t_0. If we call the dashed curve $v_1(t)$ and the solid curve $v(t)$, then we may mathematically define such a displacement by the relation

$$v(t) = v_1(t - t_0) \tag{5-33}$$

In other words, if we substitute the argument $t - t_0$ for the argument t in the expression for the dashed curve of Fig. 5-5.2, we obtain the correct expression (as a function of t) for the solid curve. The expression for the dashed curve is[7]

$$v_1(t) = V_0 e^{-t/RC} u(t) \tag{5-34}$$

thus the expression for the solid curve $v(t)$ is

$$v(t) = V_0 e^{-(t-t_0)/RC} u(t - t_0) \tag{5-35}$$

To check the validity of this expression, we note that when $t = t_0$, $v(t_0) = V_0$, as is to be expected. Similarly, one time constant later, $v(t_0 + RC) = V_0 e^{-1} = 0.368V_0$, etc.

[7]The expressions $u(t)$ and $u(t - t_0)$ are defined in Sec. 4-2.

An operation of the type defined by Eq. (5–33) is usually referred to mathematically as *shifting* or *translation*. It is readily applied to a wide variety of problems in which elements are added to or deleted from a circuit by opening or closing switches at different values of time. An example of such a situation follows.

EXAMPLE 5-5.1. The circuit shown in Fig. 5-5.3(a), at $t = 0$, has an initial condition $v(0)$ of 10 V across the capacitor, and switch *S*1 is open. At $t = 5$ s, switch *S*1 is closed. It

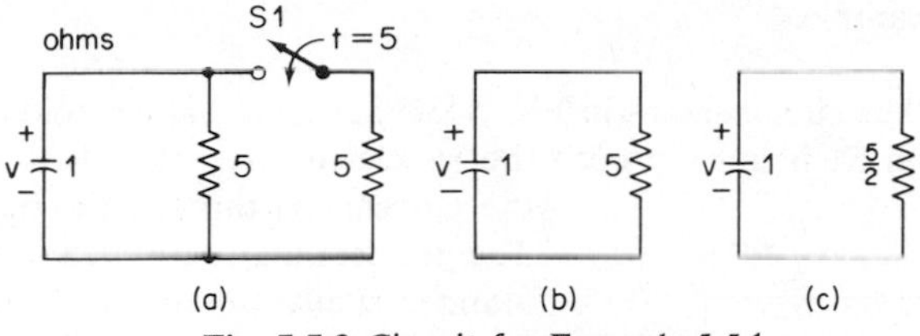

Fig. 5-5.3 Circuit for Example 5-5.1.

is desired to find the voltage $v(t)$ for all time $t \geq 0$. The problem may be attacked in two parts. For the first part, consider the circuit during the period $0 \leq t \leq 5$. For this range of time, the original circuit has the electrical appearance of the circuit shown in Fig. 5-5.3(b). The time constant for this circuit is 5 s. Thus, the solution for the voltage $v(t)$ is readily seen to be

$$v(t) = 10e^{-t/5} \qquad 0 \leq t \leq 5$$

Note that $v(5) = 10e^{-1} = 3.68$ V. We may now consider the second part of this problem, namely, the range of time $t \geq 5$, during which the switch *S*1 is closed. For this range of time, the original circuit has the electrical appearance of the one shown in Fig. 5-5.3(c). The time constant for this circuit is 2.5 s. In addition, since the voltage $v(t)$ must be continuous at the instant switch *S*1 is closed, we have the "initial" condition $v(5) = 3.68$ V. Thus, the solution for $v(t)$ is

$$v(t) = 3.68e^{-(t-5)/2.5} \qquad t \geq 5$$

A plot of the waveshape of $v(t)$ for a range of time of 0 to 10 s is shown in Fig. 5-5.4.

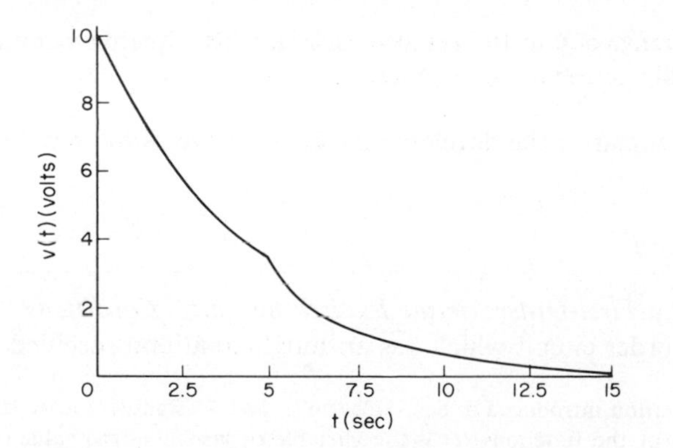

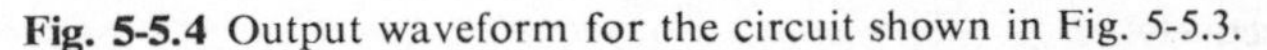

Fig. 5-5.4 Output waveform for the circuit shown in Fig. 5-5.3.

The changing of circuit parameters by means of switches illustrated in the preceding example may introduce transients into a circuit. Thus, although we know that capacitor voltages and inductor currents must be continuous in a physical network, other variables may exhibit discontinuities. To evaluate these discontinuities, and thus to determine the necessary values of initial conditions which occur when elements are switched into or out of a circuit, we may use the techniques for finding initial values of variables described in Sec. 5-4. This type of situation is illustrated by the following example.

EXAMPLE 5-5.2. The circuit shown in Fig. 5-5.5 has an initial condition of 10 V on the capacitor. The switch $S1$ is initially open, but is closed at $t = 5$ s. It is desired to find the current $i(t)$ through the indicated resistor. The problem may be solved in two parts in a manner similar to that used in the previous example. For the range of time $0 \leq t < 5$ s, the time constant of the circuit is 5 s, and the initial current through the capacitor is readily seen to be 2 A. Thus, we see that

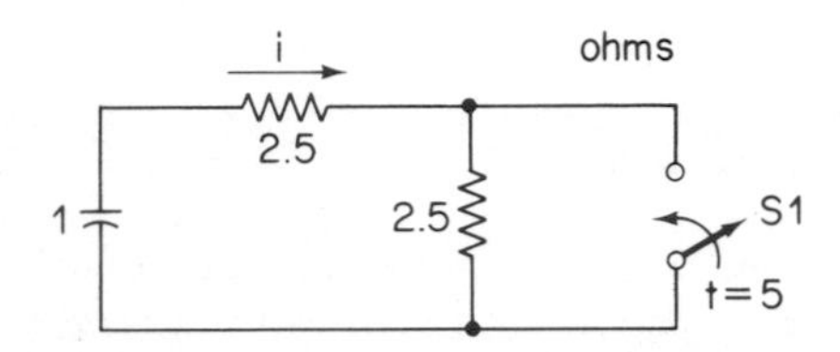

Fig. 5-5.5 Circuit for Example 5-5.2.

$$i(t) = 2e^{-t/5} \qquad 0 \leq t < 5$$

Now let us consider the circuit variables before and after the switch $S1$ is closed. For the voltage $v(t)$ across the capacitor, we may write[8]

$$v(5-) = v(5) = v(5+) = 3.68$$

For the resistor current $i(t)$, we note that $i(5-) = 0.736$ A. After the switch $S1$ is closed, however, we find that the capacitor voltage of 3.68 V now appears across a single 2.5 Ω resistor. Thus, $i(5+) = 1.472$ A. This value provides the initial condition for the second part of the problem, namely, the range of time $t > 5$. Since the time constant of the circuit for this range of time is 2.5 s, we may write

$$i(t) = 1.472e^{-(t-5)/2.5} \qquad t > 5$$

A plot of $i(t)$ for a range of 0 to 10 s is shown in Fig. 5-5.6. The discontinuous nature of this variable is readily apparent in the figure.

We may summarize the development given above as follows:

SUMMARY 5-5.1

Solution of a First-Order Circuit Excited by Initial Conditions Applied at $t \neq 0$: A first-order circuit which has an initial condition specified on a variable

[8]Using the notation introduced in Sec. 4-3, the + and − signs in the arguments are used to indicate the limit of the function $v(t)$ as the variable t approaches the value of 5 from above and below, respectively.

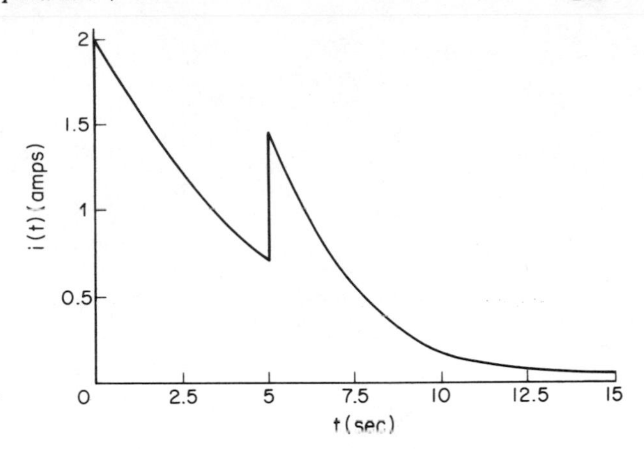

Fig. 5-5.6 Output waveform for the circuit shown in Fig. 5-5.5.

for some time $t_0 \neq 0$ will have a zero-input response of the form

$$f(t) = f(t_0)e^{-(t-t_0)/\tau} \qquad t > t_0$$

where $f(t)$ is the expression for the variable, $f(t_0)$ is the initial condition, and τ is the time constant of the circuit.

The techniques described above are readily applicable to the case where a first-order circuit is excited by sources which are "turned on" (or off) at times other than $t = 0$. The approach is illustrated by the following examples.

EXAMPLE 5-5.3. The circuit shown in Fig. 5-5.7 contains a current source with an output $i_S(t) = 3u(t - 4)$ A. Thus, for all values of $t < 4$, the source current is zero, while for values of $t > 4$, the source current is a constant 3 A. It is desired to find the current $i(t)$ in the inductor, assuming that there is no initial condition. First let us consider what would happen if the step of current was applied at $t = 0$. The techniques given in Sec. 5.4 may be directly applied in this case. Since the initial inductor current must be zero, and the final inductor current must be 3 A, the expression for the current may be directly determined from Summary 5-4.1. It has the form $3(1 - e^{-2t})u(t)$. To solve the original problem, since the circuit is linear and time-invariant, we need merely "shift" both the excitation and the response by replacing t by $t - 4$ in the above expression. Thus, we see that the answer for the original problem is

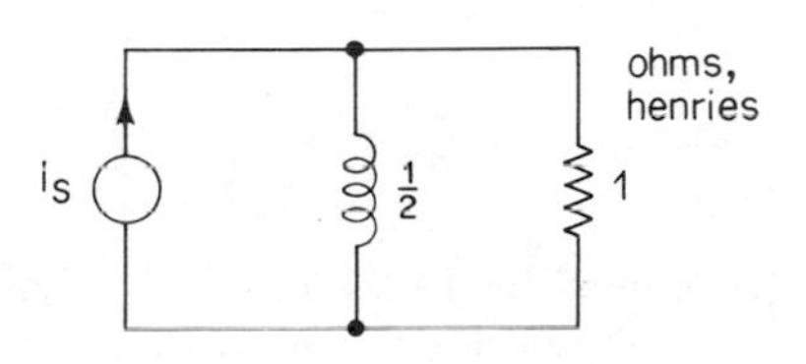

Fig. 5-5.7 Circuit for Example 5-5.3.

$$i(t) = 3(1 - e^{-2(t-4)})u(t - 4)$$

EXAMPLE 5-5.4. The circuit shown in Fig. 5-5.8(a) is excited by two sources, namely a current source which provides a constant excitation of 2 A starting at $t = 0$, and a voltage

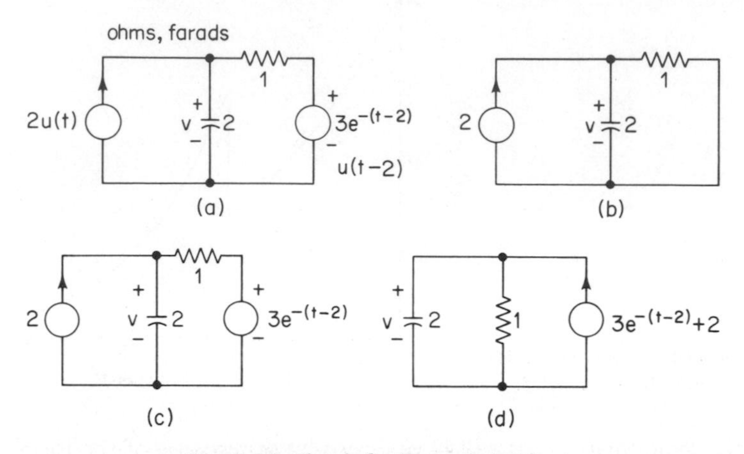

Fig. 5-5.8 Circuit for Example 5-5.4.

source which puts out zero voltage for $t \le 2$, and a voltage $3e^{-(t-2)}$ V for $t \ge 2$. To find the voltage $v(t)$ across the capacitor assuming a zero initial condition, we may consider the circuit as it appears for two different time ranges. In Fig. 5-5.8(b) the circuit is shown as it appears for the period $0 \le t \le 2$. The solution for this circuit is easily found from the techniques of Sec. 5-4. Thus, we see that

$$v(t) = 2(1 - e^{-t/2}) \qquad 0 \le t \le 2$$

From this relation we note that $v(2) = 1.264$. At this time, the second source is "turned on" by the step function term $u(t-2)$. Thus, for the range of time $t \ge 2$, the circuit appears as shown in Fig. 5-5.8(c). By making source transformations, we obtain the circuit shown in Fig. 5-5.8(d). The non-homogeneous differential equation describing the circuit (after multiplying through by the factor $\frac{1}{2}$) is

$$\frac{dv}{dt} + \frac{1}{2}v(t) = \frac{3}{2}e^{-(t-2)} + 1 = \frac{3}{2}e^{2}e^{-t} + 1 \qquad t \ge 2$$

The form of the particular solution $v_p(t)$ for this circuit is found from Table 5-2.1. It is $Ae^{-t} + B$. Solving for the constants, we find that

$$v_p(t) = -3e^{2}e^{-t} + 2$$

The complementary solution to the related homogeneous equation has the form

$$v_h(t) = Ke^{-(t-2)/2}$$

Adding the two components of the solution, and evaluating the constant K of the complementary solution so that the initial condition $v(2) = 1.264$ is satisfied, we obtain

$$v(t) = 2.264e^{-(t-2)/2} - 3e^{2}e^{-t} + 2 \qquad t \ge 2$$

This equation, together with the one found for the range of time $0 \leq t \leq 2$, specifies the solution for the capacitor voltage $v(t)$ for this problem. A plot of $v(t)$ for the range of t from 0 to 5 s is shown in Fig. 5-5.9.

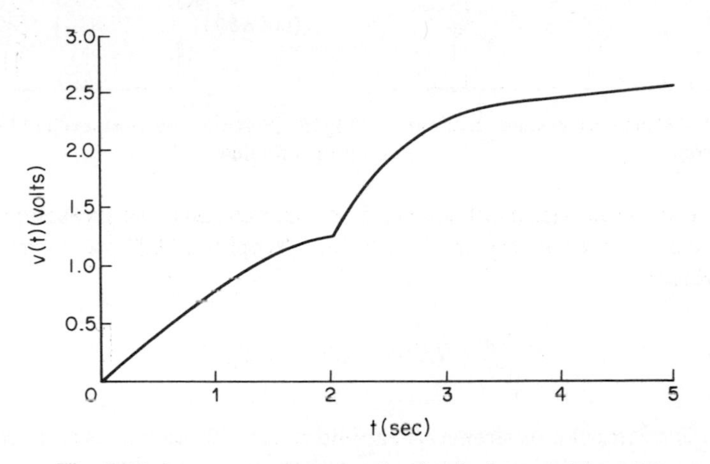

Fig. 5-5.9 Output waveform for the circuit shown in Fig. 5-5.8.

The examples given above treated cases in which the initial conditions on the circuit were zero. These results, however, are readily extended to the case where initial conditions are present and where various sources are connected to a network at times other than $t = 0$. The details of this extension are left to the reader as an exercise. We conclude this section by summarizing the development given above as follows:

SUMMARY 5-5.2

Solution of a First-Order Circuit Excited by Initial Conditions and Sources Applied at $t \neq 0$: If sources are used to excite a first-order circuit at some time $t \neq 0$, the particular solution which is valid for values of time following the application of the excitation may be found using the method outlined in Summary 5-2.2 and Table 5-2.1. The complementary solution will have the form specified in Summary 5-5.1, and will contain a constant whose value is determined so that the correct initial conditions at the time the excitation is applied are obtained.

5-6 THE USE OF IMPULSES TO GENERATE INITIAL CONDITIONS

In Sec. 4-2, we introduced the impulse function which was characterized (in the limit) by infinite magnitude and zero duration. Such a function has many applications in the investigation of the properties of networks from a theoretical viewpoint. In this section we shall present one such application, namely, the use of impulse

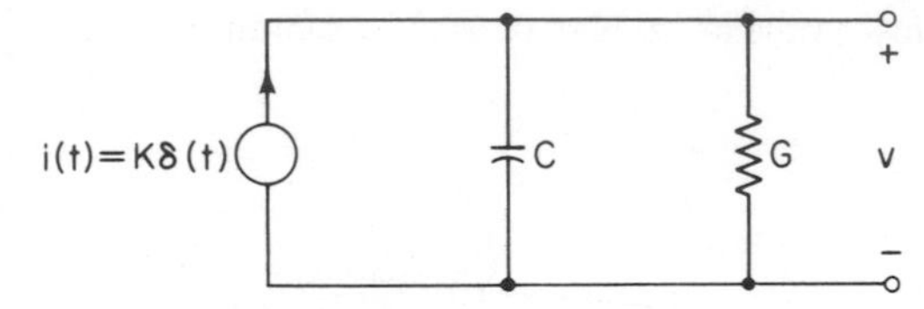

Fig. 5-6.1 An RC network excited by an impulse of current.

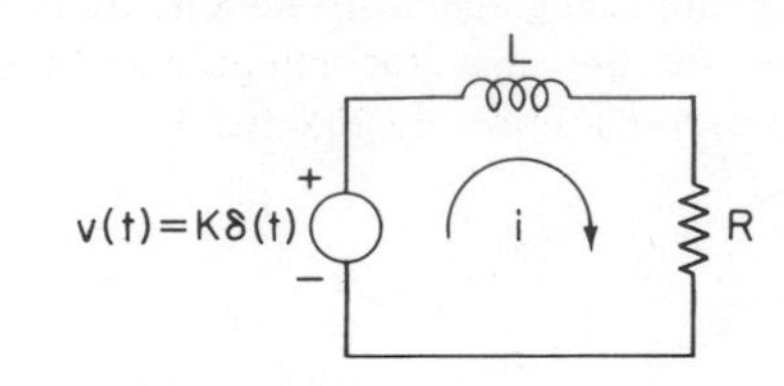

Fig. 5-6.2 An RL network excited by an impulse of voltage.

functions to establish initial conditions in first-order circuits. As an example of such a case, consider the circuit shown in Fig. 5-6.1. Applying KCL, we obtain for this circuit the relation

$$C\frac{dv}{dt} + Gv(t) = i(t) = K\delta(t) \tag{5-36}$$

where $K\delta(t)$ is an impulse of strength K applied at $t = 0$. Let us assume that there is no charge present on the capacitor for $t < 0$. If we now integrate (5-36) between the limits $-\infty$ and $0+$, we obtain

$$Cv(0+) + G\int_{-\infty}^{0+} v(\tau)\,d\tau = \int_{-\infty}^{0+} i(\tau)\,d\tau = K \tag{5-37}$$

In the left member of this equation, the integral of $v(t)$ from $-\infty$ to $0+$ must be zero, since no excitation is applied before $t = 0$. Therefore, we see that

$$v(0+) = \frac{K}{C} \tag{5-38}$$

We conclude that the impulse of current applied by the current source has established an initial condition on the previously unexcited capacitor. The response of the capacitor voltage $v(t)$ to the application of such an impulse of current at $t = 0$ is thus easily found using the methods of Sec. 5-1. We obtain

$$v(t) = \frac{K}{C}e^{-t/RC} \qquad t > 0 \tag{5-39}$$

A similar (dual) situation holds true for the circuit shown in Fig. 5-6.2, in which an impulse of voltage of strength K is applied to the circuit by the voltage source at $t = 0$. We assume that the initial condition at $t = 0-$ is zero. From KVL, we obtain for this circuit

$$L\frac{di}{dt} + Ri(t) = v(t) = K\delta(t) \tag{5-40}$$

Integrating from $-\infty$ to $0+$, we obtain

$$Li(0+) + R\int_{-\infty}^{0+} i(\tau)\,d\tau = \int_{-\infty}^{0+} v(\tau)\,d\tau = K \tag{5-41}$$

As before, the integral in the left member of this equation is zero, and we see that the impulse of voltage has produced an initial condition of current in the circuit which has the value

$$i(0+) = \frac{K}{L} \tag{5-42}$$

Thus, the solution for $i(t)$ is

$$i(t) = \frac{K}{L}e^{-Rt/L} \qquad t > 0 \tag{5-43}$$

It should be noted that in both the examples given above, the use of impulses has violated the continuity condition for the energy-storage element concerned. We may summarize the results given in the above examples as follows:

SUMMARY 5-6.1

The Use of Impulses to Generate Initial Conditions: An impulse of current applied by a current source in parallel with a capacitor will generate an initial condition of voltage on the capacitor. An impulse of voltage applied by a voltage source in series with an inductor will generate an initial condition of current in the inductor.

The response of some network variable to an impulse of excitation as illustrated in the examples given above is logically referred to as the *impulse response.* Since, from the above discussion, impulses can be thought of as establishing initial conditions in a circuit, it should be apparent that the impulse response is identical to the response which results from initial conditions which have been established by any other means. Thus, the treatment of circuits containing sources which apply impulses as well as other excitations is the same as the treatment of circuits excited by initial conditions and sources with non-impulsive outputs. This means that all the techniques of analysis which have been developed so far in this chapter can be applied to such circuits. An example of such a situation follows.

EXAMPLE 5-6.1. The circuit shown in Fig. 5-6.3 is excited by two voltage sources connected in series. Source 1 applies an impulse of strength I_0 (volts) to the circuit at $t = 0$. Source 2 applies an excitation $(10 \sin t)u(t)$ V to the circuit, i.e., an excitation of $10 \sin t$ V starting at $t = 0$. It is assumed that there is no initial condition present in the circuit for

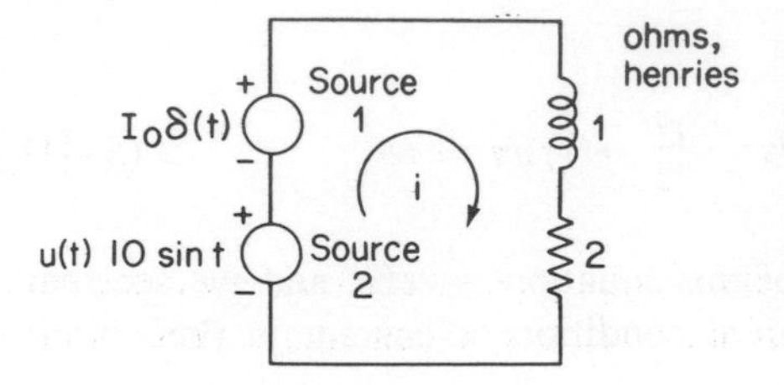

Fig. 5-6.3 Circuit for Example 5-6.1.

$t < 0$. From Eq. (5-42) we see that the effect of Source 1 is to establish an initial condition $i(0+) = I_0$ A in the inductor. The solution for the current $i(t)$ for $t > 0$ follows exactly the procedure given in Example 5-3.1. Thus we see that

$$i(t) = (2 + I_0)e^{-2t} + 4 \sin t - 2 \cos t \qquad t > 0$$

Note that we have specified this answer for the strict inequality $t > 0$, rather than the weak inequality $t \geq 0$ used in Example 5-3.1, since the conditions in the circuit discussed here are actually indeterminate at $t = 0$. In addition, we note that the two voltage sources could be combined into a single voltage source with an output voltage $v(t) = I_0\delta(t) + 10 \sin t$ V to obtain the same response for the variable $i(t)$.

It should be noted that if a voltage source (rather than a current source) is connected in parallel with a capacitor, and an impulse of voltage is applied, a quite different situation from that described above results. From the basic relation for the terminal variables of a capacitor, we know that $i(t) = C(dv/dt)$. Thus, if $v(t)$ is an impulse function, the resulting current into the capacitor is proportional to the derivative of an impulse, i.e., a doublet.[9] A similar situation results if an impulse of current is applied to an inductor. In this case, a doublet of voltage occurs across the inductor. A detailed discussion of such functions is beyond the scope of this text.

Since the impulse response described above is identical with the response found when a given network is excited only by initial conditions, its form is determined solely by the properties of the network, rather than the form of the excitation. Thus it is an intrinsic or *natural* property of the network itself. We may express this in another manner by saying that the impulse response of a given network is the result of exciting the *natural frequencies* of the network. This concept will become of great utility when we discuss the Laplace transformation in Chap. 8.

In linear circuits there is a direct relation between the impulse response of a given network and the response obtained by the application of any other waveform by the same source. We shall discuss this relation in more detail in Sec. 5-8. Here, as an introduction to the topic, consider the indefinite integral of an impulse $\delta(t)$. In Sec. 4-2 this was shown to be the unit step $u(t)$. Thus, we may write

$$u(t) = \int_{-\infty}^{t} \delta(\tau)\, d\tau \qquad (5\text{-}44)$$

Not only is the unit step the integral of the unit impulse, but, if we apply a unit step $u(t)$ as an excitation to a given network, using the same source which was used to apply an impulse, the resulting response will be the integral of the impulse response. As an example of this, consider again the circuit shown in Fig. 5-6.1. If we let $v_\delta(t)$

[9]The doublet was introduced in Sec. 4-3.

be the impulse response, then, from Eq. (5–39), we may write

$$v_\delta(t) = \frac{K}{C} e^{-t/RC} u(t) \tag{5-45}$$

where we have used the symbol $u(t)$ to avoid having to consider the limitation $t > 0$ in connection with the equation. To find the response to a unit step, we need merely integrate the impulse response given in Eq. (5–45). Let $v_u(t)$ be the step response. We obtain

$$\begin{aligned} v_u(t) &= \int_{-\infty}^{t} v_\delta(\tau)\, d\tau \\ &= \int_{-\infty}^{t} \frac{K}{C} e^{-\tau/RC} u(\tau)\, d\tau = KR(1 - e^{-t/RC})u(t) \end{aligned} \tag{5-46}$$

It is readily verified by direct solution of the problem that the expression for $v_u(t)$ given in Eq. (5–46) is the same as the voltage resulting from the excitation of the circuit by a source current $Ku(t)$, i.e., a step of magnitude K A applied at $t = 0$. A similar conclusion is readily established for the RL circuit shown in Fig. 5–6.2. We may summarize the development given above by the following:

SUMMARY 5-6.2

Relation Between the Impulse Response and the Step Response: The response of a given network variable to the application of a step input is the integral of the response of that same variable to the application of an impulse. It is assumed that the initial conditions in the network are zero before the application of the excitations.

As a further example of the importance of the impulse response, we leave the reader with the following conclusion, which will be developed further in Sec. 5–8: If the impulse response relating a given network variable to a specific source is known, then this expression may be used to find the response for any arbitrary excitation. Thus, the behavior of the network is completely characterized. For such a property to apply, of course, the network must be linear, i.e., comprised of linear elements.

5-7 NUMERICAL SOLUTION OF FIRST-ORDER DIFFERENTIAL EQUATIONS

In the preceding sections of this chapter we have discussed various methods for finding the solution of first-order differential equations. A quick review of the contents of these sections should prove to the reader that there is nothing very complicated about the procedure. Therefore, since these standard mathematical techniques of solving first-order differential equations are so straightforward, we might wonder why it is necessary to consider the use of numerical techniques to perform such a task. There are several good reasons. First of all, the numerical techniques which we shall develop here for the solution of first-order differential equations are directly applicable, with only slight modification, to the solution of higher-order differential equations and also to the solution of simultaneous sets of differential equations. The solution of such cases by standard mathematical methods, however, is considerably more complicated than the solution of the first-order case. Second, the same numerical techniques which are used to solve linear, time-invariant, first-order differential equations may be directly applied to the solution of non-linear and time-varying ones. Differential equations of these latter types are extremely difficult to solve by ordinary mathematical methods. Finally, when numerical solution techniques are used, the waveshape of the excitation function may be specified by discontinuous or even graphical relationships, rather than the explicit mathematical formulation required by classical mathematical solution techniques. Thus, there is much to be gained from a study of numerical methods for solving differential equations.

We may begin our study of numerical techniques for solving first-order differential equations by considering the circuit shown in Fig. 5-7.1. We shall assume that the initial condition for the inductor current is zero, and that the excitation voltage is a rectangular pulse of 80 V applied to the circuit at $t = 0$, and removed from the circuit at $t = 0.003$ s. The waveform of this pulse is shown in Fig. 5-7.2.

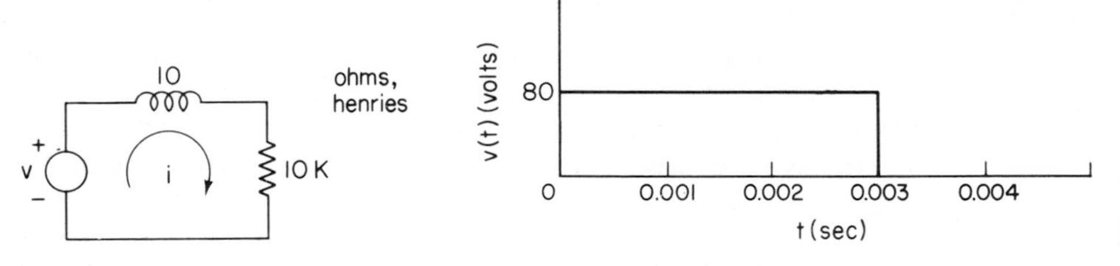

Fig. 5-7.1 A first-order *RL* circuit.

Fig. 5-7.2 Input waveform for the circuit shown in Fig. 5-7.1.

The differential equation for the inductor current $i(t)$ for the numerical values given in the figure is

$$10\frac{di}{dt} + 10^4 i(t) = v(t) \tag{5-47}$$

This may be put in the form

$$\frac{di}{dt} = -10^3 i(t) + \frac{v(t)}{10} \tag{5-48}$$

This is an expression which has the general form

$$\frac{di}{dt} = g[t, i(t)] \tag{5-49}$$

Equation (5–49) tells us that the derivative of $i(t)$ is equal to some function g, where g is a function of t and $i(t)$. For example, let us express the waveshape shown in Fig. 5–7.2 by using the unit step function defined in Sec. 4–2. Thus, we obtain

$$v(t) = 80[u(t) - u(t - 0.003)] \tag{5-50}$$

For this excitation, the differential equation of (5–48) becomes

$$i'(t) = g[t, i(t)] = -10^3 i(t) + 8[u(t) - u(t - 0.003)] \tag{5-51}$$

where, for convenience in future notation, we have used a prime to indicate differentiation with respect to t. Now let t_0 be some time at which we know an "initial" condition for $i(t)$. In other words, let $i(t_0)$ be a known quantity. Also, let t_1 be a value of time greater than t_0. We may now use the first two terms of a Taylor series to approximate $i(t_1)$. Thus, we may write

$$i(t_1) \approx i(t_0) + i'(t_0)(t_1 - t_0) \tag{5-52}$$

where $i'(t_0)$ is the derivative of $i(t)$ evaluated at $t = t_0$. We may substitute the functional relation given in (5–49) for $i'(t_0)$. Thus, we may write

$$i(t_1) \approx i(t_0) + g[t_0, i(t_0)](t_1 - t_0) \tag{5-53}$$

This relation allows us to find an approximation for $i(t_1)$. Clearly, once $i(t_1)$ is known, we could pick a new value of time t_2 and use our knowledge of $i(t_1)$ and $i'(t_1)$ to find $i(t_2)$. The process could be continued until enough values of $i(t)$ were computed to determine a plot of the function. We would expect that the more closely the adjacent values of t_i were spaced, the more accurate our results would be. Such a method of solving for a set of values of $i(t)$ which satisfy a first-order differential equation of the form given in (5–49) is called a *predictor* method since it uses known values of $i(t)$ and $i'(t)$ to predict a new value of $i(t)$. It is also referred to as a *first-order* method since it involves one evaluation of the function $g[t, i(t)]$ of (5–49) for the computation of each new value of $i(t)$. As an illustration of how this method would be applied, consider a one-mesh circuit consisting of a series connection of a voltage source with a constant output voltage of 1 V, a 1-Ω resistor, and a 1-H inductor. Applying KVL around the mesh, we obtain

$$i'(t) + i(t) = 1$$

where $i(t)$ is the mesh current. From (5–49) we find the function $g[t, i(t)]$ is given as

$$g[t, i(t)] = i'(t) = 1 - i(t)$$

Let us assume that $i(0) = 0$. We may find an approximation to $i(0.1)$ by applying (5–52) as follows

$$i(0.1) = i(0) + g[0, i(0)](0.1 - 0) = 0 + (1)(0.1) = 0.1$$

Now we may again apply (5–52) to use the value of $i(0.1)$ found above to determine $i(0.2)$. Thus, we see that

$$i(0.2) = i(0.1) + g[0.1, i(0.1)](0.2 - 0.1) = 0.1 + (0.9)(0.1) = 0.19$$

Continuing in the same manner, we find $i(0.3) = 0.271$, $i(0.4) = 0.3439$, etc. If we plot these values of $i(t)$, we find they approximate the actual solution of the differential equation which is easily shown to be $i(t) = 1 - e^{-t}$.

The first-order predictor method described above is too crude to be of much practical use. However, it does illustrate the basic principles that are involved in the numerical solution of differential equations. In order to obtain a more accurate predictor method of solving a differential equation, we note first of all that we cannot merely include more terms in the Taylor series, since the function g of Eq. (5–49) only determines di/dt, and no relation is given for the quantities d^2i/dt^2, d^3i/dt^3, etc. What we can do instead, however, is to make several evaluations of $i'(t)$ at intermediate points between t_0 and t_1, and use these values to generate an approximation to the curve describing $i(t)$ over the specified interval. For example, a fourth-order predictor method which is widely used is the Runge-Kutta method.[10] Given $i(t_0)$ and the function $g[t, i(t)]$, we may find $i(t_1)$ by computing the quantities $g^{(1)}(t_0)$, $g^{(2)}(t_0)$, $g^{(3)}(t_0)$, and $g^{(4)}(t_0)$ using the following relations

$$\begin{aligned} g^{(1)}(t_0) &= g[t_0, i(t_0)] \\ g^{(2)}(t_0) &= g\left[t_0 + \frac{\Delta t}{2}, i(t_0) + \frac{\Delta t\, g^{(1)}(t_0)}{2}\right] \\ g^{(3)}(t_0) &= g\left[t_0 + \frac{\Delta t}{2}, i(t_0) + \frac{\Delta t\, g^{(2)}(t_0)}{2}\right] \\ g^{(4)}(t_0) &= g[t_0 + \Delta t, i(t_0) + \Delta t\, g^{(3)}(t_0)] \end{aligned} \tag{5–54}$$

where

$$\Delta t = t_1 - t_0 \tag{5–55}$$

Thus, to calculate $g^{(1)}(t_0)$, we use the value t_0 for the first argument of the function g, and the value of $i(t_0)$ as the second argument. Similarly, to calculate $g^{(2)}(t_0)$ we use the value of the quantity $t_0 + \Delta t/2$ as the first argument of the function g, i.e.,

[10]To include a proof of the Runge-Kutta method at this point would lead us too far afield from our primary goal in this text. A proof of the relations given here may be found in any of the standard texts on numerical techniques. A list of some of these is given in the Bibliography at the end of this book.

we substitute this value for t in Eq. (5–49), and the value of the quantity $i(t_0) + [\Delta t\, g^{(1)}(t_0)/2]$ for the second argument of the function, i.e., we substitute this value for $i(t)$ in (5–49). Note that the quantities $g^{(i)}(t_0)$ $(i = 1, 2, 3, 4)$ must be calculated in the order given in (5–54), since the calculation for $g^{(2)}(t_0)$ requires that $g^{(1)}(t_0)$ be known, etc. Once the quantities $g^{(i)}(t_0)$ have been calculated $i(t_1)$ is found by the relation

$$i(t_1) = i(t_0) + \frac{[g^{(1)}(t_0) + 2g^{(2)}(t_0) + 2g^{(3)}(t_0) + g^{(4)}(t_0)]\Delta t}{6} \tag{5-56}$$

The relations given in Eqs. (5–54), (5–55), and (5–56) are readily programmed for the digital computer. In doing this, it is convenient to divide the interval between the initial value of time t_0 and the next value of time t_1 into separate subintervals, find the value of $i(t)$ at the first intermediate point, use this value in determining the value of $i(t)$ at the next intermediate point, etc. A flow chart illustrating the logic for doing this is shown in Fig. 5–7.3. The flow chart is readily implemented by a set of FORTRAN statements, a listing of which is given in Fig. 5–7.4. For convenience, these statements have been prepared as part of a subroutine named DFERK (for (for DiFFerential Equation solution by Runge-Kutta method). As indicated in the listing, a function subprogram G(T, A) must be prepared to specify the relations given in (5–49). A summary of the properties of the subroutine DFERK is given in Table 5–7.1 (on page 285).

To use the subroutine DFERK to obtain a plot of the solution to a given differential equation, we may use the same type of approach which was introduced with respect to the indefinite integral problem given in Sec. 4–5. Basically, this consists of calling the subroutine DFERK to determine each of the desired solution

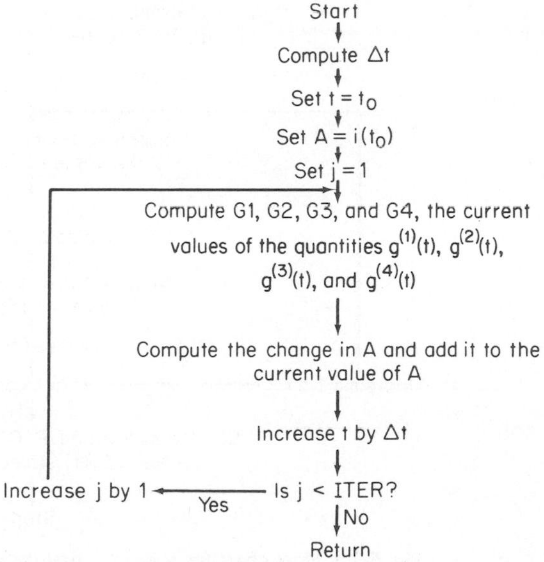

Start
Compute Δt
Set $t = t_0$
Set $A = i(t_0)$
Set $j = 1$
Compute G1, G2, G3, and G4, the current values of the quantities $g^{(1)}(t)$, $g^{(2)}(t)$, $g^{(3)}(t)$, and $g^{(4)}(t)$
Compute the change in A and add it to the current value of A
Increase t by Δt
Is j < ITER?
Yes: Increase j by 1
No: Return

Fig. 5-7.3 Flow chart for applying the Runge-Kutta algorithm.

```
      SUBROUTINE DFERK (TI, AI, TF, ITER, A)
C     SUBROUTINE FOR SOLVING A FIRST ORDER DIFFERENTIAL
C     EQUATION BY A FOURTH ORDER RUNGE-KUTTA METHOD.
C     A FUNCTION G(T,A) MUST BE PROVIDED TO DEFINE THE PROBLEM.
C          TI - INITIAL VALUE OF T
C          AI - INITIAL VALUE OF VARIABLE A(T)
C          TF - FINAL VALUE OF T
C        ITER - NUMBER OF ITERATIONS USED BETWEEN TI AND TF
C           A - FINAL VALUE OF VARIABLE A(T)
C
C     COMPUTE DT AND INITIALIZE T AND A
      HITER = ITER
      DT = (TF - TI) / HITER
      A = AI
      T = TI
C
C     PERFORM THE STEPS TO COMPUTE A(T) FOR T = TF
      DO 10 J = 1, ITER
      G1 = G(T, A)
      G2 = G(T + DT/2., A + DT*G1/2.)
      G3 = G(T + DT/2., A + DT*G2/2.)
      G4 = G(T + DT, A + DT*G3)
      A = A + (G1 + 2.*G2 + 2.*G3 + G4) * DT/6.
   10 T = T + DT
      RETURN
      END
```

Fig. 5-7.4 Listing of the subroutine DFERK.

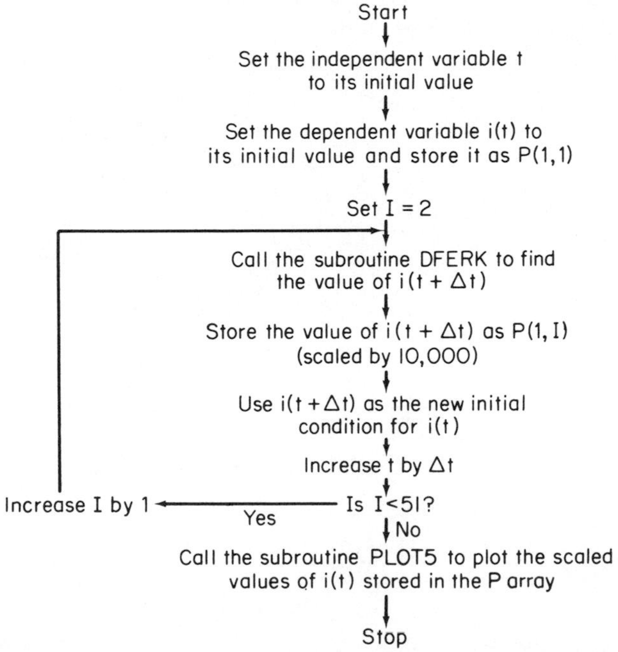

Start
↓
Set the independent variable t to its initial value
↓
Set the dependent variable i(t) to its initial value and store it as P(1,1)
↓
Set I = 2
↓
Call the subroutine DFERK to find the value of i(t + Δt)
↓
Store the value of i(t + Δt) as P(1, I) (scaled by 10,000)
↓
Use i(t + Δt) as the new initial condition for i(t)
↓
Increase t by Δt
↓
Is I<51? — Yes → Increase I by 1 (return to Call the subroutine DFERK)
↓ No
Call the subroutine PLOT5 to plot the scaled values of i(t) stored in the P array
↓
Stop

Fig. 5-7.5 Flow chart for solving a first-order differential equation.

points. The output from the subroutine is then used as the initital value of the variable for the next call of the subroutine. Simultaneously, the value of the independent variable is increased. The process is illustrated in the flow chart shown in Fig. 5-7.5. An example follows.

EXAMPLE 5-7.1. It is desired to find the current $i(t)$ over a range of time 0-0.005 s in the circuit shown in Fig. 5-7.1 when the pulse of voltage defined in (5-50) and shown in Fig. 5-7.2 is applied. It is assumed that $i(0) = 0$. The subroutine DFERK may be used to solve this problem, and the subroutine PLOT5 to display the results. A function G(T, AMPS) is

```
      FUNCTION G (T, AMPS)
C     THIS FUNCTION DEFINES THE DIFFERENTIAL
C     EQUATION OF THE FORM DI/DT=G(T,I) WHICH
C     IS USED IN EXAMPLE 5.7-1.
C              T - INDEPENDENT VARIABLE, TIME
C           AMPS - DEPENDENT VARIABLE, CURRENT
C
      IF (T .GT. 0.003) GO TO 4
      G = 8. - AMPS * 1000.
      RETURN
    4 G = - AMPS * 1000.
      RETURN
      END

C     MAIN PROGRAM FOR EXAMPLE 5.7-1
C       AMPOLD - VALUE OF CURRENT BEFORE CALLING DFERK
C       AMPNEW - VALUE OF CURRENT AFTER CALLING DFERK
C           DT - TIME INTERVAL BETWEEN PLOTTED POINTS
C            T - INDEPENDENT VARIABLE, TIME
C            P - ARRAY FOR STORING VARIABLES TO BE PLOTTED
      DIMENSION P(5,101)
C
C     INITIALIZE VARIABLES
      AMPOLD = 0.
      P(1,1) = AMPOLD
      DT = 1.E-04
      T = 0.
C
C     COMPUTE AND STORE 50 VALUES OF CURRENT
C     (SCALED BY 10000) IN P ARRAY USING TWO
C     INTERMEDIATE CALCULATIONS FOR EACH CALL
C     OF SUBROUTINE DFERK
      DO 8   I = 2, 51
      CALL DFERK (T, AMPOLD, T+DT, 2, AMPNEW)
      P(1,I) = AMPNEW * 10000.
      AMPOLD = AMPNEW
    8 T = T + DT
C
C     CALL PLOTTING SUBROUTINE TO PRODUCE PLOT
      PRINT 10
   10 FORMAT (1H1)
      CALL PLOT5 (P, 1, 51, 100)
      STOP
      END
```

Fig. 5-7.6 Listing of the program for Example 5-7.1.

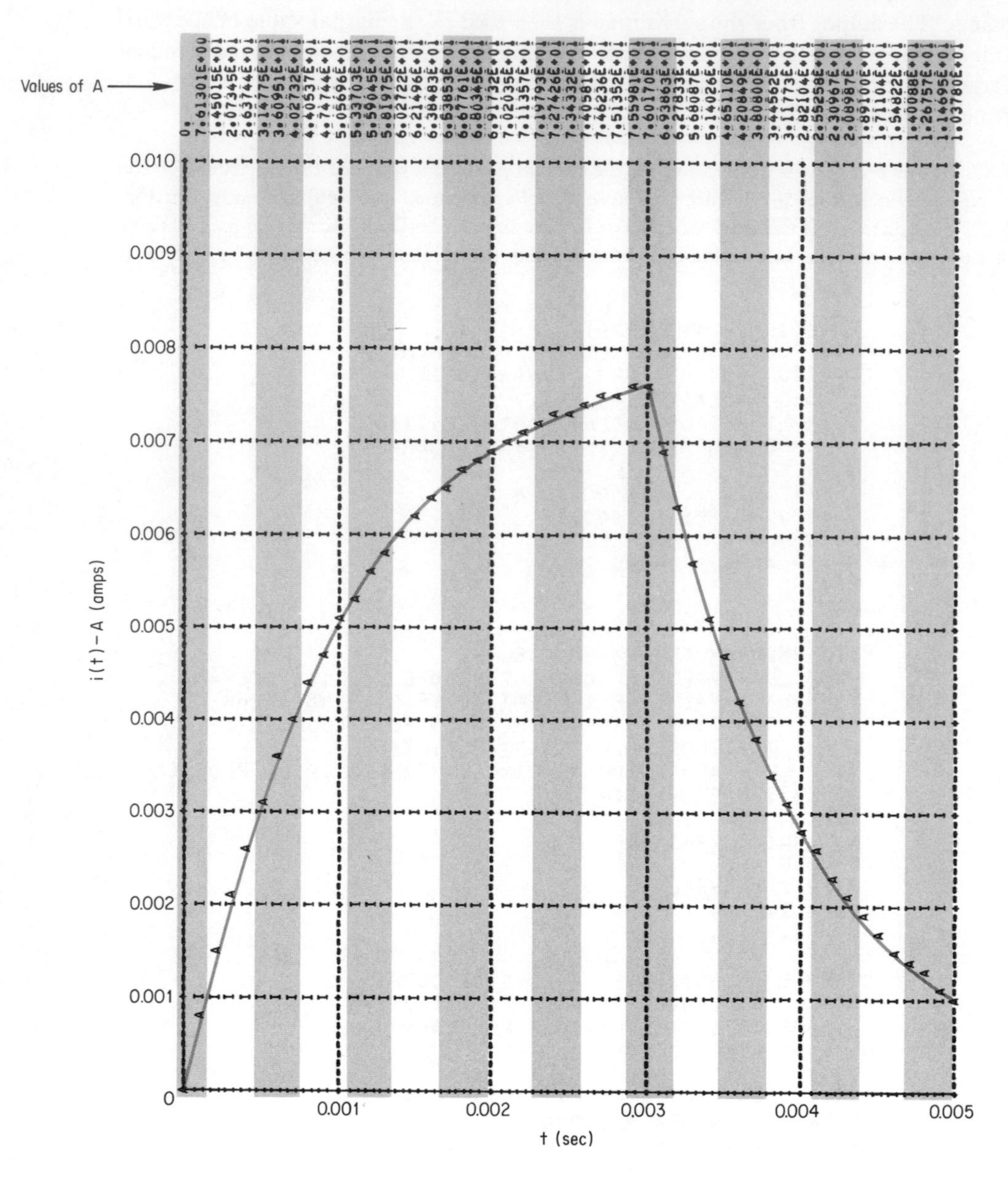

Fig. 5-7.7 Output plot for Example 5-7.1.

easily written to express the relation for the first-order differential equation given in Eq. (5-51). A listing of the function and of the main program is given in Fig. 5-7.6. A plot of the resulting output current $i(t)$ is shown in Fig. 5-7.7. The values of $i(t)$ shown in the plot are readily verified by direct (non-numerical) solution of the problem.

The differential equation solution method using the subroutine DFERK given above is readily applied to time-varying, energy-storage elements as well as to time-invariant ones. For example, for a time-varying inductor, it was shown in Sec. 4-6 that the differential equation relating the terminal variables had the form

$$v(t) = L(t)i'(t) + i(t)L'(t) \tag{5-57}$$

Rearranging the terms of this equation to put it in the form of (5-49), we obtain

$$i'(t) = g[t, i(t)] = \frac{v(t) - i(t)L'(t)}{L(t)} \tag{5-58}$$

where it is assumed that the variation of the inductance with time is given, and thus $L(t)$ and $L'(t)$ are known quantities. An example of the use of DFERK in solving a time-varying inductor problem follows.

EXAMPLE 5-7.2. A time-varying inductor is defined over a range of time from 0.05 to 2.5 s by the relation $L(t) = t$. Let the voltage applied to such an inductor be constant and equal to 4 V. Let $i(t)$ be the current through the inductor, and let $i(0.05) = 84$ A. We may use the subroutine DFERK to find the current $i(t)$ for the specified time range using (5-58) to define a function G(T, AMPS), and the subroutine PLOT5 to display the results. A listing of the function G(T, AMPS) and the main program is given in Fig. 5-7.8. A plot of the output is given in Fig. 5-7.9. It is difficult, in general, to determine explicitly the solution to a differential equation with time-varying coefficients. However, for this particular example, it is easily verified by direct substitution in (5-58) that the solution is $i(t) — 4[1 + (1/t)]$ A. A comparison of this result with the numerical values of the function plotted in Fig. 5-7.9 readily establishes the excellent agreement between the numerical results and the explicit mathematical solution.

The subroutine DFERK may also be applied to the case where it is desired to solve a circuit containing a non-linear energy-storage element. For example, consider the case of a non-linear inductor. The differential equation for such an element may be written in the form

$$v(t) = \frac{d}{dt}[L(i)i(t)] \tag{5-59}$$

Since, in the non-linear case, L is a function of $i(t)$, we obtain

$$v(t) = L(i)i'(t) + i(t)\frac{dL}{di}i'(t) \tag{5-60}$$

```
      FUNCTION G(T, AMPS)
C     THIS FUNCTION DEFINES THE DIFFERENTIAL
C     EQUATION OF THE FORM DI/DT=G(T,I) WHICH
C     IS USED IN EXAMPLE 5.7-2,
C            T - INDEPENDENT VARIABLE, TIME
C         AMPS - DEPENDENT VARIABLE, CURRENT
      G = (4. - AMPS) / T
      RETURN
      END

C     MAIN PROGRAM FOR EXAMPLE 5.7-2
C           DT - TIME INTERVAL BETWEEN PLOTTED POINTS
C            T - INDEPENDENT VARIABLE, TIME
C         AMPS - ARRAY USED TO STORE VALUES OF CURRENT
C                TO BE PLOTTED
      DIMENSION AMPS(5,101)
C
C     INITIALIZE THE VARIABLES
      AMPS(1,1) = 84.
      AMPS (1,2) = 84.
      DT = .05
      T = .05
C
C     COMPUTE AND STORE 49 VALUES OF CURRENT
C     FOR TIMES FROM .05 TO 2.5 SEC, AND STORE
C     IN AMPS ARRAY FOR PLOTTING. USE 2 INTER-
C     MEDIATE CALCULATIONS FOR EACH POINT
      DO 7   I = 2, 50
      IA = I + 1
      CALL DFERK (T, AMPS(1,I), T+DT, 2, AMPS(1,IA))
    7 T = T + DT
C
C     CALL SUBROUTINE PLOT TO PRODUCE A PLOT OF CURRENT,
      PRINT 9
    9 FORMAT (1H1)
      CALL PLOT5 (AMPS, 1, 51, 100)
      STOP
      END
```

Fig. 5-7.8 Listing of the program for Example 5-7.2.

where primes are used to indicate differentiation with respect to t. If we rearrange the above equation to put it in the form of (5–49), we obtain

$$i'(t) = g[t, i(t)] = \frac{v(t)}{L(i) + i(t)(dL/di)} \tag{5-61}$$

Since $L(i)$ and, thus, dL/di are known functions, the subroutine DFERK may be applied to solve for $i(t)$, assuming that $v(t)$ and some initial condition on $i(t)$ are known. An example follows.

EXAMPLE 5-7.3. A non-linear inductor is defined over a range of current from 0 to 0.4 A by the relation $L(i) = 3 - 6i^2$, where $i(t)$ is the current through the inductor. Let a voltage $v(t) = 6 - 144t^2$ V be applied to the inductor over the range of time from 0 to 0.2 s, and let

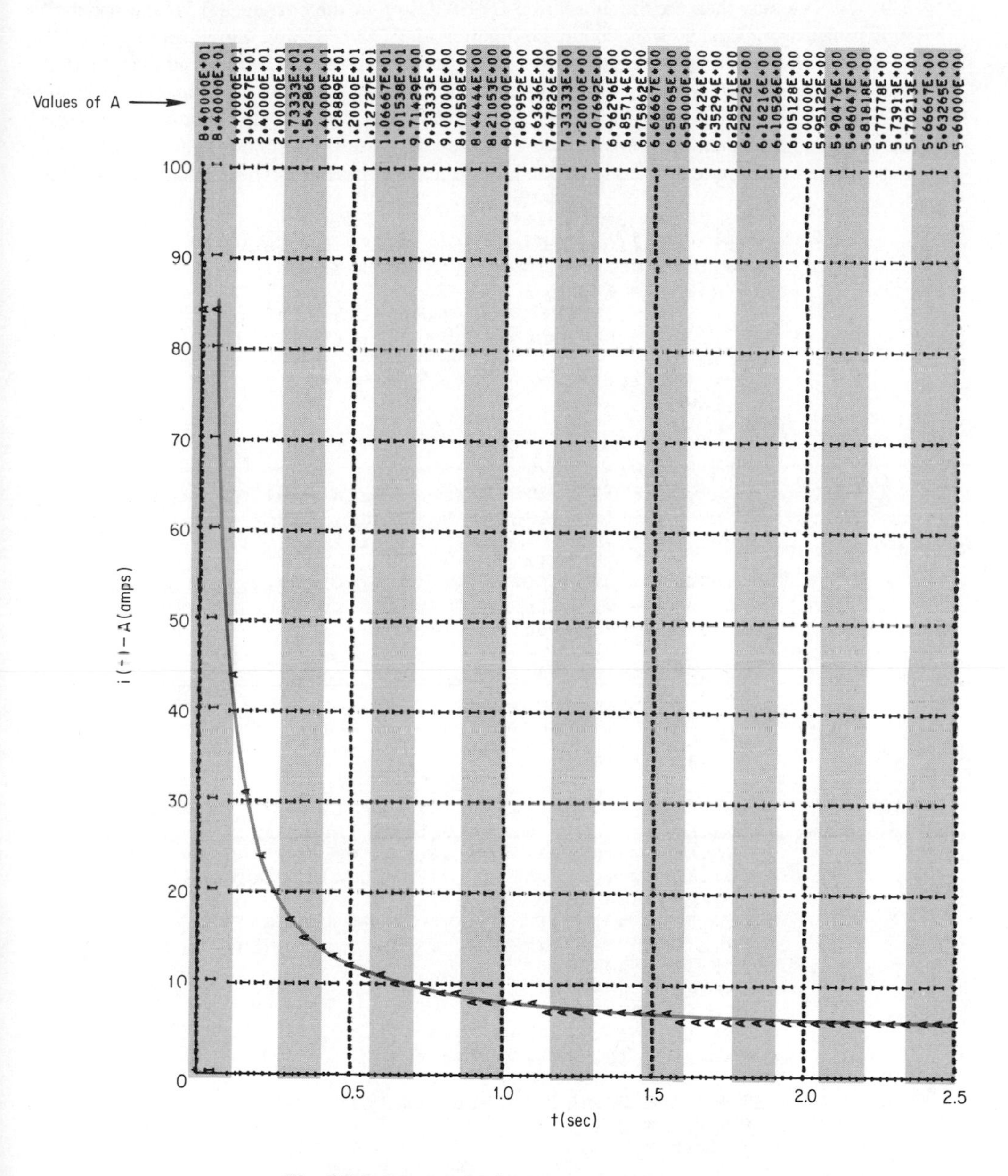

Fig. 5-7.9 Output plot for Example 5-7.2.

$i(0) = 0$. We may then use the subroutine DFERK to find the current $i(t)$ for the specified time range, using Eq. (5-61) to define the function G(T, AMPS), and the subroutine PLOT5 to display the resulting variation of $i(t)$. The flow chart that is used is similar to that shown in Fig. 5-7.5. A listing of the function G(T, AMPS) and the main program is given in Fig. 5-7.10. A plot of $i(t)$, $v(t)$, and $L[i(t)]$ is shown in Fig. 5-7.11. Although it is difficult, in general, to determine explicitly the solution to a non-linear differential equation, for this particular example it is easily verified by direct substitution in (5-61) that the solution

```
      FUNCTION G (T, I)
C     THIS FUNCTION DEFINES THE DIFFERENTIAL
C     EQUATION OF THE FORM DI/DT=G(T,I) WHICH
C     IS USED IN EXAMPLE 5.7-3.
C             T - INDEPENDENT VARIABLE, TIME
C             I - DEPENDENT VARIABLE, CURRENT
      REAL I
      G = (6.-144.*T**2) / (3.-18.*I**2)
      RETURN
      END
C
C     MAIN PROGRAM FOR EXAMPLE 5.7-3
C            DT - TIME INTERVAL BETWEEN PLOTTED POINTS
C             T - INDEPENDENT VARIABLE, TIME
C             P - ARRAY USED TO STORE VALUES OF CURRENT,
C                 INDUCTANCE, AND VOLTAGE TO BE PLOTTED
C        AMPOLD - VALUE OF CURRENT BEFORE CALLING DFERK
C        AMPNEW - VALUE OF CURRENT AFTER CALLING DFERK
      DIMENSION P(5,101)
C
C     INITIALIZE THE VARIABLES
      DT = .005
      T = 0.
      AMPOLD = 0.
      P(1,1) = AMPOLD
C
C     COMPUTE AND STORE 40 VALUES OF CURRENT IN P(1,I)
C     (SCALED BY 200), 40 VALUES OF INDUCTANCE IN P(2,I)
C     (SCALED BY 10), AND 40 VALUES OF VOLTAGE IN P(3,I)
C     (SCALED BY 10). USE 2 INTERMEDIATE CALCULATIONS
C     IN DFERK FOR EACH COMPUTATION
      DO 10  I = 2, 41
      CALL DFERK (T, AMPOLD, T+DT, 2, AMPNEW)
      P(1,I) = AMPNEW * 200.
      AMPOLD = AMPNEW
      T = T + DT
      P(2,I) = (3.-6.*AMPNEW**2) * 10.
   10 P(3,I) = (6.-144.*T**2) * 10.
C
C     STORE THE SCALED INITIAL VALUES
      P(2,1) = 30.
      P(3,1) = 60.
C
C     CALL PLOTTING SUBROUTINE TO PRODUCE A PLOT OF
C     CURRENT, INDUCTANCE, AND VOLTAGE WAVEFORMS
      PRINT 12
   12 FORMAT (1H1)
      CALL PLOT5 (P, 3, 41, 100)
      STOP
      END
```

Fig. 5-7.10 Listing of the program for Example 5-7.3.

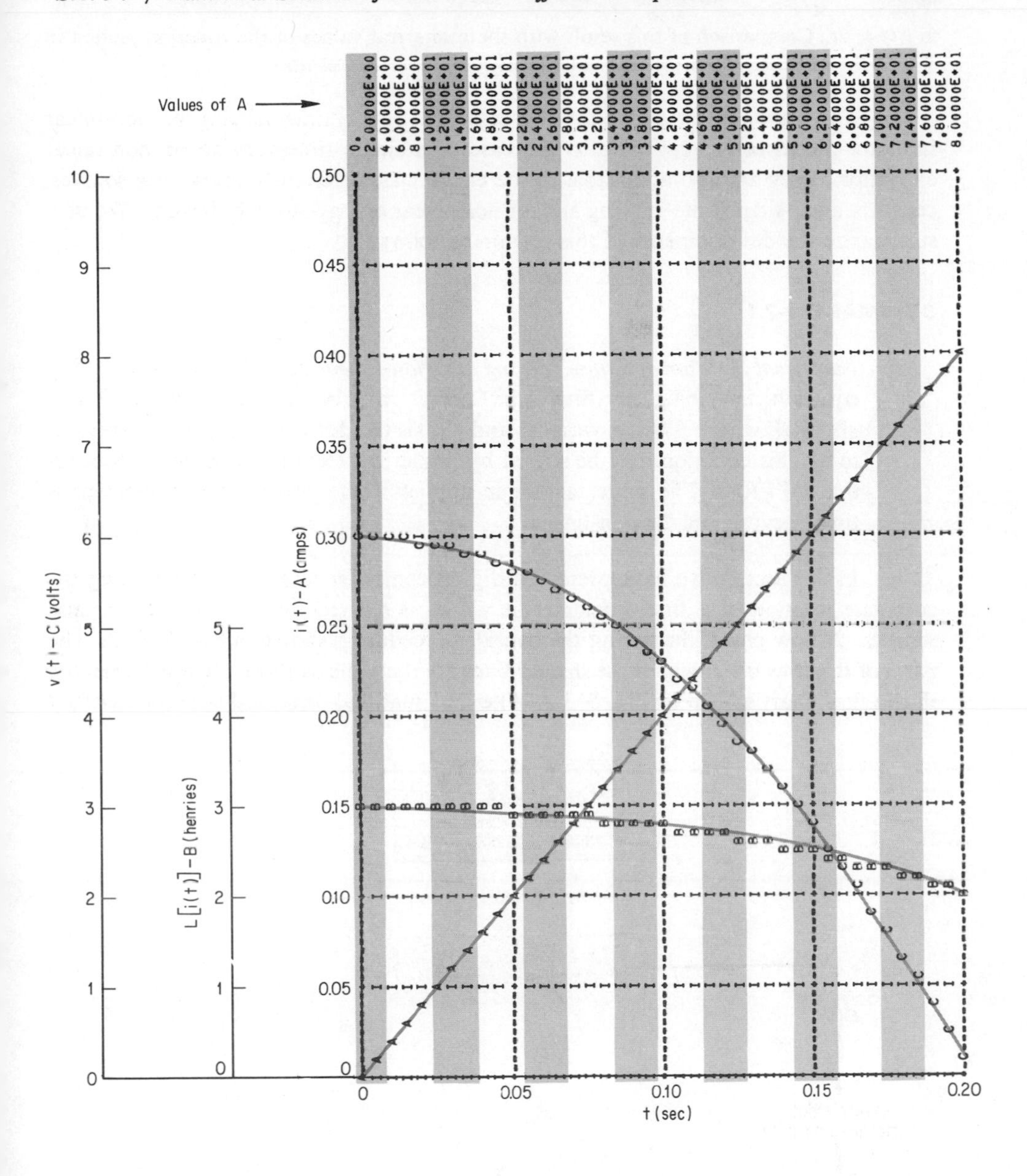

Fig. 5-7.11 Output plot for Example 5-7.3.

is $i(t) = 2t$. Comparison of this result with the numerical values of the function plotted in Fig. 5-7.11 readily establishes the validity of the numerical results.

The results given above for the case of a single time-varying or non-linear inductor are directly applicable to the case of a single time-varying or non-linear capacitor. They are also applicable to the entire class of circuits containing sources, resistors, and a single time-varying and/or non-linear energy-storage element. We may summarize the developments of this section as follows:

SUMMARY 5-7.1

Numerical Solution of Time-Varying and Non-Linear First-Order Circuits: The equation describing any first-order circuit may be put in the form $f'(t) = g[t, f(t)]$, where $f(t)$ is a variable and $f'(t)$ is the derivative of $f(t)$ with respect to t. This equation may be solved by numerical techniques using the subroutine DFERK. The same technique applies if the energy-storage element is time-varying and/or non-linear.

In this section we have presented digital computer techniques for finding the complete solution of a first-order circuit which is excited by initial conditions and sources. A flow chart illustrating the overall procedure is shown in Fig. 5-7.12. The parts of this flow chart which use shaded lines are the same as those already presented in the flow chart shown in Fig. 5-3.3. The new material presented in this flow chart

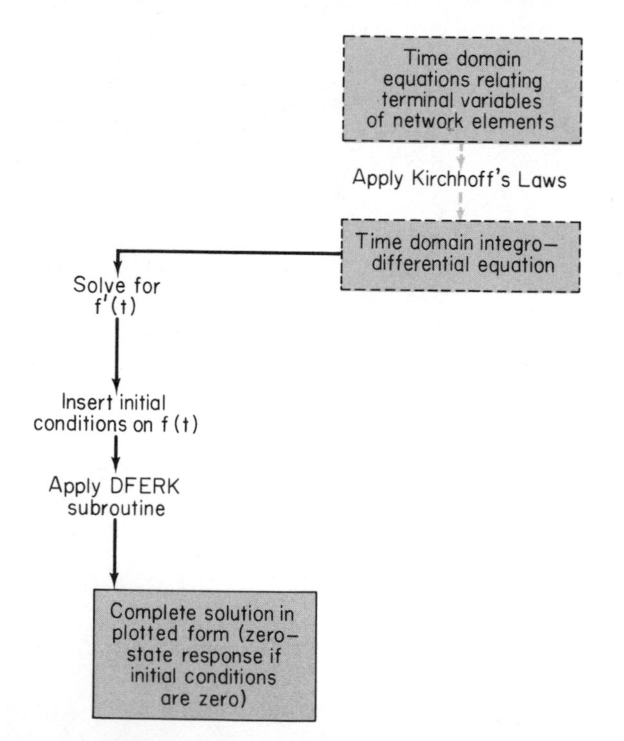

Fig. 5-7.12 Flow chart for using numerical techniques to solve a first-order circuit.

is shown with solid lines. The relative positioning of the blocks and paths have been arranged so as to permit it to be added to the flow charts previously presented in this chapter. Such an addition will be presented in Fig. 5–9.1 as part of the conclusion for the chapter.

TABLE 5-7.1 Summary of the Characteristics of the Subroutine DFERK

Identifying Statement: SUBROUTINE DFERK (TI, AI, TF, ITER, A)
Purpose: To solve the differential equation

$$i' = g(t, i)$$

and thus to find the value of $i(t)$ at the value of t specified as t_f. The process starts at t_i, a known value of t, and i_i, where $i_i = i(t_i)$.
Additional Subprogram Required: This subroutine calls a function identified by the statement

FUNCTION G(T, A)

This function is used to define the differential equation, i.e., to specify $g(t, i)$.
Input Arguments:

TI	The initial value t_i of the independent variable t.
AI	The initial value i_i of $i(t)$, i.e., $i(t_i)$.
TF	The final value t_f for which the value of $i(t)$ is desired.
ITER	The number of iterations used by the subroutine in going from t_i to t_f.

Output Argument:

A	The value of $i(t)$ at the final value of the independent variable t, i.e., $i(t_f)$.

5-8 CONVOLUTION AND SUPERPOSITION INTEGRALS

One of the limitations in the non-computer methods of analyzing first-order circuits which have been presented in this chapter is the difficulty of finding the forced response when the excitation function is defined by a complicated mathematical expression. Indeed, if the excitation is specified by a graphical waveform that has been experimentally determined, the methods are not applicable at all. In this section we shall introduce a most important general analysis method which is directly applicable to this latter case, as well as to the similar ones in which an analytic expression is given. The technique makes use of the impulse response and step response introduced in Sec. 5–6. It may be used to find the zero-state response of a network (see Summary 5–3.2) for an arbitrary waveform, even one which is graphically defined.

Let us begin our discussion by assuming that some arbitrary excitation function $g(t)$ is defined over the period $t_0 < t < t_n$ as shown in Fig. 5–8.1(a). We may approximate this function by a series of n pulses of duration Δt and of varying amplitude as shown in Fig. 5–8.1(b). Such a series of pulses may be represented analytically by defining a unit-area pulse function $p_i(t)$ by the relationship

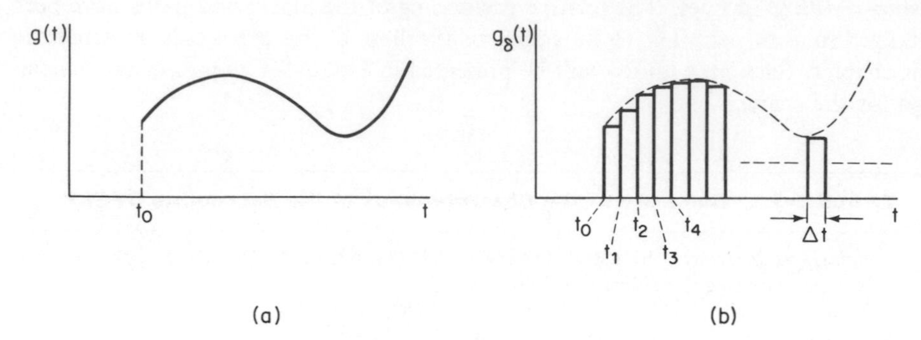

Fig. 5-8.1 An arbitrary excitation function and its approximation.

$$p_i(t) = \begin{cases} 0 & t < t_i \\ \dfrac{1}{\Delta t} & t_i < t < t_{i+1} \\ 0 & t > t_{i+1} \end{cases} \tag{5-62}$$

where $\Delta t = t_{i+1} - t_i$. If we now define the quantity $g_\delta(t)$ as the pulse representation of $g(t)$, we may write

$$g_\delta(t) = \sum_{i=0}^{n-1} p_i g(t_i)\,\Delta t \tag{5-63}$$

This expression defines a function over all values of t in terms of a set of discrete values, namely, the quantities $g(t_i)$. Since all the pulses p_i have been defined with the same width, an alternative series representation for $g_\delta(t)$ can be made which will be more useful in the following development. To do this, let us define a single rectangular pulse $p_\delta(t)$ of width Δt, beginning at $t = 0$, and having unit area. Thus, we may write

$$p_\delta(t) = \begin{cases} 0 & t < 0 \\ \dfrac{1}{\Delta t} & 0 < t < \Delta t \\ 0 & t > \Delta t \end{cases} \tag{5-64}$$

To use this function to represent any of the pulses $p_i(t)$ defined in Eq. (5-62), we need only shift $p_\delta(t)$ so that it begins at time t_i by replacing the argument t by $t - t_i$. Thus, we obtain

$$p_\delta(t - t_i) = \begin{Bmatrix} 0 & t - t_i < 0 \\ \dfrac{1}{\Delta t} & 0 < t - t_i < \Delta t \\ 0 & t - t_i > \Delta t \end{Bmatrix} = \begin{Bmatrix} 0 & t < t_i \\ \dfrac{1}{\Delta t} & t_i < t < t_i + \Delta t \\ 0 & t > t_i + \Delta t \end{Bmatrix} = p_i(t) \tag{5-65}$$

The series of pulses $g_\delta(t)$ of Eq. (5–63) may now be written

$$g_\delta(t) = \sum_{i=0}^{n-1} p_\delta(t - t_i)g(t_i)\,\Delta t \tag{5-66}$$

Let us now assume that the excitation waveform discussed above is applied as an input to a network which has a response $h_\delta(t)$ when excited by a pulse $p_\delta(t)$. If we assume that the network is linear and time-invariant, and that it is in the zero state, the response due to a shifted pulse $p_\delta(t - t_i)$ must be $h_\delta(t - t_i)$. Since homogeneity applies to linear networks, the response to a pulse $p_\delta(t - t_i)g(t_i)\,\Delta t$ must therefore be $h_\delta(t - t_i)g(t_i)\,\Delta t$, since the term $g(t_i)\,\Delta t$ only changes the amplitude of the pulse. We now have an expression for the response for any one of the pulses in the summation given in Eq. (5–66). Thus, we may apply superposition to find the response to the complete sum of pulses, i.e., to the function $g_\delta(t)$. For convenience, let us define $f_\delta(t)$ as the resulting response. From the discussion above, we see that we may write

$$f_\delta(t_n) = \sum_{i=0}^{n-1} h_\delta(t - t_i)g(t_i)\,\Delta t \tag{5-67}$$

Now let us see what happens as we increase n, the number of pulses used to represent $g(t)$ over the indicated time span from t_0 to t_n, where t_0 and t_n are fixed values. As n is increased, the quantity Δt becomes smaller, and in the limit as n approaches infinity, several items of interest occur. First of all, the pulse train $g_\delta(t)$ becomes equal to the actual input $g(t)$. Correspondingly, in the limit, $f_\delta(t)$ may be written as $f(t)$, the actual zero-state network response to the excitation $g(t)$. Now consider the pulse $p_\delta(t)$. In the limit, this simply becomes the impulse $\delta(t)$. Thus, the quantity $h_\delta(t)$ in the limit becomes the impulse response which we will designate as $h(t)$. Finally, we may note that in the limit the summation of (5–67) becomes an integral, the sequence of variables t_i becomes a continuous variable of integration which we may designate as τ, and the incremental quantity Δt becomes $d\tau$. Inserting these results in (5–67), we obtain

$$f(t) = \int_{t_0}^{t} h(t - \tau)g(\tau)\,d\tau \tag{5-68}$$

where we have replaced t_n, an arbitrary value of time which is greater than t_0, by t. This relation defines the *convolution integral*. It allows us to find the response of a network to any excitation $g(t)$ just so long as we know the impulse response $h(t)$ of the network. An example illustrating the application of the convolution integral follows.

EXAMPLE 5-8.1. A first-order *RC* network is shown in Fig. 5-8.2. The impulse response $i_\delta(t)$ of this network to an excitation $v(t) = \delta(t)$ is given as $i_\delta(t) = e^{-t}u(t)$ and has the form shown in Fig. 5-8.3(a). If a unit step $v(t) = u(t)$ is applied as input, the resulting response $i(t)$

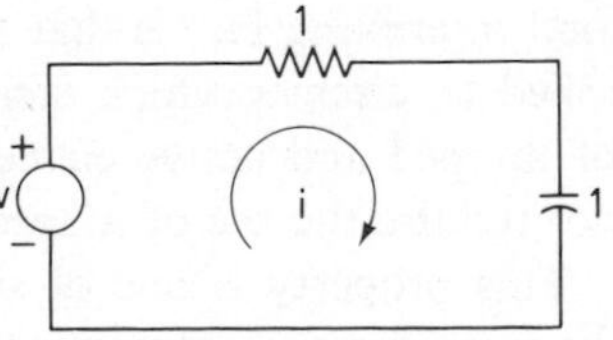

Fig. 5-8.2 A first-order *RC* network.

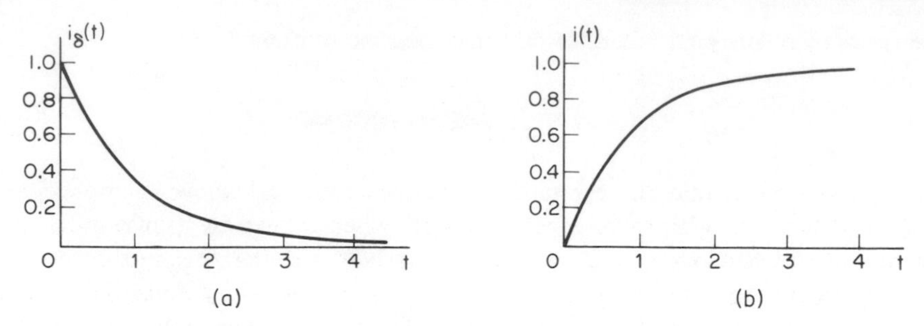

Fig. 5-8.3 Waveforms for the network shown in Fig. 5-8.2.

may be determined using Eq. (5-68). Thus, we obtain

$$i(t) = \int_0^t i_\delta(t-\tau)v(\tau)\,d\tau = \int_0^t e^{-(t-\tau)}\,d\tau = e^{-t}\int_0^t e^{\tau}\,d\tau = (1 - e^{-t})u(t)$$

A plot of the resulting output $i(t)$ is shown in Fig. 5-8.3(b).

The results given above may be summarized as follows:

SUMMARY 5-8.1

The Convolution Integral: If an arbitrary excitation $g(t)$ which is zero for $t < t_0$ is applied to a linear time-invariant network whose response to a unit impulse applied at $t = 0$ is $h(t)$, and if the network is in the zero state, the response $f(t)$ is given by the convolution integral

$$f(t) = \int_{t_0}^t h(t - \tau)g(\tau)\,d\tau$$

More specifically, this integral is said to define the convolution of two time-domain functions $h(t)$ and $g(t)$.

There are several pertinent observations that can be made regarding convolution. First of all, we note that the quantity t appears as part of the integrand in Eq. (5-68). This means that if numerical or graphical methods of integration are used, then for each new value of t at which we desire to find $f(t)$ we must perform a new integration starting from t_0. This point will be illustrated further in connection with the presentation of techniques for graphical evaluation of the convolution integral. A second interesting fact is that the convolution process indicated in Eq. (5-68) may be applied to circuits which include distributed elements as well as those consisting only of lumped and active elements. This occurs because the convolution integral does not require the use of a specific differential equation representation for a network. This property is one of significant importance since the other time-domain techniques that we present in this text are not, in general, applicable to circuits including distributed components. As a final property of the convolution integral, let us consider the case where the input function begins at $t = 0$. Letting $t_0 = 0$ in (5–68), we obtain

$$f(t) = \int_0^t h(t - \tau)g(\tau)\,d\tau \tag{5-69}$$

Now let us make a change of variable $t - \tau = x$. Thus, $\tau = t - x$, $d\tau = -dx$, and the integration limits of 0 and t become t and 0, respectively. Substituting these results in Eq. (5–69), we obtain

$$f(t) = \int_t^0 -h(x)g(t - x)\,dx = \int_0^t h(x)g(t - x)\,dx \tag{5-70}$$

Changing the variable of integration in the right member of the above equation from x to τ, we obtain

$$f(t) = \int_0^t h(\tau)g(t - \tau)\,d\tau \tag{5-71}$$

Comparing Eqs. (5–69) and (5–71), we see that the role of the impulse response $h(t)$ and the excitation function $g(t)$ is an interchangeable one in the convolution process. Such an interchange is frequently useful in evaluating convolution integrals.

The evaluation of the convolution integral for simple functions and for functions for which no explicit mathematical expression is known is conveniently done by employing graphical techniques. As an example of this, let us assume that a net-

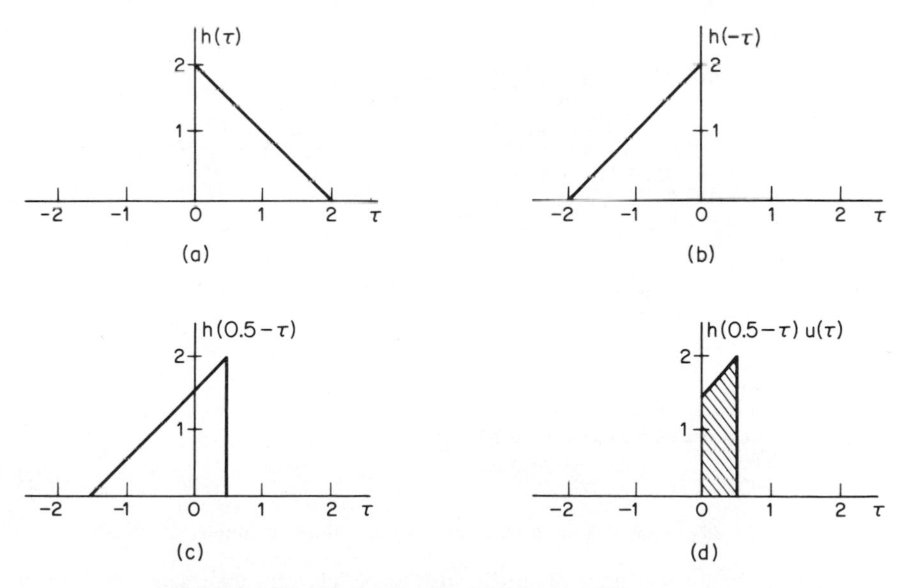

Fig. 5-8.4 Waveforms for graphical convolution example.

work has a triangular impulse response $h(\tau)$ as shown in Fig. 5–8.4(a),[11] and that the input is a unit step $u(\tau)$. First let us determine $f(0.5)$, i.e., the value of the integral given in Eq. (5–69) for $t = 0.5$. The graphical procedure is illustrated in Fig. 5–8.4. In part (b) of this figure we show $h(-\tau)$, i.e., the triangular function $h(\tau)$ in which the direction of increasing argument τ has been inverted. In part (c) we show the function $h(0.5 - \tau)$ obtained by shifting $h(-\tau)$ one-half unit to the right. In part (d) we show the product $h(0.5 - \tau)u(\tau)$. The value of the integral is equal to the shaded area. This is readily seen to be 0.875. Thus, we may write $f(0.5) = 0.875$. Repeating this process for $t = 1$, we find $f(1) = 1.5$. Similarly, $f(1.5) = 1.875$, $f(2) = 2$, and $f(t)$ for all $t > 2$ is 2. Thus, we obtain the waveform for $f(t)$ shown in Fig. 5–8.5.

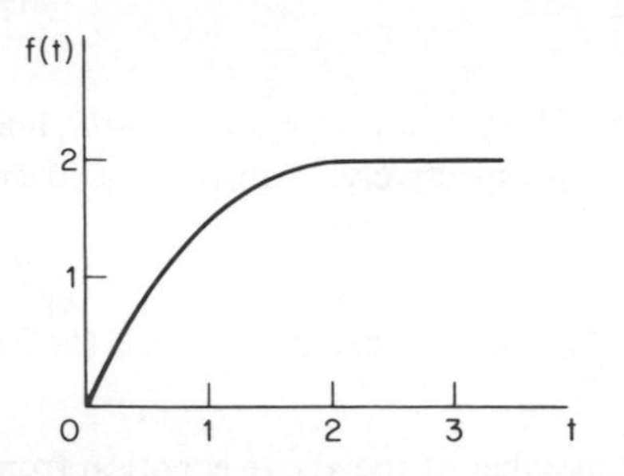

Fig. 5-8.5 Output waveform from graphical convolution example.

The process of convolution illustrated in the preceding paragraph is readily accomplished by a digital computer using the subroutine ITRPZ. To do this, we

```
      FUNCTION Y(TAU)
C     FUNCTION FOR CONVOLUTION EXAMPLE
      COMMON T
      IF(TAU.GT.2) GO TO 4
      Y = 2. - T + TAU
      RETURN
    4 Y = 2.0
      RETURN
      END

C     MAIN PROGRAM FOR CONVOLUTION EXAMPLE
C             T - INDEPENDENT VARIABLE - TIME
C            DT - INTERVAL BETWEEN PLOTTED POINTS
C             P - ARRAY USED TO STORE VALUES FOR PLOTTING
      DIMENSION P(5,101)
      COMMON T
      P(1,1) = 0.
      T = 0.05
      DT = 0.05
      DO 7   I = 2, 51
      CALL ITRPZ(0.,T,100,A)
      P(1,I) = A * 50.
    7 T = T + DT
      PRINT 9
    9 FORMAT(1H1)
      CALL PLOT5(P,1,51,100)
      STOP
      END
```

Fig. 5-8.6 Listing of program for convolution example.

[11]Such an impulse response is physically unrealizable but provides a useful mathematical example.

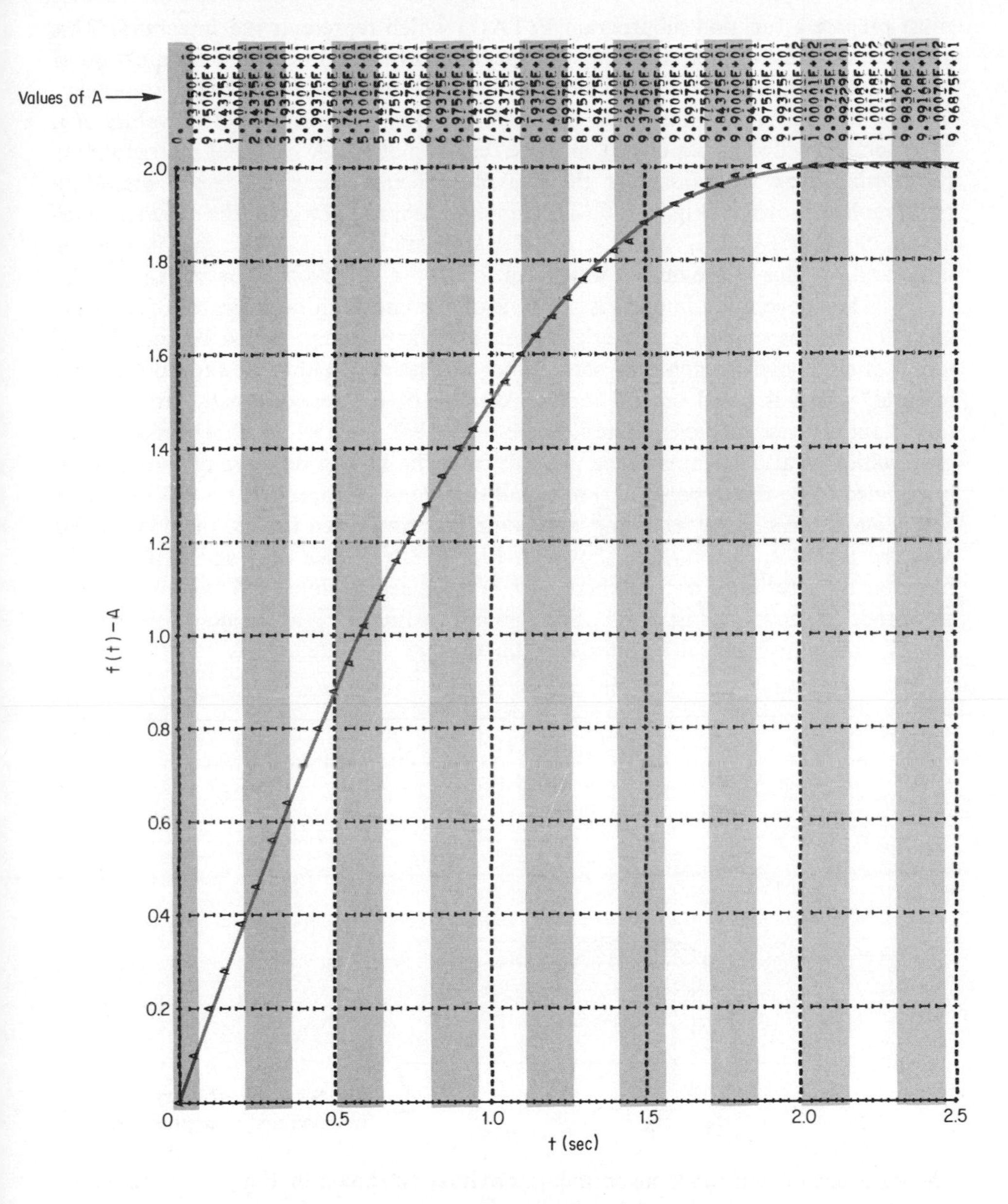

Fig. 5-8.7 Output plot for convolution example.

must prepare a function subprogram Y(TAU) which represents the integrand. This subprogram must also be supplied with the value of t at which the determination of $f(t)$ is to be made. The value can be easily supplied by using a COMMON statement. Thus, a series of integrations may be performed for a sequential set of values of t, to generate a set of values of $f(t)$, and the results plotted. A listing of a program for performing these operations for the convolution example given in the preceding paragraph is shown in Fig. 5–8.6. The function subprogram gives the required integrand. Since $h(t) = 2 - t$ for $0 < t < 2$ s, $h(t - \tau)$ is $2 - t + \tau$ as specified by the integrand. A plot of the output waveform for $0 < t < 2.5$ s is given in Fig. 5–8.7.

The convolution integral is only one of an almost limitless number of methods in which the response of a network to some arbitrary excitation can be specified in terms of its known response to some other excitation. Although the convolution integral, which is based on the impulse response of the network, is by far the most important of these methods, there is one other one, known as the superposition integral, which is also of importance. In this case, the known response of the network is assumed to be the response $w(t)$ to a unit step input $u(t)$ applied at $t = 0$. We may begin our discussion by assuming that some arbitrary input function $g(t)$ is defined over the period $t_0 < t < t_n$, as shown in Fig. 5–8.1(a). We may approximate this function by a series of n step functions $g_i(t)$ displaced in time by Δt and of varying amplitude as shown in Fig. 5–8.8. Such an approximation may be conveniently accomplished by using a shifted unit step function $u(t - t_i)$ defined as

$$u(t - t_i) = \begin{cases} 0 & t < t_i \\ 1 & t > t_i \end{cases} \tag{5–72}$$

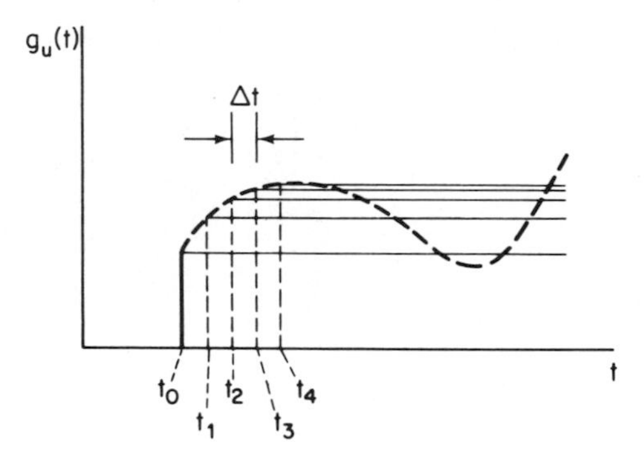

Fig. 5-8.8 An arbitrary excitation function and its approximation.

The sequence of step functions in the approximation shown in Fig. 5–8.8 may now be written explicitly by first defining the quantities g_i as

$$\begin{aligned} g_0 &= g(t_0)u(t - t_0) \\ g_1 &= [g(t_1) - g(t_0)]u(t - t_1) \\ &\vdots \\ g_i &= [g(t_i) - g(t_{i-1})]u(t - t_i) \end{aligned} \tag{5–73}$$

where the last equation holds for all i from 0 to n if we define $g(t_{-1}) = 0$. If we now define $g_u(t)$ as the step function approximation of $g(t)$, we may write

$$g_u(t) = \sum_{i=0}^{n} g_i = \sum_{i=0}^{n} [g(t_i) - g(t_{i-1})]u(t - t_i) \tag{5-74}$$

Now let us assume that the approximate excitation waveform $g_u(t)$ is applied as an input to a network which has the property that a unit step input $u(t)$ applied at $t = 0$ produces a response $w(t)$. If the network is linear, time-invariant, and in the zero state, then the response to any of the steps $g_i(t)$ defined in Eq. (5–73) is $[g(t_i) - g(t_{i-1})]w(t - t_i)$. We may now define $f_u(t)$ as the response to the series of n step functions $g_u(t)$ defined in (5–74). Applying superposition, we obtain

$$f_u(t) = \sum_{i=0}^{n} [g(t_i) - g(t_{i-1})]w(t - t_i) \tag{5-75}$$

As a preliminary step for our next development, let us multiply and divide each member of the summation in the above equation by the quantity Δt. Thus, we obtain

$$f_u(t) = \sum_{i=0}^{n} \left[\frac{g(t_i) - g(t_{i-1})}{\Delta t}\right] w(t - t_i)\, \Delta t \tag{5-76}$$

Now let us investigate what happens as we increase n, the number of step functions which approximate the input excitation $g(t)$ over the time span t_0 to t_n. As n increases, the quantity Δt becomes smaller and, in the limit as n approaches infinity, the series of step functions $g_u(t)$ becomes equal to the actual input $g(t)$. Similarly, in the limit, $f_u(t)$ becomes equal to $f(t)$, the actual zero-state network response to the excitation $g(t)$. Now consider the summation given in Eq. (5–76). In the limit, the quantity enclosed in brackets in the right member simply becomes equal to the derivative of the excitation function. In addition, the sum becomes an integral, the sequence of variables t_i becomes a continuous variable which we may designate as τ, and the incremental quantity Δt thus becomes $d\tau$. Therefore, we may write

$$f(t) = \int_{t_0}^{t} \frac{dg(\tau)}{d\tau} w(t - \tau)\, d\tau \tag{5-77}$$

where we have represented the quantity t_n, an arbitrary value of time, by t. This relation defines the *superposition integral*.[12] It enables us to find the response of the network to any excitation function $g(t)$ as long as we know the unit step response $w(t)$ of the network. An example illustrating the application of the superposition integral follows.

EXAMPLE 5-8.2. A first-order *RC* network is shown in Fig. 5-8.2. The response of this network to an input unit step $v(t) = u(t)$ was shown in Example 5-8.1 to be $i_u(t) = (1 -$

[12]This integral is also referred to as Duhamel's integral.

$e^{-t})u(t)$. We may now apply the superposition integral given in (5-77) to determine the response of this network to a unit ramp function $v(t) = tu(t)$. For such an input, the resulting response $i(t)$ may be determined as:

$$i(t) = \int_0^t i_u(t-\tau)\frac{dv(\tau)}{d\tau}\,d\tau = \int_0^t [1 - e^{-(t-\tau)}]\,d\tau$$

$$= \int_0^t d\tau - e^{-t}\int_0^t e^{\tau}\,d\tau = t - 1 + e^{-t} \qquad t > 0$$

The results given above may be summarized as follows:

SUMMARY 5-8.2

The Superposition Integral: If an arbitrary excitation $g(t)$ which is zero for $t < t_0$ is applied to a linear time-invariant network whose response to a unit step function applied at $t = 0$ is $w(t)$, and if the network is in the zero state, the response $f(t)$ is given by the superposition integral

$$f(t) = \int_{t_0}^t \frac{dg(\tau)}{d\tau} w(t-\tau)\,d\tau$$

The superposition integral may be evaluated graphically as well as analytically. The process is similar to that presented for the convolution integral in the first part of this section. Some alternative equivalent representations of the superposition integral are frequently useful in such a case. For example, if we let t_0 be 0 so that the superposition integral of Eq. (5-77) has the form

$$f(t) = \int_0^t \frac{dg(\tau)}{d\tau} w(t-\tau)\,d\tau \tag{5-78}$$

then the following form is equivalent.

$$f(t) = \int_0^t \frac{dw(\tau)}{d\tau} g(t-\tau)\,d\tau \tag{5-79}$$

Other equivalent forms of the superposition integral which may be used are

$$\begin{aligned} f(t) &= \frac{d}{dt}\int_0^t w(t-\tau)g(\tau)\,d\tau \\ f(t) &= \frac{d}{dt}\int_0^t g(t-\tau)w(\tau)\,d\tau \end{aligned} \tag{5-80}$$

The integral portions of these last two relations are similar in form to those given for the convolution integrals in Eqs. (5–69) and (5–71). Thus, we see that the superposition integral may be considered as the derivative of the convolution of an input excitation function and the step response of a given network.

We shall again refer to the results obtained in this section when we consider the Laplace transformation in Chap. 8. There we shall show that the integration process defined by the convolution and superposition integrals can be represented by a simple multiplication operation through the use of this transformation.

5-9 CONCLUSION

In this chapter we have made a detailed presentation of various methods which may be used to solve a first-order circut, i.e., one containing a single energy-storage element (a capacitor or an inductor). Here we shall briefly review the various techniques and indicate their relation to each other. The first method we discussed was the one presented in Sec. 5–1 for the case where the circuit was excited only by initial conditions. The solution procedure for this case is summarized in the flow chart shown in Fig. 5–1.6. The procedure is so simple that the student should be able to apply it almost by inspection. The procedure for solving a first-order network which is excited by sources, or by initial conditions and sources, however, becomes somewhat more complex. The approaches to be followed in these two cases are summarized in Figs. 5–2.5 and 5–3.3. An examination of the flow charts given in these two figures readily illustrates the similarity between the two cases. All of the techniques described above require that the first-order circuit be linear and time-invariant. In Sec. 5–7, however, we introduced a technique for finding the solution of a circuit which may be either time-varying or non-linear or both. The technique requires the use of the digital computer. The procedure for applying this technique is summarized in the flow chart shown in Fig. 5–7.12. To illustrate the relation between all of the above methods for the solution of a first-order circuit, in Fig. 5–9.1 we present a flow chart which combines all of the above referenced flow charts and which presents a summary of all the solution techniques. Since the techniques presented in this chapter and summarized in the flow chart shown in Fig. 5–9.1 are very similar to those which will be used to solve higher-order circuits, the flow chart shown in Fig. 5–9.1 merits careful study by the student.

A final topic which was presented in this chapter was the use of convolution and superposition integrals (in Sec. 5–8) to obtain the response of a given first-order circuit to some specified excitation. Such a technique is considerably different than the ones referred to earlier in this section since it does *not* require that an equation describing the system be known, only that the network's response to an impulse or a step excitation be given. In addition, such a response may be displayed as a plot, i.e., a mathematical expression for it is not required. Thus, and most important, it may be obtained by direct experimental measurement. The convolution and superposition techniques discussed in Sec. 5–8 are applicable to linear systems of any order.

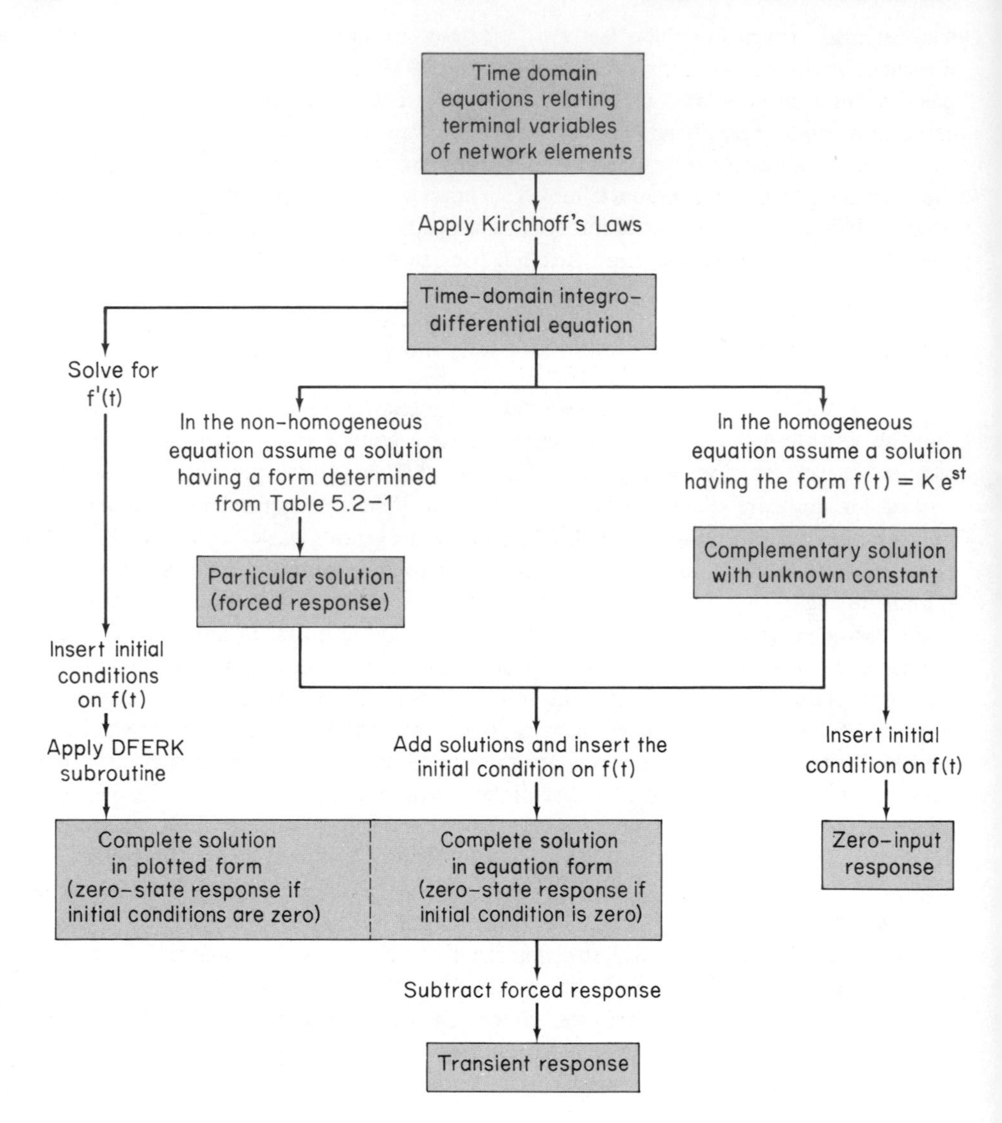

Fig. 5-9.1 General solution procedure for a first-order network.

Problems

Problem 5-1 (Sec. 5-1)

For the circuit shown in Fig. P5-1, find $i_C(t)$, $i_R(t)$, and $q(t)$ in terms of C, R, and V_0, where $q(t)$ is the charge on the capacitor and $V_0 = v(0)$.

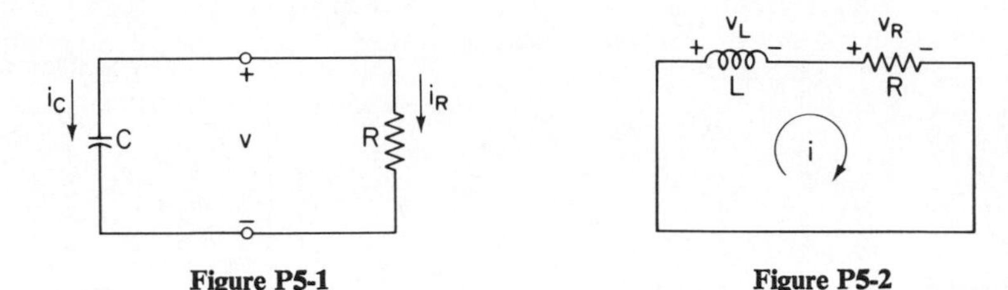

Figure P5-1 **Figure P5-2**

Problem 5–2 (*Sec. 5–1*)

For the circuit shown in Fig. P5-2, find $v_L(t)$, $v_R(t)$, and $\phi(t)$ in terms of R, L, and I_0, where $\phi(t)$ represents the flux linkages in the inductor, and $I_0 = i(0)$.

Problem 5–3 (*Sec. 5–1*)

For the circuit shown in Fig. P5-3, for $t \geq 0$, find $i(t)$, $q(t)$ (the charge in the capacitor), $v(t)$, $w_C(t)$ (the energy stored in the capacitor), and $p_R(t)$ (the power dissipated in the resistor). Assume $v(0) = 2$ V.

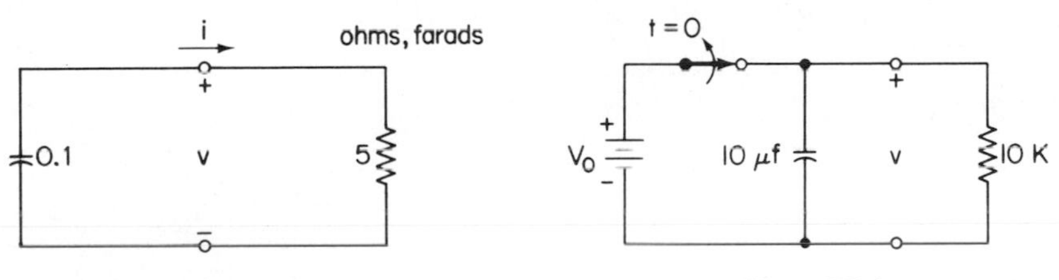

Figure P5-3

Figure P5-4

Problem 5–4 (*Sec. 5–1*)

If the switch in the circuit shown in Fig. P5-4 is opened at $t = 0$, find V_0 such that $v(0.5) = 0.3$ V.

Problem 5–5 (*Sec. 5–1*)

The switch in the circuit shown in Fig. P5-5 is closed at $t = 0$. Find V_0 such that $i(0.003) - 0.001$ A.

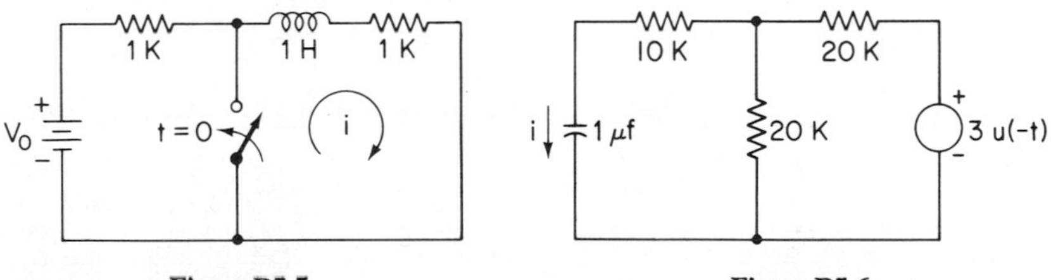

Figure P5-5 **Figure P5-6**

Problem 5–6 (*Sec. 5–1*)

For the circuit shown in Fig. P5-6, find $i(t)$ for $t > 0$.

Problem 5–7 (*Sec. 5–1*)

For the circuit shown in Fig. P5-7, find $v(t)$ for $t > 0$.

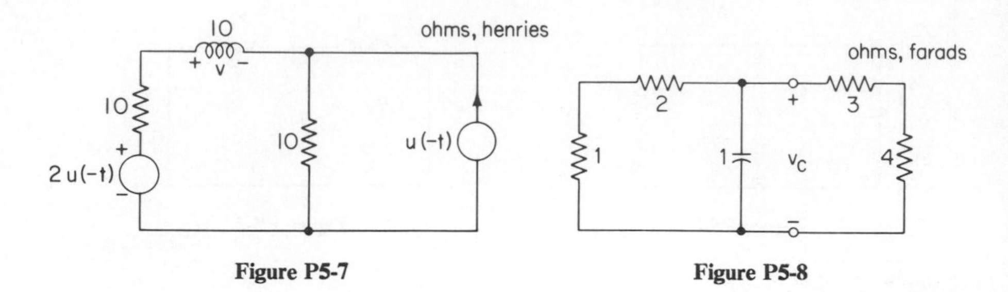

Figure P5-7 Figure P5-8

Problem 5–8 (Sec. 5–1)

For the circuit shown in Fig. P5-8, $v_C(0) = -6$ V. Find an expression for $v_C(t)$ for all $t > 0$.

Problem 5–9 (Sec. 5–1)

Show that the tangent to $f(t) = e^{-t/\tau}u(t)$ at any point t_0 intersects the t axis exactly τ seconds later as shown in Fig. P5-9.

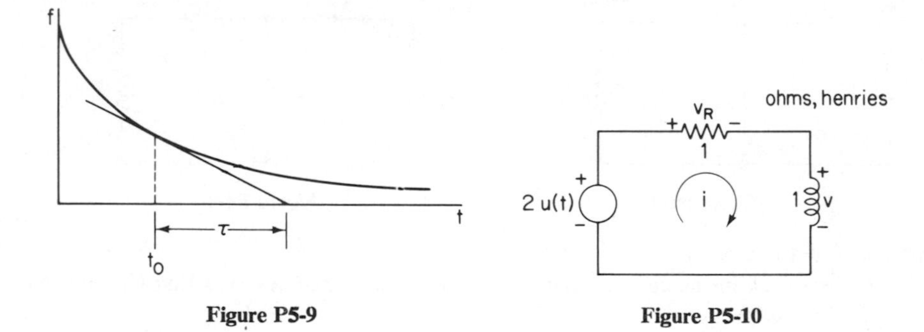

Figure P5-9 Figure P5-10

Problem 5–10 (Sec. 5–2)

Find expressions for $t \geq 0$ for the following quantities for the circuit shown in Fig. P5-10: (a) $i(t)$; (b) $\phi(t)$; (c) $v(t)$; (d) $w_L(t)$ (the energy stored in the inductor); (e) $v_R(t)$; (f) $p_L(t)$ (the power delivered to the inductor); (g) $p_R(t)$ (the power dissipated in the resistor); and (h) $p(t)$ (the power supplied by the source).

Problem 5–11 (Sec. 5–2)

For the circuit shown in Fig. P5-11, find $v(t)$ for $t \geq 0$, if $v(0) = 0$ and $i(t) = 1 +$

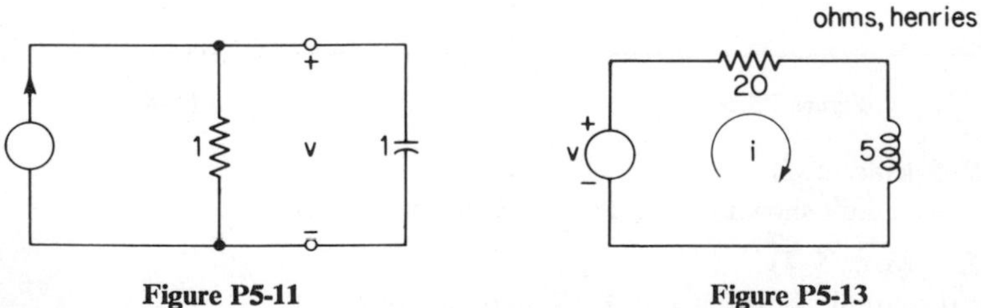

Figure P5-11 Figure P5-13

$t + t^2$ A. Then find $v(t)$ for the three cases (a) $i(t) = 1$, (b) $i(t) = t$, (c) $i(t) = t^2$. Is the sum of the last three solutions equal to the first solution? Is the sum of the *particular* solutions for the last three cases equal to the particular solution of the first case?

Problem 5–12 (*Sec. 5–2*)

Repeat Problem 5-11 for the case where the initial condition is $v(0) = 1$ V.

Problem 5–13 (*Sec. 5–2*)

In the circuit shown in Fig. P5-13, if $v(t) = 10e^{-4t} + 5$ V, find $i(t)$ for $t \geq 0$ for the initial condition $i(0) = 0$. Then find $i(t)$ for the two cases $v(t) = 10e^{-4t}$ and $v(t) = 5$. Is the sum of the last two solutions equal to the solution for the first case? Is the sum of the *particular* solutions for the last two cases equal to the particular solution for the first case?

Problem 5–14 (*Sec. 5–2*)

Repeat Problem 5-13 for the case where the initial condition is $i(0) = 1$ A.

Problem 5–15 (*Sec. 5–2*)

For the circuit shown in Fig. P5 15, find $v(t)$ for $t \geq 0$ if (a) $i(t) = \cos(t)\, u(t)$, (b) $i(t) = \sin(t)\, u(t)$ A. Assume $v(0) = 0$.

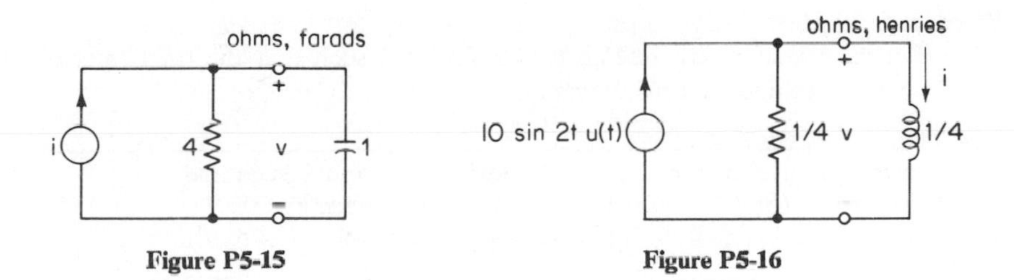

Figure P5-15 **Figure P5-16**

Problem 5–16 (*Sec. 5–2*)

In the circuit shown in Fig. P5-16, find $v(t)$ for $t \geq 0$ if $i(0) = 0$.

Problem 5–17 (*Sec. 5–2*)

In the circuit shown in Fig. P5-17, find $v(t)$ for $t \geq 0$ if the initial current in the inductor is zero.

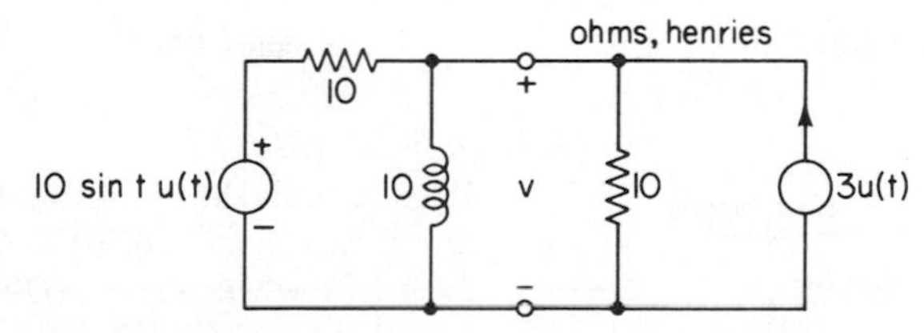

Figure P5-17

Problem 5–18 (*Sec. 5–2*)

The switch in the circuit shown in Fig. P5-18 changes from position a to position b at $t = 0$. If $i(t) = \frac{2}{3}t + 1$ A, find $v(t)$. Assume $v(0) = 0$.

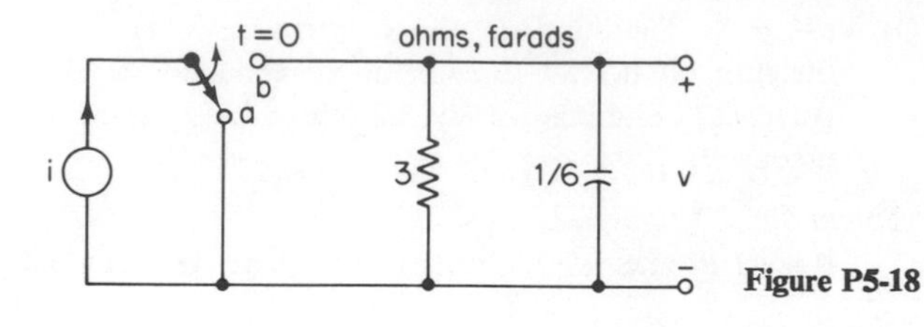

Figure P5-18

Problem 5–19 (Sec. 5–2)

In the circuit shown in Fig. P5-19, the switch closes at $t = 0$. If $v(t) = e^{-2t}$ V, find $i(t)$ for $t \geq 0$. Assume the capacitor is initially uncharged.

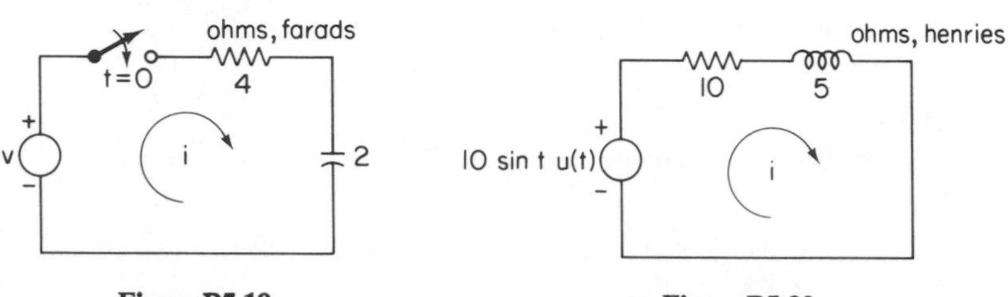

Figure P5-19

Figure P5-20

Problem 5–20 (Sec. 5–3)

For the circuit shown in Fig. P5-20, find $i(0)$ such that the *transient* part of the complete solution is equal to zero.

Problem 5–21 (Sec. 5–3)

In the circuit shown in Fig. P5-21, find R_1, R_2, and C such that:

(a) $v(t) = u(t)$, $i(t) = 0$, and $v_C(t) = 2e^{-2t} + \frac{1}{2}$ when $t \geq 0$.

(b) $i(t) = u(t)$, $v(t) = 0$, and $v_C(t) = \frac{1}{2}e^{-2t} + 2$ when $t \geq 0$.

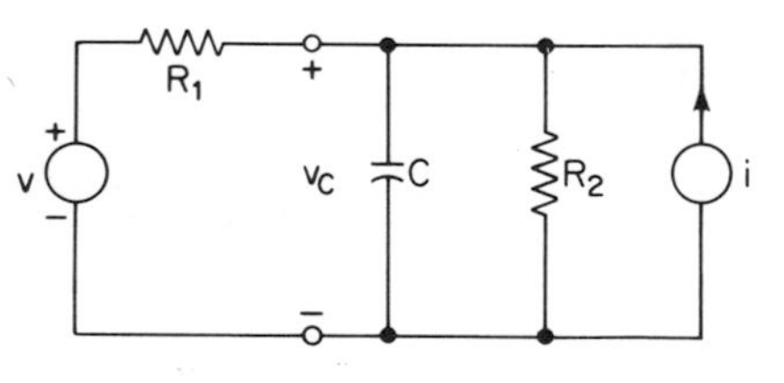

Figure P5-21

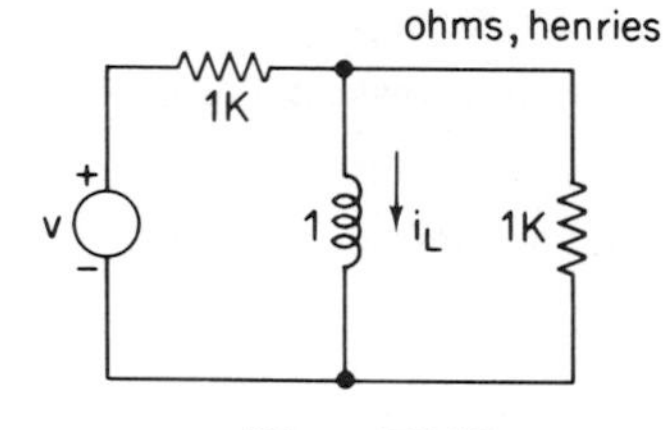

Figure P5-22

Problem 5–22 (Sec. 5–3)

In the circuit shown in Fig. P5-22, it is found that $i_L(t) = 0.001 + 0.005e^{-at}$ A for $t \geq 0$ when $v(t) = u(t)$ V. What is $i_L(t)$ if $v(t) = 2u(t)$ V?

Problem 5–23 (Sec. 5–3)

For the circuit shown in Fig. P5-23, at a given time t_0, $v_2(t_0) = 2$ V and $dv_2(t_0)/dt = -10$ V/s. Find the value of C.

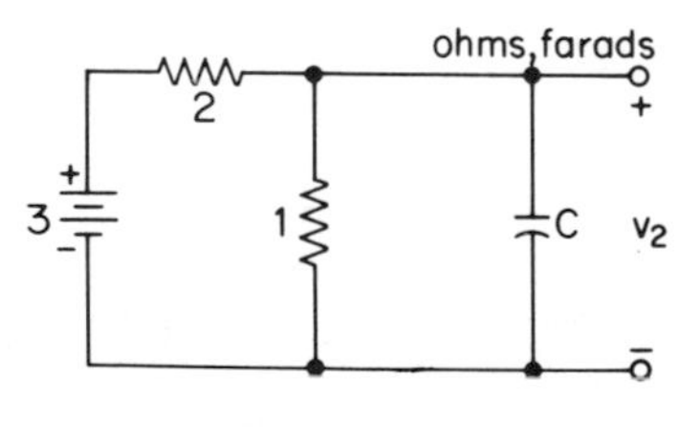

Figure P5-23

Problem 5-24 (*Sec. 5-3*)

For the circuit and the waveshape of $v_O(t)$ shown in Fig. P5-24, find $v_S(t)$.

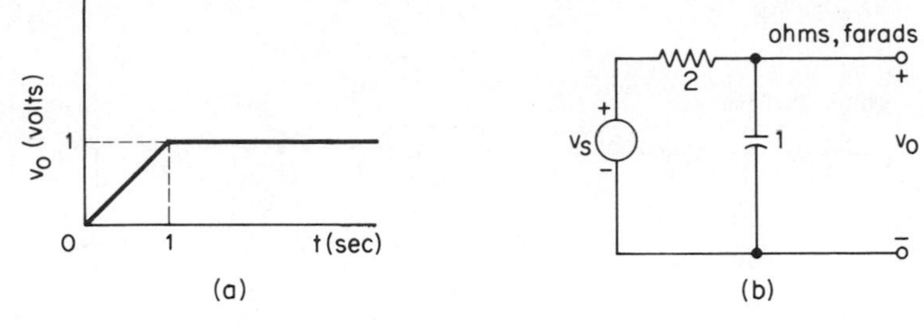

Figure P5-24

Problem 5-25 (*Sec. 5-3*)

In the circuit shown in Fig. P5-25, the switch opens at $t = 0$. Find $v(t)$ for all $t \geq 0$.

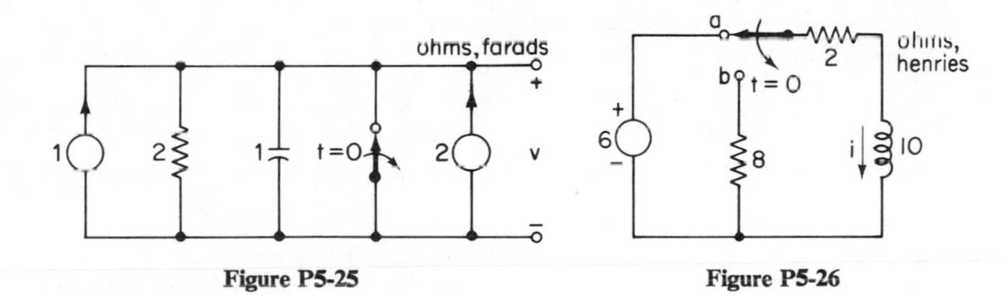

Figure P5-25 **Figure P5-26**

Problem 5-26 (*Sec. 5-3*)

In the circuit shown in Fig. P5-26, the switch has been in position *a* for a "long time." It is suddenly switched to position *b*. Find the complete solution for $i(t)$.

Problem 5-27 (*Sec. 5-3*)

In the circuit shown in Fig. P5-27, at $t = 0$ the switch is suddenly changed from position *a* to position *b*. Find the complete solution for $v(t)$.

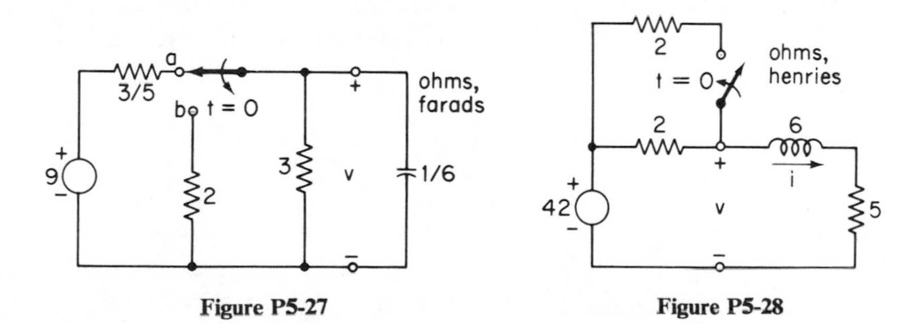

Figure P5-27 **Figure P5-28**

Problem 5-28 (*Sec. 5-3*)

In the circuit shown in Fig. P5-28, the switch closes at $t = 0$. Find $i(t)$ and $v(t)$ for $t \geq 0$.

Problem 5–29 (Sec. 5–3)

In the circuit shown in Fig. P5-29, the switch opens at $t = 0$. Find $v_1(0+)$, $v_1'(0+)$, $v_1(\infty)$, and $v_1'(\infty)$.

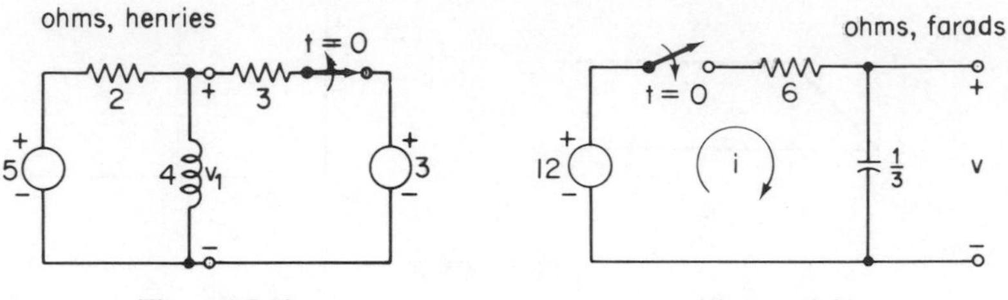

Figure P5-29 **Figure P5-30**

Problem 5–30 (Sec. 5–4)

The switch in the circuit shown in Fig. P5-30 is closed at $t = 0$. If $v(0) = 0$, find $i(0+)$, $(di/dt)(0+)$, $(d^2i/dt^2)(0+)$, $i(\infty)$, and $(di/dt)(\infty)$.

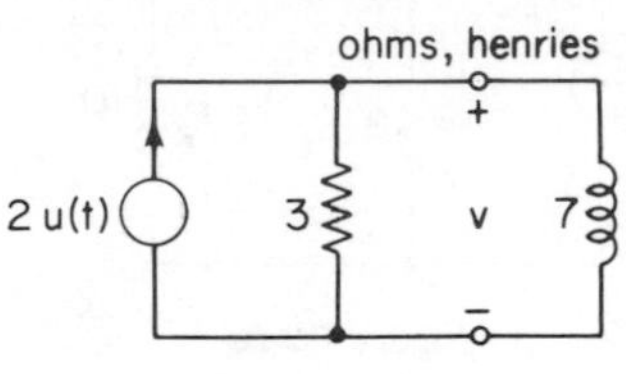

Figure P5-31

Problem 5–31 (Sec. 5–4)

In the circuit shown in Fig. P5-31, find $v(0+)$, $(dv/dt)(0+)$, $(d^2v/dt^2)(0+)$, $v(\infty)$, and $(dv/dt)(\infty)$. Assume that the initial current in the inductor is zero.

Problem 5–32 (Sec. 5–4)

In the circuit shown in Fig. P5-32, the voltage applied by the voltage source is changed from 1 V to 2 V at t equals 0. Assume that the 1-V value has been applied for a long time before t equals 0.

(a) Find an expression for $v_C(t)$, the voltage across the capacitor for $t \geq 0$.
(b) Find an expression for $i(t)$ for $t \geq 0$.

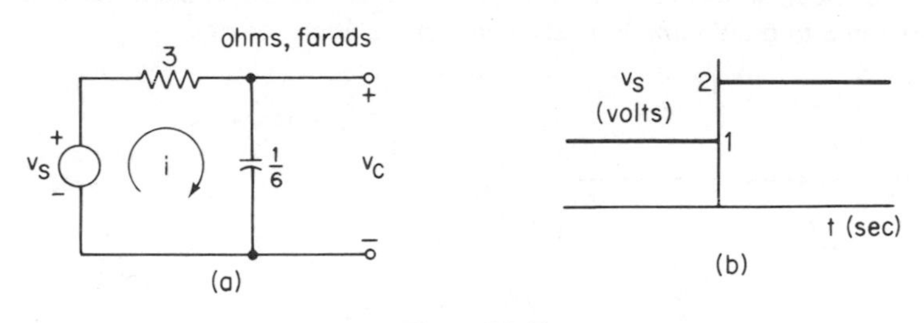

Figure P5-32

Problem 5–33 (Sec. 5–4)

For the circuit shown in Fig. P5-33, find $v_{L1}(0+)$, $v_{L2}(0+)$, $i_C(0+)$, $i_{L1}(\infty)$, $i_{L2}(\infty)$, and $v_C(\infty)$. Assume all circuit variables are zero for $t < 0$.

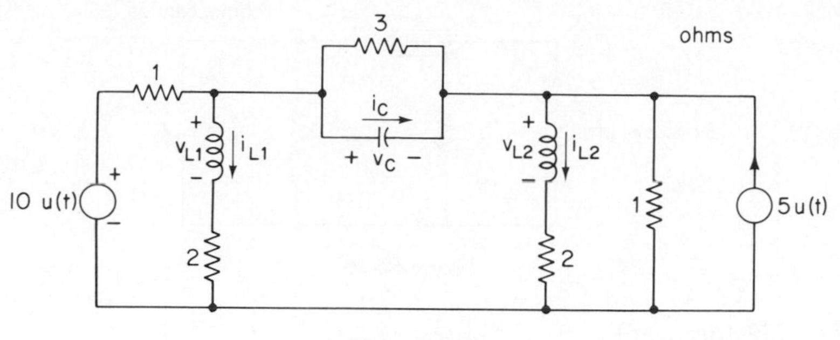

Figure P5-33

Problem 5–34 (Sec. 5–4)

In the circuit shown in Fig. P5–34, for $t < 0$ the circuit is in a steady state. The switch is closed at $t = 0$. Find $i_L(t)$ for $t \geq 0$.

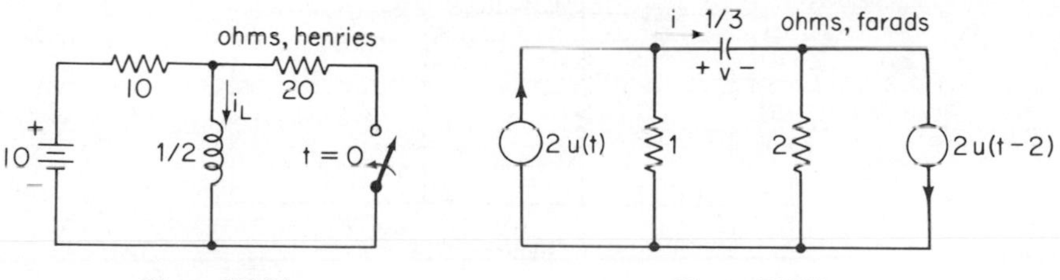

Figure P5-34 **Figure P5-35**

Problem 5–35 (Sec. 5–5)

For the circuit shown in Fig. P5–35, find $v(t)$ and $i(t)$ for $t \geq 0$. Assume $v(0) = 0$.

Problem 5–36 (Sec. 5–5)

In the circuit shown in Fig. P5–36, find $i(t)$ and $v(t)$ for $t \geq 0$. Assume $i(0) = 0$.

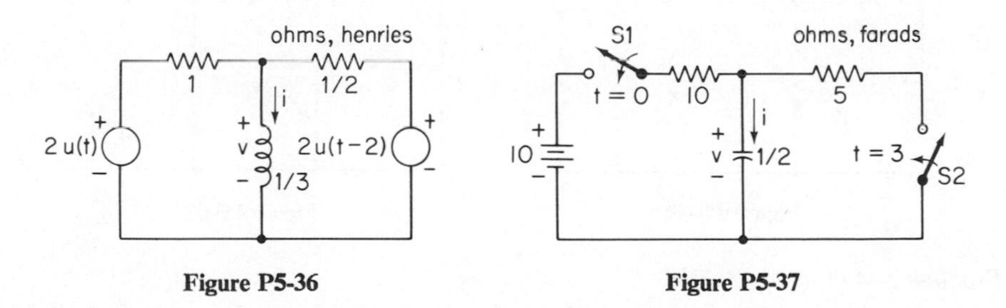

Figure P5-36 **Figure P5-37**

Problem 5–37 (Sec. 5–5)

In the circuit shown in Fig. P5–37, switch *S*1 is closed at $t = 0$ and switch *S*2 is closed at $t = 3$ s. Find $v(t)$ and $i(t)$ for $t \geq 0$. Assume $v(0) = 0$.

Problem 5–38 (Sec. 5–5)

In the circuit shown in Fig. P5–38, the switch is closed at $t = 1$ s. Find $i(t)$ and $v(t)$ for $t \geq 0$. Assume $i(0) = 0$.

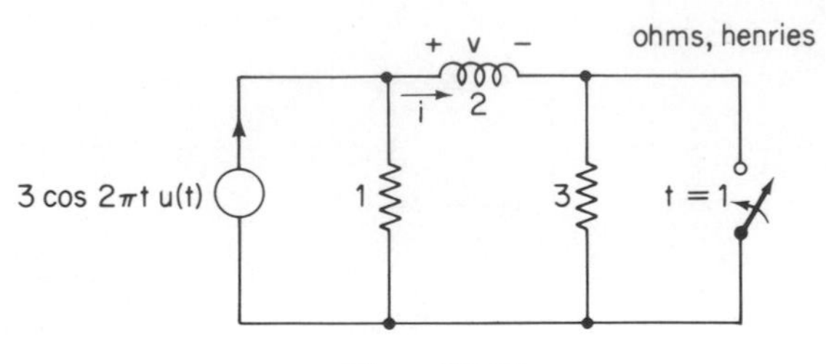

Figure P5-38

Problem 5–39 (Sec. 5–6)

For $t < 0$, the circuit shown in Fig. P5-39 is in a steady-state condition. At $t = 0$, the switch is moved from position a to position b. Draw a circuit without a switch and with a capacitor which is uncharged for $t < 0$, which will have the same solution for $v_C(t)$ for $t > 0$ as the original circuit.

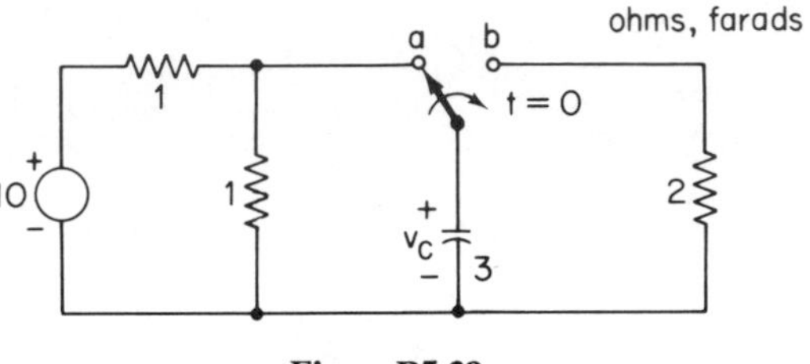

Figure P5-39

Problem 5–40 (Sec. 5–6)

For $t < 0$, the circuit shown in Fig. P5-40 is in a steady-state condition. At $t = 0$, the switch is closed. Draw a circuit without a switch, and with an inductor which is uncharged for $t < 0$, which will have the same solution for $i_L(t)$ for $t > 0$.

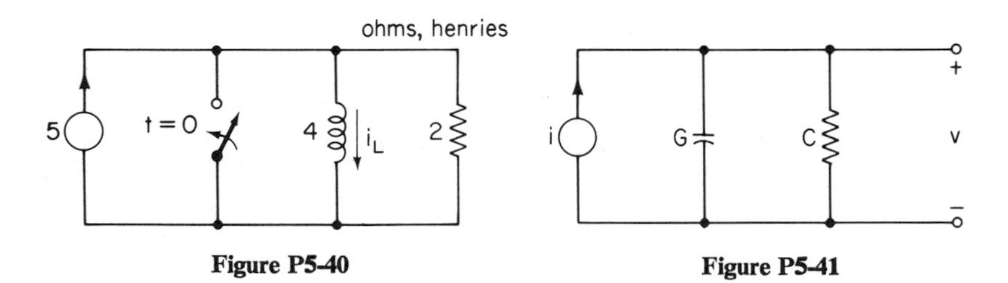

Figure P5-40

Figure P5-41

Problem 5–41 (Sec. 5–7)*[13]

For the circuit shown in Fig. P5-41, $G = 10^{-4}$ mho, $C = 1\ \mu$F, and $i(t) = 0.01u(t)$ A. Write a program using the subroutine DFERK to find $v(t)$ over the period $0 \le t \le 0.025$ s for an initial condition $v(0) = 0$. Plot 51 points. Solve for $v(t)$ by classical methods to verify your computer solution.

[13]Problems which require the use of digital computational techniques are identified by an asterisk.

*Problem 5-42** (*Sec. 5-7*)

The elements of the circuit shown in Fig. P5-41 have the values $G = 2$ mhos, $C(t) = 1 + e^{-t}$ F, and $i(t) = e^{-t}$ A. Write a program using the subroutine DFERK to solve for $v(t)$ over the period $0 \leq t \leq 5$ s for the initial condition $v(0) = 1$ V. Scale $v(t)$ by 100. Estimate a function approximating the plotted $v(t)$ and test your conclusion by inserting this function in the differential equation which describes the circuit.

*Problem 5-43** (*Sec. 5-7*)

Repeat Problem 5-42 for the case where the elements have the values $G = 1$ mho, $C(v) = 1 + v(t)$ F, and $i(t) = -2e^{-2t}$ A. Use the same initial condition and the same scaling factor.

Problem 5-44 (*Sec. 5-8*)

Apply the convolution integral to find the response $v(t)$ for the circuit shown in Fig. P5-44 for the case where the excitation $i(t)$ has the following values: (a) $e^{-t}u(t)$; (b) $\sin t[u(t) - u(t - \pi)]$; (c) $2(1 - t)[u(t) - u(t - 3)]$ A.

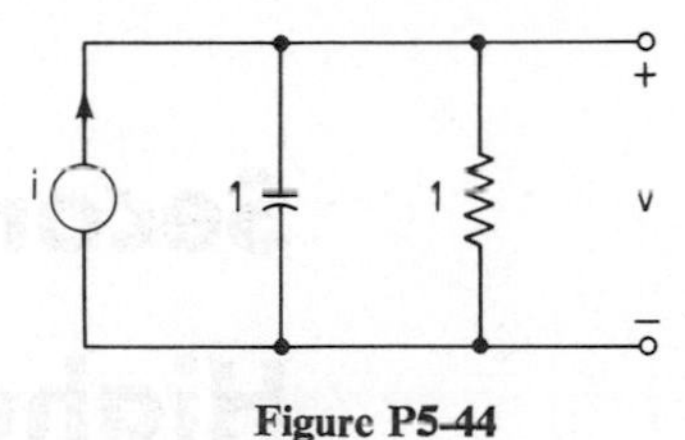

Figure P5-44

Problem 5-45 (*Sec. 5-8*)

Use graphical integration to solve part (c) of Problem 5-44.

Problem 5-46 (*Sec. 5-8*)

Repeat Problem 5-44 using the superposition integral.

Problem 5-47 (*Sec. 5-8*)

Use graphical integration to solve part (c) of Problem 5-46.

CHAPTER 6

Second-Order and Higher-Order Circuits

In Chap. 5 we studied the properties of first-order circuits, i.e., circuits containing a single energy-storage element, either an inductor or a capacitor. In this chapter we shall study the properties of second-order and higher-order circuits, i.e., circuits containing two or more energy-storage elements. First we will consider second-order circuits. Such circuits will, in general, be characterized by second-order differential equations. There are three possible types of such circuits, namely, circuits with two inductors, circuits with two capacitors, and circuits with one inductor and one capacitor. Just as was the case for the first-order circuit, the second-order circuit may also contain an arbitrary number of resistors, independent sources, and controlled sources. In our study of second-order circuits we shall find that many of the same techniques which were used to analyze first-order circuits may be applied with minor modifications to second-order circuits. The properties of the second-order circuit, however, are considerably more varied than those of the first-order circuit, as we shall discover.

6-1 EXCITATION BY INITIAL CONDITIONS—CASES I AND II

In this section we shall consider the effect of exciting a second-order circuit by initial conditions. We shall begin our discussion by considering a second-order circuit characterized by an inductor and a capacitor. Second-order circuits containing two inductors or two capacitors will be covered in a subsequent section. As an

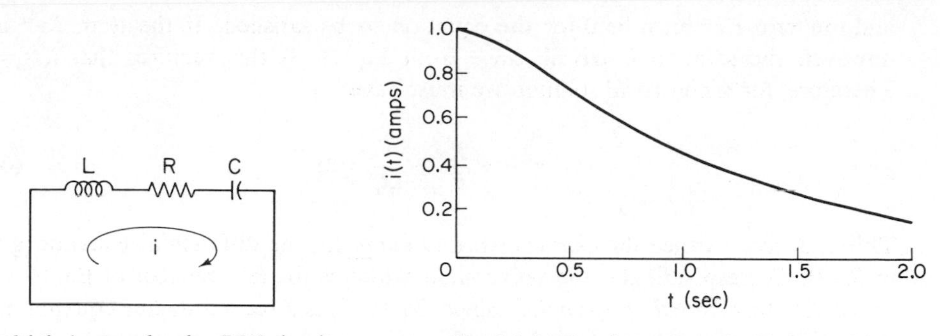

Fig.6-1.1 A second-order *RLC* circuit.

Fig. 6-1.2 Waveform of $i(t)$ for Example 6-1.1.

example of a second-order *RLC* circuit, consider the one-mesh circuit shown in Fig. 6–1.1. Applying KVL to this circuit, we may write

$$L\frac{di}{dt} + Ri(t) + \frac{1}{C}\int_{-\infty}^{t} i(\tau)\,d\tau = 0 \tag{6–1}$$

If we differentiate this equation and divide all terms by L, so as to set the leading coefficient to unity, we obtain

$$\frac{d^2i}{dt^2} + \frac{R}{L}\frac{di}{dt} + \frac{1}{LC}i(t) = 0 \tag{6–2}$$

Following the terminology conventions introduced in Sec. 5–1, we see that this equation is an ordinary, second-order, linear, homogeneous differential equation with constant coefficients, or, more briefly, *a second-order, homogeneous, differential equation.* In Sec. 5–1 we found that a simple exponential provided a solution for the first-order homogeneous differential equation. Let us try that approach here by assuming a solution for $i(t)$ of the form

$$i(t) = Ke^{st} \tag{6–3}$$

where, for the moment, K and s are undefined. Substituting this assumed solution in Eq. (6–2), we obtain

$$s^2Ke^{st} + s\frac{R}{L}Ke^{st} + \frac{1}{LC}Ke^{st} = 0 \tag{6–4}$$

This may be written in the form

$$\left(s^2 + s\frac{R}{L} + \frac{1}{LC}\right)Ke^{st} = 0 \tag{6–5}$$

In this last equation we see that the product of the polynomial $s^2 + (sR/L) + (1/LC)$

and the term Ke^{st} must be 0 for the equation to be satisfied. If the term Ke^{st} is 0, however, the solution is trivial, since from Eq. (6–3) this requires that $i(t) = 0$. Therefore, for a nontrivial solution we must have

$$s^2 + s\frac{R}{L} + \frac{1}{LC} = 0 \tag{6–6}$$

This equation is called the *characteristic equation* for the differential equation given in (6–2). Correspondingly, the polynomial which is the left member of Eq. (6–6) is called the *characteristic polynomial.* Since Eq. (6–6) is a second-degree equation in s, there will be two values of s which will satisfy it. Let these be s_1 and s_2. We find that

$$\begin{aligned} s_1 &= -\frac{R}{2L} + \sqrt{\left(\frac{R}{2L}\right)^2 - \frac{1}{LC}} \\ s_2 &= -\frac{R}{2L} - \sqrt{\left(\frac{R}{2L}\right)^2 - \frac{1}{LC}} \end{aligned} \tag{6–7}$$

There are three possible cases for the resulting values of s_1 and s_2, based on the relative values of the quantities R, L, and C. More specifically, these cases are specified by the polarity of the quantity inside the radical in Eq. (6–7), namely, $(R/2L)^2 - 1/LC$. This quantity is known as the *discriminant* of the second-degree equation. The first case (Case I) occurs when $(R/2L)^2 > 1/LC$, i.e., when the discriminant is greater than zero. For this case, the expressions given in Eq. (6–7) directly define two real roots of the characteristic equation. Since both s_1 and s_2 satisfy the original differential equation given in Eq. (6–2), the solution for $i(t)$ must be of the form

$$i(t) = K_1e^{s_1t} + K_2e^{s_2t} \tag{6–8}$$

where the constants K_1 and K_2 are determined from the initial conditions. Thus, we see that the roots of the characteristic equation are the natural frequencies of the circuit.[1] Note that in Eq. (6–8) there are *two* constants, thus *two* initial conditions must be specified. This is always the case for a second-order differential equation. Suppose that the initial conditions are given as $i(0)$ and $i'(0)$. Evaluating Eq. (6–8) for $t = 0$, we obtain

$$i(0) = K_1 + K_2 \tag{6–9}$$

This is the first of two equations needed to determine the constants K_1 and K_2. To find a second equation, we may first differentiate Eq. (6–8). We obtain

$$i'(t) = s_1K_1e^{s_1t} + s_2K_2e^{s_2t} \tag{6–10}$$

[1]See Sec. 5-1 for a definition of "natural frequency."

Evaluating this equation for $t = 0$, we obtain

$$i'(0) = s_1 K_1 + s_2 K_2 \tag{6-11}$$

Equations (6-9) and (6-11) may be written as a set of simultaneous equations, with K_1 and K_2 as unknowns. In matrix form we have

$$\begin{bmatrix} 1 & 1 \\ s_1 & s_2 \end{bmatrix} \begin{bmatrix} K_1 \\ K_2 \end{bmatrix} = \begin{bmatrix} i(0) \\ i'(0) \end{bmatrix}$$

Using the methods of Chap. 3, we find that the solution for K_1 and K_2 is

$$\begin{bmatrix} K_1 \\ K_2 \end{bmatrix} = \frac{1}{s_2 - s_1} \begin{bmatrix} s_2 & -1 \\ -s_1 & 1 \end{bmatrix} \begin{bmatrix} i(0) \\ i'(0) \end{bmatrix} \tag{6-12}$$

Thus, we have obtained the values of the unknown constants K_1 and K_2 as functions of the known quantities s_1, s_2, $i(0)$, and $i'(0)$. The process is readily illustrated by a numerical example.

EXAMPLE 6-1.1. In the series *RLC* circuit shown in Fig. 6-1.1, assume that the elements have the following values: $R = 6\ \Omega$, $L = 1$ H, $C = \frac{1}{5}$ F. From (6-7), the resulting roots of the characteristic equation are: $s_1 = -1$, $s_2 = -5$. Thus, from (6-8) the solution for $i(t)$ has the form

$$i(t) = K_1 e^{-t} + K_2 e^{-5t}$$

The values of the constants K_1 and K_2 may be expressed in terms of the initial conditions $i(0)$ and $i'(0)$. From (6-12) we see that

$$K_1 = \frac{s_2 i(0) - i'(0)}{s_2 - s_1} = \frac{5}{4} i(0) + \frac{1}{4} i'(0)$$

$$K_2 = \frac{-s_1 i(0) + i'(0)}{s_2 - s_1} = -\frac{1}{4} i(0) - \frac{1}{4} i'(0)$$

As a more specific example, suppose that the initial conditions are specified as $i(0) = 1$, $i'(0) = 0$. Then, from the above relations, $K_1 = \frac{5}{4}$, $K_2 = -\frac{1}{4}$, and the solution for $i(t)$ is

$$i(t) = \tfrac{5}{4} e^{-t} - \tfrac{1}{4} e^{-5t} \qquad t \geq 0$$

A plot of this function is shown in Fig. 6-1.2.

The process described above may be summarized as follows:

SUMMARY 6-1.1

Solution of a Second-Order Circuit Excited Only by Initial Conditions—Case I: A passive second-order circuit excited only by initial conditions is described by a second-order, homogeneous, differential equation having the form

$$\frac{d^2f}{dt^2} + a_1\frac{df}{dt} + a_0f(t) = 0 \tag{6–13}$$

where $f(t)$ is any network variable, and in which the coefficients a_1 and a_0 are positive. If $(a_1/2)^2 > a_0$, this equation has a solution of the form

$$f(t) = K_1e^{s_1t} + K_2e^{s_2t} \tag{6–14}$$

in which s_1 and s_2 are the real roots of the characteristic equation $s^2 + a_1s + a_0 = 0$, and they are not equal. The two constants K_1 and K_2 are found from the two initial conditions $f(0)$ and $f'(0)$ by the relation

$$\begin{bmatrix} K_1 \\ K_2 \end{bmatrix} = \frac{-1}{\sqrt{(a_1/2)^2 - a_0}}\begin{bmatrix} s_2 & -1 \\ -s_1 & 1 \end{bmatrix}\begin{bmatrix} f(0) \\ f'(0) \end{bmatrix} \tag{6–15}$$

There are several points of interest concerning the above development. First let us note that the roots of the characteristic equation for (6–13) are of the form

$$s_1 = \alpha + \beta \qquad s_2 = \alpha - \beta \tag{6–16}$$

where

$$\alpha = -\frac{a_1}{2} \qquad \beta = \sqrt{\left(\frac{a_1}{2}\right)^2 - a_0} \tag{6–17}$$

and where $(a_1/2)^2 - a_0$ (the discriminant) is assumed to be greater than zero. Now we may observe that:

1. From Eq. (6–17), if the coefficient a_1 is greater than zero, then α is negative.
2. From Eq. (6–17), if the coefficients a_1 and a_0 are both greater then zero, then the magnitude of β is smaller than the magnitude of α since $\beta = \sqrt{\alpha^2 - a_0}$.
3. From Eq. (6–16), if α is negative [see (1) above], and if β is smaller in magnitude than α [see (2) above], then both s_1 and s_2 are negative. In addition (for this case) they are unequal.
4. From Eq. (6–14), if s_1 and s_2 are negative [see (3) above], then $f(t)$ disappears as t approaches infinity, i.e., $f(\infty) = 0$.

Now let us make some observations relating the parity of the coefficients a_1 and a_0 to the passivity of a given second-order circuit. If such a circuit is comprised only of passive, i.e., positive-valued, resistors and energy-storage elements, then, assuming that some dissipation is present in the circuit, the energy provided by the initial conditions must eventually disappear, i.e., the variables must go to zero as t approaches infinity. From the above equations, for this case, this will only be true if a_1 and a_0 are positive. We thus conclude that for a second-order circuit, requiring the circuit to be passive guarantees that the coefficients a_1 and a_0 will be positive.

Now let us consider a second case (Case II) for the second-order, homogeneous, differential equation. We shall define this case by the condition that the discriminant be identically zero. For the circuit shown in Fig. 6–1.1, which has solutions to its characteristic equation given by (6–7), this requires that $(R/2L)^2 = 1/LC$. For such a restriction, we see that the roots of the characteristic equation are

$$s_1 = s_2 = -\frac{R}{2L} \tag{6–18}$$

Thus, this case is characterized by the fact that the characteristic equation has two equal roots. In this case, we say that the circuit has a second-order natural frequency at s_1. A solution for $i(t)$ which satisfies the original differential equation is obviously

$$i(t) = K_a e^{s_1 t} \tag{6–19}$$

where $s_1 = -R/2L$. The general solution to a second-order, homogeneous, differential equation, however, requires two arbitrary constants, not just one; therefore, we must find a second solution for $i(t)$. Let us assume that this is somewhat like the solution given in Eq. (6–19), but that it differs by having a multiplicative factor of the form $y(t)$. Thus, we assume a second solution to the differential equation which has the form

$$i(t) = y(t)e^{s_1 t} \tag{6–20}$$

Now let us see if we can find what form $y(t)$ will take. To do this, we may proceed by substituting Eq. (6–20) into the original differential equation given in Eq. (6–2). We obtain (the details are left as an exercise for the reader)

$$e^{-Rt/2L}y''(t) = 0 \tag{6–21}$$

where $y''(t)$ is the second derivative of $y(t)$ with respect to t. Since $e^{-Rt/2L}$ is obviously not equal to 0, $y''(t)$ must equal 0 for all t. Integrating the equation $y''(\mathrm{t}) = 0$ two times, we obtain

$$y(t) = K_b + K_2 t \tag{6–22}$$

where K_b and K_2 are two arbitrary constants. Substituting this result in Eq. (6–20), we obtain for the second solution for $i(t)$,

$$i(t) = K_b e^{s_1 t} + K_2 t e^{s_1 t} \tag{6–23}$$

If we now add the two solutions for $i(t)$ given in (6–19) and (6–23), we obtain

$$i(t) = K_1 e^{s_1 t} + K_2 t e^{s_1 t} \tag{6–24}$$

where we have summed the constants K_a of (6–19) and K_b of (6–23) to define a new constant K_1. This equation gives the general solution for the homogeneous differential equation given in (6–2) for the case in which the two roots of the characteristic equation are equal. If we evaluate Eq. (6–24) for $t = 0$, we obtain

$$i(0) = K_1 \tag{6–25}$$

This is the first of two equations needed to determine the constants K_1 and K_2. To find a second equation, let us first differentiate (6–24). We obtain

$$i'(t) = s_1 K_1 e^{s_1 t} + K_2 e^{s_1 t}(1 + s_1 t) \tag{6–26}$$

Evaluating this result for $t = 0$, we see that

$$i'(0) = s_1 K_1 + K_2 \tag{6–27}$$

From Eq. (6–25) and (6–27), we obtain the following expressions for K_1 and K_2.

$$K_1 = i(0) \qquad K_2 = i'(0) - s_1 i(0) \tag{6–28}$$

Thus, we have obtained expressions giving the values of the constants K_1 and K_2 in terms of the initial conditions $i(0)$ and $i'(0)$ and the natural frequency of the circuit s_1. The process is readily illustrated by a numerical example.

EXAMPLE 6-1.2. In the series *RLC* circuit shown in Fig. 6-1.1, assume that the elements have the following values: $R = 2\sqrt{5}\ \Omega$, $L = 1$ H, $C = \frac{1}{5}$ F. (These are the same values as were used in Example 6-1.1 for Case I, except for the resistor, which has been lowered in value.) The characteristic equation has a second-order root at $s_1 = -\sqrt{5}$, thus the form of the solution for the current is

$$i(t) = K_1 e^{-\sqrt{5}t} + K_2 t e^{-\sqrt{5}t}$$

If we assume that the initial conditions are $i(0) = 1$, $i'(0) = 0$, then from (6–28) we see that $K_1 = 1$, $K_2 = \sqrt{5}$. Thus, the solution for $i(t)$ is

$$i(t) = e^{-\sqrt{5}t} + \sqrt{5}\, te^{-\sqrt{5}t}$$

A plot of the resulting function $i(t)$ is given in Fig. 6-1.3.

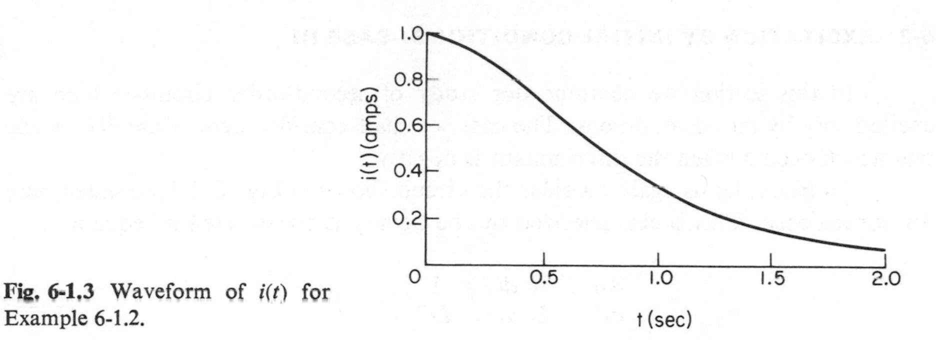

Fig. 6-1.3 Waveform of $i(t)$ for Example 6-1.2.

The process described above is easily summarized as follows:

SUMMARY 6-1.2

Solution of a Second-Order Circuit Excited Only by Initial Conditions—Case II: A passive second-order circuit excited only by initial conditions is described by a second-order, homogeneous, differential equation having the form

$$\frac{d^2f}{dt^2} + a_1\frac{df}{dt} + a_0 f(t) = 0 \tag{6-29}$$

where $f(t)$ is any network variable, and in which the coefficients a_1 and a_0 are positive. If $(a_1/2) = a_0$, this equation has a solution of the form

$$f(t) = K_1 e^{s_1 t} + K_2 t e^{s_1 t} \tag{6-30}$$

where s_1 is the root of the characteristic equation $s^2 + a_1 s + a_0 = 0$. The two constants K_1 and K_2 are found from the two initial conditions $f(0)$ and $f'(0)$ by the relations

$$K_1 = f(0) \qquad K_2 = f'(0) - s_1 f(0) \tag{6-31}$$

The conclusions made with respect to the passivity of a second-order circuit and the related positive polarity of the coefficients a_1 and a_0 for Case I also hold true for this case.

In this section we have discussed two of the three possible cases for the second-order, homogeneous, differential equation, namely, the cases where the discriminant is positive (Case I) and zero (Case II). The third case, in which the discriminant is negative, will be covered in the following section.

6-2 EXCITATION BY INITIAL CONDITIONS—CASE III

In this section we continue our study of second-order circuits which are excited only by initial conditions. The case we shall consider here (Case III) is the one which occurs when the discriminant is negative.

To begin, let us again consider the circuit shown in Fig. 6–1.1, repeated here for convenience. This is characterized by the homogeneous differential equation

$$\frac{d^2i}{dt^2} + \frac{R}{L}\frac{di}{dt} + \frac{1}{LC}i(t) = 0 \tag{6–32}$$

The characteristic equation for this differential equation is

$$s^2 + \frac{R}{L}s + \frac{1}{LC} = 0 \tag{6–33}$$

If the circuit has values of R, L, and C such that the discriminant $(R/2L)^2 - 1/LC$ is negative (Case III), the roots of this equation may be written in the form

$$\begin{aligned} s_1 &= -\frac{R}{2L} + j\sqrt{\frac{1}{LC} - \left(\frac{R}{2L}\right)^2} \\ s_2 &= -\frac{R}{2L} - j\sqrt{\frac{1}{LC} - \left(\frac{R}{2L}\right)^2} \end{aligned} \tag{6–34}$$

where $j = \sqrt{-1}$ is the usual complex operator.[2] To simplify the resulting development, let us define

$$\sigma = -\frac{R}{2L} \qquad \omega_d = \sqrt{\frac{1}{LC} - \left(\frac{R}{2L}\right)^2} \tag{6–35}$$

Thus, σ is the real part of the roots s_1 and s_2, and ω_d is the magnitude of the imaginary part of these roots. We may now write s_1 and s_2 in the form

$$s_1 = \sigma + j\omega_d \qquad s_2 = \sigma - j\omega_d \tag{6–36}$$

The solution for $i(t)$ must now have the form

$$i(t) = K_1e^{s_1t} + K_2e^{s_2t} \tag{6–37}$$

[2]A review of complex algebra is given in Appendix B.

The quantity $i(t)$ in the left member of the above equation represents a physical variable. Therefore, it must be real. The two terms in the right member, however, involve complex quantities. In order for the sum of these terms to be real, they must be complex conjugates, i.e., their real parts must be equal and their imaginary parts must be opposite in sign but equal in magnitude. From Eq. (6–36) we see that s_1 and s_2 are already complex conjugates; therefore, $e^{s_1 t}$ and $e^{s_2 t}$ will also be complex conjugates. It is necessary, therefore, that the constants K_1 and K_2 be complex conjugates to insure that $i(t)$ will be real. To emphasize this, we may define

$$K_1 = \frac{K_r}{2} - j\frac{K_i}{2} \qquad K_2 = \frac{K_r}{2} + j\frac{K_i}{2} \tag{6–38}$$

where we have introduced the factor of $\frac{1}{2}$ for convenience in future calculations. If we substitute the relations of Eqs. (6–36) and (6–38) in (6–37), we obtain

$$i(t) = \tfrac{1}{2}(K_r - jK_i)e^{\sigma t + j\omega_d t} + \tfrac{1}{2}(K_r - jK_i)e^{\sigma t - j\omega_d t} \tag{6–39}$$

Using Euler's identity,[3] we may write

$$\begin{aligned} e^{\sigma t + j\omega_d t} &= e^{\sigma t}(\cos \omega_d t + j \sin \omega_d t) \\ e^{\sigma t - j\omega_d t} &= e^{\sigma t}(\cos \omega_d t - j \sin \omega_d t) \end{aligned} \tag{6–40}$$

Substituting these relations in (6–39) and rearranging terms, we obtain

$$i(t) = e^{\sigma t}(K_r \cos \omega_d t + K_i \sin \omega_d t) \tag{6–41}$$

which is as predicted, purely real. We will show in Sec. 7–1 that such an expression may be put in the form

$$i(t) = I_0 e^{\sigma t}(\cos \omega_d t + \phi) \tag{6–42}$$

where ϕ is an angle (measured in radians). Thus, it is convenient to refer to the quantity ω_d as the *frequency of oscillation* of the circuit. For our purposes here, the form shown in Eq. (6–41) is more convenient. Equation (6–41) contains two unknown constants, namely, K_r and K_i. The first of the two equations needed to determine these constants may be found by setting $t = 0$ in (6–41). Thus, we see that

$$i(0) = K_r \tag{6–43}$$

If we differentiate Eq. (6–41), we obtain

$$i'(t) = e^{\sigma t}[K_r(\sigma \cos \omega_d t - \omega_d \sin \omega_d t) + K_i(\omega_d \cos \omega_d t + \sigma \sin \omega_d t)] \tag{6–44}$$

[3]See Appendix B.

Evaluating this expression for $t = 0$, we obtain

$$i'(0) = \sigma K_r + \omega_d K_i \tag{6-45}$$

Thus, from Eqs. (6-43) and (6-45), the solutions for K_r and K_i are readily seen to be

$$K_r = i(0) \qquad K_i = \frac{i'(0) - \sigma i(0)}{\omega_d} \tag{6-46}$$

The process described above is readily illustrated by a numerical example.

EXAMPLE 6-2.1. The circuit shown in Fig. 6-1.1 is assumed to have the following values for the components: $R = 2\,\Omega$, $L = 1$ H, $C = \frac{1}{5}$ F. (These are the same values used in Examples 6-1.1 and 6-1.2, except that the value of the resistor has been reduced.) The characteristic equation for these element values is

$$s^2 + 2s + 5 = 0$$

The roots of the characteristic equation, i.e., the natural frequencies of the network, are $s_1 = -1 + j2$ and $s_2 = -1 - j2$. Thus, $\sigma = -1$ and $\omega_d = 2$. The general form of the solution for $i(t)$ is thus

$$i(t) = e^{-t}(K_r \cos 2t + K_i \sin 2t)$$

If we assume that the initial conditions are $i(0) = 1$, $i'(0) = 0$, then, from (6-46) we see that $K_r = 1$, $K_i = \frac{1}{2}$. Thus, the solution for $i(t)$ is

$$i(t) = e^{-t}(\cos 2t + \tfrac{1}{2} \sin 2t) \qquad t \geq 0$$

A plot of the waveform of $i(t)$ is shown in Fig. 6-2.1.

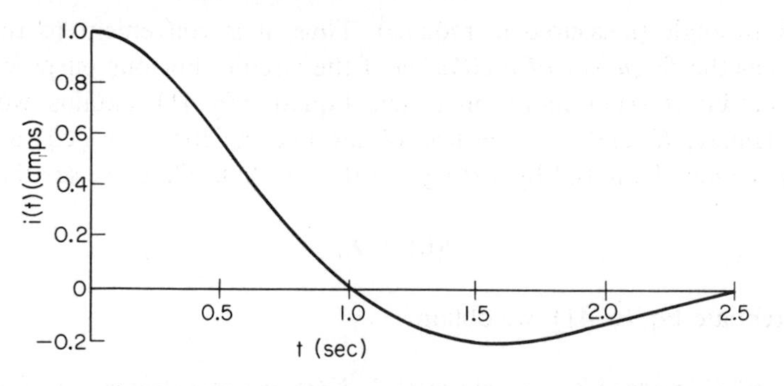

Fig. 6-2.1 Waveform of $i(t)$ for Example 6-2.1.

The process described above may be summarized as follows:

SUMMARY 6-2.1

Solution of a Second-Order Circuit Excited Only by Initial Conditions—Case III: A passive second-order circuit excited only by initial conditions is described by a second-order, homogeneous, differential equation having the form

$$\frac{d^2f}{dt^2} + a_1\frac{df}{dt} + a_0 f(t) = 0 \tag{6–47}$$

where $f(t)$ is any network variable, and in which the coefficients a_1 and a_0 are positive. If $(a_1/2)^2 < a_0$, this equation has a solution of the form

$$f(t) = e^{\sigma t}(K_1 \cos \omega_d t + K_2 \sin \omega_d t) \tag{6–48}$$

where σ and ω_d are the real and imaginary parts of the complex roots of the characteristic equation $s^2 + a_1 s + a_0 = 0$. The two constants K_1 and K_2 are found from the initial conditions $f(0)$ and $f'(0)$ using the relations

$$K_1 = f(0) \qquad K_2 = \frac{f'(0) - \sigma f(0)}{\omega_d} \tag{6–49}$$

There are several points of interest concerning the above development. First let us note that the roots of the characteristic equation are of the form

$$s_1 = \sigma + j\omega_d \qquad s_2 = \sigma - j\omega_d \tag{6–50}$$

where

$$\sigma = -\frac{a_1}{2} \qquad \omega_d = \sqrt{a_0 - \left(\frac{a_1}{2}\right)^2} \tag{6–51}$$

and where $(a_1/2)^2 - a_0$ (the discriminant) is less than zero. Now we may observe that:

1. Since $(a_1/2)^2$ is always positive, the coefficient a_0 must be positive, or this case will not apply, i.e., the discriminant will not be negative.
2. From (6–51), if the coefficient a_1 is greater than zero, then σ is negative.
3. From Eq. (6–48), the waveshape of $f(t)$ will have a sinusoidal variation with a peak magnitude or envelope proportional to $e^{\sigma t}$. If σ is negative (see 2 above), then from Eq. (6–48), $f(t)$ disappears as t approaches infinity, i.e., $f(\infty) = 0$.

An interesting subcase of Case III occurs when the coefficient a_1 of Eq. (6–47) is zero. If this is true, then from (6–51) we see that $\sigma = 0$ in the general form of the solution given in Eq. (6–48). Thus, a sinusoidal variation of the variable $f(t)$ will continue for all time. Physically, this represents a situation in which no dissipation (resistance) is present in the network, since in such a case any energy originally provided by the initial conditions as excitation for the network remains in the network indefinitely. For the network discussed in Example 6–2.1, this occurs when $R = 0$. An example of such a case follows.

EXAMPLE 6-2.2. It is assumed that the circuit shown in Fig. 6-1.1 has a value of resistance which is identically zero. In this case, the differential equation for the circuit becomes

$$\frac{d^2i}{dt^2} + \frac{1}{LC}i(t) = 0$$

The characteristic equation for this differential equation is

$$s^2 + \frac{1}{LC} = 0$$

Thus, the roots of the characteristic equation, and correspondingly, the natural frequencies of the network are purely imaginary and are equal to $\pm j\sqrt{1/LC}$. For the values $L = 1$ H and $C = \frac{1}{5}$ F (the same values used in the preceding examples), $a_1 = 0$ and $a_0 = 5$ in (6–47) and from Eq. (6–48) we see that the solution is

$$i(t) = K_1 \cos\sqrt{5}\,t + K_2 \sin\sqrt{5}\,t$$

If we assume that the initial conditions are $i(0) = 1$, $i'(0) = 0$, then $K_1 = 1$, $K_2 = 0$, and our solution is

$$i(t) = \cos\sqrt{5}\,t \qquad t \geq 0$$

The circuit described in the preceding example may be said to be *oscillating* (at a frequency of $\sqrt{5}$ rad/s). Some additional insight into this type of circuit may be gained by considering the energy relationships that exist in it. Let $w_L(t)$ be the energy stored in the inductor. Then

$$w_L(t) = \tfrac{1}{2}Li^2(t) = \tfrac{1}{2}\cos^2\sqrt{5}\,t = \tfrac{1}{4}(1 + \cos 2\sqrt{5}\,t) \tag{6–52}$$

The energy stored in the capacitor is readily found by first noting that the voltage across the capacitor $v(t)$ is the same as the voltage across the inductor, thus

$$v(t) = L\frac{di}{dt} = -\sqrt{5}\sin\sqrt{5}\,t \tag{6–53}$$

If we let $w_C(t)$ be the instantaneous value of the energy stored in the capacitor, we may write

$$w_C(t) = \tfrac{1}{2} Cv^2(t) = \tfrac{1}{2} \sin^2 \sqrt{5}\, t = \tfrac{1}{4} (1 - \cos 2\sqrt{5}\, t) \tag{6-54}$$

It is readily shown that $w_L(t) + w_C(t)$ is constant and equal to $\frac{1}{2}$. Thus, the total energy in the circuit is constant. Plots of $i(t)$, $v(t)$, $w_L(t)$, and $w_C(t)$ are shown in Fig. 6-2.2. From these plots we see that the total energy present in the circuit is transferred back and forth from the inductor to the capacitor at a frequency equal to two times the frequency at which the circuit is oscillating. Such a transfer of energy always occurs in circuits characterized by Case III, i.e., circuits in which the response has a sinusoidal component.

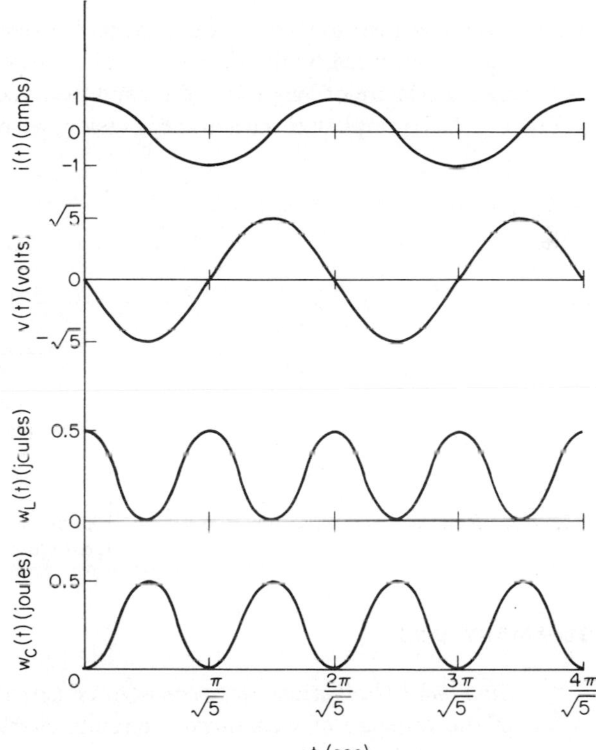

Fig. 6-2.2 Waveforms for Example 6-2.2.

From Example 6-2.2 we note that the a_0 coefficient of the characteristic equation has a special significance, namely, it is the square of the frequency (in radians per second) at which sinusoidal oscillations will occur when the dissipation of the circuit has been reduced to zero. Thus, it is determined only by the energy-storage elements of the circuit, i.e., the inductor and the capacitor. Such a frequency is referred to as the *undamped natural frequency*. Using the symbol ω_0 for this frequency, we may write

$$\omega_0 = \sqrt{a_0} \tag{6-55}$$

The undamped natural frequency ω_0 is equal to the frequency of oscillation ω_d only when the dissipation is zero, otherwise ω_d is less in value than ω_0. More explicitly, from the definitions for σ and ω_d given in Eq. (6–51), we see that we may write

$$\omega_d = \sqrt{a_0 - \left(\frac{a_1}{2}\right)^2} = \sqrt{\omega_0^2 - \sigma^2} \tag{6–56}$$

This may also be written in the form

$$\omega_0 = \sqrt{\sigma^2 + \omega_d^2} \tag{6–57}$$

Thus, it is convenient to think of the relation between the magnitudes of σ, ω_d, and ω_0 as one characterized by the sides of a right triangle in which the sides adjacent to the right angle are of length σ and ω_d and the hypotenuse is of length ω_0 as shown in Fig. 6–2.3. To emphasize the above discussion, we present the following summary:

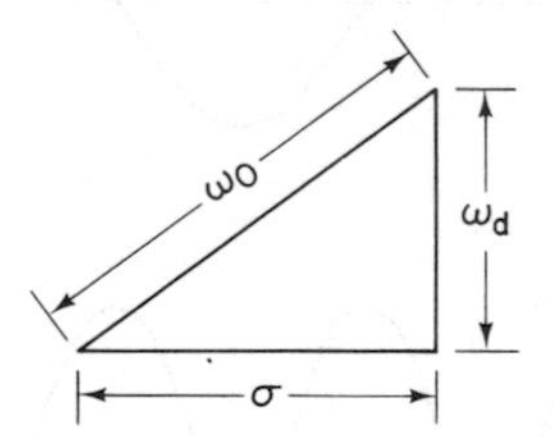

Fig. 6-2.3 Relation between ω_0, ω_d, and σ.

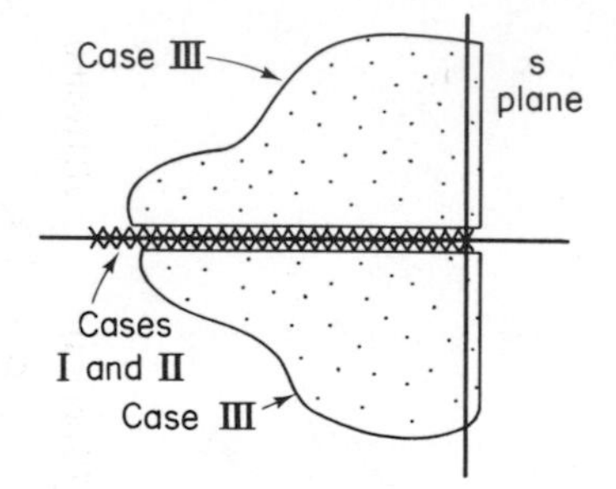

Fig. 6-2.4 Location of the zeros of the characteristic equation for Cases I, II, and III.

SUMMARY 6-2.2

Sinusoidal Oscillations in Second-Order Circuits: The sinusoidal component of the response of second-order circuits characterized by Case III (see Summary 6–2.1) has a frequency of ω_d radians per second. If the circuit contains any dissipative, i.e., resistive elements, this frequency will be lower than ω_0, the undamped natural frequency. The latter is determined only by the energy-storage elements of the circuit, i.e., the inductor and the capacitor, and is always equal to $\sqrt{a_0}$, where a_0 is defined in Eq. (6–47) as the zero-order coefficient of the differential equation.

We may correlate the separate cases for the solution of the second-order, homogeneous, differential equation presented in this and the preceding sections by indicating the permissible locations of the roots of the characteristic equation on a complex "s plane," where s is the variable of the characteristic equation, for the three cases. For passive circuits, in which the coefficients a_1 and a_0 of Eq. (6–47) are positive (and non-zero), Case I is characterized by a pair of unequal roots on the negative real axis (excluding the origin), Case II is characterized by a single second-order root on the negative real axis (excluding the origin), and Case III is characterized by complex conjugate roots in the left-half plane (and not on the imaginary axis) symmetrically located with respect to the real axis. Note that if the coefficient a_0 is zero, then one of the negative real axis zeros of Case I will be at the origin. Similarly, as has been discussed, if the coefficient a_1 is zero, the complex zeros will occur on the imaginary axis. The relative locations of the three cases are shown in Fig. 6–2.4. For example, for the network shown in Fig. 6–1.1 and used as an illustration of Cases I, II, and III in Examples 6–1.1, 6–1.2, and 6–2.1, respectively, the locus of the roots of the characteristic equation as R is varied from 6 to 0 Ω, for $L = 1$ H and $C = \frac{1}{5}$ F, is shown in Fig. 6–2.5. Note that the form of the locus for Case III is

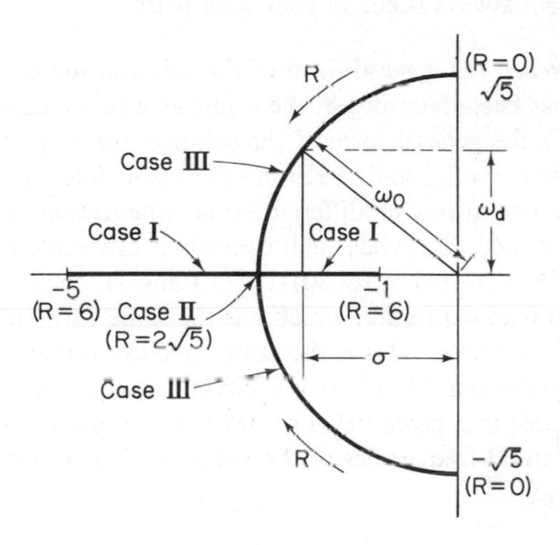

Fig. 6-2.5 Natural frequencies for the circuit of Fig. 6-1.1.

a portion of a circle centered at the origin. The radius of this circle is the value of the undamped natural frequency ω_0. For any point on this circle, the real part is σ and the imaginary part is ω_d. Thus, the right triangle relating σ, ω_d, and ω_0 may be inscribed at any point on this portion of the locus as shown in the figure. A plot of the locus of the roots of a polynomial (in this case, the *characteristic* polynomial) as a function of some parameter (in this case, R) is given the appropriate name of a *root-locus plot*. Such plots are of considerable importance in studies of control systems as well as in studies of networks. We shall see more of the root-locus plot in future chapters. It should be noted that the positions of the zeros of the characteristic equation on the s plane can be directly correlated to the form of the time response when the network is excited only by initial conditions. Thus, Cases I and II are characterized by exponential waveforms, while Case III is characterized by a sinusoidal waveform with an exponentially decreasing magnitude.

It has been pointed out that the roots of the characteristic equation are the natural frequencies of the network. This being the case, the complex s plane shown in Fig. 6–2.4 is frequently referred to as the *complex frequency plane.* We shall have considerable discussion in future chapters concerning the properties and usage of the complex frequency plane. Here we may emphasize the concept of natural frequencies and their location on the complex frequency plane by the following summary:

SUMMARY 6-2.3

Natural Frequencies of Second-Order RLC Circuits: Second-order circuits in which the two energy-storage elements are different, i.e., one is an inductor and the other a capacitor, have a characteristic equation in which the discriminant may be positive, zero, or negative. Thus, the natural frequencies may occur anywhere in the left half of the complex frequency plane (including the imaginary axis but excluding the origin). Complex natural frequencies will always occur in conjugate pairs.

When the general form of the solution for any one variable of a second-order circuit has been determined, then, just as was the case for the first-order circuit (see Sec. 5–1), the general form of the solution for any other variable can be found by appropriate application of Kirchhoff's laws. Such an application, however, involves only the operations of differentiation, integration, addition, and subtraction. It is readily verified that when such operations are made to a general solution of the form given in Eqs. (6–14) or (6–30) (Cases I and II, respectively) or Eq. (6–48) (Case III), the result is an expression which has the same form, although it will, in general, have different constants. This is the same conclusion that was reached in Summary 5–1.2 for the first-order circuit. It is a direct consequence of the fact that the behavior of any variable in a given network has a zero-input response which is determined only by the natural frequencies of the network. To emphasize this point, we present the following:

SUMMARY 6-2.4

Form of the Variables in a Second-Order Circuit Excited Only by Initial Conditions: In a second-order circuit excited only by initial conditions, all variables have a behavior characterized by the natural frequencies of the circuit. Thus, the general form of any variable $f(t)$, depending on the value of the discriminant, will be one of the following:

$$\text{Case I} \quad f(t) = K_1 e^{s_1 t} + K_2 e^{s_2 t}$$

$$\text{Case II} \quad f(t) = K_1 e^{s_1 t} + K_2 t e^{s_1 t}$$

$$\text{Case III} \quad f(t) = e^{\sigma t}(K_1 \cos \omega_d t + K_2 \sin \omega_d t)$$

The constants K_i will, in general, be different for each variable.

One other set of terms is frequently applied to second-order circuits. This has to do with the manner in which any sinusoidal oscillation of network variables disappears, i.e., is damped out. Thus, a circuit characterized by the special subcase of Case III in which the zeros of the characteristic equation are on the $j\omega$ axis is called an *undamped* circuit. Other Case III circuits are referred to as *underdamped.* Circuits characterized by Case II behavior are referred to as *critically damped* in the sense that if there is any less damping, i.e., any less amount of dissipation, then the circuit will exhibit an oscillatory component in its response. Finally, circuits characterized by Case I behavior are referred to as *overdamped.*

6-3 SECOND-ORDER CIRCUITS WITH MORE THAN ONE INDEPENDENT VARIABLE

In this section we shall extend the conclusions which have been reached to this point for the second-order one-mesh and one node-pair *RLC* circuit to circuits with greater numbers of meshes and node-pairs, and to *RL* and *RC* second-order circuits, i.e., circuits containing two inductors or two capacitors. We begin by noting that second-order *RL* and *RC* circuits must have more than one mesh or more than one pair of nodes. If this is not the case, there will be two parallel- or series-connected energy-storage elements in the circuit. These may obviously be replaced by a single equivalent energy-storage element, and the resulting circuit can then be treated as a first-order one. An example of a circuit containing two inductors, in which the inductors cannot be combined into a single inductor, is shown in Fig. 6–3.1.

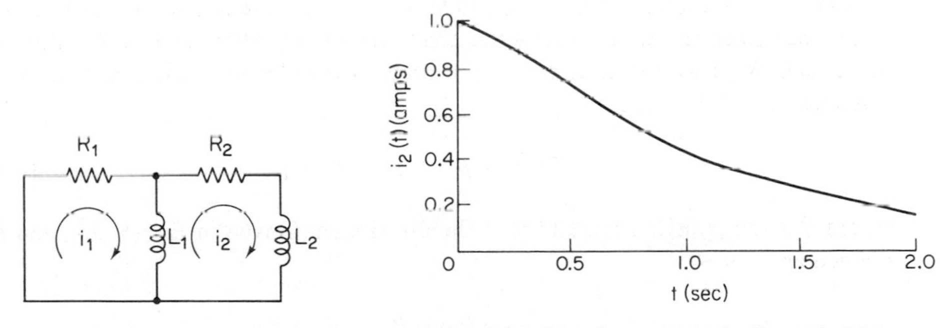

Fig. 6-3.1 Second-order *RL* circuit with two independent variables.

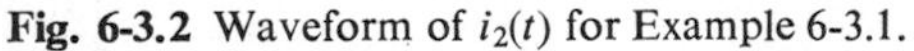

Fig. 6-3.2 Waveform of $i_2(t)$ for Example 6-3.1.

To characterize this circuit, it is necessary to write two differential equations. These are found by applying KVL around the two meshes of the circuit. Thus, we obtain

$$\begin{aligned} R_1 i_1(t) + L_1 \frac{d(i_1 - i_2)}{dt} &= 0 \\ R_2 i_2(t) + L_1 \frac{d(i_2 - i_1)}{dt} + L_2 \frac{di_2}{dt} &= 0 \end{aligned} \tag{6–58}$$

These equations may be solved by assuming solutions for $i_1(t)$ and $i_2(t)$ as follows:

$$i_1(t) = K_a e^{st} \qquad i_2(t) = K_b e^{st} \tag{6–59}$$

Substituting these values into (6–58) and putting the result into matrix format, we obtain

$$\begin{bmatrix} R_1 + sL_1 & -sL_1 \\ -sL_1 & R_2 + s(L_1 + L_2) \end{bmatrix} \begin{bmatrix} K_a e^{st} \\ K_b e^{st} \end{bmatrix} = \begin{bmatrix} 0 \\ 0 \end{bmatrix} \tag{6–60}$$

We may ignore the trivial solution $K_a e^{st} = K_b e^{st} = 0$ for this simultaneous set of equations, since from Eq. (6–59) this requires that $i_1(t) = i_2(t) = 0$, which is not very interesting. Thus, a solution to the set of equations exists only if the determinant of the square matrix in the left member of Eq. (6–60) is zero.[4] This requires

$$s^2 L_1 L_2 + s(R_1 L_1 + R_1 L_2 + R_2 L_1) + R_1 R_2 = 0 \tag{6–61}$$

This is the characteristic equation for the two-loop circuit shown in Fig. 6–3.1. The discriminant for this equation is

$$\left[\frac{R_1 + R_2}{2L_2} + \frac{R_1}{2L_1}\right]^2 - \frac{R_1 R_2}{L_1 L_2} \tag{6–62}$$

It may be shown that this discriminant is always positive (for positive-valued elements). Since the discriminant is positive, Case I, as presented in Sec. 6–1, applies. Thus, the values of s which satisfy the characteristic equation will be negative and not equal. We may define s_1 and s_2, the roots of the characteristic equation, by the relations

$$s_1 = \alpha + \beta \qquad s_2 = \alpha - \beta \tag{6–63}$$

where β is the positive square root of the discriminant given in Eq. (6–62), and α is defined as

$$\alpha = -\left[\frac{R_1 + R_2}{2L_2} + \frac{R_1}{2L_1}\right] \tag{6–64}$$

Thus, the solution for either $i_1(t)$ or $i_2(t)$ is of the form

$$K_1 e^{s_1 t} + K_2 e^{s_2 t} \tag{6–65}$$

[4]A simultaneous set of equations having the matrix form $\mathbf{Ax} = \mathbf{y}$, where the elements of $\mathbf{y}$ are *not* all zero, has a unique solution if det $\mathbf{A} \neq 0$. If the elements of $\mathbf{y}$ are all zero, however, a non-trivial solution exists only if det $\mathbf{A} = 0$. This is the case we have here.

where the constants K_1 and K_2 may be evaluated from the specified initial conditions. A numerical example follows.

EXAMPLE 6-3.1. It is desired to find the solution for $i_2(t)$ for the circuit shown in Fig. 6-3.1 for the case where the circuit elements have the values: $R_1 = 2\ \Omega$, $R_2 = 3\ \Omega$, $L_1 = L_2 = 1$ H. For these values, the square matrix given in the left member of (6-60) becomes

$$\begin{bmatrix} 2+s & -s \\ -s & 3+2s \end{bmatrix}$$

The characteristic equation is found by taking the determinant of this matrix and setting it to 0. Thus, we obtain

$$s^2 + 7s + 6 = 0$$

The roots of this equation are $s_1 = -1$, $s_2 = -6$. We may now write the general form that the solution for $i_2(t)$ (or any other network variable) must have. It is

$$i_2(t) = K_1 e^{-t} + K_2 e^{-6t}$$

Let us assume that the initial conditions are $i_2(0) = 1$ and $i_2'(0) = 0$. To use these conditions to determine expressions for K_1 and K_2, evaluating the above equation for $t = 0$, we find

$$i_2(0) = K_1 + K_2 = 1$$

A second equation involving K_1 and K_2 may be found by first differentiating the general solution for $i_2(t)$, and then evaluating it for $t = 0$. We thus obtain

$$i_2'(0) = -K_1 - 6K_2 = 0$$

Solving the two equations, we see that $K_1 = \frac{6}{5}$ and $K_2 = -\frac{1}{5}$. Thus, the solution for $i_2(t)$ is

$$i_2(t) = \tfrac{6}{5} e^{-t} - \tfrac{1}{5} e^{-6t} \qquad t > 0$$

A plot of the solution for $i_2(t)$ is shown in Fig. 6-3.2.

It was pointed out in the preceding discussion that the discriminant for the *RL* circuit shown in Fig. 6-3.1 is positive for all positive values of the network elements. This will be true for any second-order *RL* circuit, i.e., Cases II and III do not apply to circuits comprised of resistors and inductors. The same result holds true for second-order *RC* circuits, i.e., the discriminant is always positive for positive-valued circuit elements. It should be noted that even though oscillatory behavior (Case III) does not occur in such circuits, a reversal of the polarity of the variables is still possible. As an example of this, consider the circuit shown in Fig. 6–3.3. If we assume that the capacitor C_1 is initially charged and the capacitor C_2 is initially uncharged, then the first action of the circuit will be to charge the capacitor C_2. Eventually, however, the circuit will become completely discharged; thus the current $i(t)$ will first be positive, then negative. The difference between this behavior and the oscillatory behavior of *RLC* circuits is, of course, the fact that with Case III behavior, the reversals of polarity continue indefinitely and always have the same time period. Thus, although polarity reversals of variables are possible in *RL* and *RC* circuits, these circuits cannot exhibit true oscillatory behavior. These conclusions may be summarized as follows:

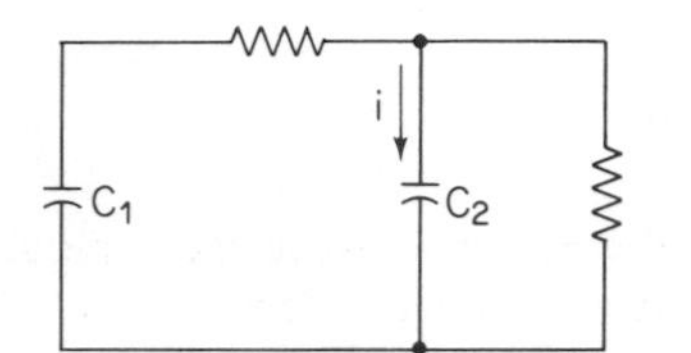

Fig. 6-3.3 Alternation of the polarity of a variable in an *RC* circuit.

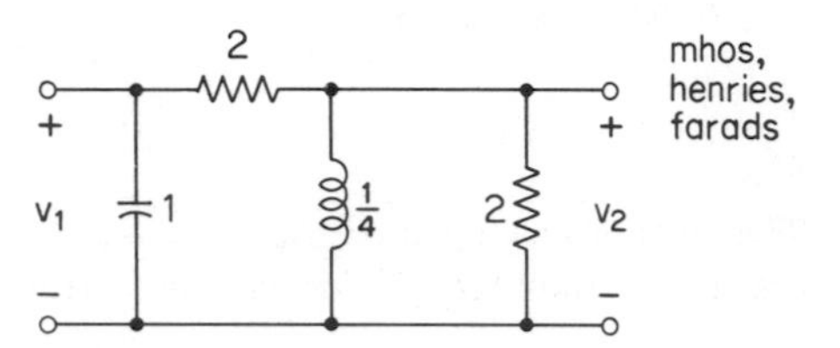

Fig. 6-3.4 A second-order *RLC* circuit with two independent variables.

SUMMARY 6-3.1

Natural Frequencies of Second-Order RL and RC Circuits: Second-order circuits consisting of two energy-storage elements of the same type always have a characteristic equation in which the discriminant is positive (Case I). Thus, the natural frequencies of such circuits are located on the negative real axis (including the origin) of the complex frequency plane and they are not equal.

In the preceding example, we showed how a set of simultaneous equations could be used to determine the characteristic equation for an *RL* (or an *RC*) second-order network. The same procedure must also be used for *RLC* circuits in which more than a single mesh or a single nodal voltage is required to define the network variables. An example follows.

EXAMPLE 6-3.2. The *RLC* network shown in Fig. 6-3.4 contains two independent nodal voltage variables $v_1(t)$ and $v_2(t)$. It is desired to determine the expression for $v_1(t)$ when the circuit has the initial conditions $v_1(0) = 1$, $v_1'(0) = 0$. We may do this by first writing the KCL equations at each of the two nodes. Thus, we obtain (note that the resistance values shown in the figure are given in units of mhos)

$$\frac{dv_1}{dt} + 2v_1(t) - 2v_2(t) = 0$$

$$4\int_{-\infty}^{t} v_2(\tau)\,d\tau + 4v_2(t) - 2v_1(t) = 0$$

If we now assume that $v_1(t) = K_a e^{st}$ and $v_2(t) = K_b e^{st}$ and substitute these expressions into the KCL equations, then, after rearranging terms, we obtain[5]

$$\begin{bmatrix} s+2 & -2 \\ -2 & \frac{4}{s}+4 \end{bmatrix} \begin{bmatrix} K_a e^{st} \\ K_b e^{st} \end{bmatrix} = \begin{bmatrix} 0 \\ 0 \end{bmatrix}$$

Setting the determinant of the square matrix in the left member of the above equation to zero yields the characteristic equation

$$s^2 + 2s + 2 = 0$$

The discriminant is negative, thus Case III applies. The natural frequencies of the network are

$$s_1 = -1 + j1 \qquad s_2 = -1 - j1$$

The general form of the solution for $v_1(t)$ is

$$v_1(t) = e^{-t}(K_1 \cos t + K_2 \sin t)$$

Evaluating this equation and its derivative at $t = 0$ to match the specified initial conditions, we find that $K_1 = K_2 = 1$. Thus, the solution for $v_1(t)$ is

$$v_1(t) = e^{-t}(\cos t + \sin t) \qquad t \geq 0$$

A plot of $v_1(t)$ is shown in Fig. 6-3.5.

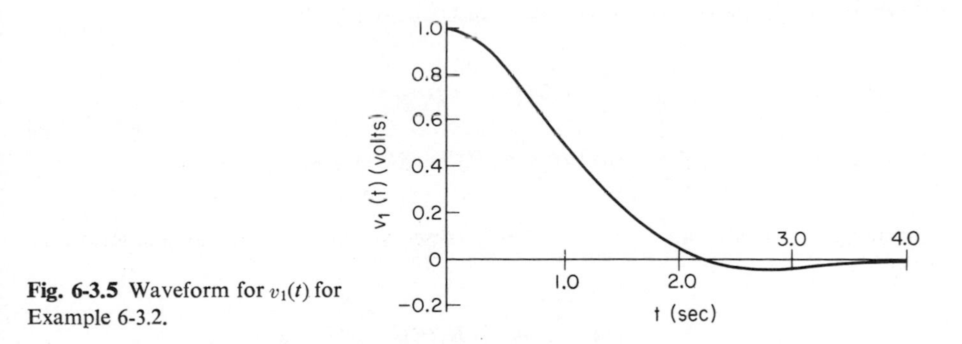

Fig. 6-3.5 Waveform for $v_1(t)$ for Example 6-3.2.

[5]It is convenient in this and the following discussion to assume that

$$\int_{-\infty}^{t} Ke^{s\tau}\,d\tau = \frac{Ke^{st}}{s}$$

This produces the same characteristic equation as would be obtained by first differentiating a given KVL or KCL integro-differential equation, and then making the substitution $K_i e^{st}$ for each of the variables. This conclusion is readily verified for this example by differentiating the KCL equation written for node 2.

The techniques described in this section for solving for the variables of second-order circuits containing two independent variables are easily extended to include the case where second-order circuits contain more than two independent variables. Such a situation would occur for a circuit containing two energy-storage elements and an arbitrary number of resistors and dependent, i.e., controlled, sources. In such a case, if the resistors and the dependent sources are replaced by equivalent resistors and/or Thevenin and Norton equivalent sources (using the techniques of Sec. 3-10), the circuit may usually be reduced to one containing only two independent variables, and the techniques of this section may be applied to determine the solution for the network variables. The process is best illustrated by an example.

EXAMPLE 6-3.3. The circuit shown in Fig. 6-3.6(a) contains three independent nodal voltage variables. In addition to the two energy-storage elements and the resistors, it also

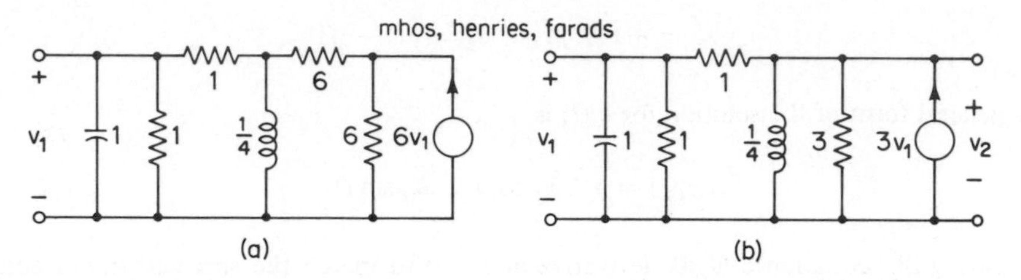

Fig. 6-3.6 A second-order *RLC* circuit with three independent variables.

contains a dependent current source whose output current depends on the voltage $v_1(t)$. It is desired to solve for the voltage $v_1(t)$ using as initial conditions $v_1(0)=1$, $v_1'(0)=0$. To do this, we may first reduce the portion of the circuit at the right which consists of two resistors and the dependent source by replacing it with a Norton equivalent circuit. The result is shown in Fig. 6-3.6(b). For this circuit, from KCL we obtain

$$\frac{dv_1}{dt} + 2v_1(t) - v_2(t) = 0$$

$$4\int_{-\infty}^{t} v_2(\tau)\,d\tau + 4v_2(t) - v_1(t) = 3v_1(t)$$

Substituting $v_1(t) = K_a e^{st}$ and $v_2(t) = K_b e^{st}$ into these equations and rearranging terms, we obtain

$$\begin{bmatrix} s+2 & -1 \\ -4 & \frac{4}{s}+4 \end{bmatrix} \begin{bmatrix} K_a e^{st} \\ K_b e^{st} \end{bmatrix} = \begin{bmatrix} 0 \\ 0 \end{bmatrix}$$

The form of these equations is similar to those found for Example 6-3.2, the only difference being in the diagonal elements of the square matrix in the left member of the equation. The characteristic equation, however, is exactly the same, namely,

$$s^2 + 2s + 2 = 0$$

Thus, since the initial conditions are the same, the details of finding the solution for $v_1(t)$ may be taken directly from that example. The result is

$$v_1(t) = e^{-t}(\cos t + \sin t)$$

It does not, of course, follow that the solutions for all the other variables of this circuit will be the same as the variables of the circuit used in Example 6-3.2.

In the discussion of second-order circuits which we have made to this point, in solving for a given variable it has been assumed that the initial conditions were given for the variable and its first derivative. In other words, if $f(t)$ was the variable, then we assumed that values for $f(0)$ and $f'(0)$ were given to permit us to determine the two arbitrary constants in the solution of the homogeneous equation. In practice, it is more convenient to be able to specify the initial conditions in terms of network variables for which the continuity conditions apply. In Secs. 4-3 and 4-6 such variables were shown to be the voltage across a capacitor and the current through an inductor. If such initial conditions are given, then, to solve the differential equation, we may use the basic circuit equations to relate these initial conditions to $f(0)$ and $f'(0)$. As an example of this type of situation, let us consider the *RLC* circuit shown in Fig. 6-3.7. The differential equation for this circuit is

$$L\frac{di}{dt} + Ri(t) + \frac{1}{C}\int_{-\infty}^{t} i(\tau)\,d\tau = 0 \tag{6-66}$$

If we assume that the values of R, L, and C are such that Case I applies, then the solution for $i(t)$ will have the form

$$i(t) = K_1 e^{s_1 t} + K_2 e^{s_2 t} \tag{6-67}$$

Now let us assume that the initial conditions are given as the current $i(0)$ through the inductor and the voltage $v(0)$ across the capacitor. To determine the two constants K_1 and K_2 in Eq. (6-67), we need to use our knowledge of $v(0)$ and $i(0)$ to find $i'(0)$. To do this, we note that

$$\frac{1}{C}\int_{-\infty}^{t} i(\tau)\,d\tau = v(t) \tag{6-68}$$

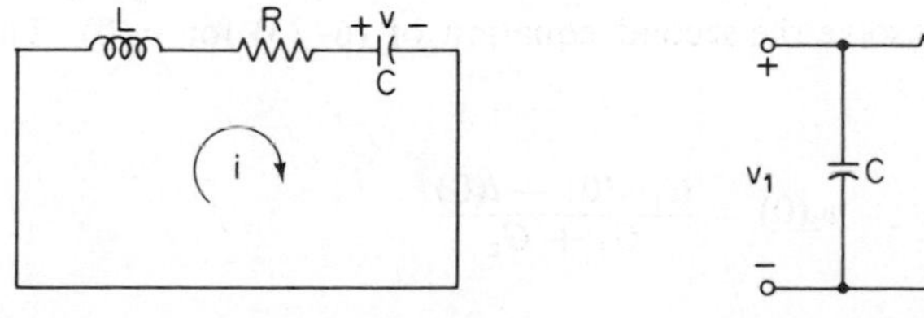

Fig. 6-3.7 A second-order *RLC* circuit.

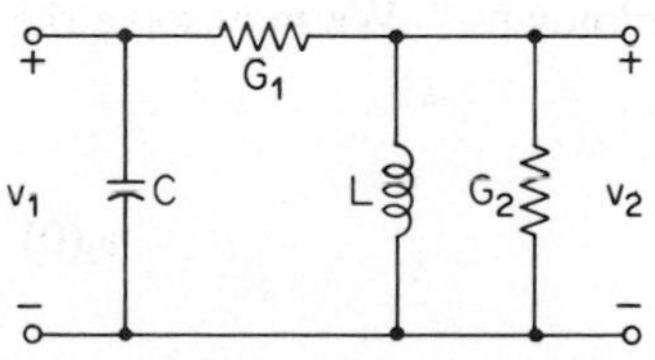

Fig. 6-3.8 A second-order *RLC* circuit.

Substituting this in (6–66) and evaluating the equation at $t = 0$, we obtain

$$Li'(0) + Ri(0) + v(0) = 0 \tag{6–69}$$

Rearranging terms, we obtain an expression for $i'(0)$ in the form

$$i'(0) = \frac{-Ri(0) - v(0)}{L} \tag{6–70}$$

where all the quantities in the right member of the equation are known. Thus, we have found an expression for the value of $i'(0)$.

The same general procedure may be applied to the case where more than one network variable is involved. For example, consider the network shown in Fig. 6–3.8. If it is desired to solve for the voltage $v_1(t)$, then we require two initial conditions, namely, $v_1(0)$ and $v_1'(0)$. To find $v_1'(0)$ in terms of the more readily specified initial condition on the value of the inductor current $i(0)$, we may proceed by first applying KCL to the nodes and writing the set of equations

$$\begin{aligned} C\frac{dv_1}{dt} + G_1 v_1(t) - G_1 v_2(t) &= 0 \\ \frac{1}{L}\int_{-\infty}^{t} v_2(\tau)\, d\tau + (G_1 + G_2)v_2(t) - G_1 v_1(t) &= 0 \end{aligned} \tag{6–71}$$

We note that the inductor current $i(t)$ may be defined as

$$i(t) = \frac{1}{L}\int_{-\infty}^{t} v_2(\tau)\, d\tau \tag{6–72}$$

Now let us substitute this relation in the set of equations of (6–71) and evaluate the resulting set of equations at $t = 0$. We obtain

$$\begin{aligned} Cv_1'(0) + G_1 v_1(0) - G_1 v_2(0) &= 0 \\ i(0) + (G_1 + G_2)v_2(0) - G_1 v_1(0) &= 0 \end{aligned} \tag{6–73}$$

The above set of equations contains two quantities which may be treated as known, i.e., the initial conditions $i(0)$ and $v_1(0)$. The quantities $v_1'(0)$ and $v_2(0)$, however, are "unknowns." We may solve the second equation of (6–73) for $v_2(0)$. Thus, we obtain

$$v_2(0) = \frac{G_1 v_1(0) - i(0)}{G_1 + G_2} \tag{6–74}$$

Substituting this value in the first equation and rearranging terms, we obtain

$$v_1'(0) = \frac{-G_1}{C} v_1(0) + \frac{G_1}{C}\left[\frac{G_1 v_1(0) - i(0)}{G_1 + G_2}\right] \tag{6-75}$$

Thus, the value of $v_1'(0)$ may be found in terms of the given initial conditions $v_1(0)$ and $i(0)$ and the values of the network elements. A similar procedure is readily applied to the problem of finding the desired initial conditions for other networks.

In the discussion of the second-order circuits which we have considered this far, we have assumed that the initial conditions were specified at $t = 0$. It is relatively easy to adapt the solution methods to the case in which initial conditions are specified for values of $t \neq 0$. The techniques are identical with those presented in Sec. 5-5. For example, assume that the values of the network elements are such that Case III applies, and thus that the solution for some network variable $f(t)$ for initial conditions applied at $t = 0$ would normally be of the form

$$f(t) = e^{\sigma t}(K_1 \cos \omega_d t + K_2 \sin \omega_d t) \qquad t \geq 0$$

For the same network, if the initial conditions were specified at t_0, i.e., if $f(t_0)$ and $f'(t_0)$ are given, then the solution will have the form

$$f(t) = e^{\sigma(t-t_0)}[K_1 \cos \omega_d(t - t_0) + K_2 \sin \omega_d(t - t_0)] \qquad t \geq t_0$$

Similar comments apply to Cases I and II, and some examples of such situations will be found in the exercises at the end of this chapter.

6-4 EXCITATION BY INITIAL CONDITIONS AND SOURCES

In the preceding sections of this chapter we have discussed the solution of second-order circuits which were excited only by initial conditions. In this section we shall extend our discussion to include the effects of excitation by sources. Thus, we shall consider the solution of *non-homogeneous* differential equations. As an example of such a situation, consider the circuit shown in Fig. 6-4.1. The KVL equation for this circuit is

$$L\frac{di}{dt} + Ri(t) + \frac{1}{C}\int_{-\infty}^{t} i(\tau)\, d\tau = v(t) \tag{6-76}$$

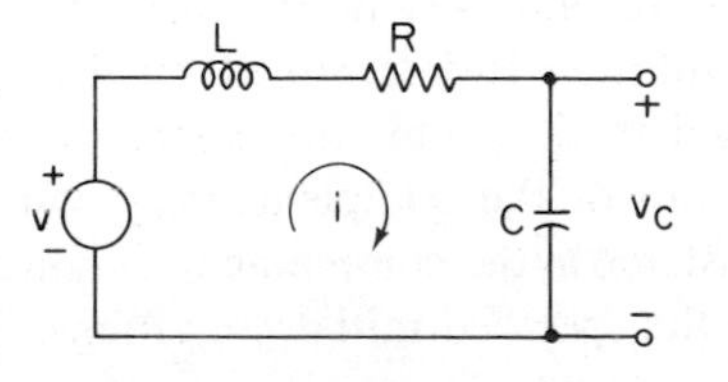

Fig. 6-4.1 A second-order *RLC* circuit.

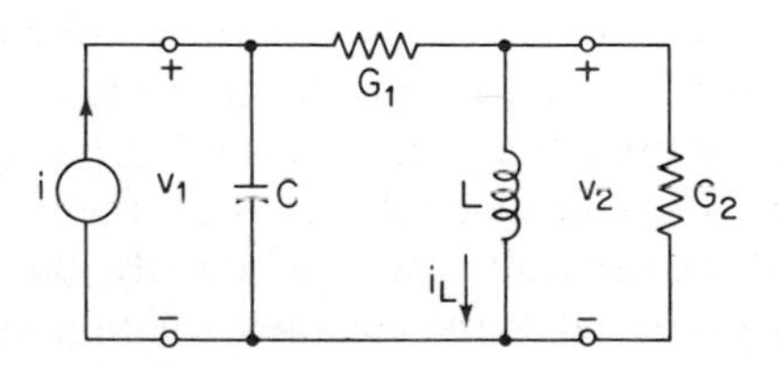

Fig. 6-4.2 A second-order *RLC* circuit with two independent variables.

This is an integro-differential equation. It is more convenient, however, to deal with differential equations than with integro-differential equations, so our first step in finding the solution of Eq. (6–76) will be to change it to a differential equation. One way of achieving this is to replace $i(t)$ in Eq. (6–76) by a different variable. To determine the variable, note that, for the network shown in Fig. 6–4.1, the voltage $v_C(t)$ across the capacitor is defined as

$$v_C(t) = \frac{1}{C}\int_{-\infty}^{t} i(\tau)\,d\tau \tag{6–77}$$

If we differentiate both sides of Eq. (6–77) two times, we obtain

$$i(t) = C\frac{dv_C}{dt} \tag{6–78}$$

and

$$\frac{di}{dt} = C\frac{d^2v_C}{dt^2} \tag{6–79}$$

Now let us substitute Eqs. (6–77), (6–78), and (6–79) into (6–76). For convenience we may also divide both members of the resulting equation by the factor LC. Thus, we obtain

$$\frac{d^2v_C}{dt^2} + \frac{R}{L}\frac{dv_C}{dt} + \frac{1}{LC}v_C(t) = \frac{1}{LC}v(t) \tag{6–80}$$

This is a non-homogeneous, second-order, differential equation. (It is also ordinary, linear, and has constant coefficients.) The techniques used for solving such a differential equation exactly parallel those presented in Sec. 5–2 for the first-order, non-homogeneous, differential equation. It was shown there that the complete solution consists of two parts. The first part is the complementary solution, i.e., the solution to the related homogeneous differential equation. As such, for a second-order circuit, it may have the form of any of the three cases described in the preceding sections of this chapter, depending on whether the discriminant of the characteristic equation is positive, zero, or negative (Cases I, II, and III, respectively). This part of the solution will contain two unspecified constants. The second part of the answer is the particular solution. The form of the particular solution is determined by the form of the excitation term. Table 5–2.1 may be used to determine the correct form to use. The complete solution is then given as the sum of the complementary solution and the particular solution. Finally, the two constants in the complementary solution are evaluated in the complete solution to match the specified initial conditions. The procedure is best illustrated by an example.

EXAMPLE 6-4.1. In the circuit shown in Fig. 6-4.1 the excitation provided by the voltage source is $v(t) = t/3$ V applied at $t = 0$. The network elements have the same values as were used in Example 6-1.1, namely, $R = 6\ \Omega$, $L = 1$ H, and $C = \frac{1}{5}$ F. For these values, the differential equation of (6-80) becomes

$$\frac{d^2 v_C}{dt^2} + 6\frac{dv_C}{dt} + 5v_C(\mathrm{t}) = \frac{5}{3}t \qquad \mathrm{t} \geq 0$$

If we let $v_h(t)$ be the complementary function, i.e., the solution to the homogeneous equation, then, following the development given in Example 6-1.1 and noting that all variables in a given circuit have the same natural behavior, we may write

$$v_h(t) = K_1 e^{-t} + K_2 e^{-5t}$$

To find the particular solution $v_p(t)$, we note that the function in the right member of the differential equation is $5t/3$. From Table 5-2.1, the form of the particular solution is seen to be

$$v_p(t) = A + Bt$$

Substituting this relation in the original differential equation and matching coefficients of like powers of t we find that $A = -\frac{2}{5}$ and $B = \frac{1}{3}$. Thus the particular solution is

$$v_p(t) = -\frac{2}{5} + \frac{t}{3}$$

The complete solution for $v_C(t)$ is now the sum of the complementary solution and the particular solution. Thus, we may write

$$v_C(t) = v_h(t) + v_p(t) = K_1 e^{-t} + K_2 e^{-5t} - \frac{2}{5} + \frac{t}{3} \qquad t \geq 0$$

To determine the initial conditions, let us assume that the circuit is unexcited for $t < 0$. Thus, $v_C(t)$ and $i(t)$ are both zero for $t < 0$. Continuity conditions, however, apply to these variables. Therefore, $i(0) = v_C(0) = 0$. In addition, since $v_C'(t)$ is proportional to $i(t)$, we see that $v_C'(0)$ is also zero. The first equation needed to determine the constants K_1 and K_2 is found by evaluating the complete solution at $t = 0$. Thus, we obtain

$$v_C(0) = K_1 + K_2 - \tfrac{2}{5} = 0$$

Differentiating the complete solution and evaluating it at $t = 0$ gives us the following second equation

$$v_C'(0) = -K_1 - 5K_2 + \tfrac{1}{3} = 0$$

Solving the two equations for K_1 and K_2 and inserting the values in the complete solution, we obtain

$$v_C(t) = \tfrac{5}{12}e^{-t} - \tfrac{1}{60}e^{-5t} - \tfrac{2}{5} + \tfrac{1}{3}t \qquad t \geq 0$$

From Eq. (6-78) we find that the solution for $i(t)$ is

$$i(t) = -\tfrac{1}{12}e^{-t} + \tfrac{1}{60}e^{-5t} + \tfrac{1}{15} \qquad t \geq 0$$

The method illustrated in the above example is summarized as follows:

SUMMARY 6-4.1

Solution of a Second-Order Circuit Excited by Initial Conditions and Sources: A second-order circuit excited by initial conditions and sources is described by a second-order, non-homogeneous, differential equation having the form

$$\frac{d^2f}{dt^2} + a_1\frac{df}{dt} + a_0 f(t) = g(t) \tag{6-81}$$

where $g(t)$ is the excitation function representing the sources and $f(t)$ is any network variable. The solution to this equation may be found as follows:

1. Find the complementary solution $f_h(t)$, i.e., the solution of the related homogeneous differential equation, using the methods of Secs. 6-1 and 6-2. This function will have two unspecified constants.
2. For the specified function $g(t)$ find the form of the particular solution $f_p(t)$ from Table 5-2.1. Substitute this expression into the original differential equation of (6-81) to find the values of any constants in $f_p(t)$.
3. Form the complete solution $f(t) = f_h(t) + f_p(t)$. The two constants in $f_h(t)$ may now be determined to match the specified initial conditions $f(0)$ and $f'(0)$.

There are other methods of changing an integro-differential equation of the form given in Eq. (6-76) to a differential equation than the change of variable technique which was used in the preceding example. One such method is simply to differentiate the entire equation. If we do this for Eq. (6-76) and divide the entire equation by L, we obtain

$$\frac{d^2i}{dt^2} + \frac{R}{L}\frac{di}{dt} + \frac{1}{LC}i(t) = \frac{1}{L}\frac{dv}{dt} \tag{6-82}$$

The techniques described in Summary 6-4.1 are readily applied to solve this equation. To illustrate the procedure, we will rework the problem presented in Example 6-4.1 as follows.

EXAMPLE 6-4.2. It is desired to find the current $i(t)$ in the circuit shown in Fig. 6-4.1 and used in Example 6-4.1 using the method given above. The differential equation given in (6-82) for the specified element values and for $v(t) = tu(t)/3$ is

$$\frac{d^2i}{dt^2} + 6\frac{di}{dt} + 5i(t) = \frac{1}{3}u(t)$$

The solution of the homogeneous equation, $i_h(t)$, has the same form as $v_h(t)$ had in Example 6-4.1, namely,

$$i_h(t) = K_1 e^{-t} + K_2 e^{-5t}$$

The particular solution $i_p(t)$ will simply be a constant. From the differential equation given above we readily find that $i_p(t) = \frac{1}{15}$. Thus, the complete solution for $i(t)$ is

$$i(t) = K_1 e^{-t} + K_2 e^{-5t} + \tfrac{1}{15} \qquad t \geq 0$$

Evaluating the solution for $i(0) = i'(0) = 0$, we find that $K_1 = -\frac{1}{12}$ and $K_2 = \frac{1}{60}$. Thus our complete solution is

$$i(t) = -\tfrac{1}{12}e^{-t} + \tfrac{1}{60}e^{-5t} + \tfrac{1}{15} \qquad t \geq 0$$

This is, of course, the same result that was obtained in Example 6-4.1.

It should be noted that the solution process illustrated in the preceding example, in which the entire integro-differential equation is differentiated, is somewhat shorter than the method illustrated in Example 6-4.1, in which a change of variable is used to convert the integro-differential equation to a differential equation. Thus, this latter method might appear to be a more desirable one. It should also be noted, however, that the latter method requires that we differentiate the expression for the excitation function. If the waveform of the excitation is discontinuous, as would be the case for a step function, then the resulting derivative will contain an impulse. The solution of a differential equation with an impulse excitation is an advanced topic which will not be treated in this chapter. However, techniques for obtaining such a solution will be covered in connection with the development of the Laplace transformation in Chap. 8.

The procedures for solving second-order differential equations excited by independent sources described in the preceding paragraphs are readily extended to the situation where the second-order circuit has more than one mesh or node-pair. For example, consider the circuit shown in Fig. 6-4.2. Applying KCL to the two nodes of this circuit, we obtain the equations

$$\begin{aligned} C\frac{dv_1}{dt} + G_1 v_1(t) - G_1 v_2(t) &= i(t) \\ -G_1 v_1(t) + (G_1 + G_2)v_2(t) + \frac{1}{L}\int_{-\infty}^{t} v_2(\tau)\,d\tau &= 0 \end{aligned} \qquad (6\text{-}83)$$

This is a simultaneous set of integro-differential equations. To begin the solution, let us change this set of equations to a set of differential equations by using the variable $i_L(t)$ in place of the variable $v_2(t)$. The necessary relations may be found from the terminal relations for the variables of the inductor, namely,

$$i_L(t) = \frac{1}{L}\int_{-\infty}^{t} v_2(\tau)\,d\tau \tag{6-84}$$

and

$$L\frac{di_L}{dt} = v_2(t) \tag{6-85}$$

Substituting these relations in (6-83), we obtain

$$\begin{aligned} C\frac{dv_1}{dt} + G_1 v_1(t) - G_1 L\frac{di_L}{dt} &= i(t) \\ -G_1 v_1(t) + L(G_1 + G_2)\frac{di_L}{dt} + i_L(t) &= 0 \end{aligned} \tag{6-86}$$

This set of equations may be solved by first assuming an exponential solution for the variables $v_1(t)$ and $i_L(t)$. Thus, if we let

$$\begin{aligned} v_1(t) &= K_a e^{st} \\ i_L(t) &= K_b e^{st} \end{aligned} \tag{6-87}$$

and substitute these values in (6-86), we obtain

$$\begin{bmatrix} sC + G_1 & -sLG_1 \\ -G_1 & sL(G_1 + G_2) + 1 \end{bmatrix} \begin{bmatrix} K_a e^{st} \\ K_b e^{st} \end{bmatrix} = \begin{bmatrix} i(t) \\ 0 \end{bmatrix} \tag{6-88}$$

The form of the complementary solution for $v_1(t)$ and $i_L(t)$ (and for any other network variable) is found by setting the determinant of the square matrix in the left member of Eq. (6-88) to zero. The roots of the resulting characteristic equation are the natural frequencies of the network. The particular solutions for the network variables $v_1(t)$ and $i_L(t)$ are found by substituting an assumed form for these variables into (6-86). The necessary form is determined from Table 5-2.1 for the specific expression given for the current source. The process is illustrated by the following example.

EXAMPLE 6-4.3. The circuit shown in Fig. 6-4.2 is assumed to have the following values for its elements: $C = 1$ F, $L = \frac{1}{4}$ H, $G_1 = G_2 = 2$ mhos. (These are the same values which were used for this circuit in Example 6-3.2.) The current source is assumed to have an excitation $i(t) = 4$ A applied at $t = 0$. The resulting set of simultaneous equations given in (6-88) may now be written

$$\begin{bmatrix} s+2 & -\frac{s}{2} \\ -2 & s+1 \end{bmatrix} \begin{bmatrix} K_a e^{st} \\ K_b e^{st} \end{bmatrix} = \begin{bmatrix} 4 \\ 0 \end{bmatrix} \qquad t > 0$$

The characteristic equation found by setting the determinant of the square matrix to zero is $s^2 + 2s + 2 = 0$. Thus, the natural frequencies of the network are located at $-1 \pm j1$, and the form of the complementary solution for either $v_1(t)$ or $i_L(t)$ is $e^{-t}(K_1 \cos t + K_2 \sin t)$, where K_1 and K_2 will, of course, be different for the two variables. The particular solution is easily found by noting from Table 5-2.1 that, for a constant-valued excitation function $i(t)$, the particular solution $v_p(t)$ for $v_1(t)$ and the particular solution $i_p(t)$ for $i_L(t)$ will both be constants. If we let $v_p(t) = A$, and $i_p(t) = B$, and substitute these values in the simultaneous differential equations of (6-86), we obtain $A = 2$, $B = 4$. Thus, the complete solution for $v_1(t)$ is of the form

$$v_1(t) = e^{-t}(K_1 \cos t + K_2 \sin t) + 2 \qquad t > 0$$

The constants K_1 and K_2 are readily determined for any specified set of initial conditions.

It should be noted that the natural frequencies found in the above example are the same as those found in Example 6–3.2. This is to be expected, since the network is identical in the two examples. Note, however, that the square matrix found in the above example is considerably different from that found in Example 6–3.2. It should especially be noted that the matrix is not symmetric.

Almost all of the techniques and conclusions derived for a first-order circuit in Secs. 5–3 to 5–6 and in Sec. 5–8 apply directly to the second-order circuit (and also to higher-order circuits). For example, in addition to separating the complete solution into a complementary solution and a particular solution, as we did in the preceding examples, we may also separate it into a zero-input response term (the portion of the solution that results from the presence of initial conditions), and a zero-state response term (the portion of the solution that is produced by excitation from sources). The general forms of such decompositions for the three cases of the second-order differential equation are tabulated in Table 6–4.1. It should be noted that the natural frequencies of the network are, in general, always excited when an input from a source or sources is applied to a network (even though the initial conditions may be zero). Thus, the familiar exponential quantities always appear in the zero-state response term as well as in the zero-input response term.[6] It should be noted that the forced response may also be referred to as the steady-state response if the excitations provided by the sources are all constant or are all sinusoidal.

Another example of the extension of the conclusions of Chap. 5 to the second-order network is the examination of linearity considerations. Thus we note that the response of any variable in a second-order circuit consisting of linear elements is linearly related to the following excitations: (1) The excitation from any initial condition when all other initial conditions are zero and when all sources are set to zero;

[6]In Case III, the term involving σ and ω_d may be considered as another way of writing a "complex exponential" term, i.e., e raised to a complex power.

and (2) the excitation from any source when all other sources are set to zero and when all initial conditions are zero. In general, the response is linearly related to any single excitation, and is not linearly related to a given excitation when other excitations are present.

TABLE 6-4.1

Case	First Decomposition of Response			Second Decomposition of Response	
	Transient Response or Complementary Solution (terms due to the natural frequencies of the network)	Forced Response or Particular Solution (terms having the form of the excitation)		Zero-Input Response (terms due to the excitation of the network by initial conditions)	Zero-State Response (terms due to the excitation of the network by sources)
I $f(t) =$	$K_1 e^{s_1 t} + K_2 e^{s_2 t}$ +	$f_p(t)$	=	$K_{1a} e^{s_1 t} + K_{2a} e^{s_2 t}$ +	$K_{1b} e^{s_1 t} + K_{2b} e^{s_2 t} + f_p(t)$
II $f(t) =$	$K_1 e^{s_1 t} + K_2 t e^{s_1 t}$ +	$f_p(t)$	=	$K_{1a} e_{s_1 t} + K_{2a} t e^{s_1 t}$ +	$K_{1b} e^{s_1 t} + K_{2b} t e^{s_1 t} + f_p(t)$
III $f(t) =$	$e^{\sigma t}(K_1 \cos \omega_d t + K_2 \sin \omega_d t)$ +	$f_p(t)$	=	$e^{\sigma t}(K_{1a} \cos \omega_d t + K_{2a} \sin \omega_d t)$ +	$e^{\sigma t}(K_{1b} \cos \omega_d t + K_{2b} \sin \omega_d t) + f_p(t)$

Note: $K_{1a} + K_{1b} = K_1$ (for all cases), $K_{2a} + K_{2b} = K_2$ (for all cases)

6-5 STATE-VARIABLE EQUATIONS FOR SECOND-ORDER CIRCUITS

In this section we shall introduce methods by means of which any second-order network may be described by a first-order matrix differential equation, i.e., by a set of simultaneous first-order differential equations. For example, consider the circuit shown in Fig. 6-5.1. Applying KVL to the two meshes of this circuit, we obtain

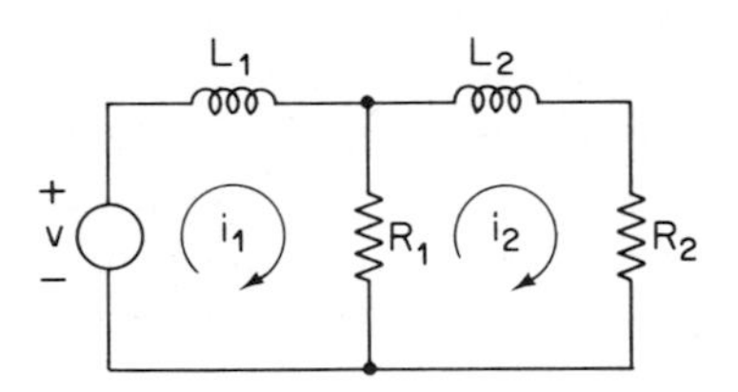

Fig. 6-5.1 A second-order *RL* circuit.

$$L_1 \frac{di_1}{dt} + R_1 i_1(t) - R_1 i_2(t) = v(t)$$
$$L_2 \frac{di_2}{dt} + (R_1 + R_2) i_2(t) - R_1 i_1(t) = 0 \tag{6-89}$$

If we divide the first equation by L_1, the second by L_2, and rearrange terms, the equations may be written in the following matrix format:

$$\begin{bmatrix} \dfrac{di_1}{dt} \\ \dfrac{di_2}{dt} \end{bmatrix} = \begin{bmatrix} -\dfrac{R_1}{L_1} & \dfrac{R_1}{L_1} \\ \dfrac{R_1}{L_2} & -\dfrac{R_1 + R_2}{L_2} \end{bmatrix} \begin{bmatrix} i_1(t) \\ i_2(t) \end{bmatrix} + \begin{bmatrix} \dfrac{v(t)}{L_1} \\ 0 \end{bmatrix} \tag{6-90}$$

This is an example of a set of simultaneous first-order differential equations having the form

$$\mathbf{f}'(t) = \mathbf{A}\mathbf{f}(t) + \mathbf{u}(t) \tag{6-91}$$

where, for this example,

$$\mathbf{f}'(t) = \begin{bmatrix} \dfrac{di_1}{dt} \\ \dfrac{di_2}{dt} \end{bmatrix} \qquad \mathbf{A} = \begin{bmatrix} -\dfrac{R_1}{L_1} & \dfrac{R_1}{L_1} \\ \dfrac{R_1}{L_2} & -\dfrac{R_1 + R_2}{L_2} \end{bmatrix} \qquad \mathbf{f}(t) = \begin{bmatrix} i_1(t) \\ i_2(t) \end{bmatrix} \qquad \mathbf{u}(t) = \begin{bmatrix} \dfrac{v(t)}{L_1} \\ 0 \end{bmatrix} \tag{6-92}$$

The relation given in (6-91) is usually referred to as a *first-order matrix differential equation.* Equations of this type play a very important role in the analysis of linear systems. In addition, they are readily written for time-varying and non-linear systems. If the equations describing a network are put in the form given in Eq. (6-91), then the network variables which are the elements of the column matrix $\mathbf{f}(t)$ are referred to as the *state variables.* Correspondingly, the equations are called the *state-variable equations*, or more simply, the *state equations.*

The state-variable formulation described above provides the key to the application of numerical techniques to the solution of a given network. Such an application will be discussed in the following section. In this section we will explore the process for writing the state equations in more detail. For the second-order circuits, which we are considering in this chapter, there are three possible ways in which the state variables may be chosen, depending on the type of energy-storage elements which are present in the circuit. The first of these ways occurs if the second-order circuit is an *RL* circuit, i.e., a circuit which has two inductors as its energy-storage elements. In this case, the state variables will be the inductor currents. The circuit described in the beginning of this section is an example of such a case. In some *RL* circuits it may happen that the conventional mesh currents do not coincide with the inductor currents. In such a case, we may use loops rather than meshes to define the independent current variables.[7] This should be done in such a manner that each loop-current variable coincides with a single inductor current. The process is illustrated by the following example.

EXAMPLE 6-5.1. It is desired to determine the state equations for the network shown in Fig. 6-5.2(a). The graph for the network is shown in Fig. 6-5.2(b). The loops formed by using a single inductor and the appropriate branches are shown in Fig. 6-5.2(c). The direction of the loop currents is, of course, arbitrary. Using KVL and the indicated loop-current reference directions, we may write the equations

[7] It should be recalled from Chap. 3 that a mesh is a restricted type of loop which has the property that no branches are enclosed by it which are not part of the loop.

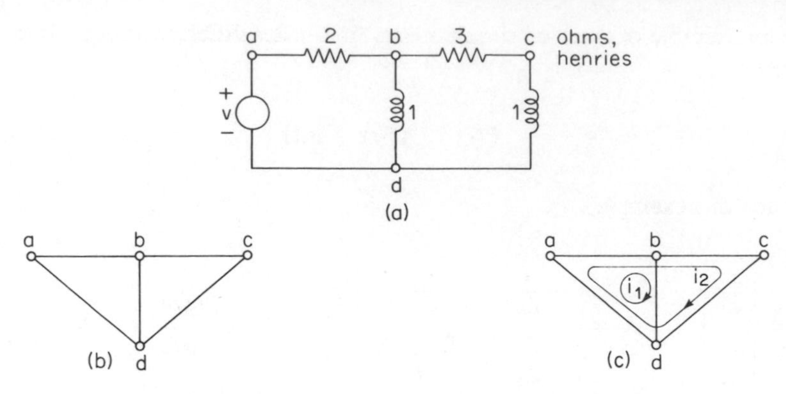

Fig. 6-5.2 Determining the state equations for a second-order *RL* network.

$$2i_1(t) + 2i_2(t) + \frac{di_1}{dt} = v(t)$$

$$2i_1(t) + 5i_2(t) + \frac{di_2}{dt} = v(t)$$

Transposing terms, we obtain

$$\begin{bmatrix} \frac{di_1}{dt} \\ \frac{di_2}{dt} \end{bmatrix} = \begin{bmatrix} -2 & -2 \\ -2 & -5 \end{bmatrix} \begin{bmatrix} i_1(t) \\ i_2(t) \end{bmatrix} + \begin{bmatrix} v(t) \\ v(t) \end{bmatrix}$$

This is the desired state-variable formulation.

A procedure similar to that described above may be used to obtain a state-variable formulation for second-order *RC* circuits, i.e., circuits which contain two capacitors as energy-storage elements. For such a network, we may select node-pair voltages in such a manner that each node-pair voltage coincides with the voltage across a capacitor. These node-pair voltage variables are then the state variables. To determine the resulting equations, we need merely draw the graph for the network and form cutsets which each contain a single capacitive branch.[8] These cutsets lead directly to a set of KCL equations which contain the capacitor voltages and their derivatives as variables. These are the state equations. The procedure is illustrated by the following example.

EXAMPLE 6-5.2. To determine the state equations for the network shown in Fig. 6-5.3(a), we first draw the circuit graph as shown in Fig. 6-5.3(b). The two branches which contain capacitors and two node-pair voltage variables are shown in Fig. 6-5.3(c). Using these vari-

[8]It may also be necessary to use equivalent sources and to replace groups of resistors by a single equivalent resistor.

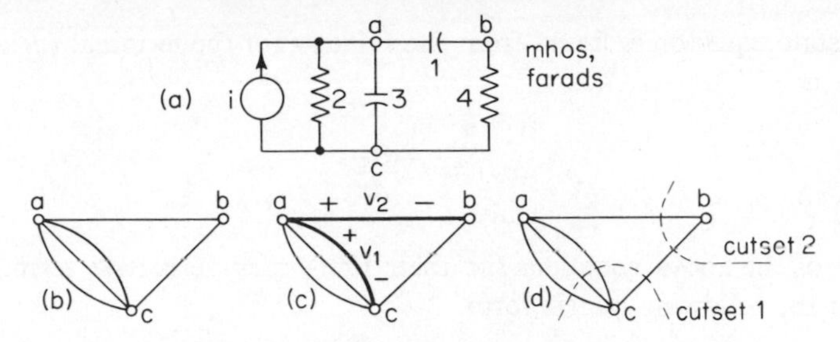

Fig. 6-5.3 Determing the state equations for a second-order *RC* network.

ables, the cutsets defining the state equations are indicated by the dashed lines in Fig. 6-5.3(d). For these cutsets, we obtain the equations

$$6v_1(t) \quad 4v_2(t) + 3\frac{dv_1}{dt} = i(t)$$

$$4v_2(t) - 4v_1(t) + \frac{dv_2}{dt} = 0$$

If we rearrange the terms of these equations, and divide the first equation by 3, the two equations may be written in the following matrix form:

$$\begin{bmatrix} \dfrac{dv_1}{dt} \\ \dfrac{dv_2}{dt} \end{bmatrix} = \begin{bmatrix} -2 & \dfrac{4}{3} \\ 4 & -4 \end{bmatrix} \begin{bmatrix} v_1(t) \\ v_2(t) \end{bmatrix} + \begin{bmatrix} \dfrac{i(t)}{3} \\ 0 \end{bmatrix}$$

This is the desired state-variable formulation.

To determine the state equations for the third type of second-order network, namely, the type in which one of the energy-storage elements is a capacitor and the other is an inductor, we may separately treat two cases. The first of these occurs when the second-order circuit consists of a single mesh or a single node-pair. In such a case, the state variables are readily chosen as the voltage across the capacitor and the current through the inductor. If the circuit contains a single mesh, as shown in Fig. 6-5.4, then the first state equation is found by applying KVL. The result is

$$v_s(t) = L\frac{di}{dt} + Ri(t) + v_C(t)$$

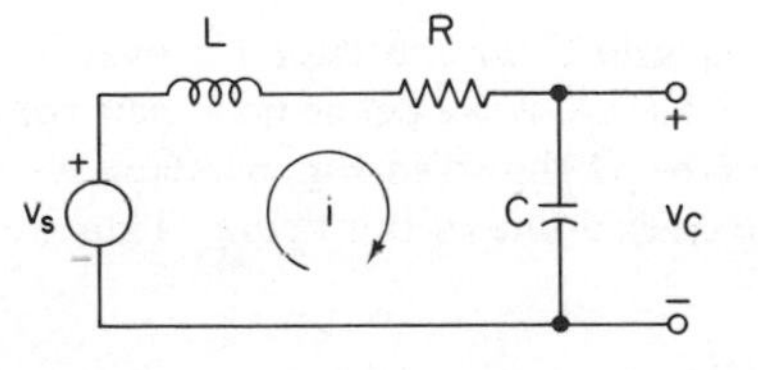

Fig. 6-5.4 A second-order *RLC* network with a single mesh.

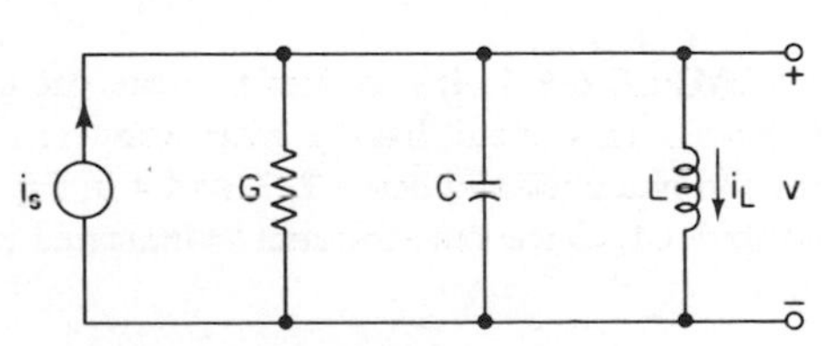

Fig. 6-5.5 A second-order *RLC* network with a single node pair.

The second state equation is found from the relation for the terminal variables of the capacitor. It is

$$i(t) = C\frac{dv_C}{dt}$$

Solving each of the above equations for their first-order derivative term, they may be written in the following matrix form.

$$\begin{bmatrix} \dfrac{di}{dt} \\ \dfrac{dv_C}{dt} \end{bmatrix} = \begin{bmatrix} -\dfrac{R}{L} & -\dfrac{1}{L} \\ \dfrac{1}{C} & 0 \end{bmatrix} \begin{bmatrix} i(t) \\ v_C(t) \end{bmatrix} + \begin{bmatrix} \dfrac{v_s(t)}{L} \\ 0 \end{bmatrix} \tag{6-93}$$

Similarly, if the circuit contains a single node-pair, as shown in Fig. 6-5.5, we may apply KCL to find one of the state equations, and we may use the terminal relation for the inductor as the other. For this circuit, after rearranging terms, we readily find the state equations to be

$$\begin{bmatrix} \dfrac{dv}{dt} \\ \dfrac{di_L}{dt} \end{bmatrix} = \begin{bmatrix} -\dfrac{G}{C} & -\dfrac{1}{C} \\ \dfrac{1}{L} & 0 \end{bmatrix} \begin{bmatrix} v(t) \\ i_L(t) \end{bmatrix} + \begin{bmatrix} \dfrac{i_S(t)}{C} \\ 0 \end{bmatrix} \tag{6-94}$$

The second case that occurs for second-order circuits, in which one of the energy-storage elements is a capacitor and the other is an inductor, is the case in which more than one variable must be used to write the network equations, i.e., the situation in which the network consists of more than a single loop or more than a single node-pair. In such a case, one of the state variables will be the current through the inductor, while the other will be the voltage across the capacitor. To achieve a state-variable formulation for such a network, we must apply KCL to a cutset which contains the capacitor but which does not contain the inductor, and apply KVL to a loop which includes the inductor but does not include the capacitor. Other KCL and KVL equations and the terminal relations for the non-energy storage elements of the circuit may then be used to eliminate any undesired variables. The procedure is illustrated by the following example.

EXAMPLE 6-5.3. It is desired to write the state equations for the circuit shown in Fig. 6-5.6(a). This circuit has the graph shown in Fig. 6-5.6(b). If we define node-pair voltages for the branches numbered 2, 3, and 4 in Fig. 6-5.6(b), as shown in Fig. 6-5.6(c), we may apply KCL to the cutset (cutset 1) indicated by the dashed line in this figure. Thus we obtain

$$\frac{1}{4}\frac{dv_1}{dt} + 2v_1(t) - 2v_2(t) = 0 \tag{6-95}$$

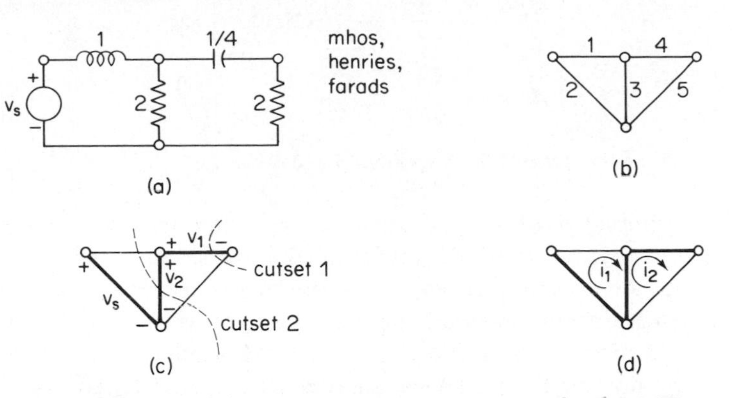

Fig. 6-5.6 Determining the state equations for a second-order *RLC* network.

Now let us assign mesh reference current directions to the two meshes shown in Fig. 6-5.6(d). Applying KVL to the mesh which contains the inductor, we may write

$$\frac{di_1}{dt} + \frac{1}{2} i_1(t) - \frac{1}{2} i_2(t) = v_s(t) \tag{6-96}$$

The two equations derived above must now be transformed so that they contain only the variables $v_1(t)$ and $i_1(t)$ (and their derivatives), and $v_s(t)$. Thus, the variables $i_2(t)$ and $v_2(t)$ must be eliminated. This is easily done by substituting other KCL and KVL relations into the original equations. For example, applying KVL around the right mesh of the circuit, we obtain

$$v_1(t) + \tfrac{1}{2} i_2(t) = \tfrac{1}{2} i_1(t) - \tfrac{1}{2} i_2(t) \tag{6-97}$$

Solving this relation for $i_2(t)$ and substituting it into (6-96), we obtain

$$\frac{di_1}{dt} + \frac{1}{4} i_1(t) + \frac{v_1(t)}{2} = v_s(t) \tag{6-98}$$

This is the first state equation. Similarly, to eliminate the variable $v_2(t)$ in (6-95), we note that the resistor in branch 3 has a terminal relation that may be written in the form

$$v_2(t) = \tfrac{1}{2} i_1(t) - \tfrac{1}{2} i_2(t) \tag{6-99}$$

If we solve (6-97) for $i_2(t)$ and substitute it into (6-99), then we have a relation giving $v_2(t)$ as a function of $i_1(t)$ and $v_1(t)$. Substituting this result in (6-95), we obtain a second state equation, namely,

$$\frac{1}{4}\frac{dv_1}{dt} + v_1(t) - \frac{1}{2} i_1(t) = 0 \tag{6-100}$$

From (6-98) and (6-100), our state equations are easily put in the form

$$\begin{bmatrix} \dfrac{di_1}{dt} \\ \dfrac{dv_1}{dt} \end{bmatrix} = \begin{bmatrix} -\dfrac{1}{4} & -\dfrac{1}{2} \\ 2 & -4 \end{bmatrix} \begin{bmatrix} i_1(t) \\ v_1(t) \end{bmatrix} + \begin{bmatrix} v_s(t) \\ 0 \end{bmatrix} \tag{6-101}$$

This is obviously of the general form specified in (6-91).

The examples given above illustrating the determination of the state equations for second-order circuits have not included any cases where the circuits contain more than two independent variables, or any cases where the circuits contain dependent sources. Such circuits, however, may be reduced to circuits similar to those discussed in this section by using techniques similar to those presented in Sec. 6-3, in which the resistors and the dependent sources are replaced by equivalent resistors and equivalent Thevenin and Norton sources.

6-6 NUMERICAL SOLUTION OF STATE EQUATIONS

In the preceding section we derived the state-equation formulation for second-order circuits. In this section we shall show how digital computational techniques can be applied to solve such circuits, i.e., we will illustrate the numerical determination of the values of the variables of a second-order network for a specified sequence of values of time.

In the last section it was shown that the equations describing a given second-order circuit could be put in the form

$$\mathbf{f}'(t) = \mathbf{A}\mathbf{f}(t) + \mathbf{u}(t) \tag{6-102}$$

where the component matrices are defined as

$$\mathbf{f}'(t) = \begin{bmatrix} \dfrac{df_1}{dt} \\ \dfrac{df_2}{dt} \end{bmatrix} \qquad \mathbf{A} = \begin{bmatrix} a_{11} & a_{12} \\ a_{21} & a_{22} \end{bmatrix} \qquad \mathbf{f}(t) = \begin{bmatrix} f_1(t) \\ f_2(t) \end{bmatrix} \qquad \mathbf{u}(t) = \begin{bmatrix} u_1(t) \\ u_2(t) \end{bmatrix} \tag{6-103}$$

and where $f_1(t)$ and $f_2(t)$ are the state variables and $u_1(t)$ and $u_2(t)$ are the excitation functions. A more general form of expression, which includes the relation given in Eq. (6-102) is

$$\mathbf{f}'(t) = \mathbf{g}[t, \mathbf{f}(t)] \tag{6-104}$$

In this expression, the notation $\mathbf{g}[t, \mathbf{f}(t)]$ represents a vector function of two arguments, namely, the scalar argument t and the vector argument $\mathbf{f}(t)$. The two separate equations represented by (6-104) may also be written in the form

$$\frac{df_1}{dt} = g_1[t, \mathbf{f}(t)]$$
$$\frac{df_2}{dt} = g_2[t, \mathbf{f}(t)] \qquad (6\text{–}105)$$

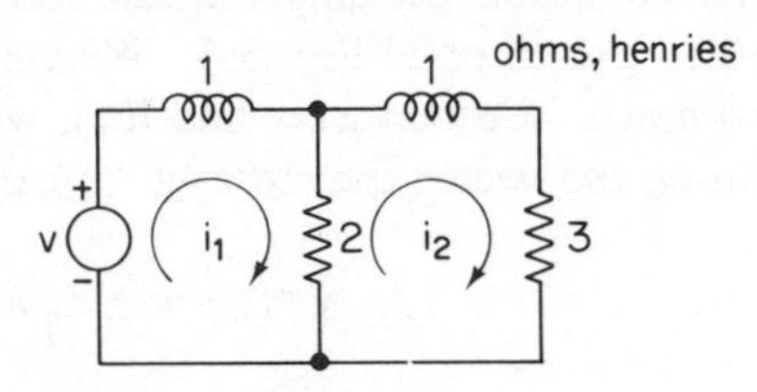

Fig. 6-6.1 A second-order *RL* network.

As an example of this type of formulation, consider the circuit shown in Fig. 6–6.1 in which the voltage source has an output of 120 $u(t)$ V. Applying KVL around each of the two meshes and rearranging terms, we obtain expressions similar to (6–105), namely,

$$\frac{di_1}{dt} = g_1[t, \mathbf{i}(t)] = -2i_1(t) + 2i_2(t) + 120u(t)$$
$$\frac{di_2}{dt} = g_2[t, \mathbf{i}(t)] = 2i_1(t) - 5i_2(t) \qquad (6\text{–}106)$$

The type of formulation illustrated above is very similar to the formulation given in (5–49). The major difference is that in this section, the equation defined is a matrix equation, i.e., a simultaneous set of differential equations, while in Sec. 5–7, the corresponding equation is a scalar equation, i.e., a single differential equation. The techniques presented in Sec. 5–7 for the solution of such a scalar differential equation, however, may be directly applied to the solution of a matrix differential equation (of first-order). For example, for the circuit shown in Fig. 6–6.1, we may define a *first-order* solution method similar to that given in Eq. (5–53). To do this let us assume that at some time t_0, $i_1(t_0)$ and $i_2(t_0)$ are known. To find the values of these variables at some other nearby value of time t_1, we may expand them in a Taylor series at t_0. From Eq. (6–106), retaining only the first two terms of the series, we obtain

$$i_1(t_1) \approx i_1(t_0) + g_1[t_0, \mathbf{i}(t_0)](t_1 - t_0)$$
$$i_2(t_1) \approx i_2(t_0) + g_2[t_0, \mathbf{i}(t_0)](t_1 - t_0) \qquad (6\text{–}107)$$

These relations may obviously be written as a single matrix equation, namely,

$$\mathbf{i}(t_1) \approx \mathbf{i}(t_0) + \mathbf{g}[t_0, \mathbf{i}(t_0)](t_1 - t_0) \qquad (6\text{–}108)$$

A comparison of this equation with (5–53) readily illustrates the fact that the first-order differential equation solution method given there for the scalar case may also be applied to a matrix differential equation.

From the above discussion it should be clear that it is easy to extend a method developed for solving *scalar*, first-order, differential equations to solve *matrix*, first-order, differential equations. All that is required is to convert to matrix form the scalar equations defining the method. We will now do this for the Runge-Kutta method presented in Sec. 5–7. We assume that $\mathbf{i}(t_0)$ and the matrix function $\mathbf{g}[t, \mathbf{i}(t)]$ are known. We may now find $\mathbf{i}(t_1)$, where $t_1 > t_0$ and is close in value to t_0, by first defining the vector quantities $\mathbf{g}^{(1)}(t_0)$, $\mathbf{g}^{(2)}(t_0)$, $\mathbf{g}^{(3)}(t_0)$, and $\mathbf{g}^{(4)}(t_0)$, using the relations

$$\begin{aligned}
\mathbf{g}^{(1)}(t_0) &= \mathbf{g}[t_0, \mathbf{i}(t_0)] \\
\mathbf{g}^{(2)}(t_0) &= \mathbf{g}\left[t_0 + \frac{\Delta t}{2}, \mathbf{i}(t_0) + \frac{\Delta t}{2}\mathbf{g}^{(1)}(t_0)\right] \\
\mathbf{g}^{(3)}(t_0) &= \mathbf{g}\left[t_0 + \frac{\Delta t}{2}, \mathbf{i}(t_0) + \frac{\Delta t}{2}\mathbf{g}^{(2)}(t_0)\right] \\
\mathbf{g}^{(4)}(t_0) &= \mathbf{g}[t_0 + \Delta t, \mathbf{i}(t_0) + \Delta t\, \mathbf{g}^{(3)}(t_0)]
\end{aligned} \tag{6–109}$$

where

$$\Delta t = t_1 - t_0 \tag{6–110}$$

Thus, to calculate $\mathbf{g}^{(1)}(t_0)$ we use the value of t_0 for the first argument of the matrix function $\mathbf{g}[t, \mathbf{i}(t)]$ and the value of $\mathbf{i}(t_0)$ for the second argument. This latter argument, of course, is a matrix (or vector) argument. The other expressions in Eq. (6–109) are similarly evaluated. Following the development of Sec. 5–7, we may now find $\mathbf{i}(t_1)$ by the relation

$$\mathbf{i}(t_1) = \mathbf{i}(t_0) + [\mathbf{g}^{(1)}(t_0) + 2\mathbf{g}^{(2)}(t_0) + 2\mathbf{g}^{(3)}(t_0) + \mathbf{g}^{(4)}(t_0)]\Delta t/6 \tag{6–111}$$

In applying the Runge-Kutta method defined by the above equations, just as was done for the scalar case in Sec. 5–7, it is usually necessary to divide the interval between t_0 and t_1 into separate subintervals, and to use the value of $\mathbf{i}(t)$ at each intermediate point as a starting value for a computation to find $\mathbf{i}(t)$ at the next intermediate point.

The procedure outlined above is readily adapted for digital computation. First of all, a subprogram must be provided which specifies the relations of the first-order, matrix, differential equation of (6–104). In Sec. 5–7 it was convenient to use a function subprogram to specify the differential equation since only a single output value was required. Here, however, we require an output quantity which is a vector. Such an output is conveniently taken as a subscripted variable. Thus, the use of a subroutine type of subprogram is appropriate. We will specify the name and argument listing of the subprogram defining the relations of Eq. (6–104) as

SUBROUTINE GN (T, A, G) (6–112)

where T is a single variable specifying the value of t, A is a one-dimensional array of variables A(I) (I = 1, 2) used as an input for the appropriate values of the variables $i_i(t)$ $(i = 1, 2)$, and G is a one-dimensional array of variables G(I) (I = 1, 2) used to output the values of the quantities $g_i(t)$ $(i = 1, 2)$. With the matrix differen-

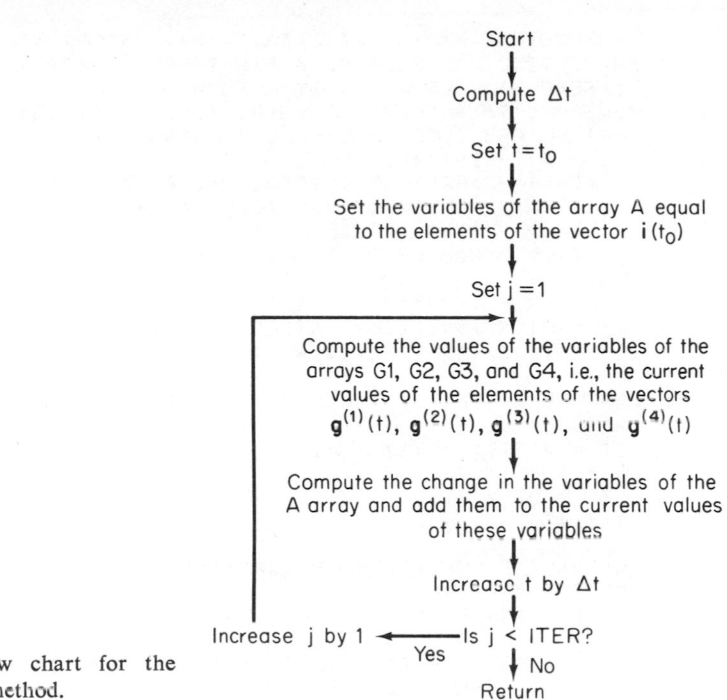

Fig. 6-6.2 Flow chart for the Runge-Kutta method.

tial equation defined by such a subroutine, we may implement Eqs. (6–109), (6–110), and (6–111) by the flow chart shown in Fig. 6–6.2 to develop a FORTRAN subroutine for solving the set of differential equations. We will use the name MXRK4 (for MatriX differential equation solution by Runge-Kutta 4th-order method) for the subroutine. A listing of the subroutine MXRK4 is given in Fig. 6–6.3, and a tabulation of its properties is given in Table 6–6.1 (on page 355). The reader should compare the flow chart and the program for MXRK4 with the flow chart and program for DFERK given in Figs. 5–7.3 and 5–7.4 to note the similarity of the two programs. Note that, although in this section we are only considering the solution of second-order differential equations, a larger dimensioning has been used in MXRK4. This is to make it usable for other applications in the following section.

To use the subroutine MXRK4 to obtain a plot of the solution of a given second-order circuit, we may use an approach similar to that used for plotting the solution of indefinite integrals in Sec. 4–5 and the solution of first-order differential equations in Sec. 5–7. This simply consists of using the values of the variables $i_1(t)$ and $i_2(t)$ at a given value of t to determine the values of these currents at some adjacent value of t. These latter values are then, in turn, used as the starting point for another application of the solution process. The general procedure is illustrated in the flow chart shown in Fig. 6–6.4. An example of such an application follows.

EXAMPLE 6-6.1. In the circuit shown in Fig. 6 6.1 the currents $i_1(t)$ and $i_2(t)$ are assumed to be zero at $t = 0$. The circuit is excited by the application of a unit step of voltage $v(t)$

```
      SUBROUTINE MXRK4 (TI, AINIT, NA, TSTOP, ITER, A)
C     SUBROUTINE FOR SOLVING A FIRST-ORDER MATRIX
C     DIFFERENTIAL EQUATION BY A FOURTH-ORDER
C     RUNGE-KUTTA METHOD.  A SUBROUTINE GN(T,A,G)
C     MUST BE PROVIDED TO DEFINE THE PROBLEM.
C           TI - INITIAL VALUE OF T
C        AINIT - ARRAY OF INITIAL VALUES OF A(I)
C           NA - NUMBER OF VARIABLES A(I)
C        TSTOP - FINAL VALUE OF T
C         ITER - NUMBER OF ITERATIONS BETWEEN TI
C                     AND TSTOP
C            A - ARRAY OF FINAL VALUES OF A(I)
      DIMENSION AINIT(30), A(30), ATEMP(30),
     1G1(10), G2(10), G3(10), G4(10)
C
C     COMPUTE DT, DT/2, AND DT/6, AND INITIALIZE T
      HITER = ITER
      DT = (TSTOP - TI) / HITER
      DT2 = DT / 2.
      DT6 = DT / 6.
      T = TI
C
C     INITIALIZE THE ARRAY OF VARIABLES A(I)
      DO 7 I = 1, NA
    7 A(I) = AINIT(I)
C
C     ENTER A DO LOOP TO COMPUTE ITER INTERMEDIATE
C     VALUES OF THE VARIABLES A(I)
      DO 23  J = 1, ITER
C
C     COMPUTE THE INTERMEDIATE ARRAYS G1, G2, G3,
C     AND G4, USING THE APPROPRIATE VALUES OF T
      CALL GN(T, A, G1)
      DO 11  I = 1, NA
   11 ATEMP(I) = A(I) + G1(I) * DT2
      T = T + DT2
      CALL GN(T, ATEMP, G2)
      DO 15  I = 1, NA
   15 ATEMP(I) = A(I) + G2(I) * DT2
      CALL GN(T, ATEMP, G3)
      DO 18  I = 1, NA
   18 ATEMP(I) = A(I) + G3(I) * DT
      T = T + DT2
      CALL GN(T, ATEMP, G4)
C
C     COMPUTE THE VALUES OF THE VARIABLES A(I) AT
C     THE JTH INTERMEDIATE POINT
      DO 22  I = 1, NA
   22 A(I) = A(I) + (G1(I)+2.*(G2(I)+G3(I))+G4(I)) * DT6
   23 CONTINUE
      RETURN
      END
```

Fig. 6-6.3 Listing of the subroutine MXRK4.

$= 120$ at $t = 0$. To apply the subroutine MXRK4 to determine the values of the mesh currents for $t \geq 0$, we must first write a subprogram identified by a statement of the form given in Eq. (6-112) to specify the relations given in Eq. (6-106). A main program is then

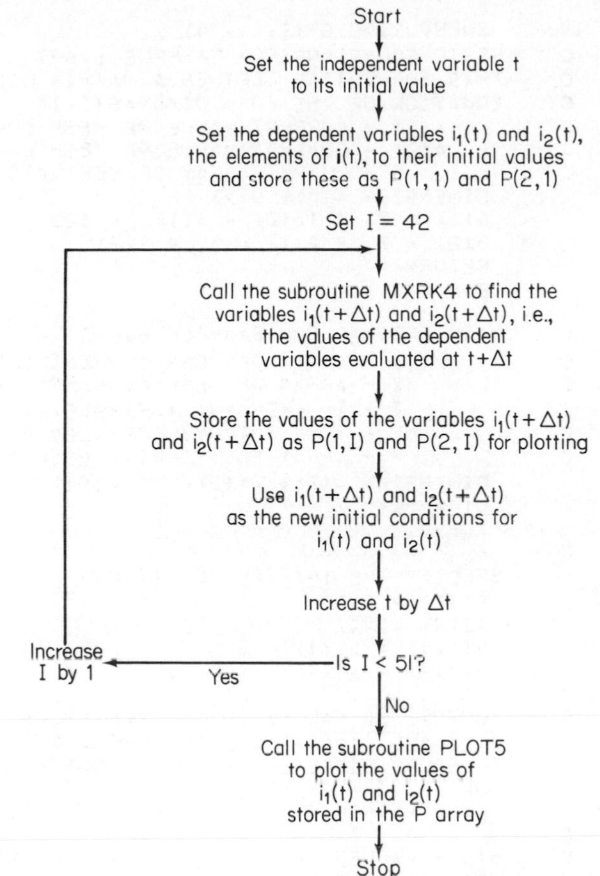

Fig. 6-6.4 Flow chart for solving first-order matrix differential equations.

easily written to compute the desired values of the variables and to store them for plotting. A listing of the main program and the subroutine GN for finding a set of fifty-one values of the mesh currents over a range of time of 0–5 s is shown in Fig. 6-6.5. A copy of the resulting plot of the variables is given in Fig. 6-6.6. Direct circuit analysis readily shows that the solutions for $i_1(t)$ and $i_2(t)$ are

$$i_1(t) = 100 - 96e^{-t} - 4e^{-6t} \qquad t \geq 0$$

$$i_2(t) = 40 - 24e^{-t} + 8e^{-6t} \qquad t \geq 0$$

The agreement of the numerical results with these expressions is readily verified.

In the example given above, it should be noted that a value of 2 has been used for the input argument ITER in the subroutine MXRK4. Thus the program separates each 0.1 interval between plotted points into two subintervals to obtain the required accuracy of the answers. The considerations which determine the proper choice of this constant are similar to those discussed in connection with the treatment

```
      SUBROUTINE GN(T, A, G)
C     STATE EQUATIONS FOR EXAMPLE 6.6-1
C     THIS SUBROUTINE DEFINES A MATRIX DIFFERENTIAL
C     EQUATION OF THE FORM DI/DT=G(T,I)
C         A(1) - CURRENT VALUE OF MESH CURRENT I1(T)
C         A(2) - CURRENT VALUE OF MESH CURRENT I2(T)
C            G - OUTPUT ARRAY OF DERIVATIVES
      DIMENSION A(2), G(2)
      G(1) = 2. * (A(2) - A(1)) + 120.
      G(2) = 2. * A(1) - 5. * A(2)
      RETURN
      END

C     MAIN PROGRAM FOR EXAMPLE 6.6-1
C            A - ARRAY OF MESH CURRENTS AFTER CALLING MXRK4
C           AI - ARRAY OF MESH CURRENTS BEFORE CALLING MXRK4
C            T - INDEPENDENT VARIABLE, TIME
C           DT - TIME INTERVAL BETWEEN PLOTTED POINTS
C            P - ARRAY FOR STORING MESH CURRENTS FOR PLOTTING
      DIMENSION A(2), AI(2), P(5,101)
      DT = 0.1
      T = 0.
C
C     SPECIFY THE INITIAL CONDITIONS
      AI(1) = 0.
      AI(2) = 0.
      P(1,1) = AI(1)
      P(2,1) = AI(2)
C
C     COMPUTE 50 VALUES OF THE VARIABLES
C     A(1) AND A(2) BY CALLING MXRK4 SUBROUTINE
C     TO SOLVE DIFFERENTIAL EQUATIONS FROM T TO T+DT
      DO 11  I = 2, 51
      CALL MXRK4 (T, AI, 2, T+DT, 2, A)
C
C     STORE VALUES OF A(1) AND A(2) FOR PLOTTING
      P(1,I) = A(1)
      P(2,I) = A(2)
C
C     RESET THE INITIAL CONDITIONS
      AI(1) = A(1)
      AI(2) = A(2)
C
C     INCREMENT THE VALUE OF T
   11 T = T + DT
C
C     CALL PLOTTING SUBROUTINE TO PLOT THE CURRENTS
      PRINT 13
   13 FORMAT (1H1)
      CALL PLOTS (P, 2, 51, 100)
      STOP
      END
```

Fig. 6-6.5 Listing of the program for Example 6-6.1.

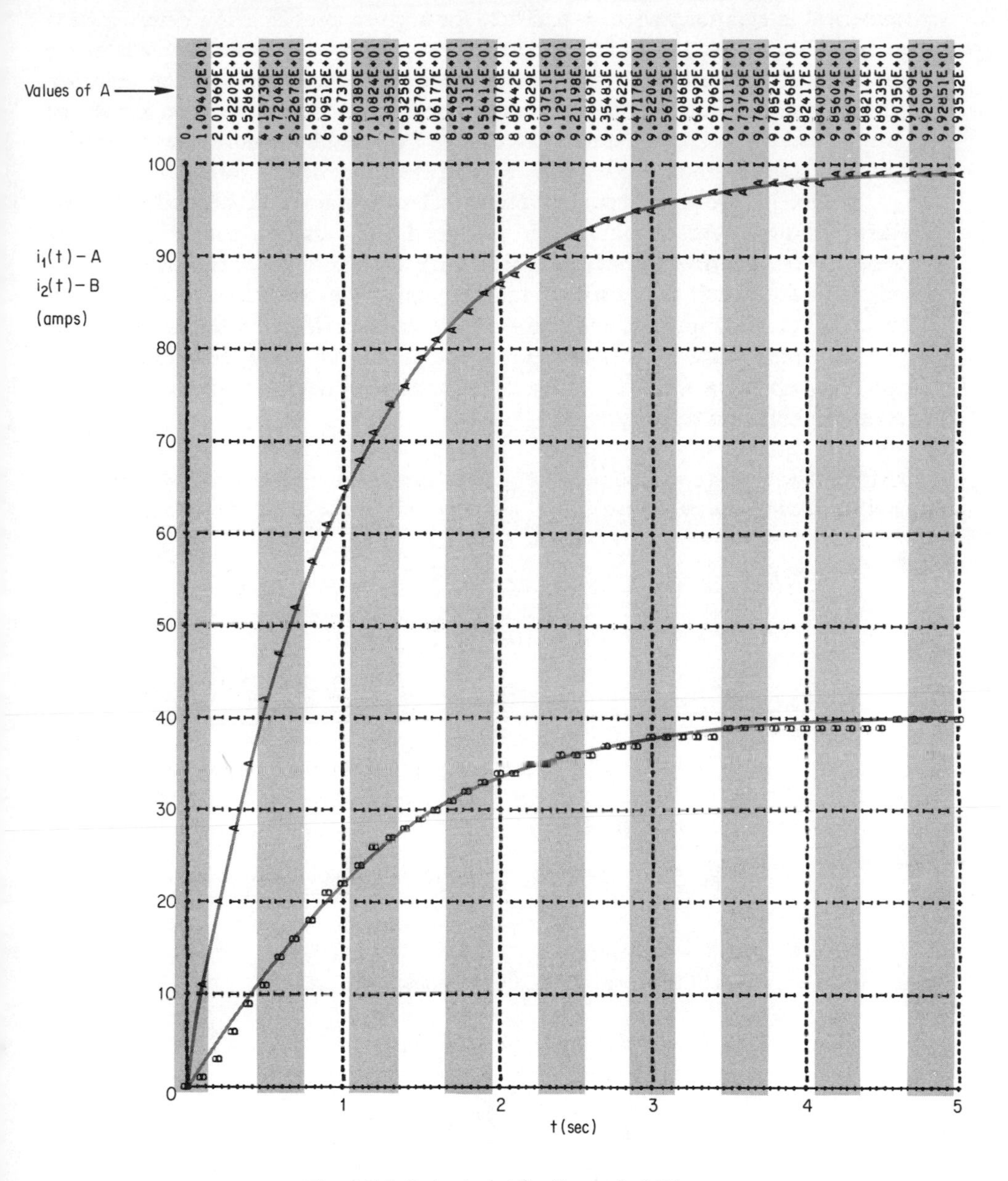

Fig. 6-6.6 Output plot for Example 6-6.1.

of numerical integration in Sec. 4–4. Thus, for a given problem, the determination of a "good" value for this constant may require that a comparison be made of the results obtained from running the problem more than once. Some of the more sophisticated methods for solving differential equations which have been developed permit the user to specify the desired accuracy of the results in advance. A detailed treatment of such methods, however, is beyond the scope of this text.

In Sec. 5–7 we pointed out that numerical techniques could be used to provide solutions for first-order circuits which contained time-varying and/or non-linear elements. The numerical techniques which have been used to obtain solutions for the second-order circuit in this section may be similarly extended, since the relation given in Eq. (6–104) includes the time-varying and non-linear cases as well as the more usual time-invariant linear one. Thus, the use of numerical techniques greatly extends our capability for the solution of second-order circuits. An example of a time-varying situation follows.

EXAMPLE 6-6.2. The circuit shown in Fig. 6–6.7 contains a time-varying inductor whose value is specified by the expression $1 + e^{-t}$. The initial conditions for the circuit are $i_1(0)$

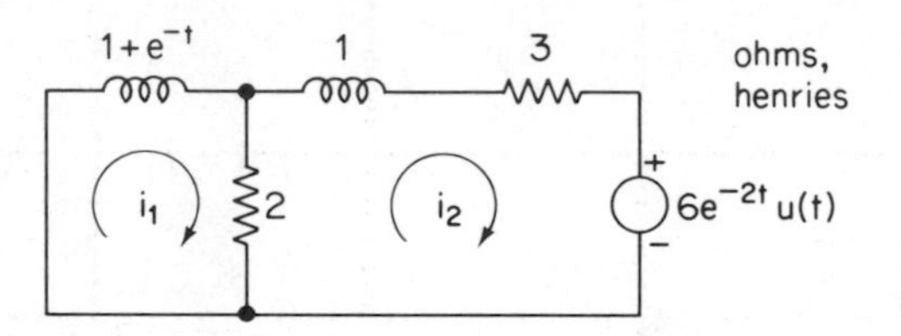

Fig. 6-6.7 Network for Example 6-6.2.

$= 2$ and $i_2(0) = -1$ A. The indicated excitation is applied by the voltage source at $t = 0$. To find the mesh currents $i_1(t)$ and $i_2(t)$ for $0 \le t \le 5$, we may begin by applying KVL around the two meshes. Thus, we obtain

$$(1 + e^{-t})i_1'(t) - e^{-t}i_1(t) + 2i_1(t) - 2i_2(t) = 0$$
$$i_2'(t) + 5i_2(t) - 2i_1(t) = -6e^{-2t}$$

Putting these equations in the form of (6–106), we obtain

$$\frac{di_1}{dt} = \frac{e^{-t} - 2i_1(t) + 2i_2(t)}{1 + e^{-t}}$$

$$\frac{di_2}{dt} = 2i_1(t) - 5i_2(t) - 6e^{-2t}$$

To determine $i_1(t)$ and $i_2(t)$ we must first write a subroutine GN which implements the above equations, and then prepare a main program which applies the subroutine MXRK4 to solve for the mesh current variables. The flow chart and main program are almost exactly the same as those used in the preceding example. A listing of the subroutine GN and the main program for this example are given in Fig. 6–6.8. A reproduction of the output plot obtained from this program is given in Fig. 6–6.9. Although, in general, it is difficult to

```
      SUBROUTINE GN(T, A, G)
C     STATE EQUATIONS FOR EXAMPLE 6.6-2
C     THIS SUBROUTINE DEFINES A MATRIX DIFFERENTIAL
C     EQUATION OF THE FORM DI/DT=G(T,I)
C        A(1) - CURRENT VALUE OF MESH CURRENT I1(T)
C        A(2) - CURRENT VALUE OF MESH CURRENT I2(T)
C           T - INDEPENDENT VARIABLE, TIME
C           G - OUTPUT ARRAY OF DERIVATIVES
      DIMENSION A(2), G(2)
      G(1) = (EXP(-T) + 2. * (A(2) - A(1)))
     1   / (1. + EXP(-T))
      G(2) = 2.*A(1) - 5.*A(2) - 6.*EXP(-2.*T)
      RETURN
      END
C     MAIN PROGRAM FOR EXAMPLE 6.6-2
C           A - ARRAY OF MESH CURRENTS AFTER CALLING MXRK4
C          AI - ARRAY OF MESH CURRENTS BEFORE CALLING MXRK4
C           T - INDEPENDENT VARIABLE, TIME
C          DT - TIME INTERVAL BETWEEN PLOTTED POINTS
C           P - ARRAY FOR STORING MESH CURRENTS FOR PLOTTING
      DIMENSION A(2), AI(2), P(5,101)
      DT = 0.1
      T = 0.
C
C     ESTABLISH THE INITIAL CONDITIONS
      AI(1) = 2.
      AI(2) = -1.
      P(1,1) = 60.
      P(2,1) = -30.
C
C     COMPUTE FIFTY VALUES OF THE VARIABLES
C     A(1) AND A(2) BY CALLING MXRK4 SUBROUTINE
C     TO SOLVE DIFFERENTIAL EQUATIONS FROM T TO T+DT
      DO 11 I = 2, 51
      CALL MXRK4 (T, AI, 2, T+DT, 2, A)
C
C     SCALE VALUES OF THE VARIABLES A(1) AND A(2)
C     AND STORE IN P ARRAY FOR PLOTTING
      P(1,I) = A(1) * 30.
      P(2,I) = A(2) * 30.
C
C     RESET INITIAL CONDITIONS
      AI(1) = A(1)
      AI(2) = A(2)
C
C     INCREMENT THE VALUE OF T
   11 T = T + DT
C
C     CALL PLOTTING SUBROUTINE TO PLOT THE CURRENTS
      PRINT 13
   13 FORMAT (1H1)
      CALL PLOTS (P, 2, 51, 60)
      STOP
      END
```

Fig. 6-6.8 Listing of the program for Example 6-6.2.

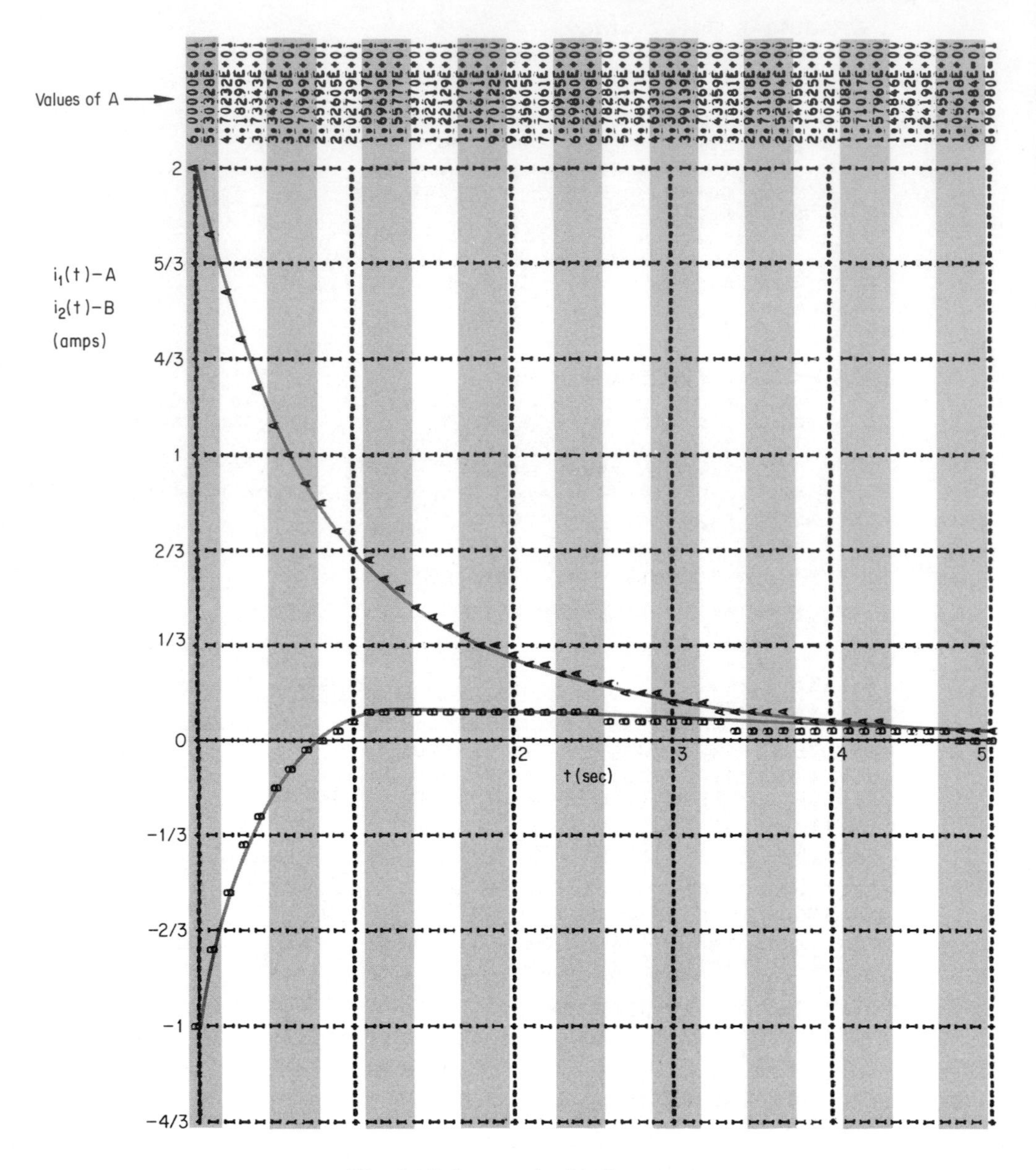

Fig. 6-6.9 Output plot for Example 6-6.2.

obtain the analytical solution of a time-varying circuit, in this case it is readily verified by direct substitution in the KVL equations that the solution is

$$i_1(t) = 2e^{-t} \qquad t \geq 0$$

$$i_2(t) = e^{-t} - 2e^{-2t} \qquad t \geq 0$$

It is easily seen that these expressions agree with the results shown in the plot in Fig. 6-6.9.

The results given above may be summarized as follows:

SUMMARY 6-6.1

Numerical Solution of Time-Varying and Non-linear Second-Order Circuits: The equation or set of equations describing any second-order circuit may be put in the form $\mathbf{f}'(t) = \mathbf{g}[t, \mathbf{f}(t)]$, where $\mathbf{f}(t)$ is a vector whose elements are

TABLE 6-6.1 Summary of the Characteristics of the Subroutine MXRK4

Identifying Statement: SUBROUTINE MXRK4 (TI, AI, NA, TSTOP, ITER, A)
Purpose: To solve the first-order matrix differential equation

$$\mathbf{i}' = \mathbf{g}[t, \mathbf{i}(t)]$$

and thus to find the elements of the column matrix $\mathbf{i}(t)$ at the value of t specified as t_{stop}, starting from t_i, a known value of t, and $\mathbf{i}_i$, where $\mathbf{i}_i = \mathbf{i}(t_i)$.
Additional Subprograms Required: This subroutine calls the subroutine identified by the statement

SUBROUTINE GN (T, A, G)

which must be used to define the differential equation, i.e., to specify $\mathbf{g}[t, \mathbf{i}(t)]$. The arguments A and G in the subroutine are both one-dimensional arrays.
Input Arguments:

- TI The initial value t_i of the independent variable t.
- AI The one-dimensional array of variables AI(I) in which are stored the initial values of the variables of the first-order differential equation, i.e., the elements of the column matrix $\mathbf{i}_i$, where $\mathbf{i}_i = \mathbf{i}(t_i)$.
- NA The number of elements in the column matrix, $\mathbf{i}(t)$, i.e., the number of variables.
- TSTOP The final value t_{stop} of t for which $\mathbf{i}(t)$ is to be found.
- ITER The number of iterations used by the subroutine in going from t_i to t_{stop}.

Output Argument:

- A The one-dimensional array of variables A(I) in which are stored the values of the first-order matrix differential equation evaluated at t_{stop}, i.e., the elements of the column matrix $\mathbf{i}(t_{\text{stop}})$.

the two state variables, and $\mathbf{f}'(t)$ is the vector whose elements are the derivatives of the state variables with respect to t. This matrix first-order differential equation may be solved by numerical techniques using the subroutine MXRK4. The same technique applies if the second-order circuit contains time-varying and/or non-linear elements.

The numerical solution of a set of first-order differential equations introduced in this section is one of the most important topics that occurs in the application of numerical techniques to the analysis of circuits and systems. Subroutines similar to the MXRK4 subroutine which we have developed here are one of the major portions of almost every modern digital computer circuit analysis program. Thus, the student who masters the concepts developed in this section will have made valuable preparation for a future study of such programs.

6-7 CIRCUITS OF HIGHER THAN SECOND ORDER

In Chap. 5 and in the preceding sections of this chapter we have developed methods for analyzing first- and second-order circuits. These methods may be directly extended to permit us to study nth-order circuits, where the value of n is restricted only by our patience in carrying out the mathematical manipulations and by the accuracy of our computations, rather than by any limit of theory. Thus, we may look upon this section as providing techniques to obtain the solution for the variables of a circuit comprised of any combination of the network elements which we have discussed to this point. Despite the magnitude of such an attainment, this section will not be an excessively long one. The reason for this is that it will consist mostly of an extension of previously developed techniques.

As a first example of a circuit of higher than second order, consider the RC network shown in Fig. 6–7.1. Let us assume that we desire to find the voltage $v_3(t)$

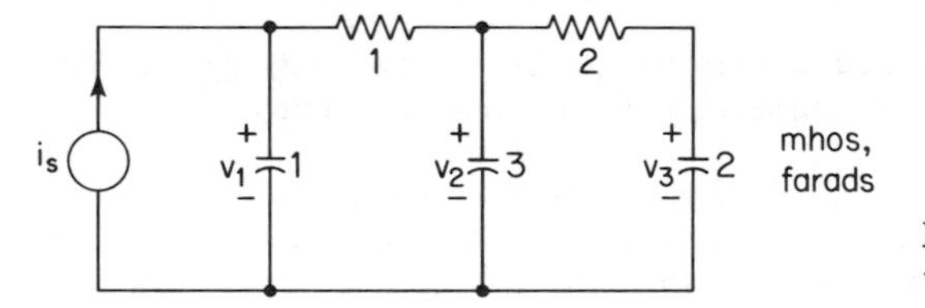

Fig. 6-7.1 A third-order RC network.

for the case where the current source is set to zero, i.e., $i_s(t) = 0$. The initial conditions will be specified as $v_3(0) = 1$, $v_3'(0) = 1$, and $v_3''(0) = 1$. The equations found by applying KCL at the three nodes are

$$\begin{aligned} v_1'(t) + v_1(t) - v_2(t) &= 0 \\ 3v_2'(t) + 3v_2(t) - v_1(t) - 2v_3(t) &= 0 \\ 2v_3'(t) + 2v_3(t) - 2v_2(t) &= 0 \end{aligned} \tag{6–113}$$

Substituting $v_1(t) = K_a e^{st}$, $v_2(t) = K_b e^{st}$, and $v_3(t) = K_c e^{st}$ in these equations, we obtain

$$\begin{bmatrix} s+1 & -1 & 0 \\ -1 & 3s+3 & -2 \\ 0 & -2 & 2s+2 \end{bmatrix} \begin{bmatrix} K_a e^{st} \\ K_b e^{st} \\ K_c e^{st} \end{bmatrix} = \begin{bmatrix} 0 \\ 0 \\ 0 \end{bmatrix} \tag{6-114}$$

Taking the determinant, we find that the characteristic equation for the network is

$$s^3 + 3s^2 + 2s = s(s+1)(s+2) = 0 \tag{6-115}$$

Thus, we see that the roots of the characteristic equation are $s_1 = 0$, $s_2 = -1$, and $s_3 = -2$. The complementary solution for $v_3(t)$ (or for any other variable in the network) is

$$v_3(t) = K_1 + K_2 e^{-t} + K_3 e^{-2t} \tag{6-116}$$

If we evaluate at $t = 0$ the equation for $v_3(t)$ given above, the equation for $v_3'(t)$, and the equation for $v_3''(t)$, we find that the constants K_1, K_2, and K_3 are related to the specified initial conditions by the equations

$$\begin{bmatrix} 1 & 1 & 1 \\ 0 & -1 & -2 \\ 0 & 1 & 4 \end{bmatrix} \begin{bmatrix} K_1 \\ K_2 \\ K_3 \end{bmatrix} = \begin{bmatrix} 1 \\ 1 \\ 1 \end{bmatrix} \tag{6-117}$$

Solving this set of equations, we obtain $K_1 = 3$, $K_2 = -3$, and $K_3 = 1$. Thus, the solution for $v_3(t)$ is

$$v_3(t) = 3 - 3e^{-t} + e^{-2t} \tag{6-118}$$

The techniques illustrated above are readily applied to an *RC* circuit of arbitrary complexity. It should be noted that although the order of the characteristic equation for an *RC* circuit cannot be greater than the number of independent network variables which is needed to write the KCL equations for the circuit, it may be less. As an example of this, in the circuit shown in Fig. 6-7.2 there are three capacitors. If we determine the characteristic equation for this circuit, we obtain

$$[s(C_1 + C_3) + (G_1 + G_3)][s(C_2 + C_3) + (G_2 + G_3)] - (sC_3 + G_3)^2 = 0 \tag{6-119}$$

This equation is obviously of second order. Thus, we see that there are only two natural frequencies and that the circuit is a second-order one.

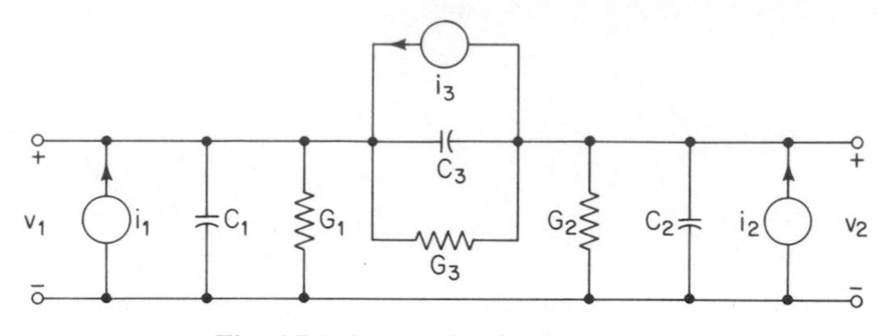

Fig. 6-7.2 A second-order *RC* network.

Now let us consider the regions in the complex s plane where the roots of the characteristic equations (i.e., the natural frequencies) of an *RC* circuit may be located. In Sec. 6–3 we showed that, for a second-order *RC* circuit, the natural frequencies were always on the negative real axis (or the origin) of the complex frequency or s plane. The same conclusion applies to nth order *RC* circuits. It may also be shown that the natural frequencies are simple.

In the above paragraphs we have discussed the solution of an nth-order RC network which is excited only by initial conditions. Only a slight extension of the method is necessary to include the case where excitation is also furnished by sources. In such a case, we must find the particular solution for the network variable that we wish to solve for. To do this we need merely substitute an assumed solution for each of the independent variables in the original KCL equations describing the circuit. The correct form of assumed solution to use is determined by the form of the excitation function, and may be taken from Table 5–2.1. The unspecified constants in the expressions given in the table must be determined separately for each of the independent variables. Once the particular solution has been determined, the complete solution may be found by evaluating the constants in the complementary solution (the part of the solution which is determined by the characteristic equation) in such a way as to match the specified initial conditions for the network. The procedure is illustrated by the following example.

EXAMPLE 6-7.1. The circuit shown in Fig. 6–7.1 is excited by a current source at the left node as illustrated. The excitation $i_s(t)$ provided by the current source is $18e^{-3t}$ A starting at t equals zero. It is desired to find an expression for the voltage $v_3(t)$. It is assumed that the initial conditions are $v_3(0) = 1$, $v_3'(0) = 0$, and $v_3''(0) = 0$. From Table 5-2.1 we find that the assumed form for each of the node voltages $v_i(t)$ $(i = 1, 2, 3)$ may be written as $A_i e^{-3t}$ $(i = 1, 2, 3)$, where the A_i are quantities whose values are to be determined. Substituting these relations into the original KCL equations of (6–113) and setting the first equation equal to $i_s(t)$, we obtain

$$\begin{bmatrix} -2 & -1 & 0 \\ -1 & -6 & -2 \\ 0 & -2 & -4 \end{bmatrix} \begin{bmatrix} A_1 \\ A_2 \\ A_3 \end{bmatrix} e^{-3t} = \begin{bmatrix} 18 \\ 0 \\ 0 \end{bmatrix} e^{-3t}$$

Since we are only interested in obtaining a solution for $v_3(t)$, we need solve only for A_3. Using determinants, we find that $A_3 = -1$. Thus, the particular solution for $v_3(t)$ is $-e^{-3t}$. The complementary solution is given by Eq. (6-116). Thus, the complete solution for $v_3(t)$ is

$$v_3(t) = K_1 + K_2 e^{-t} + K_3 e^{-2t} - e^{-3t} \qquad t \geq 0$$

Evaluating the equation for $v_3(t)$, the equation for $v_3'(t)$, and the equation for $v_3''(t)$ at $t = 0$, and inserting the specified initial conditions, we obtain the equations

$$\begin{bmatrix} 1 & 1 & 1 \\ 0 & -1 & -2 \\ 0 & 1 & 4 \end{bmatrix} \begin{bmatrix} K_1 \\ K_2 \\ K_3 \end{bmatrix} = \begin{bmatrix} 2 \\ -3 \\ 9 \end{bmatrix}$$

The solution of these equations yields the values $K_1 = 2$, $K_2 = -3$, and $K_3 = 3$. Thus, the solution for $v_3(t)$ is

$$v_3(t) = 2 - 3e^{-t} + 3e^{-2t} - e^{-3t} \qquad t \geq 0$$

There are many fairly obvious variations of the above procedure. For example, if it is desired to find a solution for some network variable which is not one of the independent variables used to determine the KCL equations, then, using the particular solutions for the independent variables, and applying KCL, KVL, and the branch terminal relations, we may determine the particular solution for any other network variable. The complete solution is then found as the sum of the particular solution and the complementary solution. Another variation in the procedure occurs for the case where more than one excitation is present, and the form of the excitations is not the same. In such a case, particular solutions for each of the excitations may be separately determined. The sum of the particular solutions and the complementary solution then gives the complete solution, and the unspecified constants of the complementary solution are evaluated to match the given initial conditions.

A procedure similar to that described in the above paragraphs may be applied to find a solution for an nth-order RL circuit. As an example of this, consider the three-mesh RL circuit shown in Fig. 6-7.3. Let us assume that it is desired to find a solution for the network variable $i_3(t)$ for the case where the network is excited only by initial conditions. Applying KVL to each of the meshes, we obtain the equations

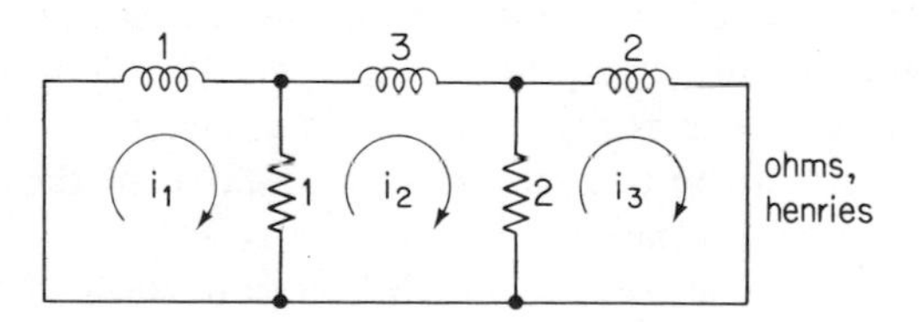

Fig. 6-7.3 A third-order *RL* network.

$$\begin{aligned} i_1'(t) + i_1(t) - i_2(t) &= 0 \\ 3i_2'(t) + 3i_2(t) - i_1(t) - 2i_3(t) &= 0 \\ 2i_3'(t) + 2i_3(t) - i_2(t) &= 0 \end{aligned} \tag{6-120}$$

Substituting the expressions $i_1(t) = K_a e^{st}$, $i_2(t) = K_b e^{st}$, and $i_3(t) = K_c e^{st}$ in the above equations, we obtain

$$\begin{bmatrix} s+1 & -1 & 0 \\ -1 & 3s+3 & -2 \\ 0 & -2 & 2s+2 \end{bmatrix} \begin{bmatrix} K_a e^{st} \\ K_b e^{st} \\ K_c e^{st} \end{bmatrix} = \begin{bmatrix} 0 \\ 0 \\ 0 \end{bmatrix} \tag{6-121}$$

Calculating the determinant, we find that the characteristic equation is

$$s^3 + 3s^2 + 2s = s(s+1)(s+2) = 0 \tag{6-122}$$

Thus we see that the expression for $i_3(t)$ may be written in the form

$$i_3(t) = K_1 + K_2 e^{-t} + K_3 e^{-2t} \tag{6-123}$$

The values of the constants K_i are readily found for any specified set of initial conditions.

The procedure illustrated above is readily applied to an *RL* circuit of arbitrary complexity. The extension of the procedure to the case where sources are present is identical with that discussed for the *RC* network. The natural frequencies will always lie on the negative real axis (or at the origin) of the complex frequency plane just as they did for the *RC* case. Although in the examples given above we used KCL to determine the network equations for an *RC* circuit and KVL to find the equations for an *RL* circuit, the procedures are readily interchanged, i.e., we may use KCL to find the equations for an *RL* circuit and KVL to find the equations for an *RC* circuit. In doing this, we need simply use the relation

$$\int K e^{st} = \frac{K}{s} e^{st} \tag{6-124}$$

Note that here it is not necessary to insert the initial condition for the integration of the exponential, since this term disappears when the equation is converted to a differential equation. Thus, in effect, we use an "operator" notation which replaces the process of integration by the operator $1/s$. Similarly, the process of differentiation may be viewed as represented by the operator s. A more rigorous justification of this process will be provided in connection with the development of the Laplace transformation in Chap. 8.

Now let us consider the determination of the solution for an *RLC* network which is of higher than second order. As an example of such a situation, consider the network shown in Fig. 6–7.4. Let us assume that we desire to find an expression for the network variable $v_2(t)$ for the case where the circuit is excited only by initial conditions, and that these are $v_2(0) = 1$, $v_2'(0) = 0$, and $v_2''(0) = 0$. We may begin by applying KCL to the network. Assuming that the output of the current source $i_s(t) = 0$, we obtain the equations

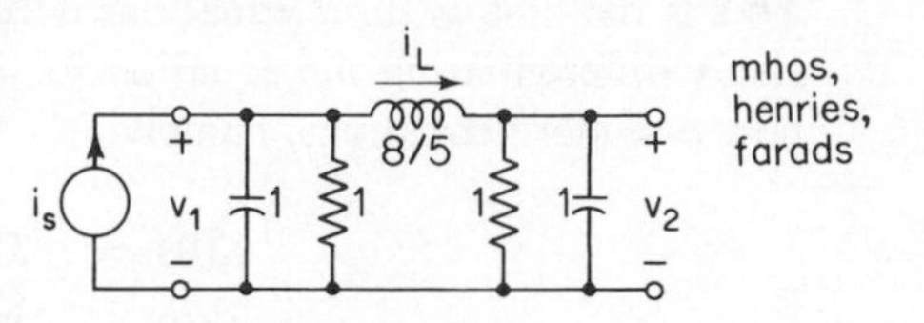

Fig. 6-7.4 A third-order *RLC* network.

$$
\begin{aligned}
v_1'(t) + v_1(t) + \tfrac{5}{8} \int [v_1(t) - v_2(t)]\, dt &= 0 \\
v_2'(t) + v_2(t) + \tfrac{5}{8} \int [v_2(t) - v_1(t)]\, dt &= 0
\end{aligned}
\tag{6–125}
$$

If we assume the exponential solutions $v_1(t) = A_1 e^{st}$ and $v_2(t) = A_2 e^{st}$ for the node voltage variables, the above equation may be written in the form

$$
\begin{bmatrix} s + 1 + \dfrac{5}{8s} & -\dfrac{5}{8s} \\ -\dfrac{5}{8s} & s + 1 + \dfrac{5}{8s} \end{bmatrix} \begin{bmatrix} A_1 e^{st} \\ A_2 e^{st} \end{bmatrix} = \begin{bmatrix} 0 \\ 0 \end{bmatrix}
\tag{6–126}
$$

The characteristic equation found by setting the determinant of the square matrix in the above relation to zero may be written as

$$
\begin{aligned}
s^3 + 2s^2 + \tfrac{9}{4}s + \tfrac{5}{4} = (s + 1)(s + \tfrac{1}{2} + j1) \\
(s + \tfrac{1}{2} - j1) = 0
\end{aligned}
\tag{6–127}
$$

Thus, the natural frequencies of the network are $s_1 = -1$, $s_2 = -\frac{1}{2} + j1$, and $s_3 = -\frac{1}{2} - j1$. The solution for $v_2(t)$ will have the form

$$
v_2(t) = K_1 e^{-t} + K_{2a} e^{[-(1/2)+j1]t} + K_{3a} e^{[-(1/2)-j1]t}
\tag{6–128}
$$

where K_{2a} and K_{3a} are complex conjugates. To facilitate computations, it is usually desirable to combine complex conjugate quantities of the type shown as the second and third terms in the right member of the above expression using the method given in Sec. 6–2. If this is done, we obtain

$$
v_2(t) = K_1 e^{-t} + e^{-(1/2)t}(K_2 \cos t + K_3 \sin t)
\tag{6–129}
$$

The advantage of this type of formulation is that whereas the constants K_{2a} and K_{3a} of (6–128) are complex, the constants K_2 and K_3 of (6–129) are real. To determine the constants K_1, K_2, and K_3, we note that

$$
v_2(0) = K_1 + K_2 = 1
\tag{6–130}
$$

This is the first of three equations needed to determine the three constants. Successively differentiating the equation for $v_2(t)$ and evaluating it at t equals zero, we obtain two more equations, namely,

$$\begin{aligned} v_2'(0) &= -K_1 - \tfrac{1}{2}K_2 + K_3 = 0 \\ v_2''(0) &= \quad K_1 - \tfrac{3}{4}K_2 - K_3 = 0 \end{aligned} \tag{6-131}$$

If we write these three equations in matrix form, we obtain

$$\begin{bmatrix} 1 & 1 & 0 \\ -1 & -\frac{1}{2} & 1 \\ 1 & -\frac{3}{4} & -1 \end{bmatrix} \begin{bmatrix} K_1 \\ K_2 \\ K_3 \end{bmatrix} = \begin{bmatrix} 1 \\ 0 \\ 0 \end{bmatrix} \tag{6-132}$$

The solution of these equations is readily found to be $K_1 = 1$, $K_2 = 0$, and $K_3 = 1$. Thus, the solution for $v_2(t)$ is

$$v_2(t) = e^{-t} + e^{-(1/2)t} \sin t \qquad t \geq 0 \tag{6-133}$$

The solution for any other variable in the network may be found by a similar process.

The procedure illustrated above is easily applied to *RLC* networks of any order. The network equations may be written by applying either KCL (if the independent variables are chosen as node voltages) or KVL (if the independent variables are chosen as mesh currents). In general, the order of an *RLC* network will be less than or equal to the number of energy-storage elements in the network. Just as was true for the second-order *RLC* case, the natural frequencies of an *n*th-order *RLC* network may be located anywhere in the left half of the complex frequency plane (including the $j\omega$ axis).

In the preceding paragraphs of this section we have discussed the solution of various types of networks of higher than second order by hand computation methods. The digital computer techniques introduced in Sec. 6–6 are also directly applicable to such networks. To apply these techniques we must first put the equations describing the network in the form of a matrix first-order differential equation, i.e., we must determine the state-variable equations using a process similar to that presented in Sec. 6–5 for the second-order case. The development of completely general procedures for doing this is a very broad subject which can easily fill a complete book.[9] Here we shall content ourselves with pointing out that, for many networks, choosing the voltages across the capacitors and the currents through the inductors as the state variables, and appropriately applying KCL and KVL, we may obtain the required set of equations. As an example of this procedure, consider the network shown in

[9]See, for example, R. A. Rohrer, *Circuit Theory: An Introduction to the State Variable Approach*, McGraw-Hill Book Company, New York, 1970.

Fig. 6–7.5. This network contains six independent energy-storage elements. As indicated above, we may choose the voltages $v_1(t)$, $v_2(t)$, and $v_3(t)$ and the currents $i_1(t)$, $i_2(t)$, and $i_3(t)$ for the variables. If we apply KCL at nodes 1 and 2, we obtain

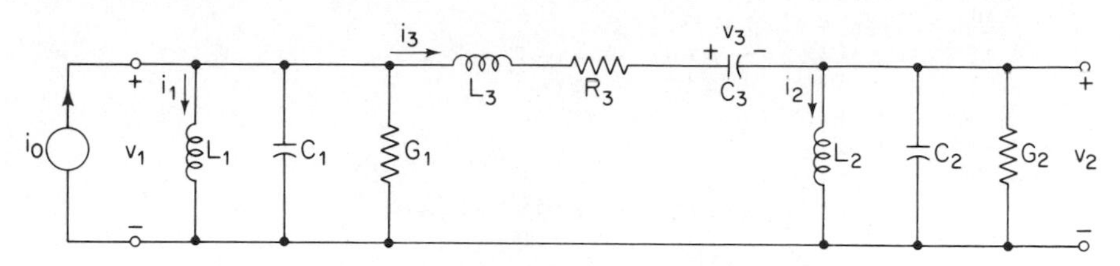

Fig. 6-7.5 A sixth-order *RLC* network.

$$\begin{aligned} C_1 v_1' + G_1 v_1 + i_1 + i_3 &= i_0 \\ C_2 v_2' + G_2 v_2 + i_2 - i_3 &= 0 \end{aligned} \tag{6–134}$$

where, for convenience, we have omitted the functional notation (t). Now let us apply KVL to the portion of the network containing L_3, R_3, and C_3. Since one end of this part of the network is at a potential of v_1 above the reference node, and the other end is at a potential of v_2 above this node, applying KVL we may write

$$L_3 i_3' + R_3 i_3 + v_3 = v_1 - v_2 \tag{6–135}$$

The equations given above provide three of the required six state-variable equations. The remaining three equations are provided by the terminal relations for L_1, L_2, and C_3. The resulting set of six equations may be written in the form:

$$\begin{bmatrix} v_1' \\ v_2' \\ v_3' \\ i_1' \\ i_2' \\ i_3' \end{bmatrix} = \begin{bmatrix} -\frac{G_1}{C_1} & 0 & 0 & -\frac{1}{C_1} & 0 & -\frac{1}{C_1} \\ 0 & -\frac{G_2}{C_2} & 0 & 0 & -\frac{1}{C_2} & \frac{1}{C_2} \\ 0 & 0 & 0 & 0 & 0 & \frac{1}{C_3} \\ \frac{1}{L_1} & 0 & 0 & 0 & 0 & 0 \\ 0 & \frac{1}{L_2} & 0 & 0 & 0 & 0 \\ \frac{1}{L_3} & -\frac{1}{L_3} & -\frac{1}{L_3} & 0 & 0 & -\frac{R_3}{L_3} \end{bmatrix} \begin{bmatrix} v_1 \\ v_2 \\ v_3 \\ i_1 \\ i_2 \\ i_3 \end{bmatrix} + \begin{bmatrix} \frac{i_0}{C_1} \\ 0 \\ 0 \\ 0 \\ 0 \\ 0 \end{bmatrix} \tag{6–136}$$

This obviously has the form of a first-order, matrix, differential equation. The solution of such a set of equations is readily performed by the digital computational techniques introduced in Sec. 6–6. Specifically, the subroutine MXRK4 may be used

```
      SUBROUTINE GN(T,A,G)
C     STATE EQUATIONS FOR SIXTH ORDER EXAMPLE
C     A(1)=V1, A(2)=V2, A(3)=V3
C     A(4)=I1, A(5)=I2, A(6)=I3
      DIMENSION A(10), G(10)
      G(1) = - A(1) - A(4) - A(6) + 1.
      G(2) = - A(2) - A(5) + A(6)
      G(3) = + A(6)
      G(4) = + A(1)
      G(5) = + A(2)
      G(6) = + A(1) - A(2) - A(3) -A(6)
      RETURN
      END

C     MAIN PROGRAM FOR SIXTH ORDER EXAMPLE
C            A - ARRAY OF STATE VARIABLES AFTER CALLING MXRK4
C           AI - ARRAY OF STATE VARIABLES BEFORE CALLING MXRK4
C            T - INDEPENDENT VARIABLE, TIME
C           DT - TIME INTERVAL BETWEEN PLOTTED POINTS
C            P - ARRAY FOR PLOTTING
      DIMENSION A(10), AI(10), P(5,101)
      DT = 0.1
      T = 0.
C
C     SPECIFY THE INITIAL CONDITIONS
      DO  2  I = 1, 6
    2 AI(I) = 0.
      P(1,1) = 0.
C
C     COMPUTE 50 VALUES OF THE VARIABLES
C     A(I) BY CALLING THE SUBROUTINE MXRK4
      DO  9  I = 2, 101
      CALL MXRK4 (T, AI, 6, T+DT, 10, A)
C
C     STORE VALUE OF V2 FOR PLOTTING
      P(1,I) = A(2) * 400.0
C
C     RESET THE INITIAL CONDITIONS
      DO  8  J = 1, 6
    8 AI(J) = A(J)
C
C     INCREMENT THE VALUE OF T
    9 T = T + DT
C
C     CALL THE PLOTTING SUBROUTINE
      PRINT 11
   11 FORMAT (1H1)
      CALL PLOT5 (P, 1, 101, 50)
      STOP
      END
```

Fig. 6-7.6 Listing of the program for solving a sixth-order *RLC* network.

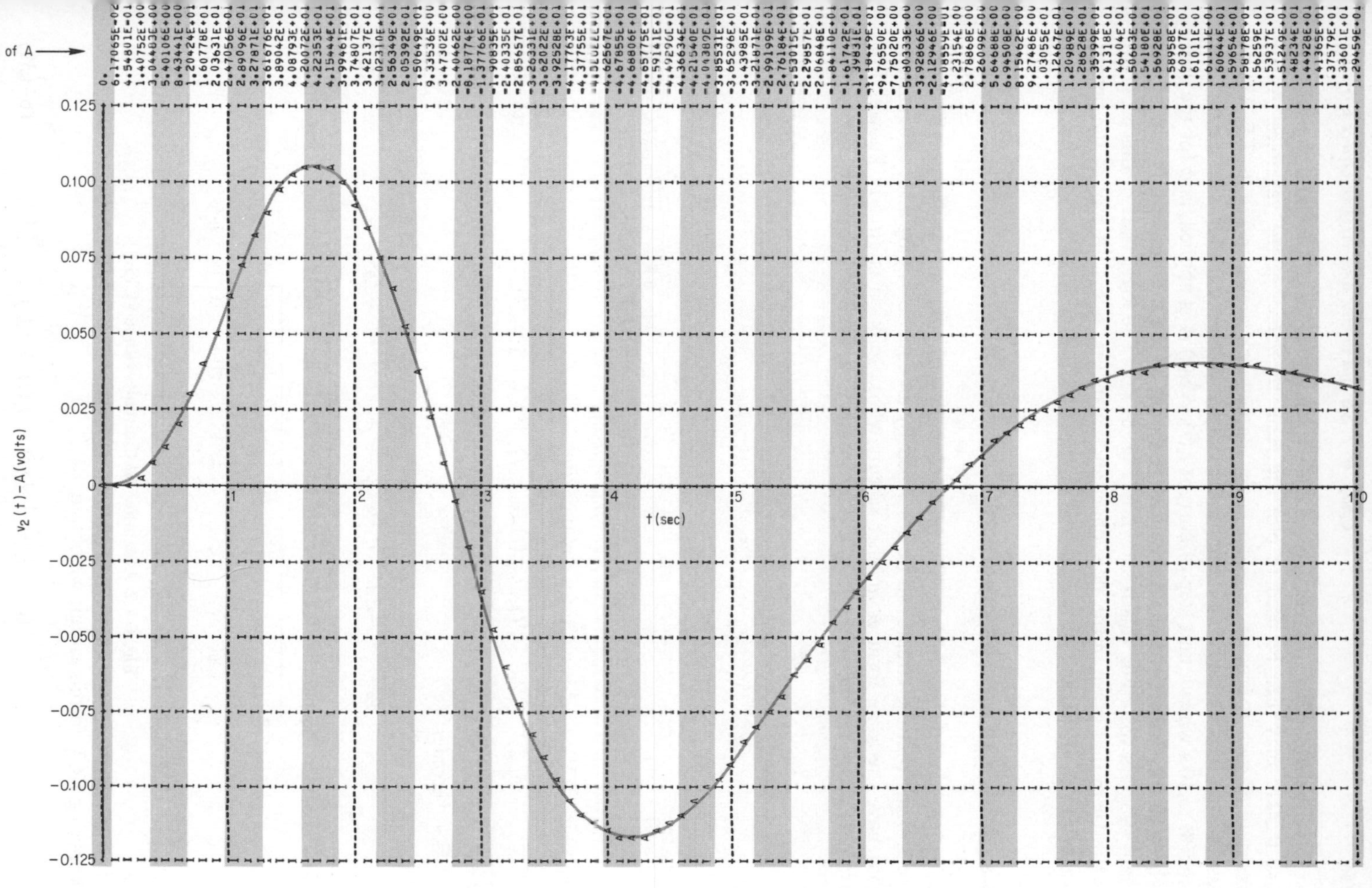

Fig. 6-7.7 Output plot for sixth-order *RLC* network.

to solve the equations at a sequence of evenly spaced values of time and the subroutine PLOT5 may then be used to plot the resulting values of the variables. As an example of such a process, consider again the network shown in Fig. 6–7.5. Let us assume that it is desired to determine the output voltage $v_2(t)$ for a range of time from 0 to 10 s when a unit step of current $i_0(t) = u(t)$ A is applied, and for the case where all the elements have unity value. A listing of the subroutine GN and of the main program for doing this is shown in Fig. 6–7.6. The logic used in this program is similar to that given in the flow chart shown in Fig. 6–6.4. A plot of the output waveform of $v_2(t)$ is shown in Fig. 6–7.7.

6-8 DUAL AND ANALOG CIRCUITS

Frequently the analysis of a given circuit configuration can be simplified by recognizing that the circuit can be described by an equation or by a set of equations that have the same form as another circuit whose solution is already known. For example, consider the series *RLC* circuit containing an ICVS shown in Fig. 6–8.1.

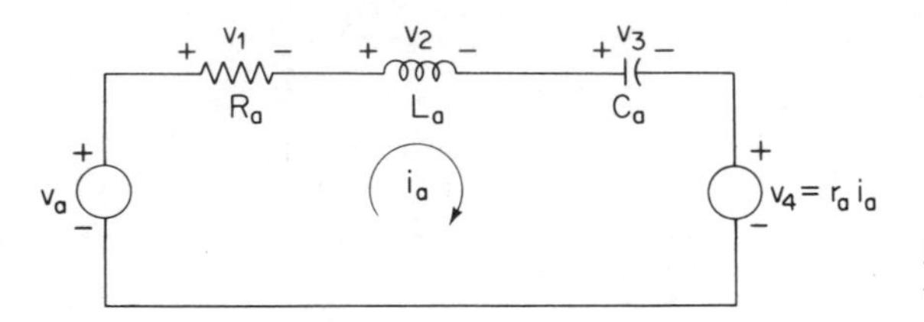

Fig. 6-8.1 A series *RLC* circuit with an ICVS.

Applying KVL to this circuit we may write

$$v_a(t) = v_1(t) + v_2(t) + v_3(t) + v_4(t) \tag{6–137}$$

After the substitution of the relations for the branch terminal variables the above relation becomes

$$v_a(t) = R_a i_a(t) + L_a \frac{di_a}{dt} + \frac{1}{C_a} \int i_a(t)\, dt + r_a i_a(t) \tag{6–138}$$

Now consider the parallel RLC circuit which includes a VCIS shown in Fig. 6–8.2.

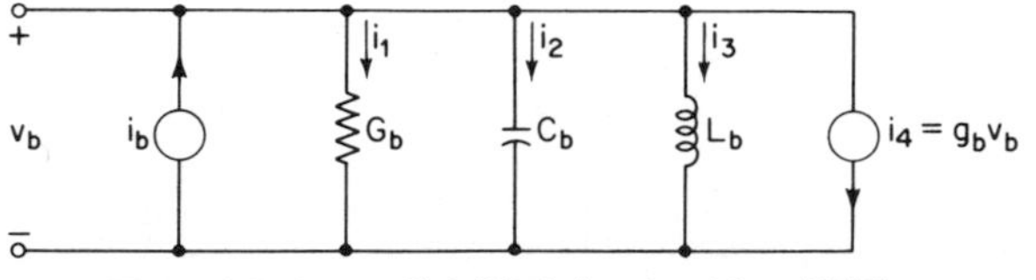

Fig. 6-8.2 A parallel *RLC* circuit with a VCIS.

Applying KCL to this circuit we obtain

$$i_b(t) = i_1(t) + i_2(t) + i_3(t) + i_4(t) \tag{6–139}$$

After the substitution of the relations for the branch terminal variables the above equation may be written

$$i_b(t) = G_b v_b(t) + C_b \frac{dv_b}{dt} + \frac{1}{L_b}\int v_b(t)\,dt + g_b v_b(t) \qquad (6\text{–}140)$$

Now let us compare Eqs. (6–137) and (6–138) (for the circuit shown in Fig. 6–8.1) with Eqs. (6–139) and (6–140) (for the circuit shown in Fig. 6–8.2). Clearly, Eqs. (6–137) and (6–139) have the same form and may be directly obtained from each other by interchanging the current and voltage variables (and the subscripts a and b). Similarly, Eqs. (6–138) and (6–140) have the same form and may be directly obtained from each other by interchanging the following quantities: (1) voltage and current, (2) resistance and conductance, and (3) inductance and capacitance; i.e., if we interchange the defining symbols v and i, R and G, and L and C (and the subscripts a and b) we may derive one set of equations from the other. In general, circuits which are described by equations having the same form, but in which the coefficients and variables are interchanged as described above, are said to be *dual* to each other. More specifically, in dual circuits each mesh which provides a KVL equation in one circuit corresponds to a node (and the branches which connect to it) which provides a KCL equation in the dual circuit and vice versa. Thus, elements in series in one circuit correspond with elements in parallel in the dual circuit. A summary of the properties of dual circuits is given in Table 6–8.1 (on page 370).

Now let us develop a general procedure for finding the dual of a given circuit. To do this we may use the following graphical construction procedure[10]:

1. In the center of each mesh of the given network place a node symbol. Number these symbols sequentially. In addition, place a ground reference node symbol outside of all the network meshes. Finally, draw a duplicate set of node symbols exterior to the network to provide the framework for the construction of the dual network.

As an example of the above step, consider the two-mesh circuit shown in Fig. 6–8.3. The addition of the node symbols is shown in Fig. 6–8.4(a). The node symbols have been redrawn in Fig. 6–8.4(b).

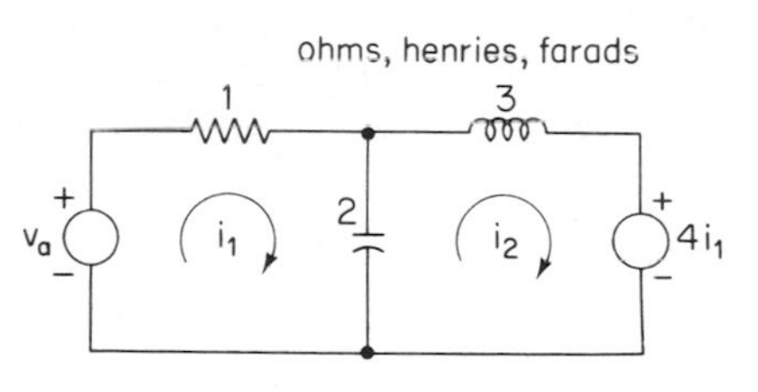

Fig. 6-8.3 A two-mesh circuit.

2. Draw lines between the node symbols in such a way that each line crosses one network element. As each line is drawn add an element to the dual network between corresponding nodes. Use the relations of Table 6–8.1 to determine the type of dual element. The reference polarity of sources in the

[10]Gardner, M.F., and Barnes, G.L., *Transients in Linear Systems*, John Wiley & Sons, Inc., New York, 1942.

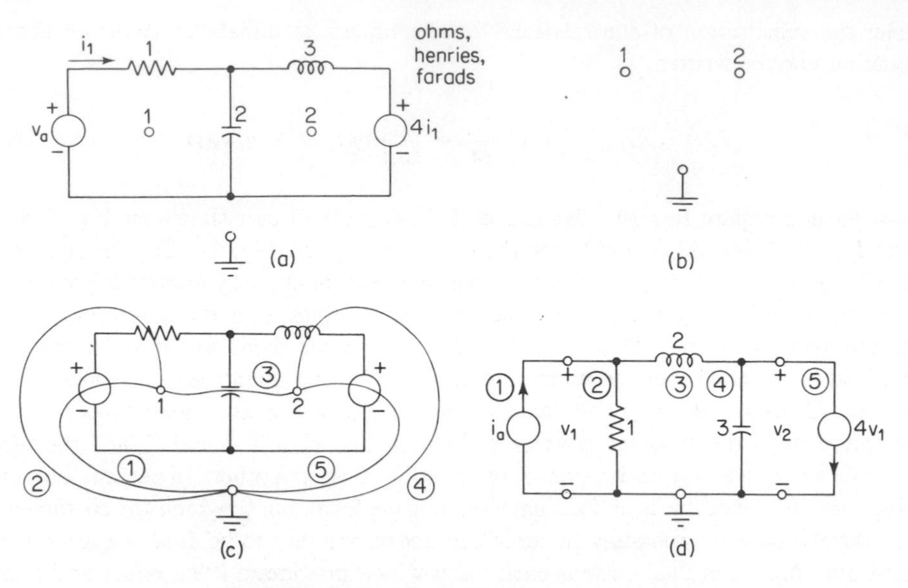

Fig. 6-8.4 Finding the dual of a two-mesh circuit.

dual network is readily observed by noting that a voltage source with a reference polarity which produces a positive mesh reference current has as its dual a current source whose reference direction is towards the numbered node symbol and away from the ground node symbol.

This step is illustrated in Figs. 6–8.4(c) and 6–8.4(d). In part (c) the lines drawn between the node symbols are shown. For convenience these have been numbered from 1 to 5. The corresponding elements in the dual network are shown in part (d) of the figure. Note that since the ICVS has a reference polarity which opposes the mesh current $i_2(t)$ shown in Fig. 6–8.3 the dual element, which is a VCIS, has a reference polarity directed away from node 2. To verify the duality of the two circuits we may write the equations for the circuit shown in Fig. 6–8.3 as

$$
\begin{aligned}
v_a(t) &= i_1(t) + \frac{1}{2}\int i_1(t)\,dt - \frac{1}{2}\int i_2(t)\,dt \\
0 &= 4i_1(t) - \frac{1}{2}\int i_1(t)\,dt + \frac{1}{2}\int i_2(t)\,dt + 3\frac{di_2}{dt}
\end{aligned}
\tag{6–141}
$$

The equations for the dual network shown in Fig. 6–8.4(d) are readily found to be

$$
\begin{aligned}
i_a(t) &= v_1(t) + \frac{1}{2}\int v_1(t)\,dt - \frac{1}{2}\int v_2(t)\,dt \\
0 &= 4v_1(t) - \frac{1}{2}\int v_1(t)\,dt + \frac{1}{2}\int v_2(t)\,dt + 3\frac{dv_2}{dt}
\end{aligned}
\tag{6–142}
$$

The similarity of the two sets of equations is readily apparent. The method described above may be applied to any network which has a planar graph. Since it is based on a transformation of the graph of the network, it is valid not only for linear time-invariant networks but also for non-linear and time-varying ones.

In the above paragraphs we saw that, given a circuit described by mesh equations we could construct a dual of the circuit in such a way that its node equations had the identical form as the mesh equations of the original circuit. Now let us treat a similar development in which we consider the use of circuits to serve as analogs for non-electrical types of systems. Specifically, we will consider mechanical systems. Elementary mechanical systems which have only translation may be considered as being composed of three types of elements, i.e., an element which is characterized only by mass and which is represented by the symbol shown in Fig. 6–8.5(a), an ele-

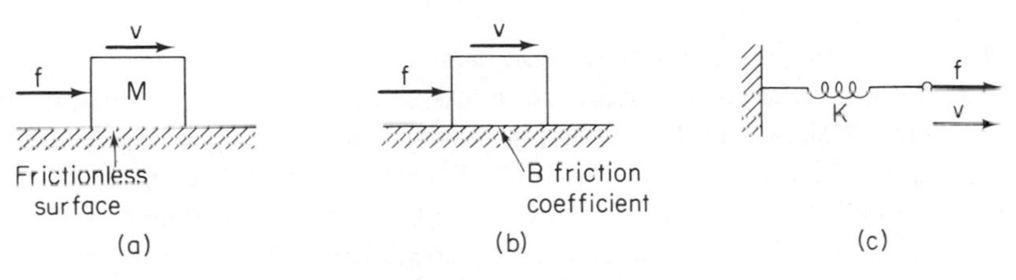

Fig. 6-8.5 Elements of a mechanical system.

ment which is characterized only by friction and which is represented by the symbol shown in Fig. 6–8.5(b), and an element which has a spring coefficient and which is represented by the symbol shown in Fig. 6–8.5(c). The defining relations for these three types of elements, using $f(t)$ for force and $v(t)$ for velocity, are

$$f(t) = M\frac{dv}{dt} \qquad \text{Mass (Newton's Law)}$$

$$f(t) = Bv(t) \qquad \text{Friction} \qquad (6\text{–}143)$$

$$f(t) = K\int_0^t v(\tau)\,d\tau + f(0) \qquad \text{Spring Coefficient}$$

where $f(0)$ is an initial condition term representing the force on the spring at $t = 0$. Now let us consider a simple mechanical system using these elements as shown in Fig. 6–8.6(a). The equation of motion for this system is found by summing the forces

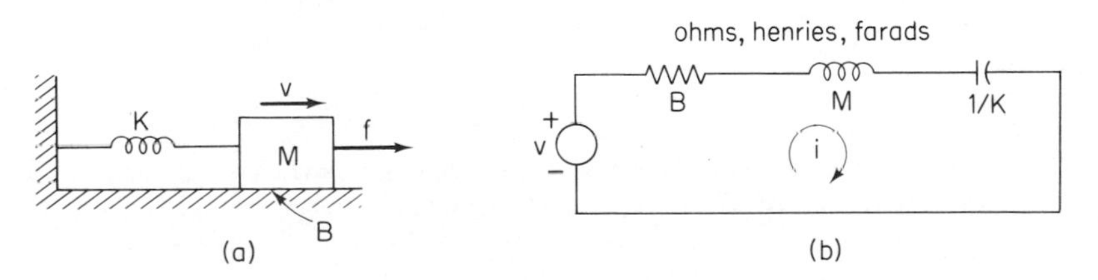

Fig. 6-8.6 A simple mechanical system.

acting on the block. This equation has the form

$$f(t) = Bv(t) + M\frac{dv}{dt} + K\int_0^t v(\tau)\,d\tau \tag{6–144}$$

where it is assumed that the spring is not in tension at $t = 0$. This equation, however, is similar to one we have frequently seen in this text. Specifically, if we apply KVL to a series *RLC* circuit comprised of a resistor of value B ohms, an inductor of value M henries and a capacitor of value $1/K$ farads as shown in Fig. 6–8.6(b), we obtain the relation

$$v(t) = Bi(t) + M\frac{di}{dt} + K\int_0^t i(\tau)\,d\tau \tag{6–145}$$

where $v(t)$ is the voltage of the source. Comparing Eqs. (6–144) and (6–145) we may conclude that the circuit shown in Fig. 6–8.6(b) will function as an analog of the mechanical circuit shown in Fig. 6–8.6(a), where, more specifically, we note that the applied voltage is the analog of the applied force, the mesh current is the analog of the velocity, and the resistance, inductance, and reciprocal capacitance are respectively the analogs of friction, mass, and spring coefficient. The different mechanical and electrical variables for such an analog are summarized in Table 6–8.2.

Analog electrical quantities are readily derived for a wide variety of types of non-electrical systems such as mechanical systems with rotational displacement, hydraulic systems, etc. In addition, engineering studies are frequently made using analogs to represent large complex systems such as manufacturing plants, airplanes, watersheds, etc. Such large scale projects are generally made using analog computers which are specifically designed to facilitate the application of basic electrical analogs to a wide range of systems.

TABLE 6-8.1 Properties of Dual Circuits

KVL Equations	*KCL Equations*
Mesh Current	Node Voltage
Voltage Source	Current Source
Series Branch	Parallel Branch
Inductor	Capacitor
Capacitor	Inductor
Resistor (ohms)	Resistor (mhos)
ICVS	VCIS
Controlling Mesh Current for an ICVS	Controlling Node Voltage for a VCIS
Output Voltage for an ICVS	Output Current for a VCIS

TABLE 6-8.2 Analogous Electrical and Mechanical Quantities

Mechanical Quantities	*Electrical Quantities*
Force	Voltage
Velocity	Current
Distance	Charge
Mass	Inductance
Friction	Resistance
Spring Coefficient	Reciprocal Capacitance

6-9 SECOND-ORDER ACTIVE *RC* CIRCUITS

In Sec. 6–3 it was pointed out that Cases II and III for the characteristic equation do not apply to networks comprised of only resistors and capacitors. Thus, we concluded that if a circuit was to be characterized by oscillatory behavior, it had to contain both an inductor and a capacitor, i.e., it had to be an *RLC* circuit. In this section we shall show that circuits comprised of resistors and capacitors *and an active element* can have a characteristic equation which may exhibit Case II or III behavior as well as Case I behavior. Such circuits are called *active RC* circuits. They form an important class of networks which are frequently used for filtering purposes in a wide variety of applications.

As an example of a second order active *RC* circuit, consider the network shown in Fig. 6–9.1. In this figure the triangle represents an ideal VCVS with a

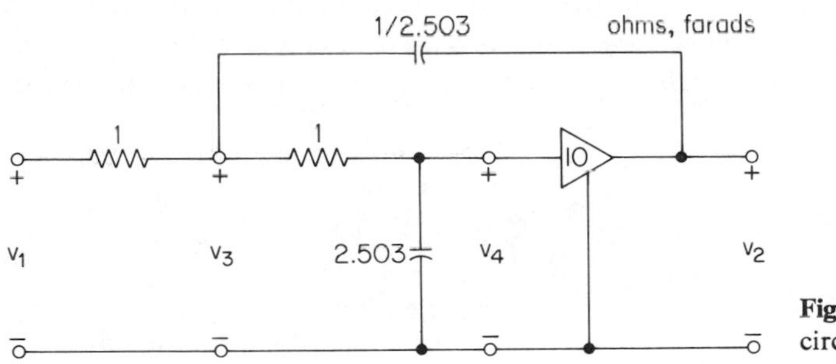

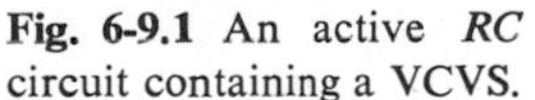
Fig. 6-9.1 An active *RC* circuit containing a VCVS.

gain of 10, i.e., a multi-terminal device whose output voltage $v_2(t)$ is equal to 10 times the input voltage $v_4(t)$. Thus, we may write

$$v_2(t) = 10v_4(t) \tag{6–146}$$

To analyze this network, i.e., to find the differential equation relating $v_2(t)$ and $v_1(t)$, let us apply KCL at nodes 3 and 4. Thus, we obtain the equations

$$
\begin{aligned}
v_1(t) - v_3(t) &= v_3(t) - v_4(t) + \frac{v_3'(t) - v_2'(t)}{2.503} \\
v_3(t) - v_4(t) &= 2.503v_4'(t)
\end{aligned}
\tag{6–147}
$$

Substituting (6–146) in the above equations, we may eliminate the variable $v_4(t)$. Thus, we obtain

$$
\begin{aligned}
v_1(t) - v_3(t) &= v_3(t) - 0.1v_2(t) + \frac{v_3'(t) - v_2'(t)}{2.503} \\
v_3(t) - 0.1v_2(t) &= 0.2503v_2'(t)
\end{aligned}
\tag{6–148}
$$

We now have two equations involving the variables $v_1(t)$, $v_2(t)$, and $v_3(t)$. Solving the second equation for $v_3(t)$, substituting the result in the first equation, and rearranging terms, we obtain

$$10v_1(t) = v_2''(t) + \sqrt{2}\, v_2'(t) + v_2(t) \tag{6–149}$$

This is the desired differential equation. The roots of the characteristic equation for (6–149) are readily seen to be

$$s = -0.707 \pm j0.707 \tag{6–150}$$

Thus, the circuit is of the type characterized by Case III in which the roots of the characteristic equation are complex. As a result, the complementary solution will have the oscillatory behavior normally associated only with passive *RLC* circuits.

As a second illustration of the use of active *RC* circuits to realize second-order differential equations which have Case II or III behavior, consider the circuit shown in Fig. 6–9.2. This circuit contains a gyrator, a device which was first discussed in

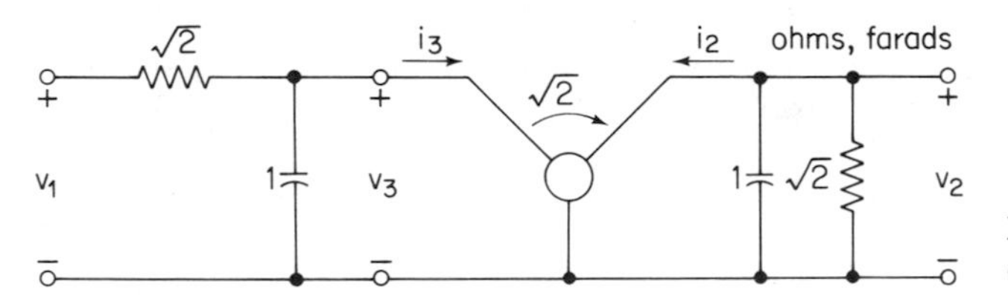

Fig. 6-9.2 An active *RC* circuit containing a gyrator.

Sec. 3.11. Adapting the defining relations given in (3–108) to the variables defined in Fig. 6–9.2, we obtain the following relations between the network variables $i_2(t)$, $v_2(t)$, $i_3(t)$, and $v_3(t)$ shown in the figure.

$$
\begin{aligned}
v_3(t) &= -\sqrt{2}\, i_2(t) \\
v_2(t) &= \sqrt{2}\, i_3(t)
\end{aligned}
\tag{6–151}
$$

Now let us analyze the circuit to find the differential equation relating $v_1(t)$ and $v_2(t)$.

Applying KCL at node 2, we obtain

$$0 = i_2(t) + v_2'(t) + \frac{v_2(t)}{\sqrt{2}} \tag{6-152}$$

Similarly, applying KCL at node 3, we obtain

$$\frac{v_1(t) - v_3(t)}{\sqrt{2}} = v_3'(t) + i_3(t) \tag{6-153}$$

Inserting the relations of (6–151) in this last result, we obtain

$$\frac{v_1(t) + \sqrt{2}\, i_2(t)}{\sqrt{2}} = -\sqrt{2}\, i_2'(t) + \frac{v_2(t)}{\sqrt{2}} \tag{6-154}$$

The variables $i_2(t)$ and $i_2'(t)$ may be eliminated from the above relation by substituting the equation given in (6–152) and its derivative. Making this substitution and rearranging terms, we obtain

$$0.5v_1(t) = v_2''(t) + \sqrt{2}\, v_2'(t) + v_2(t) \tag{6-155}$$

Thus we obtain the desired differential equation which relates $v_1(t)$ and $v_2(t)$.

It is readily observed by comparing Eqs. (6–149) and (6–155) that the circuits shown in Figs. 6–9.1 and 6–9.2 both have the same characteristic equation, i.e., the homogeneous differential equation for both networks is the same. This is a situation commonly encountered when an engineer uses active *RC* circuits to satisfy a filtering application which is characterized by a specific differential equation. Thus, network realizations which employ quite different active elements may be found to give quite similar results. One of the most interesting facets of the study of such network realizations is the development of other performance criteria, such as the dependence of the network characteristics to changes in the values of the network components, which may make one network realization preferable to another. Such studies are called *sensitivity studies.*

6-10 CONCLUSION

In this and the preceding chapter we have presented a set of general methods which may be applied to find the time-domain solution for any of the variables of a second- or higher-order network. The techniques include both conventional mathematical methods, which yield a closed-form expression for the solution, and also digital computational techniques, whose output is a plot of the waveform of the solution. The mathematical techniques may be applied only to linear time-invariant networks, whereas the digital computer approach may be directly used for non-linear and time-varying networks as well. The procedures are summarized in the flow chart given in Fig. 6-10.1. The student should study this figure carefully since it will serve as an excellent review of the material presented in this chapter. Comparison of this figure with the one presented in Fig. 5-9.1 will emphasize the differences that result between the solution methods for a first-order circuit and a second- or higher-order one.

This chapter marks the end of a fundamental division of the material contained in this book. In this and the preceding chapters we have presented many techniques concerned with the analysis of a wide range of circuits. All the network variables associated with these circuits have been specified as a function of time. Thus, we may collectively refer to these first six chapters as treating the *time-domain*

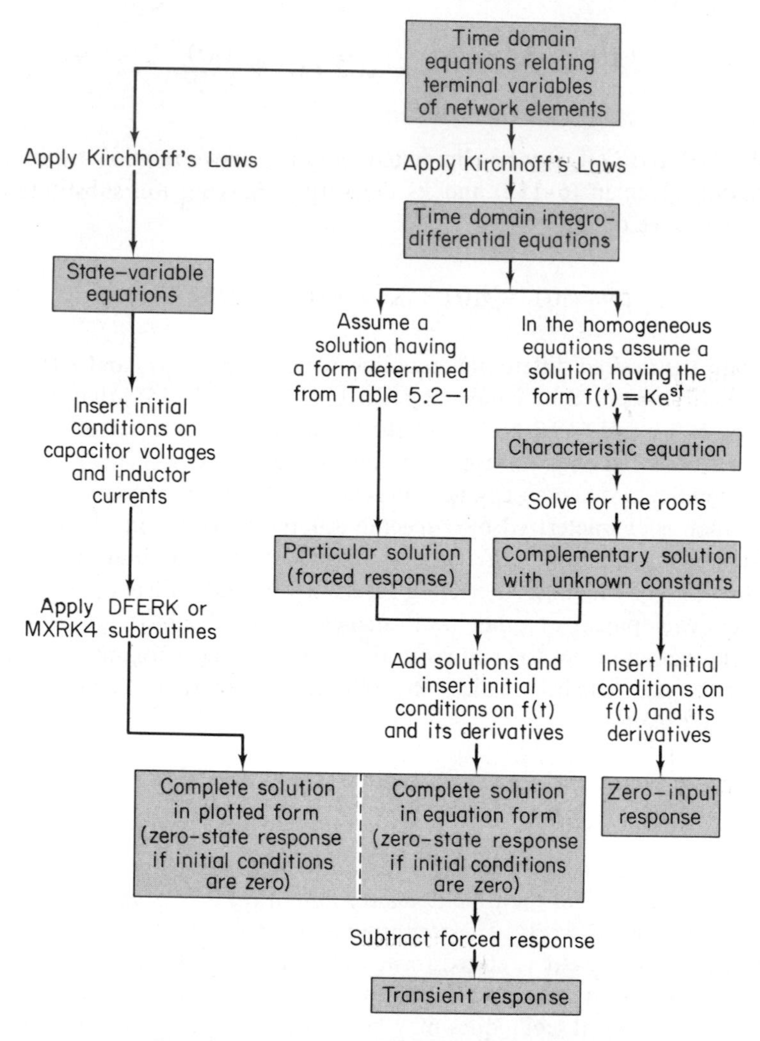

Fig. 6-10.1 General time-domain solution procedure.

analysis of circuits. In the chapters that follow we shall consider a quite different approach to network analysis. It may be categorized as the *frequency-domain* analysis of circuits. The full implications of the term "frequency-domain" will become apparent as we progress through these chapters.

Problems

Problem 6–1 (Sec. 6–1)

In the circuit shown in Fig. P6–1, find the roots of the characteristic equation [for the differential equation for $v(t)$] in terms of R, L, and C.

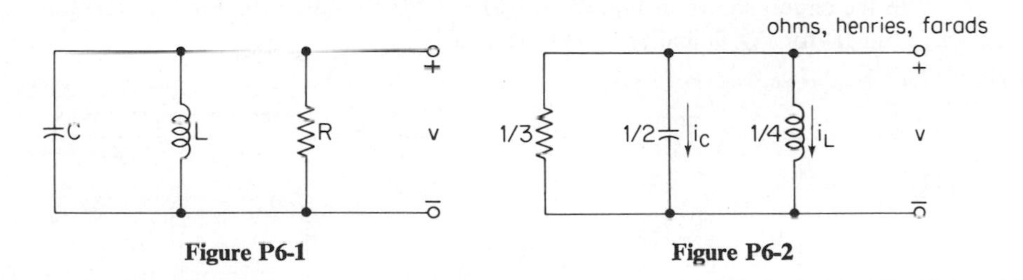

Figure P6-1 **Figure P6-2**

Problem 6–2 (Sec. 6–1)

For the circuit shown in Fig. P6–2, $v(0) = 1$ V and $v'(0) = 1$ V/s, find (a) $v(t)$ for $t \geq 0$, (b) $i_C(t)$ for $t \geq 0$, and (c) $i_L(t)$ for $t \geq 0$.

Problem 6–3 (Sec. 6–1)

In the circuit shown in Fig. P6–3, $v(0)$ equals 1 V and $v'(0)$ equals 0. Find $v(t)$ for all $t \geq 0$.

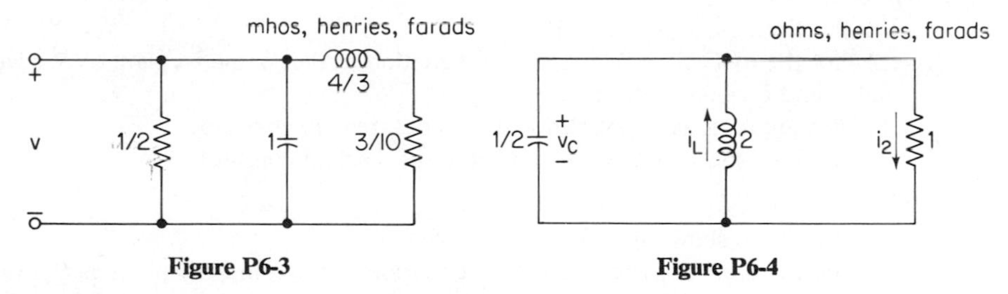

Figure P6-3 **Figure P6-4**

Problem 6–4 (Sec. 6–1)

In the circuit shown in Fig. P6–4, $i_L(0) = \frac{3}{5}$ A and $v_C(0) = 0$. Find $i_2(t)$.

Problem 6–5 (Sec. 6–1)

In the circuit shown in Fig. P6–5, find the value of R such that Case II applies, then let $v(0) = 1$ V and $v'(0) = 1$ V/s, and find $v(t)$ for $t \geq 0$, for the value of R that you have calculated.

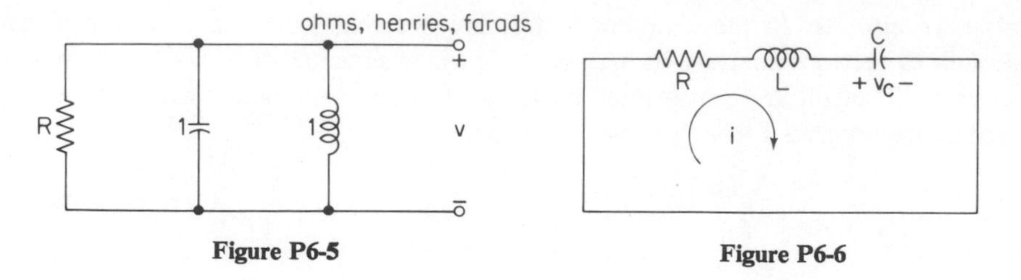

Figure P6-5

Figure P6-6

Problem 6–6 (Sec. 6–1)

For the circuit shown in Fig. P6-6, assume that Case I applies. Find a relation between $i(0)$ and $v_C(0)$ such that only the root with the smaller magnitude appears in the response. [*Hint*: Remember that any fundamental circuit variable will have the form $K_1e^{s_1t} + K_2e^{s_2t}$. Let $v_C(t)$ have this form and remember $i(t) = C\,dv_C(t)/dt$.]

Problem 6–7 (Sec. 6–2)

In the circuit shown in Fig. P6-7, $v(0) = 1$ V and $v'(0) = 0$. Find (a) $v(t)$ for $t \geq 0$, (b) $i_L(t)$ for $t \geq 0$, and (c) $i_C(t)$ for $t \geq 0$.

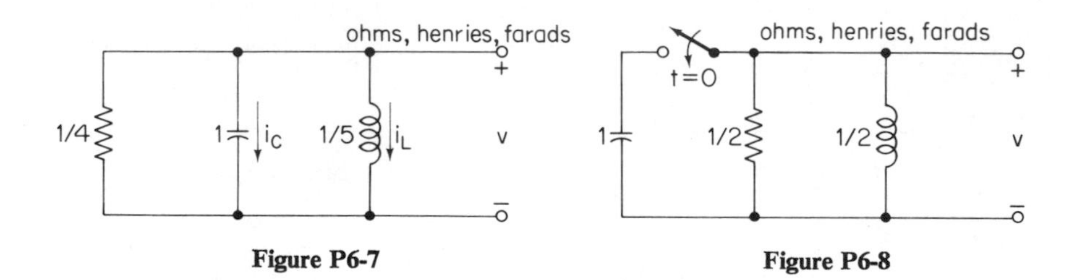

Figure P6-7

Figure P6-8

Problem 6–8 (Sec. 6–2)

In the circuit shown in Fig. P6-8, the switch closes at $t = 0$. Prior to this, the capacitor was charged to 3 V. Find $v(t)$ for all $t \geq 0$.

Problem 6–9 (Sec. 6–2)

For the circuit shown in Fig. P6-1:

(a) Plot the roots of the characteristic equation in the complex s plane as R is varied and L and C remain constant.

(b) Plot the roots as C is varied and R and L remain constant.

(c) Plot the roots as L is varied and R and C remain constant.

Problem 6–10 (Sec. 6–2)

For the circuit shown in Fig. P6-6:

(a) Plot the roots of the characteristic equation in the complex s plane as R is varied and L and C remain constant.

(b) Plot the roots as C is varied and R and L remain constant.

(c) Plot the roots as L is varied and R and C remain constant.

Problem 6–11 (Sec. 6–2)

The graph in Fig. P6-11 has the form $f(t) = Ke^{-\alpha t}\cos(\omega t)$. Find values for K, α, and ω.

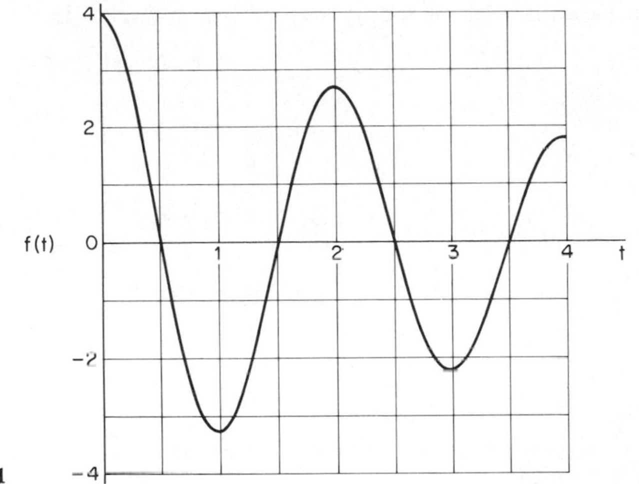

Figure P6-11

Problem 6–12 (Sec. 6–3)

For the circuit shown in Fig. P6–12, find $v(t)$ for $t \geq 0$ for the initial conditions $i_1(0) = 1$ and $i_2(0) = 0$ A.

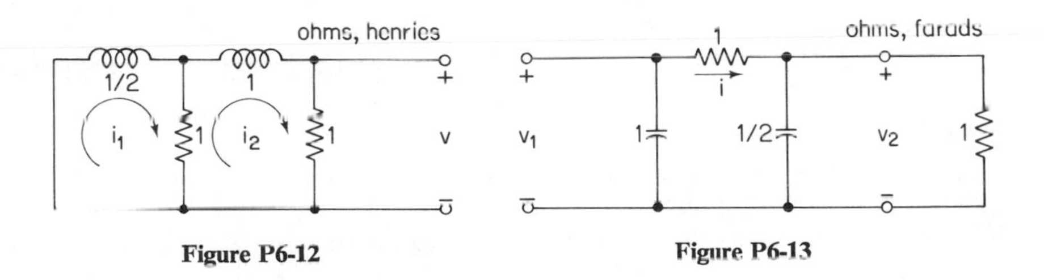

Figure P6-12 **Figure P6-13**

Problem 6–13 (Sec. 6–3)

For the circuit shown in Fig. P6–13, find $i(t)$ for $t \geq 0$ for the initial conditions $v_1(0) = 1$ and $v_2(0) = 1$ V.

Problem 6–14 (Sec. 6–3)

In the circuit shown in Fig. P6–14, find $v(t)$ for $t \geq 0$ for the initial conditions $v_C(0) = 1$ V and $i_L(0) = 1$ A.

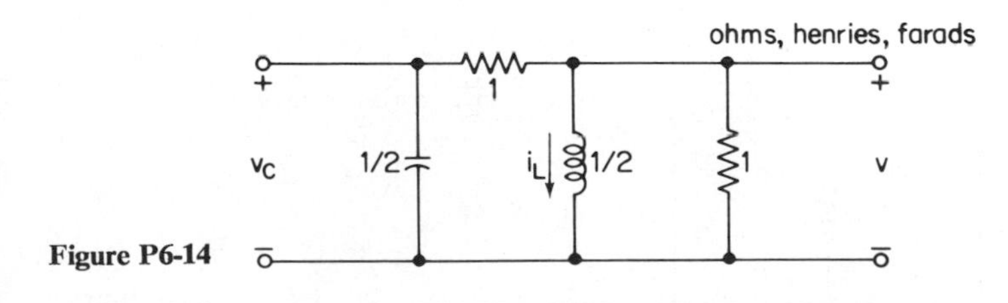

Figure P6-14

Problem 6–15 (Sec. 6–3)

Find the characteristic equation for the circuit shown in Fig. P6-15.

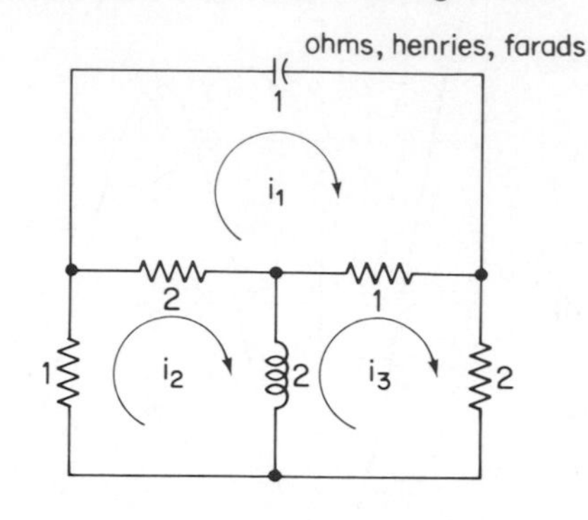

Figure P6-15

Problem 6–16 (Sec. 6–3)

For the circuit shown in Fig. P6-16, find the loci of the roots of the characteristic equation as K is varied.

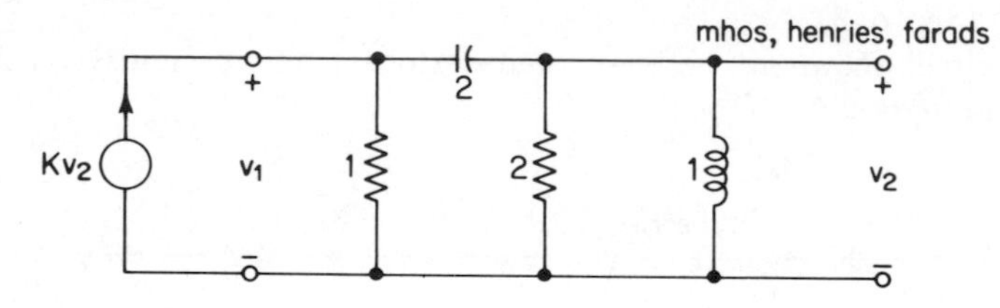

Figure P6-16

Problem 6–17 (Sec. 6–3)

Without solving the circuit, find $(di_1/dt)(0)$ and $(di_2/dt)(0)$ in the network shown in Fig. P6-17. Assume that $i_1(0) = 1$ and $i_2(0) = 2$ A.

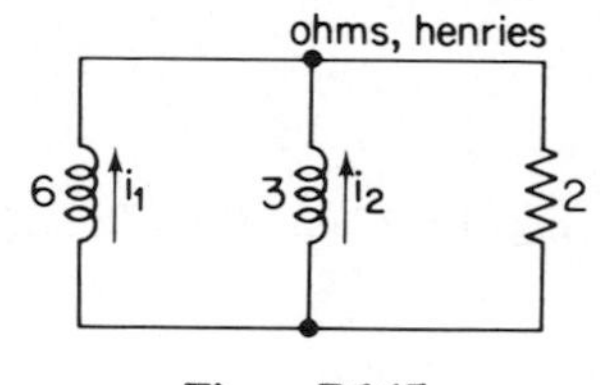

Figure P6-17

Problem 6–18 (Sec. 6–4)

In the circuit shown in Fig. P6-18, $v(0) = 0$ and $i_L(0) = 0$. Find $v(t)$, $i_C(t)$, and $i_L(t)$ for $t \geq 0$.

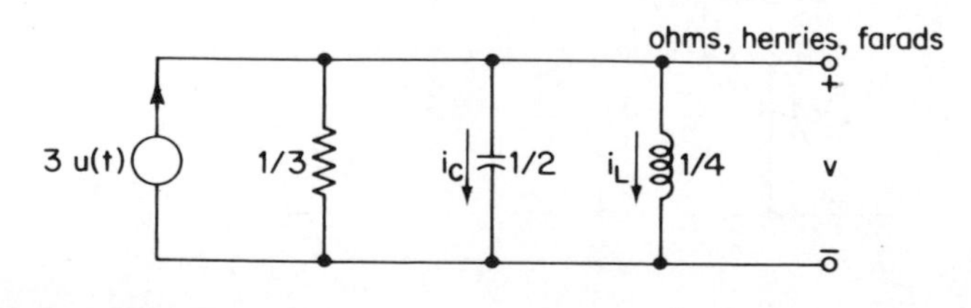

Figure P6-18

Problem 6–19 (*Sec. 6–4*)

For the circuit shown in Fig. P6-19, $v(0) = 0$ and $i_L(0) = 0$. Find $v(t)$, $i_L(t)$, and $i_C(t)$ for $t \geq 0$.

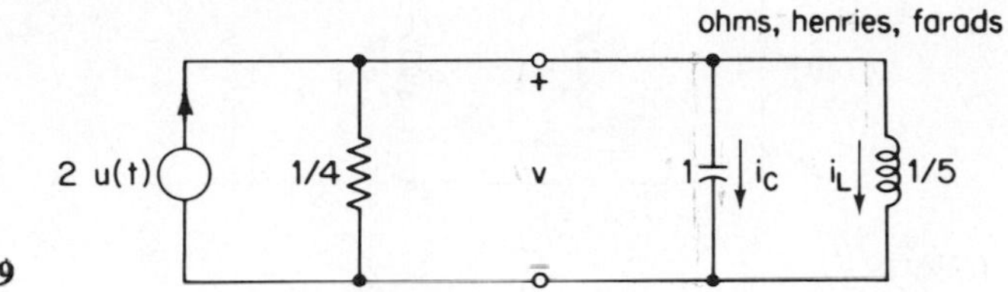

Figure P6-19

Problem 6–20 (*Sec. 6–4*)

In the circuit shown in Fig. P6-20, $v_1(0)$ equals 1 V and $i(0)$ equals 1 A. Find $v_1'(0)$, $v_2(0)$, and $v_2'(0)$.

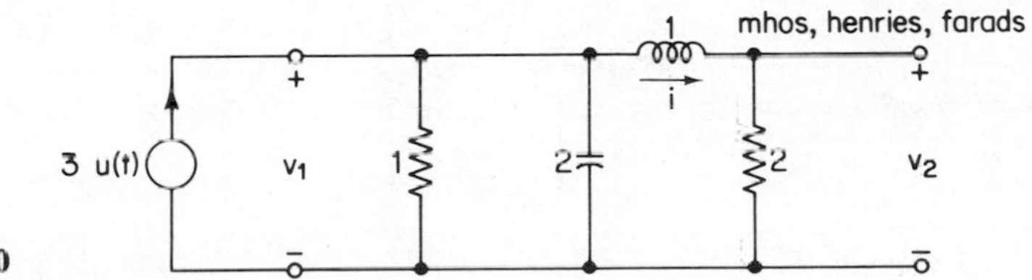

Figure P6-20

Problem 6–21 (*Sec. 6–4*)

Find the complete response for $v_2(t)$ in the circuit shown in Fig. P6-21. Assume $v_1(0) = 0$ and $v_2(0) = 0$.

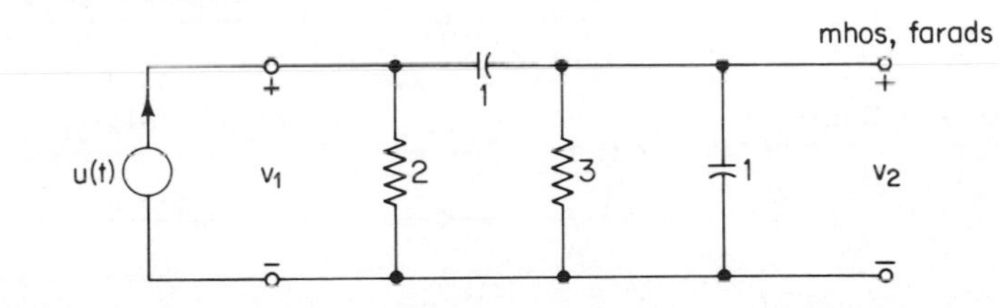

Figure P6-21

Problem 6–22 (*Sec. 6–4*)

Find $v(t)$ for $t \geq 0$ in the circuit shown in Fig. P6-22. Assume that $v_C(0) = 0$ and $i_L(0) = 0$.

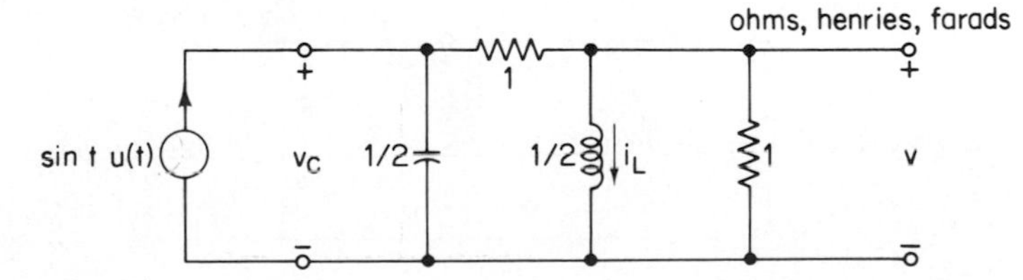

Figure P6-22

Problem 6–23 (*Sec. 6–4*)

For the circuit shown in Fig. P6-23, find $v(t)$ for $t \geq 0$ if $i_{L1}(0) = 0$ and $i_{L2}(0) = 0$.

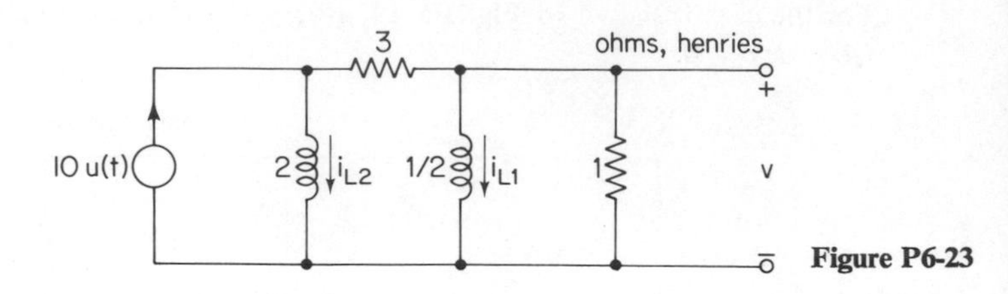

Figure P6-23

Problem 6–24 (Sec. 6–4)

For the circuit shown in Fig. P6-24, find $v(t)$ for $t \geq 0$. Assume that $v_{C1}(0) = 0$ and $v_{C2}(0) = 0$.

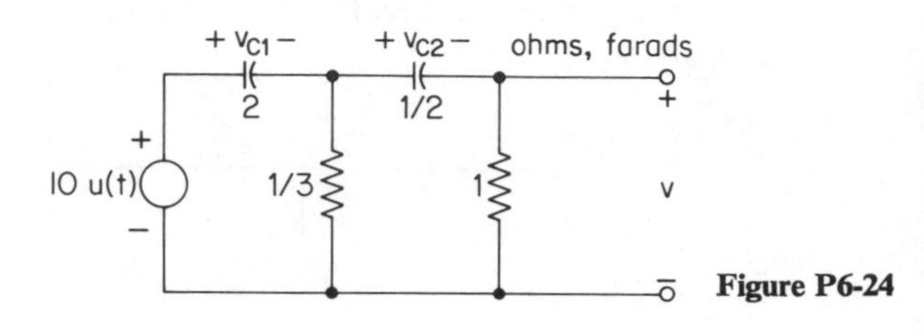

Figure P6-24

Problem 6–25 (Sec. 6–4)

In the circuit shown in Fig. P6-25, $v(0) = 0$ and $v'(0) = 0$. Find $v(t)$, $i_C(t)$, and $i_L(t)$ for $t \geq 0$.

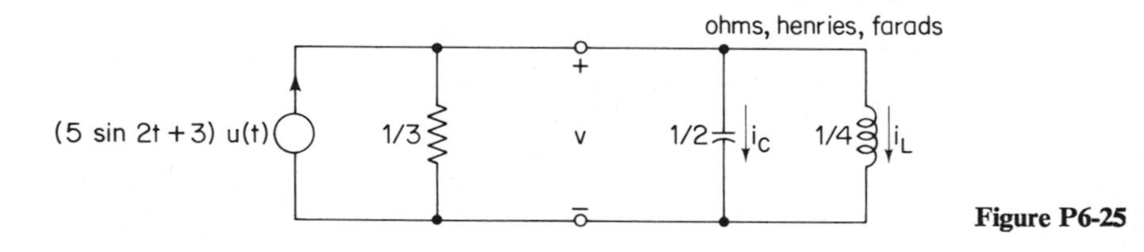

Figure P6-25

Problem 6–26 (Sec. 6–4)

In the circuit shown in Fig. P6-26, $v(0) = 0$ and $v'(0) = 0$. Find $v(t)$, $i_C(t)$, and $i_L(t)$ for $t \geq 0$.

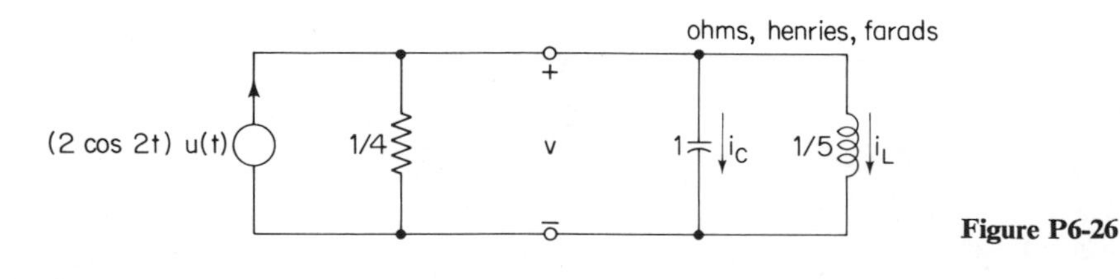

Figure P6-26

Problem 6–27 (Sec. 6–4)

In the circuit shown in Fig. P6-27, find $v(t)$ for $t \geq 0$ if $i_1(0) = 1$ A and $i_2(0) = 0$.

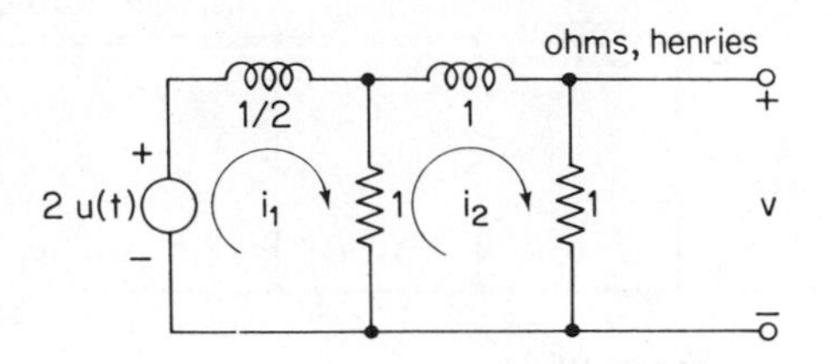

Figure P6-27

Problem 6–28 (Sec. 6–4)

In the circuit shown in Fig. P6–28, find $i(t)$ for $t \geq 0$ if $v_1(0) = 1$ and $v_2(0) = 1$ V.

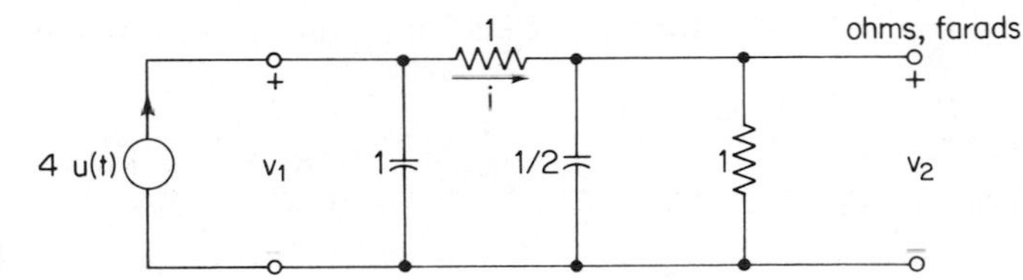

Figure P6-28

Problem 6–29 (Sec. 6–4)

In the circuit shown in Fig. P6–29, starting at t equals zero, the voltage $v(t) = -e^{-3t}$ V is applied. The current that results is $i(t) = e^{-3t} + e^{-t}\cos(2t)$ A. Find the values of the network elements R (in ohms), C (in farads), and L (in henries). Note that there is an initial condition on $i(t)$, namely, $i(0) = 2$ A.

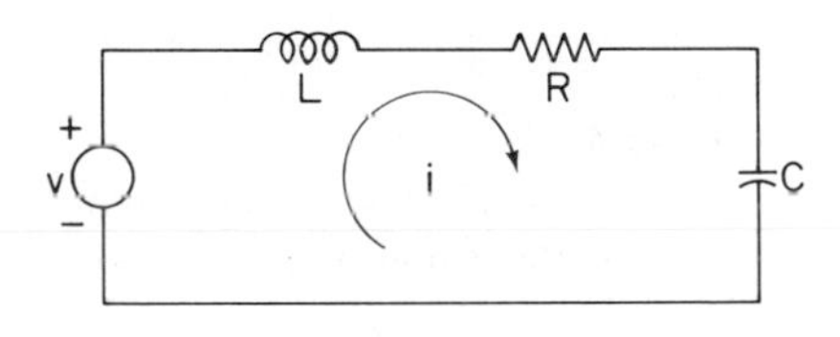

Figure P6-29

Problem 6–30 (Sec. 6–4)

The switch in the circuit shown in Fig. P6–30 moves from position a to position b at $t = 0$. The circuit is in a steady state for $t < 0$. By measurement, it is found that $i_{R2}(0+) = 2$ A and $(dv_{C1}/dt)(0+) = -1$ V/s. Find R_2 and C_1.

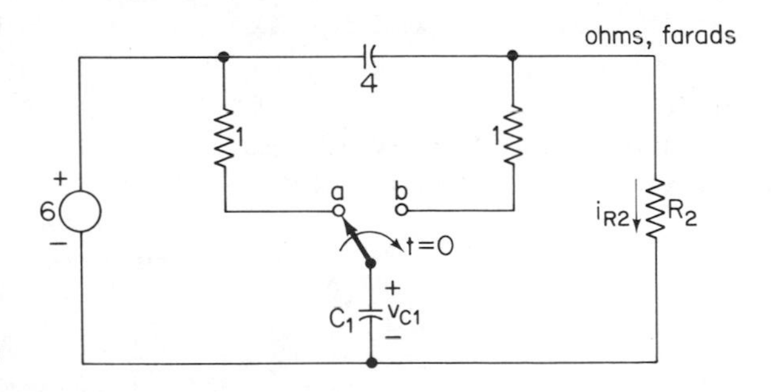

Figure P6-30

Problem 6–31 (Sec. 6–4)

In the circuit shown in Fig. P6–31, all variables are constant before the switch opens at $t = 0$. Without solving the network equations, find the values of v_1, v_2, v_1', v_2', and v_2'', all evaluated at $t = 0+$.

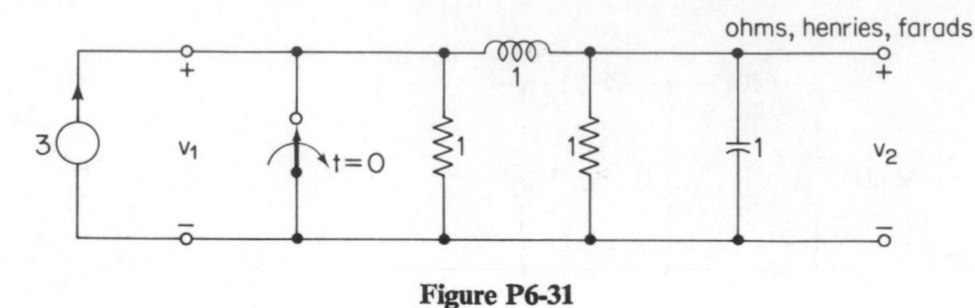

Figure P6-31

Problem 6–32 (Sec. 6–4)

In the circuit shown in Fig. P6-32, the switch is moved from a to b at $t = 0$. The circuit is in a steady state for $t < 0$. Find $v(0+)$, $(dv/dt)(0+)$, and $(d^2v/dt^2)(0+)$.

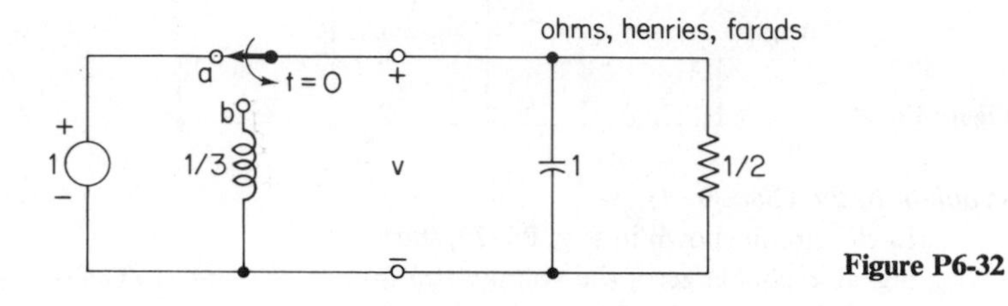

Figure P6-32

Problem 6–33 (Sec. 6–4)

The switch in the circuit shown in Fig. P6-33 closes at $t = 0$. Assume zero initial conditions and find $(di_L/dt)(0+)$, and $(di_C/dt)(0+)$.

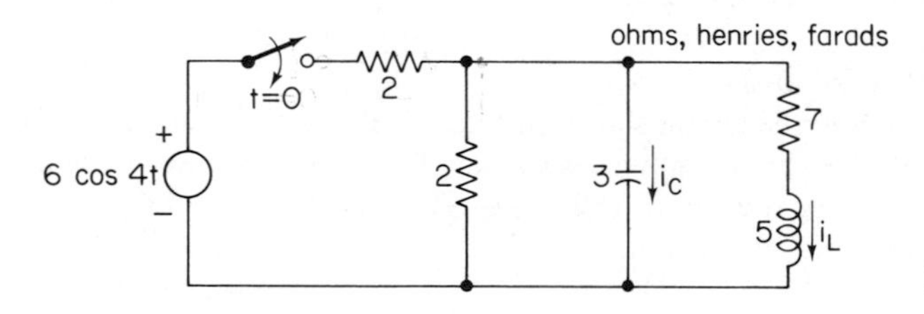

Figure P6-33

Problem 6–34 (Sec. 6–4)

In the circuit shown in Fig. P6-34, the switch opens at $t = 0$. Find $v_{sw}(0+)$ and $v_{sw}(\infty)$ (where v_{sw} is the voltage across the switch).

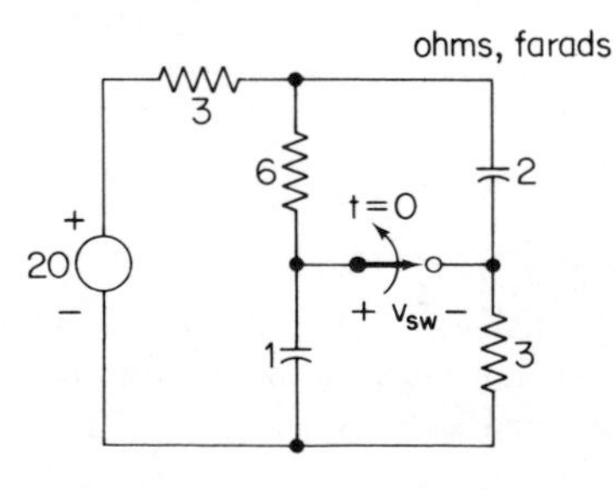

Figure P6-34

Problem 6–35 (Sec. 6–5)

Find the state-variable equations for the circuit shown in Fig. P6-35.

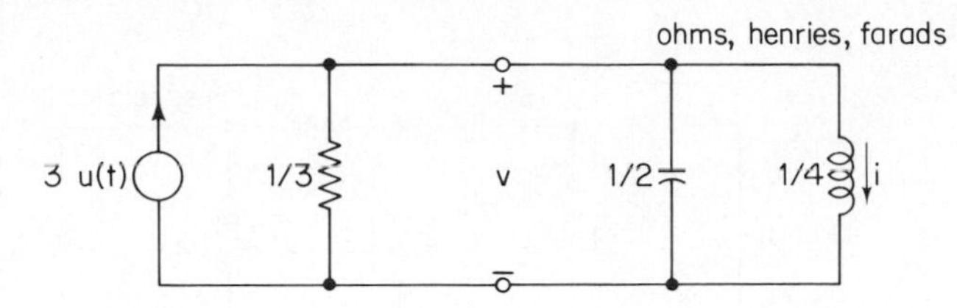

Figure P6-35

Problem 6–36 (Sec. 6–5)

Find the state-variable equations for the circuit shown in Fig. P6-36.

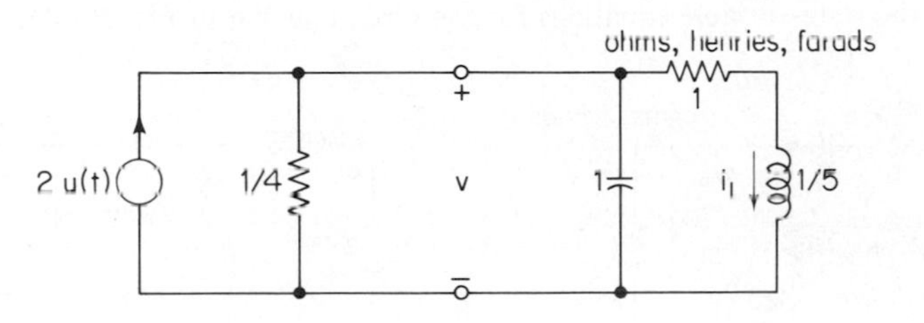

Figure P6-36

Problem 6–37 (Sec. 6–5)

Find the state-variable equations for the circuit shown in Fig. P6-37.

Problem 6–38 (Sec. 6–5)

Find the state-variable equations for the circuit shown in Fig. P6-38.

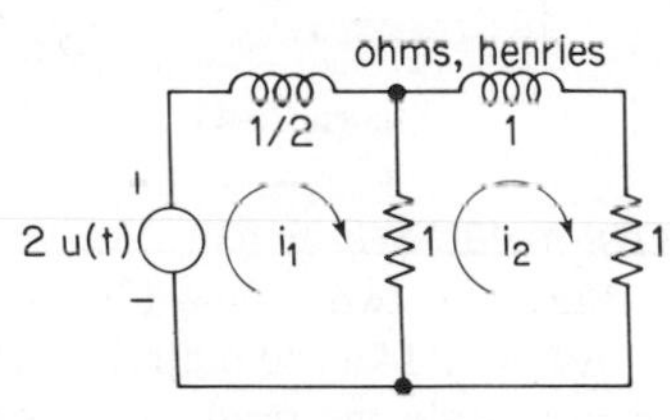

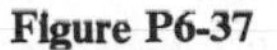

Figure P6-37

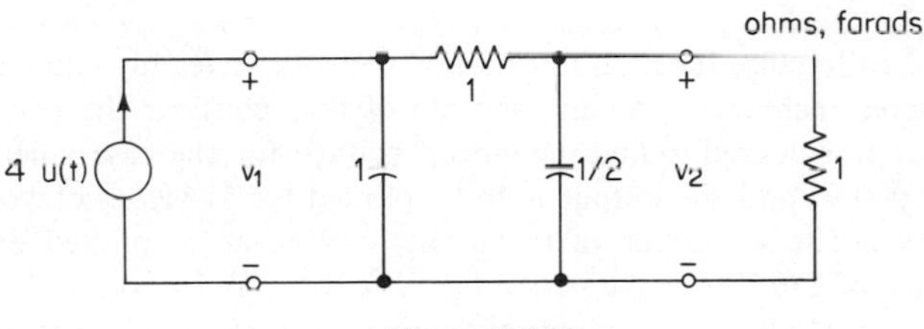

Figure P6-38

Problem 6–39 (Sec. 6–5)

Find the state-variable equations for the circuit shown in Fig. P6-39.

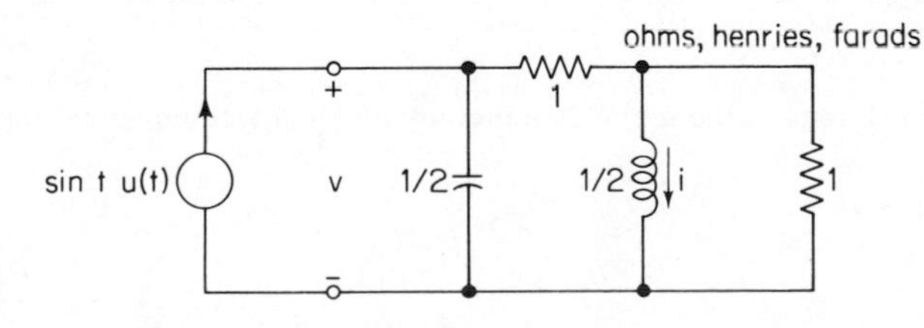

Figure P6-39

Problem 6–40 (Sec. 6–5)

Find the state-variable equations for the circuit shown in Fig. P6–40.

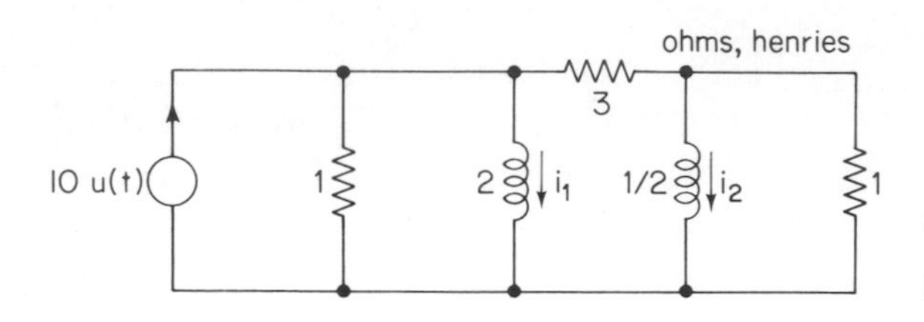

Figure P6-40

Problem 6–41 (Sec. 6–5)

Find the state-variable equations for the circuit shown in Fig. P6–41.

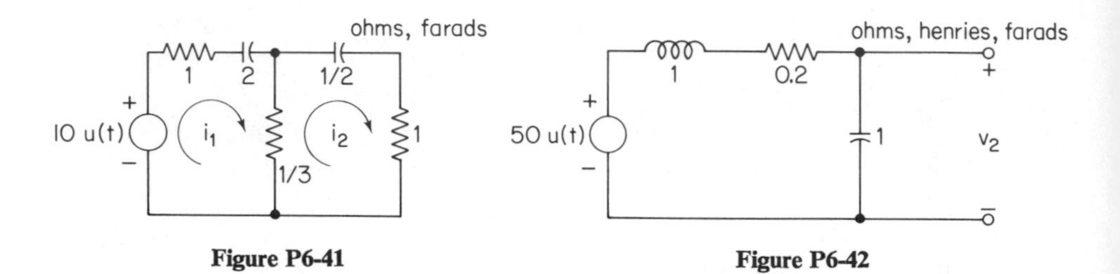

Figure P6-41

Figure P6-42

Problem 6–42(Sec. 6–6)*[11]

For the network shown in Fig. P6–42, the input excitation is a step of 50 V applied at $t = 0$. Plot the output function $v_2(t)$ for 51 values of t over the period $0 \leq t \leq 10$ s. Scale the result by 5. Assume that all the initial conditions are zero. Verify your answer by hand (non-computer) calculations.

Problem 6–43(Sec. 6–6)*

The use of an impulse function to excite a network is readily simulated by digital computational techniques. As an example of this, consider the network shown in Fig. P6–42. It is desired to find the output voltage for the case where the input is $v_1(t) = 60\,\delta(t)$ V, and the output is to be plotted for 51 values of t over the period $0 \leq t \leq 10$ s. The simulation of the input may be accomplished by letting $v_1(t)$ have a value of 150 V over the period $0 \leq t < 0.4$ s, and zero for all $t \geq 0.4$ s. Use the subroutine MXRK4 to compute the values of $v_2(t)$ and plot the output for such a simulation. Assume that all the initial conditions are zero. Verify the results by hand (non-computer) calculations.

Problem 6–44(Sec. 6–6)*

Repeat Problem 6–43 by letting $v_1(t) = 300$ V for the period $0 \leq t < 0.2$ s and zero for $t \geq 0.2$ s.

[11] Problems which require the use of digital computational techniques are identified by an asterisk.

Problem 6–45 (Sec. 6–7)

Find the characteristic equation for the circuit shown in Fig. P6-45.

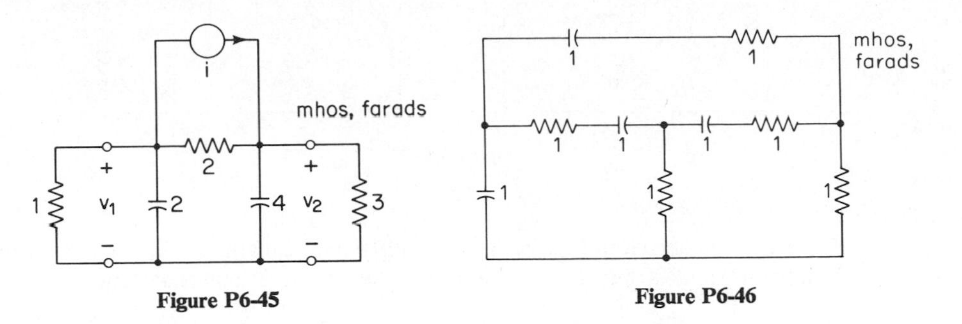

Figure P6-45 **Figure P6-46**

Problem 6–46 (Sec. 6–7)

Find the characteristic equation for the circuit shown in Fig. P6-46.

Problem 6–47 (Sec. 6–7)

For the circuit shown in Fig. P6-47, find differential equations relating (a) $v_3(t)$ and $i_s(t)$, (b) $v_2(t)$ and $i_s(t)$, and (c) $v_1(t)$ and $i_s(t)$. Assume all elements are unity-valued.

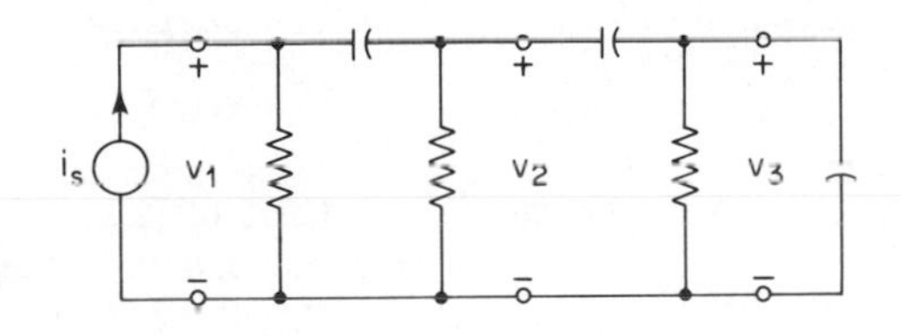

Figure P6-47

Problem 6–48 (Sec. 6–7)

Find $v_3(t)$ for $t \geq 0$ in the circuit shown in Fig. 6-7.1. Assume $v_3(0) = 1$, $v_3'(0) = 0$, $v_3''(0) = 0$, and $i_s(t) = 0$.

Problem 6–49 (Sec. 6–7)

Find the particular solution for $v_3(t)$ for $t \geq 0$ for the circuit shown in Fig. P6-49 for $i_s(t) = 2e^{-t}$ A.

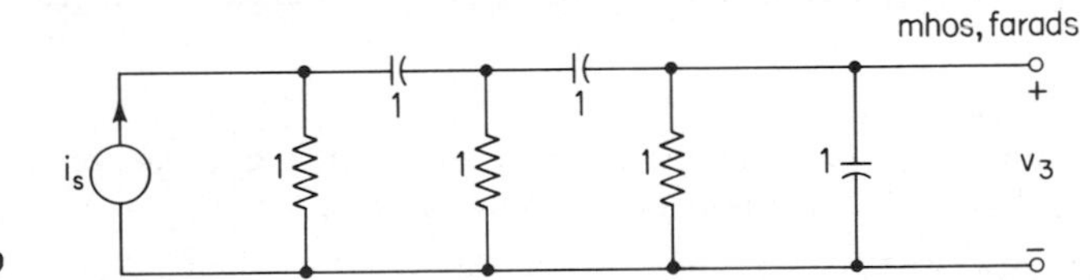

Figure P6-49

Problem 6–50 (Sec. 6–7)

Find the characteristic equation of the circuit shown in Fig. P6-50. Assume all elements are unity-valued.

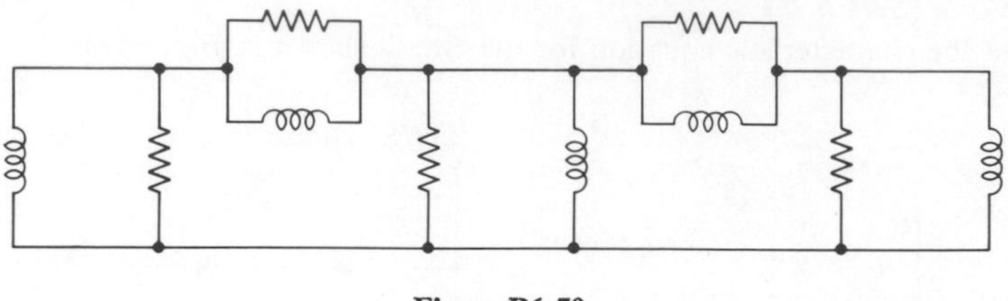

Figure P6-50

Problem 6–51 (Sec. 6–7)

For the circuit shown in Fig. P6-51, find differential equations relalting (a) $v(t)$ and $i_3(t)$, (b) $v(t)$ and $i_2(t)$, and (c) $v(t)$ and $i_1(t)$. Assume all elements are unity-valued.

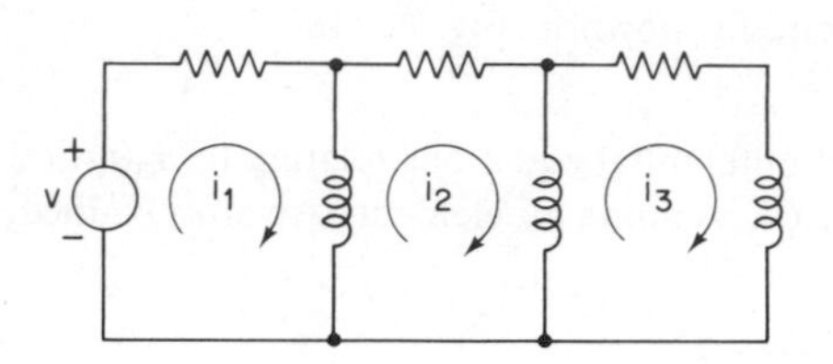

Figure P6-51

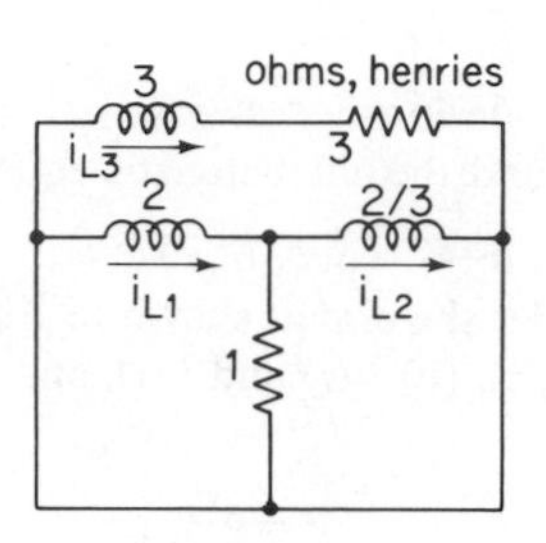

Figure P6-52

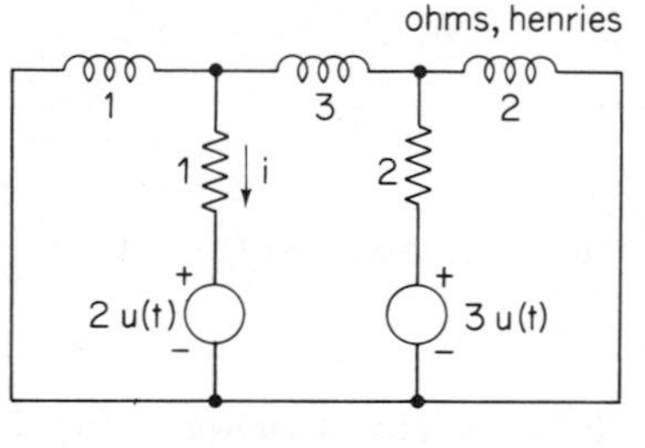

Figure P6-53

Problem 6–52 (Sec. 6–7)

In the circuit shown in Fig. P6-52, find $i_{L1}(t)$ for $t \geq 0$. Let $i_{L1}(0) = 1$, $i_{L2}(0) = -1$, and $i_{L3}(0) = 1$ A.

Problem 6–53 (Sec. 6–7)

For the circuit shown in Fig. P6-53, find $i(t)$ for $t \geq 0$. Assume that the network is unexcited for $t < 0$.

Problem 6–54 (Sec. 6–7)

Solve for $v_2(t)$ in the network shown in Fig. 6-7.4 for the following initial conditions: $v_2(0) = 0$, $v_2'(0) = 1$, $v_2''(0) = 0$. [Assume $i_s(t) = 0$.]

Problem 6–55 (Sec. 6–7)

Find a solution for $v_2(t)$ for the network shown in Fig. 6-7.4 for the case where $i_s(t) = e^{-2t}$ A, and all the initial conditions are zero.

Problem 6–56 (Sec. 6–7)

Write the state-variable equations for the *RC* network shown in Fig. P6-56. Assume that all the elements have unity value.

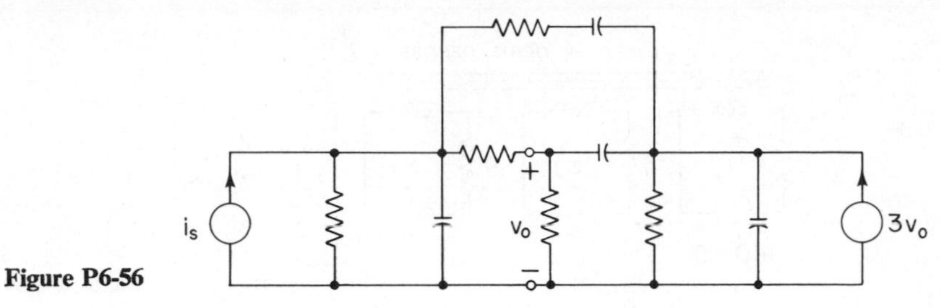

Figure P6-56

Problem 6–57 (Sec. 6–7)

Write the state-variable equations for the *RL* network shown in Fig. P6-57. Assume that all the network elements have unity value.

Problem 6–58 (Sec. 6–7)

Write the state-variable equations in matrix form for the network shown in Fig. P6-58. Assume all elements have unity value. The equations should have the form $\mathbf{x}' = \mathbf{A}\mathbf{x} + \mathbf{u}$, where $\mathbf{x}$ is defined below.

$$\mathbf{x} = \begin{bmatrix} v_1(t) \\ v_2(t) \\ i_1(t) \\ i_2(t) \\ i_3(t) \end{bmatrix}$$

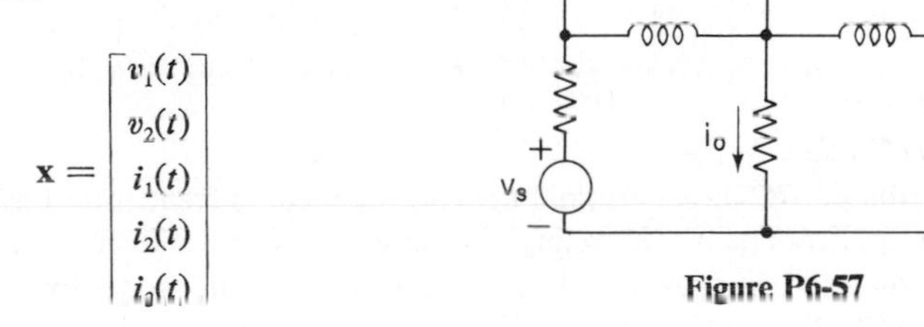

Figure P6-57

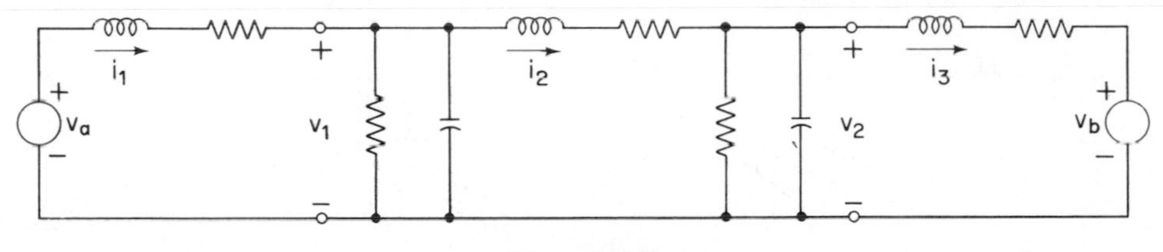

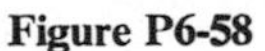

Figure P6-58

Problem 6–59 (Sec. 6–7)*

A three-winding transformer may be represented as shown in Fig. P6-59. The winding resistances are considered as lumped external resistances, and the self and mutual inductances are given by the inductance matrix shown in the figure. If a step of voltage of 140 V is applied at $t = 0$, use the subroutine MXRK4 to construct a plot of the currents in all three windings over the period $0 \le t \le 2$ s. Use 51 values of t in the plot. Assume that all initial conditions are zero.

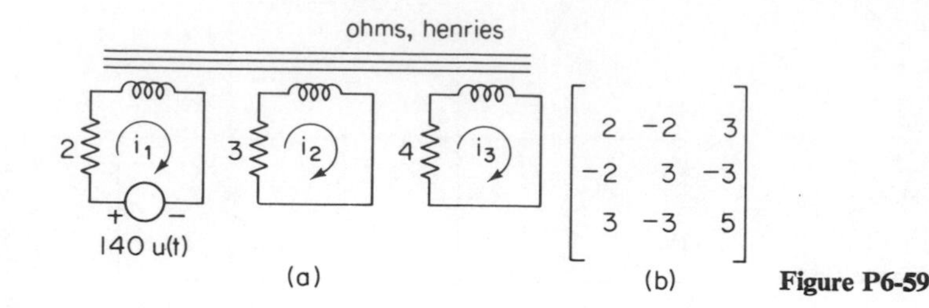

Figure P6-59

Problem 6-60 (Sec. 6-7)*

A current of 90 A is applied at $t = 0$ as an excitation to the circuit shown in Fig. P6-60. Use the subroutine MXRK4 to solve for the values of the capacitor voltages and inductor currents for 61 values of t over the period $0 \leq t \leq 6$ s. Plot $v_1(t)$ as A, $v_2(t)$ as B, and $i(t)$ as C. Assume that all the initial conditions are zero.

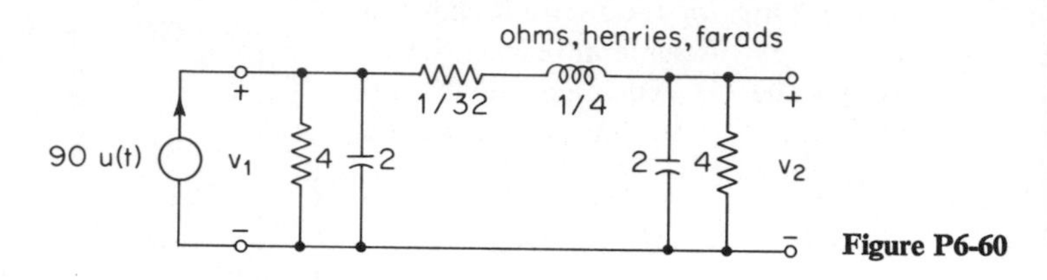

Figure P6-60

Problem 6-61 (Sec. 6-7)*

A distributed RC three-terminal network element is frequently formed in integrated circuit applications by depositing alternate layers of conducting, dielectric, and resistive material as shown in Fig. P6-61(a). A circuit model for such an element is shown in Fig. P6-61(b). Assume that ten sections are used in the model, and that each of the capacitors has a value of 0.15 F and each of the resistors has a value of

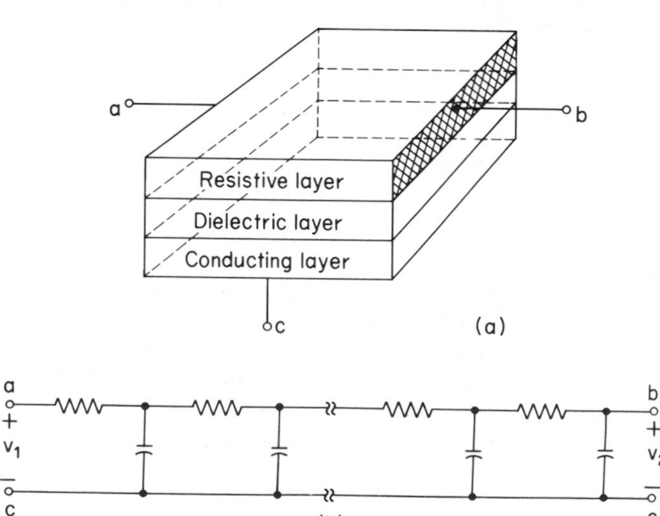

Figure P6-61

1 Ω. Use the subroutine MXRK4 to find the output voltage $v_2(t)$ for the case where $v_1(t) = u(t)$. Plot 51 values of the output voltage over the range of time from 0 to 50 s. Scale the output by 100.

Problem 6–62 (*Sec. 6–8*)

Apply the rules given in Sec. 6-8 for finding a dual network to determine the dual of the network in Fig. 6-8.4(d). Show that the resulting dual is actually the network shown in Fig. 6-8.3. That is, show that, in general, taking the dual of a dual produces the original network.

Problem 6–63 (*Sec. 6–8*)

Construct a table similar to Table 6-8.2, giving electrical and mechanical quantities which are analogous, by using a parallel *RLC* analog circuit for the mechanical system shown in Fig. 6-8.6(a).

Problem 6–64 (*Sec. 6–9*)

Find the output $v_2(t)$ for the circuit shown in Fig. 6-9.1 if a unit step $v_1(t) = u(t)$ V is applied as an input. Assume that all the initial conditions are zero.

CHAPTER 7

Sinusoidal Steady-State Analysis

In the preceding chapters we have concerned ourselves with finding the response of network variables for various excitations produced by initial conditions and sources. We have seen that the complete solution to the differential equation (or equations) describing the circuit could always be divided into two parts, namely, the complementary solution and the particular solution. In general, since the complementary solution decays exponentially as time increases, *for large values of time, only the particular solution is of interest.* This part of the complete solution always has a form similar to the form of the excitation provided by the source or sources. There are two excitation forms that occur more frequently than any others in actual practice. These are the constant excitation and the sinusoidal excitation. Methods for treating the constant excitation case were discussed in Sec. 5-4. In this chapter we shall discuss the sinusoidal case. Specifically, we shall be concerned with developing methods for finding the particular solution for network variables under the conditions that the network is excited only by sinusoids. In this case, the particular solution is frequently referred to as the *sinusoidal steady-state response.* The methods to be discussed in this chapter are of considerable importance to the electrical engineer, since sinusoids appear in many engineering applications, from the generation and transmission of electric power to satellite communication and control.

7-1 PROPERTIES OF SINUSOIDAL FUNCTIONS

Since we are planning to spend an entire chapter discussing sinusoidal steady-state response, it is important that we first explore some properties of sinusoidal functions. There are several ways in which such functions may be expressed. As a first method, consider the expression given in Table 5–2.1 for the general form that the particular solution of a network variable will have when the excitation function is a sinusoid. If we let $f(t)$ be the particular solution for the network variable, then, from the table, we may write

$$f(t) = A \cos \omega t + B \sin \omega t \tag{7–1}$$

where A and B are constants, and ω is the angular frequency in radians per second. This means of characterizing a sinusoid is particularly convenient when the method of undetermined coefficients is to be used to find the particular solution. This was the method introduced in Sec. 5-2. In this chapter, however, we shall find a different method of expressing sinusoids to be more convenient. This second method expresses the particular solution in the form

$$f(t) = F_m \cos (\omega t + \alpha) \tag{7–2}$$

where F_m is a positive quantity referred to as the *amplitude* or *peak magnitude* of the sinusoid, α is called the *argument* or *phase angle* (in radians), and ω is the *angular frequency* in radians per second. A plot of $f(t)$ is shown in Fig. 7–1.1. From this

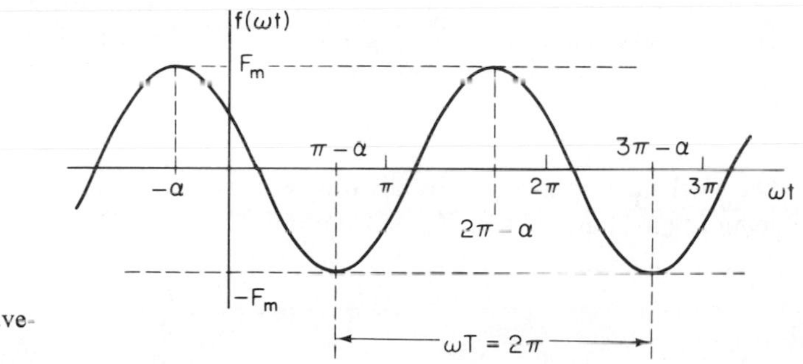

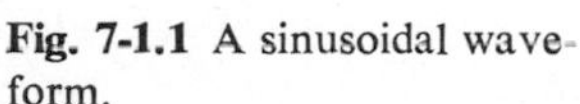
Fig. 7-1.1 A sinusoidal waveform.

figure, we see that when $\omega t = -\alpha \pm k\pi$ ($k = 0, 1, 2, 3, \ldots$), i.e., when the argument of the cosine function equals 0, $\pm\pi$, $\pm 2\pi$, $\pm 3\pi$, $\pm 4\pi$, etc., the function $f(t)$ reaches its peak magnitude, i.e., $\pm F_m$. The time between identical points on the waveform, such as the time between successive positive peak values, the time between succesive negative peak values, the time between successive zero-crossings of the abscissa (with the same slope), etc., is called the *period* T of the sinusoidal function. It is readily seen that the period $T = 2\pi/\omega$.

To see how the constants A and B of Eq. (7–1) are related to the constants F_m and α of Eq. (7–2), we may use the trigonometric identity

$$\cos (x + y) = \cos x \cos y - \sin x \sin y \tag{7–3}$$

Applying this to (7–2), we obtain

$$f(t) = F_m \cos \alpha \cos \omega t - F_m \sin \alpha \sin \omega t \tag{7-4}$$

Comparing this with the expression given in (7–1), we obtain the following relations expressing A and B in terms of F_m and α.

$$A = F_m \cos \alpha \tag{7-5}$$

$$B = -F_m \sin \alpha \tag{7-6}$$

The inverse relations which specify F_m and α in terms of A and B will now be found. We begin by squaring both members of each equation given in (7–5) and (7–6). Thus, we obtain

$$A^2 = F_m^2 \cos^2 \alpha \tag{7-7}$$

$$B^2 = F_m^2 \sin^2 \alpha \tag{7-8}$$

Adding the two equations gives

$$A^2 + B^2 = F_m^2(\cos^2 \alpha + \sin^2 \alpha) = F_m^2 \tag{7-9}$$

Thus we may write the following expression for F_m as a function of A and B:

$$F_m = \sqrt{A^2 + B^2} \tag{7-10}$$

The final relation is found by dividing corresponding members of the two equations given in (7–5) and (7–6). Thus, we obtain[1]

$$\alpha = \tan^{-1}\left(-\frac{B}{A}\right) \tag{7-11}$$

By using a simple graphical representation, the relations given above are easy to visualize, apply, and remember. To do this we need simply note that if we define an x-y plane as shown in Fig. 7–1.2(a), then a vector z as shown in the figure will be defined by the relations

$$|z| = \sqrt{x^2 + y^2} \qquad \text{Arg}\, z = \tan^{-1}\left(\frac{y}{x}\right) \tag{7-12}$$

[1]In taking the arctangent function as indicated in (7–11) and in other places in this chapter, due note must be taken of the polarity of the quantities $-B$ and A in determining the quadrant in which α lies. For example, if $-B = -1$, and $A = -1$, then $\alpha = \tan^{-1}(-1/-1) = -3\pi/4$. However, if $-B = 1$, and $A = 1$, then $\alpha = \tan^{-1}(1/1) = \pi/4$. In both cases, the argument of the arctangent function is plus one, but the angle is different.

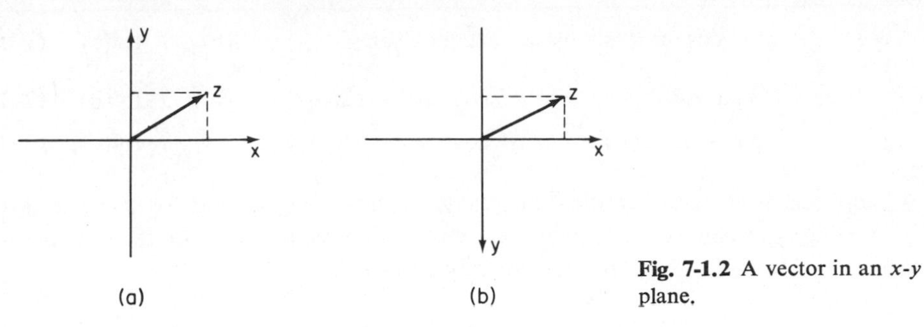

Fig. 7-1.2 A vector in an *x-y* plane.

If we now reverse the positive direction of the *y*-axis as shown in Fig. 7–1.2(b), then these relations become

$$|z| = \sqrt{x^2 + y^2} \qquad \text{Arg } z = \tan^{-1}\left(-\frac{y}{x}\right) \tag{7-13}$$

These relations have the same form as those given in (7–10) and (7–11). Thus we may define an *A-B* plane as shown in Fig. 7–1.3, in which the value of the coefficient *A* of (7–1) is plotted along the "*x*" axis, and the value of the coefficient *B* of (7–1) is plotted along the negative "*y*" axis. Any given values for *A* and *B* then define a vector. The magnitude and argument of the vector in turn define the quantities F'_m and α, respectively, in (7–2). As examples, the following equalities are illustrated in Fig. 7–1.4.[2]

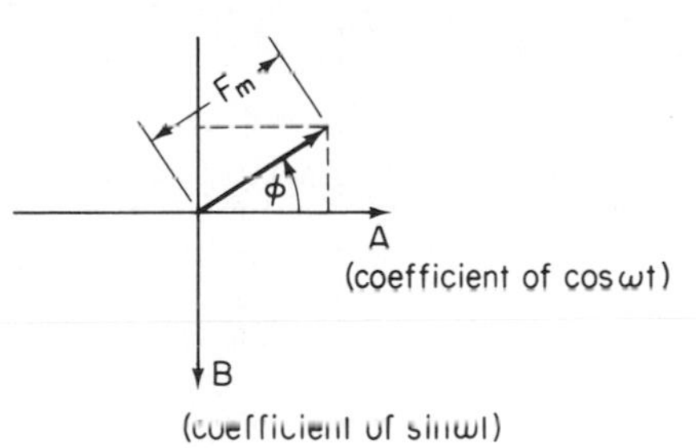

(coefficient of sin ωt)

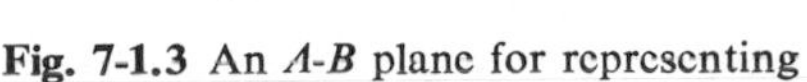

Fig. 7-1.3 An *A-B* plane for representing sinusoidal functions.

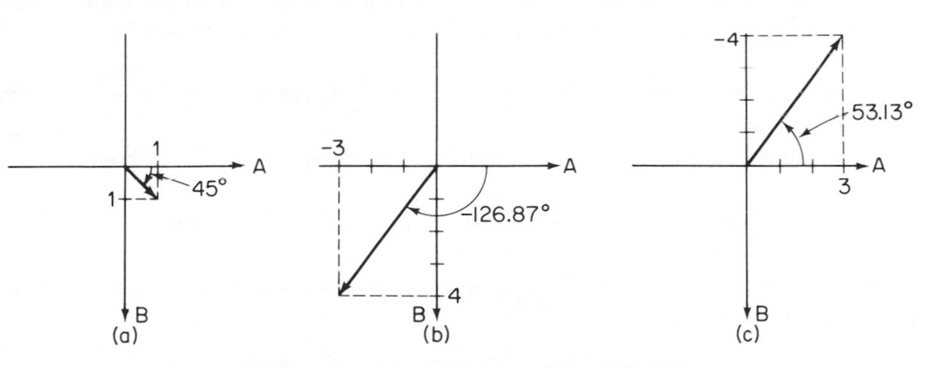

Fig. 7-1.4 Examples of the use of the *A-B* plane.

[2]To be dimensionally correct, the phase angle which is included as part of the argument of the sinusoid must be expressed in units of radians. Practically, however, for angles which are not easily expressible as simple fractions of π, we will use angles measured in degrees and identified by the superscript ° to more clearly indicate the actual value of these angles.

$$\cos \omega t + \sin \omega t = \sqrt{2} \cos\left(\omega t - \frac{\pi}{4}\right) \qquad \text{[Fig. 7-1.4(a)]} \quad (7\text{-}14)$$

$$-3 \cos \omega t + 4 \sin \omega t = 5 \cos (\omega t - 126.87^\circ) \qquad \text{[Fig. 7-1.4(b)]} \quad (7\text{-}15)$$

$$3 \cos \omega t - 4 \sin \omega t = 5 \cos (\omega t + 53.13^\circ) \qquad \text{[Fig. 7-1.4(c)]} \quad (7\text{-}16)$$

In addition, it is readily verified that the *A-B* plane may be used for computing the phase angles associated with any other type of sinusoidal expression than the positive cosine function. For example, it is readily shown that

$$\cos\left(\omega t + \frac{\pi}{6}\right) = -\sin\left(\omega t - \frac{\pi}{3}\right) = \sin\left(\omega t + \frac{2\pi}{3}\right)$$
$$= -\cos\left(\omega t - \frac{5\pi}{6}\right) \qquad (7\text{-}17)$$

The various angles are illustrated in Fig. 7-1.5.

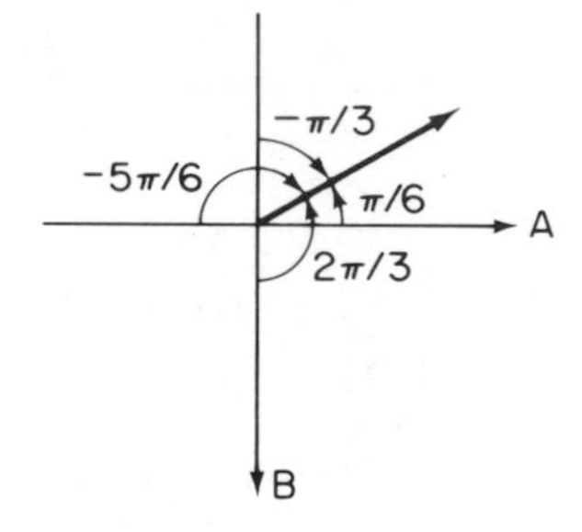

Fig. 7-1.5 Some equivalent arguments.

There are several properties of sinusoids that will be useful in succeeding sections of this chapter. One of the most important of these has to do with the addition of sinusoidal quantities. Let us consider two functions $h(t)$ and $g(t)$ defined as follows.

$$h(t) = H_m \cos (\omega t + \beta)$$
$$g(t) = G_m \cos (\omega t + \alpha) \qquad (7\text{-}18)$$

From the discussion given above we know that each of these functions may be expressed in a form similar to that given in (7-1). Let these expressions be

$$h(t) = A_h \cos \omega t + B_h \sin \omega t \qquad g(t) = A_g \cos \omega t + B_g \sin \omega t \qquad (7\text{-}19)$$

Now let us add the two functions, using the relations of (7-19). We obtain

$$h(t) + g(t) = (A_h + A_g) \cos \omega t + (B_h + B_g) \sin \omega t \qquad (7\text{-}20)$$

If we define $A = A_h + A_g$ and $B = B_h + B_g$, then the sum of the two sinusoids may be written in the form

$$h(t) + g(t) = F_m \cos (\omega t + \alpha) \qquad (7\text{-}21)$$

where F_m and α are defined by Eqs. (7-10) and (7-11). Thus, the result of the addition of the two sinusoids is another sinusoid *of the same frequency*. The *A-B* plane shown in Fig. 7-1.3 can be used to advantage when it is desired to form such a sum, since the addition of the functions is then simply reduced to the addition of the vectors representing the magnitude and phase angle of the related functions. For example, suppose that it is desired to find a single sinusoid $f(t)$ comprised of the sum of $h(t)$ and $g(t)$, where

$$\begin{aligned} h(t) &= 5\cos(\omega t + 36.87^\circ) \\ g(t) &= 3\sqrt{2}\sin(\omega t - 45^\circ) \end{aligned} \tag{7-22}$$

The vector representations of the magnitude and phase of $h(t)$ and $g(t)$ are shown in Fig. 7-1.6. The addition of these vectors yields the magnitude and phase of the sinusoid (of the same frequency) $f(t)$. Thus, we see that

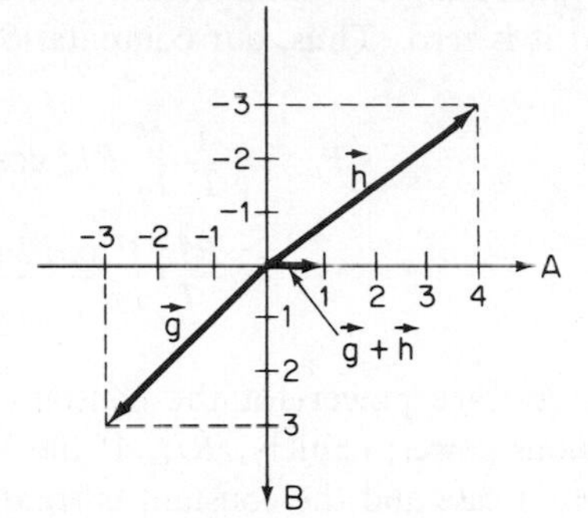

Fig. 7-1.6 Adding sinusoidal functions on the *A-B* plane.

$$f(t) = h(t) + g(t) = \cos\omega t \tag{7-23}$$

We shall find frequent application of these techniques in the remainder of this chapter.

Since sinusoidal waveforms occur with such frequency in engineering situations, it is useful to be able to characterize them according to their effects. One of the most important of such effects has to do with the power dissipated by a sinusoidal voltage or current when it is one of the variables associated with a resistive two-terminal element. To be more specific, let us see if we can make a comparison between the power dissipated in a resistor (of value R Ω) in two different situations. The first of these will be when a sinusoidal current $i(t)$ is passed through the resistor, and the second will be when a constant current of value I_0 is passed through the same resistor. For the sinusoidal case, let $i(t)$ have the form

$$i(t) = I_m\cos(\omega t + \phi) \tag{7-24}$$

The power is given by the expression

$$p(t) = Ri^2(t) = RI_m^2\cos^2(\omega t + \phi) = RI_m^2\frac{\cos(2\omega t + 2\phi) + 1}{2} \tag{7-25}$$

Correspondingly, in the constant current case, the power is given by the expression

$$p(t) = RI_0^2 \tag{7-26}$$

The expressions given in (7-25) and (7-26) cannot be directly compared, since the first fluctuates sinusoidally between the limits of 0 and RI_m^2 with a frequency of 2ω rad/s, while the second is constant. Thus, to permit a comparison between the two quantities, let us calculate the *average* value of the power over some time interval.

For the sinusoidal case, if we compute the average power over a single period, then this will be the same as the average power for any integer number of periods. The period is given by $T = 2\pi/\omega$. Since the computation is to be made over one period, the phase angle is unimportant, and for convenience of integration, we may assume that it is zero. Thus, our computation becomes

$$\begin{aligned} P_{\text{avg}} &= \frac{1}{T}\int_0^T RI_m^2 \cos^2(\omega t)\, dt = \frac{RI_m^2}{T}\int_0^T \cos^2(\omega t)\, dt \\ &= \frac{RI_m^2}{T}\int_0^T \frac{\cos 2\omega t + 1}{2} dt = \frac{RI_m^2}{2} \end{aligned} \tag{7-27}$$

The average power for the constant current case is obviously equal to the instantaneous power, namely, RI_0^2. If the values of the average power for the sinusoidal current case and the constant current case are to be the same, then we must have

$$RI_0^2 = \frac{RI_m^2}{2} \tag{7-28}$$

This requires that

$$I_0 = \frac{I_m}{\sqrt{2}} \tag{7-29}$$

Thus, a sinusoidal waveform of current with a peak magnitude of I_m will dissipate the same average power in a resistor as a constant current of value $I_m/\sqrt{2}$ A. A similar development is easily made to show that a sinusoidal waveform of voltage with a peak magnitude of V_m applied to a resistor will produce the same average power dissipation in a resistor as a constant potential of value $V_m/\sqrt{2}$ V. The details are left as an exercise to the reader.

Let us consider some of the individual steps which were made in the above development, and their effect on the function $i(t)$ of Eq. (7–24). First of all, since power considerations were involved, we *squared* the function. Second, in Eq. (7–27), we found the *mean* (or average) value of the squared function. Finally, we took the square *root* of the result. Thus, in summary, we may refer to the result as the *root-mean-square* or *rms* value of $i(t)$. This process may be summarized by the relation

$$I_{\text{rms}} = \sqrt{\frac{1}{T}\int_0^T i^2(t)\, dt} \tag{7-30}$$

where T is the length of one period of the sinusoidal variation of $i(t)$. Although our primary consideration here is with sinusoidal functions, it should be noted that the relation given in (7–30) applies equally well to the determination of the rms value of any periodic function, i.e., any function which repeats its waveform with a constant period of T s. Such a periodic function satisfies the relation

$$f(t) = f(t + T)$$

for all t.

The results given above may be summarized by pointing out that the rms value of a periodic current or voltage is equal to the value of an equivalent constant current or voltage which will produce the same average power dissipation in a given resistor.

As an example of the use of rms values, note that the familiar distribution of household electricity is predominantly done using sinusoidal variables, and the value of the voltage is specified in terms of rms units. Thus, a "110-V outlet" provides a sinusoidal voltage which varies between the (peak) limits of $\pm 110 \times \sqrt{2}$ or about ± 155.6 V. The frequency of such a sinusoid is usually 60 Hz (cycles per/second), and this is $60 \times 2\pi$ or about 377 rad/s. Thus, the sinusoidal voltage which is provided by most electric power companies may be characterized as $155.6 \cos(377t + \phi)$ V, where ϕ is the phase angle.

7-2 THE EXPONENTIAL FUNCTION

Let us begin our study of sinusoidal steady-state response by considering the circuit shown in Fig. 7-2.1. Applying KCL to this circuit, if $i(t)$ is a sinusoid, we may write

$$C\frac{dv}{dt} + Gv(t) = i(t) = I_m \cos(\omega t + \phi) \quad (7\text{-}31)$$

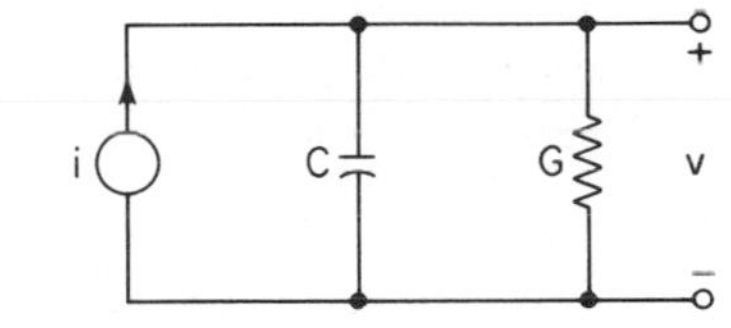

Fig. 7-2.1 A simple RC network.

where I_m is the positive peak magnitude of the excitation sinusoid and ϕ is the phase angle. From Table 5-2.1 we know that the particular solution for $v(t)$ will also be a sinusoid. From the discussion of the preceding section, such a particular solution may be written in the form

$$v(t) = V_m \cos(\omega t + \theta) \quad (7\text{-}32)$$

Instead of directly substituting this expression into Eq. (7-31) to determine the value of the constants V_m and θ, let us first convert the expression given in (7-32) to the exponential form for the cosine function. Thus, we may write

$$v(t) = \frac{V_m}{2}\left(e^{j(\omega t+\theta)} + e^{-j(\omega t+\theta)}\right) = \frac{V_m e^{j\theta}}{2} e^{j\omega t} + \frac{V_m e^{-j\theta}}{2} e^{-j\omega t} \quad (7\text{-}33)$$

The coefficients of the exponential functions in the right member of (7-33) have equal magnitudes but opposite angles, thus they are complex conjugates. To simplify our notation we may use a script letter $\mathscr{V}$ to represent the numerator of these complex quantities. Thus, we may write

$$\mathscr{V} = V_m e^{j\theta} \qquad \mathscr{V}^* = V_m e^{-j\theta} \tag{7-34}$$

where the superscript asterisk is used to indicate the complex conjugate. The node voltage $v(t)$ may now be written

$$v(t) = \frac{\mathscr{V}}{2} e^{j\omega t} + \frac{\mathscr{V}^*}{2} e^{-j\omega t} \tag{7-35}$$

In a similar fashion, we may express the excitation current $i(t)$ as

$$i(t) = \frac{I_m}{2} [e^{j(\omega t + \phi)} + e^{-j(\omega t + \phi)}] = \frac{I_m e^{j\phi}}{2} e^{j\omega t} + \frac{I_m e^{-j\phi}}{2} e^{-j\omega t} \tag{7-36}$$

If we define

$$\mathscr{I} = I_m e^{j\phi} \qquad \mathscr{I}^* = I_m e^{-j\phi} \tag{7-37}$$

then (7–36) may be put in the form

$$i(t) = \frac{\mathscr{I}}{2} e^{j\omega t} + \frac{\mathscr{I}^*}{2} e^{-j\omega t} \tag{7-38}$$

Now let us substitute the expressions for $v(t)$ and $i(t)$ given in (7–35) and (7–38) in the original differential equation of (7–31). After rearranging terms, we obtain

$$[(G + j\omega C)\mathscr{V} - \mathscr{I}]\frac{e^{j\omega t}}{2} + [(G - j\omega C)\mathscr{V}^* - \mathscr{I}^*]\frac{e^{-j\omega t}}{2} = 0 \tag{7-39}$$

Since the exponentials are functions of t, the only way in which the equation given above will be satisfied *for all values of t* is for the terms multiplying the exponentials to be zero. Thus, we see that

$$\mathscr{V} = \frac{\mathscr{I}}{G + j\omega C}$$
$$\mathscr{V}^* = \frac{\mathscr{I}^*}{G - j\omega C} \tag{7-40}$$

These equations provide the information that we need to determine the particular solution of the network. For the moment, let us only use the first equation of (7–40). Thus, we may write

$$\mathscr{V} = V_m e^{j\theta} = \frac{\mathscr{I}}{G + j\omega C} = \frac{I_m e^{j\phi}}{|G + j\omega C| e^{j \tan^{-1}(\omega C/G)}}$$
$$= \frac{I_m}{\sqrt{G^2 + (\omega C)^2}} e^{j[\phi - \tan^{-1}(\omega C/G)]} \tag{7-41}$$

Thus we conclude that the separate relations for the magnitude and phase of $v(t)$ of Eq. (7–32) are

$$V_m = \frac{I_m}{\sqrt{G^2 + (\omega C)^2}} \qquad \theta = \phi - \tan^{-1}\left(\frac{\omega C}{G}\right) \tag{7–42}$$

When these terms are substituted in Eq. (7–32), the particular solution for $v(t)$ is completely defined. An example follows.

EXAMPLE 7-2.1. The elements shown in the circuit diagram in Fig. 7-2.1 are specified as having the values $G = 2$ mhos and $C = 1$ F. It is desired to find the sinusoidal steady-state response for $v(t)$ under the conditions that $i(t) = 5\cos[2t - (\pi/4)]$. From (7-37), $\mathscr{I} = 5e^{-j\pi/4}$ and, from Eq. (7-40), we obtain

$$\mathscr{V} = \frac{\mathscr{I}}{G + j\omega C} = \frac{5e^{-j\pi/4}}{2 + j2} = \frac{5e^{-j\pi/4}}{2\sqrt{2}\, e^{j\pi/4}} = \frac{5}{2\sqrt{2}} e^{-j\pi/2}$$

Thus, our solution for $v(t)$ is

$$v(t) = \frac{5}{2\sqrt{2}} \cos\left(2t - \frac{\pi}{2}\right) = \frac{5}{2\sqrt{2}} \sin 2t$$

It should be noted that the most laborious part of the procedure given above is the determination of the equation of the form of (7–40) relating the excitation to the response. Thus, even though the procedure for finding $v(t)$ as illustrated in the numerical example is simple, the overall process may appear more complicated than the method of undetermined coefficients presented in Chaps. 5 and 6. We will shortly show, however, that this approach may be applied very simply to arbitrary network configurations, and thus it is of considerable value.

Now let us return to consider again the two equations of (7–40). It is easily seen that if we take the conjugate of both sides of the first equation, we obtain the second. Thus, the second equation is always satisfied if the first one is satisfied. Therefore, we see that once the first equation is known, finding the second equation does not given us any new information concerning the network. The first equation, however, is the equation associated with the exponential term $e^{j\omega t}$ (the *positive* exponential term) in the expression for the excitation current $i(t)$ and the expression for the particular solution for $v(t)$. Thus, this equation can be generated if we adopt the mathematical artifice of assuming that the excitation has the form[3]

[3]Note that the quantities given for $i(t)$ and $v(t)$ in Eqs. (7–43) and (7–44), respectively, cannot actually be equal to physical variables, since they are *complex* quantities, while the physical variables of current and voltage must, of course, be *real*. However, there is no reason why, in an equation such as that shown in (7–31), we cannot mathematically make a substitution of the form indicated in Eqs. (7–43) and (7–44). This is what we imply by the words "mathematical artifice." In the next section we shall use a more sophisticated approach which is acceptable both from a mathematical viewpoint and a physical viewpoint.

$$i(t) = \frac{\mathscr{I}}{2} e^{j\omega t} \tag{7-43}$$

where $\mathscr{I}$ is defined in Eq. (7-37). Similarly, we may assume that the particular solution for the node voltage has the form

$$v(t) = \frac{\mathscr{V}}{2} e^{j\omega t} \tag{7-44}$$

where $\mathscr{V}$ is defined in (7-34). Finally, substituting Eqs. (7-43) and (7-44) into the basic differential equation defined by the left members of (7-31), we obtain the identical relation given as the first equation of (7-40). The effort involved in obtaining this result, however, is exactly half of that required for the original determination of Eq. (7-40), a considerable computational saving.

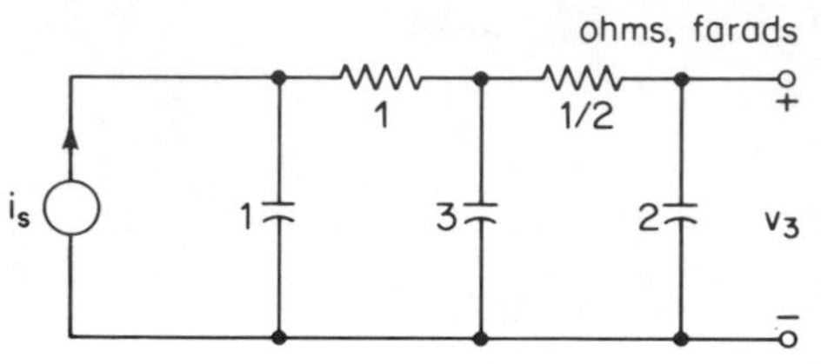

Fig. 7-2.2 A third-order *RC* network.

It is readily shown that the simplified procedure outlined above may be applied to a differential equation of any order. As an example of this, consider the circuit shown in Fig. 7-2.2. The differential equation relating the excitation $i_s(t)$ and the node voltage $v_3(t)$ for this network may be shown to be

$$i_s(t) = 3v_3''' + 9v'' + 6v_3' \tag{7-45}$$

Let us now assume that $i_s(t)$ is a sinusoid. Using the original method presented above, we may use the notation

$$i_s(t) = I_s \cos(\omega t + \phi) = \frac{\mathscr{I}_s}{2} e^{j\omega t} + \frac{\mathscr{I}_s^*}{2} e^{-j\omega t} \tag{7-46}$$

where

$$\mathscr{I}_s = I_s e^{j\phi} \tag{7-47}$$

The particular solution for $v_3(t)$, i.e., the sinusoidal steady-state response, will also have the form of a sinusoid with the same frequency. Thus, we define

$$v_3(t) = V_3 \cos(\omega t + \theta) = \frac{\mathscr{V}_3}{2} e^{j\omega t} + \frac{\mathscr{V}_3^*}{2} e^{-j\omega t} \tag{7-48}$$

where

$$\mathscr{V}_3 = V_3 e^{j\theta} \tag{7-49}$$

Substituting the right members of (7–46) and (7–48) into the differential equation of (7–45) and rearranging terms, we obtain the equation

$$\{[3(j\omega)^3 + 9(j\omega)^2 + 6(j\omega)]\mathscr{V}_3 - \mathscr{I}_s\}e^{j\omega t} \\ + \{[3(-j\omega)^3 + 9(-j\omega)^2 + 6(-j\omega)]\mathscr{V}_3^* - \mathscr{I}_s^*\}e^{-j\omega t} = 0 \qquad (7\text{–}50)$$

The left member of the above equation will be zero for all values of t only if the coefficients of the exponential terms are zero. Thus, we may write

$$\mathscr{V}_3 = \frac{\mathscr{I}_s}{3(j\omega)^3 + 9(j\omega)^2 + 6(j\omega)}$$

$$\mathscr{V}_3^* = \frac{\mathscr{I}_s^*}{3(-j\omega)^3 + 9(-j\omega)^2 + 6(-j\omega)} \qquad (7\text{–}51)$$

Again we see that the equations are conjugates; thus, for a given value of ω either of these equations may be used to determine the (complex) value of $\mathscr{V}_3$. Thus, using Eqs. (7–48) and (7–49), we may find the steady-state solution for $v_3(t)$. To simplify the solution procedure, we need merely change the form of the excitation given in (7–46) and the form of the particular solution given in (7–48) to the reduced forms

$$i_s(t) = \frac{\mathscr{I}_s}{2}e^{j\omega t} \qquad v_3(t) = \frac{\mathscr{V}_3}{2}e^{j\omega t} \qquad (7\text{–}52)$$

Substituting these forms into the basic differential equation of (7–45) is readily shown to produce the first equation of (7–50), and thus it may be used to find the steady-state solution for $v_3(t)$.

It should be noted that the procedure given is also easily applied if the network is described by an integral equation or an integro-differential equation, since in such a case we need merely differentiate the given expression to obtain a purely differential equation. For example, if we apply KVL to a circuit consisting of a series-connected voltage source, a resistor, and a capacitor, we obtain

$$v(t) = Ri(t) + \frac{1}{C}\int_{-\infty}^{t} i(\tau)\,d\tau \qquad (7\text{–}53)$$

Differentiating this equation, we obtain

$$\frac{dv}{dt} = R\frac{di}{dt} + \frac{1}{C}i(t) \qquad (7\text{–}54)$$

From this point, substitutions similar to those described above may be used to find the sinusoidal steady-state response for $i(t)$ for a given sinusoidal excitation $v(t)$. The details are left to the reader.

The procedure described above is readily summarized and generalized as follows:

SUMMARY 7-2.1

Sinusoidal Steady-State Response for a Circuit Described by a Single Differential Equation: If a sinusoidal excitation variable $g(t) = G_m \cos(\omega t + \phi)$ (and/or its derivatives) is related to a response variable $f(t)$ and its derivatives by a single differential equation of the form

$$a_0 f(t) + a_1 \frac{df}{dt} + a_2 \frac{d^2 f}{dt^2} + \cdots = b_0 g(t) + b_1 \frac{dg}{dt} + b_2 \frac{d^2 g}{dt^2} + \cdots \quad (7\text{-}55)$$

then the particular solution for the response will have the form $f(t) = F_m \cos(\omega t + \theta)$. The quantities F_m and θ may be found by making the following substitutions in the differential equation:

$$g(t) = \mathscr{G} e^{j\omega t} \qquad f(t) = \mathscr{F} e^{j\omega t}$$

where $\mathscr{G} = G_m e^{j\phi}$, and $\mathscr{F} = F_m e^{j\theta}$. Thus, we obtain

$$\mathscr{F} = \frac{b_0 + b_1(j\omega) + b_2(j\omega)^2 + \cdots}{a_0 + a_1(j\omega) + a_2(j\omega)^2 + \cdots} \mathscr{G}$$

The resulting expression for $f(t)$ is called the sinusoidal steady-state response.

7-3 PHASORS AND SINUSOIDAL STEADY-STATE RESPONSE

In the preceding section we formulated a method for finding the sinusoidal steady-state response of a given network by assuming that the excitation and response variables could be expressed as quantities involving a positive exponential term and having the form

$$\mathscr{F} e^{j\omega t} \quad (7\text{-}56)$$

where

$$\mathscr{F} = F_m e^{j\phi} \quad (7\text{-}57)$$

and the quantities F_m and ϕ are, respectively, the amplitude and phase of the sinusoidally varying quantity $f(t)$, which has the form

$$f(t) = F_m \cos(\omega t + \phi) \quad (7\text{-}58$$

A more rigorous justification of the method may be presented by noting that, from (7–56) and (7–57), we may write

$$\begin{aligned}\mathscr{F} e^{j\omega t} &= F_m e^{j\phi} e^{j\omega t} = F_m e^{j(\omega t + \phi)} \\ &= F_m[\cos(\omega t + \phi) + j \sin(\omega t + \phi)]\end{aligned} \tag{7–59}$$

Comparing Eqs. (7–58) and (7–59), we now may write

$$f(t) = \text{Re}(\mathscr{F} e^{j\omega t}) \tag{7–60}$$

where "Re" stands for "the real part of." Thus, we see that, using the formal relation given in Eq. (7–60), we may use the complex quantity $\mathscr{F} e^{j\omega t}$ as defined by Eq. (7–59) to represent the sinusoidal function $F_m \cos(\omega t + \phi)$. Such a complex quantity is called a *phasor*.[4] A phasor has a definite representation on the complex plane, i.e., the plane on which the value of the real part of a complex quantity is plotted in the "x" direction and on which the value of the imaginary part of a complex quantity is plotted in the "y" direction. Such a representation is shown in Fig. 7–3.1

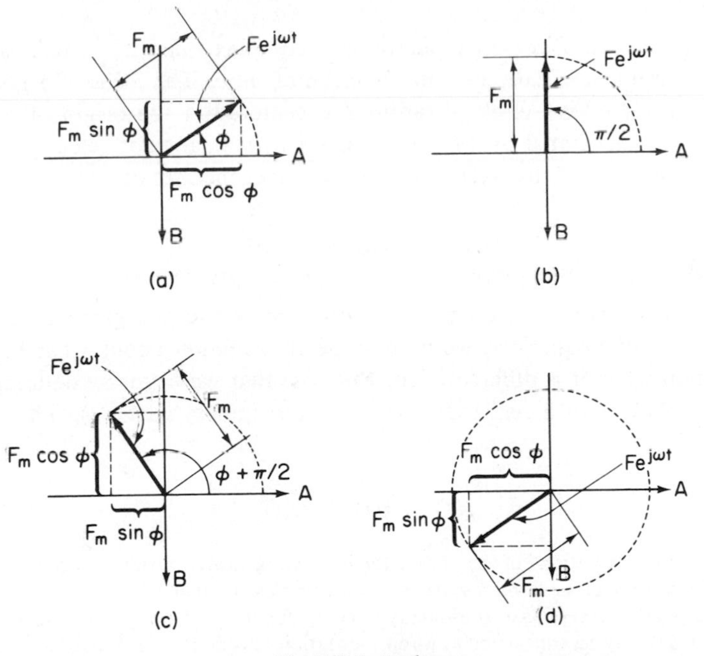

Fig. 7-3.1 A phasor.

[4]It is customary to use the same units for the phasor as were used for the original sinusoidally varying quantity. Thus, a phasor used to represent a sinusoidally varying voltage will be treated as having units of volts. A phasor used to represent a sinusoidally varying current will be treated as having units of amperes, etc.

for various values of ωt. For example, at $t = 0$, the real part of $\mathcal{F}e^{j\omega t}$ is $F_m \cos \phi$ and the imaginary part is $F_m \sin \phi$. These two values determine a vector of magnitude F_m and of angle ϕ as shown in Fig. 7–3.1(a). Since, from Eq. (7–60), $f(t)$ is the real part of $\mathcal{F}e^{j\omega t}$, it is equal to the component of $\mathcal{F}e^{j\omega t}$ that lies along the real axis, i.e., it is equal to the projection of $\mathcal{F}e^{j\omega t}$ on the real axis. Thus, when t equals zero, $f(0)$ is equal to Re $\mathcal{F}$ or $F_m \cos \phi$, as shown in Fig. 7–3.1(a). Now let us consider the situation where $\omega t = (\pi/2) - \phi$, i.e., the case where $t = [(\pi/2) - \phi)]/\omega$. In this case, the real part of $\mathcal{F}e^{j\omega t}$ is zero, and the imaginary part is F_m. Thus, for this value of t the phasor $\mathcal{F}e^{j\omega t}$ appears as shown in Fig. 7–3.1(b). The corresponding projection on the real axis tells us that $f(t)$ evaluated for $t = [(\pi/2) - \phi]/\omega$ equals zero, as is also apparent from Eq. (7–58). Continuing in the same fashion, we see that when $\omega t = \pi/2$, i.e., when $t = \pi/2\omega$, the phasor $\mathcal{F}e^{j\omega t}$ appears as shown in Fig. 7–3.1(c), namely, it has a magnitude of F_m, and an angle of $(\pi/2) + \phi$. Thus, $f(\pi/2\omega) = -F_m \sin \phi$. Similarly, when $t = \pi/\omega$, the phasor will have the value shown in Fig. 7–3.1(d), etc. Considering all the above, we note that the locus of $\mathcal{F}e^{j\omega t}$ is a circle of radius F_m centered at the origin of the complex plane.[5] The preceding discussion may be summarized as follows:

SUMMARY 7-3.1

Definition of a Phasor: A phasor is a complex quantity which is used to represent a sinusoidally varying function of time. The locus of a phasor such as $\mathcal{F} = F_m e^{j\omega t}$ is a circle of radius F_m, centered at the origin of a complex plane, and generated by a rotating vector with an angular velocity of ω rad/s. The projection of this vector on the real axis generates the sinusoidal function

$$f(t) = F_m \cos (\omega t + \phi)^6$$

In order to discuss the use of functions such as the one given in Eq. (7–60) in solving differential equations, we need some information about what happens to such a function when it is differentiated. Suppose that we begin by defining a complex function $g(t)$, with a real part $x(t)$ and an imaginary part $y(t)$. Thus, we may write

$$g(t) = x(t) + jy(t) \tag{7–61}$$

[5]Note that the "rotation" of the phasor for increasing values of time is what distinguishes it from a stationary *vector*, i.e. a representation of a complex number.

[6]Since a phasor represents a sinusoidally varying function of time, a quantity such as is given in Eq. (7–57) may be considered as a representation of a sinusoid evaluated at t equals zero, while a quantity of the form given in Eq. (7–56) may be considered as a representation of a sinusoid at all values of time. For convenience, we shall refer to expressions of the form of both Eqs. (7–56) and (7–57) as phasors, even though in the latter case there is no explicit dependence on t.

Now let us consider a function $f(t)$ defined as

$$f(t) = \text{Re}\,[g(t)] = \text{Re}\,[x(t) + jy(t)] = x(t) \tag{7-62}$$

If we desire to differentiate $f(t)$, then clearly we may write

$$\frac{df}{dt} = \frac{d}{dt}\,\text{Re}\,[g(t)] = \frac{dx}{dt} \tag{7-63}$$

Let us now investigate what happens if we perform the operations in the opposite order, i.e., if we first differentiate $g(t)$ and then take the real part of the result. If we do this, we may write

$$\text{Re}\left[\frac{dg}{dt}\right] = \text{Re}\left[\frac{dx}{dt} + j\frac{dy}{dt}\right] = \frac{dx}{dt} \tag{7-64}$$

Comparing Eqs. (7-63) and (7-64), we see that

$$\frac{d}{dt}\,\text{Re}\,[g(t)] = \text{Re}\left[\frac{dg}{dt}\right] \tag{7-65}$$

Thus we may conclude that if a complex function of t is differentiated, and then its real part is extracted, the result is the same as if the order of the two operations had been reversed, i.e., as if first the real part had been taken and then the result differentiated. A similar result may be shown to hold with respect to integration.

A second item of information that we shall require before we can discuss the use of relations of the type given in (7-60) is the relative implications of an equality between two complex functions having the form given in Eq. (7-56), and an equality between the real parts of such functions. Let the two complex functions be $g_1(t)$ and $g_2(t)$, where

$$\begin{aligned} g_1(t) &= \mathscr{G}_1 e^{j\omega t} = (G_{1r} + jG_{1i})e^{j\omega t} \\ g_2(t) &= \mathscr{G}_2 e^{j\omega t} = (G_{2r} + jG_{2i})e^{j\omega t} \end{aligned} \tag{7-66}$$

As indicated, G_{1r} and G_{1i} are the real and imaginary parts of the complex quantity $\mathscr{G}_1$. Similarly, G_{2r} and G_{2i} are the real and imaginary parts of $\mathscr{G}_2$. It should be apparent that, if we specify $g_1(t) = g_2(t)$ for all t, then, by definition of an equality for complex functions, it must hold true that $\text{Re}\,[g_1(t)] = \text{Re}\,[g_2(t)]$. However, here we are more interested in the reverse situation. That is, for functions of the type given in Eq. (7-66), if the real parts are specified as being equal, what can we conclude about the functions themselves? To find the answer to this question, let us first evaluate the functions of Eq. (7-66) for t equals 0. We obtain

$$g_1(0) = G_{1r} + jG_{1i} \qquad g_2(0) = G_{2r} + jG_{2i} \tag{7-67}$$

If we require that $\text{Re}\,[g_1(t)] = \text{Re}\,[g_2(t)]$ for all t, then, for the particular value $t = 0$, we conclude from (7–67), that $G_{1r} = G_{2r}$. Now let us evaluate the relations of (7–66) for $t = \pi/2\omega$. For this value of t, $e^{j\omega t} = j$; thus, from (7–66) we see that

$$g_1\left(\frac{\pi}{2\omega}\right) = -G_{1i} + jG_{1r} \qquad g_2\left(\frac{\pi}{2\omega}\right) = -G_{2i} + jG_{2r} \tag{7-68}$$

From Eq. (7–68) we see that requiring $\text{Re}\,[g_1(t)] = \text{Re}\,[g_2(t)]$ for this value of t implies that $G_{1i} = G_{2i}$. Since we have shown that the real and the imaginary parts of G_1 and G_2 are separately equal, the complex quantities themselves are obviously equal, thus, $g_1(t)$ and $g_2(t)$ must also be equal.

We have now developed sufficient background to present a more rigorous justification of the method for the determination of the sinusoidal steady-state response presented in the previous section. Let us again consider the network shown in Fig. 7–2.1. The differential equation describing this circuit is

$$C\frac{dv}{dt} + Gv(t) = i(t) \tag{7-69}$$

If the excitation is a sinusoid of magnitude I_m, phase ϕ, and frequency ω, then the sinusoid may be represented by the phasor defined as $\mathscr{I}e^{j\omega t}$, where

$$\mathscr{I} = I_m e^{j\phi} \tag{7-70}$$

Thus, following the form of (7–60), we may write

$$i(t) = \text{Re}\,(\mathscr{I}e^{j\omega t}) \tag{7-71}$$

Similarly, since we know that the response for the voltage will be a sinusoid, let us assume that it will have an amplitude V_m and a phase θ. Then we may represent the voltage sinusoid by a phasor $\mathscr{V}e^{j\omega t}$, where

$$\mathscr{V} = V_m e^{j\theta} \tag{7-72}$$

Thus, the expression for $v(t)$ becomes

$$v(t) = \text{Re}\,(\mathscr{V}e^{j\omega t}) \tag{7-73}$$

Now let us substitute Eqs. (7–71) and (7–73) into the differential equation of (7–69). We obtain

$$C\frac{d}{dt}\text{Re}\,(\mathscr{V}e^{j\omega t}) + G\,\text{Re}\,(\mathscr{V}e^{j\omega t}) = \text{Re}\,(\mathscr{I}e^{j\omega t}) \tag{7-74}$$

From the result given above, we may interchange the operations of differentiation and real part extraction in the first term of the left member of Eq. (7–74). Doing this and performing the differentiation, we obtain

$$C\,\mathrm{Re}\,(j\omega\mathscr{V}e^{j\omega t}) + G\,\mathrm{Re}\,(\mathscr{V}e^{j\omega t}) = \mathrm{Re}\,(\mathscr{I}e^{j\omega t}) \tag{7–75}$$

In the left member of Eq. (7–75), the constants may be moved inside the Re operation. Also, since the sum of two real parts is equal to the real part of the sum of the two corresponding complex quantities, the preceding equation may be rewritten in the form

$$\mathrm{Re}\,[(j\omega C\mathscr{V} + G\mathscr{V})e^{j\omega t}] = \mathrm{Re}\,(\mathscr{I}e^{j\omega t}) \tag{7–76}$$

From the preceding discussion, we know that the equality of the real parts given above implies the equality of the entire functions. Thus, we may write

$$(j\omega C\mathscr{V} + G\mathscr{V})e^{j\omega t} = \mathscr{I}e^{j\omega t} \tag{7–77}$$

If we multiply both sides of the above by $e^{-j\omega t}$, we find that the solution for the phasor $\mathscr{V}$ is

$$\mathscr{V} = \frac{\mathscr{I}}{G + j\omega C} \tag{7–78}$$

which is the same result as was obtained in the last section. We may obviously use the value of $\mathscr{V}$ found in Eq. (7–78) to specify the value of $v(t)$ using Eq. (7–73).

The procedure given above is readily generalized. Assume that some excitation $g(t)$ and some response $f(t)$ are related by a differential equation of the form

$$a_1 f(t) + a_2\frac{df}{dt} + a_3\frac{d^2f}{dt^2} + \cdots + a_{n+1}\frac{d^nf}{dt^n} = b_1 g(t) + b_2\frac{dg}{dt} + b_3\frac{d^2g}{dt^2} + \cdots + b_{m+1}\frac{d^mg}{dt^m} \tag{7–79}$$

If $g(t)$ is sinusoidal, then we may use the phasor $\mathscr{G}e^{j\omega t}$ to represent it, and the phasor $\mathscr{F}e^{j\omega t}$ may then be used to represent the sinusoidal steady-state component of the response $f(t)$ (where $\mathscr{G}$ and $\mathscr{F}$ are complex constants). Substituting the relations

$$f(t) = \mathrm{Re}\,(\mathscr{F}e^{j\omega t}) \qquad g(t) = \mathrm{Re}\,(\mathscr{G}e^{j\omega t}) \tag{7–80}$$

into (7–79) and interchanging the operations of differentiation and real part extraction, we obtain

$$\begin{aligned}&\operatorname{Re}\{[a_1 + a_2(j\omega) + a_3(j\omega)^2 + \cdots + a_{n+1}(j\omega)^n]\mathscr{F}e^{j\omega t}\}\\ &= \operatorname{Re}\{[b_1 + b_2(j\omega) + b_3(j\omega)^2 + \cdots + b_{m+1}(j\omega)^m]\mathscr{G}e^{j\omega t}\}\end{aligned} \tag{7-81}$$

Equating the complex functions, multiplying both sides of the result by $e^{-j\omega t}$, and rearranging terms, we obtain

$$\mathscr{F} = \frac{b_1 + b_2(j\omega) + b_3(j\omega)^2 + \cdots + b_{m+1}(j\omega)^m}{a_1 + a_2(j\omega) + a_3(j\omega)^2 + \cdots + a_{n+1}(j\omega)^n}\mathscr{G} \tag{7-82}$$

Thus, the response phasor $\mathscr{F}$ is defined in terms of the excitation phasor $\mathscr{G}$ and the coefficients of the original differential equation. Comparing Eqs. (7-79) and (7-82), we obtain a most important conclusion, namely, that we may directly write the relation between an excitation phasor $\mathscr{G}$ and a response phasor $\mathscr{F}$ by simply making the substitutions

$$\begin{aligned}\frac{d^i f}{dt^i} &= (j\omega)^i\mathscr{F} \qquad i = 0, 1, 2, \ldots, n\\ \frac{d^i g}{dt^i} &= (j\omega)^i\mathscr{G} \qquad i = 0, 1, 2, \ldots, m\end{aligned} \tag{7-83}$$

Obviously, once the relation between the phasors is known, we may find the corresponding relation between the sinusoidal functions $g(t)$ and $f(t)$ which are represented by the phasors. The process is illustrated by the following example.

EXAMPLE 7-3.1. The excitation $i_1(t)$ and the node voltage $v_2(t)$ of the network shown in Fig. 7-3.2 may be shown to be related by the differential equation

$$\frac{5}{4}v_2(t) + \frac{9}{4}\frac{dv_2}{dt} + 2\frac{d^2v_2}{dt^2} + \frac{d^3v_2}{dt^3} = \frac{5}{8}i_1(t) + \frac{di_1}{dt} + \frac{d^2i_1}{dt^2}$$

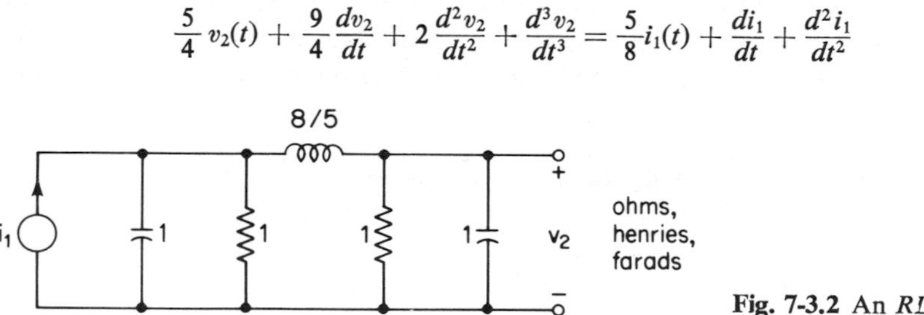

Fig. 7-3.2 An *RLC* network.

If $i_1(t)$ is a sinusoid, then we may use a phasor $\mathscr{I}_1e^{j\omega t}$ to represent it. Similarly, the sinusoidal steady-state component of $v_2(t)$ may be represented by the phasor $\mathscr{V}_2e^{j\omega t}$. From the development given above, we see that the relation between $\mathscr{V}_2$ and $\mathscr{I}_1$ may be directly written as

$$\frac{\mathscr{V}_2}{\mathscr{I}_1} = \frac{\frac{5}{8} + (j\omega) + (j\omega)^2}{\frac{5}{4} + \frac{9}{4}(j\omega) + 2(j\omega)^2 + (j\omega)^3}$$

This is more conveniently written by grouping the real and the imaginary terms together. Thus, we may write

$$\frac{\mathscr{V}_2}{\mathscr{I}_1} = \frac{\frac{5}{8} - \omega^2 + j\omega}{\frac{5}{4} - 2\omega^2 + j(\frac{9}{4}\omega - \omega^3)}$$

The above equation relates $\mathscr{V}_2$ and $\mathscr{I}_1$ at any value of ω. For example, if $i_1(t) = 10 \cos [t + (\pi/6)]$, then $\omega = 1$ rad/s, and $\mathscr{I}_1 = 10e^{j\pi/6}$. From the above equation, we see that

$$\mathscr{V}_2 = \frac{-3 + j8}{-6 + j10} 10e^{j\pi/6} = 7.36e^{j19.60^\circ}$$

Therefore, the sinusoidal steady-state value of $v_2(t)$ is

$$v_2(t) = 7.36 \cos (t + 19.60^\circ)$$

A procedure of the type illustrated in the preceding example is referred to as the *phasor method* of determining the sinusoidal steady-state response. We may summarize the material presented above as follows:

SUMMARY 7-3.1

Use of Phasors in Solving for the Sinusoidal Steady-State Response of a Circuit: If a given circuit is described by a differential equation having the form

$$a_1 f(t) + a_2 \frac{df}{dt} + a_3 \frac{d^2 f}{dt^2} + \cdots + a_{n+1} \frac{d^n f}{dt^n}$$
$$= b_1 g(t) + b_2 \frac{dg}{dt} + b_3 \frac{d^2 g}{dt^2} + \cdots + b_{m+1} \frac{d^m g}{dt^m}$$

which relates some response variable $f(t)$ to a sinusoidal excitation variable $g(t)$ with magnitude G_m and phase ϕ, we may find the sinusoidal steady-state component of $f(t)$ by substituting $\mathscr{G} = G_m e^{j\phi}$ in the following:

$$\mathscr{F} = \frac{b_1 + b_2(j\omega) + b_3(j\omega)^2 + \cdots + b_{m+1}(j\omega)^m}{a_1 + a_2(j\omega) + a_3(j\omega)^2 + \cdots + a_{n+1}(j\omega)^n} \mathscr{G}$$

The magnitude and phase of the phasor $\mathscr{F}$ are the amplitude and phase of the sinusoidal steady-state component of $f(t)$, respectively.

7-4 PHASOR SOLUTIONS OF SIMULTANEOUS SETS OF INTEGRO-DIFFERENTIAL EQUATIONS

The procedure described in the preceding section for using phasors to determine the sinusoidal steady-state solution for a response variable in a *single* differential equation is readily applied to a simultaneous set of differential equations and/or integro-differential equations. For example, consider the network shown in Fig. 7-4.1.

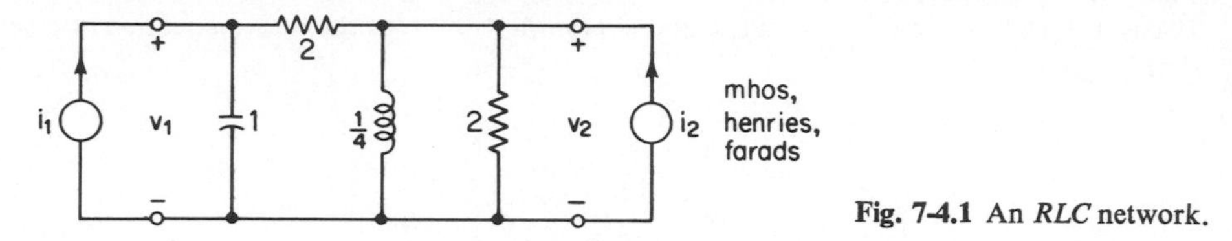

Fig. 7-4.1 An *RLC* network.

Applying KCL to the two nodes of this network, we obtain the equations

$$\begin{aligned} &\frac{dv_1}{dt} + 2v_1(t) - 2v_2(t) = i_1(t) \\ &4\int_{-\infty}^{t} v_2(\tau)\,d\tau + 4v_2(t) - 2v_1(t) = i_2(t) \end{aligned} \tag{7-84}$$

If we desire to find the sinusoidal steady-state response for $v_1(t)$ and $v_2(t)$, we may begin by differentiating the second of the equations given in Eq. (7-84) to convert it to a differential equation rather than an integro-differential equation. Thus, the set of equations becomes

$$\begin{aligned} &\frac{dv_1}{dt} + 2v_1(t) - 2v_2(t) = i_1(t) \\ &4v_2(t) + 4\frac{dv_2}{dt} - 2\frac{dv_1}{dt} = \frac{di_2}{dt} \end{aligned} \tag{7-85}$$

Now let us consider the case where the excitation to the network is purely sinusoidal, i.e., where the source currents $i_1(t)$ and $i_2(t)$ have the form

$$i_1(t) = I_1 \cos(\omega t + \phi_1) \qquad i_2(t) = I_2 \cos(\omega t + \phi_2) \tag{7-86}$$

It is assumed that the magnitudes I_1 and I_2 and the phase angles ϕ_1 and ϕ_2 are known. The two magnitudes may, of course, be different, as may the two phase angles. The frequency ω, however, is assumed to be the same for the two excitations. The particular solution for the node voltages, i.e., the steady-state solution for these network variables, will also have a sinusoidal form. Thus, we may write

$$v_1(t) = V_1 \cos(\omega t + \theta_1) \qquad v_2(t) = V_2 \cos(\omega t + \theta_2) \tag{7-87}$$

where the values of V_1, V_2, θ_1, and θ_2 are to be found. To determine these quantities, we need simply use the phasors $\mathscr{I}_1 e^{j\omega t}$ and $\mathscr{I}_2 e^{j\omega t}$ to represent the two excitation variables $i_1(t)$ and $i_2(t)$, and the phasors $\mathscr{V}_1 e^{j\omega t}$ and $\mathscr{V}_2 e^{j\omega t}$ to represent the two response variables $v_1(t)$ and $v_2(t)$. Thus, we may write

$$\begin{aligned} i_1(t) &= \text{Re}\,(\mathscr{I}_1 e^{j\omega t}) \qquad & i_2(t) &= \text{Re}\,(\mathscr{I}_2 e^{j\omega t}) \\ v_1(t) &= \text{Re}\,(\mathscr{V}_1 e^{j\omega t}) \qquad & v_2(t) &= \text{Re}\,(\mathscr{V}_2 e^{j\omega t}) \end{aligned} \tag{7-88}$$

where the complex constants $\mathscr{I}_1$, $\mathscr{I}_2$, $\mathscr{V}_1$, and $\mathscr{V}_2$ are defined as

$$\begin{aligned} \mathscr{I}_1 &= I_1 e^{j\phi_1} \qquad & \mathscr{I}_2 &= I_2 e^{j\phi_2} \\ \mathscr{V}_1 &= V_1 e^{j\theta_1} \qquad & \mathscr{V}_2 &= V_2 e^{j\theta_2} \end{aligned} \tag{7-89}$$

Substituting the relations of Eq. (7-88) into the differential equations given in (7-85) and interchanging the order of differentiation and real part extraction, we obtain

$$\text{Re}\left\{\begin{bmatrix} 2 + j\omega & -2 \\ -j2\omega & 4 + j4\omega \end{bmatrix}\begin{bmatrix} \mathscr{V}_1 e^{j\omega t} \\ \mathscr{V}_2 e^{j\omega t} \end{bmatrix}\right\} = \text{Re}\left\{\begin{bmatrix} \mathscr{I}_1 e^{j\omega t} \\ j\omega \mathscr{I}_2 e^{j\omega t} \end{bmatrix}\right\} \tag{7-90}$$

Equating the entire complex matrices of (7-90) and multiplying both sides by $e^{-j\omega t}$, we obtain

$$\begin{bmatrix} 2 + j\omega & -2 \\ -j2\omega & 4 + j4\omega \end{bmatrix}\begin{bmatrix} \mathscr{V}_1 \\ \mathscr{V}_2 \end{bmatrix} = \begin{bmatrix} \mathscr{I}_1 \\ j\omega \mathscr{I}_2 \end{bmatrix} \tag{7-91}$$

Note that in the right member of (7-91), the term $\mathscr{I}_2$ is multiplied by $j\omega$, since this term represents the derivative of $i_2(t)$ with respect to time. If we divide the entire second equation of (7-91) by $j\omega$, the set of equations becomes

$$\begin{bmatrix} 2 + j\omega & -2 \\ -2 & 4 + \dfrac{4}{j\omega} \end{bmatrix}\begin{bmatrix} \mathscr{V}_1 \\ \mathscr{V}_2 \end{bmatrix} = \begin{bmatrix} \mathscr{I}_1 \\ \mathscr{I}_2 \end{bmatrix} \tag{7-92}$$

If we now compare the left members of (7-92) with the original equations of (7-84), we see that the integral operation in the original integro-differential equations produces a factor $1/j\omega$ in the equations relating the phasors, and the derivative operation produces a factor of $j\omega$. This is easily shown to be always true. Thus, in general, phasor methods may be directly applied to integro-differential equations, without the necessity of differentiating these equations so as to convert them to the form of purely differential equations.

The relations given in (7-92) are a simultaneous set of equations in which all the elements are complex and are functions of $j\omega$. Solving such a set of equations follows exactly the procedures which were given in Chap. 3, since the methods developed there apply irrespective of whether the matrices have elements which are real or complex, or whether the elements are constant or are functions of some variable. Thus, we see that the determinant of the square matrix in Eq. (7-92) is

$$\frac{4(j\omega)^2 + 8(j\omega) + 8}{j\omega} \tag{7-93}$$

Using the relations given in Sec. 3-1 for the solution of a set of two simultaneous equations, the relations of (7-92) become

$$\begin{bmatrix} \mathscr{V}_1 \\ \mathscr{V}_2 \end{bmatrix} = \frac{j\omega}{4(j\omega)^2 + 8(j\omega) + 8} \begin{bmatrix} 4 + \dfrac{4}{j\omega} & 2 \\ 2 & 2 + j\omega \end{bmatrix} \begin{bmatrix} \mathscr{I}_1 \\ \mathscr{I}_2 \end{bmatrix} \tag{7-94}$$

This set of equations provides the solution for the network. That is, given any specified values for $\mathscr{I}_1$ and $\mathscr{I}_2$, we may determine the corresponding values of $\mathscr{V}_1$ and $\mathscr{V}_2$, and, using Eqs. (7-87) and (7-89), we obtain the expressions for the sinusoidal steady-state components of $v_1(t)$ and $v_2(t)$. The procedure is illustrated by the following example.

EXAMPLE 7-4.1. The network shown in Fig. 7-4.1 is excited as follows.

$$i_1(t) = \cos t \qquad i_2(t) = 2\cos\left(t + \frac{\pi}{2}\right)$$

It is desired to find the sinusoidal steady-state response for $v_1(t)$ and $v_2(t)$. To do this, we first note from (7-86) and (7-89) that $\mathscr{I}_1 = 1e^{j0}$, $\mathscr{I}_2 = 2e^{j\pi/2}$, and $\omega = 1$. Substituting these values in (7-94), we obtain

$$\begin{bmatrix} \mathscr{V}_1 \\ \mathscr{V}_2 \end{bmatrix} = \frac{j1}{4 + j8} \begin{bmatrix} 4 - j4 & 2 \\ 2 & 2 + j1 \end{bmatrix} \begin{bmatrix} 1e^{j0} \\ 2e^{j\pi/2} \end{bmatrix}$$

Performing the indicated multiplication, we find that

$$\mathscr{V}_1 = \frac{j1}{1 + j2} = \frac{1}{\sqrt{5}} e^{j26.57^\circ} \qquad \mathscr{V}_2 = \frac{-1}{1 + j2} = \frac{1}{\sqrt{5}} e^{j116.57^\circ}$$

The corresponding solutions for the quantities $v_1(t)$ and $v_2(t)$ are

$$v_1(t) = \frac{1}{\sqrt{5}} \cos(t + 26.57^\circ) \qquad v_2(t) = \frac{1}{\sqrt{5}} \cos(t + 116.57^\circ)$$

These are the sinusoidal steady-state values of the nodal voltages, i.e., the particular solutions for these network variables.

The procedure outlined above may be generalized by considering a simultaneous set of n integro-differential equations relating a set of response variables $f_i(t)$ $(i = 1, 2, \ldots, n)$ to a set of excitation variables $g_i(t)$ $(i = 1, 2, \ldots, n)$. Let us assume that the equations may be written in the form (for convenience of representation here we will assume that $n = 3$)

$$\begin{bmatrix} D_{11} & D_{12} & D_{13} \\ D_{21} & D_{22} & D_{23} \\ D_{31} & D_{32} & D_{33} \end{bmatrix} \begin{bmatrix} f_1(t) \\ f_2(t) \\ f_3(t) \end{bmatrix} = \begin{bmatrix} g_1(t) \\ g_2(t) \\ g_3(t) \end{bmatrix} \tag{7-95}$$

where the quantities D_{ij} are symbolic representations for integro-differential operators having the form[7]

$$D_{ij} = a_{ij}^{(1)} \int dt + a_{ij}^{(2)} + a_{ij}^{(3)} \frac{d}{dt} \tag{7-96}$$

The relations of (7-95) may be written using matrix notation by defining

$$\mathbf{D} = \begin{bmatrix} D_{11} & D_{12} & D_{13} \\ D_{21} & D_{22} & D_{23} \\ D_{31} & D_{32} & D_{33} \end{bmatrix} \qquad \mathbf{f}(t) = \begin{bmatrix} f_1(t) \\ f_2(t) \\ f_3(t) \end{bmatrix} \qquad \mathbf{g}(t) = \begin{bmatrix} g_1(t) \\ g_2(t) \\ g_3(t) \end{bmatrix} \tag{7-97}$$

Thus, (7-95) may be written in the form

$$\mathbf{Df}(t) - \mathbf{g}(t) \tag{7-98}$$

Let us now assume that the functions $g_i(t)$ are sinusoids, having the form

$$g_i(t) = G_i \cos(\omega t + \phi_i) \tag{7-99}$$

We may represent these functions by phasors of the form $\mathcal{G}_i e^{j\omega t}$, where

$$\mathcal{G}_i = G_i e^{j\phi_i} \tag{7-100}$$

Thus, we may write

$$g_i(t) = \mathrm{Re}\,(\mathcal{G}_i e^{j\omega t}) \tag{7-101}$$

Similarly, for any response variables $f_i(t)$, the particular solutions will have the form

[7]The student may wonder why the integro-differential operators defined in (7-96) do not include higher-order derivative or integral terms. Although there is no mathematical reason for not including such cases, we shall show in the next section that the three operations indicated, namely, simple integration, multiplication by a constant, and first-order differentiation, are sufficient to completely characterize any network situation.

$$f_i(t) = F_i \cos(\omega t + \theta_i) \tag{7-102}$$

Thus, if we define the phasors $\mathscr{F}_i e^{j\omega t}$, where

$$\mathscr{F}_i = F_i e^{j\theta_i} \tag{7-103}$$

the response variables may be written in the form

$$f_i(t) = \text{Re}\,(\mathscr{F}_i e^{j\omega t}) \tag{7-104}$$

For convenience, we may now define the matrices **F** and **G** by the relations

$$\mathbf{F} = \begin{bmatrix} \mathscr{F}_1 \\ \mathscr{F}_2 \\ \mathscr{F}_3 \end{bmatrix} \qquad \mathbf{G} = \begin{bmatrix} \mathscr{G}_1 \\ \mathscr{G}_2 \\ \mathscr{G}_3 \end{bmatrix} \tag{7-105}$$

Thus, the relations of (7–95) and (7–98) may be written in matrix form as

$$\mathbf{D}\,\text{Re}\,(\mathbf{F}e^{j\omega t}) = \text{Re}\,(\mathbf{G}e^{j\omega t}) \tag{7-106}$$

As before, we may interchange the operations of differentiation (or integration) and taking the real part, thus we may write the set of equations in the form

$$\text{Re}\,(\mathbf{EF}e^{j\omega t}) = \text{Re}\,(\mathbf{G}e^{j\omega t}) \tag{7-107}$$

where **E** is defined as the square matrix with elements e_{ij}, which, from (7–96), have the form

$$e_{ij} = \frac{a_{ij}^{(1)}}{j\omega} + a_{ij}^{(2)} + j\omega a_{ij}^{(3)} \tag{7-108}$$

If we now equate the complex values of both sides of (7–107) and multiply by $e^{-j\omega t}$, we obtain a result which may be written in the form

$$\mathbf{EF} = \mathbf{G} \tag{7-109}$$

Thus, the excitation phasors $\mathscr{G}_i$ are related to the response phasors $\mathscr{F}_i$ by the matrix **E** whose elements are functions of $j\omega$. As a result, once the quantities $\mathscr{G}_i$ have been specified, then the quantities $\mathscr{F}_i$ are easily found by solving the set of simultaneous equations using any of the procedures given in Chap. 3. An example follows.

EXAMPLE 7–4.2. The network shown in Fig. 7-4.2 is excited by three voltage sources having the values

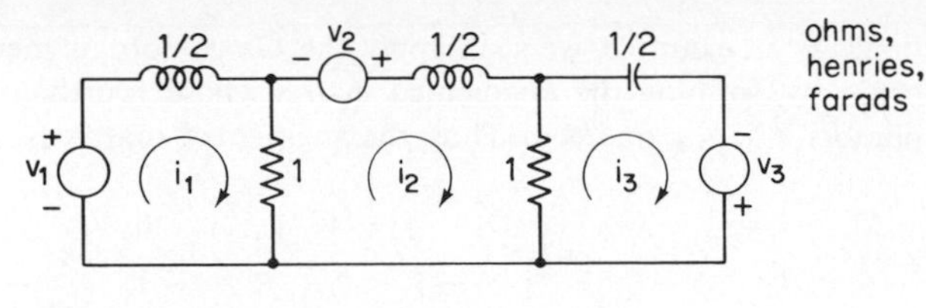

Fig. 7-4.2 Network for Example 7-4.2.

$$v_1(t) = 4 \cos 2t$$

$$v_2(t) = -6 \cos 2t$$

$$v_3(t) = 2\sqrt{2} \cos\left(2t - \frac{\pi}{4}\right)$$

It is desired to find the sinusoidal steady-state components of the mesh currents $i_1(t)$, $i_2(t)$, and $i_3(t)$. To do this, we may first apply KVL to the three meshes to find the integro-differential equations describing the network. These are

$$\frac{1}{2}\frac{di_1}{dt} + i_1(t) - i_2(t) = v_1(t)$$

$$\frac{1}{2}\frac{di_2}{dt} + 2i_2(t) - i_1(t) - i_3(t) = v_2(t)$$

$$2\int i_3(t)\,dt + i_3(t) - i_2(t) = v_3(t)$$

The voltage phasors are readily found to be[8]

$$\mathscr{V}_1 = 4/\underline{0^\circ} = 4 + j0$$

$$\mathscr{V}_2 = 6/\underline{180^\circ} = -6 + j0$$

$$\mathscr{V}_3 = 2\sqrt{2}/\underline{-45^\circ} = 2 - j2$$

Applying the method described in this section, we may directly write the equations relating the voltage phasors $\mathscr{V}_1$, $\mathscr{V}_2$, and $\mathscr{V}_3$ to the current phasors $\mathscr{I}_1$, $\mathscr{I}_2$, and $\mathscr{I}_3$ by an equation of the general form of Eq. (7-109). Thus, we obtain (for $\omega = 2$)

$$\begin{bmatrix} 1 + j1 & -1 & 0 \\ -1 & 2 + j1 & -1 \\ 0 & -1 & 1 - j1 \end{bmatrix} \begin{bmatrix} \mathscr{I}_1 \\ \mathscr{I}_2 \\ \mathscr{I}_3 \end{bmatrix} = \begin{bmatrix} 4 + j0 \\ -6 + j0 \\ 2 - j2 \end{bmatrix}$$

This is a set of simultaneous equations with complex coefficients which may be solved for $\mathscr{I}_1$, $\mathscr{I}_2$, and $\mathscr{I}_3$. Any of the methods given in Chap. 3 may be used to determine the solu-

[8]Here we use an alternative representation for the phase of a complex quantity which is frequently more convenient than the exponential format and which is often found in the literature. It is defined by the relation

$$Ae^{j\alpha} = A/\underline{\alpha}$$

tion. As an example, we shall apply the Gauss-Jordan method described in Sec. 3-4. We begin by forming the augmented matrix whose fourth column contains the values of the phasors $\mathscr{V}_1$, $\mathscr{V}_2$, and $\mathscr{V}_3$. Thus, the augmented matrix is

$$\begin{bmatrix} 1+j1 & -1 & 0 & 4 \\ -1 & 2+j1 & -1 & -6 \\ 0 & -1 & 1-j1 & 2-j2 \end{bmatrix}$$

The following steps use row operations to convert the three columns on the left of the augmented matrix given above to an identity matrix. First we multiply the first row by $1/(1+j1)$. We obtain

$$\begin{bmatrix} 1 & -\frac{1}{2}+j\frac{1}{2} & 0 & 2-j2 \\ -1 & 2+j1 & -1 & -6 \\ 0 & -1 & 1-j1 & 2-j2 \end{bmatrix}$$

The next step is to replace the second row by the sum of the first and second rows. Doing this, we obtain

$$\begin{bmatrix} 1 & -\frac{1}{2}+j\frac{1}{2} & 0 & 2-j2 \\ 0 & \frac{3}{2}+j\frac{3}{2} & -1 & -4-j2 \\ 0 & -1 & 1-j1 & 2-j2 \end{bmatrix}$$

Now let us multiply the second row by $1/(\frac{3}{2}+j\frac{3}{2})$. We obtain

$$\begin{bmatrix} 1 & -\frac{1}{2}+j\frac{1}{2} & 0 & 2-j2 \\ 0 & 1 & -\frac{1}{3}+j\frac{1}{3} & -2+j\frac{2}{3} \\ 0 & -1 & 1-j1 & 2-j2 \end{bmatrix}$$

If we now replace the third row by its sum with the second row, and replace the first row by its sum with the second row multiplied by $\frac{1}{2}-j\frac{1}{2}$, we obtain

$$\begin{bmatrix} 1 & 0 & j\frac{1}{3} & \frac{4}{3}-j\frac{2}{3} \\ 0 & 1 & -\frac{1}{3}+j\frac{1}{3} & -2+j\frac{2}{3} \\ 0 & 0 & \frac{2}{3}-j\frac{2}{3} & -j\frac{4}{3} \end{bmatrix}$$

Now let us multiply the third row by $1/(\frac{2}{3}-j\frac{2}{3})$. We obtain

$$\begin{bmatrix} 1 & 0 & j\frac{1}{3} & \frac{4}{3} - j\frac{2}{3} \\ 0 & 1 & -\frac{1}{3} + j\frac{1}{3} & -2 + j\frac{2}{3} \\ 0 & 0 & 1 & 1 - j1 \end{bmatrix}$$

The final steps are: (1) to multiply the third row by $-j\frac{1}{3}$ and add it to the first row using the resultant sum to replace the first row, and (2) to multiply the third row by $\frac{1}{3} - j\frac{1}{3}$ and add it to the second row using the sum to replace the second row. Performing these operations, we obtain

$$\begin{bmatrix} 1 & 0 & 0 & 1 - j1 \\ 0 & 1 & 0 & -2 \\ 0 & 0 & 1 & 1 - j1 \end{bmatrix}$$

Since the first three columns of the augmented matrix have been converted to an identity matrix, the last column of the above array gives the solution to the set of simultaneous equations. Thus, we may write the following values for the mesh current phasors.

$$\mathscr{I}_1 = 1 - j1 = \sqrt{2}\underline{/-45^\circ}$$
$$\mathscr{I}_2 = -2 = 2\underline{/180^\circ}$$
$$\mathscr{I}_3 = 1 - j1 = \sqrt{2}\underline{/-45^\circ}$$

The corresponding values for the sinusoidal steady-state components of the mesh currents are

$$i_1(t) = \sqrt{2}\cos(2t - 45^\circ)$$
$$i_2(t) = -2\cos 2t$$
$$i_3(t) = \sqrt{2}\cos(2t - 45^\circ)$$

Thus, we have found the solution for the network. The validity of this solution is readily demonstrated by substituting the steady-state components of the mesh currents into the original integro-differential equations and verifying that they do indeed satisfy these equations. This is left as an exercise for the reader.

The procedure illustrated above may be summarized as follows:

SUMMARY 7-4.1

Sinusoidal Steady-State Response for a Circuit Described by a Set of Integro-Differential Equations: If a set of excitation variables $g_i(t) = G_i \cos(\omega t + \phi_i)$ $(i = 1, 2, \ldots, n)$ are related to a set of response variables $f_i(t)$ $(i = 1, 2, \ldots, n)$ by a set of n simultaneous integro-differential equations, then the particular solutions for the response variables will have the form $f_i(t) = F_i \cos(\omega t + \theta_i)$. If we define the phasors $\mathscr{G}_i = G_i e^{j\phi_i}$ and $\mathscr{F}_i = F_i e^{j\theta_i}$, then the relations between the excitation phasors $\mathscr{G}_i$ and the response phasors $\mathscr{F}_i$ may be found by solving the complex set of simultaneous equations $\mathbf{EF} = \mathbf{G}$, where $\mathbf{F}$ is the column matrix with elements $\mathscr{F}_i$, $\mathbf{G}$ is the column matrix with elements $\mathscr{G}_i$, and $\mathbf{E}$ is the matrix formed directly from the original integro-differential equations by replacing the operation of differentiation with a multiplicative factor $j\omega$ and the operation of integration by the multiplicative factor $1/j\omega$. Once the quantities $\mathscr{F}_i$ have been determined, the corresponding expressions for the quantities $f_i(t)$ are referred to as the sinusoidal steady-state responses for these variables.

It is readily apparent that the procedure outlined above requires considerable computational effort, even for the case where only three simultaneous equations are to be solved. As a result of this conclusion, we will find that it is advantageous to employ digital computational methods for circuits with considerably fewer numbers of independent variables than would be the case for a resistance network. In a subsequent section we shall treat some methods for doing this. In the following section we shall present some important additional procedures for determining the elements of the matrix $\mathbf{E}$ which relates the excitation phasors to the response phasors.

7-5 IMPEDANCE AND ADMITTANCE

All the techniques of network analysis which we have developed to this point in this text are based on the use of KVL and KCL to determine a set of network equations. In this section we shall begin an investigation of the way in which these techniques may be extended to include the sinusoidal steady-state case. For example, consider the simple one-mesh circuit shown in Fig. 7-5.1. Applying KVL, we may write

$$v_L(t) + v_C(t) + v_R(t) = v_S(t) \tag{7-110}$$

Now let us consider what happens to (7-110) when only the sinusoidal steady-state response is desired. We shall use the phasors $\mathscr{V}_L$, $\mathscr{V}_C$, $\mathscr{V}_R$, and $\mathscr{V}_S$ to represent the sinusoidally varying variables in the circuit. Thus, following the procedure given in the preceding section, we may write

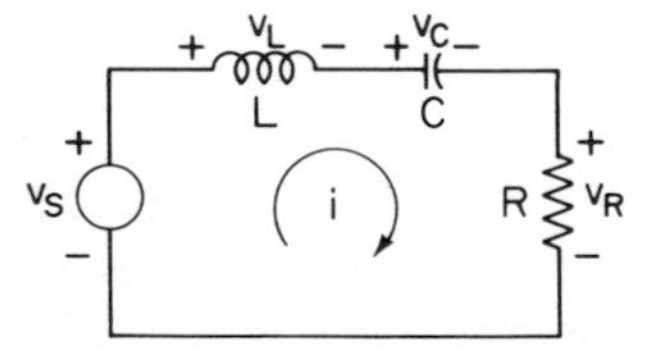

Fig. 7-5.1 A simple one-mesh *RLC* circuit.

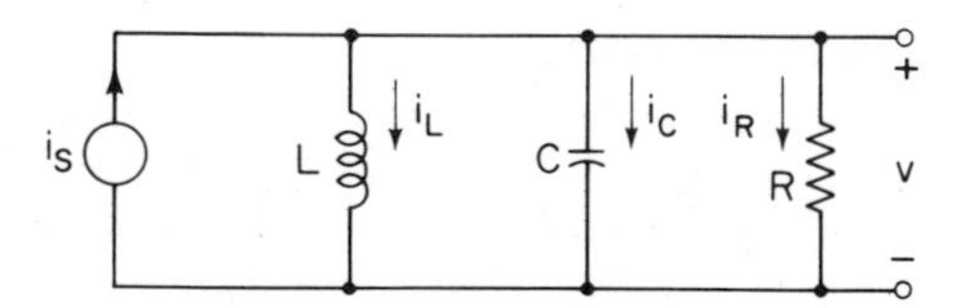

Fig. 7-5.2 A simple one-node *RLC* circuit.

$$v_L(t) = \text{Re}\,(\mathscr{V}_L e^{j\omega t}) \qquad v_C(t) = \text{Re}\,(\mathscr{V}_C e^{j\omega t})$$
$$v_R(t) = \text{Re}\,(\mathscr{V}_R e^{j\omega t}) \qquad v_S(t) = \text{Re}\,(\mathscr{V}_S e^{j\omega t}) \tag{7-111}$$

If we substitute the relations of (7–111) into (7–110), we obtain

$$\text{Re}\,(\mathscr{V}_L e^{j\omega t}) + \text{Re}\,(\mathscr{V}_C e^{j\omega t}) + \text{Re}\,(\mathscr{V}_R e^{j\omega t}) = \text{Re}\,(\mathscr{V}_S e^{j\omega t}) \tag{7-112}$$

Since the sum of a group of real parts of complex quantities is equal to the real part of the sum of the complex quantities, we may write Eq. (7–112) in the form

$$\text{Re}\,[(\mathscr{V}_L + \mathscr{V}_C + \mathscr{V}_R)e^{j\omega t}] = \text{Re}\,(\mathscr{V}_S e^{j\omega t}) \tag{7-113}$$

As was shown in the last section, this implies the equality of the complex members of both sides of the equation. Thus we may write

$$\mathscr{V}_L + \mathscr{V}_C + \mathscr{V}_R = \mathscr{V}_S \tag{7-114}$$

Comparing Eqs. (7–114) and (7–110), we see that the phasors $\mathscr{V}_L$, $\mathscr{V}_C$, $\mathscr{V}_R$, and $\mathscr{V}_S$ satisfy the same KVL relation that the sinusoidally varying quantities $v_L(t)$, $v_C(t)$, $v_R(t)$, and $v_S(t)$ satisfy. This conclusion is readily generalized. Thus we may say that when a set of voltage variables are related by KVL, the phasors which represent these voltage variables also satisfy an identical relation. The same conclusion holds true when KCL is applied. For example, consider the circuit shown in Fig. 7-5.2. Applying KCL, we may write

$$i_L(t) + i_C(t) + i_R(t) = i_S(t) \tag{7-115}$$

If we now assume that the current variables are all sinusoidal functions, then, under steady-state conditions, they may be represented by a set of phasors $\mathscr{I}_L$, $\mathscr{I}_C$, $\mathscr{I}_R$, and $\mathscr{I}_S$. The relations between the phasors and the original currents are

$$i_L(t) = \text{Re}\,(\mathscr{I}_L e^{j\omega t}) \qquad i_C(t) = \text{Re}\,(\mathscr{I}_C e^{j\omega t})$$
$$i_R(t) = \text{Re}\,(\mathscr{I}_R e^{j\omega t}) \qquad i_S(t) = \text{Re}\,(\mathscr{I}_S e^{j\omega t}) \tag{7-116}$$

Substituting the relations of Eq (7–116) into (7–115) and making manipulations identical with those which were made for the KVL example, we obtain

$$\mathscr{I}_L + \mathscr{I}_C + \mathscr{I}_R = \mathscr{I}_S \tag{7-117}$$

Thus, we see that phasors will also satisfy relations identical with the KCL equations that relate the variables that the phasors represent. These conclusions are easily generalized, and may be summarized as follows:

SUMMARY 7-5.1

Phasors and Kirchhoff's Laws: If a set of sinusoidally varying quantities are related by a group of KVL and KCL equations, the set of phasors which represent these sinusoidally varying quantities are related by an identical group of equations. Thus, we may say that Kirchhoff's laws may be directly applied to phasors.[9]

Now let us consider again the circuit shown in Fig. 7-5.1. The equations relating the terminal variables of each of the passive elements shown in this circuit are

$$v_L(t) = L\frac{di}{dt} \qquad v_C(t) = \frac{1}{C}\int i(t)\,dt \qquad v_R(t) = Ri(t) \tag{7-118}$$

Now let us define a current phasor $\mathscr{I}$ to represent the variable $i(t)$ of (7-118). The defining relation is

$$i(t) = \text{Re}\,(\mathscr{I}e^{j\omega t}) \tag{7-119}$$

If we substitute the relations of (7-111) and (7-119) in (7-118), we obtain the expressions

$$\begin{aligned} \text{Re}\,(\mathscr{V}_L e^{j\omega t}) &= L\frac{d}{dt}\text{Re}\,(\mathscr{I}e^{j\omega t}) \\ \text{Re}\,(\mathscr{V}_C e^{j\omega t}) &= \frac{1}{C}\int \text{Re}\,(\mathscr{I}e^{j\omega t})\,dt \\ \text{Re}\,(\mathscr{V}_R e^{j\omega t}) &= R\,\text{Re}\,(\mathscr{I}e^{j\omega t}) \end{aligned} \tag{7-120}$$

Since the operations of differentiation and taking the real part, and the operations of integration and taking the real part may be interchanged, if we perform the indicated differentiation and integration, we obtain

$$\begin{aligned} \text{Re}\,(\mathscr{V}_L e^{j\omega t}) &= \text{Re}\,(j\omega L\mathscr{I}e^{j\omega t}) \\ \text{Re}\,(\mathscr{V}_C e^{j\omega t}) &= \text{Re}\left(\frac{1}{j\omega C}\mathscr{I}e^{j\omega t}\right) \\ \text{Re}\,(\mathscr{V}_R e^{j\omega t}) &= \text{Re}\,(R\mathscr{I}e^{j\omega t}) \end{aligned} \tag{7-121}$$

Equating the complex members of the above and multiplying both sides of each relation by $e^{-j\omega t}$, we obtain

$$\mathscr{V}_L = j\omega L\mathscr{I} \qquad \mathscr{V}_C = \frac{1}{j\omega C}\mathscr{I} \qquad \mathscr{V}_R = R\mathscr{I} \tag{7-122}$$

[9]Note that due account must be taken of the reference polarity of phasors when applying KVL and KCL. These reference polarities are identical with the reference polarities of the variables that the phasors represent.

The expressions given in (7–122) are the phasor equivalents of the terminal relations given in (7–118). Comparing the two sets of equations, we see that the operation of differentiation has been replaced by multiplication by the factor $j\omega$, and the operation of integration has been replaced by multiplication by the factor $1/j\omega$. Thus, the *integral and differential relations* of Eq. (7–118) are replaced by the *algebraic relations* of Eq. (7–122) when phasors are used to represent the sinusoidal network variables. Let us now express the relations of (7–122) in the form

$$\frac{\mathscr{V}_L}{\mathscr{I}} = j\omega L \qquad \frac{\mathscr{V}_C}{\mathscr{I}} = \frac{1}{j\omega C} \qquad \frac{\mathscr{V}_R}{\mathscr{I}} = R \tag{7–123}$$

The left members of the expressions given in Eq. (7–123) each consist of the ratio of a voltage phasor to a current phasor. Thus, the right members of these expressions, namely, the expressions $j\omega L$, $1/j\omega C$, and R must have the same dimensional units. The name given to such quantities is *impedance*, and the units are *ohms*. When an impedance is purely real, it is referred to as a *resistive* impedance. When it is purely imaginary, it is referred to as a *reactive* impedance. An impedance which is neither purely real nor purely imaginary is simply referred to as a *complex* impedance. In general, we may use the letter Z to refer to an impedance and the letters R and X to refer to its *resistive part* and its *reactive part*, respectively. Thus, in general, we may write

$$Z = R + jX \tag{7–124}$$

For example, considering the first relation given in (7–123), we see that an inductor has an impedance which is purely imaginary and has the value $j\omega L$. Thus, the reactance of an inductor is ωL. Obviously, this reactance will always be positive (for a positive-valued L). From the second relation given in (7–123), we see that a capacitor has an impedance which is purely imaginary and which has the value $-j/\omega C$. Thus, the reactance of a capacitor equals $-1/\omega C$, and it is always negative. Finally, a resistor has an impedance which is purely real and has a value of R.

Impedance, since it is a complex quantity, specifies not only the relation between the magnitudes of the voltage and current phasors, but also the relation between their phase angles. For example, from the first relation of Eq. (7–122), we see that the phase angle of $\mathscr{I}$ plus the phase angle of the impedance $j\omega L$ (which is 90°) must equal the phase angle of $\mathscr{V}_L$. We conclude that the phase angle of $\mathscr{V}_L$ will always be 90° greater than the phase angle of $\mathscr{I}$. Thus, we may say that *for an inductor, under conditions of sinusoidal steady-state excitation, the voltage sinusoid leads the current sinusoid by 90°*. Alternatively, we may say that the current sinusoid lags the voltage sinusoid by 90°. It is frequently useful to express such relations by drawing a *phasor diagram* on the $A - B$ plane introduced in Sec. 7–1. Thus, for the variables $\mathscr{V}_L$ and $\mathscr{I}$ of the first relation of Eq. (7–122), the phasor diagram would appear as shown in Fig. 7–5.3(a). The relative position of the two vectors, i.e., the 90° angle, will be the same for all values of frequency ω and all (positive) values of L, although

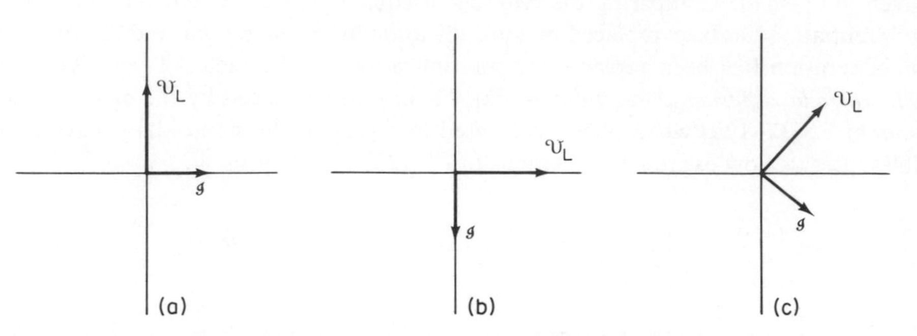

Fig. 7-5.3 Phasor diagrams for an inductor.

the relative lengths of the $\mathscr{V}_L$ and $\mathscr{I}$ vectors will be different. A phasor diagram of the type drawn in Fig. 7–5.3(a) only shows the *relative phase between phasors*. The actual phase angle is of little importance. Thus the phasor diagrams of Figs. 7–5.3(b) and 7–5.3(c) have exactly the same significance as the one shown in Fig. 7–5.3(a), namely, they show that the voltage across an inductor leads the current through it by 90° under conditions of sinusoidal steady-state excitation.

A similar situation, as defined by the second relation given in (7–122), holds for a capacitor. Since the impedance of a capacitor is $-j/\omega C$, its phase angle will always be $-90°$. Thus, *for a capacitor under conditions of sinusoidal steady-state excitation, the voltage sinusoid lags the current sinusoid by 90°* (or the current leads the voltage by 90°). Some phasor diagrams for the voltage and current phasors of a capacitor are given in Fig. 7–5.4. Finally, for a resistor, since the impedance of a

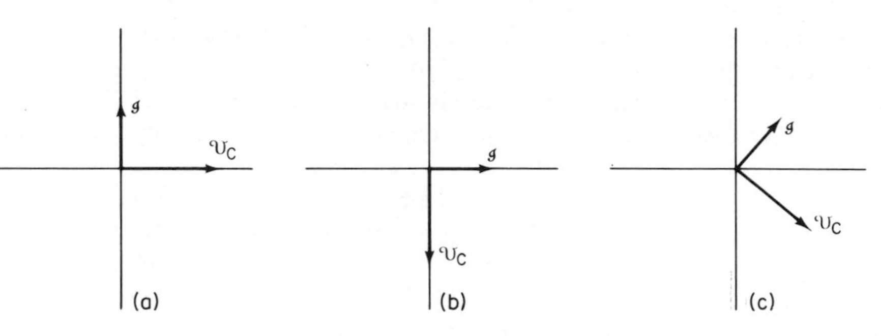

Fig. 7-5.4 Phasor diagrams for a capacitor.

resistor has a phase angle of zero degrees, we may say that *the voltage and current sinusoids for a resistor are in phase*, i.e., there is no phase lead or lag, under conditions of sinusoidal steady-state excitation. Some phasor diagrams for the voltage and current variables of a resistor are shown in Fig. 7–5.5. We may summarize the results given above as follows:

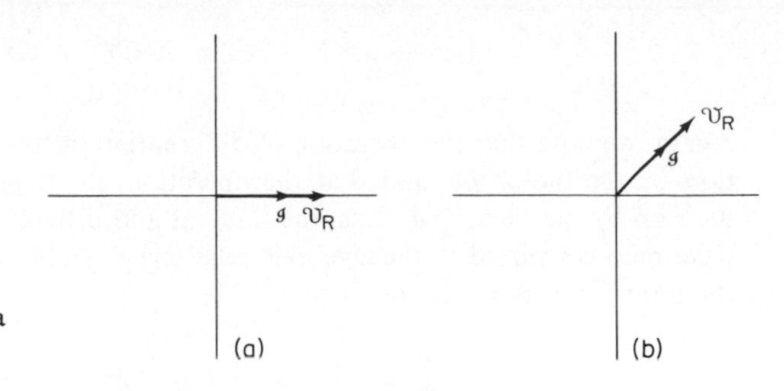

Fig. 7-5.5 Phasor diagrams for a resistor.

SUMMARY 7-5.2

Definition of Impedance: If voltage and current phasors are defined for the sinusoidal voltage and current variables of a two-terminal network element, then the expression relating the voltage phasor to the current phasor is defined as the impedance of the element. The real part of the impedance is called the resistance of the element, and the imaginary part is called the reactance. An inductor has a positive reactance which is proportional to frequency, and zero resistance. A capacitor has a negative reactance which is inversely proportional to frequency, and zero resistance. Under sinusoidal steady-state conditions, the voltage variable of an inductor leads the current variable by 90°, while the current variable of a capacitor leads the voltage variable by 90°. For a resistor, the voltage and current variables are in phase.[10]

We may develop a set of conclusions similar to those given above by starting with the circuit shown in Fig. 7-5.2. The equations relating the terminal variables are

$$i_L(t) = \frac{1}{L}\int v(t)\,dt \qquad i_C(t) = C\frac{dv}{dt} \qquad i_R(t) = Gv(t) \tag{7-125}$$

Now let us define a voltage phasor $\mathscr{V}$ to represent the variable $v(t)$ under conditions of sinusoidal steady-state. The defining relation is

$$v(t) = \mathrm{Re}\,(\mathscr{V} e^{j\omega t}) \tag{7-126}$$

If we substitute the relations of (7–116) and (7–126) in (7–125) and reduce the equations in a manner identical with that used in (7–120) through (7–122), we obtain the relations

[10]The conclusions given here are based on the requirement that the relative reference polarities for the voltage and current phasors are the associated ones normally used for a two-terminal element, i.e., the reference arrow for the current phasor enters the terminal marked by the positive reference sign of the voltage phasor.

$$\mathscr{I}_L = \frac{1}{j\omega L}\mathscr{V} \qquad \mathscr{I}_C = j\omega C\mathscr{V} \qquad \mathscr{I}_R = G\mathscr{V} \tag{7-127}$$

Again, we note that the operation of differentiation has been replaced by multiplication by the factor $j\omega$, and that the operation of integration has been replaced by division by the factor $j\omega$, thus, the integral and differential relations of Eq. (7–125) have been converted to the algebraic relations given in Eq. (7–127). Now let us write the latter expressions in the form

$$\frac{\mathscr{I}_L}{\mathscr{V}} = \frac{1}{j\omega L} \qquad \frac{\mathscr{I}_C}{\mathscr{V}} = j\omega C \qquad \frac{\mathscr{I}_R}{\mathscr{V}} = G \tag{7-128}$$

In these expressions, the left members all consist of the ratio of a current phasor to a voltage phasor. Thus, the expressions on the right, namely, $1/j\omega L$, $j\omega C$, and G, must have the same dimensional units. The name given to these quantities is *admittance*, and they have the units of mhos. When an admittance is purely real, it is referred to as a *conductive admittance.* When it is purely imaginary, it is referred to as a *susceptive admittance.* In general, we use the latter Y to refer to an admittance. The letters G and B are frequently used to refer to the *conductive part* and the *susceptive part* of an admittance, respectively. Thus, in general, an admittance may be written as follows:

$$Y = G + jB \tag{7-129}$$

where G and B are real and Y is complex. For example, considering the first relation given in Eq. (7–128), we see that an inductor has an admittance which is purely imaginary and equals $-j/\omega L$. The susceptance of an inductor is $-1/\omega L$ and it is always negative. Similarly, a capacitor has an admittance of $j\omega C$. Its susceptance is ωC, and it is always positive. Finally, a resistor has an admittance which is purely real and which is equal to the value G. Comparing these results with those determined for the impedance of the inductor, the resistor, and the capacitor, we see that the admittance of these elements is the reciprocal of their impedance, and vice versa. The conclusions reached above with respect to the leading and lagging of the sinusoidal voltage and current variables across inductors and capacitors are, of course, exactly the same whether they are derived from the relations of Eq. (7–127) or those of (7–122).

The conclusions given above may be summarized as follows:

SUMMARY 7-5.3

Definition of Admittance: If voltage and current phasors are defined for the sinusoidal voltage and current variables of a two-terminal network element, then the expression relating the current phasor to the voltage phasor is defined as the admittance of the element. The real part of the admittance is referred to as the conductance of the element, and the imaginary part is called the susceptance. The admittance is the reciprocal of the impedance for a two-terminal network element.

A summary of the properties of the impedance and admittance of the resistor, the inductor, and the capacitor are given in Table 7–5.1.

TABLE 7-5.1 Impedance and Admittance of Passive Two-Terminal Network Elements

Element	Impedance	Resistance	Reactance	Admittance	Conductance	Susceptance
R	$Z_R = R$	$R_R = R$	$X_R = 0$	$Y_R = \frac{1}{R}$	$G_R = \frac{1}{R}$	$B_R = 0$
L	$Z_L = j\omega L$	$R_L = 0$	$X_L = \omega L$	$Y_L = \frac{1}{j\omega L}$	$G_L = 0$	$B_L = \frac{-1}{\omega L}$
C	$Z_C = \frac{1}{j\omega C}$	$R_C = 0$	$X_C = \frac{-1}{\omega C}$	$Y_C = j\omega C$	$G_C = 0$	$B_C = \omega C$

In the preceding paragraphs of this section we have defined the quantities of impedance and admittance, and we have pointed out that, for a given two-terminal element, these quantities are reciprocally valued. Thus, if we know the impedance, we obviously know the admittance and vice versa. Frequently, it will be convenient in our future discussions to refer to the concept implied by impedance and admittance without specifically choosing one or the other of these quantities. To do this we shall use the word *immittance*, which is formed from the words *im*pedance and ad*mittance*. Thus, the immittance of a two-terminal network element may be either the ratio of a voltage phasor to a current phasor or the (reciprocal) ratio of a current phasor to a voltage phasor. The connections between the terms immittance, impedance, admittance, resistance, reactance, conductance, and susceptance, as used in sinusoidal steady-state phasor analysis, are summarized in Table 7–5.2.

TABLE 7-5.2 Terminology for Phasor Analysis

Immittance—Impedance or Admittance

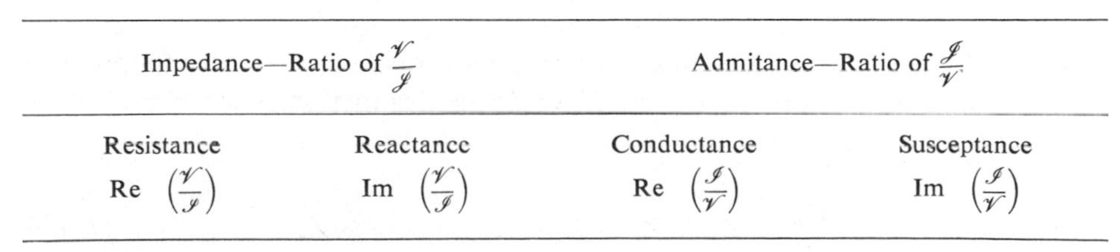

Impedance—Ratio of $\frac{\mathcal{V}}{\mathcal{I}}$		Admitance—Ratio of $\frac{\mathcal{I}}{\mathcal{V}}$	
Resistance	Reactance	Conductance	Susceptance
$\mathrm{Re}\left(\frac{\mathcal{V}}{\mathcal{I}}\right)$	$\mathrm{Im}\left(\frac{\mathcal{V}}{\mathcal{I}}\right)$	$\mathrm{Re}\left(\frac{\mathcal{I}}{\mathcal{V}}\right)$	$\mathrm{Im}\left(\frac{\mathcal{I}}{\mathcal{V}}\right)$

Now let us investigate the immittance of two terminal networks formed by interconnecting resistors, capacitors, and inductors. As a first example, consider the circuit shown in Fig. 7–5.1. If we substitute the relations of Eq. (7–122) in the phasor KVL equation given in (7–124), we obtain

$$j\omega L\mathscr{I} + \frac{1}{j\omega C}\mathscr{I} + R\mathscr{I} = \mathscr{V}_S \tag{7–130}$$

This may be written in the form

$$\frac{\mathscr{V}_S}{\mathscr{I}} = R + j\omega L + \frac{1}{j\omega C} \tag{7–131}$$

Thus, we see that if we consider the entire circuit (as seen from the terminals of the voltage source) as a two-terminal network element, the impedance of this circuit is equal to the sum of the separate impedances of each of the individual network elements. To see that this is true in general, consider the circuit consisting of a series connection of impedances Z_i $(i = 1, 2, \ldots, n)$ as shown in Fig. 7–5.6. Voltage and

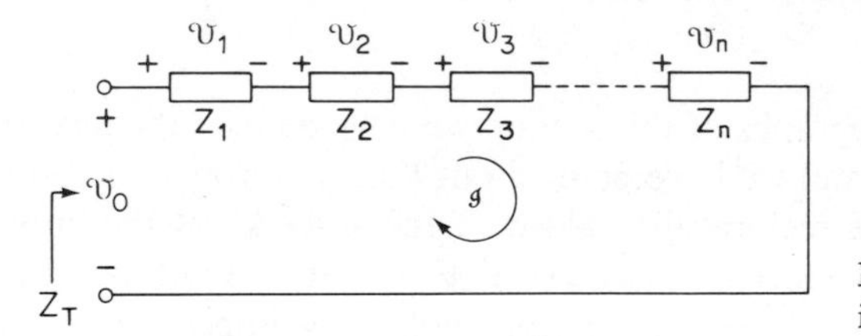

Fig. 7-5.6 A series connection of impedances.

current phasors are defined as indicated in the figure. Applying KVL to the voltage phasors (see Summary 7–5.1), we may write

$$\mathscr{V}_1 + \mathscr{V}_2 + \mathscr{V}_3 + \cdots + \mathscr{V}_n = \mathscr{V}_0 \tag{7–132}$$

If we now substitute the branch relations for each of the individual impedances, we may rewrite (7–132) as follows

$$\begin{aligned} &Z_1\mathscr{I} + Z_2\mathscr{I} + Z_3\mathscr{I} + \cdots + Z_n\mathscr{I} \\ &\qquad = (Z_1 + Z_2 + Z_3 + \cdots + Z_n)\mathscr{I} = \mathscr{V}_0 \end{aligned} \tag{7–133}$$

The impedance of the series connection of the individual impedances is simply equal to the ratio of the phasors $\mathscr{V}_0$ and $\mathscr{I}$. If we define this total impedance as Z_T, then, from (7–133), we may write

$$\frac{\mathscr{V}_0}{\mathscr{I}} = Z_T = Z_1 + Z_2 + Z_3 + \cdots + Z_n \tag{7–134}$$

Thus, we see that the total impedance is the sum of the individual impedances when the elements are connected in series.

A similar development is possible with admittances. To see this, let us substitute the relations of (7–127) in the phasor KCL equation for Fig. 7–5.2 given in (7–117). We obtain

$$\frac{1}{j\omega L}\mathscr{V} + j\omega C\mathscr{V} + G\mathscr{V} = \mathscr{I}_S \qquad (7\text{–}135)$$

This may be rewritten in the form

$$\frac{\mathscr{I}_S}{\mathscr{V}} = \frac{1}{j\omega L} + j\omega C + G \qquad (7\text{–}136)$$

Thus, if we consider the entire circuit as a single two-terminal element with phasor variables $\mathscr{I}_S$ and $\mathscr{V}$, we see that the admittance of the circuit is given as the sum of the individual admittances which are connected in parallel. This conclusion is easily generalized. To see this, consider the circuit consisting of a parallel connection of admittances Y_i $(i = 1, 2, \ldots, n)$ as shown in Fig. 7–5.7. Applying KCL to the current phasors indicated in the figure, we obtain

$$\mathscr{I}_1 + \mathscr{I}_2 + \mathscr{I}_3 + \cdots + \mathscr{I}_n = \mathscr{I}_0 \qquad (7\text{–}137)$$

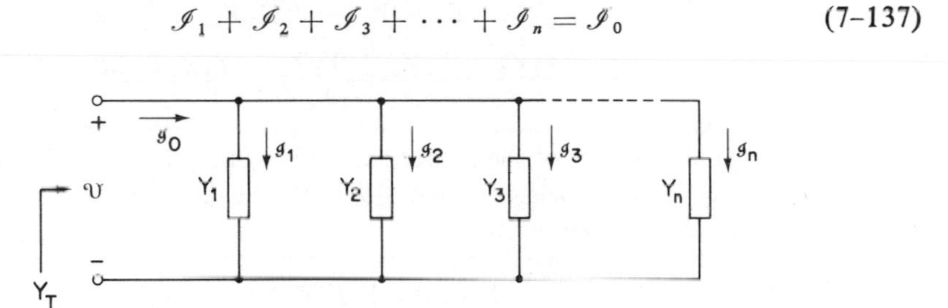

Fig. 7-5-7 A parallel connection of admittances.

If we now substitute the branch relations for each of the admittances in (7–137), we obtain

$$\begin{aligned} &Y_1\mathscr{V} + Y_2\mathscr{V} + Y_3\mathscr{V} + \cdots + Y_n\mathscr{V} \\ &\qquad = (Y_1 + Y_2 + Y_3 + \cdots + Y_n)\mathscr{V} = \mathscr{I}_0 \end{aligned} \qquad (7\text{–}138)$$

The total admittance Y_T of the entire network is simply the ratio of the phasors $\mathscr{I}_0$ and $\mathscr{V}$. Thus, we see that

$$\frac{\mathscr{I}_0}{\mathscr{V}} = Y_T = Y_1 + Y_2 + Y_3 + \cdots + Y_n \qquad (7\text{–}139)$$

We conclude that the total admittance of a number of two-terminal network elements connected in parallel is simply the sum of the admittances of the individual elements.

The developments given above are readily extended to other network configurations. For example, consider the ladder network shown in Fig. 7-5.8. If we define Z_a as the total impedance of the two series branches labeled Y_1 and Z_2, then we see that

$$Z_a = \frac{1}{Y_1} + Z_2 = \frac{1 + Y_1 Z_2}{Y_1} \tag{7-140}$$

If we now define Y_b as the admittance of the branches labeled Y_1, Z_2, and Y_3, we see that the branch labeled Y_3 may be considered as being connected in parallel with the series connection of the branches labeled Y_1 and Z_2. The impedance of these two branches has already been labeled Z_a. Thus, we may write

$$Y_b = \frac{1}{Z_a} + Y_3 = \frac{Y_1}{1 + Y_1 Z_2} + Y_3 = \frac{Y_1 + Y_3 + Y_1 Z_2 Y_3}{1 + Y_1 Z_2} \tag{7-141}$$

Continuing in this fashion, we may define Z_c as the impedance of the series connection of the branch labeled Z_4 with the other branches on its right. Thus, we may write

$$\begin{aligned} Z_c &= \frac{1}{Y_b} + Z_4 = \frac{1 + Y_1 Z_2}{Y_1 + Y_3 + Y_1 Z_2 Y_3} + Z_4 \\ &= \frac{1 + Y_1 Z_4 + Y_3 Z_4 + Y_1 Z_2 + Y_1 Z_2 Y_3 Z_4}{Y_1 + Y_3 + Y_1 Z_2 Y_3} \end{aligned} \tag{7-142}$$

Finally, the total admittance of the ladder network Y_T may be found by noting that the branch labeled Y_5 is connected in parallel with the impedance Z_c. Thus, their admittances add, and we may write

$$\begin{aligned} Y_T &= \frac{1}{Z_c} + Y_5 = \frac{Y_1 + Y_3 + Y_1 Z_2 Y_3}{1 + Y_1 Z_4 + Y_3 Z_4 + Y_1 Z_2 + Y_1 Z_2 Y_3 Z_4} + Y_5 \\ &= \frac{\begin{aligned} Y_1 + Y_3 + Y_5 + Y_1 Z_4 Y_5 + Y_3 Z_4 Y_5 + Y_1 Z_2 Y_5 \\ + Y_1 Z_2 Y_3 + Y_1 Z_2 Y_3 Z_4 Y_5 \end{aligned}}{1 + Y_1 Z_4 + Y_3 Z_4 + Y_1 Z_2 + Y_1 Z_2 Y_3 Z_4} \end{aligned} \tag{7-143}$$

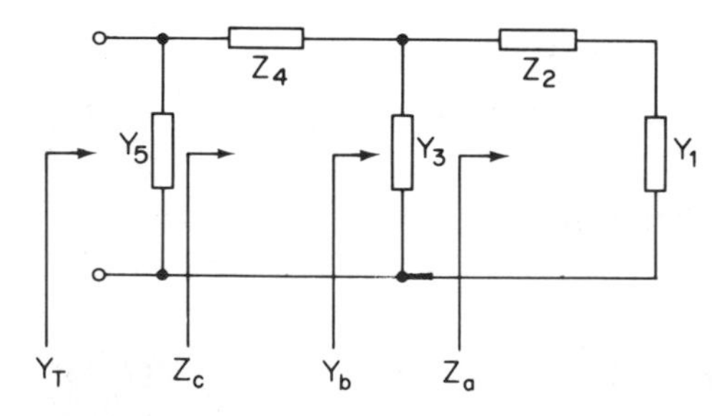

Fig. 7-5.8 A ladder network of immittances.

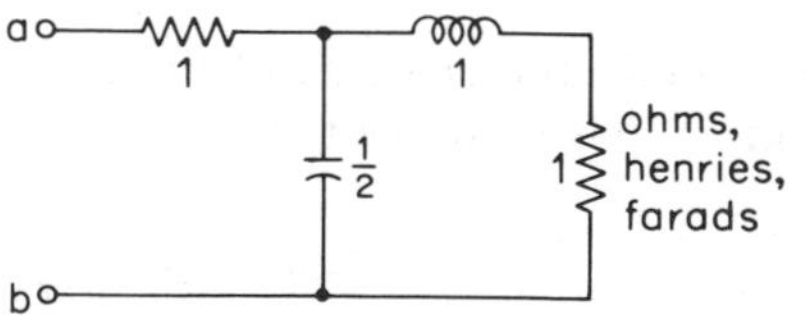

Fig. 7-5.9 Network for Example 7-5.1.

The relation given in (7–143) may be expressed more simply as a continued fraction. As such, it may be written

$$Y_T = Y_5 + \cfrac{1}{Z_4 + \cfrac{1}{Y_3 + \cfrac{1}{Z_2 + \cfrac{1}{Y_1}}}} \qquad (7\text{–}144)$$

An example of the determination of the immittance of a ladder network follows.

EXAMPLE 7-5.1. It is desired to find the impedance Z_0 of the ladder network shown in Fig. 7-5.9 at the terminals $a - b$. This is easily done by using a continued fraction similar to that given in Eq. (7-144). Thus, we may write

$$Z_0 = 1 + \cfrac{1}{j\cfrac{\omega}{2} + \cfrac{1}{j\omega + \cfrac{1}{1}}}$$

Clearing the fractions we obtain

$$Z_0 = \frac{2 - (\omega^2/2) + j(3\omega/2)}{1 - (\omega^2/2) + j(\omega/2)}$$

To separate the real part and the imaginary part of the above expression, we need merely multiply the numerator and the dominator by the conjugate of the dominator. Doing this and rearranging the terms, we obtain

$$Z_0 = \frac{2 - (3\omega^2/4) + (\omega^4/4) + j[(\omega/2) - (\omega^3/2)]}{1 - (3\omega^2/4) + (\omega^4/4)}$$

Thus, the resistive component of the impedance Z_0 is $[(2 - (3\omega^2/4) + (\omega^4/4)]/[1 - (3\omega^2/4) + (\omega^4/4)]$ and the reactive component is $[(\omega/2) - (\omega^3/2)]/[1 - (3\omega^2/4) + (\omega^4/4)]$.

We may make some interesting observations about the basic concept introduced in this section by studying the results of the example given above. First of all, we may note that the expression for the real or resistive part of the impedance is affected not only by the values of the resistive elements of the circuit, but also by the values of the inductor and the capacitor. As such, the resistive component of the impedance is a function of ω. Thus, in general, we must be careful to differentiate between the *resistance of the resistors* which are included in a given circuit (and which are clearly not functions of ω), and the *resistive portion of an impedance* which, except for the simplest of circuits, will be a function of frequency. Thus, if we let R stand for the real part of an impedance, we may write it more explicitly as $R(\omega)$, indicating its functional dependence.

In a similar manner, from the example given above, we may note that the reactive portion of an impedance is also a function of frequency. Thus, if we let X stand for such a reactive part, we may indicate the functional dependence by writing $X(\omega)$. Now let us consider the impedance itself. It is obviously comprised of algebraic combinations of terms which have three possible forms: (1) terms which are constant; (2) terms which are proportional to $j\omega$; and (3) terms which are proportional to $1/j\omega$. Thus, if we let Z stand for impedance, we may more explicitly write $Z(j\omega)$ to indicate the dependence of Z on the variable $j\omega$. Considering all the above, we may rewrite our general expression for impedance given in Eq. (7–124) as

$$Z(j\omega) = R(\omega) + jX(\omega) \tag{7–145}$$

A similar development is easily applied to the general expression for admittance given in Eq. (7–129). Thus, in general, we may write

$$Y(j\omega) = G(\omega) + jB(\omega) \tag{7–146}$$

Now let us consider what happens when we evaluate an expression of the type given in Eq. (7–145) at a particular value of ω. In such a case, we simply obtain a complex number for the impedance, since the dependence on ω has been eliminated by substituting a particular value of this variable. Such an impedance is equivalent (at that specific frequency) to a series connection of a single resistor and a single inductor or capacitor, As an example of this, consider the circuit shown in Fig. 7–5.9. If we evaluate the expression for Z_0 given in Example 7–5.1 with $\omega = 2$ rad/s, we obtain $Z_0 = \frac{3}{2} - j\frac{3}{2}$. We may construct an equivalent circuit which has the same impedance (at $\omega = 2$ rad/s) by using a single resistor and capacitor as shown in Fig. 7–5.10(a). Similarly, if we evaluate the impedance of the circuit of Fig. 7–5.9 at $\omega = \frac{1}{2}$ rad/s, we obtain $Z_0 = \frac{117}{53} + j\frac{12}{53}$. Thus, the given circuit has the same impedance as the equivalent circuit shown in Fig. 7–5.10(b) at the specified frequency. It should be noted that, depending on whether the reactive component is positive or negative at a given frequency, the equivalent circuit may include either an inductor

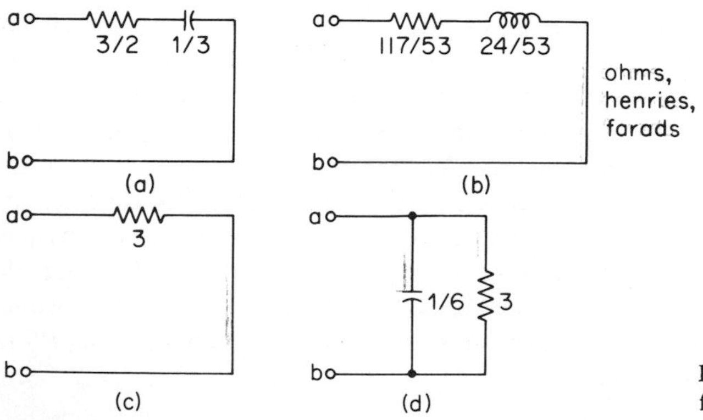

Fig. 7-5.10 Equivalent circuits for the network of Fig. 7-5.9.

or a capacitor. It is also possible for the reactive component to be zero. For example, if we evaluate the impedance Z_0 at $\omega = 1$ rad/s, we obtain $Z_0 = 3 + j0$. Thus, at this frequency, the circuit shown in Fig. 7-5.9 has exactly the same properties as an equivalent circuit consisting of a single resistor of value $3\ \Omega$, as shown in Fig. 7-5.10(c). The circuits shown in Fig. 7-5.10 are, of course, only equivalent with respect to sinusoidal steady-state conditions. More specifically, they are not equivalent if we are also considering transient behavior. Conclusions similar to those given above may be made on an admittance basis. In this case, we may derive an equivalent circuit consisting of a resistor in parallel with a single inductor or capacitor to represent any two-terminal admittance at a given frequency. As an example of this, since the impedance of the circuit shown in Fig. 7-5.9 has been shown to be $\frac{3}{2} - j\frac{3}{2}$ at $\omega = 2$ rad/s, its admittance at that same frequency is obviously equal to $1/(\frac{3}{2} - j\frac{3}{2}) = \frac{1}{3} + j\frac{1}{3}$. At the specified frequency, this is the same admittance as that of the equivalent circuit shown in Fig. 7-5.10(d). Thus, at $\omega = 2$ rad/s the circuits shown in Figs. 7-5.9, 7-5.10(a), and 7-5.10(d) are all equivalent.

We may summarize the discussion given above as follows:

SUMMARY 7-5.4

Equivalence of Circuits Under Conditions of Sinusoidal Steady-State Excitation: Any two-terminal immittance formed by an arbitrary number of resistors, capacitors, and inductors may be replaced at a specified value of frequency by an equivalent two-terminal immittance consisting either of a series connection of a resistor with a capacitor or inductor, or a parallel connection of a resistor with a capacitor or inductor.

7-6 MESH AND NODE ANALYSIS WITH IMPEDANCES AND ADMITTANCES

In this section we will extend the concepts which have been presented in the preceding sections so as to develop general methods of phasor analysis for networks which are under conditions of sinusoidal steady-state excitation. The methods are very similar to those for resistance networks which were presented in Chap. 3. A review of that chapter will prove very worthwhile in enabling the reader to understand the material which follows.

We may begin our study of general phasor analysis methods by considering the circuit shown in Fig. 7-6.1. Let us assume that the circuit is to be analyzed

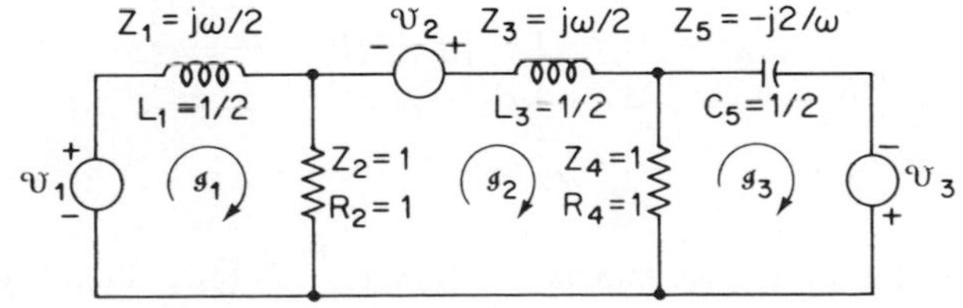

Fig. 7-6.1 An *RLC* network analyzed by applying KVL.

under sinusoidal steady-state conditions, and that the values given for the various elements in the figure are impedances. Three mesh-current phasors $\mathscr{I}_1$, $\mathscr{I}_2$, and $\mathscr{I}_3$ are defined as indicated. If we apply KVL directly to the phasor quantities and the impedances indicated in the figure, we obtain the following (phasor) mesh equations:

$$\begin{aligned} \mathscr{V}_1 &= j\frac{\omega}{2}\mathscr{I}_1 + (\mathscr{I}_1 - \mathscr{I}_2) \\ \mathscr{V}_2 &= j\frac{\omega}{2}\mathscr{I}_2 + (\mathscr{I}_2 - \mathscr{I}_1) + (\mathscr{I}_2 - \mathscr{I}_3) \\ \mathscr{V}_3 &= -j\frac{2}{\omega}\mathscr{I}_3 + (\mathscr{I}_3 - \mathscr{I}_2) \end{aligned} \tag{7–147}$$

These equations may be more conveniently written in matrix form as

$$\begin{bmatrix} \mathscr{V}_1 \\ \mathscr{V}_2 \\ \mathscr{V}_3 \end{bmatrix} = \begin{bmatrix} 1 + j\frac{\omega}{2} & -1 & 0 \\ -1 & 2 + j\frac{\omega}{2} & -1 \\ 0 & -1 & 1 - j\frac{2}{\omega} \end{bmatrix} \begin{bmatrix} \mathscr{I}_1 \\ \mathscr{I}_2 \\ \mathscr{I}_3 \end{bmatrix} \tag{7–148}$$

Note that the elements of the square matrix given in Eq. (7–148) each relate a voltage phasor to a current phasor, thus they have the dimensions of impedance. We will refer to such a matrix as an *impedance matrix*. If we use the symbol $\mathbf{Z}(j\omega)$ for this matrix, then we may consider (7–148) as an example of a general matrix equation having the form

$$\mathbf{V} = \mathbf{Z}(j\omega)\mathbf{I} \tag{7–149}$$

where $\mathbf{V}$ is a column matrix, each element of which is a phasor $\mathscr{V}_i$, and $\mathbf{I}$ is a column matrix, each element of which is a phasor $\mathscr{I}_i$. A set of equations having the form of Eq. (7–149) will be referred to as a set of *mesh impedance equations*. It should be noted that the above determination of the mesh impedance equations by the use of the phasor version of KVL yields exactly the same set of equations as were obtained by the substitution of expressions of the form $v_i(t) = \text{Re}\,(\mathscr{V}_i e^{j\omega t})$ and $i_i(t) = \text{Re}\,(\mathscr{I}_i e^{j\omega t})$ into the original integro-differential equations describing the network. As an example of this, note that if we let $\omega = 2$ rad/s in (7–148), the equations become

$$\mathbf{V} = \begin{bmatrix} \mathscr{V}_1 \\ \mathscr{V}_2 \\ \mathscr{V}_3 \end{bmatrix} = \begin{bmatrix} 1 + j1 & -1 & 0 \\ -1 & 2 + j1 & -1 \\ 0 & -1 & 1 - j1 \end{bmatrix} \begin{bmatrix} \mathscr{I}_1 \\ \mathscr{I}_2 \\ \mathscr{I}_2 \end{bmatrix} = \mathbf{Z}(j2)\mathbf{I} \tag{7–150}$$

The impedance matrix $\mathbf{Z}(j2)$ of Eq. (7–150) is readily seen to be identical with the square matrix obtained when the substitutions referred to above were made in Sec. 7–4 in the integro-differential equations describing the network shown in Fig. 7–4.2, which is identical with the network shown in Fig. 7–6.1.

As another example of the determination of the mesh impedance equations, consider the network shown in Fig. 7–6.2. This network consists of a set of six impedances $Z_i(j\omega)$. Applying KVL to the three meshes defined by the phasor mesh current variables $\mathscr{I}_i$ we obtain

$$\begin{aligned} \mathscr{V}_1 - \mathscr{V}_3 &= Z_1\mathscr{I}_1 + Z_3(\mathscr{I}_1 - \mathscr{I}_2) + Z_2(\mathscr{I}_1 - \mathscr{I}_3) \\ -\mathscr{V}_2 &= Z_5\mathscr{I}_2 + Z_3(\mathscr{I}_2 - \mathscr{I}_1) + Z_4(\mathscr{I}_2 - \mathscr{I}_3) \\ \mathscr{V}_3 &= Z_6\mathscr{I}_3 + Z_2(\mathscr{I}_3 - \mathscr{I}_1) + Z_4(\mathscr{I}_3 - \mathscr{I}_2) \end{aligned} \tag{7–151}$$

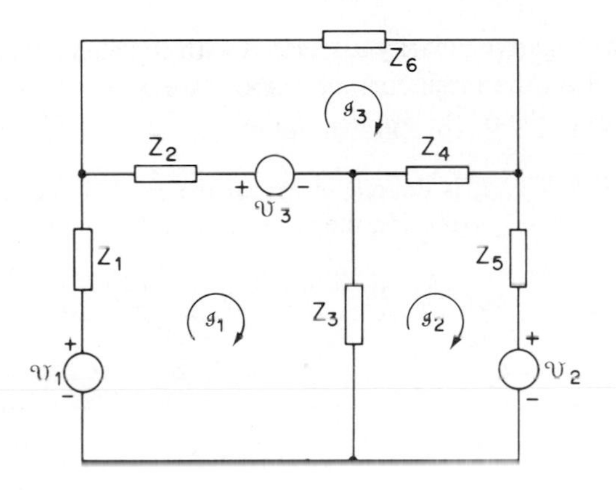

Fig. 7-6.2 A network analyzed using KVL.

Putting these equations in matrix format, we may write

$$\begin{bmatrix} \mathscr{V}_1 - \mathscr{V}_3 \\ -\mathscr{V}_2 \\ \mathscr{V}_3 \end{bmatrix} = \begin{bmatrix} Z_1 + Z_2 + Z_3 & -Z_3 & -Z_2 \\ -Z_3 & Z_3 + Z_4 + Z_5 & -Z_4 \\ -Z_2 & -Z_4 & Z_2 + Z_4 + Z_6 \end{bmatrix} \begin{bmatrix} \mathscr{I}_1 \\ \mathscr{I}_2 \\ \mathscr{I}_3 \end{bmatrix} \tag{7–152}$$

If we examine the elements of the matrices of the mesh impedance equations derived for the above examples, we readily see that we may formulate a set of rules for directly determining the expressions for these elements. These rules exactly parallel the rules given in Sec. 3–2 for determining the elements of the matrix equation $\mathbf{v}(t) = \mathbf{Ri}(t)$ for a resistance network. They may be summarized as follows:

SUMMARY 7-6.1

Rules for Forming the Mesh Impedance Equations: The elements of the matrices of the mesh impedance equation given in (7–149) may be found as follows:

$\mathscr{V}_i$ The summation of the phasors defining all the voltage sources in the ith mesh, with sources whose reference polarity constitutes a voltage rise (in the positive reference direction of the ith mesh current) treated as positive, and with sources whose reference polarity constitutes a voltage drop treated as negative.

$Z_{ii}(j\omega)$ The summation of the values of all the impedances which are located in the ith mesh.

$Z_{ij}(j\omega)$ $(i \neq j)$ The negative of the sum of the values of any impedances which are common to the ith and the jth meshes.[11]

$\mathscr{I}_j$ The mesh current phasors which are the unknown variables for the network.

The negative polarity associated with the elements $Z_{ij}(j\omega)$ $(i \neq j)$ only occurs, of course, if all the mesh current phasors are given the same positive reference direction. An example of the application of these rules follows.

EXAMPLE 7-6.1. It is desired to obtain the mesh impedance equations for the circuit shown in Fig. 7-6.3. Applying the rules given in Summary 7-6.1, we obtain

$$\begin{bmatrix} \mathscr{V}_1 \\ -\mathscr{V}_2 \\ 0 \end{bmatrix} = \begin{bmatrix} R_1 + j\omega L_1 - \dfrac{j}{\omega C_1} & -R_1 + \dfrac{j}{\omega C_1} & -j\omega L_1 \\ -R_1 + \dfrac{j}{\omega C_1} & R_1 + R_2 + R_4 - \dfrac{j}{\omega C_1} & -R_2 \\ -j\omega L_1 & -R_2 & R_2 + R_3 + j\omega L_1 - \dfrac{j}{\omega C_2} \end{bmatrix} \begin{bmatrix} \mathscr{I}_1 \\ \mathscr{I}_2 \\ \mathscr{I}_3 \end{bmatrix}$$

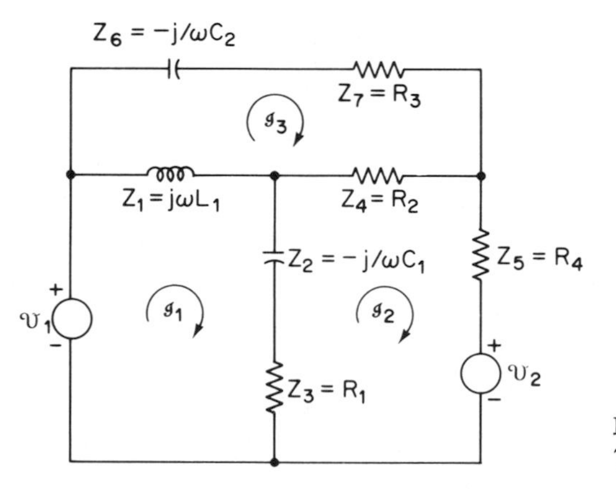

Fig. 7-6.3 Circuit for Example 7-6.1.

In the above development, it should be noted that the elements of the impedance matrix $\mathbf{Z}(j\omega)$ are determined by three types of network elements: resistors, which produce terms which are not functions of $j\omega$; inductors, which produce terms

[11]The quantity $Z_{ii}(j\omega)$ is sometimes called the *self impedance* of the ith mesh, while the quantity $Z_{ij}(j\omega)$ $(i \neq j)$ is called the impedance *mutual* to the ith and jth meshes.

which are proportional to $j\omega$; and capacitors, which produce terms which are inversely proportional to $j\omega$. Thus, if we define a *resistance matrix* **R**, an *inductance matrix* **L**, and an *elastance* (*reciprocal capacitance*) matrix **S**, we may write the impedance matrix in the form

$$\mathbf{Z}(j\omega) = \mathbf{R} + j\omega\mathbf{L} + \frac{1}{j\omega}\mathbf{S} = \mathbf{R} + j\left(\omega\mathbf{L} - \frac{1}{\omega}\mathbf{S}\right) = \mathbf{R} + j\mathbf{X}(\omega) \qquad (7\text{–}153)$$

where $\mathbf{X}(\omega)$ is referred to as the *reactance matrix*. Note that in this form, the resistance matrix is not a function of ω, although the reactance matrix, as indicated, is a function of this variable. The elements of the **R**, **L**, and **S** matrices may be found by application of rules directly paralleling those given in Summary 7–6.1. For example, for the network shown in Fig. 7–6.1, these matrices are readily shown to be

$$\mathbf{R} = \begin{bmatrix} 1 & -1 & 0 \\ -1 & 2 & -1 \\ 0 & -1 & 1 \end{bmatrix} \qquad \mathbf{L} = \begin{bmatrix} \frac{1}{2} & 0 & 0 \\ 0 & \frac{1}{2} & 0 \\ 0 & 0 & 0 \end{bmatrix} \qquad \mathbf{S} = \begin{bmatrix} 0 & 0 & 0 \\ 0 & 0 & 0 \\ 0 & 0 & 2 \end{bmatrix} \qquad (7\text{–}154)$$

It is readily verified that, for networks comprised of resistors, inductors, and capacitors, the matrices $\mathbf{Z}(j\omega)$, **R**, **L**, and **S** will always be symmetric.

The operations involved in forming the impedance matrix from the values of the impedances and the numbers of the meshes in which these impedances are located may easily be implemented for the digital computer for cases in which numerical values are known for the element values and for the frequency variable. If the numerical values of the impedances are stored in a complex one-dimensional array IMPED and the corresponding values of the mesh numbers in which the impedances are located are stored in one-dimensional arrays MESH1 and MESH2, then, if there are NZ impedances and N meshes, a subroutine ZMESH may be defined for computing the effect that these impedances have in determining the elements of a two-dimensional complex array Z which contains the values of the elements of the impedance matrix. The identifying statement for the subroutine will have the form

```
SUBROUTINE ZMESH (N, NZ, IMPED, MESH1, MESH2, Z)
```

The logic for the algorithm is very similar to that used for the subroutine RMESH defined in Sec. 3–2. The only significant difference is that a FORTRAN *type declaration* must be used to define the complex arrays. Just as was the case in the subroutine RMESH, a value of zero must be specified for any element of MESH2 corresponding with an impedance which occurs only in a single mesh. A listing of the subroutine ZMESH is given in Fig. 7–6.4 and summary of its properties is given in Table 7–6.1 (on page 443). An example of its application follows.

EXAMPLE 7-6.2. As an example of the use of the subroutine ZMESH in forming the impedance matrix for a given network, let us consider the network used in Example 7-6.1

```
      SUBROUTINE ZMESH (N, NZ, IMPED, MESH1, MESH2, Z)
C     SUBROUTINE TO COMPUTE ELEMENTS OF IMPEDANCE MATRIX Z
C     FROM THE VALUES OF THE IMPEDANCES AND THE NUMBERS OF
C     THE MESHES IN WHICH THESE IMPEDANCES ARE LOCATED.
C     EACH OF THE MESHES MUST HAVE THE SAME REFERENCE
C     DIRECTION, AND NO IMPEDANCE CAN BE PRESENT IN MORE
C     THAN TWO MESHES
C          N - NUMBER OF MESHES
C         NZ - NUMBER OF IMPEDANCES
C      IMPED - ARRAY OF IMPEDANCE VALUES
C      MESH1 - FIRST ARRAY OF MESH NUMBERS
C      MESH2 - SECOND ARRAY OF MESH NUMBERS (IF IMPEDANCE
C                  OCCURS ONLY IN A SINGLE MESH, THE
C                  CORRESPONDING ELEMENT IN MESH2 IS ZERO)
C          Z - OUTPUT IMPEDANCE MATRIX
      DIMENSION MESH1(20), MESH2(20)
      COMPLEX IMPED(20), Z(10,10)
C
C     SET THE ELEMENTS OF THE IMPEDANCE MATRIX Z TO ZERO
      DO 3  I = 1, N
      DO 3  J = 1, N
    3 Z(I,J) = (0.,0.)
C
C     MODIFY THE ELEMENTS OF THE IMPEDANCE MATRIX Z FOR
C     EACH OF THE IMPEDANCES
      DO 11 I = 1, NZ
      J = MESH1(I)
      K = MESH2(I)
C
C     TEST TO DETERMINE WHETHER THE IMPEDANCE OCCURS
C     ONLY IN A SINGLE MESH
      IF (K) 11, 11, 8
    8 Z(K,K) = Z(K,K) + IMPED(I)
      Z(J,K) = Z(J,K) - IMPED(I)
      Z(K,J) = Z(J,K)
   11 Z(J,J) = Z(J,J) + IMPED(I)
      RETURN
      END
```

Fig. 7-6.4 Listing of the subroutine ZMESH.

and shown in Fig. 7-6.3. Let us assume that ω is 5 rad/s, and that the elements of the network have the values and impedances given below.

Element	*Value*	*Impedance*
R_1	1 Ω	$1 + j0\ \Omega$
R_2	2 Ω	$2 + j0\ \Omega$
R_3	3 Ω	$3 + j0\ \Omega$
R_4	4 Ω	$4 + j0\ \Omega$
L_1	0.3 H	$0 + j1.5\ \Omega$
C_1	0.1 F	$0 - j2\ \Omega$
C_2	0.5 F	$0 - j4\ \Omega$

We may now easily write a main program to read values for the number of meshes and the number of impedances, and the mesh numbers and the value of the impedance for each element. To determine the impedance matrix, it is merely necessary for the main program to call the subroutine ZMESH and to print the resulting values of the elements of the **Z** matrix. A listing of such a main program is given in Fig. 7-6.5. The program is similar to the one used in Example 3-2.4, the only significant difference being the use of complex variables rather than real variables. A listing of the input data and the resulting output is given in Fig. 7-6.6. It is readily verified that the elements of the **Z** matrix as computed by the program agree with those given in Example 7-6.1 when the specified numerical values are inserted in the relations given there.

A development similar to that given in the preceding paragraphs is readily made if it is desired to write a set of equations relating the dependent and independent variables of a network using quantities of admittance rather than impedance. The procedure is quite similar to that given for the determination of the node equations for a resistance network in Sec. 3-7. As an example of the procedure, consider the network shown in Fig. 7-6.7. If we apply KCL to the two nodes of this network, then, specifying the branch currents in terms of the branch admittances and the phasor node voltages, we may write

```
C     MAIN PROGRAM FOR EXAMPLE 7.6-2
C            N - NUMBER OF MESHES
C           NZ - NUMBER OF IMPEDANCES
C          RES - ARRAY OF IMPEDANCE VALUES
C        MESH1 - FIRST ARRAY OF MESH NUMBERS
C        MESH2 - SECOND ARRAY OF MESH NUMBERS
C            Z - IMPEDANCE MATRIX
      DIMENSION MESH1(20), MESH2(20)
      COMPLEX IMPED(20), Z(10,10)
C
C     READ INPUT DATA
      READ 2, N, NZ
    2 FORMAT (2I2)
      READ 4, (MESH1(I), MESH2(I), IMPED(I), I = 1, NZ)
    4 FORMAT (2I1, 8X, 2E10.0)
C
C     CALL THE SUBROUTINE ZMESH TO CONSTRUCT
C     THE IMPEDANCE MATRIX Z
      CALL ZMESH (N, NZ, IMPED, MESH1, MESH2, Z)
C
C     PRINT A LISTING OF THE IMPEDANCE MATRIX Z
      PRINT 7
    7 FORMAT  (9X,34H* * * *  IMPEDANCE MATRIX  * * * */)
      DO 11  I = 1, N
      PRINT 10, (Z(I,J), J = 1, N)
   10 FORMAT (3(1X, F5.2, 1X, 2H+J, F5.2, 5X))
   11 CONTINUE
      STOP
      END
```

Fig. 7-6.5 Listing of the program for Example 7-6.2.

```
INPUT DATA FOR EXAMPLE 7.6-2

0307
12          1.0          0.0
23          2.0          0.0
30          3.0          0.0
20          4.0          0.0
13          0.0          1.5
12          0.0         -2.0
30          0.0         -4.0


OUTPUT DATA FOR EXAMPLE 7.6-2

          * * * *   IMPEDANCE MATRIX   * * * *

  1.00 +J -.50        -1.00 +J 2.00         0.00 +J-1.50
 -1.00 +J 2.00         7.00 +J-2.00        -2.00 +J 0.00
  0.00 +J-1.50        -2.00 +J 0.00         5.00 +J-2.50
```

Fig. 7-6.6 Input and output data for Example 7-6.2.

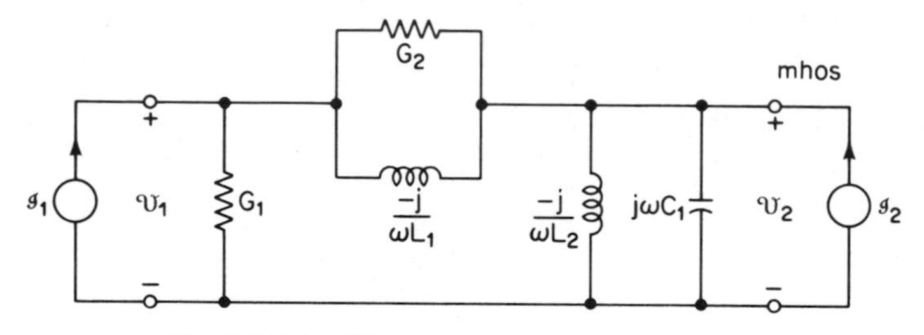

Fig. 7-6.7 An *RLC* network analyzed using KCL.

$$
\begin{aligned}
\mathscr{I}_1 &= G_1\mathscr{V}_1 - \frac{j}{\omega L_1}(\mathscr{V}_1 - \mathscr{V}_2) + G_2(\mathscr{V}_1 - \mathscr{V}_2) \\
\mathscr{I}_2 &= \frac{-j}{\omega L_2}\mathscr{V}_2 + j\omega C_1\mathscr{V}_2 - \frac{j}{\omega L_1}(\mathscr{V}_2 - \mathscr{V}_1) + G_2(\mathscr{V}_2 - \mathscr{V}_1)
\end{aligned}
\tag{7–155}
$$

Rearranging the terms and putting the equations in matrix format, we obtain

$$
\begin{bmatrix} \mathscr{I}_1 \\ \mathscr{I}_2 \end{bmatrix} = \begin{bmatrix} G_1 + G_2 - \dfrac{j}{\omega L_1} & -G_2 + \dfrac{j}{\omega L_1} \\ -G_2 + \dfrac{j}{\omega L_1} & G_2 - \dfrac{j}{\omega L_1} - \dfrac{j}{\omega L_2} + j\omega C_1 \end{bmatrix} \begin{bmatrix} \mathscr{V}_1 \\ \mathscr{V}_2 \end{bmatrix}
\tag{7–156}
$$

Note that the elements of the square matrix given in (7–156) each relate a voltage phasor to a current phasor; thus, these elements each have the dimensions of admittance. It is therefore logical to refer to the square matrix as an *admittance matrix.* If we use the designation $\mathbf{Y}(j\omega)$ for this matrix, then Eq. (7–156) may be written in the general form

$$\mathbf{I} = \mathbf{Y}(j\omega)\mathbf{V} \tag{7–157}$$

where **I** and **V** are column matrices whose elements are the phasors $\mathscr{I}_i$ and $\mathscr{V}_i$, respectively. A set of equations having the form of (7–157) will be referred to as a set of *node admittance equations.*

As another example of the determination of the node admittance equations, consider the network shown in Fig. 7-6.8. Applying KCL to each of the nodes of this figure, we obtain the equations

$$\begin{aligned}\mathscr{I}_1 + \mathscr{I}_2 &= Y_1\mathscr{V}_1 + Y_2(\mathscr{V}_1 - \mathscr{V}_2) + Y_6(\mathscr{V}_1 - \mathscr{V}_3)\\ 0 &= Y_3\mathscr{V}_2 + Y_2(\mathscr{V}_2 - \mathscr{V}_1) + Y_4(\mathscr{V}_2 - \mathscr{V}_3)\\ \mathscr{I}_3 - \mathscr{I}_2 &= Y_5\mathscr{V}_3 + Y_4(\mathscr{V}_3 - \mathscr{V}_2) + Y_6(\mathscr{V}_3 - \mathscr{V}_1)\end{aligned} \tag{7–158}$$

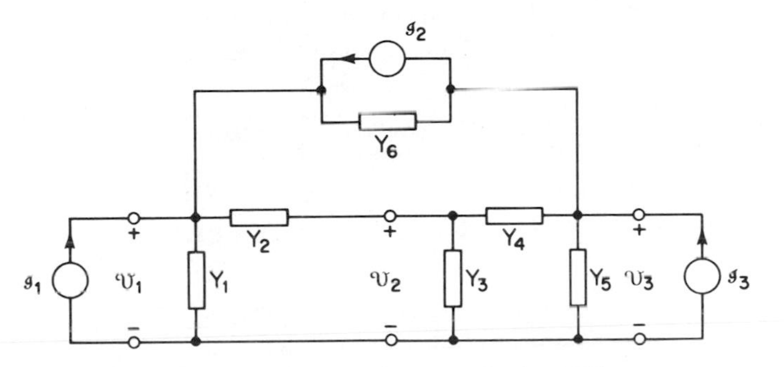

Fig. 7-6.8 A network analyzed using KCL.

Putting these in the form of (7–157), we obtain

$$\begin{bmatrix}\mathscr{I}_1 + \mathscr{I}_2\\ 0\\ \mathscr{I}_3 - \mathscr{I}_2\end{bmatrix} = \begin{bmatrix}Y_1 + Y_2 + Y_6 & -Y_2 & -Y_6\\ -Y_2 & Y_2 + Y_3 + Y_4 & -Y_4\\ -Y_6 & -Y_4 & Y_4 + Y_5 + Y_6\end{bmatrix}\begin{bmatrix}\mathscr{V}_1\\ \mathscr{V}_2\\ \mathscr{V}_3\end{bmatrix} \tag{7–159}$$

If we examine the elements of the matrices of the node admittance equations for the above examples, we readily note that we may formulate a set of rules for directly determining the expressions for these elements. These rules exactly parallel the rules given in Sec. 3–7 for determining the elements of the matrix equation $\mathbf{i}(t) = \mathbf{Gv}(t)$ for a resistance network. They may be summarized as follows:

SUMMARY 7-6.2

Rules for Forming the Node Admittance Equations: The elements of the matrices of the node admittance equation given in (7–157) may be found as follows:

$\mathscr{I}_i$ The summation of the phasors representing the current sources connected to the ith node. Sources whose positive reference direction is toward the node are treated as positive, and those whose reference direction is away from the node are treated as negative.

$Y_{ii}(j\omega)$ The summation of the values of all the admittances connected to the ith node.

$Y_{ij}(j\omega)$ $(i \neq j)$ The negative of the sum of the values of any admittances which are connected between the ith and jth nodes.[12]

$\mathscr{V}_i$ The node voltage phasors which are the unknown variables for the network.

An example of the application of these rules follows.

EXAMPLE 7-6.3. It is desired to obtain the node admittance equations for the circuit shown in Fig. 7-6.9. Applying the rules given in Summary 7-6.2, we obtain

$$\begin{bmatrix} \mathscr{I}_1 \\ \mathscr{I}_2 \\ -\mathscr{I}_2 \\ \mathscr{I}_3 \end{bmatrix} = \begin{bmatrix} j4 & j2 & -j4 & 0 \\ j2 & 14+j10 & -j7 & -8 \\ -j4 & -j7 & 19 & -10+j11 \\ 0 & -8 & -10+j11 & 30+j2 \end{bmatrix} \begin{bmatrix} \mathscr{V}_1 \\ \mathscr{V}_2 \\ \mathscr{V}_3 \\ \mathscr{V}_4 \end{bmatrix}$$

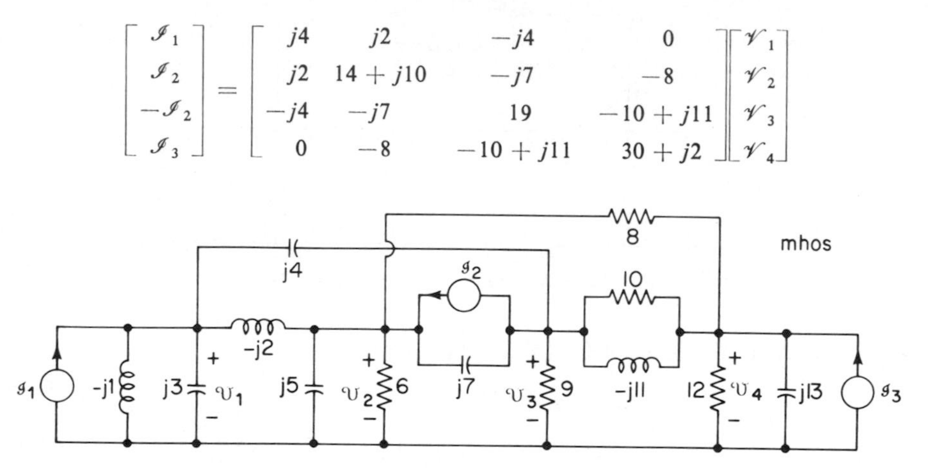

Fig. 7-6.9 Network for Example 7-6.3.

In a manner similar to that done for the impedance matrix, we may write the admittance matrix $\mathbf{Y}(j\omega)$ in terms of a *conductance matrix* $\mathbf{G}$, a *reciprocal inductance matrix* $\mathbf{\Gamma}$, and a capacitance matrix $\mathbf{C}$. The expression is

$$\mathbf{Y}(j\omega) = \mathbf{G} + j\omega\mathbf{C} + \frac{1}{j\omega}\mathbf{\Gamma} = \mathbf{G} + j\left(\omega\mathbf{C} - \frac{1}{\omega}\mathbf{\Gamma}\right) = \mathbf{G} + j\mathbf{B}(\omega) \qquad (7\text{-}160)$$

where $\mathbf{B}$ is referred to as the *susceptance matrix*. The elements of the matrices $\mathbf{G}$, $\mathbf{C}$, and $\mathbf{\Gamma}$ are readily found by applying rules directly paralleling those given in Summary 7-6.2.

[12]The quantity $Y_{ij}(j\omega)$ is sometimes called the *self admittance* of the ith node, while the quantity $Y_{ij}(j\omega)$ $(i \neq j)$ is called the *mutual admittance* between the ith and jth nodes.

Since the rules for forming the elements of the admittance matrix in terms of the component admittances and the nodes to which these admittances are connected are identical to the rules for forming the elements of the impedance matrix in terms of the component impedances and the meshes in which these impedances are located, the subroutine ZMESH defined earlier in this section may be used without modification to make a digital computation of the admittance matrix. In such a case, the variable N specifies the number of independent nodes, the variable NZ specifies the number of admittances, the complex values of the admittances are stored in the one-dimensional complex array IMPED, the numbers of the nodes to which the admittances are connected are stored in the one-dimensional arrays MESH1 and MESH2 (if an admittance is connected to ground, the corresponding element of MESH2 is set to zero), and the resulting admittance matrix is computed and stored in the elements of the two-dimensional array Z. If desired, of course, in the main program which calls ZMESH the variable names may be changed to ones more symbolic of an admittance formulation. Such a change, however, is certainly not required, since the program given in Fig. 7-6.5 may be directly applied to the computation of an admittance matrix with only a trivial change in the Hollerith statements.

Practically any of the other analysis techniques which were introduced in Chapter 3 as aids in analyzing resistance networks may be applied to the phasor analysis of *RLC* networks. For example, a non-ideal source consisting of the series connection of a voltage source and an impedance may be converted to an equivalent non-ideal source consisting of the parallel connection of a current source and an admittance. The general form of the two sources is shown in Fig. 7-6.10. It is easily

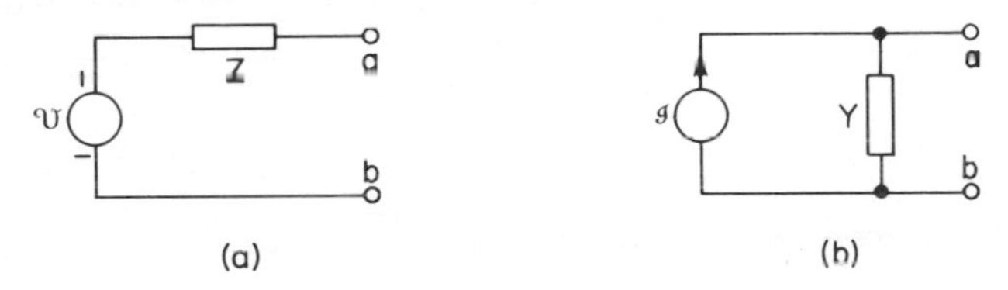

Fig. 7-6.10 Non-ideal sources.

shown that the sources are equivalent in terms of their behavior at the external terminals $a - b$ under the conditions that

$$Y(j\omega) = \frac{1}{Z(j\omega)} \qquad \mathscr{V} = \mathscr{I} Z(j\omega) \tag{7-161}$$

As an example of such an equivalence, the two non-ideal sources shown in Figs. 7-6.11(a) and 7-6.11(b) are equivalent. In addition, since the series impedance $1 + j1\ \Omega$ is equal to an admittance of $1/(1 + j1)$ or $\frac{1}{2} - j\frac{1}{2}$ mho, the immittance itself may also be realized by a parallel connection of elements as shown in Figs. 7-6.11(c) and 7-6.11(d). Such an equivalence is only valid, of course, at the specified frequency.

Another parallel between sinusoidal steady-state networks and resistance networks lies in the fact that any collection of sources and two-terminal elements may be replaced, at a given pair of terminals *and at a given frequency*, by a Thevenin or Norton equivalent circuit consisting of a single source plus an immittance, the latter

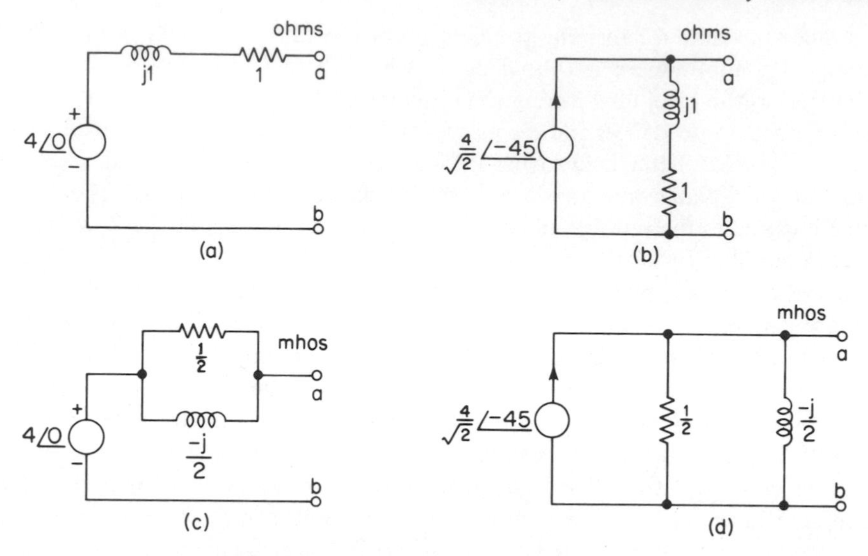

Fig. 7-6.11 An example of non-ideal sources.

in turn consisting of a single resistor and an inductor or capacitor. The techniques for determining the value of the source and the value of the immittance are similar to those given in Sec. 3-9 for a resistance network. Thus, if a Thevenin equivalent circuit is to be found, it is necessary to determine the open-circuit voltage at the

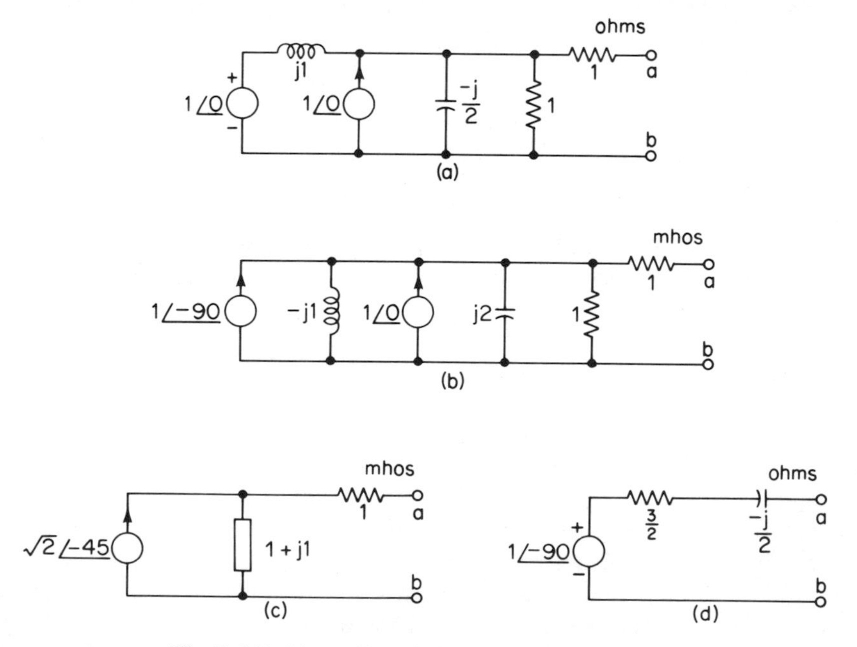

Fig. 7-6.12 Thevenin and Norton equivalent circuits.

specified pair of terminals, and the impedance seen at those terminals when all *independent* sources are set to zero. As an example of this, consider the circuit shown in Fig. 7–6.12(a). To determine the open-circuit voltage at terminals $a - b$ it is convenient to first convert the non-ideal source consisting of a voltage source and a series inductor to a current source and a parallel inductor as shown in Fig. 7–6.12(b). Combining the two parallel current sources and adding the admittances of the paralleled passive elements, we obtain the circuit shown in Fig. 7–6.12(c). From this figure we see that the open-circuit voltage phasor is $-j1$. The shunt admittance is readily converted to a series impedance of value $\frac{1}{2} - j\frac{1}{2}\,\Omega$. Thus, the Thevenin equivalent circuit has the form shown in Fig. 7–6.12(d). A Norton equivalent for the circuit shown in Fig. 7–6.12(a) is readily found by determining the short-circuit current at the terminals $a - b$, and the admittance at these terminals when the independent sources in the network are set to zero. Alternatively, the Norton equivalent may be

TABLE 7-6.1 Summary of the Characteristics of the Subroutine ZMESH

Identifying Statement: SUBROUTINE ZMESH (N, NZ, IMPED, MESH1, MESH2, Z)

Purpose: To form the complex impedance matrix **Z** (or the complex admittance matrix **Y**) which represents a network under conditions of sinusoidal steady-state excitation.

Additional Subprograms Required: None

Input Arguments:

- N The number of meshes (or the number of nodes not including the reference or ground node) in the network.
- NZ The number of complex impedances (or admittances) that are in the network.
- IMPED The one-dimensional array of complex variables IMPED(I) in which are stored the values of the complex impedances (or admittances).
- MESH1 The one-dimensional array of integer variables MESH1(I) in which are stored the values of the first mesh (or the first node) in which the Ith impedance (or admittance) is connected (this must not be the reference or ground node).
- MESH2 The one-dimensional array of integer variables MESH2(I) in which are stored the values of the second mesh (or the second node) in which the Ith impedance (or admittance) is connected. If the Ith impedance occurs only in a single mesh (if one terminal of the Ith admittance is connected to the ground or reference node), the variable MESH2(I) must be set equal to zero.

Output Argument:

- Z The two-dimensional array of complex variables Z(I, J) in which are stored the elements z_{ij} (or y_{id}) of the complex impedance (admittance) matrix **Z** (or **Y**).

Note: The variables of the subroutine are dimensioned as follows: IMPED(20), MESH1(20), MESH2(20), Z(10, 10).

found directly from the Thevenin equivalent circuit given in Fig. 7-6.12(d), using the rules for conversion of non-ideal sources described above.

It should be apparent from the above discussion that, using the techniques of source conversion, adding voltage sources in parallel, and adding current sources in series as discussed in connection with resistance networks in Sec. 3-8, it is possible as a preliminary step to finding the solution of the network (i.e., of determining the value of some variable or variables) to write either a set of mesh impedance equations *or* a set of node admittance equations for a given network. The choice of which formulation to use is determined by two main factors, namely, computational efficiency and convenience. Computational efficiency is mainly concerned with the number of simultaneous equations which are to be solved. The number of equations for the two types of formulation can be found as follows: If n is the number of nodes, and b the number of branches, there will be $b - (n - 1)$ mesh impedance equations and $n - 1$ node admittance equations. The "convenience" factor is somewhat harder to evaluate. For example, for a network with only voltage sources, with all elements specified in terms of their impedances, and in which it is desired to find the value of some network current, it may be more convenient to use a mesh impedance formulation than a node admittance formulation. This avoids the necessity of transforming sources, taking the reciprocal of the impedance values to find the admittances, etc., even though the number of mesh impedance equations is greater than the corresponding number of node equations. This conclusion would be even more applicable if digital computer methods are to be used to obtain the solution. Other examples where the convenience factor dominates are easily found.

7-7 DIGITAL COMPUTER TECHNIQUES FOR SOLVING THE IMPEDANCE AND ADMITTANCE EQUATIONS

In the preceding section we discussed some general techniques for writing the impedance and admittance equations for quite arbitrary network configurations. In this section we shall discuss methods for the solution of such sets of equations. There are two general cases which may be considered. The first of these is when the frequency variable ω is retained as an explicit variable, i.e., when the mesh impedance matrix or the node admittance matrix have the form $\mathbf{Z}(j\omega)$ or $\mathbf{Y}(j\omega)$, respectively. The second case occurs when it is desired to find a solution to the impedance or admittance equations at some specific value of ω, i.e., when ω is given some numerical value. Let this numerical value be ω_0. The matrices that we must deal with in this case may be designated as $\mathbf{Z}(j\omega_0)$ or $\mathbf{Y}(j\omega_0)$. We shall see that there are significant differences between these two cases, the most significant of which is that digital computation techniques may be applied to the second case, but not, in general, to the first case.

Now let us consider the first case in more detail. For simplicity, we shall consider a set of impedance equations, since a similar development is easily made for the admittance case. Thus, our problem will be the solution of a set of equations having the form

$$\mathbf{V} = \mathbf{Z}(j\omega)\mathbf{I} \tag{7-162}$$

where **V** and **I** are column matrices whose elements are the phasors representing equivalent mesh voltage sources and mesh currents, respectively. The solution procedure consists of finding the inverse of the matrix $\mathbf{Z}(j\omega)$, i.e., putting the relations of Eq. (7–162) in the form

$$\mathbf{I} = \mathbf{Z}(j\omega)^{-1}\mathbf{V} \tag{7-163}$$

This may be done using the methods given in Chap. 3. The only difference is that, at each step of the simultaneous equation solution procedure or the matrix inversion procedure, the elements of the arrays which must be manipulated will be functions of $j\omega$, rather than being purely numerical as they were in Chap. 3. In general, these elements will also be rational functions, i.e., ratios of polynomials in the variable $j\omega$. One example of such a situation was given in Sec. 7–4, in which the inverse of a 2×2 matrix with elements which were functions of $j\omega$ was found. As another example of the process, consider the network shown in Fig. 7–7.1. Using the techniques of the preceding section, the impedance matrix is readily found to be

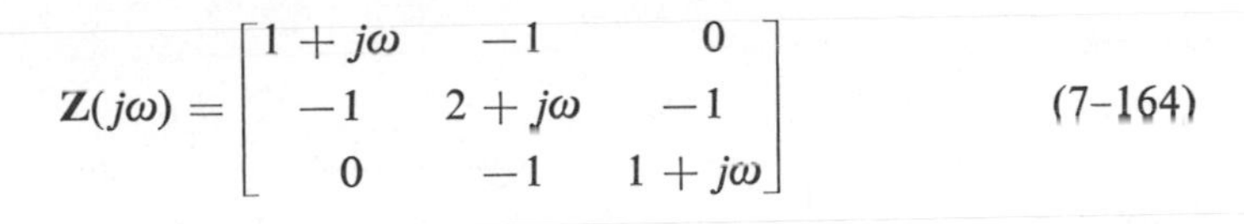

$$\mathbf{Z}(j\omega) = \begin{bmatrix} 1 + j\omega & -1 & 0 \\ -1 & 2 + j\omega & -1 \\ 0 & -1 & 1 + j\omega \end{bmatrix} \tag{7-164}$$

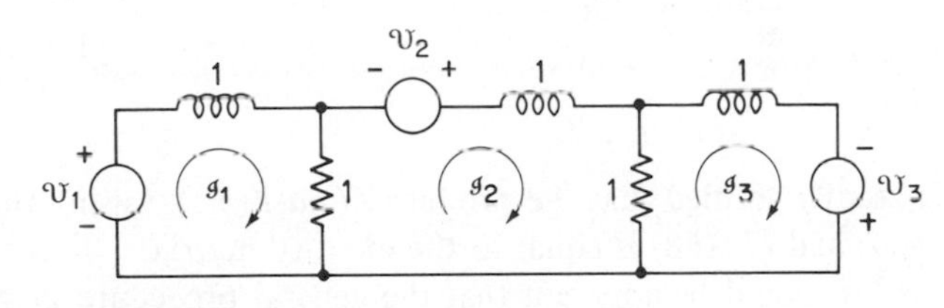

Fig. 7-7.1 A three-mesh *RL* circuit.

To find the inverse matrix, we first find the determinant. Following the rules given in Appendix A, we obtain

$$\begin{aligned} |\mathbf{Z}(j\omega)| &= (1 + j\omega)(2 + j\omega)(1 + j\omega) - (1 + j\omega) - (1 + j\omega) \\ &= -4\omega^2 + j(3\omega - \omega^3) \end{aligned} \tag{7-165}$$

If we let the quantities Z_{ij} be the cofactors of $\mathbf{Z}(j\omega)$, then these cofactors are readily found as follows:

$$
\begin{aligned}
Z_{11} &= \begin{vmatrix} 2+j\omega & -1 \\ -1 & 1+j\omega \end{vmatrix} = 1-\omega^2+j3\omega \\
Z_{12} &= -\begin{vmatrix} -1 & -1 \\ 0 & 1+j\omega \end{vmatrix} = 1+j\omega = Z_{21} \\
Z_{13} &= \begin{vmatrix} -1 & 2+j\omega \\ 0 & -1 \end{vmatrix} = 1 = Z_{31} \\
Z_{22} &= \begin{vmatrix} 1+j\omega & 0 \\ 0 & 1+j\omega \end{vmatrix} = 1-\omega^2+j2\omega \\
Z_{23} &= -\begin{vmatrix} 1+j\omega & -1 \\ 0 & -1 \end{vmatrix} = 1+j\omega = Z_{32} \\
Z_{33} &= \begin{vmatrix} 1+j\omega & -1 \\ -1 & 2+j\omega \end{vmatrix} = 1-\omega^2+j3\omega
\end{aligned}
\tag{7-166}
$$

Substituting the above values in the general relation for the inverse of a 3×3 matrix,

$$
\mathbf{Z}(j\omega)^{-1} = \frac{1}{\det \mathbf{Z}(j\omega)} \begin{bmatrix} Z_{11} & Z_{21} & Z_{31} \\ Z_{12} & Z_{22} & Z_{32} \\ Z_{13} & Z_{23} & Z_{33} \end{bmatrix} \tag{7-167}
$$

we obtain

$$
\mathbf{Z}(j\omega)^{-1} = \frac{1}{-4\omega^2 + j(3\omega - \omega^3)} \begin{bmatrix} 1-\omega^2+j3\omega & 1+j\omega & 1 \\ 1+j\omega & 1-\omega^2+j2\omega & 1+j\omega \\ 1 & 1+j\omega & 1-\omega^2+j3\omega \end{bmatrix} \tag{7-168}
$$

It is readily verified that the product $\mathbf{Z}(j\omega)\mathbf{Z}(j\omega)^{-1}$ using the matrices given in Eqs. (7-164) and (7-168) is equal to the identity matrix.

It should be apparent that the general procedure given above becomes quite involved for networks having three or more independent variables, i.e., networks having three or more simultaneous equations in their impedance or admittance formulation. As a matter of fact, the network used in the example given above for a 3×3 impedance matrix was purposely selected so that the elements of the impedance matrix were polynomials rather than rational functions (ratios of polynomials). In the latter (and more usual) case, the computations would be considerably more complex. Since hand computations are so difficult, we now turn our attentions to the use of digital computational techniques.

In general, the implementation of algebraic manipulations of polynomials and functions of the type encountered above is cumbersome and difficult to achieve on a digital computer. The situation conceptually parallels the one discussed in connection with the introduction of numerical integration techniques in Sec. 4–5. There it was pointed out that, although it is not feasible to use numerical techniques to produce closed-form expressions for integrals, an alternative approach well suited to digital computers is to apply numerical integration techniques over successive small ranges of the independent variable and plot the resulting values of the integral. This is similar to the philosophy that we shall follow in applying numerical techniques to the solution of immittance equations. Thus, we may begin our study of such an application by considering the second case formulated at the beginning of this section. For convenience, let us use an impedance formulation since the identical techniques apply to the admittance case. The general set of equations to be considered may be written in the form

$$\mathbf{V} = \mathbf{Z}(j\omega_0)\mathbf{I} \tag{7–169}$$

where ω_0 is given some specific numerical value. The first subcase occurs when specific numerical values are also given for the elements of the $\mathbf{V}$ matrix. In this situation, the solution procedure consists of determining specific numerical values for the phasor elements of the matrix $\mathbf{I}$. The second and more general subcase, which includes the first, is the determination of the general solution to the set of equations given in (7–169) for unspecified excitation voltage phasors, i.e., for a general matrix $\mathbf{V}$. In this case, the solution procedure consists of finding the inverse of $\mathbf{Z}(j\omega_0)$, i.e., determining the elements of the square matrix $\mathbf{Z}(j\omega_0)^{-1}$, where this matrix is defined by the relation

$$\mathbf{I} = \mathbf{Z}(j\omega_0)^{-1}\mathbf{V} \tag{7–170}$$

Both of the subcases defined above are characterized by the fact that the elements of $\mathbf{Z}(j\omega_0)$ are complex, i.e., that they have the form $a + jb$, where the numerical values of a and b are known. The numerical techniques for solving either of these subcases are identical with the techniques presented in Chap. 3, except that the procedures must be applied to matrices with complex elements rather than real elements. An example of the application of one of these procedures (the Gauss-Jordan method for the solution of a set of simultaneous equations) to the first subcase has already been given in connection with Example 7–4.2. To generalize this approach and to provide a means for implementing it on the digital computer, we need only take the subroutine GJSEQ described in Sec. 3–4, and use a FORTRAN *type declaration* to specify that the variables are complex. Let us name the resulting subroutine CGJSEQ (for Complex Gauss-Jordan Simultaneous EQuation solution). If we consider that such a subroutine will be used to solve the equations defined by (7–169), then the argument listing may logically be given as (V, Z, AMPS, N), where V is the (known) one-dimensional complex array of voltage phasors, Z is the two-dimensional complex array defining the matrix $\mathbf{Z}(j\omega_0)$, AMPS is the (unknown) complex output array of mesh current phasors (the elements of $\mathbf{I}$), and N is the number of equations in the set. Thus, the identification statement for the subroutine will be

SUBROUTINE CGJSEQ (V, Z, AMPS, N) (7–171)

A listing of a subroutine for performing such a computation is given in Fig. 7-7.2. It is readily noted that this subroutine is almost identical with the subroutine GJSEQ described in Sec. 3-4. The only major difference is that a test has been added in the subroutine to make certain that the value of a given main diagonal element is not close to zero before using this value as a divisor for the other elements of the row.

```
      SUBROUTINE CGJSEQ (V, Z, AMPS, N)
C     SUBROUTINE TO SOLVE THE SET OF SIMULTANEOUS EQUATIONS V=Z*AMPS
C     SUBROUTINE USES GAUSS-JORDAN REDUCTION, Z MATRIX IS PRESERVED
C            V - COLUMN MATRIX OF KNOWN COMPLEX VARIABLES
C            Z - SQUARE NON-SINGULAR COMPLEX MATRIX
C         AMPS - OUTPUT COMPLEX COLUMN MATRIX
C            N - NUMBER OF EQUATIONS
      COMPLEX V(10), Z(10,10), AMPS(10), ZA(10,11), ALFA, BETA
C
C     ENTER Z MATRIX IN ZA ARRAY AND ENTER V AS
C     N+1TH COLUMN OF ZA ARRAY
      DO 23  I = 1, N
      DO 22  J = 1, N
   22 ZA(I,J) = Z(I,J)
   23 ZA(I,N+1) = V(I)
C
C     REDUCE MATRIX ZA TO SOLVE SIMULTANEOUS EQUATIONS.
C     NP IS NUMBER OF COLUMNS IN AUGMENTED MATRIX
      NP = N + 1
      DO 12  I = 1, N
C
C     SET MAIN DIAGONAL ELEMENTS TO UNITY, AFTER
C     TESTING DIAGONAL ELEMENT FOR ZERO VALUE
      ALFA = ZA(I,I)
      IF (CABS(ALFA).LT.0.0001) GO TO 33
      DO  5  J = I, NP
    5 ZA(I,J) = ZA(I,J) / ALFA
C
C     SET ELEMENTS OF ITH COLUMN TO ZERO
      DO 11  K = 1, N
      IF (K - I) 8, 11, 8
    8 BETA = ZA(K,I)
      DO 10  J = I, NP
   10 ZA(K,J) = ZA(K,J) - BETA * ZA(I,J)
   11 CONTINUE
   12 CONTINUE
C
C     SET OUTPUT MATRIX AMPS EQUAL TO LAST COLUMN OF MATRIX ZA
      DO 31  I = 1, N
   31 AMPS(I) = ZA(I,NP)
      RETURN
C
C     PRINT ERROR MESSAGE IF METHOD FAILS
   33 PRINT 34
   34 FORMAT (1H0, 29HMAIN DIAGONAL ELEMENT IS ZERO)
      RETURN
      END
```

Fig. 7-7.2 Listing of the subroutine CGJSEQ.

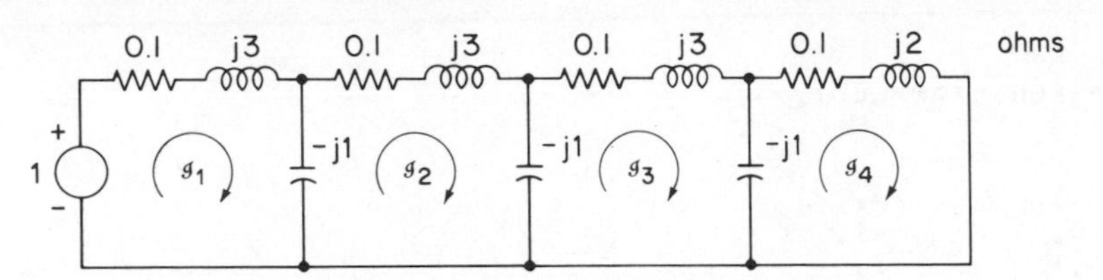

Fig. 7-7.3 Network for Example 7-7.1.

This test is necessary since reactances may be positive or negative and, therefore, it is possible for the impedance of a given main diagonal matrix element to be zero at some frequency and for the impedance matrix to be singular at that frequency. The only other difference between CGJSEQ and GJSEQ is the use of the variable Z in place of the variable R (a trivial difference), and the type declaration specifying that the arrays V, Z, and AMPS, and the variables ALFA and BETA, are complex. A summary of the properties of CGJSEQ is given in Table 7-7.1 (on page 454). An example of the use of this subroutine follows.

EXAMPLE 7-7.1. It is desired to find the phasor mesh currents for the network shown in Fig. 7-7.3, under conditions of sinusoidal steady-state excitation. All the elements are specified in terms of their impedances. We may use the subroutine ZMESH described in the preceding section to find the impedance matrix and the subroutine CGJSEQ to solve the resulting set of simultaneous equations. A listing of the main program, the input data, and the output data is given in Fig. 7-7.4. A separate program has been used to determine the validity of the computed mesh current phasors by inserting these values in the original equations and determining the resulting mesh voltage phasors. It is readily observed that excellent agreement with the original values is reached.

The second subcase that we will consider is the problem of finding the inverse of a given matrix $\mathbf{Z}(j\omega_0)$. This problem may be solved by using techniques similar to those used to find the inverse of a matrix composed of real elements as discussed in Sec. 3-5. In that section, the Gauss-Jordan procedure was extended to define an algorithm for finding the inverse of a matrix. The procedure was implemented in the subroutine MXINV. We need only modify the subroutine by changing the pertinent variables from real to complex to adapt this subroutine so that it may be used to find the inverse of a complex matrix. If we use the name CMXINV (for Complex MatriX INVersion) for this subroutine, it may be identified by a statement of the form

$$\text{SUBROUTINE CMXINV (Z, N, ZI)} \tag{7-172}$$

where Z is the two-dimensional complex (input) array containing the original matrix, N is the order of the matrix, and ZI is the two-dimensional complex (output) array containing the inverse matrix. The only significant difference between this subroutine and the subroutine MXINV is the fact that each main diagonal element is tested to make certain that it is not zero (or close to that value) before dividing the elements

```
INPUT DATA FOR EXAMPLE 7.7-1

 4 7
10          .1          3.
12          0.          -1.
20          .1          3.
23          0.          -1.
30          .1          3.
34          0.          -1.
40          .1          2.
1.0
```

```
OUTPUT DATA FOR EXAMPLE 7.7-1

                        * * * *  Z  MATRIX  * * * *

   .10  +J 2.00        0.00  +J 1.00        0.00  +J 0.00        0.00  +J 0.00
  0.00  +J 1.00         .10  +J 1.00        0.00  +J 1.00        0.00  +J 0.00
  0.00  +J 0.00        0.00  +J 1.00         .10  +J 1.00        0.00  +J 1.00
  0.00  +J 0.00        0.00  +J 0.00        0.00  +J 1.00         .10  +J 1.00


             * * * *  MESH  CURRENT  PHASORS  * * * *

I(1)= 7.3357892E-02+J-4.9628498E-01
I(2)=-9.7087287E-02+J-9.4259502E-05
I(3)= 2.3738821E-02+J 4.8667051E-01
I(4)= 2.4681416E-02+J-4.8420236E-01
```

```
TEST OF RESULTS OF EXAMPLE 7.7-1

          * * * *  COMPUTED  MESH  VOLTAGE  PHASORS  * * * *

V(1)= 1.0000000E+00+J-2.6645353E-15
V(2)=-5.3290705E-15+J-4.4408921E-16
V(3)=-3.5527137E-15+J-1.1102230E-16
V(4)= 0.            +J 1.5543122E-15
```

Fig. 7-7.4 Listing of the program, input data, and output data for Example 7-7.1.

of the corresponding row by the diagonal element. A listing of the subroutine CMXINV is given in Fig. 7–7.5. A summary of the properties of the subroutine is given in Table 7–7.2 (on page 457). An example of the use of the subroutine follows.

EXAMPLE 7–7.2. For the network shown in Fig. 7-7.3 with the mesh impedance equations determined in Example 7-7.1, it is desired to find the inverse of the impedance matrix. A listing of a main program and the necessary input data for doing this are given in Fig. 7-7.6. The resulting output data are also shown in this figure. By using a separate program,

```
C     MAIN PROGRAM FOR EXAMPLE 7.7-1
C            N - NUMBER OF MESHES
C           NZ - NUMBER OF IMPEDANCES
C        IMPED - ARRAY OF COMPLEX IMPEDANCE VALUES
C        MESH1 - FIRST ARRAY OF MESH NUMBERS
C        MESH2 - SECOND ARRAY OF MESH NUMBERS
C            V - ARRAY OF KNOWN MESH VOLTAGE PHASORS
C            Z - MESH IMPEDANCE MATRIX
C         AMPS - OUTPUT ARRAY OF MESH CURRENT PHASORS
      DIMENSION MESH1(20), MESH2(20)
      COMPLEX V(10), Z(10,10), AMPS(10), IMPED(20)
C
C     READ INPUT DATA
      READ 2, N, NZ
    2 FORMAT (2I2)
      READ 4, (MESH1(I), MESH2(I), IMPED(I), I = 1, NZ)
    4 FORMAT (2I1, 8X, 2E10.0)
      READ 6, (V(I), I = 1, N)
    6 FORMAT (8E10.0)
C
C     CALL THE SUBROUTINE ZMESH TO CONSTRUCT
C     THE IMPEDANCE MATRIX Z
      CALL ZMESH (N, NZ, IMPED, MESH1, MESH2, Z)
C
C     PRINT A LISTING OF THE VALUES OF THE
C     ELEMENTS OF THE IMPEDANCE MATRIX Z
      PRINT 9
    9 FORMAT (///, 23X, 27H* * * *  Z  MATRIX  * * * */)
      DO 13 J = 1, N
      PRINT 12, (Z(J,I), I = 1, N)
   12 FORMAT (4(1X, F5.2, 1X, 2H+J, F5.2, 5X))
   13 CONTINUE
C
C     CALL THE SUBROUTINE CGJSEQ TO SOLVE
C     THE SET OF SIMULTANEOUS EQUATIONS
      CALL CGJSEQ (V, Z, AMPS, N)
C
C     PRINT AN OUTPUT LISTING OF THE VALUES
C     OF THE MESH CURRENT PHASORS FOR THE NETWORK
      PRINT 16
   16 FORMAT (///, 16X, 40H* * * *  MESH  CURRENT  PHASORS  * * * */)
      PRINT 18, (I, AMPS(I), I = 1, N)
   18 FORMAT (1X, 2HI(, I1, 2H)=, E14.7,2H+J, E14.7)
      STOP
      END
```

Fig. 7-7.4 (b)

it is readily verified that the product of the original matrix and the inverse matrix produces a matrix whose elements very closely approximate those of the identity matrix. The results of such a computation are also given in the figure. The details of this verification are left to the reader as an exercise.

Situations where it is desired to use digital computational techniques to simulate the first case mentioned at the beginning of this section, i.e., cases in which the solution of a set of immittance equations is desired for a range of values of ω, are easily handled by the general approach referred to earlier—that is, by determining the numerical values of the elements of the immittance matrix at each of a sequence of

```
      SUBROUTINE CMXINV (Z, N, ZI)
C     SUBROUTINE TO FIND THE INVERSE OF A COMPLEX MATRIX Z
C     USING GAUSS-JORDAN REDUCTION.  Z MATRIX IS PRESERVED
C            Z - COMPLEX MATRIX WHICH IS TO BE INVERTED
C            N - ORDER OF MATRIX
C           ZI - OUTPUT INVERSE MATRIX
      COMPLEX Z(10,10), ZA(10,20), ZI(10,10), ALFA, BETA
C
C     ENTER Z ARRAY INTO ZA ARRAY AND SET
C     LAST N COLUMNS OF ZA ARRAY TO IDENTITY MATRIX
      DO 26  I = 1, N
      DO 24  J = 1, N
      ZA(I,J) = Z(I,J)
      NJ = N + J
   24 ZA(I,NJ) = (0.,0.)
      NI = N + I
   26 ZA(I,NI) = (1.,0.)
C
C     REDUCE MATRIX ZA SO THAT FIRST N COLUMNS
C     ARE SET EQUAL TO THE IDENTITY MATRIX
      NP = 2 * N
      DO 12  I = 1, N
C
C     SET MAIN DIAGONAL ELEMENTS TO UNITY AFTER
C     TESTING FOR ZERO VALUE
      ALFA = ZA(I,I)
      IF (CABS(ALFA).LT.0.0001) GO TO 35
      DO  5  J = I, NP
    5 ZA(I,J) = ZA(I,J) / ALFA
C
C     SET ELEMENTS OF ITH COLUMN TO ZERO
      DO 11  K = 1, N
      IF (K - I) 8, 11, 8
    8 BETA = ZA(K,I)
      DO 10  J = I, NP
   10 ZA(K,J) = ZA(K,J) - BETA * ZA(I,J)
   11 CONTINUE
   12 CONTINUE
C
C     SET INVERSE MATRIX ZI EQUAL TO LAST
C     N COLUMNS OF ZA ARRAY
      DO 33  J = 1, N
      JN = J + N
      DO 33  I = 1, N
   33 ZI(I,J) = ZA(I,JN)
      RETURN
C
C     PRINT ERROR MESSAGE IF METHOD FAILS
   35 PRINT 36
   36 FORMAT (1H0, 29HMAIN DIAGONAL ELEMENT IS ZERO/)
      END
```

Fig. 7-7.5 Listing of the subroutine CMXINV.

```
C     MAIN PROGRAM FOR EXAMPLE 7.7-2
C            Z - MATRIX TO BE INVERTED
C            N - ORDER OF MATRIX
C           ZI - INVERSE MATRIX
      COMPLEX Z(10,10), ZI(10,10)
      N = 4
C
C     READ THE ELEMENTS OF THE MATRIX TO BE INVERTED
      READ  3, ((Z(I,J), J = 1, N), I = 1, N)
    3 FORMAT (4(1X, F5.2, 3X, F5.2, 5X))
C
C     CALL THE SUBROUTINE CMXINV TO COMPUTE THE INVERSE
      CALL CMXINV (Z, N, ZI)
C
C     PRINT AN OUTPUT LISTING OF THE INVERSE MATRIX
      PRINT 6
    6 FORMAT (///, 6X, 23HINVERSE MATRIX ELEMENTS/)
      DO 11 I = 1, N
      DO 9 J = 1, N
    9 PRINT 10, I, J, ZI(I,J)
   10 FORMAT (1X, 3HZI(, 2I1, 2H)=, E12.5, 2H+J, E12.5)
   11 PRINT 12
   12 FORMAT (1H0)
      STOP
      END
```

```
INPUT DATA FOR EXAMPLE 7.7-2

  .10 +J 2.00      0.00 +J 1.00      0.00 +J 0.00      0.00 +J 0.00
 0.00 +J 1.00       .10 +J 1.00      0.00 +J 1.00      0.00 +J 0.00
 0.00 +J 0.00      0.00 +J 1.00       .10 +J 1.00      0.00 +J 1.00
 0.00 +J 0.00      0.00 +J 0.00      0.00 +J 1.00       .10 +J 1.00
```

```
OUTPUT DATA FOR EXAMPLE 7.7-2              TEST OF RESULTS OF EXAMPLE 7.7-2

     INVERSE MATRIX ELEMENTS                         MATRIX TIMES INVERSE

ZI(11)= 7.33579E-02+J-4.96285E-01         Z*ZI(11)= 1.00000E+00+J-2.66454E-15
ZI(12)=-9.70873E-02+J-9.42595E-05         Z*ZI(12)=-9.38138E-15+J 1.15463E-14
ZI(13)= 2.37388E-02+J 4.86671E-01         Z*ZI(13)= 3.55271E-15+J-1.77636E-15
ZI(14)= 2.46814E-02+J-4.84202E-01         Z*ZI(14)=-3.55271E-15+J-1.77636E-15
ZI(21)=-9.70873E-02+J-9.42595E-05         Z*ZI(21)=-5.32907E-15+J-4.44089E-16
ZI(22)= 1.94184E-01+J-9.52021E-03         Z*ZI(22)= 1.00000E+00+J 5.32907E-15
ZI(23)=-9.61447E-02+J-9.70967E-01         Z*ZI(23)= 3.55271E-15+J-1.77636E-15
ZI(24)=-9.42595E-04+J 9.70873E-01         Z*ZI(24)=-1.77636E-15+J-2.22045E-15
ZI(31)= 2.37388E-02+J 4.86671E-01         Z*ZI(31)=-3.55271E-15+J-1.11022E-16
ZI(32)=-9.61447E-02+J-9.70967E-01         Z*ZI(32)=-3.55271E-15+J-3.09822E-15
ZI(33)= 1.69503E-01+J 4.74682E-01         Z*ZI(33)= 1.00000E+00+J-4.44089E-16
ZI(34)=-1.20826E-01+J-4.86765E-01         Z*ZI(34)= 7.10543E-15+J-8.88178E-16
ZI(41)= 2.46814E-02+J-4.84202E-01         Z*ZI(41)= 0.         +J 1.55431E-15
ZI(42)=-9.42595E-04+J 9.70873E-01         Z*ZI(42)= 7.10543E-15+J-4.88498E-15
ZI(43)=-1.20826E-01+J-4.86765E-01         Z*ZI(43)=-1.77636E-15+J 0.
ZI(44)= 1.70445E-01+J-4.96191E-01         Z*ZI(44)= 1.00000E+00+J-8.88178E-16
```

Fig. 7-7.6 Listing of the program, input data, and output data for Example 7-7.2.

TABLE 7-7.1 Summary of the Characteristics of the Subroutine CGJSEQ

Identifying Statement: SUBROUTINE CGJSEQ (V, Z, A, N)
Purpose: To solve a set of simultaneous equations having the form

$$\mathbf{V} = \mathbf{ZA}$$

i.e., given the elements of the matrices **V** and **Z**, to find the elements of **A**. The quantities v_i, the elements of **V**, z_{ij}, the elements of **Z**, and a_i, the elements of **A** are specified as complex quantities. Thus, this subroutine is ideally suited to application to phasor problems in ac network theory.
Additional Subprograms Required: None.
Input Arguments:

V The one-dimensional complex array of variables V(I) in which are stored the complex values of the known quantities v_i, the elements of the matrix **V**. For sinusoidal steady-state phasor problems, these variables represent the sum of the phasor output voltages of the voltage sources located in the ith mesh of the network.

Z The two-dimensional complex array of variables Z(I, J) in which are stored the complex values of the coefficients z_{ij} of the square matrix **Z**. For sinusoidal steady-state phasor problems, these quantities represent the values of the self and mutual complex impedances.

N The order of the matrix **Z**, i.e., the number of simultaneous equations to be solved.

Output Argument:

A The one-dimensional complex array of variables A(I) in which are stored the complex values of the unknown variables a_i of the matrix **A**. For sinusoidal steady-state phasor problems, these variables represent the phasor values of the mesh current in the ith mesh.

Notes:

1. The dimensions of the variables used in this subroutine are V(10), Z(10, 10), and A(10).
2. The subroutine uses a Gauss-Jordan reduction algorithm similar to that used in the subroutine GJSEQ. The original **Z** matrix is preserved.
3. If the algorithm fails, i.e., if an attempt is made to divide the elements of any row by a very small number, an error message is printed.

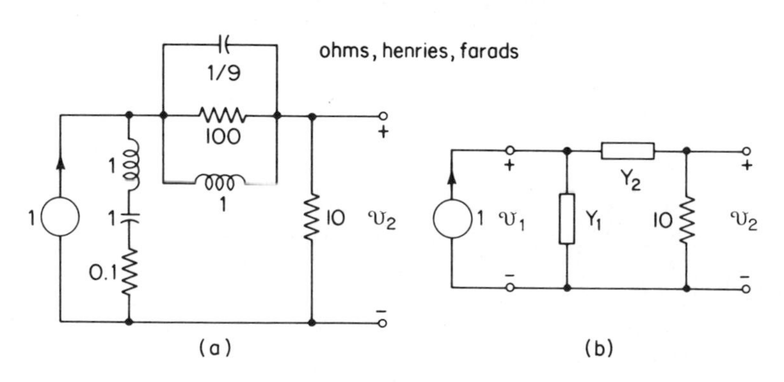

Fig. 7-7.7 Network for Example 7-7.3.

```
C     MAIN PROGRAM FOR EXAMPLE 7.7-3
C          AMPS - ARRAY OF PHASOR CURRENT EXCITATIONS
C             Y - COMPLEX NODE ADMITTANCE MATRIX
C             V - OUTPUT ARRAY OF PHASOR NODE VOLTAGES
C         OMEGA - FREQUENCY VARIABLE
      COMPLEX AMPS(10), Y(10,10), V(10), Y1, Y2, OMJ
      DIMENSION P(5,101)
C
C     INITIALIZE THE VARIABLES
      OMEGA = 0.1
      AMPS(1)=(10.,0.)
      AMPS(2)=(0.,0.)
      P(1,1) = 101.0
C
C     COMPUTE 50 VALUES OF THE PHASOR NODE
C     VOLTAGES BY USING CGJSEQ SUBROUTINE
C     TO SOLVE SIMULTANEOUS EQUATIONS FOR 50
C     VALUES OF FREQUENCY.  STORE THE MAGNITUDE
C     OFTHE PHASOR V2 FOR PLOTTING
      DO 16  I = 2, 51
      OM2 = OMEGA * OMEGA
      OMJ = CMPLX (0.0, OMEGA)
      Y1 = OMJ / (1.-OM2+0.1*OMJ)
      Y2 = (1.-OM2/9.+.01*OMJ) / OMJ
      Y(1,1) = Y1 + Y2
      Y(1,2) = - Y2
      Y(2,1) = Y(1,2)
      Y(2,2) = Y2 + 0.1
      CALL CGJSEQ (AMPS, Y, V, 2)
      P(1,I) = CABS(V(2))
      OMEGA = OMEGA + 0.1
   16 CONTINUE
C
C     CALL THE PLOTTING SUBROUTINE TO
C     PLOT THE MAGNITUDE OF THE PHASOR V2
      PRINT 18
   18 FORMAT (1H1)
      CALL PLOTS (P, 1, 51, 100)
      STOP
      END
```

Fig. 7-7.8 Listing of the program for Example 7-7.3.

values of ω, solving the simultaneous equations or finding the inverse matrix for each of these cases, and, if desired, plotting the results. An example of this procedure follows.

EXAMPLE 7–7.3. It is desired to find the voltage phasor $\mathscr{V}_2$ for the network shown in Fig. 7-7.7(a) for a range of values of ω from 0 to 5 rad/s. It is assumed that the input phasor is kept constant at unity magnitude and zero phase angle for all frequencies. As a first solution step, let us redraw the circuit using the two admittances Y_1 and Y_2 as shown in Fig. 7-7.7(b). The admittance equations for this circuit are

$$\begin{bmatrix} 1 \\ 0 \end{bmatrix} = \begin{bmatrix} Y_1 + Y_2 & -Y_2 \\ -Y_2 & Y_2 + 0.1 \end{bmatrix} \begin{bmatrix} \mathscr{V}_1 \\ \mathscr{V}_2 \end{bmatrix}$$

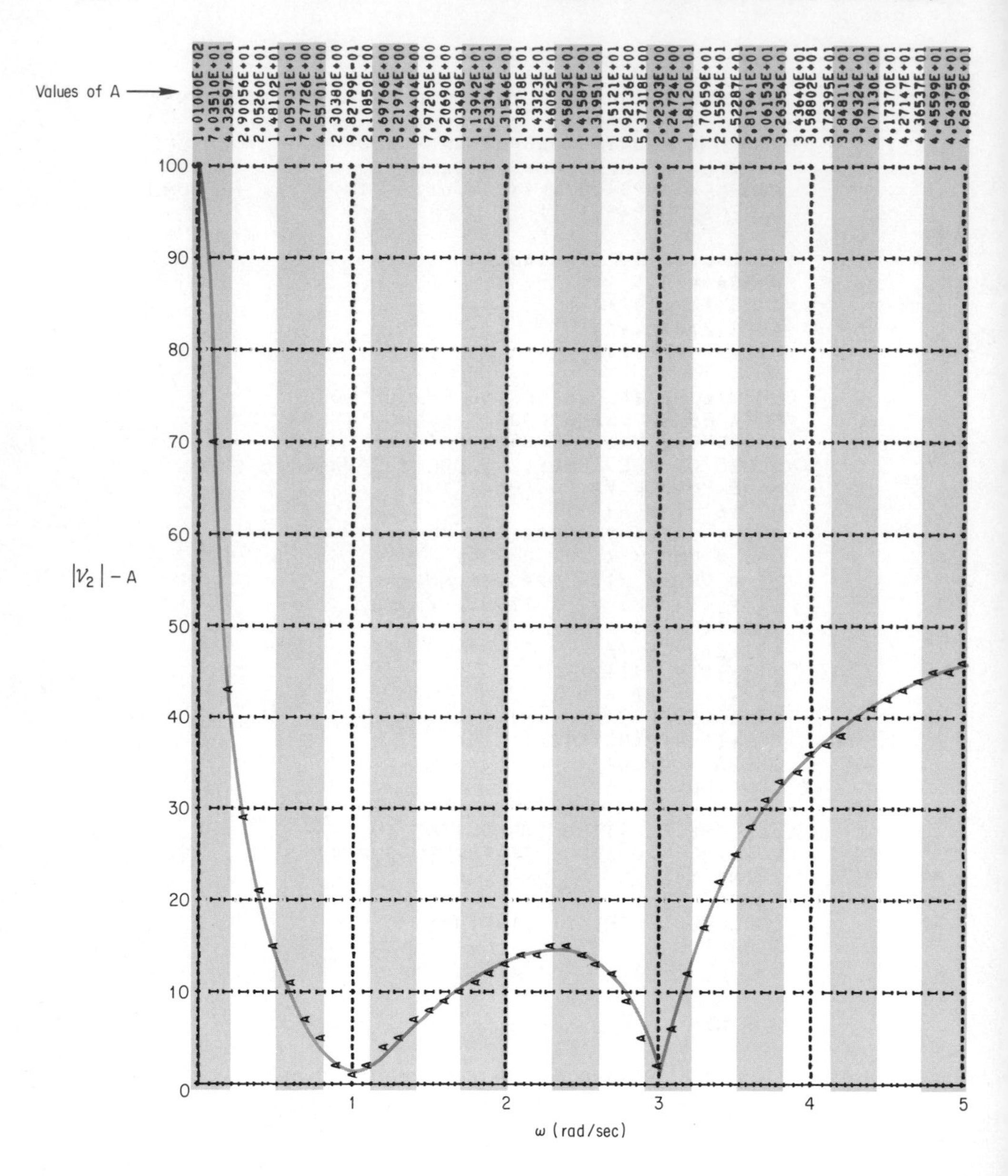

Fig. 7-7.9 Output plot for Example 7-7.3.

TABLE 7-7.2 Characteristics of the Subroutine CMXINV

Identifying Statement: SUBROUTINE CMXINV (Z, N, ZI)
Purpose: To find the complex elements of a matrix which is the inverse of a non-singular square matrix **Z** of order n which has complex elements.
Additional Subprograms Required: None.
Input Arguments:
- Z The two-dimensional array of complex variables Z(I, J) in which are stored the elements z_{ij} of the non-singular complex matrix **Z** which is to be inverted.
- N The order n of the matrix **Z**.

Output Argument:
- ZI The two-dimensional array of complex variables ZI(I, J) in which are stored the values of the elements of the complex matrix $\mathbf{Z}^{-1}$.

Notes:
1. The variables of the subroutine are dimensioned as follows: Z(10, 10), ZI(10, 10).
2. The main diagonal elements of the matrix **Z** must be non-zero.
3. The original **Z** matrix is preserved.

From the original circuit we may readily find expressions for the values of Y_1 and Y_2 as functions of ω. These are

$$Y_1 = \frac{1}{(1/j\omega) + j\omega + 0.1} = \frac{j\omega}{1 - \omega^2 + 0.1j\omega}$$

$$Y_2 = \frac{j\omega}{9} + \frac{1}{j\omega} + 0.01 = \frac{1 - (\omega^2/9) + 0.01j\omega}{j\omega}$$

The relations given above are readily implemented in a FORTRAN program, which may be set up to calculate the elements of the admittance matrix for 51 values of ω from 0 to 5 rad/s. The subroutine CGJSEQ may then be used to calculate the node voltage phasors, and the subroutine PLOT5 to display the results. A listing of a main program for accomplishing this is given in Fig. 7-7.8. For simplicity, only the magnitude of the phasor $\mathscr{V}_2$ as a function of ω has been plotted. A plot of the output is shown in Fig. 7-7.9.

7-8 NETWORK FUNCTIONS

In Sec. 7-5 we introduced the concept of immittance, as embodied in impedance and admittance. These latter quantities were defined, respectively, as the ratio of a voltage phasor to a current phasor, and the ratio of a current phasor to a voltage phasor, with the added conditions that the voltage and current phasors were both defined at the same pair of terminals, and that their relative reference polarities followed the usual associated convention for two-terminal elements. It was also pointed out that impedance was reciprocal to admittance, i.e., once one of these quantities was known, the other could be easily found by simple inversion. The general form of an immittance was shown to be that of a rational function, i.e., a ratio of polynomials which were functions of the variable $j\omega$. In this section we shall show that impedance and admittance are two examples of a more general class of rational functions which are referred to as *network functions*. More specifically, we may define a network function as the ratio of a response phasor to an excitation phasor. In general, it is not necessary that these phasors be located at the same pair of terminals (as they must be to satisfy the definitions of immittances which we have used previously). As an example of a network function, consider the circuit shown in Fig. 7-8.1. Let us assume that we desire to find the network function defined as the ratio of the response phasor $\mathscr{I}_2$ to the excitation phasor $\mathscr{V}_1$. We may begin by writing the phasor KVL equations around the two meshes. Using matrix notation, we obtain

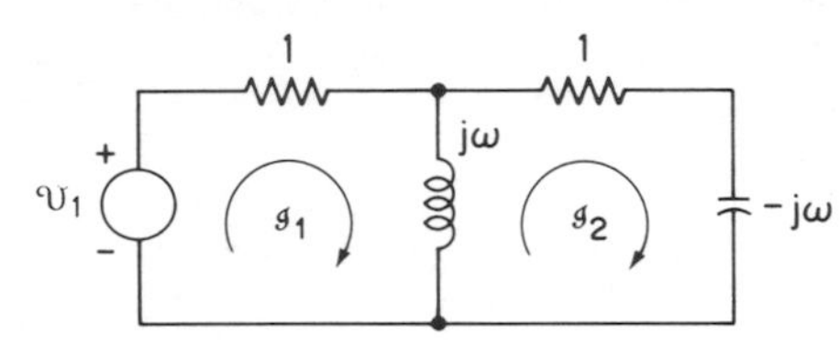

Fig. 7-8.1 An *RLC* network.

$$\begin{bmatrix} \mathscr{V}_1 \\ 0 \end{bmatrix} = \begin{bmatrix} 1 + j\omega & -j\omega \\ -j\omega & 1 + j\left(\omega - \frac{1}{\omega}\right) \end{bmatrix} \begin{bmatrix} \mathscr{I}_1 \\ \mathscr{I}_2 \end{bmatrix} \tag{7-173}$$

To determine $\mathscr{I}_2$ as a function of $\mathscr{V}_1$, the inverse set of equations is required. Using the techniques of Sec. 3-1, we obtain

$$\begin{bmatrix} \mathscr{I}_1 \\ \mathscr{I}_2 \end{bmatrix} = \frac{\omega}{2\omega + j(2\omega^2 - 1)} \begin{bmatrix} 1 + j\left(\omega - \frac{1}{\omega}\right) & j\omega \\ j\omega & 1 + j\omega \end{bmatrix} \begin{bmatrix} \mathscr{V}_1 \\ 0 \end{bmatrix}$$

The desired network function is easily seen to be

$$\frac{\mathscr{I}_2}{\mathscr{V}_1} = \frac{j(\omega)^2}{2\omega + j(2\omega^2 - 1)} \tag{7-174}$$

Since it represents the ratio of a current phasor to a voltage phasor, this network function has the dimensions of admittance (units of mhos). The function, however, does not satisfy the definition previously given for an admittance, since the excitation and response phasors are located at different points in the network. We shall use the term *transfer admittance* to define such a network function. The type of admittance defined earlier in this chapter (referring to two phasors located at the *same pair* of network terminals) may be called a *driving-point admittance* to emphasize its character. In a similar manner, we may define a *transfer impedance* as the ratio of a voltage response phasor to a current excitation phasor, where the phasors are *not* located at the same terminal pair. A network function relating a voltage phasor to a current phasor at the same pair of terminals is called a *driving-point impedance*. The term *transfer immittance* is used to collectively refer to transfer admittances and transfer impedances. Similarly, the term *driving-point immittance* is used to collectively refer to driving-point impedances and admittances.

A second class of transfer functions consists of *dimensionless transfer functions.* The first of these is defined as the ratio of a voltage response phasor to a voltage excitation phasor. Such a network function is called a *voltage transfer function.* Similarly, we may define a *current transfer function* as the ratio of a current response phasor to a current excitation phasor. The various network functions described above are summarized in Table 7–8.1. Note that they are all defined as the ratio of response to excitation, and *not* as the ratio of excitation to response. It is possible to also define network functions relating charge and/or flux linkages to voltage, current, charge, etc. In practice, however, these are rarely used.

TABLE 7-8.1 Types of Network Functions

Driving-Point Functions		Transfer Functions			
Immittance Functions		Immittance Functions		Dimensionless Functions	
Driving-Point Impedance	Driving-Point Admittance	Transfer Impedance	Transfer Admittance	Transfer Voltage Ratio	Transfer Current Ratio
$\frac{\mathcal{V}}{\mathcal{I}}$	$\frac{\mathcal{I}}{\mathcal{V}}$	$\frac{\mathcal{V}}{\mathcal{I}}$	$\frac{\mathcal{I}}{\mathcal{V}}$	$\frac{\mathcal{V}_2}{\mathcal{V}_1}$	$\frac{\mathcal{I}_2}{\mathcal{V}_1}$

In analyzing the properties of a given network function, it is frequency advantageous to make a plot of its characteristics as the frequency variable ω is varied over some range. The range usually starts from 0 rad/s and goes to some value of frequency which is large enough so that all the significant variations of the network function have been included. Such a plot shows the *frequency response* of the network function. Since the value of the network function is complex, two loci are generally required to display its frequency response. In this text we shall usually use the magnitude and the phase as the quantities plotted against frequency.[13]

[13]In studying some properties of networks, it is also useful to plot the real part and the imaginary part of the network function.

As an introduction to the study of the frequency response of network functions, let us consider the *RC* network shown in Fig. 7–8.2(a). The network function that relates the response voltage phasor $\mathscr{V}_2$ to the excitation voltage phasor $\mathscr{V}_1$ is

$$\frac{\mathscr{V}_2}{\mathscr{V}_1} = \frac{1}{1 + j\omega RC} \tag{7–175}$$

Sketches of the magnitude and phase of this network function as ω is varied are shown in Figs. 7–8.2(b) and 7–8.2(c). Note that in the magnitude curve shown in Fig. 7–8.2(b), the magnitude is unity at $\omega = 0$, but that it approaches zero as ω approaches ∞. Such a function is referred to as a *low-pass function*, since the region where its magnitude is large is lower in frequency than the region where its magnitude is small and, thus, more of any low-frequency sinusoidal excitations which are applied are "passed" than are high-frequency ones. Since the function is only of first-order in $j\omega$, it may also be referred to as a *first-order low-pass function.* Note

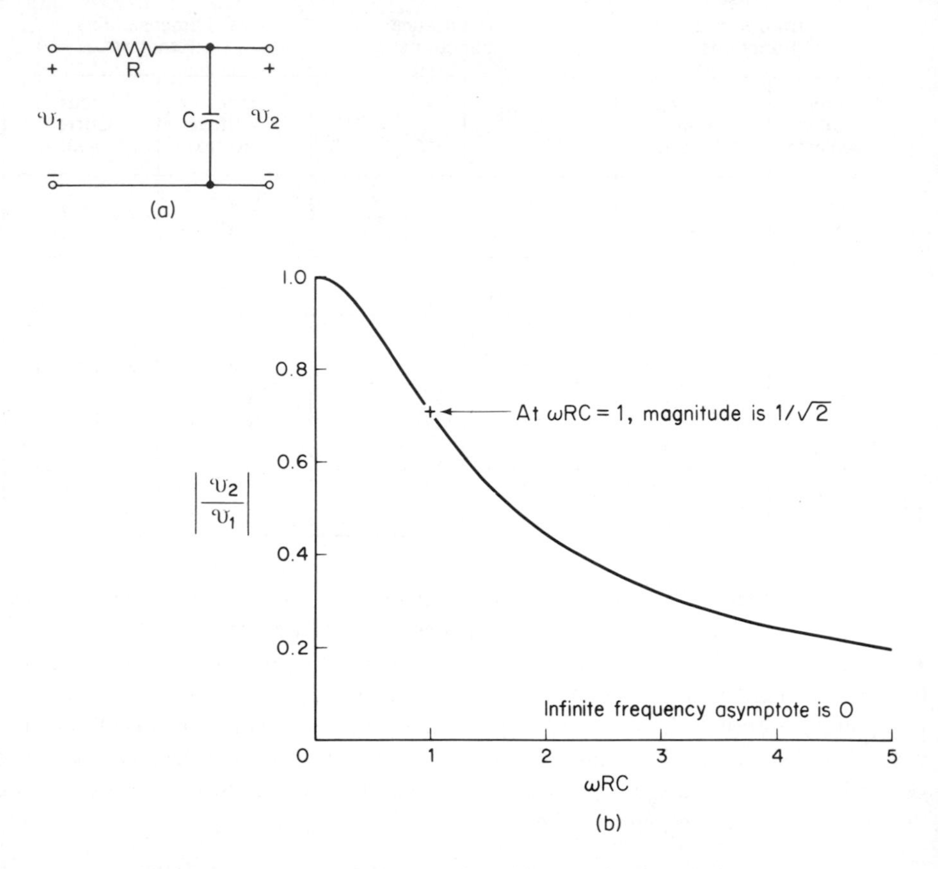

Fig. 7-8.2 A low-pass *RC* network and its magnitude and phase plots.

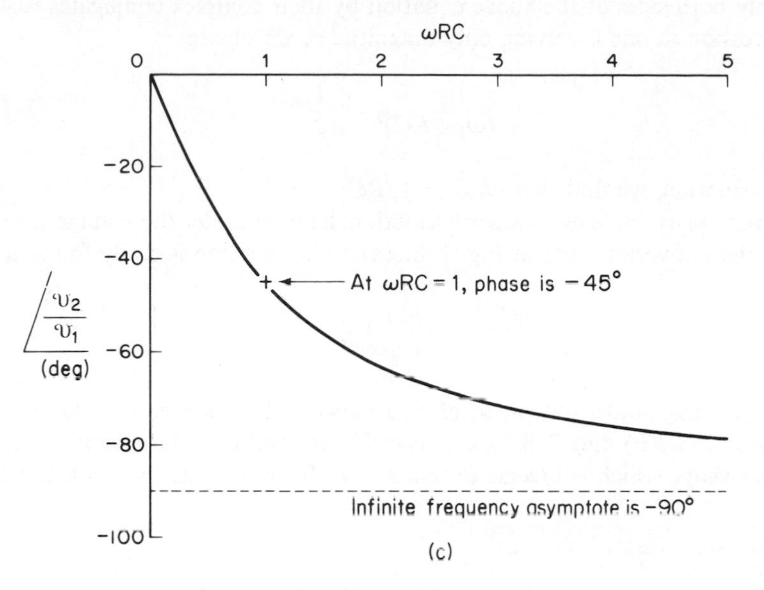

Fig. 7-8.2 Continued

that the phase of the network function as shown in Fig. 7–8.2(c) varies from 0° (at $\omega = 0$) to $-90°$ (when ω becomes very large). Thus, the response sinusoid will lag behind the excitation sinusoid by a phase angle of 0 to 90°, the exact angle depending on the value of ω. Considering this, it is customary to refer to a network of this type as a *lag network*. In general, for a network with a low-pass characteristic, each reactive element such as an inductor or a capacitor will contribute 90° to the total phase lag provided by the network.

One of the important characteristics of low-pass functions of the type given in Eq. (7–175) is the frequency at which the magnitude characteristic decreases to 0.707 of its zero frequency value. The frequency at which this occurs is called the *3-dB frequency*, where dB is an abbreviation for *decibel*, a logarithmic measure of the magnitude of the network function which is defined by the relation

$$\text{dB} = 20 \log |N(j\omega)| \tag{7–176}$$

where $N(j\omega)$ is the network function and log is the logarithm to the base 10. For a transfer function, when the magnitude drops to 0.707 of its zero frequency value, the power being transmitted drops to the square of this, i.e., one-half. Thus the 3-dB frequency is also referred to as the *half-power frequency*. The 3-dB frequency for the function given in Eq. (7–175) is readily found by noting that the magnitude at $\omega = 0$ rad/s is unity. Thus, at the 3-dB frequency $\omega_{3\,\text{dB}}$, we have

$$\frac{1}{1 + j\omega_{3\,\text{dB}}RC} = 0.707 = \frac{1}{\sqrt{2}}$$

If we multiply both sides of the above equation by their complex conjugates to convert the expression to one involving only magnitudes, we obtain

$$\frac{1}{1+(\omega_{3\text{ dB}}RC)^2} = \frac{1}{2} \tag{7-177}$$

Solving this equation, we find that $\omega_{3\text{ dB}} = 1/RC$.

As a second example of a network function, let us consider the voltage transfer function for the network shown in Fig. 7-8.3(a). The function is easily found to be

$$\frac{\mathcal{V}_2}{\mathcal{V}_1} = \frac{j\omega RC}{1 + j\omega RC} \tag{7-178}$$

Sketches of the magnitude and phase of this network function as ω is varied are shown in Figs. 7-8.3(b) and 7-8.3(c). It is easily verified that the magnitude characteristic has a shape which is inverse to that shown for a low-pass network function

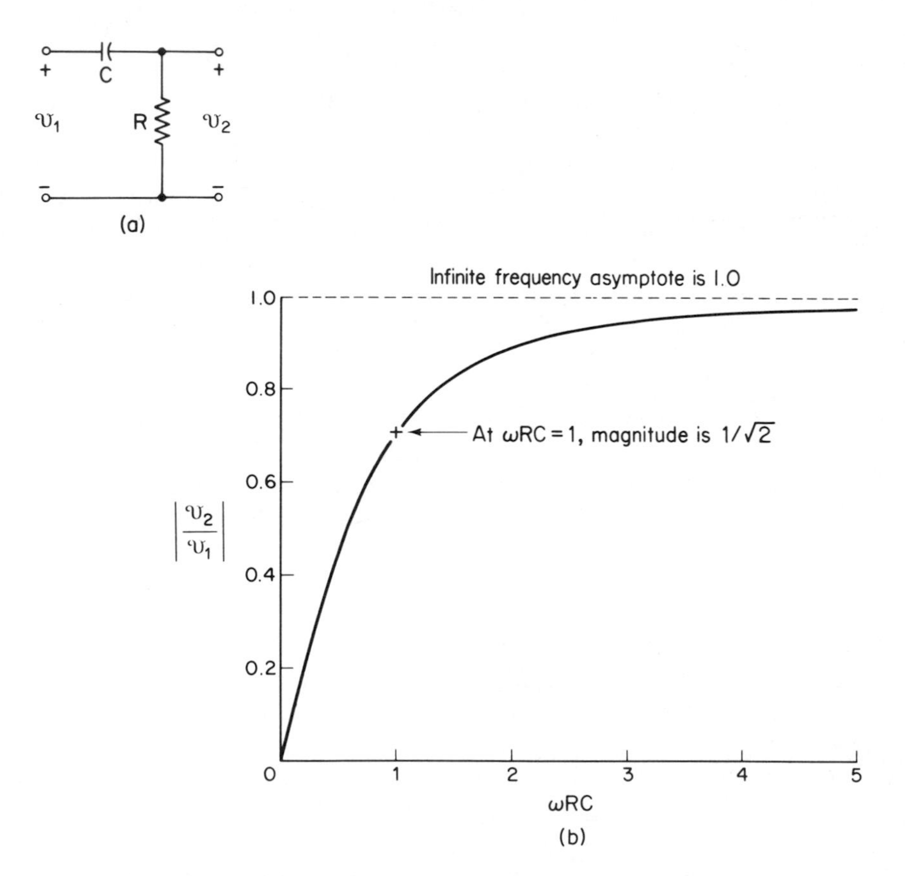

Fig. 7-8.3 A high-pass *RC* network and its magnitude and phase plots.

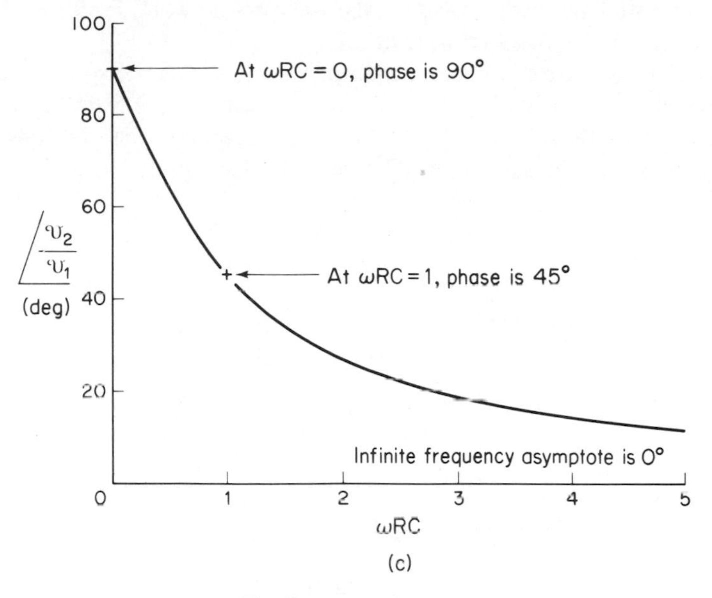

Fig. 7-8.3 Continued

in Fig. 7–8.2(b). Since this network function provides most of its transmission in the high-frequency range, and since it has one reactive element, it is appropriately referred to as a *first-order high-pass network function.* Note that the phase of the network function as shown in Fig. 7–8.3(c) varies from 90° (at $\omega = 0$) to 0° (for the range where ω becomes very large). Therefore, the response sinusoid will lead the excitation sinusoid by a phase angle of 90° to 0°, depending on the frequency. Thus, the network is appropriately called a *lead network.* In general, each reactive element, such as an inductor or a capacitor, in a high-pass network will contribute 90° to the total phase shift of the network function. It should be noted that *causality,* i.e., the principle that an output cannot precede the input that produces it, naturally precludes the output sinusoid from "leading" the input sinusoid. Thus, a lead network such as the one shown in Fig. 7–8.3(a) actually produces a negative phase which ranges from −270° to −360° rather than a positive phase going from +90° to 0°. However, since individual cycles of the sinusoidal excitation lose their identity during sinusoidal steady-state conditions, it is customary to refer to such a situation by using the term lead rather than lag.

In a manner similar to that used for low-pass functions, high-pass functions are frequently described in terms of the frequency at which the magnitude characteristic decreases to 0.707 of its infinite frequency value. The frequency at which this occurs is also referred to as the 3-dB frequency. For the high-pass network shown in Fig. 7–8.3(a), let us write the network function in the form

$$\frac{\mathscr{V}_2}{\mathscr{V}_1} = \frac{1}{1 + (1/j\omega RC)} \tag{7-179}$$

A development similar to that made for the low-pass network readily shows that the 3-dB frequency for this function is $1/RC$.

As a slightly more complicated example of a low-pass network, let us consider the network shown in Fig. 7-8.4(a). A second-order voltage transfer function may be defined by using the voltage phasor $\mathscr{V}_C$ as the response quantity, and the input voltage phasor $\mathscr{V}_1$ as the excitation. The network function is

$$\frac{\mathscr{V}_C}{\mathscr{V}_1} = \frac{1}{1 - \omega^2 LC + j\omega RC} \tag{7-180}$$

For convenience in emphasizing the properties of such a network function, let us choose unity values for L and C. Thus the network function may be written[14]

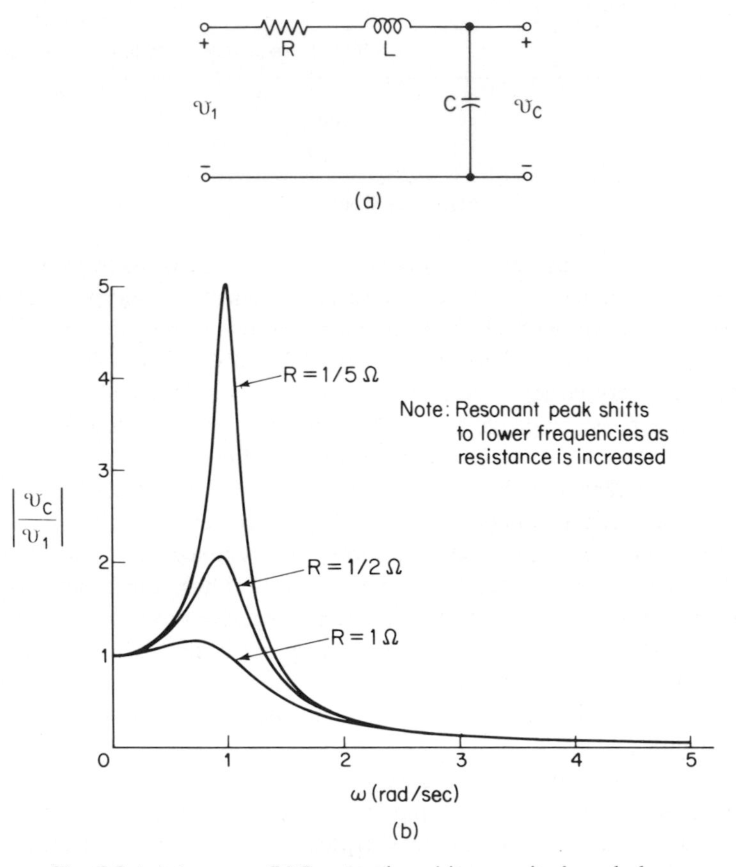

Fig. 7-8.4 A low-pass *RLC* network and its magnitude and phase plots.

[14]We will show in Sec. 8-8 that such a choice of element values, in effect, constitutes a *frequency and impedance normalization* of the network's frequency characteristics, and thus it does not affect the general properties of the network function.

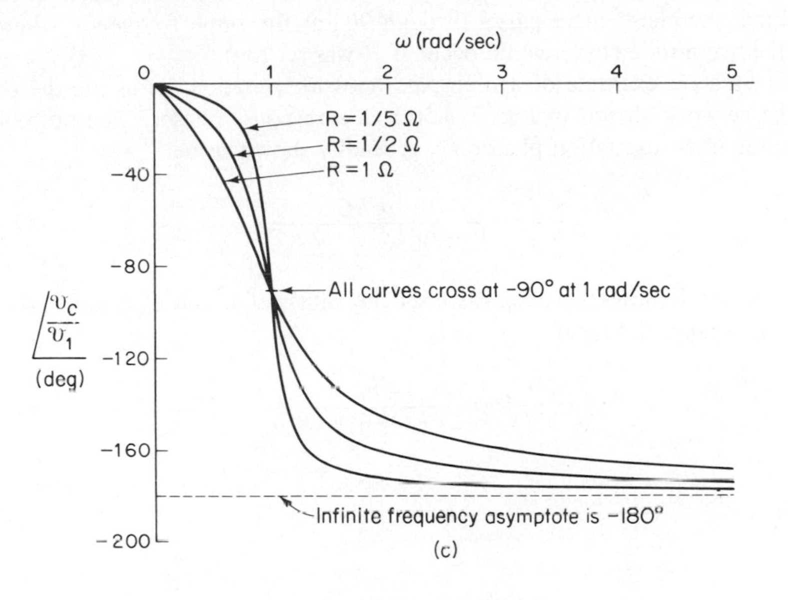

Fig. 7-8.4 Continued

$$\frac{\mathscr{V}_C}{\mathscr{V}_1} = \frac{1}{1 - \omega^2 + j\omega R} \tag{7-181}$$

The magnitude and phase of the function defined in Eq. (7–181) are shown for several values of R in Figs. 7–8.4(b) and 7–8.4(c). There are several points of interest that may be observed in these frequency response curves. First it should be noted that the magnitude is non-zero at $\omega = 0$, but that it approaches zero as ω approaches ∞. This behavior is readily verified from Eq. (7–181), since the function equals unity when $\omega = 0$, and its magnitude is inversely proportional to ω^2 for large values of ω. Thus the function is a low-pass function. Next, note that the phase or argument of the denominator polynomial goes from 0° at $\omega = 0$, to 90° when $\omega \approx 1$ or, more generally, when $\omega \approx 1/\sqrt{LC}$, and to $-180°$ when ω becomes very large. Thus, the phase of the entire network function goes from 0° to $-180°$. This result is consistent with the statement made for the first-order low-pass network, namely, that each reactive element in a low-pass network contributes 90° to the total phase shift of the function. A further examination of the various curves shown in Figs. 7–8.4(b) and 7–8.4(c) shows that as the value R of the resistor is decreased, the magnitude curve develops a peak at $\omega = \sqrt{1 - (R^2/2)}$ rad/s. As R approaches 0, the frequency approaches 1 rad/s. In the more general case, where L and C are not set to unity, this peak appears at

$$\omega = \frac{1}{\sqrt{LC}}\left[1 - \frac{R^2C}{2L}\right]^{1/2}$$

In this case, as R approaches 0, the frequency approaches $1/\sqrt{LC}$ rad/s. It should be noted that the phase curve passes through 90° at the same frequency where the peak of the magnitude curve would occur if R was set to 0.

As a second example of a high-pass network function, let us use the phasor $\mathscr{V}_L$ of the network shown in Fig. 7-8.5(a) as a response phasor. The ratio of this phasor to the input excitation phasor $\mathscr{V}_1$ is readily shown to be

$$\frac{\mathscr{V}_L}{\mathscr{V}_1} = \frac{-\omega^2 LC}{1 - \omega^2 LC + j\omega RC} \tag{7-182}$$

As was done for the low-pass case, if we set the values of L and C to unity, the network function takes the form

$$\frac{\mathscr{V}_L}{\mathscr{V}_1} = \frac{-\omega^2}{1 - \omega^2 + j\omega R} \tag{7-183}$$

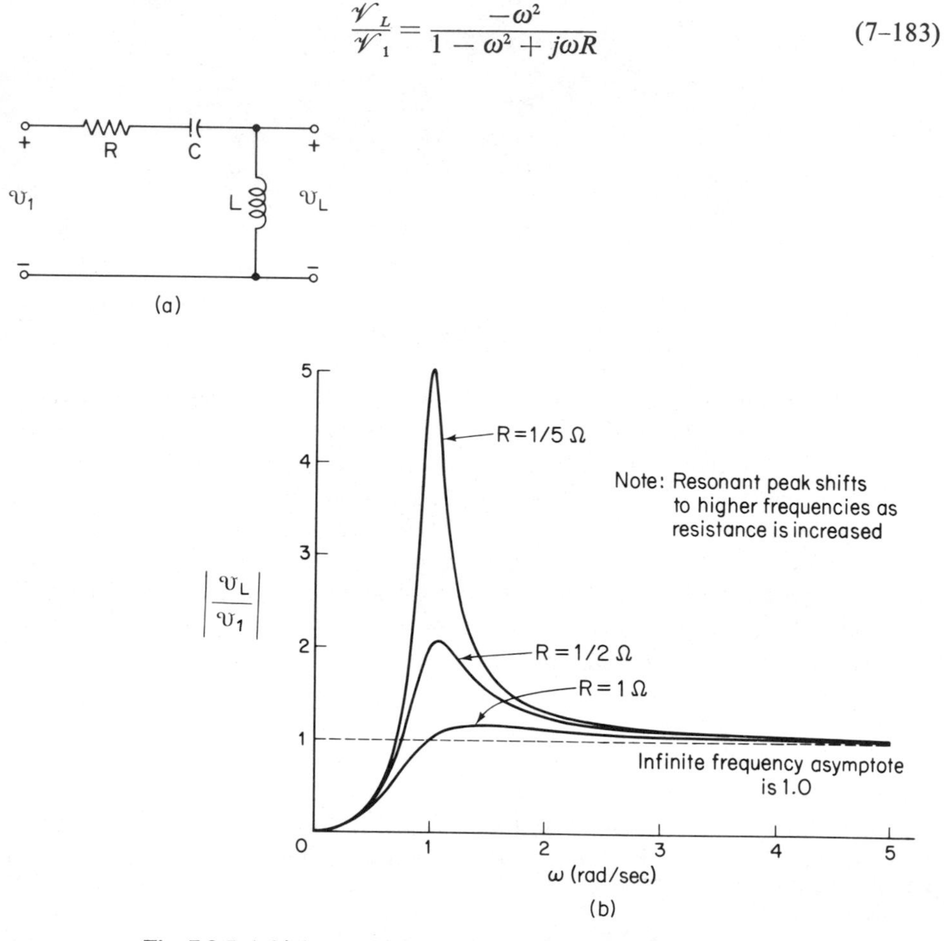

Fig. 7-8.5 A high-pass *RLC* network and its magnitude and phase plots.

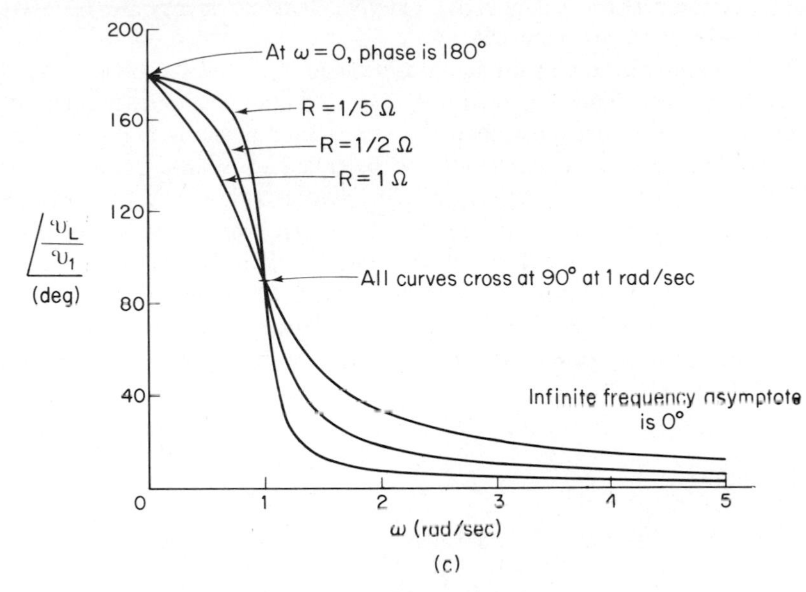

Fig. 7-8.5 Continued

The magnitude and phase of the network function defined by (7–183) are plotted for several values of R in Figs. 7–8.5(b) and 7–8.5(c). An examination of these curves shows that the two reactive elements produce a total of 180° of phase shift in the network, and that the peak of the magnitude curve which appears for low values of resistance occurs when the phase shift is 90°.

As an example of another class of network functions, we will use the phasor $\mathscr{V}_R$ of the network shown in Fig. 7–8.6(a) as the response phasor. The ratio of this to the excitation phasor $\mathscr{V}_1$ is readily shown to be

$$\frac{\mathscr{V}_R}{\mathscr{V}_1} = \frac{j\omega RC}{1 - \omega^2 LC + j\omega RC} \tag{7–184}$$

If we set the values of L and C to unity, the network function becomes

$$\frac{\mathscr{V}_R}{\mathscr{V}_1} = \frac{j\omega R}{1 - \omega^2 + j\omega R} \tag{7–185}$$

The magnitude and phase of the network function defined by (7–185) are plotted for several values of R in Figs. 7–8.6(b) and 7–8.6(c). The magnitude characteristic in these plots approaches zero for every low and very high values of ω and reaches a maximum when $\omega = 1$ rad/s. Thus, most of the response for this network function occurs for the band of frequencies which are near 1 rad/s (more generally for ω near $1/\sqrt{LC}$). A network function which has a magnitude curve of this type is referred to as a *band-pass network function*. The network is called a *band-pass network*. In this case, since the network contains two reactive elements, it may also be referred to as a *second-order* band-pass network.

The peaking of the magnitude characteristic of Fig. 7-8.6(b) is also referred to as a *resonant* effect. Thus, the circuit of Fig. 7-8.6(a) is frequently referred to as a *resonant circuit*. Since all the elements are connected in series, and since there are two reactive elements, this circuit may also be called a *second-order series-resonant* circuit. The frequency at which the magnitude characteristic of Fig. 7-8.6(b) reaches its peak value is called the *center frequency* or the *resonant frequency* of the band-pass network. On each side of this center frequency, there is a frequency at which the magnitude characteristic has a value which is 0.707 of its value at the center frequency. These frequencies are referred to as the *upper and lower 3-dB frequencies.* The difference between these frequencies is called the *bandwidth* of the function. If we let B be the bandwidth, then[15]

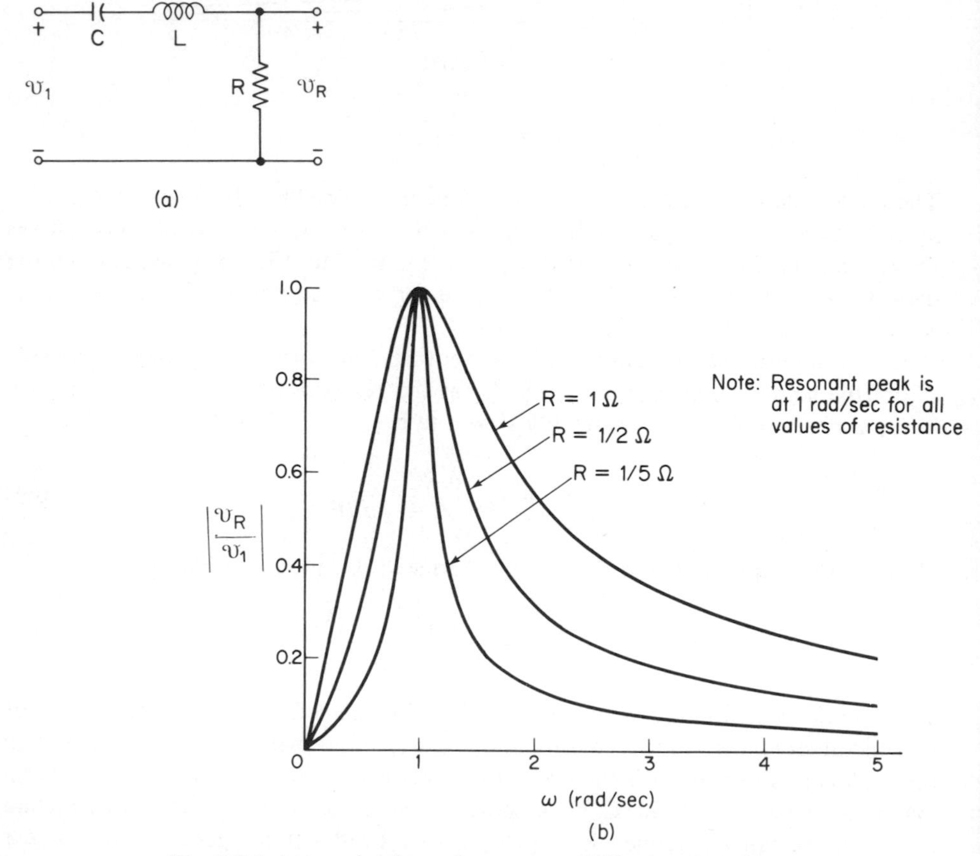

Fig. 7-8.6 A second-order series-resonant *RLC* network and its magnitude and phase plots.

[15]For a *low-pass* network the bandwidth is defined as being equal to the frequency at which the magnitude is down 3 dB from its zero frequency value.

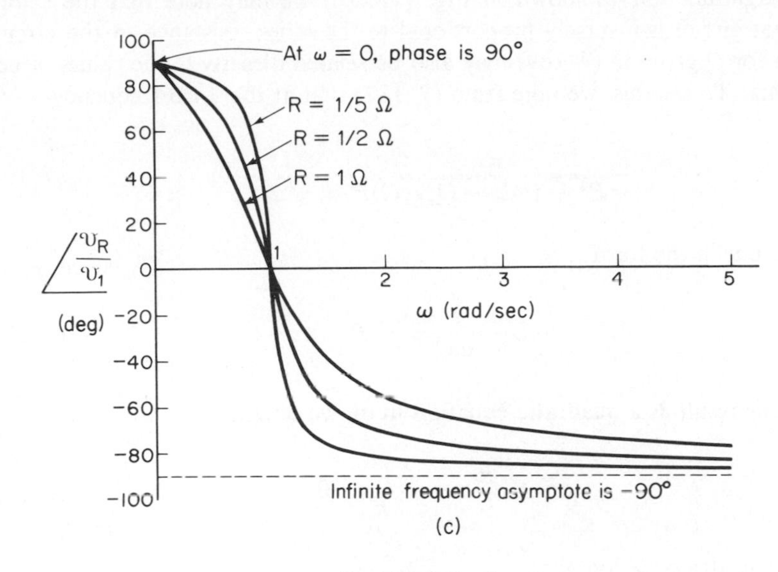

Fig. 7-8.6 Continued

$$B = \omega_{\text{upper 3 dB}} - \omega_{\text{lower 3 dB}} \tag{7-186}$$

The resonant effect in a second-order network is brought about by a cancellation of the imaginary components in the denominator of the network function. This is the result of the fact that reactances may be positive or negative in value. To see how such a cancellation comes about, let us write the band-pass network function given in (7-184) in the following form.

$$\frac{\mathcal{V}_R}{\mathcal{V}_1} = \frac{R}{R + j[\omega L - (1/\omega C)]} \tag{7-187}$$

The magnitude of this function has its maximum value when the imaginary term in the denominator becomes zero. If we let ω_0 be the frequency at which this occurs, then

$$\omega_0 L - \frac{1}{\omega_0 C} = 0 \qquad \omega_0 = \frac{1}{\sqrt{LC}} \tag{7-188}$$

Thus, ω_0 is the resonant frequency for this circuit. When ω has this value, the circuit is said to be in *resonance*.

One measure of the sharpness of the resonant peak in a band-pass circuit is the *quality factor*, commonly designated as Q. This may be defined as

$$Q = \frac{\text{Resonant frequency}}{\text{Bandwidth}} = \frac{\omega_0}{B} \tag{7-189}$$

From the magnitude curves shown in Fig. 7–8.6(b) we may note that the Q of a series-resonant circuit is inversely proportional to the series resistance in the circuit. The relation for Q given in (7–189) may also be related directly to the values of network elements. To see this, we note from (7–187) that at the 3-dB frequency

$$\frac{R}{\sqrt{R^2 + [\omega L - (1/\omega C)]^2}} = \frac{1}{\sqrt{2}} \tag{7–190}$$

This may be put in the form

$$\omega L - \frac{1}{\omega C} = \pm R \tag{7–191}$$

Expressing this result as a quadratic equation in ω, we obtain

$$\omega^2 \pm \frac{R}{L}\,\omega - \frac{1}{LC} = 0 \tag{7–192}$$

The solutions to this equation are

$$\omega = \pm\frac{R}{2L} \pm \sqrt{\left(\frac{R}{2L}\right)^2 + \frac{1}{LC}} \tag{7–193}$$

Since ω must be positive, we may identify the upper and lower 3-dB frequencies as

$$\omega_{\text{upper 3 dB}} = \frac{R}{2L} + \sqrt{\left(\frac{R}{2L}\right)^2 + \frac{1}{LC}} \tag{7–194}$$

$$\omega_{\text{lower 3 dB}} = -\frac{R}{2L} + \sqrt{\left(\frac{R}{2L}\right)^2 + \frac{1}{LC}} \tag{7–195}$$

Thus, the bandwidth may be expressed

$$B = \omega_{\text{upper 3 dB}} - \omega_{\text{lower 3 dB}} = \frac{R}{L} \tag{7–196}$$

Substituting this result in (7–189), we obtain the following expression for the Q of this network function.

$$Q = \frac{\omega_0 L}{R} \tag{7–197}$$

The process of resonance may be illustrated in more detail by the use of phasor diagrams. To make the example more concrete, let us use the numerical values $L = 1$, $C = 1$, and $R = 0.5$ for the circuit shown in Fig. 7–8.6(a). Since only the relative position of the phasors will be of interest to us, let us choose the reference angle for $\mathscr{V}_R$ as zero degrees. Similarly, since we are primarily concerned with the ratio of $\mathscr{V}_R$ to $\mathscr{V}_1$ rather than the actual values of these phasors, let us set the magnitude of $\mathscr{V}_R$ to unity. Thus, we may draw the phasor for $\mathscr{V}_R$ as shown in the phasor diagram in Fig. 7–8.7(a). The phasor $\mathscr{I}$ is easily found from $\mathscr{V}_R$, using the relation $\mathscr{I} = \mathscr{V}_R/R$. Thus, $\mathscr{I} = 2\angle 0°$ and it may be added to the phasor diagram as shown in Fig.

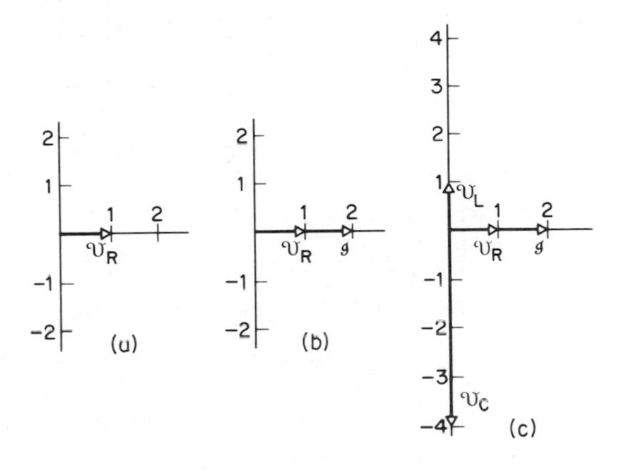

Fig. 7-8.7 Phasor diagrams for the network shown in Fig. 7-8.6(a).

7–8.7(b). Next, we may compute the phasors $\mathscr{V}_C = \mathscr{I}/j\omega C$, and $\mathscr{V}_L = j\omega L\mathscr{I}$. Let us make our first determination of these quantities at $\omega = \frac{1}{2}$ rad/s. We find that $\mathscr{V}_C = 4\angle -90°$ and $\mathscr{V}_L = 1\angle 90°$. If these quantities are added to the phasor diagram, the result is shown in Fig. 7–8.7(c). We may now reposition the phasors on the phasor diagram, in exactly the same way that vectors are positioned on a vector diagram, to indicate the sum $\mathscr{V}_R + \mathscr{V}_C + \mathscr{V}_L$ of the phasors. From KVL, this quantity is equal to $\mathscr{V}_1$. The construction is shown in Fig. 7–8.8(a). From the figure it is easily seen that $\mathscr{V}_R/\mathscr{V}_1 = 1/\sqrt{10}\angle -72.6°$. This conclusion is readily verified if the network function given in (7–185) is evaluated at $\omega = 0.5$ rad/s, and $R = 0.5$. Now let us construct a phasor diagram to show the relation of the phasors when $\omega = 1$ rad/s. In this case, again assuming that $\mathscr{V}_R = 1\angle 0°$, we find that $\mathscr{I} = 2\angle 0°$, $\mathscr{V}_C = 2\angle -90°$, and $\mathscr{V}_L = 2\angle 90°$. The resulting phasor diagram is shown in Fig. 7–8.8(b). Obviously, $\mathscr{V}_R/\mathscr{V}_1 = 1\angle 0°$, which may be verified from (7–185). As a final example, consider the phasor diagram for $\omega = 2$ rad/s, still assuming that $\mathscr{V}_R = 1\angle 0°$. In this case, $\mathscr{V}_C = 1\angle -90°$, and $\mathscr{V}_L = 4\angle 90°$. The phasor diagram appears as shown in Fig. 7–8.8(c). From the figure we see that $\mathscr{V}_R/\mathscr{V}_1 = 1/\sqrt{10}\angle 72.6°$. Thus, this is the value of the network function at $\omega = 2$ rad/s, as may readily be verified from Eq. (7–185). The three phasor diagrams shown in Fig. 7–8.8 illustrate typical conditions in a series-resonant circuit at frequencies below resonance, at resonance, and above resonance. Since the response phasor is maintained at unity magnitude, the magnitude of the network function is inversely proportional to the length of the phasor $\mathscr{V}_1$. From the diagrams, we readily see that the

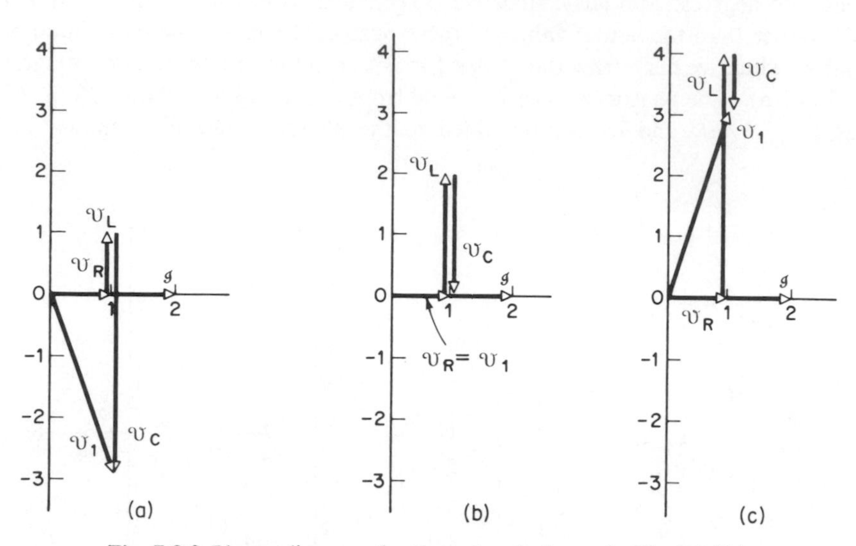

Fig. 7-8.8 Phasor diagrams for the network shown in Fig. 7-8.6(a).

minimum magnitude of this phasor occurs at the resonant frequency. In addition, we note that at frequencies below resonance, the circuit acts as an *RC* circuit, in that the current $\mathcal{I}$ leads the applied voltage $\mathcal{V}_1$. Similarly, at frequencies above resonance, the applied voltage $\mathcal{V}_1$ leads the current $\mathcal{I}$ and, thus, the circuit behaves like an *RL* circuit, exhibiting the phase lead characteristic typical of an inductor. Finally, at resonance, the circuit behaves like a pure resistance, since the applied voltage $\mathcal{V}_1$ is in phase with the current $\mathcal{I}$ flowing through the circuit.

The above discussion may be summarized as follows:

SUMMARY 7-8.1

Properties of a Second-Order Series-Resonant Circuit: A circuit consisting of a series connection of a resistor, a capacitor, and an inductor is called a second-order series-resonant circuit. The resonant frequency is $\omega_0 = 1/\sqrt{LC}$ rad/s. Below resonance, the circuit acts like an *RC* network. Above resonance, the circuit acts like an *RL* circuit. At resonance, the circuit acts like a resistor.

It should be noted that at resonance, the voltages across the capacitor and the inductor are equal and opposite. In high-Q circuits, i.e., ones in which the resistance is low, these voltages can become very large, and can actually be far larger than the applied voltage. As an example of this, suppose that a series-resonant circuit in which $R = 100\ \Omega$, $L = 1$ H, and $C = 7.04\ \mu$F (7.04×10^{-6} F) is connected to an electrical wall outlet which provides 110 V (rms) at 60 Hz (377 rad/s). The magnitude of the the voltage phasor is 155.6 V (see Sec. 8–1). Substitution of the values for L and C

in Eq. (7–188) shows that the circuit is resonant at the applied frequency, namely, 377 rad/s. The magnitude of the current phasor is thus determined only by the resistance. It equals 155.6/100 or 1.556 A. The magnitude of the phasor for the voltage appearing across the inductor is $\mathcal{I}\omega L$, i.e., 1.556 × 377 × 1 or 586 V! The magnitude of the phasor voltage across the capacitor is also 586 V. Note that these voltages are so much higher than the applied voltage that they may present a dangerous shock hazard. A number of people who failed to realize that, in a series-resonant circuit, the voltages across component network elements may be many times larger than the applied voltage are now learning circuit theory in "heavenly" universities. It is hoped that the readers of this text will not join them prematurely.

It is also possible to define a second-order *parallel-resonant circuit*. As an example of a network function for such a circuit, consider the driving-point impedance of the circuit shown in Fig. 7–8.9. In terms of the (phasor) variables defined in this circuit, we may write

$$\frac{\mathcal{V}}{\mathcal{I}_1} = \frac{1}{G + j[\omega C - (1/\omega L)]} \tag{7–198}$$

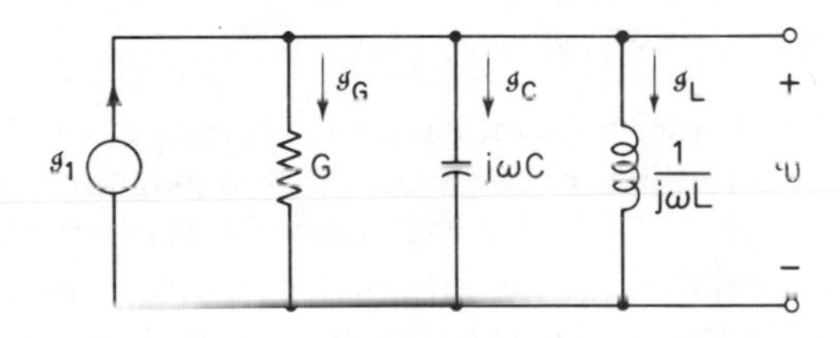

Fig. 7-8.9 A second-order parallel resonant circuit.

Resonance occurs in this circuit when the imaginary part of the denominator (the susceptance) goes to zero. Thus, we see that the resonant frequency ω_0 is defined by

$$\omega_0 C - \frac{1}{\omega_0 L} = 0 \qquad \omega_0 = \frac{1}{\sqrt{LC}} \tag{7–199}$$

This is the same relation as was determined for the series-resonant circuit. Phasor diagrams for the parallel-resonant circuit for frequencies below resonance, at resonance, and above resonance are shown in Fig. 7–8.10. For convenience in drawing these phasor diagrams, it has been assumed that $\mathcal{I}_R$, the current through the resistor, has the same magnitude in each of the three cases. From the diagram we may conclude that below resonance, a parallel-resonant circuit is inductive in behavior, since the voltage across the circuit leads the applied current. Similarly, for frequencies above the resonant frequency, the circuit acts in a capacitive manner, in the sense that the current leads the voltage. Finally, at resonance, the circuit acts as a pure conductance, i.e., the applied current is in phase with the voltage across the three parallel circuit elements.

The discussion given above may be summarized as follows:

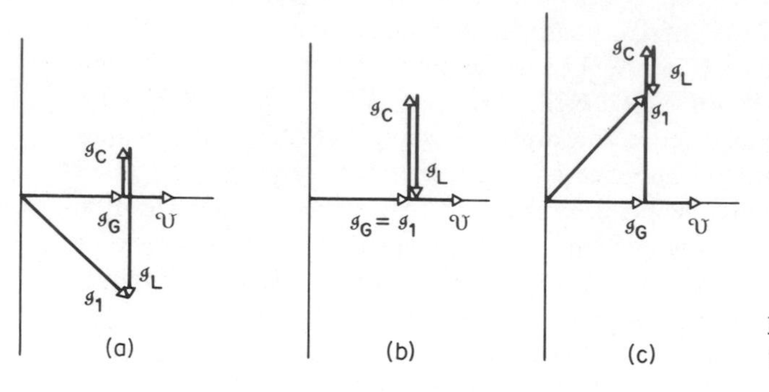

Fig. 7-8.10 Phasor diagrams for the circuit shown in Fig. 7-8.9.

SUMMARY 7-8.2

Properties of a Second-Order Parallel-Resonant Circuit: A circuit consisting of a parallel connection of a resistor, a capacitor, and an inductor is called a second-order parallel-resonant circuit. The resonant frequency is $\omega_0 = 1/\sqrt{LC}$ rad/s. Below resonance, the circuit acts like an *RL* circuit. Above resonance, the circuit acts like an *RC* circuit. At resonance, the circuit acts like a resistor.

In a parallel-resonant circuit at resonance, conditions dual to those described for the series-resonant circuit exist, i.e., the magnitude of the phasor currents $\mathscr{I}_C$ and $\mathscr{I}_L$ shown in Fig. 7–8.10(b) may become considerably larger than the magnitude of the input current $\mathscr{I}_1$. These large currents are sometimes referred to as *circulating currents* since they circulate in the loop formed by the inductor and capacitor. Unless proper consideration is given to the magnitude of these currents, they may become large enough to destroy circuit elements.

7-9 FREQUENCY RESPONSE PLOTS

In the preceding section we introduced the concept of a network function and showed how it could be used to relate a pair of input and output phasors. To illustrate the properties of such functions, magnitude and phase plots were used to graphically display the manner in which a given network function was affected by its sinusoidal frequency variable ω. More specifically, if a network function was defined as $N(j\omega)$, we used a magnitude plot to display $|N(j\omega)|$ as a function of ω and a phase plot to display Arg $N(j\omega)$ vs. ω. In this section we shall consider another useful graph which may be used for the display of magnitude and phase information. This is one in which logarithmic scales are used for $|N(j\omega)|$ and for the variable ω. Such graphs are called Bode plots (or Bode diagrams). Obviously, such graphs contain the same information which is present in the (non-logarithmic) plots previously discussed. We shall see, however, that the logarithmic plots are considerably easier to construct. As a matter of fact, we shall find that they can be approximated to a surprising degree of accuracy by using only straight-line segments.

To begin our discussion of Bode plots, let us consider a network function $N(j\omega)$ which is characterized in terms of its magnitude and phase components. Thus, we may write

$$N(j\omega) = |N(j\omega)|e^{j \operatorname{Arg} N(j\omega)} \tag{7-200}$$

If we now take the natural logarithm of both sides of Eq. (7-200) and note that the logarithm of a product of terms is the sum of the logarithms of the individual terms, we obtain

$$\ln [N(j\omega)] = \ln |N(j\omega)| + \ln [e^{j \operatorname{Arg} N(j\omega)}] \tag{7-201}$$

Since $\ln (e^a) = e^{\ln (a)} = a$, the above expression may be put in the form

$$\ln [N(j\omega)] = \ln |N(j\omega)| + j \operatorname{Arg} N(j\omega) \tag{7-202}$$

The significance of this result is that by taking the natural logarithm of the network function $N(j\omega)$, we convert the function to a sum of two quantities, namely, a real part which is a function only of the magnitude of $N(j\omega)$, and an imaginary part which is a function only of its phase. In the relation given in Eq. (7-202), the unit for $\ln |N(j\omega)|$ is the *neper* and the unit for the $\operatorname{Arg} N(j\omega)$ is the *radian*. To convert these quantities to the more usual *dB* and *degree* units we may use the following relations:

Magnitude in decibels (dB) = 8.6589 × magnitude in nepers

Argument in degrees = 57.2958 × argument in radians

To see how the use of logarithmic quantities simplifies the graphical representation of network functions, let us consider a simple network function $N_z(j\omega)$ having the form

$$N_z(j\omega) = 1 + \frac{j\omega}{z} \tag{7-203}$$

Taking the natural logarithm of both sides of this, we obtain

$$\ln [N_z(j\omega)] = \ln \left|1 + \frac{j\omega}{z}\right| + j \tan^{-1} \frac{\omega}{z} \tag{7-204}$$

The first term on the right-hand side of this relation is concerned only with the magnitude of the network function. Thus, it may equally well be expressed in dB. If we define such a magnitude as $M_z(\omega)$, we may write

$$M_z(\omega) = 8.6589 \ln\left|1 + \frac{j\omega}{z}\right| = 20 \log\left|1 + \frac{j\omega}{z}\right| \tag{7-205}$$

This term is obviously a function of the variable ω. However, for values of ω such that $\omega/z \ll 1$, i.e., such that $\omega \ll z$, this term becomes

$$M_z(\omega) \Big|_{\omega \ll z} \approx 20 \log 1 = 0 \text{ dB} \tag{7-206}$$

Thus, for such a range of ω the term $M_z(\omega)$ may be plotted as a straight line at 0 dB on a dB vs. log ω plot as shown in Fig. 7-9.1(a). For values of ω such that $\omega \gg z$, the term $M_z(\omega)$ given in (7-205) may be approximated as

$$M_z(\omega) \Big|_{\omega \gg z} \approx 20 \log \frac{\omega}{z} = 20(\log \omega - \log z) \text{ dB} \tag{7-207}$$

On a dB vs. log ω plot this is simply a straight line with a slope of 20. It intersects the 0-dB line (which represents $M_z(\omega)$ for small values of ω) when $\log \omega = \log z$, i.e., when $\omega = z$. Thus, it may be characterized as shown in Fig. 7-9.1(b). Since

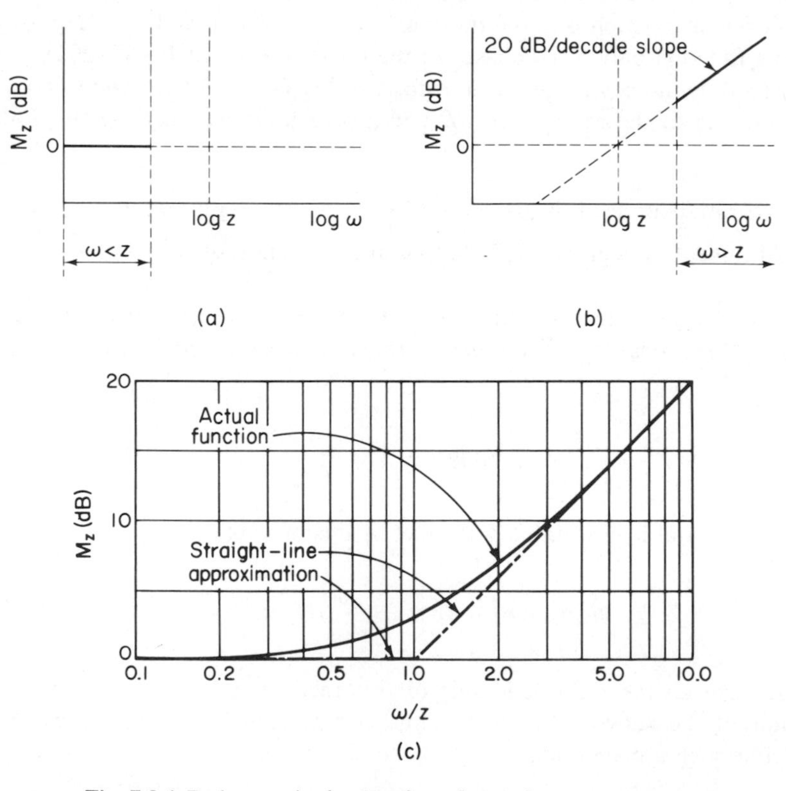

Fig. 7-9.1 Bode magnitude plots for a first-order numerator term.

the slope of this line is 20, in any region for which this characterization is valid, a unit change in log ω such as log ω going from 1 to 2 or from 9 to 10, or from 9.7 to 10.7, etc., must produce a resulting change in $M_z(\omega)$ of 20 dB. The corresponding change in ω for such a unit change in log ω can be determined by letting ω_1 and ω_2 be any two frequencies the difference of whose logarithms is unity. Thus, we may write (assuming $\omega_2 > \omega_1$)

$$\log \omega_2 - \log \omega_1 = \log \frac{\omega_2}{\omega_1} = 1 \tag{7-208}$$

Thus, the ratio ω_2/ω_1 must equal 10. A pair of frequencies ω_2 and ω_1 such that their ratio is 10 is said to encompass a *decade* of frequency. Thus the solid line shown in Fig. 7-9.1(b) is frequently referred to as having a slope of 20 dB/decade of frequency or, more concisely, simply as a *20 dB/decade* slope. Another frequently used expression for the slope of the line shown in Fig. 7-9.1(b) is *6 dB/octave* where an *octave* is defined as the frequency range encompassed by two frequencies, one of which is twice the value of the other. If we extend the two straight-line characteristics shown in Figs. 7-9.1(a) and (b) until they intersect as shown in Fig. 7-9.1(c), we obtain an asymptotic straight-line approximation for the quantity $M_z(\omega)$ defined in (7-205). The actual plot of this function is also shown in the figure. It is readily observed that the maximum error in the asymptotic approximation is 3 dB and that this occurs at a frequency $\omega = z$. In Fig. 7-9.1(c) it is important to note that values of ω itself rather than values of log ω are given on the abscissa scale. From the values of ω, however, we see that the scale is obviously logarithmic. Such a convention is usually considerably more convenient than using values of log ω for the scale.

Now let us consider the imaginary part of ln $[N_z(j\omega)]$, i.e., the part which gives the phase of $N_z(j\omega)$. From (7-202) and (7-204) we have

$$\operatorname{Arg} N_z(j\omega) = \tan^{-1} \frac{\omega}{z} \tag{7-209}$$

A straight-line asymptotic approximation which is frequently used to plot this function on a degree vs. log ω plot consists of a horizontal straight line at zero degrees (for $\omega \ll z$) and a second straight line at $90°$ (for $\omega \gg z$). The two horizontal lines are joined by a straight line with a slope of $45°$ per decade passing through the point (log z, $45°$). Such a straight-line phase approximation is shown in Fig. 7-9.2(a). A portion of the actual phase curve as defined in Eq. (7-209) is shown in Fig. 7-9.2(b). The maximum error in the straight-line approximation is about $6°$. An example of the use of the technique described above follows.

EXAMPLE 7-9.1. As an example of the use of Bode plots to approximate the magnitude and phase characteristics of a simple network function, consider the driving-point admittance $Y(j\omega)$ for the network shown in Fig. 7-9.3. For this network

$$Y(j\omega) = 3(j\omega + 10)$$

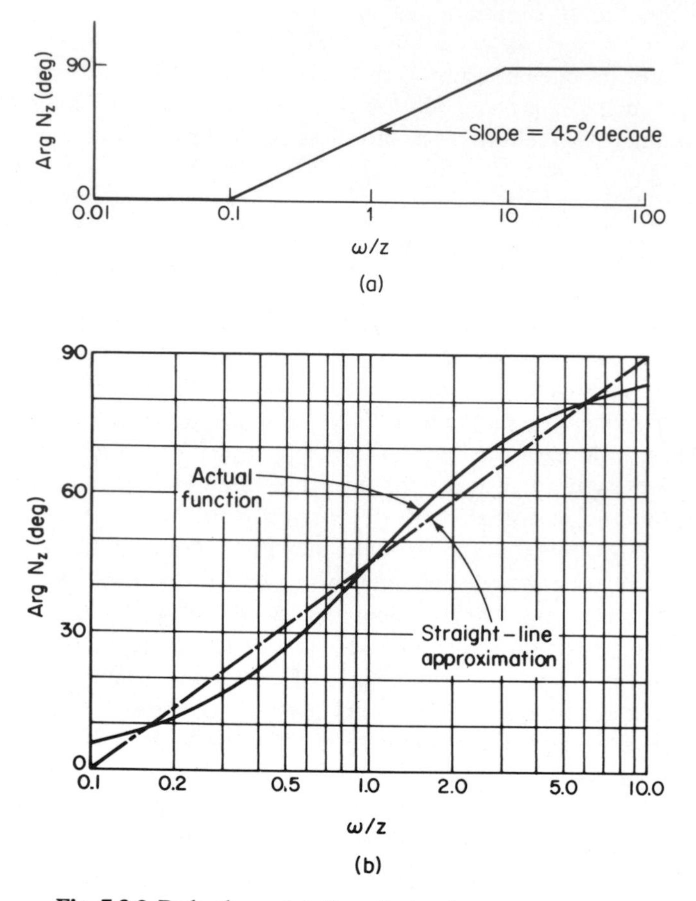

Fig. 7-9.2 Bode phase plots for a first-order numerator term.

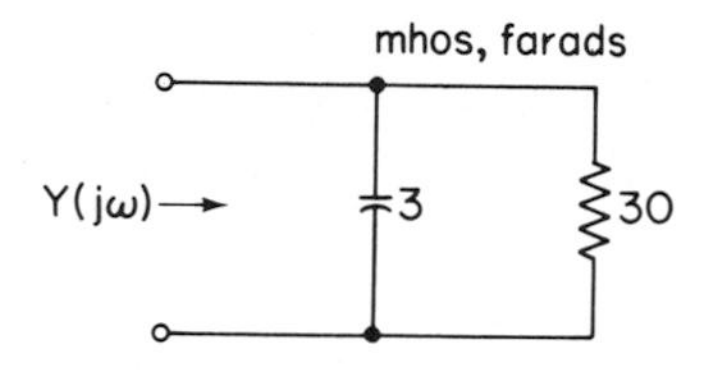

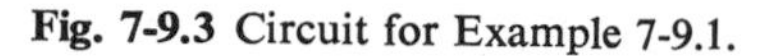

Fig. 7-9.3 Circuit for Example 7-9.1.

Putting the term in parentheses in a form similar to that given in Eq. (7-203), we obtain

$$Y(j\omega) = 30\left(1 + \frac{j\omega}{10}\right)$$

The magnitude in dB may be expressed as

$$20\,|\log Y(j\omega)| = 20 \log 30 + 20 \log \left|1 + \frac{j\omega}{10}\right| = 3.854 + 20 \log \left|1 + \frac{j\omega}{10}\right|$$

Using the method described above to obtain the asymptotic characteristics for the second term of the right member of the equation and adding the constant 3.854, we obtain the Bode magnitude plot shown in Fig. 7-9.4(a). Since the phase of the multiplicative constant is zero, the phase of the network function $Y(j\omega)$ is simply that of the term $[1 + (j\omega/10)]$.

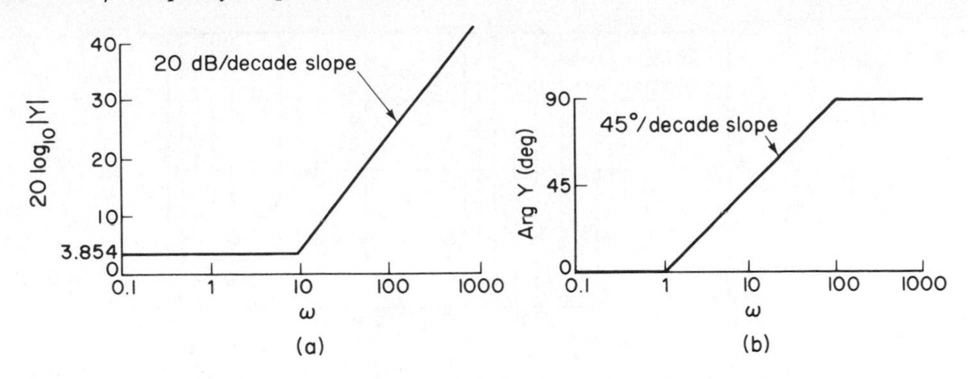

Fig. 7-9.4 Bode magnitude and phase plots for Example 7-9.1.

Thus, applying the techniques described above, we obtain the Bode phase plot shown in Fig. 7-9.4(b).

An analysis similar to that given in the preceding paragraphs may be used to construct the Bode diagrams for a network function $N_p(j\omega)$ having the form

$$N_p(j\omega) = \frac{1}{1 + (j\omega/p)} \tag{7-210}$$

To see this, let us again take the natural logarithm of both members of the equation. We obtain

$$\ln\,[N_p(j\omega)] = \ln\left|\frac{1}{1 + (j\omega/p)}\right| - j\tan^{-1}\frac{\omega}{p} \tag{7-211}$$

Since the logarithm of the reciprocal of a quantity is equal to the negative of the logarithm of the quantity, the above relation may be put in the form

$$\ln\,[N_p(j\omega)] = -\ln\left|1 + \frac{j\omega}{p}\right| - j\tan^{-1}\frac{\omega}{p} \tag{7-212}$$

It is readily observed that this is simply the negative of the expression given in Eq. (7-204) (with the substitution of p for z). Thus, the asymptotic magnitude and phase Bode plots will be as shown in Fig. 7-9.5. These are simply the negative of those shown in Figs. 7-9.1(c) and 7-9.2(b). The actual plots are also shown. It is readily seen that the magnitude characteristic again has a maximum error of 3 dB and the phase characteristic has a maximum error of 6°. Thus, we see that an asymptotic straight-line approximation for the dB vs. log ω plot for $|N_p(j\omega)|$ may be constructed by drawing a horizontal 0-dB line from the ordinate to the value log p and continuing this with a straight line with a *minus* 20 dB/decade slope. An asymptotic straight-line plot of degrees vs. log ω for Arg $N_p(j\omega)$ may be constructed by drawing a

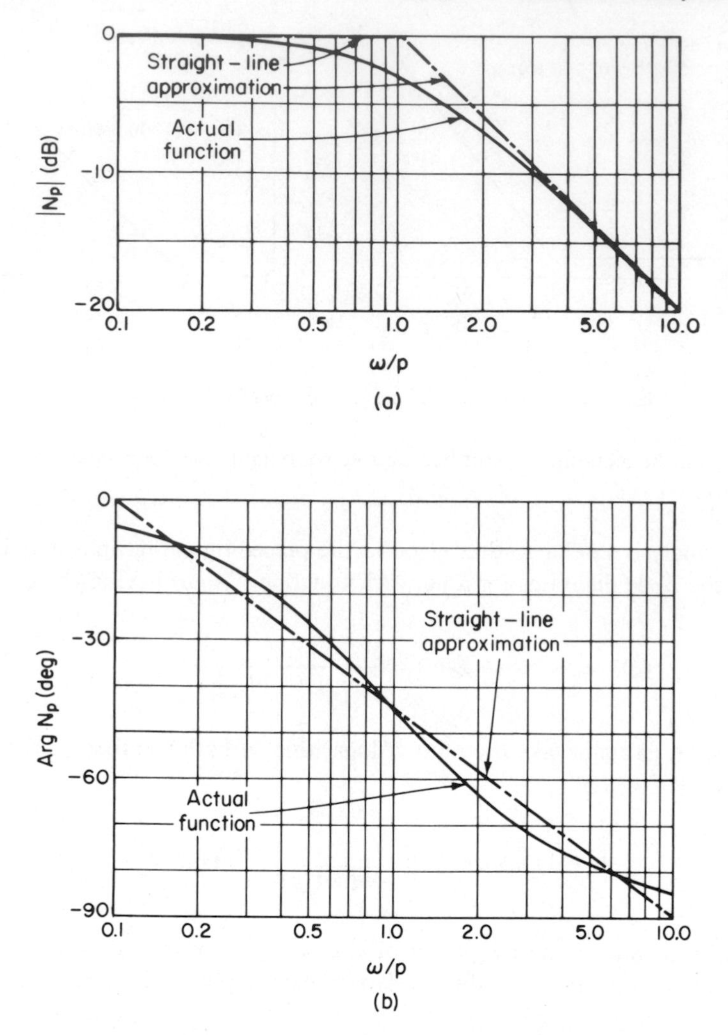

Fig. 7-9.5 Bode plots for a first-order denominator term.

straight line with a slope of $-45°$/decade through the point (log p, $45°$), and extending this line with $0°$ and $-90°$ horizontal asymptotes.

In the preceding paragraphs we have discussed the construction of Bode plots for two types of simple network functions, i.e., those with a first-order factor in either the numerator or the denominator. There is one other type of first-order network function that also occurs frequently. This is one having the form $j\omega$ (or $1/j\omega$). The magnitude Bode plot for such a function simply consists of a straight line which crosses the 0-dB abscissa at $\log \omega = 0$, i.e., at $\omega = 1$, and which has a slope of $+20$ dB/decade (-20 dB/decade for the function $1/j\omega$). The Bode phase plot simply consists of a horizontal line at $+90°$ ($-90°$ for the function $1/j\omega$).

The results developed in this section to this point are readily extended to provide a means of constructing Bode plots for any function consisting of a product of numerator factors having the form shown in Eq. (7–203) and denominator factors of the form shown in Eq. (7–210). To see this, consider the following network function:

$$N(j\omega) = \frac{A(j\omega)}{B(j\omega)} = K\frac{[1 + (j\omega/z_1)][1 + (j\omega/z_2)] \cdots [1 + (j\omega/z_m)]}{[1 + (j\omega/p_1)][1 + (j\omega/p_2)] \cdots [1 + (j\omega/p_n)]} \qquad (7\text{–}213)$$

If we take the logarithm of both members of this equation, we obtain

$$\begin{aligned}
\ln [N(j\omega)] = \ln K &+ \ln\left|1 + \frac{j\omega}{z_1}\right| + \ln\left|1 + \frac{j\omega}{z_2}\right| \\
&+ \cdots + \ln\left|1 + \frac{j\omega}{z_m}\right| + \ln\left|\frac{1}{1 + (j\omega/p_1)}\right| \\
&+ \ln\left|\frac{1}{1 + (j\omega/p_2)}\right| + \cdots + \ln\left|\frac{1}{1 + (j\omega/p_n)}\right| \\
&+ j\Big[\tan^{-1}\frac{\omega}{z_1} + \tan^{-1}\frac{\omega}{z_2} + \cdots + \tan^{-1}\frac{\omega}{z_m} \\
&- \tan^{-1}\frac{\omega}{p_1} - \tan^{-1}\frac{\omega}{p_2} - \cdots - \tan^{-1}\frac{\omega}{p_n}\Big] \qquad (7\text{–}214)
\end{aligned}$$

If we now define $|N(j\omega)|$ in dB as $M(\omega)$, we may write

$$\begin{aligned}
M(\omega) = 20 \log |N(j\omega)| = 20 \log K &+ 20 \log\left|1 + \frac{j\omega}{z_1}\right| \\
&+ 20 \log\left|1 + \frac{j\omega}{z_2}\right| + \cdots + 20 \log\left|1 + \frac{j\omega}{z_m}\right| \\
&+ 20 \log\left|\frac{1}{1 + (j\omega/p_1)}\right| + 20 \log\left|\frac{1}{1 + (j\omega/p_2)}\right| \\
&+ \cdots + 20 \log\left|\frac{1}{1 + (j\omega/p_n)}\right| \qquad (7\text{–}215)
\end{aligned}$$

Thus we see that an asymptotic plot of $M(\omega)$ can be made by summing the separate asymptotic plots of factors having the form $20 \log |1 + (j\omega/z_i)|$ and $20 \log |1/[1 + (j\omega/p_i)]|$. Since these latter plots simply consist of straight lines, such a summation is readily done graphically. In a similar manner, if we investigate the behavior of $\text{Arg}\, N(j\omega)$, we may write

$$\begin{aligned}
\text{Arg}\, N(j\omega) = \tan^{-1}\frac{\omega}{z_1} &+ \tan^{-1}\frac{\omega}{z_2} + \cdots + \tan^{-1}\frac{\omega}{z_m} - \tan^{-1}\frac{\omega}{p_1} \\
&- \tan^{-1}\frac{\omega}{p_2} - \cdots - \tan^{-1}\frac{\omega}{p_n} \qquad (7\text{–}216)
\end{aligned}$$

From this result we again see that the addition of separate Bode phase plots for the numerator and denominator factors of $N(j\omega)$ gives us the resulting overall phase plot for the network function. An example follows.

EXAMPLE 7-9.2. It is desired to construct Bode plots for the network function

$$N(j\omega) = 10\frac{1 + (j\omega/10)}{[1 + (j\omega/1)][1 + (j\omega/100)]}$$

The separate magnitude and phase plots for the factors $1 + (j\omega/10)$, $1/[1 + (j\omega/1)]$, and $1/[1 + (j\omega/100)]$ are shown in Figs. 7-9.6(a) and 7-9.7(a). The addition of these plots and the inclusion of the constant term $20 \log 10 = 20$ for the constant multiplier of $N(j\omega)$

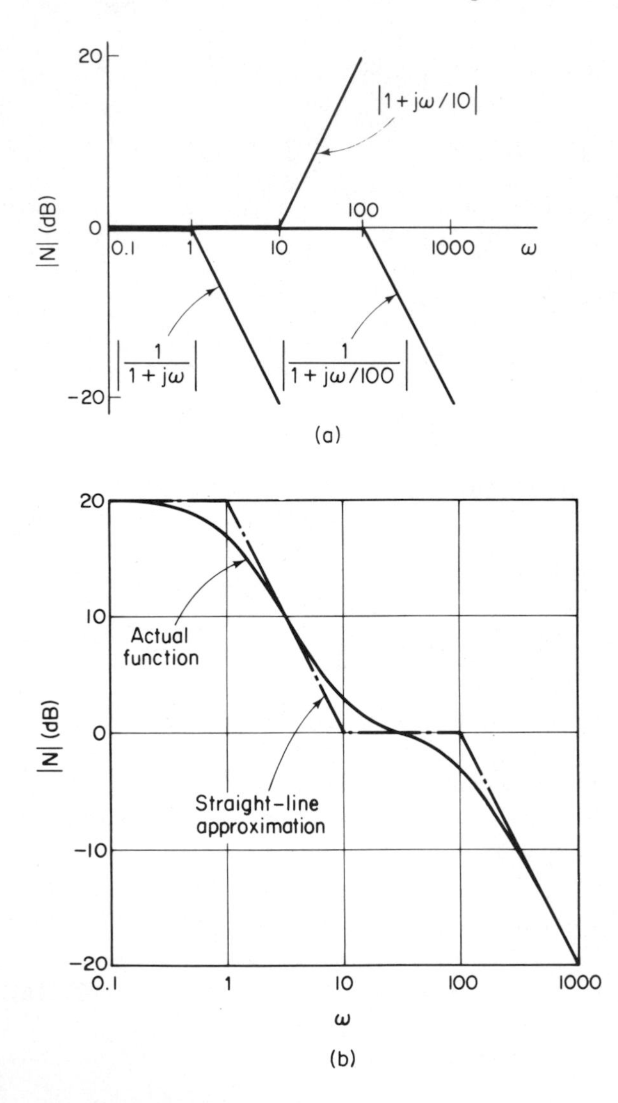

Fig. 7-9.6 Bode magnitude plots for Example 7-9.2.

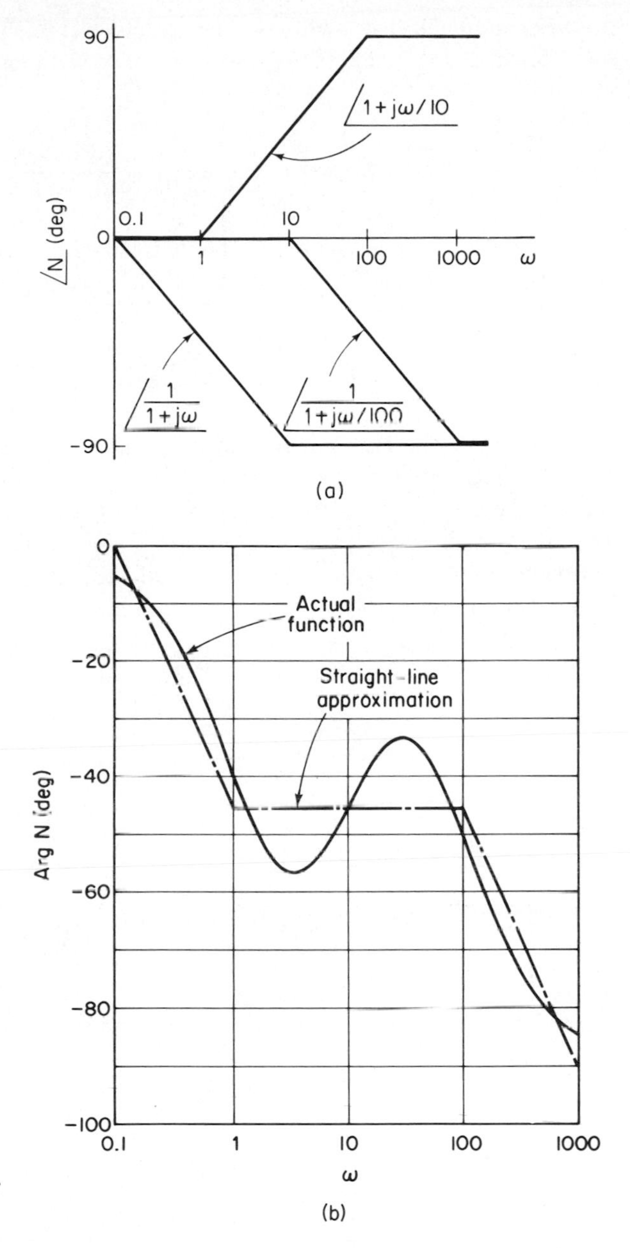

Fig. 7-9.7 Bode phase plots for Example 7-9.2.

gives the Bode magnitude and phase plots shown in Figs. 7-9.6(b) and 7-9.7(b). The actual magnitude and phase curves are also shown in these plots for comparison.

In the preceding paragraphs we showed how to construct Bode plots for network functions consisting of first-order factors in the numerator and denominator. Now let us consider the effect of second-order factors which have complex conjugate zeros. For simplicity, we will consider a network function $N_c(j\omega)$ consisting of a second-order denominator factor. Such a network function may be written in the form

$$N_c(j\omega) = \frac{1}{1 + 2\zeta(j\omega/\omega_0) + (j\omega/\omega_0)^2} \tag{7-217}$$

where ζ is called the damping factor and ω_0 is called the undamped natural frequency. Taking the natural log of both sides of (7-217), we obtain

$$\ln [N_c(j\omega)] = \ln \left| \frac{1}{1 + 2\zeta(j\omega/\omega_0) + (j\omega/\omega_0)^2} \right| - j \tan^{-1} \left(\frac{2\zeta\omega/\omega_0}{1 - (\omega/\omega_0)^2} \right) \tag{7-218}$$

If we investigate the real and imaginary parts of the above function in a manner similar to that done for the first-order factors, we see that a plot of 20 log $|N_c(j\omega)|$ has a low-frequency asymptote at 0 dB, and a high-frequency asymptotic locus with a slope of -40 dB/decade (-12 dB/octave). The break point is at $\omega = \omega_0$, and is independent of the value of ζ. At the break point, however, we obtain from the actual equation

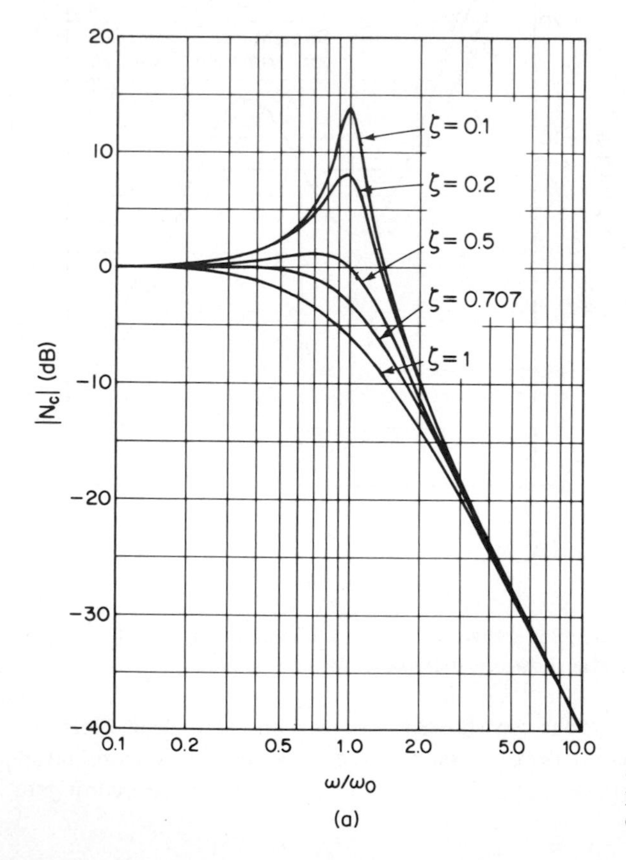

Fig. 7-9.8 Bode magnitude and phase plots for a second-order denominator factor.

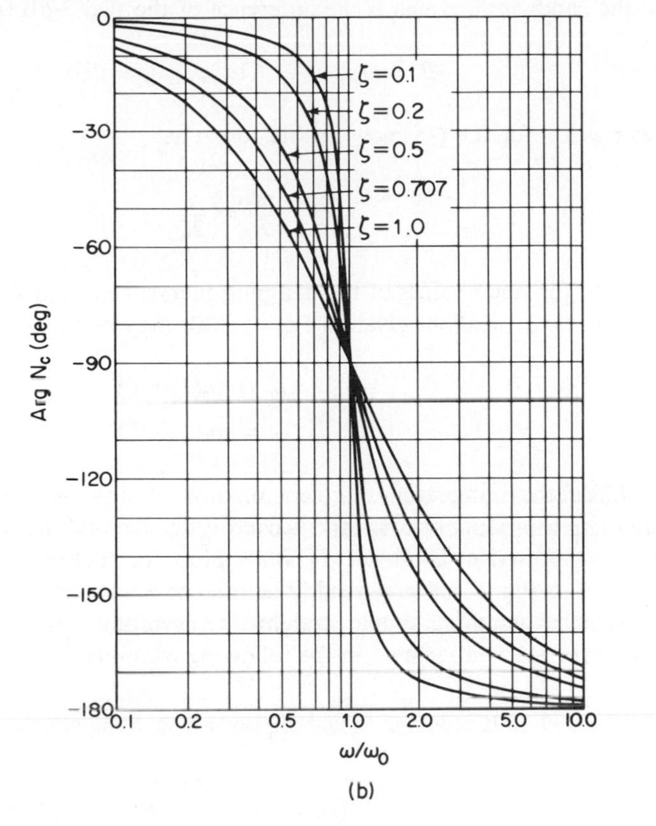

Fig. 7-9.8 Continued

$$20 \log |N_c(j\omega)| = -20 \log 2\zeta \qquad (7\text{–}219)$$

Since the asymptotic curve always has a value of 0 dB at this point, we see that the error of the approximation for the magnitude increases as ζ increases. These results are readily observed in the plots of 20 log $|N_c(j\omega)|$ vs. log ω for various values of ζ shown in Fig. 7–9.8(a). Similar considerations are readily established for the phase. From (7–218), we see that the low-frequency phase approaches 0° while the high-frequency phase equals $-180°$. The phase at $\omega = \omega_0$ is $-90°$ and it is independent of ζ. A series of plots of the actual phase for various values of ζ is shown in Fig. 7–9.8(b). These magnitude and phase plots may also be applied to network functions that have second-order numerator factors by simply reversing the sign of the ordinate scales.

The damping factor and undamped natural frequency ω_0 introduced in the preceding paragraph may be directly related to the network Q which was defined in Sec. 7–8. To see this, let us solve for the 3-dB frequencies of the network function $N_c(j\omega)$ given in Eq. (7–217). Selecting only the positive solutions, we obtain

$$\begin{aligned} \omega_{\text{upper 3 dB}} &= \omega_0(\sqrt{\zeta^2 + 1} + \zeta) \\ \omega_{\text{lower 3 dB}} &= \omega_0(\sqrt{\zeta^2 + 1} - \zeta) \end{aligned} \qquad (7\text{–}220)$$

Thus, the bandwidth, which is the difference of the two 3-dB frequencies, is

$$B = \omega_{\text{upper 3 dB}} - \omega_{\text{lower 3 dB}} = 2\zeta\omega_0 \tag{7-221}$$

The expression for the Q is now readily found as

$$Q = \frac{\omega_0}{B} = \frac{1}{2\zeta} \tag{7-222}$$

Frequently, for small values of the damping factor, i.e., high values of Q, simplified versions of the expressions given in Eq. (7–220) may be used. These are

$$\begin{aligned} \omega_{\text{upper 3 dB}} &= \omega_0(1 + \zeta) \\ \omega_{\text{lower 3 dB}} &= \omega_0(1 - \zeta) \end{aligned} \tag{7-223}$$

Since the numerator and denominator of any network function may be factored into a product of first- and second-order factors, the techniques introduced above may be used to construct the Bode plots for such a factored function. The plots shown in Fig. 7–9.8 may readily be used to apply corrections to the asymptotic plots made by assuming simple straight-line asymptotes for second-degree factors. Such a correction is illustrated in the following example.

EXAMPLE 7–9.3. It is desired to find the Bode magnitude plot for the network function

$$N(j\omega) = \frac{j\omega}{(1 + j\omega)[1 + (0.4j\omega/100) + (j\omega/100)^2]}$$

The separate asymptotic Bode plots for the numerator and for the two denominator factors are shown in Fig. 7-9.9(a). It should be noted that the second-order denominator factor has an asymptote of -40 dB/decade for high frequencies, while the first-order factor has an asymptotic slope of -20 dB/decade. The sum of these loci is given in Fig. 7-9.9(b). A correction to the asymptotic straight-line approximation for the second-order denominator factor may be obtained from the curves of Fig. 7-9.8(a) for the case where $\zeta = 0.2$. This is readily apparent in the locus giving the actual magnitude of $N(j\omega)$, which is also shown in Fig. 7-9.9(b).

There is one other type of graphical representation which is frequently used to display the dependence of network functions on the sinusoidal frequency variable ω. This representation consists of a two-dimensional plot in which the real part of the network function is displayed along the abscissa or x axis and the imaginary part is displayed along the ordinate or y axis. The locus which represents a given network function is generated by evaluating the function for a range of frequency starting from zero and going to infinity. The result is called a *Nyquist* plot. As an example of such a plot, consider the network shown in Fig. 7–9.10. If we define $N(j\omega)$ as the open-circuit voltage transfer function, then $N(j\omega) = 1/[1 + (j\omega/p)]$. This may

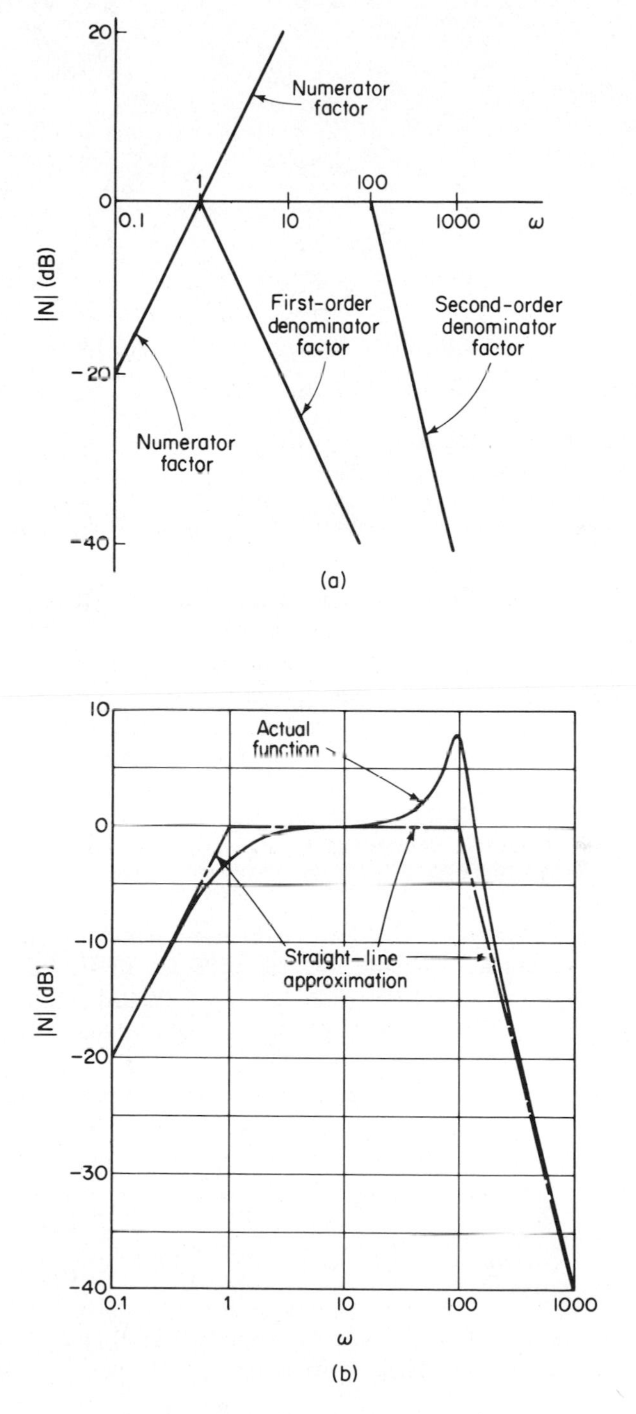

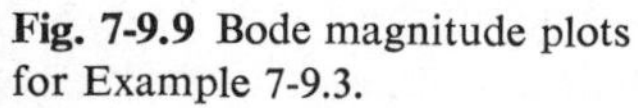
Fig. 7-9.9 Bode magnitude plots for Example 7-9.3.

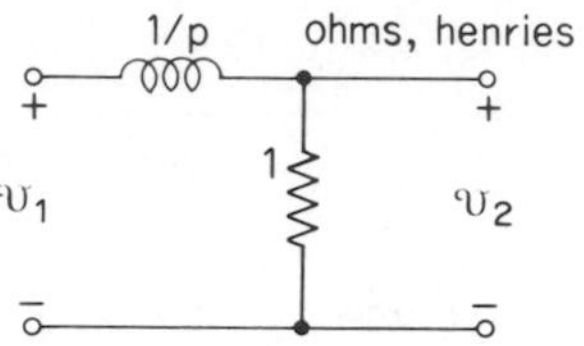

1/p ohms, henries

Fig. 7-9.10 An *RL* network.

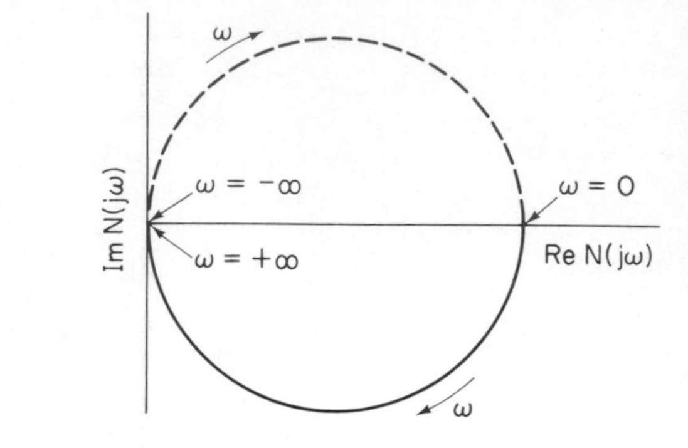

Fig. 7-9.11 Nyquist plot for the network shown in Fig. 7-9.10.

be written in the form

$$N(j\omega) = \frac{1}{1 + (\omega/p)^2} - j\frac{\omega/p}{1 + (\omega/p)^2} \tag{7-224}$$

A plot of this function for a range of values of ω from zero to infinity is shown by the solid line in Fig. 7-9.11. Also shown in this figure is a dashed line which represents the locus for a range of negative values of ω from $-\infty$ to 0. Nyquist diagrams have considerable use in studies of the stability of active networks and control systems. Such applications are covered in detail in books on modern control theory.

7-10 DIGITAL COMPUTER TECHNIQUES FOR PLOTTING FREQUENCY RESPONSE

The use of the digital computer is especially attractive when it is desired to make frequency response plots, since the multitude of computations required by such displays are, in general, quite laborious. In this section, we shall present techniques for making three types of plots: (1) magnitude and phase plots with linear scales, (2) magnitude and phase plots using logarithmic scales (Bode plots), and (3) Nyquist plots. A fundamental computation which must be made in constructing any of these plots is the determination of the magnitude and the phase (or the determination of the real and imaginary parts) of a network function when evaluated at a specific value of its frequency variable. Thus, given a network function defined as

$$N(j\omega) = \frac{A(j\omega)}{B(j\omega)} = \frac{a_1 + a_2(j\omega) + a_3(j\omega)^2 + \cdots + a_{m+1}(j\omega)^m}{b_1 + b_2(j\omega) + b_3(j\omega)^2 + \cdots + b_{n+1}(j\omega)^n} \tag{7-225}$$

we desire to find the following quantities: Re $[N(j\omega_0)]$, Im $[N(j\omega_0)]$, Arg $N(j\omega_0)$, and $|N(j\omega_0)|$, where ω_0 is some specified value of ω. Such a computation is then repeated for a sequence of other values of ω_0. A convenient way of performing such a computation is to separately determine the complex values of the numerator and denominator polynomials $A(j\omega_0)$ and $B(j\omega_0)$. In doing this, let us here anticipate a slightly more complicated situation (which will be of use to us in the following chapter) in which we desire to determine

$$A(p_0) = a_1 + a_2 p_0 + a_3(p_0)^2 + \cdots + a_{m+1}(p_0)^m \qquad (7\text{-}226)$$

where p_0 is a *complex* number. Such an implementation may be applied to the problem considered here by simply letting $p_0 = 0 + j\omega_0$. To program the relation given in Eq. (7-226), it is more efficient to rewrite the equation in a form which does not require raising the quantity p_0 to the various powers indicated in each term. As an example of this form, consider the case where $A(p_0)$ is of third degree. We may write

$$A(p_0) = a_1 + a_2 p_0 + a_3(p_0)^2 + a_4(p_0)^3 = a_1 + p_0[a_2 + p_0(a_3 + p_0 a_4)] \qquad (7\text{-}227)$$

It is easily verified that when the terms in the right member of the above equation are multiplied out, the equality is satisfied. More generally, this new form for writing a polynomial may be shown to be

$$A(p_0) = a_1 + p_0(a_2 + p_0\{a_3 + p_0[a_4 + \cdots + p_0(a_m + p_0 a_{m+1}) \cdots]\}) \qquad (7\text{-}228)$$

It is readily verified that the expressions given in the right members of Eqs. (7-226) and (7-228) are equal.

The operations shown in Eq. (7-228) are readily programmed as a digital computer algorithm. To see this, let us use the term "Value" as an intermediate variable used in the evaluation process. The algorithm for computing $A(p_0)$ may then be expressed as follows:

1. Set "Value" equal to a_{m+1}.
2. Multiply "Value" by p_0 and add a_m to it. Redefine "Value" as being equal to this result.
3. Multiply "Value" by p_0 and add a_{m-1} to it. Call the result "Value."

---

i. Multiply "Value" by p_0 and add a_{m+2-i} to it. Call the result "Value."

---

m. Multiply "Value" by p_0 and add a_2 to it. Call the result "Value."

$m + 1$. Multiply "Value" by p_0 and add a_1 to it. The result is $A(p_0)$.

The algorithm described above is shown in the form of a flow chart in Fig. 7-10.1. The operations are readily programmed in a subroutine which may be given the name CVALPL (for Complex eVALuation of a PoLynomial) and identified by the statement

SUBROUTINE CVALPL (M, A, P, VALUE) (7-229)

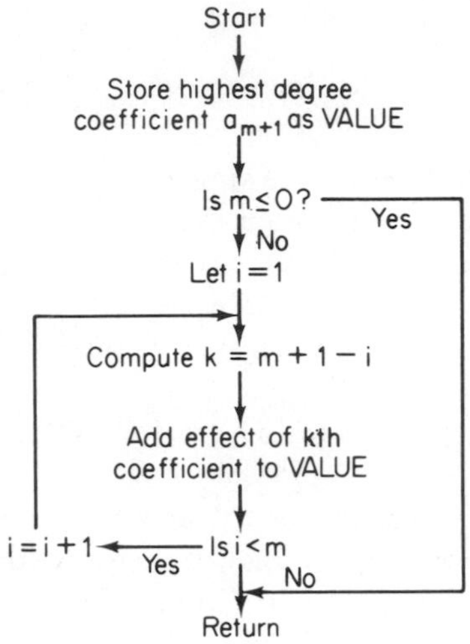

Fig. 7-10.1 Flow chart for the subroutine CVALPL.

where M is the degree of the polynomial, A is a single-subscripted real array in which are stored the values of the coefficients of the polynomial $A(p_0)$, P is the value of complex frequency p_0 at which it is desired to evaluate the polynomial, and VALUE is the resulting complex value of $A(p_0)$. A listing of the subroutine CVALPL is shown in Fig. 7-10.2. The characteristics of the subroutine are summarized in Table 7-10.1 (on page 499). To use the subroutine CVALPL to determine the magnitude of a network function, we will use another subroutine SSS (for Sinusoidal Steady-State evaluation) which calls the subroutine CVALPL twice, once to determine the value of the numerator polynomial $A(j\omega_0)$ and a second time to determine the value of the denominator polynomial $B(j\omega_0)$, and then divides the two values to obtain the desired value of the network function. A suitable identifying statement for such a subroutine is

```
      SUBROUTINE CVALPL (M, A, P, VALUE)
C     SUBROUTINE FOR DETERMINING THE VALUE OF A POLYNOMIAL
C     A(1) + A(2)*P + A(3)*P**2 + ... + A(M+1)*P**M
C     FOR A COMPLEX VALUE OF ITS VARIABLE
C          M - DEGREE OF POLYNOMIAL
C          A - ARRAY OF POLYNOMIAL COEFFICIENTS
C                 (IN ASCENDING ORDER)
C          P - COMPLEX VALUE OF FREQUENCY VARIABLE
C      VALUE - OUTPUT COMPLEX VALUE OF THE POLYNOMIAL
      DIMENSION A(10)
      COMPLEX B(10), P, VALUE
      MC = M + 1
C
C     STORE COEFFICIENTS AS COMPLEX VARIABLES
      DO 3  I = 1, MC
    3 B(I) = CMPLX(A(I), 0.)
C
C     STORE VALUE OF HIGHEST DEGREE COEFFICIENT
      VALUE = B(MC)
      IF (M.LE.0) RETURN
C
C     ACCUMULATE EFFECT OF OTHER COEFFICIENTS
      DO 8  I = 1, M
      K = MC - I
    8 VALUE = VALUE * P + B(K)
      RETURN
      END
```

Fig. 7-10.2 Listing of the subroutine CVALPL.

SUBROUTINE SSS (M, A, N, B, OMEGA, VREAL, VIMAG, VPHASE, VMAG)

(7-230)

where M is the degree of the numerator polynomial $A(j\omega)$ of the network function and A is a single-subscripted real array containing the coefficients of the polynomial, N is the degree of the denominator polynomial $B(j\omega)$ and B is a single-subscripted real array containing the polynomial coefficients. OMEGA is the real value of ω at which the network function is to be evaluated. The output arguments VREAL, VIMAG, VPHASE, and VMAG give the values of the real part, the imaginary part, the phase (in degrees), and the magnitude of $N(j\omega)$, respectively. A listing of the

```
      SUBROUTINE SSS(M, A, N, B, OMEGA, VREAL, VIMAG, VPHASE, VMAG)
C     SUBROUTINE FOR EVALUATING A NETWORK FUNCTION GIVING THE
C     SINUSOIDAL STEADY-STATE RELATION BETWEEN AN INPUT PHASOR
C     AND A RESPONSE PHASOR
C            M - DEGREE OF NUMERATOR POLYNOMIAL A(JW)
C            A - ARRAY OF NUMERATOR POLYNOMIAL COEFFICIENTS
C            N - DEGREE OF DENOMINATOR POLYNOMIAL B(JW)
C            B - ARRAY OF DENOMINATOR POLYNOMIAL COEFFICIENTS
C        OMEGA - VALUE OF W AT WHICH THE NETWORK FUNCTION
C                   IS TO BE EVALUATED
C        VREAL - THE REAL PART OF THE NETWORK FUNCTION
C        VIMAG - THE IMAGINARY PART OF THE NETWORK FUNCTION
C       VPHASE - THE PHASE OF THE NETWORK FUNCTION (IN DEGREES)
C         VMAG - THE MAGNITUDE OF THE NETWORK FUNCTION
      DIMENSION A(10), B(10)
      COMPLEX P, VA, VB, V
C
C     DEFINE A COMPLEX VARIABLE WHOSE IMAGINARY PART IS OMEGA
      P = CMPLX(0.0, OMEGA)
C
C     CALL THE SUBROUTINE CVALPL TO EVALUATE THE NUMERATOR
C     AND DENOMINATOR POLYNOMIALS
      CALL CVALPL(M, A, P, VA)
      CALL CVALPL(N, B, P, VB)
C
C     TEST TO MAKE CERTAIN THAT THE DENOMINATOR IS NOT ZERO
      IF (CABS(VB).LE.1.E-08) GO TO 12
C
C     COMPUTE THE VARIOUS OUTPUT QUANTITIES
      V = VA / VB
      VREAL = REAL(V)
      VIMAG = AIMAG(V)
      VMAG = CABS(V)
      IF (VMAG.LT.1.E-08) GO TO 15
      VPHASE = ATAN2(VIMAG, VREAL) * 57.2958
      RETURN
   12 PRINT 13, OMEGA
   13 FORMAT (31HOMAGNITUDE IS INFINITE, OMEGA =,E10.3)
      RETURN
   15 PRINT 16, OMEGA
   16 FORMAT (32HOPHASE IS INDETERMINATE, OMEGA =,E10.3/)
      RETURN
      END
```

Fig. 7-10.3 Listing of the subroutine SSS.

subroutine SSS is given in Fig. 7-10.3.[16] The characteristics of the subroutine are summarized in Table 7-10.2 (on page 500). An example of the use of this subroutine follows.

EXAMPLE 7-10.1. As an example of the use of the subroutine SSS to construct the magnitude and phase plots of a network function, consider the network shown in Fig. 7-10.4. In this figure, the triangle represents an ideal VCVS with a gain of 100. The open-circuit voltage transfer function for this network is readily shown to be

$$\frac{\mathscr{V}_2}{\mathscr{V}_1} = N(j\omega) = \frac{200j\omega}{(j\omega)^2 + 2j\omega + 400}$$

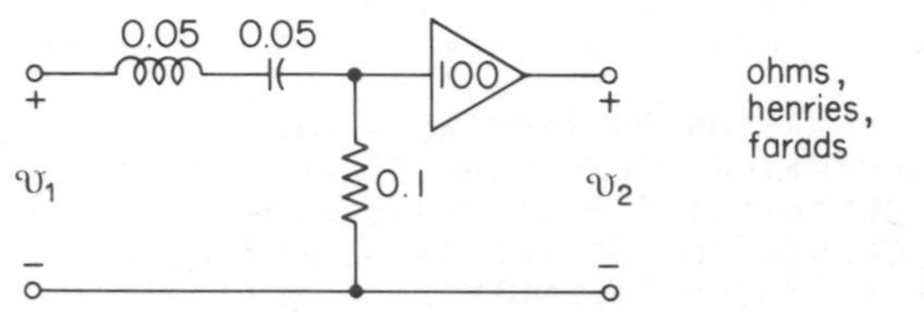

Fig. 7-10.4 A simple active network.

The determination of the magnitude and phase plots over a range of frequencies from 0-50 rad/s is accomplished by the digital computer program given in Fig. 7-10.5. The resulting magnitude and phase output plots are shown in Fig. 7-10.6.

In the preceding paragraphs we discussed the use of a digital computer program to realize magnitude and phase plots as a function of a linear scale in ω. These techniques are readily extended to permit the construction of Bode plots which have a logarithmic frequency scale. The only additional subprogram required is one which generates logarithmically spaced values of the frequency variable. To implement such a generation, we may use a FORTRAN function. Such a function must be supplied with the following information: (1) an index n indicating that the function is currently generating the nth logarithmically spaced value of frequency that is required. This index corresponds with the value of the line of the plot in which the data are to be graphically presented; thus, n is set to zero for the first value of ω used in computing the value of the network function. (2) A number 1_o giving the logarithm of the first frequency (corresponding to $n = 0$) that is required. For example, to start a plot at 0.1 rad/s, $1_o = -1$. (3) A number n_d giving the number of points desired per decade of frequency. For example, $n_d = 5$ means that five values of frequency will be generated for any range such as $0.1 < \omega \leq 1$ or $10 < \omega \leq 100$, etc. Note that the first of the inequalities in these expressions is a weak inequality,

[16]The subroutine SSS uses a two-argument arctangent function ATAN2 to compute the phase (taking into account the correct quadrant) over a range of $-\pi$ to π radians. If only a single argument arctangent function is available in the user's computer installation, an auxiliary program must be constructed to correctly determine the quadrant of the angle. Details of such a program may be found in Fig. 9.3 in *Digital Computations in Basic Circuit Theory*, L. P. Huelsman, McGraw-Hill Book Company, New York, 1968.

```
C     MAIN PROGRAM FOR EXAMPLE 7.10-1
C             A - ARRAY OF COEFFICIENTS OF NUMERATOR POLYNOMIAL
C             B - ARRAY OF COEFFICIENTS OF DENOMINATOR POLYNOMIAL
C          PMAG - TWO-DIMENSIONAL ARRAY FOR STORING VALUES OF
C                 THE NETWORK FUNCTION MAGNITUDE FOR PLOTTING
C           PHS - TWO-DIMENSIONAL ARRAY FOR STORING VALUES OF
C                 THE NETWORK FUNCTION PHASE FOR PLOTTING
      DIMENSION A(2), B(3), PMAG(5,101), PHS(5,101)
C
C     STORE THE VALUES OF THE COEFFICIENTS OF THE NETWORK FUNCTION
C     AND THE ZERO-FREQUENCY VALUES OF THE MAGNITUDE AND PHASE
      A(1) = 0.
      A(2) = 200.
      B(1) = 400.
      B(2) = 2.
      B(3) = 1.
      PMAG(1,1) = 0.0
      PHS(1,1) = 90. / 2.
      OMEGA = 1.
C
C     COMPUTE AND STORE THE MAGNITUDE AND PHASE AT EACH FREQUENCY
      DO 10 I = 2, 51
      CALL SSS(1, A, 2, B, OMEGA, VR, VI, PHASE, PMAG(1,I))
      PHS(1,I) = PHASE / 2.
   10 OMEGA = OMEGA + 1.0
C
C     CALL THE PLOTTING SUBROUTINE TO PLOT THE MAGNITUDE AND PHASE
      PRINT 12
   12 FORMAT(1H1)
      CALL PLOT5(PMAG, 1, 51, 100)
      PRINT 12
      CALL PLOT5(PHS, 1, 51, 50)
      STOP
      END
```

Fig. 7-10.5 Listing of the program for Example 7-10.1.

since for $n_d = 5$, six values of frequency will be generated over the ranges specified as $0.1 \leq \omega \leq 1$ and $10 \leq \omega \leq 100$, etc. If we call such a function XL, it will have an identifying statement of the form

$$\text{FUNCTION XL (N, LO, ND)} \tag{7-231}$$

where the arguments correspond with the quantities n, 1_o, and n_d described above. A listing of such a function is given in Fig. 7-10.7. It should be noted that the index N is automatically increased by one each time the function is called. Thus, this input argument must be a variable (not a constant) in the calling statement for the function. The input arguments LO and ND, however, can be specified as constants. To use the function XL in connection with the plotting subroutine PLOT5, a value of N = 0 should be established in the program prior to the first call for the function XL. A summary of the characteristics of the function are given in Table 7-10.3 (on page 500). An example of its use follows.

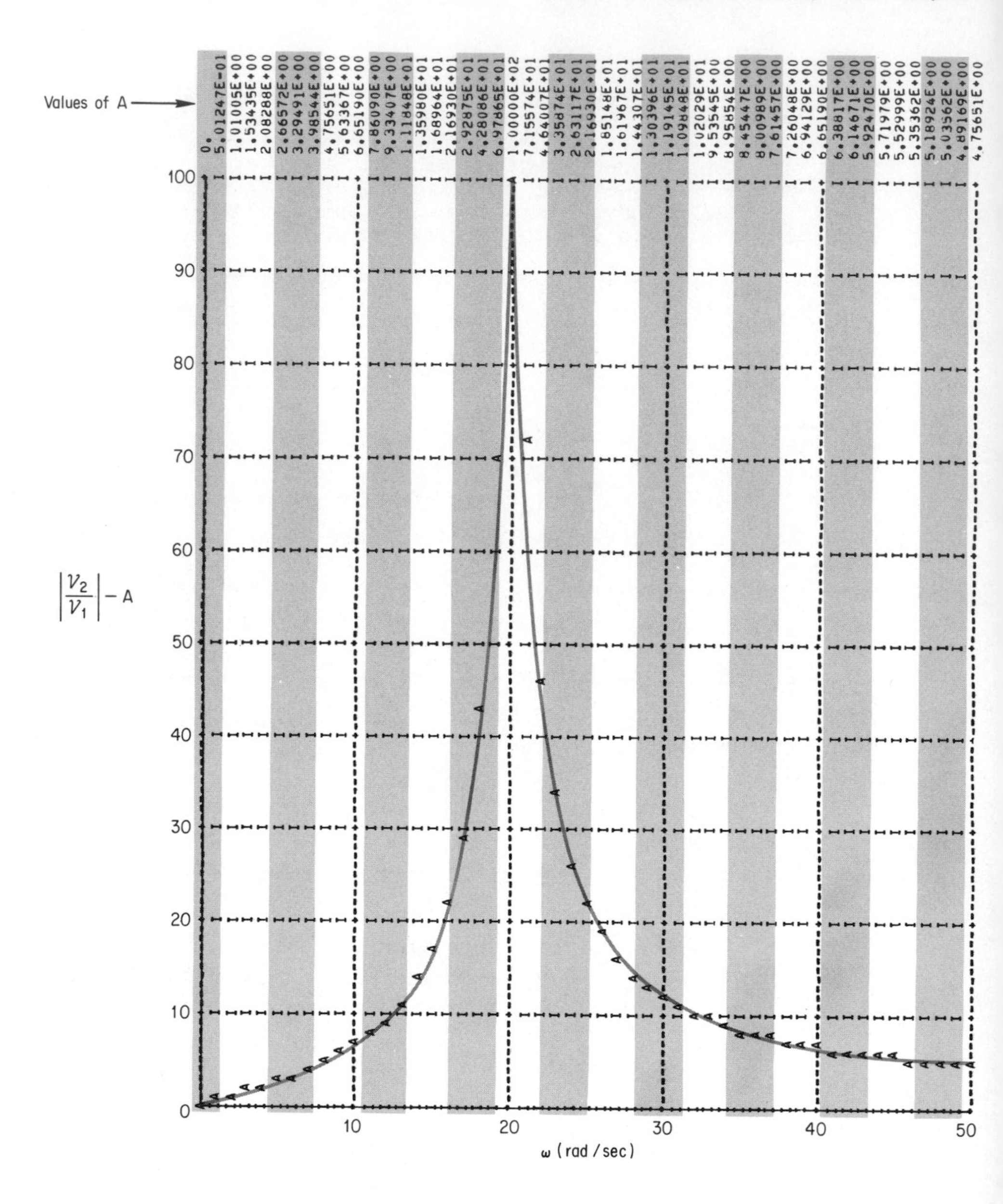

Fig. 7-10.6 Output plots for Example 7-10.1.

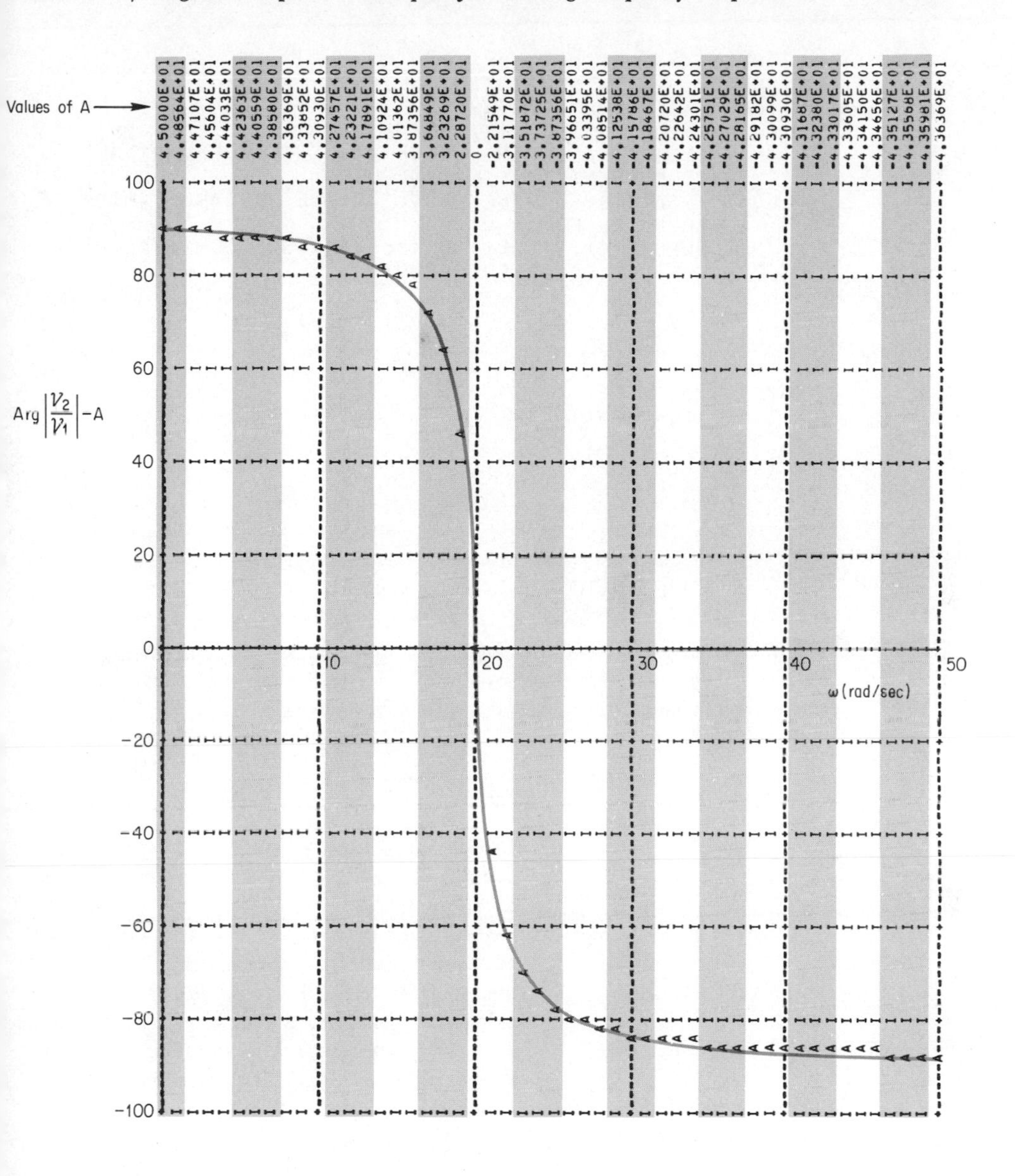

Fig. 7-10.6 Continued

```
      FUNCTION XL(N, LO, ND)
C     FUNCTION FOR GENERATING LOGARITHMICALLY SPACED
C     VALUES OF A FREQUENCY VARIABLE
C            XL - VALUE OF THE VARIABLE
C             N - INDEX INDICATING WHICH VALUE OF THE VARIABLE
C                 IS CURRENTLY BEING COMPUTED
C            LO - LOGARITHM OF THE VALUE OF THE VARIABLE WHICH IS
C                 COMPUTED FOR N = 0
C            ND - NUMBER OF VALUES OF THE VARIABLE COMPUTED
C                 PER DECADE OF FREQUENCY
      AN = N
      ALO = LO
      AND = ND
      AA = AN / AND + ALO
      IF (AA) 6, 6, 8
    6 XL = 1. / 10.**(-AA)
      GO TO 9
    8 XL = 10.**AA
    9 N = N + 1
      RETURN
      END
```

Fig. 7-10.7 Listing of the function XL.

EXAMPLE 7–10.2. As an example of the use of the function XL in constructing Bode plots for sinusoidal steady-state frequency response, consider the network shown in Fig. 7-10.8. The open-circuit voltage transfer function for this network is

$$\frac{\mathscr{V}_2}{\mathscr{V}_1} = N(j\omega) = \frac{1}{1 + (0.2j\omega/10^4) + [(j\omega)^2/10^8]}$$

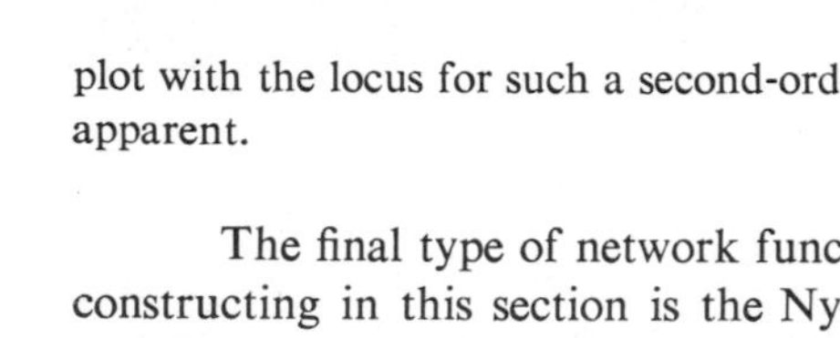

Fig. 7-10.8 Network for Example 7-10.2.

In comparing this network function with the general form given in (7-217), we see that the damping factor $\zeta = 0.1$ and the undamped natural frequency $\omega_0 = 10^4$. To construct a Bode magnitude plot over six decades of frequency from 10 to 10^7 rad/s with 10 frequency values per decade, the program shown in Fig. 7-10.9 may be used. The resulting output is shown in Fig. 7-10.10. The agreement of this plot with the locus for such a second-order network function shown in Fig. 7-9.8 is readily apparent.

The final type of network function characterization which we shall consider constructing in this section is the Nyquist plot. In such a plot, values of frequency are chosen in an increasing sequence. However, the values of the real and imaginary parts of the network function which are plotted along the x axis and y axis, respectively, may, in general, increase or decrease in value as the frequency is increased. Thus, the values of the points which are to be plotted are not, in general, generated in an ascending order with respect to either of the rectangular axes. Such a plot is frequently referred to an x-y plot. It is obviously not practical to use a plotting

```
C     MAIN PROGRAM FOR EXAMPLE 7.10-2
C             A - ARRAY OF COEFFICIENTS OF NUMERATOR POLYNOMIAL
C             B - ARRAY OF COEFFICIENTS OF DENOMINATOR POLYNOMIAL
C           PLT - TWO-DIMENSIONAL ARRAY FOR PLOTTING
      DIMENSION PLT(5,101), A(5), B(5)
C
C     INITIALIZE THE VALUES OF THE POLYNOMIAL COEFFICIENTS
C     AND THE INDEX FOR THE LOGARITHMIC FUNCTION XL
      A(1) = 1.
      B(1) = 1.
      B(2) = 0.2E-04
      B(3) = 1.0E-08
      N = 0
C
C     COMPUTE AND STORE FOR PLOTTING 61 VALUES OF THE
C     MAGNITUDE OF THE NETWORK FUNCTION
      DO 9  I = 1, 61
      OMEGA = XL(N, 3, 20)
      CALL SSS(0, A, 2, B, OMEGA, VR, VI, VPH, VMAG)
    9 PLT(1,I) = 20. * ALOG10(VMAG)
C
C     CALL THE PLOTTING SUBROUTINE
      PRINT 11
   11 FORMAT (1H1)
      CALL PLOT5(PLT, 1, 61, 20)
      STOP
      END
```

Fig. 7-10.9 Listing of the program for Example 7-10.2.

subroutine such as PLOT5 to plot such data, since that subroutine requires that the values of the independent variable be arranged in increasing order and that they be evenly spaced. Therefore, we introduce here a second type of plotting subroutine specifically designed to construct *x*-*y* plots. The details of the operation of this subroutine are described more fully in Appendix C. Here we shall only consider its use. The subroutine is named XYPLTS. It is identified by the statement

SUBROUTINE XYPLTS (NDP, X, Y, NSCLX, NSCLY, NNPX) (7-232)

where NDP is the number of points which are to be plotted, X and Y are single-subscripted arrays containing the X and Y coordinates of the points which are to be plotted, NSCLX is the maximum value desired on the x axis, NSCLY is the maximum value desired on the y axis, and NNPX is the total desired range of values on the x axis. The plot is generated vertically downward on the printer page in the direction of the increasing x-axis coordinate. The minimum value which is realized along the x axis is NSCLX − NNPX. The range of the x axis which is specified is independent of the dimensioning of any of the variables and it may be made as large as required for a given problem. The minimum value of the y axis is NSCLY − 100. An automatic scaling option is provided which may be called by setting the argument NNPX to the value 999. In this case, a grid 90 units by 100 units is produced and the data is automatically scaled so that the range of the values of the X(I) quantities

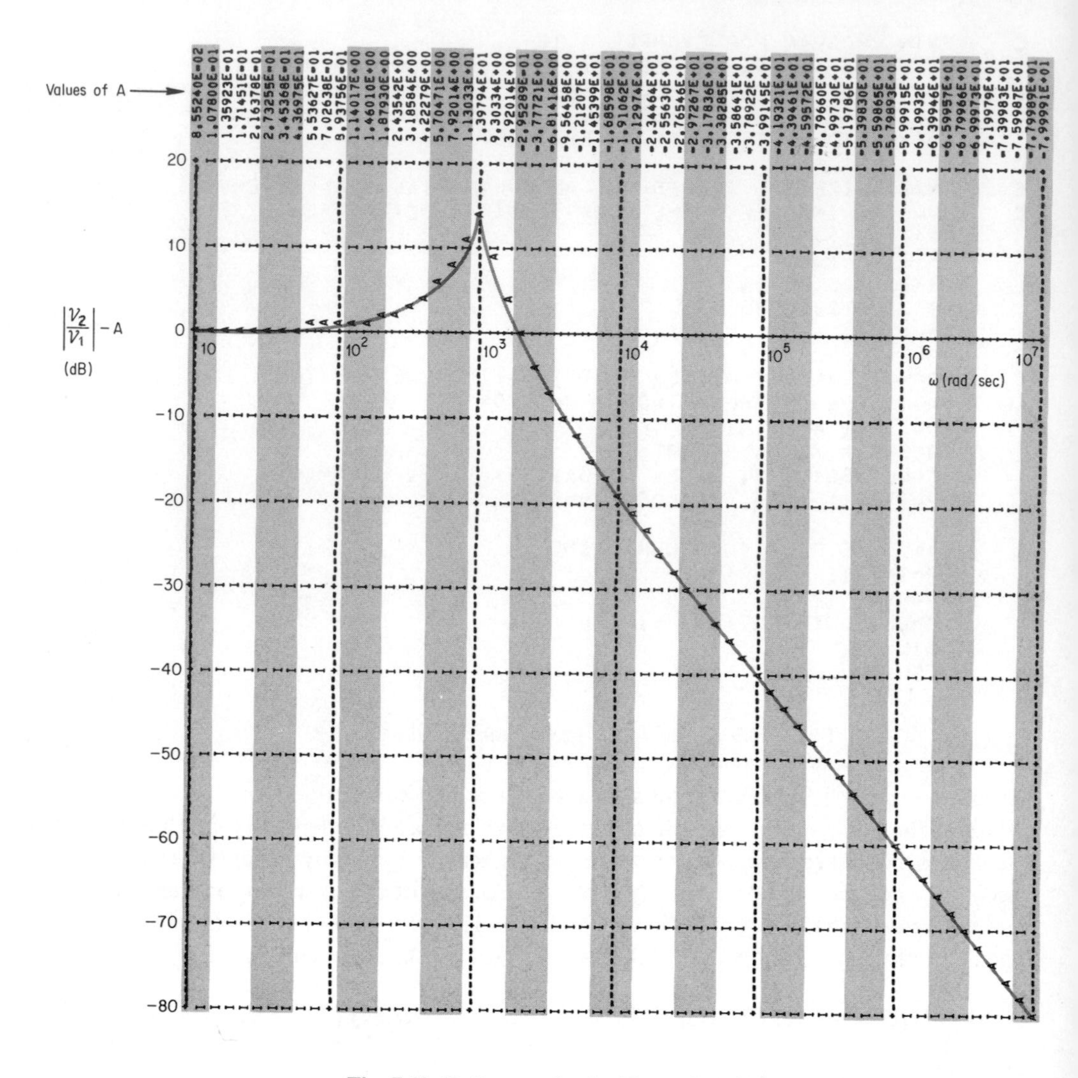

Fig. 7-10.10 Output plot for Example 7-10.2.

or the Y(I) quantities (whichever is larger) fills the entire available range. A summary of the characteristics of the subroutine XYPLTS is given in Table 7-10.4 (on page 501).

The x-y plotting subroutine XYPLTS is readily used in connection with the subroutine SSS to produce a Nyquist plot. One problem associated with the construction of such plots is the difficulty of choosing values of ω such that the plotted points are uniformly spaced. This problem occurs because in the typical Nyquist plot the frequency variable covers a range from zero to some very large value. Obviously, any linear incrementation of frequency which may be satisfactory for low values of ω will not be satisfactory for high values. To partially solve such a problem we may also include a geometric factor for determining the interval between adjacent frequency values. Thus, we may include in the program a statement of the form

$$\text{OMEGA} = (\text{OMEGA} + \text{A}) * \text{B} \tag{7-233}$$

where the constant A will primarily determine the point spacing in the low-frequency range and the constant B will determine the spacing in the high-frequency range. The constants A and B are determined by trial and error. A more sophisticated approach is to have the program itself compute the difference between adjacent points and compute the change in ω so that the points are equally spaced. Such a development is left to the interested reader as a project. An example of the digital computer construction of a Nyquist plot follows.

EXAMPLE 7-10.3. As an example of the use of the subroutines XYPLTS and SSS to construct a Nyquist plot, consider the network shown in Fig. 7-10.11. The network function for the transfer admittance of this network may be shown to be

---

TABLE 7-10.1 Summary of the Characteristics of the Subroutine CVALPL

Identifying Statement: SUBROUTINE CVALPL (M, A, P, VALUE)

Purpose: To determine the complex value of a polynomial $A(s)$ having the form

$$A(s) = a_1 + a_2 s + a_3 s^2 + \cdots + a_m s^{m-1} + a_{m+1} s^m$$

where the argument s is set to some complex value p_0, i.e., to find $A(p_0)$.

Additional Subroutines Required: None.

Input Arguments:

M The degree m of the polynomial $A(s)$.

A The one-dimensional array of variables A(I) in which are stored the values of the coefficients a_i of the polynomial $A(s)$.

P The complex value p_0 of the variable s at which the polynomial $A(s)$ is to be evaluated.

Output Argument:

VALUE The value of the polynomial $A(s)$ when s is set equal to the complex value p_0.

Note: The array A of the subroutine is dimensioned A(10).

TABLE 7-10.2 Summary of the Characteristics of the Subroutine SSS

Identifying Statement: SUBROUTINE SSS (M, A, N, B, OMEGA, VREAL, VIMAG, VPHASE, VMAG)

Purpose: To determine the real part, imaginary part, phase, and magnitude of a network function $N(j\omega)$ having the form

$$N(j\omega) = \frac{A(j\omega)}{B(j\omega)} = \frac{a_1 + a_2 j\omega + a_3(j\omega)^2 + \cdots + a_{m+1}(j\omega)^m}{b_1 + b_2 j\omega + b_3(j\omega)^2 + \cdots + b_{n+1}(j\omega)^n}$$

which determines the ratio of a response phasor to an excitation phasor under sinusoidal steady-state conditions.

Additional Subprograms Required: This subroutine calls the subroutine CVALPL.

Input Arguments:

M The degree m of the numerator polynomial $A(j\omega)$.

A The one-dimensional array of variables A(I) in which are stored the coefficients a_i of the numerator polynomial $A(j\omega)$ of the rational function $N(j\omega)$.

N The degree n of the denominator polynomial $B(j\omega)$.

B The one-dimensional array of variables B(I) in which are stored the coefficients b_i of the denominator polynomial $B(j\omega)$ of the rational function $N(j\omega)$.

OMEGA The frequency ω in radians per second at which the polynomial is to be evaluated.

Output Arguments:

VREAL The value of the real part of $N(j\omega)$.

VIMAG The value of the imaginary part of $N(j\omega)$.

VPHASE The value (in degrees) of the phase of $N(j\omega)$.

VMAG The value of the magnitude of $N(j\omega)$.

Notes:

1. If $N(j\omega)$ is infinite, a statement to that effect is printed and any previously computed values of VREAL, VIMAG, VPHASE, and VMAG are given as output.

2. The variables of the subroutine are dimensioned as follows: A(10), B(10).

TABLE 7-10.3 Summary of the Characteristics of the Function XL

Identifying Statement: FUNCTION XL (N, LO, ND)

Purpose: To generate a sequence of logarithmically spaced numbers for use in plots with a logarithmic scale for the frequency variable.

Additional Subprograms Required: None.

Input Arguments:

N The index n giving the number of times the function has been called. The index is automatically advanced by 1 each time the function is called.

LO The logarithm (to the base 10) of the value of the function when the index $n = 0$.

ND The number of intermediate values that the function should generate for each decade of output values.

Output Value:

XL The logarithmically spaced value generated by the function.

Note: For use in connection with the plotting subroutine PLOT5, the argument N should be set to zero before the first time the function XL is called by the main program.

TABLE 7-10.4 Summary of the Characteristics of the Subroutine XYPLTS

Identifying Statement: SUBROUTINE XYPLTS (NDP, X, Y, NSCLX, NSCLY, NNPX)

Purpose: To plot n_d pairs of numbers x_i and y_i, that is, to make an x-y plot of the data points (x_i, y_i).

Additional Subprograms Required: None.

Input Arguments:

NDP The number n_d of data points to be plotted, i.e., the number of values of x_i and y_i.

X The one-dimensional array of variables X(I) in which are stored the values of the quantities x_i.

Y The one-dimensional array of variables Y(I) in which are stored the values of the quantities y_i.

NSCLX The scale factor giving the maximum value of the x_i which can be plotted along the abscissa.

NSCLY The scale factor giving the maximum value of the y_i which can be plotted along the ordinate.

NNPX The total range of the values of the x_i which can be plotted along the abscissa. For confining the plot to a single page of printed output, NNPX = 90.

Output: The program provides a plot of the data points (x_i, y_i) for a range of values of y_i from NSCLY − 100 to NSCLY along the ordinate, and for a range of values of x_i from NSCLX − NNPX to NSCLX along the abscissa. Coordinate lines are printed every 10 units in both the ordinate direction and the abscissa direction. The direction of increasing values on the abscissa is vertically downward on the printed page.

Notes:

1. The dimensioning of the variables is as follows: X(200), Y(200).

2. An automatic scaling option may be used by setting the argument NNPX to 999. In this case, a plot 90 × 100 units is produced in which the data are scaled so that the range of the x_i or the y_i quantities (whichever is larger) encompasses the full range of its scale.

$$\frac{\mathscr{I}_2}{\mathscr{V}_1} = \frac{50}{1 + 2j\omega + 2(j\omega)^2 + (j\omega)^3}$$

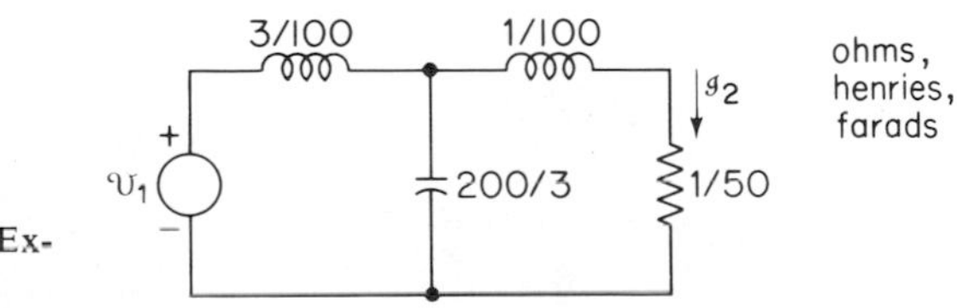

Fig. 7-10.11 Network for Example 7-10.3.

A program for determining the Nyquist plot for this function is given in Fig. 7-10.12. The resulting output is shown in Fig. 7-10.13(a) (for the unscaled data) and Fig. 7-10.13(b) (for the automatically scaled data).

```
C     MAIN PROGRAM FOR EXAMPLE 7.10-3
C             A - ARRAY OF NUMERATOR POLYNOMIAL COEFFICIENTS
C             B - ARRAY OF DENOMINATOR POLYNOMIAL COEFFICIENTS
C             X - ARRAY FOR STORING REAL PART OF NETWORK FUNCTION
C             Y - ARRAY FOR STORING IMAGINARY PART OF NETWORK FUNCTION
C             L - INDEX FOR THE NUMBER OF THE POINT BEING COMPUTED
      DIMENSION A(4), B(4), X(200), Y(200)
C
C     INITIALIZE THE VALUES OF THE NUMERATOR AND DENOMINATOR
C     POLYNOMIAL COEFFICIENTS
      A(1) = 50.
      B(1) = 1.
      B(2) = 2.
      B(3) = 2.
      B(4) = 1.
      OMEGA = 0.
      L = 1
C
C     COMPUTE POINTS UNTIL THE UPPER VALUE OF THE FREQUENCY
C     OR THE LIMITS OF THE DIMENSIONING HAVE BEEN EXCEEDED
    8 CALL SSS(0, A, 3, B, OMEGA, X(L), Y(L), VPHS, VMAG)
      IF (OMEGA.GT.5) GO TO 14
      IF (L.GT.200) GO TO 14
      L = L + 1
C
C     INCREASE THE FREQUENCY BY AN ARITHMETIC AND A GEOMETRIC FACTOR
      OMEGA = (OMEGA + 0.03) * 1.005
      GO TO 8
C
C     CALL THE PLOTTING SUBROUTINE WITHOUT SCALING AND WITH SCALING
   14 PRINT 15
   15 FORMAT(1H1)
      CALL XYPLTS(L, X, Y, 50, 10, 90)
      PRINT 15
      CALL XYPLTS(L, X, Y, 50, 10, 999)
      STOP
      END
```

Fig. 7-10.12 Program for Example 7-10.3.

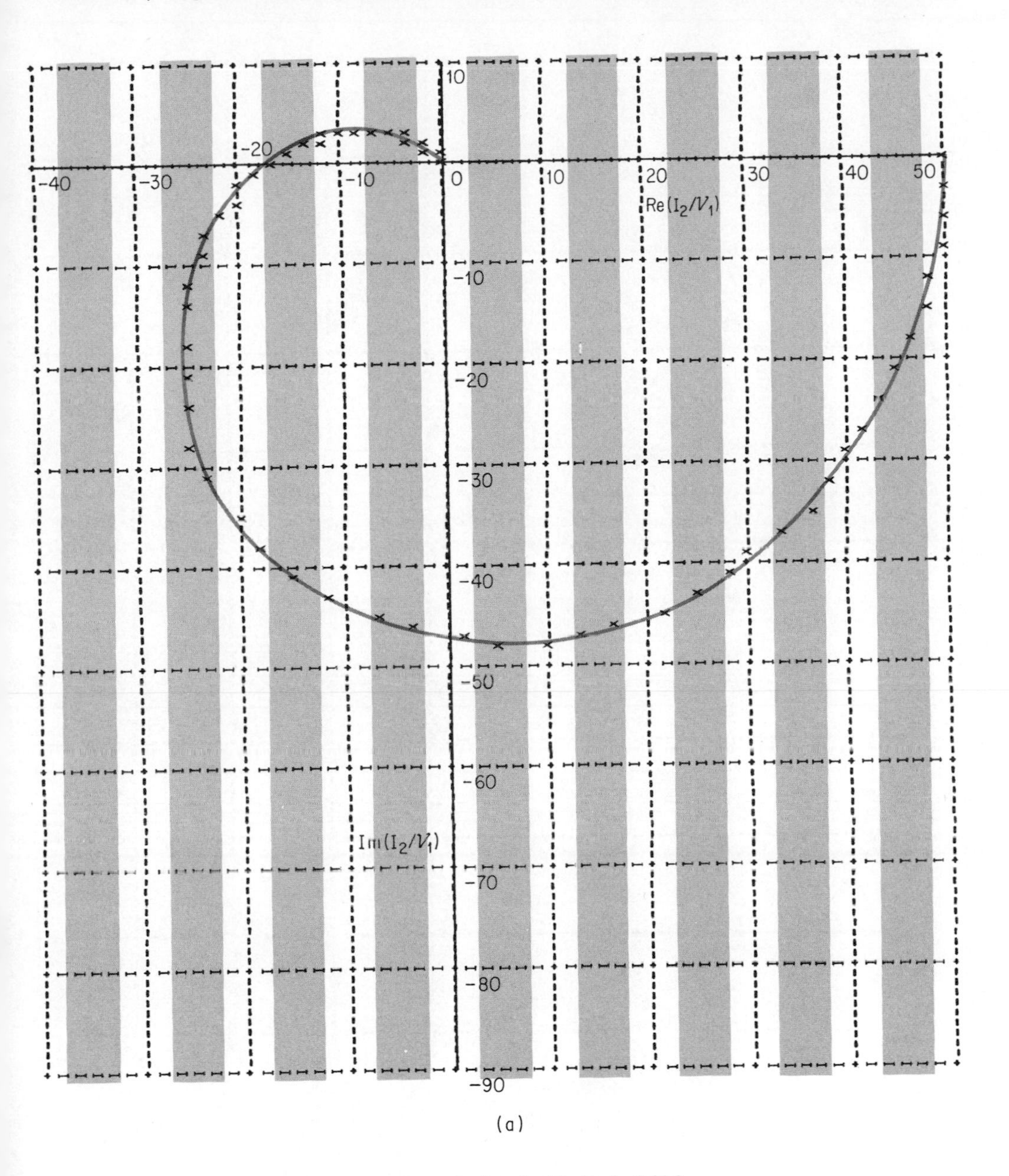

(a)

Fig. 7-10.13 Output plots for Example 7-10.3.

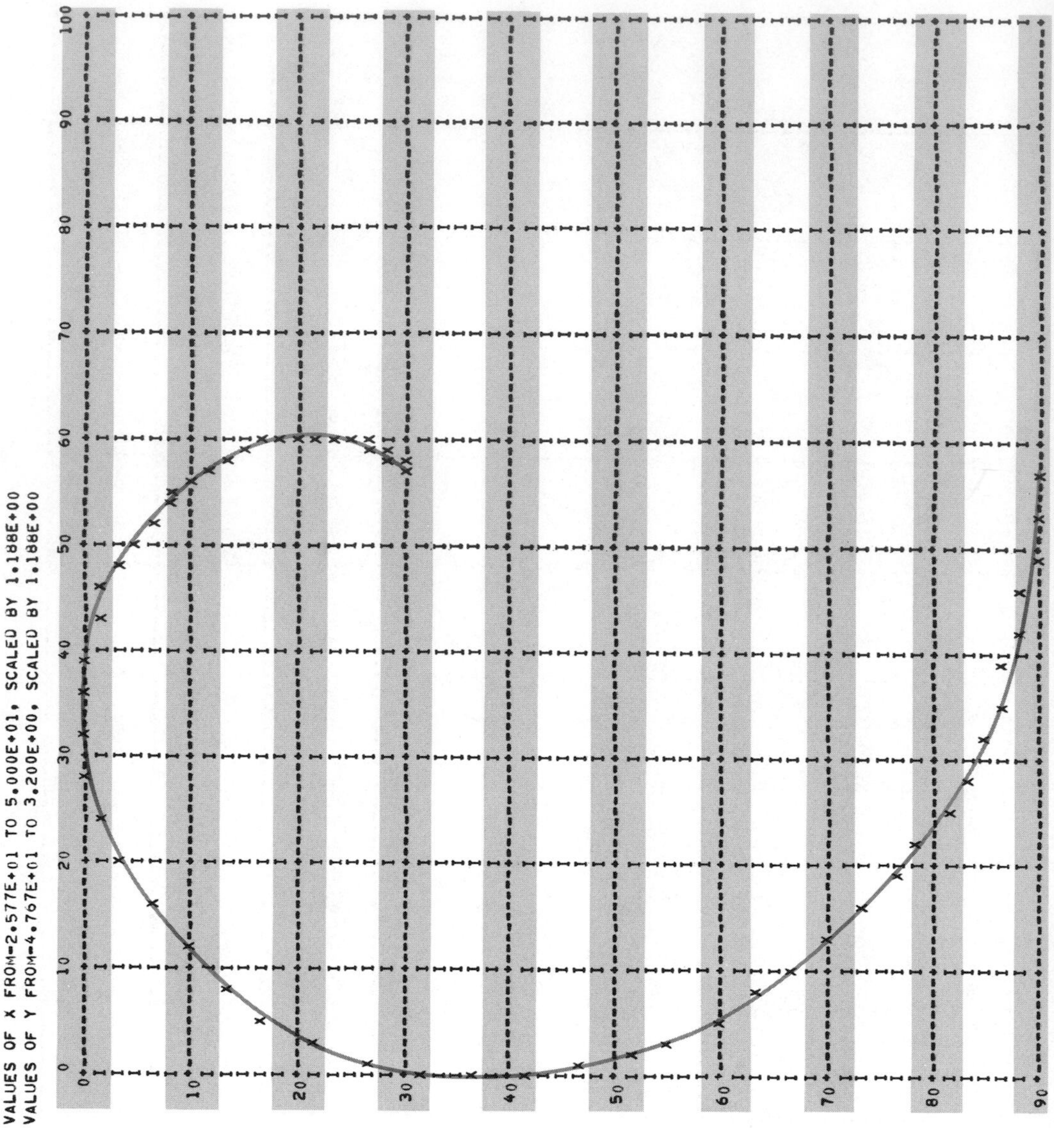

(b)

Fig. 7-10.13 Continued

7-11 SINUSOIDAL STEADY-STATE POWER

In Sec. 7–1 we introduced the use of rms values to characterize sinusoidally varying quantities. Specifically, we showed that a sinusoidal function $f(t)$ having the form

$$f(t) = F_m \cos(\omega t + \phi) \tag{7-234}$$

where F_m is the peak value of $f(t)$, has an rms value $F_{\text{rms}} = F_m/\sqrt{2}$. The definition of an rms value was shown to be such that a sinusoidal steady-state current with an rms value I_{rms} would dissipate the same average power in a given resistor as a constant current with a magnitude equal to I_{rms}. In this section we shall delve more deeply into the use of rms values for voltage and current and show how these may be used to determine the manner in which power is dissipated and stored in circuits under sinusoidal steady-state conditions.

To begin our discussion, consider a two-terminal network with voltage and current variables defined using the usual associated reference directions as shown in Fig. 7–11.1. We shall also assume that there are no independent sources in the network. Let us now define a network function giving the driving-point impedance of the network as

$$Z(j\omega) = R(\omega) + jX(\omega) = |Z|e^{j\theta} = |Z|\cos\theta + j|Z|\sin\theta \tag{7-235}$$

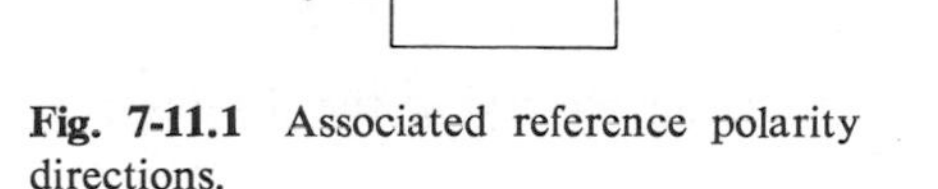

Fig. 7-11.1 Associated reference polarity directions.

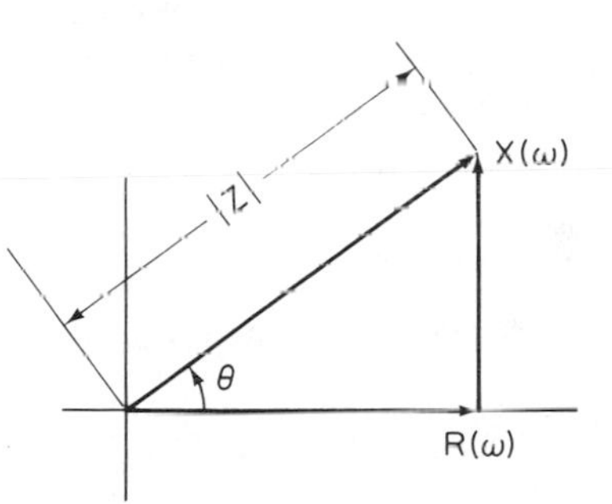

Fig. 7-11.2 An impedance triangle.

where $R(\omega)$ is the real part of $Z(j\omega)$ and $X(\omega)$ is the imaginary part. The relation between $R(\omega)$, $X(\omega)$, $|Z|$, and θ may be represented by a vector diagram as shown in Fig. 7–11.2. Such a diagram is sometimes called an *impedance triangle*. Now let us assume that a sinusoidal voltage is applied as an excitation to the network shown in Fig. 7–11.1. If we let this voltage be

$$v(t) = V_m \cos(\omega t + \alpha) = \sqrt{2}\, V_{\text{rms}} \cos(\omega t + \alpha) \tag{7-236}$$

then the sinusoidal steady-state current that results may be found by applying the phasor methods derived in this chapter. Thus, we may write

$$i(t) = I_m \cos(\omega t + \alpha - \theta) = \sqrt{2}\, I_{\text{rms}} \cos(\omega t + \alpha - \theta) \tag{7-237}$$

where

$$I_m = \frac{V_m}{|Z|} \qquad I_{\text{rms}} = \frac{V_{\text{rms}}}{|Z|} \tag{7-238}$$

The instantaneous power supplied to this network is found by multiplying (7–236) and (7–237). Thus, since $\cos a \cos b = \frac{1}{2}\cos(a-b) + \frac{1}{2}\cos(a+b)$, we may write

$$p(t) = v(t)i(t) = V_{\text{rms}} I_{\text{rms}}[\cos\theta + \cos(2\omega t + 2\alpha - \theta)] \tag{7-239}$$

Now let us find the average value of the instantaneous power $p(t)$. This is done by integrating $p(t)$ in Eq. (7–239) over one period of the sinusoidal excitation and dividing the result by the value of the period. The average value of the second term inside the bracket in the right member of Eq. (7–239) is zero. Thus, we find that

$$P_{\text{avg}} = \int_0^T p(t)\, dt = V_{\text{rms}} I_{\text{rms}} \cos\theta \tag{7-240}$$

where T is the period. From this expression, we see that the average power is always less than or equal to the product of the rms values of the voltage and the current. To indicate the ratio of these quantities, we may define the *power factor* (abbreviated PF) as follows:

$$\text{PF} = \cos\theta = \frac{V_{\text{rms}} I_{\text{rms}}}{P_{\text{avg}}} \tag{7-241}$$

Using (7–238) to eliminate V_{rms} in (7–240) and applying (7–235), we may also write

$$P_{\text{avg}} = I_{\text{rms}}^2 |Z| \cos\theta = I_{\text{rms}}^2 R(\omega) \tag{7-242}$$

In a similar manner, if we define the network function specifying the driving-point admittance of the network shown in Fig. 7–11.1 as

$$Y(j\omega) = G(\omega) + jB(\omega) = |Y|e^{j\phi} = |Y|\cos\phi + j|Y|\sin\phi \tag{7-243}$$

then since $\phi = -\theta$ and $|Y| = 1/|Z|$, using (7–238) to eliminate I_{rms} in (7–240) we obtain

$$P_{\text{avg}} = \frac{V_{\text{rms}}^2}{|Z|} \cos\theta = V_{\text{rms}}^2 G(\omega) \tag{7-244}$$

It should be noted that $G(\omega) = 1/R(\omega)$ only if $X(\omega)$, and thus $B(\omega)$, is zero.

To more clearly indicate the relation between instantaneous power and average power, let us consider three cases in detail. The first of these is the case for which the imaginary part of $Z(j\omega)$ is zero, i.e., the phase angle θ is zero and the power factor is unity. Typical waveforms of the voltage, current, and power variables for this case are shown in Fig. 7–11.3(a). It is readily observed that the instantaneous power is always positive, similarly, from Eq. (7–240) we see that the average power P_{avg} is simply the product $V_{rms}\, I_{rms}$. Now let us consider a second case defined by letting $R(\omega) = X(\omega)$. Thus, $\theta = 45°$ and the power factor is 0.707. Typical plots of the waveforms of the voltage, current, and power are shown for this case in Fig. 7–11.3(b). In this case, although the instantaneous power $p(t)$ goes positive and negative, the average power P_{avg} is clearly positive and, from Eq. (7–240), it has the value $V_{rms}\, I_{rms} \cos 45°$. As a third case, let us see what happens if the real part of the driving-point impedance $Z(j\omega)$ is zero. In this case, $\theta = 90°$ and, from Eq. (7–240), the average power is zero. Typical waveforms of voltage, current, and power for this case are shown in Fig. 7–11.3(c). From the waveforms in this figure, we may note that the *instantaneous* power has quite large values for some instants of time. The *average* value of the power, however, is zero. The positive and negative values of

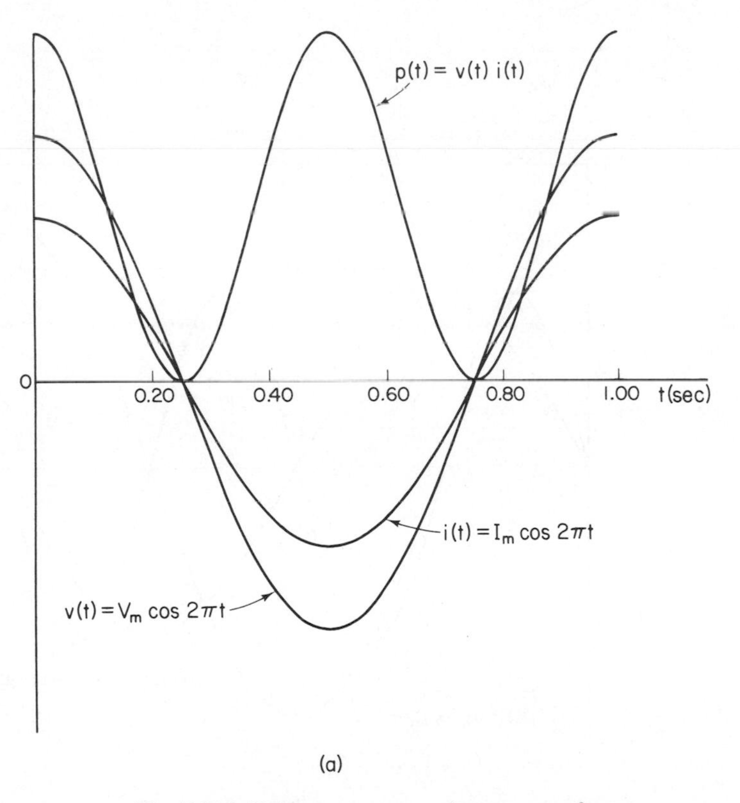

Fig. 7-11.3 Voltage, current, and power waveforms: (a) $\theta = 0°$; (b) $\theta = 45°$; (c) $\theta = 90°$.

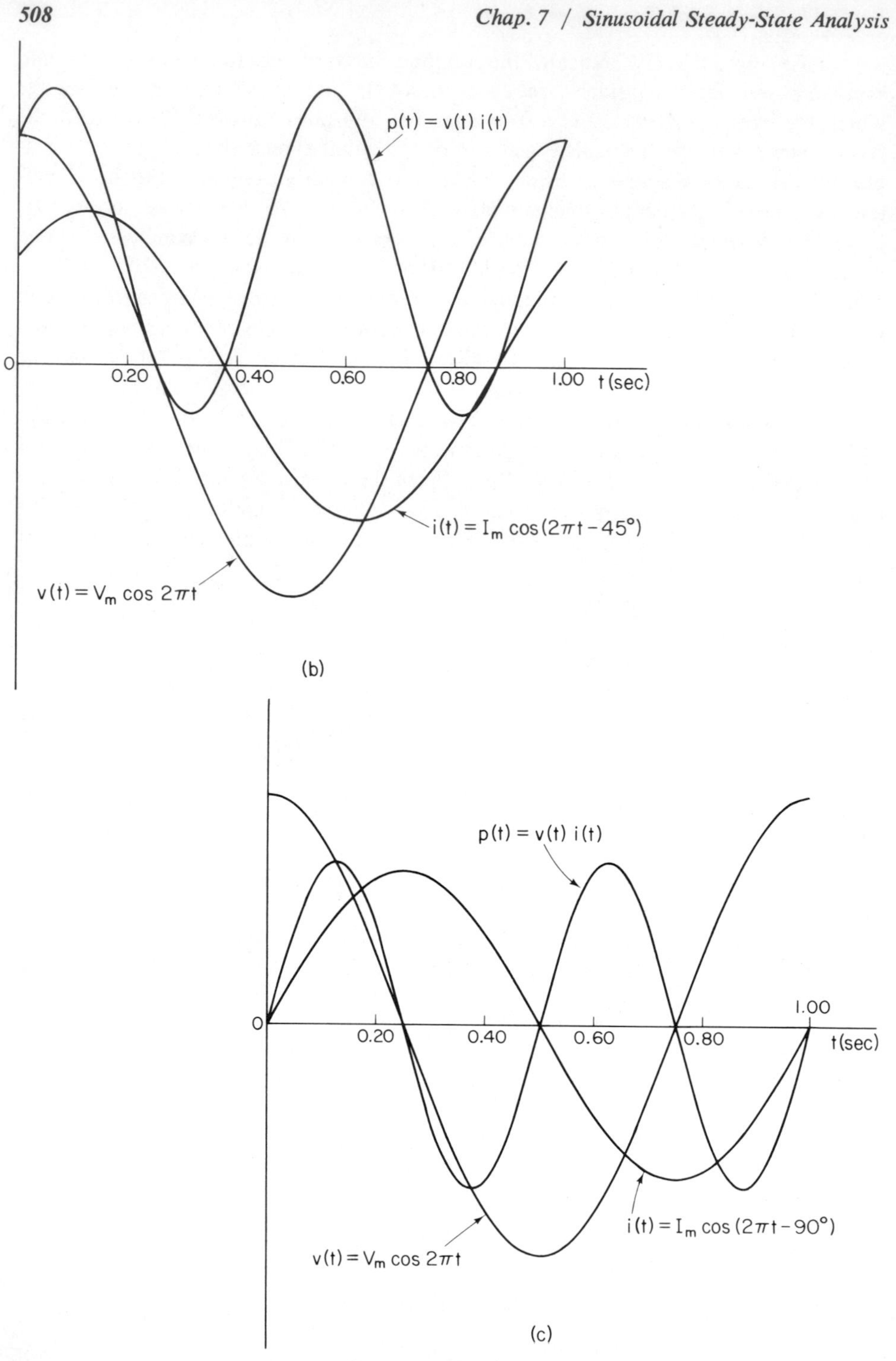

Fig. 7-11.3 Continued

the instantaneous power represent the transfer of energy back and forth between the source which excites the circuit and the elements of the circuit itself. In other words, during the portions of the cycle where $p(t)$ is positive, energy is stored in the electric and magnetic fields associated with the capacitive and inductive elements, respectively. During portions of the applied cycle where $p(t)$ is negative, this energy is delivered back to the source. The power which oscillates back and forth between source and network in this manner is frequently spoken of as being *reactive power*. Although it is not dissipated in the network, i.e., it is not turned into heat, it does represent an actual flow of current into and out of the network. Thus, in designing power systems it is important to provide sufficient wiring capacity to permit the flow of such reactive currents. This is especially important since all "real-world" power lines will have some resistance present in them. Thus, even though the reactive power does not dissipate energy in the load, in practice it will actually produce energy losses in the power lines themselves. To more clearly indicate the component of the instantaneous power which is reactive, we may use the relation

$$\cos(\alpha + \beta) = \cos\alpha \cos\beta + \sin\alpha \sin\beta$$

to write Eq. (7-239) in the following form.

$$p(t) = V_{\text{rms}} I_{\text{rms}} \cos\theta\,[1 + \cos(2\omega t + 2\alpha)] + V_{\text{rms}} I_{\text{rms}} \sin\theta\,[\sin(2\omega t + 2\alpha)] \tag{7-245}$$

When $\theta = 90°$, the instantaneous power is obviously determined only by the second term in the right member of the above. On the other hand, for values of θ in the range $0 < \theta < 90°$, the magnitude of this second term is $V_{\text{rms}}I_{\text{rms}} \sin\theta$. Thus, we may define a reactive average power represented by this term. Specifically, using the symbol Q for average *reactive power*, we may write[17]

$$Q = V_{\text{rms}} I_{\text{rms}} \sin\theta \tag{7-246}$$

The units of Q are VARS (*V*olt *A*mperes *R*eactive). It should be noted that for an inductive circuit, θ is greater than zero and, therefore, Q is positive; for a capacitive circuit, θ is less than zero and Q will be negative. Thus, in general reactive power can be positive or negative. The relation between P, the average (or real) power, and Q, the reactive power, is simply that of a right triangle. To illustrate this, we may define a *complex power* W by the relation

$$W = V_{\text{rms}} I_{\text{rms}}\, e^{j\theta} = V_{\text{rms}} I_{\text{rms}} (\cos\theta + j\sin\theta) = P + jQ \tag{7-247}$$

The units for W are VA (*V*olt *A*mperes). The right-triangle relation may be emphasized by drawing a power triangle as shown in Fig. 7-11.4.

[17]Note that the symbol Q as used here does *not* refer to the Q of a resonant circuit.

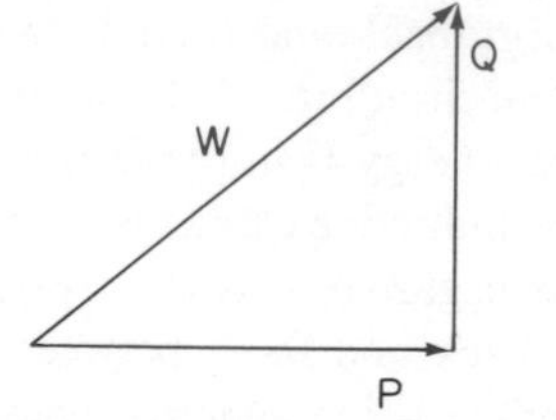

Fig. 7-11.4 A power triangle.

Complex power may be conveniently expressed in terms of the phasors used to represent the voltage and current variables. If we let $\mathscr{V}$ and $\mathscr{I}$ be the phasors for these variables, where[18]

$$\mathscr{V} = V_{\text{rms}}e^{j\alpha} \qquad \mathscr{I} = I_{\text{rms}}e^{j\beta} \tag{7-248}$$

Then, if these phasors are the terminal variables associated with the driving-point impedance $Z(j\omega) = |Z|e^{j\theta}$ we may write

$$\mathscr{V} = V_{\text{rms}}\, e^{j\alpha} = |Z|\, e^{j\theta}\, I_{\text{rms}}\, e^{j\beta} = |Z|\, I_{\text{rms}}\, e^{j(\beta+\theta)} \tag{7-249}$$

Equating the arguments of the members of the above equation, we see that

$$\alpha = \beta + \theta \tag{7-250}$$

Now let us multiply both sides of Eq. (7-249) by $\mathscr{I}^* = I_{\text{rms}}\, e^{-j\beta}$. Thus, we obtain

$$\mathscr{V}\mathscr{I}^* = |Z|\, I_{\text{rms}}^2\, e^{j\theta} \tag{7-251}$$

Substituting for one of the I_{rms} quantities from (7-238), we obtain

$$\mathscr{V}\mathscr{I}^* = V_{\text{rms}}\, I_{\text{rms}}\, e^{j\theta} = W \tag{7-252}$$

The results given above may be summarized as follows:

SUMMARY 7-11.1

Sinusoidal Steady-State Power: If $\mathscr{V}$ and $\mathscr{I}$ are the phasors representing the rms values of sinusoidal steady-state voltage and current variables with associated reference directions at a pair of terminals of a network, the complex power W (in volt amperes) delivered to the pair of terminals is defined as $\mathscr{V}\mathscr{I}^*$. The real part of W is the average power P (in watts) transmitted to the network and the imaginary part of W is the reactive power Q (in vars).

Now let us investigate the power supplied to two two-terminal impedances connected in parallel and excited by a voltage source. Such a connection is shown in Fig. 7-11.5(a). In terms of the phasors defined in the figure, we may write

$$W = \mathscr{V}\mathscr{I}^* = \mathscr{V}(\mathscr{I}_1^* + \mathscr{I}_2^*) = \mathscr{V}\mathscr{I}_1^* + \mathscr{V}\mathscr{I}_2^* \tag{7-253}$$

[18]For convenience, here we define the phasors in terms of the rms value of the sinusoidal functions which they represent rather than the peak magnitude. Such a convention is usually followed when considering sinusoidal steady-state power.

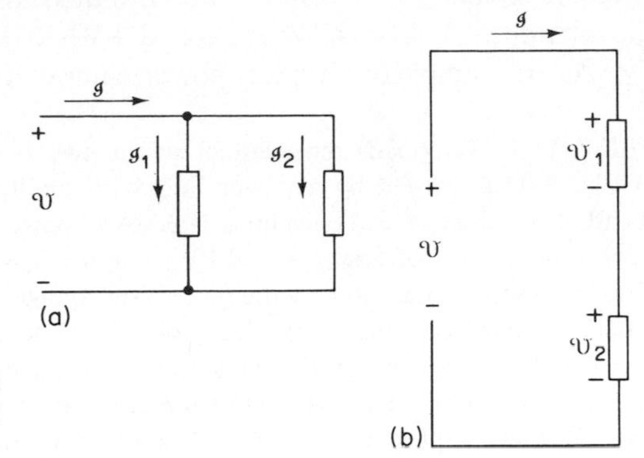

Fig. 7-11.5 Series and parallel connections of two two-terminal impedances.

where W is the total complex power supplied to both networks. If we define the component power to network 1 as $P_1 + jQ_1$, and that to network 2 as $P_2 + jQ_2$, we may write, from (7–253),

$$W = (P_1 + jQ_1) + (P_2 + jQ_2) = P_1 + P_2 + j(Q_1 + Q_2) \qquad (7\text{–}254)$$

Thus, we conclude that the total real power supplied to the parallel connection of the two networks is the sum of the powers consumed by the individual networks. Similarly the total reactive power is the sum of the individual reactive powers. Now let us consider a similar situation in which two networks are connected in series as shown in Fig. 7–11.5(b). For this case, the total complex power W is

$$W = \mathcal{V}\mathcal{I}^* = \mathcal{V}_1\mathcal{I}^* + \mathcal{V}_2\mathcal{I}^* \qquad (7\text{–}255)$$

Again we see that the total real and reactive powers are simply given as the sum of the real and reactive powers of the individual networks. The conclusion given above for parallel and series connected networks may be readily shown to apply to any arbitrary interconnection of two-terminal networks. As such, we may refer to these results as the *conservation of complex power*. These results may be summarized as follows:

SUMMARY 7-11.2

Conservation of Complex Power: If in a set of two-terminal networks N_i the real and reactive powers are P_i and Q_i, respectively, the total real and reactive power for *any* interconnection of the N_i is the sum of the P_i and the sum of the Q_i, respectively. Thus, the total complex power for such an interconnection of networks is the sum of the complex powers of the component networks.

It should be noted that the units of voltamperes, watts, and vars used for complex power, real power, and reactive power, respectively, are too small for many practical power computations. Units which are more frequently used in practice are kVA (kilovoltamperes), kW (kilowatts), and kVAR (kilovoltamperes reactive), respectively. As an example of a typical power computation, consider the following.

EXAMPLE 7-11.1. Two loads represented by the two two-terminal networks labeled A and B in Fig. 7-11.6 are fed by a power line with an impedance $Z = 0.1 + j0.1\ \Omega$ as shown. Load A is inductive and consumes 10 kVA of power at 0.6 PF and load B is capacitive and consumes 5 kVA of power at 0.8 PF. The voltage $\mathscr{V}$ at the loads is 100 V rms. It is desired to find the rms value of the generator voltage $\mathscr{V}_g$. To solve this problem, we may first determine the complex power for each of the loads. For load A the real power is $10 \times 0.6 = 6$ kW. Using a power triangle similar to the one shown in Fig. 7-11.4, we find that the reactive power is $+8$ kVAR. Thus the complex power is $6 + j8$ kVA. Similarly, for network B, the complex power is $4 - j3$ kVA. Summing these powers, we find that the total complex load power is $10 + j5$ kVA, thus $|W| = \sqrt{125}$ kVA. Since the total load is inductive, the load voltage leads the total load current by $\tan^{-1}(\frac{5}{10})$. If we assume a reference angle of 0° for the load voltage phasor, then, using rms quantities, the load voltage phasor is $100\angle 0°$ V or $0.1\angle 0°$ kV. Thus, from (7-252), the total current phasor $\mathscr{I}$ may be computed as follows:

$$\mathscr{I}^* = \frac{W}{\mathscr{V}} = \frac{10 + j5}{0.1} = 100 + j50 \qquad \mathscr{I} = 100 - j50 \text{ A} \tag{7-256}$$

It should be noted that the current lags the voltage, as must be the case for an inductive load. The phasor voltage drop caused by the power line impedance Z is now found as $\mathscr{I}Z = (100 - j50) \times (0.1 + j0.1) = 15 + j5$ V. Thus the generator voltage $\mathscr{V}_g = \mathscr{I}Z + \mathscr{V} = (15 + j5) + 100 = 115 + j5$ V, and the magnitude of $\mathscr{V}_g$ is very nearly equal to 115 V rms. We see that the generator voltage in this case is 15 per cent greater than the load voltage.

In the example given above we noted that the generator voltage was approximately 15 per cent higher than the load voltage. Such a situation is usually very undesirable since, if the loads A and B shown in Fig. 7-11.6 are disconnected, the voltage at the load terminals will rise from 100 V to 115 V. If some relatively low-wattage electronic apparatus should remain connected to these load terminals, and if the rated voltage for this equipment was 100 V, such a rise in voltage might seriously affect the operation of the equipment (or even cause it to smoke). Thus, it is desirable to minimize the change in voltage that occurs when a large load is connected to or

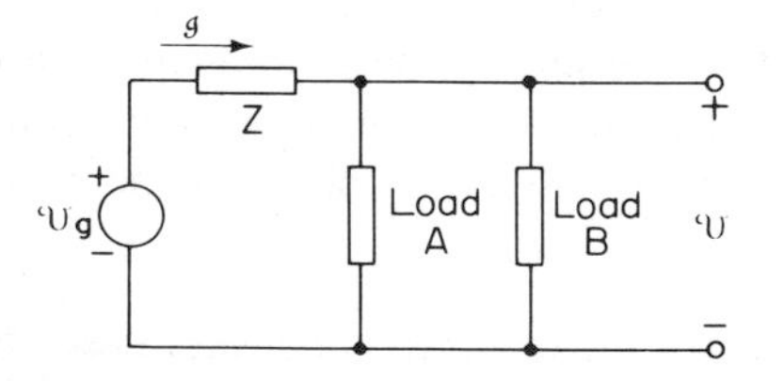

Fig. 7-11.6 Network for Example 7-11.1.

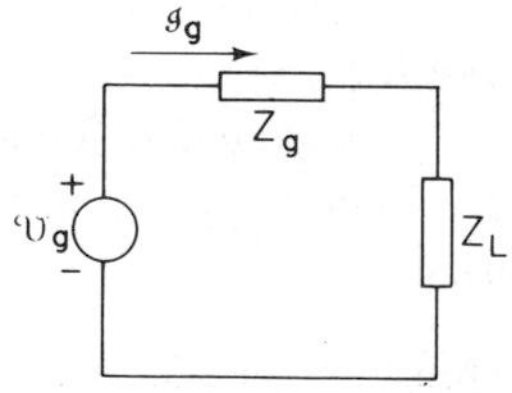

Fig. 7-11.7 A non-ideal source and a load.

removed from a given power line. One way of doing this for loads which are highly reactive is to raise the power factor of the load. Such an approach is treated in the following example.

EXAMPLE 7–11.2. As an example of power factor correction, consider the addition of a third load to the two loads used in the preceding example. We will choose this load so that the resulting power factor of all the loads is unity. From the computations made in the previous example, we see that the power factor correcting load must have a complex power $W = 0 - j5$ kVA, i.e., it must consume zero real power and have a reactive power of -5 kVAR. Physically, such a load simply consists of a capacitor. With the addition of this compensating load, the total complex load power W is now $10 + j0$. The total phasor current to the loads is thus $\mathscr{I} = 100 + j0$ A and the generator voltage is $\mathscr{V}_g = \mathscr{I}Z + \mathscr{V} = (10 + j10) + 100 = 110 + j10$ V. Thus, the magnitude of the generator voltage is now very nearly equal to 110 V rms and the regulation has been reduced from 15 per cent to 10 per cent.

Load factor correction techniques similar to the one illustrated in the preceding example are greatly encouraged by power companies to the extent that for large industrial users, low power factor rates, which are usually considerably higher than high power factor rates, are levied. As a result, it is frequently economically attractive for large companies to install devices whose sole function is power factor correction. Such corrective techniques may range in size from fluorescent light ballasts specially designed for high power factor all the way to large rotating machines called synchronous condensers where sole purpose is the simulation of a capacitive load.

It is frequently desirable to maximize the power transferred to a given load by adjusting the load itself or by inserting some network between the generator and the load. Such a maximization is important in a broad range of applications that are encountered in communication systems and electronic circuits. One approach to the general problem is illustrated in Fig. 7–11.7. In this figure, a voltage generator with phasor output voltage $\mathscr{V}_g$ and series impedence $Z_g = R_g + jX_g$ drives a load impedance $Z_L = R_L + jX_L$. Such a generator circuit may be considered as the Thevenin equivalent for a broad range of network configurations. Now let us assume that R_L and X_L are free to be varied such that the power delivered by the generator to the load is maximized. The phasor current $\mathscr{I}_g$ is readily found to be

$$\mathscr{I}_g = \frac{\mathscr{V}_g}{Z_g + Z_L} = \frac{\mathscr{V}_g}{(R_g + R_L) + j(X_g + X_L)} \tag{7–257}$$

The average power delivered to the load is given as

$$P_L = |\mathscr{I}_g|^2 R_L = \frac{|\mathscr{V}_g|^2 R_L}{(R_g + R_L)^2 + (X_g + X_L)^2} \tag{7–258}$$

Let us first consider the case where R_L is a constant and find a value of X_L to maximize the power P_L. Taking the partial derivative of the expression in Eq. (7–258) with respect to X_L we obtain

$$\frac{\partial P_L}{\partial X_L} = |\mathscr{V}_g|^2 R_L \left[\frac{-2(X_g + X_L)}{[(R_g + R_L)^2 + (X_g + X_L)^2]^2} \right] \tag{7-259}$$

Setting this derivative to zero, we obtain the relation $X_g = -X_L$. Substituting this value in Eq. (7–258) we obtain

$$P_L|_{X_L=-X_g} = \frac{|\mathscr{V}_g|^2 R_L}{(R_g + R_L)^2} \tag{7-260}$$

To determine the effect R_L has on power maximization, we may now differentiate this result again, this time with respect to R_L, and set the result to zero. Doing this, we find the maximum occurs when $R_G = R_L$. Considering both of the above results, we see that the maximum power transfer is obtained when

$$Z_L = Z_g^* \tag{7-261}$$

Thus, maximum power transfer occurs *when the source impedance and the load impedance are conjugates.*

Another example of the maximization of power transfer occurs when both the generator impedance and the load impedance are fixed. In this situation a method which is frequently used to maximize the transfer of power is the insertion of a transformer between the generator and the load. The turns ratio of the transformer is chosen such that the desired maximum is obtained. As an example of such a situation, consider the network shown in Fig. 7–11.8 in which an ideal transformer of turns ratio $1:n$ is inserted between a source impedance Z_g and a load impedance Z_L. The driving-point impedance seen at the left end of the transformer $Z_0 = Z_L/n^2$. If the phase of Z_L is θ, then

$$Z_0 = |Z_0|e^{j\theta} = |Z_0| \cos\theta + j|Z_0| \sin\theta \tag{7-262}$$

Since the transformer is a lossless device, the power dissipated in Z_0 is the same as that dissipated in Z_L. Thus, we may write

$$P_L = |\mathscr{I}_g|^2 |Z_0| \cos\theta = \frac{|\mathscr{V}_g|^2 |Z_0| \cos\theta}{(R_g + |Z_0|\cos\theta)^2 + (X_g + |Z_0|\sin\theta)^2} \tag{7-263}$$

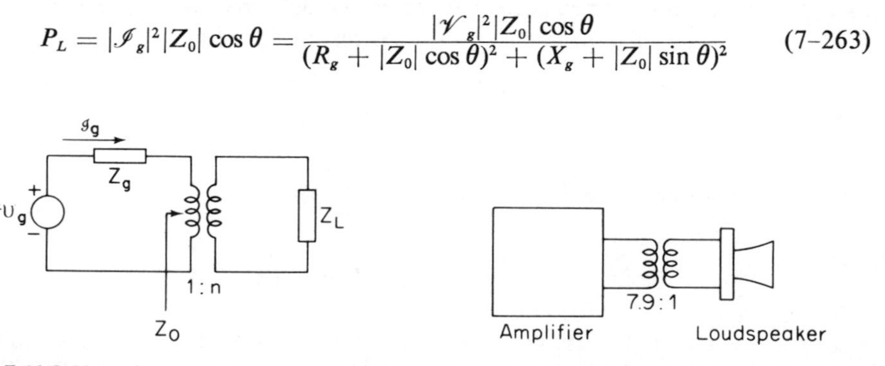

Fig. 7-11.8 Use of a transformer to maximize power transfer.

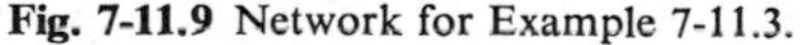

Fig. 7-11.9 Network for Example 7-11.3.

Since $|Z_0|$ is adjusted by changing the value of n, the turns ratio, let us now take the derivative of P_L with respect to $|Z_0|$. Doing this and setting the result to zero, we obtain

$$|Z_0| = |Z_g| \tag{7-264}$$

Thus, since $|Z_0| = |Z_L|/n^2$, we obtain

$$n = \sqrt{\frac{|Z_L|}{|Z_g|}} \tag{7-265}$$

as the value of n which provides maximum power transfer. An example follows.

EXAMPLE 7-11.3. As an example of the use of a transformer to maximize power transfer, let us consider an audio amplifier which has an output impedance of 500 Ω. It is desired to connect the amplifier to a loudspeaker with a coil impedance of 8 Ω. To achieve the desired maximum power transfer, we may insert a transformer between the amplifier and the loudspeaker with the turns ratio $n = \sqrt{8/500}$, i.e., approximately 7.9 : 1 as shown in Fig. 7-11.9.

7-12 THREE-PHASE CIRCUITS

In this section we shall show how the phasor methods developed to analyze circuits operating under sinusoidal steady-state conditions can be applied to the study of three-phase ac circuits. Such circuits have wide use in the generation, transmission, and distribution of electric power since they provide constant power rather than sinusoidally varying power. Thus, three-phase circuits provide an important example of the application of circuit-analysis techniques to an everyday engineering situation. An ideal three-phase power generator may be visualized as consisting of three independent voltage sources, each generating a sinusoidal voltage displaced in phase by 120° from the other two. For example, consider the three sources with outputs $v_a(t)$, $v_b(t)$, and $v_c(t)$ shown in Fig. 7-12.1(a). For clarity we have indicated expressions for both the time-domain and the phasor representations for these sources. It should be noted that the magnitude and frequency of the output of each source is the same,

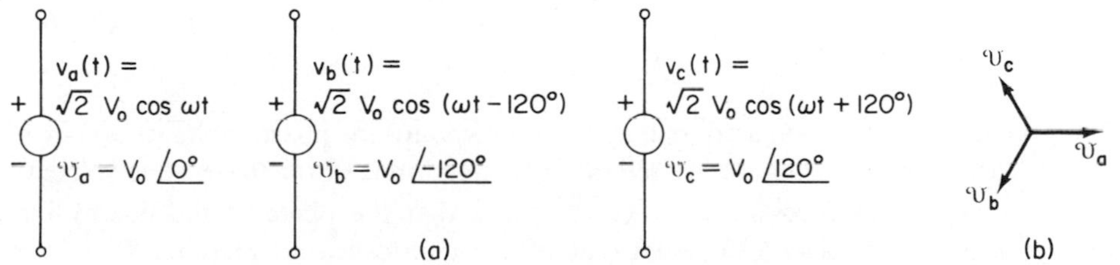

Fig. 7-12.1 Three sources and their phasor diagram.

namely, $\sqrt{2}V_o$V and ω rad/s. In addition, it should be noted that the phasors are defined using rms quantities for their magnitudes. A phasor diagram for the three phasor voltages $\mathscr{V}_a$, $\mathscr{V}_b$, and $\mathscr{V}_c$ is shown in Fig. 7–12.1(b). Now let us consider the case where the negative reference terminals of the three sources shown in Fig. 7–12.1(a) are connected to a common point as shown in Fig. 7–12.2(a). The common point

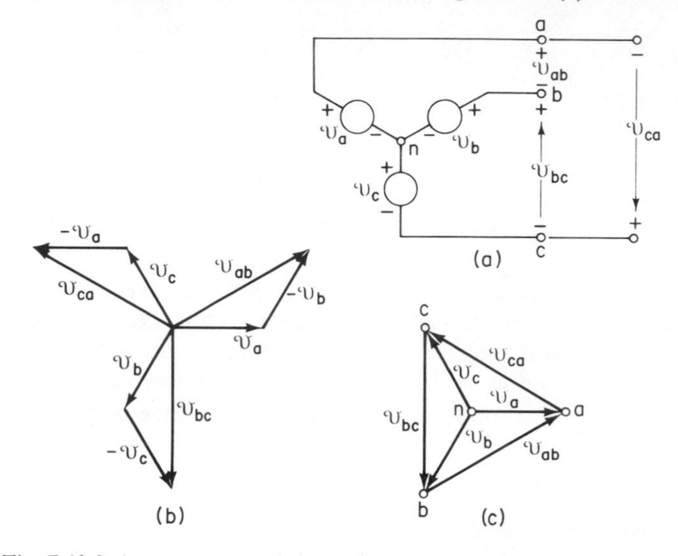

Fig. 7-12.2 A wye-connected three-phase source and its phasor diagrams.

is identified by the letter n. It is called a *neutral*. Such a network is called a *wye-connected* three-phase generator. The relative orientation of the three voltage sources with respect to terminals a, b, and c which is shown in Fig. 7–12.2(a) is called *abc* rotation. If the outputs of voltage sources $\mathscr{V}_b$ and $\mathscr{V}_c$ were interchanged, the result would be referred to as *acb* rotation. Since the terminals labeled a, b, and c are usually connected to a set of three transmission lines which carry power to some load remotely located from the generator, it is convenient to refer to the source voltages shown in Fig. 7–12.2(a) as *line-to-neutral* voltages. We may also define a set of *line-to-line* voltages $\mathscr{V}_{ab}$, $\mathscr{V}_{bc}$, and $\mathscr{V}_{ca}$ as shown in Fig. 7–12.2(a). The phasor representations for these voltages are readily found in terms of those for the line-to-neutral voltages as follows.

$$\begin{aligned} \mathscr{V}_{ab} &= \mathscr{V}_a - \mathscr{V}_b \\ \mathscr{V}_{bc} &= \mathscr{V}_b - \mathscr{V}_c \\ \mathscr{V}_{ca} &= \mathscr{V}_c - \mathscr{V}_a \end{aligned} \tag{7–266}$$

The graphical construction used to determine the line-to-line phasor voltages is shown in 7–12.2(b). From this figure, we see that the magnitude of the line-to-line voltages may be computed as $2 \cos 30° \; V_o = \sqrt{3}\,V_o$ and that the phase of the line-to-line voltage phasors is displaced 30° from that of the line-to-neutral phasors. Thus, we may write

$$\mathscr{V}_{ab} = \sqrt{3}\,V_o\underline{/30^\circ} \qquad v_{ab}(t) = \sqrt{6}\,V_o \cos(\omega t + 30^\circ)$$
$$\mathscr{V}_{bc} = \sqrt{3}\,V_o\underline{/-90^\circ} \qquad v_{bc}(t) = \sqrt{6}\,V_o \cos(\omega t - 90^\circ) \qquad (7\text{–}267)$$
$$\mathscr{V}_{ca} = \sqrt{3}\,V_o\underline{/150^\circ} \qquad v_{ca}(t) = \sqrt{6}\,V_o \cos(\omega t + 150^\circ)$$

For brevity, the line-to-neutral voltages are frequently referred to simply as *phase* voltages and the line-to-line voltages simply as *line* voltages. It is somewhat easier to visualize the relation between the phase and the line voltages if the phasors for the latter are removed from the origin and drawn as shown in Fig. 7–12.2(c). From this latter phasor diagram, we see that if KVL is applied to the line voltages we obtain

$$\mathscr{V}_{ab} + \mathscr{V}_{bc} + \mathscr{V}_{ca} = 0 \qquad (7\text{–}268)$$

Frequently, three-phase voltages are generated by connecting voltage sources directly to the lines rather than to a neutral point. Such a connection is shown in Fig. 7–12.3(a). It is called a *delta connection*. A three-phase transformer bank is one

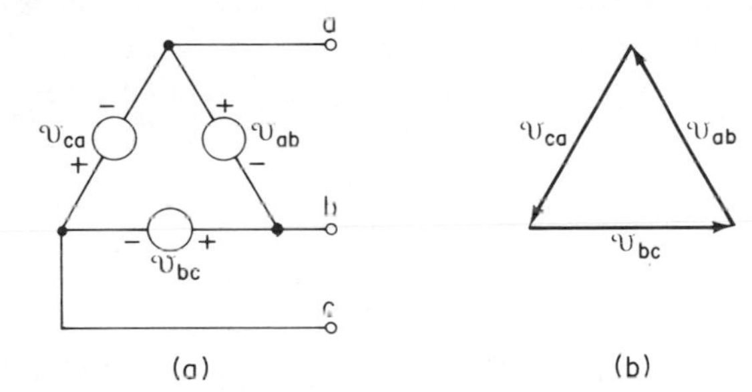

Fig. 7-12.3 A delta-connected three-phase source and its phasor diagram.

means of providing such a connection which is frequently encountered. The phasor diagram for this connection is shown in Fig. 7–12.3(b). For such a connection no neutral point is available. It should be noted that if any of the sources shown in Fig. 7–12.3(a) are reversed, the results could be disastrous. In this case, the sum of the phasor voltages no longer equals zero. Thus there will be a net non-zero voltage around the loop and, since the impedance of the loop is that of the voltage sources, namely, zero, in theory infinite current will flow. In practice, if there is no circuit breaker to protect the circuit, the heat generated by the resulting large currents will rapidly produce smoke and may even cause the transformer to explode.

As a first example of the type of computations which are normally encountered for three-phase circuits, let us consider the case of a *balanced* wye-connected load, i.e., a load consisting of three equal impedances $Z(j\omega)$, all of which have a common neutral terminal. Such a load, driven by a wye-connected three-phase generator, is shown in Fig. 7–12.4(a). If we assume that the phase angle of $Z(j\omega)$ is θ, then the phasor diagram (using $\mathscr{V}_a$ as the reference) for the circuit is as shown in Fig. 7–12.4(b). From this phasor diagram we see that the sum of the line currents $\mathscr{I}_a$, $\mathscr{I}_b$, and $\mathscr{I}_c$ is zero. Thus, applying KCL to the circuit shown in Fig. 7–12.4(a), we see that the

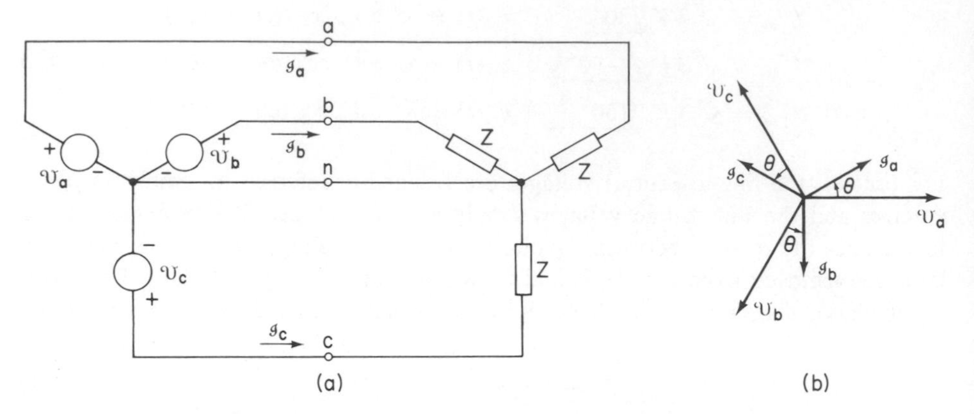

Fig. 7-12.4 A balanced wye-connected load and its phasor diagram.

current in the neutral wire must be zero. Thus, for a balanced wye-connected load the neutral connection may be omitted without any change occurring in the circuit variables. If we let V_P be the rms magnitude of the phase voltage, and V_L be the rms magnitude of the line voltage, then, as we have previously shown, $V_L = \sqrt{3}\, V_P$. The average real and reactive power delivered to each phase of the wye-connected load may be designated as P_P and Q_P, respectively. Thus, applying the results of the preceding section, we may write

$$P_P = V_P I \cos\theta = \left(\frac{V_L}{\sqrt{3}}\right) I \cos\theta \tag{7-269a}$$

$$Q_P = V_P I \sin\theta = \left(\frac{V_L}{\sqrt{3}}\right) I \sin\theta \tag{7-269b}$$

where I is the rms value of the line current. Since there are three phases, we may make a total power analysis of the circuit as follows.

$$P_{\text{total}} = 3P_P = 3V_P I \cos\theta = \sqrt{3}\, V_L I \cos\theta \tag{7-270a}$$

$$Q_{\text{total}} = 3Q_P = 3V_P I \sin\theta = \sqrt{3}\, V_L I \sin\theta \tag{7-270b}$$

$$|W_{\text{total}}| = (P^2_{\text{total}} + Q^2_{\text{total}})^{1/2} = \sqrt{3}\, V_L I \tag{7-270c}$$

$$\text{PF} = \frac{P_{\text{total}}}{|W_{\text{total}}|} = \cos\theta \tag{7-270d}$$

As an example of a typical problem involving a balanced wye-connected load, consider the following.

EXAMPLE 7-12.1. A three-phase voltage generator with line voltages of 120 V rms is used to drive a balanced wye-connected load consisting of three impedances $Z_L = 4 + j3\Omega$ through a set of power lines, each of which has an impedance $Z_G = 0.1 + j0.1\Omega$ as shown in Fig. 7-12.5. It is desired to find the line voltages at the load and the power delivered to

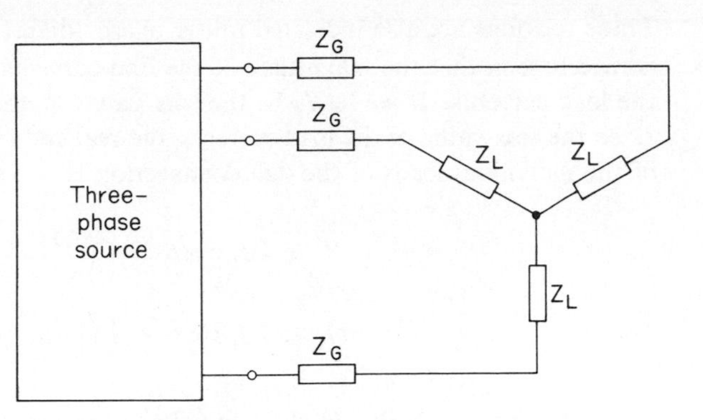

Fig. 7-12.5 Network for Example 7-12.1.

the load. To do this we first find I, the rms magnitude of the line currents. Since the load is balanced, such a determination may be made on the basis of phase (line-to-neutral) quantities even though no neutral connection is present. Thus, we see that

$$I = \frac{V_P}{|Z_G + Z_L|} = \frac{120/\sqrt{3}}{|0.1 + j0.1 + 4 + j3|} = 10.82$$

The line voltage at the load is $\sqrt{3}\, I|Z_L| = \sqrt{3} \times 10.82 \times 5 = 93.7$ V rms. The total power delivered to the load is $3I^2 \operatorname{Re} Z_L = 3 \times (10.82)^2 \times 4 = 1404$ W.

A second example of a three-phase power system frequently encountered in practice is one in which a balanced delta-connected load is used. Such a system is shown in Fig. 7-12.6(a). Since the system is balanced, the phasor load currents $\mathscr{I}_1$, $\mathscr{I}_2$, and $\mathscr{I}_3$ must be separated 120° in phase from each other. Thus, arbitrarily using $\mathscr{I}_1$ as reference and assuming *abc* rotation, the load currents may be represented as shown in Fig. 7-12.6(b). The individual phasor line currents $\mathscr{I}_a$, $\mathscr{I}_b$, and $\mathscr{I}_c$ defined in the figure are now related to the phasor load currents as follows:

$$\begin{aligned} \mathscr{I}_a &= \mathscr{I}_1 - \mathscr{I}_3 \\ \mathscr{I}_b &= \mathscr{I}_2 - \mathscr{I}_1 \\ \mathscr{I}_c &= \mathscr{I}_3 - \mathscr{I}_2 \end{aligned} \tag{7-271}$$

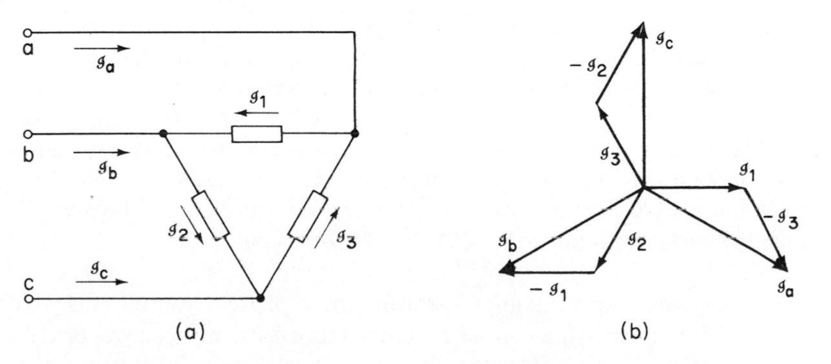

Fig. 7-12.6 A balanced delta-connected load and its phasor diagram.

These relations are also indicated in the phasor diagram shown in Fig. 7–12.6(b). It is readily seen that the magnitude of the line currents is $\sqrt{3}$ times the magnitude of the load currents. If we let I_P be the rms value of the phase (or line) currents and I_L be the rms value of the load currents, the real and reactive power delivered to each of the individual loads of the delta connection is

$$P_L = VI_L \cos\theta = V\left(\frac{I_P}{\sqrt{3}}\right)\cos\theta \tag{7–272a}$$

$$Q_L = VI_L \sin\theta = V\left(\frac{I_P}{\sqrt{3}}\right)\sin\theta \tag{7–272b}$$

where V is the rms value of the line voltage. Summing these quantities to determine the total power delivered to the load, we find

$$P_{\text{total}} = 3P_L = \sqrt{3}\,VI_P\cos\theta \tag{7–273a}$$

$$Q_{\text{total}} = 3Q_L = \sqrt{3}\,VI_P\sin\theta \tag{7–273b}$$

$$|W_{\text{total}}| = (P_{\text{total}}^2 + Q_{\text{total}}^2)^{1/2} = \sqrt{3}\,VI_P \tag{7–273c}$$

$$\text{PF} = \frac{P_{\text{total}}}{|W_{\text{total}}|} = \cos\theta \tag{7–273d}$$

It should be noted that these relations are identical with those given for the balanced wye-connected load.

In solving problems in which the power lines feeding power to a delta-connected load have some non-zero impedance, it is usually convenient to transform the delta-connected load to a wye-connected load. This can be accomplished using relations similar to those given for making delta-wye transformations of resistance networks in Sec. 3–8. Extending these relations, for the balanced case, we obtain

$$Z_\Delta = 3Z_Y \tag{7–274}$$

where Z_Δ and Z_Y are the load impedances in the delta- and wye-connected cases, respectively. An example of the use of such a transformation follows.

EXAMPLE 7–12.2. It is desired to find a balanced delta-connected load which will create the same circuit conditions in the power lines as those found in connection with the preceding example in which the individual impedances of the wye-connected load had the value $Z_Y = 4 + j3 = 5\underline{/36.9^\circ}$. Applying (7–274), we see that $Z_\Delta = 12 + j9$. Thus, for the circuit shown in Fig. 7–12.7, using the results of the preceding example, we see that the line currents are 10.82 A rms and that the line voltages at the load are 93.7 V rms. The power delivered to the total load consisting of the three impedances Z_Δ is still 1404 W. The currents in the individual loads are $10.82/\sqrt{3} = 6.26$ A rms.

The one remaining situation illustrating three-phase power circuits which will be treated in this section is the case of an unbalanced load, either wye- or delta-con-

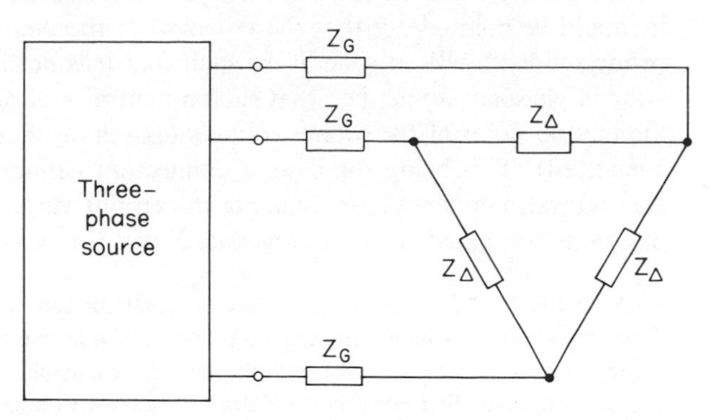

Fig. 7-12.7 Network for Example 7-12.7.

nected. Frequently delta-wye or wye-delta transformations may be used to simplify the resulting analysis. For the unbalanced case, the delta-to-wye transformations are readily shown to be

$$
\begin{aligned}
Z_a &= \frac{Z_1 Z_3}{Z_1 + Z_2 + Z_3} \\
Z_b &= \frac{Z_1 Z_2}{Z_1 + Z_2 + Z_3} \\
Z_c &= \frac{Z_2 Z_3}{Z_1 + Z_2 + Z_3}
\end{aligned}
\tag{7–275}
$$

where the component impedances are indicated in Fig. 7–12.8. Similarly, the wye-to-delta relations are

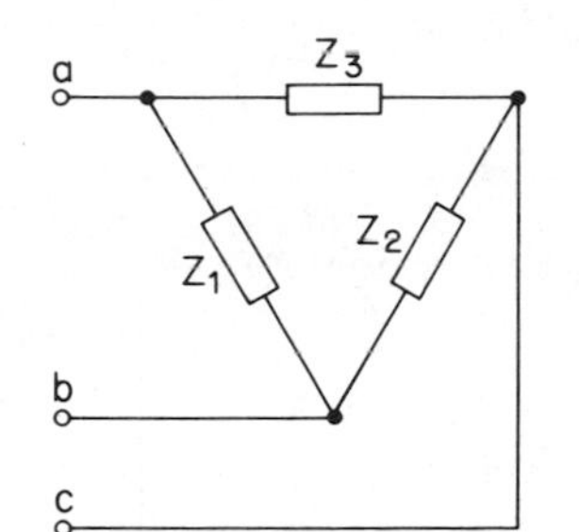

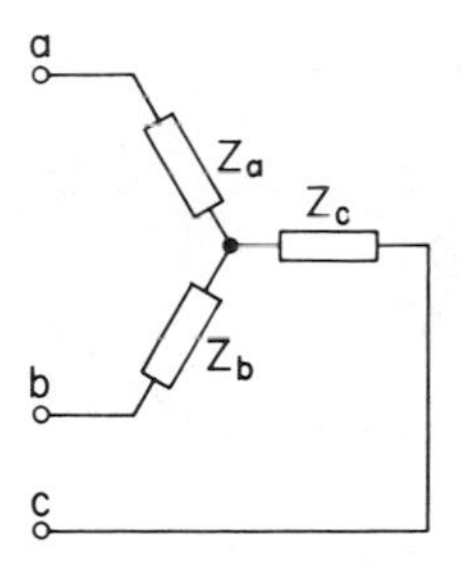

Fig. 7-12.8 Wye and delta networks.

$$
\begin{aligned}
Z_1 &= \frac{Z_a Z_b + Z_a Z_c + Z_b Z_c}{Z_c} \\
Z_2 &= \frac{Z_a Z_b + Z_a Z_c + Z_b Z_c}{Z_a} \\
Z_3 &= \frac{Z_a Z_b + Z_a Z_c + Z_b Z_c}{Z_b}
\end{aligned}
\tag{7–276}
$$

In using these relations to realize a wye-connected load from a delta-connected load, it should be pointed out that the potential of the neutral point in the resulting wye-connected load will, in general, be such that it is not located at the center of the line voltage phasor triangle, i.e., that such a neutral will not be at the same potential as the neutral point of the source voltages (assuming that the source voltages are wye-connected). This being the case, a connection cannot be made between the source and neutral points without changing the circuit variables. An example of the computations associated with an unbalanced load follows.

EXAMPLE 7-12.3. The delta-connected load shown in Fig. 7-12.9(a) is excited by a three-phase set of 200-V rms line voltages with *abc* rotation (generated by three line-to-neutral sources). The line voltage phasor triangle is shown in Fig. 7-12.9(b). To analyze this circuit, let us first convert the delta of load impedances to a wye. Applying the relations given in Eq. (7-275), we find that

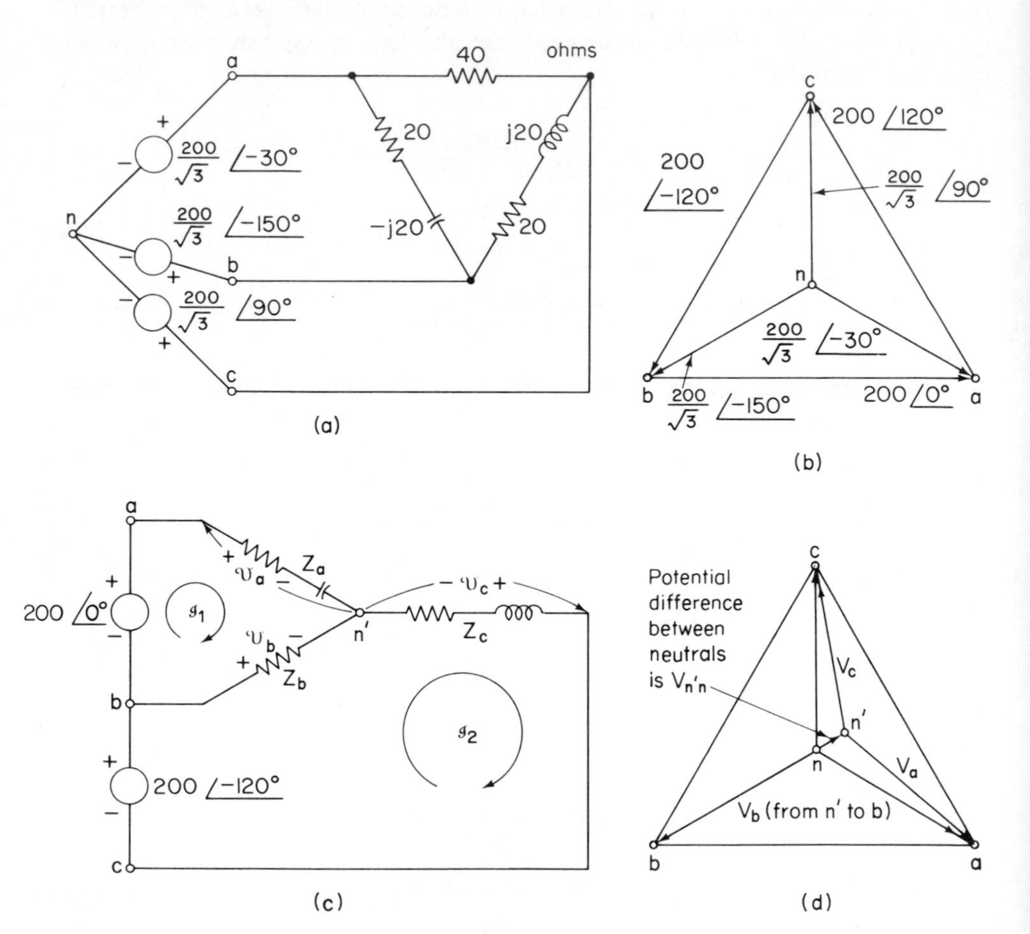

Fig. 7-12.9 Network and phasor diagram for Example 7-12.3.

$$Z_a = 10 - j10$$
$$Z_b = 10$$
$$Z_c = 10 + j10$$

The resulting wye of impedances is shown in Fig. 7-12.9(c). As indicated in this figure, this circuit can be analyzed as a two-mesh circuit with the indicated phasor voltage sources. Defining phasor mesh currents $\mathscr{I}_1$ and $\mathscr{I}_2$ as shown in the figure, the mesh equations in matrix form are

$$\begin{bmatrix} 200 \\ -173.2 - j100 \end{bmatrix} = \begin{bmatrix} 20 - j10 & -10 \\ -10 & 20 + j10 \end{bmatrix} \begin{bmatrix} \mathscr{I}_1 \\ \mathscr{I}_2 \end{bmatrix}$$

Solving for the phasor mesh currents, we obtain $\mathscr{I}_1 = 7.53\underline{/5.1^\circ}$ and $\mathscr{I}_2 = 7.53\underline{/-125^\circ}$. The line-to-neutral or phase voltages for the load may thus be computed as

$$\mathscr{V}_a = \mathscr{I}_1 Z_a = 106.5\underline{/-39.9^\circ}$$
$$\mathscr{V}_b = (\mathscr{I}_2 - \mathscr{I}_1)Z_b = 136.6\underline{/-150^\circ}$$
$$\mathscr{V}_c = -\mathscr{I}_2 Z_c = 106.5\underline{/99.9^\circ}$$

Drawing these phasors inside the line-to-line phasor voltage triangle given in Fig. 7-12.9(b), we obtain the phasor diagram shown in Fig. 7-12.9(d). Thus, we see that the magnitude of the potential between the neutral of the load and the neutral of the generator (assuming a wye-connected generator) is 21.13 V, i.e., that the two points are at different potentials.

7-13 ACTIVE RC CIRCUITS

In Sec. 7-8 various types of network functions were introduced. Three types which were covered in detail were the second-order voltage transfer functions for the low-pass, high-pass, and band-pass cases. Passive *RLC* network realizations for these functions were analyzed and described. Frequently, however, such passive network realizations are undesirable because they require the use of inductors. There are many disadvantages to such a use. For example, if a filter is being designed for an airborne or satellite application, the relatively large weight and size of inductors as compared to other network components is very undesirable. Another disadvantage is that inductors, in general, have a smaller signal range over which they exhibit a linear behavior than do resistors or capacitors. Finally, it is not possible to construct inductors by the fabrication techniques used in making integrated circuits. Thus, network realizations which require inductors, in general, cannot be realized in integrated form. From the above discussion we may conclude that it is highly desirable in many applications to replace *RLC* circuits with some other class of circuits in which inductors are not present. This is readily done by using resistors and capacitors and some type of active element. The resulting circuits are called *active RC* circuits.

Not only are they able to provide many more types of network functions than their passive *RLC* counterparts, but, in addition, they can be used to provide gain. This further increases their attractiveness. Some examples of active *RC* networks were previously introduced in Sec. 6–9 in connection with the discussion of second-order circuits (in the time domain) given there. Here we shall give some further examples of active *RC* circuits to show how these may be used to provide high-pass, low-pass, and band-pass network functions.

As a first example of active *RC* circuits, let us consider the case where a VCVS is used as the active element to produce various transfer functions. In Fig. 7–13.1 an active *RC* circuit which has a low-pass second-order characteristic is shown. The triangle represents an ideal VCVS. The voltage transfer function of this circuit is

$$\frac{\mathscr{V}_2}{\mathscr{V}_1} = \frac{K}{1 - \omega^2 + j\omega(3 - K)} \tag{7-277}$$

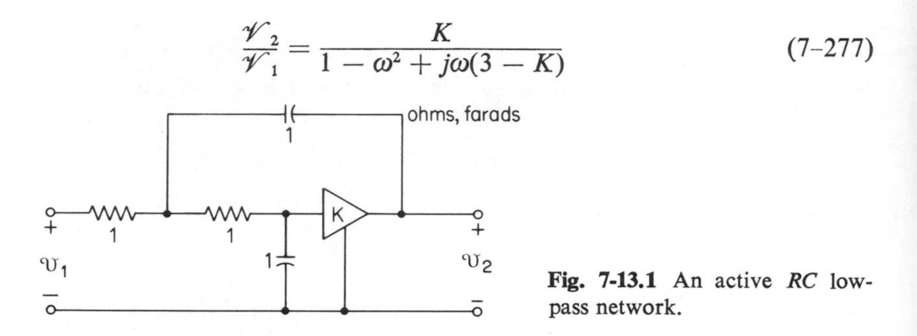

Fig. 7-13.1 An active *RC* low-pass network.

where *K* is the gain of the VCVS. The magnitude characteristics for the voltage transfer function for three different values of *K* are shown in Fig. 7–13.2. By way of comparison, a passive *RLC* low-pass circuit is shown in Fig. 7–13.3. This is the same circuit which was originally presented in Fig. 7–8.4(a). The voltage transfer function for this network is

$$\frac{\mathscr{V}_2}{\mathscr{V}_1} = \frac{1}{1 - \omega^2 + j\omega R} \tag{7-278}$$

Comparing (7–277) and (7–278), we see that the denominators of the two functions will be identical if

$$3 - K = R \tag{7-279}$$

In addition, we see that the numerators differ only by the multiplicative constant *K*. Thus, we might expect that the shape of the magnitude characteristics for the two networks will be very similar. Corresponding to the values of *K* of $\frac{14}{5}$, $\frac{5}{2}$, and 2 used in Fig. 7–13.2, we have from (7–279) values for *R* of $\frac{1}{5}$, $\frac{1}{2}$, and 1. Examination of Fig. 7–8.4(b), which gives the magnitude curves for the network of Fig. 7–13.3 for these values of *R*, verifies the expected similarity of the two sets of curves. The only difference between the two sets is due to the different numerator constants. Since these

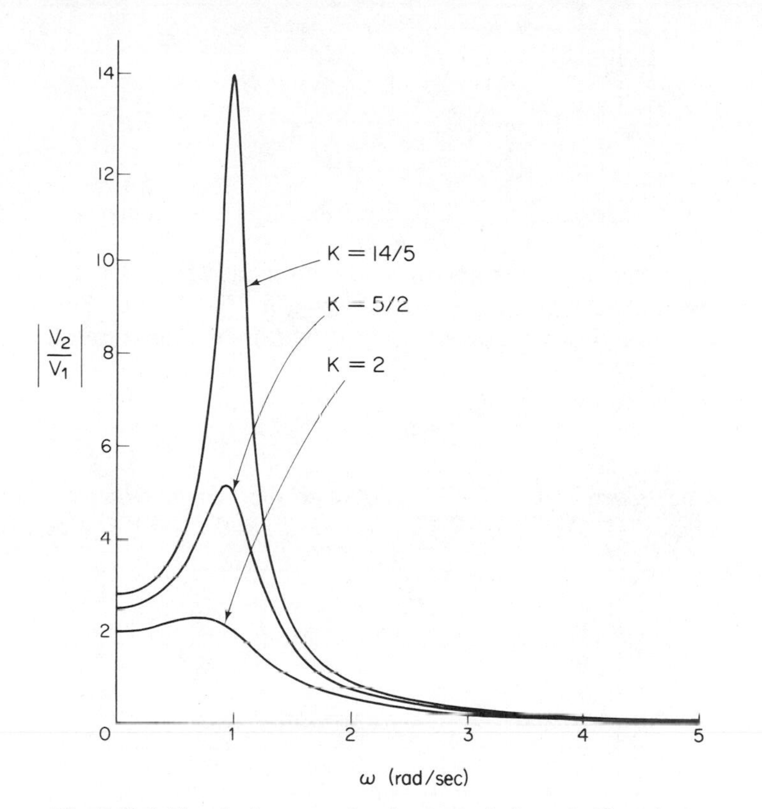

Fig. 7-13.2 Magnitude curves for the network shown in Fig. 7-13.1.

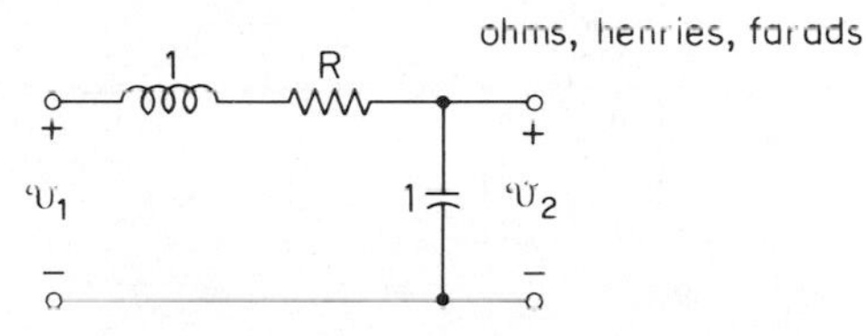

Fig. 7-13.3 A passive *RLC* low-pass network.

constants do not affect the phase determination, the phase curves shown in Fig. 7–8.4(c) apply to the network of Fig. 7–13.1 (for the corresponding values of K), as well as to the network of Fig. 7–13.3. It should be noted that the VCVS shown in the circuit in Fig. 7–13.1 is readily realized using an operational amplifier and two resistors as described in Sec. 3.11.

A second example of an active *RC* circuit which uses a VCVS as its active element is shown in Fig. 7–13.4. This is a high-pass network. It has the voltage transfer function

$$\frac{\mathscr{V}_2}{\mathscr{V}_1} = \frac{-K\omega^2}{1 - \omega^2 + j\omega(3 - K)} \tag{7–280}$$

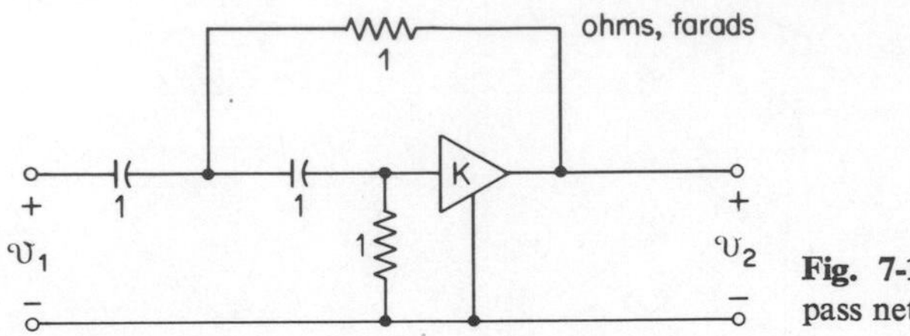

Fig. 7-13.4 An active *RC* high-pass network.

Magnitude characteristics for this network are shown in Fig. 7–13.5. By way of comparison, a passive *RLC* high-pass circuit is shown in Fig. 7–13.6. This is the same circuit which was originally presented in Fig. 7–8.5(a). Its voltage transfer function is

$$\frac{\mathscr{V}_2}{\mathscr{V}_1} = \frac{-\omega^2}{1 - \omega^2 + j\omega R} \tag{7-281}$$

The equivalence given in Eq. (7–279) applies to the denominators of the network functions in Eqs. (7–280) and (7–281) just as it does for the low-pass functions in Eqs.

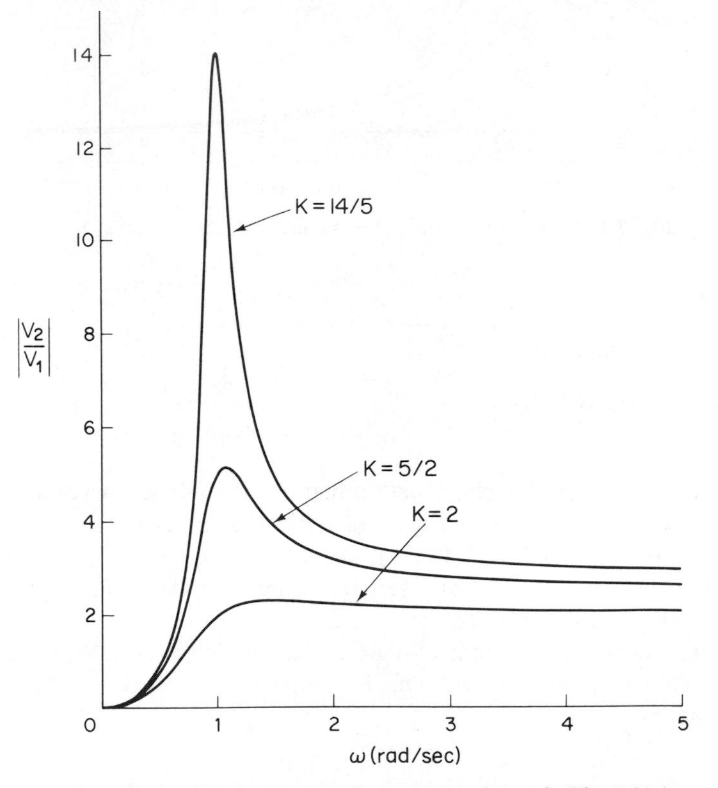

Fig. 7-13.5 Magnitude curves for the network shown in Fig. 7-13.4.

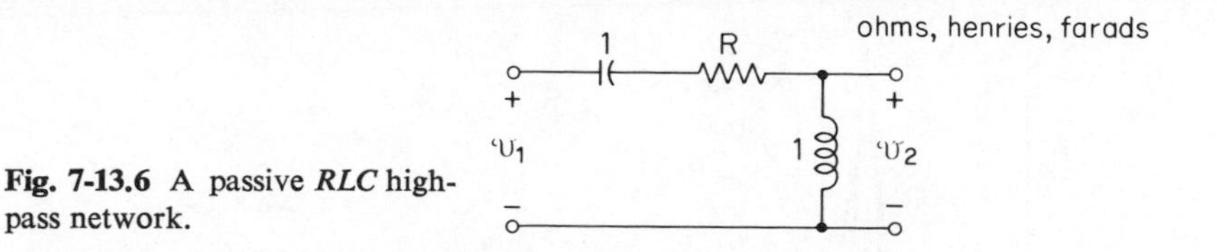

Fig. 7-13.6 A passive *RLC* high-pass network.

(7–277) and (7–278). In addition, the numerators again differ by the constant *K*. Thus, the magnitude curves of Fig. 7–13.5 differ only by a constant from those given in Fig. 7–8.5(b) for the passive *RLC* circuit shown in Fig. 7–13.6. In addition, the phase curves given in Fig. 7–8.5(c) apply to both of the high-pass networks.

A third example of an active *RC* circuit which uses a VCVS as its active element is the band-pass network shown in Fig. 7–13.7. This network has the voltage transfer function

$$\frac{\mathcal{V}_2}{\mathcal{V}_1} = \frac{j\omega K/2}{1 - \omega^2 + j\omega[3 - (K/2)]} \tag{7–282}$$

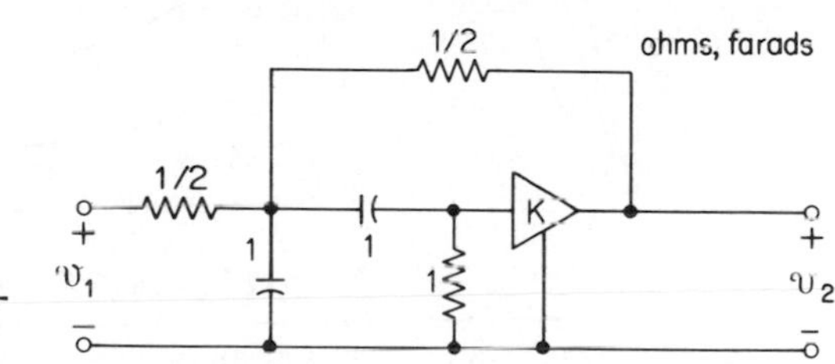

Fig. 7-13.7 An active *RC* band-pass network.

Magnitude characteristics for this network for several values of *K* are shown in Fig. 7–13.8. The behavior of this network may be compared with the passive *RLC* network shown in Fig. 7–13.9. This is the same circuit as was shown in Fig. 7–8.6(a). Its voltage transfer function is

$$\frac{\mathcal{V}_2}{\mathcal{V}_1} = \frac{j\omega R}{1 - \omega^2 + j\omega R} \tag{7–283}$$

Comparing (7–282) and (7–283) we see that the denominators will be the same if

$$3 - \frac{K}{2} = R \tag{7–284}$$

Thus, the three values for *K* of $\frac{28}{5}$, 5, and 4 used for the loci of Fig. 7–13.8 correspond with values of *R* of $\frac{1}{5}$, $\frac{1}{2}$, and 1 which were used in the magnitude curves shown in Fig. 7–8.6(b) for the passive *RLC* circuit. The numerators of the two network functions differ only by a constant, thus the magnitude curves shown in Fig. 7–13.8 are very similar to those shown in Fig. 7–8.6(b). In addition, the phase curves shown in Fig. 7–8.6(c) apply to both of the band-pass network configurations.

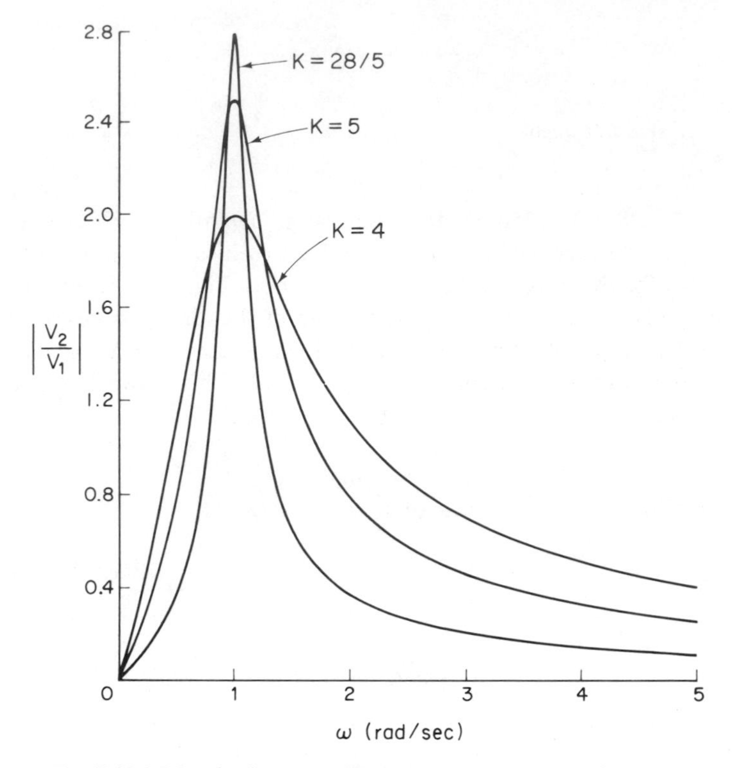

Fig. 7-13.8 Magnitude curves for the network shown in Fig. 7-13.7.

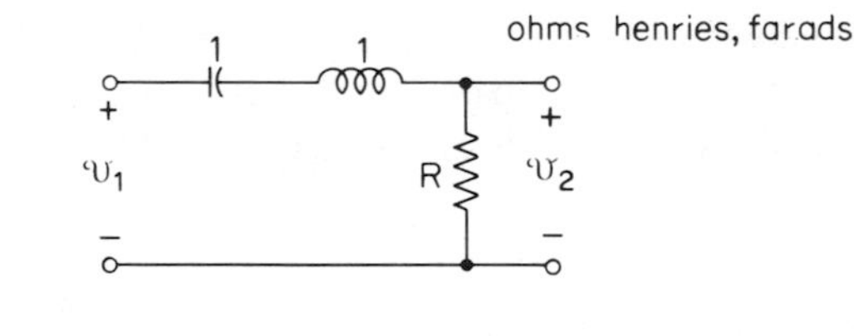

Fig. 7-13.9 A passive *RLC* band-pass network.

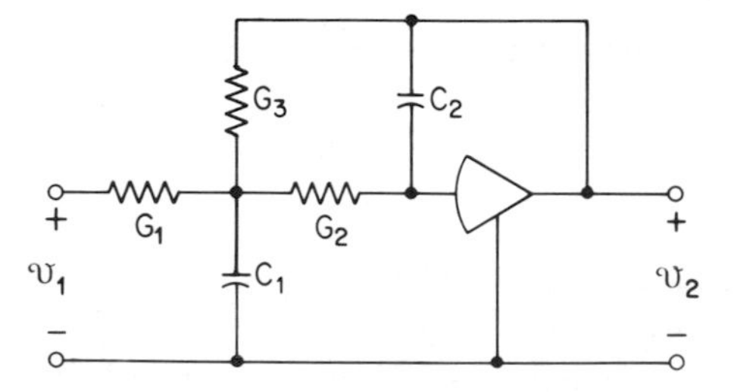

Fig. 7-13.10 An active *RC* low-pass network.

In the preceding paragraphs we have considered three active *RC* circuits which use VCVS's as the active element. Now let us consider some circuits which use other types of active elements. One such element which is frequently used is the operational amplifier. This device was introduced in Sec. 3.11. A useful low-pass network cofiguration using an operational amplifier is shown in Fig. 7–13.10. The voltage transfer function for this network is

$$\frac{\mathscr{V}_2}{\mathscr{V}_1} = \frac{-G_1G_2}{G_2G_3 - \omega^2C_1C_2 + j\omega C_1(G_1 + G_2 + G_3)} \tag{7-285}$$

The determination of the values for the elements necessary to realize a specific transfer function requires the solution of a set of non-linear equations. One such solution is given by the following procedure. We first assume that the network function has the form

$$\frac{\mathscr{V}_2}{\mathscr{V}_1} = \frac{-H}{1 - \omega^2 + j\omega a} \tag{7-286}$$

where $H > 0$. If C_2 is chosen to have some convenient value, then we may compute the other elements as follows:

$$\begin{aligned} C_1 &= \frac{4C_2(H+1)}{a^2} \\ G_1 &= \frac{2HC_2}{a} \\ G_2 &= \frac{2C_2(H+1)}{a} \\ G_3 &= \frac{2C_2}{a} \end{aligned} \tag{7-287}$$

As an example of the use of these relations, consider a design to realize the network function

$$\frac{\mathscr{V}_2}{\mathscr{V}_1} = \frac{-10}{1 - \omega^2 + j\omega\sqrt{2}} \tag{7-288}$$

Comparing this with the general form given in Eq. (7–286), we see that $H = 10$ and $a = \sqrt{2}$. Choosing $C_2 = 0.1$ F, we obtain the values $C_1 = 2.2$ F, $G_1 = \sqrt{2}$ mhos, $G_2 = 1.1\sqrt{2}$ mhos, and $G_3 = \sqrt{2}/10$ mhos.

A more complex network which uses operational amplifiers as active elements is shown in Fig. 7–13.11. This configuration has the advantage that it may be used to realize very high-Q network functions. In addition, the resulting characteristic is extremely stable, i.e., the magnitude and phase characteristics produced by the network are relatively unaffected by changes in the values of the elements or changes in the gain of the operational amplifiers. A further advantage of this circuit is that it may be used to produce low-pass, high-pass, or band-pass transfer functions depending on the point from which the output is taken. In terms of the variables defined in the figure, the voltage transfer functions for this circuit may be shown to be

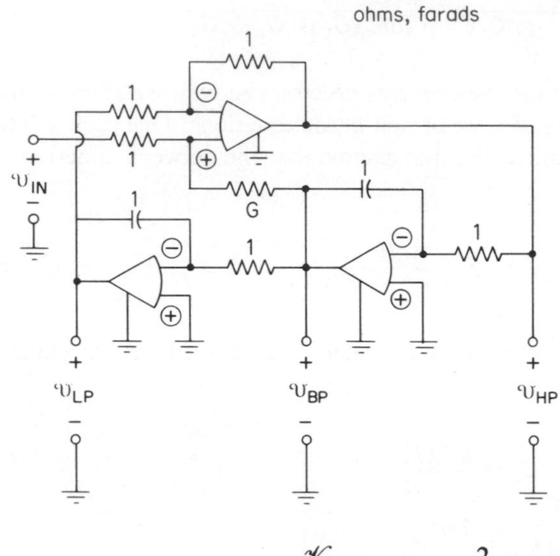

Fig. 7-13.11 An active *RC* network using operational amplifiers.

$$\frac{\mathcal{V}_{LP}}{\mathcal{V}_{IN}} = \frac{2}{1 - \omega^2 + j2\omega G}$$
$$\frac{\mathcal{V}_{HP}}{\mathcal{V}_{IN}} = \frac{-2\omega^2}{1 - \omega^2 + j2\omega G} \tag{7-289}$$
$$\frac{\mathcal{V}_{BP}}{\mathcal{V}_{IN}} = \frac{-j2\omega}{1 - \omega^2 + j2\omega G}$$

These relations only hold for the case where $G \ll 1$.[19]

We shall conclude our discussion of the use of active *RC* circuits to realize second-order network functions by presenting two realizations which are somewhat less well-known than the ones presented above. The first of these is a band-pass circuit using an NIC (see Sec. 3.11) as an active element. The network configuration is shown in Fig. 7-13.12. The voltage transfer function for this network is

$$\frac{\mathcal{V}_2}{\mathcal{V}_1} = \frac{-j\omega K}{1 - \omega^2 + j\omega(2 - K)} \tag{7-290}$$

where K is the gain of the NIC. The band-pass nature of this characteristic is readily apparent.

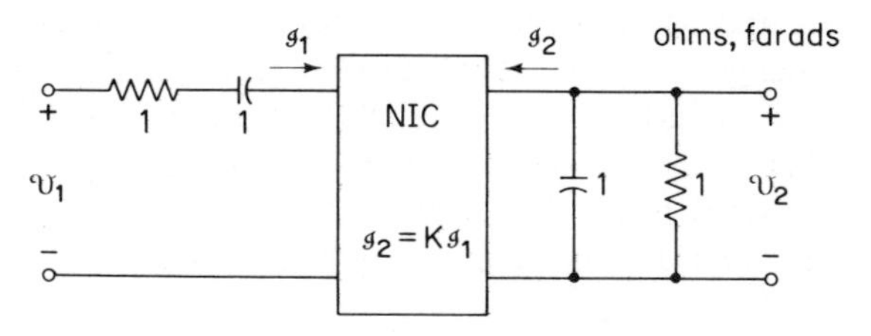

Fig. 7-13.12 An active *RC* band-pass network using an NIC.

[19]A more exact expression for the voltage transfer functions may be found in L. P. Huelsman, *Theory and Design of Active RC Circuits*, McGraw-Hill Book Company, New York, 1968.

A second active *RC* circuit which is less well-known than the ones using VCVS's but which has some interesting properties is a band-pass network which uses a gyrator (see Sec. 3.11) as an active element. The actual network configuration is shown in Fig. 7-13.13. The voltage transfer function for the circuit is

$$\frac{\mathscr{V}_2}{\mathscr{V}_1} = \frac{j\omega G}{1 + G^2 - \omega^2 + j2\omega} \tag{7-291}$$

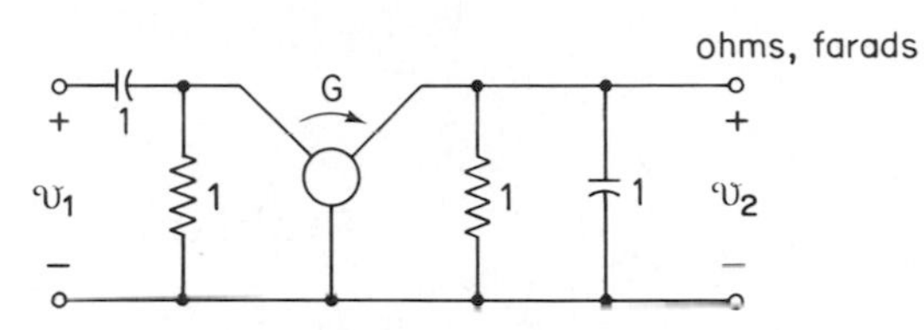

Fig. 7-13.13 An active *RC* band-pass network using a gyrator.

where G is the gyration conductance (the reciprocal of the gyration resistance). This circuit has the advantage that it will remain stable for all values of the gyration conductance.

In this section we have presented some examples of active *RC* circuits which may be used to realize some of the more frequently encountered network functions. This presentation can only be considered as an introduction to a rapidly growing area of network theory, namely, the use of resistors, capacitors, and active elements to provide circuits capable of realizing a broad range of network functions. More information on this subject may be found in the references given in the Bibliography at the end of this book.

7-14 CONCLUSION

This chapter represents a significant departure from the topics and the techniques presented in the first six chapters of this book. The earlier chapters were primarily concerned with the analysis of circuits in the time domain. In this chapter, however, we have been concerned with the behavior of networks in the frequency domain, i.e., under conditions of sinusoidal steady-state excitation. The solution techniques presented in this chapter relate directly to the time-domain material presented in Chaps. 1–6 since they provide a systematic and practical method for determining the particular solution, i.e., the forced response, for the case where the network excitation is sinusoidal. A review of the examples in Chaps. 5 and 6 will readily show that the mathematics connected with the determination of particular solutions is considerably simplified by using the phasor methods introduced in this chapter. In addition to its relation to time-domain network analysis, however, the sinusoidal steady-state analysis techniques presented here have considerable additional significance since they directly relate to the experimental determination of the properties of a given circuit. This is, of course, due to the fact that sinusoidal quantities are so easily generated and measured in practice. Thus, in many cases such quantities provide a definition of network performance which is more closely related to actual usage than are corresponding time-domain results.

The frequency-domain analysis of networks, however, is not limited only to the sinusoidal steady-state studies presented in this chapter. In the following chapter we shall see that the concept of a frequency domain may be extended to include all the analysis results which were obtained in the time-domain studies made in Chaps. 1–6. In addition, we shall find that these general frequency-domain solution techniques are frequently easier and more convenient to apply than the time-domain ones. Thus, collectively, they represent an analysis approach which has major importance.

Problems

Problem 7–1 (Sec. 7–1)

Put the following expressions in the form $F_m \cos(\omega t + \phi)$ and in the form $F_m \sin(\omega t + \theta)$.

(a) $\sqrt{3}\cos\omega t + \sin\omega t$

(b) $-2\cos\omega t + 4\sin\omega t$

(c) $2\cos\omega t - 5\sin\omega t$

(d) $-\cos\omega t - \sin\omega t$

Problem 7–2 (Sec. 7–1)

Put the following expressions into the form $A\cos\omega t + B\sin\omega t$.

(a) $(2/\sqrt{3})\cos(\omega t + 30°)$

(b) $5\sin(\omega t - 53.13°)$

(c) $-2\cos(\omega t + 45°)$

(d) $7\sin(\omega t + 160°)$

Problem 7–3 (Sec. 7–1)

If $\cos[(\pi/2) - \omega t] + \cos(\omega t) = F_m\cos(\omega t - \alpha)$, find α and F_m.

Problem 7–4 (Sec. 7–1)

Find the rms values of the periodic functions shown in Fig. P7–4.

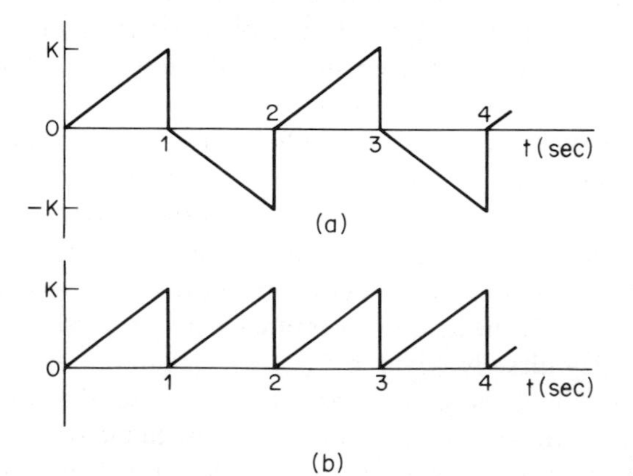

Figure P7-4

Problem 7–5 (Sec. 7–1)

Find the rms values of the periodic functions shown in Fig. P7-5.

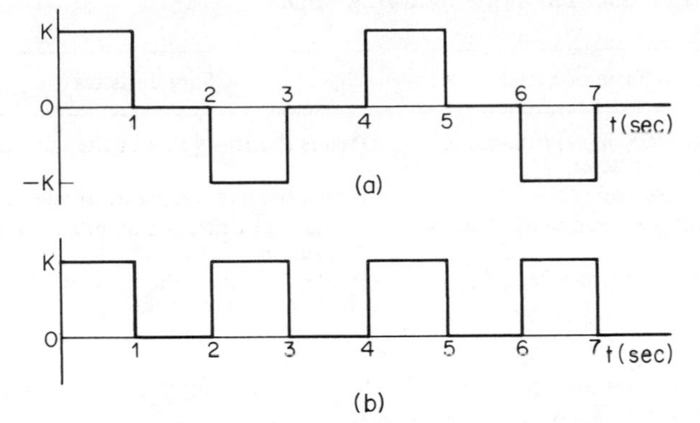

Figure P7-5

Problem 7–6 (Sec. 7–1)

Find the average value and the root-mean-square value of the voltage waveform shown in Fig. P7-6.

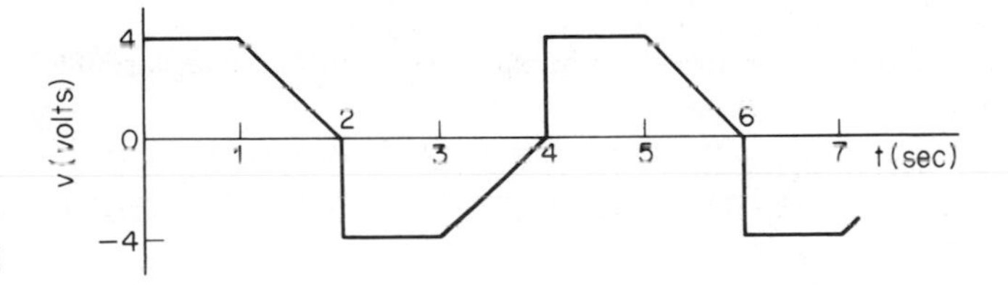

Figure P7-6

Problem 7–7 (Sec. 7–1)

Find the average value and the root-mean-square value of the waveform shown in Fig. P7-7.

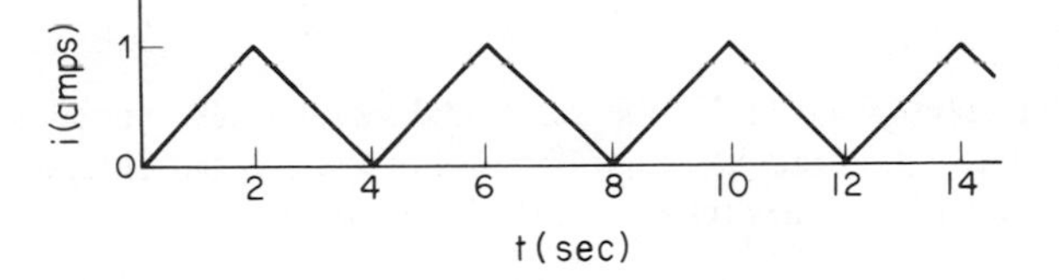

Figure P7-7

Problem 7–8 (Sec. 7–1)

Find the average value and the root-mean-square value of the current waveform shown in Fig. P7-8.

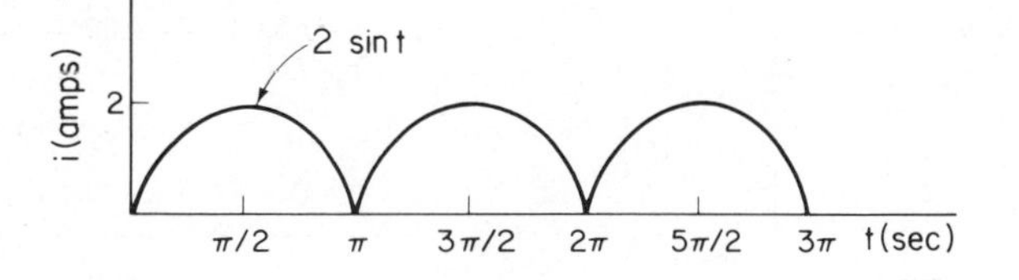

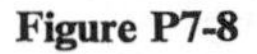

Figure P7-8

Problem 7–9 (Sec. 7–1)

The characteristics of ammeters which are used to measure current fall into four categories as summarized in the following table.

Category	Type of Meter	Characteristics
1	Thermocouple, iron-vane, dynamometer	Reads the rms value of the current.
2	D'Arsonval (dc)	Reads the average value of the current.
3	Full-wave rectifier	Reads 1.11 times the average value of the magnitude of the current.
4	Half-wave rectifier	Reads 2.22 times the average value of the positive portion of the current waveform.

Assume that the waveshapes shown in the following figures represent current and find the values that each type of meter will read for each of the waveshapes: (a) Fig. P7-4; (b) Fig. P7-5; (c) Fig. P7-6; (d) Fig. P7-7.

Problem 7–10 (Sec. 7–2)

Use the technique described in Summary 7-2.1 to find the particular solution for the circuit given in Problem 5-16.

Problem 7–11 (Sec. 7–2)

Use the technique described in Summary 7-2.1 to find the particular solution for the circuit given in Problem 5-17.

Problem 7–12 (Sec. 7–2)

Use the technique described in Summary 7-2.1 to find the particular solution for $v(t)$ for the circuit given in Problem 6-26.

Problem 7–13 (Sec. 7–3)

Show by graphical methods that the sum of two phasors $\mathscr{F}e^{j\omega t}$ and $\mathscr{F}^*e^{-j\omega t}$ where $\mathscr{F} = F_o e^{j\phi}$ produces a real sinusoidal function.

Problem 7–14 (Sec. 7–3)

What is the locus of a phasor $\mathscr{F}e^{-j\omega t}$ where $\mathscr{F} = 3e^{j(\pi/4)}$?

Problem 7–15 (Sec. 7–3)

Prove that the solution given for the sinusoidal steady-state component of $v_2(t)$ in Example 7-3.1 is indeed the particular solution of the original differential equation given in the example by substituting it into that equation.

Problem 7–16 (Sec. 7–3)

In the circuit shown in Fig. P7-16, find the steady-state component of $i(t)$ if $v(t) = 2 \sin t$ V.

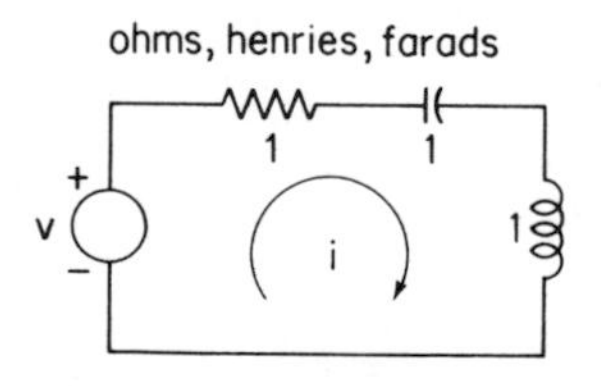

Figure P7-16

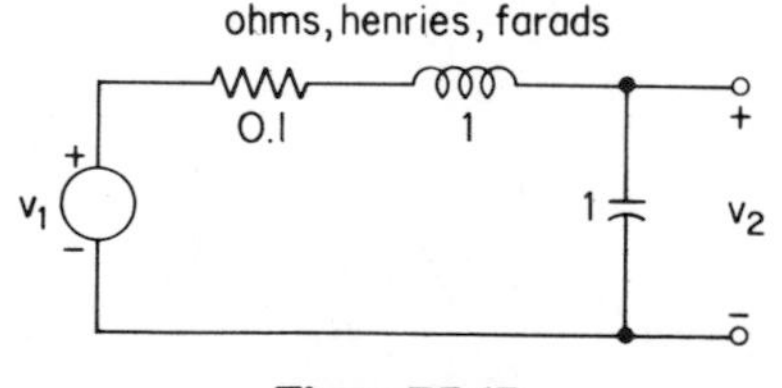

Figure P7-17

Problem 7–17 (Sec. 7–3)

In the circuit shown in Fig. P7-17 the input voltage $v_1(t) = \cos t$ V. Find the sinusoidal steady-state component of the output voltage $v_2(t)$.

Problem 7–18 (Sec. 7–3)

The circuit shown in Fig. P7-18 is in a steady-state condition at 1 rad/s. Draw a phasor diagram for the phasor quantities $\mathscr{I}_0$, $\mathscr{I}_G$, $\mathscr{I}_L$, $\mathscr{I}_C$, and $\mathscr{V}$.

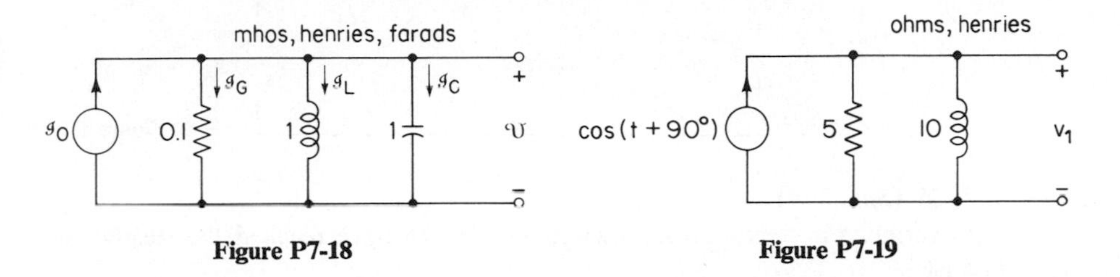

Figure P7-18 **Figure P7-19**

Problem 7–19 (Sec. 7–3)

For the network shown in Fig. P7-19, assume all variables have reached the steady state. Find $v_1(t)$.

Problem 7–20 (Sec. 7–3)

When a current $i(t) = 8\cos t - 11\sin t$ A is applied to the circuit shown in Fig. P7-20, it is found that $v(t) = \sin t + 2\cos t$ V for steady-state conditions. Find the values of L, $i_R(t)$, $i_L(t)$, and $i_C(t)$.

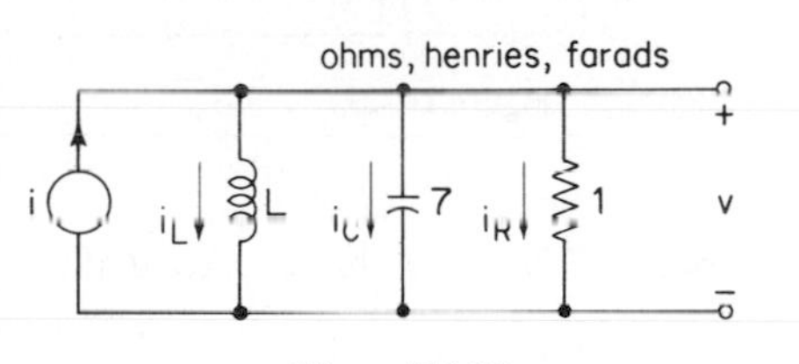

Figure P7-20

Problem 7–21 (Sec. 7–4)

Prove that the expressions computed for $v_1(t)$ and $v_2(t)$ in Example 7-4.1 are particular solutions by substituting them into the original set of differential equations given in (7-84).

Problem 7–22 (Sec. 7–4)

Apply the extended Gauss-Jordan procedure to find the inverse of the following matrix.

$$\mathbf{A} = \begin{bmatrix} 1 + j1 & -1 & 0 \\ -1 & 2 + j1 & -1 \\ 0 & -1 & 1 - j1 \end{bmatrix}$$

Problem 7–23 (Sec. 7–4)

Use cofactors to find the inverse of the matrix given in Problem 7-22.

Problem 7–24 (Sec. 7–4)

Find the particular solutions for $f_1(t)$ and $f_2(t)$ if $g_1 = \cos t$ and $g_2(t) = u(t)$ in the equations that follow:

$$3f_1' - 2f_2' + 2f_1 - 2f_2 = g_1$$
$$f_1' + 2f_1 + 2f_2' - f_2 = g_2$$

Problem 7–25 (*Sec. 7–4*)

In the network shown in Fig. P7-25, the two voltage sources have the values $v_1(t) = \cos 2t$ and $v_2(t) = \sqrt{2}\cos(2t - 45°)$ V. Find the sinusoidal steady-state components of $i_1(t)$ and $i_2(t)$.

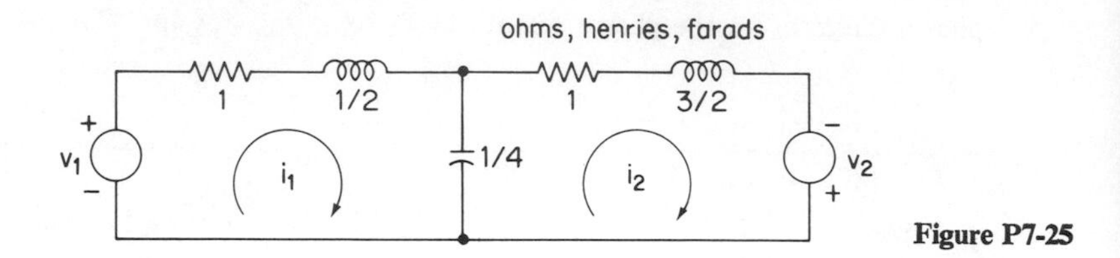

Figure P7-25

Problem 7–26 (*Sec. 7–4*)

If all variables in the network shown in Fig. P7-26 have reached the steady state, find $v_2(t)$.

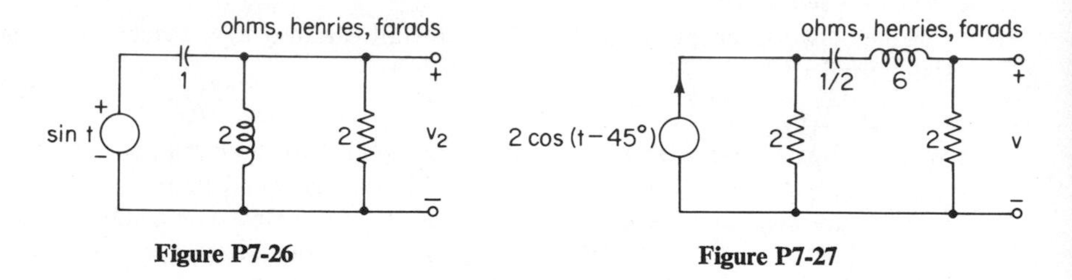

Figure P7-26 **Figure P7-27**

Problem 7–27 (*Sec. 7–4*)

Assuming that all variables in the network shown in Fig. P7-27 have reached the steady state, find $v(t)$.

Problem 7–28 (*Sec. 7–4*)

Assume that all the voltages and currents in the network shown in Fig. P7-28 have reached the steady state. Find $i(t)$.

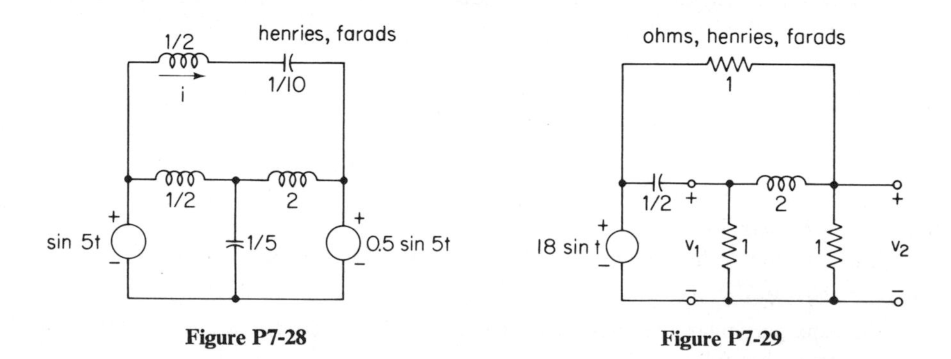

Figure P7-28 **Figure P7-29**

Problem 7–29 (*Sec. 7–4*)

In the circuit shown in Fig. P7-29, find $v_1(t)$ and $v_2(t)$ for steady-state conditions.

Problem 7–30 (Sec. 7–4)

Assuming steady-state conditions in the circuit shown in Fig. P7–30, find $i_2(t)$ for the case $K = 1$.

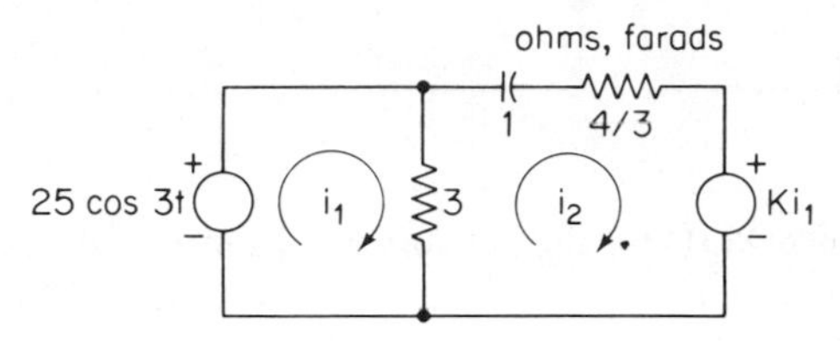

Figure P7-30

Problem 7–31 (Sec. 7–5)

Find expressions for the input impedance $Z(j\omega)$ for each of the circuits shown in Fig. P7–31.

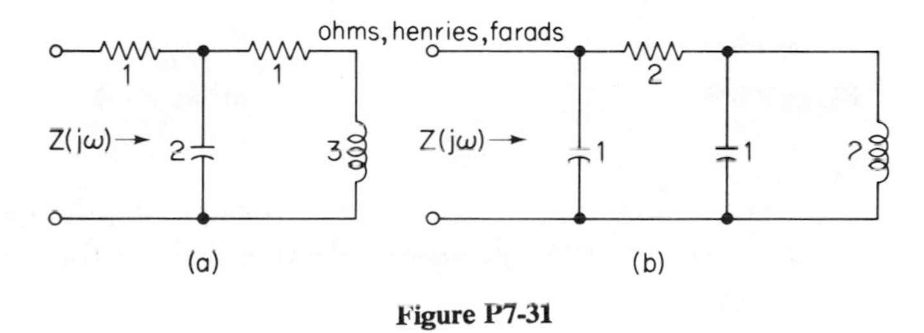

Figure P7-31

Problem 7–32 (Sec. 7–5)

Find expressions for the input admittance $Y(j\omega)$ for each of the circuits shown in Fig. P7–32.

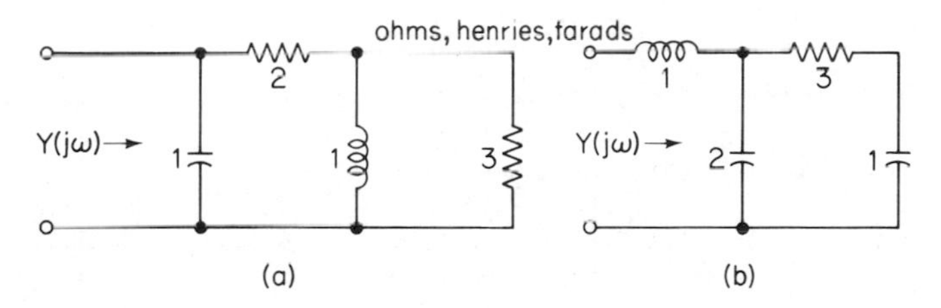

Figure P7-32

Problem 7–33 (Sec. 7–5)

For each of the circuits shown in Fig. P7–31, find two two-element circuits which have the same driving-point impedance at $\omega = 2$ rad/s.

Problem 7–34 (Sec. 7–5)

For each of the circuits shown in Fig. P7–32, find two two-element circuits which have the same driving-point admittance at $\omega = 2$ rad/s.

Problem 7–35 (Sec. 7–5)

Find the driving-point impedance $Z(j\omega)$ for the network containing an ideal transformer shown in Fig. P7–35.

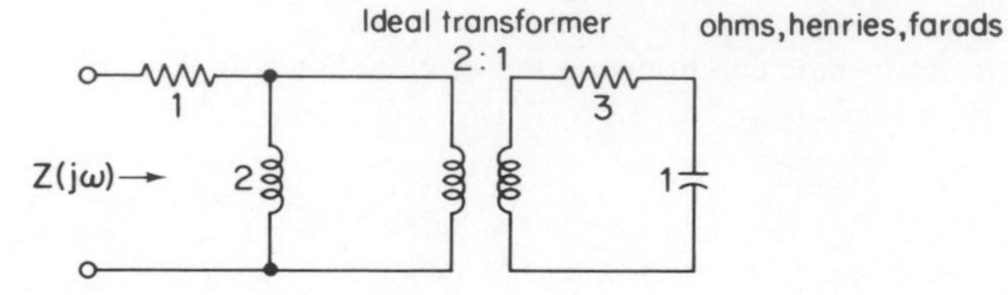

Figure P7-35

Problem 7–36 (Sec. 7–5)

Find the input admittance $Y(j1)$ for the network shown in Fig. P7-36.

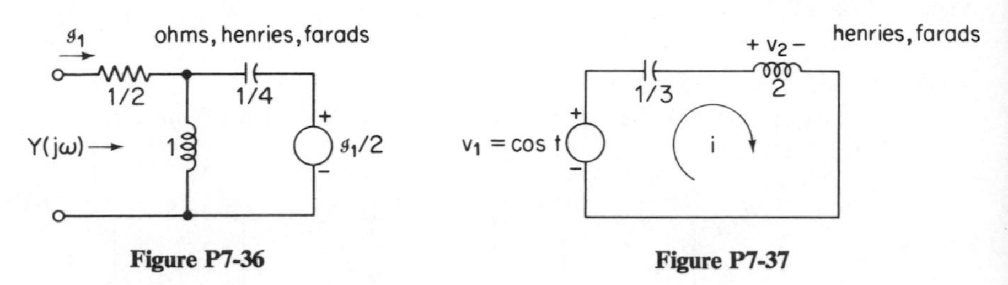

Figure P7-36

Figure P7-37

Problem 7–37 (Sec. 7–5)

Draw a phasor diagram showing the magnitude and phase relationships between the variables $i(t)$, $v_1(t)$, and $v_2(t)$ in the network shown in Fig. P7-37.

Problem 7–38 (Sec. 7–6)

Write, but do not solve, the mesh impedance equations in matrix form for the network shown in Fig. P7-38 when (a) $\omega = 1$ rad/s, (b) $\omega = 2$ rad/s.

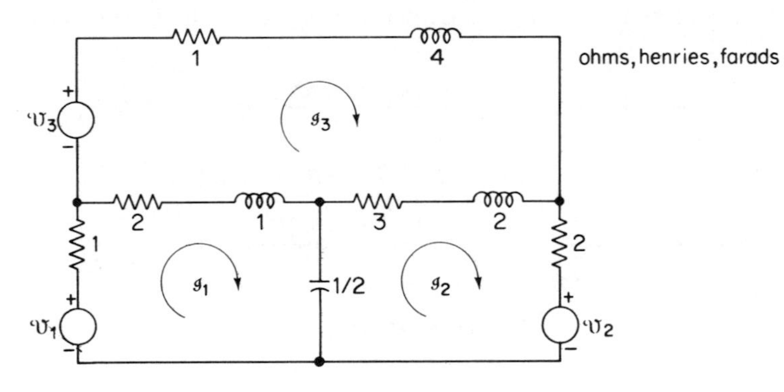

Figure P7-38

Problem 7–39 (Sec. 7–6)

Find the resistance matrix **R**, the inductance matrix **L**, the elastance matrix **S**, and the reactance matrix **X** for the network shown in Fig. P7-38.

Problem 7–40 (Sec. 7–6)

Write, but do not solve, the node admittance equations in matrix form for the network shown in Fig. P7-40 when (a) $\omega = 1$ rad/s, (b) $\omega = 2$ rad/s. Treat the network as a three-node network.

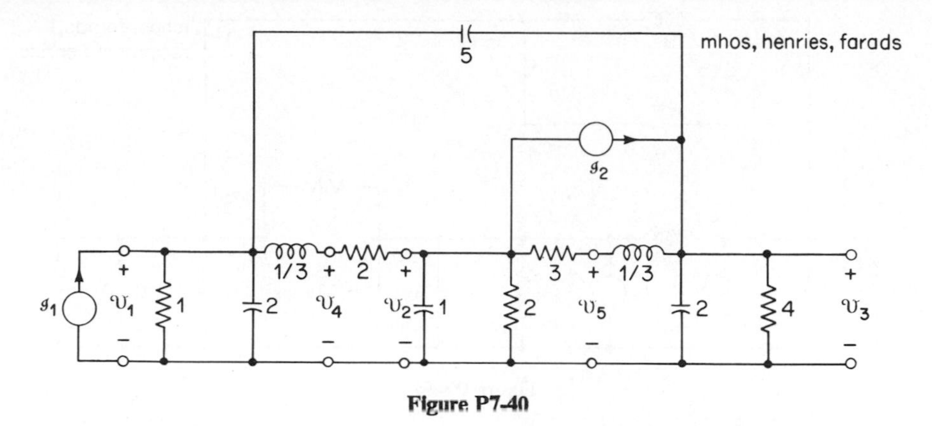

Figure P7-40

Problem 7–41 (*Sec. 7–6*)

Find the conductance matrix **G**, the reciprocal inductance matrix **Γ**, the capacitance matrix **C**, and the susceptance matrix **B** for the network shown in Fig. P7-40. Treat the network as a five-node network.

Problem 7–42 (*Sec. 7–6*)

Write, but do not solve, the mesh impedance equations in matrix form for the network shown in Fig. P7-42.

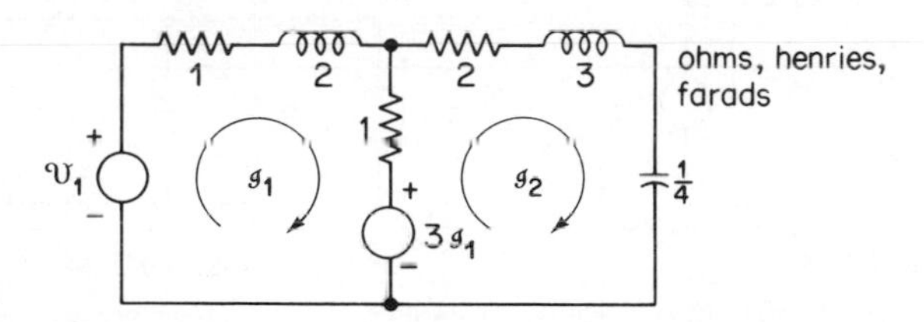

Figure P7-42

Problem 7–43 (*Sec. 7–6*)

Find, but do not solve, the node admittance matrix for the network shown in Fig. P7-43. Treat the network as having three nodes.

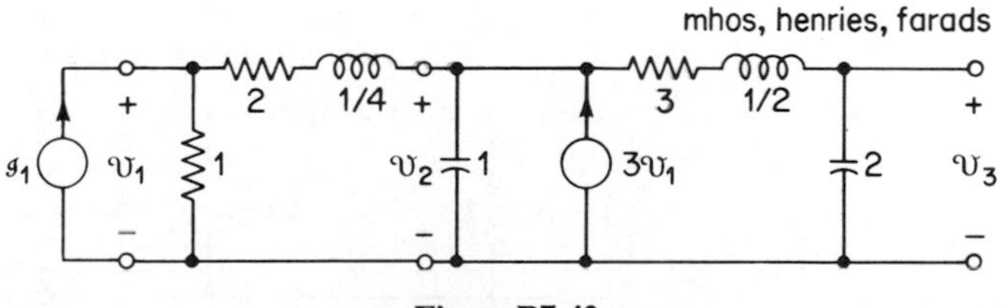

Figure P7-43

Problem 7–44 (*Sec. 7–6*)

Find the admittance matrix for the network shown in Fig. P7-44 under the conditions that the angular frequency is 1 rad/s.

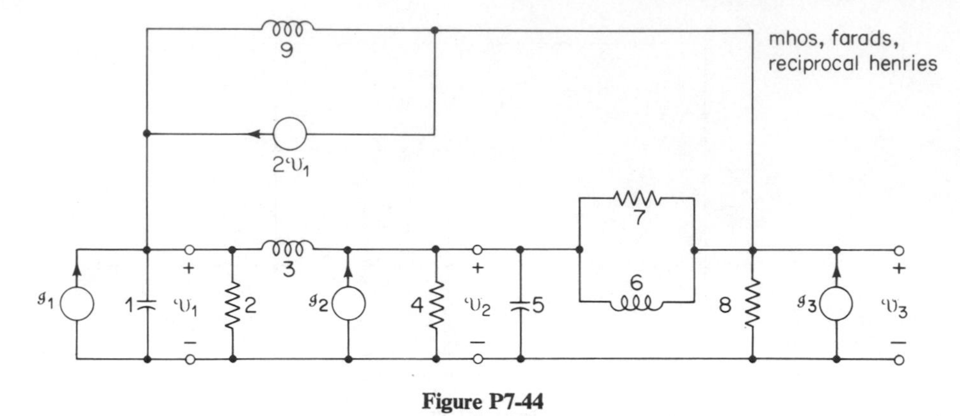

Figure P7-44

Problem 7–45 (Sec. 7–6)

Write a set of matrix equations describing the network shown in Fig. P7-45.

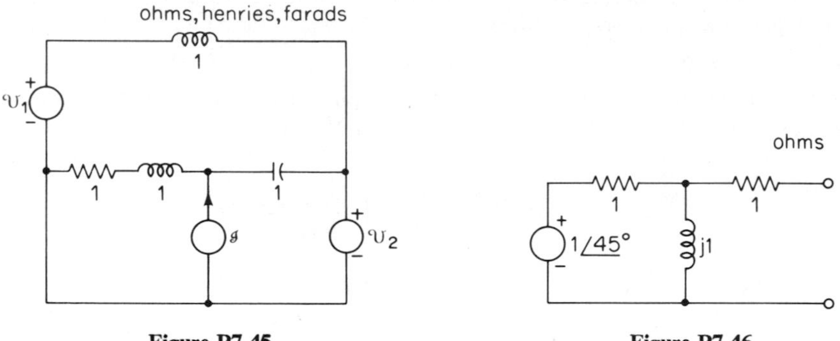

Figure P7-45

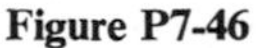

Figure P7-46

Problem 7–46 (Sec. 7–6)

Find a Thevenin equivalent circuit for the network shown in Fig. P7-46.

Problem 7–47 (Sec. 7–6)

Find Thevenin and Norton equivalent circuits for the network shown in Fig. P7-47 at the terminals *a-b* when $\omega = 2$ rad/s.

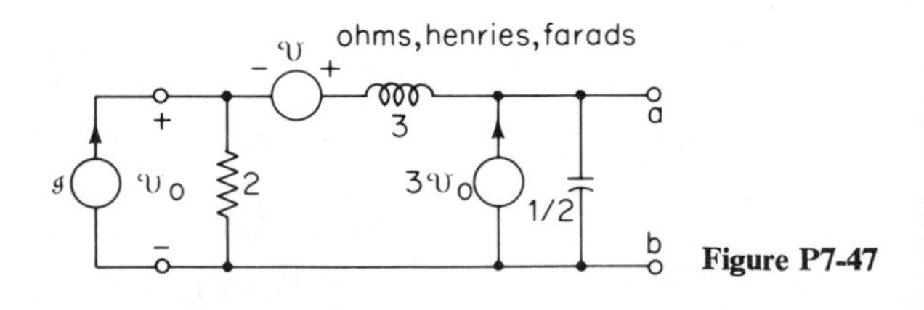

Figure P7-47

Problem 7–48 (Sec. 7–7)

The circuit shown in Fig. P7-48 is defined in terms of the complex impedance of the component elements and the value of the input voltage phasor. Find the values of the mesh current phasors.

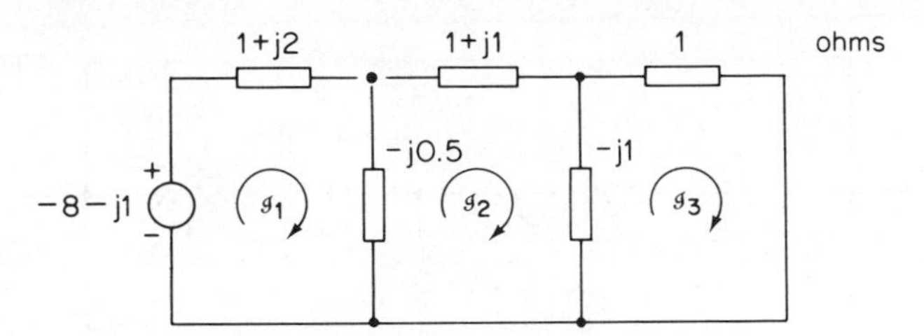

Figure P7-48

*Problem 7–49** (*Sec. 7–7*)[20]

Verify the data given in the last section of Fig. 7-7.4(b) by adding the necessary statements to the FORTRAN program given in Fig. 7-7.4(a).

*Problem 7–50** (*Sec. 7–7*)

Use the subroutines ZMESH and CGJSEQ to solve for the mesh current phasors $\mathscr{I}_1$, $\mathscr{I}_2$, and $\mathscr{I}_3$ in the network shown in Fig. P7-50. Check the values obtained by inserting them in the network equations.

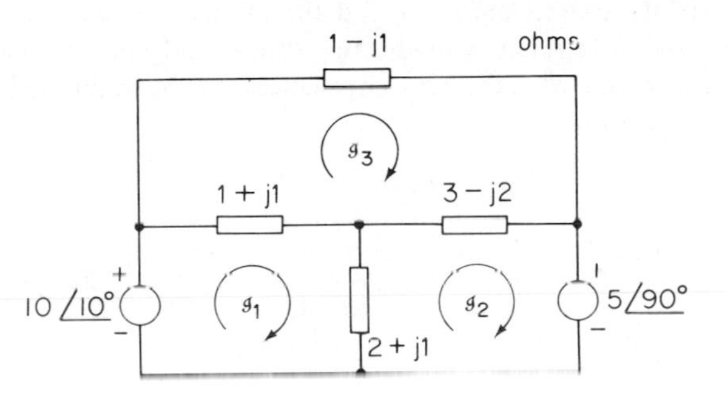

Figure P7-50

*Problem 7–51** (*Sec. 7–7*)

Use the subroutines ZMESH and CGJSEQ to find the node voltage phasors $\mathscr{V}_1$, $\mathscr{V}_2$, and $\mathscr{V}_3$ for the network shown in Fig. P7-51. Check the values obtained by inserting them in the actual network equations.

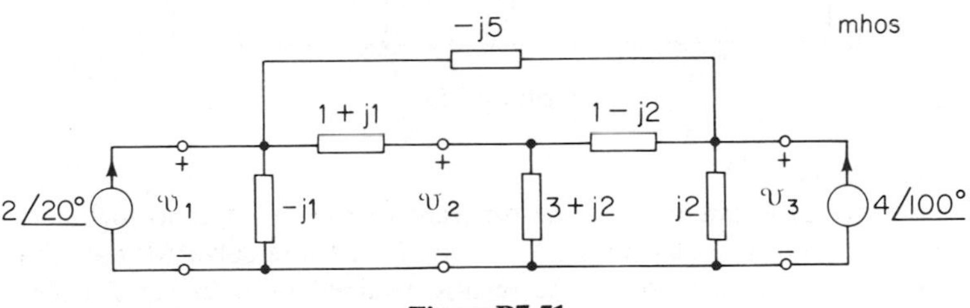

Figure P7-51

*Problem 7–52** (*Sec. 7–7*)

Use the subroutines ZMESH and CGJSEQ to find the sinusoidal steady-state component of the voltage $v_0(t)$ in the network shown in Fig. P7-52.

[20]Problems which require the use of digital computational techniques are identified by an asterisk.

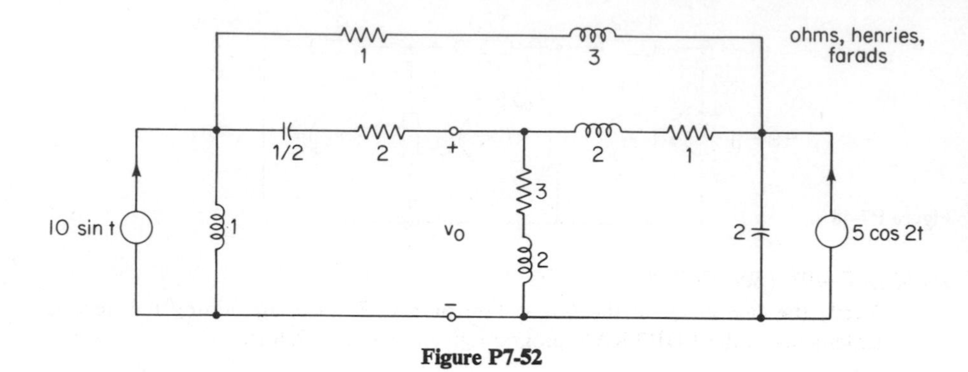

Figure P7-52

*Problem 7–53** (*Sec. 7–7*)

Two sinusoidal independent sources, one a voltage source, the other a current source, are used to excite the network shown in Fig. P7-53. Use the digital computer subroutines ZMESH and CGJSEQ to find the phasor voltage $\mathscr{V}_0$. (*Hint:* Add a second current source in series with the one shown and convert these sources to voltage sources in series with the two impedances on the right of the figure, then solve for the four mesh currents.)

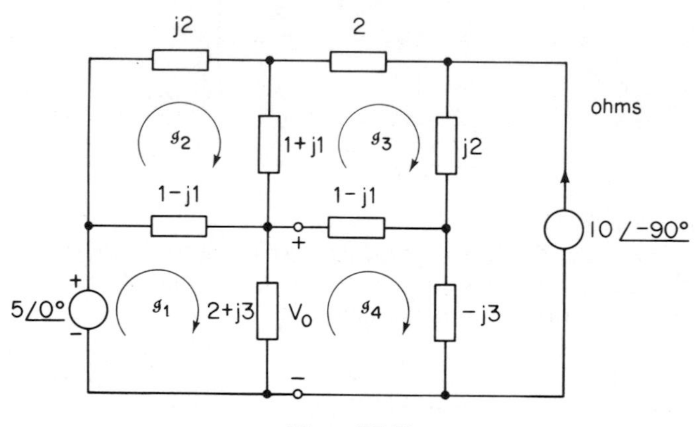

Figure P7-53

*Problem 7–54** (*Sec. 7–7*)

Add the necessary statements to the program in Fig. 7-7.4(a) to include the effect of the ICVS shown in the network in Fig. P7-54 and solve for the mesh current phasors of this network. Verify the results obtained by inserting the phasor mesh current values into the actual network equations.

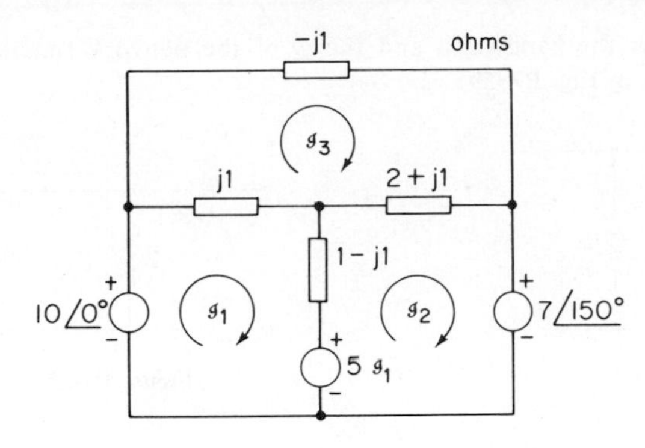

Figure P7-54

Problem 7–55 (Sec. 7–8)

Find the network function $\mathscr{I}/\mathscr{V}$, where $\mathscr{I}$ and $\mathscr{V}$ are the phasors defined in Fig. P7-55.

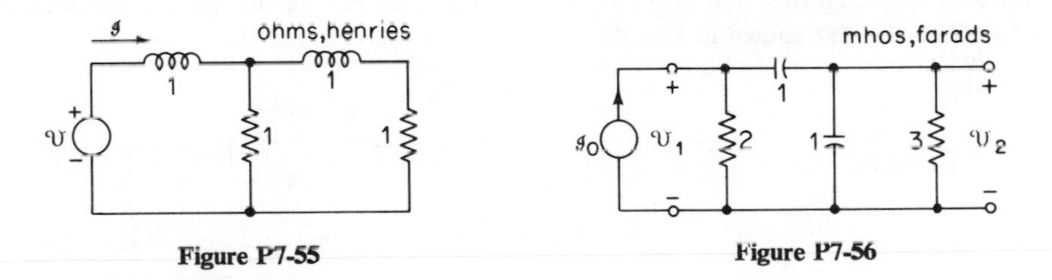

Figure P7-55 **Figure P7-56**

Problem 7–56 (Sec. 7–8)

Find the network function $\mathscr{V}_2/\mathscr{I}_0$, where $\mathscr{V}_2$ and $\mathscr{I}_0$ are the phasors defined in Fig. P7-56.

Problem 7–57 (Sec. 7–8)

Find the network function $\mathscr{V}_2/\mathscr{V}_1$, where $\mathscr{V}_2$ and $\mathscr{V}_1$ are the phasors defined in Fig. P7-57.

Problem 7–58 (Sec. 7–8)

Find the network function $\mathscr{V}_2/\mathscr{V}_1$, where $\mathscr{V}_2$ and $\mathscr{V}_1$ are the phasors defined in the network shown in Fig. P7-58.

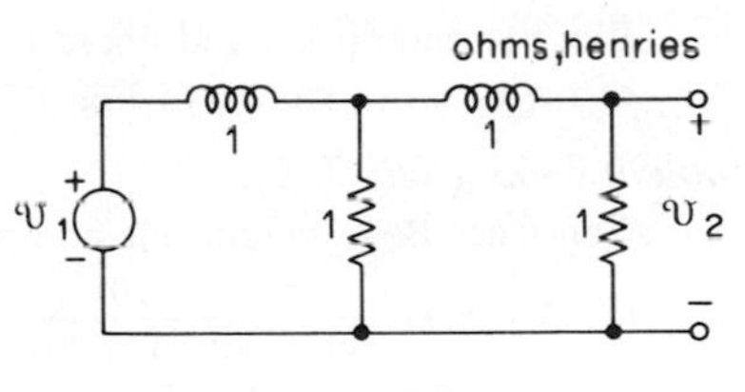

Figure P7-57

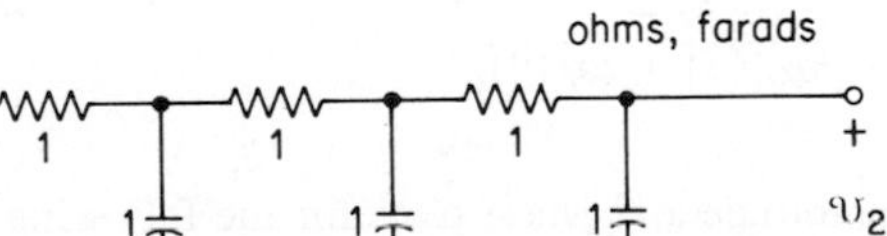

Figure P7-58

Problem 7–59 (Sec. 7–8)

Find expressions for the bandwidth and the Q of the network function $\mathscr{V}/\mathscr{I}$ for the network shown in Fig. P7-59.

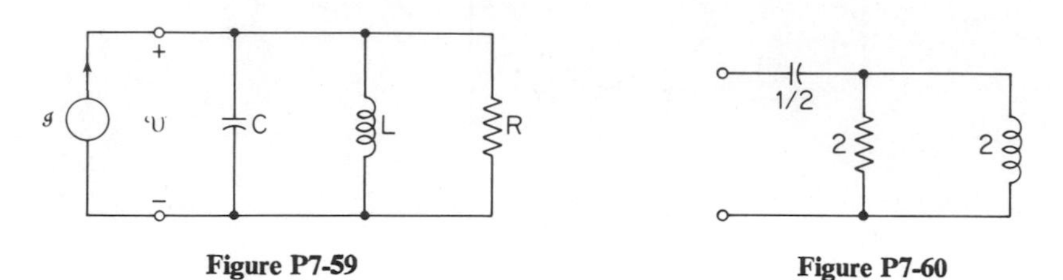

Figure P7-59 **Figure P7-60**

Problem 7–60 (Sec. 7–8)

Plot the magnitude and phase as a function of ω for the driving-point impedance of the network shown in Fig. P7-60.

Problem 7–61 (Sec. 7–8)

Plot the magnitude and phase as a function of ωRC for the voltage transfer function of the network shown in Fig. P7-61.

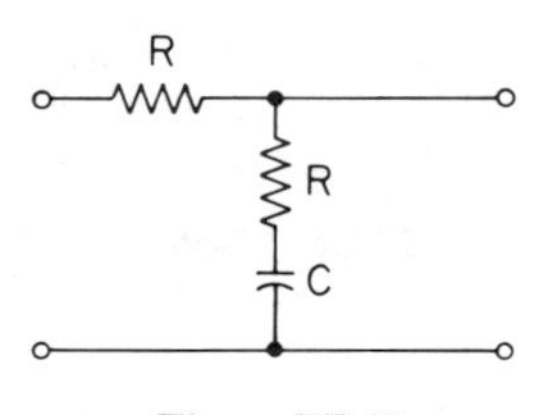

Figure P7-61

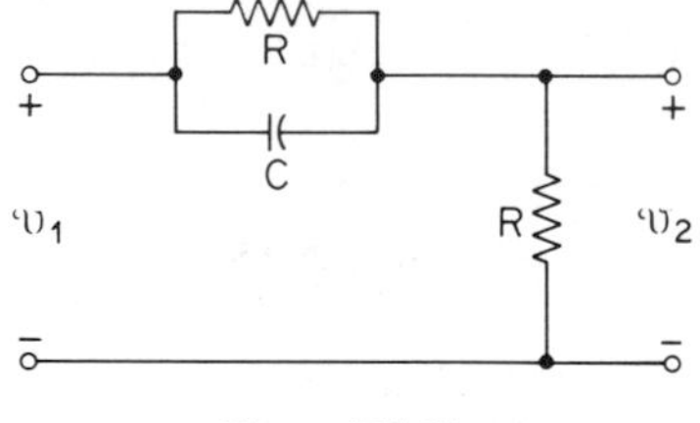

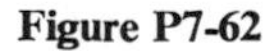

Figure P7-62

Problem 7–62 (Sec. 7–8)

Plot the magnitude and phase as a function of ωRC for the voltage transfer function of the network shown in Fig. P7-62.

Problem 7–63 (Sec. 7–9)

Construct Bode magnitude and phase plots for the following network functions:

(a) $N_a(j\omega) = \dfrac{10}{(1 + j\omega)[1 + (j\omega/10)]}$

(b) $N_b(j\omega) = \dfrac{10j\omega}{(1 + j\omega)[1 + (j\omega/10)]}$

(c) $N_c(j\omega) = \dfrac{10(j\omega)^2}{(1 + j\omega)[1 + (j\omega/10)]}$

Problem 7–64 (Sec. 7–9)

Construct Bode magnitude and phase plots for the following network function:

$$N(j\omega) = \frac{10[1 + j\omega + (j\omega)^2]}{[1 + (j\omega/0.1)][1 + (j\omega/10)]}$$

Problem 7–65 (Sec. 7–9)

Construct Nyquist plots for the networks shown in Figs. P7-60, P7-61, and P7-62.

*Problem 7-66** (*Sec. 7-10*)

Use a linear frequency scale to plot the magnitude and phase of the open-circuit voltage transfer function for the network shown in Fig. 7-10.4, for values of the resistor of 0.2, 0.5, and 1 Ω. Use 51 values of frequency covering the range 0-50 rad/s. Put all three magnitude loci on one plot and all three phase loci on another.

*Problem 7-67** (*Sec. 7-10*)

Make a plot of the magnitude of the open-circuit voltage transfer function of the network shown in Fig. P7-67 for 51 values of frequency from 0 to 2.5 rad/s. Scale the magnitude by 70 in the plot. Verify the fact that the network function will have zeros of transmission at the frequencies at which the impedance of the series *LC* branch is infinite and at which the impedance of the shunt *LC* branch is zero.

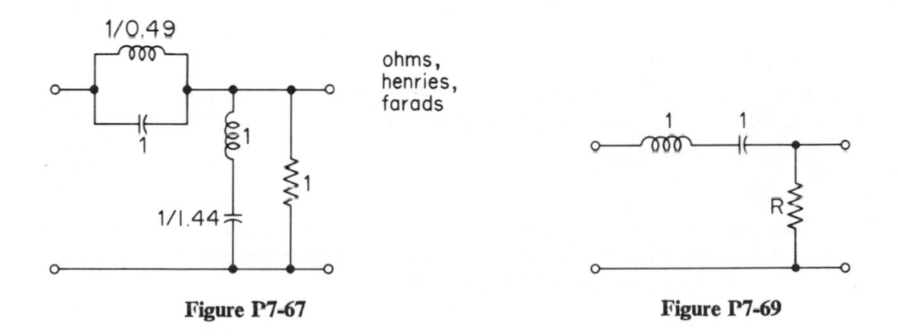

Figure P7-67

Figure P7-69

*Problem 7-68** (*Sec. 7-10*)

Use the subroutine SSS and the function XL to verify the Bode magnitude plot given in connection with Example 7-9.3. In addition, construct the phase plot for the function. Use a range of frequency from 0.1 to 1000 rad/s and use 15 points per decade.

*Problem 7-69** (*Sec. 7-10*)

Construct Bode magnitude plots for the open-circuit voltage transfer function for the network shown in Fig. P7-69 for values of 0.1, 1, and 10 Ω. Use a range of frequency from 0.01 to 100 rad/s with 15 values of frequency per decade. Put all three loci on the same plot.

*Problem 7-70** (*Sec. 7-10*)

Repeat Problem 7-69 for the network shown in Fig. P7-70.

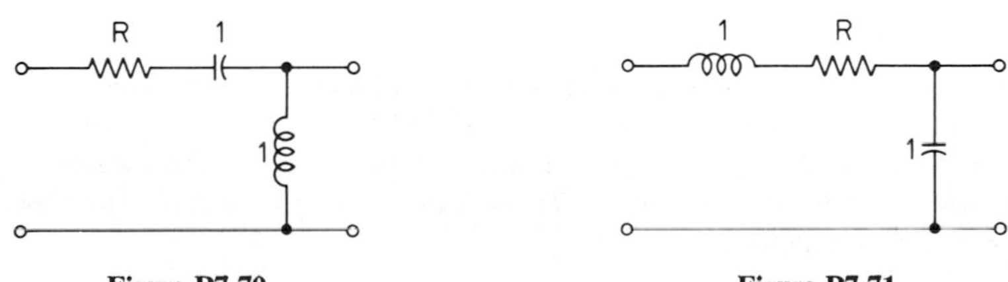

Figure P7-70

Figure P7-71

*Problem 7-71** (*Sec. 7-10*)

Repeat Problem 7-69 for the network shown in Fig. P7-71.

Problem 7–72 (Sec. 7–10)*

Make a Nyquist plot for the open-circuit voltage transfer function for the network shown in Fig. P7–72 for $K = 2$ (the triangle represents an ideal VCVS with gain K). Use 0.05 for A and 1.02 for B in (7–233) to increase ω. Compute and plot 70 points starting with $\omega = 0$.

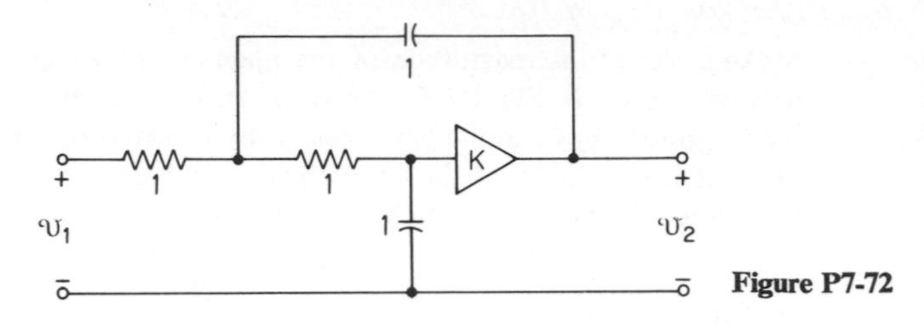

Figure P7-72

Problem 7–73 (Sec. 7–10)*

Make a Nyquist plot of the open-circuit voltage transfer function of the network shown in Fig. P7–67. Use $A = 0.03$ and $B = 1.05$ in (7–233) to increase ω. Compute 100 points starting with $\omega = 0$.

Problem 7–74 (Sec. 7–10)*

Make a Nyquist plot for the network function given in Example 7–9.3. Use $A =$ 0.09 and $B = 1.04$ in (7–233) to increase the value of ω. Compute 120 points starting with $\omega = 0$. Scale the plotted points by 25.

Problem 7–75 (Sec. 7–11)

The square symbol (with four terminals) shown in Fig. P7–75 represents a wattmeter, a device used to measure average power in circuits under sinusoidal steady-state conditions. As shown, two of the terminals are connected in series with a load Z_L (to measure the current), and two are connected across the load (to measure the voltage). Determine the readings that the wattmeter, ammeter, and voltmeter will indicate.

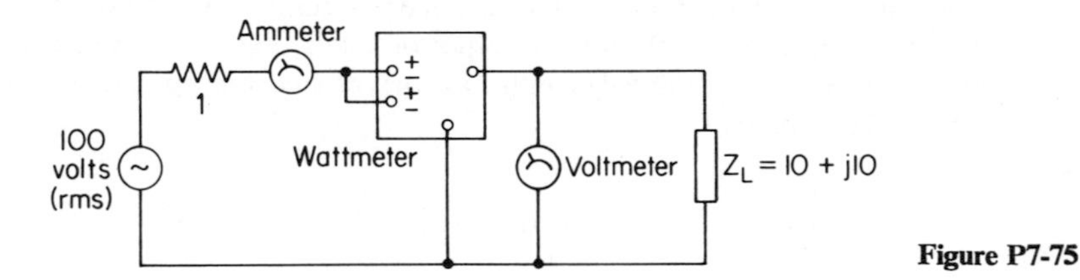

Figure P7-75

Problem 7–76 (Sec. 7–11)

In the circuit shown in Fig. P7–76, a load impedance Z_L is connected to a generator by an ideal transformer with a turns ratio of 1 : 2. The circuit variables are measured by two (rms reading) ammeters and two wattmeters. Find the reading of each of the four meters. (See Problem 7–75 for a discussion of the convention used in designating a wattmeter.)

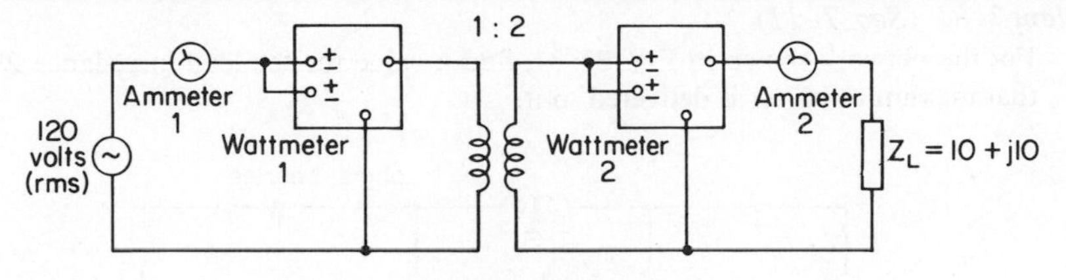

Figure P7-76

Problem 7–77 (Sec. 7–11)

In the circuit shown in Fig. P7-77, a set of rms meter readings for the voltage and current variables are given. Use these values to find the impedance Z_1. Assume that the overall power factor is positive and that Z_2 is purely resistive.

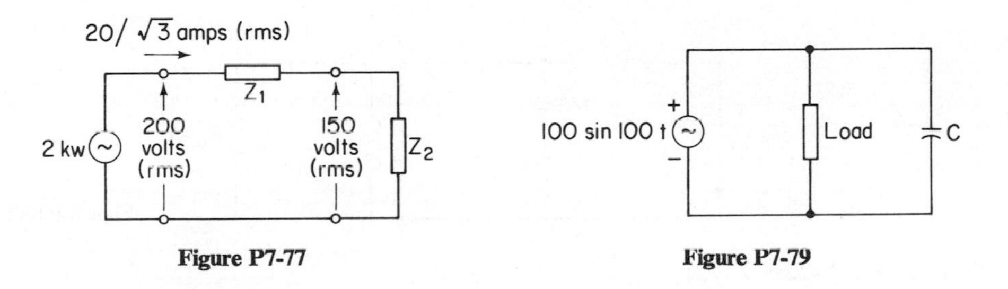

Figure P7-77 **Figure P7-79**

Problem 7–78 (Sec. 7–11)

Prove that the conservation of complex power applies in the circuits shown in the following figures: (a) Fig. P7-75; (b) Fig. P7-77.

Problem 7–79 (Sec. 7–11)

The real and reactive powers consumed by the load shown in Fig. P7-79 are 6 W and +8 VAR, respectively (without considering the capacitor). Find the value of capacitance *C* which will make the source see a unity power factor load.

Problem 7–80 (Sec. 7–11)

In the network shown in Fig. P7-80, load *A* consumes 10 kW of power at 0.6 PF lagging, and load *B* consumes 8 kW with a power factor of 0.8 lagging. Find a value for the resistor such that the overall power factor will be 0.9. Is this an effective way of achieving power factor correction? Why, or why not?

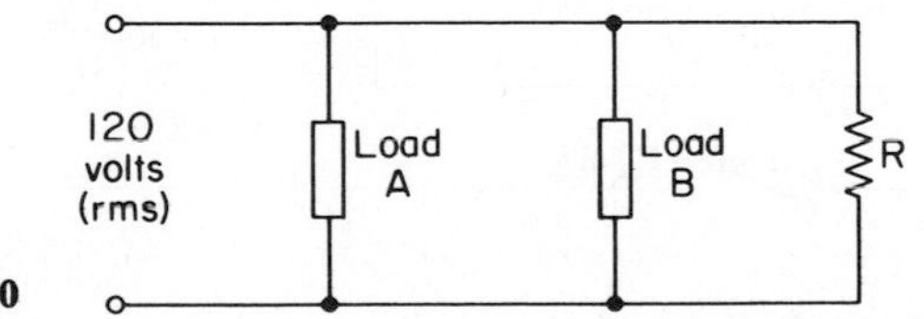

Figure P7-80

Problem 7–81 (*Sec. 7–11*)

For the network shown in Fig. P7-81, find a value for the load impedance Z_L such that maximum power is delivered to it.

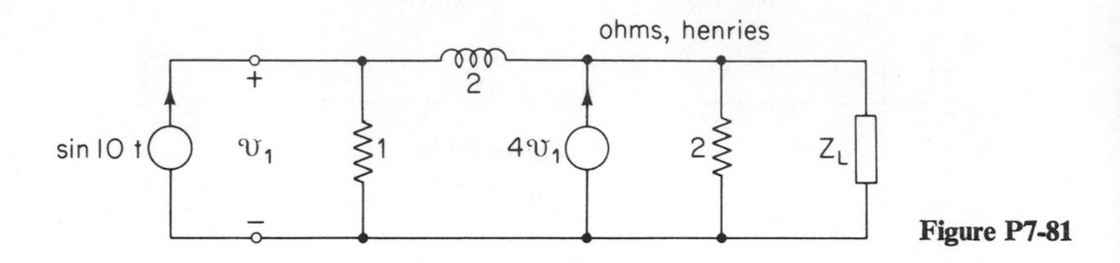

Figure P7-81

Problem 7–82 (*Sec. 7–11*)

For the circuit shown in Fig. P7-82, find the value of the turns ratio n of the ideal transformer such that maximum power is delivered to the load resistor R_L.

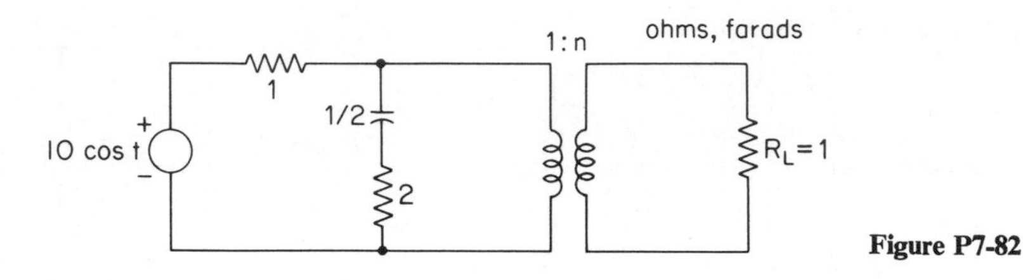

Figure P7-82

Problem 7–83 (*Sec. 7–12*)

Draw the phasor diagram for the line and load voltages and the line currents for the three-phase circuit shown in Fig. P7-83. What is the total power delivered to the load?

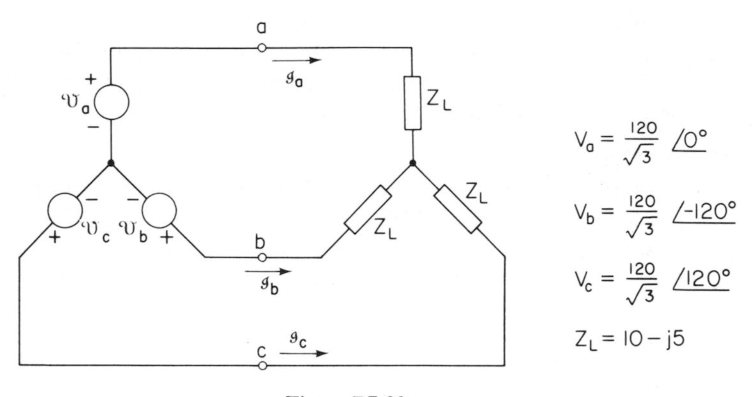

Figure P7-83

Problem 7–84 (*Sec. 7–12*)

Draw the phasor diagram for the line and phase voltages and the line and load currents for the three-phase circuit shown in Fig. P7-84. What is the total power delivered to the load?

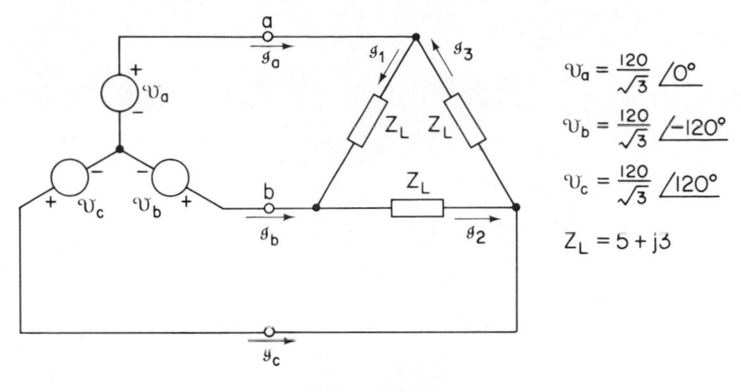

Figure P7-84

Problem 7–85 (*Sec. 7–12*)

Solve for the line currents for the three-phase circuit shown in Fig. P7-85. Draw the phasor diagram. Find the phasor $V_{nn'}$ which represents the voltage between the neutral n of the three-phase generator and the common point n' of the wye of impedances.

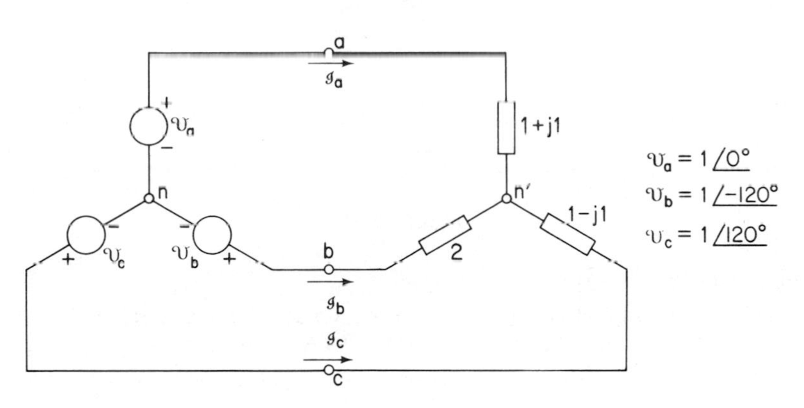

Figure P7-85

Problem 7–86 (*Sec. 7–12*)

For the three-terminal circuits shown in Fig. P7-86, find delta equivalents for the wye circuits, and wye equivalents for the delta circuits.

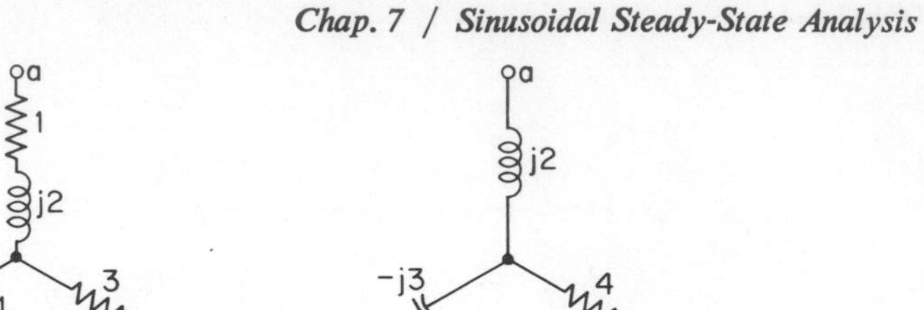

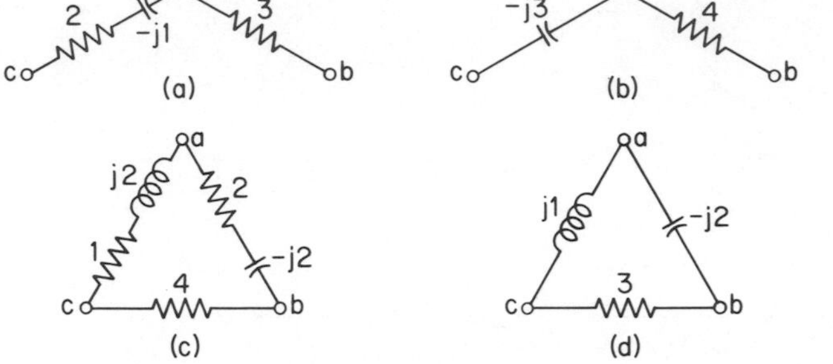

Figure P7-86

Problem 7–87 (*Sec. 7–12*)

Solve for the line and load currents for the three-phase circuit containing three impedances in a delta connection shown in Fig. P7-87.

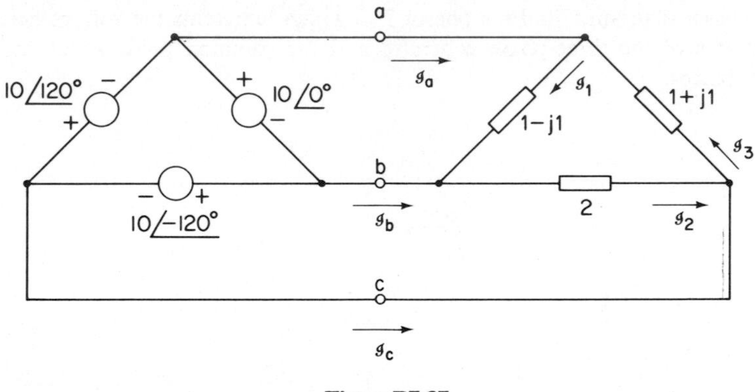

Figure P7-87

Problem 7–88 (*Sec. 7–13*)

Find the open-circuit voltage transfer function for the active *RC* circuit shown in Fig. P7-88. Assume that the gain of the VCVS (represented by a triangle in the figure) is negative. Identify the circuit as low-pass, high-pass, or band-pass.

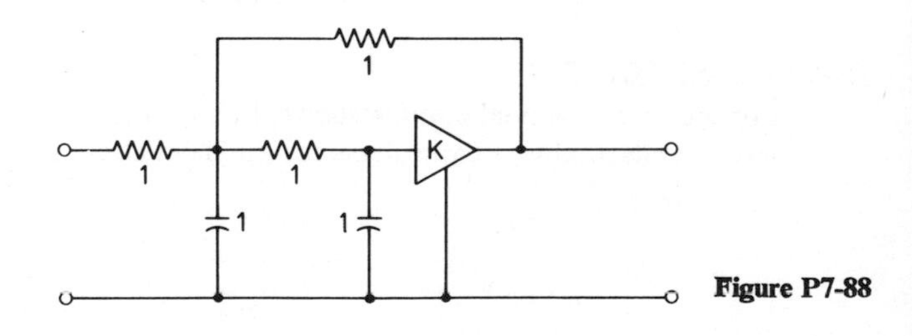

Figure P7-88

Problem 7–89 (*Sec. 7–13*)

Find the open-circuit voltage transfer function for the active *RC* circuit shown in Fig. P7–89. Assume that the gain of the VCVS (represented by a triangle in the figure) is negative. Identify the circuit as low-pass, high-pass, or band-pass.

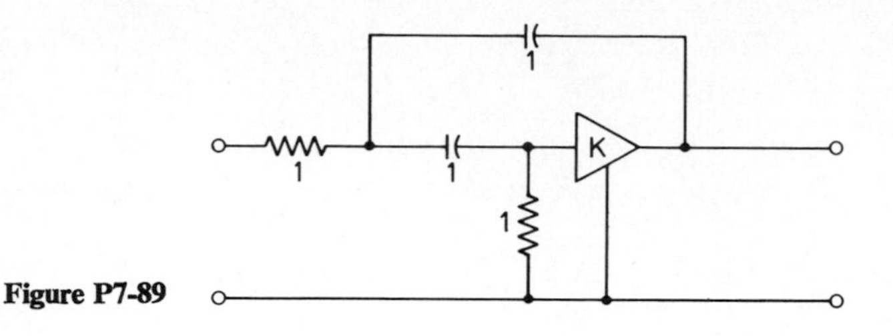

Figure P7-89

CHAPTER 8

The Laplace Transformation

In the preceding chapters of this book we have been concerned with the formulation and solution of sets of network equations relating variables which are functions of time. If the network is comprised only of resistors, such sets of equations are algebraic. More generally, however, when inductors and capacitors are present, sets of integro-differential equations result. It has been shown that the solution of such sets of equations requires the development of many specialized techniques and skills. There exists another quite different approach to finding the solution of sets of integro-differential equations than the ones we have studied so far. This is the use of transformed network variables, i.e., network variables which have been submitted to some mathematical process so that the equations which involve these variables are more amenable to solution. There are many such mathematical transformations which are useful in engineering applications. In this book we shall only study the simplest, most basic, and most used of such transformations, the *Laplace transformation.* We shall see that it has the property of transforming integro-differential equations into algebraic equations, thus providing a valuable tool for the analysis of networks.

8-1 DEVELOPING THE LAPLACE TRANSFORMATION

Let us see if we can develop the Laplace transformation by simple reasoning. Our goal will be to "invent" some mathematical relationship that transforms a function $f(t)$ of a real variable t into a function of some other variable. Let us use F for the transformed function and s for the new variable. The process may be indicated symbolically as[1]

$$f(t) \longrightarrow F(s) \tag{8-1}$$

where the arrow represents the process of transformation. In describing such a process, it is common to refer to functions of t, where t is a real variable, as being in the t (or time) domain, and the resulting transformed function as being in the s domain. Since at this point we don't know anything about the mathematical nature of the variable s, we must be alert for the possibility that it will be complex rather than real. This will actually be shown to be the case. Now let us assume that the purpose of our transformation is to convert the operation of differentiation in the time domain to multiplication by the transformed variable in the s domain. This result may be indicated symbolically as

$$\frac{d}{dt}f(t) \longrightarrow sF(s) \tag{8-2}$$

where $F(s)$ is defined by (8–1). This result is preliminary, since it is possible that the transformation of the differentiated function will produce terms other than the term $sF(s)$; however, this point we may defer until later. Now let us specify the form that the transformation must have. Let us assume that we are interested only in positive values of t, i.e., that all the functions $f(t)$ that we shall consider are zero for $t < 0$. For the transformation process to accurately represent a function $f(t)$ for all positive values of t, the transformation must take into account all values that $f(t)$ takes on. One way of doing this is by integrating $f(t)$ over the range from zero to infinity. To ensure that the effect of any singularity functions (such as impulses) which are applied at $t = 0$ are included, we will specify the lower limit of integration as $0-$ rather than 0.[2] In evaluating such a definite integral, the variable of integration, namely, t, is eliminated from the resulting expression. Therefore, to have such an integration produce a function $F(s)$, a term containing the variable s must be added to the integrand. In general, such a term must be a function of both t and s. Thus, if we use the letter h to represent the function, we may write it in the form $h(t, s)$. The actual form of $h(t, s)$ is yet to be determined. However, as a preliminary step, we have now specified the general form that our transformation will have. Thus, we may write

$$F(s) = \int_{0-}^{\infty} f(t)h(t, s)\,dt \tag{8-3}$$

To determine $h(t, s)$ (this is sometimes referred to as the *kernel* of the transformation

[1]The symbol p is also widely used as the variable for the transformed function.

[2]The impulse function was briefly introduced in Sec. 4-2. A more detailed treatment of this function will be given later in this chapter. The quantity $0-$ is defined in Sec. 4-3.

integral), we must satisfy the specified condition that differentiation is transformed to multiplication. Thus, from (8–2) and (8–3), we may write

$$sF(s) = \int_{0-}^{\infty} f'(t)h(t, s)\, dt \tag{8-4}$$

where $f'(t) = df/dt$. Let us integrate the right member of (8–4) using integration by parts. The general rule for integration by parts for two functions $w(t)$ and $u(t)$ is

$$\int_{a}^{b} w'(t)u(t)\, dt = w(t)u(t)\Big|_{a}^{b} - \int_{a}^{b} w(t)u'(t)\, dt \tag{8-5}$$

where the prime indicates differentiation with respect to the independent variable t. Thus, if we let $w'(t) = f'(t)$, and let $u(t) = h(t, s)$, we may write

$$\int_{0-}^{\infty} f'(t)h(t, s)\, dt = f(t)h(t, s)\Big|_{0-}^{\infty} - \int_{0-}^{\infty} f(t)h'(t, s)\, dt \tag{8-6}$$

The first term in the right member of (8–6) only involves the initial and final values of t, namely, zero and infinity. Thus, this term is unlikely to affect the transformation, which is necessarily based on the entire range of values of $f(t)$. For the moment, therefore, let us ignore this term. Thus, (8–6) may be written as

$$s\int_{0-}^{\infty} f(t)h(t, s)\, dt = -\int_{0-}^{\infty} f(t)h'(t, s)\, dt \tag{8-7}$$

where we have used (8–3) and (8–4) to modify the form of the left member of (8–6). For this equation to be satisfied, we require

$$sh(t, s) = -h'(t, s) = -\frac{dh(t, s)}{dt} \tag{8-8}$$

This equation is easily solved by separating variables. Thus, we may write

$$\frac{dh(t, s)}{h(t, s)} = -s\, dt \tag{8-9}$$

Integrating both sides of the above, we obtain

$$\ln h(t, s) = -st \tag{8-10}$$

If we now raise both sides of (8–10) to the exponential power, we obtain

$$h(t, s) = e^{-st} \tag{8-11}$$

This is our desired expression for $h(t, s)$. Thus, from (8–3) we may define the Laplace transformation by the relation

$$\mathscr{L}[f(t)] = F(s) = \int_{0-}^{\infty} f(t)e^{-st}\,dt \tag{8–12}$$

where the term $\mathscr{L}[f(t)]$ may be read as "the Laplace transformation of $f(t)$," and $F(s)$ is called the *Laplace transform* of $f(t)$ or, more simply, the transform of $f(t)$. In general, s is a complex variable which may take on any value. In actually performing the integration specified in (8–12), however, Re (s) (the real part of s) must be somewhat restricted. As an example of such a restriction, consider the determination of the Laplace transform of the unit step function $u(t)$. If we define $s = \sigma + j\omega$, then from (8–12) we may write

$$\mathscr{L}[u(t)] = F(s) = \int_{0-}^{\infty} u(t)e^{-st}\,dt = \int_{0-}^{\infty} e^{-\sigma t}(\cos\omega t - j\sin\omega t)\,dt \tag{8–13}$$

If σ (the real part of s) is negative, then, since the sine and cosine terms do not approach a given value as t approaches infinity, the improper integral is not convergent and cannot be evaluated. The same is true if $\sigma = 0$. On the other hand, if σ is positive, then the integrand approaches zero as t goes to infinity and the integral exists. Specifically, we may write

$$\mathscr{L}[u(t)] = \int_{0-}^{\infty} e^{-st}\,dt = -\frac{1}{s}e^{-st}\bigg|_{0-}^{\infty} = \frac{1}{s} \qquad \text{Re}\, s > 0 \tag{8–14}$$

Restrictions on Re s similar to that used in determining the transform of the unit step function in (8–14) are readily defined for any of the functions that one normally encounters in network theory applications. Specifically, for a function $f(t)$ to be transformable, it is sufficient to require that

$$\int_{0-}^{\infty} |f(t)|e^{-\sigma_0 t}\,dt < \infty \tag{8–15}$$

for some real positive value of σ_0. It should be noted that even if a function does not satisfy (8–15), it may still be transformed if we suitably restrict the range of our interest in the function. For example, the function e^{t^2} is not transformable; however, a function $f(t)$ defined as

$$f(t) = \begin{cases} e^{t^2} & t < t_0 \\ 0 & t > t_0 \end{cases} \tag{8–16}$$

is transformable, and may be used in place of e^{t^2} by simply restricting the range of time over which the function is considered, to be $0 \le t \le t_0$ where t_0 may be selected to have any positive finite value.

At this point, it should be noted that there are differences, both conceptual and practical, between the *Laplace transformation*, i.e., the integral defined by (8–12), and the resulting *Laplace transform*, i.e., the function $F(s)$. For example, consider the unit step function $u(t)$ which was found in (8–14) to have the transform $1/s$. The defining integral transformation has been shown to exist only for values of s satisfying the relation $\text{Re}\, s > 0$. The resulting transform, however, obviously has a value for all values of s (including those in the left half-plane) except $s = 0$. Mathematically, we may justify the use of the transform for ranges of values of s for which the original transformation integral was not defined by a process referred to as *analytic continuation.* A treatment of the details of this procedure may be found in any standard text on complex variable theory.

Now let us determine whether the transformation defined by (8–12) actually accomplishes the desired goal, i.e., whether it transforms the operation of differentiation into the operation of multiplication. To find this out we may proceed by applying the transformation to the derivative of $f(t)$ with respect to t. Using (8–12) and integrating by parts, we obtain

$$\mathscr{L}[f'(t)] = \int_{0-}^{\infty} f'(t)e^{-st}\,dt = f(t)e^{-st}\Big|_{0-}^{\infty} - \int_{0-}^{\infty} - sf(t)e^{-st}\,dt \tag{8–17}$$

In evaluating the first term in the right member of the above equation at its upper (infinite) limit, if we assume that $f(t)$ satisfies (8–15), then[3]

$$\lim_{t\to\infty} f(t)e^{-st} = 0 \tag{8–18}$$

Almost all the functions with which we shall deal may be shown to have this property. Evaluating this same term at its lower limit, we simply obtain $-f(0-)$. Now consider the last term of the right member of (8–17). Since the integration is with respect to t, the term $-s$ may be moved outside the integral. From (8–12), the remaining integral is simply equal to $F(s)$. From the above, we see that we may simplify (8–17) to the form

$$\mathscr{L}[f'(t)] = sF(s) - f(0-) \tag{8–19}$$

Thus, we see that the Laplace transformation as defined by (8–12) has the property that the operation of differentiation (in the t domain) is transformed into the operation of multiplication by s (in the s domain), and that the initial condition is introduced explicitly by the transformation.

Now let us consider what properties the variable s must have. It is always true that the coefficient of an exponential term must be dimensionless, therefore, from

[3]Some restrictions must also be placed on $\text{Re}\, s$ to correctly define the convergence of the function in the limit. In general, these are the same as those required to ensure that the Laplace transformation of $f(t)$ exists.

(8–12) we see that the quantity $-st$ must be dimensionless. Since t has the dimensions of time, s must have the dimensions of *reciprocal* time, i.e., *frequency*. In later developments in this chapter, we shall show that, in general, s will be a complex quantity. Thus, we may appropriately refer to s as the *complex frequency variable*. The relationship between the complex frequency variable s introduced here and the frequency variable used in the last chapter will be made clearer in a later section of this chapter.

The development given above may be summarized as follows:

SUMMARY 8-1.1

The Laplace Transformation: The Laplace transformation is defined by the relation

$$F(s) = \mathscr{L}[f(t)] = \int_{0-}^{\infty} f(t)e^{-st}\,dt$$

where $f(t)$ is a time-domain function of the real variable t and $F(s)$ is a frequency-domain function of the complex variable s. The transformation has the property that

$$\mathscr{L}\left[\frac{df}{dt}\right] = sF(s) - f(0-)$$

where $F(s)$ is defined above.

In the following section we will investigate some basic properties of the Laplace transformation. It should be noted that the transformed quantities that result from the Laplace transformation process are not "physical" network variables, i.e., if $v(t)$ is some voltage (which may be physically measured by connecting a voltmeter across the terminals at which the voltage is defined), then the transformed quantity $V(s) = \mathscr{L}[v(t)]$ is *not* a physical quantity. That is, there is no voltmeter yet built which will measure such a term. Despite this fact, it is sometimes found convenient to use the same units in referring to $V(s)$ as were used for $v(t)$. It must be strongly emphasized that such use *is only a convenient way of "remembering" what units were originally associated with the time-domain quantity*, and that such a use does not imply any physical properties of the transformed quantity.

8-2 PROPERTIES OF THE LAPLACE TRANSFORMATION

In the last section we defined the Laplace transformation which transformed a function $f(t)$ of a real variable t into a function $F(s)$ of a complex variable s by the relation

$$F(s) = \int_{0-}^{\infty} f(t)e^{-st}\,dt \tag{8-20}$$

The transformation was shown to have the property that a derivative operation on an untransformed function was changed into an algebraic operation on a transformed function. Thus, we derived the relation

$$\mathscr{L}\left[\frac{df}{dt}\right] = sF(s) - f(0-) \tag{8-21}$$

In this section we shall develop some additional properties of the Laplace transformation. These properties will be of considerable use when we begin applying the transformation to the solution of network problems later in this chapter.

One of the most important properties of the Laplace transformation is that it is a *linear* transformation. Thus, it satisfies the property of *homogeneity*, i.e., the Laplace transformation of $K\,f(t)$ is K times the Laplace transformation of $f(t)$. To see this, we may write

$$\mathscr{L}[Kf(t)] = \int_{0-}^{\infty} Kf(t)e^{-st}\,dt = K\int_{0-}^{\infty} f(t)e^{-st}\,dt = KF(s) \tag{8-22}$$

For this result to hold, of course, the quantity K cannot be a function of either t or $f(t)$. Since the Laplace transformation is linear, the property of *superposition* also applies, namely, the Laplace transformation of the function $f_1(t) + f_2(t)$ is the sum of the separate transformations of $f_1(t)$ and $f_2(t)$. To see this, we may write

$$\begin{aligned}\mathscr{L}[f_1(t) + f_2(t)] &= \int_{0-}^{\infty} [f_1(t) + f_2(t)]e^{-st}\,dt \\ &= \int_{0-}^{\infty} f_1(t)e^{-st}\,dt + \int_{0-}^{\infty} f_2(t)e^{-st}\,dt = \mathscr{L}[f_1(t)] + \mathscr{L}[f_2(t)]\end{aligned} \tag{8-23}$$

These linearity considerations are easily generalized as follows:

SUMMARY 8-2.1

Linearity of the Laplace Transformation: If the Laplace transforms of a set of functions $f_i(t)$ of the real variable t are the functions $F_i(s)$, where s is the complex frequency variable, then

$$\mathscr{L}\left[\sum_{i=1}^{n} a_i f_i(t)\right] = \sum_{i=1}^{n} a_i F_i(s)$$

where the quantities a_i are real constants.

As a consequence of the linearity of the Laplace transformation, it should be noted that in a given network situation each variable is transformed separately. As an example of this, consider the equation relating the terminal variables of a capacitor. This has the form

$$i(t) = C\frac{dv}{dt} \tag{8-24}$$

Now let us define transforms of the time-domain functions $i(t)$ and $v(t)$ by the relations

$$\begin{aligned} \mathscr{L}[i(t)] &= \int_{0-}^{\infty} i(t)e^{-st}\,dt = I(s) \\ \mathscr{L}[v(t)] &= \int_{0-}^{\infty} v(t)e^{-st}\,dt = V(s) \end{aligned} \tag{8-25}$$

For the general case, where specific expressions are not given for $i(t)$ and $v(t)$, we may use the symbolic forms $I(s)$ and $V(s)$ to represent the transformed variables. Thus, from (8–20) and (8–21), the transform of the equation given in (8–24) may be written

$$I(s) = sCV(s) - Cv(0-) \tag{8-26}$$

Now let us consider a second example of the application of the linearity properties of the Laplace transformation. In Fig. 8–2.1 a parallel RC circuit is shown. The time-domain equation for this circuit is

$$i(t) = C\frac{dv}{dt} + Gv(t) \tag{8-27}$$

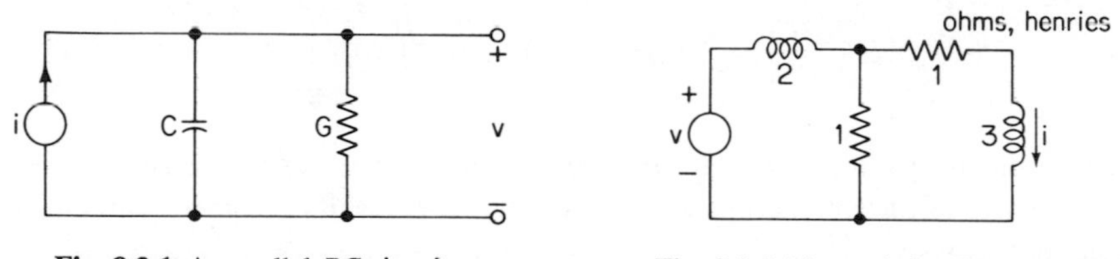

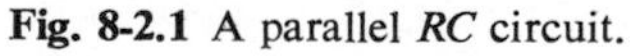

Fig. 8-2.1 A parallel *RC* circuit.

Fig. 8-2.2 Network for Example 8-2.1.

A problem usually associated with such a circuit is to find $v(t)$ when $i(t)$ is known. To do this let us transform the equation. Following the rules given above, we obtain

$$\begin{aligned} I(s) &= sCV(s) - Cv(0-) + GV(s) \\ &= (sC + G)V(s) - Cv(0-) \end{aligned} \tag{8-28}$$

The original differential equation given in (8–27) may be thought of as having two unknown quantities, namely, dv/dt and $v(t)$ [it is assumed that $i(t)$ is known]. In the transformed equation given in (8–28), however, there is only one unknown quantity, namely, $V(s)$, since the initial condition term $v(0-)$ is assumed to be known. Thus, by transformation we have reduced the number of unknown quantities.

As a second property of the Laplace transformation, let us investigate what happens when higher-order derivative functions are transformed. We will start with the second derivative. To do this, we define $g(t) = df/dt$. Then we may write the second derivative of $f(t)$ in the form

$$\frac{d^2f}{dt^2} = \frac{d}{dt}\frac{df}{dt} = \frac{dg}{dt} \tag{8–29}$$

Now let us take the Laplace transformation of the right member of the above equation. From (8–21) we may write

$$\mathcal{L}\left[\frac{dg}{dt}\right] = s\mathcal{L}[g(t)] - g(0-) \tag{8–30}$$

Now let us substitute the relation $g(t) = df/dt$ in (8–30). We obtain

$$\mathcal{L}\left[\frac{d^2f}{dt^2}\right] = s\mathcal{L}\left[\frac{df}{dt}\right] - \frac{df}{dt}(0-) \tag{8–31}$$

where $df(0-)/dt$ is the value of the derivative of $f(t)$ evaluated at $t = 0-$. We may again use (8–21) to evaluate the term $\mathcal{L}[df/dt]$ in the above equation. Thus, we find

$$\mathcal{L}\left[\frac{d^2f}{dt^2}\right] = s[sF(s) - f(0-)] - \frac{df}{dt}(0-) \tag{8–32}$$

This may be put in the form

$$\mathcal{L}\left[\frac{d^2f}{dt^2}\right] = s^2F(s) - sf(0-) - \frac{df}{dt}(0-) \tag{8–33}$$

The development given above is easily extended to treat a derivative of any order. The details are left as an exercise for the reader. The results obtained may be summarized as follows:

SUMMARY 8-2.2

The Laplace Transform of Derivatives: If the Laplace transform of a function $f(t)$ of a real variable t is a function $F(s)$ of a complex variable s, then the transform of the nth derivative of $f(t)$ is given as

$$\mathscr{L}\left[\frac{d^n f}{dt^n}\right] = s^n F(s) - s^{n-1} f(0-) - s^{n-2}\frac{df}{dt}(0-) - \cdots - \frac{d^{n-1}}{dt^{n-1}} f(0-)$$

The development given above makes it possible for us to obtain the Laplace transform of differential equations of any order. An example follows.

EXAMPLE 8-2.1. The circuit shown in Fig. 8-2.2 may be analyzed using the techniques given in Chap. 6 to find a differential equation relating the input voltage $v(t)$ and the current $i(t)$. The differential equation is

$$v(t) = 6\frac{d^2 i}{dt^2} + 7\frac{di}{dt} + i(t) \tag{8-34}$$

Applying the results of Summary 8-2.2 and using the transformed variables $V(s)$ and $I(s)$, we may write

$$V(s) = 6\left[s^2 I(s) - si(0-) - \frac{d}{dt}i(0-)\right] + 7[sI(s) - i(0-)] + I(s)$$

Rearranging the terms of the equation, we may write

$$V(s) = (6s^2 + 7s + 1)I(s) - (6s + 7)i(0-) - 6\frac{d}{dt}i(0-) \tag{8-35}$$

This is the Laplace transform of the original differential equation of (8-34).

The Laplace transform of the differential equation which was found in the example given above illustrates some important points concerning the use of the Laplace transformation to transform differential equations. In general, the transformed equation will consist of two parts. The first part, as illustrated by the first term in the right member of Eq. (8-35), consists of a polynomial in s which multiplies the transformed dependent variable $I(s)$. Comparing Eqs. (8-34) and (8-35) we see that the coefficients of this polynomial are the same as those of the original differential equation, and that the power of s which multiplies each term corresponds with the order of the individual derivatives. Thus, this polynomial is identical with the one that is obtained when the trial solution $f(t) = Ke^{st}$ is substituted in the original differential equation and the common term Ke^{st} is factored out. Because this is so, this polynomial is identical to the *characteristic polynomial* which was introduced in Sec. 6-1. The second part of the transformed equation, i.e., the remaining terms in the right member of Eq. (8-35), is a function of the coefficients, the complex frequency variable s, and the initial conditions. Thus, the entire transformed equation includes the differential equation and the initial conditions. These conclusions are easily shown to be true in general, and may be summarized as follows:

SUMMARY 8-2.3

The Laplace Transform of a Differential Equation: If a general differential equation which relates some excitation function $g(t)$ and some response function $f(t)$ has the form

$$a_1 f(t) + a_2 \frac{df}{dt} + a_3 \frac{d^2f}{dt_2} + \cdots + a_{n+1} \frac{d^nf}{dt^n}$$
$$= b_1 g(t) + b_2 \frac{dg}{dt} + b_3 \frac{d^2g}{dt^2} + \cdots + b_{m+1} \frac{d^mg}{dt^m}$$

then the Laplace transform of this equation will have the form

$$(a_1 + a_2 s + a_3 s^2 + \cdots + a_{n+1} s^n)F(s) + F_{\text{i.c.}}$$
$$= (b_1 + b_2 s + b_3 s^2 + \cdots + b_{m+1} s^m)G(s)$$

where $F(s)$ and $G(s)$ are the Laplace transforms of $f(t)$ and $g(t)$, respectively, and where $F_{\text{i.c.}}$ is a function of the initial conditions $f(0-)$, $df(0-)/dt, \ldots,$ the coefficients a_i $(i = 1, 2, \ldots)$, and s. The polynomial which multiplies $F(s)$ is the characteristic polynomial.

It should be noted that, in the above development, the results apply for the case where the quantities a_i and b_i are not functions of t, i.e., the equation (and thus the network represented by it) must be time-invariant. In addition, no terms of the type $f^2(t)$, $(df/dt)^2$, etc. can be present, i.e., the equation must be linear. Thus the Laplace transformation is only applicable to linear time-invariant networks.

In a manner similar to that used in the last section for finding the transform of a derivative function, we may find the transform of an integral function. From (8-20) we may write

$$\mathcal{L}\left[\int_{0-}^{t} f(\tau)\, d\tau\right] = \int_{0-}^{\infty} \left[\int_{0-}^{t} f(\tau)\, d\tau\right] e^{-st}\, dt \tag{8-36}$$

where, for clarity, we have used τ as the variable of integration in the integrand. The integration is easily carried out by parts, using (8-5) and letting

$$u(t) = \int_{0-}^{t} f(\tau)\, d\tau \qquad u'(t) = f(t)\, dt$$
$$w'(t) = e^{-st}\, dt \qquad w(t) = -\frac{e^{-st}}{s} \tag{8-37}$$

Thus, we obtain

$$\mathcal{L}\left[\int_{0-}^{t} f(\tau)\, d\tau\right] = -\frac{e^{-st}}{s} \int_{0-}^{t} f(\tau)\, d\tau \Big|_{0-}^{\infty} + \frac{1}{s} \int_{0-}^{\infty} f(t) e^{-st}\, dt \tag{8-38}$$

Now let us consider the first term of the right member of this equation. Evaluating the term at the upper limit of infinity, we obtain zero. The reasoning is similar to that used in connection with Eq. (8–14). Evaluating the same term at its lower limit, we again obtain zero, since the integral vanishes when t is set to zero and thus the entire first term is zero. Now let us consider the second term of the right member of Eq. (8–38). The integral term here is the same as the one given in Eq. (8–20), thus, it simply defines the transformed function $F(s)$. Summarizing the above discussion, we may conclude

$$\mathscr{L}\left[\int_{0-}^{t} f(\tau)\,d\tau\right] = \frac{F(s)}{s} \tag{8–39}$$

As an example of the application of this result, let us consider the integral relation for the voltage and current variables of a capacitor. This is[4]

$$v(t) = \frac{1}{C}\int_{0-}^{t} i(\tau)\,d\tau + v(0-) \tag{8–40}$$

The right member of this equation contains a constant, namely, $v(0-)$. Therefore, before we can take the Laplace transformation of this equation, we must determine how to take the transformation of a constant. Let K be such a constant. Since, in applying the Laplace transformation, it is assumed that all functions are zero for $t < 0$, such a constant must be defined in connection with the unit step operator $u(t)$, i.e., we must find the transform of $Ku(t)$. From (8–14) we see that this is simply K/s. Now let us return to the problem of transforming the equation given in (8–40) From (8–39) we may write

$$V(s) = \frac{1}{sC}I(s) + \frac{v(0-)}{s} \tag{8–41}$$

where $V(s)$ is the transform of $v(t)$, and $I(s)$ is the transform of $i(t)$. This conclusion may be combined with the preceding results obtained for differential equations to obtain the solution of integro-differential equations. As an example of such a situation, consider the network shown in Fig. 8–2.3. If the switch S is closed at $t = 0$, then the excitation applied to the circuit is $V_0 u(t)$. Thus, from KVL, we may write

$$V_0 u(t) = Ri(t) + L\frac{di}{dt} + \frac{1}{C}\int_{0-}^{t} i(\tau)\,d\tau + v_C(0-) \tag{8–42}$$

[4]In this relation we have written $v(0-)$ to be consistent with our notation throughout this treatment of the Laplace transformation. Actually, since the continuity condition applies to the voltage across the capacitor, we might as well have written $v(0)$, except for the fact that frequently in Laplace transform analysis of networks, impulses may be used to establish initial conditions, thus, integral functions may be discontinuous. More will be said about this in a later section.

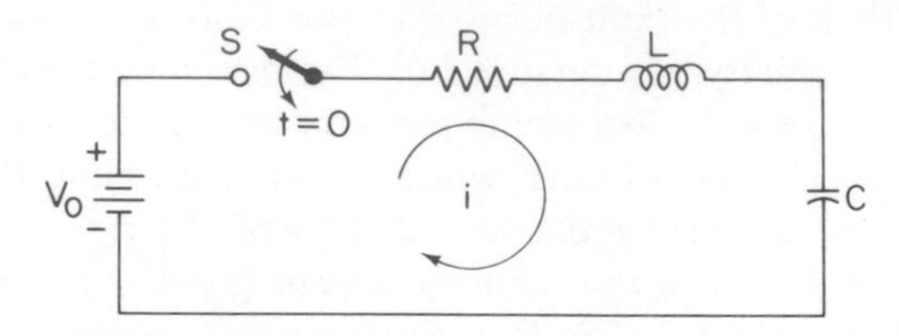

Fig. 8-2.3 A series *RLC* network.

Transforming the equation and rearranging terms, we obtain

$$\frac{V_0}{s} = \left(R + sL + \frac{1}{sC}\right) I(s) + \frac{v_C(0-)}{s} - Li(0-) \tag{8-43}$$

It is readily noted that the first term in the right member of this equation is the transformed equivalent of the right member of the original integro-differential equation with multiplication by the variable s replacing differentiation, and with division by s replacing integration. The second and third terms in the right member are functions of the initial conditions on the various elements of the network. The results given in this example are readily shown to be true in general. Thus we may summarize the above discussion as follows:

SUMMARY 8-2.4

The Laplace Transform of an Integro-Differential Equation: If the Laplace transformation is applied to an integro-differential equation, the result is a transformed equation, each member of which consists of two terms. One of these terms will be identical in form with the original equation, except that differentiation of any order is replaced by multiplication of the transformed quantity by the complex frequency variable s raised to the same order, and that integration is similarly replaced by division by s. The second term of each member will be a function of the initial conditions in the network (and also of s).

In this section we have developed the transformations for the elementary operations of integration and differentiation, and showed how these may be used to transform integro-differential equations describing networks. Later in this chapter we shall see how to use these transformed relations to actually solve for the unknown network variable, i.e., to retransform the equation in the complex frequency variable s back to an equation in the time domain in such a manner as to find an expression in t for some unknown network variable. Before doing this, however, we must develop some additional transformations for various functions of time. This will be done in the next section.

8-3 LAPLACE TRANSFORMS OF FUNCTIONS OF TIME

In the first section of this chapter we derived the Laplace transform of one simple function of time, namely, the step function or constant-valued (for $t > 0$) function. Specifically, we showed that

$$\mathscr{L}[u(t)] = \frac{1}{s} \tag{8-44}$$

A pair of terms such as those given in (8-44), namely, the terms $u(t)$ and $1/s$, are referred to as a *transform pair*. In this section we shall derive several such transform pairs, corresponding with the time-domain functions most frequently encountered in network theory. We shall assume that all such functions are zero for $t < 0$. Thus, the functions will all be shown as being multiplied by $u(t)$.

As a first example of such a transform pair, let us consider the Laplace transformation of an impulse function. Just as we did in Sec. 4-2, let us first define an approximation $\delta_\epsilon(t)$ to the impulse function by the plot shown in Fig. 8-3.1. This waveform has the property

$$\int_{-\infty}^{\infty} \delta_\epsilon(t)\, dt = 1 \tag{8-45}$$

In the limit, as the quantity ϵ shown in Fig. 8-3.1 approaches zero, the resulting function may be defined as the impulse function $\delta(t)$. Thus, we may write for the impulse

$$\int_{-\infty}^{\infty} \delta(t)\, dt = 1 \tag{8-46}$$

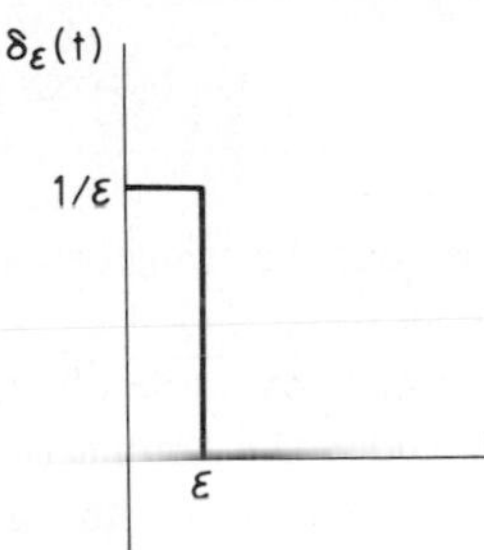

Fig. 8-3.1 An approximation to the impulse function.

Now let us investigate an integral over the range of $-\infty$ to ∞ of the *product* of the approximation to the impulse $\delta_\epsilon(t)$ and some other function $f(t)$. Using the definition of $\delta_\epsilon(t)$ given in Fig. 8-3.1, we obtain

$$\int_{-\infty}^{\infty} \delta_\epsilon(t) f(t)\, dt = \int_0^\epsilon \frac{1}{\epsilon} f(t)\, dt = \frac{1}{\epsilon}\int_0^\epsilon f(t)\, dt \tag{8-47}$$

As ϵ becomes very small, the value of $f(t)$ over the range $0 \le t \le \epsilon$ becomes essentially constant, and may be approximated as $f(0)$. Thus, we write

$$\int_{-\infty}^{\infty} \delta_\epsilon(t) f(t)\, dt = \frac{1}{\epsilon} f(0) \int_0^\epsilon dt = f(0) \tag{8-48}$$

In the limit as ϵ approaches zero, the above relation may be written in terms of $\delta(t)$ rather than $\delta_\epsilon(t)$. Thus, we obtain

$$\int_{-\infty}^{\infty} \delta(t)f(t)\,dt = f(0) \tag{8-49}$$

This result is sometimes referred to as the *sampling property* of a unit impulse, since it samples or selects a single value of the function $f(t)$, namely $f(0)$, from the continuum of values which constitute such a function.[5] The result given above is readily extended to impulses occurring at other values of time. Thus, in general, we may write

$$\int_{-\infty}^{\infty} \delta(t-a)f(t)\,dt = f(a) \tag{8-50}$$

Now let us apply the result given in Eq. (8–49) to the determination of the Laplace transform of an impulse. From the general definition of the transformation, we may write

$$\mathscr{L}[\delta(t)] = \int_{0-}^{\infty} \delta(t)e^{-st}\,dt = e^{-st}\Big|_{t=0-} = 1 \tag{8-51}$$

Thus, we see that the transformation of a unit impulse occurring at $t = 0$ is simply the constant unity.

As another example of the transformation of a frequently encountered time function, consider the simple exponential function $e^{-at}u(t)$ for $a \geq 0$. The Laplace transformation of this function may be found as follows.

$$\mathscr{L}[e^{-at}u(t)] = \int_{0-}^{\infty} e^{-at}e^{-st}\,dt = \int_{0-}^{\infty} e^{-(s+a)t}\,dt = \frac{-1}{s+a}e^{-(s+a)t}\Big|_{0-}^{\infty} \tag{8-52}$$

In evaluating the right member of (8–52), we note that the exponential term evaluated at $t = \infty$ is zero.[6] At the lower limit, namely $t = 0-$, the exponential term is 1. Thus, we may define a transform pair relating the function $e^{-at}u(t)$ for $a \geq 0$, and its transform, by the relation

$$\mathscr{L}[e^{-at}u(t)] = \frac{1}{s+a} \tag{8-53}$$

It is readily shown that this transform pair also exists for $a < 0$. It may also be shown that, when $a = 0$, this transform pair is the same as the one given in Eq. (8–44).

[5]It is possible to rigorously justify the result given in Eq. (8–49). Such a development, however, requires the use of an area of mathematics called the theory of distributions, which is beyond the scope of this text.

[6]Convergence of the integral also requires that $\operatorname{Re} s > -a$.

Now let us consider the transformation of the imaginary exponential function $e^{j\omega t}u(t)$. The Laplace transform of this function may be found as follows:

$$\mathscr{L}[e^{j\omega t}u(t)] = \int_{0-}^{\infty} e^{j\omega t}e^{-st}\,dt = \int_{0-}^{\infty} e^{-(s-j\omega)t}\,dt = \frac{-1}{s-j\omega}e^{-(s-j\omega)t}\Big|_{0-}^{\infty} \tag{8-54}$$

In the evaluation of the right member of the above equation, we again find that the term disappears when evaluated at the upper limit of $t = \infty$. Thus, our transform pair becomes

$$\mathscr{L}[e^{j\omega t}u(t)] = \frac{1}{s-j\omega} \tag{8-55}$$

By a similar process, we may show that the transformation of the exponential $e^{-j\omega t}u(t)$ is defined as

$$\mathscr{L}[e^{-j\omega t}u(t)] = \frac{1}{s+j\omega} \tag{8-56}$$

We may use the transform pairs defined in (8-55) and (8-56) to determine the transformation of the trigonometric functions $\sin \omega t$ and $\cos \omega t$. From Euler's identity (see Appendix B) we know that $\cos \omega t = (e^{j\omega t} + e^{-j\omega t})/2$. Thus, from (8-55) and (8-56), we may write

$$\mathscr{L}[\cos \omega t\, u(t)] = \frac{1}{2}\left[\frac{1}{s-j\omega} + \frac{1}{s+j\omega}\right] = \frac{s}{s^2+\omega^2} \tag{8-57}$$

Similarly, since $\sin \omega t = (e^{j\omega t} - e^{-j\omega t})/2j$, we may define the transform pair for the sine function as

$$\mathscr{L}[\sin \omega t\, u(t)] = \frac{1}{2j}\left[\frac{1}{s-j\omega} - \frac{1}{s+j\omega}\right] = \frac{\omega}{s^2+\omega^2} \tag{8-58}$$

It has already been shown that the Laplace transform of the derivative of a function of time is

$$\mathscr{L}\left[\frac{df}{dt}\right] = sF(s) - f(0-) \tag{8-59}$$

where $F(s) = \mathscr{L}[f(t)]$. The importance of choosing $f(0-)$ rather than $f(0)$ for the point at which the initial condition term is evaluated can be appreciated in terms of the two transforms derived above. To see this, first let us consider the transform of the *derivative* of the function $\sin \omega t\, u(t)$. Applying the formula for differentiation, we may write

$$\mathscr{L}\left[\frac{d}{dt}\sin\omega t\,u(t)\right] = s\mathscr{L}[\sin\omega t\,u(t)] - \sin\omega t\,u(t)\bigg|_{t=0-}$$

$$= s\frac{\omega}{s^2+\omega^2} - 0 = \frac{s\omega}{s^2+\omega^2} \qquad (8\text{–}60)$$

To verify this result we note that the derivative of $\sin\omega t\,u(t)$ is $\omega\cos\omega t\,u(t)$. Thus, directly transforming the differentiated time-domain function, we obtain

$$\mathscr{L}[\omega\cos\omega t\,u(t)] = \frac{s\omega}{s^2+\omega^2} \qquad (8\text{–}61)$$

This is the same as the result obtained in Eq. (8–60). An examination of the above steps will show no significant difference whether we use 0, 0−, or 0+ as the value of t at which the initial condition term of (8–59) is evaluated. Now let us consider the opposite problem, namely, the transform of the derivative of the function $\cos\omega t\,u(t)$. Applying the formula for differentiation, we obtain

$$\mathscr{L}\left[\frac{d}{dt}\cos\omega t\,u(t)\right] = s\mathscr{L}[\cos\omega t\,u(t)] - \cos\omega t\,u(t)\bigg|_{t=0-}$$

$$= s\frac{s}{s^2+\omega^2} - 0 = \frac{s^2}{s^2+\omega^2} \qquad (8\text{–}62)$$

Note that if we had used 0+ as the value of t to be used in evaluating the initial condition term, we would obtain $-\omega^2/(s^2+\omega^2)$ for the transform of the derivative of $\cos\omega t\,u(t)$. To show that this would be incorrect, let us actually differentiate the time-domain function $\cos\omega t\,u(t)$. We obtain

$$\frac{d}{dt}[\cos\omega t\,u(t)] = -\omega\sin\omega t\,u(t) + \cos\omega t\,\delta(t) \qquad (8\text{–}63)$$

Since the impulse function is non-zero only at $t = 0$, and since the cosine function evaluated at $t = 0$ is unity, the term $\cos\omega t\,\delta(t)$ at the right of the above equation simply equals $\delta(t)$. Using this value, and taking the transform of the right member of (8–63), we obtain

$$\mathscr{L}[-\omega\sin\omega t\,u(t) + \delta(t)] = \frac{-\omega^2}{s^2+\omega^2} + 1 = \frac{s^2}{s^2+\omega^2} \qquad (8\text{–}64)$$

This is the same result as was obtained from (8–62), thus it was necessary that 0− rather than 0 or 0+ be used in evaluating the initial condition term in Eq. (8–62). In summarizing the two developments given above, we see from the first example that when we transform the derivative of the sine function, since the function is zero at $t = 0$, it makes no difference whether we use 0, 0−, or 0+ when we evaluate the initial condition term. From the second example, we see that when we transform the derivative of the cosine function, in order to have the formula for the transform of a derivative as applied in (8–62) agree with the transform of the differentiated function as given in (8–64), we must use the value $0-$ in evaluating the initial condition term. This conclusion applies to any function which is discontinuous at $t = 0$. Thus, we must use the value $0-$ throughout our discussion of the Laplace transformation for such cases.

Now let us consider the *ramp function* $Ktu(t)$ (when K equals unity, this function may be called a *unit* ramp function). One way of finding the transform of this function would be to proceed in the same fashion we have followed above, i.e., to evaluate

$$\mathscr{L}[Ktu(t)] = \int_{0-}^{\infty} Kte^{-st}\,dt \tag{8–65}$$

Rather than attempting to evaluate the integral in the right member of (8–65), however, let us investigate a simpler procedure which may be used to determine the transform of the ramp function. In the preceding section we showed that the transform of the integral of a function of time was equal to the transform of the function divided by s. It is easy to see that the ramp function $Ktu(t)$ is simply the integral of the step function $Ku(t)$. Since the transform of the step function is K/s, the transform of the ramp is simply K/s^2. Thus, we may write directly

$$\mathscr{L}[Ktu(t)] = \frac{K}{s^2} \tag{8–66}$$

This procedure is easily applied to determine the transform of the general class of functions $Kt^n u(t)$. The details are left as an exercise for the reader.

A very useful transformation is one which treats the *shifted unit step function* $u(t - a)$. This represents a unit step applied at $t = a$. We need only concern ourselves with values of $a > 0$, since the Laplace transformation is only defined for positive values of t. Thus, a unit step which begins at negative values of time is no different (when transformed) than a unit step starting at $t = 0$. The Laplace transformation of the shifted step function is defined by

$$\mathscr{L}[u(t-a)] = \int_{0-}^{\infty} u(t-a)e^{-st}\,dt = \int_{a-}^{\infty} e^{-st}\,dt = \left.\frac{-e^{-st}}{s}\right|_{a-}^{\infty} \tag{8–67}$$

The right member of the above, evaluated at $t = \infty$, is zero. Therefore, we may write

$$\mathscr{L}[u(t-a)] = e^{-as}\frac{1}{s} \tag{8–68}$$

If we examine the transform pair given in Eq. (8–68), we find that the transformed expression is the same as that given in (8–44) for a unit step applied in $t = 0$, namely,

$1/s$, multiplied by the additional factor e^{-as}. Thus, the effect of shifting the unit step from $t = 0$ to $t = a$ is simply to multiply the transform of the unit step applied at $t = 0$ by e^{-as}. Many different functions may be constructed from combinations of shifted step functions. An example follows.

EXAMPLE 8-3.1. We may use a step function and a shifted step function to define a pulse of magnitude K starting at $t = 0$ and terminating at $t = a$. Such a pulse has been designated $f(t)$ in Fig. 8-3.2(a). The pulse may be considered as the summation of two step functions: the first, of strength K, starting at $t = 0$ as shown in Fig. 8-3.2(b), and the second, of strength $-K$, starting at $t = a$ as shown in Fig. 8-3.2(c). Thus, we may write

$$f(t) = K[u(t) - u(t - a)]$$

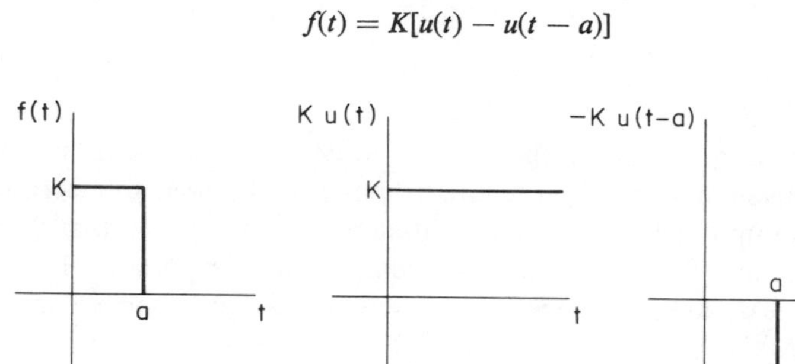

Fig. 8-3.2 Waveforms for Example 8-3.1.

From the discussion given above, the Laplace transforms of the two step functions are readily found to be

$$\mathscr{L}[Ku(t)] = \frac{K}{s} \qquad \mathscr{L}[-Ku(t - a)] = \frac{-Ke^{-as}}{s}$$

Thus, since the transform of a sum of terms is equal to the sum of the transforms (see Summary 8-2.1), we may write

$$\mathscr{L}[f(t)] = \frac{K(1 - e^{-as})}{s}$$

where $f(t)$ is the pulse defined in the time domain in Fig. 8-3.2(a). The term $u(t) - u(t - a)$ used in this example is sometimes referred to as a *gating function*, since, when used to multiply a given function of time, the result is a function which is zero for all values of time less than 0 and greater than a. Thus, this term provides a "gate" through which only those values of the function which exist for the range $0 \leq t \leq a$ are permitted to pass.

Now let us show that the results obtained above, namely, that shifting the time scale by a quantity a for a unit step corresponds with multiplication of the transform of the unit step by the quantity e^{-as}, applies to time functions in general. To see this, consider an arbitrary function $f(t)u(t)$. Now let us assume that we desire to find the transform of a function which has a waveform identical with that of $f(t)u(t)$, but which has been shifted to the right along the real t axis by the distance a, where $a \geq 0$. Such a function may be described as $f(t-a)u(t-a)$. To begin, let us write the general integral expression for the Laplace transformation of a function $f(t)u(t)$ using a "dummy" variable of integration τ. The expression is

$$\mathscr{L}[f(t)u(t)] = F(s) = \int_{0-}^{\infty} f(\tau)e^{-s\tau}\, d\tau \tag{8-69}$$

Since the variable of integration is arbitrary, let us define a new variable by the relation $\tau = t - a$. The upper limit of integration remains ∞, but the lower limit of integration becomes $a-$. Finally, $d\tau$ becomes dt. Thus, Eq. (8-69) may be written in the form

$$\mathscr{L}[f(t)u(t)] = F(s) = \int_{a-}^{\infty} f(t-a)e^{-s(t-a)}\, dt = \int_{a-}^{\infty} f(t-a)e^{-st}e^{as}\, dt \tag{8-70}$$

The lower limit of integration may be changed to $0-$ by including the term $u(t-a)$ inside the integral. Also, since e^{as} is not a function of t, it may be moved outside the integral. Making these changes, we obtain

$$\mathscr{L}[f(t)u(t)] = e^{as}\int_{0-}^{\infty} f(t-a)u(t-a)e^{-st}\, dt \tag{8-71}$$

The integral term in the right member of the above equation is simply the Laplace transformation of $f(t-a)u(t-a)$. Thus, multiplying both sides of the above equation by e^{-as}, we obtain

$$\mathscr{L}[f(t-a)u(t-a)] = e^{-as}\mathscr{L}[f(t)u(t)] \tag{8-72}$$

Thus, we see that the transformation of a function $f(t)u(t)$ which is delayed to begin at $t = a$ is simply the transformation of the original function multiplied by e^{-as}. Thus, we have generalized the result originally obtained for the shifted unit step function in Eq. (8-68).

The result derived above is called the *shifting theorem* or, since it provides a translation of the real variable t, the *real translation theorem.* It is of considerable application in determining the Laplace transform of functions for cases where the direct application of the defining integral for the transformation might be difficult. An example follows.

EXAMPLE 8-3.2. It is desired to find the Laplace transform for a single sinusoidal pulse defined as $f(t)u(t)$ in Fig. 8-3.3(a). Such a pulse is readily seen to be the result of adding two sinusoidal waves of period $2a$, one starting at $t = 0$, as shown in Fig. 8-3.3(b), the other starting at $t = a$, as shown in Fig. 8-3.3(c). Thus, we may write

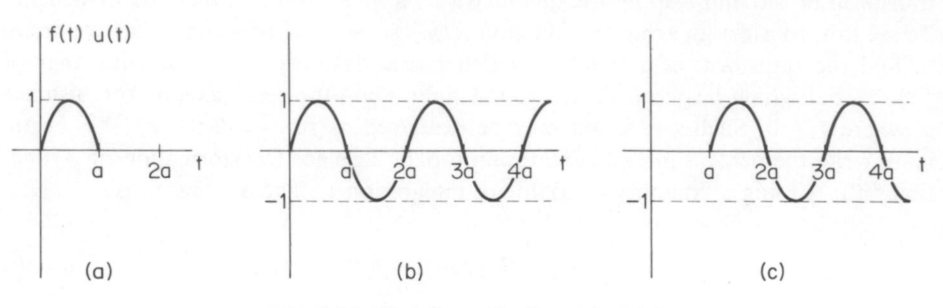

Fig. 8-3.3 Waveforms for Example 8-3.2.

$$f(t)u(t) = \sin \frac{\pi t}{a} u(t) + \sin \frac{\pi(t-a)}{a} u(t-a)$$

From (8-58) we see that the transform of $\sin (\pi t/a)u(t)$ is

$$\mathscr{L}\left[\sin \frac{\pi t}{a} u(t)\right] = \frac{\pi/a}{s^2 + (\pi/a)^2}$$

Thus, using (8-72), we may write

$$\mathscr{L}[f(t)u(t)] = \frac{(1 + e^{-as})\pi/a}{s^2 + (\pi/a)^2}$$

The shifting theorem discussed above may be summarized as follows:

SUMMARY 8-3.1

The Shifting Theorem: If a function $f(t)u(t)$ which is zero for $t < 0$ has a Laplace transform $F(s)$, then for a positive real value of a, a function $f(t - a)u(t - a)$ which is zero for $t < a$ and which corresponds to the original function shifted in time by a seconds, has a Laplace transform $e^{-as}F(s)$.

An extension of the shifting theorem is readily made to functions which are periodic for $t > 0$. If we let $f(t)$ be such a function, and $f_1(t)$ be a function which describes *only the first cycle* of the periodic function and is zero for all other values of t, then, if the period of $f(t)$ is T, we may write $f(t)$ in the form

$$f(t) = f_1(t) + f_1(t - T)u(t - T) + f_1(t - 2T)u(t - 2T) + \cdots \quad (8\text{-}73)$$

Here $f_1(t - T)u(t - T)$ is the first cycle of $f(t)$ shifted one period, thus it is the second cycle of $f(t)$; similarly, $f_1(t - 2T)u(t - 2T)$ is the third cycle, etc. If we now take the Laplace transform of Eq. (8-73), simultaneously applying the shifting theorem given in Summary 8-3.1 above, we obtain

$$\mathscr{L}[f(t)] = F(s) = F_1(s)(1 + e^{-Ts} + e^{-2Ts} + \cdots) \quad (8\text{-}74)$$

where $F_1(s)$ is the Laplace transform of $f_1(t)$. Equation (8–74) may be written in the form

$$\mathscr{L}[f(t)] = F(s) = \frac{F_1(s)}{1 - e^{-Ts}} \tag{8–75}$$

where the form of the denominator term in the right member of (8–75) is easily related to the term in parentheses in the right member of Eq. (8–74) by simple division. An example of the application of this result follows.

EXAMPLE 8-3.3. It is desired to obtain the Laplace transform of the periodic pulse train with period b shown as $f(t)$ in Fig. 8-3.4. From the result of Example 8-3.1, we know that the transform of the first cycle of this pulse train is simply the transform of the pulse defined over $0 < t < a$. Thus, following the terminology given preceding Eq. (8–73), we may write

$$F_1(s) = \frac{1 - e^{-as}}{s}$$

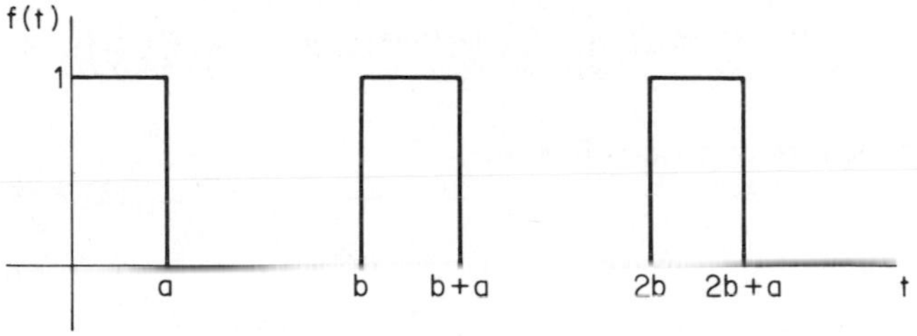

Fig. 8-3.4 Waveform for Example 8-3.3.

Applying Eq. (8–75), we see that the transform of $f(t)$ is given as

$$\mathscr{L}[f(t)] = F(s) = \frac{1 - e^{-as}}{s(1 - e^{-bs})}$$

Note that when $b = a$, then $F(s) = 1/s$, i.e., the function becomes a unit step.

Another important property of the Laplace transformation follows from the transform of the term $e^{-at}f(t)u(t)$. For this, we may write

$$\mathscr{L}[e^{-at}f(t)u(t)] = \int_{0-}^{\infty} e^{-at}f(t)e^{-st}\,dt = \int_{0-}^{\infty} f(t)e^{-(s+a)t}\,dt = F(s+a) \tag{8–76}$$

Thus, the result of multiplying a time function by e^{-at} is a translation of value a in the complex frequency variable s. This result is a very useful one in permitting us to readily extend our list of transform pairs. For example, from Eq. (8–58), since the Laplace transformation of $\sin bt\,u(t)$ is $b/(s^2 + b^2)$, we see that the transform of

$e^{-at} \sin bt\, u(t)$ is simply found to be $b/[(s+a)^2 + b^2]$. Other useful results are readily obtained. We may summarize the above discussion as follows:

SUMMARY 8-3.2

Complex Frequency Variable Translation Theorem: The Laplace transformation of a function $f(t)u(t)$ which is multiplied by the term e^{-at} is equal to $F(s+a)$, where $F(s)$ is the transform of the function $f(t)u(t)$ alone.

One other procedure involving the Laplace transform that is of interest in studies of circuit theory is the transformation of functions in the time domain which have been subjected to a time scaling. For our purposes here, we may define "scaling" as multiplying all values of the variable t by some positive constant a. Thus, given an unscaled function $f(t)u(t)$, let us investigate the Laplace transformation of a time-scaled function $f(at)u(at) = f(at)u(t)$. We may write

$$\begin{aligned} \mathscr{L}[f(at)u(t)] &= \int_{0-}^{\infty} f(at)e^{-st}\,dt = \int_{0-}^{\infty} f(at)e^{-s(at)/a}\frac{d(at)}{a} \\ &= \frac{1}{a}\int_{0-}^{\infty} f(at)e^{-at(s/a)}\,d(at) = \frac{1}{a}F\left(\frac{s}{a}\right) \end{aligned} \tag{8-77}$$

This result may be summarized as follows:

SUMMARY 8-3.3

Scaling Theorem: If a function $f(t)u(t)$ has its real variable t multiplied by a positive constant a, then the Laplace transform of the function will have its complex frequency variable s multiplied by $1/a$. In addition, the entire transform is multiplied by $1/a$.

The result given above is quite reasonable when one considers that, in general, frequency and time are reciprocal quantities and, thus, multiplying time by a scale factor should correspondingly divide frequency by the same factor.

The various Laplace transform pairs which have been derived in this chapter, as well as some other ones which are commonly encountered, are listed for reference in Table 8-3.1. The properties of the Laplace transformation which have been developed in this chapter are listed in Table 8-3.2. These tables will be of use to the reader when applications of the Laplace transformation are discussed in future sections of this chapter.

TABLE 8-3.1 Basic Laplace Transform Pairs

Number	$f(t)$†	$F(s)$
1	$\delta(t)$	1
2	$\delta^{(n)}(t)$‡	s^n
3	$u(t)$	$\frac{1}{s}$
4	t	$\frac{1}{s^2}$
5	t^n	$\frac{n!}{s^{n+1}}$
6	$e^{-at} \quad a > 0$	$\frac{1}{s+a}$
7	$te^{-at} \quad a > 0$	$\frac{1}{(s+a)^2}$
8	$\sin \omega t$	$\frac{\omega}{s^2+\omega^2}$
9	$\cos \omega t$	$\frac{s}{s^2+\omega^2}$
10	$\sin(\omega t + \theta)$	$\frac{s \sin\theta + \omega\cos\theta}{s^2+\omega^2}$
11	$\cos(\omega t + \theta)$	$\frac{s \cos\theta - \omega\sin\theta}{s^2+\omega^2}$
12	$e^{-at}\sin \omega t \quad a > 0$	$\frac{\omega}{(s+a)^2+\omega^2}$
13	$e^{-at}\cos \omega t \quad a > 0$	$\frac{s+a}{(s+a)^2+\omega^2}$
14	$e^{-at}\sin(\omega t+\theta) \quad a > 0$	$\frac{(s+a)\sin\theta + \omega\cos\theta}{(s+a)^2+\omega^2}$
15	$e^{-at}\cos(\omega t+\theta) \quad a > 0$	$\frac{(s+a)\cos\theta - \omega\sin\theta}{(s+a)^2+\omega^2}$
16	$2\lvert K\rvert e^{-at}\cos(\omega t + \text{Arg } K) \quad a > 0$	$\frac{K}{s+a-j\omega} + \frac{K^*}{s+a+j\omega}$

†All of the functions $f(t)$ are assumed to be zero for $t < 0$, i.e., they may be considered as being multiplied by $u(t)$.

‡The function $\delta^{(n)}(t)$ is defined as $d^n\delta(t)/dt^n$, i.e., the nth derivative of $\delta(t)$. Thus $\delta^{(1)}(t)$ is a doublet, $\delta^{(2)}(t)$ is a triplet, etc.

TABLE 8-3.2 Basic Laplace Transform Operations

Number	$f(t)$†	$F(s)$
1	$f_1(t) \pm f_2(t)$	$F_1(s) \pm F_2(s)$
2	$Kf(t)$	$KF(s)$
3	$\frac{d}{dt} f(t)$	$sF(s) - f(0-)$
4	$\frac{d^2}{dt^2} f(t)$	$s^2 F(s) - sf(0-) - \frac{df}{dt}(0-)$
5	$\frac{d^3}{dt^3} f(t)$	$s^3 F(s) - s^2 f(0-) - s\frac{df}{dt}(0-) - \frac{d^2 f}{dt^2}(0-)$
6	$g(t) = \int_0^t f(\tau)\, d\tau + g(0-)$	$\frac{F(s)}{s} + \frac{g(0-)}{s}$
7	$u(t-a)$	e^{-as}
8	$f(t-a)u(t-a)$	$e^{-as} F(s)$
9	$e^{-at} f(t)$	$F(s+a)$
10	$f(at)$	$\frac{1}{a} F\left(\frac{s}{a}\right)$
11	$tf(t)$	$-\frac{d}{ds} F(s)$
12	$t^n f(t)$	$(-1)^n \frac{d^n}{ds^n} F(s)$
13‡	$\lim_{t \to 0} f(t)$	$\lim_{s \to \infty} sF(s)$
14‡	$\lim_{t \to \infty} f(t)$	$\lim_{s \to 0} sF(s)$

†All of the functions $f(t)$ are assumed to be zero for $t < 0$, i.e., they may be considered as being multiplied by $u(t)$.

‡This operation is defined in Sec. 8-4.

8-4 THE INVERSE LAPLACE TRANSFORMATION—PARTIAL-FRACTION EXPANSIONS

Let us summarize what we have accomplished so far in this chapter. In Sec. 8-1 we developed an expression for a mathematical transformation which changed functions of a real variable t into functions of a complex frequency variable s, and which had the property that the operations of differentiation and integration in the time domain were changed to algebraic operations in the frequency domain. In Sec. 8-2 we illustrated how integro-differential time-domain equations could be transformed into algebraic frequency-domain equations involving transformed variables and in Sec. 8-3 we showed how specific time functions could be similarly transformed. We are now in a position to attack our final goal, namely, reclaiming the time-domain solutions for the actual physical network variables from the transformed equations. The mathematical process by means of which this is accomplished is referred to as an *inverse transformation.* Such a process will be the major subject of this section.

The *inverse Laplace transformation* is defined by a complex *inversion integral* which may be applied to a function $F(s)$ to generate a function $f(t)$. The integral itself is somewhat formidable, and its rigorous application involves a knowledge of complex variable theory that is beyond the scope of our text.[7] Fortunately, however, for our applications, there is no need to become involved in such a complicated process. The reason for this is the *uniqueness* property of the Laplace transformation. This may be defined as follows:

SUMMARY 8-4.1

Uniqueness of the Laplace Transformation: For a given $F(s)$, there is one and only one function $f(t)$, defined for $t \geq 0$, which is related to $F(s)$ by the Laplace transformation.[8] Because of this uniqueness, we may write

$$\mathscr{L}[f(t)] = F(s) \qquad \mathscr{L}^{-1}[F(s)] = f(t) \tag{8-78}$$

where $\mathscr{L}^{-1}[F(s)]$ may be read as "the inverse transformation of $F(s)$."

As a most important application of the conclusion given above, consider the transform pairs given in Table 8-3.1. These pairs were originally derived "from left to right," i.e., given the time-domain functions $f(t)$, we found the corresponding functions $F(s)$. Since the Laplace transformation of these functions is unique, we may also use this table of transform pairs "from right to left," i.e., given any of the tabulated frequency-domain functions $F(s)$ or any linear combination of them, we may directly write the corresponding functions $f(t)$ in the time domain, thus obtaining the desired inverse transform. An example of this procedure follows.

EXAMPLE 8-4.1. As an example of the procedure for finding the inverse transform, consider the following function:

[7]The complex inversion integral defining the inverse Laplace transformation is defined as

$$f(t) = \frac{1}{2\pi j}\int_{\sigma_0 - j\infty}^{\sigma_0 + j\infty} F(s)e^{st}\, ds$$

where the path of integration is defined as the *Bromwich path*, and is chosen so as to achieve the necessary convergence of the integral.

[8]It should be noted that this uniqueness only applies to the Laplace transformation which is defined as an integral with limits of integration from 0− to infinity. This transformation is sometimes referred to as the *single-sided* Laplace transformation. Another integral transformation, the *double-sided* Laplace transformation, has limits of integration from minus infinity to infinity. Such a transformation may be applied to functions which have a non-zero value for $t < 0$. In such a case, the uniqueness property described here does not apply.

$$F(s) = \frac{3s - 8}{s^2 + 4} = \frac{3(s)}{s^2 + 4} - \frac{4(2)}{s^2 + 4}$$

Here in the right member, we have separated the function into two parts which have the general form of items 8 and 9 in Table 8-3.1. From the table, we may write

$$\begin{aligned} f(t) &= \mathcal{L}^{-1}[F(s)] = \mathcal{L}^{-1}\left[\frac{3(s)}{s^2 + 4} - \frac{4(2)}{s^2 + 4}\right] = 3\cos 2t - 4\sin 2t \\ &= 5\cos(2t + 53.13°) \end{aligned}$$

where the last result is obtained by the methods outlined in Sec. 7-1.

From the above example, it should be apparent that a simple means of determining the inverse transform of a complicated function is to rewrite the function as a sum of terms, with each term having the form of one of the transforms given in Table 8-3.1. We shall now describe a general procedure for writing a rational function in such a way. The first step of this procedure is to evaluate the behavior of the function $F(s)$ in the limit as s approaches infinity. There are three possible situations that may occur. These are:

$$\begin{aligned} &1.\ \lim_{s \to \infty} F(s) = Ks^m \qquad (m = 1, 2, 3, \dots) \\ &2.\ \lim_{s \to \infty} F(s) = K \\ &3.\ \lim_{s \to \infty} F(s) = \frac{K}{s^m} \qquad (m = 1, 2, 3, \dots) \end{aligned} \tag{8-79}$$

Before describing the above situations, let us introduce the concept of a point in the complex frequency plane at which the magnitude of the function $F(s)$ goes to infinity. Such a point is called a *pole* of $F(s)$. Thus, we may describe the first situation defined in Eq. (8-79) by saying that $F(s)$ has an *mth-order pole at infinity*. This requires that the degree of the numerator polynomial of $F(s)$ be m greater than the degree of the denominator polynomial. An example of such a function is

$$F(s) = \frac{2s^4 + 1}{s^3 + 3s^2 + 4s + 1} \qquad \lim_{s \to \infty} F(s) = 2s \tag{8-80}$$

For this function, we say that $F(s)$ has a first-order pole at infinity. To describe the second situation listed in Eq. (8-79), we may say that $F(s)$ is constant at s equals infinity. This requires that the degree of the numerator and denominator polynomials be equal. An example of such a function is

$$F(s) = \frac{3s^2 + 2s + 1}{s^2 + 4s + 4} \qquad \lim_{s \to \infty} F(s) = 3 \tag{8-81}$$

To describe the third situation listed in Eq. (8-79), we may introduce the concept of a *zero* to indicate a place in the complex frequency plane at which the magnitude of $F(s)$ goes to zero. Thus, we may describe this case by saying that $F(s)$ has an *mth-order zero at infinity*. This requires that the degree of the numerator of $F(s)$ be m lower than the degree of the denominator. An example of a function which has a second-order zero at infinity is

$$F(s) = \frac{7}{s^2 + 3s + 2} \qquad \lim_{s \to \infty} F(s) = \frac{7}{s^2} \tag{8-82}$$

In evaluating the behavior of a given function at infinity, if the first or second situation listed in Eq. (8–79) occurs, then the first step in putting $F(s)$ in a form such that the inverse transform is easily determined is to separate the terms defining the behavior at infinity from the rest of the rational function. This is simply done by long division. For example, consider

$$F(s) = \frac{s^3 + 5s^2 + 11s + 9}{s^2 + 3s + 2} = s + 2 + \frac{3s + 5}{s^2 + 3s + 2} \tag{8-83}$$

where the right member is obtained as follows:

$$\begin{array}{r|l|l} s^2 + 3s + 2 & s^3 + 5s^2 + 11s + 9 & s + 2 \\ & s^3 + 3s^2 + 2s & \\ \hline & 2s^2 + 9s + 9 & \\ & 2s^2 + 6s + 4 & \\ \hline & 3s + 5 \quad \text{Remainder} & \end{array} \tag{8-84}$$

Considering (8–83), we see that we may write from Table 8–3.1

$$f(t) = \mathscr{L}^{-1}[F(s)] = \delta^{(1)}(t) + 2\delta(t) + \mathscr{L}^{-1}\left[\frac{3s + 5}{s^2 + 3s + 2}\right] \tag{8-85}$$

where the inverse transform of the last term in the right member is yet to be found. The results given above are easily generalized as follows:

SUMMARY 8-4.2

The Inverse Transform of the Infinite Frequency Characteristics of a Rational Function: A rational function $f(s) = C(s)/B(s)$, in which the degree of the numerator polynomial $C(s)$ is m greater ($m = 0, 1, 2, \ldots$) than the degree of the denominator polynomial $B(s)$, may be written in the form

$$F(s) = g_1 + g_2 s + g_3 s^2 + \cdots + g_{m+1} s^m + \frac{A(s)}{B(s)}$$

where the degree of $A(s)$ is less than the degree of $B(s)$. The inverse transform of this expression is given as

$$f(t) = \mathscr{L}^{-1}[F(s)] = g_1\delta(t) + g_2\delta^{(1)}(t) + g_3\delta^{(2)}(t) + \cdots + g_{m+1}\delta^{(m)}(t) + \mathscr{L}^{-1}\left[\frac{A(s)}{B(s)}\right]$$

where the quantities $\delta(t)$, $\delta^{(1)}(t)$, $\delta^{(2)}(t)$, ... are, respectively, an impulse, a doublet, a triplet, ... occurring at $t = 0$.

In the above paragraphs we have seen how to transform the infinite frequency characteristics of a rational function into time-domain behavior. In most network situations, the degree of the numerator will never be more than 1 greater than the degree of the denominator; thus, only the impulse and the doublet will normally appear in the time function. Now let us consider the inverse transform of the remaining portion of the rational function, i.e., the part in which the degree of the numerator is less than the degree of the denominator.[9] Let $F(s) = A(s)/B(s)$ be such a function, and let us assume that any non-unity leading coefficient of the denominator polynomial $B(s)$ has been moved to the numerator. If we now factor both the numerator and the denominator polynomials of $F(s)$, we may write the function in the form

$$F(s) = \frac{A(s)}{B(s)} = \frac{K(s-z_1)(s-z_2)(s-z_3)\cdots(s-z_m)}{(s-p_1)(s-p_2)(s-p_3)\cdots(s-p_n)} = \frac{K\prod_{i=1}^{m}(s-z_i)}{\prod_{i=1}^{n}(s-p_i)} \qquad m < n \tag{8-86}$$

The quantities p_i represent locations in the finite s plane where the magnitude of $F(s)$ goes to infinity. Thuse, they are referred to as the *poles of* $F(s)$. Similarly, the quantities z_i represent places in the finite complex plane where the magnitude of $F(s)$ is zero. They are called the *zeros of* $F(s)$. The constant K is sometimes referred to as the overall multiplicative constant or, more simply, as the *gain constant*. It should be apparent that the number of finite zeros of $F(s)$ plus the number of zeros at infinity is equal to the number of finite poles of $F(s)$, i.e., *the total number of poles of a rational function equals the total number of zeros.*

There are three cases which we can consider for functions which are of the general type shown in Eq. (8-86). The first of these is the case in which all the poles of $F(s)$ are real and simple. Thus, in this case, all the quantities p_i in Eq. (8-86) will be real numbers, and no two will have the same value. Now let us assume that we can find a method for writing $F(s)$ in the form

[9]Such a function is sometimes called a *proper* rational function.

$$F(s) = \frac{K_1}{s - p_1} + \frac{K_2}{s - p_2} + \frac{K_3}{s - p_3} + \cdots + \frac{K_n}{s - p_n} = \sum_{i=1}^{n} \frac{K_i}{s - p_i} \tag{8-87}$$

If this can be done, then, from the transform pairs given in Table 8-3.1 we see that the inverse transform may be immediately written as

$$\begin{aligned} f(t) = \mathscr{L}^{-1}[F(s)] &= K_1 e^{p_1 t} + K_2 e^{p_2 t} + K_3 e^{p_3 t} + \cdots + K_n e^{p_n t} \\ &= \sum_{i=1}^{n} K_i e^{p_i t} \end{aligned} \tag{8-88}$$

An expansion of a rational function $F(s)$ in the form indicated in Eq. (8-87) is called a *partial-fraction expansion.* The constants associated with each of the terms of the partial-fraction expansion are called the *residues.* More specifically, K_i is called *the residue of* $F(s)$ *at its pole at* p_i. If $F(s)$ has only two or three poles, then the method of *undetermined coefficients* is easily used to find the values of the K_i. As an example of this, consider the function

$$F(s) = \frac{3s + 5}{s^2 + 3s + 2} = \frac{3s + 5}{(s + 1)(s + 2)} \tag{8-89}$$

The poles are $p_1 = -1$, and $p_2 = -2$. Putting $F(s)$ in the form of a partial-fraction expansion, we obtain

$$F(s) = \frac{K_1}{s + 1} + \frac{K_2}{s + 2} = \frac{K_1(s + 2) + K_2(s + 1)}{(s + 1)(s + 2)} = \frac{s(K_1 + K_2) + (2K_1 + K_2)}{s^2 + 3s + 2} \tag{8-90}$$

If we now compare the numerators of the right members of (8-89) and (8-90) and equate corresponding coefficients of s, we obtain the equations

$$K_1 + K_2 = 3 \qquad 2K_1 + K_2 = 5$$

Solving these equations, we find that $K_1 = 2$, $K_2 = 1$. Thus, $F(s)$ of (8-89) may be written in the form

$$F(s) = \frac{3s + 5}{s^2 + 3s + 2} = \frac{2}{s + 1} + \frac{1}{s + 2} \tag{8-91}$$

The inverse transform of $F(s)$ of Eq. (8-91) may be found by taking the inverse transform of each term in the right member. Thus, from Table 8-3.1 we obtain

$$f(t) = \mathscr{L}^{-1}[F(s)] = [2e^{-t} + e^{-2t}]u(t) \tag{8-92}$$

The method of undetermined coefficients outlined above for finding the residues K_i requires the solution of a set of simultaneous equations of order n, where

n is the number of poles. Such an approach becomes unwieldy for values of n greater than 2 or 3. Thus, it is desirable to consider an alternative method for finding the residues. To see how this alternative approach works, let us again consider the function $F(s)$ defined in Eq. (8-86). If we multiply $F(s)$ by the factor $(s - p_1)$, we obtain

$$(s - p_1)F(s) = \frac{(s - p_1)K\prod_{i=1}^{m}(s - z_i)}{\prod_{i=1}^{n}(s - p_i)} = \frac{K\prod_{i=1}^{m}(s - z_1)}{\prod_{i=2}^{n}(s - p_i)} \tag{8-93}$$

Note that in the right member of (8-93), we have cancelled the factor $(s - p_1)$ in the numerator and denominator, thus the index of the product symbol in the denominator goes from 2 to n, rather than from 1 to n. Now let us also multiply the partial-fraction expansion form of $F(s)$ as given in Eq. (8-87) by the factor $(s - p_1)$. Doing this, we may write

$$(s - p_1)F(s) = K_1 + (s - p_1)\left[\frac{K_2}{s - p_2} + \frac{K_3}{s - p_3} + \cdots + \frac{K_n}{s - p_n}\right] \tag{8-94}$$

Now consider what happens if we evaluate Eqs. (8-93) and (8-94) at $s = p_1$. From the right member of Eq. (8-93), we see that even though p_1 is a pole of $F(s)$ [and therefore $F(p_1)$ is infinite], the product $(s - p_1)F(s)$ will be finite when evaluated at $s = p_1$, since the factor $(s - p_1)$ cancels in the numerator and the denominator. If we now evaluate Eq. (8-94) at $s = p_1$, the right member is simply K_1. Combining these results, we obtain

$$(s - p_1)F(s)\big|_{s=p_1} = K_1 \tag{8-95}$$

This procedure is easily extended to determine the other residues, since, if we multiply $F(s)$ by the factor $(s - p_i)$, we see that the residue K_i is determined by the expression

$$K_i = (s - p_i)F(s)\big|_{s=p_i} \tag{8-96}$$

Thus, we may apply this procedure to find the residue at any pole p_i. An example follows.

EXAMPLE 8-4.2. It is desired to find the partial-fraction expansion and the inverse transform of the following function.

$$F(s) = \frac{6s^2 + 25s + 23}{s^3 + 6s^2 + 11s + 6} = \frac{6s^2 + 25s + 23}{(s + 1)(s + 2)(s + 3)}$$

The partial-fraction expansion for this function will have the form

$$F(s) = \frac{K_1}{s + 1} + \frac{K_2}{s + 2} + \frac{K_3}{s + 3}$$

The residues K_1, K_2, and K_3 may be found using Eq. (8-96) as follows:

$$K_1 = (s+1)F(s)\Big|_{s=-1} = \frac{6s^2+25s+23}{(s+2)(s+3)}\Big|_{s=-1} = 2$$

$$K_2 = (s+2)F(s)\Big|_{s=-2} = \frac{6s^2+25s+23}{(s+1)(s+3)}\Big|_{s=-2} = 3$$

$$K_3 = (s+3)F(s)\Big|_{s=-3} = \frac{6s^2+25s+23}{(s+1)(s+2)}\Big|_{s=-3} = 1$$

Thus, we may write $F(s)$ as the following partial-fraction expansion.

$$F(s) = \frac{2}{s+1} + \frac{3}{s+2} + \frac{1}{s+3}$$

Since, from Table 8-3.1, the inverse transform of a term of the type $k/(s+a)$ is $ke^{-at}u(t)$, the inverse transform of $F(s)$ is readily seen to be

$$f(t) = \mathscr{L}^{-1}[F(s)] = [2e^{-t} + 3e^{-2t} + e^{-3t}]u(t)$$

The results given above are summarized as follows:

SUMMARY 8-4.3

The Inverse Transform of a Rational Function Containing Only Simple Real Poles: A rational function $F(s)$ which contains only n simple real poles, and in which the degree of the numerator polynomial is less than the degree of the denominator polynomial may always be expressed as a partial-fraction expansion having the form

$$F(s) = \sum_{i=1}^{n} \frac{K_i}{s - p_i}$$

where the K_i are the residues and the p_i are the poles. The residues are found as

$$K_i = (s - p_i)F(s)\big|_{s=p_i}$$

The inverse transform is given as

$$f(t) = \mathscr{L}^{-1}[F(s)] = \sum_{i=1}^{n} K_i e^{p_i t}$$

The second case which we shall discuss for determining a partial-fraction expansion of a function $F(s)$ in which the degree of the numerator polynomial is less than the degree of the denominator polynomial is the one in which the function has only real poles, but in which one or more of these poles is non-simple, i.e., of higher order than first. For simplicity, let us assume that there is only one such pole, located at $s = p_0$, and that it is of order n ($n = 2, 3, 4, \dots$). We shall also assume that there are m other poles, and that these are all simple. Such a function may be written in the form

$$F(s) = \frac{A(s)}{(s - p_0)^n \prod_{i=1}^{m} (s - p_i)} \tag{8-97}$$

where it is assumed that the degree of the numerator polynomial is less than $n + m$, i.e., the degree of the denominator polynomial. Now let us assume that we can find a method for writing such a function in the form

$$\begin{aligned} F(s) &= \frac{K_{01}}{s - p_0} + \frac{K_{02}}{(s - p_0)^2} + \frac{K_{03}}{(s - p_0)^3} + \cdots + \frac{K_{0n}}{(s - p_0)^n} \\ &\quad + \frac{K_1}{s - p_1} + \frac{K_2}{s - p_2} + \cdots + \frac{K_m}{s - p_m} \\ &= \sum_{i=1}^{n} \frac{K_{0i}}{(s - p_0)^i} + \sum_{i=1}^{m} \frac{K_i}{s - p_i} \end{aligned} \tag{8-98}$$

If this can be done, then, from the transform pairs given in Table 8-3.1, we see that the inverse transform may be immediately written as

$$\begin{aligned} f(t) &= \mathscr{L}^{-1}[F(s)] = K_{01}e^{p_0t} + K_{02}te^{p_0t} + \frac{K_{03}t^2e^{p_0t}}{2} + \cdots + \frac{K_{0n}t^{n-1}e^{p_0t}}{(n-1)!} \\ &\quad + K_1e^{p_1t} + K_2e^{p_2t} + \cdots + K_me^{p_mt} \\ &= e^{p_0t} \sum_{i=1}^{n} \frac{K_{0i}t^{i-1}}{(i-1)!} + \sum_{i=1}^{m} K_ie^{p_it} \end{aligned} \tag{8-99}$$

In the partial-fraction expansion given in Eq. (8-98), the determination of the residues K_i associated with the m simple poles is accomplished by the procedure given in Summary 8-4.3. Such a determination presents no new problems. The determination of the constants K_{0i} associated with the various powers of the non-simple pole, however, requires a slight modification of the basic technique.[10] To see this, consider the function

$$F(s) = \frac{4s^2 + 11s + 9}{s^3 + 4s^2 + 5s + 2} = \frac{4s^2 + 11s + 9}{(s + 1)^2(s + 2)} \tag{8-100}$$

The partial-fraction expansion for this function will have the form

[10]Of the constants K_{0i} ($i = 1, 2, 3, \dots, n$) of Eq. (8-98), only K_{01} is formally called the residue.

$$F(s) = \frac{K_{01}}{s+1} + \frac{K_{02}}{(s+1)^2} + \frac{K_1}{s+2} \tag{8-101}$$

Let us first see how to find the constant K_{02} associated with the highest degree term representing the second-order pole at $s = -1$. If we multiply both sides of Eq. (8-101) by the denominator of the term containing K_{02}, i.e., the term $(s+1)^2$, and rearrange terms, we obtain

$$(s+1)^2 F(s) = K_{02} + K_{01}(s+1) + \frac{K_1(s+1)^2}{s+2} \tag{8-102}$$

If we now evaluate both sides of (8-102) at $s = -1$, the last two terms in the right member of Eq. (8-102) will disappear, since they are multiplied by zero. Since the left member also contains a term $(s+1)^2$ in the denominator of $F(s)$, however, this member will not be zero. Therefore, we may write

$$K_{02} = (s+1)^2 F(s)\Big|_{s=-1} = \frac{4s^2 + 11s + 9}{s+2}\Big|_{s=-1} = 2 \tag{8-103}$$

Now let us consider how the term K_{01} may be found. As a first step, let us differentiate both sides of Eq. (8-102). Thus, we obtain

$$\frac{d}{ds}[(s+1)^2 F(s)] = K_{01} + K_1\left[\frac{2(s+1)}{s+2} + (s+1)^2 \frac{d}{ds}\frac{1}{s+2}\right] \tag{8-104}$$

Note that in the left member of the above, the factor $(s+1)^2$ will be cancelled by an identical factor in the denominator of $F(s)$ *before* performing the differentiation. Now let us evaluate both sides of Eq. (8-104) at $s = -1$. All the terms in the right member go to zero except the quantity K_{01}. Thus, we may write

$$K_{01} = \frac{d}{ds}[(s+1)^2 F(s)]\Big|_{s=-1} = \frac{d}{ds}\left[\frac{4s^2 + 11s + 9}{s+2}\right]\Bigg|_{s=-1}$$
$$= \left[\frac{8s+11}{s+2} - \frac{4s^2 + 11s + 9}{(s+2)^2}\right]\Bigg|_{s=-1} = 1 \tag{8-105}$$

Thus, we have found K_{01} and K_{02} of Eq. (8-101). The residue K_1 is easily shown to be 3, therefore, we may write

$$F(s) = \frac{1}{s+1} + \frac{2}{(s+1)^2} + \frac{3}{s+2} \tag{8-106}$$

From Table 8-3.1 we find that the inverse transform of this function is

$$f(t) = \mathscr{L}^{-1}[F(s)] = [e^{-t} + 2te^{-t} + 3e^{-2t}]u(t) \tag{8-107}$$

The results given above are easily generalized as follows:

SUMMARY 8-4.4

The Inverse Transform of a Rational Function Containing Non-Simple Real Poles: If a rational function $F(s)$ contains a non-simple real pole of order n at $s = p_0$, then there will be a sum of terms of the form $K_{0i}/(s - p_0)^i$ ($i = 1, 2, \ldots, n$) in the partial-fraction expansion for the function. The multiplicative constants K_{0i} of these terms are found by the relation[11]

$$K_{0i} = \frac{1}{(n-i)!} \frac{d^{n-i}}{ds^{n-i}} (s - p_0)^n F(s) \bigg|_{s=p_0}$$

where the constants are most easily evaluated in the order $i = n, n-1, n-2, \ldots, 2, 1$. There will also be a series of terms in the inverse transform corresponding with such a non-simple pole. These will have the form $e^{p_0 t} K_{0i} t^{i-1}/(i-1)!$

The third case which we shall discuss for determining a partial-fraction expansion of a rational function $F(s)$ is the case in which the function contains a pair of complex conjugate poles. For simplicity, we shall consider that the order of the poles is simple and that they are located at p_0 and p_0^*. Thus, we may write $F(s)$ in the form

$$F(s) = \frac{A(s)}{(s - p_0)(s - p_0^*)Q(s)} \tag{8-108}$$

where the degree of the numerator polynomial $A(s)$ is less than the degree of the denominator [the degree of $Q(s)$ plus 2], and where $Q(s)$ contains all the other poles of $F(s)$. Now let us separate the partial-fraction expansion terms involving the complex conjugate poles from the rest of the function. Thus, we may write

$$F(s) = \frac{K_1}{s - p_0} + \frac{K_2}{s - p_0^*} + \frac{R(s)}{Q(s)} \tag{8-109}$$

where $R(s)/Q(s)$ is the partial-fraction expansion for all the other poles of $F(s)$. The residues K_1 and K_2 are easily found as follows:

[11]Many alternative methods have been proposed for finding the terms in the partial-fraction expansion for poles which are of higher order than the first. A sampling of some of the literature on this subject may be found in the following.

Moad, M. F., "On partial-fraction expansion with multiple poles through derivatives," *Proc. IEEE*, Vol. 57, No. 11, pp. 2056–2958, November, 1969.

Karni, S., "Easy partial-fraction expansion with multiple poles," *Proc. IEEE*, Vol. 57, No. 2, pp. 231–232, February, 1969.

Rao, K. R., and N. Ahmed, "Recursive techniques for obtaining the partial-fraction expansion of a rational function," *IEEE Trans. on Education*, Vol. E-11, No. 2, pp. 152–154, June, 1968.

$$K_1 = (s - p_0)F(s)|_{s=p_0}$$
$$K_2 = (s - p_0^*)F(s)|_{s=p_0^*} \tag{8-110}$$

From these equations we may note that, since the evaluation of the right members is made at a complex value of the argument, the residues K_1 and K_2 will, in general, be complex. In addition, we may note that $K_2 = K_1^*$. Let us define the real and imaginary parts of K_1 and p_0 by the expressions

$$K_1 = k_r + jk_i \qquad p_0 = \sigma_0 + j\omega_0 \tag{8-111}$$

The related conjugate terms then are given as

$$K_2 = K_1^* = k_r - jk_i \qquad p_0^* = \sigma_0 - j\omega_0 \tag{8-112}$$

Substituting the expressions of (8–111) and (8–112) in the general form for $F(s)$ given in (8–109) and rearranging terms, we obtain

$$F(s) = \frac{2k_r s - 2k_r\sigma_0 - 2k_i\omega_0}{(s - \sigma_0)^2 + \omega_0^2} + \frac{R(s)}{Q(s)} \tag{8-113}$$

We conclude that in the partial-fraction expansion the terms produced by a simple pair of complex conjugate poles with complex residues may be combined to form an expression of the form $(as + b)/[(s - \sigma_0)^2 + \omega_0^2]$ where the coefficients a, b, σ_0, and ω_0 are all real. Frequently, for low-order rational functions, the coefficients a and b can be found directly, using the method of undetermined coefficients. This is usually easier than using the complex arithmetic required when determining the residues directly by Eq. (8–110). As an example of this, consider the function

$$F(s) = \frac{4s^2 + 7s + 13}{(s^2 + 2s + 5)(s + 2)} = \frac{as + b}{s^2 + 2s + 5} + \frac{K}{s + 2} \tag{8-114}$$

The residue K for the pole at -2 is easily shown to be 3. Using this value and equating the numerator of both sides of Eq. (8–114), we obtain

$$\begin{aligned} 4s^2 + 7s + 13 &= (as + b)(s + 2) + 3(s^2 + 2s + 5) \\ &= s^2(a + 3) + s(2a + b + 6) + (2b + 15) \end{aligned} \tag{8-115}$$

Equating the coefficients of like powers of s, we find that $a = 1$ and $b = -1$.[12]

[12]It should be noted that, since there are three coefficients of the various powers of s (including s^0), performing this operation provides three equations, with only two unknowns to be solved for. The reason for this is that there are really three unknowns, namely, a, b, and K as given in Eq. (8–114). We have already inserted the value of one of the unknowns, namely K, into the expression. Thus, any two of the three equations may be used to find a and b. In general, it is easier to find the residues for real poles directly, and thus to reduce the order of the set of simultaneous equations which must be solved to find the coefficients associated with complex conjugate pairs of poles.

Thus, we see that

$$F(s) = \frac{s-1}{s^2+2s+5} + \frac{3}{s+2} = \frac{s-1}{(s+1)^2+(2)^2} + \frac{3}{s+2} \tag{8-116}$$

The form used for the complex conjugate pole pair in the denominator of the right member of the above equation provides the key to one of the simplest methods of taking the inverse transform of such a term. To see this, let us use the transforms of the sine and cosine terms, multiplied by an exponential as given in Table 8-3.1. These are

$$\begin{aligned} \mathscr{L}[e^{-\sigma t} \sin \omega t \, u(t)] &= \frac{\omega}{(s+\sigma)^2+\omega^2} \\ \mathscr{L}[e^{-\sigma t} \cos \omega t \, u(t)] &= \frac{s+\sigma}{(s+\sigma)^2+\omega^2} \end{aligned} \tag{8-117}$$

Now let us apply these transform pairs to the first term of the right member of Eq. (8-116). In doing this, we must first separate this term into the sum of two terms, such that each is separately transformable, using the transform pairs of Eq. (8-117). Thus, we must find constants k_1 and k_2 such that

$$\frac{k_1(s+1)}{(s+1)^2+(2)^2} + \frac{k_2(2)}{(s+1)^2+(2)^2} = \frac{s-1}{(s+1)^2+(2)^2} \tag{8-118}$$

Equating coefficients in the numerators of the left and right members of Eq. (8-118), we find that $k_1 = 1$ and $k_2 = -1$. Thus we may write $F(s)$ of (8-116) in the form

$$F(s) = \frac{s+1}{(s+1)^2+(2)^2} + \frac{-1(2)}{(s+1)^2+(2)^2} + \frac{3}{s+2} \tag{8-119}$$

The inverse transforms of the first two terms in the right member of this equation are readily found from Eq. (8-117). Thus, the entire inverse transform for the function $F(s)$ given in (8-116) is

$$\begin{aligned} f(t) = \mathscr{L}^{-1}[F(s)] &= [e^{-t} \cos 2t - e^{-t} \sin 2t + 3e^{-2t}]u(t) \\ &= [\sqrt{2}\, e^{-t} \cos (2t + 45°) + 3e^{-2t}]u(t) \end{aligned} \tag{8-120}$$

where the second form of $f(t)$, in which the two sinusoidal terms have been combined, is formed by following the methods given in Sec. 7-1. The results given above are easily generalized as follows:

SUMMARY 8-4.5

The Inverse Transform of Complex Conjugate Pole Pairs: If a rational function $F(s)$ contains a pair of complex conjugate simple poles located at $s = -\sigma_0 \pm j\omega_0$, then there will be a term of the form $(as + b)/[(s + \sigma_0)^2 + \omega_0^2]$ in the partial-fraction expansion for the function. Such a term may always be modified to have the form

$$\frac{k_1(s+\sigma_0)}{(s+\sigma_0)^2+\omega_0^2} + \frac{k_2\omega_0}{(s+\sigma_0)^2+\omega_0^2}$$

The inverse transform of this pair of terms is

$$e^{-\sigma_0 t}(k_1 \cos \omega_0 t + k_2 \sin \omega_0 t)u(t)$$

The above results are easily modified to treat the case where multiple-order complex conjugate poles are encountered. The method is similar to that described in Summary 8-4.3, except that it is necessary to use complex arithmetic. The details are left to the reader as an exercise.

An alternative formulation for the inverse transformation of the partial-fraction expansion of a pair of complex conjugate poles may be obtained by expressing the residues associated with these poles in a polar notation. Thus, in Eq. (8-109), if we let $K_1 = Ke^{j\phi}$ and $K_2 = K_1^* = Ke^{-j\phi}$, where K is the (non-zero) magnitude of the residue, it is easily shown that the inverse transformation of the first two terms of the function $F(s)$ defined in (8-109) is

$$f(t) = 2Ke^{\sigma_0 t} \cos(\omega_0 t + \phi) \tag{8-121}$$

where σ_0 and ω_0 are defined in Eq. (8-111). The details of determining this result are left as an exercise for the reader.

Frequently, it is useful to be able to determine the initial value of an inverse Laplace transform without actually performing the inversion process. To see how this may be done, let us first consider the expression for the Laplace transform of the derivative of a time function $f(t)$. This has been shown to be

$$\int_{0-}^{\infty} f'(t)e^{-st}\,dt = sF(s) - f(0-) \tag{8-122}$$

Now let us modify this result by dividing the integral in the left member of the above expression into two parts as follows:

$$\int_{0-}^{0+} f'(t)e^{-st}\,dt + \int_{0+}^{\infty} f'(t)e^{-st}\,dt = sF(s) - f(0-) \tag{8-123}$$

The first integral in the left member of the above is taken over a range of time from $0-$ to $0+$. In this range, the quantity e^{-st} is simply equal to unity. Thus, this first integral may simply be evaluated as follows:

$$\int_{0-}^{0+} f'(t)e^{-st}\,dt = \int_{0-}^{0+} f'(t)\,dt = f(0+) - f(0-) \tag{8-124}$$

Now let us insert the above result in Eq. (8–123) and take the limit of both sides of the equation as s approaches infinity. We obtain

$$f(0+) - f(0-) + \lim_{s\to\infty} \int_{0+}^{\infty} f'(t)e^{-st}\,dt = \lim_{s\to\infty} sF(s) - f(0-) \tag{8-125}$$

Since the limits of integration are from $0+$ to infinity in the integral in the left member of the above expression, the variable of integration t is always positive. Thus, as s approaches infinity, the integrand approaches 0 [we require that $f'(t)$ be transformable for this result to hold]. As a result, the value of the integral is also zero and, adding $f(0-)$ to both sides of (8–125), we obtain

$$f(0+) = \lim_{s\to\infty} sF(s) \tag{8-126}$$

This result may be summarized as follows:

SUMMARY 8-4.6

The Initial-Value Theorem: The initial value $f(0+)$ of the inverse transform $f(t)$ of a function $F(s)$ may be found by evaluating the quantity $sF(s)$ in the limit as s approaches infinity.[13]

It is also possible to directly determine the value of $f(t)$ as t approaches *infinity* from the Laplace transform without actually performing the inversion process. To see this, we may let s approach zero in the expression for the Laplace transformation of a derivative. This may be written

$$\lim_{s\to 0} \int_{0-}^{\infty} \left[\frac{df}{dt}\right] e^{-st}\,dt = \lim_{s\to 0}\, [sF(s) - f(0-)] \tag{8-127}$$

The left member of this equation may be rewritten as follows:

[13]The result given above assumes that $f(t)$ does not have any impulses or derivatives of impulses at the origin. If it does, the result of Summary 8-4.6 must be modified. A discussion of this case may be found in P. M. Chirlian's *Basic Circuit Theory*, Chapter 5, McGraw-Hill Book Company, New York, 1969.

$$\lim_{s\to 0}\int_{0-}^{\infty}\left[\frac{df}{dt}\right]e^{-st}\,dt = \int_{0-}^{\infty}\left[\frac{df}{dt}\right]dt = \lim_{t\to\infty}\int_{0-}^{t}\left[\frac{df}{d\tau}\right]d\tau = \lim_{t\to\infty}\,[f(t) - f(0-)] \tag{8-128}$$

Equating the right members of (8–127) and (8–128), since $f(0-)$ is not a function of t or s, it may be subtracted from both members. Thus, we obtain

$$\lim_{t\to\infty} f(t) = \lim_{s\to 0} sF(s) \tag{8-129}$$

This result requires that the function $sF(s)$ have no poles on the imaginary axis or in the right half-plane, since otherwise the final value of the related time function $f(t)$ is indeterminate. For example, if $sF(s) = s/(s^2 + 1)$ (which has poles on the $j\omega$ axis), the relation of Eq. (8–129) does not apply. This is clearly true, since $F(s) = 1/(s^2 + 1)$ and, thus, $f(t) = \sin t$, which is indeterminate for $t = \infty$. The result given above may be summarized as follows:

SUMMARY 8-4.7

The Final-Value Theorem: It $F(s)$ is the transform of a function $f(t)$, the final value of $f(t)$ as t approaches infinity may be found by evaluating the quantity $sF(s)$ as s approaches zero provided that $sF(s)$ has no $j\omega$ axis or right half-plane poles.

In the preceding sections of this chapter we have shown how an integro-differential equation in the time domain may be transformed into an algebraic equation in the frequency domain using the Laplace transformation. This procedure, coupled with the techniques for finding the inverse transformation by means of the partial-fraction expansion, provides a complete method for the analysis of networks based on the use of the Laplace transformation. A flow chart illustrating the general procedure is given in Fig. 8–4.1. This figure summarizes the various steps involved in the process and the various starting and terminal points for the procedure. The flow chart given in the figure should be carefully compared with the one given in Fig. 6–8.1, which summarizes the time-domain solution procedures discussed in Chaps. 5 and 6. Such a comparison will provide considerable insight into the manner in which the two methods complement each other.

The general method which has been presented in the preceding sections of this chapter and which is outlined in Fig. 8–4.1 is only one of the ways in which the Laplace transformation can be used to solve network analysis problems. An even more useful method is to make use of a concept which may be called the "transformed network." This approach will be covered in more detail in Sec. 8–6. Before undertaking such a discussion, however, let us first investigate some ways in which digital computational techniques can be applied to implement the inverse Laplace transformation procedures. This is the topic of the following section.

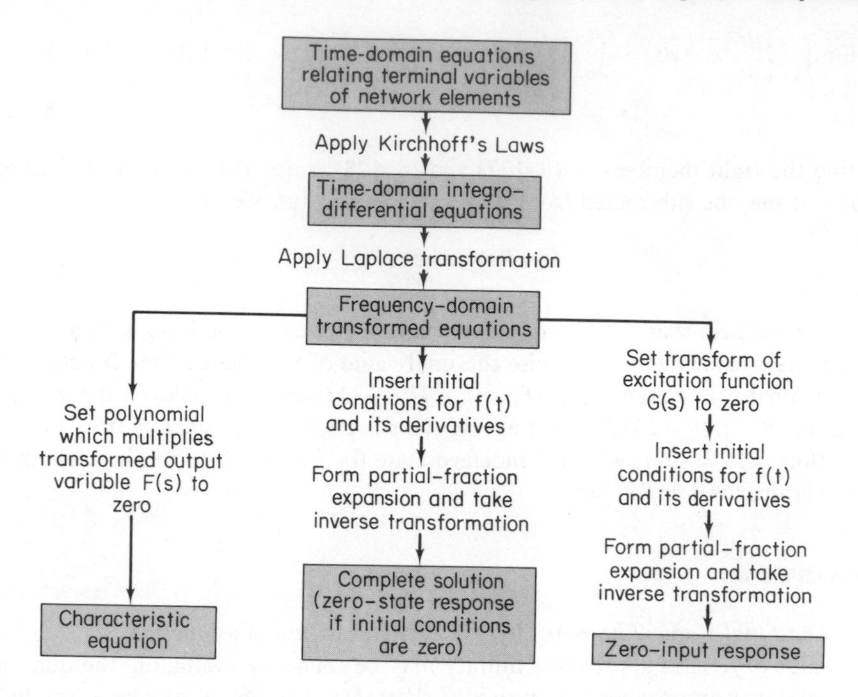

Fig. 8-4.1 Flow chart of Laplace transformation solution procedure.

8-5 DIGITAL COMPUTER METHODS FOR TAKING THE INVERSE LAPLACE TRANSFORMATION

In the last section we presented various techniques for finding the inverse Laplace transform of a rational function $F(s)$, where s is the complex frequency variable. In this section we shall show how the various operations of finding the partial-fraction expansion and evaluating the inverse transform may be implemented by digital computational procedures.

Let us consider a general rational function $F(s)$ having the form

$$F(s) = \frac{C(s)}{B(s)} = \frac{c_1 + c_2 s + c_3 s^2 + \cdots + c_m s^{m-1} + c_{m+1} s^m}{b_1 + b_2 s + b_3 s^2 + \cdots + b_n s^{n-1} + b_{n+1} s^n} \tag{8-130}$$

where the coefficients c_i and b_i are real numbers. In general, such a function will be the transform of some response variable. Thus, it will represent the product of a network function and some transformed excitation variable. The basic problem to be implemented is to determine the inverse transform $f(t)$. The first step in accomplishing this is to check the infinite frequency behavior of the function. If $m \geq n$, then infinite frequency terms corresponding with an impulse, a doublet, etc. in the time domain must be removed. This is simply done by long division. The first step of such a process may be shown as follows:

$$
\begin{array}{r|l|l}
b_{n+1}s^n + b_n s^{n-1} + \cdots & c_{m+1}s^m + c_m s^{m-1} + \cdots & \dfrac{c_{m+1}}{b_{n+1}} s^{m-n} + \cdots \\
& c_{m+1}s^m + \dfrac{c_{m+1}}{b_{n+1}} b_n s^{m-1} + \cdots & \\
\hline
& \left[c_m - \dfrac{c_{m+1}}{b_{n+1}} b_n\right] s^{m-1} + \cdots &
\end{array}
$$

Thus, from this first step, we obtain a result of the form

$$
F(s) = \frac{C(s)}{B(s)} = \frac{c_{m+1}}{b_{n+1}} s^{m-n} + \frac{a_m s^{m-1} + a_{m-1}s^{m-2} + \cdots}{B(s)} \tag{8-131}
$$

where $a_m = c_m - (c_{m+1}b_n/b_{n+1})$, etc. The polynomial $A(s) = a_m s^{m-1} + a_{m-1}s^{m-2} + \cdots$ may simply be considered as a new denominator of degree $m - 1$, and, if $(m - 1) \geq n$, then the long division process may be continued by letting $C(s) = A(s)$ and finding a new remainder polynomial $A(s)$. The process is continued until a numerator polynomial $A(s)$ of degree $n - 1$ is obtained. The algorithm is easily implemented by a digital computer subroutine. Let the name of such a subprogram be PDIV (for Polynomial DIVision). If we use a single-subscripted array C to store the coefficients of the numerator polynomial, an array B to store the coefficients of the denominator polynomial, an array G to store the coefficients of the infinite behavior terms, and an array A to store the coefficients of the polynomial $A(s)$, then the subroutine may be identified by a statement of the form

SUBROUTINE PDIV (M, C, N, B, KI, G, KR, A) (8-132)

where M and N are the degree of the numerator polynomial of $F(s)$ and the degree of the denominator polynomial, respectively, KI is the difference between M and N, and KR = N − 1. A flow chart and a listing for the subroutine are shown in Figs. 8-5.1 and 8-5.2, respectively. A summary of the characteristics of the subroutine is given in Table 8-5.1 (on page 610). It should be noted that the subroutine, in effect, takes a rational function $F(s)$ of the form given in Eq. (8-130), where the degree of the numerator polynomial $C(s)$ may be greater than that of $B(s)$, and puts it in the form

$$
F(s) = \frac{C(s)}{B(s)} = G(s) + \frac{A(s)}{B(s)} \tag{8-133}
$$

where the degree of $A(s)$ is less than the degree of $B(s)$. From this point on, we shall consider that the subroutine PDIV has been used and, thus, that the form of $F(s)$ is $A(s)/B(s)$, where m is the degree of $A(s)$, n is the degree of $B(s)$, and $m < n$.

The next step in finding the inverse transform of $F(s)$ is to determine the poles of the function $F(s)$, i.e., the roots of the polynomial $B(s)$. Such a root-finding problem is one frequently encountered in many mathematical and scientific situations. As a result, sophisticated computer programs for finding these roots have been de-

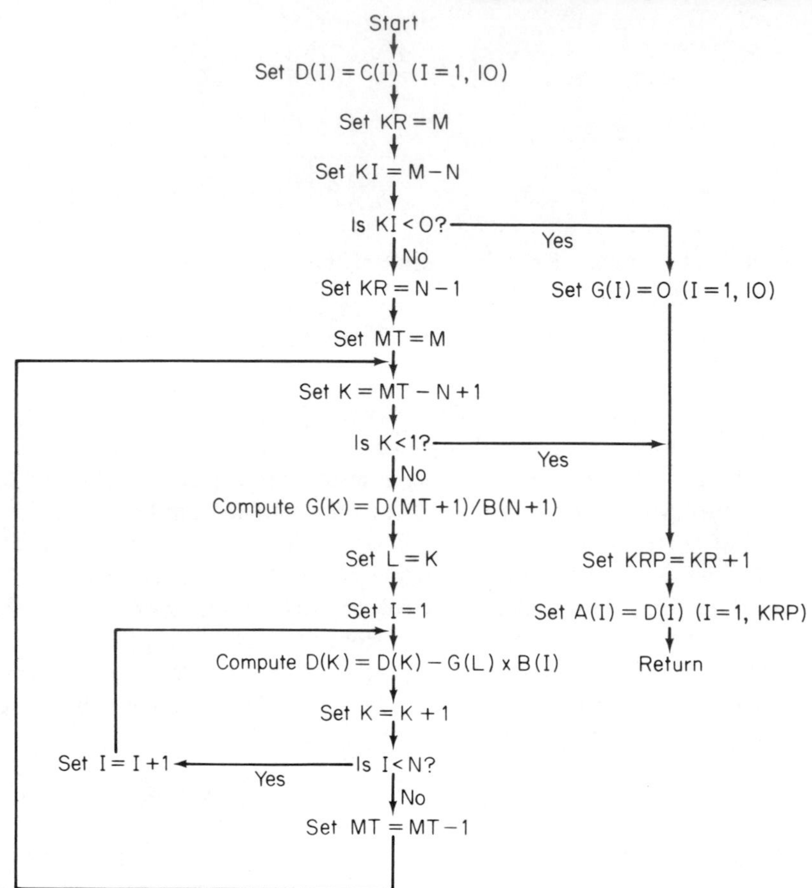

Fig. 8-5.1 Flow chart for the subroutine PDIV.

veloped and are usually available in the library of programs maintained at any computer center. We shall not attempt to match the capabilities of such large-scale computational tools here. However, in order to make the software package accompanying this text as complete as possible, a simple root-solving subroutine is included. This subroutine compiles efficiently, and should prove adequate for problems of the scope considered here. The program is designed to find the zeros of a polynomial $B(s)$, where

$$B(s) = b_1 + b_2 s + b_3 s^2 + \cdots + b_n s^{n-1} + b_{n+1} s^n \tag{8-134}$$

The subroutine is called ROOT and it is identified by a statement of the form

$$\text{SUBROUTINE ROOT (N, B, P)} \tag{8-135}$$

where N is the degree of the polynomial $B(s)$, B is the single-subscripted array containing the coefficients of the polynomial, and P is a single-subscripted complex array

```
      SUBROUTINE PDIV (M, C, N, B, KI, G, KR, A)
C     SUBROUTINE TO REMOVE SINGULARITIES AT INFINITY FROM A RATIO OF
C     POLYNOMIALS C(S)/B(S). THIS RATIO IS EXPRESSED AS G(S) + A(S)/B(S)
C     WHERE ALL POLYNOMIALS HAVE THE FORM A(1)+A(2)*S+A(3)*S**2+...
C     THE ORIGINAL POLYNOMIALS ARE ALL PRESERVED
C            M - DEGREE OF NUMERATOR POLYNOMIAL C(S)
C            C - ARRAY OF NUMERATOR POLYNOMIAL COEFFICIENTS
C            N - DEGREE OF DENOMINATOR POLYNOMIAL B(S)
C            B - ARRAY OF DENOMINATOR POLYNOMIAL COEFFICIENTS
C           KI - DEGREE OF QUOTIENT POLYNOMIAL G(S)  (= M - N)
C            G - OUTPUT ARRAY OF QUOTIENT POLYNOMIAL COEFFICIENTS
C           KR - DEGREE OF REMAINDER POLYNOMIAL A(S)  (= N - 1)
C            A - OUTPUT ARRAY OF REMAINDER POLYNOMIAL COEFFICIENTS
      DIMENSION C(10), B(10), G(10), A(10), D(10)
C
C     STORE THE COEFFICIENTS OF C(S) IN THE D ARRAY
      DO 2    I = 1, 10
    2 D(I) = C(I)
      KR = M
      KI = M - N
      IF (KI.LT.0) GO TO 17
      KR = N - 1
      MT = M
    8 K = MT - N + 1
      IF (K.LT.1) GO TO 19
C
C     COMPUTE THE KTH COEFFICIENT OF THE POLYNOMIAL G(S)
      G(K) = D(MT+1) / B(N+1)
      L = K
C
C     COMPUTE THE COEFFICIENTS OF THE REDUCED POLYNOMIAL D
      DO 14   I = 1, N
      D(K) = D(K) - G(L) * B(I)
   14 K = K + 1
      MT = MT - 1
      GO TO 8
C
C     IF THE NUMERATOR IS LESS IN DEGREE THAN THE DENOMINATOR
C     SET THE COEFFICIENTS OF THE G ARRAY TO ZERO
   17 DO 18   I = 1, 10
   18 G(I) = 0.0
      KI = 0
   19 KRP = KR + 1
C
C     STORE THE COEFFICIENTS OF THE REDUCED POLYNOMIAL D IN
C     THE A ARRAY AS THE REMAINDER POLYNOMIAL
      DO 21   I = 1, KRP
   21 A(I) = D(I)
      RETURN
      END
```

Fig. 8-5.2 Listing for the subroutine PDIV.

containing the roots of the polynomial. A detailed description of the operation of this subroutine would take us too far from our primary goal in this section. However, a brief description, together with a listing of the subroutine, is given in Appendix C. A summary of the characteristics of the subroutine ROOT is given in Table 8–5.2 (on page 611).

In the preceding paragraphs, as the first steps in determining the inverse transform of the function $F(s)$ of (8–130), we have discussed the problem of removing poles and constants at infinity from the function and the problem of factoring the denominator polynomial $B(s)$, i.e., of finding the poles of the function $F(s)$. The next problem which we shall consider is that of finding the residues of $F(s)$ at its various finite poles. First let us consider the problem of finding the residues for *simple* poles. We may develop an alternative expression for the one given in the preceding section by first considering a factored form for the denominator of the function $F(s)$. Thus, we may write

$$F(s) = \frac{A(s)}{B(s)} = \frac{A(s)}{b_{n+1}\prod_{i=1}^{n}(s - p_i)} \tag{8–136}$$

where n is the degree of the denominator polynomial $B(s)$, i.e., it is the number of finite poles which $F(s)$ possesses, and where the p_i are the locations of the poles. These latter may be real or complex, but it is assumed that they are simple. If we let K_1 be the residue for the pole at p_1, then, from Eq. (8–96), we may write

$$K_1 = \left.\frac{A(s)}{b_{n+1}\prod_{i=2}^{n}(s - p_i)}\right|_{s=p_1} \tag{8–137}$$

Now let us consider what happens if we differentiate the denominator polynomial $B(s)$, using the form given in the right member of Eq. (8–136). Performing this operation, we obtain

$$\begin{aligned}\frac{d}{ds}B(s) = b_{n+1}\Bigg[&\prod_{i=2}^{n}(s - p_i) + (s - p_1)\prod_{i=3}^{n}(s - p_i) \\ &+ (s - p_1)(s - p_2)\prod_{i=4}^{n}(s - p_i) + \cdots + \prod_{i=1}^{n-1}(s - p_i)\Bigg]\end{aligned} \tag{8–138}$$

Note that in this expression there is a term $(s - p_1)$ in all the terms inside the brackets *except the first term.* Thus, if we evaluate both members of Eq. (8–138) at $s = p_1$, all the terms except the first go to zero, and we may write

$$\left.\frac{d}{ds}B(s)\right|_{s=p_1} = b_{n+1}\left.\prod_{i=2}^{n}(s - p_i)\right|_{s=p_1} \tag{8–139}$$

The right member of this expression, however, is identical with the denominator of the right member of Eq. (8–137). Thus, we conclude that we may write the expression for the residue K_1 of a simple pole located at p_1 in the form

$$K_1 = \left.\frac{A(s)}{(d/ds)B(s)}\right|_{s=p_1} \tag{8–140}$$

This conclusion is easily extended to the general case. The results are summarized as follows:

SUMMARY 8-5.1

Finding the Residue of a Simple Pole: For a rational function $F(s) = A(s)/B(s)$, the residue K_i of a simple pole located at p_i is

$$K_i = \frac{A(s)}{(d/ds)B(s)}\bigg|_{s=p_i} \tag{8-141}$$

where K_i will be real if p_i is real, and complex if p_i is complex.

To implement the relation given in Eq. (8-141) on the digital computer, two operations are required. These are the operation of differentiation of a polynomial and the operation of evaluating a polynomial at a specific value of its argument. Let us consider the differentiation operation first. Let $A(s)$ be a polynomial of degree n which has the form

$$A(s) = a_1 + a_2 s + a_3 s^2 + a_4 s^3 + \cdots + a_n s^{n-1} + a_{n+1} s^n \tag{8-142}$$

If we let $B(s)$ be the derivative of $A(s)$ with respect to s, then we may write

$$\begin{aligned} B(s) &= b_1 + b_2 s + b_3 s^2 + \cdots + b_n s^{n-1} \\ &= a_2 + 2a_3 s + 3a_4 s^2 + \cdots + n a_{n+1} s^{n-1} \end{aligned} \tag{8-143}$$

Thus, we see that the coefficients b_i are defined by the relation

$$b_i = i a_{i+1} \qquad (i = 1, 2, 3, \ldots, n) \tag{8-144}$$

These relations are easily implemented by a digital computer subroutine. Let DIFPL (for DIFerentiation of a PoLynomial) be the name of the subroutine, let A be a single-subscripted array with elements A(I) containing the coefficients of the original polynomial, and let N be the degree of the polynomial. Similarly, let B be an array containing the coefficients of the resulting differentiated polynomial, and M be its degree. The subroutine may be identified by a FORTRAN statement of the form

SUBROUTINE DIFPL (N, A, M, B) (8-145)

A flow chart for the logic of such a subroutine is given in Fig. 8-5.3. A listing of a subroutine which will perform such an operation is given in Fig. 8-5.4. The subroutine's characteristics are summarized in Table 8-5.3 (on page 611).

The second operation listed above, namely, the evaluation of a polynomial at a given value of its complex argument, has already been implemented for the digital

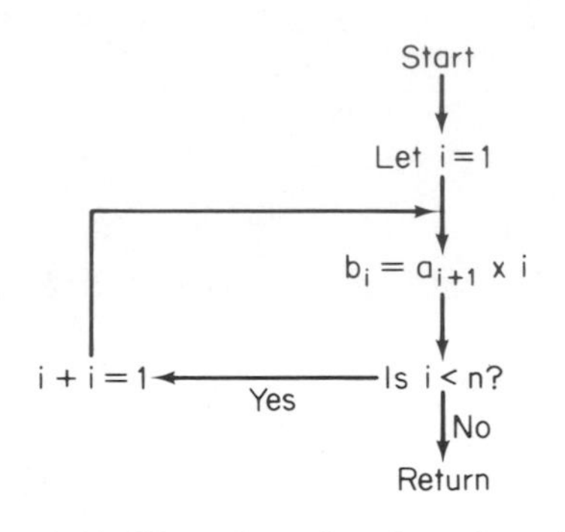

Fig. 8-5.3 Flow chart for the subroutine DIFPL.

computer in connection with the determination of sinusoidal steady-state response given in Sec. 7–10. This is the function of the subroutine CVALPL, whose characteristics are summarized in Table 7–10.1.

We may readily apply the subroutines DIFPL and CVALPL described above to the problem of finding the residue of a given function $F(s)$ at a simple pole as defined in Eq. (8–141). The process is illustrated by the flow chart shown in Fig. 8–5.5. Before describing the implementation of these operations in a digital computer program, let us investigate the more complex problem of finding the constants in a partial-fraction expansion for the terms associated with a *second*-order pole. This is done in the following paragraphs.

```
      SUBROUTINE DIFPL (N, A, M, B)
C     SUBROUTINE FOR DIFFERENTIATING A POLYNOMIAL OF THE FORM
C     A(1) + A(2)*P + A(3)*P**2 + ... + A(N+1)*P**N
C     TO OBTAIN THE RESULTING (DIFFERENTIATED POLYNOMIAL
C     B(1) + B(2)*P + B(3)*P**2 + ... + A(M+1)*P**M
C           N - DEGREE OF INPUT POLYNOMIAL
C           A - ARRAY OF COEFFICIENTS OF INPUT POLYNOMIAL
C           M - DEGREE OF DIFFERENTIATED POLYNOMIAL (=N-1)
C           B - OUTPUT ARRAY OF COEFFICIENTS OF DIFFERENTIATED
C                   POLYNOMIAL
      DIMENSION A(10), B(10)
      DO 3  I = 1, N
      C = I
    3 B(I) = A(I+1) * C
      M = N - 1
      RETURN
      END
```

Fig. 8-5.4 Listing of the subroutine DIFPL.

Although it is possible to use the algorithm given in Summary 8–4.4 to determine the constants in the partial-fraction expansion for a function with a second-order pole, in general, it is difficult to develop a digital computer program which implements this approach. Therefore, instead, we shall use a method based on the location of the *zeros* of the rational function being expanded.[14] For this purpose, we may write the function in the form

[14]This method was originally presented by P. A. Payne in a manuscript entitled "The Graphical Evaluation of System Time Response."

$$F(s) = \frac{A(s)}{B(s)} = K\frac{\prod_{j=1}^{m}(s - z_j)}{\prod_{i=1}^{m}(s - p_i)}$$

$$= K\frac{\prod_{j=1}^{m}(s - z_j)}{(s - p_1)^2\prod_{i=3}^{n}(s - p_i)} \qquad (8\text{-}146)$$

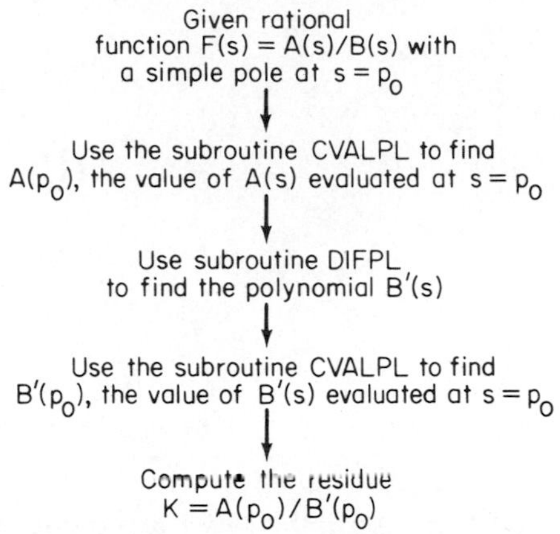

In the above expression, K is a multiplicative constant, the quantities z_i are the zeros of $F(s)$, and the quantities p_i are the poles. In addition, as indicated in the right member of the expression, we have assumed that there is a second-order pole located at $s = p_1$. The expression for the partial-fraction expansion of $F(s)$ may now be written in the form

Given rational function F(s) = A(s)/B(s) with a simple pole at s = p_0

↓

Use the subroutine CVALPL to find A(p_0), the value of A(s) evaluated at s = p_0

↓

Use subroutine DIFPL to find the polynomial B′(s)

↓

Use the subroutine CVALPL to find B′(p_0), the value of B′(s) evaluated at s = p_0

↓

Compute the residue K = A(p_0)/B′(p_0)

Fig. 8-5.5 Flow chart for finding the residue of a simple pole.

$$F(s) = \frac{K_{11}}{s - p_1} + \frac{K_{12}}{(s - p_1)^2} + \frac{C(s)}{D(s)} \qquad (8\text{-}147)$$

where $D(s)$ is the remainder of the polynomial $B(s)$ after the factor representing the second-order pole at p_1 has been removed. Thus, the function $C(s)/D(s)$ represents the remaining terms in the partial-fraction expansion. From the discussion given in Sec. 8-4, we may determine the value of the constant K_{12} as follows:

$$K_{12} = \left.\frac{A(s)}{D(s)}\right|_{s=p_1} \qquad (8\text{-}148)$$

To use this expression, however, we must first find $D(s)$. However, removal of the second-order factor $(s - p_1)^2$ from $B(s)$, which must be done in order to determine $D(s)$, is somewhat difficult to program. This is especially true for the case of complex roots. To avoid such an operation, we may extend the result given in Summary 8-5.1 as an alternative approach. Let us begin this extension by differentiating the denominator of the right member of Eq. (8-146). Thus, we obtain

$$\frac{d}{ds}B(s) = 2(s - p_1)\prod_{i=3}^{n}(s - p_i) + (s - p_1)^2\prod_{i=4}^{n}(s - p_i)$$
$$+ (s - p_1)^2(s - p_3)\prod_{i=5}^{n}(s - p_i) + \cdots \qquad (8\text{-}149)$$

Now let us differentiate this result a second time. Thus, we obtain

$$\frac{d^2}{ds^2}B(s) = 2\prod_{i=3}^{n}(s-p_i) + 2(s-p_1)\prod_{i=4}^{n}(s-p_i)$$
$$+ 2(s-p_1)(s-p_3)\prod_{i=5}^{n}(s-p_i) + \cdots$$
$$+ 2(s-p_1)\prod_{i=4}^{n}(s-p_1)^2\prod_{i=5}^{n}(s-p_i)$$
$$+ (s-p_1)^2(s-p_4)\prod_{i=6}^{n}(s-p_i) + \cdots$$
$$+ 2(s-p_1)(s-p_3)\prod_{i=5}^{n}(s-p_i) + (s-p_1)^2\prod_{i=5}^{n}(s-p_i)$$
$$+ (s-p_1)^2(s-p_3)\prod_{i=6}^{n}(s-p_i) + \cdots \tag{8-150}$$

At first glance, this latter expression appears rather formidable. Each group of quantities, however, simply represents the result of the term-by-term differentiation of each of the quantities shown in the right member of Eq. (8–149). If we now let $s = p_1$ in Eq. (8–150), *every term except the first disappears.* The result may be written as follows

$$\left.\frac{d^2}{ds^2}B(s)\right|_{s=p_1} = 2\prod_{i=3}^{n}(s-p_i) \tag{8-151}$$

The right member of this last equation, however, is simply equal to two times the value of the denominator polynomial of Eq. (8–148), i.e., it equals $2D(p_1)$. Thus, Eq. (8–148) may be written in the form

$$K_{12} = \left.\frac{2A(s)}{(d^2/ds^2)B(s)}\right|_{s=p_1} \tag{8-152}$$

This relation provides an easily programmed algorithm for determining the constant K_{12} associated with a second-order pole at p_1. It is readily implemented using the subroutines DIFPL and CVALPL already defined.

Now let us consider the determination of the residue K_{11} of (8–147). As we originally mentioned, the use of the relation given in Summary 8–4.4 is not particularly convenient. Instead, we will use the expression

$$K_{11} = K_{12}\left[\sum_{j=1}^{m}\frac{1}{s-z_j} - \sum_{i=3}^{n}\frac{1}{s-p_i}\right] \tag{8-153}$$

where the quantities z_j and p_i are defined in Eq. (8–146). To avoid detracting from the main argument being presented in this section, a discussion of the proof of this result has been placed in Appendix C. The overall logic described above for finding the constants associated with the first- and second-order terms of the partial-fraction

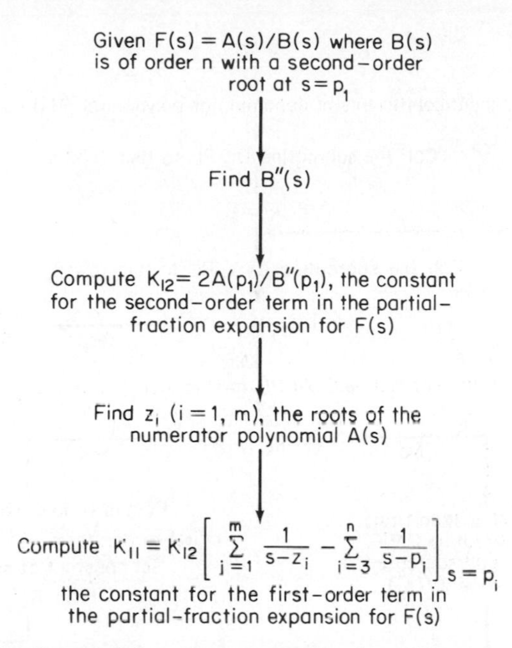

Fig. 8-5.6 Simplified flow chart for finding the residues of second-order poles.

expansion for a second-order pole is summarized in the flow chart shown in Fig. 8-5.6.

We may now implement the general logic defined in the flow charts shown in Figs. 8-5.5 and 8-5.6 (for first- and second-order poles, respectively) into a detailed flow chart suitable for digital computer implementation as a subroutine. Such a flow chart is shown in Fig. 8-5.7. The first step of the procedure is to store the coefficients of the polynomial $B(s)$ (which are stored in the B array) in a second array C, with elements C(I). This is necessary so that operations may be made on this array without destroying the original set of coefficients, which thus are still available for use in other portions of the program which calls this subroutine. In defining the variables used in the flow chart (and in the subroutine), the suffix D has been used to indicate differentiation. Thus, the quantities CD(I) are the coefficients of the derivative of the polynomial whose coefficients are stored as C(I). Similarly, the quantities CDD(I) are the coefficients of the second derivative of the polynomial whose coefficients are stored as C(I). In a similar fashion, the suffix V has been used to indicate the complex value. Thus, CV is the complex value of the polynomial whose coefficients are stored as C(I) when evaluated at the pole location, and CDV is the complex value of the derivative of the polynomial, etc. In Fig. 8-5.8, a listing is given of a subroutine which implements the logic defined in the flow chart of Fig. 8-5.7. The subroutine is named PFEXSD (for Partial Fraction EXpansion for Simple and Double poles), and the identifying statement is

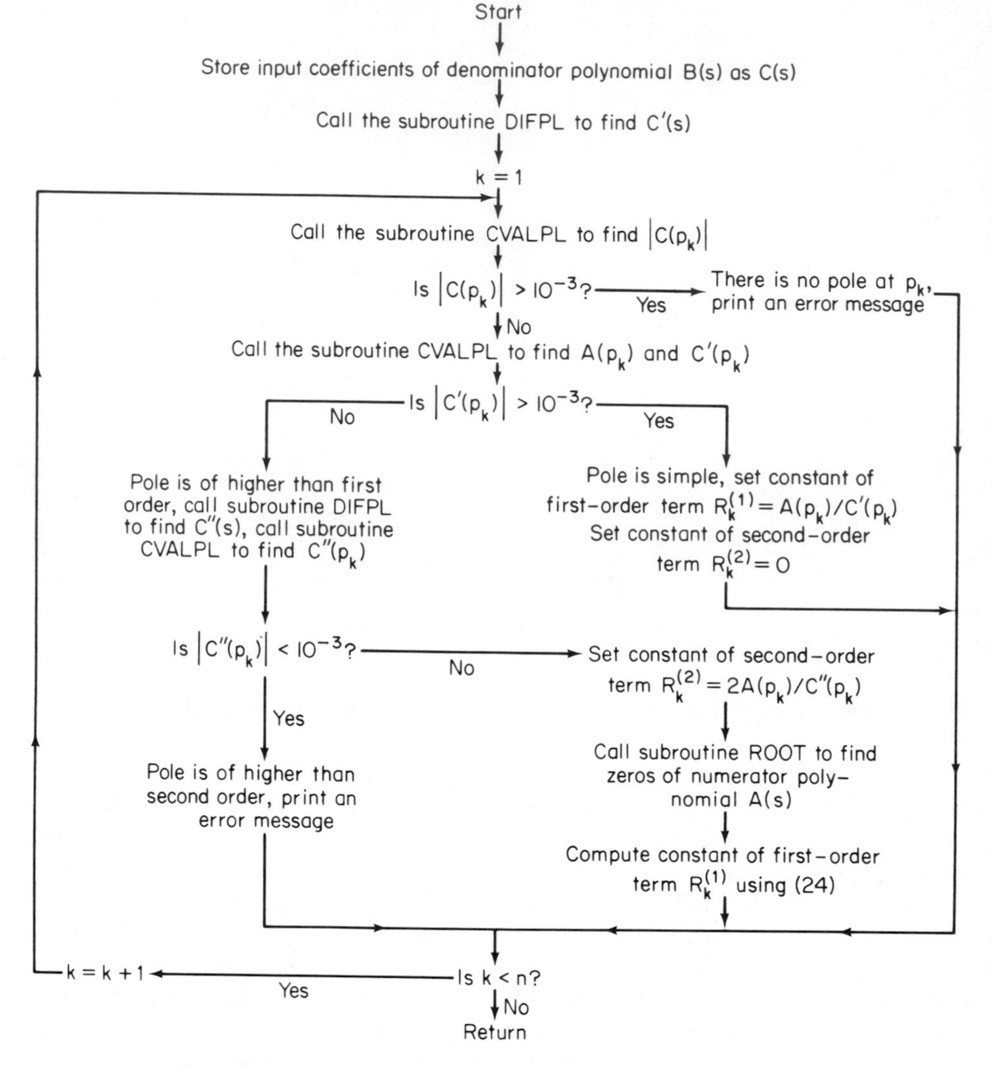

Fig. 8-5.7 Flow chart for finding the residues of first- and second-order poles.

$$\text{SUBROUTINE PFEXSD (M, A, N, B, P, R1, R2)} \tag{8–154}$$

where the coefficients of the numerator polynomial $A(s)$ of degree M are stored as the quantities A(I), the coefficients of the denominator polynomial $B(s)$ of degree N are stored as B(I), P is the complex array of the pole locations of the function, and R1 and R2 are the complex arrays of the coefficients of the partial-fraction expansion terms. A summary of the characteristics of the subroutine PFEXSD is given in Table 8–5.4 (on page 612).

```
      SUBROUTINE PFEXSD (M, A, N, B, P, R1, R2)
C     SUBROUTINE FOR FINDING THE COEFFICIENTS IN THE PARTIAL
C     FRACTION EXPANSION FOR SIMPLE AND DOUBLE POLES OF A
C     RATIONAL FUNCTION F(S)=A(S)/B(S) WHERE THE POLYNOMIALS
C     HAVE THE FORM A(1) + A(2)*S + A(3)*S**2+ ...
C            M - DEGREE OF NUMERATOR POLYNOMIAL A(S)
C            A - ARRAY OF COEFFICIENTS OF POLYNOMIAL A(S)
C            N - DEGREE OF DENOMINATOR POLYNOMIAL B(S)
C            B - ARRAY OF COEFFICIENTS OF POLYNOMIAL B(S)
C            P - ARRAY OF POLES FOR WHICH RESIDUES ARE TO BE FOUND
C           R1 - RESIDUE OF A SIMPLE POLE, COEFFICIENT OF FIRST
C                          DEGREE TERM FOR A DOUBLE POLE
C           R2 - RESIDUE FOR A DOUBLE POLE (SET TO ZERO FOR A
C                          SIMPLE POLE)
C     THE COEFFICIENTS OF THE POLYNOMIALS A(S) AND B(S) ARE PRESERVED
      COMPLEX P(10), PA(10), R1(10), R2(10), CV, AV, CDV, CDDV, SA, SB
      DIMENSION A(10), B(10), C(10), CD(10), CDD(10)
C
C     CONSTRUCT THE POLYNOMIAL C(S) EQUAL TO B(S) AND FIND
C     ITS DERIVATIVE
      NP = N + 1
      DO 3  I = 1, NP
    3 C(I) = B(I)
      CALL DIFPL (N, C, NM, CD)
C
C     PERFORM THE PARTIAL-FRACTION EXPANSION PROCEDURE N TIMES
      DO 49  K = 1, N
C
C     TEST TO MAKE CERTAIN THAT S = P(K) IS A ZERO OF C(S)
   17 CALL CVALPL (N, C, P(K), CV)
      IF (CABS(CV).GT.0.001)  GO TO 42
      CALL CVALPL (M, A, P(K), AV)
C
C     TEST THE DERIVATIVE OF C(S) TO DETERMINE THE ORDER
C     OF THE ZERO AT S = P(K)
      CALL CVALPL (NM, CD, P(K), CDV)
      IF (CABS(CDV).LT.0.001)  GO TO 25
C
C     IF THE ZERO OF C(S) IS SIMPLE, COMPUTE THE RESIDUE
C     R1, SET R2 TO ZERO, AND RETURN
      R1(K) = AV / CDV
      R2(K) = (0.0,0.0)
      GO TO 49
C
C     TEST TO MAKE CERTAIN THAT THE ZERO OF C(S) IS NOT
C     OF HIGHER DEGREE THAN SECOND
   25 CALL DIFPL (NM, CD, NC, CDD)
      CALL CVALPL (NC, CDD, P(K), CDDV)
      IF (CABS(CDDV).LT.0.001)  GO TO 45
C
C     IF THE ZERO OF C(S) IS DOUBLE, COMPUTE THE COEFFICIENTS R1 AND R2
C     FOR THE TERMS IN THE PARTIAL FRACTION EXPANSION
      R2(K) = 2. * AV / CDDV
      PRK = REAL(P(K))
      PIK = AIMAG(P(K))
      SA = (0.,0.)
      IF (M.EQ.0)  GO TO 34
      CALL ROOT (M, A, PA)
      DO 33  I = 1, M
```

Fig. 8-5.8 Listing of the subroutine PFEXSD.

```
33 SA = SA + (1.,0.) / (P(K) - PA(I))
34 SB = (0.,0.)
   DO 39 I = 1, N
   IF (ABS(PRK - REAL(P(I))).GT.0.001) GO TO 38
   IF (ABS(PIK - AIMAG(P(I))).LT.0.001) GO TO 39
38 SB = SB + (1.,0.) / (P(K) - P(I))
39 CONTINUE
   R1(K) = R2(K) * (SA - SB)
   GO TO 49
C
C  PRINT ANY ERROR MESSAGES THAT ARE REQUIRED
42 PRINT 43
43 FORMAT (/40H0FUNCTION F(S) DOES NOT HAVE A POLE AT P/)
   GO TO 47
45 PRINT 46
46 FORMAT (/42H0POLE OF F(S) IS OF GREATER THAN 2ND ORDER/)
47 PRINT 48, P(K)
48 FORMAT (4H P =,E11.3,3H +J,E11.3/)
49 CONTINUE
   RETURN
   END
```

Fig. 8-5.8 Continued

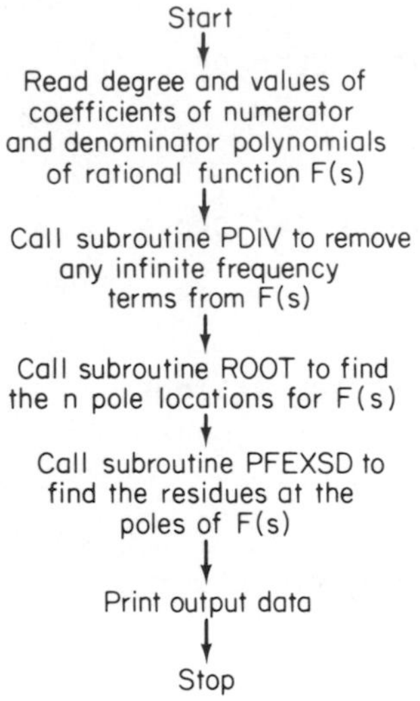

Start
↓
Read degree and values of coefficients of numerator and denominator polynomials of rational function F(s)
↓
Call subroutine PDIV to remove any infinite frequency terms from F(s)
↓
Call subroutine ROOT to find the n pole locations for F(s)
↓
Call subroutine PFEXSD to find the residues at the poles of F(s)
↓
Print output data
↓
Stop

Fig. 8-5.9 Flow chart for a program for finding a partial-fraction expansion.

The various subroutines described so far in this section are easily assembled into a program for finding the partial-fraction expansion for an arbitrary function $F(s)$. A flow chart for such a program is shown in Fig. 8–5.9. After data have been read in, the subroutine PDIV is used to determine the constants associated with any singularity functions such as impulses, doublets, etc., that may be present. Then the subroutine ROOT is used to find the locations of the poles. Finally, the subroutine PFEXSD is used to determine the residues of these poles. A listing of a program which implements the logic of the flow chart is shown in Fig. 8–5.10(a). As an example of the application of this program, consider the function

$$F(s) = \frac{2 + 19s + 60s^2 + 82s^3 + 64s^4 - 30s^5 - 8s^6 - s^7}{8 + 28s + 44s^2 + 40s^3 + 22s^4 + 7s^5 + s^6} \tag{8–155}$$

Listings of the input data and the output data for this function are given in Fig. 8–5.10(b). From this data, we see that $F(s)$ of Eq. (8–155) may be written in the form

```
C      MAIN PROGRAM FOR FINDING PARTIAL FRACTION EXPANSION
C                 M - DEGREE OF NUMERATOR POLYNOMIAL
C                 N - DEGREE OF DENOMINATOR POLYNOMIAL
C                 C - ARRAY OF NUMERATOR POLYNOMIAL COEFFICIENTS
C                 B - ARRAY OF DENOMINATOR POLYNOMIAL COEFFICIENTS
C                 P - COMPLEX ARRAY OF POLE LOCATIONS
C                R1 - COMPLEX ARRAY OF RESIDUES FOR 1ST-ORDER POLES
C                R2 - COMPLEX ARRAY OF CONSTANTS FOR 2ND-ORDER POLES
       DIMENSION C(10), B(10), G(10), A(10)
       COMPLEX P(10), R1(10), R2(10)
C
C      READ THE DATA FOR THE NUMERATOR AND DENOMINATOR POLYNOMIALS
       READ  2, M, N
     2 FORMAT (2I2)
       MP = M + 1
       READ  5, (C(I), I = 1, MP)
     5 FORMAT (8E10.0)
       NP = N + 1
       READ  5, (B(I), I = 1, NP)
C
C      CALL SUBROUTINE PDIV TO REMOVE THE INFINITE FREQUENCY TERMS
       CALL PDIV (M, C, N, B, KI, G, KR, A)
       IF (KI.LT.0) GO TO 13
       KIP = KI + 1
       PRINT 12, (G(I), I = 1, KIP)
    12 FORMAT (/17H INFINITE TERMS =, 5E13.5)
C
C      CALL SUBROUTINE ROOT TO FIND THE POLE LOCATIONS
    13 CALL ROOT (N, B, P)
    14 PRINT 15
    15 FORMAT (/7X,13HPOLE LOCATION,7X,25HRESIDUE (OF 1ST DEG TERM)
      1,2X,25HRESIDUE (OF 2ND DEG TERM))
C
C      CALL SUBROUTINE PFEXSD TO FIND THE RESIDUES
       CALL PFEXSD (KR, A, N, B, P, R1, R2)
C
C      PRINT THE OUTPUT
       DO 20 I = 1, N
       PRINT 19, P(I), R1(I), R2(I)
    19 FORMAT (1X,3(E11.4,2H+J,E11.4,2X))
    20 CONTINUE
       STOP
       END
```

```
INPUT DATA FOR PARTIAL-FRACTION EXPANSION PROBLEM

 7 6
2.0       19.0      60.0      82.0      64.0      30.0      8.0       1.0
8.        28.       44.       40.       22.       7.        1.
```

```
OUTPUT DATA FOR PARTIAL-FRACTION EXPANSION PROBLEM

 INFINITE TERMS =  1.00000E+00  1.00000E+00

       POLE LOCATION         RESIDUE (OF 1ST DEG TERM)  RESIDUE (OF 2ND DEG TERM)
-2.0000E+00+J 0.             1.0000E+00+J 0.            0.         +J 0.
-1.0000E+00+J 0.             2.0000E+00+J 0.            0.         +J 0.
-1.0000E+00+J-1.0000E+00    -9.9999E-01+J-2.0000E+00    5.0000E-01+J-2.5000E-01
-1.0000E+00+J 1.0000E+00    -9.9999E-01+J 2.0000E+00    5.0000E-01+J 2.5000E-01
-1.0000E+00+J-1.0000E+00    -9.9999E-01+J-2.0000E+00    5.0000E-01+J-2.5000E-01
-1.0000E+00+J 1.0000E+00    -9.9999E-01+J 2.0000E+00    5.0000E-01+J 2.5000E-01
```

Fig. 8-5.10 Listing of a program for finding a partial-fraction expansion.

$$F(s) = \frac{-1 - j2}{s + 1 + j1} + \frac{-1 + j2}{s + 1 - j1} + \frac{(\frac{1}{2}) - j(\frac{1}{4})}{(s + 1 + j1)^2} + \frac{(\frac{1}{2}) + j(\frac{1}{4})}{(s + 1 - j1)^2}$$
$$+ \frac{2}{s + 1} - \frac{2}{s + 2} + s + 1 \tag{8-156}$$

Thus the function contains a pole at infinity, a second-order complex conjugate pole, and two real poles.

As a final operation which may conveniently be implemented by digital computational techniques, let us consider the problem of determining the time-domain waveshape of a function for which the partial-fraction expansion is known. To do this, suppose we have a function $F(s)$ which has m distinct poles of first or second order located at $s = p_i$ $(i = 1, 2, \ldots, m)$ and which has a partial-fraction expansion of the form

$$F(s) = \sum_{i=1}^{m} F_i(s) = \sum_{i=1}^{m} \left[\frac{R_i^{(1)}}{s - p_i} + \frac{R_i^{(2)}}{(s - p_i)^2} \right] \tag{8-157}$$

where the quantities $R_i^{(1)}$ and $R_i^{(2)}$ are the constants associated with the first- and second-order terms for the ith pole. Our problem now is to find $f(t) = L^{-1}[F(s)]$. In computing the inverse transformation of such a function, the individual terms given in the summation in Eq. (8-157) fall into one of four cases. The first of these may be defined as the case where p_i is real and simple. In this case, $R_i^{(2)}$ is zero and the corresponding time-domain term $f_i(t) = L^{-1}[F_i(s)]$ will be

$$f_i(t) = R_i^{(1)} e^{p_i t} \tag{8-158}$$

Since, for this case, p_i has been defined as being real, $R_i^{(1)}$ will also be real. Thus, an expression which is equivalent to the one given above is

$$f_i(t) = \text{Re}\,(R_i^{(1)}) e^{t\,\text{Re}(p_i)} \tag{8-159}$$

This latter form will be more convenient for the digital computer implementation which will be discussed shortly. The second case to be considered may be defined as the one in which p_i is real and of second order. In this case, $R_i^{(2)}$ will be non-zero and real. Thus, the time-domain term in the inverse transformation corresponding with this pole may be written

$$f_i(t) = \text{Re}\,(R_i^{(1)}) e^{t\,\text{Re}(p_i)} + \text{t}\,\text{Re}\,(R_i^{(2)}) e^{t\,\text{Re}(p_i)} \tag{8-160}$$

The third case to be considered is the one in which p_i is simple and complex, i.e., it has a non-zero imaginary part. In such a case, there will be two conjugate terms in the partial-fraction expansion. If we let the conjugate poles be p_i and p_{i+1}, where $p_{i+1} = p_i^*$, then the time-domain behavior $f_i(t) + f_{i+1}(t) = \mathscr{L}^{-1}[F_i(s) + F_{i+1}(s)]$ is given as

$$f_i(t) + f_{i+1}(t) = R_i^{(1)}e^{p_i t} + R_i^{*(1)}e^{p_i^* t} \tag{8-161}$$

An equivalent form in which this equation may be written is

$$f_i(t) + f_{i+1}(t) = 2\,\mathrm{Re}\,(R_i^{(1)}e^{p_i t}) \tag{8-162}$$

The final case to bc considered is one in which p_i is of second order and also complex. In such a case, if we let $p_{i+1} = p_i^*$, the corresponding time-domain terms for the second-order pole and its conjugate may be expressed as follows

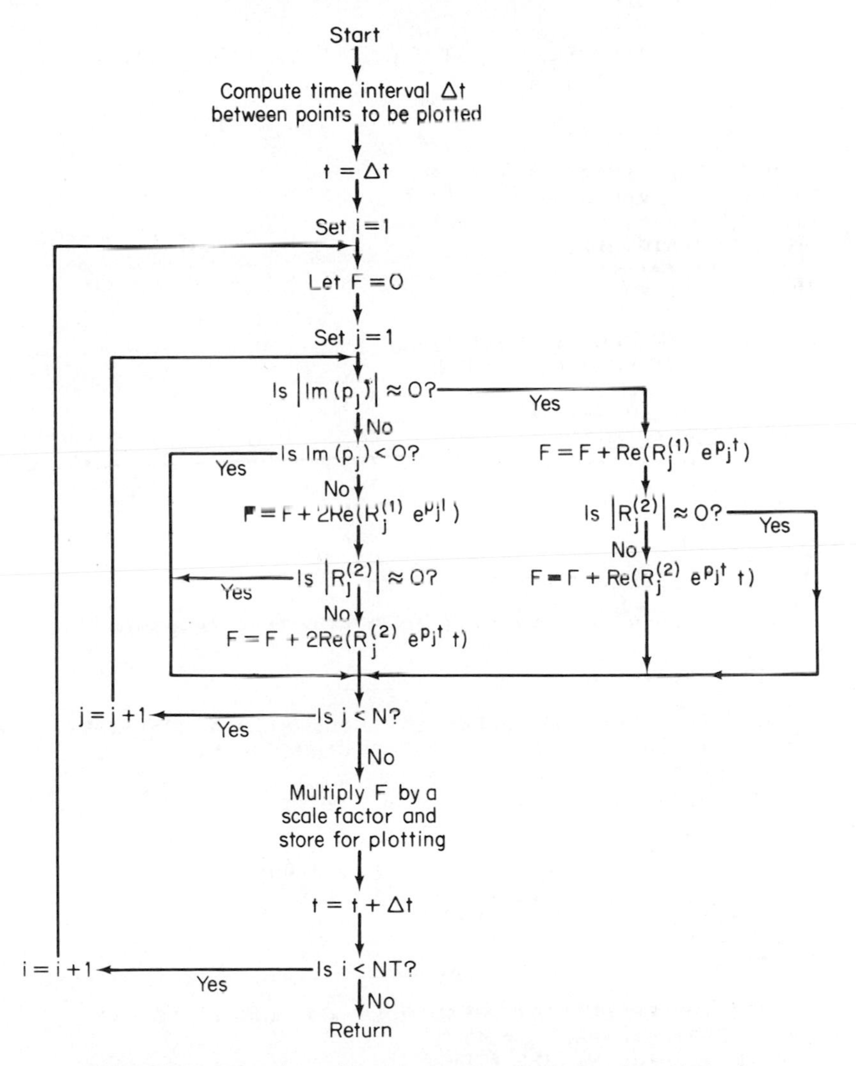

Fig. 8-5.11 Flow chart of a program for computing the values of an inverse transformation.

```
      SUBROUTINE ILPX (NT, TR, N, P, R1, R2, NA, A, SCALE)
C     SUBROUTINE FOR MAKING A PLOT OF THE INVERSE LAPLACE
C     TRANSFORM OF A PARTIAL FRACTION EXPANSION
C             NT - NUMBER OF VALUES OF T AT WHICH F(T) IS TO BE PLOTTED
C             TR - RANGE OF T OVER WHICH F(T) IS TO BE PLOTTED
C              N - NUMBER OF FINITE POLES IN F(S)
C              P - COMPLEX ARRAY OF POLE LOCATIONS
C             R1 - COMPLEX ARRAY OF RESIDUES FOR POLES
C             R2 - COMPLEX ARRAY OF CONSTANTS FOR SECOND-ORDER POLES
      DIMENSION A(5,101)
      COMPLEX P(10), R1(10), R2(10), TC, EPT
      XNT = NT
      DT = TR / XNT
      T = 0.
C
C     FOR EACH OF THE DESIRED VALUES OF T COMPUTE F(T)
      DO 32  I = 1, NT
      F = 0.
      TC = CMPLX(T, 0.0)
C
C     COMPUTE THE EFFECT OF EACH POLE AT
C     THE CURRENT VALUE OF T
      DO 30  J = 1, N
      PI = AIMAG(P(J))
      PR = REAL(P(J))
      IF(J.EQ.1) GO TO 17
C
C     TEST TO MAKE CERTAIN THIS POLE
C     IS DIFFERENT FROM ALL OTHERS
      DO 16 K = 2,J
      PKI = AIMAG(P(K-1))
      PKR = REAL(P(K-1))
      IF  (ABS(PR-PKR).GT.1.E-05) GO TO 16
      IF  (ABS(PI-PKI).LT.1.E-05) GO TO 30
   16 CONTINUE
C
C     TEST TO SEE IF THE POLE IS PURELY REAL
   17 IF  (ABS(PI).LT.1.E-05) GO TO 26
C
C     IF THE POLE IS COMPLEX TEST TO SEE IF ITS IMAGINARY
C     PART IS NEGATIVE
      IF  (PI.LT.0.0) GO TO 30
C
C     COMPUTE THE EFFECT OF A COMPLEX POLE WITH A POSITIVE
C     IMAGINARY PART AND ADD IT TO THE VALUE OF F(T)
      EPT = CEXP(TC * P(J))
      FR = REAL(R1(J) * EPT)
      F = F + 2. * FR
C
C     TEST TO SEE IF THE POLE IS OF SECOND ORDER
      IF  (CABS(R2(J)).LT.1.E-05) GO TO 30
      FR = REAL(R2(J) * TC * EPT)
      F = F + 2. * FR
      GO TO 30
C
C     COMPUTE THE EFFECT OF A REAL POLE AND ADD IT TO F(T)
   26 ER =  EXP(REAL(P(J)) * T)
      F =  F +  REAL(R1(J)) * ER
C
```

Fig. 8-5.12 Listing of the subroutine ILPX.

```
C     TEST TO SEE IF THE POLE IS OF SECOND ORDER
      IF (CABS(R2(J)).LT.1.E-05) GO TO 30
      F =  F + REAL (R2(J)) * ER * T
   30 CONTINUE
C
C     SCALE THE VALUE OF F(T) AND STORE IT FOR PLOTTING
      A(NA,I) = F * SCALE
   32 T = T + DT
      RETURN
      END
```

Fig. 8-5.12 Continued

```
C     MAIN PROGRAM FOR PLOTTING INVERSE LAPLACE TRANSFORM
C             N - DEGREE OF POLYNOMIAL
C             P - COMPLEX 1-D ARRAY OF POLE LOCATIONS
C            R1 - COMPLEX 1-D ARRAY OF RESIDUES OF FIRST DEGREE POLES
C            R2 - COMPLEX 1-D ARRAY OF CONSTANTS FOR 2ND DEGREE POLES
C             A - ARRAY USED TO STORE VALUES OF F(T) FOR PLOTTING
      DIMENSION A(5,101)
      COMPLEX P(10), R1(10), R2(10)
C
C     READ THE INPUT DATA FOR THE POLES AND THE RESIDUES
    1 READ 2, N
    2 FORMAT (I2)
      READ 4, (P(I), R1(I), R2(I), I = 1, N)
    4 FORMAT (6E10.0)
C
C     CALL SUBROUTINE ILPX TO COMPUTE THE VALUES OF F(T) AND
C     STORE THEM FOR PLOTTING
      CALL ILPX(61, 2.5, N, P, R1, R2, 1, A, 1.0)
      PRINT 7
    7 FORMAT (1H1)
C
C     CALL PLOTTING SUBROUTINE TO PLOT THE WAVEFORM OF F(T)
      CALL PLOT5(A, 1, 61, 50)
    9 STOP
      END

INPUT DATA
 2
-1.       +10.      0.0       -25.0     0.0       0.0
-1.       -10.      0.0       +25.0     0.0       0.0
```

Fig. 8-5.13 Main program for computing an inverse transformation.

$$f_i(t) + f_{i+1}(t) = R_i^{(1)} e^{p_i t} + R_i^{*(1)} e^{p_i^* t} + tR_i^{(2)} e^{p_i t} + tR_i^{*(2)} e^{p_i^* t} \qquad (8\text{–}163)$$

This relation may also be written in the form

$$f_i(t) + f_{i+1}(t) = 2 \operatorname{Re} (R_i^{(1)} e^{p_i t} + tR_i^{(2)} e^{p_i t}) \qquad (8\text{–}164)$$

The relations covering the four cases found above are readily implemented as a digital computer subroutine. Such a subroutine must perform a sequence of computations for a specified equally-spaced set of values of the variable t and store the results for plotting. A flow chart of the logic for such an operation is shown in Fig. 8–5.11. The relations used are those given in Eqs. (8–159), (8–160), (8–162), and (8–164). To implement this logic, the subroutine must be supplied with the following input arguments: NT, the number of values of t at which the inverse transformation is to be evaluated; TR, the total time range encompassed; N, the number of poles; the complex arrays P, R1, and R2 containing the pole locations and the values of the residues associated with the first- and second-degree terms in the partial-fraction expansion; NA, the location in the plotting array in which the data are to be stored; and SCALE, the scale factor to be used in preparing the data for plotting. If no scaling is desired, this input variable must be set to unity. The output argument of

---

TABLE 8-5.1 Summary of the Characteristics of the Subroutine PDIV

Identifying Statement: SUBROUTINE PDIV (M, C, N, B, KI, G, KR, A)

Purpose: To reduce a given quotient of polynomials $F(s)$ so it is expressed as a simple, proper, rational fraction plus a series of terms representing the behavior at infinity. Thus, it computes the coefficients of the polynomials $G(s)$ and $A(s)$ in the relation

$$F(s) = \frac{C(s)}{B(s)} = G(s) + \frac{A(s)}{B(s)}$$

where $C(s)$ and $B(s)$ are given, and all polynomials have the form

$$b_1 + b_2 s + b_3 s^2 + \cdots + b_n s^{n-1} + b_{n+1} s^n$$

Additional Subroutines Required: None.

Input Arguments:

- M The degree of the original numerator polynomial $C(s)$.
- C The one-dimensional array of variables C(I) in which are stored the values of the coefficients c_i of the polynomial $C(s)$.
- N The degree of the denominator polynomial $B(s)$.
- B The one-dimensional array of variables B(I) in which are stored the values of the coefficients b_i of the polynomial $B(s)$.

Output Arguments:

- KI The degree of the infinite frequency polynomial $G(s)$.
- G The one-dimensional array of variables G(I) in which are stored the values of the coefficients g_i of the polynomial $G(s)$ representing the behavior of the rational function $F(s)$ at infinity.
- KR The degree of the remainder polynomial $A(s)$.
- A The one-dimensional array of variables A(I) in which are stored the values of the coefficients a_i of the remainder polynomial $A(s)$.

Notes:

1. The values of the coefficients of $C(s)$ and $B(s)$, the input polynomials, are preserved invariant.

2. If the degree of the polynomial $C(s)$ is less than the degree of the polynomial $B(s)$, the coefficients of the polynomial $G(s)$ are set to zero and the coefficients of $A(s)$ are set equal to those of $C(s)$.

3. The variables of this subroutine are dimensioned as follows: C(10), B(10), G(10), A(10).

TABLE 8-5.2 Summary of the Characteristics of the Subroutine ROOT

Identifying Statement: SUBROUTINE ROOT (N, B, P)
Purpose: To find the roots of a polynomial $B(s)$ of degree n having the form

$$B(s) = b_1 + b_2 s + b_3 s^2 + \cdots + b_n s^{n-1} + b_{n+1} s^n$$

i.e., to find the quantities p_i where

$$B(s) = K \prod_{i=1}^{n} (s - p_i)$$

Additional Subroutines Required: None.
Input Arguments:
- N The degree of polynomial $B(s)$.
- B The one-dimensional array of variables B(I) in which are stored the values of the coefficients b_i of the polynomial $B(s)$.

Output Argument:
- P The complex one-dimensional array of variables P(I) in which are stored the values of the roots p_i of the polynomial $B(s)$.

Notes:
1. The values of the coefficients of the polynomial $B(s)$ are preserved invariant in the B array.
2. The subroutine uses a Lin-Bairstow method for determining the roots.
3. The variables of this subroutine are dimensioned as follows: B(10), P(10).

TABLE 8-5.3 Summary of the Characteristics of the Subroutine DIFPL

Identifying Statement: SUBROUTINE DIFPL (N, A, M, B)
Purpose: To differentiate a polynomial $A(s)$ of the form

$$A(s) = a_1 + a_2 s + a_3 s^2 + \cdots + a_{n+1} s^n$$

The resulting polynomial will be $A'(s)$, where

$$A'(s) = b_1 + b_2 s + b_3 s^2 + \cdots + b_{m+1} s^m = B(s)$$

Additional Subprograms Required: None.
Input Arguments:
- N The degree n of the polynomial $A(s)$.
- A The one-dimensional array of variables A(I) in which are stored the values of the coefficients a_i defining the polonomial $A(s)$.

Output Arguments:
- M The degree of the polynomial $A'(s) = B(s)$.
- B The one-dimensional array of variables B(I) in which are stored the values of the coefficients b_i of the polynomial $A'(s) = B(s)$.

Note: The variables of this subroutine are dimensioned as follows: A(10), B(10).

the subroutine is A, the usual plotting array. If we call the subroutine ILPX (for Inverse LaPlace Transformation), the identifying statement will have the form

SUBROUTINE ILPX (NT, TR, N, P, R1, R2, NA, A, SCALE)

A listing of such a subroutine is given in Fig. 8–5.12. A comparison of the FORTRAN statements with the logic of the flow chart given in Fig. 8–5.11 will readily show how the subroutine operates. A summary of the characteristics of the subroutine is shown in Table 8–5.5 (on page 614). As an example of the use of this subroutine, consider finding the inverse transform of the function

$$F(s) = \frac{500}{(s+1)^2 + 100} = \frac{-j25}{s+1-j10} + \frac{j25}{s+1+j10}$$

Let us assume it is desired to plot the resulting inverse transform over a range of $0 \leq t \leq 2.5$ s. A listing of a main program for plotting 61 points over this range is shown in Fig. 8–5.13. The resulting output plot is shown in Fig. 8–5.14.

TABLE 8-5.4 Summary of the Characteristics of the Subroutine PFEXSD

Identifying Statement: SUBROUTINE PFEXSD (M, A, N, B, P, R1, R2)
Purpose: To find the residues for a rational function $F(s)$ at its simple or double poles located at $s = p_i$, i.e., to find the coefficients K_{i1} and K_{i2} where

$$F(s) = \frac{A(s)}{B(s)} = \frac{a_1 + a_2 s + a_3 s^2 + \cdots + a_{m+1} s^m}{b_1 + b_2 s + b_3 s^2 + \cdots + b_{n+1} s^n}$$

$$= \sum_i \left| \frac{K_{i1}}{s - p_i} + \frac{K_{i2}}{(s-p_i)^2} \right|$$

Additional Subprograms Required: This subroutine calls the subroutines DIFPL, CVALPL, and ROOT.
Input Arguments:
- M The degree of the numerator polynomial $A(s)$.
- A The one-dimensional array of variables A(I) in which are stored the values of the coefficients a_i of the numerator polynomial $A(s)$.
- N The degree of the denominator polynomial $B(s)$.
- B The one-dimensional array of variables B(I) in which are stored the values of the coefficients b_i of the denominator polynomial $B(s)$.
- P The one-dimensional complex array of variables P(I) in which are stored the values of the locations of the poles p_i of the rational function $F(s)$.

Output Arguments:
- R1 The one-dimensional complex array of variables R1(I) in which are stored the values of the residues K_{i1} for a first-order pole at p_i or the complex value of the quantity K_{i1} for a pole p_i which is of second order.
- R2 The one-dimensional complex array of variables R2(I) in which are stored the values of the residues K_{i2} for a second-order pole located at p_i (this quantity is set to zero if the pole is of first order).

Note: The variables of the subroutine are dimensioned as follows: A(10), B(10), P(10), R1(10), R2(10).

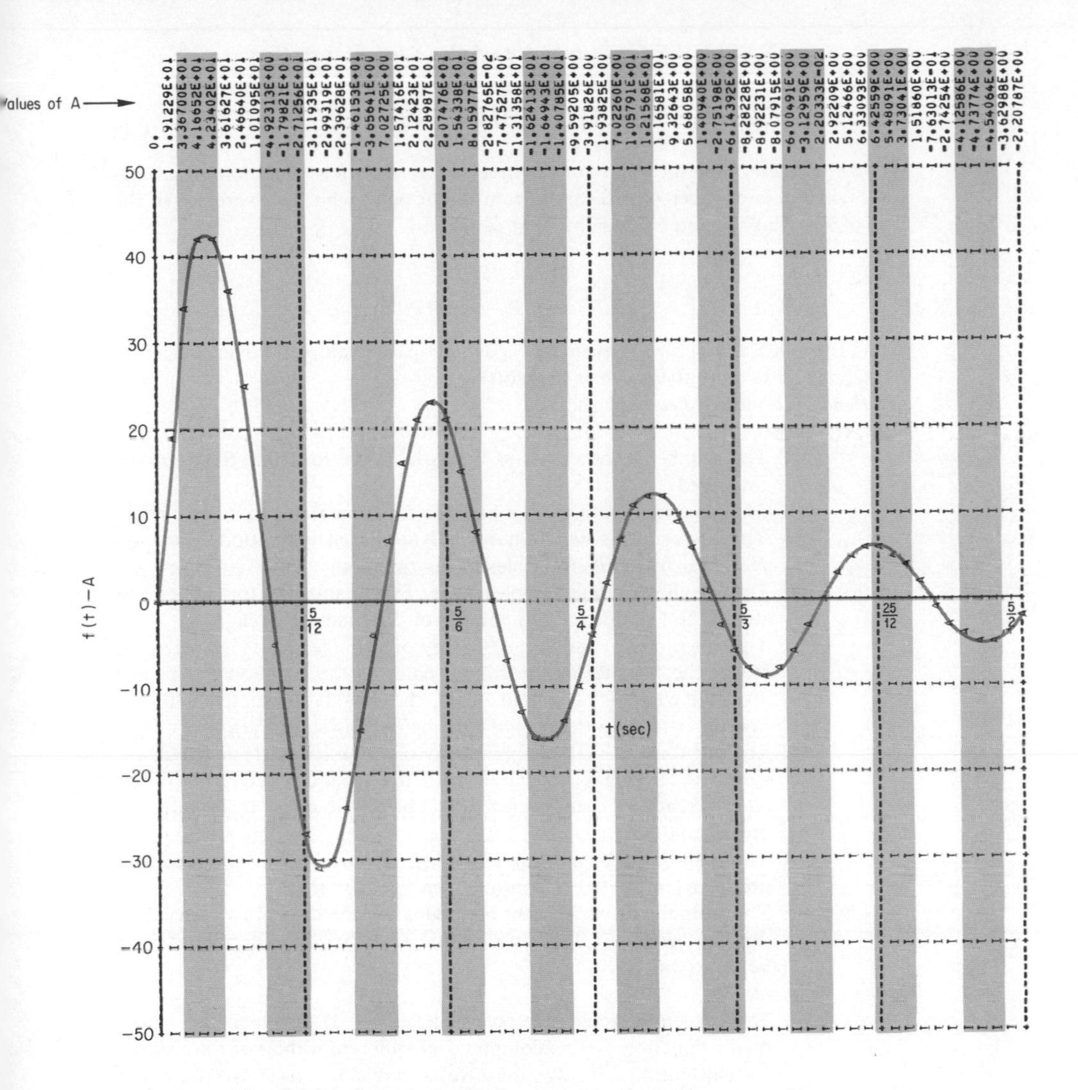

Fig. 8-5.14 Output plot for inverse transformation example.

TABLE 8-5.5 Summary of the Characteristics of the Subroutine ILPX

Identifying Statement: SUBROUTINE ILPX (NT, TR, N, P, R1, R2, NA, A, SCALE)

Purpose: To determine the values of the inverse Laplace transformation of a function $F(s)$ containing a set of first- and second-order poles which is expressed in the form of a partial-fraction expansion. That is, given

$$F(s) = \sum_i \left[\frac{R_i^{(1)}}{s - p_i} + \frac{R_i^{(2)}}{(s - p_i)^2}\right]$$

to find the value of $f(t)$ for a sequence of n_t evenly spaced values of t over the period $0 \leq t \leq t_r$ and to store these values for plotting.

Additional Subroutines Required: None.

Input Arguments:

- NT The number n_t of values of t at which the function $f(t)$ is to be evaluated.
- TR The maximum value t_r at which the function $f(t)$ is to be evaluated.
- N The degree of the denominator polynomial of the rational function $F(s)$, i.e., the number of poles of the function.
- P The one-dimensional complex array of variables P(I) in which are stored the locations of the poles p_i of the function $F(s)$.
- R1 The one-dimensional complex array of variables R1(I) in which are stored the residues of the simple poles of the function $F(s)$ or, for any pole which is of second order, the residue associated with the first-order term in the partial-fraction expansion of $F(s)$.
- R2 The one-dimensional complex array of variables R2(I) in which are stored the residues associated with the second-order terms in the partial-fraction expansion for $F(s)$. For simple poles, these variables are set to $0 + j0$.
- NA The position in the plotting array A(I, J) in which the computed data are to be stored, i.e., the value given to the index I.
- SCALE The scale factor which is to be used when the data are stored in the plotting array A(I, J). If no scaling is desired, this argument must be set to unity.

Output Argument:

- A The two-dimensional array of variables A(I, J) in which the values of the function $f(t)$ as computed at different values of t are stored for plotting. Specifically, these values are stored as A(NA, J), where J has a range from 1 to NT.

Note: The variables of the subroutine are dimensioned as follows: P(10), R1(10), R2(10), A(5, 101).

8-6 IMPEDANCE AND ADMITTANCE—NETWORK FUNCTIONS

In this section we shall apply the methods for finding transforms and inverse transforms developed in the preceding sections of this chapter to the general problem of analyzing a network. As a simple introduction, consider a capacitor excited by a voltage source as shown in Fig. 8 6.1(a). The terminal variables are related by the equation

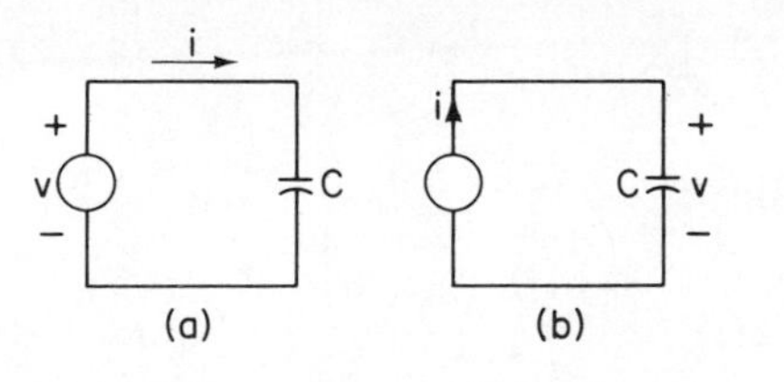

Fig. 8-6.1 A capacitor excited by sources.

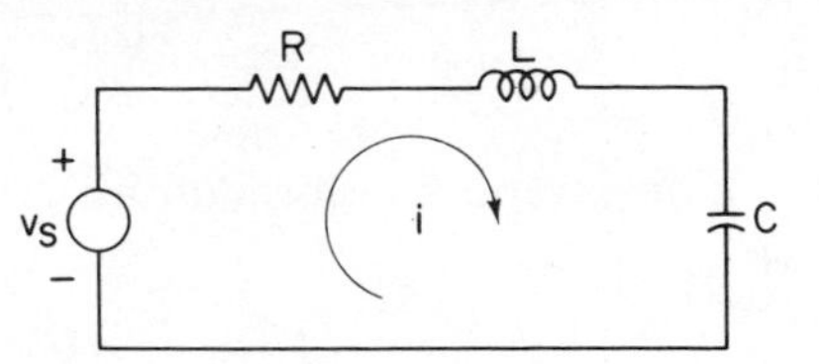

Fig. 8-6.2 A series *RLC* circuit.

$$i(t) = C\frac{dv}{dt} \tag{8–165}$$

Transforming this equation, we obtain

$$I(s) = CsV(s) - Cv(0-) \tag{8–166}$$

Now let us assume that the excitation is a step of magnitude K volts applied at $t = 0$. Thus, $v(t) = Ku(t)$, and $V(s) = K/s$. If there is no initial condition on the capacitor, then $v(0-) = 0$, and from (8–166) we obtain

$$I(s) = sC\left(\frac{K}{s}\right) = KC \tag{8–167}$$

Taking the inverse transform of (8–167), we find that

$$i(t) = KC\delta(t) \tag{8–168}$$

Thus, we find that the resulting current is an impulse occurring at $t = 0$. This is the same result that was obtained in Sec. 4–3. More generally, if $v(0-)$ is not zero, then the resulting current will be

$$i(t) = C[K - v(0-)]\,\delta(t) \tag{8–169}$$

Now consider the case where a capacitor is excited by a current source as shown in Fig. 8–6.1(b). The defining relation for the terminal variables is

$$v(t) = \frac{1}{C}\int_{0-}^{t} i(\tau)\,d\tau + v(0-) \tag{8–170}$$

Transforming this equation, we obtain

$$V(s) = \frac{1}{sC}I(s) + \frac{v(0-)}{s} \tag{8–171}$$

If we now apply a step of current $i(t) = Ku(t)$ A, then $I(s) = K/s$, and, if $v(0-) = 0$, we obtain

$$V(s) = \frac{1}{sC}\frac{K}{s} = \frac{K}{s^2C} \tag{8-172}$$

Taking the inverse transform of this, we obtain

$$v(t) = \frac{Kt}{C}u(t) \tag{8-173}$$

Thus, the voltage is a ramp of slope K/C starting at $t = 0$. More generally, if $v(0-)$ is not zero, then

$$v(t) = \left[\frac{Kt}{C} + v(0-)\right]u(t) \tag{8-174}$$

This function represents the sum of a ramp of slope K/C and a step of value $v(0-)$, i.e., a ramp of slope K/C starting from a value of $v(0-)$ rather than from a value of 0.

The developments given above are easily extended to more complex networks. For example, consider the RLC network shown in Fig. 8-6.2. The integro-differential equation found by applying KVL to this circuit may be written in the form

$$v_S(t) = L\frac{di}{dt} + \frac{1}{C}\int_{0-}^{t} i(\tau)\,d\tau + v_C(0-) + v_R(t) \tag{8-175}$$

Transforming this equation, we obtain

$$V_S(s) = L[sI(s) - i(0-)] + \frac{1}{sC}I(s) + \frac{v_C(0-)}{s} + V_R(s) \tag{8-176}$$

Now let us assume that we desire to solve for $V_R(s)$. The variable $I(s)$ may be eliminated from Eq. (8-176) by substituting the transformed form of Ohm's law, namely, $I(s) = V_R(s)/R$. Making this substitution and rearranging terms, we obtain

$$V_S(s) = \left[\frac{s^2 + s(R/L) + (1/LC)}{s(R/L)}\right]V_R(s) - Li(0-) + \frac{v_C(0-)}{s} \tag{8-177}$$

Solving this equation for $V_R(s)$, we obtain

$$V_R(s) = \frac{s(R/L)V_S(s)}{s^2 + s(R/L) + (1/LC)} + \frac{sRi(0-) - v_C(0-)(R/L)}{s^2 + s(R/L) + (1/LC)} \tag{8-178}$$

Thus, we have found a transformed equation which explicitly gives $V_R(s)$, i.e., the transform of the response variable $v_R(t)$, in terms of the transform of the excitation variable $v_S(t)$, and the initial conditions on the inductor current $i(0-)$ and the capacitor voltage $v_C(0-)$. Specifying these quantities will permit us to determine $V_R(s)$ explicitly, and, by taking the inverse transform, we are thus able to find $v_R(t)$. Note that in Eq. (8–178) we have written the expression for $V_R(s)$ in such a way as to separate the term involving $V_S(s)$ from the term involving the initial conditions. Thus, the inverse transform of the first term in the right member of Eq. (8–178) will give the *zero-state response* for the network, i.e., the response that results from an excitation which is applied when all the initial conditions are zero. The poles of this term come from two places: (1) the network equation [the zeros of the quantity $s^2 + s(R/L) + (1/LC)$]; and (2) the poles of the function $V_S(s)$. Thus, the resulting inverse transform will have terms which are the result of the element values of the network, as well as terms which are the result of the particular form of $v_S(t)$. The situation will be identical with that shown in the "Zero-State Response" column of the "Second Decomposition of Response" section of Table 6–4.1. Whether Case I, II, or III applies will depend on whether the zeros of $s^2 + s(R/L) + (1/LC)$ are real and simple, real and double, or complex, respectively.

Now let us consider the second term in the right member of Eq. (8–178). The inverse transform of this term will give the *zero-input response* for the network, i.e., the response that results from initial conditions when the excitation $v_S(t)$ is zero. Thus, the inverse transform of this term will have elements whose basic behavior is the result of the network elements, and whose magnitude and/or phase is the result of the relative magnitude of the two initial conditions. This situation will be identical with that shown in the "Zero-Input Response" column of Table 6–4.1.

As has been discussed above, the expression for $V_R(s)$ given in Eq. (8–178) has been written in such a form as to agree with the entries entitled "Second Decomposition of Response" in Table 6–4.1. It is also possible to put the transformed expression given in Eq. (8–178) in a form so that the resulting inverse transform has the general appearance given in the columns entitled "First Decomposition of Response" in Table 6–4.1. To do this, let us define $V_S(s) = N(s)/D(s)$. Equation (8–178) may now be written in the form

$$V_R(s) = \frac{as + b}{s^2 + s(R/L) + (1/LC)} + \frac{P(s)}{D(s)} \tag{8–179}$$

where the coefficients a and b, and the polynomial $P(s)$ are readily found from Eq. (8–178). The inverse transform of the first term in the right member of Eq. (8–179) will be the transient response, i.e., the complementary solution for the network. The inverse transform of the second term will be the forced response, i.e., the particular solution. Thus, these terms give the transformed equivalent of the entries in the first two columns of Table 6–4.1. A numerical example illustrating the principles described above follows.

EXAMPLE 8 6.1. For the *RLC* network shown in Fig. 8-6.2, the two-terminal elements have the values $R = 1\ \Omega$, $C = 1$ F, and $L = 0.5$ H. It is desired to find the output voltage $v_R(t)$ under the conditions that a unit ramp $v_S(t) = tu(t)$ V is applied as an excitation. It is assumed that the initial conditions in the network are zero. Since the transform of $tu(t)$ is $1/s^2$, we may substitute the given element values, directly in (8-178) to obtain an expression for $V_R(s)$. This is

$$V_R(s) = \frac{2}{s(s^2 + 2s + 2)} = \frac{K_1}{s} + \frac{as + b}{(s + 1)^2 + (1)^2}$$

Using the methods given in Sec. 8-4, we find that $K_1 = 1, a = -1$, and $b = -2$. Thus, we may write

$$V_R(s) = \frac{1}{s} - \frac{s + 2}{(s + 1)^2 + (1)^2} = \frac{1}{s} - \frac{s + 1}{(s + 1)^2 + (1)^2} - \frac{1}{(s + 1)^2 + (1)^2}$$

The inverse transform of each of the terms in the right member of the above equation is easily found using the methods of Sec. 8-3. Thus, we find that

$$V_R(t) = [1 - e^{-t}(\cos t + \sin t)]u(t) = [1 - \sqrt{2}\, e^{-t} \cos (t - 45^\circ)]u(t)$$

is the desired response. The first term inside the brackets is readily seen to be the particular response, corresponding with the form of the particular response given as $A + Bt$ in Table 5-2.1. In this case, $B = 0$, since the zero at the origin of the network function [the term s in the numerator of the left term in the right member of (8-178)] has cancelled one of the poles at the origin of the transform $1/s^2$ of $v_S(t)$. The second term inside the brackets is the transient response, the form of which is due to the natural frequencies of the network.

The fact that the transformed expression for the response variable includes the initial conditions explicitly makes it easy not only to observe the effect of these initial conditions, but also to select them in such a way as to achieve some desired effect on the response. An example of such an application follows.

EXAMPLE 8-6.2. For the *RLC* network and the excitation used in Example 8-6.1, it is desired to specify a set of non-zero initial conditions in such a way that the phase of the sinusoidal term in the response $v_R(t)$ is zero, i.e., in such a way that the inverse transform of the term representing the complex pole pair is of the form $Ke^{-t} \cos t$. To do this, using the results found in Example 8-6.1, we may write the general expression for the response given in (8-178) as

$$V_R(s) = \frac{1}{s} - \frac{s + 2}{(s + 1)^2 + (1)^2} + \frac{si(0-) - 2v_C(0-)}{(s + 1)^2 + (1)^2}$$

It is readily observed that if we specify $i(0-) = 0$, and $v_C(0-) = -1/2$, then

$$V_R(s) = \frac{1}{s} - \frac{s + 1}{(s + 1)^2 + (1)^2}$$

Taking the inverse transform, we obtain

$$v_R(t) = (1 - e^{-t}\cos t)u(t)$$

as was desired.

The techniques illustrated above are readily applied to the solution of networks containing more than a single independent variable. As an example of this, consider the network shown in Fig. 8-6.3. In addition to the two independent sources specified as $i_1(t)$ and $i_2(t)$, this circuit also has a dependent VCIS connected to the second node and controlled by the voltage at the first node. The time-domain equations for this circuit are easily written by applying KCL at the two nodes. Thus, we obtain

$$\begin{aligned} i_1(t) &= \frac{dv_1}{dt} + 2v_1(t) - v_2(t) \qquad t \geq 0 \\ 3v_1(t) + i_2(t) &= 4\int_{0-}^{t} v_2(\tau)\,d\tau + i_L(0-) + 4v_2(t) - v_1(t) \qquad t \geq 0 \end{aligned} \tag{8-180}$$

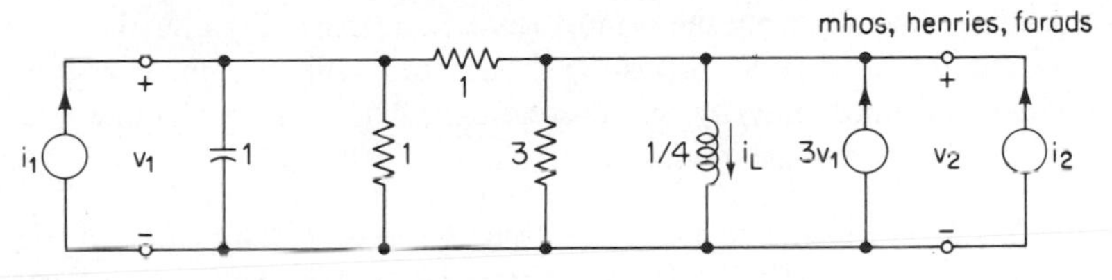

Fig. 8-6.3 A circuit with an independent source.

Combining the $v_1(t)$ terms in the left and right members of the second equation and transforming both of the equations, we obtain

$$\begin{aligned} I_1(s) &= sV_1(s) - v_1(0\;\;) + 2V_1(s) - V_2(s) \\ I_2(s) &= \frac{4}{s}V_2(s) + \frac{i_L(0-)}{s} + 4V_2(s) - 4V_1(s) \end{aligned} \tag{8-181}$$

Now let us write the equations in matrix form. If we move the initial condition terms to the left member at the same time, we obtain

$$\begin{bmatrix} I_1(s) + v_1(0-) \\ I_2(s) - \dfrac{i_L(0-)}{s} \end{bmatrix} = \begin{bmatrix} s+2 & -1 \\ -4 & \dfrac{4}{s} + 4 \end{bmatrix} \begin{bmatrix} V_1(s) \\ V_2(s) \end{bmatrix} \tag{8-182}$$

To solve for such a set of equations, we need merely multiply both members of the equation by the inverse of the square matrix. The inverse is readily found using the techniques given in Sec. 3-1. Thus, we obtain

$$\begin{bmatrix} V_1(s) \\ V_2(s) \end{bmatrix} = \frac{s/4}{s^2 + 2s + 2} \begin{bmatrix} \dfrac{4}{s} + 4 & 1 \\ 4 & s + 2 \end{bmatrix} \begin{bmatrix} I_1(s) + v_1(0-) \\ I_2(s) - \dfrac{i_L(0-)}{s} \end{bmatrix} \tag{8-183}$$

Let us assume that we desire to find an expression for the transform of $V_2(s)$. From the matrix equation given above, we may write

$$V_2(s) = \frac{sI_1(s)}{s^2 + 2s + 2} + \frac{(s/4)(s + 2)I_2(s)}{s^2 + 2s + 2} + \frac{sv_1(0-) - i_L(0-)(s + 2)/4}{s^2 + 2s + 2} \tag{8-184}$$

Obviously, if desired, a similar expression can be written for $V_1(s)$. In the right member of the expression given above, we have indicated three separate terms. The first term is the transform of the zero-state response for the case in which only $i_1(t)$ is non-zero. The second term is the transform of the zero-state response for the case in which only $i_2(t)$ is non-zero. It is readily observed that superposition applies, namely, the transform of the zero-state response for the case for which both $i_1(t)$ and $i_2(t)$ are non-zero is simply the sum of these two terms. Finally, the third term in the right member of Eq. (8–184) is the transform of the zero-input response, i.e., the response determined only by initial conditions. An example of the use of Eq. (8–184) to find the output $v_2(t)$ follows.

EXAMPLE 8-6.3. For the network shown in Fig. 8-6.3, it is desired to find $v_2(t)$ when excitation currents $i_1(t) = tu(t)$ and $i_2(t) = e^{-2t}u(t)$ A are applied. The initial conditions are specified as $v_1(0-) = \frac{1}{4}$ V and $i_L(0-) = 0$. The transforms of $i_1(t)$ and $i_2(t)$ are $1/s^2$ and $1/(s + 2)$, respectively. Inserting these values in (8–184), we obtain

$$V_2(s) = \frac{1/s}{s^2 + 2s + 2} + \frac{(s/4)(s + 2)/(s + 2)}{s^2 + 2s + 2} + \frac{s/4}{s^2 + 2s + 2}$$

Combining terms and expressing the result in the form of a partial-fraction expansion, we obtain

$$V_2(s) = \frac{(s^2/2) + 1}{s(s^2 + 2s + 2)} = \frac{K_1}{s} + \frac{as + b}{s^2 + 2s + 2}$$

Using the techniques given in Sec. 8-4, we find that $K_1 = \frac{1}{2}$, $a = 0$, and $b = -1$. Thus, we may write

$$V_2(s) = \frac{\frac{1}{2}}{s} - \frac{1}{(s + 1)^2 + (1)^2}$$

Taking the inverse transform, we find that

$$v_2(t) = [\tfrac{1}{2} - e^{-t} \sin t]u(t)$$

The developments given above are quite general, in the sense that they may be used to determine the *complete* response (i.e., the sum of the zero-state response and the zero-input response) of an arbitrary network. In many network situations, however, the initial conditions may be specified as zero. In such a situation, it is only necessary to find the zero-state response. In this case, the determination of the transformed network equations can be considerably simplified. To see this, let us consider the transformed equation for the terminal variables of a capacitor as given in Eq. (8–166) for the case where the initial condition $v(0-)$ is zero. Since, as shown in Fig. 8–6.1(a), the excitation is assumed to be a voltage source, the resulting relation may be written in the form

$$\frac{I(s)}{V(s)} = \frac{\text{Response}}{\text{Excitation}} = sC \tag{8–185}$$

In a manner similar to that used when we considered phasors in Sec. 7–8, we may define an expression of this type which relates a transformed response variable to a transformed excitation variable as a *network function*. In the case given in Eq. (8–185), since the network function is the ratio of a transformed current variable to a transformed voltage variable, and since both variables are defined at the same pair of terminals, the expression sC is called the *driving-point admittance*, or, more simply, the *admittance* of a capacitor of value C farads. Note that if we let $s = j\omega$, then we obtain the expression for the admittance of a capacitor as used in phasor analysis in Sec. 7–5.

Now consider a capacitor excited by a current source as shown in Fig. 8–6.1(b). If we again consider the case where the initial condition is zero, i.e., the zero-state case, then Eq. (8–171) may be written in the form

$$\frac{V(s)}{I(s)} = \frac{\text{Response}}{\text{Excitation}} = \frac{1}{sC} \tag{8–186}$$

Since this network function is the ratio of a transformed voltage variable to a transformed current variable, it defines the *driving-point impedance* of a capacitor. It is readily observed from Eqs. (8–185) and (8–186) that driving-point impedances and admittances are reciprocal quantities. Just as was the case for the sinusoidal steady state, we may use the term *immittance* to collectively refer to impedance and admittance.[15]

The immittance of other network elements is easily determined by transforming the time-domain equations relating their terminal behavior, and by assuming zero initial conditions. For example, for an inductor excited by a current source, the time-domain and frequency-domain relations are

[15]The terms *network function, impedance, admittance,* etc. used in this section are identical with those used for the sinusoidal steady-state case in Sec. 7–8. Such a usage follows common practice, the differentiation between the two usages being provided by the functional notation (s) or $(j\omega)$.

$$v(t) = L\frac{di}{dt} \qquad V(s) = L[sI(s) - i(0-)] \tag{8-187}$$

Similarly, for an inductor excited by a voltage source, we obtain

$$i(t) = \frac{1}{L}\int_{0-}^{t} v(\tau)\,d\tau + i(0-) \qquad I(s) = \frac{1}{sL}V(s) + \frac{i(0-)}{s} \tag{8-188}$$

Thus, we see that the driving-point impedance of an inductor is sL; correspondingly, its admittance is $1/sL$. It is easily shown also that the impedance of a resistor is R, and its admittance is $1/R$. Note that there is a significant difference between immittances as defined in the complex frequency domain in this chapter and those defined for sinusoidal steady-state conditions in Chap. 7. The immittances defined here may be evaluated for any value of the complex frequency variable s. Thus, the impedance of an inductance, which is sL, will be real when s takes on real values, imaginary when s takes on imaginary values, and complex when s takes on complex values. However, the impedance of an inductor under sinusoidal steady-state conditions, which is $j\omega L$, is always imaginary for all values of ω, since ω is always real.

All the terminology derived with respect to network functions for sinusoidal steady-state conditions in Sec. 7-8 is directly applicable to the network functions which have been defined in this section as a ratio of transformed variables. Thus, in addition to the driving-point immittances introduced above, we may consider *transfer immittances* and *dimensionless transfer ratios*. Table 7-8.1 may thus be used as a basis for defining network functions in the complex frequency variable as well as a basis for defining network functions under sinusoidal steady-state conditions. This conclusion may be summarized as follows:

SUMMARY 8-6.1

Network Functions for Transformed Network Variables: Let $E(s)$ be the transform of some excitation variable $e(t)$ for a given network. Similarly, let $R(s)$ be the transform of some response variable $r(t)$. If we transform the integro-differential equation relating $e(t)$ and $r(t)$, and if all the initial conditions are set to zero, then the remaining equation may be used to define a network function $N(s)$ relating the transformed response variable to the transformed excitation variable. Specifically, $N(s) = R(s)/E(s)$. Such a network function may be a driving-point immittance, a transfer immittance, or a dimensionless transfer ratio.

The concept of a network function developed in the preceding discussion applies equally well to circuits in which mutual inductance and/or active elements are present. Such cases are illustrated in the following examples.

EXAMPLE 8-6.4. It is desired to find the network function relating the transformed excitation voltage $V_1(s) = \mathscr{L}[v_1(t)]$ and the transformed output voltage $V_2(s) = \mathscr{L}[v_2(t)]$ in the network shown in Fig. 8-6.4. The time-domain loop equations for this circuit are

$$v_1(t) = R_1 i_1(t) + L_1 \frac{di_1}{dt} + M \frac{di_2}{dt}$$

$$0 = M \frac{di_1}{dt} + L_2 \frac{di_2}{dt} + R_2 i_2(t)$$

The Laplace transformation of these relations yields the following matrix equation:

$$\begin{bmatrix} V_1(s) \\ 0 \end{bmatrix} = \begin{bmatrix} R_1 + sL_1 & M \\ M & R_2 + sL_2 \end{bmatrix} \begin{bmatrix} I_1(s) \\ I_2(s) \end{bmatrix} - \begin{bmatrix} L_1 i_1(0-) + Mi_2(0-) \\ L_2 i_2(0-) + Mi_1(0-) \end{bmatrix}$$

The network function $V_2(s)/V_1(s)$ is readily found by setting the initial conditions to zero, inverting the square matrix, solving for $I_2(s)$, and applying the relation $V_2(s) = R_2 I_2(s)$. The details are left as an exercise for the reader.

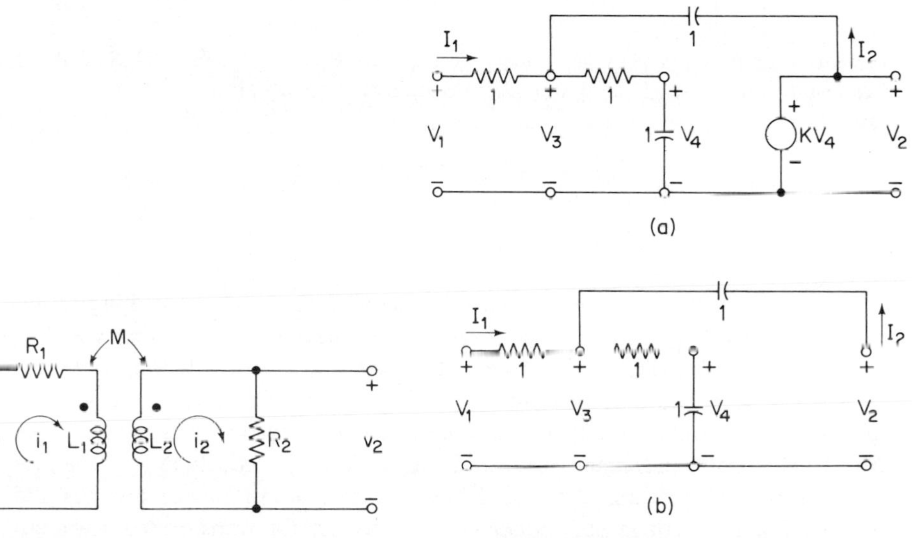

Fig. 8-6.4 Network for Example 8-6.4.

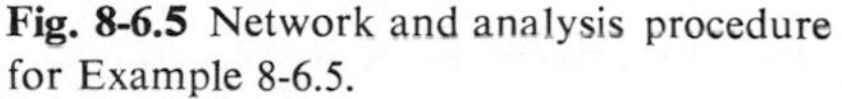

Fig. 8-6.5 Network and analysis procedure for Example 8-6.5.

EXAMPLE 8-6.5. It is desired to find the network transfer function relating the transformed excitation voltage $V_1(s)$ and the transformed output voltage $V_2(s)$ in the network shown in Fig. 8-6.5(a) in which an active element modeled by an ideal VCVS is used.

A convenient way to analyze this network is to first find the admittance matrix for a reduced network containing only the passive elements which were in the original circuit. For this example, such a reduced network may be described in matrix form as follows:

$$\begin{bmatrix} I_1(s) \\ I_2(s) \\ 0 \\ 0 \end{bmatrix} = \begin{bmatrix} 1 & 0 & -1 & 0 \\ 0 & s & -s & 0 \\ -1 & -s & s+2 & -1 \\ 0 & 0 & -1 & s+1 \end{bmatrix} \begin{bmatrix} V_1(s) \\ V_2(s) \\ V_3(s) \\ V_4(s) \end{bmatrix}$$

For convenience, let us explicitly write out the last two equations contained in the above matrix formulation. These are

$$0 = -V_1(s) - sV_2(s) + (s + 2)V_3(s) - V_4(s)$$
$$0 = -V_3(s) + (s + 1)V_4(s)$$

Now let us consider what changes must be made in these equations when we replace the VCVS in the circuit. Since this device relates the voltages $V_2(s)$ and $V_4(s)$, we may eliminate the variable $V_4(s)$ in the above equations by substituting $V_4(s) = V_2(s)/K$. Thus, we obtain

$$0 = -V_1(s) + \left(-s - \frac{1}{K}\right)V_2(s) + (s + 2)V_3(s)$$
$$0 = \left[\frac{s + 1}{K}\right]V_2(s) - V_3(s)$$

From the second equation given above, we see that $V_3(s) = V_2(s)(s + 1)/K$. Substituting this result in the first equation and rearranging terms, we obtain

$$\frac{V_2(s)}{V_1(s)} = \frac{K}{s^2 + (3 - K)s + 1}$$

This is the desired voltage transfer function.

There is a very close tie between the network functions defined in this chapter and those defined in Chap. 7. In this chapter, we have seen in Summary 8-2.4 that, when initial conditions are not considered, network functions which relate transformed network variables may be found from the original equations by replacing differentiation by multiplication by s and integration by division by s. These, however, are the identical replacements used in Chap. 7. We may conclude that the network functions (of $j\omega$) determined for sinusoidal steady-state or phasor analysis and the network functions (of s) determined in this chapter for transformed variables are identical if we set $j\omega = s$.

As an example of this, consider the series RLC network shown in Fig. 8-6.2. From Eq. (8-178), for the zero-state case we may write

$$\frac{V_R(s)}{V_S(s)} = \frac{\text{Response}}{\text{Excitation}} = \frac{sRC}{s^2LC + sRC + 1} \tag{8-189}$$

If we substitute $s = j\omega$ in the above equation, we obtain

$$\left.\frac{V_R(s)}{V_S(s)}\right|_{s=j\omega} = \frac{j\omega RC}{-\omega^2LC + j\omega RC + 1} \tag{8-190}$$

The right member of this expression is identical with the one given in Eq. (7–184). Thus, we see that by letting $s = j\omega$ in a network function in the complex frequency variable s, we obtain the network function which relates excitation and response phasors for the sinusoidal steady-state case. We may conclude that, if we know the network function which relates a transformed response variable to a transformed excitation variable (and which may thus be used to determine the zero-state response for any excitation), letting $s = j\omega$ yields the network function relating the corresponding response phasor to the excitation phasor (which may thus be used to determine the particular solution for sinusoidal steady-state conditions). Note that the phasor method only determines the *particular solution component* of the zero-state response, and not the transient component. The transform network function method, however, determines *both* components of the zero-state response, namely, the particular solution *and the transient solution.* Furthermore it accomplishes this for any input excitation function, not merely for the sinusoidal case. These differences may be observed in the following example.

EXAMPLE 8-6.6. The excitation $v_S(t) = \sqrt{2}\cos(2t + 45^\circ)u(t) = [\cos 2t - \sin 2t]u(t)$ V is applied to the simple *RLC* network shown in Fig. 8-6.2 and used in Example 8-6.1. If $v_R(t)$ is considered as the response variable, then the transfer function for the transformed variables is

$$\frac{V_R(s)}{V_S(s)} = \frac{2s}{s^2 + 2s + 2}$$

To find the sinusoidal steady-state response using phasors, we may define $\mathscr{V}_R$ as the response phasor and $\mathscr{V}_S$ as the excitation phasor. Substituting $s = j\omega$ in the network function given above, we obtain

$$\frac{\mathscr{V}_R}{\mathscr{V}_S} = \frac{j2\omega}{2 - \omega^2 + j2\omega}$$

The angular frequency of the sinusoidal excitation, however, is 2 rad/s. Substituting $\omega = 2$ in the preceding relation, we obtain

$$\frac{\mathscr{V}_R}{\mathscr{V}_S} = \frac{j4}{-2 + j4} = \frac{4 - j2}{5}$$

The phasor representation for $v_S(t)$ is $1 + j1$. Inserting this value in the last relation, we find that the phasor representation for $\mathscr{V}_R$ is $(6 + j2)/5$. Thus, the sinusoidal steady-state component of $v_R(t)$ is $\frac{6}{5}\cos 2t - \frac{2}{5}\sin 2t$. Now let us compare this with the result obtained using transform techniques. Using Table 8-3.1, we find that the transform of $v_S(t)$ is

$$V_S(s) = \frac{s}{s^2 + 4} - \frac{2}{s^2 + 4} = \frac{s - 2}{s^2 + 4}$$

Thus, using the transfer function given at the beginning of this example, we find that the transform for $V_R(s)$ is

$$V_R(s) = \frac{2s}{s^2 + 2s + 2}\,\frac{s - 2}{s^2 + 4}$$

The inverse transform of this expression can be easily found if it is put in the form

$$V_R(s) = \frac{as + b}{s^2 + 2s + 2} + \frac{cs + d}{s^2 + 4}$$

Equating coefficients in the last two expressions, we may readily find values for a, b, c, and d. Inserting these values, the expression for $V_R(s)$ may be written

$$\begin{aligned} V_R(s) &= \frac{-\frac{6}{5}s + \frac{2}{5}}{s^2 + 2s + 2} + \frac{\frac{6}{5}s - \frac{4}{5}}{s^2 + 4} \\ &= \frac{-\frac{6}{5}(s + 1)}{(s + 1)^2 + (1)^2} + \frac{\frac{8}{5}}{(s + 1)^2 + (1)^2} + \frac{\frac{6}{5}s}{s^2 + 4} - \frac{\frac{2}{5}2}{s^2 + 4} \end{aligned}$$

where, in the last expression, the terms have been decomposed in such a way as to facilitate taking the inverse transform. From the entries in Table 8-3.1, we readily find that the inverse transform is

$$V_R(t) = \{e^{-t}[-\tfrac{6}{5}\cos t + \tfrac{8}{5}\sin t] + \tfrac{6}{5}\cos 2t - \tfrac{2}{5}\sin 2t\}u(t)$$

The last two terms in the right member of the above equation are, of course, identical with the sinusoidal steady-state solution found earlier in this example by the use of phasor methods. Thus, they represent the particular solution component of the zero-state response. The first two terms of the right member of the expression for $v_R(t)$, however, cannot be found by the phasor method, since they represent the transient component of the zero-state solution. Thus, we see that the transformed network function provides us with a means of readily finding the entire zero-state response of a network, namely, both the particular solution and the transient response.

The results given above are readily summarized as follows:

SUMMARY 8-6.2

The Relation Between Network Functions Used in Sinusoidal Steady-State Analysis and Network Functions Used in Transform Analysis: Consider a network in which a response variable $r(t)$ and an excitation variable $e(t)$ have been defined. Let $\mathscr{R}$ and $\mathscr{E}$ be the phasor representations for these variables, and let $R(s)$ and $E(s)$ be the Laplace transforms of these variables. If we define the network functions $N_1(j\omega)$ and $N_2(s)$ by the relations

$$N_1(j\omega) = \frac{\mathscr{R}}{\mathscr{E}} \qquad N_2(s) = \frac{R(s)}{E(s)}$$

respectively, then $N_1(j\omega)$ and $N_2(s)$ are related as follows

$$N_1(j\omega)|_{j\omega=s} = N_2(s) \qquad N_2(s)|_{s=j\omega} = N_1(j\omega)$$

Thus, we see that the functions N_1 and N_2 are identical.

The concepts of impedance and admittance, as developed for phasors, are identical with those developed for transformed variables. Therefore, all the manipulations made with respect to impedances and admittances for networks under conditions of sinusoidal steady-state excitations (such as the addition of impedances for elements connected in series and the addition of admittances for network elements connected in parallel) apply to networks described by transformed equations. In the next section we shall see how to apply these rules to directly write the transformed equations and determine any desired network function.

In this and the preceding sections of this chapter we have developed a powerful network analysis tool, the Laplace transformation. This transformation defines a complex frequency domain in which the integral and differential operations which relate the actual time-domain network variables are replaced by the algebraic operations of division and multiplication. Several methods for applying the Laplace transform to the analysis of networks have been developed. One of these methods is summarized in the flow chart shown in Fig. 8-6.6. Starting in the upper left corner of this flow chart, we have indicated that by transforming the time-domain equations relating the terminal variables of network elements and applying Kirchhoff's laws, we obtain frequency-domain equations with the initial condition terms specified in terms of the terminal variables. Correspondingly, we may first apply Kirchhoff's laws in the time domain to obtain a set of simultaneous integro-differential equations and then apply the Laplace transformation. In this case, we obtain frequency-domain equations in which the initial conditions are specified in terms of the node voltage variables or the mesh current variables (depending on whether KCL or KVL has been used). In either case, the sets of frequency-domain equations may be solved for some desired response variable and any of the various time-domain response components may be obtained. Examples of the use of the procedure given in the flow chart may be found in Examples 8-6.1, 8-6.2, and 8-6.3. It is strongly recommended that the student restudy these examples and correlate each step of the solution procedure with a corresponding step in the flow chart. It should be noted that the various elements of the flow chart have been physically oriented so as to make it possible to merge this flow chart with some other ones which will be developed in later sections of this chapter.

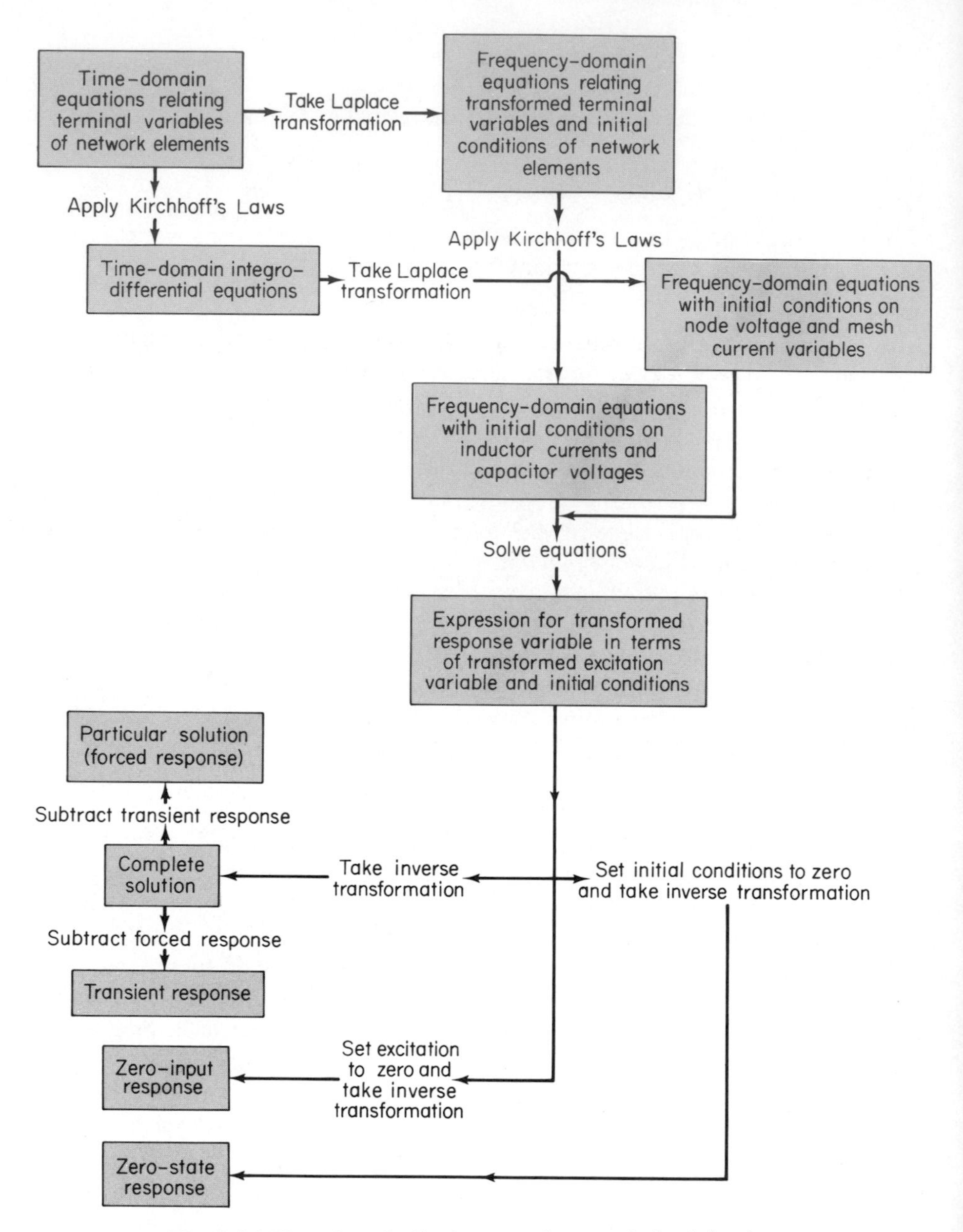

Fig. 8-6.6 Flow chart for Laplace transform analysis of circuits.

8-7 MESH AND NODE EQUATIONS FOR TRANSFORMED NETWORKS

In the preceding section we showed how to apply transform methods to find the solution of networks by transforming the time-domain equations describing the network variables. Starting from these time-domain equations, we showed that if the initial conditions were specified as zero, we could then define network functions specifying a ratio of transformed network variables. Frequently, it is more convenient to determine the network functions directly from the network itself, rather than from the time-domain equations. In order to do this, in this section we will develop the concept of a *transformed network*. Such a concept is one of the most important and useful results of Laplace transformation theory.

To begin our study of transformed networks, consider the series connection of two-terminal network elements shown in Fig. 8–7.1(a). For this connection, applying KVL, we may write

$$v_0(t) = v_1(t) + v_2(t) + v_3(t) + \cdots + v_n(t) \tag{8–191}$$

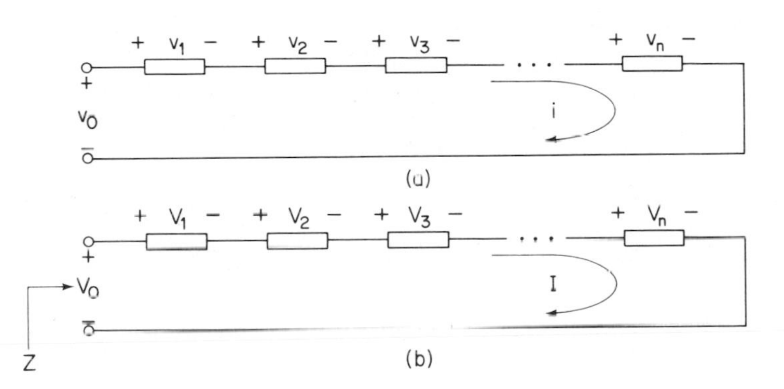

Fig. 8-7.1 A series connection of two-terminal network elements.

If we transform both sides of Eq. (8–191), we obtain from Summary 8–2.1,

$$V_0(s) = V_1(s) + V_2(s) + V_3(s) + \cdots + V_n(s) \tag{8–192}$$

where $V_i(s) = \mathscr{L}[v_i(t)]$. We conclude that transformed voltage variables satisfy KVL. In a similar fashion, we may show that transformed current variables satisfy KCL. These conclusions are similar to those summarized in Summary 7–5.1 with respect to phasors. Thus, we may redraw the circuit of Fig. 8–7.1(a) as shown in Fig. 8–7.1(b), where the reference polarities of the transformed voltage variables $V_i(s)$ follow those of the original variables $v_i(t)$. Now let $I(s)$ be the transform of $i(t)$. *If we assume that the initial conditions are zero*, we may express each of the transformed voltages in terms of the transformed current $I(s)$ and the individual impedances $Z_i(s)$ by the relations $V_i(s) = Z_i(s)I(s)$. Substituting these relations in Eq. (8–192) and rearranging terms, we obtain

$$V_0(s) = [Z_1(s) + Z_2(s) + Z_3(s) + \cdots + Z_n(s)]I(s) \tag{8–193}$$

Since the network function defined as $V_0(s)/I(s)$ is the driving-point impedance of the entire circuit, we conclude that *the impedance of a series connection of impe-*

dances is the sum of the individual impedances. In a similar fashion, we may show that KCL is satisfied by a set of transformed current variables, and that *the admittance of a parallel connection of admittances is equal to the sum of the individual admittances.* These conclusions are similar to those reached with respect to phasor immittances in Sec. 7.5. Considering the results given above, we see that we may define a *transformed network*, in which all the variables are the transforms of the original time-domain network variables, and in which the elements are defined in terms of their immittances. In such a transformed network, all the usual rules followed for forming network equations may be applied directly to the elements of the network. As an example of this, consider the network with time-domain variables shown in Fig. 8-7.2(a) and the corresponding transformed network shown in Fig. 8-7.2(b). Applying the voltage-divider law directly to the impedances of the transformed network, we may define the network function

$$\frac{V_L(s)}{V_S(s)} = \frac{sL}{R + (1/sC) + sL} \tag{8-194}$$

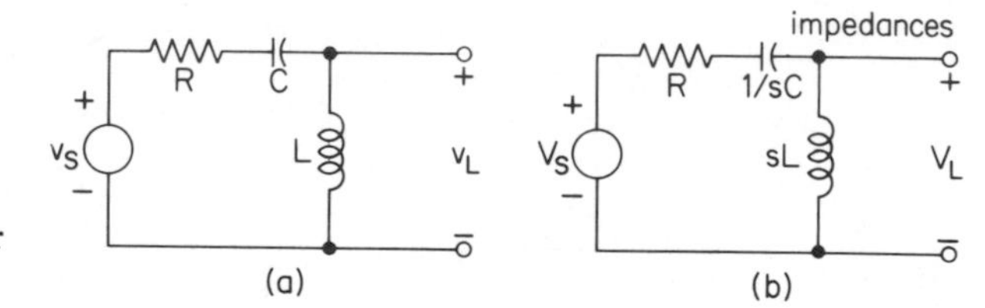

Fig. 8-7.2 A voltage-divider network.

Thus, using the techniques developed in the preceding section, we may directly find the zero-state response for $v_L(t)$ for any arbitrary excitation $v_S(t)$ by inserting the transformed expression for the latter in Eq. (8-194) and finding the inverse transform for $V_L(s)$. These conclusions may be summarized as follows:

SUMMARY 8-7.1

> *The Transformed Network:* If a network configuration is specified in the time domain, and if it is in the zero state, a transformed network may be defined in the complex-frequency domain by replacing all time-domain variables by their equivalent transformed variables and by specifying all the network elements in terms of their immittances. KVL and KCL may be directly applied to the transformed network variables. Thus, elements in series have their impedances add, and elements connected in parallel have their admittances add.

Just as was true for the phasor case, it should be noted that when specifying the element values for the transformed network, some identification must be made to indicate whether impedances or admittances are being specified.

The transformed network may be used to actually find the integro-differential equation relating a response and an excitation variable in a given network. To illustrate this approach, consider the network shown in Fig. 8–7.2(b). Let us assume that it is desired to find the integro-differential equation relating $v_L(t)$ and $v_S(t)$. To begin, let us put the network function relating $V_L(s)$ and $V_S(s)$ given in (8–194) in the following form.

$$sLV_S(s) = \left[R + \frac{1}{sC} + sL\right]V_L(s) \tag{8–195}$$

Since multiplication by s is the transformed equivalent of differentiation, and since division by s is similarly related to integration, the original integro-differential equation must have the form

$$L\frac{d}{dt}v_S(t) = Rv_L(t) + \frac{1}{C}\int_{0-}^{t} v_L(\tau)\,d\tau + L\frac{d}{dt}v_L(t) \tag{8–196}$$

In general, it is easier to apply this method than it is to derive the integro-differential equation directly in the time domain. For example, to find a relation between $v_S(t)$ and $v_L(t)$ for the network shown in Fig. 8–7.2(a), we must first apply KVL and write

$$v_S(t) = Ri(t) + \frac{1}{C}\int_{0-}^{t} i(\tau)\,d\tau + v_L(t) \tag{8–197}$$

Next we must make the substitution $i = (1/L)\int v_L(t)\,dt$. Thus, we obtain

$$v_S(t) = \frac{R}{L}\int_{0-}^{t} v_L(\tau)\,d\tau + \frac{1}{C}\int_{0-}^{t}\left[\frac{1}{L}\int_{0-}^{t} v_L(\tau)\,d\tau\right]dt + v_L(t) \tag{8–198}$$

Differentiating Eq. (8–198) to eliminate the double integral and multiplying through by L, we obtain

$$L\frac{d}{dt}v_S(t) = Rv_L(t) + \frac{1}{C}\int_{0-}^{t} v_L(\tau)\,d\tau + L\frac{d}{dt}v_L(t) \tag{8–199}$$

which corresponds directly with Eq. (8–196). The ease with which the original integro-differential equation relating a pair of network variables can be found from the transformed network is even more apparent when networks with more than a single independent variable are considered. An example follows.

EXAMPLE 8-7.1. It is desired to determine the integro-differential equation relating $v(t)$ and $i_1(t)$ in the network shown in Fig. 8–7.3(a). To do this, we may first draw the transformed network as shown in Fig. 8–7.3(b). Applying KVL around the two meshes, we obtain

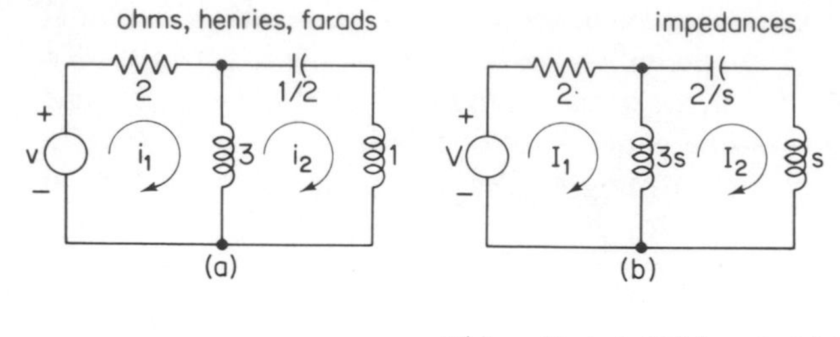

Fig. 8-7.3 Network for Example 8-7.1.

$$V(s) = (2 + 3s)I_1(s) - 3sI_2(s)$$

$$0 = -3sI_1(s) + \left(4s + \frac{2}{s}\right)I_2(s)$$

In matrix form, these equations may be written

$$\begin{bmatrix} V(s) \\ 0 \end{bmatrix} \begin{bmatrix} 2 + 3s & -3s \\ -3s & 4s + \frac{2}{s} \end{bmatrix} \begin{bmatrix} I_1(s) \\ I_2(s) \end{bmatrix}$$

Solving these equations using the method of Sec. 3-1, we obtain

$$\begin{bmatrix} I_1(s) \\ I_2(s) \end{bmatrix} = \frac{s}{3s^3 + 8s^2 + 6s + 4} \begin{bmatrix} 4s + \frac{2}{s} & 3s \\ 3s & 2 + 3s \end{bmatrix} \begin{bmatrix} V(s) \\ 0 \end{bmatrix}$$

Thus, the network function relating $I_1(s)$ and $V(s)$ is

$$\frac{I_1(s)}{V(s)} = \frac{4s^2 + 2}{3s^3 + 8s^2 + 6s + 4}$$

The original differential equation is easily seen to be

$$4v''(t) + 2v(t) = 3i_1'''(t) + 8i_1''(t) + 6i_1'(t) + 4i_1(t)$$

where the primes indicate differentiation with respect to t. To appreciate the simplicity and generality of the transformed network approach, the reader should compare the process of finding the differential equation using the transformed network in the preceding example with the process of solving the original time-domain network for the desired differential equation.

The results given above may be summarized as follows:

SUMMARY 8-7.2

Finding the Integro-Differential Equations for a Time-Domain Network from the Network Function: If $R(s)$ and $E(s)$ are the transforms of a response variable $r(t)$ and an excitation variable $e(t)$ respectively, and, for the zero-state case, are related by the network function

$$\frac{R(s)}{E(s)} = \frac{a_0 + a_1 s + a_2 s^2 + \cdots}{b_0 + b_1 s + b_2 s^2 + \cdots}$$

where the a_i and the b_i are real coefficients, then the differential equation relating $r(t)$ and $e(t)$ is

$$a_0 e(t) + a_1 e'(t) + a_2 e''(t) + \cdots = b_0 r(t) + b_1 r'(t) + b_2 r''(t) + \cdots$$

where the primes indicate differentiation with respect to t.

The basic techniques which were introduced in Chap. 3 for writing mesh and node equations for resistance networks, and which were used in Chap. 7 to write similar equations for complex immittances, may be directly applied to determine the network equations for a transformed network. Thus, in general, the transformed *mesh impedance equations* will have the form

$$\mathbf{V}(s) = \mathbf{Z}(s)\mathbf{I}(s) \tag{8-200}$$

where the rules for finding the elements of the component matrices are summarized as follows:

SUMMARY 8-7.3

Rules for Forming the Mesh Impedance Equations for a Transformed Network: The elements of the matrices of Eq. (8-200) defining the mesh impedance equations for a transformed network may be found as follows:

- $V_i(s)$ The sum of the transforms of all the excitation voltage sources which are present in the ith mesh, with sources whose reference polarity represents a rise when traversing the mesh in the positive reference current direction treated as positive.
- $z_{ij}(s)(i = j)$ The sum of the transformed impedances which are present in the ith mesh.
- $z_{ij}(s)(i \neq j)$ The negative of the sum of all the transformed impedances which are common to the ith and jth meshes.
- $I_i(s)$ The transform of the ith mesh current.

These rules assume that all the mesh currents have the same reference direction, and that no element is traversed by more than two mesh currents.

An example of the application of these rules may be found in the matrix expression given in Example 8-7.1.

A similar development may be easily made to show that node equations may be written by inspection for the transformed network. Such a set of transformed *node admittance equations* has the general form

$$\mathbf{I}(s) = \mathbf{Y}(s)\mathbf{V}(s) \tag{8-201}$$

where the elements of the component matrices may be found by the following rules:

SUMMARY 8-7.4

Rules for Forming the Node Admittance Equations for a Transformed Network: The elements of the matrices of (8-201) defining the node admittance equations for a transformed network may be found as follows:

$I_i(s)$ The sum of the transforms of all the excitation current sources which are connected to the ith node, with sources whose reference polarity is into the node treated as positive.

$y_{ij}(s)(i = j)$ The sum of all the transformed admittances which are connected to the ith node.

$y_{ij}(s)(i \neq j)$ The negative of the sum of any transformed admittances which are connected between nodes i and j.

$V_i(s)$ The transform of the ith node voltage.

These rules assume that the negative reference polarity of all the transformed node voltages are located at a common ground terminal.

An example of the application of these rules to directly determine the node equations follows.

EXAMPLE 8-7.2. It is desired to write the node admittance equations for the network shown in Fig. 8-7.4. Note that this network contains a VCIS, i.e., a dependent voltage-controlled current source as well as two independent current sources. For convenience, assume that all the elements have unity value. Following the rules given above, we obtain

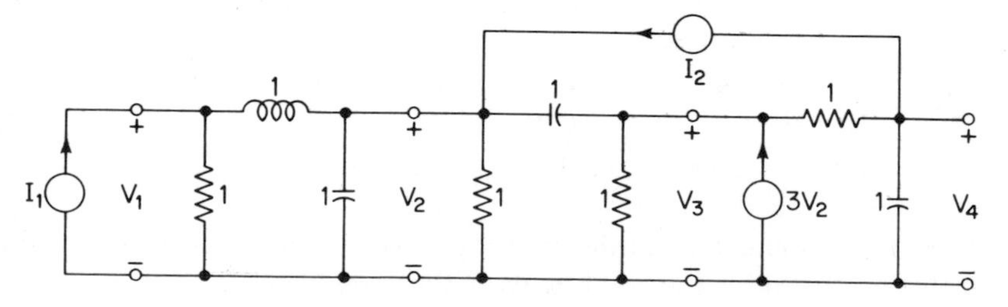

Fig. 8-7.4 Network for Example 8-7.2.

$$\begin{bmatrix} I_1(s) \\ I_2(s) \\ 3V_2(s) \\ -I_2(s) \end{bmatrix} = \begin{bmatrix} 1+\frac{1}{s} & -\frac{1}{s} & 0 & 0 \\ -\frac{1}{s} & 2s+1+\frac{1}{s} & -s & 0 \\ 0 & -s & s+2 & -1 \\ 0 & 0 & -1 & s+1 \end{bmatrix} \begin{bmatrix} V_1(s) \\ V_2(s) \\ V_3(s) \\ V_4(s) \end{bmatrix}$$

Obviously, the term $3V_2(s)$ in the left member of the equation, which is due to the presence of the VCIS, can be moved into the square matrix, thus changing the "32" element from $-s$ to $-(s + 3)$.

The techniques which have been presented in this section for using transformed networks to directly determine network functions, and thus to permit the solution for the zero-state response of time-domain network variables, are easily adapted to include the case in which it is desired to retain the simplicity of the transformed network concept but also to include the effects of initial conditions. As an introduction to the procedure for accomplishing this, consider the network consisting of an uncharged capacitor and a shunt current source as shown in Fig. 8-7.5. The relation for the terminal variables is

$$V(s) = \frac{1}{sC} I(s) \tag{8-202}$$

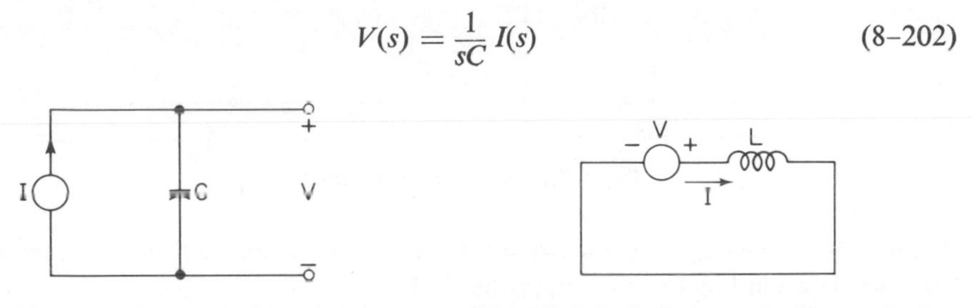

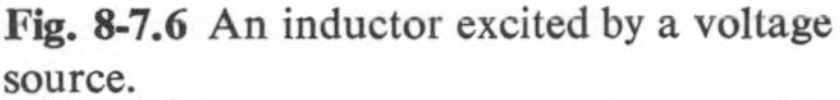

Fig. 8-7.5 A capacitor excited by a current source.

Fig. 8-7.6 An inductor excited by a voltage source.

Now let the current source provide an excitation which is an impulse function of strength KC, i.e., let $i(t) = KC\delta(t)$. Then $I(s) = KC$, and $V(s) = K/s$. Thus, at $t = 0+$, the capacitor voltage is K volts, and we have established an initial condition on the originally uncharged capacitor. A similar procedure may be used to establish an initial condition on an inductor. Consider the series connection of an uncharged inductor and a voltage source shown in Fig. 8-7.6. The relation for the terminal variables is

$$I(s) = \frac{1}{sL} V(s) \tag{8-203}$$

If the voltage source now provides an impulse of strength KL at $t = 0$, then we obtain $I(s) = K/s$. Thus, at $t = 0+$, the inductor current has the value K A, and an initial condition has been established through the use of an external source. The use of impulse generators to establish initial conditions described above is further illustrated by the following example.

EXAMPLE 8-7.3. For the transformed network shown in Fig. 8-7.7(a), it is desired to use the concept of a transformed network to find the transformed response for $I_2(s)$ for the case in which the initial conditions on the inductor currents $i_1(0-)$ and $i_2(0-)$ are not zero. The first step is to redraw the network inserting voltage sources in series with the inductors as shown in Fig. 8-7.7(b). As indicated in the figure, the transformed output of the voltage source which has been placed in series with the 3-H inductor in the first mesh in $3K_1$ (an impulse of strength $3K_1$), while the transformed output of the voltage source which has been placed in series with the $\frac{1}{2}$-H inductor in the second mesh is $K_2/2$. The mesh impedance equations for the network may now be written by applying KVL. Thus, we obtain

$$\begin{bmatrix} V_1(s) + 3K_1 \\ \dfrac{K_2}{2} \end{bmatrix} = \begin{bmatrix} 3s+1 & -1 \\ -1 & \dfrac{s}{2}+1 \end{bmatrix} \begin{bmatrix} I_1(s) \\ I_2(s) \end{bmatrix}$$

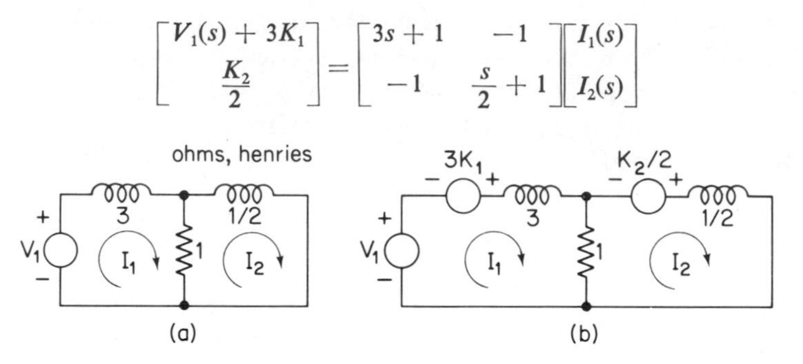

Fig. 8-7.7 Network for Example 8-7.3.

Before solving these equations, let us first write the time-domain equations for the original network shown in Fig. 8-7.7(a). Applying KVL, we obtain

$$v_1(t) = 3\frac{di_1}{dt} + i_1(t) - i_2(t)$$

$$0 = \frac{1}{2}\frac{di_2}{dt} + i_2(t) - i_1(t)$$

Transforming these equations, we see that we may write

$$V_1(s) = 3sI_1(s) - 3i_1(0-) + I_1(s) - I_2(s)$$
$$0 = \tfrac{1}{2}sI_2(s) - \tfrac{1}{2}i_2(0-) + I_2(s) - I_1(s)$$

Rearranging the terms and putting the equations in matrix form, we obtain

$$\begin{bmatrix} V_1(s) + 3i_1(0-) \\ \dfrac{i_2(0-)}{2} \end{bmatrix} = \begin{bmatrix} 3s+1 & -1 \\ -1 & \dfrac{s}{2}+1 \end{bmatrix} \begin{bmatrix} I_1(s) \\ I_2(s) \end{bmatrix}$$

Comparing these equations with those derived directly from the transformed network with the impulse generators shown in Fig. 8-7.7(b), we conclude that if we set K_1 and K_2 equal to the desired initial conditions $i_1(0-)$ and $i_2(0-)$, respectively, then the equations are identical for all values of $t > 0$. Thus, either set of equations may be solved to give an expression for the transformed current $I_2(s)$. The result is

$$I_2(s) = \frac{2V_1(s) + 6K_1 + K_2(3s + 1)}{3s^2 + 7s} = \frac{2V_1(s) + 6i_1(0-) + i_2(0-)(3s + 1)}{3s^2 + 7s}$$

The results given above may be summarized as follows:

SUMMARY 8-7.5

Using Impulses to Establish Initial Conditions in a Transformed Network: Non-zero initial conditions may be established in a transformed network by shunting capacitors with current sources and by placing voltage sources in series with inductors. If the outputs of these sources are all impulses occurring at $t = 0$, then their only effect is to establish initial conditions on the network elements, and they have no effect on the network at any other time.

A second method for establishing initial conditions in the capacitors and inductors of a transformed network may be developed from a consideration of the transform of the integral relations for the terminal variables of these elements. For example, for capacitors, we may write

$$V(s) = \frac{1}{sC}I(s) + \frac{v(0-)}{s} \tag{8-204}$$

Thus, a voltage source in series with an uncharged capacitor may be used as an equivalent circuit for a capacitor with a non-zero initial condition of voltage. The output of the voltage source must be a step function of magnitude $v(0-)$ starting at $t = 0$. Similarly, for an inductor, the transform of the integral expression for the terminal variables is

$$I(s) = \frac{1}{sL}V(s) + \frac{i(0-)}{s} \tag{8-205}$$

Thus, a current source in shunt with an uncharged inductor may be used as an equivalent circuit for an inductor with a non-zero initial condition of current. The output of the current source must be a step function of magnitude $i(0-)$ starting at $t = 0$. It is important to observe the difference between the approach based on the use of transformed integral relations described above and the approach given in Summary 8-7.5. The method described in the summary was based on the use of impulses. Thus, the only function of the added sources was to establish the initial conditions on the uncharged network elements. For all other values of time, the sources had no effect on the circuit; thus, they could be ignored. In the method described in this paragraph, however, the uncharged element *remains initially uncharged.* Thus, the effect of the sources must be considered for all values of time. This approach may be summarized as follows:

SUMMARY 8-7.6

Using Step Functions to Establish Initial Conditions in a Transformed Network: Non-zero initial conditions may be established in a transformed network by connecting voltage sources in series with capacitors and by shunting inductors with current sources. If the output of these sources are step functions starting at $t = 0$, then each combination of a source and an uncharged reactive element is equivalent to a reactive element which has a non-zero initial condition.

In deciding which of the above methods to apply in analyzing a given network, several factors must be taken into consideration. For example, if there is already some independent excitation voltage source in series with a reactive element, then the simplest approach is usually to use another series voltage source (with an impulse output for an inductor and a step output for a capacitor) to establish the initial condition. Similarly, if there is already an independent excitation current source in shunt with a reactive element, it is usually most convenient to use an additional shunt current source (with an impulse output for a capacitor and a step output for an inductor) to establish the initial condition.

As an example of the use of step and impulse functions to establish initial conditions, consider the following.

EXAMPLE 8-7.4. The network shown in Fig. 8-7.8(a) is at rest with the switch S1 in the open position. The current that flows through the first of the coupled set of coils (after all the transients have disappeared) is 5 A. Since the energy associated with this current is stored in the magnetic field, and since this magnetic field links both windings, it is more meaningful to use a T equivalent circuit for the set of coupled coils (see Sec. 4-8). With such a modification, the circuit may be redrawn as shown in Fig. 8-7.8(b). In this circuit, two of the inductors in the T have an initial current and one does not. We may now add two shunt current sources each with a step output current of 5 A starting at $t = 0$ to the network as shown in Fig. 8-7.8(c). With such a representation, all three of the inductors may be treated as being uncharged. Thus, the current sources provide the initial conditions. If the switch $S1$ is closed at $t = 0$, the transformed network representation for $t > 0$ is as shown in Fig. 8-7.8(d). In this figure the initial conditions on all three of the inductors are zero. An alternative transformed representation using impulse generators is shown in Fig. 8-7.8(e). It is left as an exercise to the reader to verify that the network equations which result from the transformed networks shown in Figs. 8-7.8(d) and (e) both have the same solution.

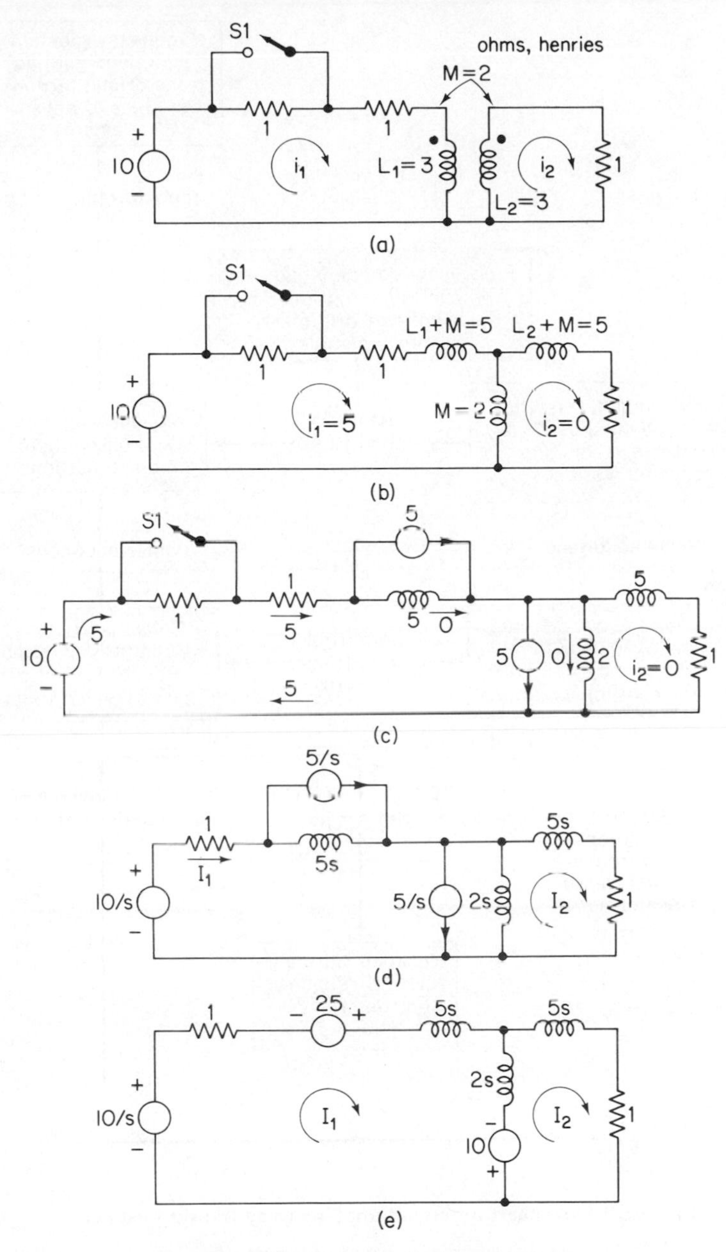

Fig. 8-7.8 Network for Example 8-7.4.

The various cases of generating initial conditions which are most conveniently applied to the mesh and node analysis situations are summarized in Table 8-7.1. It should be noted that the two types of excitation used to generate initial conditions for either an inductor or a capacitor are easily related to each other by the use of

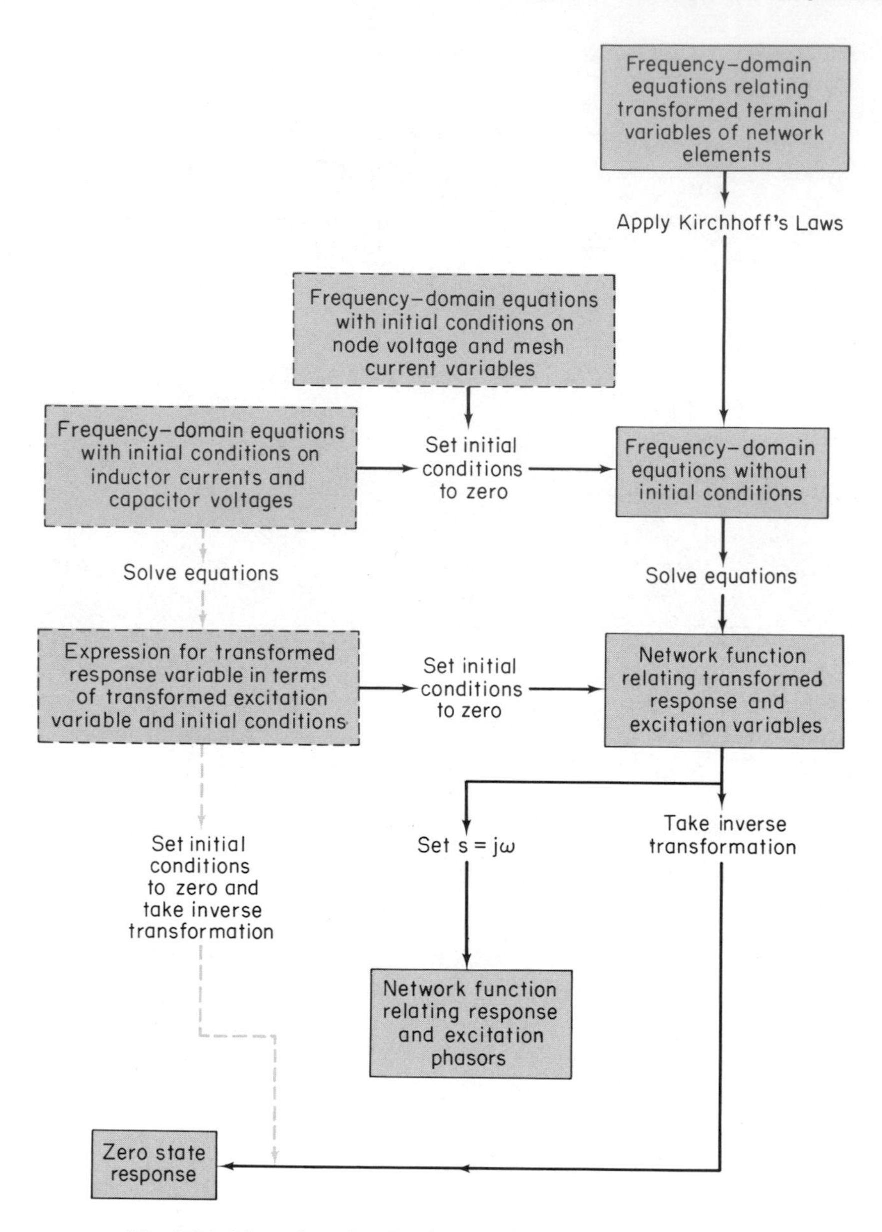

Fig. 8-7.9 Flow chart for circuit analysis using transformed networks.

Thevenin and/or Norton equivalent circuits. The proof of this is left to the reader as an exercise. In the next section we shall see that the use of voltage sources in series with inductors and current sources in shunt with capacitors to establish initial conditions is an example of a more general principle which makes it possible to add

TABLE 8-7.1 Generating Initial Conditions in a Transformed Network

Type of Analysis	Type of Element	
	Capacitor	Inductor
Mesh	series voltage source with step output $v(0)u(t)$	series voltage source with impulse output $i(0)L\delta(t)$
Node	shunt current source with impulse output $v(0)C\delta(t)$	shunt current source with step output $i(0)u(t)$

sources to networks in such a way that the natural frequencies of the network are unchanged.

In this section we have introduced the concept of a transformed network and showed how it may be used in circuit analysis. A flow chart illustrating the general process is shown in Fig. 8–7.9. The main flow followed by the process is illustrated using solid lines starting at the upper right corner of the chart. In addition, some shaded lines have been used to include portions of the flow chart given in Fig. 8–6.6 which summarizes the time-domain equation approach. The interrelationships between the two methods are readily apparent. It is strongly recommended that the student carefully study both of these flow charts as an aid to his understanding of the material covered to this point in this chapter.

8-8 NATURAL FREQUENCIES OF NETWORKS

In the preceding sections of this chapter we defined the concept of a transformed network, and (in Summaries 8–7.3 and 8–7.4) we showed how the mesh impedance and node admittance equations could be directly written from the transformed network. The solution of such sets of equations follows the general rules for the solution of a set of simultaneous equations given in Secs. 3–3 and 3–4. For example, consider the mesh impedance equations. In matrix notation these will have the form

$$\mathbf{V}(s) = \mathbf{Z}(s)\mathbf{I}(s) \tag{8–206}$$

In expanded form, such a set of equations may be written

$$\begin{bmatrix} V_1(s) \\ V_2(s) \\ \cdot \\ \cdot \\ \cdot \\ V_n(s) \end{bmatrix} = \begin{bmatrix} z_{11}(s) & z_{12}(s) & \cdots & z_{1n}(s) \\ z_{21}(s) & z_{22}(s) & \cdots & z_{2n}(s) \\ \cdot & \cdot & \cdots & \cdot \\ \cdot & \cdot & \cdots & \cdot \\ \cdot & \cdot & \cdots & \cdot \\ z_{n1}(s) & z_{n2}(s) & \cdots & z_{nn}(s) \end{bmatrix} \begin{bmatrix} I_1(s) \\ I_2(s) \\ \cdot \\ \cdot \\ \cdot \\ I_n(s) \end{bmatrix} \tag{8–207}$$

where the elements of the various matrices are defined in Summary 8–7.3, and where the quantities $V_i(s)$ may include the effects of impulse generators which have been used to establish initial conditions, as well as other types of excitations. The solution to this set of equations will have the form

$$\begin{bmatrix} I_1(s) \\ I_2(s) \\ \cdot \\ \cdot \\ \cdot \\ I_n(s) \end{bmatrix} = \frac{1}{\det \mathbf{Z}(s)} \begin{bmatrix} Z_{11}(s) & Z_{21}(s) & \cdots & Z_{n1}(s) \\ Z_{12}(s) & Z_{22}(s) & \cdots & Z_{n2}(s) \\ \cdot & \cdot & \cdots & \cdot \\ \cdot & \cdot & \cdots & \cdot \\ \cdot & \cdot & \cdots & \cdot \\ Z_{1n}(s) & Z_{2n}(s) & \cdots & Z_{nn}(s) \end{bmatrix} \begin{bmatrix} V_1(s) \\ V_2(s) \\ \cdot \\ \cdot \\ \cdot \\ V_n(s) \end{bmatrix} \tag{8–208}$$

where the quantities $Z_{ij}(s)$ are the cofactors of the $\mathbf{Z}(s)$ matrix. Note that, as specified by Cramer's rule (see Sec. 3–3), the "12" element in the inverse matrix is the $Z_{21}(s)$ cofactor, etc. The linear nature of the transformed network is readily apparent from the equations given in (8–208). For example, consider the transform for the mesh current $I_1(s)$. From (8–208), this may be written as

$$I_1(s) = \frac{Z_{11}(s)}{\det \mathbf{Z}(s)} V_1(s) + \frac{Z_{21}(s)}{\det \mathbf{Z}(s)} V_2(s) + \cdots + \frac{Z_{n1}(s)}{\det \mathbf{Z}(s)} V_n(s) \tag{8–209}$$

From this equation we may note that the transformed response $I_1(s)$ is given as the sum of the effects produced by the individual excitations $V_1(s), V_2(s), \ldots, V_n(s)$. Thus, the principle of *superposition* applies. Similarly, if all the excitations except one, say $V_j(s)$, are zero, then multiplying this excitation by a factor k must necessarily multiply the response $I_1(s)$ by the same factor. Thus, the principle of *homogeneity* applies. These two properties, of course, define linearity.

Another very significant property of linear networks is readily observed from Eq. (8–209). This is the fact that the quantity det $\mathbf{Z}(s)$ appears in the denominator of each term in the right member. Thus, the zeros of the determinant of $\mathbf{Z}(s)$ are the poles of all network functions which have $I_1(s)$ as a response variable, independent of which variable is the excitation variable.

The conclusions reached above for the transformed mesh current $I_1(s)$ are readily extended to apply to any mesh current. To do this, we need merely note that the expression for the jth transformed mesh current $I_j(s)$ is readily shown to be

$$I_j(s) = \frac{Z_{1j}(s)}{\det \mathbf{Z}(s)} V_1(s) + \frac{Z_{2j}(s)}{\det \mathbf{Z}(s)} V_2(s) + \cdots + \frac{Z_{nj}(s)}{\det \mathbf{Z}(s)} V_n(s) \tag{8–210}$$

Comparing the form of this equation with that given for $I_1(s)$ in Eq. (8–209), we see that all the network functions which have mesh currents as their response variables will also have the zeros of det $\mathbf{Z}(s)$ as their poles. Thus, we may conclude that *the zeros of the determinant of the mesh impedance matrix* $\mathbf{Z}(s)$ *are the natural frequencies of the network.*

A development similar to that given above may be used to show that if a transformed network is described by a set of node admittance equations having the form

$$\mathbf{I}(s) = \mathbf{Y}(s)\mathbf{V}(s) \tag{8-211}$$

where the elements of the various matrices are defined in Summary 8-7.4, then the solution for the jth transformed node voltage variable will be

$$V_j(s) = \frac{Y_{1j}(s)}{\det \mathbf{Y}(s)} I_1(s) + \frac{Y_{2j}(s)}{\det \mathbf{Y}(s)} I_2(s) + \cdots + \frac{Y_{nj}(s)}{\det \mathbf{Y}(s)} I_n(s) \tag{8-212}$$

where the quantities $Y_{ij}(s)$ are the cofactors of the matrix $\mathbf{Y}(s)$. Thus, we see that the poles of all network functions which have the transformed node voltages as their response variables are the zeros of det $\mathbf{Y}(s)$. Thus, we may conclude *that the zeros of the determinant of the node admittance matrix* $\mathbf{Y}(s)$ *are the natural frequencies of the network.* The above conclusions, of course, assume that none of the zeros of the determinants are cancelled by corresponding zeros in the cofactors. Such a case is sometimes referred to as a *degenerate* case. The conclusions given above may be summarized as follows:

SUMMARY 8-8.1

Natural Frequencies of a Network: The natural frequencies of a network are the zeros of the determinant of the matrix $\mathbf{Z}(s)$ formed by writing the mesh impedance equations for the transformed network. The natural frequencies of a network are also the zeros of the determinant of the matrix $\mathbf{Y}(s)$ formed by writing the node admittance equations for the network. It is possible for some of these natural frequencies to be cancelled by a corresponding zero of a cofactor.

The natural frequencies of a network are determined by several factors: the type of network elements, the values of elements, and the topology used to interconnect them. The natural frequencies specify the general form of the zero-input response of the network, i.e., the general form of the time-domain behavior of the network variables which results from excitation by initial conditions. Thus, the addition (or deletion) of independent sources to a network will leave its natural frequencies invariant if these sources are added in such a way so as not to change the topology of the network that results when all the independent sources are set to zero. Specifically, this is true for independent *current* sources which are connected in *parallel* with a given element or elements, and independent *voltage* sources which are connected in *series* with a given element. The latter connection is made by breaking one of the leads to the given element and connecting the two ends of the broken lead to the voltage source. An example of such a situation occurred in connection with the discussion leading to Summary 8–7.5, in which series voltage sources and shunt current sources were used in the transformed network to establish initial conditions on inductors and capacitors, respectively. As another example of this conclusion, consider the transformed network shown in Fig. 8–8.1(a). It is claimed that the natural frequencies of this network are located at $s = -1 \pm j1$. To verify this claim, let us add independent current sources $I_1(s)$ and $I_2(s)$ and define transformed node voltages $V_1(s)$ and $V_2(s)$ as shown in Fig. 8–8.1(b). Applying KCL to this network, we obtain the following node admittance equations:

$$\begin{bmatrix} I_1(s) \\ I_2(s) \end{bmatrix} = \begin{bmatrix} s+2 & -2 \\ -2 & \frac{4}{s}+4 \end{bmatrix} \begin{bmatrix} V_1(s) \\ V_2(s) \end{bmatrix} \tag{8–213}$$

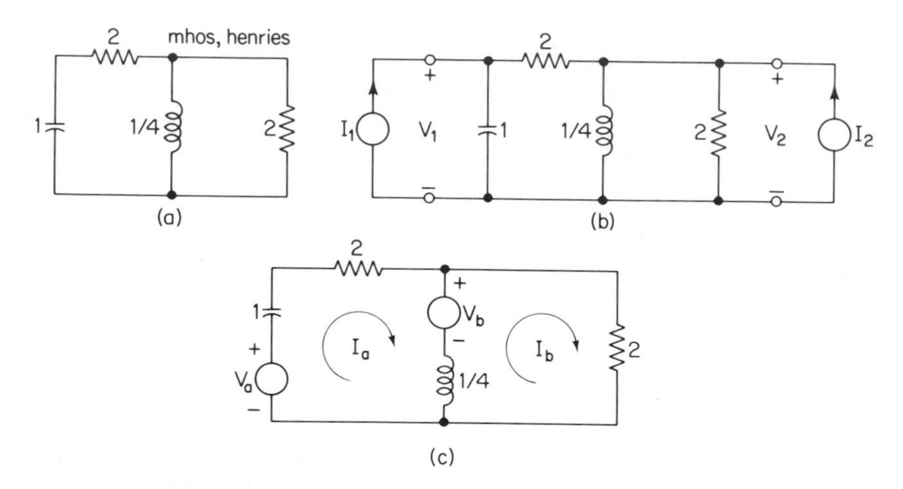

Fig. 8-8.1 Networks with identical natural frequencies.

Setting the determinant of the admittance matrix of Eq. (8–213) to zero, we obtain

$$s^2 + 2s + 2 = (s + 1 + j1)(s + 1 - j1) = 0 \tag{8–214}$$

We conclude that the natural frequencies of the network shown in Fig. 8–8.1(a) are the values of s which satisfy Eq. (8–214), namely, $s = -1 \pm j1$, as originally claimed. The network natural frequencies, of course, will be the same whether mesh or node analysis is used. As an example of this, consider the circuit shown in Fig. 8–8.1(c). This is the original transformed network as shown in Fig. 8–8.1(a), which has been modified by inserting independent voltage sources $V_a(s)$ and $V_b(s)$ in series with the 1-F capacitor and the $\frac{1}{4}$-H inductor. Defining mesh currents $I_a(s)$ and $I_b(s)$ as shown and applying KVL, we obtain the mesh impedance equations

$$\begin{bmatrix} V_a(s) - V_b(s) \\ V_b(s) \end{bmatrix} = \begin{bmatrix} \frac{1}{s} + \frac{s}{4} + \frac{1}{2} & -\frac{s}{4} \\ -\frac{s}{4} & \frac{s}{4} + \frac{1}{2} \end{bmatrix} \begin{bmatrix} I_a(s) \\ I_b(s) \end{bmatrix} \tag{8-215}$$

Setting the determinant of the impedance matrix of Eq. (8–215) to zero, we again obtain Eq. (8–214), verifying our claim that the natural frequencies of the network are located at $s = -1 \pm j1$. From the above example, we may conclude that the natural frequencies of a network are independent of the type of excitation (i.e., whether by independent current sources or independent voltage sources), the number of such excitation sources, and their location in the network, as long as the network is not changed when such sources are set to zero. Thus, for example, all the networks shown in Fig. 8–8.1 have the same natural frequencies. As a counter example, consider the network shown in Fig. 8–8.2(a). This network does not have the same natural frequencies as the ones shown in Fig. 8–8.1, since the capacitor and resistor which are in series with the current source do not affect the natural frequencies of the network. As an example of this, let us determine the transfer function $V_2(s)/I(s)$. This may be done directly by noting that the impedance $Z(s)$ of the parallel combination of the inductor and the resistor is

$$Z(s) = \frac{1}{(4/s) + 2} = \frac{s}{2s + 4} = \frac{s}{2(s + 2)} \tag{8-216}$$

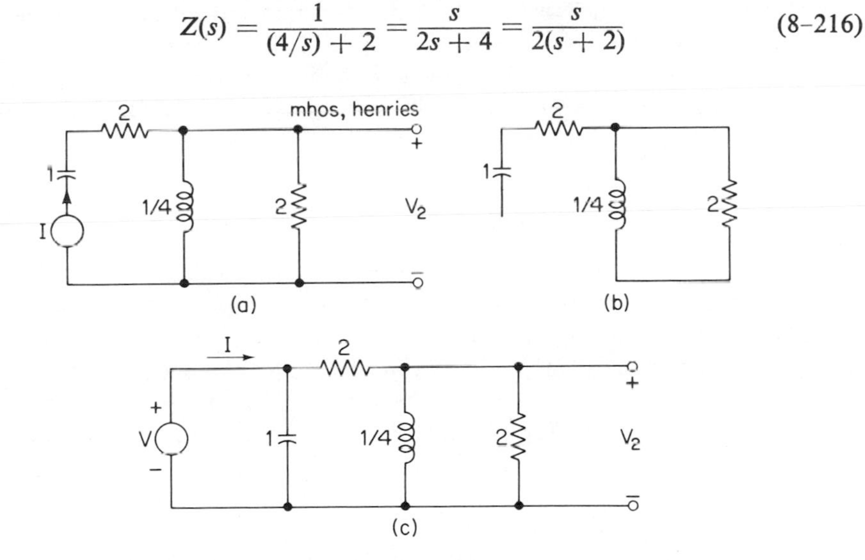

Fig. 8-8.2 Networks which do not have identical natural frequencies.

Since $V_2(s)$ is simply $Z(s)$ times $I(s)$, we may directly write

$$\frac{V_2(s)}{I(s)} = \frac{s}{2(s + 2)} \tag{8-217}$$

thus, the network has a natural frequency at $s = -2$. This result is easily generalized to show that elements which are *in series with a current source*, such as the resistor and capacitor shown in Fig. 8–8.2(a), will not affect the natural frequencies of a network. This is done by noting that when the current source is set to zero and replaced by an open circuit, as shown in Fig. 8–8.2(b), such elements are completely superfluous as far as making any contribution to the properties of the zero-input network response, since no current flows through them. It should be noted, however, that such elements, namely, those which are in series with a current source, will always have an effect on the natural frequencies of the driving-point impedance of the network as seen at the terminals of the current source. For example, for the network shown in Fig. 8–8.2(a), the driving-point impedance $Z_S(s)$ seen at the terminals of the current source is the sum of the impedance $Z(s)$ of Eq. (8–216) and the impedance of the series-connected resistor and capacitor. Thus, we may write

$$Z_S(s) = \frac{V_1(s)}{I(s)} = \frac{1}{s} + \frac{1}{2} + \frac{s}{2(s+2)} = \frac{s^2 + 2s + 2}{s(s+2)} \tag{8–218}$$

Thus, the driving-point impedance has a pole at $s = 0$, which is not one of the natural frequencies of the network. It should be noted that driving-point functions are independent of the type of source which is used to excite the network at the given pair of terminals at which the driving-point function is defined. As an example of this, consider again the network shown in Fig. 8–8.1(c). The relation for $I_a(s)/V_a(s)$ [when $V_b(s)$ is set to zero] defines a driving-point admittance at the same pair of terminals and for the same network as the driving-point impedance $Z_S(s)$ defined in Eq. (8–218). Thus, solving Eq. (8–215) for $I_a(s)/V_a(s)$, we obtain

$$\frac{I_a(s)}{V_a(s)} = \frac{s(s+2)}{s^2 + 2s + 2} \tag{8–219}$$

This is, of course, the reciprocal of the expression obtained in (8–218).

As another example of using an excitation source in such a way as to change the natural frequencies of a network, consider the circuit shown in Fig. 8–8.2(c). The capacitor which is in parallel with the voltage source will not affect the natural frequencies of the network. It is readily shown, for example, that the voltage transfer function $V_2(s)/V(s)$ is

$$\frac{V_2(s)}{V(s)} = \frac{s}{2(s+1)} \tag{8–220}$$

Thus, the network has a single natural frequency at $s = -1$ independent of the capacitor value. The capacitor, of course, does affect the driving-point functions of the network. For example, we may readily show that the driving-point admittance at the terminals of the voltage source is

$$\frac{I(s)}{V(s)} = \frac{s^2 + 2s + 2}{s + 1} \tag{8-221}$$

Thus, the driving-point admittance has an additional pole at s equals infinity which is not a natural frequency of the network. The admittance is, of course, the reciprocal of the driving-point impedance, found by exciting the network at the same pair of terminals with a current source. That is, it is the reciprocal of the network function $V_1(s)/I_1(s)$ for the network shown in Fig. 8-8.1(b), and thus the numerator of Eq. (8-221) is the same as the left member of (8-214).

The results given above are easily generalized. Let a network be excited at a pair of terminals by a current source as shown in Fig. 8-8.3(a), and let the driving-point impedance have the form

$$\frac{V(s)}{I(s)} = \frac{A(s)}{B(s)} \tag{8-222}$$

where $A(s)$ and $B(s)$ are polynomials with real coefficients. In general, the zeros of $B(s)$ will *include* the natural frequencies of the network, and, if there are no elements in series with the current source, then *all the zeros* of $B(s)$ will be the natural frequencies of the network. If the current source is set to zero, the given network

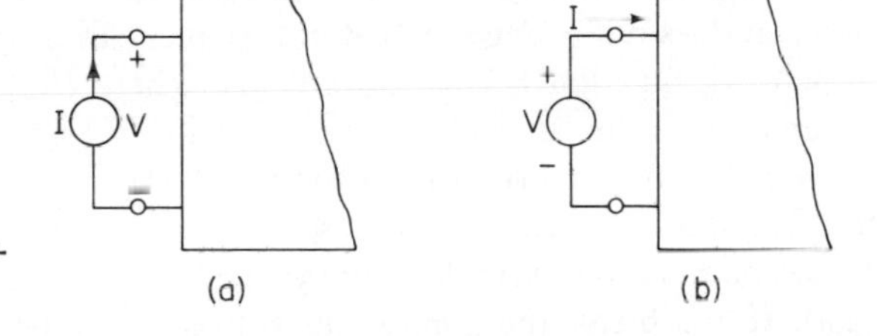

Fig. 8-8.3 Open-circuit and short-circuit natural frequencies.

terminals are effectively open-circuited, thus these natural frequencies are frequently referred to as the *open-circuit natural frequencies* of the network. It should be noted that the term "open-circuit" refers to the condition at the indicated pair of terminals. Now let the same network be excited by a voltage source as shown in Fig. 8 8.3(b). Since the driving-point admittance must always be the reciprocal of the driving-point impedance, we see that

$$\frac{I(s)}{V(s)} = \frac{B(s)}{A(s)} \tag{8-223}$$

In this case, the zeros of $A(s)$ will *include* the natural frequencies of the network, and, if there are no elements in shunt with the voltage source, then *all the zeros* of $A(s)$ will be the natural frequencies of the network. Since, if the voltage source is set to zero, the given network terminals are effectively short-circuited, the network natural frequencies in this case are referred to as the *short-circuit natural frequencies* of the network with respect to the given terminal pair. The results given above may be summarized as follows:

SUMMARY 8-8.2

The Effect of Independent Excitation Sources on the Natural Frequencies of a Network: The addition or deletion of current sources in parallel with network elements, or voltage sources in series with network elements, does not change the natural frequencies of the network. Correspondingly, network elements which are in series with current sources or in parallel with voltage sources have no effect on the natural frequencies of a network, although they do affect the driving-point immittance as seen from the terminals of such sources.

The discussion given above concerning the effect of independent voltage and current sources on the natural frequencies of a network provides the key to some methods of determining the natural frequencies which may be easier to apply than the methods given in Summary 8–8.1. To see this, consider a network in which the independent sources have been set to zero. Now let us break some connection in the network and insert a voltage source across the terminals of the break. Such a procedure is illustrated in Figs. 8–8.1(a) and (c). Obviously, from Summary 8–8.2, such an insertion does not change the natural frequencies of the network. Now let us define a network function $1/Z(s) = I(s)/V(s)$, where $V(s)$ is the output of the voltage source and $I(s)$ is the current flowing out of it. We see that the natural frequencies [i.e., the poles of the network function $1/Z(s)$] are the zeros of $Z(s)$. Thus, these natural frequencies can be found by setting the impedance to zero and solving for the values of s. As an example of this procedure, consider the network shown in Fig. 8–8.1(a). If we break the connection between the 1-F capacitor and the 2-mho resistor, the input $Z(s)$ seen at this break is

$$Z(s) = \frac{1}{2} + \frac{1}{s} + \frac{1}{(4/s) + 2} = \frac{s^2 + 2s + 2}{s(s + 2)} \tag{8–224}$$

Setting $Z(s) = 0$, we find the natural frequencies are located at $s = -1 \pm j1$, as was previously shown. The dual situation is also readily verified, i.e., setting the immittance $Y(s)$ seen at any pair of terminals to zero and solving for s also gives the natural frequencies of the network.

In the preceding paragraphs of this section we have made some observations on the natural frequencies of networks and the manner in which these are related to driving-point functions. Now let us investigate some characteristics of *transfer* functions. Specifically, let us determine the effect on the natural frequencies of a given network when we interchange the point at which an excitation is applied to a network and the point at which a response is measured. Since there are two types of excitations (independent voltage and current sources) and two types of response (voltage and current), there are four possible cases. Let the first case involve a voltage excitation source and a current response (a transfer admittance). Thus, we have the situation shown in Fig. 8-8.4(a). The interchange of the points of excitation and response produces the situation shown in Fig. 8-8.4(b). In both of these circuits, setting the source to zero produces the network shown in Fig. 8-8.4(c). Thus, we may conclude that the natural frequencies of the network are unchanged when the points of excitation and response are interchanged. As a second case, consider the use of a current excitation source and a voltage response (a transfer impedance) as shown in Fig. 8-8.5(a). Interchanging these quantities yields the situation shown in Fig. 8-8.5(b). In both of these circuits, setting the source to zero yields the network shown in Fig. 8-8.5(c). Thus, in this case, the natural frequencies are again unaffected by the interchange. Let the third case be the one involving the use of a voltage excitation source and a voltage response (a transfer voltage ratio) as shown in Fig. 8-8.6(a). Interchanging these quantities produces the situation shown in Fig. 8-8.6(b). Comparing these two figures, we see that setting the sources to zero in the two figures produces different terminations on the network terminals. Thus, we conclude that the interchange of source and excitation necessarily produces a different

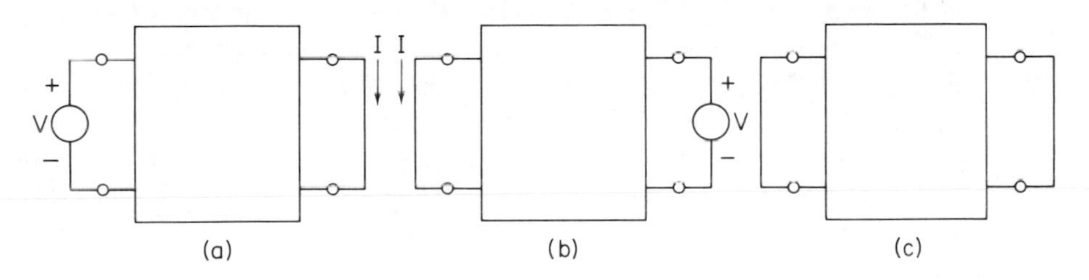

Fig. 8-8.4 Network connections with the same natural frequencies.

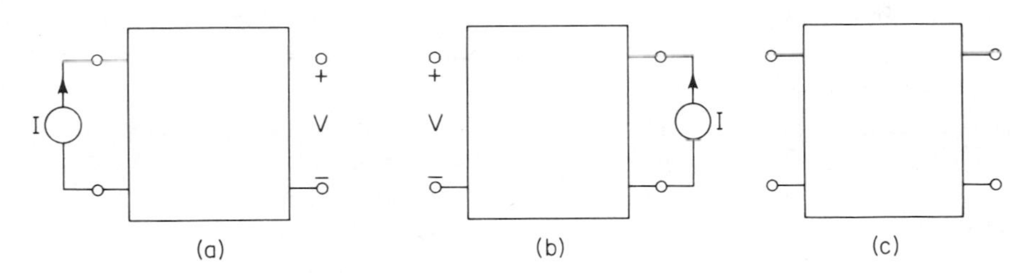

Fig. 8-8.5 Network connections with the same natural frequencies.

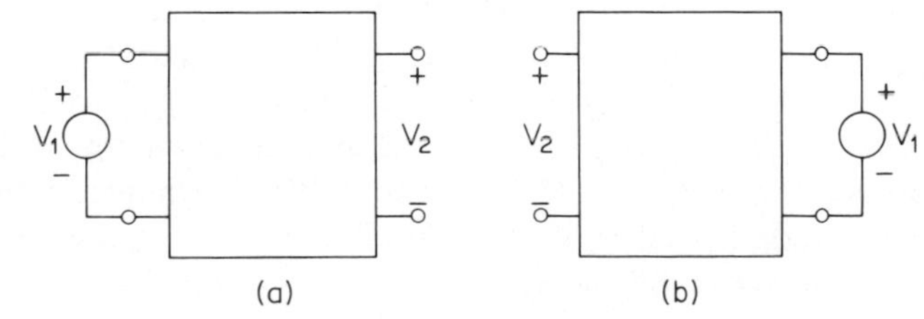

Fig. 8-8.6 Network connections with different natural frequencies.

set of network natural frequencies in this case. The final case, consisting of a current excitation source and a current response (a transfer current ratio), is shown in Fig. 8–8.7(a). The interchange of variables produces the situation shown in Fig. 8–8.7(b). Again we see that the two networks will have different natural frequencies. These conclusions may be summarized as follows:

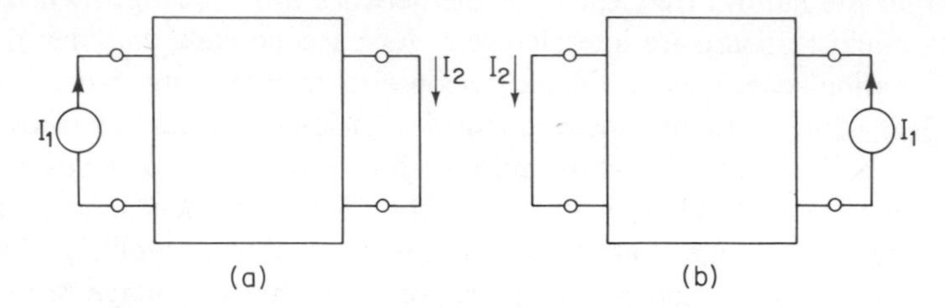

Fig. 8-8.7 Network connections with different natural frequencies.

SUMMARY 8-8.3

Interchange of Excitation and Response: If a network function as defined for a given network is a transfer admittance or a transfer impedance, the interchange of the points at which the excitation is applied and the response is measured will not affect the natural frequencies of the network. If a network function is a voltage transfer ratio or a current transfer ratio, however, the interchange of excitation and response will change the natural frequencies of the network.

In the first two cases we defined above, namely, the transfer immittance cases, we have seen that the natural frequencies, i.e., the *poles* of the network functions, are not affected when the points of excitation and response are interchanged. Now let us consider the *zeros* of the network functions for these two cases. For each case, there are two possibilities. The first is that the zeros will also remain unchanged. If such a result is true for any point of excitation and any point of response (excluding driving-point situations), then the network is said to be *reciprocal*. If there are one or more transfer functions for which the zeros are different when such an interchange is made, then the network is correspondingly said to be *non-reciprocal*. A direct result of the reciprocal or non-reciprocal nature of a network is seen in the mesh impedance formulation given in Eq. (8–207). If the matrix $\mathbf{Z}(s)$ is symmetric, i.e., if $z_{ij}(s) = z_{ji}(s)$, then the inverse matrix given in Eq. (8–208) will also be symmetric. Thus, the effect on the ith mesh current $I_i(s)$ due to a voltage source $V_j(s)$ in the jth mesh, i.e., $Z_{ji}(s)/\det \mathbf{Z}(s)$, will be the same as the effect on the jth mesh current $I_j(s)$ produced by a voltage source $V_i(s)$ in the ith mesh, i.e., $Z_{ij}(s)/\det \mathbf{Z}(s)$. Thus, *if* $\mathbf{Z}(s)$ *is symmetric, the network is reciprocal.* A similar conclusion holds for the node admittance matrix given in Eq. (8–211). That is, if $\mathbf{Y}(s)$ is symmetric, the network is reciprocal. The converse is readily shown to hold, namely, if $\mathbf{Z}(s)$ is non-symmetric, the network is non-reciprocal. Similarly, if $\mathbf{Y}(s)$ is non-symmetric, the network is non-reciprocal.

This result is summarized as follows:

SUMMARY 8-8.4

Reciprocity: A reciprocal network is one in which the interchange of excitation and response in any transfer immittance leaves the given transfer immittance invariant. A non-reciprocal network is one in which this is not true. Reciprocal networks have symmetric mesh impedance and node admittance matrices.

If the various example networks which have been analyzed in this chapter are examined, it is fairly simple to make some general observations concerning the type of networks which are reciprocal and those which are non-reciprocal. In general, networks comprised only of two-terminal elements such as resistors, capacitors, and inductors will always be reciprocal, while networks containing controlled sources will usually be non-reciprocal.[16] This is true independent of whether the two-terminal elements are active or passive.

When we first introduced resistors (in Sec. 2–1) and capacitors and inductors (in Secs. 4–1 and 4–6, respectively), we pointed out that in the analysis of circuits containing these elements it would be convenient to restrict our attention to elements which had small values (say from approximately 0 to 10) so as to simplify the resulting numerical computations. We now demonstrate that such a restriction is indeed a trivial one, since, by a process called *normalization*, we may translate such circuits to ones which have any range of element values that we may desire. There are two basic types of normalizations, namely, frequency normalizations and impedance normalizations. Both of these will be treated in the following paragraphs.

The first type of normalization that we shall consider is *frequency* normalization. Such a normalization can be viewed as a change of the complex frequency variable. Thus, we may define a normalized frequency variable p by the relation

$$p = k_f s \tag{8–225}$$

where s is the usual complex frequency variable defined by the Laplace transformation and k_f is a frequency normalization constant.

Now let us consider how such a substitution affects the poles and zeros of a network function. A factor of the form $(s - s_0)$ defining a singular point (a pole or a zero) at $s = s_0$, becomes a factor $[(p/k_f) - s_0] = (p - k_f s_0)/k_f$ representing a singular point at $p = k_f s_0$. Thus all the singular points will be relocated radially, along lines passing through the origin, by the transformation. They will be either moved closer to the origin or farther away, depending on whether k_f is less than or

[16] If the controlling variable *and* the output variable are both defined at the same branch, then the controlled source acts as a single two-terminal element, and the network containing such elements is reciprocal.

greater than unity. Obviously, a network which has a certain behavior at a point $j\omega_0$ in the s plane will have the same behavior at the point $jk_f\omega_0$ in the p plane. As an example of such a frequency normalization, consider the network function

$$\frac{V_2}{V_1}(s) = \frac{2s}{s^2 + 2s + 2} \tag{8-226}$$

If, as a "sample" point, we evaluate this network function at $s = j\sqrt{2}$, we find that it has a value of unity. In addition, it is readily seen that the poles of this network function are at $s = -1 \pm j1$. Now let us apply the transformation $p = 10^3 s$ to the network function. We obtain

$$\frac{V_2}{V_1}(p) = \frac{2 \times 10^{-3} p}{10^{-6}p^2 + 2 \times 10^{-3}p + 2} = \frac{2 \times 10^3 p}{p^2 + 2 \times 10^3 p + 2 \times 10^6} \tag{8-227}$$

The poles of this network function are at $p = -10^3 \pm j10^3$. Clearly, the poles have been shifted by the factor 10^3. In addition, if we evaluate the network function at $p = j\sqrt{2} \times 10^3$, we find that it also has a value of unity just as the original network function had at $s = j\sqrt{2}$.

In the preceding paragraph we showed how a frequency normalization can be applied to a network function to shift the position of the critical frequencies. Identical results can be obtained by applying the frequency normalization directly to the elements of a network which realizes the network function. For example, consider the network shown in Fig. 8-8.8(a). It is readily verified that the voltage

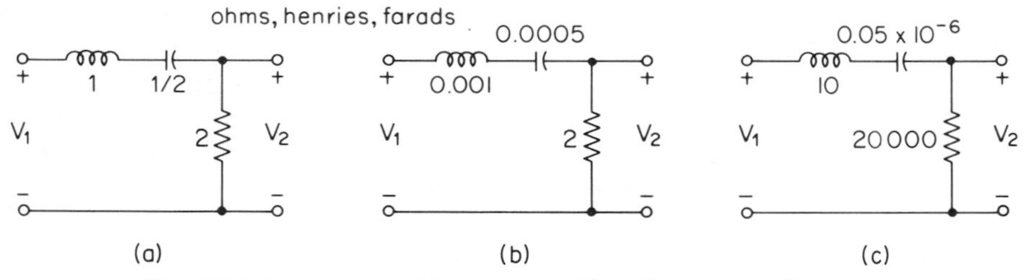

Fig. 8-8.8 An example of frequency and impedance normalization.

transfer function for this network is as given in (8-226). If we apply the frequency normalization $p = 10^3 s$ directly to the network elements, then the 1-H inductor $[Z(s) = s]$ becomes a 1-mH inductor $[Z(p) = 10^{-3}p]$ and the $\frac{1}{2}$-F capacitor $[Y(s) = 0.5s]$ becomes a $\frac{1}{2}$-mF capacitor $[Y(p) = 0.5 \times 10^{-3}p]$. Since the resistor is not a function of frequency, its value stays constant. Thus, the frequency normalized network has the form shown in Fig. 8-8.8(b). It is readily verified that this network has the voltage transfer function given in (8-227). Thus, we see that the normalization process may be applied either to the individual elements of a network or to the network function describing the network with identical results.

The second type of normalization that we shall consider here is *impedance* normalization. This is a type of normalization which changes the values of the elements of a given network without changing the location of any of the critical frequencies. If we define the impedance and admittance of the elements of a network by $Z(s)$ and $Y(s)$, respectively, then the normalized impedance and admittance $Z_n(s)$ and $Y_n(s)$ are given as

$$Z_n(s) = k_z Z(s) \qquad Y_n(s) = \frac{Y(s)}{k_z} \tag{8-228}$$

where k_z is the impedance normalization constant. From the relations given above, we see that the values of all the resistors and inductors are multiplied by k_z, while the values of the capacitors are divided by k_z. As an example of an impedance normalization, let us again consider the network shown in Fig. 8-8.8(b). Let us assume that it is desired to retain the (frequency normalized) pole locations, but in addition, to impedance normalize the network so that the resistor has a value of 10 kΩ. From Eq. (8-228) we see that $k_z = 10^4$. Thus, the 1-mH inductor in the network shown in Fig. 8-8.8(b) becomes a 10-H inductor, the $\frac{1}{2}$-mF capacitor becomes a 0.05-μF capacitor, and the 2-Ω resistor becomes a 20-kΩ resistor. The resulting impedance normalized network is shown in Fig. 8-8.8(c). It is easily verified that the voltage transfer function for this network is given by Eq. (8-227), i.e., it is the same as the one for the network shown in Fig. 8-8.8(b). Thus, we see that the impedance normalization, although it has changed the values of the network elements, has not changed the locations of the critical frequencies. It should be noted that if the network function which is being considered has dimensions of impedance (rather than being dimensionless, as a voltage transfer function is), then an impedance normalization will multiply the network function by the impedance normalization constant k_z. Similarly, if the network function has the dimensions of admittance, the result of an impedance normalization will be to multiply the network function by $1/k_z$. In both cases, however, the location of all the critical frequencies will still remain invariant. It should also be noted that frequency and impedance normalizations may be applied in any order. Thus, if we first impedance normalize the network shown in Fig. 8-8.8(a) using $k_z = 10^4$ and then frequency normalize the resulting network using $k_f = 10^3$, the final result is still the same as that shown in Fig. 8-8.8(c).

8-9 LOCATIONS OF CRITICAL FREQUENCIES OF NETWORK FUNCTIONS

In the preceding sections of this chapter we have discussed two types of critical frequencies for network functions. These are *poles*, which are defined as points in the complex frequency plane (or at infinity) where the magnitude of a given network function goes to infinity, and *zeros*, which are places where the magnitude goes to zero. One very useful way of visualizing such critical frequencies of a network function is by drawing a three-dimensional representation in which the real and the imaginary axes of the complex frequency plane are plotted in two directions, and the magnitude of the network function is plotted in the third direction. In such a representation, the locus of all points defined by the network function will generate

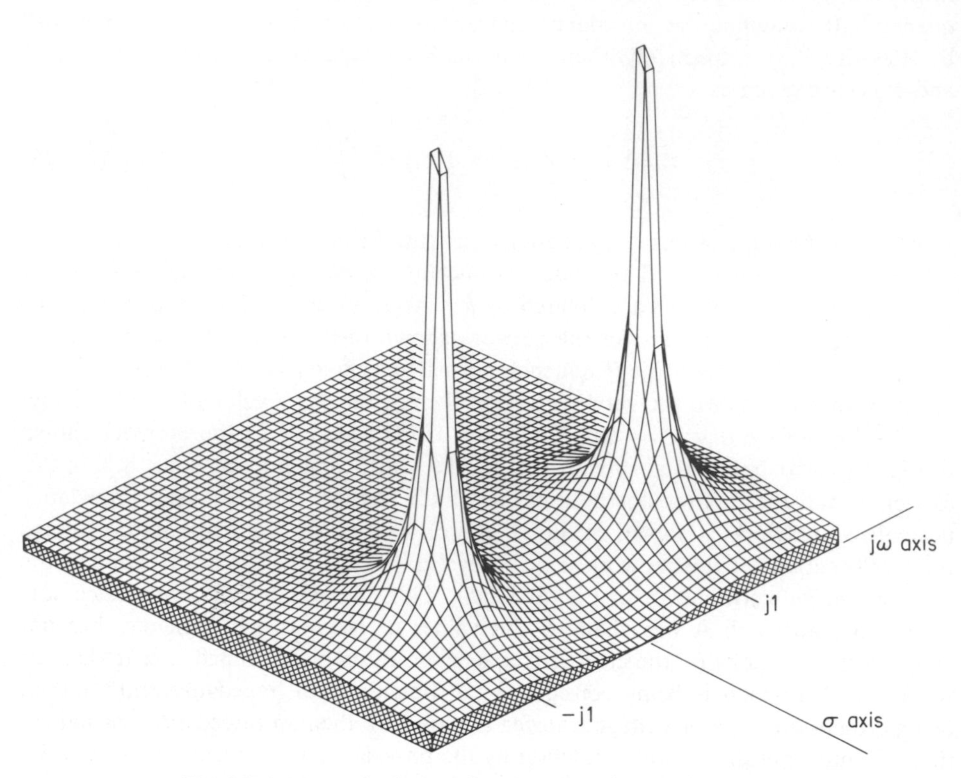

Fig. 8-9.1 A magnitude function representation on the complex plane.

a three-dimensional surface specifying the magnitude of the function at every point on the complex plane. Such a surface is illustrated in Fig. 8–9.1 for the network function $F(s)$, where

$$F(s) = \frac{s}{(s + 1 + j1)(s + 1 - j1)} \tag{8–229}$$

From the figure, we see that as the points $-1 \pm j1$ are approached on the s plane, the magnitude surface goes to infinity, thus visually as well as mathematically producing a "pole." Similarly, at the point $s = 0$, the magnitude surface goes to zero. In this section we shall present a summary of the basic principles which limit the locations of the critical frequencies, i.e., the poles and zeros, for various classes of network functions, and for various types of networks. Because of the complexity of the underlying mathematics, our treatment will be non-rigorous. The results, however, are quite interesting, and well worth including as part of our study of the properties of networks.

To begin our study of the locations of the critical frequencies of network functions, let us define a network function $F(s)$ as

$$F(s) = \frac{R(s)}{E(s)} = \frac{A(s)}{B(s)} = \frac{a_0 + a_1 s + a_2 s^2 + \ldots + a_{m+1} s^m}{b_0 + b_1 s + b_2 s^2 + \ldots + b_{n+1} s^n} \qquad (8\text{–}230)$$

where $R(s)$ and $E(s)$ are rational functions representing the transformed response and excitation, and $A(s)$ and $B(s)$ are polynomials. If we let the excitation to the network be a unit impulse, then, for all $t > 0$, there is no excitation to the network. Thus, the resulting response is the zero-input response, i.e., the response which results from the initial conditions which are generated by the excitation impulse. The situation is similar to that described in Summary 8–7.5. For this case, $E(s)$ is the transform of a unit impulse, i.e., $E(s) = 1$. Thus, we may write

$$F(s) = R(s) = \frac{A(s)}{B(s)} \qquad (8\text{–}231)$$

We see that, *for a unit impulse excitation, the network function equals* $R(s)$, *the transform of the response variable.* The resulting response is appropriately referred to as the *impulse response.* Thus, we may determine all the properties of the network function by examining the properties of $R(s)$ for an applied impulse of excitation. As an example of the determination of such a property, let us consider the bounds on network variables. For a passive network under zero-input conditions, the total energy stored in the network will either be constant (for a lossless or *LC* network) or decreasing (if dissipation, i.e., resistance, is present). Thus the zero-input response for any network variable must be bounded. Thus, pole locations which represent time-domain behavior which grows indefinitely (and thus is not bounded) are not possible for such a passive network. For example, poles which are in the right half-plane (RHP) represent a time-domain positive exponential behavior. Thus, they violate the requirement for boundedness and they are not possible for passive networks. Similarly, poles on the $j\omega$ axis of higher than first order represent non-bounded time-domain behavior, thus they will not occur in passive networks. Simple poles on the $j\omega$ axis, however, represent bounded behavior and thus they are permissible. Poles in the left half-plane (LHP) of any order are permitted. For example, for a pole of order m on the negative real axis at $s = -a$ $(a > 0)$, the corresponding time-domain behavior is $Kt^{m-1}e^{-at}$. By applying l'Hospital's rule, it is easy to show that

$$\lim_{t \to \infty} Kt^{m-1}e^{-at} = 0 \qquad a > 0 \qquad (8\text{–}232)$$

for all finite m. This is due to the fact that the negative exponential term dominates for large t. It may similarly be shown that *network function poles of multiple order anywhere in the LHP* represent bounded time-domain zero-input behavior. A sum-

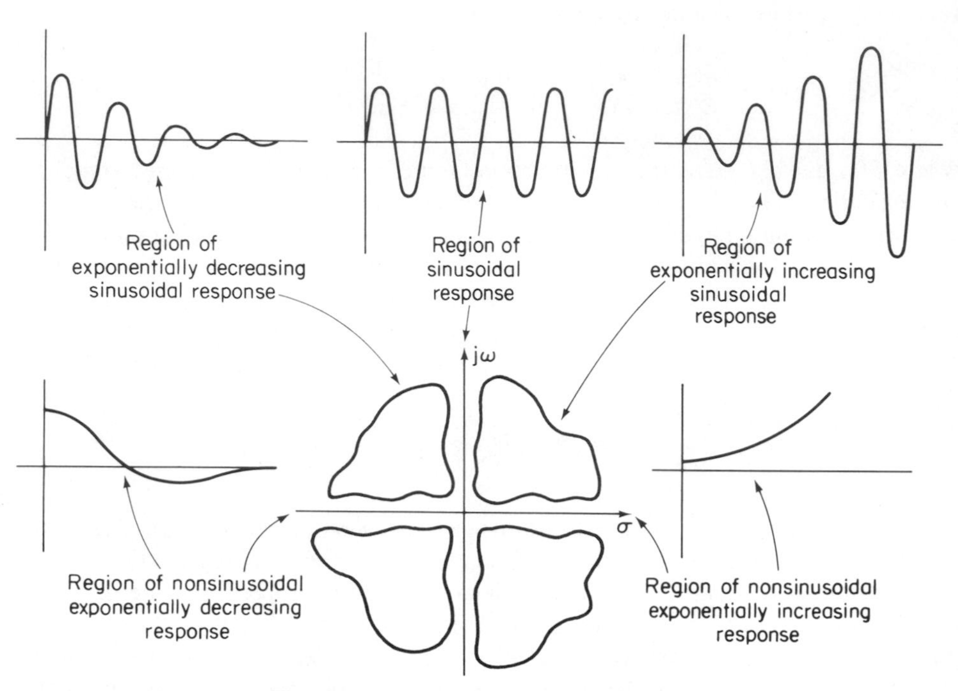

Fig. 8-9.2 Impulse response characteristics.

mary of the impulse response characteristics for various pole locations on the complex frequency plane is given in Fig. 8–9.2. From this figure and the discussion given above, we conclude the following:

SUMMARY 8-9.1

Natural Frequencies of Passive Networks: The natural frequencies of a network are the poles of the transform of the impulse response. These are the same as the poles of the network function. The poles of network functions for passive networks must be in the LHP or on the $j\omega$ axis. Any poles on the $j\omega$ axis must be simple.

In general, no simple conclusion can be made concerning the poles of *active* networks, since it is equally possible for these to be in the RHP or the LHP. As an example of this, consider the network shown in Fig. 8–9.3. This network contains passive resistors and capacitors and a VCVS. Let us investigate the possible locations for the poles of the voltage transfer function of this network. In Example 8–6.5 the voltage transfer function for this network was shown to be

$$\frac{V_2(s)}{V_1(s)} = \frac{K}{s^2 + s(3 - K) + 1} \tag{8–233}$$

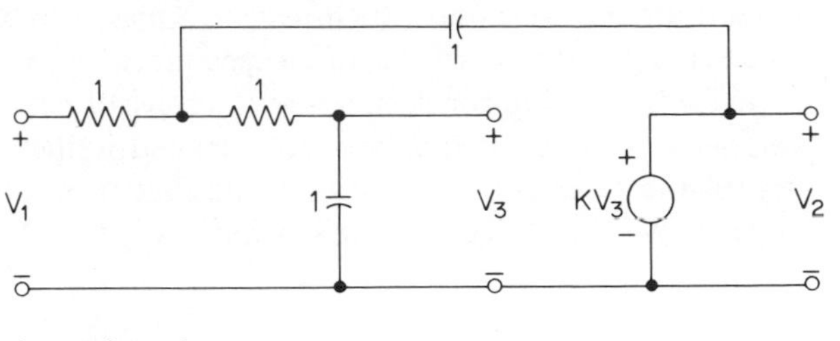

Fig. 8-9.3 An active *RC* network.

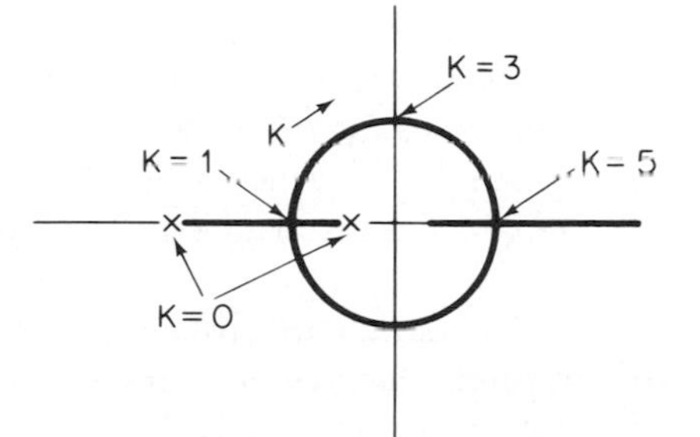

Fig. 8-9.4 A root-locus plot.

Obviously, from this equation, the poles of the voltage transfer function, i.e., the natural frequencies of the network, are a function of the value of K, i.e., the gain of the VCVS. A plot of the locus of these poles as K is varied is given in Fig. 8-9.4. Such a plot is frequently referred to as a *root-locus plot*, since it shows the locus of the roots of the denominator polynomial of the network function $V_2(s)/V_1(s)$ as they are affected by the value of K. From the figure we note that if $0 \leq K \leq 3$, the poles of the voltage transfer function are in the LHP or on the $j\omega$ axis. For this range of K, the impulse response of the network is bounded, and the network is said to be *stable*. For values of $K > 3$, however, the poles are in the RHP, and the impulse response is unbounded. For such values of K, the network is said to be *unstable*. Thus, we may conclude that active networks may have stable or unstable characteristics, depending on the particular configuration and on the specific values of the elements. On the other hand, passive networks, i.e., networks comprised only of passive network elements, are always stable.

Now let us investigate some other restrictions on the poles of various network functions. First, let us consider networks comprised only of inductors and capacitors, i.e., *LC* networks. Since there is no resistance in such circuits, there is no dissipation of any energy present in the circuit. Thus, the variables of such a circuit will oscillate sinusoidally as energy is transferred back and forth between the capacitors and inductors of the circuit. The amplitude of each sinusoidal oscillation component will remain fixed. The frequency-domain location for the poles of the network function corresponding with such time-domain behavior is the $j\omega$ axis. Thus, we conclude that *LC* networks have their poles restricted to the $j\omega$ axis. For boundedness of the response, such poles must be simple. Now let us consider *RC* circuits, i.e., circuits comprised only of resistors and capacitors. Since there is only one type of energy-storage element in such a circuit, i.e., capacitors, there can be no sinusoidal component in the zero-input response. In addition, since there is dissipation in the circuit, the values of all variables must eventually go to zero. The location of the poles in the complex frequency plane that correspond with such time-domain behavior is the negative-real axis. We conclude that *RC* networks have their poles only on the negative-real axis. Although not required by boundedness, it can be shown that these poles must also be simple. Such networks may also have a simple pole at the origin. The driving-point impedance of a capacitor is an example of this latter case. Similar logic leads us to the conclusion that the poles of network functions for *RL* networks, i.e., networks comprised only of resistors and inductors, will have poles only on the negative-real axis (or at the origin), and that these poles will be simple. The results given above may be summarized as follows:

SUMMARY 8-9.2

Critical Frequencies of LC, RC, and RL Networks: The poles of network functions for *LC* networks are restricted to the $j\omega$ axis (and the origin) of the complex frequency plane, while those for *RC* and *RL* networks are restricted to the negative-real axis (and the origin). All the poles are simple.

The restrictions given above apply to any type of network function, i.e., driving-point functions *or* transfer functions. If we now restrict our attention to driving-point functions, we may make some observations about the zeros of the three classes of networks discussed above. Since the zeros of driving-point functions become the poles when the excitation is changed from a voltage source to a current source or vice versa, the restrictions given in Summary 8–9.2 apply equally well to the zeros of driving-point functions. Now let us see if we can determine some other properties for such zeros. To begin our study, consider the *RC* driving-point impedance function $Z(s)$ of the cascade of shunt-connected resistors and capacitors shown in Fig. 8–9.5. This is easily seen to be

$$Z(s) = \frac{1}{sC_1 + G_1} + \frac{1}{sC_2 + G_2} + \dots + \frac{1}{sC_n + G_n} \tag{8–234}$$

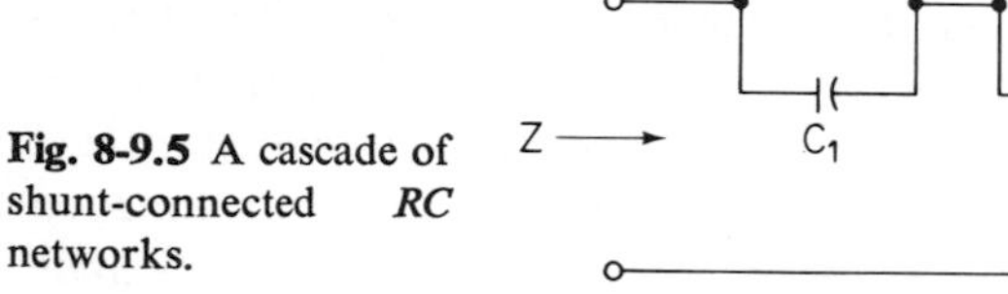

Fig. 8-9.5 A cascade of shunt-connected *RC* networks.

This expression may be rewritten as

$$Z(s) = \frac{1/C_1}{s + (G_1/C_1)} + \frac{1/C_2}{s + (G_2/C_2)} + \dots + \frac{1/C_n}{s + (G_n/C_n)} \tag{8–235}$$

This last expression is easily recognized as a partial-fraction expansion for $Z(s)$. The quantities $1/C_i$ are the residues, and these are clearly positive for a passive net-

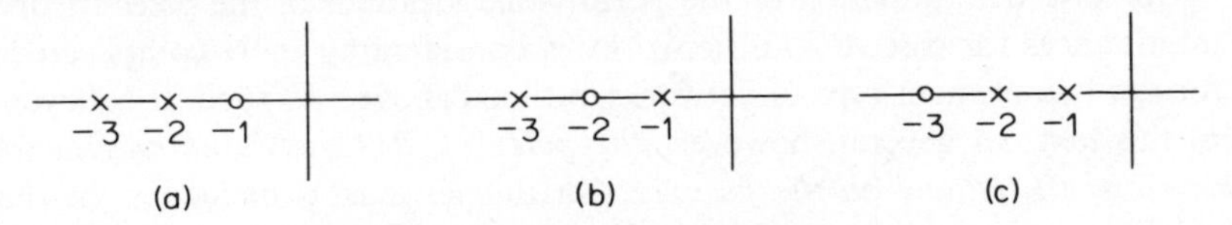

Fig. 8-9.6 Different pole-zero alternations.

work with positive-valued capacitors. Let us now consider the way in which some different alternations of poles and zeros along the negative-real axis affect the signs of the residues at nearby poles. In Fig. 8–9.6, three different such alternations are shown using **X**'s to represent the pole positions, and circles to represent the locations of the zeros. If we assume that the multiplicative constants of the network functions are unity, then these pole-zero patterns define three different network functions. The functions and the corresponding partial-fraction expansions are given in Table 8–9.1.

TABLE 8-9.1 Some Pole-Zero Alternations and Their Partial-Fraction Expansions

Figure	*Network Function*	*Partial-Fraction Expansion*
8–9.6(a)	$\dfrac{s+1}{(s+2)(s+3)}$	$\dfrac{-1}{s+2}+\dfrac{2}{s+3}$
8–9.6(b)	$\dfrac{s+2}{(s+1)(s+3)}$	$\dfrac{\frac{1}{2}}{s+1}+\dfrac{\frac{1}{2}}{s+3}$
8–9.6(c)	$\dfrac{s+3}{(s+1)(s+2)}$	$\dfrac{2}{s+1}+\dfrac{-1}{s+2}$

An examination of the entries of this table shows that only an *alternation* of the poles and zeros, as shown in Fig. 8–9.6(b), provides positive residues and, thus, corresponds with Eq. (8–235) and has a passive *RC* realization of the form shown in Fig. 8–9.5. Although our development here has treated only a simple numerical *RC* example, it may be shown that the alternation of poles and zeros along the negative-real or $j\omega$ axes is a necessary condition for the driving-point immittances for any *RC*, *RL*, or *LC* network. These results may be summarized as follows:

SUMMARY 8-9.3

Alternation of Poles and Zeros in Driving-Point Functions: The poles and zeros of the driving-point immittances of *RC* and *RL* networks must alternate along the negative-real axis. Those of *LC* networks must alternate along the $j\omega$ axis.

A precise determination of the permissible locations of the poles of driving-point immittances for passive *RLC* networks is considerably more complicated than those for the two-element type of networks outlined above. As such, it is beyond the scope of this text. In general, however, *the poles of RLC networks may be located anywhere in the LHP and on the $j\omega$ axis.* On the $j\omega$ axis, boundedness of the network response requires that they be simple.

Now let us consider the permissible locations for the poles and zeros of *transfer functions.* The poles of such functions are, of course, the natural frequencies of the network. Thus, they are determined by the same restrictions given above for the poles of driving-point functions. The zeros, however, are almost completely unrestricted, and, in general, they may occur in either the LHP or the RHP and they need not necessarily be simple. Certain network configurations do, however, place restrictions on the location of the zeros. As an example of such a network configuration, consider the voltage divider shown in Fig. 8–9.7(a). The network function for the voltage transfer ratio is

$$\frac{V_2(s)}{V_1(s)} = \frac{Z_2(s)}{Z_1(s) + Z_2(s)} \tag{8–236}$$

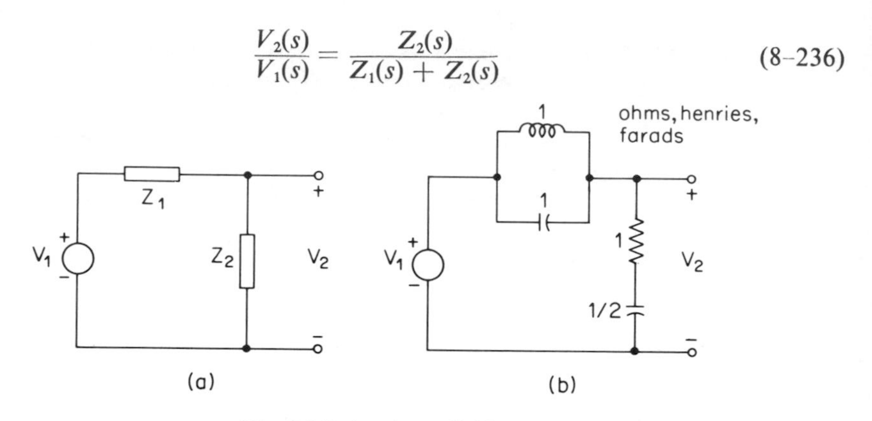

Fig. 8-9.7 A voltage divider.

where $Z_1(s)$ is called a *series* element (it is in series with the source), and $Z_2(s)$ is called a *shunt* element (it is in shunt with the output terminals). As a more specific example, consider the network shown in Fig. 8–9.7(b). The impedance $Z_1(s)$ of the series *LC* element is $s/(s^2 + 1)$ and the impedance $Z_2(s)$ of the shunt *RC* element is $(s + 2)/s$. From (8–236), the network function for the voltage transfer ratio is

$$\frac{V_2(s)}{V_1(s)} = \frac{(s + 2)/s}{[s/(s^2 + 1)] + [(s + 2)/s]} \tag{8–237}$$

Now let us consider what happens if the excitation $v_1(t)$ is chosen so that its transform has a pole at the identical frequency at which the impedance $Z_2(s)$ of the shunt *RC* element *has a zero*, namely, at $s = -2$. For this, from Table 8–3.1, the excitation will have the form $v_1(t) = Ke^{-2t}$ and its transform will be $V_1(s) = K/(s + 2)$. Substituting this expression in Eq. (8–237), we obtain

$$V_2(s) = \frac{K/s}{[s/(s^2+1)] + [(s+2)/s]} = \frac{K(s^2+1)}{s^3+3s^2+s+2} \tag{8-238}$$

From Eq. (8-238) we see that there is no pole of $V_2(s)$ at $s = -2$. Thus, even though the excitation $v_1(t)$ was of the form Ke^{-2t}, *there will not be an* e^{-2t} *term present in the response for* $v_2(t)$. We may summarize this result by saying that since the network function of Eq. (8-237) has a zero at $s = -2$, exciting the network at this complex frequency mathematically cancels the factor $(s + 2)$ in the numerator and denominator of the expression for $V_2(s)$. From a physical viewpoint, we may say that when a network is excited at a complex frequency at which it has a zero, then this excitation does not appear in the response. Even though the excitation frequency does not appear in the response, there will, of course, be other response terms due to the other natural frequencies of the network. These other natural frequencies are, of course, excited by the applied excitation function.

A similar situation occurs if the network shown in Fig. 8-9.7(b) is excited at some frequency at which the impedance $Z_1(s)$ of the series LC element *has a pole*, i.e., at $s = +j1$. Such an excitation occurs if we let $v_1(t) = K \sin t$; thus, $V_1(s) = K/(s^2 + 1)$. Substituting this expression in (8-237), we obtain

$$V_2(s) = \frac{K(s+2)/[s(s^2+1)]}{[s/(s^2+1)] + [(s+2)/s]} = \frac{K(s+2)}{s^3+3s^2+s+2} \tag{8-239}$$

From Eq. (8-239) we see that there are no poles of $V_2(s)$ at $s = \pm j1$, even though the transformed excitation function has poles there. This is because the impedance of $Z_1(s)$ also has poles at $s = \pm j1$.

In the example discussed above, we showed that the zero of the driving-point impedance $Z_2(s)$ of the shunt element and the pole of the driving-point impedance $Z_1(s)$ of the series element of the network shown in Fig. 8-9.7(b) both produced a zero in the resulting network voltage transfer function. This conclusion is easily generalized. To see this, consider again the voltage divider shown in Fig. 8-9.7(a). Suppose that the driving-point impedance of $Z_2(s)$ has a zero at $s = p_0$. Thus, we may write

$$Z_2(s) = \frac{(s-p_0)A(s)}{B(s)} \tag{8-240}$$

where $A(s)$ and $B(s)$ are arbitrary polynomials, and $B(p_0) \neq 0$. The voltage transfer function for the network may now be written as

$$\frac{V_2(s)}{V_1(s)} = \frac{(s-p_0)A(s)/B(s)}{Z_1(s) + (s-p_0)A(s)/B(s)} = \frac{(s-p_0)A(s)}{Z_1(s)B(s) + (s-p_0)A(s)} \tag{8-241}$$

Evaluating the network function at $s = p_0$, we see that since $Z_2(p_0) = 0$, the network function itself is also equal to 0. Now let us assume that $V_1(s)$ has a pole at p_0. In this case, it may be written in the form

$$V_1(s) = \frac{C(s)}{(s - p_0)D(s)} \tag{8-242}$$

where $C(s)$ and $D(s)$ are arbitrary polynomials. The response $V_2(s)$ must now have the form

$$\begin{aligned} V_2(s) &= \frac{(s - p_0)A(s)}{Z_1(s)B(s) + (s - p_0)A(s)} \frac{C(s)}{(s - p_0)D(s)} \\ &= \frac{A(s)C(s)}{[Z_1(s)B(s) + (s - p_0)A(s)]D(s)} \end{aligned} \tag{8-243}$$

From Eq. (8-243) we see that $V_2(s)$ does not have a pole at p_0. Thus, we may conclude that, in general, when $Z_2(s)$ has a zero at some complex frequency, any excitation which is applied at that frequency produces no output voltage (at that frequency). In effect, we can think of the impedance $Z_2(s)$ as being a "short circuit" at that value of *complex* frequency, and thus shorting such an excitation frequency to ground. The input will, of course, excite the other natural frequencies of the network. Now let us consider a different case for the same network. For this case, let us assume that $Z_1(s)$ has a pole at $s = p_0$. Thus, we may write

$$Z_1(s) = \frac{A(s)}{(s - p_0)B(s)} \tag{8-244}$$

where $A(s)$ and $B(s)$ are arbitary polynomials, and $A(p_0) \neq 0$. The voltage transfer function will now have the form

$$\frac{V_2(s)}{V_1(s)} = \frac{(s - p_0)B(s)Z_2(s)}{(s - p_0)B(s)Z_2(s) + A(s)} \tag{8-245}$$

If we evaluate this expression at $s = p_0$, then the network function itself will be equal to 0. Again we may conclude that if $V_1(s)$ contains an excitation frequency at p_0, there will be no output frequency p_0 present in $V_2(s)$. In this case, we may think of $Z_1(s)$ as being an "open circuit" at the complex frequency p_0; thus, it blocks the transmission of any input at that frequency.

The conclusions given above for a voltage divider are easily generalized to include a ladder network of the form shown in Fig. 8-9.8. As such they apply to

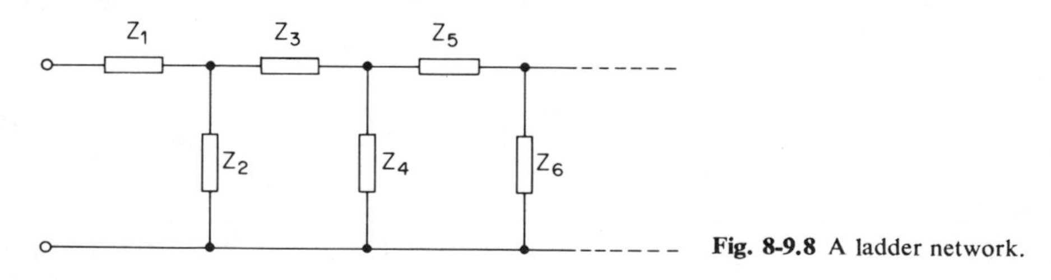

Fig. 8-9.8 A ladder network.

any transfer function defined from one end of the network to the other. At any value of complex frequency at which any of the impedances $Z_1(s)$, $Z_3(s)$, $Z_5(s)$, etc., has a pole, there will be no transmission, i.e., the transfer function will have a zero. Similarly, at any value of complex frequency at which any of the impedances $Z_2(s)$, $Z_4(s)$, $Z_6(s)$, etc., has a zero, there will also be a zero in the transfer function. Since *RLC* networks can only have poles and zeros in the LHP, we may conclude that *the transfer functions of ladder networks can only have zeros in the LHP and on the $j\omega$ axis.* Such conclusions make it easy to directly evaluate some of the characteristics of frequently encountered ladder networks. For example, consider the network shown in Fig. 8-9.9. The poles of the series impedances (the inductors) are all at $s = \infty$.

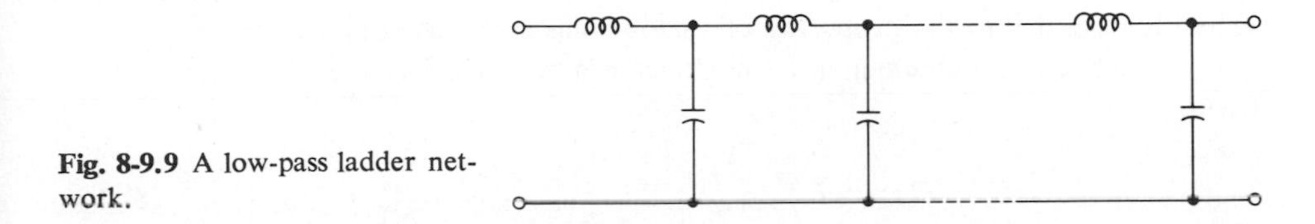

Fig. 8-9.9 A low-pass ladder network.

Similarly, the zeros of the shunt impedances (the capacitors) are at $s = \infty$. Thus, the transmission zeros of the network will all be at $s = \infty$. We conclude that the network is a low-pass network. Similar reasoning shows that the network shown in Fig. 8-9.10 is a high-pass network. These results may be summarized as follows:

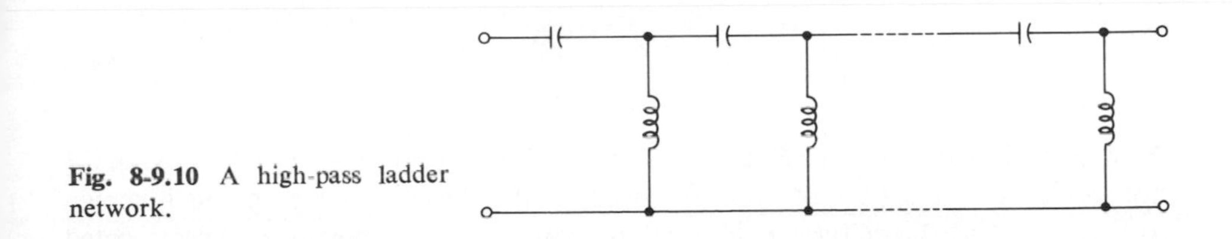

Fig. 8-9.10 A high-pass ladder network.

SUMMARY 8-9.4

Transmission Zeros of Ladder Networks: The zeros of transfer functions for ladder networks will occur at the complex frequencies at which the driving-point impedances of any of the series elements have poles and at the complex frequencies at which the driving-point impedances of any of the shunt elements have zeros. Thus the zeros can only occur in the LHP or on the $j\omega$ axis.[17]

[17]For a shunt element to produce a zero of transmission at a given frequency, the impedance of the remainder of the network as seen from the element must *not* also be zero at that frequency. Similarly, for a series element to produce a zero of transmission, at a given frequency, the impedance of the rest of the network *not* be infinite at that frequency.

8-10 POLES AND ZEROS AND FREQUENCY RESPONSE

In Summary 8-6.2 we showed that there is a direct relation between the type of network function which relates *transformed excitation and response variables*, and the type of network function which relates *excitation and response phasors*. This latter network function, of course, determines the sinusoidal steady-state response of a given network. Specifically, we showed that the two network functions are related by a simple argument change, namely, the substitution of $j\omega$ for s and vice versa. In this section we shall illustrate a graphical method of relating these types of network functions, i.e., of determining the sinusoidal steady-state response of a network directly from the locations of the poles and zeros in the complex frequency plane. The method has the advantage that it is readily visualized. In addition, it provides valuable insight into the properties of various classes of network functions.

To begin our discussion, let us consider a network function $F(s)$ written in the form

$$F(s) = K\frac{(s - z_1)(s - z_2)\cdots(s - z_m)}{(s - p_1)(s - p_2)\cdots(s - p_n)} \tag{8-246}$$

where the quantities z_i and p_i are the zeros and the poles of the network function, respectively, and where K is a multiplicative constant. Let us now consider what happens when we evaluate the function $F(s)$ by letting $s = s_0$, where s_0 is some specific value of s, i.e., some specific point on the complex frequency plane. For such a situation, Eq. (8-246) may be written

$$F(s_0) = K\frac{(s_0 - z_1)(s_0 - z_2)\cdots(s_0 - z_m)}{(s_0 - p_1)(s_0 - p_2)\cdots(s_0 - p_n)} \tag{8-247}$$

Now let us consider a typical numerator term of the form $(s_0 - z_i)$ where s_0 and z_i are specific points on the complex frequency plane. Such a pair of points is shown in Fig. 8-10.1(a). Since these points are complex quantities, they may be represented by vectors as shown in Fig. 8-10.1(b). Now let us construct a vector W going from z_i to s_0 as shown in Fig. 8-10.1(c). By simple vector addition, we see that $s_0 = W + z_i$ or $W = s_0 - z_i$. Thus, we may conclude that the complex quantity $(s_0 - z_i)$ is equal to the (complex) value of a vector drawn from the point z_i to the point s_0 on the complex plane. The actual magnitude and phase of such a vector are, of course, more readily visualized if the vector is shifted to the origin as shown in Fig. 8-10.1(d). Thus, we have developed a graphical means of evaluating numerator factors of Eq. (8-247) having the form $(s_0 - z_i)$. Obviously, denominator factors of Eq. (8-247) which have the form $(s_0 - p_i)$ may be evaluated in the same manner.

The graphical evaluation technique described above is readily applied to determine the magnitude and phase of a given network function at a specified point on the complex plane. As an example, consider the network function

$$F(s) = 3\frac{s(s+1)}{s^2 + 2s + 2} = 3\frac{s(s+1)}{(s + 1 - j1)(s + 1 + j1)} \tag{8-248}$$

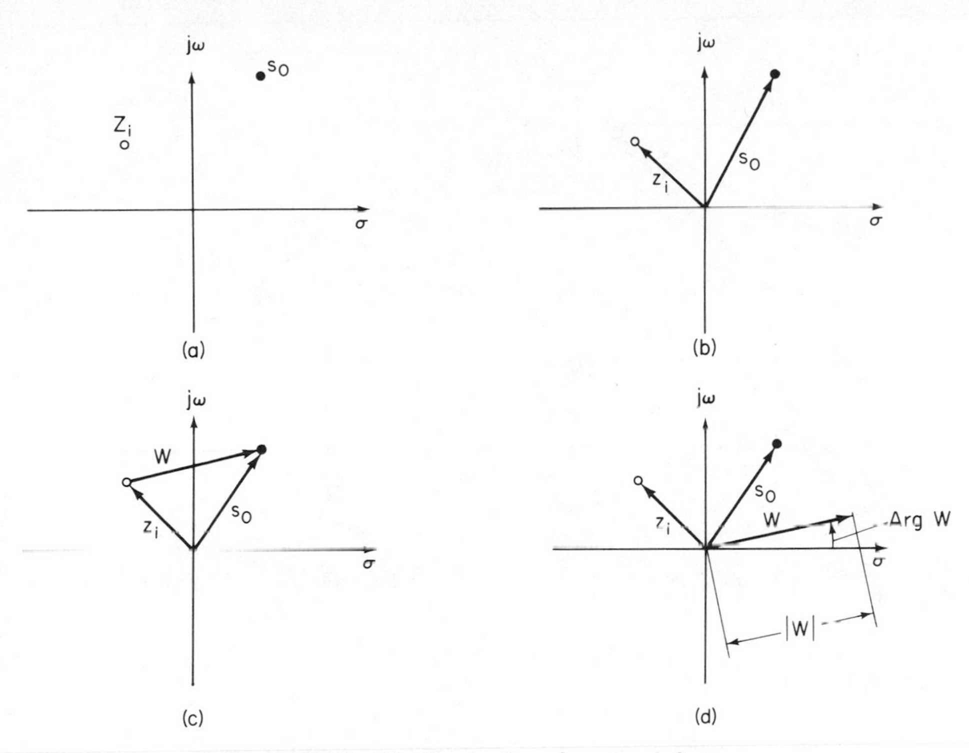

Fig. 8-10.1 Graphical evaluation of a network factor.

The zeros of this function are $z_1 = 0$ and $z_2 = -1$, and the poles are $p_1 = -1 + j1$ and $p_2 = -1 - j1$. In addition, the multiplicative constant is 3. Now let us suppose that it is desired to evaluate $F(s)$ of Eq. (8-248) for $s = s_0$, where $s_0 = 1 + j1$. We may do this by the process shown in Fig. 8-10.2. The first numerator factor s, evaluated at $s = s_0$, is simply s_0, i.e., the vector W_1 of length $\sqrt{2}$ and angle 45° shown in the figure. The second numerator factor $(s + 1)$, evaluated at s_0, is the vector W_2. This has length $\sqrt{5}$ and angle 26.57°. Similarly, the denominator factors may be represented by the vectors U_1 and U_2 shown in the figure. In terms of these four vectors, we see that

$$F(1 + j1) = 3\frac{W_1 W_2}{U_1 U_2} \tag{8-249}$$

Inserting the values of the magnitude and phase of the individual vectors, we may write

$$F(1 + j1) = 3\frac{\sqrt{2}\, e^{j45^\circ} \sqrt{5}\, e^{j26.57^\circ}}{2e^{j0^\circ} \sqrt{8}\, e^{j45^\circ}} = \frac{3\sqrt{5}}{4}\, e^{j26.57^\circ} \tag{8-250}$$

The results given above are easily generalized as follows:

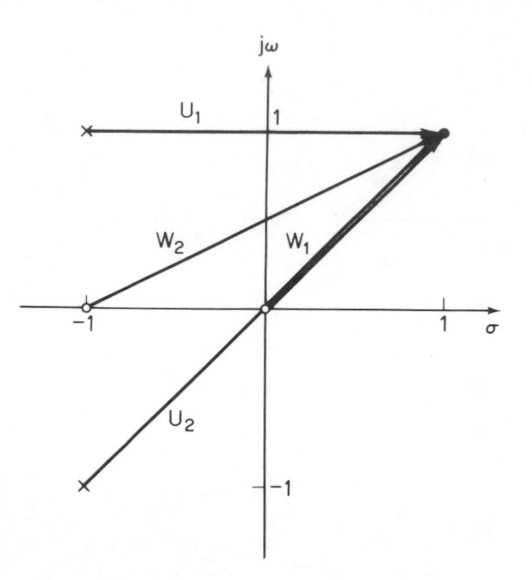

Fig. 8-10.2 Graphical evaluation of a network function.

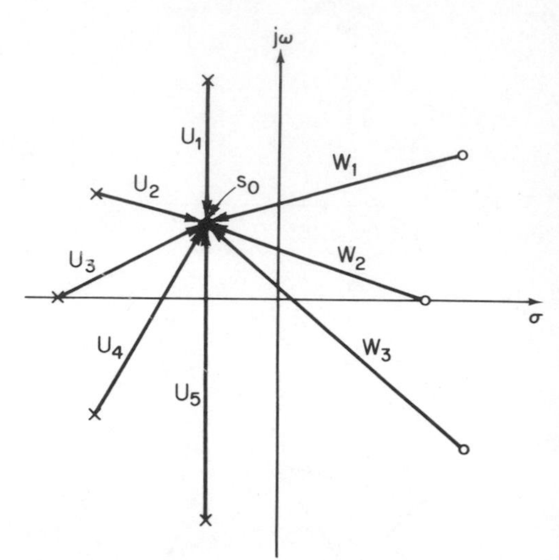

Fig. 8-10.3 Graphical evaluation of a network function.

SUMMARY 8-10.1

Graphical Evaluation of Network Functions: A network function $F(s)$ having the form

$$F(s) = K\frac{(s - z_1)(s - z_2)\cdots(s - z_m)}{(s - p_1)(s - p_2)\cdots(s - p_n)}$$

may be evaluated at a specific value of its argument $s = s_0$ by constructing a set of vectors with values W_i from the various zeros z_i of the function to the point s_0, and by constructing a set of vectors with values U_i from the various poles p_i to the point s_0 as shown in Fig. 8-10.3. In terms of these vectors, we may write

$$F(s_0) = K\frac{\prod_{i=1}^{m} W_i}{\prod_{i=1}^{n} U_i}$$

There are several reasons why the procedure described above is a useful one. First of all, it is easily implemented by graphical means; thus, for complicated network functions, a scale and protractor may be conveniently used to implement the evaluation process. Second, it provides a means of evaluating a network function while it is in factored form, without the necessity for multiplying out the various factors. Finally, it provides an easy means for obtaining insight into how the relative locations of the poles and zeros of a network function influence the properties of the function in the frequency domain and thus indirectly in the time domain. As an example of such an application, consider the network function

$$F(s) = \frac{s + 1.5}{(s^2 + 6s + 18)(s^2 + 1.6s + 0.68)(s + 0.8)} \tag{8-251}$$

The locations of the poles and zeros of this function are shown in Fig. 8-10.4(a). Its partial-fraction expansion will have the form

$$F(s) = \frac{K_1}{s + 3 - j3} + \frac{K_1^*}{s + 3 + j3} + \frac{K_2}{s + 0.8 - j0.2} + \frac{K_2^*}{s + 0.8 + j0.2} + \frac{K_3}{s + 0.8} \tag{8-252}$$

Now let us see how graphical techniques may be used to determine the various residues. First we may note that, as shown in Summary 8-4.3, the value of K_1 is given as

$$K_1 = \frac{s + 1.5}{(s + 3 + j3)(s + 0.8 - j0.2)(s + 0.8 + j0.2)(s + 0.8)}\bigg|_{s=-3+j3} \tag{8-253}$$

The right member of (8-253) may be evaluated graphically from the plot of the pole and zero locations as shown in Fig. 8-10.4(b). Thus, using the vectors defined in this figure, we may write

$$K_1 = \frac{W_1}{U_1 U_2 U_3 U_4} \tag{8-254}$$

Examining this result, we conclude that the residue of the pole at $-3 + j3$ is found by deleting the pole at $-3 + j3$ from the original network function and evaluating the remaining function at $s = -3 + j3$. For simplicity, let us first compute only the magnitudes of the residues. Using the values given in Fig. 8-10.4(b), we see that

$$|K_1| = \frac{3.354}{6.0 \times 3.562 \times 3.883 \times 3.722} = 0.01071 \tag{8-255}$$

Similarly, the value of the magnitude of K_2, the residue of the pole at $-0.8 + j0.2$, may be computed using the vectors shown in Fig. 8-10.4(c). Thus, we obtain

$$|K_2| = \frac{0.728}{3.562 \times 3.883 \times 0.4 \times 0.2} = 0.6579 \tag{8-256}$$

For this residue, note that since the poles at -0.8 and $-0.8 - j0.2$ are quire close to the pole at $-0.8 + j0.2$ at which the residue is being determined, the quantities

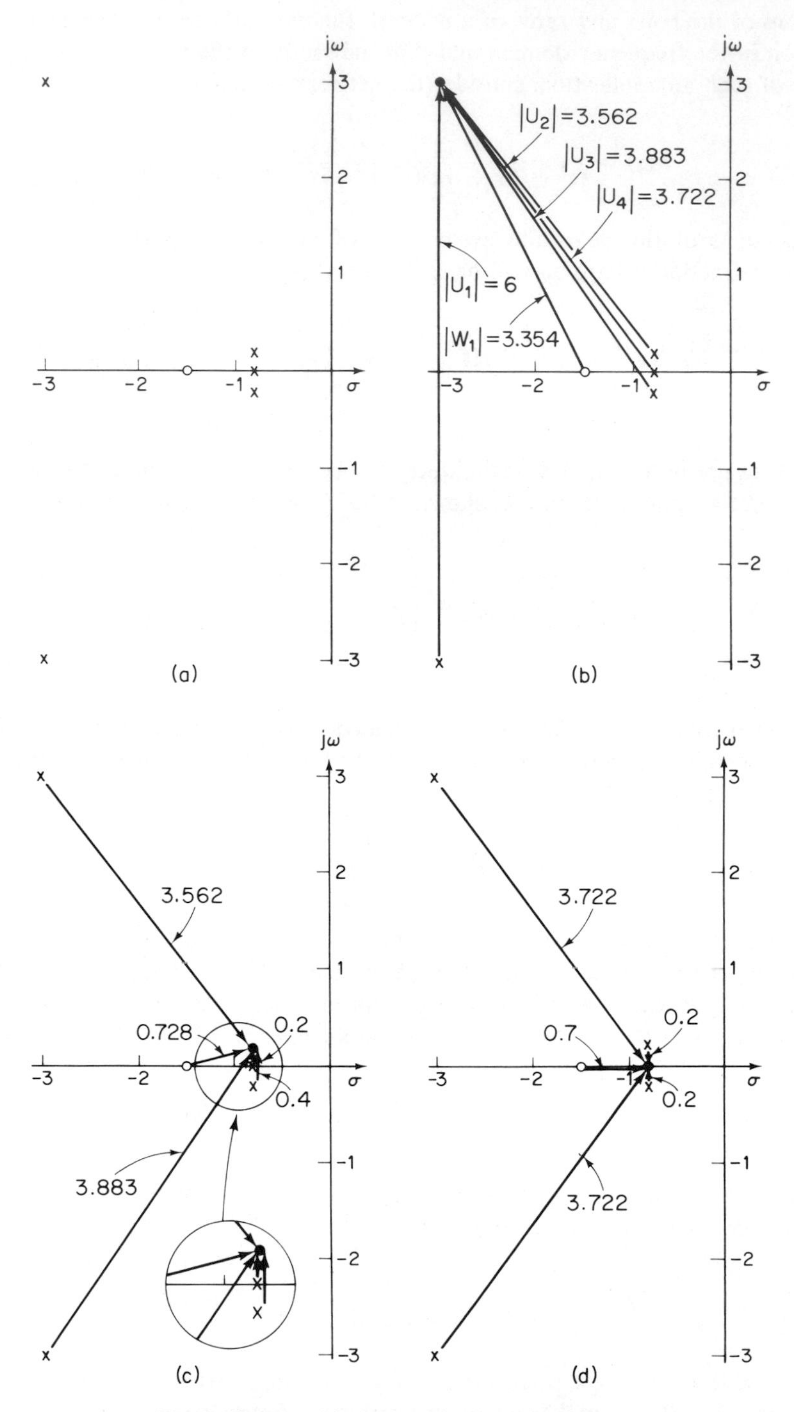

Fig. 8-10.4 Graphical determination of residues.

in the denominator representing the magnitudes of these distances are small. The result is that the magnitude of the residue for this pole is considerably larger than the magnitude of the residue K_1 for the pole at $-3 + j3$, since there are no other poles which are close to the latter one. Following this reasoning further, we may expect the residue K_3 for the pole at -0.8 also to be large. Using the vectors shown in Fig. 8-10.4(d), we find the magnitude of K_3 to be

$$|K_3| = \frac{0.7}{3.722 \times 3.722 \times 0.2 \times 0.2} = 1.263 \tag{8-257}$$

which, as expected, is also considerably larger than K_1. Since the relative magnitudes of K_2 and K_3 are so much larger than the magnitude of K_1, only a small inaccuracy will occur in our answer if we do not consider the effects of the first two terms in the partial-fraction expansion given in Eq. (8-252). Thus, we may write

$$F(s) \approx \frac{K_2}{s + 0.8 - j0.2} + \frac{K_2^*}{s + 0.8 + j0.2} + \frac{K_3}{s + 0.8} \tag{8-258}$$

and we have simplified the amount of effort required in determining the residues and in finding the resulting inverse transform. Note that although we have eliminated the terms involving the poles at $-3 + j3$ from the partial-fraction expansion, these poles must, of course, be considered in finding the values of the residues K_2 and K_3.

The result obtained above, namely, poles which are close together will have larger residues than poles which are spaced far apart, is readily shown to be true in general. Many other interesting conclusions can be obtained concerning the properties of residues based on the use of graphical methods in evaluating network functions. However, we shall leave these to more advanced texts.

The graphical evaluation techniques introduced above may be directly applied to the evaluation of a network function which relates a pair of excitation and response phasors. Since such a function is simply the network function relating the transformed network variables with the substitution of $s = j\omega$, we need simply set $s_0 = j\omega_0$ in the discussion given above in order to determine the ratio between response and excitation phasors at any point ω_0 on the $j\omega$ axis, i.e., for any sinusoidal frequency of ω_0 rad/s. As an example of the insight produced by such a procedure, consider the network function

$$F(s) = \frac{10s}{(s + 0.1 - j1)(s + 0.1 + j1)} \tag{8-259}$$

The poles and zeros of the network function and the vectors used to evaluate $F(j\omega)$ are shown in Fig. 8-10.5(a). Note that as the value of ω at which the evaluation is made approaches 1, the length of the vector U_1 approaches its minimum value. Since the magnitude of $F(j\omega)$ may be specified as

$$|F(j\omega)| = \frac{10|W_1|}{|U_1||U_2|} \tag{8-260}$$

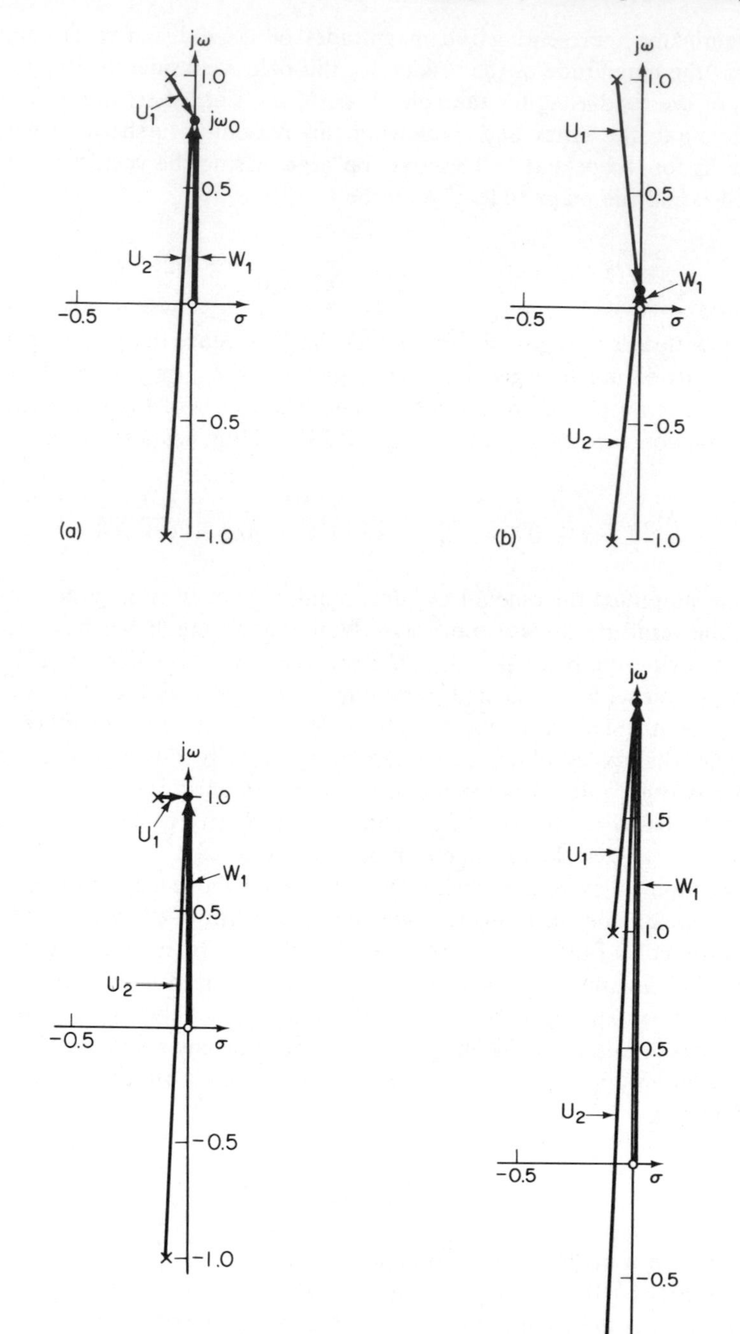

Fig. 8-10.5 Graphical evaluation of frequency response.

we see that such a *minimum* value in the denominator of $F(j\omega)$ will produce a *maximum* in the magnitude of the network function. Thus, in general, we may conclude that when a pole of a network function is close to the $j\omega$ axis, the values of ω in the vicinity of such a pole will produce a large magnitude in the network function. This is simply another way of looking at the phenomenon of resonance which was introduced in Sec. 7–8. Similar conclusions are readily made for the phase of the network function. For example, when a pole is close to the $j\omega$ axis, then the angle of the vector from that pole to the $j\omega$ axis will change rapidly for small changes in the value of $j\omega_0$ in the vicinity of the pole. Thus, the phase of the network function changes rapidly as a function of ω for values of ω which are close to such a pole. As an example of this, for the network function given in Eq. (8–259) we see that the argument of $F(j\omega)$ may be written in terms of the vectors defined in Fig. 8–10.5(a) as

$$\text{Arg } F(j\omega) = \frac{\text{Arg } W_1}{\text{Arg } U_1 + \text{Arg } U_2} \tag{8-261}$$

For values of ω which are very small, as shown in Fig. 8–10.5(b) the values of Arg U_1 and Arg U_2 are very nearly equal and opposite; thus, Arg $F(j\omega) \approx$ Arg $W_1 = 90°$. As we increase ω, the phase decreases until, in the vicinity of $\omega = 1$ rad/s, there will be a frequency at which Arg U_1 is very small, but at which Arg U_1 + Arg $U_2 = 90°$. Since Arg W_1 is always 90°, at such a frequency Arg $F(j\omega) = 0$. This situation is shown in Fig. 8–10.5(c). In the vicinity of this frequency, the phase changes very rapidly as a function of ω. Finally, for large values of ω, Arg U_1 and Arg U_2 are each approximately 90°, and Arg W_1 is 90°; therefore, Arg $F(j\omega)$ approaches $-90°$. This situation is shown in Fig. 8–10.5(d). The overall magnitude and phase curves will be similar to those shown in Figs. 7–8.6(b) and 7–8.6(c). Observations similar to those made above for poles may be made for zeros located on or near the $j\omega$ axis. These results may be summarized as follows:

SUMMARY 8-10.2

Use of Poles and Zeros in Determination of Frequency Response: A network function $F(s)$ having the form

$$F(s) = K\frac{(s - z_1)(s - z_2)\cdots(s - z_m)}{(s - p_1)(s - p_2)\cdots(s - p_n)}$$

may be evaluated for sinusoidal steady-state behavior for any value of frequency ω_0 by constructing a set of vectors with complex values W_i from the various zeros z_i to the point $j\omega_0$, and by constructing a set of vectors of values U_i from the various poles p_i to the same point. The process is illustrated in Fig. 8–10.6. In terms of these vectors, we may write

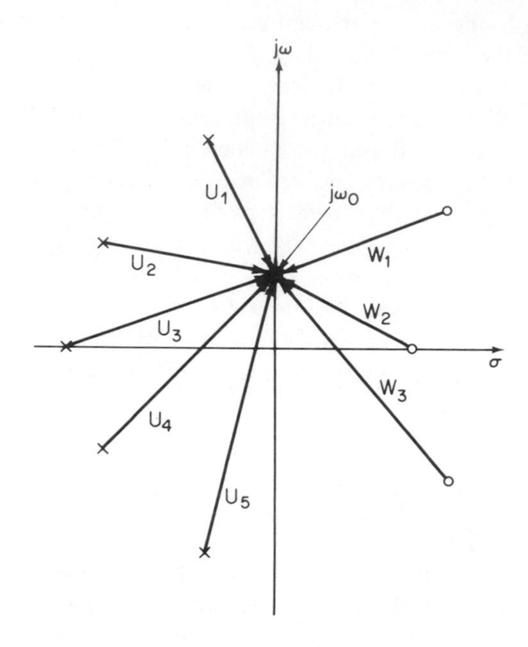

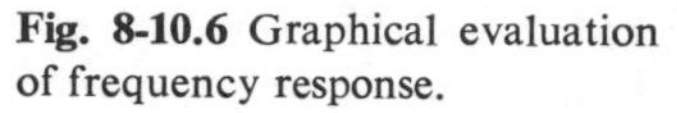
Fig. 8-10.6 Graphical evaluation of frequency response.

$$F(j\omega) = K \frac{\prod_{i=1}^{m} W_i}{\prod_{i=1}^{n} U_i}$$

The results given above may be used to determine many useful results concerning the sinusoidal steady-state response of various networks. For example, in the limit as ω approaches infinity, the vectors from all the finite poles and zeros will approach angles of 90°. Thus, the phase of the network function in this limit is simply equal to the number of finite zeros minus the number of finite poles times 90°. As another example, it is readily shown that for values of frequency above and below a zero on the $j\omega$ axis, the phase must differ by 180°.

As a final topic in this section we now show that there is a direct relation between the quality factor Q introduced in Sec. 7-8 and the pole positions of a given network function. As an example of this relation, consider the band-pass network function

$$N(s) = \frac{Ks}{s^2 + as + b} \tag{8-262}$$

Evaluating this for $s = j\omega$, we obtain

$$N(j\omega) = \frac{j\omega K}{b - \omega^2 + j\omega a} \tag{8-263}$$

Dividing the numerator and denominator by $j\omega$, this may be written in the form

$$N(j\omega) = \frac{K}{(b/j\omega) + j\omega + a} = \frac{K}{a + j[\omega - (b/\omega)]} \tag{8-264}$$

At resonance the denominator must have its smallest value. This occurs when the imaginary part of the denominator is zero. Thus, letting ω_0 be the resonant frequency, we see that the following relation must be satisfied:

$$\omega_0 - \frac{b}{\omega_0} = 0 \tag{8-265}$$

This requires that

$$\omega_0 = \sqrt{b} \tag{8-266}$$

At resonance, i.e., when $s = j\omega_0$, the network function is real and has the value

$$N(j\omega_0) = \frac{K}{a} \tag{8-267}$$

Now let us compute the value of the frequencies at which the magnitude of the network function $|N(j\omega)|$ is 0.707 of its value at resonance, i.e., the 3-dB frequencies. From Eqs. (8-264) and (8-267), these values of frequency ω must satisfy the relation

$$\frac{K}{\sqrt{[\omega - (b/\omega)]^2 + a^2}} = \frac{K}{\sqrt{2}\,a} \tag{8-268}$$

Squaring both members of the above equation, we obtain

$$\left(\omega - \frac{b}{\omega}\right)^2 + a^2 = 2a^2 \tag{8-269}$$

Rearranging terms and taking the square root of both members of the above, we obtain

$$\omega - \frac{b}{\omega} = \pm a \tag{8-270}$$

This may be written as a quadratic equation having the form

$$\omega^2 \pm a\omega - b = 0 \tag{8-271}$$

The roots of this equation are

$$\omega = \pm\frac{a}{2} \pm \sqrt{\left(\frac{a}{2}\right)^2 + b} \tag{8-272}$$

Choosing the appropriate combination of signs, we see that

$$\omega_{\text{upper 3 dB}} = \frac{a}{2} + \sqrt{\left(\frac{a}{2}\right)^2 + b}$$
$$\omega_{\text{lower 3 dB}} = -\frac{a}{2} + \sqrt{\left(\frac{a}{2}\right)^2 + b} \tag{8-273}$$

Thus, we find that the bandwidth B is

$$B = \omega_{\text{upper 3 dB}} - \omega_{\text{lower 3 dB}} = a \tag{8-274}$$

In Sec. 7–8 we showed that Q was defined as

$$Q = \frac{\omega_0}{B} \tag{8-275}$$

Inserting the relations of Eqs. (8–266) and (8–274) in the above, we obtain the following expression for Q in terms of the coefficients of the network function given in Eq. (8–262). This is

$$Q = \frac{\sqrt{b}}{a} \tag{8-276}$$

Thus, we conclude that for a second-order band-pass network function having the form given in Eq. (8–262), the Q is simply expressed as the ratio of the square root of the zero-degree coefficient to the first-degree coefficient. Since the square root of the zero-degree coefficient is equal to the magnitude of a vector drawn from the origin to the pole position and since the first-degree coefficient is -2 times the real part of the pole position, we see that an alternative definition for Q for a complex conjugate pair of poles located at $s = p_0$ and $s = p_0^*$ is

$$Q = \left|\frac{p_0}{2\,\text{Re}\,p_0}\right| \tag{8-277}$$

The relations given in Eqs. (8–276) and (8–277) provide a simple means of calculating Q in network functions which are characterized by a second-degree denominator polynomial.

8-11 THEVENIN AND NORTON EQUIVALENT TRANSFORMED NETWORKS

To this point in this chapter we have introduced the Laplace transformation and showed how it may be used to solve for the values of the variables of transformed networks. Many of the techniques for forming the equations describing the transformed network which we have introduced in this chapter are practically identical with those which were introduced in Chap. 3, where we discussed the formulation of network equations for resistance networks. For example, in Sec. 8-7, we showed that mesh impedance and node admittance equations could be written for transformed networks by following the same rules for the determination of the elements of the matrices that were used in the resistance network case. The modification of almost all the other techniques for analyzing resistance networks given in Chap. 3 (such as converting voltage sources to current sources and vice versa, adding parallel voltage sources and series current sources, etc.) so that they apply to the transformed network is a trivial task, since the only major change in the techniques is to substitute the word impedance for the word resistance, and the word admittance for the word conductance. As an example of the modification of one such technique, consider the development of a Thevenin equivalent circuit for a transformed network. As an example of this procedure, we will use the network shown in Fig. 8-11.1. The circuit consists of an independent voltage source with an output $V_S(s) = 3s/(s^2 + 4)$ [thus, the time-domain excitation is $v_S(t) = 3 \cos 2t u(t)$], a $\frac{1}{9}$-H inductor, and a 1-F capacitor. The Thevenin equivalent circuit will, of course, consist of an independent voltage source and a series two-terminal impedance. To determine the components of the Thevenin equivalent circuit, following the procedure outlined in Sec. 3-9, we first find the transformed open-circuit output voltage $V_0(s)$. This is easily done by considering the capacitor and the inductor as elements of a voltage divider. Thus, we may write

$$V_0(s) = \frac{1/s}{(s/9) + (1/s)} V_S(s) = \frac{9}{s^2 + 9} \frac{3s}{s^2 + 4} \tag{8-278}$$

The techniques developed in Sec. 8-4 for determining a partial-fraction expansion are readily applied to change the product form of the right member of (8-278) into a sum form. Thus, we may write

$$V_0(s) = \frac{-27s/5}{s^2 + 9} + \frac{27s/5}{s^2 + 4} \tag{8-279}$$

Considering this result, and recalling that the open-circuit voltage specifies the value

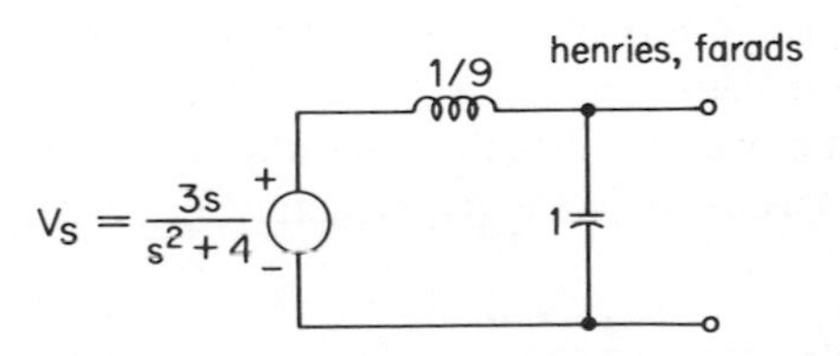

Fig. 8-11.1 A simple transformed network.

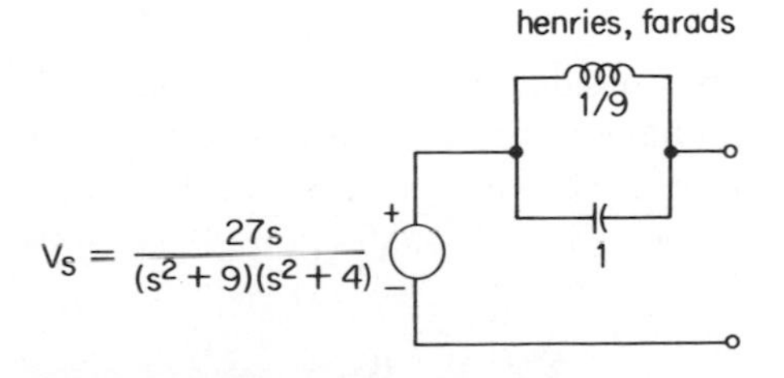

Fig. 8-11.2 Thevenin equivalent for the transformed network shown in Fig. 8-11.1.

of the voltage source in the equivalent Thevenin circuit, we see that the time-domain equivalent voltage source will have an output $v_0(t) = (27/5)\,[-\cos 3t + \cos 2t]\,u(t)$. This may be modeled in the transformed network as a single voltage source with an output as given by the right member of Eq. (8–278). Now let us find the second part of the equivalent circuit, the two-terminal impedance. This is found by setting the output of the independent source $V_S(s)$ to zero and determining the impedance at the terminals of the network. For the network shown in Fig. 8–11.1 this is simply the impedance of the parallel combination of the inductor and the capacitor. Thus, the Thevenin equivalent of the transformed network shown in Fig. 8–11.1 may be drawn as shown in Fig. 8–11.2.

To verify the validity of the Thevenin equivalent circuit derived above, let us determine the expression for the current $I(s)$ which flows in a 1-Ω resistor that is connected to the terminals of the original circuit shown in Fig. 8–11.1. The connection is illustrated in Fig. 8–11.3(a). For this circuit, we may find $I(s)$ by writing the mesh equations. Thus, we obtain

$$\begin{bmatrix} \dfrac{3s}{s^2+4} \\ 0 \end{bmatrix} = \begin{bmatrix} \dfrac{s}{9} + \dfrac{1}{s} & -\dfrac{1}{s} \\ -\dfrac{1}{s} & 1 + \dfrac{1}{s} \end{bmatrix} \begin{bmatrix} I_1(s) \\ I(s) \end{bmatrix} \tag{8–280}$$

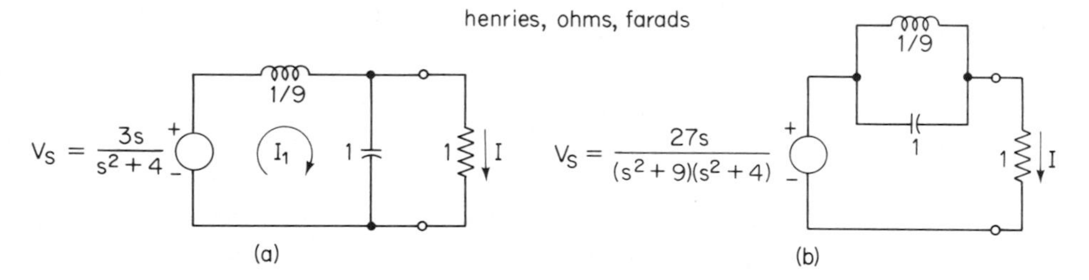

Fig. 8-11.3 Testing the circuits of Figs. 8-11.1 and 8-11.2 using a one-ohm termination.

Solving these equations, we find

$$I(s) = \frac{27s}{(s^2+4)(s^2+s+9)} \tag{8–281}$$

Now consider the problem of finding the current $I(s)$ in a 1-Ω resistor used as a termination for the Thevenin equivalent circuit shown in Fig. 8–11.2. The configuration is shown in Fig. 8–11.3(b). In this case, we find that the total impedance of the mesh is $Z(s) = (s^2 + s + 9)/(s^2 + 9)$. Thus, the current $I(s)$ is simply given as

$$I(s) = \frac{V_0(s)}{Z(s)} = \frac{27s}{(s^2+4)(s^2+9)}\,\frac{s^2+9}{s^2+s+9}$$

$$= \frac{27s}{(s^2+4)(s^2+s+9)} \tag{8–282}$$

Therefore, as is to be expected, the same results are obtained in the two cases. The procedure outlined above for finding the Thevenin equivalent circuit for a transformed network is summarized as follows:

SUMMARY 8-11.1

The Thevenin Equivalent Circuit for a Transformed Network: A transformed network representation for an actual time-domain circuit in which the initial conditions are zero may be converted at a given pair of terminals to a transformed Thevenin equivalent network consisting of a transformed voltage source in series with a two-terminal impedance. The output of the transformed Thevenin voltage source is simply the transformed voltage seen at the pair of terminals when these are open-circuited. The Thevenin impedance is the impedance of the network seen between the pair of terminals when all the independent sources have been set to zero.

By a similar development, we may extend the procedure for finding a Norton equivalent circuit given in Sec. 3-9 to include transformed networks. The transformed equivalent circuit will consist of an independent current source and a shunt two-terminal admittance. As an example, consider finding a Norton equivalent circuit for the network shown in Fig. 8-11.1. First we will find the transformed current that will flow in a short circuit placed across the terminals of the network. This gives the value for the equivalent transformed Norton current generator. Letting $I_0(s)$ be the expression for this current, we find from Fig. 8-11.4(a) that

$$I_0(s) = \frac{3s}{s^2+4}\frac{9}{s} = \frac{27}{s^2+4} \tag{8-283}$$

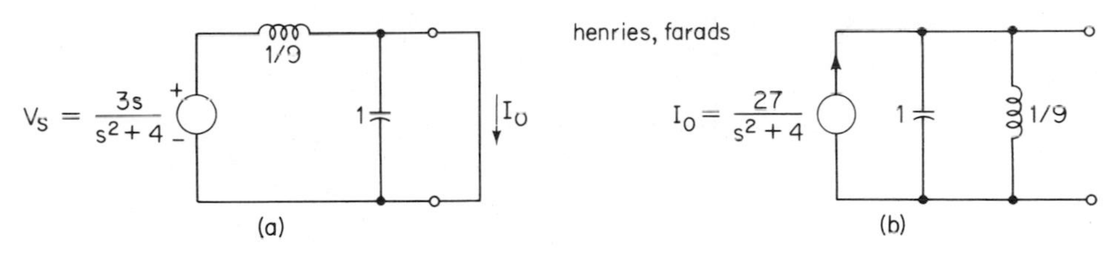

Fig. 8-11.4 Finding a Norton equivalent circuit.

The equivalent admittance is simply the parallel connection of the inductor and the capacitor, i.e., the admittance seen at the specified pair of terminals when the independent source is set to zero. Thus, we obtain an equivalent Norton circuit as shown in Fig. 8-11.4(b). These results may be summarized as follows:

SUMMARY 8-11.2

The Norton Equivalent Circuit for a Transformed Network: A transformed network representation for an actual time-domain circuit in which the initial conditions are zero may be converted at a given pair of terminals to an equivalent transformed Norton circuit consisting of a transformed current source in shunt with a two-terminal admittance. The output of the Norton current source is simply the transformed current that flows through a short circuit placed across the pair of terminals. The Norton admittance is the admittance of the network seen between the pair of terminals when all the independent sources have been set to zero. Its value is the reciprocal of the value of the Thevenin impedance.

In practice, a circuit realization for a Thevenin impedance (or a Norton admittance) is easily found directly from the transformed network without any computation. This is so because it is simply the portion of the original transformed circuit connected to the specified terminal pair which remains after the independent transformed sources have been set to zero, i.e., after voltage sources have been replaced by a short circuit and current sources by an open circuit. An example of such a situation follows.

EXAMPLE 8-11.1. The transformed network shown in Fig. 8-11.5(a) contains two independent sources and one (dependent) VCIS. It is desired to find a Thevenin equivalent circuit for the original network as seen at the pair of terminals labeled *a-b*. To aid in such a determination, we may first convert the branch consisting of a series-connected voltage source and an inductance to a Norton equivalent. The resulting network is shown in Fig. 8-11.5(b). For this network, applying KCL, we obtain the node admittance equations

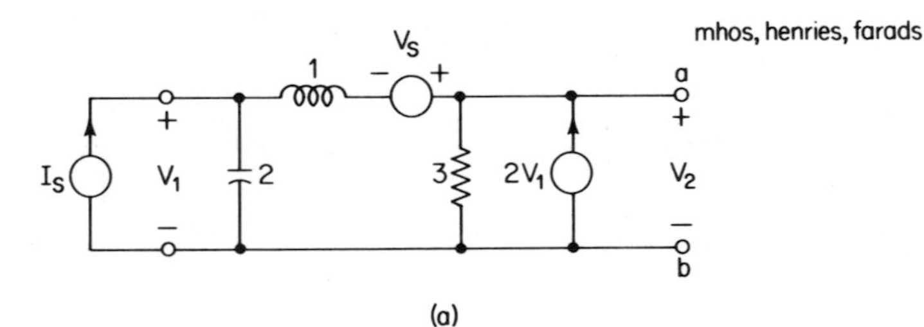

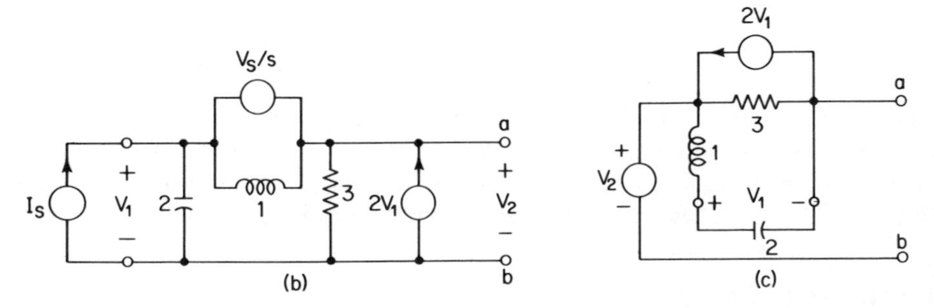

Fig. 8-11.5 Network for Example 8-11.1.

$$\begin{bmatrix} I_S - \dfrac{V_S}{s} \\ \dfrac{V_S}{s} \end{bmatrix} = \begin{bmatrix} 2s + \dfrac{1}{s} & -\dfrac{1}{s} \\ -2 - \dfrac{1}{s} & 3 + \dfrac{1}{s} \end{bmatrix} \begin{bmatrix} V_1(s) \\ V_2(s) \end{bmatrix}$$

where the effect of the controlled source has been included in the node admittance matrix. Solving for $V_2(s)$, we obtain

$$V_2(s) = \frac{I_S(s)(2s + 1) - 2(s - 1)V_S(s)}{6s^2 + 2s + 1}$$

This is the value of the transformed Thevenin voltage source. The Thevenin impedance may simply be represented by the original network in which the independent sources have been set to zero. Thus, the complete Thevenin equivalent circuit is as shown in Fig. 8-11.5(c).

The procedures given above for finding the Thevenin or Norton equivalent of a transformed network in which all the initial conditions have been assumed to be zero are readily modified for the case where the effect of non-zero initial conditions is to be considered. This may be done by applying the results of Summaries 8-7.5 and 8-7.6, in which it is shown that impulse and step generators may be used to establish initial conditions in a network which otherwise would be in the zero state. An example follows.

EXAMPLE 8-11.2. For the network shown in Fig. 8-11.6(a), it is desired to use the concept of a transformed network to find the Thevenin equivalent circuit as seen at terminals *a-b* for the case where the initial conditions $v_1(0+)$ and $i_L(0+)$ are not zero. The first step is to redraw the network inserting a current source in shunt across the capacitor and a voltage source in series with the inductor as shown in Fig. 8-11.6(b). As indicated in the figure, the output of the additional transformed current source is K_1 (an impulse of strength K_1) and the output of the new transformed voltage source is $K_2/4$. In order to put the network in a form more amenable to node analysis, we may convert the branch consisting of the inductor in series with the voltage source to a Norton equivalent circuit consisting of the inductor in shunt with a current source. The current source will have an output which is equal to the current which would flow through a short circuit placed across the terminals of the branch (when it is removed from the network). This is easily shown to be $-K_2/s$ with the indicated reference polarity. Thus, the circuit appears as shown in Fig. 8-11.6(c). The network equations for this circuit are

$$\begin{bmatrix} I_1(s) + K_1 \\ I_2(s) - \dfrac{K_2}{s} \end{bmatrix} = \begin{bmatrix} s + 2 & -1 \\ -4 & \dfrac{4}{s} + 4 \end{bmatrix} \begin{bmatrix} V_1(s) \\ V_2(s) \end{bmatrix}$$

where the term resulting from the VCIS has been moved to the right member of the equation. If we make the substitutions $K_1 = v_1(0+)$ and $K_2 = i_L(0+)$, then the relations given

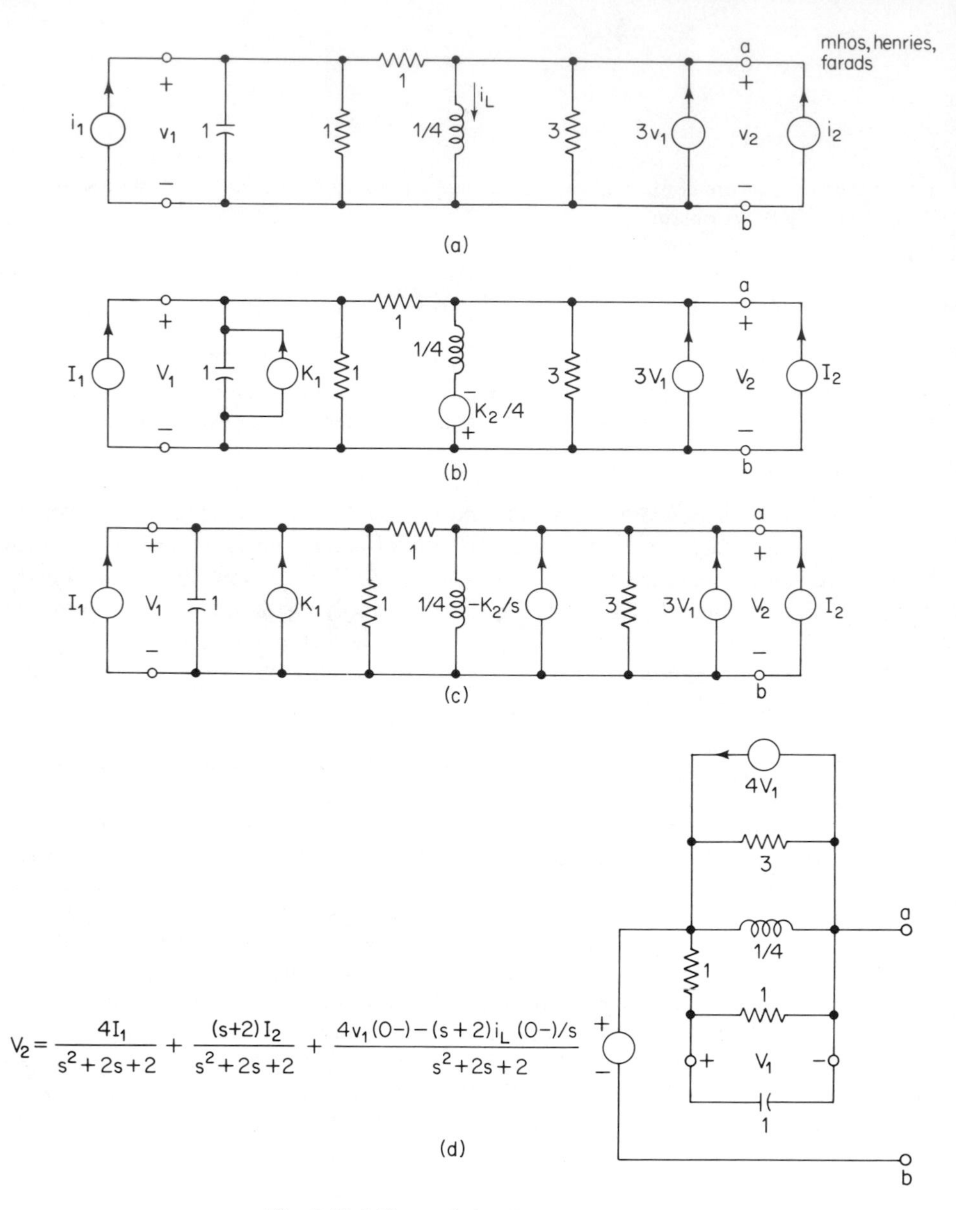

Fig. 8-11.6 Network for Example 8-11.2.

above are identical with those given in Eq. (8-182), which were derived directly from the time-domain equations for the same network (assuming that the initial conditions are the same at 0+ as they are at 0−). Thus, the solution for $V_2(s)$ yields the value of the equivalent Thevenin voltage generator. The resulting Thevenin equivalent circuit is shown in Fig. 8-11.6(d).

In the preceding examples we have shown how the original network itself may be used to realize the Thevenin or Norton immittance when an equivalent network is being constructed. Frequently, it is possible to replace circuits such as those shown in Figs. 8–11.5(c) and 8–11.6(d) with simpler networks which have the same driving-point immittance, and which are, therefore, equally as valid as the Thevenin or Norton immittance. The reason for this is that, in general, there is no unique realization for a specified driving-point admittance at a given pair of terminals. As an almost trivial example of this statement, note that if a driving-point impedance of $3s$ is desired at a pair of terminals, it may be realized by a single 3-H inductor as well as by a 1-H and a 2-H inductor connected in series, etc. A sizeable area of network theory is devoted to the problem of finding network realizations for specified driving-point immittances, and also for other network functions. This is called network *synthesis*. The titles of several books which treat this subject may be found in the Bibliography at the end of this text.

8-12 CONCLUSION

In this chapter we have introduced one of the most useful and powerful techniques that exists for the analysis of lumped linear time-invariant circuits, namely, the complex frequency plane as defined by the Laplace transformation. As a summary of the various techniques for applying Laplace transformation methods to the analysis of networks and a comparison of those techniques with the time-domain methods introduced in Chaps. 5 and 6, we present on a single flow chart in Fig. 8–12.1 the most significant operations of both time-domain and frequency-domain analysis. At first glance, this flow chart will probably appear somewhat formidable to the reader. This shouldn't be too surprising when you consider that it summarizes several hundred pages of text material! A more detailed study, however, will show that the chart is actually easier to understand than might at first be apparent. One reason for this is that all the individual blocks of this flow chart have already been presented as portions of smaller flow charts in earlier parts of the text. In addition, when these earlier flow charts were presented, the blocks and paths were given the same relative positions and orientation as are used in this flow chart. Thus, as a first step towards understanding the flow chart shown in Fig. 8–12.1, it is strongly recommended that the student review the following figures:

Fig. 5–1.6 Flow chart for solving a first-order circuit excited only by initial conditions.

Fig. 5–2.5 Flow chart for solving a first-order circuit excited only by sources.

Fig. 5–3.3 Flow chart for solving a first-order circuit excited by sources and initial conditions.

Fig. 6–10.1 General time-domain solution procedure.

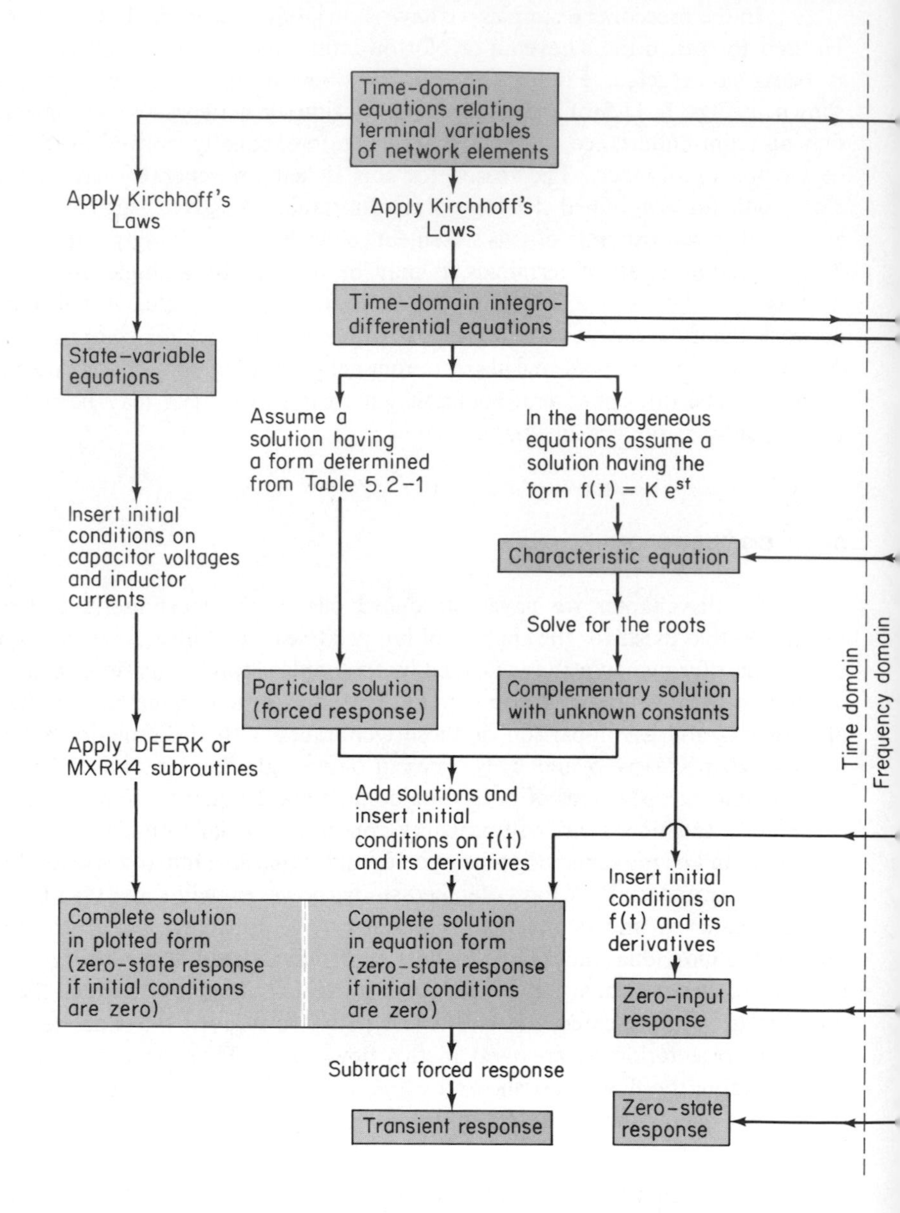

Fig. 8-12.1 Flow chart for time-domain and frequency-domain circuit analysis.

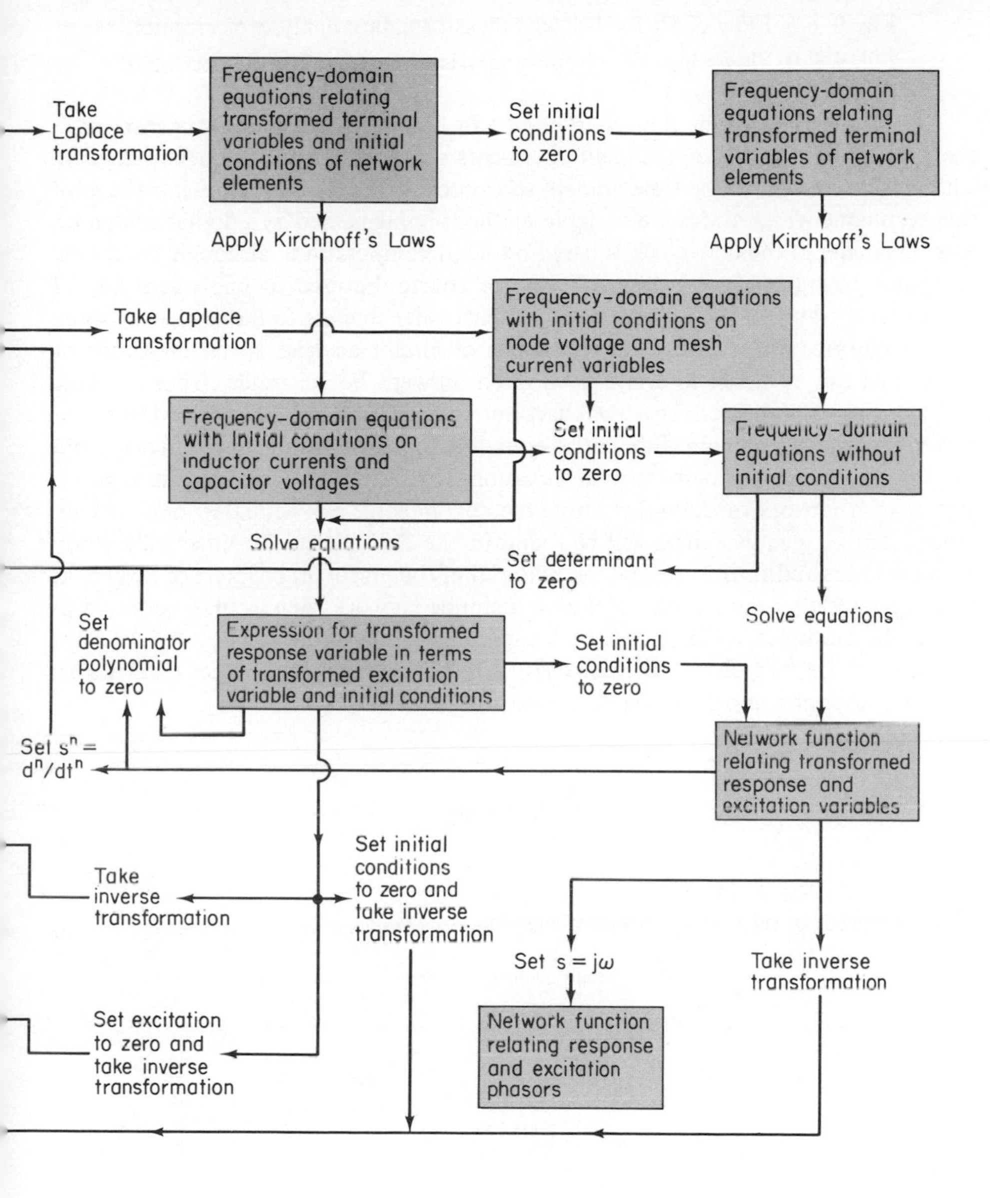

Fig. 8-12.1 Continued

The left half of the flow chart shown in Fig. 8–12.1 represents operations in the time domain and the right half represents operations in the frequency domain. The far left column of the time-domain section of the flow chart represents the solution technique which uses state-variable methods implemented by a digital computer. The remainder of the flow chart is based on hand computations, although the digital computer programs given in Sec. 8–5 can, of course, be used to implement any of the Laplace transformation operations. Of particular interest to the student wishing a thorough grasp of the basic fundamentals of circuit analysis is the multitude of routes that can be taken to arrive at a given answer. For example, from the flow chart it is readily apparent that the characteristic equation can be obtained from the homogeneous time-domain differential equation [by substituting $f(t) = Ke^{st}$], from the frequency-domain mesh or node equations (by setting the determinant to zero), or from the network function (by setting the denominator polynomial to zero). Many other interesting equivalences will be found in the flow chart. To effectively study the flow chart and thus to review the material of the first eight chapters of this book, it is suggested that the student start with a simple network such as may be found in any of the examples in Chap. 5, 6, or 8 and actually carry through each of the steps indicated in the flow chart. The successful completion of such a project indicates an excellent level of comprehension of the text material.

Problems

Problem 8–1 (Sec. 8–2)

For each of the circuits shown in Fig. P8-1:

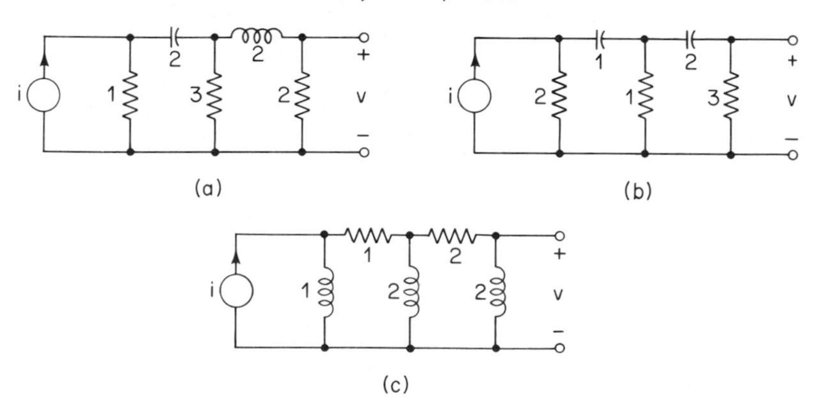

Figure P8-1

(a) Find the differential equation relating the output voltage $v(t)$ and the input current $i(t)$.
(b) Find an expression for $V(s)$, the Laplace transform of $v(t)$ in terms of $I(s)$, the Laplace transform of $i(t)$, and the initial conditions on $v(t)$.
(c) Identify the characteristic polynomial.

Problem 8-2 (Sec. 8-2)

For each of the circuits shown in Fig. P8-2:
(a) Find the differential equation relating the output voltage $v_2(t)$ and the input voltage $v_1(t)$.
(b) Find an expression for $V_2(s)$, the Laplace transform of $v_2(t)$, in terms of $V_1(s)$, the Laplace transform of $v_1(t)$, and the initial conditions on $v_2(t)$.
(c) Identify the characteristic polynomial.

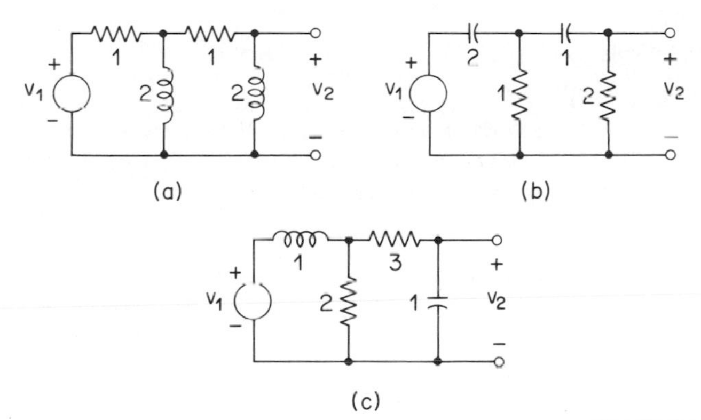

Figure P8-2

Problem 8-3 (Sec. 8-2)

Derive the expression for the Laplace transform of the nth order derivative of a time-domain function $f(t)$ given in Summary 8-2.2.

Problem 8-4 (Sec. 8-3)

Find the Laplace transform of each of the functions shown in Fig. P4-14.

Problem 8-5 (Sec. 8-3)

Find the Laplace transform of each of the functions shown in Fig. P4-15. Assume that the functions are zero for $t < 0$.

Problem 8-6 (Sec. 8-3)

Derive the expression for the Laplace transform of the function $Kt^n u(t)$, where n is a positive integer, given in Table 8-3.1.

Problem 8-7 (Sec. 8-4)

Find the inverse Laplace transform of the frequency-domain functions given below.

(a) $\dfrac{3s + 2}{s^2 + 1}$ (b) $\dfrac{1}{(s + 1)^2}$

(c) $\dfrac{3s^2 + 2s + 2}{(s + 2)^2(s + 3)}$ (d) $\dfrac{s - 1}{s^2 + 2s + 2}$

(e) $\dfrac{4s^2 - 3s + 5}{s(s^2 + 2s + 5)}$ (f) $\dfrac{3s^3 + 2s^2 + s + 1}{(s^2 + 2s + 2)^2}$

Problem 8-8 (Sec. 8-4)

Find the inverse Laplace transform of the frequency-domain functions given below.

(a) $\dfrac{s^3}{s + 1}$ (b) $\dfrac{s^3 + s^2 + 3s + 2}{s(s^2 + s + 1)}$

(c) $\dfrac{s^4 + 1}{(s + 1)^3}$ (d) $\dfrac{s}{(s^2 + 1)^2}$

Problem 8-9 (Sec. 8-4)

Find the initial values for the inverse Laplace transform of each of the functions given in Problem 8-7 by applying the results of Summary 8-4.6.

Problem 8-10 (Sec. 8-4)

Find the final values for the inverse Laplace transform of each of the functions given in Problems 8-7 and 8-8 by applying the results of Summary 8-4.7.

Problem 8-11 (Sec. 8-4)

Derive the result given in (8-121).

Problem 8-12(Sec. 8-5)*[18]

Use digital computer techniques to make a plot of the values of the residues K_1 and K_2 for the function $F(s)$ given below. Use a as the independent variable over the range $0 < a \leq 0.98$. Use 49 discrete values of a.

$$F(s) = \frac{1}{(s + 1)(s + a)} = \frac{K_1}{s + 1} + \frac{K_2}{s + a}$$

Problem 8-13(Sec. 8-5)*

Use digital computer techniques to find a partial-fraction expansion for the following functions.

(a) $\dfrac{1.0}{s^4 + 4s^3 + 8s^2 + 8s + 4}$

(b) $\dfrac{10s^4 - 20}{s^5 + 6s^4 + 15s^3 + 10s^2 + 14s + 4}$

Problem 8-14(Sec. 8-5)*

Find the partial-fraction expansion and plot the inverse Laplace transform over the time period from 0 to 2.5 s for the function $F(s)$ given below. Use 51 points on the plot.

$$F(s) = \frac{25s^3 - 900}{s^4 + s^3 + 36s^2 + 36s}$$

Problem 8-15(Sec. 8-5)*

Determine the response to a step function $90u(t)$ for each of the network functions given below. $F_1(s)$ is a low-pass function satisfying a Bessel or maximally-flat delay (in the time-domain) characteristic. $F_2(s)$ is a low-pass function satisfying a Butter-

[18]Problems which require the use of digital computational techniques are identified by an asterisk.

worth or maximally-flat magnitude (as a function of a sinusoidal frequency) characteristic. Plot the responses as a function of time over a range from 0 to 10 s. Put both characteristics on the same plot using 51 values of time.

$$F_1(s) = \frac{15}{s^3 + 6s^2 + 15s + 15} \qquad F_2(s) = \frac{1}{s^3 + 2s^2 + 2s + 1}$$

Problem 8-16 (*Sec. 8-6*)

For the network used in Example 8-6.1, find the voltage $v_R(t)$ across the resistor (with the positive reference polarity on the left), for the case where the input voltage $v_S(t)$ is 10 cos $t\ u(t)$ V and the initial conditions are zero.

Problem 8-17 (*Sec. 8-6*)

Use the Laplace transform technique to find $v_2(t)$ in the network shown in Fig. P8-17. Assume that all initial conditions are zero and that $i(t) = 2e^{-t}u(t)$ A.

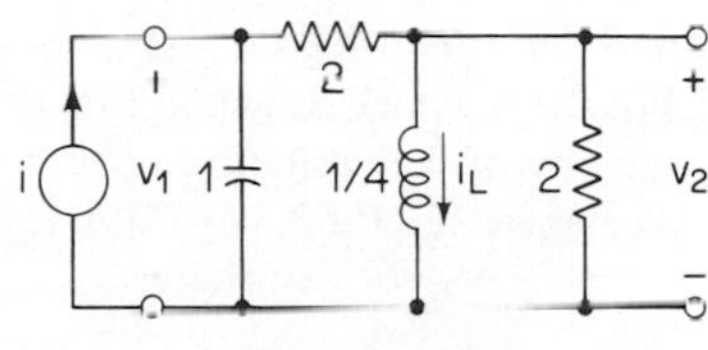

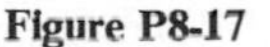
Figure P8-17

Problem 8-18 (*Sec. 8-6*)

Use the Laplace transform technique to find an expression for $v_2(t)$ for the network shown in Fig. P8-17 if $i(t) = u(t)$ A and if the initial conditions are $v_1(0-) = 1$ V, $i_L(0-) = 2$ A.

Problem 8-19 (*Sec. 8-6*)

Find the initial conditions $v_1(0-)$ and $i_L(0-)$ such that the output $v_2(t) = e^{-t} \cos t$ V in the circuit shown in Fig. P8-17. Let $i(t) = u(t)$ A.

Problem 8-20 (*Sec. 8-6*)

Find the driving point immittance of each of the networks shown in Fig. P8-20.

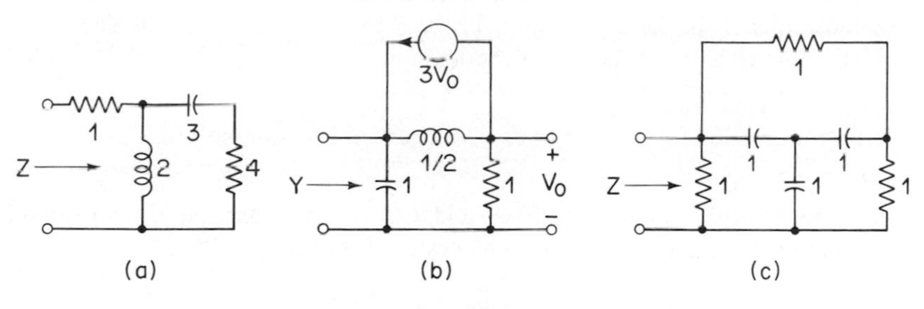

Figure P8-20

Problem 8-21 (*Sec. 8-6*)

In the network shown in Fig. P8-21, if the initial conditions are zero, find an expression for $v_2(t)$ for the following cases: (a) when the excitation voltage $v_1(t)$ is sin $t\ u(t)$ V, and (b) when the excitation voltage $v_1(t)$ is sin $2t\ u(t)$ V.

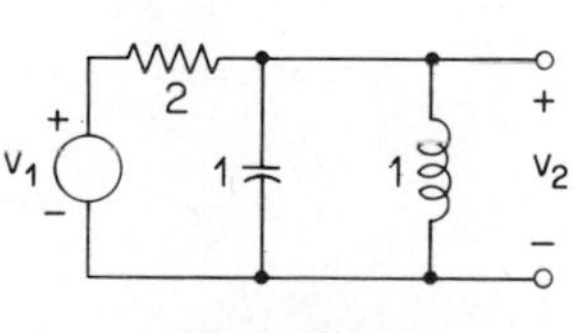

Figure P8-21

Problem 8–22 (*Sec. 8–6*)

Find an expression for the Laplace transform $V_C(s)$ of the voltage $v_C(t)$ which appears across the capacitance (with the positive reference polarity at the top) in Fig. 8-6.2 in terms of the Laplace transform $V_S(s)$ of the excitation voltage $v_S(t)$ and the initial conditions. Identify the zero-state response component and the zero-input response component of your answer.

Problem 8–23 (*Sec. 8–6*)

Find an expression for the Laplace transform $V_L(s)$ of the voltage $v_L(t)$ which appears across the inductance (with the positive reference polarity at the left) in Fig. 8-6.2 in terms of the Laplace transform $V_S(s)$ of the excitation voltage $v_S(t)$ and the initial conditions. Identify the zero-state response component and the zero-input response component of your answer.

Problem 8–24 (*Sec. 8–6*)

For the network shown in Fig. P8-24, find the expressions for the following variables in terms of the excitation current $I_1(s)$ and the initial conditions on the energy-storage elements: (a) $V_1(s)$, (b) $V_2(s)$, (c) $V_L(s)$, (d) $I_C(s)$.

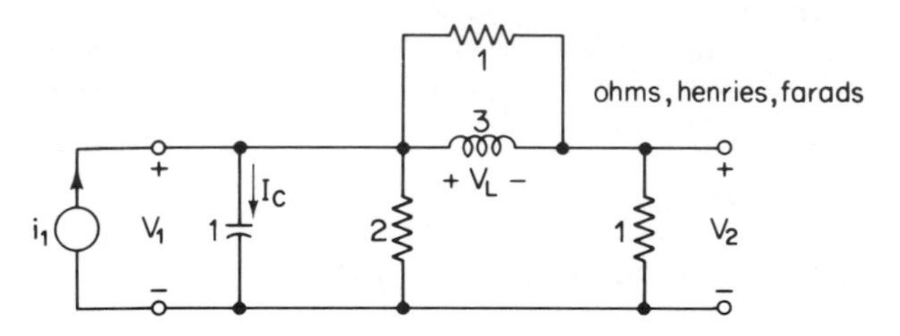

Figure P8-24

Problem 8–25 (*Sec. 8–6*)

For the *RLC* network and the values used in Example 8-6.1, find a set of initial conditions such that the sinusoidal term in the response of $v_R(t)$, the voltage across the resistor, is of the form $Ke^{-t}\sin t$, where $K > 0$. The positive reference polarity $v_R(t)$ is assumed to be at the left of the resistor.

Problem 8–26 (*Sec. 8–6*)

Verify the solution for the constants *a*, *b*, *c*, and *d* in Example 8-6.6.

Problem 8–27 (*Sec. 8–6*)

Find a set of initial conditions $v_1(0-)$ and $i_L(0-)$ such that there is no sinusoidal oscillation in the output for $v_2(t)$ in the circuit shown in Fig. P8-27.

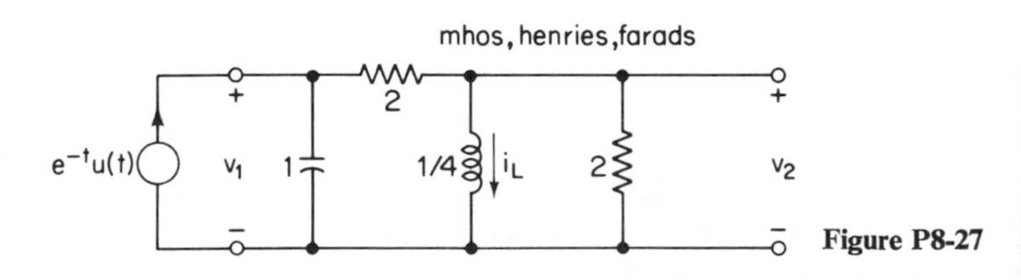

Figure P8-27

Problem 8–28 (*Sec. 8–7*)

Write the mesh equations for the transformed network shown in Fig. P8-28.

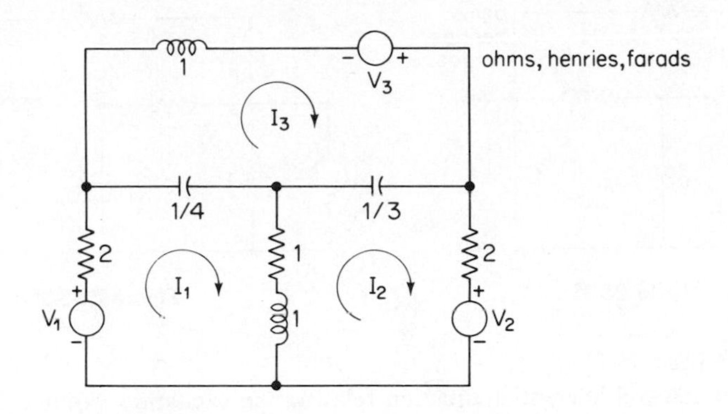

Figure P8-28

Problem 8–29 (*Sec. 8–7*)

Write the node equations for the transformed network shown in Fig. P8-29.

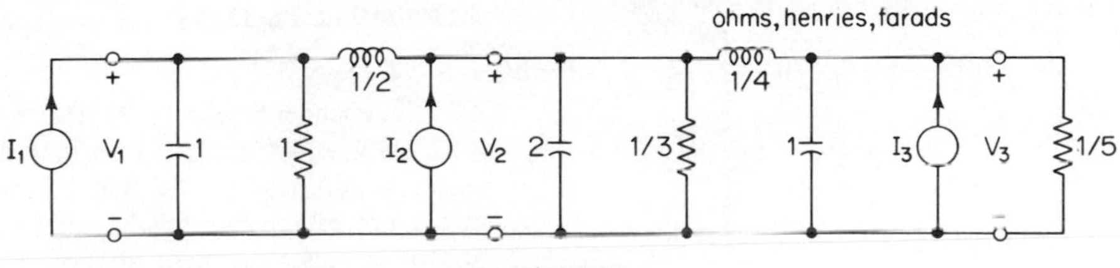

Figure P8-29

Problem 8–30 (*Sec. 8–7*)

Write the network equations (transformed) in the matrix format $\mathbf{I} = \mathbf{YV}$ for the network shown in Fig. P8-30. Assume all initial conditions are zero and all elements have unity value. *Do not solve* the equations.

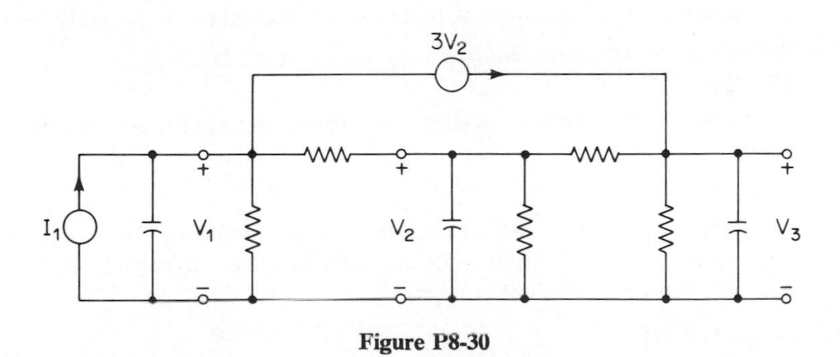

Figure P8-30

Problem 8–31 (*Sec. 8–7*)

Find the integro-differential equation relating the input quantity $v_1(t)$ and the output quantity $v_2(t)$ for the network shown in Fig. P8-31.

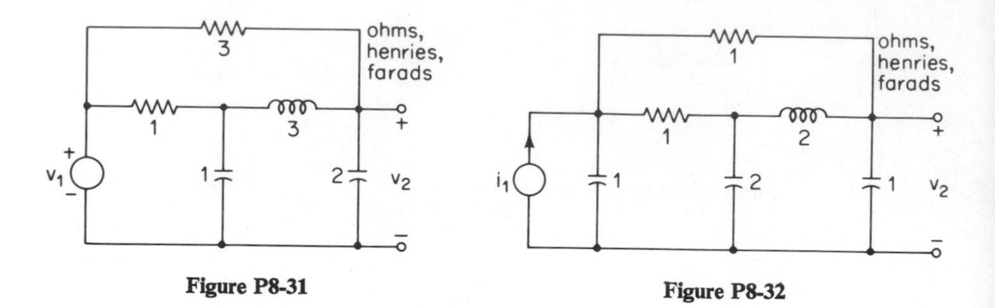

Figure P8-31

Figure P8-32

Problem 8–32 (*Sec. 8–7*)

Find the integro-differential equation relating the excitation variable $i_1(t)$ and the response variable $v_2(t)$ in the network shown in Fig. P8-32.

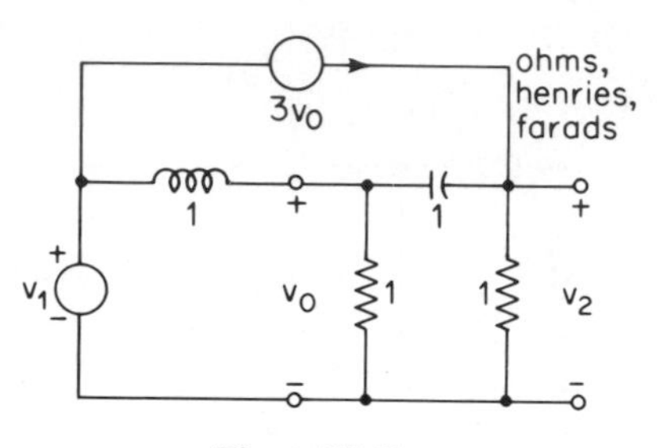

Figure P8-33

Problem 8–33 (*Sec. 8–7*)

Find the differential equation which relates $v_1(t)$ and $v_2(t)$ in the network shown in Fig. P8-33.

Problem 8–34 (*Sec. 8–7*)

Add independent sources to the network shown in Fig. P8-31 so that the initial conditions on the reactive elements are established by the specification of impulses on these sources, then write the mesh equations for the network.

Problem 8–35 (*Sec. 8–7*)

Repeat Problem 8-34 for the network shown in Fig. P8-32. Use node equations instead of mesh equations.

Problem 8–36 (*Sec. 8–7*)

Repeat Problem 8-34 using sources with a step function output to generate the initial conditions.

Problem 8–37 (*Sec. 8–7*)

Repeat Problem 8-35 using sources with a step function output to generate the initial conditions.

Problem 8–38 (*Sec. 8–7*)

Show that the two circuits used to establish initial conditions on a capacitor as described in Summaries 8-7.5 and 8-7.6 are directly related to each other by using Thevenin and Norton equivalent circuits.

Problem 8–39 (*Sec. 8–8*)

Find the natural frequencies of each of the networks shown in Fig. P8-39 by setting the determinant of the mesh impedance matrix **Z** to zero.

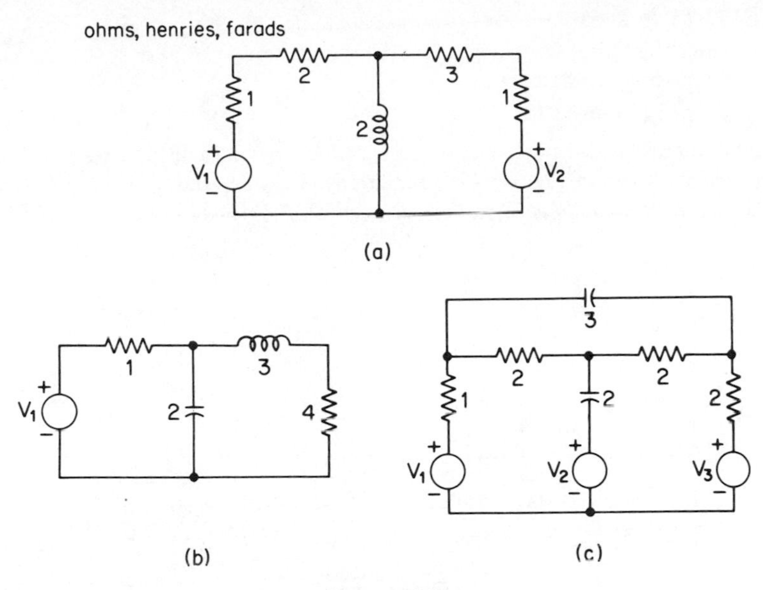

Figure P8-39

Problem 8–40 (Sec. 8–8)

Find the natural frequencies of each of the networks shown in Fig. P8-40 by setting the determinant of the node admittance matrix **Y** to zero.

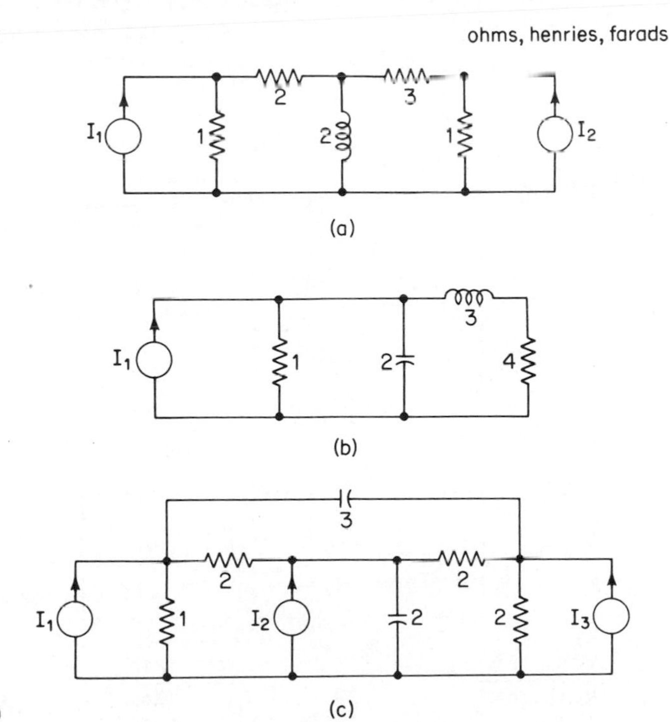

Figure P8-40

Problem 8–41 (*Sec. 8–8*)

Compare the natural frequencies found in Problems 8-39 and 8-40. Comment on any similarities or differences.

Problem 8–42 (*Sec. 8–8*)

A set of network transfer functions are defined by making two tests on the same network as shown in Fig. P8-42. Predict any equalities that will exist between the various polynomials $A(s)$, $B(s)$, $C(s)$, $D(s)$, $E(s)$, $F(s)$, $G(s)$, and $H(s)$.

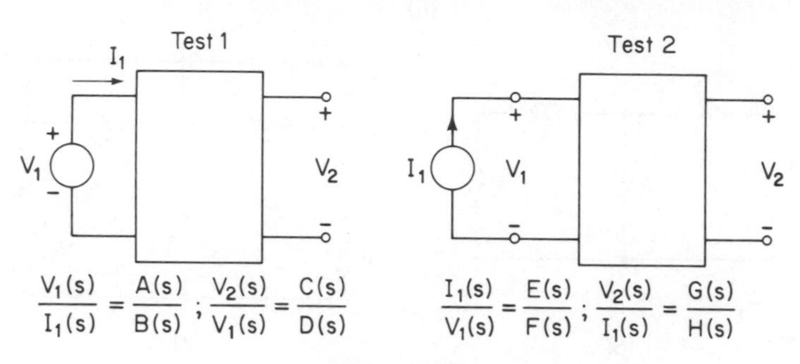

Figure P8-42

Problem 8–43 (*Sec. 8–8*)

As a physical example of the conclusions reached in Problem 8-42, find the polynomials $A(s)$, $B(s)$, $C(s)$, $D(s)$, $E(s)$, $F(s)$, $G(s)$, and $H(s)$ defined in that problem for the network shown in Fig. P8-43.

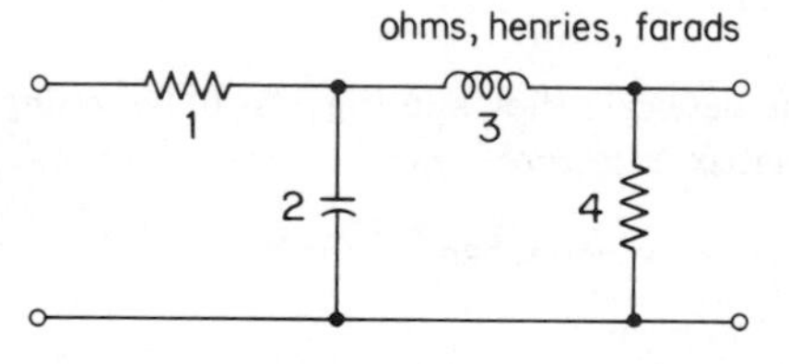

Figure P8-43

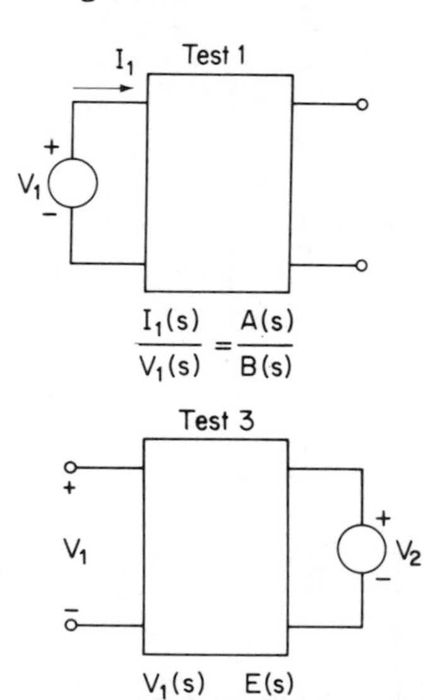

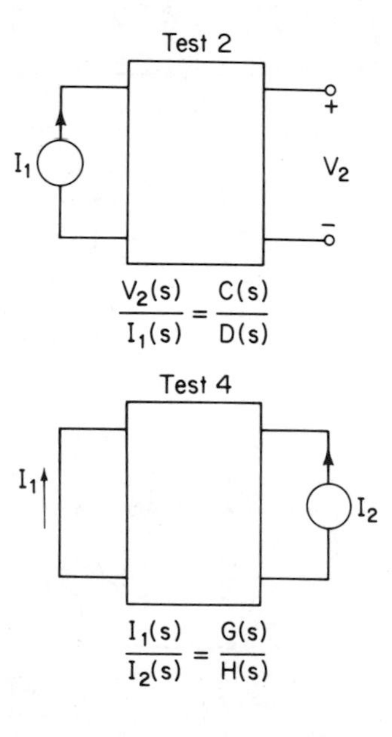

Figure P8-44

Problem 8-44 (*Sec. 8-8*)

A series of four tests are made on a two two-terminal-pair network as shown in Fig. P8-44. If the network is reciprocal, find any equalities that will exist between the various polynomials $A(s)$, $B(s)$, $C(s)$, $D(s)$, $E(s)$, $F(s)$, $G(s)$, and $H(s)$.

Problem 8-45 (*Sec. 8-8*)

As a physical example of the conclusions reached in Problem 8-44, find the polynomials $A(s)$, $B(s)$, $C(s)$, $D(s)$, $E(s)$, $F(s)$, $G(s)$, and $H(s)$ defined in that problem for the network shown in Fig. P8-43.

Problem 8-46 (*Sec. 8-8*)

Find any relations that exist between any of the polynomials $A(s)$, $B(s)$, $C(s)$, $D(s)$, $E(s)$, and $F(s)$ in the three network functions defined by the three tests on the same network shown in Fig. P8-46.

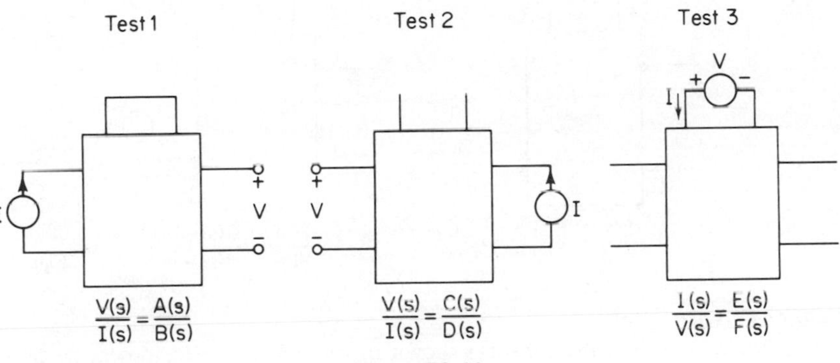

Figure P8-46

Problem 8-47 (*Sec. 8-8*)

Find any relations that may exist between the various polynomials $A(s)$, $B(s)$, $C(s)$, $D(s)$, $E(s)$, $F(s)$, $G(s)$, and $H(s)$ which define the different network functions for the network shown in Fig. P8-47.

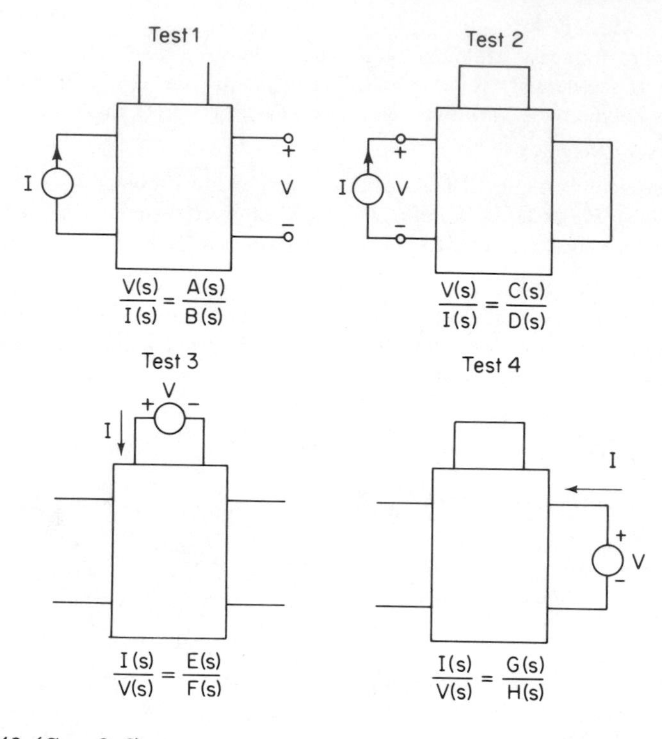

Figure P8-47

Problem 8–48 (Sec. 8–8)

(a) Predict which of the networks shown in Fig. P8-48 are reciprocal by observing the form and the components of the network.

(b) Find the network functions $Z_{12}(s) = I_1(s)/V_2(s)$ [with $I_2(s) = 0$] and $Z_{21}(s) = I_2(s)/V_1(s)$ [with $I_1(s) = 0$] or the network functions $Y_{12}(s) = V_1(s)/I_2(s)$ [with $V_2(s) = 0$] and $Y_{21}(s) = V_2(s)/I_1(s)$ [with $V_2(s) = 0$] for each of the networks shown in the figure to verify the conclusions reached in part (a).

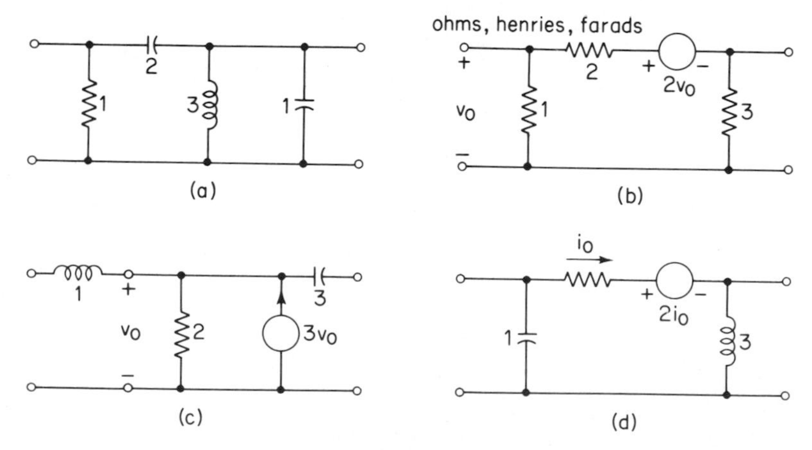

Figure P8-48

Problem 8–49 (Sec. 8–9)

The networks shown in Fig. P8-49 are to be operated as open-circuit voltage dividers. Without determining the actual network function defining the voltage transfer ratio for these networks, determine the location of any transmission zeros.

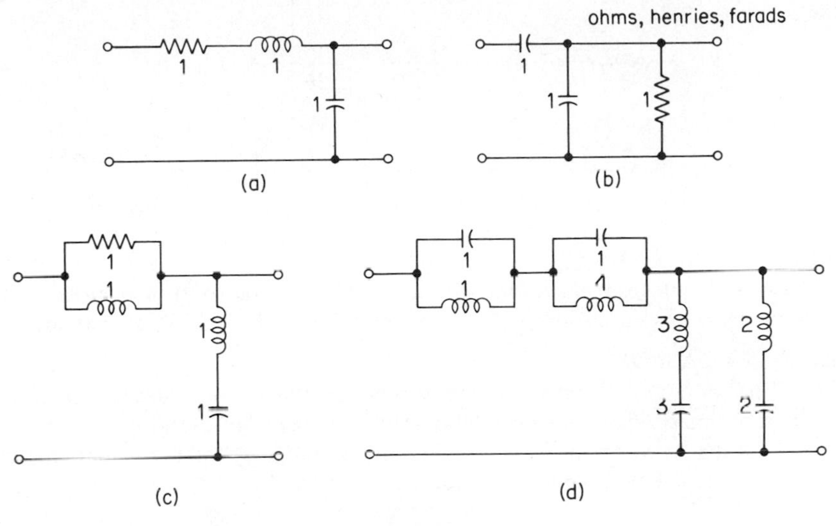

Figure P8-49

Problem 8–50 (Sec. 8–9)

Find the location and multiplicity of any zeros of transmission which are present in the ladder network shown in Fig. P8-50.

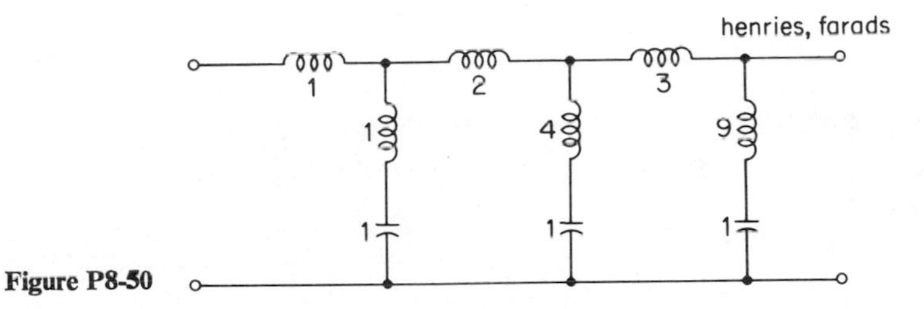

Figure P8-50

Problem 8–51 (Sec. 8–9)

Find the network functions $V_2(s)/V_1(s)$ for the active networks shown in Fig. P8-51. (The triangles represent VCVS's.)

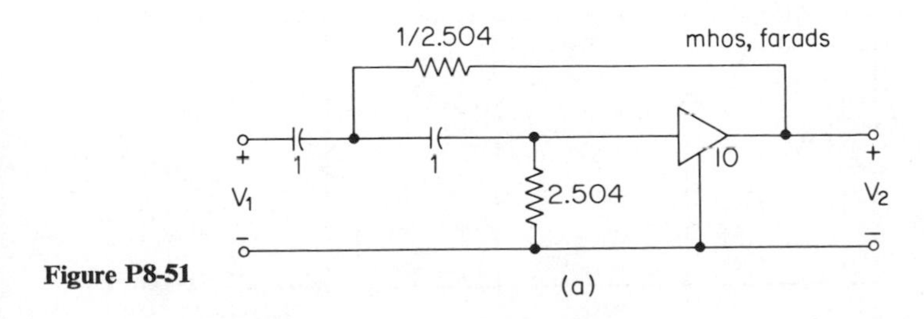

Figure P8-51

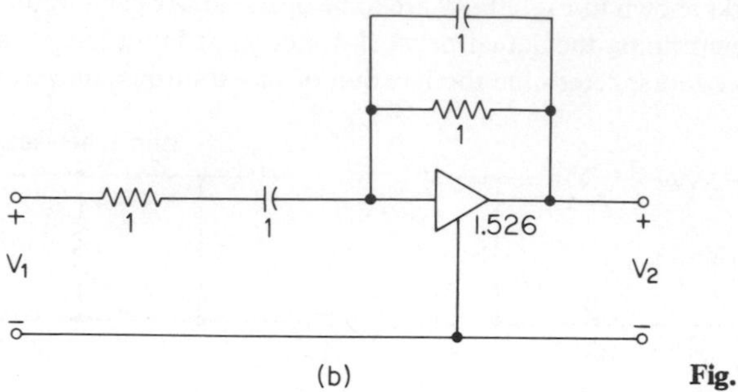

Fig. P8-51. Continued

Problem 8–52 (Sec. 8–9)

Draw the root-locus plot showing the variation of the poles of the network functions for the networks shown in Fig. P8-51 as the gains of the VCVS's are varied.

Problem 8–53 (Sec. 8–9)

(a) Find values of L and C for the network shown in Fig. P8-53 such that the circuit has no steady-state sinusoidal output component [of $v_2(t)$].

(b) Find the time-domain expression for the output $v_2(t)$ for the values chosen in part (a). Assume that the initial conditions are zero.

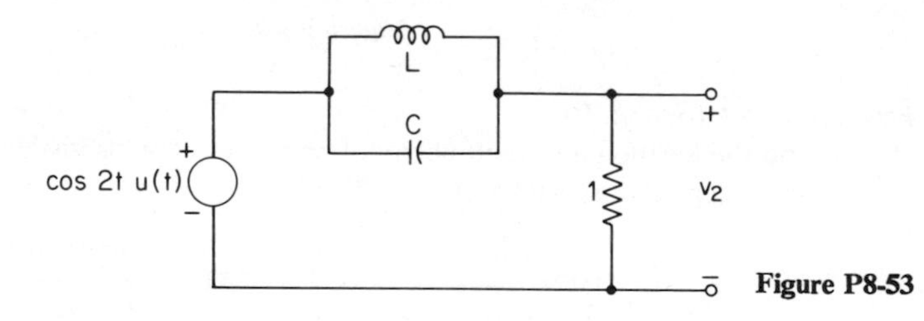

Figure P8-53

Problem 8–54 (Sec. 8–9)

The circuit shown in Fig. P8-54 is to be used to realize the open-circuit voltage transfer function

$$\frac{V_2}{V_1} = \frac{(s^2 + 1)(s^2 + 4)}{as^4 + bs^3 + cs^2 + ds + 8}$$

Find values for the elements L_1, L_2, C_1, C_2, and R.

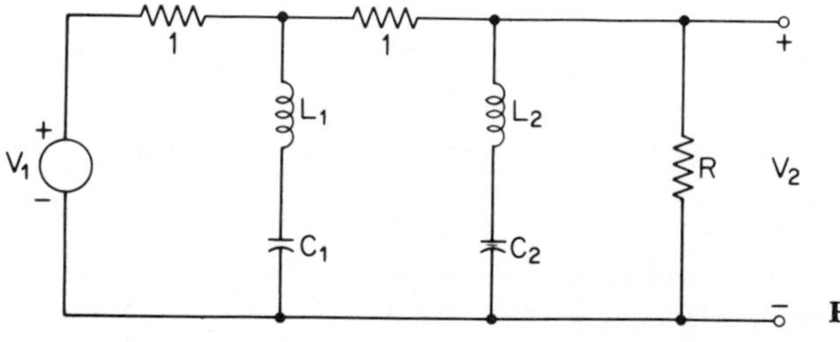

Figure P8-54

Problem 8–55 (Sec. 8–10)

Draw approximate sketches of the magnitude and phase of each of the pole-zero configurations shown in Fig. P8-55.

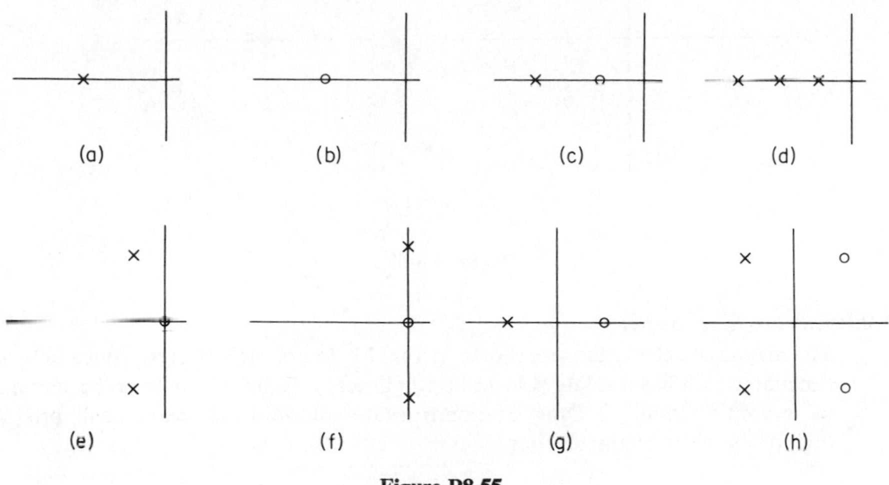

Figure P8-55

Problem 8–56 (Sec. 8–10)

Sketch the magnitude and the phase response curves for each of the pole-zero configurations shown in Fig. P8-56. Give numerical values for significant points on your plots. Assume all network functions have unity gain constants.

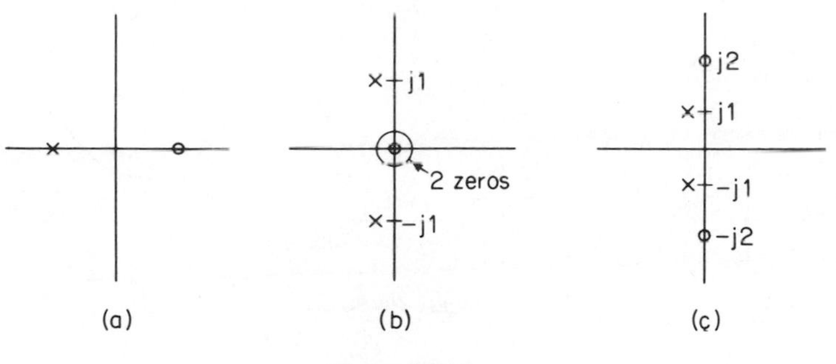

Figure P8-56

Problem 8–57 (Sec. 8–10)

Sketch magnitude and phase plots for each of the pole-zero configurations shown in Fig. P8-57. Assume that the gain constant is unity and indicate significant values on the ordinates and abscissas of your plots.

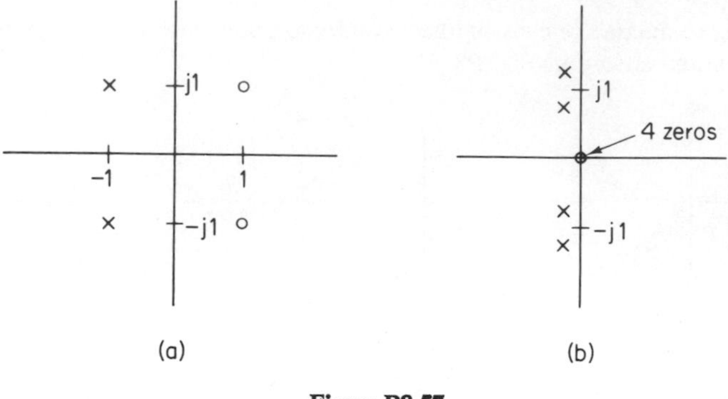

Figure P8-57

Problem 8–58 (Sec. 8–10)

The magnitude characteristics shown in Fig. P8-58 are each derived from a network function which has a single pole and a single zero. (Either the pole or the zero may be located at infinity.) Draw an approximate pole-zero plot which will produce each of the given characteristics.

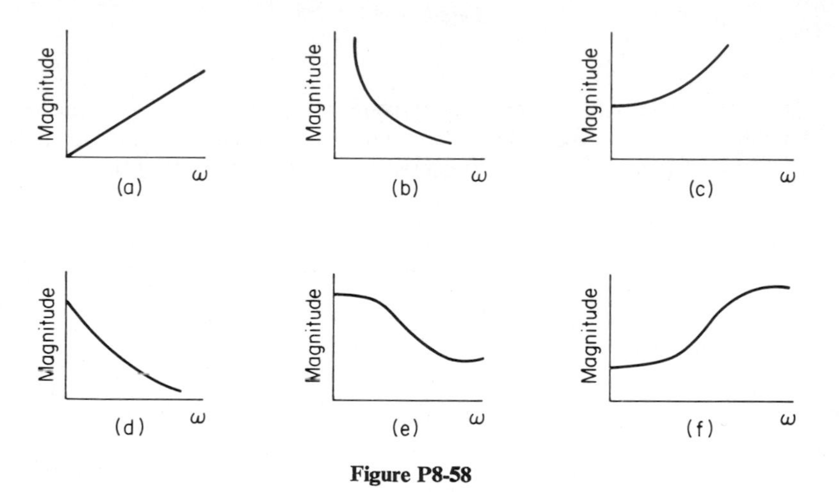

Figure P8-58

Problem 8–59 (Sec. 8–10)

The phase characteristics shown in Fig. P8-59 are each derived from a network function which has a single pole and a single zero. (Either the pole or the zero may be located at infinity.) Draw an approximate pole-zero plot which will produce each of the given characteristics.

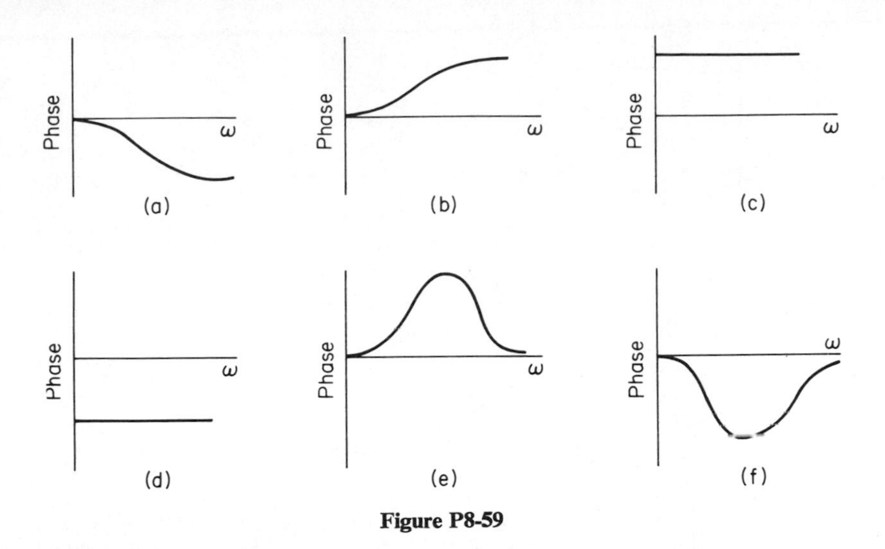

Figure P8-59

Problem 8–60 (Sec. 8–10)

The magnitude characteristics shown in Fig. P8-60 are each derived from a network function which has two complex conjugate poles and two zeros. The latter may only be located at the origin or at infinity. Draw the pole-zero plot which will produce each of the characteristics.

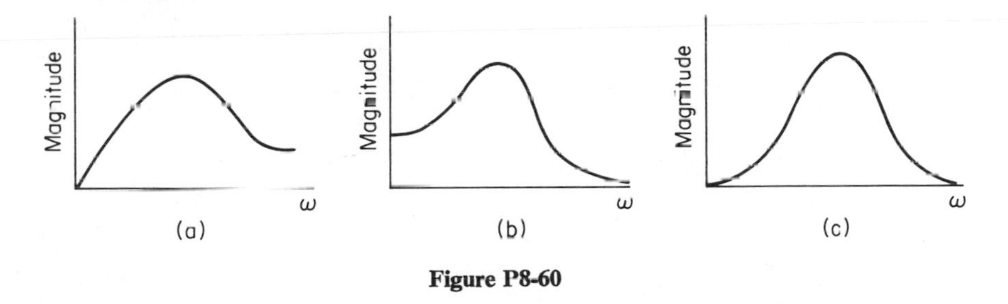

Figure P8-60

Problem 8–61 (Sec. 8–10)

The phase characteristics shown in Fig. P8-61 are each derived from a network function which has two complex conjugate poles and two zeros. The latter may be located only at the origin or at infinity. Draw the pole-zero plot which will produce each of the characteristics.

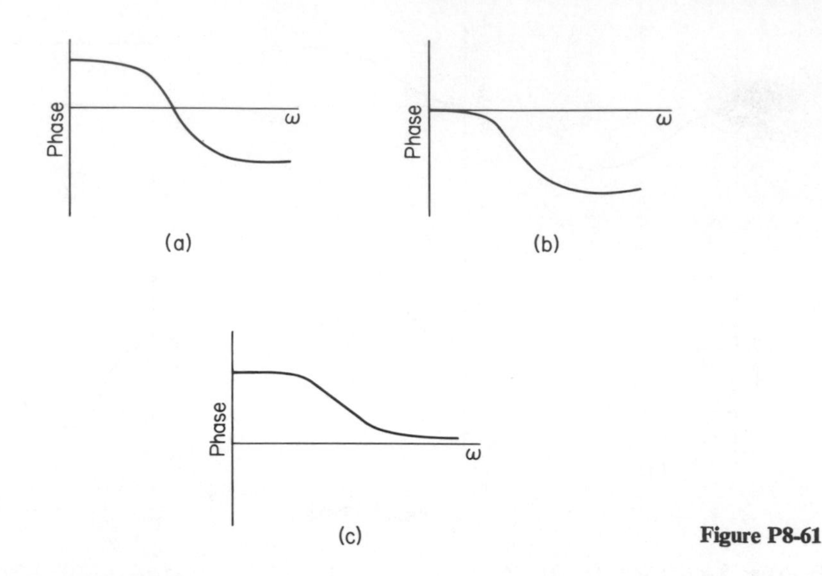

Figure P8-61

Problem 8–62 (Sec. 8–11)

Find Thevenin and Norton equivalent circuits for each of the transformed networks shown in Fig. P8-62. Assume that all the initial conditions are zero.

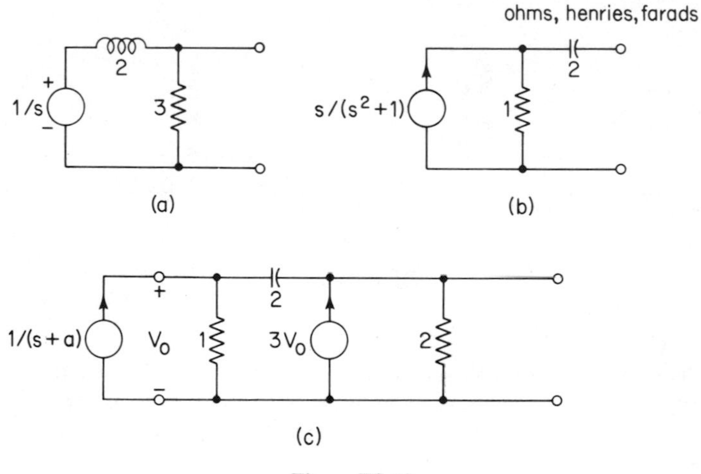

Figure P8-62

Problem 8–63 (Sec. 8–11)

Test the validity of the Thevenin and Norton circuits found in Problem 8-62 by calculating the transformed current that will flow in a 1-Ω resistor connected across the output terminals of the original circuit, the Thevenin equivalent circuit, and the Norton equivalent circuit.

Problem 8-64 (*Sec. 8-11*)

Repeat Problem 8-62 for the circuit shown in Fig. P8-62(a) under the assumption that the reactive element shown has a non-zero initial condition.

Problem 8-65 (*Sec. 8-11*)

Repeat Problem 8-63 for the circuit shown in Fig. P8-62(a) under the assumption that the reactive element shown has a non-zero initial condition.

CHAPTER 9

Two-Port Network Parameters

One of the most frequent tasks to which networks are applied is the process of shaping (or filtering) some electrical signal information. In order to accomplish this filtering process, it is necessary that a specific pair of terminals in the network be designated as the place at which an excitation or input is to be applied and another pair of terminals be designated as the place where the response or output, i.e., the filtered signal, is present. A frequently-used name applied to a pair of terminals used for input or output purposes is a *port*. Thus, the input terminal-pair (this is customarily drawn at the left side of the network schematic) may be referred to as the *input port* and the output terminal-pair (this is usually drawn at the right side of the network schematic) is called the *output port*. Thus, in a typical filtering application, the network may be referred to as a *two-port network*. In this chapter we shall investigate some ways of characterizing two-port networks. Before we do this, however, we must consider some of the more general details that apply to all networks which have two ports. This is done in the following section.

9-1 ONE-PORT AND TWO-PORT NETWORKS

Before we investigate some of the properties of the general two-port network, let us first consider a simpler case, namely, the one-port network. It should be noted here that all the developments which follow refer to *transformed networks* (as defined in Sec. 8-7). Thus, all the variables and all the network functions will be functions of the complex frequency variable. Now consider the simple inductor shown in Fig. 9–1.1(a). The two terminals of this element constitute a port, thus this network may be thought of as a one-port network. More generally, *any* two-terminal network element may be considered to be a one-port network. In Fig. 9–1.1(b) a more complicated network is shown, in which only two terminals are indicated as being of interest. Thus, in this case, we may also describe the network as a one-port network. The description, of course, refers to the fact that we plan to make use only of the indicated two terminals. More generally, we may draw a one-port network as shown in Fig. 9–1.1(c), indicating that only two of all the network's terminals are to be

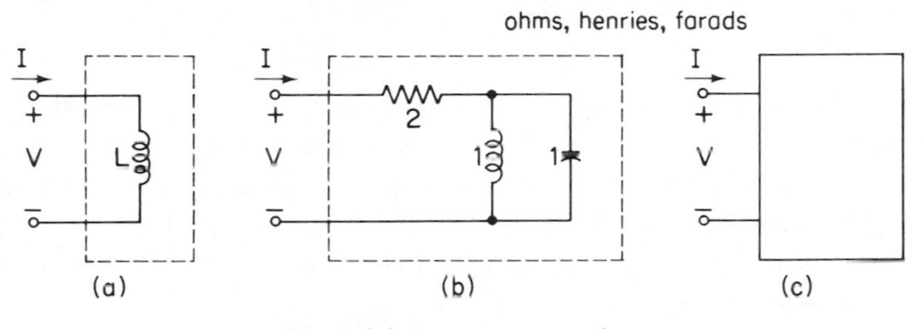

Fig. 9-1.1 One-port networks.

made available for excitation and measurement of response. Each of the networks shown in Fig. 9–1.1 has two variables defined at the port. The relative reference directions for these two variables follow the convention for associated reference directions developed in Sec. 1–2, i.e., the positive current reference direction is *into* the positive voltage reference terminal. The network itself will, of course, impose a relation between these two variables. For example, for the network shown in Fig. 9–1.1(a), we may write

$$V(s) = sLI(s) \tag{9–1}$$

Similarly, for the network shown in Fig. 9–1.1(b), applying routine network analysis, we may write

$$V(s) = \left[\frac{2s^2 + s + 2}{s^2 + 1}\right] I(s) \tag{9–2}$$

Here the expression within the brackets is the impedance of the network as seen at the port. More generally, we may simply indicate the fact that the network has some driving-point impedance $Z(s)$. Thus, for the network shown in Fig. 9–1.1(c), we may write the expression

$$V(s) = Z(s)I(s) \tag{9–3}$$

relating the variables defined at the port.

In the three equations given above we have written the relations defining the variables of a single port so that both members of the equations have the dimensions of voltage. We can obviously invert all of these equations so that both members have the dimensions of current. In this case, the relations of Eqs. (9–1), (9–2), and (9–3) become

$$I(s) = \frac{1}{sL} V(s) \tag{9–4a}$$

$$I(s) = \frac{s^2 + 1}{2s^2 + s + 2} V(s) \tag{9–4b}$$

$$I(s) = Y(s)V(s) \tag{9–4c}$$

respectively. In these last relations, the quantity multiplying the variable $V(s)$ will have the dimensions of admittance. We may conclude that the relation between the port variables of any one-port network is simply determined by specifying either the impedance or the admittance of the specific network at that port. Thus, we see that in a one-port network, two network functions are defined. One of these is the driving-point impedance $Z(s)$, where

$$Z(s) = \frac{V(s)}{I(s)} \tag{9–5}$$

in which (since network functions are defined as response/excitation) it is assumed that the excitation is provided by a current source as shown in Fig. 9–1.2. The other network function defined for a one-port network is $Y(s)$, where

$$Y(s) = \frac{I(s)}{V(s)} \tag{9–6}$$

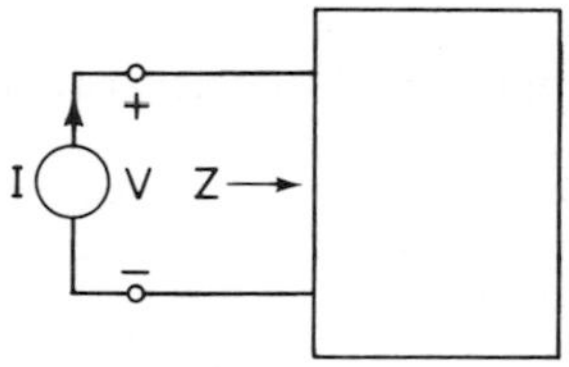

Fig. 9-1.2 A one-port network excited by a current source.

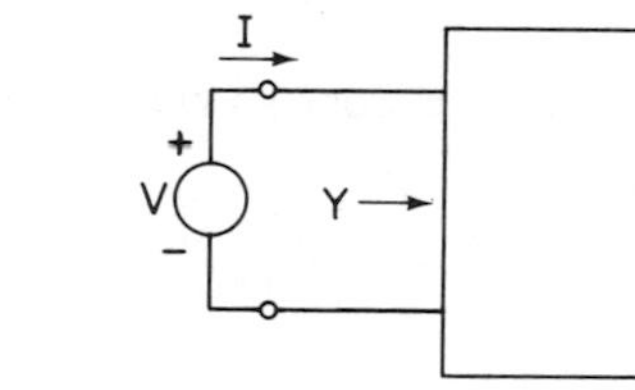

Fig. 9-1.3 A one-port network excited by a voltage source.

and the excitation is provided by a voltage source as shown in Fig. 9–1.3. Obviously, $Y(s)$ of Eq. (9–6) equals the reciprocal of $Z(s)$ of Eq. (9–5).

The description of a one-port network given above assumes that *there are no independent sources* in the network, since, if such sources are present, a voltage may appear across the open-circuited terminals of the network. Thus, if independent sources are present, Eq. (9–3) will have the general form

$$V(s) = Z(s)I(s) + V_o(s) \tag{9-7}$$

where the term $V_o(s)$ represents the effects of the independent sources, i.e., when $I(s)$ is zero, $V(s) = V_o(s)$. Another effect which is observed if independent sources are present in a one-port network is that a current may flow through a short circuit placed across the terminals of the port. In this case, Eq. (9-4c) must have the form

$$I(s) = Y(s)V(s) + I_o(s) \tag{9-8}$$

where the term $I_o(s)$ represents the effects of the independent sources. The above discussion may be summarized as follows:

SUMMARY 9-1.1

Definition of a One-Port Network: A one-port network is any network which does not contain independent sources and in which only a single terminal pair is to be used for the application of an excitation and the measurement of a response. It has the following properties:

1. There are two port variables, $V(s)$ and $I(s)$, with the relative reference polarities shown in Fig. 9-1.1(c).
2. There is one equation which relates the variables.
3. The equation may be written in either of two forms; thus, it defines two network functions $Y(s)$ and $Z(s)$ which are reciprocal to each other.

Now let us see how the situation changes when we have a two-port network. At each of the ports we shall now define a voltage variable and a current variable as shown in Fig. 9-1.4. For convenience, we have used the subscript 1 to refer to the variables at the input port (at the left) and the subscript 2 to refer to the variables at the output port (at the right). At both of these ports the variables are defined so that their relative reference directions obey the usual convention. Since we now have four variables rather than the two used to describe the one-port network, we now require two equations to relate these variables. These equations will have the general form

$$\begin{aligned} U_1(s) &= k_{11}(s)W_1(s) + k_{12}(s)W_2(s) \\ U_2(s) &= k_{21}(s)W_1(s) + k_{22}(s)W_2(s) \end{aligned} \tag{9-9}$$

where the quantities $U_1(s)$, $U_2(s)$, $W_1(s)$, and $W_2(s)$ may be any of the voltage and current variables $V_1(s)$, $V_2(s)$, $I_1(s)$, or $I_2(s)$, and the $k_{ij}(s)$ are the network functions which relate these variables. These latter are frequently referred to as *network parameters.*

Now let us consider the various ways in which we can select two variables for the left members $U_1(s)$ and $U_2(s)$ of Eq. (9–9). If we investigate the number of possible combinations in which two quantities may be selected from a total group of four, we find that there are six possibilities. Thus, there will be six different possible sets of coefficients $k_{ij}(s)$. A tabulation of these six possible combinations is given in Table 9–1.1. In this table, it will be noted that the fifth and sixth sets of network parameters have one variable, namely, $I_2(s)$, defined with a minus sign. The reason for this will be made clear in a following section. Each of the sets of parameters listed in the table has very specific properties that make it different from any other combination and make it the best one to apply to a certain class of specific network situations. These properties will be discussed in more detail in the sections that follow. Here we may note that any one of these six sets of parameters (if it exists) has the property that it completely characterizes the network. We shall find that if a given set of parameters for a network is known, then any other parameter set which exists may be found directly from the known set.

TABLE 9-1.1. The Six Sets of Network Parameters

Case	$U_1(s)$	$U_2(s)$	$W_1(s)$	$W_2(s)$
1	$V_1(s)$	$V_2(s)$	$I_1(s)$	$I_2(s)$
2	$I_1(s)$	$I_2(s)$	$V_1(s)$	$V_2(s)$
3	$I_1(s)$	$V_2(s)$	$V_1(s)$	$I_2(s)$
4	$V_1(s)$	$I_2(s)$	$I_1(s)$	$V_2(s)$
5	$V_1(s)$	$I_1(s)$	$V_2(s)$	$-I_2(s)$
6	$V_2(s)$	$I_2(s)$	$V_1(s)$	$-I_1(s)$

The general two-port network configuration shown in Fig. 9–1.4 has four individual terminals. Thus, it is theoretically possible to define four current variables rather than the two shown in the figure. To eliminate this possibility, it is customary to add the requirement to a network port that the current into one of the port terminals is, at every instant of time, equal to the current out of the other terminal of the port. This is called the *port current requirement* and it is illustrated in Figure 9–1.5. Most of the situations to which two-port networks are applied automatically satisfy this requirement. For example, if a port is open-circuited, i.e., if nothing is

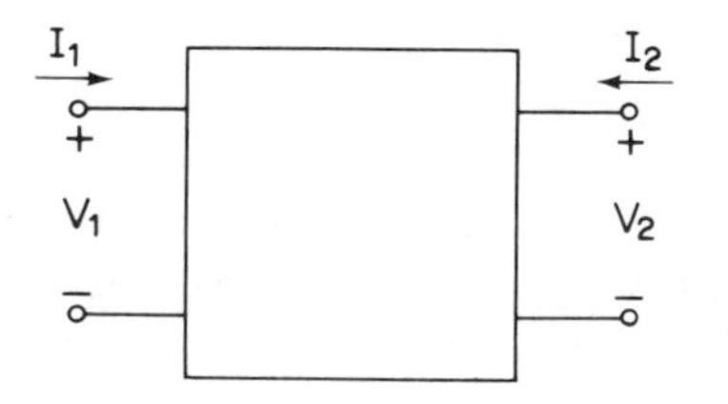

Fig. 9-1.4 Variables for a two-port network.

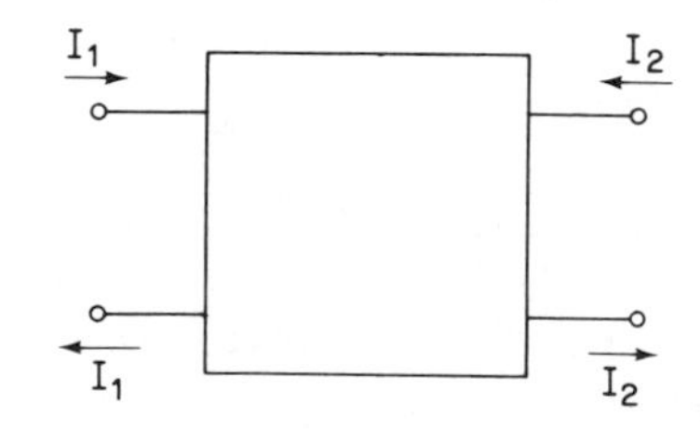

Fig. 9-1.5 The port current requirement.

connected to it, then obviously the currents at the two terminals defining the port are zero. Thus, the requirement is satisfied. Now consider the case where *any two-terminal element* (including a short circuit) is connected to the port. Since the currents into and out of the two-terminal element must obviously be equal, the port current requirement is always satisfied in this case. The requirement is also satisfied for the case where a group of two-port networks is connected in cascade as shown in Fig. 9-1.6. To see this, consider the first network in the cascade. The restriction

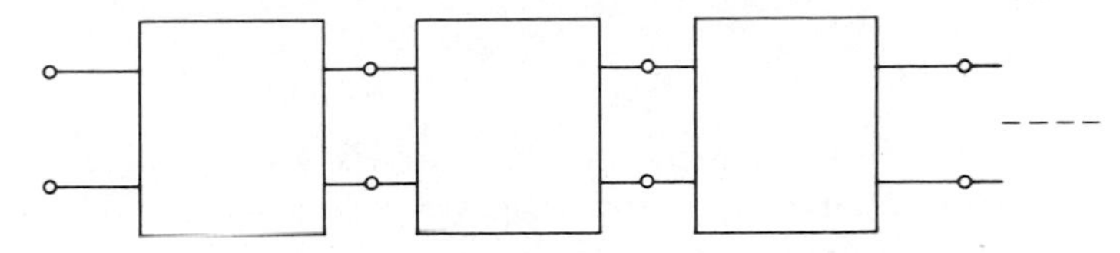

Fig. 9-1.6 A cascade connection of two-port networks.

on the currents is obviously satisfied at the input port of this network. Thus, the sum of the terminal currents at this port is zero. KCL now requires that the sum of the other two terminal currents also be zero, thus the port current requirement is satisfied. Obviously, since the requirement is satisfied at the *output* port of the first network, it must also be satisfied at the *input* port of the second network since the ports are directly connected. Continuing in this fashion, we conclude that the restrictions on the port currents are satisfied for all the networks in the cascade. A network configuration in which the port condition may not be satisfied is shown in Fig. 9-1.7. Depending upon the elements of the B network, the currents at the terminals of each of the two ports of the A network may be different. This type of a configuration will not be frequently encountered in this text.

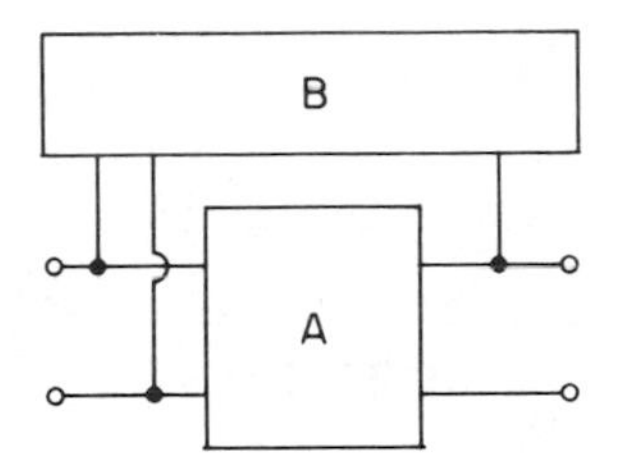

Fig. 9-1.7 A network interconnection which may not satisfy the port current requirement.

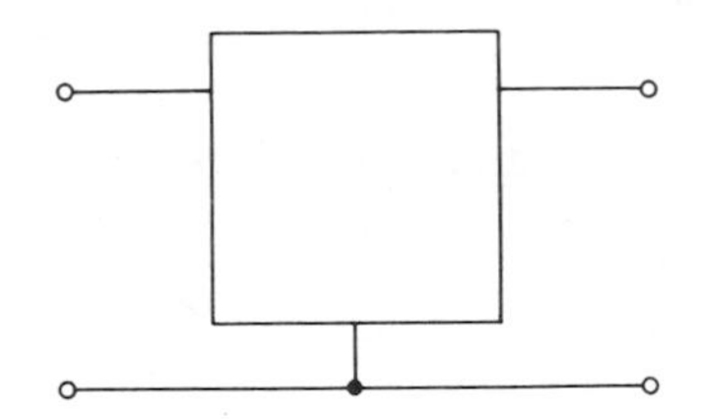

Fig. 9-1.8 A three-terminal network.

By far the most important subclass of two-port networks is the one in which the minus reference terminals of the input and output ports are at the same potential, i.e., a network in which a short circuit exists in the network between these terminals. Such a network can be drawn with the short circuit external to the box representing the network as shown in Fig. 9-1.8. This type of a network configuration is appropriately referred to as a *three-terminal* network as well as a two-port network. Since the only modification required to change a three-terminal network to a two-port network is to connect an extra wire to the third terminal, all the developments that we will make in this chapter regarding two-port networks apply equally well to three-terminal networks.

In this section we have discussed some general considerations that are applicable to one- and two-port networks. In the following sections of this chapter we shall present a detailed treatment of each of the six possible types of two-port network parameters.

9-2 *z* PARAMETERS

In the last section we showed that there are six ways in which the voltage and current variables of a two-port network may be selected so as to define a set of network functions. The various cases are shown in Table 9-1.1. In this section we shall consider the first of these cases, namely, we shall choose $V_1(s)$ and $V_2(s)$ for the quantities $U_1(s)$ and $U_2(s)$ shown in Eq. (9-9), and $I_1(s)$ and $I_2(s)$ as the quantities $W_1(s)$ and $W_2(s)$. Since the left members of the resulting set of equations have the dimensions of voltage, all the terms in the right members of these equations must also have the dimensions of voltage. Therefore, since the quantities $W_1(s)$ and $W_2(s)$ have the dimensions of current, this requires that coefficients $k_{ij}(s)$ have the dimensions of impedance. Thus, the equations may be written in the form

$$\begin{aligned} V_1(s) &= z_{11}(s)I_1(s) + z_{12}(s)I_2(s) \\ V_2(s) &= z_{21}(s)I_1(s) + z_{22}(s)I_2(s) \end{aligned} \tag{9-10}$$

where the quantities $z_{ij}(s)$ are called the *z parameters*. More succinctly, we may write the equations given above in matrix form as

$$\mathbf{V}(s) = \mathbf{Z}(s)\mathbf{I}(s) \tag{9-11}$$

where the individual matrices are identified by the following expression:

$$\mathbf{V}(s) = \begin{bmatrix} V_1(s) \\ V_2(s) \end{bmatrix} = \begin{bmatrix} z_{11}(s) & z_{12}(s) \\ z_{21}(s) & z_{22}(s) \end{bmatrix} \begin{bmatrix} I_1(s) \\ I_2(s) \end{bmatrix} = \mathbf{Z}(s)\mathbf{I}(s) \tag{9-12}$$

The matrix $\mathbf{Z}(s)$ is referred to as the *z-parameter matrix*.

For a given network the *z* parameters may be found by applying a set of test excitations to the network. To see this, consider the first equation given in Eq. (9-10). If we let the output port (port 2) of the network be open-circuited, then the variable $I_2(s)$ must be zero. Thus, the equation will have the form

$$V_1(s) = z_{11}(s)I_1(s)\big|_{I_2(s)=0} \tag{9-13}$$

The quantity $z_{11}(s)$ is a network function. Thus, recalling that network functions are defined as response/excitation, (9–13) implies the use of a current excitation (rather than a voltage excitation) at the input port (port 1) of the network. Thus, the parameter $z_{11}(s)$ gives the driving-point impedance at port 1 when port 2 is open-circuited. More precisely, $z_{11}(s)$ is defined by the relation

$$z_{11}(s) = \left.\frac{V_1(s)}{I_1(s)}\right|_{I_2(s)=0} \tag{9–14}$$

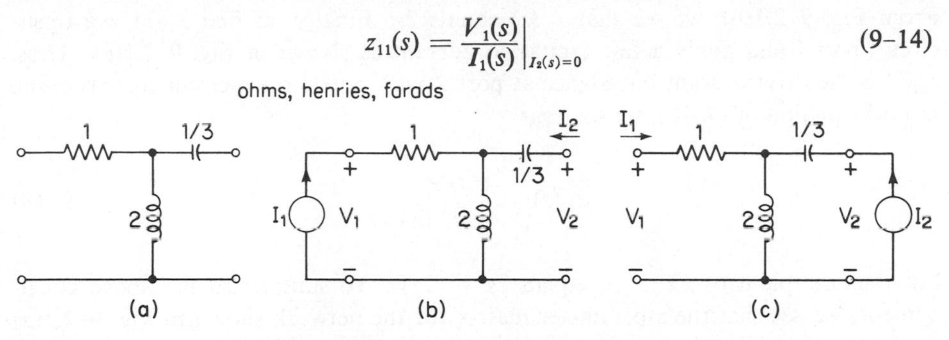

Fig. 9-2.1 Tests for the z parameters.

As an example of finding this parameter, consider the network shown in Fig. 9–2.1(a). If we apply a current source of value $I_1(s)$ at the input port, as shown in Fig. 9–2.1(b), the resulting impedance at this port is $2s + 1$. Thus, for this network $z_{11}(s)$ equals $2s + 1$.

A similar testing procedure may be applied to find the parameter $z_{12}(s)$. In this case, we must let port 1 be open-circuited, thus, $I_1(s) = 0$, and the first equation of Eq. (9–10) has the form

$$V_1(s) = z_{12}(s)I_2(s)|_{I_1(s)=0} \tag{9–15}$$

The quantity $z_{12}(s)$ thus defines the transfer impedance relating a current excitation at port 2 to a voltage response at port 1 (when port 1 is open-circuited). Thus, we may write

$$z_{12}(s) = \left.\frac{V_1(s)}{I_2(s)}\right|_{I_1(s)=0} \tag{9–16}$$

An example of the determination of this parameter for the network shown in Fig. 9–2.1(a) is shown in Fig. 9–2.1(c). From this figure, we see that $V_1 = 2sI_2(s)$. Thus, $z_{12}(s)$ equals $2s$.

Test excitations similar to those described above may be applied so that the second equation of Eq. (9–10) may be used to find the parameters $z_{21}(s)$ and $z_{22}(s)$. For example, to find $z_{21}(s)$, we open-circuit port 2 and apply a current excitation at port 1. Thus, for the example network shown in Fig. 9–2.1(a), we may use the testing configuration shown in Fig. 9–2.1(b). The quantity $z_{21}(s)$ is now seen to be the transfer impedance from port 1 to port 2 when port 2 is open-circuited. Thus, in general, it is defined as

$$z_{21}(s) = \left.\frac{V_2(s)}{I_1(s)}\right|_{I_2(s)=0} \tag{9-17}$$

From Fig. 9-2.1(b), we see that $z_{21}(s)$ equals $2s$. Finally, to find $z_{22}(s)$, we open-circuit port 1 and apply a test excitation current as shown in Fig. 9-2.1(c). Thus, $z_{22}(s)$ is the driving-point impedance at port 2 with port 1 open-circuited. From the second equation of (9-10), we see that

$$z_{22}(s) = \left.\frac{V_2(s)}{I_2(s)}\right|_{I_1(s)=0} \tag{9-18}$$

For the example network, $z_{22}(s)$ equals $2s + (3/s)$. To summarize the above development, we see that the z-parameter matrix for the network shown in Fig. 9-2.1(a) is

$$\mathbf{Z}(s) = \begin{bmatrix} 2s+1 & 2s \\ 2s & 2s + \dfrac{3}{s} \end{bmatrix} \tag{9-19}$$

In the applications of the test excitations described above, we note that in every case the port at which the response is measured is *open-circuited*. In addition, since the impedance of the current sources used to apply the test excitations is infinite, the port to which such an excitation is applied also sees an infinite impedance termination. Thus, from an impedance viewpoint, the excitation port may also be considered as being open-circuited, although, of course, an excitation current does flow into it. Considering the above, we see that another name which is also applicable to the z parameters is the *open-circuit impedance parameters*.

All the z parameters described above are defined as a ratio of voltage to current. In determining these parameters, rather than considering the excitation current as a variable and writing it explicitly, as was done above, it is frequently convenient to simply set the output of the excitation current source equal to unity. Since we are considering transformed variables, this implies that the physical current is set equal to an impulse [the inverse transform of 1 is $\delta(t)$]. As was pointed out in Sec. 8-9, when the transformed excitation is set to unity, *the network function is identically equal to the response*. Thus, if we set the excitation currents to unity, the resulting expressions for the response voltages directly specify the z parameters. This is usually a convenient technique for finding the z parameters. An example of this procedure follows.

EXAMPLE 9-2.1. It is desired to find the z parameters for the network shown in Fig. 9-2.2(a). First, let us set up a testing condition which will permit us to determine the

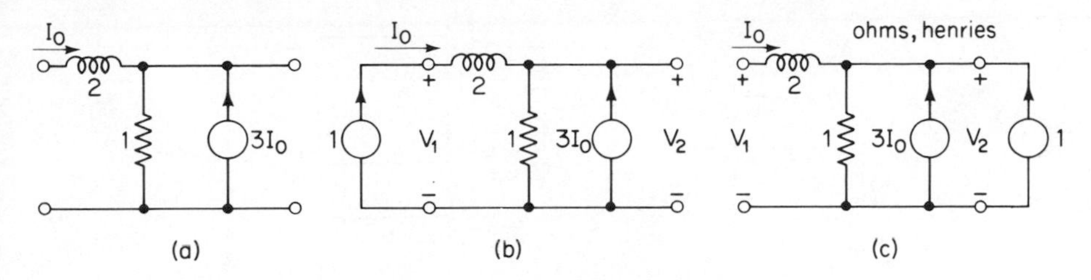

Fig. 9-2.2 Tests for the z parameters.

parameters $z_{11}(s)$ and $z_{21}(s)$. This is shown in Fig. 9-2.2(b). Note that the transformed current excitation applied at port 1 is unity. Since the port current $I_1(s)$ is also equal to the controlling current for the ICIS, the current through the resistor will be 4 while the current through the inductor is 1. Thus, applying KVL, $V_1(s) = 2s + 4$ and $V_2(s) = 4$. Since, for unity excitation, the response $V_1(s)$ equals the network function $z_{11}(s)$ and the response $V_2(s)$ equals $z_{21}(s)$, we conclude that $z_{11}(s)$ equals $2s + 4$ and $z_{21}(s)$ equals 4. A similar procedure is used to find the remaining z parameters, namely, $z_{12}(s)$ and $z_{22}(s)$. For this case, we apply an excitation current source of unity value at port 2 as shown in Fig. 9-2.2(c). Since the controlling current for the ICIS is zero, in this case we find that $V_1(s) = V_2(s) = 1$ and thus, $z_{12}(s) = z_{22}(s) = 1$. The z-parameter matrix for the network shown in Fig. 9-2.2(a) is thus

$$\begin{bmatrix} 2s + 4 & 1 \\ 4 & 1 \end{bmatrix}$$

The results given above may be summarized as follows:

SUMMARY 9-2.1

The z Parameters of a Two-Port Network: The z parameters of a two-port network are the functions $z_{ij}(s)$ which specify the port voltages as functions of the port currents. The exact relations are

$$\begin{bmatrix} V_1(s) \\ V_2(s) \end{bmatrix} = \begin{bmatrix} z_{11}(s) & z_{12}(s) \\ z_{21}(s) & z_{22}(s) \end{bmatrix} \begin{bmatrix} I_1(s) \\ I_2(s) \end{bmatrix}$$

Two of the four z parameters may be found by applying a unit test excitation current at port 1 and measuring the resulting voltage response at both ports. The other two z parameters may be found by similarly applying a unit test excitation current at the other port. The various test conditions used to define each parameter are summarized in Table 9-2.1.

Now let us investigate some properties of the z parameters. In Sec. 8-8 we showed that if the points of excitation and response are interchanged, the transfer impedance of a *reciprocal* network will be unchanged, while the transfer impedance

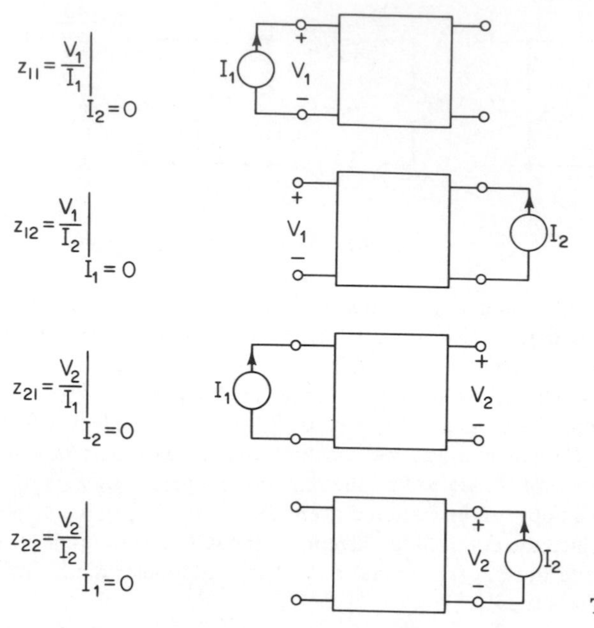

Table 9-2.1

of a *non-reciprocal* network will be modified. Such an interchange of excitation and response is exactly what is done in determining the parameters $z_{12}(s)$ and $z_{21}(s)$ as shown in Table 9–2.1. We may conclude that a reciprocal network will, in general, have a *symmetric* z-parameter matrix, i.e., $z_{12}(s)$ will equal $z_{21}(s)$, and that a non-reciprocal network will have a non-symmetric z-parameter matrix. As an example of this the parameter matrix given in (9–19) for the network shown in Fig. 9–2.1(a) is symmetric. This is to be expected since the network is comprised only of two-terminal network elements and such networks are always reciprocal. An example of a non-symmetric z-parameter matrix is given in Example 9–2.1. This result again could be anticipated, since the network contains an active element, i.e., an ICIS.

An important property of the z parameters is that if, for a given network, the z parameters are known, we may find any other desired network function directly from them. For example, consider the determination of the open-circuit voltage transfer function $V_2(s)/V_1(s)$. By the term "open-circuit" here, we imply that the response port, i.e., the port where $V_2(s)$ is measured, is open-circuited. To find this voltage transfer function, let us rewrite the equations of (9–10) with the added restriction that $I_2(s) = 0$. Thus, we obtain

$$\begin{aligned} V_1(s) &= z_{11}(s)I_1(s) \\ V_2(s) &= z_{21}(s)I_1(s) \end{aligned} \tag{9–20}$$

Taking the ratio of these two equations and inverting the result, we obtain

$$\frac{V_2(s)}{V_1(s)} = \frac{z_{21}(s)}{z_{11}(s)} \tag{9-21}$$

This is the desired voltage transfer function expressed in terms of the z parameters. In a similar fashion we may obtain the short-circuit current transfer function $I_2(s)/I_1(s)$. By the term "short-circuit" here, we imply that the response port, i.e., the one at which $I_2(s)$ is measured, is short-circuited. To find this function, we need simply write the second equation of (9–10) with the added restriction that $V_2(s) = 0$. Thus, we obtain

$$0 = z_{21}(s)I_1(s) + z_{22}(s)I_2(s) \tag{9-22}$$

Rearranging the terms of the above equation, we see that

$$\frac{I_2(s)}{I_1(s)} = \frac{-z_{21}(s)}{z_{22}(s)} \tag{9-23}$$

This is the desired current transfer function. As a third example of using the z parameters to find other network functions, consider the problem of finding the input impedance of a given two-port network when the output port is short-circuited. In this case, the relation between the variables $I_2(s)$ and $I_1(s)$ is given by Eq. (9–23). Substituting this result in the first equation of (9–10) to eliminate $I_2(s)$, we obtain

$$\frac{V_1(s)}{I_1(s)} = z_{11}(s) - \frac{z_{12}(s)z_{21}(s)}{z_{22}(s)} \tag{9-24}$$

This is the short-circuit driving-point impedance of a two-port network. A procedure similar to those given above may be used to find expressions for any other network function which is defined for a two-port network in terms of the z parameters of the two-port network.

In many network configurations we can simplify the determination of the z parameters by separating the network into component parts which are simpler to analyze than the original network. As an example of this, consider the case where two-port networks are connected as shown in Fig. 9–2.3. This type of a connection is frequently referred to as a *series* connection of two-port networks. For the A network and the voltage and current variables shown in the figure, we may define a set of z parameters $z_{ij}^a(s)$ by the relations

$$\mathbf{V}^a(s) = \begin{bmatrix} V_1^a(s) \\ V_2^a(s) \end{bmatrix} = \begin{bmatrix} z_{11}^a(s) & z_{12}^a(s) \\ z_{21}^a(s) & z_{22}^a(s) \end{bmatrix} \begin{bmatrix} I_1^a(s) \\ I_2^a(s) \end{bmatrix} = \mathbf{Z}^a(s)\mathbf{I}^a(s) \tag{9-25}$$

Similarly, for the B network we may define a set of parameters $z_{ij}^b(s)$ by the relations

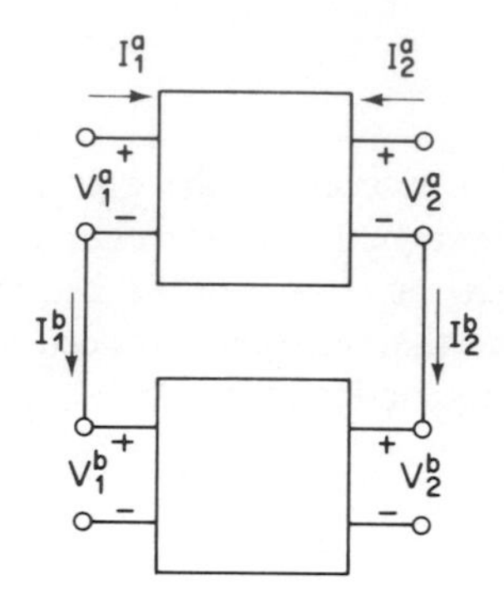

Fig. 9-2.3 A series connection of two two-port networks.

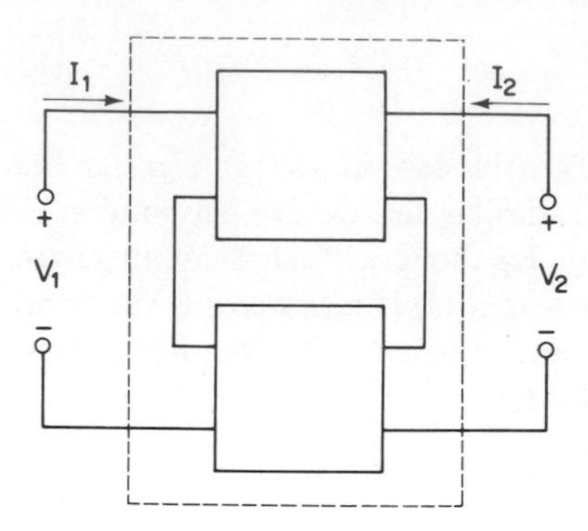

Fig. 9-2.4 A series connection of two two-port networks.

$$\mathbf{V}^b(s) = \begin{bmatrix} V_1^b(s) \\ V_2^b(s) \end{bmatrix} = \begin{bmatrix} z_{11}^b(s) & z_{12}^b(s) \\ z_{21}^b(s) & z_{22}^b(s) \end{bmatrix} \begin{bmatrix} I_1^b(s) \\ I_2^b(s) \end{bmatrix} = \mathbf{Z}^b(s)\mathbf{I}^b(s) \tag{9-26}$$

The series connection of the separate two-port networks labeled A and B can be considered as still another two-port network as shown in Fig. 9–2.4. Using the voltage and current variables shown in the figure, we may define a third set of parameters $z_{ij}(s)$ by the relations

$$\mathbf{V}(s) = \begin{bmatrix} V_1(s) \\ V_2(s) \end{bmatrix} = \begin{bmatrix} z_{11}(s) & z_{12}(s) \\ z_{21}(s) & z_{22}(s) \end{bmatrix} \begin{bmatrix} I_1(s) \\ I_2(s) \end{bmatrix} = \mathbf{Z}(s)\mathbf{I}(s) \tag{9-27}$$

Applying KVL, we see that the sum of the port voltages of the A and B networks equals the port voltages of the overall network. Similarly, it is readily noted that the port currents at the input and output ports are the same for all three networks. Thus, the following relations must hold between the three sets of voltage and current variables for the various two-port networks.

$$\mathbf{V}(s) = \begin{bmatrix} V_1(s) \\ V_2(s) \end{bmatrix} = \begin{bmatrix} V_1^a(s) + V_1^b(s) \\ V_2^a(s) + V_2^b(s) \end{bmatrix} = \mathbf{V}^a(s) + \mathbf{V}^b(s) \tag{9-28}$$

$$\mathbf{I}(s) = \begin{bmatrix} I_1(s) \\ I_2(s) \end{bmatrix} = \begin{bmatrix} I_1^a(s) \\ I_2^a(s) \end{bmatrix} = \begin{bmatrix} I_1^b(s) \\ I_2^b(s) \end{bmatrix} = \mathbf{I}^a(s) = \mathbf{I}^b(s) \tag{9-29}$$

Combining the results from Eqs. (9–27), (9–28), and (9–29), we obtain

$$\begin{aligned} \mathbf{V}(s) &= \mathbf{V}^a(s) + \mathbf{V}^b(s) = \mathbf{Z}^a(s)\mathbf{I}^a(s) + \mathbf{Z}^b(s)\mathbf{I}^b(s) \\ &= [\mathbf{Z}^a(s) + \mathbf{Z}^b(s)]\mathbf{I}(s) = \mathbf{Z}(s)\mathbf{I}(s) \end{aligned} \tag{9-30}$$

Thus, we may conclude that, for a two-port network which is formed as a series connection of two separate two-port networks, the z parameters of the overall network may be found by adding the z parameters of the component networks. An example follows.

EXAMPLE 9-2.2. In Fig. 9-2.5(a), a network consisting of three resistors and an ICVS is shown. To simplify the determination of the z parameters of this network it may be separated into two component networks connected in series as shown in parts (b) and (c) of the figure. For the network shown in Fig. 9-2.5(b), using the techniques developed in this section, the z-parameter matrix is readily shown to be

$$\begin{bmatrix} 3 & 0 \\ K & 2 \end{bmatrix}$$

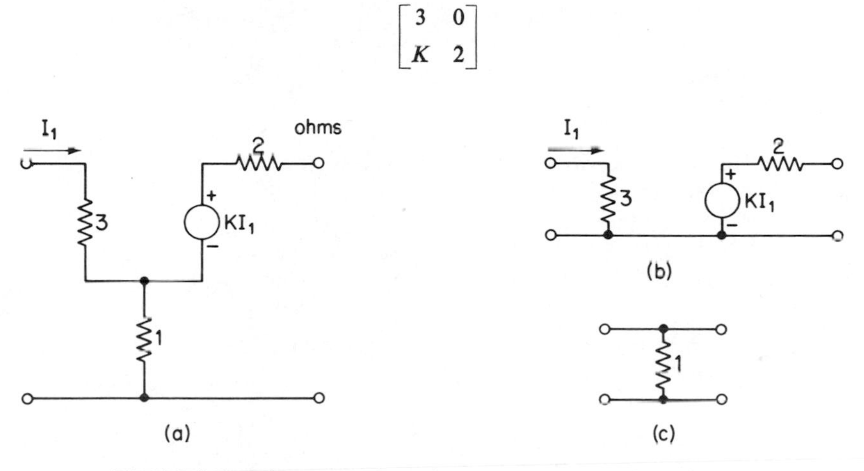

Fig. 9-2.5 A series connection of two-port networks in which the z parameters add.

For the network shown in Fig. 9-2.5(c), the z-parameter matrix is

$$\begin{bmatrix} 1 & 1 \\ 1 & 1 \end{bmatrix}$$

If we now add the two z-parameter matrices given above, we obtain

$$\begin{bmatrix} 4 & 1 \\ 1+K & 3 \end{bmatrix}$$

These are the z parameters for the overall network shown in Fig. 9-2.5(a).

It should be noted that, in making a series connection of two two-port networks, the result given above, namely, that the z parameters add, does not apply if the port current requirement given in Sec. 9-1 is violated. As an example of this, consider the network shown in Fig. 9-2.6(a) [with the test current $I_1(s)$ applied]. This may be considered as a series connection of the two component networks shown in parts (b) and (c) of the figure. For the network shown in Fig. 9-2.6(b), the z parameters are

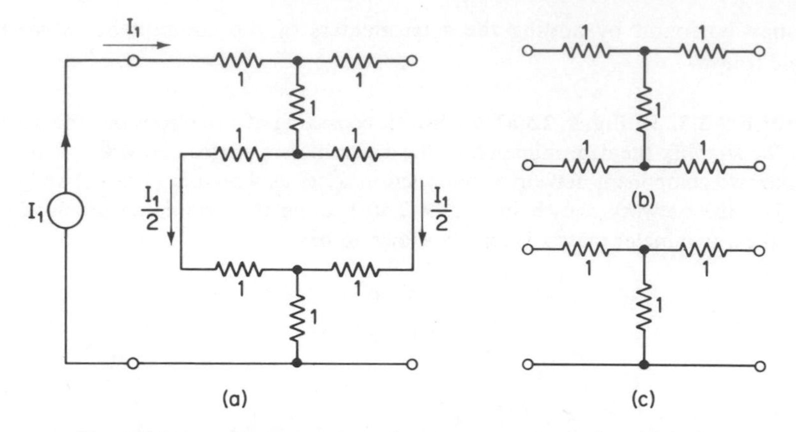

Fig. 9-2.6 A series connection of two-port networks in which the z parameters do NOT add.

$$\begin{bmatrix} 3 & 1 \\ 1 & 3 \end{bmatrix} \tag{9-31}$$

For the network shown in Fig. 9–2.6(c), the z parameters are

$$\begin{bmatrix} 2 & 1 \\ 1 & 2 \end{bmatrix} \tag{9-32}$$

The series connection of the two networks, shown in Fig. 9–2.6(a), however, has z parameters

$$\begin{bmatrix} 4 & 3 \\ 3 & 4 \end{bmatrix} \tag{9-33}$$

Obviously, in this case, the z parameters of the overall network are not equal to the sum of the z parameters of the component networks. To see why this is so, consider the path of the test current $I_1(s)$ shown in Fig. 9–2.6(a). At the indicated interior node this current branches into two equal currents $I_1(s)/2$ as shown. Thus, for the upper two-port component network, the current into one of the terminals of its port 1 is $I_1(s)$, and the current out of the other terminal of this port is $I_1(s)/2$. This clearly violates the port current condition, and, as a result, it invalidates the z parameters for the component networks.

The results given above may be summarized as follows:

SUMMARY 9-2.2

Properties of the z Parameters of a Two-Port Network: The four z parameters of a given two-port network completely determine the properties of the network in the sense that they may be used to find any other network function that is defined for the two-port network. If two two-port networks are connected in series, the sum of their individual z parameters gives the z parameters for the resulting overall two-port network (unless the connection violates the port current requirement on the port currents of the component two-port networks).

In the preceding paragraphs descriptions of various techniques for finding the z parameters of a specific network have been given. Now let us consider the inverse problem: Given a set of z parameters, how do we find a two-port network which has such a set of parameters? This problem is basically a synthesis problem, and there is, in general, no unique solution to synthesis problems. There are, however, some useful network forms which may be used as circuit representations for a given set of z parameters. One of the most common of these forms is the "**T**" network representation. This is applicable to all reciprocal networks, i.e., ones in which $z_{12}(s)$ is equal to $z_{21}(s)$. To see how this **T** representation may be developed, let us consider first of all the **T** configuration of two-terminal impedances shown in Fig. 9-2.7. The z-parameter matrix for this network is readily shown to be

$$\begin{bmatrix} z_{11}(s) & z_{12}(s) \\ z_{12}(s) & z_{22}(s) \end{bmatrix} = \begin{bmatrix} Z_a(s) + Z_c(s) & Z_c(s) \\ Z_c(s) & Z_b(s) + Z_c(s) \end{bmatrix}$$

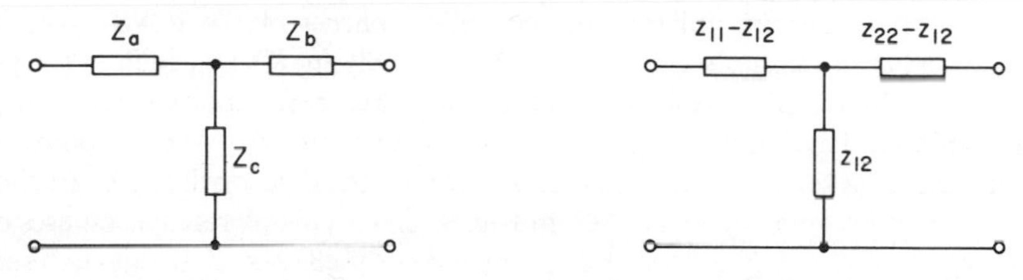

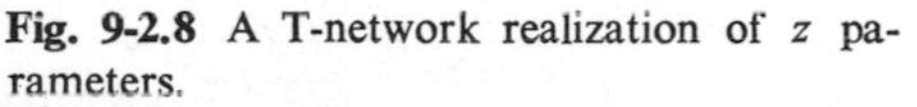

Fig. 9-2.7 A T network.

Fig. 9-2.8 A T-network realization of z parameters.

Equating like terms in the above equation, we find the relations giving the network elements $Z_a(s)$, $Z_b(s)$, and $Z_c(s)$ as functions of the network parameters are

$$\begin{aligned} Z_a(s) &= z_{11}(s) - z_{12}(s) \\ Z_b(s) &= z_{22}(s) - z_{12}(s) \\ Z_c(s) &= z_{12}(s) \end{aligned} \tag{9-34}$$

Thus, the **T** network may be redrawn with the expressions specifying the two-terminal impedances given directly in terms of the z parameters as shown in Fig. 9-2.8. As an example of finding a network representation for a set of z parameters, consider the z-parameter matrix

$$\begin{bmatrix} 7 & 3 \\ 3 & 5 \end{bmatrix} \tag{9-35}$$

Applying the relations given in (9–34), we see that the **T** configuration for a two-port network which realizes these parameters is as shown in Fig. 9–2.9. It should be noted that when the relations given in Eq. (9–34) are applied to the case where the z parameters are complicated functions of the complex frequency variable, due to the subtraction of rational functions required in the right members of these equations, the network functions describing the impedances $Z_a(s)$, $Z_b(s)$, and $Z_c(s)$ may be difficult to synthesize.

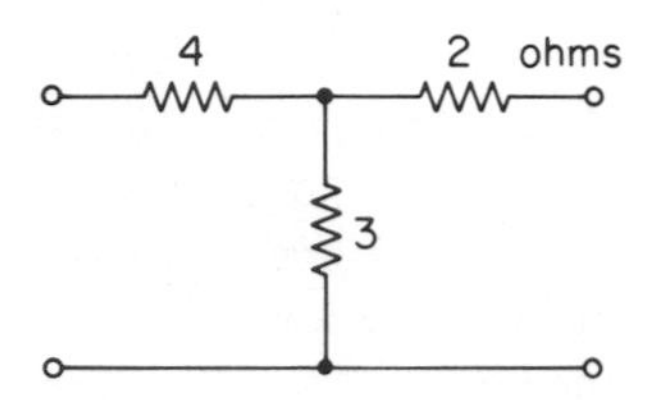

Fig. 9-2.9 A T network realizing the z parameters of (9-35).

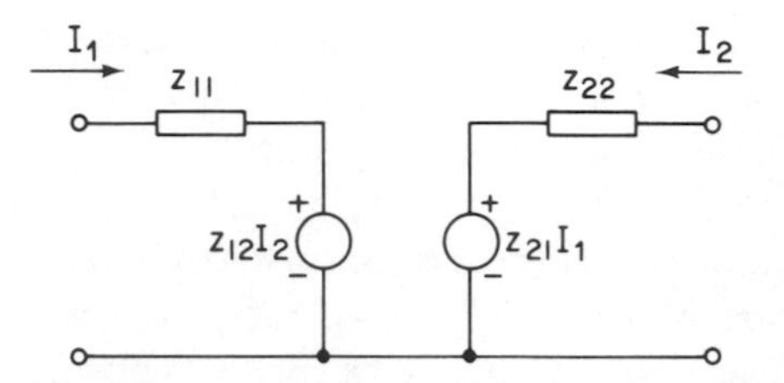

Fig. 9-2.10 A two-controlled-source realization of a non-reciprocal set of z parameters.

In order to realize non-symmetric z-parameter matrices, we must use network representations which are non-reciprocal. One of the most useful representations for such a set of z parameters utilizes two controlled sources of the ICVS type. The representation has the form shown in Fig. 9–2.10. Applying KVL at each of the ports shown in this figure, it is readily verified that this representation satisfies the original equations defining the z parameters given in Eq. (9–10). Another equivalent representation for a set of z parameters for a non-reciprocal network is the one shown in Fig. 9–2.11. This representation uses only a single controlled source. If this representation is applied to a reciprocal network in which $z_{21}(s) = z_{12}(s)$, it will simply appear as the **T** configuration shown in Fig. 9–2.8.

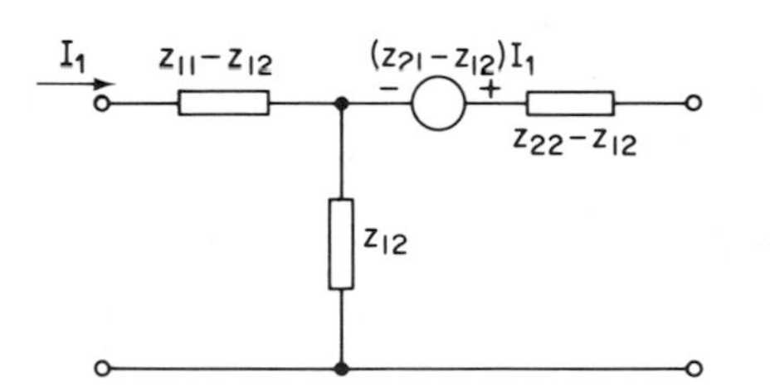

Fig. 9-2.11 A one-controlled-source realization of a non-reciprocal set of z parameters.

In this section we have investigated the use of z parameters to describe two-port networks. In the following sections we shall make a similar investigation of the other types of parameters which are categorized in Table 9–1.1.

9-3 *y* PARAMETERS

In this section we shall discuss the second of the six ways in which the voltage and current variables of a two-port network may be selected so as to define a set of network functions. The case is the second one shown in Table 9–1.1, namely, we will choose $I_1(s)$ and $I_2(s)$ for the quantities $U_1(s)$ and $U_2(s)$ shown in (9–9) and $V_1(s)$ and $V_2(s)$ as the quantities $W_1(s)$ and $W_2(s)$. For this case, the left members of the resulting set of equations have the dimensions of current. Therefore, all the terms in the right members of these equations must also have the dimensions of current. Therefore, the coefficients $k_{ij}(s)$ of (9–9) must have the dimensions of admittance. Thus, the equations may be written in the form

$$\begin{aligned} I_1(s) &= y_{11}(s)V_1(s) + y_{12}(s)V_2(s) \\ I_2(s) &= y_{21}(s)V_1(s) + y_{22}(s)V_2(s) \end{aligned} \tag{9–36}$$

where the quantities $y_{ij}(s)$ are called the *y parameters*. The equations given above may also be written in matrix form as

$$\mathbf{I}(s) = \mathbf{Y}(s)\mathbf{V}(s) \tag{9–37}$$

where the individual matrices are identified by the following expression.

$$\mathbf{I}(s) = \begin{bmatrix} I_1(s) \\ I_2(s) \end{bmatrix} = \begin{bmatrix} y_{11}(s) & y_{12}(s) \\ y_{21}(s) & y_{22}(s) \end{bmatrix} \begin{bmatrix} V_1(s) \\ V_2(s) \end{bmatrix} = \mathbf{Y}(s)\mathbf{V}(s) \tag{9–38}$$

The matrix $\mathbf{Y}(s)$ is referred to as the *y-parameter matrix.*

For a given network, the y parameters may be found in a manner similar to that used for the z parameters in the previous section. In this case, however, the test excitations will be applied by voltage sources rather than by current sources. For example, consider the first equation given in (9–36). If we short-circuit port 2, then the variable $V_2(s)$ must be 0. Thus, the network function $y_{11}(s)$ is seen to be the driving-point admittance at port 1 when port 2 is short-circuited, and it is defined by the relation

$$y_{11}(s) = \left.\frac{I_1(s)}{V_1(s)}\right|_{V_2(s)=0} \tag{9–39}$$

In a similar manner, if we consider the second equation of (9–36) and the same terminating condition, namely, $V_2(s) = 0$, then the network function $y_{21}(s)$ which is the transfer admittance from port 1 to port 2 is defined by the relation

$$y_{21}(s) = \left.\frac{I_2(s)}{V_1(s)}\right|_{V_2(s)=0} \tag{9–40}$$

As an example of finding these two parameters, consider the network shown in Fig. 9–3.1(a). If we apply a voltage source $V_1(s)$ at port 1 and short-circuit the terminals of port 2 as shown in Fig. 9–3.1(b), then we see from the figure that $I_1(s) = (2s^2 + 1)V_1(s)/2s$ and $I_2(s) = -V_1(s)/2s$. Thus, for this network $y_{11}(s) = (2s^2 + 1)/$

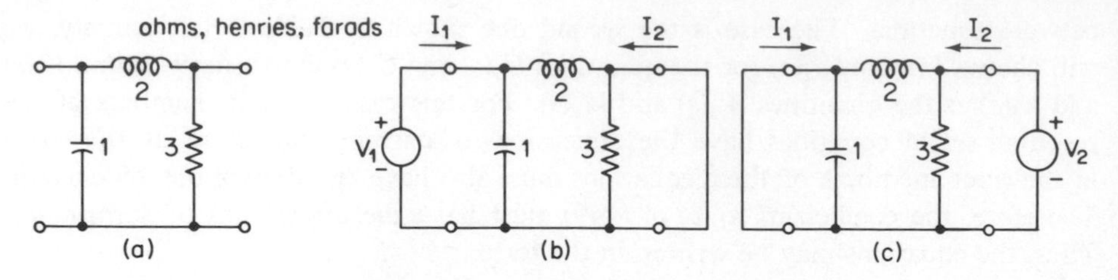

Fig. 9-3.1 Tests for the y parameters.

$2s$ and $y_{21}(s) = -1/2s$. Note that the minus sign in the expression for $y_{21}(s)$ occurs because for a positive $V_1(s)$, the actual current direction in the short circuit flows opposite to the assumed positive reference direction for $I_2(s)$.

A similar testing procedure may be applied to find the parameters $y_{12}(s)$ and $y_{22}(s)$. In this case, let us short-circuit port 1. Thus, the variable $V_1(s)$ must equal 0. From the first equation of (9–36), we find that $y_{12}(s)$, the transfer admittance from port 2 to port 1, is defined by the relation

$$y_{12}(s) = \left.\frac{I_1(s)}{V_2(s)}\right|_{V_1(s)=0} \tag{9–41}$$

Similarly, from the second equation of (9–36) the parameter $y_{22}(s)$, the driving-point admittance at port 2 when port 1 is short-circuited, is defined by the relation

$$y_{22}(s) = \left.\frac{I_2(s)}{V_2(s)}\right|_{V_1(s)=0} \tag{9–42}$$

As an example of the determination of these last two parameters, the network shown in Fig. 9–3.1(a) may be excited by a voltage source at port 2 with port 1 short-circuited as shown in Fig. 9–3.1(c). From this figure, we see that $I_1(s) = -V_2(s)/2s$ and $I_2(s) = (2s + 3)V_2(s)/6s$. Thus, we conclude that $y_{12}(s) = -1/2s$ and $y_{22}(s) = (2s + 3)/6s$. Combining the above results, we may write the y-parameter matrix for the network shown in Fig. 9–3.1(a) as

$$\mathbf{Y}(s) = \begin{bmatrix} \dfrac{2s^2 + 1}{2s} & -\dfrac{1}{2s} \\ -\dfrac{1}{2s} & \dfrac{(2s/3) + 1}{2s} \end{bmatrix} \tag{9–43}$$

In the applications of the test excitations described above, we note that in each case the port at which the response is measured is short-circuited. Also, since the internal impedance of the voltage sources used to apply the test excitations is zero, the port to which these excitations are applied in effect sees a zero impedance termination. Thus, from an impedance standpoint, the excitation port may also be considered as being short-circuited, although, of course, there will be a voltage present across its terminals. Considering the above, we see that another equally descriptive name which may be applied to the y parameters is *short-circuit admittance parameters.* The same remarks which were made in the preceding section with respect to the use of unit-valued excitations to directly determine the network parameters also applies to the y-parameter case. An example of this procedure follows.

EXAMPLE 9-3.1. It is desired to find the y parameters for the network shown in Fig. 9-3.2(a). In order to find $y_{11}(s)$ and $y_{21}(s)$, we may apply a unit-valued test excitation voltage at port 1 and a short circuit at port 2, as shown in Fig. 9-3.2(b). For this excitation

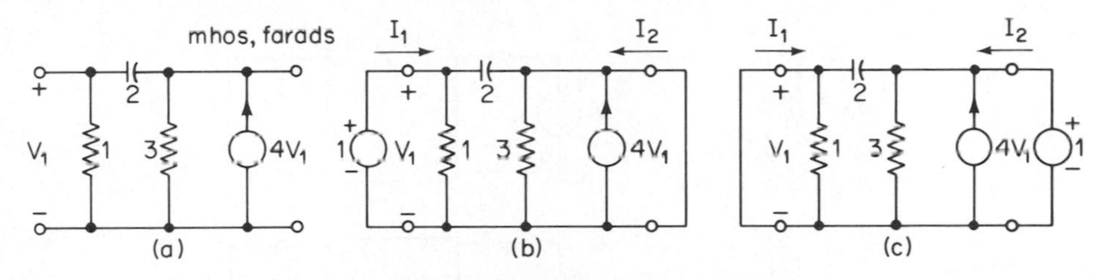

Fig. 9-3.2 Tests for the y parameters.

the (transformed) output of the VCIS is 4. Routine circuit analysis shows that $I_1(s) = y_{11}(s) = 2s + 1$ and $I_2(s) = y_{21}(s) = -2s - 4$. To find the remaining y parameters, we may apply a unit-valued test excitation voltage at port 2 and a short circuit at port 1 as shown in Fig. 9-3.2(c). In this case, since $V_1(s) = 0$, the output of the VCIS is zero. It is easily shown that $I_1(s) = y_{12}(s) = -2s$ and $I_2(s) = y_{22}(s) = 2s + 3$. Thus, the y parameters for the network shown in Fig. 9-3.2(a) are

$$\mathbf{Y}(s) = \begin{bmatrix} 2s + 1 & -2s \\ -2s - 4 & 2s + 3 \end{bmatrix}$$

The results given above may be summarized as follows:

SUMMARY 9-3.1

The y Parameters of a Two-Port Network: The y parameters of a two-port network are the network functions $y_{ij}(s)$ which specify the port currents as functions of the port voltages as follows:

$$\begin{bmatrix} I_1(s) \\ I_2(s) \end{bmatrix} = \begin{bmatrix} y_{11}(s) & y_{12}(s) \\ y_{21}(s) & y_{22}(s) \end{bmatrix} \begin{bmatrix} V_1(s) \\ V_2(s) \end{bmatrix}$$

Two of the four y parameters may be found by applying a unit test excitation voltage at port 1, short-circuiting port 2, and measuring the resulting current response at the two ports. The other two y parameters may be found by similarly applying a unit test excitation voltage at port 2 and short-circuiting port 1. The various excitations used to determine the y parameters are summarized in Table 9-3.1.

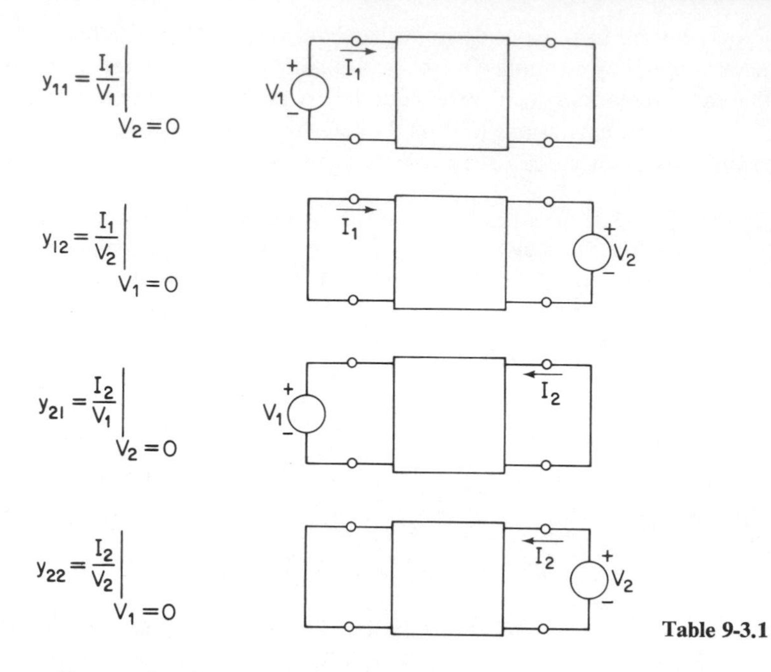

Table 9-3.1

In the preceding section it was shown that the z parameters for a two-port network were characterized by a matrix equation having the form

$$\mathbf{V}(s) = \mathbf{Z}(s)\mathbf{I}(s) \tag{9-44}$$

where $\mathbf{Z}(s)$ is the z-parameter matrix. Now let us pre-multiply both sides of this expression by the y-parameter matrix $\mathbf{Y}(s)$. We obtain

$$\mathbf{Y}(s)\mathbf{V}(s) = \mathbf{Y}(s)\mathbf{Z}(s)\mathbf{I}(s) \tag{9-45}$$

The left member of this equation, however, from (9–37), simply defines the column matrix $\mathbf{I}(s)$. Thus, considering the right member of the equation, we see that the matrix product $\mathbf{Y}(s)\mathbf{Z}(s)$ can only be equal to the identity matrix. We conclude that the y-parameter matrix is the inverse of the z-parameter matrix, namely, that the relations

$$\mathbf{Y}(s) = [\mathbf{Z}(s)]^{-1} \qquad \mathbf{Z}(s) = [\mathbf{Y}(s)]^{-1} \tag{9-46}$$

are satisfied. Since this is true, we may directly apply the relations for the inverse of a second-order matrix developed in Sec. 3–1 to determine the elements of the y-parameter matrix in terms of the elements of the z-parameter matrix. Thus, we may write

$$\begin{bmatrix} y_{11}(s) & y_{12}(s) \\ y_{21}(s) & y_{22}(s) \end{bmatrix} = \frac{1}{\det \mathbf{Z}(s)} \begin{bmatrix} z_{22}(s) & -z_{12}(s) \\ -z_{21}(s) & z_{11}(s) \end{bmatrix} \tag{9-47}$$

Similarly, the elements of the z-parameter matrix may be found from the elements of the y-parameter matrix by the relations

$$\begin{bmatrix} z_{11}(s) & z_{12}(s) \\ z_{21}(s) & z_{22}(s) \end{bmatrix} = \frac{1}{\det \mathbf{Y}(s)} \begin{bmatrix} y_{22}(s) & -y_{12}(s) \\ -y_{21}(s) & y_{11}(s) \end{bmatrix} \tag{9-48}$$

Since the inverse of a symmetric matrix is still a symmetric matrix, the conclusions relating symmetricity of the parameter matrix and reciprocity of the network made for the z parameters also apply to the y parameters; namely, a reciprocal network will have a symmetric y-parameter matrix and the y-parameter matrix for a non-reciprocal network will not be symmetric. It is also easily shown that, as was the case for the z parameters, once the y parameters for a network are known, any other network function may be expressed in terms of them. The procedures used for determining the other network functions are similar to those presented for the z parameters.

In many network configurations, we can simplify the determination of the y parameters by separating the network configuration into component parts which are simpler to analyze than the original network. For example, consider the case where two two-port networks are connected as shown in Fig. 9–3.3. This type of a connection is called a *parallel* connection of two two-port networks. For the A network and the voltage and current variables shown in the figure, we may define a set of y parameters $y_{ij}^a(s)$ by the relations

$$\mathbf{I}^a(s) = \begin{bmatrix} I_1^a(s) \\ I_2^a(s) \end{bmatrix} = \begin{bmatrix} y_{11}^a(s) & y_{12}^a(s) \\ y_{21}^a(s) & y_{22}^a(s) \end{bmatrix} \begin{bmatrix} V_1^a(s) \\ V_2^a(s) \end{bmatrix} = \mathbf{Y}^a(s)\mathbf{V}^a(s) \tag{9-49}$$

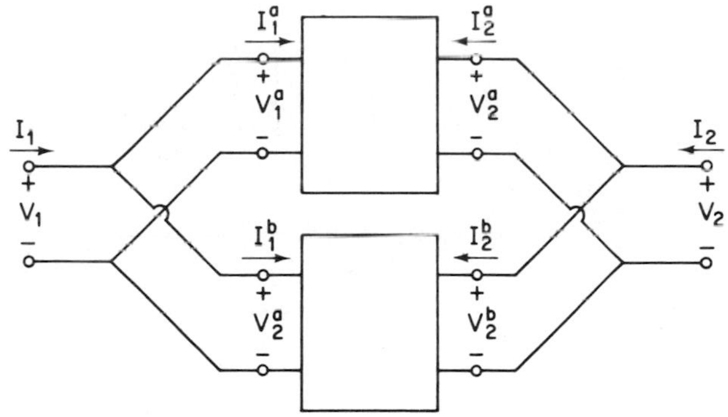

Fig. 9-3.3 A parallel connection of two two-port networks.

Similarly, for the B network we may define a set of parameters $y_{ij}^b(s)$ by the relations

$$\mathbf{I}^b(s) = \begin{bmatrix} I_1^b(s) \\ I_2^b(s) \end{bmatrix} = \begin{bmatrix} y_{11}^b(s) & y_{12}^b(s) \\ y_{21}^b(s) & y_{22}^b(s) \end{bmatrix} \begin{bmatrix} V_1^b(s) \\ V_2^b(s) \end{bmatrix} = \mathbf{Y}^b(s)\mathbf{V}^b(s) \tag{9-50}$$

The parallel connection of the separate two-port networks A and B can be considered as still another two-port network as shown in Fig. 9-3.3. The voltage and current variables specified in the figure for this overall two-port network define a third set of parameters $y_{ij}(s)$ by the relations

$$\mathbf{I}(s) = \begin{bmatrix} I_1(s) \\ I_2(s) \end{bmatrix} = \begin{bmatrix} y_{11}(s) & y_{12}(s) \\ y_{21}(s) & y_{22}(s) \end{bmatrix} \begin{bmatrix} V_1(s) \\ V_2(s) \end{bmatrix} = \mathbf{Y}(s)\mathbf{V}(s) \tag{9-51}$$

Applying KCL, we see that the sum of the port currents of the A and B networks equals the port currents of the overall network. Similarly, it is readily noted that the port voltages at the input and output ports are the same for all three networks. Thus, by a proof analogous to that used for the z-parameter case, we can show that the y parameters of the overall network may be found by adding the y parameters of the component networks. This result is expressed by the following relation:

$$\mathbf{Y}(s) = \mathbf{Y}^a(s) + \mathbf{Y}^b(s) \tag{9-52}$$

An example follows.

EXAMPLE 9-3.2. As an example of a situation in which the y parameters of component paralleled networks may be added to find the parameters of an overall two-port network, consider the network shown in Fig. 9-3.4(a). This may be treated as the parallel combination of the network shown in Fig. 9-3.4(b) with a y-parameter matrix

$$\begin{bmatrix} 1.05s + 0.1 & -0.05s \\ -0.05s & 0.55s + 0.2 \end{bmatrix}$$

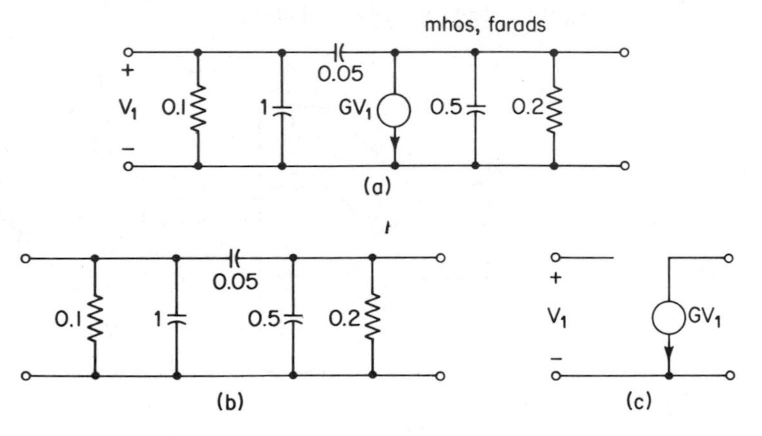

Fig. 9-3.4 A parallel connection of two-port networks in which the y parameters add.

and the network shown in Fig. 9-3.4(c) with a y-parameter matrix

$$\begin{bmatrix} 0 & 0 \\ G & 0 \end{bmatrix}$$

The sum of the two matrices given above is the y-parameter matrix for the overall two-port network shown in Fig. 9-3.4(a). It is

$$\begin{bmatrix} 1.05s + 0.1 & -0.05s \\ G - 0.05s & 0.55s + 0.2 \end{bmatrix}$$

The same caution which was presented in the preceding section regarding the possible violation of the port current requirement by the interconnection of certain types of networks also applies to the y-parameter case. It may be shown, however, that when *three-terminal networks (treated as two-port networks) are connected in parallel, the port current requirement is automatically satisfied.* Thus, the y parameters of three-terminal networks may always be added, when such networks are connected in parallel, to determine the y parameters of the resulting two-port network. The results given above may be summarized as follows:

SUMMARY 9-3.2

Properties of the y Parameters of a Two-Port Network: The four y parameters of a two-port network have the following properties:

1. They completely determine the characteristics of the network in the sense that they may be used to find any other network function which is defined for the network.
2. The y-parameter matrix is the inverse of the z-parameter matrix.
3. If two two-port networks are connected in parallel, the sum of their individual y parameters gives the y parameters for the resulting overall two-port network (unless the connection violates the port current requirement on the port currents of the component two-port networks).
4. When three-terminal networks are connected in parallel, the port current requirement is always satisfied.

In the preceding paragraphs a description of various techniques for finding the y parameters of a specific network have been discussed. Now let us consider the inverse problem. Given a set of y parameters, how do we find a two-port network which realizes such a set of parameters? One of the most common of the network forms for doing this is the "π" network representation. This is applicable to all reciprocal networks, i.e., networks for which $y_{12}(s)$ is equal to $y_{21}(s)$. The π representation is developed in a manner similar to that used for the determination of the **T** representation for the z parameters. To see this, consider the π configuration of two-terminal admittances shown in Fig. 9-3.5. The y-parameter matrix for this network is readily shown to be

$$\begin{bmatrix} y_{11}(s) & y_{12}(s) \\ y_{12}(s) & y_{22}(s) \end{bmatrix} = \begin{bmatrix} Y_a(s) + Y_c(s) & -Y_c(s) \\ -Y_c(s) & Y_b(s) + Y_c(s) \end{bmatrix} \tag{9-53}$$

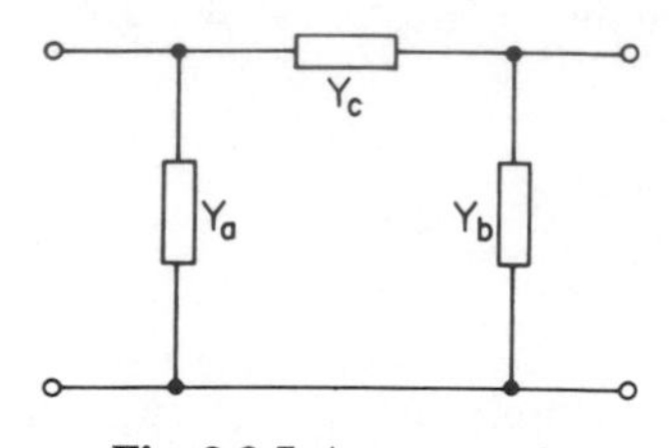

Fig. 9-3.5 A π network.

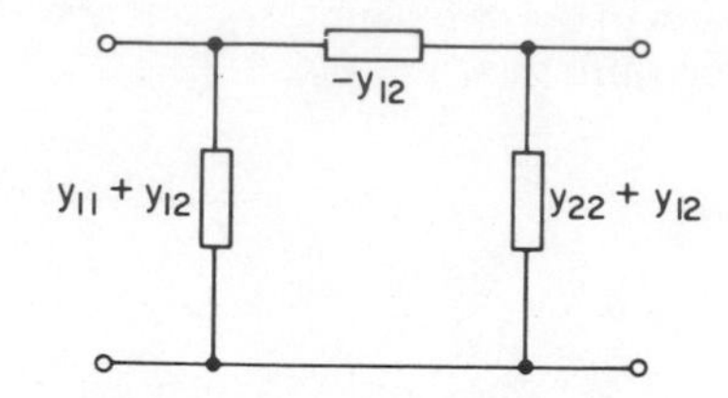

Fig. 9-3.6 A π network realization of y parameters.

Equating like terms in the above equation, we find the relations giving the network elements $Y_a(s)$, $Y_b(s)$, and $Y_c(s)$ as functions of the network parameters are

$$\begin{aligned} Y_a(s) &= y_{11}(s) + y_{12}(s) \\ Y_b(s) &= y_{22}(s) + y_{12}(s) \\ Y_c(s) &= -y_{12}(s) \end{aligned} \tag{9-54}$$

Thus, the π network may be redrawn with the expressions specifying the two-terminal admittances given directly in terms of the y parameters as shown in Fig. 9–3.6. As an example of finding a network representation for a set of y parameters, consider the y-parameter matrix

$$\begin{bmatrix} 3s + 2 & -1 \\ -1 & 2s + 3 \end{bmatrix} \tag{9-55}$$

Applying the relations given in (9–54), we see that the π configuration for a two-port network which realizes these parameters is as shown in Fig. 9–3.7. It should be noted that, in general, the synthesis of the two-terminal admittances may not be as simple as it is in this example if the expressions for the $y_{ij}(s)$ are complicated functions.

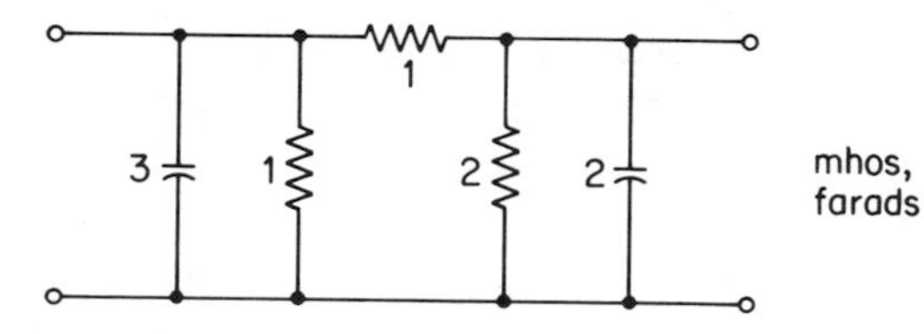

Fig. 9-3.7 A network realizing the y parameters of Eq. (9-55).

In order to realize non-symmetrical y-parameter matrices, one of the most useful representations is one utilizing two controlled sources of the VCIS type. This representation has the form shown in Fig. 9–3.8. If we apply KCL at each of the ports shown in this figure, it is readily verified that this representation satisfies the original equations defining the y parameters given in (9–36). Another equivalent representation for a set of y parameters for a non-reciprocal network is the one shown in Fig. 9–3.9. This representation utilizes only a single controlled source. For a

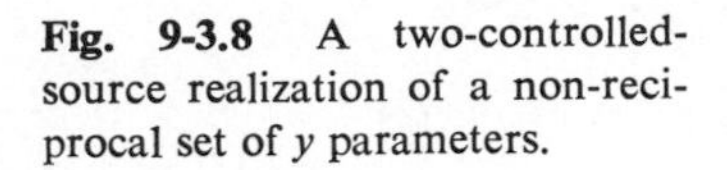
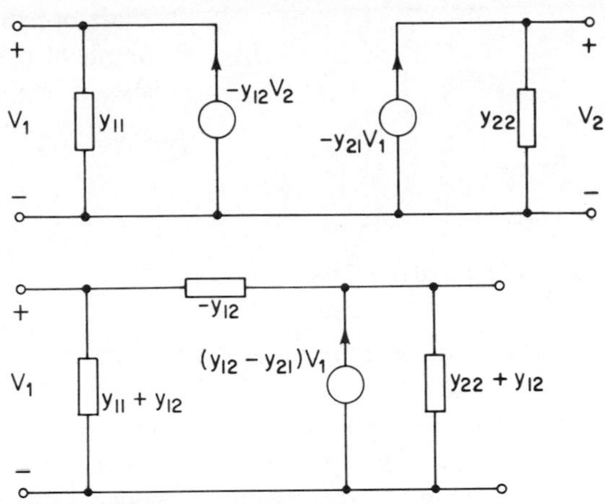

Fig. 9-3.8 A two-controlled-source realization of a non-reciprocal set of y parameters.

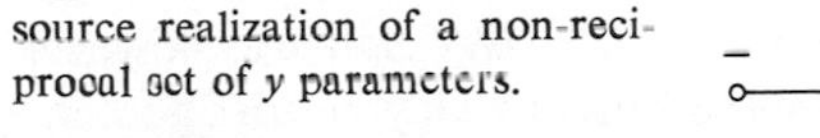

Fig. 9-3.9 A one-controlled-source realization of a non-reciprocal set of y parameters.

reciprocal network, i.e., for the case where $y_{12}(s) = y_{21}(s)$, this representation simply becomes the π configuration shown in Fig. 9-3.6.

The z- and y-parameter sets may not both exist for a given network configuration. For example, consider a two-port network consisting of a single series element as shown in Fig. 9-3.10. The y-parameter matrix for this network is readily shown to be

$$\begin{bmatrix} Y(s) & -Y(s) \\ -Y(s) & Y(s) \end{bmatrix} \tag{9-56}$$

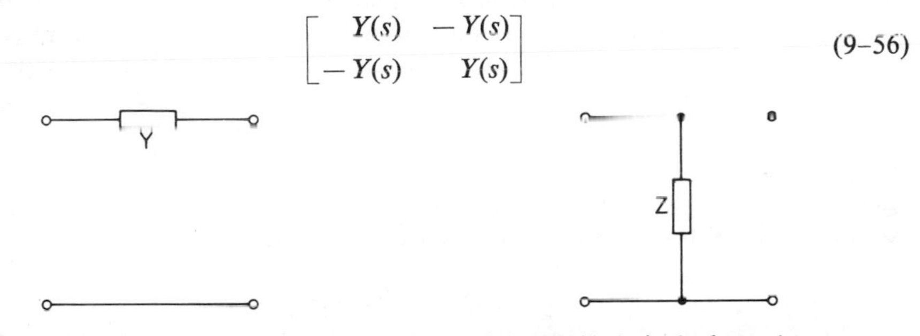

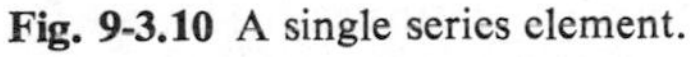

Fig. 9-3.10 A single series element.

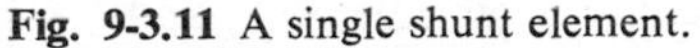

Fig. 9-3.11 A single shunt element.

Clearly, the determinant of this matrix equals zero. Therefore, the inverse matrix, and consequently the z parameters as determined by (9-48), do not exist. As another example of a network which does not have both z and y parameters, consider a two-port network consisting of a single shunt element as shown in Fig. 9-3.11. The z-parameter matrix for this network is

$$\begin{bmatrix} Z(s) & Z(s) \\ Z(s) & Z(s) \end{bmatrix} \tag{9-57}$$

Again, since the matrix is singular, the inverse matrix, and therefore the y parameters as determined by (9-47), do not exist for this network. For some networks, neither

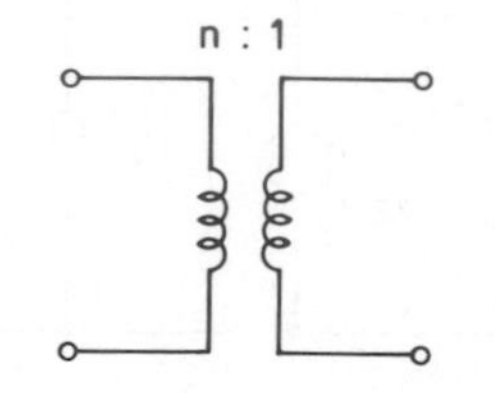

Fig. 9-3.12 An ideal transformer.

the z nor the y parameters exist. As an example of this, consider the ideal transformer shown in Fig. 9-3.12. The equations defining this two-port network are

$$V_1(s) = nV_2(s)$$

$$I_1(s) = -\frac{1}{n} I_2(s)$$

Since these equations require that the voltage at one port be specified as a function of the voltage at the other port and similarly that the current be expressed as a function of the other port current, neither the z parameters nor the y parameters can be used to describe the characteristics of this network. We shall see, however, that some of the other representations which we shall discuss may be used to characterize it. In general, except for some very unusual degenerate networks, there will be at least one parameter set which will exist for any given network configuration.

9-4 HYBRID PARAMETERS

In the last two sections we discussed the use of z and y parameters to represent two-port networks. These parameters are characterized by the fact that they have the dimensions of impedance and admittance, respectively. Thus, collectively they may be referred to as *immittance parameters*. In this section we shall introduce two more sets of parameters which are quite different in nature. These are the third and fourth types given in the list in Table 9-1.1 of the six possible parameter sets for a two-port network. They are named the g parameters and the h parameters and are collectively referred to as *hybrid parameters*. The name is appropriate because the individual elements of both of these types of parameters do not have a common dimension.

The first set of hybrid parameters is defined by the third case in Table 9-1.1, namely, the variables $I_1(s)$ and $V_2(s)$ are chosen for the quantities $U_1(s)$ and $U_2(s)$ in Eq. (9-9), and the variables $V_1(s)$ and $I_2(s)$ are chosen for the quantities $W_1(s)$ and $W_2(s)$. The equations have the form

$$\begin{aligned} I_1(s) &= g_{11}(s)V_1(s) + g_{12}(s)I_2(s) \\ V_2(s) &= g_{21}(s)V_1(s) + g_{22}(s)I_2(s) \end{aligned} \tag{9-58}$$

where the quantities $g_{ij}(s)$ are called the *g parameters*. In the first equation, the left member has the dimensions of current. Therefore, in the right member, $g_{11}(s)$ must have the dimensions of admittance and $g_{12}(s)$ must be dimensionless. In the second equation, the left member has the dimensions of voltage; therefore, in the right member $g_{21}(s)$ must be dimensionless and $g_{22}(s)$ must have the dimensions of impedance. The matrix representation for the g parameters will have the form

$$\begin{bmatrix} I_1(s) \\ V_2(s) \end{bmatrix} = \begin{bmatrix} g_{11}(s) & g_{12}(s) \\ g_{21}(s) & g_{22}(s) \end{bmatrix} \begin{bmatrix} V_1(s) \\ I_2(s) \end{bmatrix} = \mathbf{G}(s) \begin{bmatrix} V_1(s) \\ I_2(s) \end{bmatrix} \tag{9-59}$$

where the matrix $\mathbf{G}(s)$ is called the *g-parameter matrix*.

Test excitations similar to those defined for the z and y parameters may be used to determine the g parameters. For example, if we open-circuit port 2, thus setting $I_2(s) = 0$, and apply a voltage test excitation at port 1, from (9-58) we obtain the relations

$$I_1(s) = g_{11}(s)V_1(s) \tag{9-60a}$$

$$V_2(s) = g_{21}(s)V_1(s) \tag{9-60b}$$

Thus, $g_{11}(s)$ is the open-circuit driving-point admittance at port 1 of the network and $g_{21}(s)$ is the open-circuit voltage transfer function from port 1 to port 2. As an example of the determination of these parameters, consider the network shown in Fig. 9-4.1(a). The test conditions are shown in Fig. 9-4.1(b). From this figure, it is easily seen that $g_{11}(s) = (2s^2 + 3s + 1)/(2s + 3)$, and $g_{21}(s) = 3/(2s + 3)$. In a similar manner, we may find the remaining two g parameters by short-circuiting port 1, thus setting $V_1(s) = 0$, and applying a test excitation current at port 2. From Eq. (9-58), the relations defining the parameters are

$$I_1(s) = g_{12}(s)I_2(s) \tag{9-61a}$$

$$V_2(s) = g_{22}(s)I_2(s) \tag{9-61b}$$

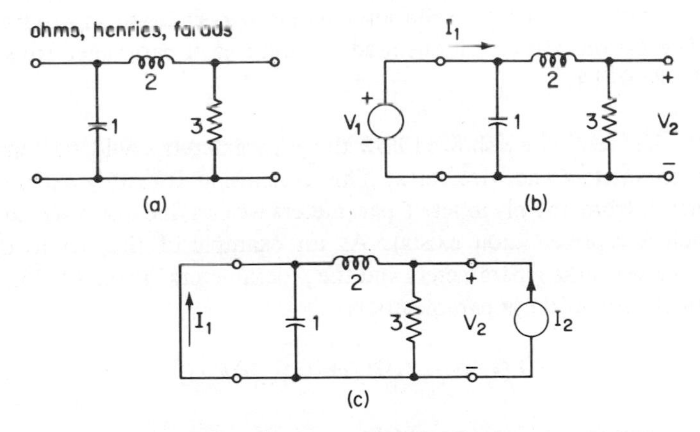

Fig. 9-4.1 Tests for the g parameters.

Thus, $g_{12}(s)$ is the short-circuit current transfer ratio from port 2 to port 1 and $g_{22}(s)$ is the short-circuit driving-point impedance at port 2. For the network shown in Fig. 9-4.1(a), the test conditions are shown in Fig. 9-4.1(c). From the figure we find that $g_{12}(s) = -3/(2s + 3)$. Note that the minus sign occurs because for a positive current $I_2(s)$ the actual current flows opposite to the assumed positive reference direction for $I_1(s)$. From Fig. 9–4.1(c) we also see that $g_{22}(s) = 6s/(2s + 3)$. Thus, the g-parameter matrix for the example network shown in Fig. 9–4.1(a) is

$$\mathbf{G}(s) = \begin{bmatrix} \dfrac{2s^2 + 3s + 1}{2s + 3} & \dfrac{-3}{2s + 3} \\ \dfrac{3}{2s + 3} & \dfrac{6s}{2s + 3} \end{bmatrix} \tag{9–62}$$

It should be apparent that we may use unit input excitations to determine the g parameters just as we did in the case of z and y parameters. The results given above may be summarized as follows:

SUMMARY 9-4.1

The g Parameters of a Two-Port Network: The g parameters of a two-port network are the network functions $g_{ij}(s)$ which specify the current at port 1 and the voltage at port 2 as functions of the other port variables. The exact relations are

$$\begin{bmatrix} I_1(s) \\ V_2(s) \end{bmatrix} = \begin{bmatrix} g_{11}(s) & g_{12}(s) \\ g_{21}(s) & g_{22}(s) \end{bmatrix} \begin{bmatrix} V_1(s) \\ I_2(s) \end{bmatrix}$$

Two of the four g parameters may be found by applying a unit test excitation voltage at port 1 and open-circuiting port 2. The other two may be found by applying a unit test excitation current at port 2 and short-circuiting port 1. The various test conditions used to define each parameter are summarized in Table 9–4.1.

In the last section we showed how the y parameters could be found directly from the z parameters and vice versa. This conclusion is readily extended to show that, in general, from any given set of parameters we can find any other set of parameters (if such a representation exists). As an example of this, let us develop the relationship between the g parameters and the y parameters. From (9–36), the second equation which defines the y parameters is

$$I_2(s) = y_{21}(s)V_1(s) + y_{22}(s)V_2(s) \tag{9–63}$$

Solving this equation for $V_2(s)$, we obtain

$$V_2(s) = \frac{1}{y_{22}(s)} I_2(s) - \frac{y_{21}(s)}{y_{22}(s)} V_1(s) \tag{9–64}$$

Comparing this equation with (9–58), we see that it has the same form as the second equation for the g parameters, namely, it expresses $V_2(s)$ as a function of $V_1(s)$ and

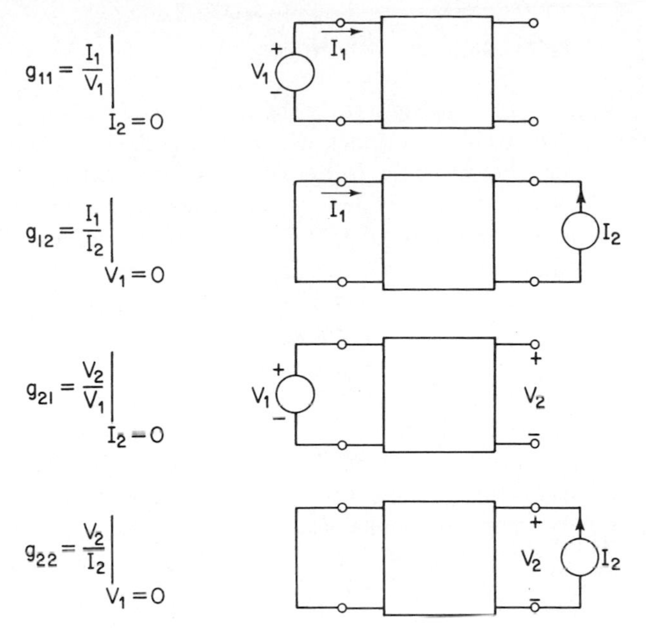

Table 9-4.1

$I_2(s)$. To derive the first g-parameter equation in terms of the y parameters, let us consider the first equation for the y parameters given in (9–36). This is

$$I_1(s) = y_{11}(s)V_1(s) + y_{12}(s)V_2(s) \tag{9–65}$$

If we substitute (9–64) in this equation and rearrange terms, we obtain

$$I_1(s) = \frac{\det \mathbf{Y}(s)}{y_{22}(s)} V_1(s) + \frac{y_{12}(s)}{y_{22}(s)} I_2(s) \tag{9–66}$$

where $\det \mathbf{Y}(s) = y_{11}(s)y_{22}(s) - y_{12}(s)y_{21}(s)$. This equation has the form of the first equation of the g parameters, namely, it expresses $I_1(s)$ as a function of $V_1(s)$ and $I_2(s)$. We may summarize the results given in (9–64) and (9–66) by the matrix equality

$$\begin{bmatrix} g_{11}(s) & g_{12}(s) \\ g_{21}(s) & g_{22}(s) \end{bmatrix} = \frac{1}{y_{22}(s)} \begin{bmatrix} \det \mathbf{Y}(s) & y_{12}(s) \\ -y_{21}(s) & 1 \end{bmatrix} \tag{9–67}$$

Thus, we have developed the relations giving the g parameters as functions of the y parameters. A series of steps similar to the above may be used to obtain the inverse relations, namely, to specify the y parameter in terms of the g parameters. The results of these steps is the matrix equality

$$\begin{bmatrix} y_{11}(s) & y_{12}(s) \\ y_{21}(s) & y_{22}(s) \end{bmatrix} = \frac{1}{g_{22}(s)} \begin{bmatrix} \det \mathbf{G}(s) & g_{12}(s) \\ -g_{21}(s) & 1 \end{bmatrix} \tag{9-68}$$

where $\det \mathbf{G}(s) = g_{11}(s)g_{22}(s) - g_{12}(s)g_{21}(s)$. It should be noted that, implicit in Eqs. (9–67) and (9–68), are the conditions under which it is possible to find one set of parameters from another. For example, in Eq. (9–67) we see that every element of the right member of the equation is divided by $y_{22}(s)$. We may conclude that, for a two-port network, if the y parameters exist, the g parameters will also exist if and only if $y_{22}(s) \neq 0$. Similarly, from Eq. (9–68), if a set of g parameters for a two-port network exists, the y parameters will exist if and only if $g_{22}(s) \neq 0$.

By an extension of the method given above, it is readily verified that all the parameter sets are related. Thus, any properties of a two-port network which are expressed in terms of one set of parameters may be easily related to another set of parameters. For example, it has been shown that a reciprocal network is one in which $y_{12}(s)$ equals $y_{21}(s)$. From Eq. (9–68) we see that the equivalent equality in terms of the g parameters is that $g_{12}(s)/g_{22}(s) = -g_{21}(s)/g_{22}(s)$. Thus, the condition for reciprocity in term of the g parameters is simply

$$g_{12}(s) = -g_{21}(s) \tag{9-69}$$

By a similar substitution, any desired network function may be found in terms of the g parameters. For example, for the open-circuit voltage transfer function, we readily see that

$$\frac{V_2(s)}{V_1(s)} = \frac{-y_{21}(s)}{y_{22}(s)} = g_{21}(s) \tag{9-70}$$

Just as was the case for the z and y parameters, it is possible to specify an interconnection of two-port networks such that the g parameters add. Such an interconnection, however, is seldom encountered in practice. The details concerning it are left as an exercise for the reader.

The second set of hybrid parameters, which is given as the fourth case in Table 9–1.1, is the one in which $U_1(s)$ in Eq. (9–9) is set equal to $V_1(s)$ and $U_2(s)$ is set equal to $I_2(s)$. Similarly, $W_1(s)$ and $W_2(s)$ are set equal to $I_1(s)$ and $V_2(s)$, respectively. The resulting equations have the form

$$\begin{aligned} V_1(s) &= h_{11}(s)I_1(s) + h_{12}(s)V_2(s) \\ I_2(s) &= h_{21}(s)I_1(s) + h_{22}(s)V_2(s) \end{aligned} \tag{9-71}$$

where the quantities $h_{ij}(s)$ are called the *h parameters*. The left member of the first equation given in Eq. (9–71) has the dimensions of voltage. Thus, we see that $h_{11}(s)$ has the dimensions of impedance and $h_{12}(s)$ is dimensionless. Similarly, since the left member of the second equation has the dimensions of current, $h_{21}(s)$ must be dimensionless and $h_{22}(s)$ must have the dimensions of admittance. If the equations are written in matrix form, we obtain

$$\begin{bmatrix} V_1(s) \\ I_2(s) \end{bmatrix} = \begin{bmatrix} h_{11}(s) & h_{12}(s) \\ h_{21}(s) & h_{22}(s) \end{bmatrix} \begin{bmatrix} I_1(s) \\ V_2(s) \end{bmatrix} = \mathbf{H}(s) \begin{bmatrix} I_1(s) \\ V_2(s) \end{bmatrix} \tag{9-72}$$

where the square matrix $\mathbf{H}(s)$ is referred to as the *h-parameter matrix*. If we compare the form of the equation in (9–72) with that given in (9–59), we see that the *h*-parameter matrix and the *g*-parameter matrix are related by

$$\mathbf{H} = \mathbf{G}^{-1} \qquad \mathbf{G} = \mathbf{H}^{-1} \tag{9-73}$$

The application of the test excitations to determine the *h* parameters is similar to that given for the *g* parameters. The results are summarized as follows:

SUMMARY 9-4.2

The h Parameters of a Two-Port Network: The *h* parameters of a two-port network are the network functions $h_{ij}(s)$ which specify the voltage variable at port 1 and the current variable at port 2 as functions of the other port variables. The exact relations are:

$$\begin{bmatrix} V_1(s) \\ I_2(s) \end{bmatrix} = \begin{bmatrix} h_{11}(s) & h_{12}(s) \\ h_{21}(s) & h_{22}(s) \end{bmatrix} \begin{bmatrix} I_1(s) \\ V_2(s) \end{bmatrix}$$

Two of the four *h* parameters may be found by applying a unit test excitation current at port 1 with port 2 short-circuited. The other two *h* parameters may be found by applying a unit test excitation voltage at port 2 with port 1 open-circuited. The various test conditions used to define each of the *h* parameters are summarized in Table 9–4.2.

Remarks similar to those made for the *g* parameter also apply to the *h* parameters, namely, reciprocity implies that $h_{12}(s)$ equals $-h_{21}(s)$, and, in general, any other set of parameters or any network function may be expressed directly in terms of the *h* parameters. A summary of the interrelations between the various parameter sets is given in Table 9–7.1.

Just as was the case for the *z* and *y* parameters, there are several network representations which may be used to model a given set of hybrid parameters. As an example, let us again consider the *g* parameters. The first equation of (9–58) says that $I_1(s)$ is equal to the sum of a term which is proportional to $V_1(s)$ (such a term may be physically represented by a shunt admittance) and a term which is proportional to $I_2(s)$ (which may be represented by a controlled source). Similarly, from

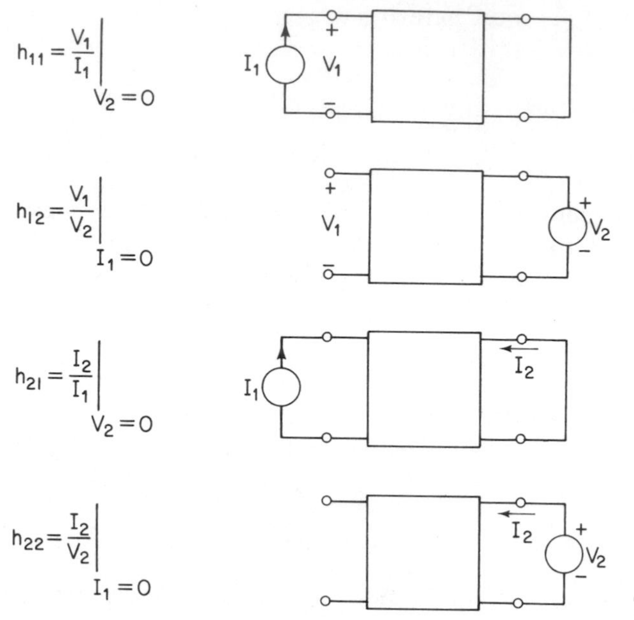

Table 9-4.2

the second equation of (9–58) we see that $V_2(s)$ is equal to the sum of a term which is proportional to $V_1(s)$ (represented by a VCVS) and a term which is proportional to $I_2(s)$ (which is represented by a series impedance). Combining these results, we see that the set of equations given in Eq. (9–58) may be modeled by the network representation shown in Fig. 9–4.2. If $g_{12}(s)$ equals zero, this model appears as shown in Fig. 9–4.3. Thus, there is no transmission from port 2 to port 1 and the network may be considered as a *unilateral* one. A representation of the type shown in Fig. 9–4.3 is useful for modeling many active devices. An example follows.

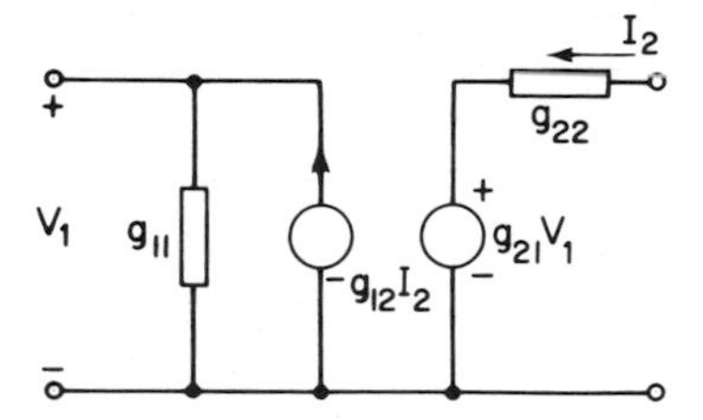

Fig. 9-4.2 A two-controlled-source realization of a set of g parameters.

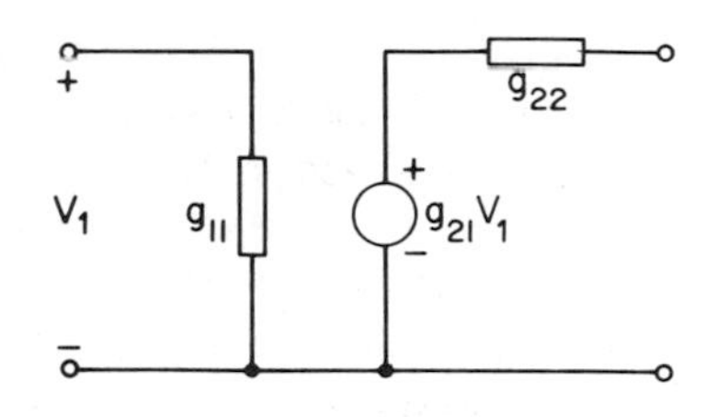

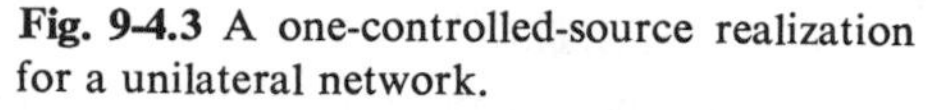
Fig. 9-4.3 A one-controlled-source realization for a unilateral network.

EXAMPLE 9-4.1. As an example of the use of g parameters to model an active element, consider a unilateral voltage amplifier which is specified as having an input resistance of 100 kΩ, and a no-load low-frequency gain of 100 which decreases at a rate of 6 dB/octave starting at 1000 rad/s. In addition, suppose that the no-load output voltage at any frequency is halved if a 1-kΩ load is added. Such a device may be directly modeled using the g parameters. Since it is unilateral, we conclude that $g_{12}(s)$ equals 0. The parameter $g_{11}(s)$

equals 1/100,000, namely, the reciprocal of the input impedance. To meet the no-load specifications the transfer admittance $g_{21}(s)$ must have a pole at $s = -1000$. In addition, $g_{21}(0)$ must equal 100. Thus, $g_{21}(s)$ equals $10^5/(s + 10^3)$. Finally, the network load characteristic specifies $g_{22}(s)$ equals $10^3\ \Omega$, since this "internal" impedance must equal the external impedance which causes the output voltage to be halved. Thus, the amplifier, as specified, may be modeled by the following set of g parameters.

$$\begin{bmatrix} 10^{-5} & 0 \\ \dfrac{10^5}{s + 10^3} & 10^3 \end{bmatrix}$$

The g-parameter model for this amplifier is shown in Fig. 9-4.4.

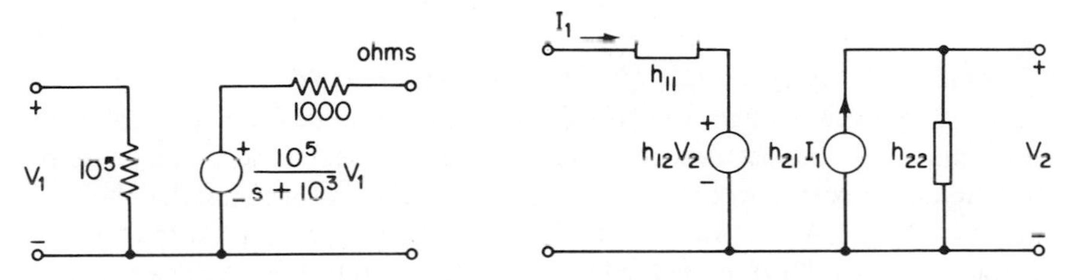

Fig. 9-4.4 Network realization for Example 9-4.1.

Fig. 9-4.5 A two-controlled-source realization of a set of h parameters.

A network model which may be used to represent a set of h parameters is easily derived following logic similar to that given above. It serves as a useful model for the case where the output of an active element has the basic characteristics of a current source. Thus, it is common to use h parameters to represent transistors. The model is shown in Fig. 9-4.5.

9-5 *ABCD* PARAMETERS

All the types of two-port network parameters which have been introduced so far in this chapter have had the property that the quantities $U_1(s)$ and $U_2(s)$ shown in Eq. (9-9) have included one variable from each of the two ports. Thus, for the z parameters these quantities have included a voltage variable from each port; similarly, for the y parameters they have included a current variable from each port; finally, for the hybrid parameters they have included a voltage variable from one port and a current variable from another. In this section we shall introduce two sets of network parameters which are quite different in that the quantities $U_1(s)$ and $U_2(s)$ will include both the voltage *and* the current variable from a single port. The first of these parameter sets covers the situation shown as the fifth case in Table 9-1.1. Namely, the variables $V_1(s)$ and $I_1(s)$ are used as the quantities $U_1(s)$ and $U_2(s)$ and the variables $V_2(s)$ and $-I_2(s)$ are used as the quantities $W_1(s)$ and $W_2(s)$. Note that this last assignment of variable includes a minus sign. The reason for this will become apparent later in this section. For the coefficients k_{ij} in Eq. (9-9), we shall use the quantities $A(s)$, $B(s)$, $C(s)$, and $D(s)$. Thus, the equations for this case may be written in the form

$$\begin{aligned} V_1(s) &= A(s)V_2(s) - B(s)I_2(s) \\ I_1(s) &= C(s)V_2(s) - D(s)I_2(s) \end{aligned} \tag{9-74}$$

The quantities $A(s)$, $B(s)$, $C(s)$, and $D(s)$ defined for the above equations are collectively referred to as the *ABCD parameters.* For convenience of notation, we may write these in the form

$$\begin{bmatrix} V_1(s) \\ I_1(s) \end{bmatrix} = \begin{bmatrix} A(s) & B(s) \\ C(s) & D(s) \end{bmatrix} \begin{bmatrix} V_2(s) \\ -I_2(s) \end{bmatrix} = \mathbf{A}(s) \begin{bmatrix} V_2(s) \\ -I_2(s) \end{bmatrix} \tag{9-75}$$

where the matrix $\mathbf{A}(s)$ is called the *ABCD parameter matrix.*

For a given network the *ABCD* parameters may be found by applying a set of test excitations to the network in a manner similar to that done for the other parameters which have been introduced in this chapter. To see this, consider the first equation given in (9–74). If we let port 2 of the network be open-circuited, then the variable $I_2(s)$ must be zero. For such a termination, the equation becomes

$$V_1(s) = A(s)V_2(s) \tag{9-76}$$

To set up a test situation which may be used to implement this equation, we note that $V_2(s)$ *cannot* be used as an excitation variable, since we have already postulated that $I_2(s) = 0$. Therefore, we conclude that $V_1(s)$ must be the excitation variable and $V_2(s)$ must be the response variable. Since we have followed the practice of writing network functions as the ratio of response to excitation (rather than the ratio of excitation to response), it is more consistent to actually define the parameter $A(s)$ by the following expression:

$$\frac{1}{A(s)} = \left.\frac{V_2(s)}{V_1(s)}\right|_{I_2(s)=0} \tag{9-77}$$

Thus, we see that the parameter $A(s)$ is the reciprocal of the open-circuit voltage transfer ratio from port 1 to port 2 of a two-port network. As an example of the determination of this parameter, consider the network consisting of a single shunt element $Y(s)$ as shown in Fig. 9–5.1. If a test voltage excitation is applied at port 1 and port 2 is open-circuited, as shown in Fig. 9–5.2, then $V_1(s) = V_2(s)$ and, from Eq. (9–77), we conclude that $A(s) = 1$. Now consider the determination of the $B(s)$ parameter. If we set $V_2(s)$ to zero by short-

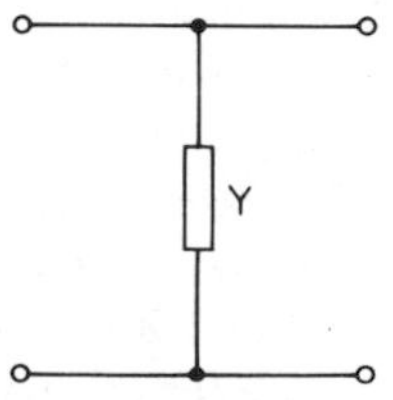

Y

Fig. 9-5.1 A single shunt element.

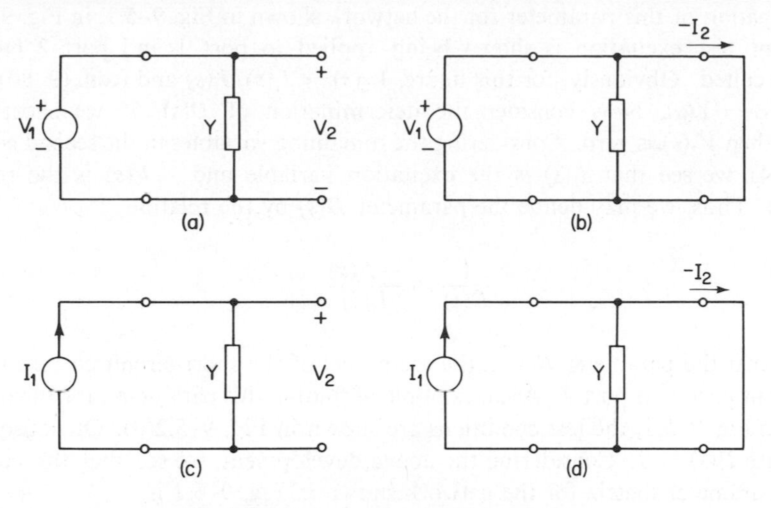

Fig. 9-5.2 Tests for the *ABCD* parameters.

circuiting port 2, the first equation of (9–74) becomes

$$V_1(s) = -B(s)I_2(s) \tag{9–78}$$

Following the logic given above, we see that $V_1(s)$ is the excitation variable and $-I_2(s)$ is the response variable. Thus, the parameter $B(s)$ is defined by the relation

$$\frac{1}{B(s)} = \left.\frac{-I_2(s)}{V_1(s)}\right|_{V_2(s)=0} \tag{9–79}$$

We conclude that $B(s)$ is the reciprocal of the short-circuit transfer impedance from port 1 to port 2. As an example of finding this parameter, again consider the network shown in Fig. 9–5.1. The test conditions with a short circuit on port 2 are shown in Fig. 9–5.2(b). Note that, since the short circuit is effectively applied across the voltage source, and since the impedance of a voltage source is zero, the magnitude of $I_2(s)$ will be infinite for any non-zero voltage excitation. Therefore, the magnitude of $1/B(s)$ is also infinite, and we see that the parameter $B(s) = 0$.

Now consider the second equation of (9–74). If we open-circuit port 2 to set $I_2(s)$ to zero, then, considering the remaining variables in this equation, we see that $I_1(s)$ is the excitation variable and $V_2(s)$ is the response variable. Following the procedure used above, we may define the network function $C(s)$ by the relation

$$\frac{1}{C(s)} = \left.\frac{V_2(s)}{I_1(s)}\right|_{I_2(s)=0} \tag{9–80}$$

From the above we see that the network parameter $C(s)$ is the reciprocal of the open-circuit current transfer admittance from port 1 to port 2. As an example of the determination of this parameter for the network shown in Fig. 9–5.1, in Fig. 9–5.2(c) a current test excitation is shown being applied to port 1 and port 2 has been open-circuited. Obviously, for this figure, $V_2(s) = I_1(s)/Y(s)$ and from (9–80) we see that $C(s) = Y(s)$. Now consider the determination of $D(s)$. If we short-circuit port 2, then $V_2(s)$ is zero. Considering the remaining variables in the second equation of (9–74), we see that $I_1(s)$ is the excitation variable and $-I_2(s)$ is the response variable. Thus, we may define the parameter $D(s)$ by the relation

$$\frac{1}{D(s)} = \left.\frac{-I_2(s)}{I_1(s)}\right|_{V_2(s)=0} \tag{9–81}$$

We see that the parameter $D(s)$ is the reciprocal of the short-circuit current transfer ratio from port 1 to port 2. As an example of finding this parameter, for the network shown in Fig. 9–5.1, the test conditions are shown in Fig. 9–5.2(d). Obviously, from this figure $D(s) = 1$. Considering the above development, we see that the complete *ABCD* parameter matrix for the network shown in Fig. 9–5.1 is

$$\begin{bmatrix} 1 & 0 \\ Y(s) & 1 \end{bmatrix} \tag{9–82}$$

In all the descriptions of the *ABCD* parameters given above, we see that each parameter defines a *transfer* function rather than a driving-point function. Thus, all these parameters are concerned with the *transmission* of a signal from port 1 to port 2 for various types of excitation and termination. Considering this, we see that another name which is equally applicable to these parameters (and one which is frequently used) is *transmission parameters.*

All the *ABCD* parameters described above are readily determined by applying unity-valued excitations following the methods used in previous sections. Thus, we may summarize the discussion given above as follows:

SUMMARY 9-5.1

The *ABCD Parameters of a Two-Port Network:* The *ABCD* parameters of a two-port network are the quantities $A(s)$, $B(s)$, $C(s)$, and $D(s)$ which specify the variables at port 1 as functions of the variables at port 2. The exact relations are

$$\begin{bmatrix} V_1(s) \\ I_1(s) \end{bmatrix} = \begin{bmatrix} A(s) & B(s) \\ C(s) & D(s) \end{bmatrix} \begin{bmatrix} V_2(s) \\ -I_2(s) \end{bmatrix}$$

Two of the four *ABCD* parameters may be found by applying a unit test excitation voltage at port 1 and measuring the resulting open-circuit and short-circuit responses at port 2. The other two *ABCD* parameters may be found by applying a unit test excitation current at port 1 and similarly measuring the responses at port 2. The parameters are the reciprocal of the network functions thus determined. The various test conditions used to define the parameters are summarized in Table 9–5.1.

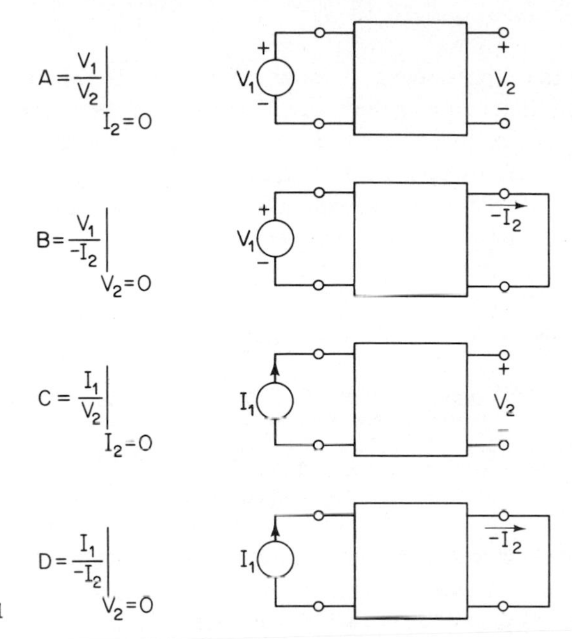

Table 9-5.1

Now let us investigate some properties of the *ABCD* parameters. In general, these are similar to the properties of the other parameter sets discussed in previous sections. For example, any network function which is defined for a two-port network may be expressed in terms of the *ABCD* parameters. Similarly, any other set of parameters may be expressed in terms of the *ABCD* parameters. As an example of such a determination, if we solve the second equation of (9–74) for the variable $V_2(s)$, we obtain

$$V_2(s) = \frac{1}{C(s)} I_1(s) + \frac{D(s)}{C(s)} I_2(s) \tag{9–83}$$

Substituting this result in the first equation of (9–74) and rearranging terms, we obtain

$$V_1(s) = \frac{A(s)}{C(s)} I_1(s) + \left[\frac{A(s)D(s)}{C(s)} - B(s)\right] I_2(s) \tag{9–84}$$

The two equations given above may be written in the following form:

$$\begin{bmatrix} V_1(s) \\ V_2(s) \end{bmatrix} = \frac{1}{C(s)} \begin{bmatrix} A(s) & \det \mathbf{A}(s) \\ 1 & D(s) \end{bmatrix} \begin{bmatrix} I_1(s) \\ I_2(s) \end{bmatrix} \tag{9–85}$$

Comparing this with (9–12), we see that we may write the following matrix equality:

$$\begin{bmatrix} z_{11}(s) & z_{12}(s) \\ z_{21}(s) & z_{22}(s) \end{bmatrix} = \frac{1}{C(s)} \begin{bmatrix} A(s) & \det \mathbf{A}(s) \\ 1 & D(s) \end{bmatrix} \tag{9–86}$$

Thus, we have defined the z parameters in terms of the *ABCD* parameters. Other expressions determining the relations between various parameter sets are easily derived. They are summarized in Table 9–7.1. As another example of the properties of the *ABCD* parameters, let us consider the restriction that reciprocity places on these parameters. In Sec. 9–2 it was shown that a reciprocal network is characterized as one in which the z-parameter matrix is symmetric, i.e., $z_{12}(s) = z_{21}(s)$. From Eq. (9–86) we see that this requires that

$$\det \mathbf{A}(s) = A(s)D(s) - B(s)C(s) = 1 \tag{9–87}$$

We conclude that the determinant of the *ABCD* parameter matrix is unity for a reciprocal network.

In previous sections we showed how the z-parameter matrices of component two-port networks are added when such networks are connected in series and considered as a new two-port network, and how the y-parameter matrices similarly add when networks are connected in parallel. Thus, we may anticipate that there is also some connection of two-port networks such that the *ABCD* parameters of the component networks may be used to find the *ABCD* parameters of some overall two-port network. There is, indeed, such a connection; however, it involves matrix *multiplication* rather than matrix *addition.* To see how this comes about, consider the connection of two two-port networks shown in Fig. 9–5.3. This is called a *cascade*

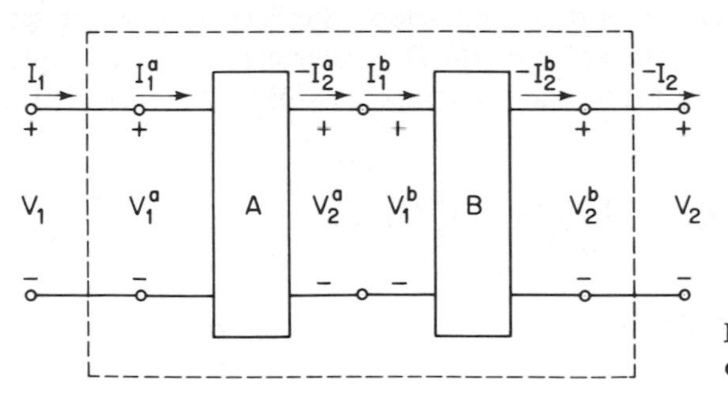

Fig. 9-5.3 A cascade connection of two two-port networks.

connection of networks. For the A network and the voltage and current variables shown in the figure, we may define a set of *ABCD* parameters $A^a(s)$, $B^a(s)$, $C^a(s)$, and $D^a(s)$ by the relations

$$\begin{bmatrix} V_1^a(s) \\ I_1^a(s) \end{bmatrix} = \begin{bmatrix} A^a(s) & B^a(s) \\ C^a(s) & D^a(s) \end{bmatrix} \begin{bmatrix} V_2^a(s) \\ -I_2^a(s) \end{bmatrix} \tag{9–88}$$

Similarly, for the B network we may define a set of parameters $A^b(s)$, $B^b(s)$, $C^b(s)$, and $D^b(s)$ by the relations

$$\begin{bmatrix} V_1^b(s) \\ I_1^b(s) \end{bmatrix} = \begin{bmatrix} A^b(s) & B^b(s) \\ C^b(s) & D^b(s) \end{bmatrix} \begin{bmatrix} V_2^b(s) \\ -I_2^b(s) \end{bmatrix} \tag{9-89}$$

The quantities $V_1^b(s)$ and $I_1^b(s)$ are obviously equal to the quantities $V_2^a(s)$ and $-I_2^a(s)$, respectively. Thus, we may combine Eqs. (9–88) and (9–89) to obtain

$$\begin{bmatrix} V_1^a(s) \\ I_1^a(s) \end{bmatrix} = \begin{bmatrix} A^a(s) & B^a(s) \\ C^a(s) & D^a(s) \end{bmatrix} \begin{bmatrix} A^b(s) & B^b(s) \\ C^b(s) & D^b(s) \end{bmatrix} \begin{bmatrix} V_2^b(s) \\ -I_2^b(s) \end{bmatrix} \tag{9-90}$$

This result, of course, occurs since the port 2 current variable is defined with a minus sign. Now let us consider the overall two-port network formed as the cascade of networks A and B with port variables defined as shown in Fig. 9–5.3. Obviously, the variables at port 1 of the overall network are the same as those at port 1 of network A. Similarly, the variables at port 2 of the overall network are the same as those at port 2 of network B. Thus, we may write Eq. (9–90) in terms of the variables of the overall network. We obtain

$$\begin{bmatrix} V_1(s) \\ I_1(s) \end{bmatrix} = \begin{bmatrix} A^a(s) & B^a(s) \\ C^a(s) & D^a(s) \end{bmatrix} \begin{bmatrix} A^b(s) & B^b(s) \\ C^b(s) & D^b(s) \end{bmatrix} \begin{bmatrix} V_2(s) \\ -I_2(s) \end{bmatrix} \tag{9-91}$$

The product of the two component *ABCD* parameter matrices clearly defines a new second-order matrix which gives the *ABCD* parameters of the overall network formed by the cascade connection of the two component networks. Note that since matrix multiplication is not commutative, i.e., in general, if $\mathbf{E}(s)$ and $\mathbf{F}(s)$ are two square matrices, the product $\mathbf{E}(s)\mathbf{F}(s)$ is not equal to the product $\mathbf{F}(s)\mathbf{E}(s)$, the component parameter matrices must be multiplied in the same left-to-right order as the component networks appear in the cascade connection.

As an example of the application of the results given above, let us determine the transmission parameters of the "L" network shown in Fig. 9–5.4(a). This may be considered as a cascade connection of two component networks A and B as shown in Fig. 9–5.4(b). For the A network, the *ABCD* parameter matrix has already been

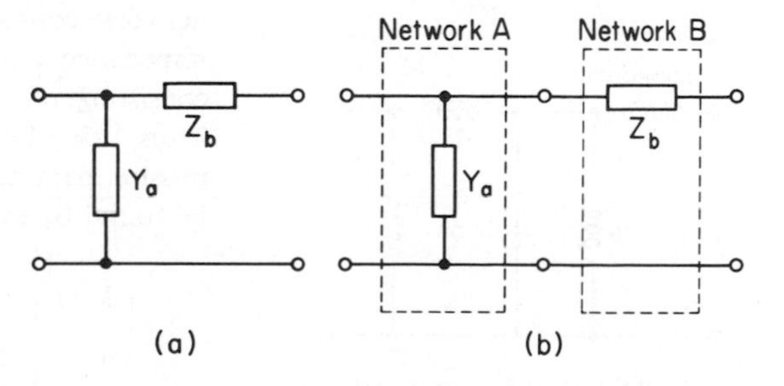

Fig. 9-5.4 A shunt element cascaded with a series element.

given in Eq. (9–82) [with $Y(s)$ replaced by $Y_a(s)$]. For the B network, the *ABCD* parameter matrix is readily shown to be

$$\begin{bmatrix} 1 & Z_b(s) \\ 0 & 1 \end{bmatrix} \tag{9–92}$$

Thus, the *ABCD* parameters for the overall network are defined by the equation

$$\begin{bmatrix} 1 & 0 \\ Y_a(s) & 1 \end{bmatrix}\begin{bmatrix} 1 & Z_b(s) \\ 0 & 1 \end{bmatrix} = \begin{bmatrix} 1 & Z_b(s) \\ Y_a(s) & 1 + Y_a(s)Z_b(s) \end{bmatrix} \tag{9–93}$$

The results given above are easily generalized to cover a cascade of an arbitrary number of two-port networks as shown in Fig. 9–5.5. For such a cascade, the trans-

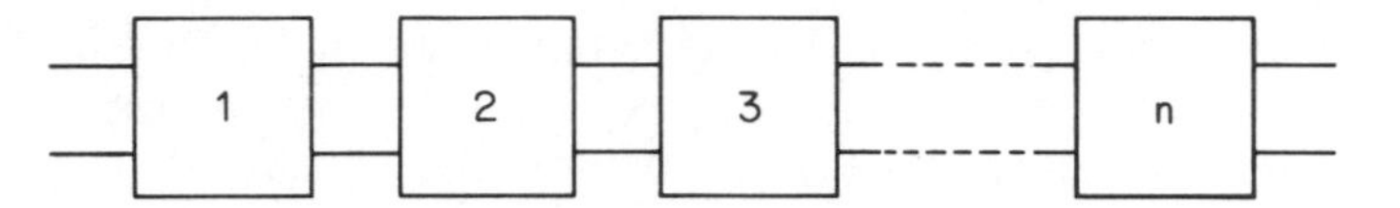

Fig. 9-5.5 A cascade of n two-port networks.

mission parameters are found by multiplying the transmission matrices of the individual component networks in the proper order. Thus, if $\mathbf{A}^{(i)}(s)$ $(i = 1, 2, \ldots, n)$ is the transmission parameter matrix for the ith component network, we may write

$$\mathbf{A}(s) = \mathbf{A}^{(1)}(s)\mathbf{A}^{(2)}(s)\mathbf{A}^{(3)}(s) \cdots \mathbf{A}^{(n)}(s) \tag{9–94}$$

where $\mathbf{A}(s)$ is the transmission parameter matrix for the overall two-port network. Note that in a cascade connection of two-port networks such as is shown in Fig. 9–5.5, we need not concern ourselves with whether or not the port current requirement is satisfied, since, as was shown in Sec. 9–1, this requirement is always satisfied for a cascade connection of networks. The results of Eqs. (9–82) and (9–92) provide a a simple procedure for finding the *ABCD* parameters of a ladder network. An example follows.

EXAMPLE 9-5.1. The ladder network shown in Fig. 9-5.6 may be considered as a ladder network consisting of two series branches of impedance 1 and $4s$, and two shunt branches consisting of admittances $(6s^2 + 1)/2s$ and 1. Thus, using Eqs. (9-82) and (9-92), the transmission parameters for the entire network may be found by evaluating the matrix product

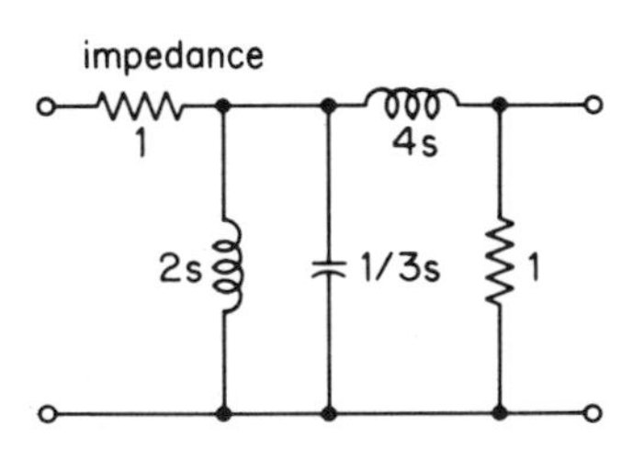

Fig. 9-5.6 A ladder network.

$$\begin{bmatrix} 1 & 1 \\ 0 & 1 \end{bmatrix}\begin{bmatrix} 1 & 0 \\ \dfrac{6s^2+1}{2s} & 1 \end{bmatrix}\begin{bmatrix} 1 & 4s \\ 0 & 1 \end{bmatrix}\begin{bmatrix} 1 & 0 \\ 1 & 1 \end{bmatrix}$$

Performing the matrix multiplication, we obtain

$$\begin{bmatrix} \dfrac{24s^3 + 14s^2 + 8s + 1}{2s} & \dfrac{24s^3 + 8s^2 + 6s}{2s} \\ \dfrac{24s^3 + 6s^2 + 6s + 1}{2s} & \dfrac{24s^3 + 6s}{2s} \end{bmatrix}$$

The results given above may be summarized as follows:

SUMMARY 9-5.2

Properties of the ABCD Parameters of a Two-Port Network: The four *ABCD* parameters of a given two-port network completely determine the properties of the network in the sense that they may be used to find any other network function that is defined for the two-port network. If a set of two-port networks are connected in cascade, the product of their individual *ABCD* parameter matrices, in the same order from left to right in which the component networks are cascaded, gives the *ABCD* parameters for the resulting overall two-port network.

The transmission parameters of a two-port network may be modeled by a network consisting of three controlled sources and one impedance as shown in Fig. 9-5.7. Applying KVL at port 1 and KCL at port 2, it is readily shown that this network satisfies the relations of (9-82). As an example of the use of this model, consider the ideal transformer shown in Fig. 9-5.8(a). It is easily shown that the *ABCD* parameters for this two-port network are

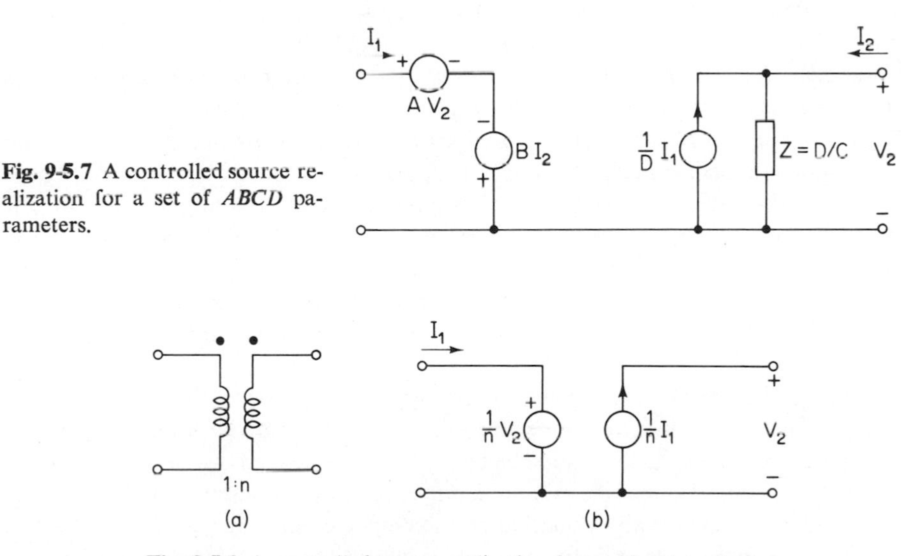

Fig. 9-5.7 A controlled source realization for a set of *ABCD* parameters.

Fig. 9-5.8 A controlled source realization for an ideal transformer.

$$\begin{bmatrix} \dfrac{1}{n} & 0 \\ 0 & n \end{bmatrix} \tag{9-95}$$

Thus, the ideal transformer may be modeled with two controlled sources as shown in Fig. 9-5.8(b).

The last remaining set of network parameters for a two-port network to be considered is defined as the sixth case in Table 9-1.1. This parameter set is usually referred to as the $\mathscr{ABCD}$ *parameters* (read "script" ABCD parameters). It is defined by the matrix equation

$$\begin{bmatrix} V_2(s) \\ I_2(s) \end{bmatrix} = \begin{bmatrix} \mathscr{A}(s) & \mathscr{B}(s) \\ \mathscr{C}(s) & \mathscr{D}(s) \end{bmatrix} \begin{bmatrix} V_1(s) \\ -I_1(s) \end{bmatrix} \tag{9-96}$$

where the square matrix is called the $\mathscr{ABCD}$ *parameter matrix* or the *inverse transmission parameter matrix.* Its characteristics differ only slightly from those of the *ABCD* parameters. As a result, it is used very infrequently.

9-6 APPLICATIONS OF TWO-PORT NETWORK PARAMETERS

In the preceding sections of this chapter we have defined various sets of network parameters and illustrated how they may be applied to characterize a wide range of two-port networks. Additional applications of these parameters can be made in many typical circuit situations. One such application frequently encountered is the use of a two-port network to serve as a signal transmitting device. As such, it is usually excited by some non-ideal signal source, and transmits the signal to some load impedance. In addition to transmitting the signal, the two-port network is frequently called upon to provide isolation between the excitation and the load, as well as to amplify, filter, or modify the signal in some way. The basic configuration is shown in Fig 9-6.1. In this figure, a box is used to represent the two-port network. The load and source impedances $Z_L(s)$ and $Z_S(s)$ are sometimes called terminations. There are many network functions which are of interest for such a configuration. In this section we shall develop specific equations for several of these. In determining the network functions, we shall assume that the two-port network is specified in terms of one of the parameter sets defined in the previous sections. Expressions for the desired network function in terms of other parameter sets are readily obtained using the relations given in Table 9-7.1.

The first network function for the configuration shown in Fig. 9-6.1 that we shall determine is $Z_i(s)$, i.e., the impedance seen looking into the input port of the two-port network when a load impedance $Z_L(s)$ is connected to the output port. Such a network function specifies the impedance seen by the source which drives the network. Thus, it permits an evaluation of the degree of impedance match which is obtained at the input port and thus (as was discussed in Sec. 7-11) an evaluation of

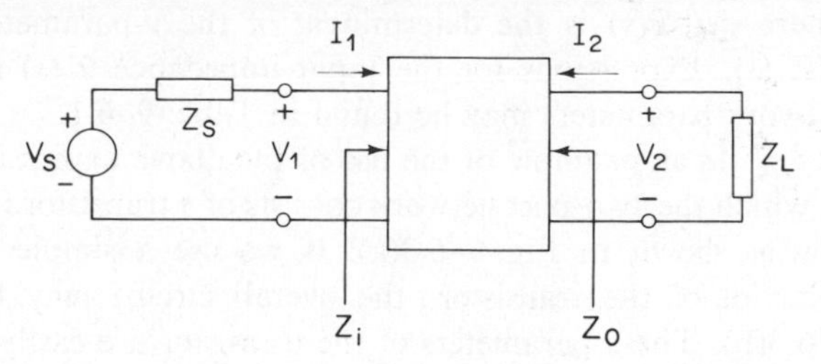

Fig. 9-6.1 A two-port network with terminating impedances.

the efficiency of the power transfer. In defining the input admittance $Z_i(s)$, it is assumed that the effect of the source impedance $Z_S(s)$ is not considered. Thus, the actual test to determine $Z_i(s)$ is as shown in Fig. 9-6.2. Comparing this circuit with the one shown in Fig. 9-6.1, we see that the arrow used in Fig. 9-6.1 to define $Z_i(s)$ may be interpreted as implying that all circuit components behind the arrowhead are ignored.

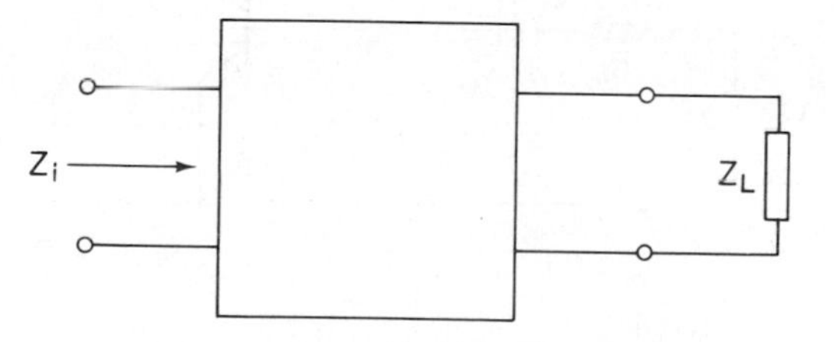

Fig. 9-6.2 Definition of Z_i.

To determine $Z_i(s)$, we first note that the presence of the load impedance $Z_L(s)$ requires that the transformed variables $V_2(s)$ and $I_2(s)$ satisfy the relation

$$V_2(s) = -Z_L(s)I_2(s) \tag{9-97}$$

Now consider the first equation defining the y parameters of a two-port network. This is

$$I_1(s) = y_{11}(s)V_1(s) + y_{12}(s)V_2(s) \tag{9-98}$$

If we substitute (9-97) in (9-98), we may solve for the relation between $I_1(s)$ and $V_1(s)$ which gives us the reciprocal of the input impedance. Thus, we obtain

$$\frac{1}{Z_i(s)} = \frac{I_1(s)}{V_1(s)} = y_{11}(s) - \frac{y_{12}(s)y_{21}(s)}{y_{22}(s) + [1/Z_L(s)]} \tag{9-99}$$

Inverting this relation, we obtain the desired expression for the input impedance in terms of the y parameters of the two-port network

$$Z_i(s) = \frac{y_{22}(s) + Y_L(s)}{\det \mathbf{Y}(s) + y_{11}(s)Y_L(s)} \tag{9-100}$$

where det $\mathbf{Y}(s)$ is the determinant of the y-parameter matrix and where $Y_L(s) = 1/Z_L(s)$. Expressions for the input impedance $Z_i(s)$ in terms of the other two-port network parameters may be found in Table 9–6.1.

As an example of the use of the above expression for $Z_i(s)$, consider a circuit in which the two-port network consists of a transistor in a common emitter configuration as shown in Fig. 9–6.3(a). If we use a simple **T** model for the small-signal behavior of the transistor, the overall circuit may be redrawn as shown in Fig. 9–6.3(b). The z parameters of the transistor are easily determined to be

$$\begin{bmatrix} r_e + r_b & r_e \\ r_e - \alpha r_c & r_e + r_c(1 - \alpha) \end{bmatrix} \tag{9–101}$$

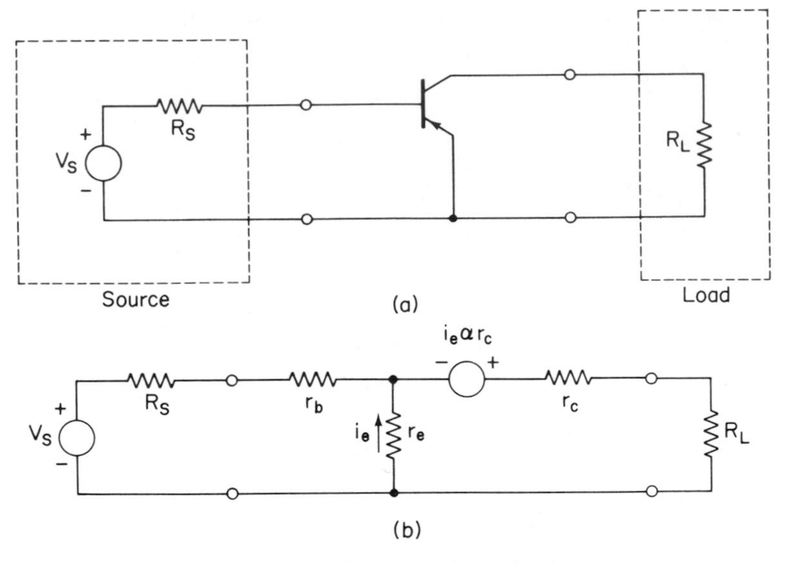

Fig. 9-6.3 A simple transistor circuit.

Substituting the values $r_e = 25\ \Omega$, $r_b = 100\ \Omega$, $r_c = 10^6\ \Omega$, and $\alpha = 0.99$, the approximate z parameters become

$$\begin{bmatrix} 125 & 25 \\ -10^6 & 10^4 \end{bmatrix} \tag{9–102}$$

Applying the relations given in Table 9–6.1, we readily find that the input impedance $Z_i(s)$ with a load impedance $Z_L(s)$ of 1000 Ω is 2390 Ω. This is the impedance with which the non-ideal source is loaded.

The second network function which we shall consider is $Z_0(s)$, i.e., the impedance seen looking into the output terminals of a two-port network when a source which has an internal impedance $Z_S(s)$ is connected to the input port. Such a network

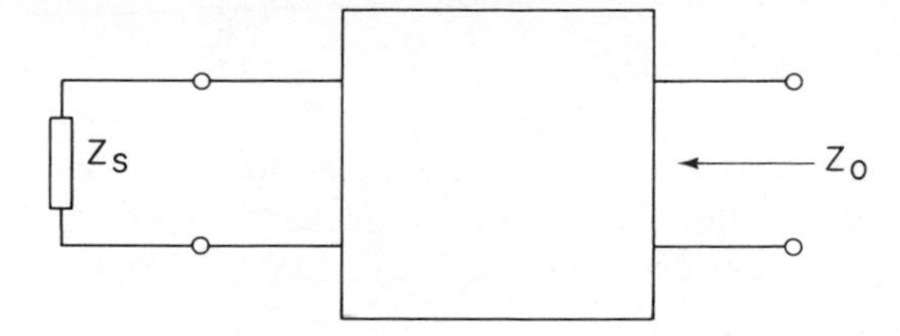

Fig. 9-6.4 Definition of Z_0.

function gives us the impedance seen by a circuit connected to the output port. Thus, it permits us to evaluate the impedance match at the output port. The test for determining $Z_0(s)$ is shown in Fig. 9–6.4. In this figure it should be noted that the source voltage $V_S(s)$ has been set to zero. This is the test condition implied by the arrow defining $Z_0(s)$ shown in Fig. 9–6.1. In the determination of the network function $Z_0(s)$, we note from Fig. 9–6.4 that the transformed input variables $V_1(s)$ and $I_1(s)$ are related as follows:

$$V_1(s) = -Z_S(s)I_1(s) \tag{9–103}$$

Now let us consider the second equation defining the y parameters of a two-port network. This is

$$I_2(s) = y_{21}(s)V_1(s) + y_{22}(s)V_2(s) \tag{9–104}$$

Substituting (9–103) into (9–104) and solving for the ratio of $I_2(s)$ and $V_2(s)$, we obtain

$$\frac{1}{Z_0(s)} = \frac{I_2(s)}{V_2(s)} = y_{22}(s) - \frac{y_{12}(s)y_{21}(s)}{y_{11}(s) + [1/Z_S(s)]} \tag{9–105}$$

The reciprocal of this relation gives us the expression for the desired network function

$$Z_0(s) = \frac{y_{11}(s) + Y_S(s)}{\det \mathbf{Y}(s) + y_{22}(s)Y_S(s)} \tag{9–106}$$

where $Y_S(s) = 1/Z_S(s)$. Expressions for the output impedance $Z_0(s)$ in terms of other sets of two-port network parameters are readily obtained and are summarized in Table 9–6.1. As an example of the determination of $Z_0(s)$, let us consider again the circuit shown in Fig. 9–6.3(b). Using the z parameters for the transistor given in (9–102) and the expression given in Table 9–6.1, we find that the output impedance $Z_0(s)$ for a source impedance $Z_S(s)$ of 600 Ω is 44.5 kΩ. Thus any network connected to the output terminals of this circuit will be driven by an impedance of 44.5 kΩ.

The third network function that we shall consider is the current transfer function $A_i(s)$ for the terminated circuit. This network function is defined in terms of the variables shown Fig. 9–6.1 as

$$A_i(s) = \frac{-I_2(s)}{I_1(s)} \tag{9-107}$$

It should be noted that, for convenience, we define the output current as $-I_2(s)$ where $I_2(s)$ is the conventional two-port output current variable. To determine this network function, let us consider the second equation defining the z parameters for a two-port network. This is

$$V_2(s) = z_{11}(s)I_1(s) + z_{12}I_2(s) \tag{9-108}$$

Inserting the relation of Eq. (9–97) and rearranging terms, we obtain

$$A_i(s) = \frac{z_{21}(s)}{z_{22}(s) + Z_L(s)} \tag{9-109}$$

This is the desired network function for the current gain of the terminated two-port network. Alternative expressions for $A_i(s)$ in terms of other sets of two-port network parameters are given in Table 9–6.1. As an example of the determination of $A_i(s)$, consider again the circuit shown in Fig. 9–6.3(b). Using the z parameters given in Eq. (9–102) and a load impedance $Z_L(s)$ of 1000 Ω, we find $A_i(s) = -190$.

The final network function that we shall consider in this section is the voltage transfer function $A_v(s)$. This network function is defined in terms of the variables shown in Fig. 9–6.1 as

$$A_v(s) = \frac{V_2(s)}{V_1(s)} \tag{9-110}$$

To determine this network function, consider the second equation for the y parameters of a two-port network. This is

$$I_2(s) = y_{21}(s)V_1(s) + y_{22}(s)V_2(s) \tag{9-111}$$

Inserting the relation of Eq. (9–97) and rearranging terms, we obtain

$$A_v(s) = \frac{-y_{21}(s)}{y_{22}(s) + Y_L(s)} \tag{9-112}$$

where $Y_L(s) = 1/Z_L(s)$. This is the desired network function for the voltage gain of the terminated network. Alternative expressions for $A_v(s)$ in terms of other sets of two-port network parameters are given in Table 9–6.1. As an example of the determination of this network function, consider again the circuit shown in Fig. 9–6.3(b). Using the z parameters given in Eq. (9–102) and the expression for $A_v(s)$ in terms of the z parameters given in Table 9–6.1, and assuming a load impedance $Z_L(s)$ of 1000 Ω, we find $A_v(s) = -38$.

TABLE 9-6.1 Two-Port Network Properties

	Z	**Y**	**G**	**H**	**A**	$\mathcal{A}$
Z_i	$\dfrac{\det \mathbf{Z} + z_{11}Z_L}{z_{22} + Z_L}$	$\dfrac{y_{22} + Y_L}{\det \mathbf{Y} + y_{11}Y_L}$	$\dfrac{g_{22} + Z_L}{\det \mathbf{G} + g_{11}Z_L}$	$\dfrac{\det \mathbf{H} + h_{11}Y_L}{h_{22} + Y_L}$	$\dfrac{AZ_L + B}{CZ_L + D}$	$\dfrac{\mathcal{D}Z_L + \mathcal{B}}{\mathcal{C}Z_L + \mathcal{A}}$
Z_0	$\dfrac{\det \mathbf{Z} + z_{22}Z_S}{z_{11} + Z_S}$	$\dfrac{y_{11} + Y_S}{\det \mathbf{Y} + y_{22}Y_S}$	$\dfrac{\det \mathbf{G} + g_{22}Y_S}{g_{11} + Y_S}$	$\dfrac{h_{11} + Z_S}{\det \mathbf{H} + h_{22}Z_S}$	$\dfrac{DZ_S + B}{CZ_S + A}$	$\dfrac{\mathcal{A}Z_S + \mathcal{B}}{\mathcal{C}Z_S + \mathcal{D}}$
$A_i = -\dfrac{I_2}{I_1}$	$\dfrac{z_{21}}{z_{22} + Z_L}$	$\dfrac{-y_{21}Y_L}{\det \mathbf{Y} + y_{11}Y_L}$	$\dfrac{g_{21}}{\det \mathbf{G} + g_{11}Z_L}$	$\dfrac{-h_{21}Y_L}{h_{22} + Y_L}$	$\dfrac{1}{D + CZ_L}$	$\dfrac{\det \mathcal{A}}{\mathcal{A} + \mathcal{C}Z_L}$
$A_v = \dfrac{V_2}{V_1}$	$\dfrac{z_{21}Z_L}{\det \mathbf{Z} + z_{11}Z_L}$	$\dfrac{-y_{21}}{y_{zz} + Y_L}$	$\dfrac{g_{21}Z_L}{g_{22} + Z_L}$	$\dfrac{-h_{21}}{\det \mathbf{H} + h_{11}Y_L}$	$\dfrac{Z_L}{B + AZ_L}$	$\dfrac{\det \mathcal{A}}{\mathcal{B} + \mathcal{D}Z_L}$

9-7 CONCLUSION

When the Laplace transformation was introduced in Chap. 8, we employed the concept of a transformed network and the network functions describing it as a tool for the analysis of lumped linear time-invariant circuits. In this chapter we have seen an additional application of such an approach, namely, the use of a set of four parameters, each of which is a network function, to characterize a given two-port network. Such a characterization provides a convenient technique for decomposing a given system into simpler subcomponents as an aid to its analysis. Such a characterization also provides a means of defining the properties of a network in a manner which is independent of the planned use of the network or the type of termination which may be connected to it. Network parameters are widely used to define the properties of transistors and other electronic devices, and a set of such parameters will usually be found on the manufacturer's specification sheets for such devices.

TABLE 9-7.1 Two-Port

	Z		**Y**		**G**	
Z	z_{11}	z_{12}	$\frac{y_{22}}{\det \mathbf{Y}}$	$-\frac{y_{12}}{\det \mathbf{Y}}$	$\frac{1}{g_{11}}$	$-\frac{g_{12}}{g_{11}}$
	z_{21}	z_{22}	$-\frac{y_{21}}{\det \mathbf{Y}}$	$\frac{y_{11}}{\det \mathbf{Y}}$	$\frac{g_{21}}{g_{11}}$	$\frac{\det \mathbf{G}}{g_{11}}$
Y	$\frac{z_{22}}{\det \mathbf{Z}}$	$-\frac{z_{12}}{\det \mathbf{Z}}$	y_{11}	y_{12}	$\frac{\det \mathbf{G}}{g_{22}}$	$\frac{g_{12}}{g_{22}}$
	$-\frac{z_{21}}{\det \mathbf{Z}}$	$\frac{z_{11}}{\det \mathbf{Z}}$	y_{21}	y_{22}	$-\frac{g_{21}}{g_{22}}$	$\frac{1}{g_{22}}$
G	$\frac{1}{z_{11}}$	$-\frac{z_{12}}{z_{11}}$	$\frac{\det \mathbf{Y}}{y_{22}}$	$\frac{y_{12}}{y_{22}}$	g_{11}	g_{12}
	$\frac{z_{21}}{z_{11}}$	$\frac{\det \mathbf{Z}}{z_{11}}$	$-\frac{y_{21}}{y_{22}}$	$\frac{1}{y_{22}}$	g_{21}	g_{22}
H	$\frac{\det \mathbf{Z}}{z_{22}}$	$\frac{z_{12}}{z_{22}}$	$\frac{1}{y_{11}}$	$-\frac{y_{12}}{y_{11}}$	$\frac{g_{22}}{\det \mathbf{G}}$	$-\frac{g_{12}}{\det \mathbf{G}}$
	$-\frac{z_{21}}{z_{22}}$	$\frac{1}{z_{22}}$	$\frac{y_{21}}{y_{11}}$	$\frac{\det \mathbf{Y}}{y_{11}}$	$-\frac{g_{21}}{\det \mathbf{G}}$	$\frac{g_{11}}{\det \mathbf{G}}$
A	$\frac{z_{11}}{z_{21}}$	$\frac{\det \mathbf{Z}}{z_{21}}$	$-\frac{y_{22}}{y_{21}}$	$-\frac{1}{y_{21}}$	$\frac{1}{g_{21}}$	$\frac{g_{22}}{g_{21}}$
	$\frac{1}{z_{21}}$	$\frac{z_{22}}{z_{21}}$	$-\frac{\det \mathbf{Y}}{y_{21}}$	$-\frac{y_{11}}{y_{21}}$	$\frac{g_{11}}{g_{21}}$	$\frac{\det \mathbf{G}}{g_{21}}$
$\mathscr{A}$	$\frac{z_{22}}{z_{12}}$	$\frac{\det \mathbf{Z}}{z_{12}}$	$-\frac{y_{11}}{y_{12}}$	$\frac{1}{y_{12}}$	$-\frac{\det \mathbf{G}}{g_{12}}$	$-\frac{g_{22}}{g_{12}}$
	$\frac{1}{z_{12}}$	$\frac{z_{11}}{z_{12}}$	$-\frac{\det \mathbf{Y}}{y_{12}}$	$\frac{y_{22}}{y_{12}}$	$-\frac{g_{11}}{g_{12}}$	$-\frac{1}{g_{12}}$

Parameter Relations

$\mathbf{H}$		$\mathbf{A}$		$\mathcal{A}$	
$\frac{\det \mathbf{H}}{h_{22}}$	$\frac{h_{12}}{h_{22}}$	$\frac{A}{C}$	$\frac{\det \mathbf{A}}{C}$	$\frac{\mathcal{D}}{\mathcal{C}}$	$\frac{1}{\mathcal{C}}$
$-\frac{h_{21}}{h_{22}}$	$\frac{1}{h_{22}}$	$\frac{1}{C}$	$\frac{D}{C}$	$\frac{\det \mathcal{A}}{\mathcal{C}}$	$\frac{\mathcal{A}}{\mathcal{C}}$
$\frac{1}{h_{11}}$	$-\frac{h_{12}}{h_{11}}$	$\frac{D}{B}$	$-\frac{\det \mathbf{A}}{B}$	$\frac{\mathcal{A}}{\mathcal{B}}$	$-\frac{1}{\mathcal{B}}$
$\frac{h_{21}}{h_{11}}$	$\frac{\det \mathbf{H}}{h_{11}}$	$-\frac{1}{B}$	$\frac{A}{B}$	$-\frac{\det \mathcal{A}}{\mathcal{B}}$	$\frac{\mathcal{D}}{\mathcal{B}}$
$\frac{h_{22}}{\det \mathbf{H}}$	$-\frac{h_{12}}{\det \mathbf{H}}$	$\frac{C}{A}$	$-\frac{\det \mathbf{A}}{A}$	$\frac{\mathcal{C}}{\mathcal{D}}$	$-\frac{1}{\mathcal{D}}$
$-\frac{h_{21}}{\det \mathbf{H}}$	$\frac{h_{11}}{\det \mathbf{H}}$	$\frac{1}{A}$	$\frac{B}{A}$	$\frac{\det \mathcal{A}}{\mathcal{D}}$	$\frac{\mathcal{B}}{\mathcal{D}}$
h_{11}	h_{12}	$\frac{B}{D}$	$\frac{\det \mathbf{A}}{D}$	$\frac{\mathcal{B}}{\mathcal{A}}$	$\frac{1}{\mathcal{A}}$
h_{21}	h_{22}	$-\frac{1}{D}$	$\frac{C}{D}$	$\frac{\det \mathcal{A}}{\mathcal{A}}$	$\frac{\mathcal{C}}{\mathcal{A}}$
$-\frac{\det \mathbf{H}}{h_{21}}$	$-\frac{h_{11}}{h_{21}}$	A	B	$\frac{\mathcal{D}}{\det \mathcal{A}}$	$\frac{\mathcal{B}}{\det \mathcal{A}}$
$-\frac{h_{22}}{h_{21}}$	$-\frac{1}{h_{21}}$	C	D	$\frac{\mathcal{C}}{\det \mathcal{A}}$	$\frac{\mathcal{A}}{\det \mathcal{A}}$
$\frac{1}{h_{12}}$	$\frac{h_{11}}{h_{12}}$	$\frac{D}{\det \mathbf{A}}$	$\frac{B}{\det \mathbf{A}}$	$\mathcal{A}$	$\mathcal{B}$
$\frac{h_{22}}{h_{12}}$	$\frac{\det \mathbf{H}}{h_{12}}$	$\frac{C}{\det \mathbf{A}}$	$\frac{A}{\det \mathbf{A}}$	$\mathcal{C}$	$\mathcal{D}$

Problems

Problem 9-1 (*Sec. 9-2*)

Find the z parameters for each of the two-port networks shown in Fig. P9-1.

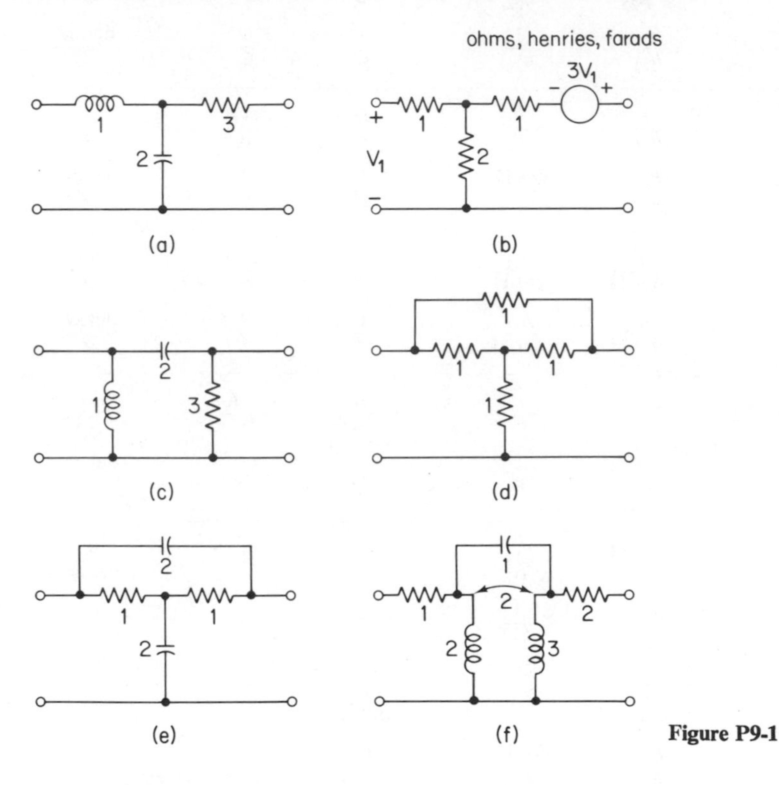

Figure P9-1

Problem 9-2 (*Sec. 9-2*)

Find the z parameters of the two-port network shown in Fig. P9-2.

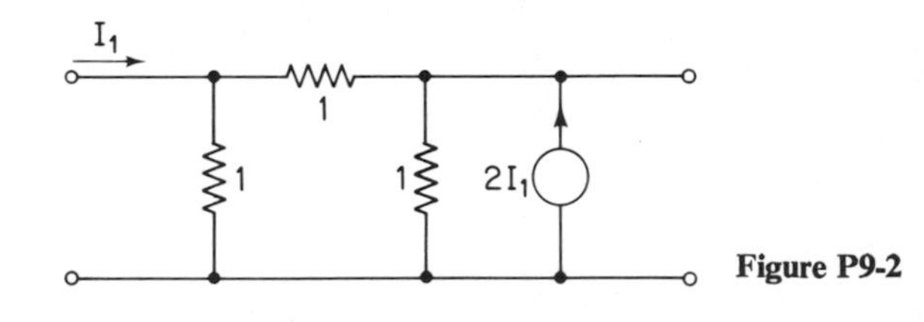

Figure P9-2

Problem 9-3 (*Sec. 9-2*)

Find the z parameters for the networks shown in Fig. P9-3 by (a) applying the usual testing conditions and (b) separating the network into two component subnetworks connected in series and finding the z parameters of each subnetwork separately.

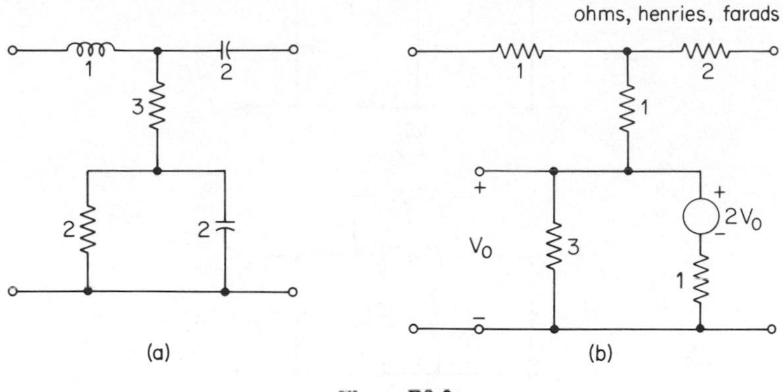

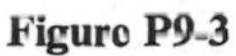
Figure P9-3

Problem 9–4 (Sec. 9–2)

Find a network which realizes each of the following sets of z parameters:

$$\text{(a)}\ \begin{bmatrix} 3 & 1 \\ 1 & 2 \end{bmatrix};\quad \text{(b)}\ \begin{bmatrix} 1+\dfrac{4}{s} & \dfrac{2}{s} \\ \dfrac{2}{s} & 3+\dfrac{2}{s} \end{bmatrix};\quad \text{(c)}\ \begin{bmatrix} 3 & 2 \\ -4 & 4 \end{bmatrix}.$$

Problem 9–5 (Sec. 9–3)

Find the y parameters for each of the two-port networks shown in Fig. P9-5.

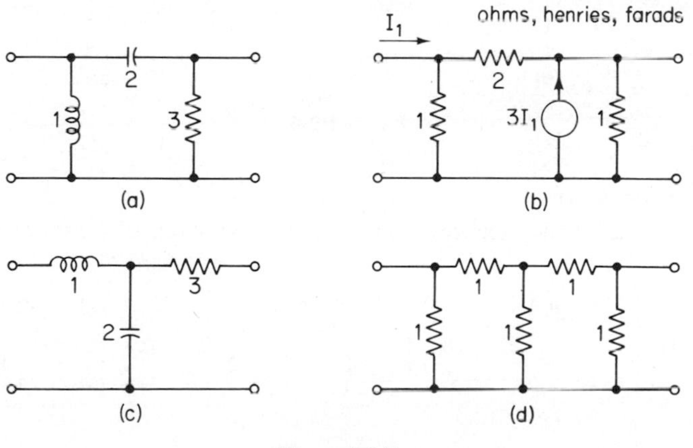

Figure P9-5

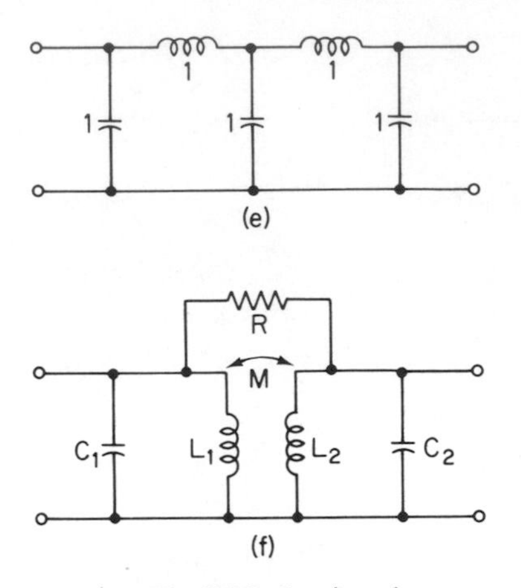

Fig. P9-5 Continued

Problem 9–6 (Sec. 9–3)

Find the y parameters for each of the networks shown in Fig. P9-6 by (a) applying the usual test conditions and (b) separating the network into two component subnetworks connected in parallel and finding the y parameters of each subnetwork separately.

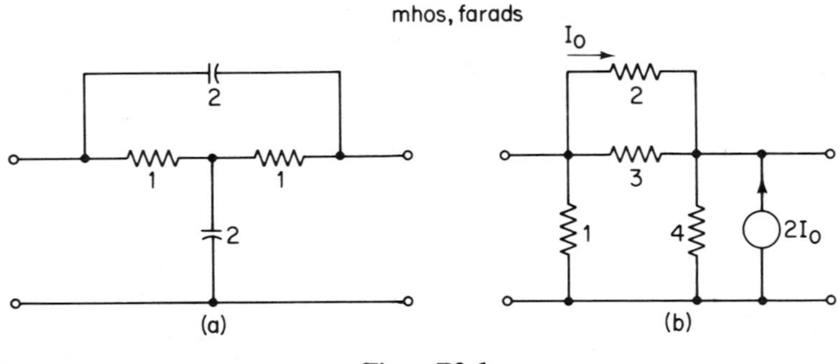

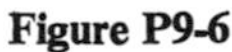

Figure P9-6

Problem 9–7 (Sec. 9–3)

Find a network which realizes each of the following sets of y parameters:

$$\text{(a)}\begin{bmatrix} 3 & -1 \\ -1 & 2 \end{bmatrix}; \quad \text{(b)}\begin{bmatrix} 1+\dfrac{4}{s} & -\dfrac{2}{s} \\ -\dfrac{2}{s} & 3+\dfrac{2}{s} \end{bmatrix}; \quad \text{(c)}\begin{bmatrix} 3 & -2 \\ 4 & 4 \end{bmatrix}.$$

Problem 9–8 (Sec. 9–3)

Show that the product of the z-parameter matrix for the network shown in Fig. P9-1(a) and the y-parameter matrix for the network shown in Fig. P9-5(c) yields the identity matrix.

Problem 9–9 (Sec. 9–3)

Show that the product of the z-parameter matrix for the network shown in Fig. P9-1(c) and the y-parameter matrix for the network shown in Fig. P9-5(a) gives the identity matrix.

Problem 9–10 (Sec. 9–3)

To find the y parameters of the network shown in Fig. P9-10(a), a student decides to find the y parameters of the two networks shown in parts (b) and (c) of the figure separately and add the result. Is his reasoning correct? Test your conclusion by performing the analysis.

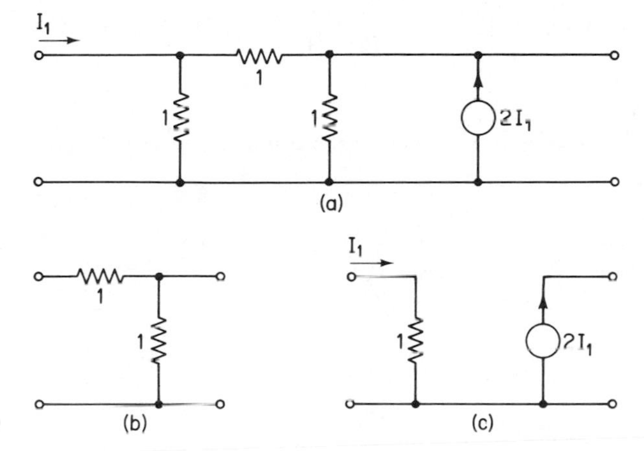

Figure P9-10

Problem 9–11 (Sec. 9–4)

Find the g parameters for each of the two-port networks shown in Fig. P9-11.

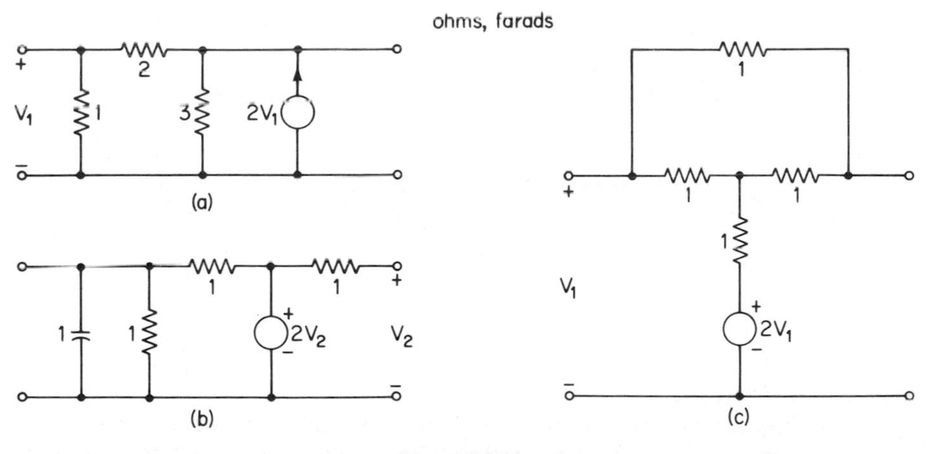

Figure P9-11

Problem 9–12 (Sec. 9–4)

Show an interconnection of two two-port networks such that their g parameters add. Use blocks to represent the two-port networks and assume that the port condition is not violated.

Problem 9–13 (Sec. 9–4)

Find the h parameters for each of the two-port networks shown in Fig. P9-11.

Problem 9–14 (Sec. 9–4)

Show an interconnection of two two-port networks such that their h parameters add. Use blocks to represent the two-port networks and assume that the port condition is not violated.

Problem 9–15 (Sec. 9–4)

Find a network which realizes the following g parameters.

$$\begin{bmatrix} s+4 & 0 \\ 10 & 4 \end{bmatrix}$$

Problem 9–16 (Sec. 9–5)

Find the transmission parameters for each of the two-port networks shown in Fig. P9-16.

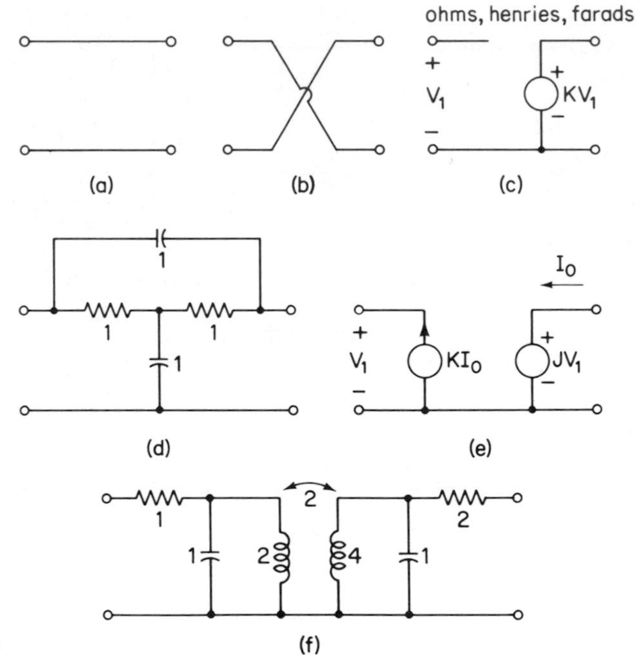

Figure P9-16

Problem 9–17 (Sec. 9–5)

Find the y parameters of the network shown in Fig. P9-17 by first finding the transmission parameters of the transformer and the series resistors, converting these into y parameters, and then adding the y parameters of the capacitor. The transformer is assumed to be ideal. Test the result by finding the y parameters directly using the usual test conditions.

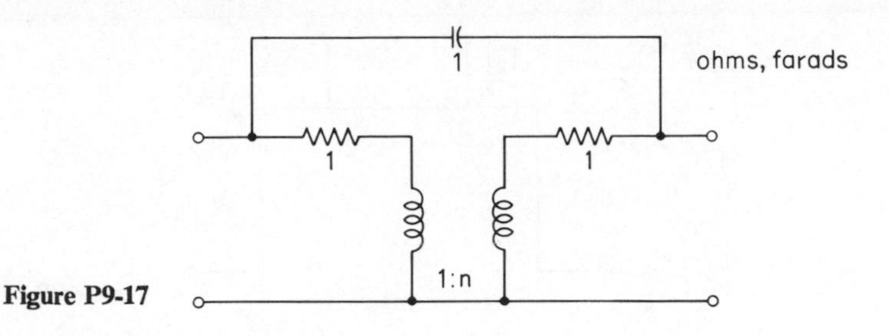

Figure P9-17

Problem 9–18 (*Sec. 9–5*)

Find the z parameters of the network shown in Fig. P9-18 by first finding the transmission parameters of each element, then using these to find the overall transmission parameters of the cascade of elements, and finally converting the result to z parameters. Check your answer by directly applying the usual tests to determine the z parameters of the network.

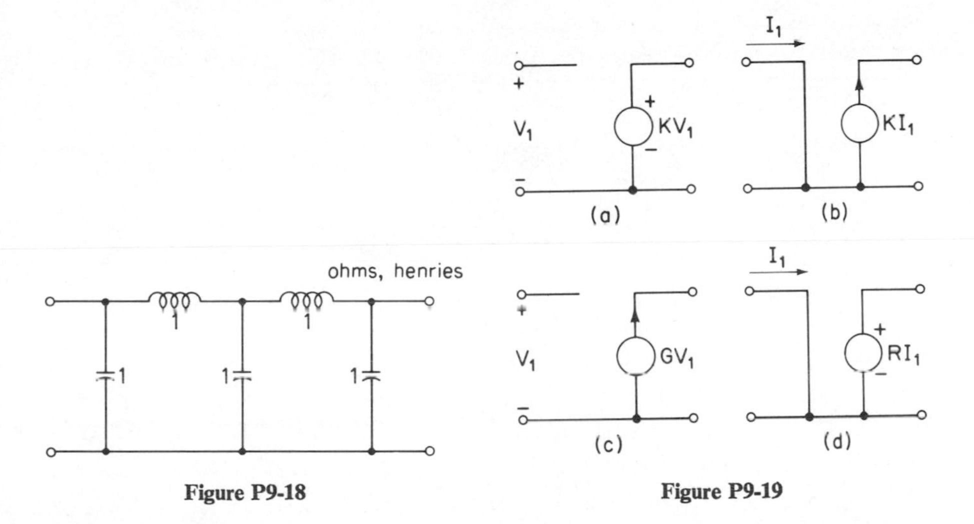

Figure P9-18

Figure P9-19

Problem 9–19 (*Sec. 9–5*)

Find all possible sets of parameters for each of the active devices shown in Fig. P9-19.

Problem 9–20 (*Sec. 9–5*)

Two three-port networks with variables defined as shown in Fig. P9-20 are each specified by a set of parameters having the variables indicated below. Show an interconnection of the ports of the two networks such that their individual parameters add to form the parameters of the overall three-port network.

$$\begin{bmatrix} V_1 \\ I_2 \\ V_3 \end{bmatrix} = [\text{Set of parameters}] \begin{bmatrix} I_1 \\ V_2 \\ I_3 \end{bmatrix}$$

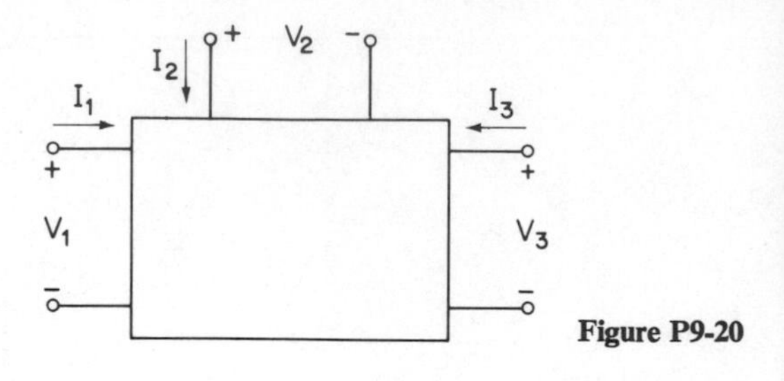

Figure P9-20

Problem 9-21 (Sec. 9-6)

Find the network functions $Z_i(s)$, $Z_0(s)$, $A_i(s)$, and $A_v(s)$ for the circuit shown in Fig. 9-6.3(a), with the transistor in a common base configuration rather than a common emitter configuration. Use the transistor model shown in Fig. 9-6.3(b), and assume that $r_b = 100\ \Omega$, $r_e = 25\ \Omega$, $r_c = 10^6\ \Omega$, Z_S $100\ \Omega$, $\alpha = 0.99$ and $Z_L = 10^4\ \Omega$.

CHAPTER 10

The Fourier Series and the Fourier Integral

In Chap. 7 we discussed the determination of the steady-state response of networks which were excited only by sinusoidal functions. As part of our development, we introduced the concept of a network function and showed that the product of an input sinusoidal function (represented as a phasor) and the network function (a function of $j\omega$) gave us the output or response sinusoid (also represented as a phasor). A most important extension of this approach allows similar techniques to be applied to find the steady-state response of a network which is excited by periodic functions *which are not sinusoids*. Basically, the method consists of using a series of sinusoids to represent the non-sinusoidal periodic input and output waveforms. The basic concepts of such an approach can also be extended to cover non-periodic functions. In this context, the ideas of this chapter provide an important link between the sinusoidal steady-state material introduced in Chap. 7 and the concepts of a transformed network presented in Chap 8. The details of this connection will become apparent in the material which follows.

10-1 NON-SINUSOIDAL PERIODIC FUNCTIONS

In this section we shall introduce the concept of using a series of sinusoidal functions to represent a non-sinusoidal periodic function. First of all, let us review some terminology associated with a sinusoidal function. If we let $f(t)$ be a sinusoidal waveform, then we may write

$$f(t) = F_0 \sin(\omega_0 t + \phi) \tag{10-1}$$

where F_0 is the peak value or magnitude of the sinusoidal function, ω_0 is the frequency of oscillation in radians per second, and ϕ is the phase angle (or argument) in radians. Two other quantities of interest which are used in connection with such a sinusoidal function are f_0, the frequency of oscillation in hertz (where $f_0 = \omega_0/2\pi$), and T, the period in seconds (where $T = 1/f_0$). For purposes of sinusoidal steady-state analysis it is assumed that a function such as is given by Eq. (10-1) is defined over all t, i.e., over the interval $-\infty \leq t \leq \infty$. A sinusoidal function of this type has the property that

$$f(t) = f\left(t \pm \frac{2\pi}{\omega_0}\right) = f(t \pm T) \tag{10-2}$$

for all t.

Let us now see how a series of sinusoidal functions as described above can be utilized to describe non-sinusoidal periodic functions. Such non-sinusoidal functions are of considerable importance since they are frequently encountered in network applications. For example, the square wave shown in Fig. 10-1.1(a) is available as an

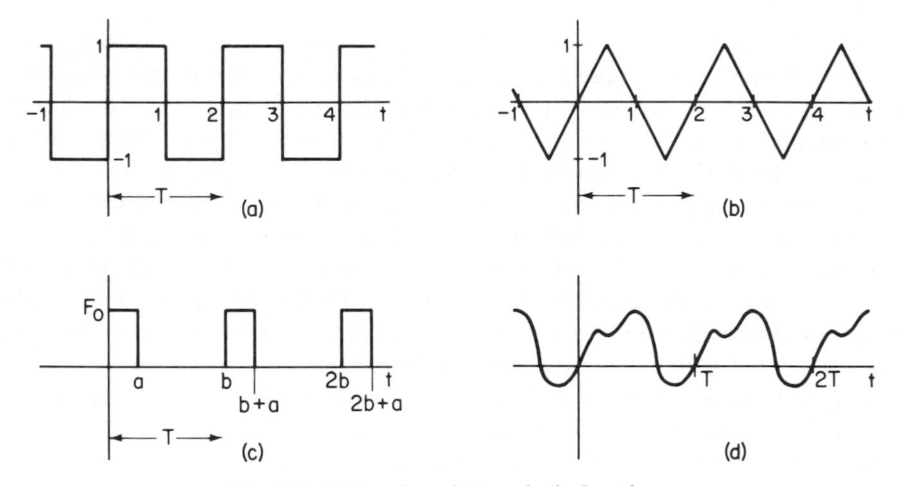

Fig. 10-1.1 Non-sinusoidal periodic functions.

output from most signal generators used to test electronic circuits. Other examples of non-sinusoidal periodic functions are the triangular wave shown in Fig. 10-1.1(b), the train of pulses shown in Fig. 10-1.1(c), and the irregular waveform shown in Fig. 10-1.1(d). In general, any function that satisfies the relation

$$f(t) = f(t + T) \tag{10-3}$$

for a given value of T, and for $-\infty \leq t \leq \infty$ is said to be a *periodic function* with a period of T seconds. To see how (10–3) applies, consider the square wave shown in Fig. 10–1.1(a). The period of this square wave is 2. Thus, if we let $f(t)$ designate this function and if we choose $t = 1.1$ as a "test" value of t, we see that $f(1.1) = -1$. From the figure we also see that $f(3.1) = -1$, $f(5.1) = -1$, etc., as required by (10–3).

To see how a square wave such as is shown in Fig. 10–1.1(a) may be treated as a series of sinusoids, let us start with a sinusoid $f_1(t)$ which has the same period, namely $T = 2$. Thus, $\omega_0 = 2\pi/T = \pi$. In addition, let $f_1(t)$ have a peak magnitude of $4/\pi$. Thus, we may write

$$f_1(t) = \frac{4}{\pi} \sin \pi t \tag{10–4}$$

This function is plotted in Fig. 10–1.2(a). It obviously doesn't look very much like a square wave. To this sinusoid let us now add a second sinusoid with a frequency three times that of $f_1(t)$ and a magnitude $\frac{1}{3}$ as great. In the next section we shall see how this second sinusoid was chosen. Let us call the resulting sum $f_2(t)$. It may be expressed as

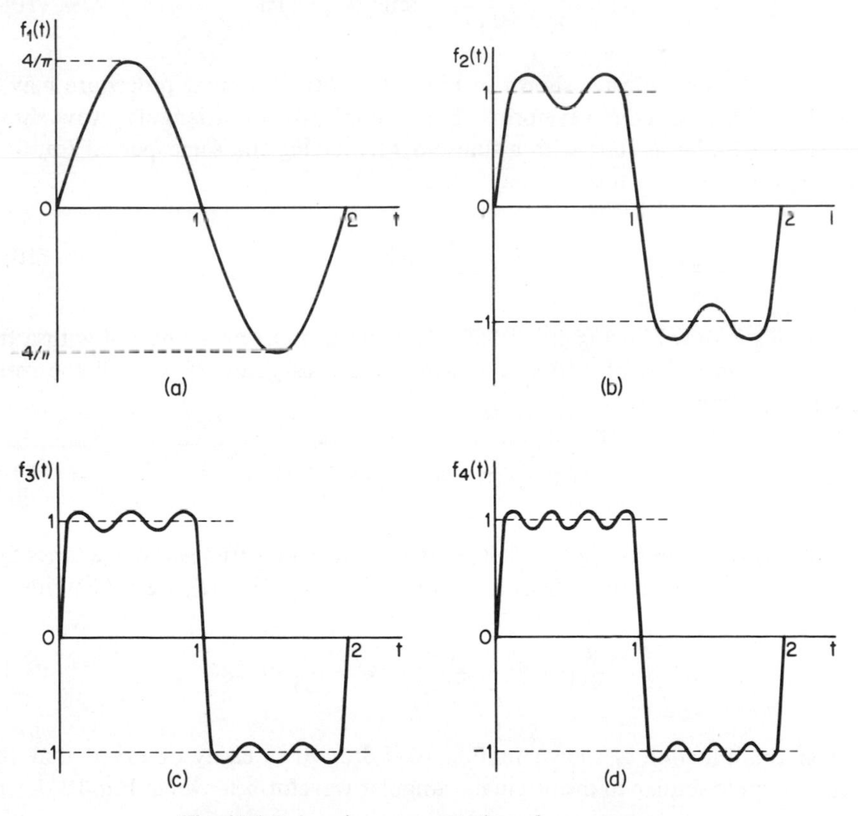

Fig. 10-1.2 A series representation of a square wave.

$$f_2(t) = \frac{4}{\pi}\left(\sin \pi t + \frac{1}{3} \sin 3\pi t\right) \tag{10-5}$$

A plot of this function is shown in Fig. 10–1.2(b). Now let us add a third sinusoid which has a frequency five times as great as $f_1(t)$ and a magnitude $\frac{1}{5}$ as large. If we call the result $f_3(t)$, we may write

$$f_3(t) = \frac{4}{\pi}\left(\sin \pi t + \frac{1}{3} \sin 3\pi t + \frac{1}{5} \sin 5\pi t\right) \tag{10-6}$$

A plot of this function is shown in Fig. 10–1.2(c). Continuing in this fashion by adding a sinusoid of frequency seven times as large as $f_1(t)$ and a magnitude $\frac{1}{7}$ as large, we obtain the function $f_4(t)$ shown in Fig. 10–1.2(d). This latter function obviously begins to approach the general appearance of a square wave. In the next section we shall show that in the limit as the number of sinusoidal terms becomes infinite, their sum generates a waveform identical with the one shown in Fig. 10–1.1(a). Thus, we may write

$$f(t) = \frac{4}{\pi} \sum_{k=1}^{\infty} \frac{1}{2k - 1} \sin (2k - 1)\pi t \tag{10-7}$$

where $f(t)$ is the square wave shown in Fig. 10–1.1(a). A similar procedure may be followed for other periodic waveforms. For example, for the triangular wave shown in Fig. 10–1.1(b), let us start with a sinusoid $f_1(t)$ having the same period ($\omega_0 = \pi$) and a magnitude $8/\pi^2$. Thus, we may write

$$f_1(t) = \frac{8}{\pi^2} \sin \pi t \tag{10-8}$$

A plot of this is shown in Fig. 10–1.3(a). Now let us *subtract* a sinusoid with a frequency three times that of $f_1(t)$ and a magnitude $\frac{1}{9}$ as great. If we call the result $f_2(t)$, we may write

$$f_2(t) = \frac{8}{\pi^2}\left(\sin \pi t - \frac{1}{9} \sin 3\pi t\right) \tag{10-9}$$

A plot of this is shown in Fig. 10–1.3(b). If we now add a frequency five times that of $f_1(t)$ and with a magnitude $\frac{1}{25}$ as large and call the result $f_3(t)$, we may write

$$f_3(t) = \frac{8}{\pi^2}\left(\sin \pi t - \frac{1}{9} \sin 3\pi t + \frac{1}{25} \sin 5\pi t\right) \tag{10-10}$$

A plot of this function is shown in Fig. 10–1.3(c). It is easily observed that this waveform is quite similar to the original triangular waveform shown in Fig. 10–1.1(b).

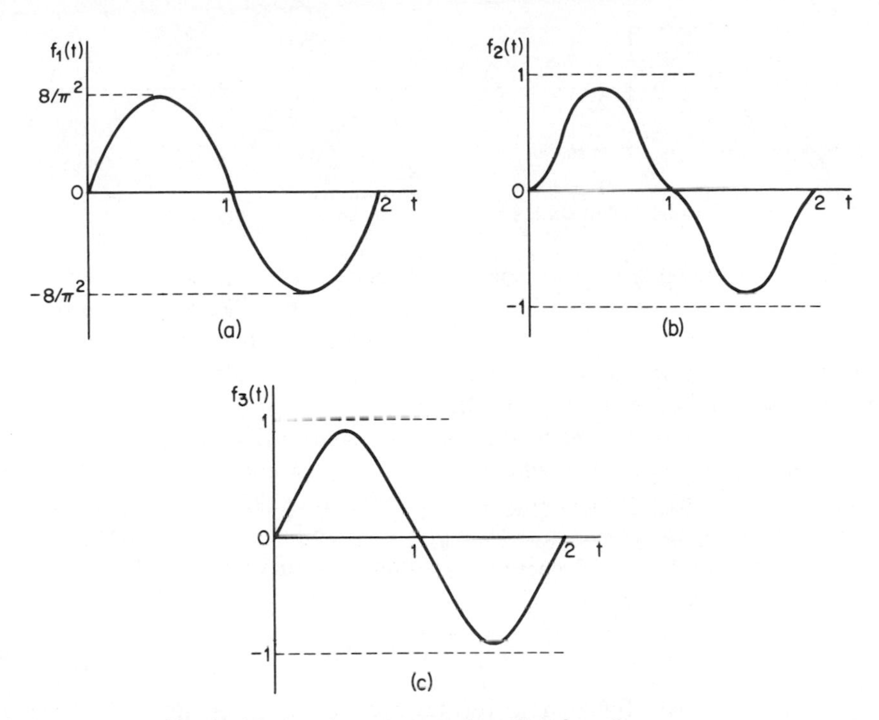

Fig. 10-1.3 A series representation of a triangular wave.

In the above discussion we have demonstrated that we can add and subtract sinusoidal functions in such a way as to produce functions whose waveforms approach those of non-sinusoidal periodic functions. In addition, we have shown that the more sinusoidal terms which we add, the closer the resemblance becomes. Thus, we have used a series representation of sinusoids to represent non-sinusoidal periodic functions. Such series are called *Fourier series*. In the next section we shall show how to find the elements of such series, i.e., how to find the magnitude, frequency, and phase of the component sinusoidal terms. Before proceeding to that task, however, we will determine some general properties of periodic functions.

The examples of using a series of sinusoidal functions to represent a non-sinusoidal periodic function which were given above all used sine functions in the resulting expressions. Other periodic non-sinusoidal functions may require the use of cosine functions. For example, consider the square wave shown in Fig. 10–1.4(a). We will see shortly that such a function may be represented by a Fourier series similar to (10–7) in which the sine functions are replaced by a cosine functions. Thus, for the function $f(t)$ shown in this figure, we may write

$$f(t) = \frac{4}{\pi}\left(\cos \pi t - \frac{1}{3}\cos 3\pi t + \frac{1}{5}\cos 5\pi t - \cdots\right) \tag{10–11}$$

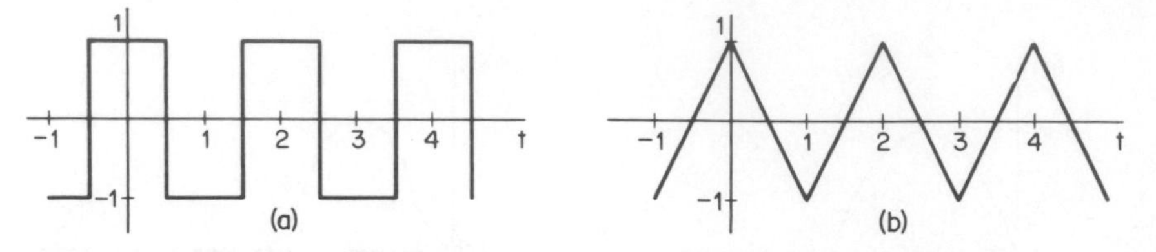

Fig. 10-1.4 Waveforms represented by series of cosine terms.

Similarly, a series which involves only cosine terms may be written for the triangular wave shown in Fig. 10–1.4(b). The above examples which use either sine or cosine functions are actually special cases of a more general class of functions which require the use of both sine and cosine functions for their representaton. These will be covered in more detail in the following section.

The functions expressed by the sine and cosine series given above are examples of two special types of functions which are frequently encountered in treating periodic functions. To emphasize their properties, we will define these types more fully in the following paragraphs. The definitions apply to all periodic functions (sinusoidal and non-sinusoidal) and have to do with the symmetry of the function. The first special type is defined by the relation

$$f(t) = f(-t) \tag{10–12}$$

where $f(t)$ is a periodic function. If this relation holds for all values of t, then $f(t)$ is said to be an *even function.* As an example of such a function, consider the square wave shown in Fig. 10–1.4(a). It is readily shown that this function satisfies Eq. (10–12). Similarly, the triangular wave shown in Fig. 10–1.4(b) is readily observed to be an even function. One of the most well-known even functions is the cosine function itself. We shall make reference to its even properties frequently in the sections that follow. In general, a function whose waveform is symmetrical with respect to the ordinate is an even function. Frequently, such a function is referred to as being "symmetrical about the origin." It should be noted that even functions may have a non-zero average value.

The second case that we shall be concerned with regarding the symmetry of periodic functions is defined by the relations

$$f(t) = -f(-t) \tag{10–13}$$

where $f(t)$ is a periodic function. If this relation holds for all t, then $f(t)$ is said to be an *odd function.* As an example of such a function, consider the square wave shown in Fig. 10–1.1(a). It is readily shown that this waveform satisfies Eq. (10–13) and, thus, is an odd function. Similarly, the triangular wave shown in Fig. 10–1.1(b) is an odd function. Another important odd function which we will use frequently in the sections that follow is the sine function. A function which satisfies the defining relation given in (10–13) is frequently said to be "skew-symmetric about the origin." Unlike even functions, the average value of an odd function is always zero.

We may now develop some properties which have to do with even and odd functions which will be of importance in the sections that follow. As a first property, let us consider the sum of a series of even functions $f_i^{(e)}(t)$ $(i = 1, 2, \ldots, n)$. If we let the sum of such a series of functions be $f(t)$, we may write

$$f(t) = \sum_{i=1}^{n} f_i^{(e)}(t) \tag{10-14}$$

Now let us replace the argument t by the argument $-t$ in the above equation. Since the functions in the right member have been defined as being even, from Eq. (10-12) such a replacement does not change the value of the right member. Therefore, the left member of the equation must also be unchanged, i.e., $f(-t) = f(t)$. We may conclude that the sum of a series of even functions is also an even function. As a second property of even and odd functions, let us consider the sum of a series of odd functions $f_i^{(o)}(t)$ $(i = 1, 2, \ldots, n)$. If we let the sum be designated as $f(t)$, we may write

$$f(t) = \sum_{i=1}^{n} f_i^{(o)}(t) \tag{10-15}$$

Now let us again replace the argument t by $-t$. Since the functions in the right member are odd, from Eq. (10-13) such a replacement changes the sign of the right member. Therefore the sign of the left member must be correspondingly changed, i.e., $f(-t) = -f(t)$. We conclude that the sum of a series of odd functions is also an odd function.

Now let us evaluate the situation that occurs when even and odd functions are multiplied. First let us consider the product of two even functions. Let $f_1^{(e)}$ and $f_2^{(e)}(t)$ be two even functions and let $f(t)$ be their product. Thus, we may write

$$f(t) = f_1^{(e)}(t) \times f_2^{(e)}(t) \tag{10-16}$$

If we replace t by $-t$ in both members of the above equation, we obtain

$$f(-t) = f_1^{(e)}(-t) \times f_2^{(e)}(-t) \tag{10-17}$$

Now let us apply Eq. (10-12) to the even functions in the right member of the above. We obtain

$$f(-t) = f_1^{(e)}(t) \times f_2^{(e)}(t) \tag{10-18}$$

Comparing Eqs. (10-16) and (10-18), we conclude that the product of two even functions is also an even function. In a similar manner, we may show that the product of two odd functions is an even function. To do this, let $f_1^{(o)}(t)$ and $f_2^{(o)}(t)$ be odd functions and let $f(t)$ be their product. Thus, we may write

$$f(t) = f_1^{(o)}(t) \times f_2^{(o)}(t) \tag{10-19}$$

If we let $t = -t$ in the above equation and use (10–13), we see that

$$f(-t) = f_1^{(o)}(-t) \times f_2^{(o)}(-t) = -f_1^{(o)}(t) \times -f_2^{(o)}(t) = f(t) \qquad (10\text{–}20)$$

We conclude that the product of two odd functions is also an even function. Finally, let us consider the case where an even function $f^{(e)}(t)$ and an odd function $f^{(o)}(t)$ are multiplied to produce a function $f(t)$. Thus, we may write

$$f(t) = f^{(e)}(t) \times f^{(o)}(t) \qquad (10\text{–}21)$$

Changing t to $-t$ in both members in the above equation and applying (10–12) and (10–13), we obtain

$$f(-t) = f^{(e)}(-t) \times f^{(o)}(-t) = -f^{(e)}(t) \times f^{(o)}(t) = -f(t) \qquad (10\text{–}22)$$

Thus, we conclude that the product of an even function and an odd function is an odd function.

The results given above may be summarized as follows:

SUMMARY 10-1.1

Sums and Products of Even and Odd Functions: An even periodic function is one which is symmetric about the origin, i.e., one in which $f(t) = f(-t)$. An odd periodic function is one which is skew-symmetric about the origin, i.e., one in which $f(t) = -f(-t)$. Sums and products of such functions satisfy the following characterizations.

even + even = even
odd + odd = odd
even × even = even
odd × odd = even
odd × even = odd

The one case of combining even and odd functions not covered in the above discussion is the case in which an even function is added to an odd function. In general, such a sum will be neither even nor odd. To see this, let $f^{(e)}(t)$ be an even function and $f^{(o)}(t)$ be an odd function. If we let $f(t)$ be their sum, we may write

$$f(t) = f^{(e)}(t) + f^{(o)}(t) \qquad (10\text{–}23)$$

Now let us replace the argument t by $-t$ in the above. We obtain

$$f(-t) = f^{(e)}(-t) + f^{(o)}(-t) \qquad (10\text{–}24)$$

Substituting from the defining relations for even and odd functions given in Eqs. (10–12) and (10–13), this expression may be rewritten

$$f(-t) = f^{(e)}(t) - f^{(o)}(t) \tag{10–25}$$

If we now add Eqs. (10–23) and (10–25) and rearrange terms, we obtain

$$f^{(e)}(t) = \tfrac{1}{2}[f(t) + f(-t)] \tag{10–26}$$

Similarly, subtracting (10–23) and (10–25) and rearranging terms, we obtain

$$f^{(o)} = \tfrac{1}{2}[f(t) - f(-t)] \tag{10–27}$$

From the results of Eqs. (10–26) and (10–27) given above, we obtain a most important result: If we have any arbitrary function $f(t)$ which does not have even or odd symmetricity and apply Eq. (10–26), we obtain an even function $f^{(e)}(t)$ which is the *even part of* $f(t)$. Similarly, applying Eq. (10–27) to $f(t)$, we obtain an odd function $f^{(o)}(t)$ which is called the *odd part of* $f(t)$. Thus, we may conclude that an arbitrary periodic function may always be decomposed into an even and an odd part. It is readily shown that the addition of such parts gives the original function. An example follows.

EXAMPLE 10-1.1. As an example of the determination of the even and odd parts of an arbitrary periodic function, consider the periodic sequence of rectangular pulses labeled $f(t)$ in Fig. 10-1.5(a). The corresponding function $f(-t)$ is shown in Fig. 10-1.5(b). Adding these two functions and dividing by two to satisfy (10-26), we obtain a function $f^{(e)}(t)$ as show in Fig. 10-1.5(c). Obviously this function has even symmetricity. Similarly, applying (10-27) we obtain the odd function $f^{(o)}(t)$ shown in Fig. 10-1.5(d). The sum of the even part and the odd part of $f(t)$ obviously gives us the original function $f(t)$ shown in Fig. 10-1.5(a).

The symmetry aspects of periodic functions discussed above are sometimes referred to as the properties of *full-wave symmetry*. Another type of symmetry which is frequently found to be useful in discussing periodic functions is *half-wave symmetry*. There are two types, odd and even. *Odd half-wave symmetry* is defined by the relation

$$f(t) = -f\left(t + \frac{T}{2}\right) \tag{10–28}$$

for all t over the range $-\infty \leq t \leq \infty$, where T is the period. Thus, the shape of the waveform in the second half of any given period is the negative of the waveform in the first half. As an example of odd half-wave symmetry, consider the periodic function shown in Fig. 10–1.6. It is readily observed that this waveform satisfies the relation given in Eq. (10–28). The other type of half-wave symmetry that might be considered is *even half-wave symmetry*. This would be defined by the relation

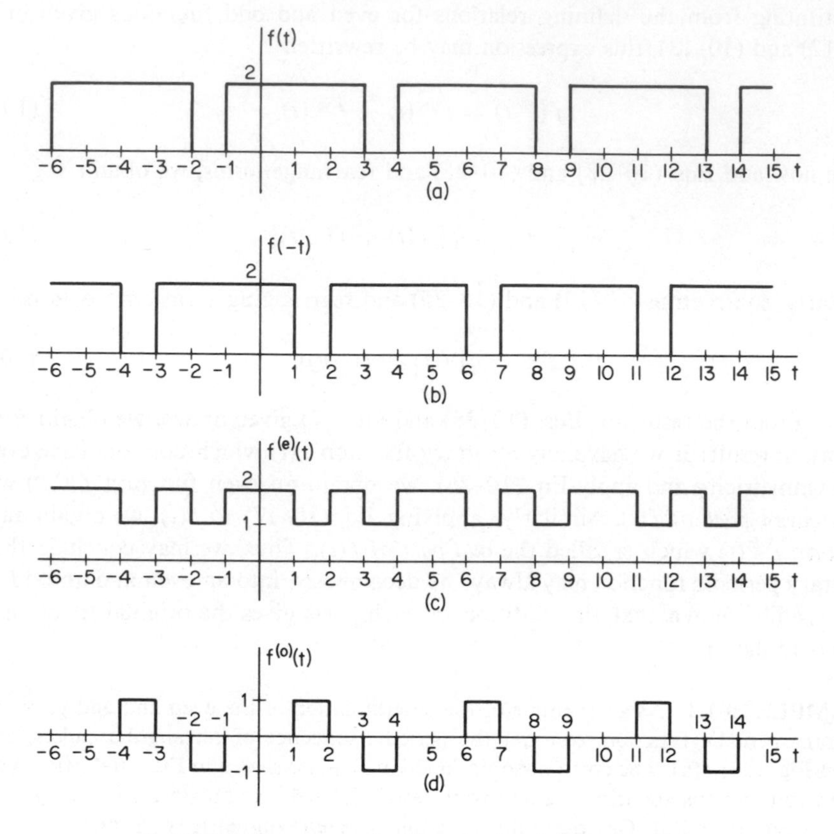

Fig. 10-1.5 The even and odd parts of a periodic function.

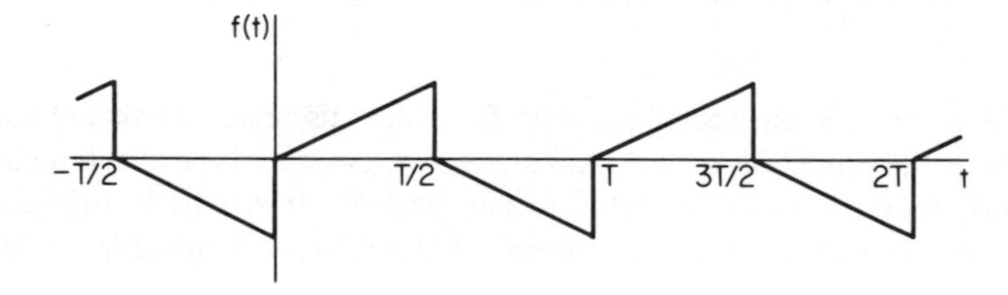

Fig. 10-1.6 A periodic function with half-wave symmetry.

$$f(t) = f\left(t + \frac{T}{2}\right) \tag{10–29}$$

where T is the period. Actually, this is not a special case, since, from Eq. (10–2) this relation merely defines a function with a period $T/2$ rather than a period T. Henceforth, when we refer to half-wave symmetry, we shall imply *odd* half-wave symmetry as defined by Eq. (10–28).

10-2 THE FOURIER SERIES

In the first section of this chapter we presented some specific examples of how series of sinusoidal terms called Fourier series could be formed such that in the limit as the number of terms was increased the waveform of the series approached that of a specified non-sinusoidal periodic function. In this section, we shall investigate the general form of such a Fourier series and show how to determine it for any arbitrary periodic non-sinusoidal function.

Let us assume $f(t)$ is an arbitrary periodic function of period T. To avoid any possibility of trying to obtain a Fourier series for "weird" functions, we shall place some restrictions on $f(t)$. First, let us assume that it has only a finite number of maxima or minima in any one period, and that these maxima and minima are finite in magnitude. Second, let us assume that there are only a finite number of jumps or discontinuities in any one period. Finally, let us assume that the integral of the absolute value of the function taken over one period is also finite, i.e.,

$$\int_0^T |f(t)|\, dt < \infty \tag{10–30}$$

These conditions are called the *Dirichlet conditions*, and they are sufficient to ensure that the Fourier series of $f(t)$ exists. An example of a function which does not satisfy the Dirichlet conditions is $\sin(1/t)$. In practice, all the functions that are normally encountered satisfy these conditions. For such a function it may be shown that the Fourier series has the form

$$\begin{aligned} f(t) &= a_0 + a_1 \cos \omega_0 t + a_2 \cos 2\omega_0 t + a_3 \cos 3\omega_0 t + \cdots \\ &\quad + b_1 \sin \omega_0 t + b_2 \sin 2\omega_0 t + b_3 \sin 3\omega_0 t + \cdots \\ &= a_0 + \sum_{i=1}^{\infty} (a_i \cos i\omega_0 t + b_i \sin i\omega_0 t) \end{aligned} \tag{10–31}$$

In this equation ω_0 $(= 2\pi/T)$ is called the *fundamental frequency* in radians per second of the periodic function and each of the quantities $i\omega_0$ $(= i2\pi/T)$ is referred to as the *ith harmonic* of the fundamental frequency. These quantities can, of course, also be expressed in hertz, in which case f_0 $(= 1/T)$ is the fundamental frequency and if_0 $(= i/T)$ is the ith harmonic of the fundamental. The quantities a_i and b_i are called the *Fourier coefficients*. In this section we shall see how they may be determined. Before doing this, however, let us review some properties of the functions $\cos n\omega_0 t$ and $\sin m\omega_0 t$ (where n and m are positive integers). We may first note that the integral covering one period of the fundamental of a sine or a cosine function or any of its harmonics is zero. Thus, we may write

$$\begin{aligned} \int_0^T \cos n\omega_0 t\, dt &= 0 \\ \int_0^T \sin m\omega_0 t\, dt &= 0 \end{aligned} \tag{10–32}$$

These relations are easily seen to be true, since, over one period, the integrands generate equal positive- and negative-valued areas. Now let us note that over one period of the fundamental the integral of either the sine squared or the cosine squared of the fundamental or any harmonic is equal to $T/2$. Thus, we may write

$$\begin{aligned} \int_0^T \cos^2 n\omega_0 t\, dt &= T/2 \\ \int_0^T \sin^2 m\omega_0 t\, dt &= T/2 \end{aligned} \tag{10-33}$$

As a final property of the integrals of sine and cosine functions, we note that the integrals of products of sine and cosine terms over one period, as given in the following relations, are zero

$$\begin{aligned} \int_0^T \cos n\omega_0 t \sin m\omega_0 t\, dt &= 0 \qquad \text{all } m, n \\ \int_0^T \cos n\omega_0 t \cos m\omega_0 t\, dt &= 0 \qquad m \neq n \\ \int_0^T \sin n\omega_0 t \sin m\omega_0 t\, dt &= 0 \qquad m \neq n \end{aligned} \tag{10-34}$$

The proofs of the relations given above may be found in any basic text on integral calculus. In the expressions given in Eqs. (10–32), (10–33), and (10–34), it should be noted that, although the limits of integration are 0 and T, in general, the theorems hold for any limits of integration which define one complete period. Thus, we may replace the lower limit of the integrals by t_0 and the upper limit by $t_0 + T$ in any of the above relations without changing the results. The relations of Eq. (10–34) may be collectively summarized by saying that the functions $\cos n\omega_0 t$ and $\sin m\omega_0 t$ are *orthogonal* with respect to integration from t_0 to $t_0 + T$, i.e., integration over one period.

Using the relations given in Eqs. (10–32), (10–33), and (10–34), we may now present a general process for determining the Fourier coefficients a_i and b_i. As an example of this procedure, let us first consider the details of obtaining the coefficient a_0. If we integrate both sides of (10–31) from 0 to T, we obtain

$$\int_0^T f(t)\, dt = \int_0^T a_0\, dt + \int_0^T \sum_{i=1}^{\infty} (a_i \cos i\omega_0 t + b_i \sin i\omega_0 t)\, dt \tag{10-35}$$

It may be shown that the operations of integration and summation indicated in the right term of the right member of (10–35) can be interchanged. Doing this, we may write

$$\int_0^T f(t)\, dt = \int_0^T a_0\, dt + \sum_{i=1}^{\infty} \left[\int_0^T a_i \cos i\omega_0 t\, dt + \int_0^T b_i \sin i\omega_0 t\, dt \right] \tag{10-36}$$

Integrating the first term in the right member of Eq. (10–36), we obtain a_0T. Integrating the terms in the summation in the right member, we find that all these terms are zero. Thus, we have derived an integral relation which only involves the original function $f(t)$ and the coefficient a_0, and which may be written in the form

$$a_0 = \frac{1}{T}\int_0^T f(t)\,dt \tag{10–37}$$

From this relation, we see that a_0 gives the average value of the periodic function $f(t)$ over any single period. Obviously, since the function is periodic, this average value will be the same for any other period and we may conclude that a_0 is the average value of the non-sinusoidal periodic function $f(t)$. As an example of determining such a coefficient, consider the square wave shown in Fig. 10–2.1. The period T is obviously 2 s and the integral of the function over one period (the area under one square pulse) is 2. Applying (10–37), we conclude that $a_0 = 1$.

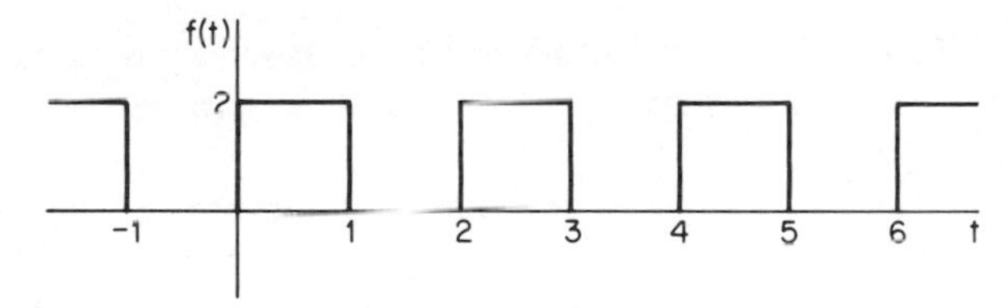

Fig. 10-2.1 A square wave.

In determining the other Fourier coefficients a procedure quite similar to the one described above is followed. The only major difference is that we must multiply both sides of Eq. (10–31) by an appropriate sinusoidal function before integrating. Thus, the orthogonality relations of Eq. (10–34) may be used to isolate any single Fourier coefficients. As an example of this procedure, consider the determination of the coefficient a_1. Multiplying both members of (10–31) by the term $\cos \omega_0 t$ and integrating from 0 to T, we obtain

$$\int_0^T f(t)\cos \omega_0 t\,dt = \int_0^T a_0 \cos \omega_0 t\,dt + \int_0^T \cos \omega_0 t \sum_{i=1}^{\infty} (a_i \cos i\omega_0 t + b_i \sin i\omega_0 t)\,dt \tag{10–38}$$

Again the operations of integration and summation may be interchanged. Doing this and applying the orthogonality relations of (10–34), all terms in the right member of the above equation except the term involving the coefficient a_1 become zero Thus, we obtain

$$\int_0^T f(t)\cos \omega_0 t\,dt = \int_0^T a_1 \cos^2 \omega_0 t\,dt = a_1\frac{T}{2} \tag{10–39}$$

Rearranging the terms, we see that

$$a_1 = \frac{2}{T}\int_0^T f(t) \cos \omega_0 t \, dt \tag{10–40}$$

We conclude that the relation given above may be applied to any periodic function to find the Fourier coefficient a_1. As an example of such an application, let us determine the coefficient a_1 for the square wave shown in Fig. 10–2.1. From Eq. (10–40) we obtain

$$a_1 = \frac{2}{2}\int_0^1 2 \cos \pi t \, dt = \frac{2 \sin \pi t}{\pi}\bigg|_0^1 = 0 \tag{10–41}$$

The approach used above to determine a_1 is easily extended to provide a general method for obtaining any of the Fourier coefficients a_i. The result is

$$a_i = \frac{2}{T}\int_0^T f(t) \cos i\omega_0 t \, dt \qquad i = 1, 2, 3, \ldots \tag{10–42}$$

This result is easily verified using the orthogonality relations of Eq. (10–34). Applying this relation to the square wave shown in Fig. 10–2.1, we find that all the coefficients a_i are zero.

Now let us consider the determination of the coefficients associated with the sine terms in the Fourier series. To determine the coefficient b_1 we may multiply both members of Eq. (10–31) by the factor $\sin \omega_0 t$ and integrate from 0 to T. Thus, we obtain

$$\int_0^T f(t) \sin \omega_0 t \, dt = \int_0^T a_0 \sin \omega_0 t \, dt + \int_0^T \sin \omega_0 t \sum_{i=1}^{\infty} (a_i \cos i\omega_0 t + b_i \sin i\omega_0 t) \, dt \tag{10–43}$$

Interchanging the operations of integration and summation and applying the orthogonality relations of Eq. (10–34), all the terms in the right member except the one involving b_1 become zero. Thus, we conclude that

$$b_1 = \frac{2}{T}\int_0^T f(t) \sin \omega_0 t \, dt \tag{10–44}$$

Extending this procedure, we obtain the following general relation for determining the coefficients b_i of a Fourier series.

$$b_i = \frac{2}{T}\int_0^T f(t) \sin i\omega_0 t \, dt \tag{10–45}$$

As an example of the application of this relation, consider the square wave shown in Fig. 10–2.1. For the coefficient b_1 we obtain

$$b_1 = \frac{2}{2}\int_0^1 2 \sin \pi t \, dt = -\frac{2 \cos \pi t}{\pi}\bigg|_0^1 = \frac{4}{\pi} \tag{10-46}$$

Similarly, for the other coefficients b_i we find that

$$b_i = \begin{cases} 0 & i \text{ even} \\ 4/i\pi & i \text{ odd} \end{cases} \tag{10-47}$$

Comparing the results given above for the determination of the coefficients a_i and b_i of the square wave shown in Fig. 10-2.1, we find that this square wave may be expressed as the following Fourier series

$$f(t) = 1 + \frac{4}{\pi}\left(\sin \pi t + \frac{1}{3} \sin 3\pi t + \frac{1}{5} \sin 5\pi t + \cdots\right) \tag{10-48}$$

Thus, we have verified (with the addition of a constant term) the assumed form given for the series representation of a square wave in Eq. (10-7). If we apply the procedure given above to the square wave shown in Fig. 10-2.2, then we obtain

$$f(t) = 1 + \frac{4}{\pi}\left(\cos \pi t - \frac{1}{3} \cos 3\pi t + \frac{1}{5} \cos 5\pi t - \cdots\right) \tag{10-49}$$

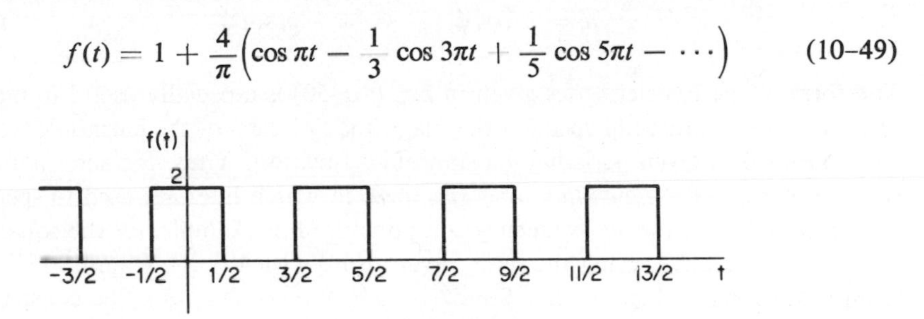

Fig. 10-2.2 A square wave.

This is similar to the result obtained in Eq. (10-11). The results given above may be summarized as follows:

SUMMARY 10-2.1

The Fourier Series for Periodic Functions: Any periodic function $f(t)$ which satisfies the Dirichlet conditions may be expressed as a Fourier series having the form

$$f(t) = a_0 + \sum_{i=1}^{\infty} (a_i \cos i\omega_0 t + b_i \sin i\omega_0 t)$$

The quantities a_i and b_i are called the Fourier coefficients and are found using the relations summarized in Table 10-3.1.

The general expression for a Fourier series given in Eq. (10–31) is only one of several forms which may be used to express such a series. As an example of another form, let us recall from Sec. 7–1 that when two sinusoids having the same frequency are added, the result is another sinusoid of the same frequency. Thus, each of the terms in the summation in the right member of Eq. (10–31) may be expressed as a single sinusoidal term which includes a phase angle. An alternative form for a Fourier series may thus be written

$$f(t) = c_0 + \sum_{i=1}^{\infty} c_i \cos(i\omega_0 t + \phi_i) \tag{10–50}$$

where

$$c_0 = a_0 \qquad c_i = \sqrt{a_i^2 + b_i^2} \qquad \phi_i = -\tan^{-1}\frac{b_i}{a_i} \tag{10–51}$$

The inverse relations for the quantities a_i and b_i in terms of c_i and ϕ_i are

$$a_i = c_i \cos\phi_i \qquad b_i = -c_i \sin\phi_i \tag{10–52}$$

The form of the Fourier series given in Eq. (10–50) is especially useful in many applications since it directly specifies the magnitude of each of the harmonic frequency components of a given periodic non-sinusoidal function. Thus, for such a function, we may construct a *magnitude line spectrum* in which lines are used to specify the magnitude of these various frequency components. For example, for the square wave shown in Fig. 10–2.1, using the series given in Eq. (10–48), the magnitude line spectrum is as shown in Fig. 10–2.3. Similarly, a *phase spectrum* may be constructed in which lines are used to specify the value of the phase for the various frequency components.

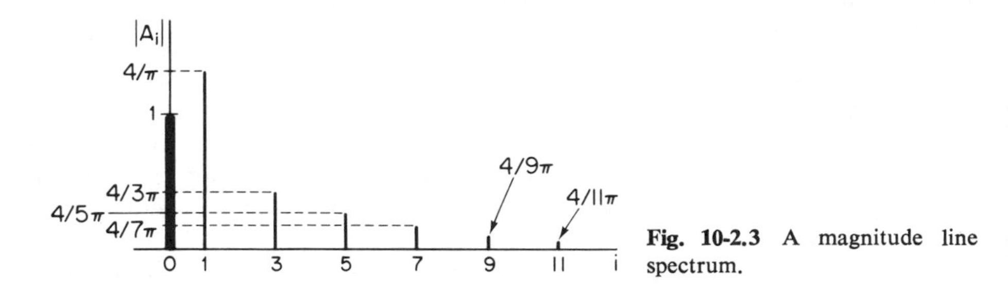

Fig. 10-2.3 A magnitude line spectrum.

Now let us consider the effect of changing the reference time $t = 0$ on a Fourier series. As an example of such a change of reference, consider the square waves shown in Figs. 10–2.1 and 10–2.2. The waveform of these square waves is identical but the reference point $t = 0$ on the t axis has been shifted by $\frac{1}{4}$ of the period. The result of this is to change the form of the series from one involving only sine terms as given in Eq. (10–48) to one involving only cosine terms as given in Eq. (10–49). Comparing these two series and using the representation form given in Eq. (10–50), we see that the magnitude coefficients c_i for the two series are the same, thus they have the same magnitude line spectrum. In considering the effect on the phase angles ϕ_i when a series such as (10–48) is written in the form of Eq. (10–50), we note that a change of θ radians in the phase of the fundamental frequency produces a change of $n\theta$ radians in the phase of the nth harmonic. This conclusion may be seen more clearly by considering Fig. 10–2.4. In this figure, a fundamental sinusoidal function $f_1(t)$ and a third harmonic sinusoidal function $f_2(t)$ are shown. If we choose point 1 as the reference point for $t = 0$, then clearly $f_1(t)$ is a sine function and $f_2(t)$ is a sine function with a frequency three times as great, i.e., a third harmonic frequency. On the other hand, if we use point 2 as a reference point for $t = 0$, then we have changed the phase of $f_1(t)$ by $\frac{1}{4}$ of its period. Correspondingly, we have changed the phase of the function $f_2(t)$ by $\frac{3}{4}$ of one of its periods. Applying this result to the square wave shown in Fig. 10–2.1 with the Fourier series given in Eq. (10–48), we see that we may write such a series in the form

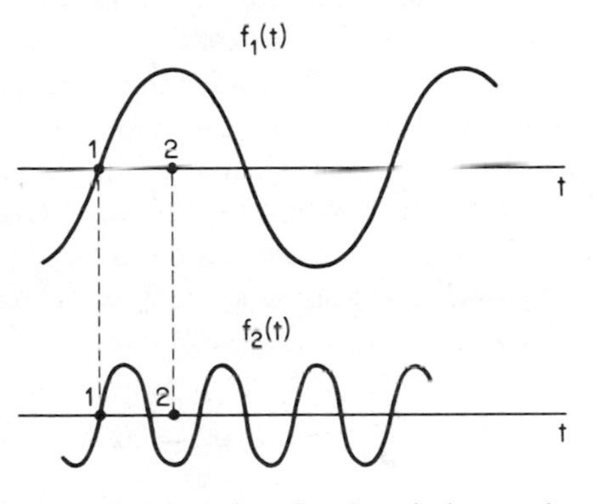

Fig. 10-2.4 Relations for phase in harmonic waveforms.

$$f(t) = 1 + \frac{4}{\pi} \sum_{i=1,3,5}^{\infty} \frac{1}{i} \cos\left(i\pi t - \frac{i\pi}{2}\right) \tag{10–53}$$

Comparing this series with the one given in Eq. (10–49) for the square wave shown in Fig. 10–2.2, we see that the difference between the two square waves is simply one of phase. The results given above may be summarized as follows:

SUMMARY 10-2.2

Magnitude and Phase Representation of a Fourier Series: A Fourier series for any function $f(t)$ may be written in the form

$$f(t) = c_0 + \sum_{i=1}^{\infty} c_i \cos\,(i\omega_0 t + \phi_i)$$

where

$$c_0 = a_0 \qquad c_i = \sqrt{a_i^2 + b_i^2} \qquad \phi = -\tan^{-1}\left(\frac{b_i}{a_i}\right)$$

and the coefficients a_i and b_i are defined in Summary 10–2.1. Changing the reference point $t = 0$ which has been chosen for such a function does not affect the values of the magnitude coefficients c_1 although it will change the phase angles ϕ_i.

An interesting result of the properties described above is that an even function with a zero average value may be changed to an odd function by choosing a different point for the reference time. The two square waves shown in Figs. 10–1.1(a) and 10–1.4(a) illustrate this.

In the preceding paragraphs we have discussed the effect of shifting a periodic function along the horizontal (or t) axis, i.e., selecting different reference points for $t = 0$. Now let us consider the effect of shifting a function along the vertical axis. This simply adds a constant term to all values of the function. Thus, it affects only the term a_0 in the Fourier series. As an example of this conclusion, consider the square wave shown in Fig. 10–2.5. This differs from the square wave shown in Fig. 10–2.2 only in that it has an average value which is easily seen to be 4. Thus, we may obtain its Fourier series directly from Eq. (10–48) by changing the constant term to a value of 4. Thus, we obtain

$$f(t) = 4 + \frac{4}{\pi}\left(\cos \pi t - \frac{1}{3}\cos 3\pi t + \frac{1}{5}\cos 5\pi t - \cdots\right) \tag{10–54}$$

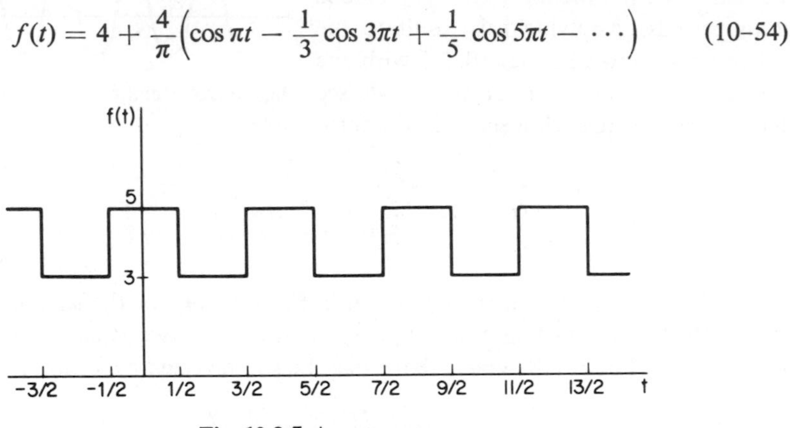

Fig. 10-2.5 A square wave.

As a final topic for this section, in the following paragraphs we consider the question of how well the Fourier series performs its task of representing a given non-sinusoidal periodic function. Such information is necessarily of great interest, especially with respect to the problem of evaluating the accuracy of the representation as a function of the number of terms which are used in the series. To investigate this, suppose we let $f(t)$ be a periodic function and let $f_n(t)$ be its Fourier series which includes only the first n harmonic terms. The truncation error caused by not including the remaining terms may be defined as $e_n(t)$, where

$$e_n(t) = f(t) - f_n(t) \tag{10–55}$$

In evaluating this error, it is usually helpful to use the function $e_n(t)$ to define some scalar error constant from which the time dependence has been eliminated. One of the most frequently used means of doing this is to define a mean-square error E_n by the relation

$$E_n = \int_0^T e_n^2(t)\,dt \tag{10-56}$$

This expression in effect sums the time dependent error $e_n(t)$ over an entire period. In addition, the use of the square relation weights the larger errors more heavily than the smaller ones and also prevents any false minimization that might occur as a result of positive and negative polarity errors tending to cancel. It may be shown that a Fourier series with coefficients a_i and b_i as determined by Eqs. (10–37), (10–42), and (10–45) is optimum in the sense that there are no other values of the coefficients that will give a lower value of E_n for any specified value of n.[1] This property of the Fourier series is referred to as the *least squares error property*. It is one of the reasons that the Fourier series is such a valuable tool for the representation of non-sinusoidal periodic waveforms.

In general, the type of function for which the Fourier series representation is the poorest is a periodic function in which discontinuities occur. It will be recalled that the Dirichlet conditions permit a finite number of such discontinuities. The first problem associated with a discontinuity is the question of what value the series will assign to the function at the time at which the discontinuity occurs. For example, suppose a function $f(t)$ has a discontinuity at $t = t_0$. This requires that $f(t_0 -)$ is not equal to $f(t_0 +)$.[2] The truncated series $f_n(t)$, being composed of a sum of continuous terms, obviously cannot be discontinuous at $t = t_0$. Instead, it may be shown that it provides the value $f(t_0)$ which satisfies the relation

$$f(t_0) = \frac{f(t_0+) + f(t_0-)}{2} \tag{10-57}$$

i.e., the series provides the average of the values of the function which precede and follow the discontinuity. As higher numbers of terms are added to the series $f_n(t)$, it will tend to overshoot the values of $f(t_0+)$ and $f(t_0-)$ in its effort to approximate a discontinuity. A good example of this is a square wave. The Fourier series for such a wave produces a characteristic overshoot and damped oscillatory decay which approaches 8.95 per cent of the total magnitude of the discontinuity. This phenomenon is illustrated in Fig. 10–2.6. As n is increased, the frequency of oscillation increases; however, the magnitude of the overshoot still remains the same. This tendency of a truncated Fourier series $f_n(t)$ to exhibit a damped oscillatory overshoot in its repre-

[1]See P. M. Chirlian, *Basic Network Theory*, Sec. 11.3, McGraw-Hill Book Company, New York, 1969, for a development which indicates the proof of this statement.

[2]The symbols t_0- and t_0+ are explained in Sec. 4-3.

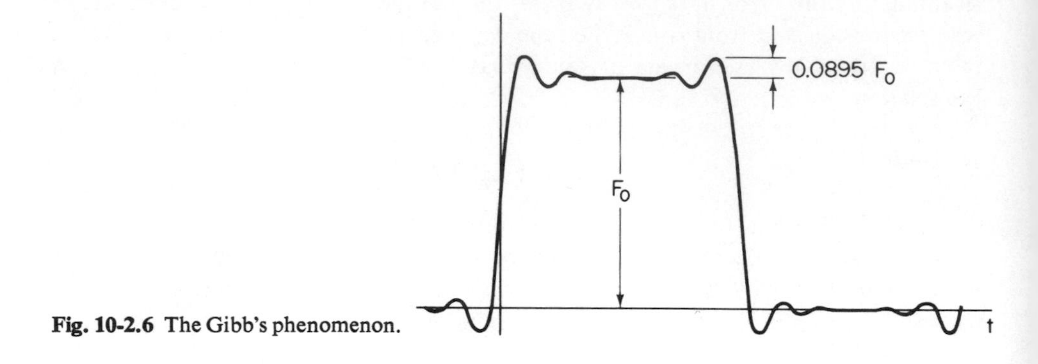

Fig. 10-2.6 The Gibb's phenomenon.

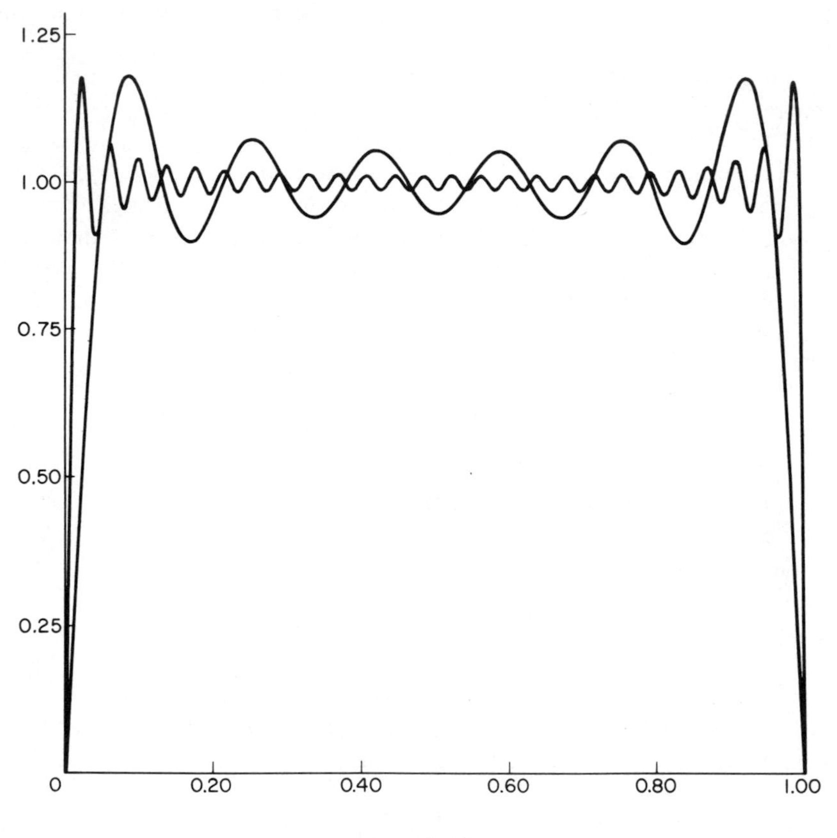

Fig. 10-2.7 The Gibb's phenomenon.

sentation of a discontinuity in a function $f(t)$ is called the *Gibb's phenomenon.* A more specific example of this phenomenon is shown in Fig. 10–2.7. Here we show a computer-plotted reconstruction of the first half-cycle of a square wave with a peak magnitude of plus and minus unity, and a period of two seconds. One of the loci shows the waveform that results when terms through the eleventh harmonic are used. The second locus shows the results of using terms through the fifty-first harmonic. It is readily observed that an overshoot of $2 \times 0.0895 = 0.179$ occurs which is independent of the number of terms used in the series. (The "2" occurs since the magnitude of the discontinuity is from -1 to $+1$, i.e., a total discontinuity of 2.)

10-3 HOW THE FORM OF A PERIODIC FUNCTION AFFECTS THE COEFFICIENTS OF THE FOURIER SERIES

In this section we shall discuss some of the most significant ways in which the form of a non-sinusoidal periodic function affects the various coefficients a_i and b_i of the Fourier series as defined by Eq. (10–31). We shall investigate two major types of such phenomena. The first of these is the effect that the various symmetry properties discussed in Sec. 10–1 have on the coefficients. The second is the effect that discontinuities in the function being represented have on the coefficients.

To begin, let $f(t)$ be an arbitrary periodic function with an even part $f^{(e)}(t)$ and an odd part $f^{(o)}(t)$. Now let us use Eq. (10–42) to find the coefficients a_i. Since the functions $f^{(e)}(t)$ and $f^{(o)}(t)$ specify symmetry properties with respect to the origin, we will change the limits of integration in the integral defining the coefficients a_i from zero and T to $-T/2$ and $T/2$, respectively, so that the integration interval is symmetrical with respect to the origin. Thus, we obtain

$$a_i = \frac{2}{T}\left[\int_{-T/2}^{T/2} f^{(e)}(t) \cos i\omega_0 t \, dt + \int_{-T/2}^{T/2} f^{(o)}(t) \cos i\omega_0 t \, dt\right] \tag{10–58}$$

In the right member of the above equation the first integrand contains a product of even functions; therefore, from Summary 10–1.1 it must be an even function, i.e., one in which $f(t) = f(-t)$. Thus, the integral may be evaluated by integrating over the limits zero to $T/2$ and multiplying the result by two. The second integrand contains the product of an even and an odd function. Therefore, the integrand is an odd function in which $f(t) = -f(-t)$. The integrand of such a function taken over limits which are symmetrical with respect to the origin must be zero. Considering the above, we see that the expression for the coefficients a_i may be written

$$a_i = \frac{4}{T}\int_{0}^{T/2} f^{(e)} \cos i\omega_0 t \, dt \tag{10–59}$$

where $f^{(e)}(t)$ is the even part of some periodic function $f(t)$. A most important conclusion of the above is that only the even part of an arbitrary periodic function $f(t)$ determines the coefficients a_i. Thus, the Fourier series of a purely odd function will contain no cosine terms. It should be noted that since an even function can have a non-zero average value, there may also be a constant a_0 as determined by Eq. (10–37) for such a case. As an example of this conclusion, consider the expression given in Eq. (10–54) for the square wave shown in Fig. 10–2.5. The series obviously contains only cosine terms plus a constant term and the square wave is clearly an even function.

Now let us make a development similar to the above for the Fourier sine coefficients b_i. Using Eq. (10–45) and changing the limits of integration, we may write

$$b_i = \frac{2}{T}\left[\int_{-T/2}^{T/2} f^{(e)}(t) \sin i\omega_0 t \, dt + \int_{-T/2}^{T/2} f^{(o)}(t) \sin i\omega_0 t \, dt\right] \tag{10–60}$$

In the right member of the above equation the first integrand contains a product of an even and an odd function. Therefore, it is odd and the resulting integral is zero. Similarly, the second integrand contains the product of two odd functions; therefore, it is even and its integral may be evaluated by taking the integral from zero to $T/2$ and multiplying by two. As a result, the expression for determining the coefficients b_i may be written in the form

$$b_i = \frac{4}{T}\int_0^{T/2} f^{(o)}(t) \sin i\omega_0 t \, dt \tag{10–61}$$

where $f^{(o)}(t)$ is the odd part of some periodic function $f(t)$. We conclude that the Fourier series of an even function will contain no sine terms. As an example of this, consider the Fourier series given in Eq. (10–7) for the square wave shown in Fig. 10–1.1(a). The series obviously contains only sine terms and the square wave clearly has odd symmetry. Since the average value of an odd function is zero, the coefficient a_0 will always be zero for such a function.

Now let us consider the effect of half-wave symmetry as defined by Eq. (10–28) on the Fourier coefficients. An example of a function with half-wave symmetry is shown by the solid line in Fig. 10–3.1(a). Also, on this figure we have indicated by a dashed line a cosine wave with the same period as the original function. The expression used to determine the Fourier coefficients a_i ($i = 1, 2, \ldots$) is

$$a_i = \frac{2}{T}\int_{-T/2}^{T/2} f(t) \cos i\omega_0 t \, dt \tag{10–62}$$

Thus, the integrand for the constant a_1 is determined by the product of the dashed and solid functions shown in the figure. Now consider the portion of the integral over the limits $-T/2$ to zero. Suppose we reverse the functions indicated in the figure over this interval. The result of doing this is shown in Fig. 10–3.1(b). The value of the integral over this interval is unchanged since the total area is still the same. From Fig. 10–3.1(b), however, we see that over the entire period from $-T/2$ to $T/2$ both the dashed line and the solid line define odd functions. Therefore, the

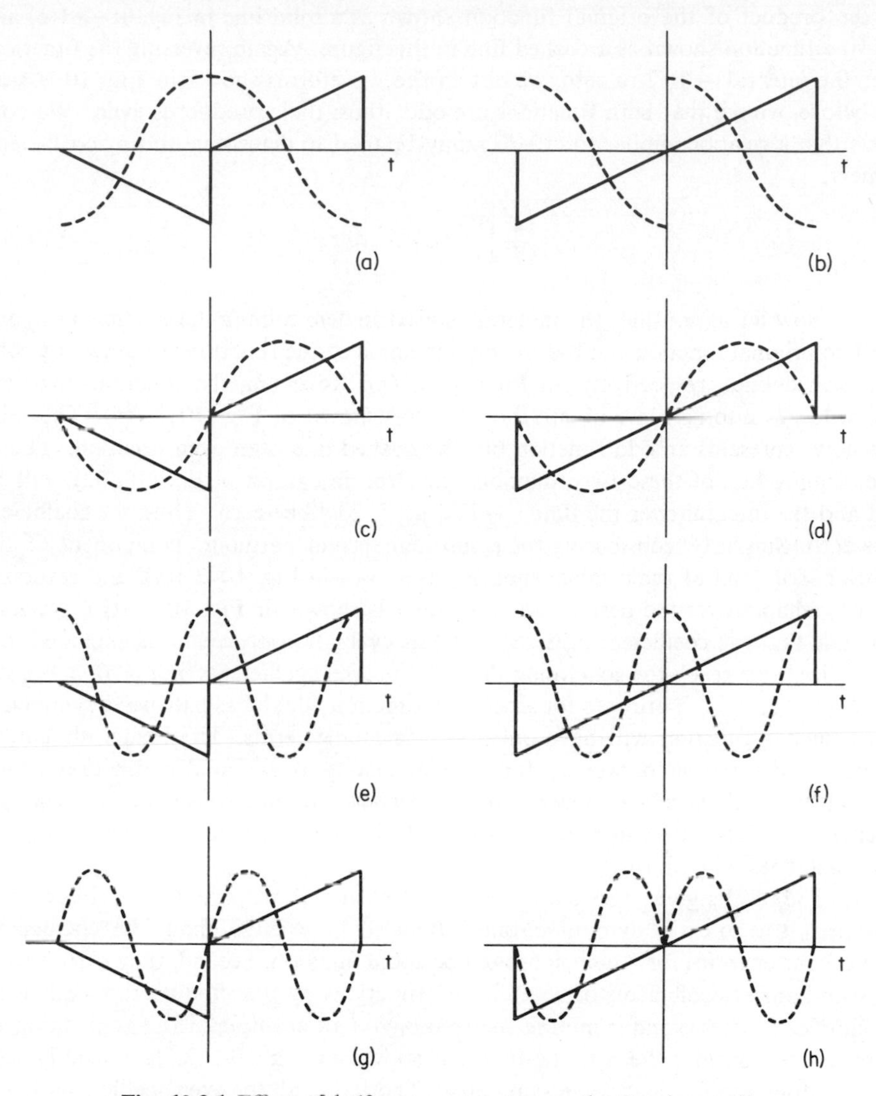

Fig. 10-3.1 Effect of half-wave symmetry on the Fourier coefficients.

product of these two, namely, the integrand of Eq. (10–62), is even and an expression for the coefficient a_1 may be derived similar in form to the one given in (10–59), namely,

$$a_1 = \frac{4}{T}\int_0^{T/2} f(t) \cos \omega_0 t \, dt \tag{10–63}$$

The same logic may be used to find the coefficient b_1. In this case, the integrand will be the product of the original function shown as a solid line in Fig. 10–3.1(c) and the sine function shown as a dashed line in this figure. Again reversing the functions over the interval $-T/2$ to zero, we obtain the waveforms shown in Fig. 10–3.1(d). As before, we see that both functions are odd; thus, their product is even. We conclude that a relation similar to (10–61) may be used to determine the b_1 coefficient, namely,

$$b_1 = \frac{4}{T}\int_0^{T/2} f(t) \sin \omega_0 t \, dt \tag{10–64}$$

Now let us consider the integrands used in determining the coefficients a_2 and b_2. The original function and the second harmonic cosine function are shown by solid and dashed lines, respectively, in Fig. 10–3.1(e). Reversing the functions over the interval $-T/2$ to zero, we obtain the situation shown in Fig. 10–3.1(f). The solid line now represents an odd function but the dashed line is an even function. Therefore, the product of these two functions, i.e., the integrand of Eq. (10–62), will be odd and the integral over the limits $-T/2$ to $T/2$ will be zero. Thus, the coefficient a_2 is zero. Similarly, considering the sinusoidal second harmonic component of the Fourier series and of the original function as shown in Fig. 10–3.1(g), and reversing the waveshapes over the period $-T/2$ to zero as shown in Fig. 10–3.1(h), we may conclude that the coefficient b_2 is zero. If this evaluation procedure is extended and generalized, we reach the conclusion that all the even coefficients a_i $(i = 0, 2, 4, \ldots)$ and b_i $(i = 2, 4, \ldots)$ are zero for a periodic function which has half-wave symmetry. Thus, such a function will have only odd harmonic terms. The reader should be careful to distinguish between a function of *odd symmetry* and a function which contains only *odd harmonic terms*. The former has only sine terms in its series expansion. The letter will, in general, have both sine and cosine terms but only the odd harmonic ones.

The results given above are useful in two major ways. First, by inspection, they permit us to easily determine some advanced knowledge about the coefficients of the Fourier series for a non-sinusoidal periodic function. Second, they permit us to find the Fourier coefficients of complicated functions by first finding the coefficients for simpler functions and summing the results. As an example of the application of these results, consider the periodic function shown in Fig. 10–3.2. It is readily seen that this function has half-wave symmetry. Therefore, all the even coefficients of the Fourier series must be zero. If we now split the function into its even and odd parts, the even part $f^{(e)}(t)$ is a square wave indentical with the one shown in Fig. 10–1.4(a). Its Fourier series is given in Eq. (10–11), namely,

$$f^{(e)}(t) = \frac{4}{\pi}\left(\cos \pi t - \frac{1}{3}\cos 3\pi t + \frac{1}{5}\cos 5\pi t - \cdots\right) \tag{10–65}$$

Similarly, the odd part $f^{(o)}(t)$ is a triangular wave identical with the one shown in Fig. 10–1.1(b), which has the Fourier series [see (10–10)]

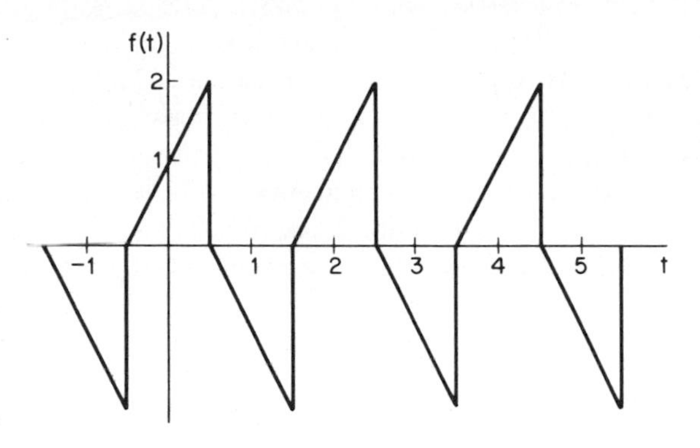

Fig. 10-3.2 A periodic function.

$$f^{(o)}(t) = \frac{8}{\pi^2}\left(\sin \pi t - \frac{1}{9}\sin 3\pi t + \frac{1}{25}\sin 5\pi t - \cdots\right) \tag{10-66}$$

The Fourier series of the function given in Fig. 10-3.2 is now simply found by summing the two series. Thus, we obtain

$$f(t) = \frac{4}{\pi}\cos \pi t + \frac{8}{\pi^2}\sin \pi t - \frac{4}{3\pi}\cos 3\pi t - \frac{8}{9\pi^2}\sin 3\pi t + \cdots \tag{10-67}$$

The results given above may be summarized as follows:

SUMMARY 10-3.1

Effect of Waveform Symmetry on the Fourier Series Coefficients: The various waveform symmetries which a periodic function may have affect the Fourier series coefficients in the following ways:

1. An even function will have a Fourier series consisting only of cosine terms and possibly a constant. Thus, the coefficients b_i $(i = 1, 2, \ldots)$ of the general series given in Eq. (10-31) are zero.
2. An odd function will have a Fourier series consisting only of sine terms. Thus, the coefficients a_i $(i = 0, 1, 2, \ldots)$ are zero.
3. A function with half-wave symmetry will have only odd harmonic terms. Thus, the coefficients a_i $(i = 0, 2, 4, \ldots)$ and b_i $(i = 2, 4, \ldots)$ are zero. A summary of the various symmetry relations and the equations for finding the Fourier coefficients are given in Tables 10-3.1 and 10-3.2.

The second relationship between the form of a non-sinusoidal periodic function and the resulting coefficients of the Fourier series that we shall investigate in this section is the effect of discontinuities in the function. At the end of Sec. 10-2, we showed that the Fourier series for a function containing a discontinuity will produce an overshoot referred to as the Gibbs phenomenon. Therefore, we might very well expect that such a function will require large numbers of high-frequency harmonics in its Fourier series representation to achieve reasonable accuracy. Correspondingly, we might expect that functions which are well behaved in the sense that they are not discontinuous might require considerably fewer terms. Such a conjecture is indeed well founded. In fact, it may be shown that the number of terms required in a Fourier series for a given accuracy of representation has a direct relation to the number of times the function being considered may be differentiated without producing any discontinuities. As an example of this consider three periodic functions which have a cycle as shown in Fig. 10–3.3. In part (a) of this figure, we show a function which has a parabolic behavior. It is defined by the relations

$$f_a(t) = \begin{cases} \dfrac{F_0 t^2}{2} - \dfrac{F_0 T}{4} t & 0 \le t \le \dfrac{T}{2} \\[2ex] \dfrac{3F_0 T}{4} t - \dfrac{F_0 t^2}{2} - \dfrac{F_0 t^2}{4} & \dfrac{T}{2} \le t \le T \end{cases} \tag{10–68}$$

TABLE 10-3.1 Expressions For Fourier Coefficients

Type of symmetry	Condition for symmetry	a_0	a_i $(i = 1, 2, 3, \ldots)$	b_i $(i = 1, 2, 3, \ldots)$
none	—	$\frac{1}{T}\int_0^T f(t)\,dt$	$\frac{2}{T}\int_0^T f(t)\cos i\omega_0 t\,dt$	$\frac{2}{T}\int_0^T f(t)\sin i\omega_0 t\,dt$
even	$f(t) = f(-t)$	$\frac{2}{T}\int_0^{T/2} f(t)\,dt$	$\frac{4}{T}\int_0^{T/2} f(t)\cos i\omega_0 t\,dt$	0
odd	$f(t) = -f(-t)$	0	0	$\frac{4}{T}\int_0^{T/2} f(t)\sin i\omega_0 t\,dt$

TABLE 10-3.2 Summary of Symmetry Properties

Coefficient	Description	Symmetry of $f(t)$: Even $f(t) = f(-t)$	Odd $f(t) = -f(-t)$	Half-wave $f(t) = -f\left(t + \frac{2}{T}\right)$
a_0	dc term	may be non-zero	zero	zero
$a_i\ (i = 1, 3, \ldots)$	odd harmonic cosine terms	may be non-zero	zero	may be non-zero
$a_i\ (i = 2, 4, \ldots)$	even harmonic cosine terms	may be non-zero	zero	zero
$b_i\ (i = 1, 3, \ldots)$	odd harmonic sine terms	zero	may be non-zero	may be non-zero
$b_i\ (i = 2, 4, \ldots)$	even harmonic sine terms	zero	may be non-zero	zero

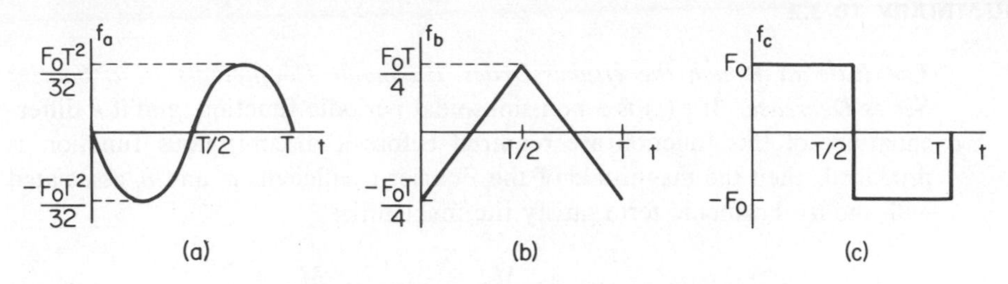

Fig. 10-3.3 Periodic functions related by differentiation.

If we differentiate this parabolic function, we obtain the relations

$$f_b(t) = \begin{cases} F_0t - F_0\dfrac{T}{4} & 0 \leq t \leq \dfrac{T}{2} \\ \dfrac{3F_0T}{4} - F_0t & \dfrac{T}{2} \leq t \leq T \end{cases} \tag{10-69}$$

These equations define the triangular periodic function shown in Fig. 10-3.3(b). Similarly, a second differentiation of the parabolic waveform produces the relations

$$f_c(t) = \begin{cases} F_0 & 0 < t < \dfrac{T}{2} \\ -F_0 & \dfrac{T}{2} < t < T \end{cases} \tag{10-70}$$

These relations define a discontinuous function, namely, the square wave shown in Fig. 10-3.3(c). Now let us consider the Fourier series for the three functions. We have already encountered the square and triangular waveforms in previous discussions. From series similar to those given in Eqs. (10-7) and (10-10), we see that for the square wave the magnitude of the non-zero coefficients decreases in direct proportion to $1/n$, where n is the number of the harmonic term which the coefficient multiplies. The magnitudes of the coefficients for the continuous triangular wave (which if differentiated *once* leads to the discontinuous square wave), however, are seen to be proportional to $1/n^2$. Now consider the parabolic function shown in Fig. 10-3.3(a). This function must be differentiated *twice* before a discontinuous function (a square wave) results. If we determine the Fourier series for this function, we find that the coefficients of the various terms decrease in direct proportion to $1/n^3$. Comparing the results for the three waveforms given in Fig. 10-3.3, we may conclude that the more times a function must be differentiated before discontinuities are produced, the more rapidly the magnitudes of the coefficients of the Fourier series decrease. This conclusion is an example of a general law regarding the manner in which the magnitudes of the Fourier coefficients decrease as determined by the nature of the function which is being represented. Such a general law may be summarized as follows:

SUMMARY 10-3.2

The Rate at Which the Higher-Order Harmonic Coefficients of a Fourier Series Decrease: If $f(t)$ is a non-sinusoidal periodic function, and if k differentiations of this function are required before a discontinuous function is produced, then the magnitude of the Fourier coefficients a_i and b_i associated with the ith harmonic term satisfy the inequalities

$$|a_i| \leq \frac{M}{i^{k+1}} \qquad |b_i| \leq \frac{M}{i^{k+1}} \tag{10–71}$$

where the value of the constant M depends on the specific function $f(t)$ being analyzed.[3]

The relations given above are of considerable value in anticipating the form of the magnitude spectrum of Fourier coefficients that will characterize a given periodic function. They are also of value in helping the engineer to anticipate the form of a periodic function from its magnitude spectrum. For example, from Eq. (10–71) it is readily verified that any *continuous function* will have terms in its Fourier series whose magnitude decreases at least as rapidly as $1/n^2$. Such a conclusion provides valuable information on the number of terms which must be included when a Fourier series is used to analyze a network for a given input waveform. Such an application will be considered in the following section.

10-4 NETWORK ANALYSIS USING THE FOURIER SERIES

In the preceding sections of this chapter we have described the use of the Fourier series in the representation of non-sinusoidal periodic functions. Now let us see how these representations may be used to make a steady-state analysis of a network excited by such a function. Note that here we say "steady-state" rather than "*sinusoidal* steady-state" since we are considering *non*-sinusoidal periodic wave forms. The basic technique is simply to determine the network function (as a function of $j\omega$) by the methods described in Chap 7. The Fourier series representation of the non-sinusoidal periodic waveform may then be treated as a series of separate sinusoidal excitations at different frequencies, each of which produces its own sinusoidal output. Because of the linear nature of the network, the superposition of the input waveforms (which yields the Fourier series representation of the non-sinusoidal periodic input function) produces a corresponding superposition of the output waveforms. This latter gives the desired response of the network. As an example of the technique, consider the network shown in Fig. 10–4.1(a). As indicated, the excitation voltage $v_s(t)$ is a square wave with a period $T = 4\pi$. Thus, the fundamental frequency is 0.5 rad/s.

[3]R. J. Schwarz and B. Friedland, *Linear Systems*, p. 144, McGraw-Hill Book Company, New York, 1965.

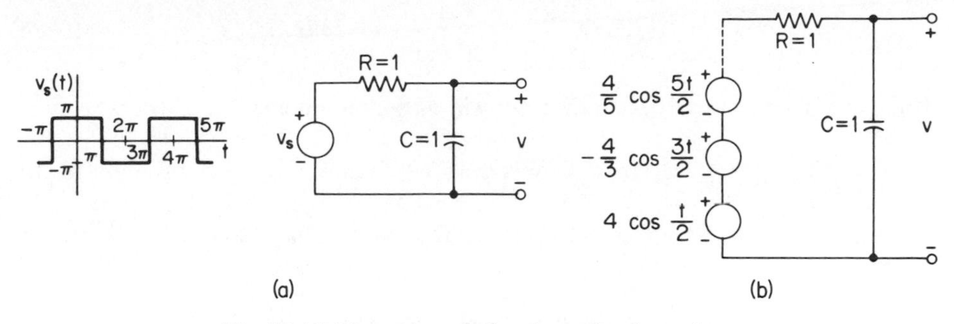

Fig. 10-4.1 Network analysis using a Fourier series.

For simplicity, let us consider only the first three harmonic terms. Thus, the Fourier series of the input waveform is readily found to be [see Eq. (10–11)]

$$v_s(t) = 4(\cos \tfrac{1}{2}t - \tfrac{1}{3}\cos \tfrac{3}{2}t + \tfrac{1}{5}\cos \tfrac{5}{2}t) \qquad (10\text{–}72)$$

Since the right member of the above is a summation of terms all of which have the dimensions of voltage, the input waveform may also be represented as a series connection of voltage sources each of which has an output voltage at one of the sinusoidal frequencies represented in the Fourier series. Such a representation is shown in Fig. 10–4.1(b). Now let us use phasor notation for the network. Using the methods of Chap. 7, we find that the network function $N(j\omega)$ for this network is

$$N(j\omega) = \frac{\mathscr{V}_2}{\mathscr{V}_1} = \frac{1}{1 + j\omega RC} \qquad (10\text{–}73)$$

where $\mathscr{V}_2$ and $\mathscr{V}_1$ are phasors. For the first or fundamental frequency [as applied by the voltage source whose output is $v_1(t)$] we may define the phasor $\mathscr{V}_s^{(1)} = 4\underline{/0^\circ}$. Since the frequency of this source is 0.5 rad/s, we may write

$$\mathscr{V}_0^{(1)} = N(j0.5)\mathscr{V}_s^{(1)} = \frac{4}{1 + j0.5} = 3.576\underline{/-26.57^\circ} \qquad (10\text{–}74)$$

where $\mathscr{V}_0^{(1)}$ is the phasor component of the output produced by this first harmonic term. Thus, the steady-state component of this output in the time domain may be expressed as $v_0^{(1)}(t)$, where

$$v_0^{(1)}(t) = 3.576 \cos(\tfrac{1}{2}t - 26.57^\circ) \qquad (10\text{–}75)$$

Continuing in this manner, we may compute the effect of the second voltage source. Its frequency is 1.5 rad/s and its phasor representation is $\mathscr{V}_s^{(2)} = \frac{4}{3}\underline{/180^\circ}$. If we let its effect on the output be the phasor $\mathscr{V}_0^{(2)}$, we may write

$$\mathscr{V}_0^{(2)} = N(j1.5)\mathscr{V}_s^{(2)} = \frac{-\frac{4}{3}}{1+j1.5} = 0.7392\underline{/123.7^\circ} \tag{10-76}$$

The corresponding steady-state time-domain output component is $v_0^{(2)}(t)$, where

$$v_0^{(2)}(t) = 0.7392 \cos\left(\tfrac{3}{2}t + 123.7^\circ\right) \tag{10-77}$$

Continuing, the output $v_0^{(3)}(t)$ produced by the third voltage source $v_0^{(3)}(t)$ may be shown to be

$$v_0^{(3)}(t) = 0.2972 \cos\left(\tfrac{5}{2}t - 68.2^\circ\right) \tag{10-78}$$

The process described above obviously may be continued for as many terms of the Fourier series as may be required. The steady-state output waveform $v_0(t)$ is then readily found by summing the individual output terms $v_0^{(i)}(t)$. For this example, using only the terms given in Eqs. (10–75), (10–77), and (10–78), we obtain

$$v_0(t) = 3.576 \cos\left(\tfrac{1}{2}t - 26.57^\circ\right) + 0.7392 \cos\left(\tfrac{3}{2}t + 123.7^\circ\right) + 0.2972 \cos\left(\tfrac{5}{2}t - 68.2^\circ\right) \tag{10-79}$$

A sketch of this output waveform is shown in Fig. 10–4.2. It should be noted that this waveform represents only the steady-state component, i.e., the particular solution, of the network response to the applied square wave input. If the complete response is desired, then the time-domain techniques described in Chaps. 5 and 6 must also be applied. The procedure illustrated in the preceding paragraphs may be summarized as follows:

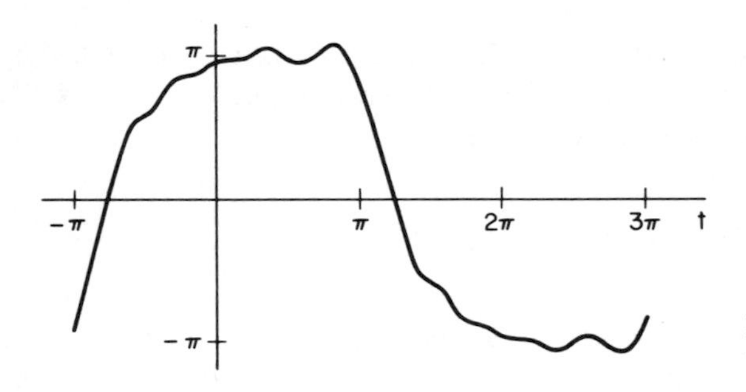

Fig. 10-4.2 Output waveform for the circuit shown in Fig. 10-4.1.

SUMMARY 10-4.1

Analysis of a Network Excited by a Non-Sinusoidal Periodic Waveform: If a non-sinusoidal periodic waveform $f(t)$ with fundamental frequency ω_0 and Fourier series

$$f(t) = c_0 + \sum_{k=1}^{\infty} c_k \cos(k\omega_0 t + \alpha_k)$$

is applied as an excitation to a linear network, the output function $g(t)$ will be described by a non-sinusoidal periodic waveform with Fourier series

$$g(t) = d_0 + \sum_{k=1}^{\infty} d_k \cos (k\omega_0 t + \beta_k)$$

The quantities d_0, d_k, and β_k are determined by the relations

$$\begin{aligned} d_0 &= N(j0)c_0 \\ d_k &= |N(jk\omega_0)|c_k \qquad (k = 1, 2, \dots) \\ \beta_k &= \underline{/N(jk\omega_0)} + \alpha_k \qquad (k = 1, 2, \dots) \end{aligned}$$

where $N(j\omega)$ is the network function relating the output and input phasor.

Fourier series techniques are of special use in evaluating the steady-state performance of filter circuits in which it is desired to select or eliminate a certain frequency component from a waveform which may contain many harmonic components. An example of such an application follows.

EXAMPLE 10-4.1. The circuit shown in Fig. 10-4.3(a) consists of a full-wave bridge rectifier operated from a secondary transformer winding. The other transformer winding is connected to a sinusoidal voltage source of 60 Hz. The rectifier portion of the circuit is

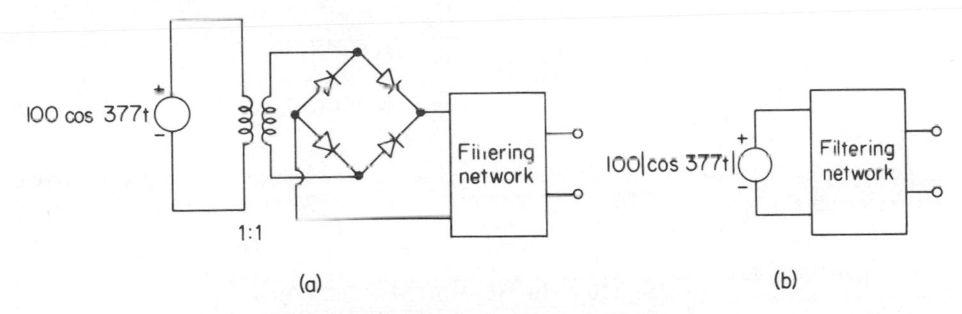

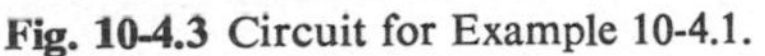
Fig. 10-4.3 Circuit for Example 10-4.1.

followed by a filtering network. This is a common configuration used in electronic circuits to provide a filtered or nearly-constant value of dc voltage from a sinusoidally varying ac supply. The full-wave rectifier portion of the circuit may be modeled by a single voltage source whose output is a rectified sine-wave voltage; thus, the circuit may be represented as shown in Fig. 10-4.3(b).[4] Such a waveform has a second half-cycle which is identical to its first half-cycle. Thus, for convenience, it may be considered as a periodic wave having twice the applied frequency. The first few terms of the Fourier series for such a waveform may be shown to be

[4]It should be noted that this model is only valid if the input current to the filter does not reverse in polarity. Thus, the final result of the example must be checked to make certain that this condition is satisfied.

$$v_s(t) = \frac{200}{\pi}(1 - \tfrac{2}{3}\cos 754t - \tfrac{2}{15}\cos 1508t - \tfrac{2}{35}\cos 2262t + \cdots)$$

Since the first harmonic component of this series (which is the second harmonic component of the source frequency) is the largest of the harmonic terms in the series, it is desirable to design the filtering network to reduce this term as much as possible. To do this we will use the network shown in Fig. 10-4.4. The sinusoidal steady-state network function for this circuit is readily found to be

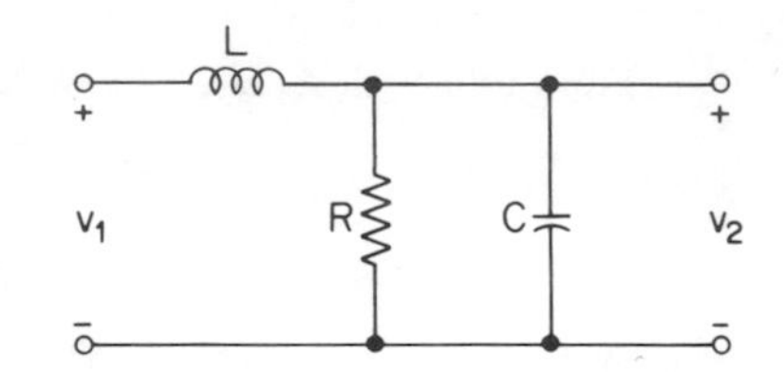

Fig. 10-4.4 Filtering network for Example 10-4.1.

$$N(j\omega) = \frac{\mathscr{V}_2}{\mathscr{V}_1} = \frac{1}{1 - \omega^2 LC + (j\omega L/R)}$$

Let us assume that circuit considerations require that the value of the resistor be fixed at 2 kΩ and the value of the capacitor at 20 μF. We may now choose the value of inductance so as to make the total peak-to-peak second-harmonic output voltage variation at 754 rad/s be some small precentage of the dc output voltage. If we write the output voltage $v_0(t)$ in the form

$$v_0(t) = V_0^{(0)} + V_0^{(1)} \cos 754t + V_0^{(2)} \cos 1508t + \cdots$$

and choose the percentage as 5 per cent, we may write

$$0.05 = \frac{2V_0^{(1)}}{V_0^{(0)}} = \frac{2\left(\frac{400}{3\pi}\right)|N(j754)|}{\left(\frac{200}{\pi}\right)|N(j0)|}$$

In determining $|N(j754)|$, if we substitute the values for the resistor and capacitor given above, we obtain

$$|N(j754)| = \left|\frac{1}{1 - (754)^2(20 \times 10^{-6})L + j(754 \times 0.0005)L}\right| \approx \left|\frac{1}{1 - 11.38L}\right|$$

where we have assumed that the magnitude of the imaginary term in the denominator is smaller than the real term. Since $N(j0)$ is clearly 1, we find that L is 2.433 H. Evaluating the network function at the first, second, and third harmonics of the series (the second, fourth, and sixth harmonics of the original sinusoidal waveform), we obtain

$$N(j754) = 0.0375\underline{/-178.0^\circ}$$
$$N(j1508) = 0.00912\underline{/-179.0^\circ}$$
$$N(j2262) = 0.00403\underline{/-179.4^\circ}$$

We may now apply phasor methods to determine the output phasors $\mathscr{V}_0^{(1)}$, $\mathscr{V}_0^{(2)}$, and $\mathscr{V}_0^{(3)}$ for the first, second, and third harmonic components. Thus, we obtain

$$\mathscr{V}_0^{(1)} = N(j754)\left[\frac{400}{3\pi}\angle 180^\circ\right] = 1.59\angle 2.0^\circ$$

$$\mathscr{V}_0^{(2)} = N(j1508)\left[\frac{400}{15\pi}\angle 180^\circ\right] = 0.0774\angle 1.0^\circ$$

$$\mathscr{V}_0^{(3)} = N(j2262)\left[\frac{400}{35\pi}\angle 180^\circ\right] = 0.0147\angle 0.6^\circ$$

Thus, the time-domain expression for the sinusoidal output waveform becomes

$$V_0(t) = \frac{200}{\pi} + 1.59\cos(754t + 2.0^\circ) + 0.0774\cos(1508t + 1.0^\circ) + 0.0147\cos(2262t + 0.6^\circ)$$

The usefulness of such a filtering circuit in transforming the pulsating rectified sinusoidal input wave into a relatively constant dc voltage is readily apparent.

One of the most useful aspects of the treatment of non-sinusoidal periodic waveforms using the Fourier series given above is that it provides a ready tool for making computations of power dissipation and transmission through a given network. As a preliminary step to a discussion of this topic, since we have already demonstrated the utility of rms quantities in power considerations concerning sinusoidal signals, let us first consider the rms value of a *non*-sinusoidal periodic waveform. To do this, let $f(t)$ be a nonsinusoidal periodic waveform represented by the magnitude and phase form of the Fourier series. Thus, we may write

$$f(t) = c_0 + c_1\cos(\omega_0 t + \phi_1) + c_2\cos(2\omega_0 t + \phi_2) + \cdots \tag{10-80}$$

The general relation for the rms value of a periodic function $f(t)$ with period T as developed in Sec. 7-1 is

$$F_{\text{rms}} = \sqrt{\frac{1}{T}\int_0^T f^2(t)\,dt} \tag{10-81}$$

Substituting the expression for the Fourier series of $f(t)$ as given in Eq. (10-80) into this last expression, we obtain

$$F_{\text{rms}} = \sqrt{\frac{1}{T}\int_0^T [c_0 + c_1\cos(\omega_0 t + \phi_1) + c_2\cos(2\omega_0 t + \phi_2) + \cdots]^2\,dt} \tag{10-82}$$

In this expression the bracketed integrand containing the infinite summation is squared. We may indicate the first terms of the resulting integrand produced when the squared expression is expanded as follows:

$$c_0^2 + c_0 c_1\cos(\omega_0 t + \phi_1) + \cdots + c_1^2\cos^2(\omega_0 t + \phi_1) + c_1\cos(\omega_0 t + \phi_1)\,c_2\cos(2\omega_0 t + \phi_2) + \cdots \tag{10-83}$$

If we now apply the orthogonality relations of (10–32), (10–33), and (10–34), we find that when the integration indicated in (10–82) is performed, all the product terms involving components of different harmonic frequencies disappear. Thus, we may write the expression for the rms value in the form

$$F_{\text{rms}} = \sqrt{\frac{1}{T}\int_0^T \left[c_0^2 + \sum_{k=1}^{\infty} c_k^2 \cos^2(k\omega_0 t + \phi_k)\right] dt} \tag{10–84}$$

If we interchange the operations of integration and summation in the above equation, we obtain

$$F_{\text{rms}} = \sqrt{\frac{1}{T}\int_0^T c_0^2\, dt + \sum_{k=1}^{\infty} \frac{1}{T}\int_0^T c_k^2 \cos^2(k\omega_0 t + \phi_k)\, dt} \tag{10–85}$$

The separate integrals of this expression, however, simply define the rms values of the separate harmonic components of the Fourier series. Thus, in general, we may write

$$F_{\text{rms}} = \sqrt{c_0^2 + \sum_{k=1}^{\infty} F_k^2} \tag{10–86}$$

where the quantity F_k^2 is the square of the rms value of the kth harmonic component of the Fourier series $f(t)$ and where c_0^2 is the square of the constant component of the Fourier series. This constant, of course, is numerically equal to its own rms value. We may summarize the results given as follows:

SUMMARY 10-4.2

The rms Value of a Non-Sinusoidal Periodic Function: The rms or effective value of a non-sinusoidal periodic function is equal to the square root of the sum composed of the squares of the rms values of the various Fourier series harmonic components and the square of the constant (or average) term.

Now let us consider the power which appears at a pair of terminals at which the voltage and current waveforms are non-sinusoidal but periodic. Defining the voltage and current variables as $v(t)$ and $i(t)$ we may write the Fourier series for these quantities as

$$\begin{aligned} v(t) &= v_0 + \sum_{k=1}^{\infty} v_k \cos(k\omega_0 t + \phi_k) \\ i(t) &= i_0 + \sum_{k=1}^{\infty} i_k \cos(k\omega_0 t + \theta_k) \end{aligned} \tag{10–87}$$

Rather than treating the instantaneous power $p(t) = v(t)i(t)$ (which is a function of t), let us define an average power P_{avg} by the relation

$$P_{\text{avg}} = \frac{1}{T}\int_0^T p(t)\,dt = \frac{1}{T}\int_0^T v(t)i(t)\,dt \tag{10-88}$$

Substituting the expressions for the Fourier series for $v(t)$ and $i(t)$ given in (10–87), we obtain

$$P_{\text{avg}} = \frac{1}{T}\int_0^T \left[v_0 + \sum_{k=1}^{\infty} v_k \cos(k\omega_0 t + \phi_k)\right]\left[i_0 + \sum_{k=1}^{\infty} i_k \cos(k\omega_0 t + \theta_k)\right] dt \tag{10-89}$$

If we again apply the orthogonality relations of Sec. 10–2 to the integrand terms that result from multiplying the bracketed terms in Eq. (10–89), we again find that the product terms involving different harmonic frequencies disappear. Thus, the expression may be simplified as follows.

$$P_{\text{avg}} = \frac{1}{T}\int_0^T v_0 i_0\,dt + \sum_{k=1}^{\infty}\left[\frac{1}{T}\int_0^T v_k \cos(k\omega_0 t + \phi_k) i_k \cos(k\omega_0 t + \theta_k)\,dt\right] \tag{10-90}$$

The separate integrals of this expression, however, simply define the average power dissipated at each of the various harmonic frequency components of the Fourier series. Thus, in general, we may write

$$P_{\text{avg}} = P_0 + \sum_{k=1}^{\infty} P_k \tag{10-91}$$

where P_0 is the average dc power, and the quantity P_k is the average of the power being dissipated at the kth harmonic frequency component in the Fourier series. The results given above may be summarized as follows:

SUMMARY 10-4.3

Average Power at a Pair of Terminals at Which the Variables are Periodic but Non-Sinusoidal: The average power dissipated at a pair of terminals of a network at which the variables are periodic but non-sinusoidal may be found by summing the average powers found at each of the harmonic component frequencies (including dc) of the Fourier series representation for the variables.

The conclusions given above with respect to average power are readily extended to the case where complex power is considered.

10-5 USE OF NUMERICAL TECHNIQUES IN FOURIER SERIES ANALYSIS

The methods for applying Fourier series to the analysis of networks excited by non-sinusoidal periodic waveforms that were presented in the preceding sections of this chapter clearly require considerable numerical computations. Such tasks can advantageously be relegated to the digital computer in many cases. In this section we shall describe some means for accomplishing this. Specifically, we shall consider three ways in which digital computational techniques may be advantageously used. These are: (1) the determination of the Fourier series coefficients by numerical integration; (2) the use of digital computational processes in reconstructing the waveform of a function when the Fourier series coefficients are known; and (3) the techniques for determining the output waveform resulting from the application of a non-sinusoidal periodic function as an excitation for a linear time-invariant network.

Let us begin our discussion of the use of numerical techniques in Fourier series analysis methods by considering the problem of determining the coefficients of Fourier series. For a series having the form

$$f(t) = a_0 + \sum_{h=1}^{\infty} (a_h \cos h\omega_0 t + b_h \sin h\omega_0 t) \tag{10-92}$$

it was shown in Sec. 10-2 that the coefficients are determined by the integral equations

$$\begin{aligned} a_0 &= \frac{1}{T}\int_0^T f(t)\,dt \\ a_h &= \frac{2}{T}\int_0^T f(t)\cos h\omega_0 t\,dt \\ b_h &= \frac{2}{T}\int_0^T f(t)\sin h\omega_0 t\,dt \end{aligned} \tag{10-93}$$

where $T = 2\pi/\omega_0$ is the period. We have seen in Chap. 4, however, that numerical integration is readily implemented by digital computational techniques. In that chapter we introduced the subroutine ITRPZ for performing such integrations. A summary of its characteristics is given in Table 4-4.1. To apply this subroutine, we must define an integrand function Y(T) which is called by ITRPZ. For our application here, this function must be so constructed that, in addition to being supplied with the value of the variable of integration T, it must also be supplied with the value of ω_0 (which will be different for each problem) and the value of h (which will be different for each integration in a given problem). Thus, two more input arguments must be provided for the function. Since we cannot change the argument list of the function Y(T) (without rewriting the subroutine ITRPZ) we will introduce these additional arguments by using a *common* statement. The reader who is unfamiliar with the use of such statements should consult any of the standard texts on FORTRAN. If we let the variable OM represent ω_0 and the variable H represent h in Eq. (10-93), the common statement will have the from

$$\text{COMMON OM, H, K} \tag{10–94}$$

where K is an indicator which is used to specify the use of the cosine term or the sine term in the integrand. The quantity $f(t)$ for which we wish to determine the Fourier series may now be specified by a separate function FCN(T) which is called by the function Y(T). Thus, we may specify a function Y(T) which may be used as the integrand in any of the relations of Eq. (10–93) for any arbitrary non-sinusoidal periodic function $f(t)$. A flow chart of the logic used to accomplish this is given in Fig. 10–5.1 together with a listing of the function Y(T). To actually find the Fourier

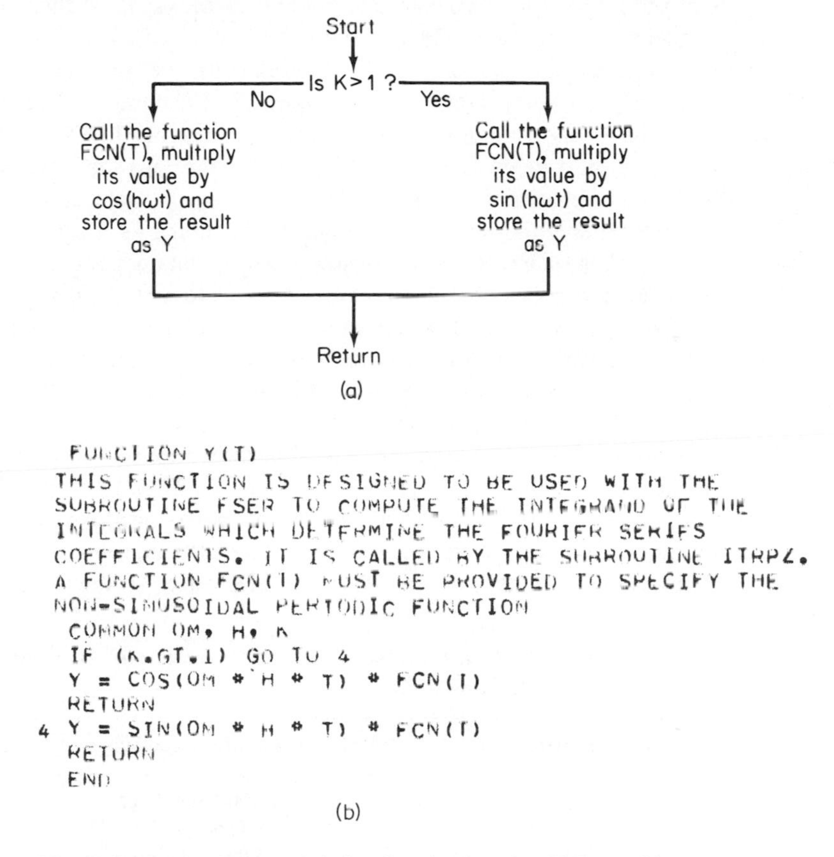

Fig. 10-5.1 Flow chart and listing for the function Y(T) used in determining Fourier series coefficients.

series coefficients of Eq. (10–93) we may now specify a subroutine FSER (for Fourier SERies). The purpose of this subroutine will be to call the subroutine ITRPZ in such a way as to perform the various integrations specified by the relations of Eq. (10–93). In considering the argument list which such a function requires, we note that it must be supplied with a constant specifying the number of coefficients desired. We will use the argument NF to indicate the highest harmonic term to be computed. An input argument is also needcd to specify the period of the non-sinusoidal periodic function. We will use the variable PERIOD for this purpose. A final input argument is used to specify the number of iterations to be used by the integration subroutine ITRPZ. We will use the variable ITER for this. The comments made in Sec. 4–4 with respect to the proper choice of value for this variable apply equally well here. It should be noted, however, that the integrand for the higher-order harmonic terms will be more rapidly varying than that for the lower-order terms. Thus, the integration for the higher-order terms might logically be expected to require higher values of the variable ITER. To offset this requirement, however, we note that in general the magnitude of the higher-order harmonic terms will be lower; thus, their determination can be made less accurate without affecting the total accuracy of the series representation. Considering both of the above comments, we may conclude that in many problems a good compromise between numerical accuracy and computation time may be achieved by using the same value of the variable ITER for the determination of all the coefficients. This is the approach which we shall use here.

Another consideration which is of interest with respect to the choice of a value for ITER occurs when functions with discontinuities (such as a square wave) are analyzed. In general, in such cases, better accuracy will be obtained if the time divisions used by the integration subroutine do not exactly coincide with the discontinuity. As an example of this, consider the determination of the coefficient a_1 (which multiplies the first harmonic cosine term) for a square wave which has odd symmetry. From basic symmetry considerations we know that the actual value of such a coefficient should be zero. In Table 10–5.1 are shown the results obtained from a determination of this coefficient with the subroutine FSER using various values for the input argument ITER. It is readily observed that by using 101 iterations (in which case, the time divisions used in the integration process do not coincide with the discontinuity), considerably better accuracy is obtained than the case where 200 iterations are used. Thus, a lower number of iterations provides better accuracy in this case.

TABLE 10-5.1 Effect of Various Numbers of Iterations in Determining a Fourier Coefficient

Number of iterations	Value of the coefficient a_1
100	1.00008444E–02
101	8.44800814E–07
200	5.00084460E–03
201	8.44699000E–07

As output, the subroutine FSER is designed to compute the variable A0 (the value of the coefficient a_0), the one-dimensional array A [containing the variables A(I) which give the values of the coefficients a_h], and the one-dimensional array B [with variables B(I) which give the values of the coefficients b_h]. In addition, it is convenient to have the subroutine simultaneously compute the coefficients of the magnitude and phase form of the Fourier series. This has the form

$$b(t) = c_0 + \sum_{h=1}^{\infty} c_h \cos (h\omega_0 t + \phi_h) \tag{10-95}$$

For this purpose we may specify the output one-dimensional arrays C and PH for the coefficients c_h and ϕ_h (in radians), respectively. Considering all of the above, the identifying statement for the subroutine FSER will have the form

SUBROUTINE FSER (NF, PERIOD, ITER, A0, A, B, C, PH) (10-96)

The flow chart and listing for the subroutine FSER is given in Fig. 10-5-2. A sum-

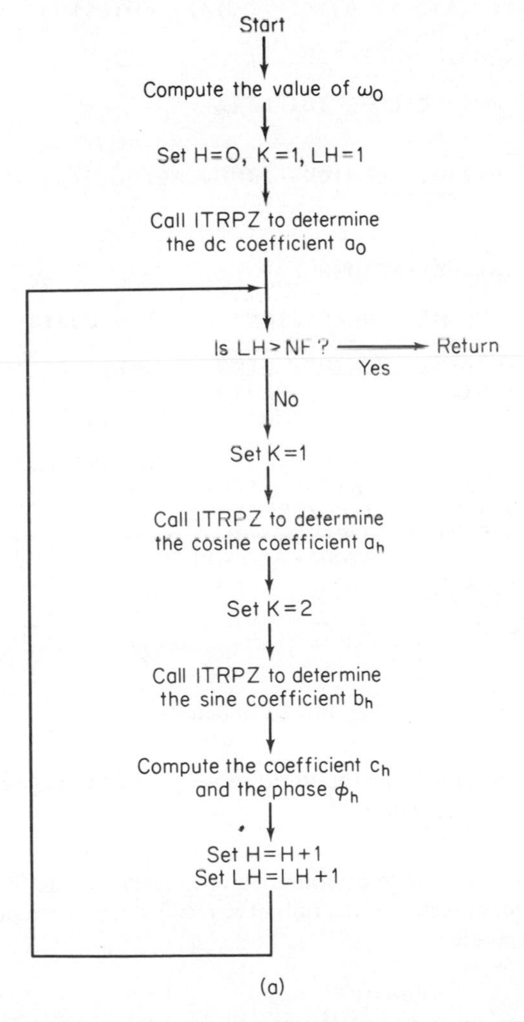

Fig. 10-5.2 Flow chart and listing for the subroutine FSER.

```
      SUBROUTINE FSER(NF, PERIOD, ITER, A0, A, B, C, PH)
C     SUBROUTINE FOR DETERMINING THE COEFFICIENTS OF THE
C     TERMS IN THE FOURIER SERIES FOR A PERIODIC
C     NON-SINUSOIDAL FUNCTION SPECIFIED AS FCN(T).  THIS
C     SUBROUTINE MUST BE USED WITH A SPECIAL FUNCTION Y(T)
C     WHICH IS CALLED BY THE SUBROUTINE ITRPZ
C           NF - FINAL HARMONIC TERM TO BE EVALUATED
C       PERIOD - TIME FOR ONE COMPLETE CYCLE
C         ITER - NUMBER OF ITERATIONS USED BY SUBROUTINE
C                ITRPZ FOR EACH COEFFICIENT
C           A0 - OUTPUT ARGUMENT, VALUE OF DC COEFFICIENT
C            A - OUTPUT ARRAY, VALUES OF COSINE COEFFICIENTS
C            B - OUTPUT ARRAY, VALUES OF SINE COEFFICIENTS
C            C - OUTPUT ARRAY, COEFFICIENT MAGNITUDES
C           PH - OUTPUT ARRAY (PHASE, RADIANS)
      DIMENSION A(10), B(10), C(10), PH(10)
      COMMON OM, H, K
      OM = 6.28318 / PERIOD
C
C     COMPUTE THE DC COEFFICIENT A0
      H = 0.
      K = 1
      CALL ITRPZ(0., PERIOD, ITER, A0)
      A0 = A0 / PERIOD
      H = 1.
      LH = 1
    9 IF (LH.GT.NF) RETURN
C
C     COMPUTE THE HTH COEFFICIENT FOR THE COSINE TERMS
      K = 1
      CALL ITRPZ(0., PERIOD, ITER, A(LH))
      A(LH) = A(LH) * 2.0 / PERIOD
      K = 2
C
C     COMPUTE THE HTH COEFFICIENT FOR THE SINE TERMS
      CALL ITRPZ(0., PERIOD, ITER, B(LH))
      B(LH) = B(LH) * 2.0 / PERIOD
      C(LH) = SQRT(A(LH)**2 + B(LH)**2)
      PH(LH) = ATAN2(-B(LH), A(LH))
      H = H + 1.
      LH = LH + 1
      GO TO 9
      END
```

(b)

Fig. 10-5.2 Continued

mary of its characteristics may be found in Table 10–5.2 (on page 808). An example of the use of the subroutine follows.

EXAMPLE 10-5.1. It is desired to determine the coefficients of the first three harmonic components of the Fourier series for the full-wave rectified sine wave shown in Fig. 10-5.3 and defined by the expression

$$f(t) = \frac{\pi}{2}\left|\sin t\right|$$

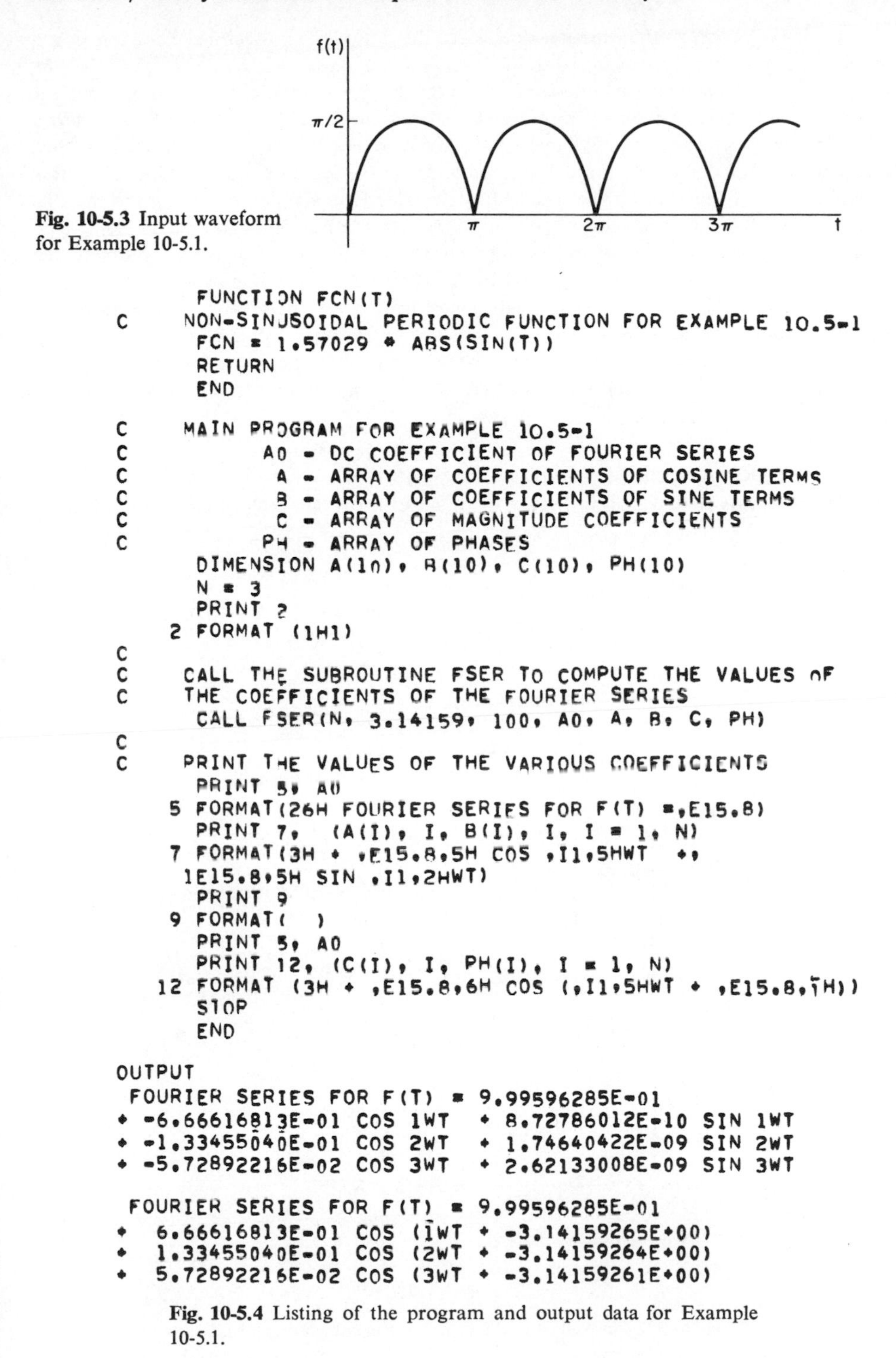

Fig. 10-5.3 Input waveform for Example 10-5.1.

```
      FUNCTION FCN(T)
C     NON-SINUSOIDAL PERIODIC FUNCTION FOR EXAMPLE 10.5-1
      FCN = 1.57029 * ABS(SIN(T))
      RETURN
      END

C     MAIN PROGRAM FOR EXAMPLE 10.5-1
C           A0 - DC COEFFICIENT OF FOURIER SERIES
C            A - ARRAY OF COEFFICIENTS OF COSINE TERMS
C            B - ARRAY OF COEFFICIENTS OF SINE TERMS
C            C - ARRAY OF MAGNITUDE COEFFICIENTS
C           PH - ARRAY OF PHASES
      DIMENSION A(10), B(10), C(10), PH(10)
      N = 3
      PRINT 2
    2 FORMAT (1H1)
C
C     CALL THE SUBROUTINE FSER TO COMPUTE THE VALUES OF
C     THE COEFFICIENTS OF THE FOURIER SERIES
      CALL FSER(N, 3.14159, 100, A0, A, B, C, PH)
C
C     PRINT THE VALUES OF THE VARIOUS COEFFICIENTS
      PRINT 5, A0
    5 FORMAT(26H FOURIER SERIES FOR F(T) =,E15.8)
      PRINT 7,  (A(I), I, B(I), I, I = 1, N)
    7 FORMAT(3H + ,E15.8,5H COS ,I1,5HWT  +,
     1E15.8,5H SIN ,I1,2HWT)
      PRINT 9
    9 FORMAT(  )
      PRINT 5, A0
      PRINT 12, (C(I), I, PH(I), I = 1, N)
   12 FORMAT (3H + ,E15.8,6H COS (,I1,5HWT + ,E15.8,1H))
      STOP
      END

OUTPUT
 FOURIER SERIES FOR F(T) = 9.99596285E-01
+ -6.66616813E-01 COS 1WT  + 8.72786012E-10 SIN 1WT
+ -1.33455040E-01 COS 2WT  + 1.74640422E-09 SIN 2WT
+ -5.72892216E-02 COS 3WT  + 2.62133008E-09 SIN 3WT

 FOURIER SERIES FOR F(T) = 9.99596285E-01
+   6.66616813E-01 COS (1WT + -3.14159265E+00)
+   1.33455040E-01 COS (2WT + -3.14159264E+00)
+   5.72892216E-02 COS (3WT + -3.14159261E+00)
```

Fig. 10-5.4 Listing of the program and output data for Example 10-5.1.

Although the period of the function sin t is 2π, the period of the rectified function is π s. Thus, we expect that only even harmonic terms such as cos $2t$, sin $2t$, cos $4t$, sin $4t$, cos $8t$, etc., and the term a_0 may be non-zero in the Fourier series. In addition, we note that the function is even. Thus, we may expect that the coefficients associated with the sine terms of all frequencies will be zero. We may now proceed to use the subroutine FSER described above to find the coefficients of the Fourier series by numerical integration. This subroutine requires that the function $f(t)$ be specified by the function FCN(T), where T is the variable of integration. A listing of such a function for this problem is given in Fig. 10-5.4. Since there are three harmonic terms to be computed, we will set NF = 3 in the argument list for the subroutine FSER. The argument PERIOD is set to the value of π. We shall use a value of 100 for the constant ITER used in performing the integration. A listing of the main program together with the output data is shown in Fig. 10-5.4. It is readily verified that the values of the coefficients are similar to those given in Example 10-4.1 for the same waveform and that the values of the coefficients of the sine terms are very small, especially when compared to those of the cosine terms.

The second application of digital computational techniques to Fourier series analysis methods that we shall discuss in this section is the construction and plotting

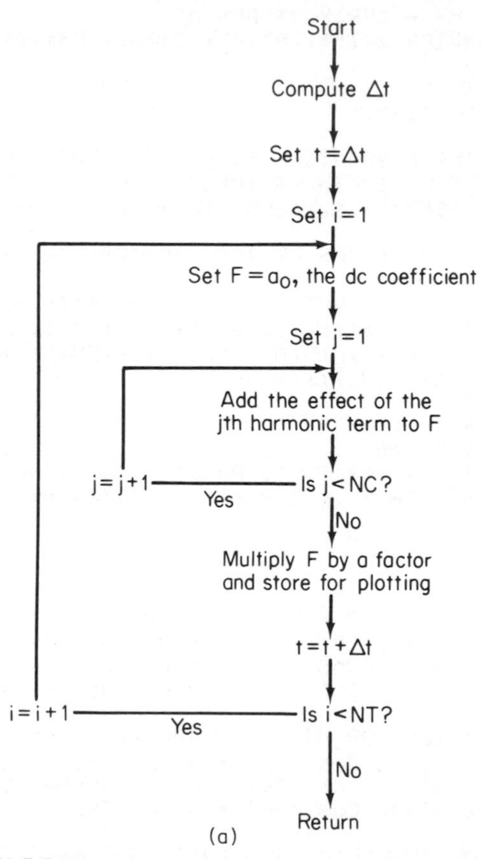

Fig. 10-5.5 Flow chart and listing for the subroutine FPLT.

```
      SUBROUTINE FPLT(NT, PERIOD, A0, NC, C, PH, NP, P, SCALE)
C           NT - NUMBER OF TIME VALUES TO BE PLOTTED
C       PERIOD - TIME FOR ONE PERIOD OF FUNCTION
C           A0 - CONSTANT FOURIER COEFFICIENT
C           NC - NUMBER OF HARMONIC TERMS TO BE USED
C            C - ARRAY OF FOURIER COEFFICIENT MAGNITUDES
C           PH - ARRAY OF FOURIER COEFFICIENT PHASES
C           NP - STORAGE LOCATION IN PLOTTING ARRAY
C            P - OUTPUT ARRAY FOR USE IN SUBROUTINE PLOT5
C        SCALE - SCALING FACTOR FOR PLOTTED DATA
      DIMENSION C(10), PH(10), P(5,101)
      RAD = 6.28318 / PERIOD
      TNT = NT - 1
      DT = PERIOD / TNT
      T = 0.0
C
C     MAKE COMPUTATIONS AT NT DIFFERENT VALUES OF TIME
C     AND STORE THE RESULTS FOR PLOTTING
      DO 11 I = 1, NT
      F = A0
C
C     DETERMINE THE TOTAL EFFECT OF NC HARMONIC TERMS
C     AT EACH VALUE OF TIME
      DO 9  J = 1, NC
      X = J
    9 F = F + C(J) * COS(RAD * X * T + PH(J))
      P(NP,I) = F * SCALE
   11 T = T + DT
      RETURN
      END
```

(b)

Fig. 10-5.5 Continued

of a waveform for which the Fourier coefficients are known. To accomplish this we need a computational package which can be supplied with the values of the Fourier coefficients and the value of the period of the periodic waveform and which will then calculate the value of the function at each of a number of time increments over the specified period. Such an operation is easily implemented by a subroutine. In its argument list, we will use the variable A0 for the coefficient a_0 and the one-dimensional arrays C and PH for the coefficients c_i and ϕ_i, respectively. Thus, we assume that the Fourier series is given in the magnitude and phase form shown in Eq. (10-95). The argument NC will be used to specify the number of coefficients c_i and ϕ_i which are to be used, and the argument PERIOD will be used for the period. The number of values of t at which an evaluation of the function is made, and thus the resulting number of points which are plotted, is determined by the argument NT. The computed data are stored in the output array P which is dimensioned (5,101) for use with the plotting subroutine PLOT5. The argument SCALE is used to scale the data at the time they are stored in the P array. Finally, to permit the possibility of preparing multiple plots, we will use the argument NP to determine the first index of the array P in which the data are to be stored. Thus, the scaled values of the function to be plotted will be stored in P(NP, I) for values of I from 1 to NT. If we use the name FPLT for this subroutine, it may be identified by the statement

$$\text{SUBROUTINE FPLT (NT, PERIOD, A0, NC, C, PH, NP, P, SCALE)} \quad (10\text{–}97)$$

A listing and a flow chart for such a subroutine is given in Fig. 10–5.5. Its characteristics are summarized in Table 10–5.3 (on page 809). An example of its use follows.

EXAMPLE 10-5.2 It is desired to plot the Fourier series representation of the function $f(t)$ shown in Fig. 10-5.6 for two different cases. The first case is defined by terminating the Fourier series for the function after the second harmonic term, and the second case by terminating the series after the sixth harmonic term. To do this, we will use the subroutine FSER to determine the coefficients of the Fourier series and the subroutine FPLT to reconstruct the function. A flow chart and a listing of the main program and the function subprogram FCN defining the function $f(t)$ is given in Fig. 10-5.7. A plot of the output

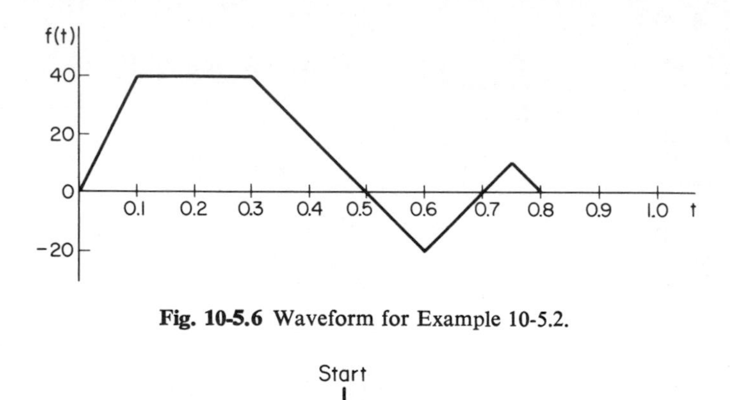

Fig. 10-5.6 Waveform for Example 10-5.2.

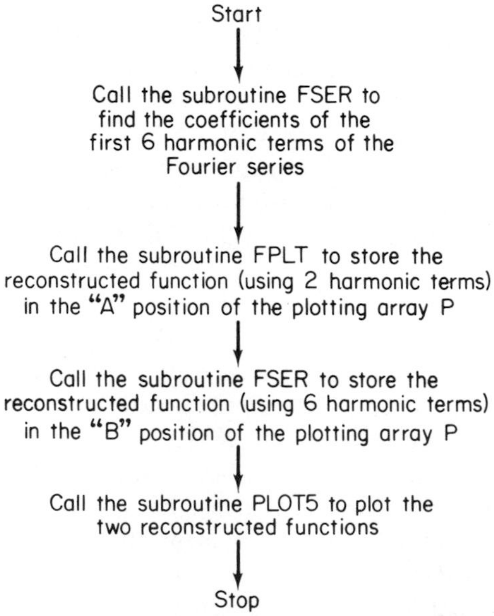

Start

↓

Call the subroutine FSER to find the coefficients of the first 6 harmonic terms of the Fourier series

↓

Call the subroutine FPLT to store the reconstructed function (using 2 harmonic terms) in the "A" position of the plotting array P

↓

Call the subroutine FSER to store the reconstructed function (using 6 harmonic terms) in the "B" position of the plotting array P

↓

Call the subroutine PLOT5 to plot the two reconstructed functions

↓

Stop

Fig. 10-5.7 Flow chart and program listing for Example 10-5.2.

```
      FUNCTION FCN(T)
C     FUNCTION FOR FOURIER ANALYSIS AND PLOTTING
      IF (T.GT.0.1) GO TO 4
      FCN = 400. * T
      RETURN
    4 IF (T.GT.0.3) GO TO 7
      FCN = 40.
      RETURN
    7 IF (T.GT.0.6) GO TO 10
      FCN = 100. - 200. * T
      RETURN
   10 IF (T.GT.0.75) GO TO 13
      FCN = -140. + 200. * T
      RETURN
   13 IF (T.GT.0.8) GO TO 18
      FCN = 160. - 200. * T
      RETURN
   18 FCN = 0.
      RETURN
      END

C     PROGRAM FOR FOURIER ANALYSIS AND PLOTTING
C             A0 - CONSTANT FOURIER COEFFICIENT
C              C - ARRAY OF FOURIER COEFFICIENT MAGNITUDES
C             PH - ARRAY OF FOURIER COEFFICIENT PHASES
C              P - TWO-DIMENSIONAL ARRAY FOR USE IN PLOTTING
      DIMENSION A(10), B(10), C(10), PH(10), P(5,101)
      CALL FSER(6, 1.0, 100, A0, A, B, C, PH)
      CALL FPLT(101, 1.0, A0, 2, C, PH, 1, P, 1.0)
      CALL FPLT(101, 1.0, A0, 6, C, PH, 2, P, 1.0)
      PRINT 5
    5 FORMAT(1H1)
      CALL PLOT5(P, 2, 101, 70)
      STOP
      END
```

Fig. 10-5.7 Continued

waveforms for the two cases is given in Fig. 10-5.8. The manner in which the series converges to approach the original function as higher-order harmonic terms are added is clearly indicated.

The third application of digital computational techniques to Fourier series methods of network analysis that we will treat in this section is the determination of the output waveform for a given network when a non-sinusoidal periodic function is applied as an input. To accomplish this, three steps are required. First, we may use the subroutine FSER to determine the values of the Fourier coefficients for the periodic function. Second, we may use the subroutine SSS whose properties are described in Table 7-10.2 to determine the magnitude and phase of the network func-

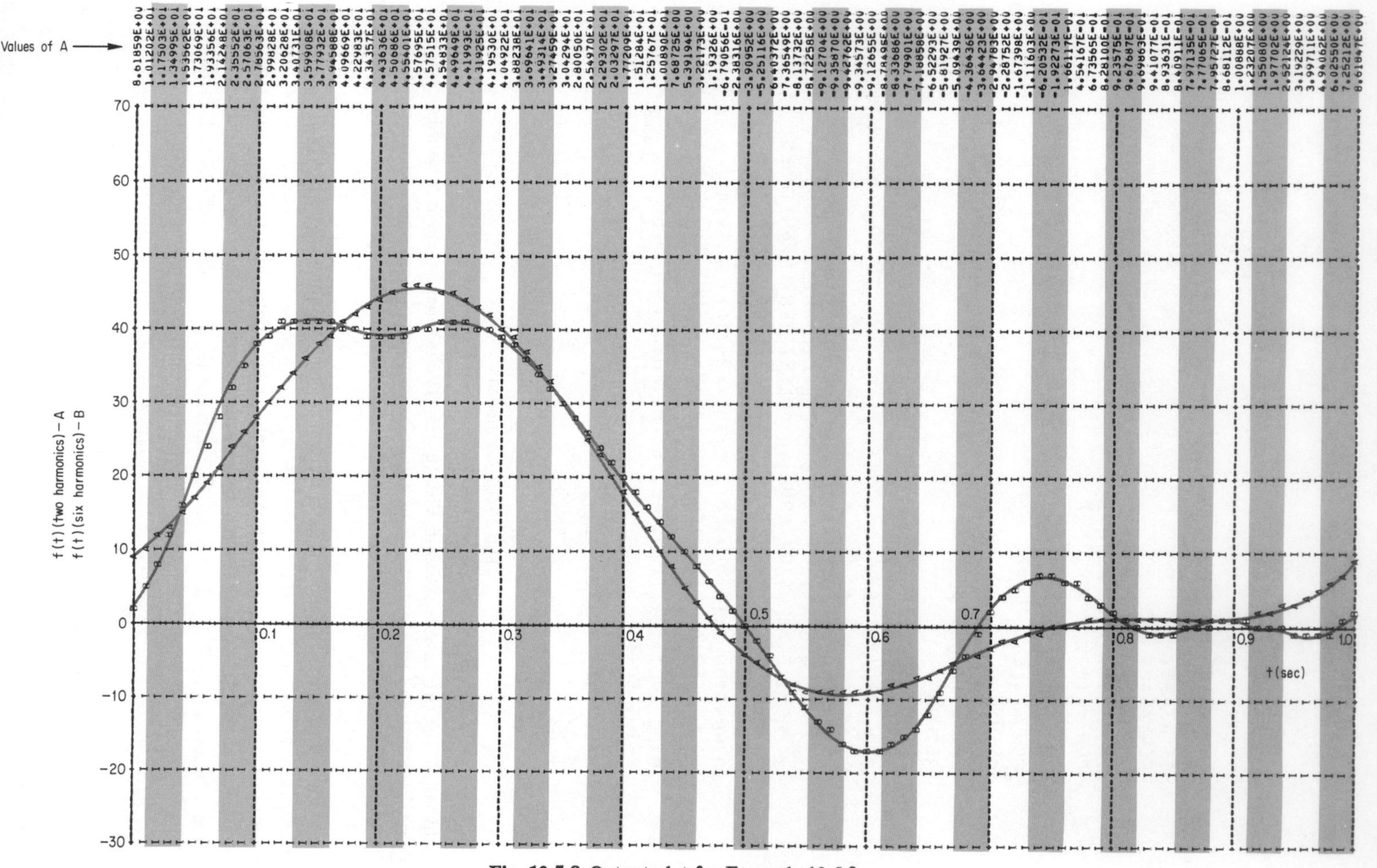

Fig. 10-5.8 Output plot for Example 10-5.2.

tion at each of the frequencies of the harmonic terms in the Fourier series. These values may then be used to modify the Fourier coefficients describing the input waveform. Finally, we may use the subroutine FPLT to plot the resulting output waveform. The method is illustrated in the following example.

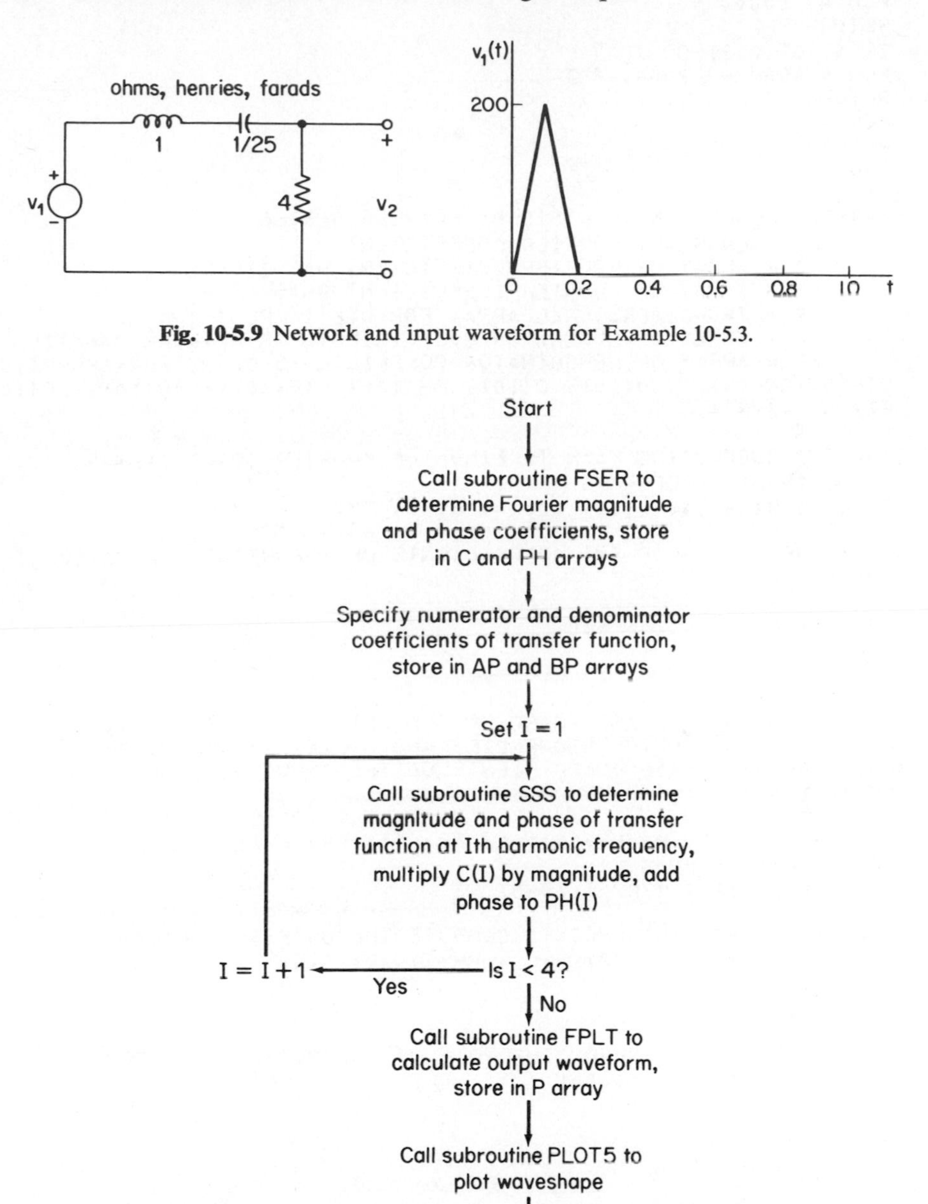

Fig. 10-5.9 Network and input waveform for Example 10-5.3.

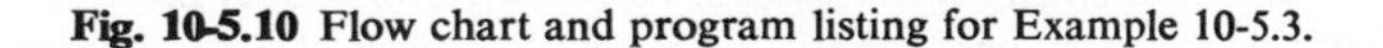

Fig. 10-5.10 Flow chart and program listing for Example 10-5.3.

```
      FUNCTION FCN(T)
C     NON-SINUSOIDAL PERIODIC INPUT FUNCTION FOR
C     FOURIER SERIES NETWORK ANALYSIS EXAMPLE
      IF (T.GT.0.1) GO TO 4
      FCN =  2000. * T
      RETURN
    4 IF (T.GT.0.2) GO TO 7
      FCN = 400. -  2000. * T
      RETURN
    7 FCN = 0.
      RETURN
      END

C     EXAMPLE OF NETWORK ANALYSIS BY FOURIER SERIES
C           A0 - CONSTANT FOURIER COEFFICIENT
C            C - ARRAY OF FOURIER COEFFICIENT MAGNITUDES
C           PH = ARRAY OF FOURIER COEFFICIENT PHASES
C            P = TWO-DIMENSIONAL ARRAY FOR USE IN PLOTTING
C           AP - ARRAY OF NUMERATOR COEFFICIENTS OF NETWORK FUNCTION
C           BP = ARRAY OF DENOMINATOR COEFFICIENTS OF NETWORK FUNCTION
      DIMENSION A(10), B(10), C(10), PH(10), P(5,101), AP(10), BP(10)
      PI2 = 6.28318
C
C     CALL THE SUBROUTINE FSER TO FIND THE FOURIER COEFFICIENTS
C     OF THE INPUT WAVEFORM
      CALL FSER(4, 1.0, 100, A0, A, B, C, PH)
C
C     STORE THE VALUES OF THE COEFFICIENTS OF THE NETWORK FUNCTION
      AP(1)  = 0.
      AP(2)  = 4.
      BP(1)  = PI2 * PI2 + 4.0
      BP(2)  = 4.
      BP(3)  = 1.
C
C     FIND THE MAGNITUDE AND PHASE OF THE NETWORK FUNCTION AT
C     EACH OF THE HARMONIC FREQUENCIES AND MODIFY THE FOURIER
C     MAGNITUDE AND PHASE COEFFICIENTS OF THE INPUT WAVEFORM
      DO 12 I = 1, 4
      X = I
      CALL SSS(1, AP, 2, BP, PI2*X, FR, FI, FP, FMAG)
      C(I) = C(I) * FMAG
   12 PH(I) = PH(I) + (FP/57.2958)
C
C     CALL THE SUBROUTINE FPLT TO COMPUTE THE OUTPUT WAVEFORM
      CALL FPLT(51, 1.0, A0, 4, C, PH, 1, P, 1.0)
      PRINT 15
   15 FORMAT(1H1)
C
C     CALL THE PLOTTING SUBROUTINE TO PLOT THE OUTPUT WAVEFORM
      CALL PLOT5(P, 1, 51, 80)
      STOP
      END
```

(b)

Fig. 10-5.10 Continued

EXAMPLE 10-5.3 It is desired to find the output waveform $v_2(t)$ for the network shown in Fig. 10-5.9 when the non-sinusoidal periodic input waveform $v_1(t)$ shown in the figure is applied as an input. A flow chart of the logic used to accomplish this is shown in Fig.

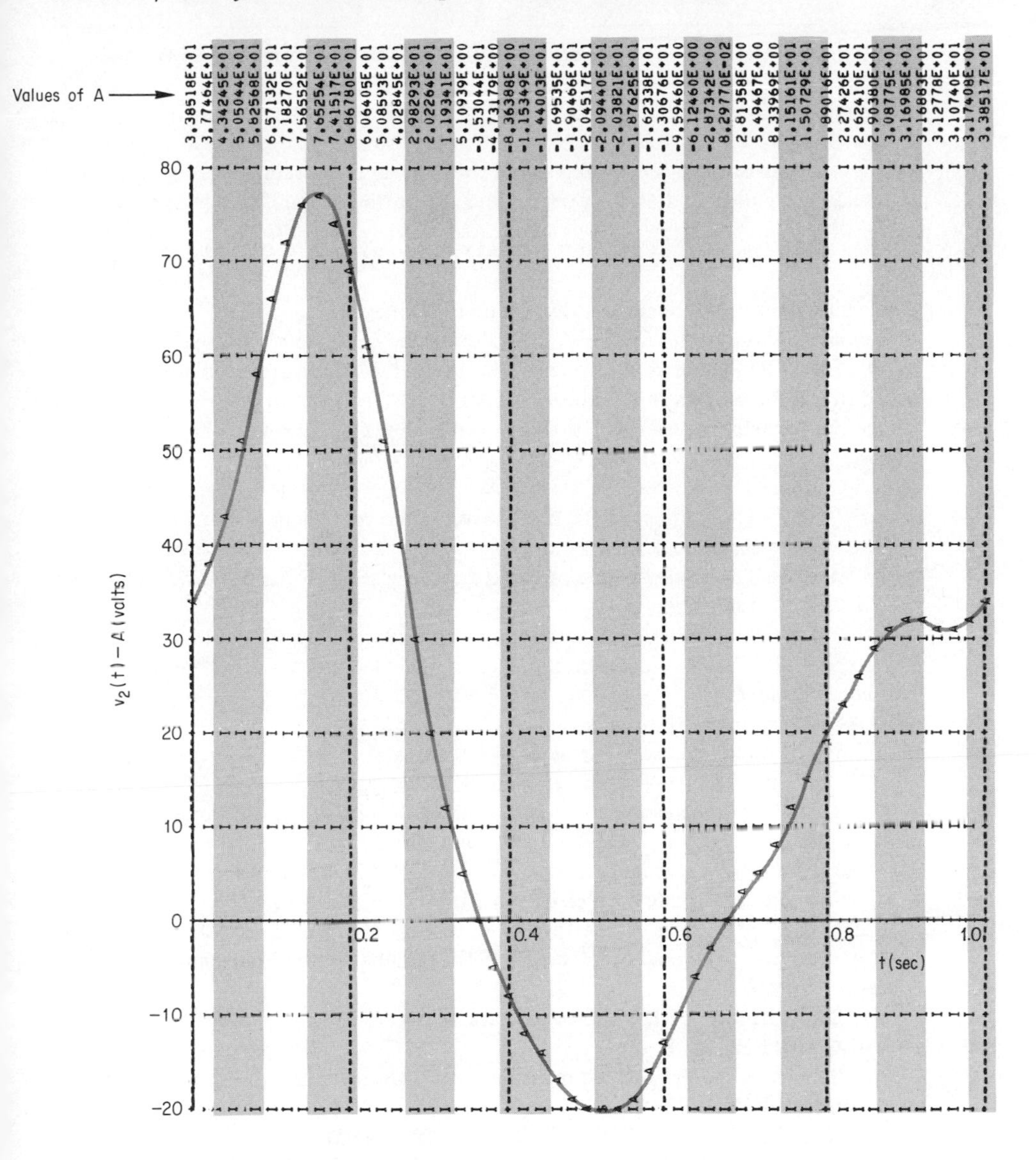

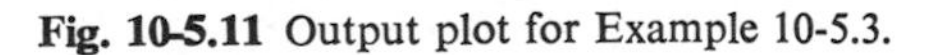
Fig. 10-5.11 Output plot for Example 10-5.3.

TABLE 10-5.2 Characteristics of the Subroutine FSER

Identifying Statement: SUBROUTINE FSER (NF, PERIOD, ITER, A0, A, B, C, PH)

Purpose: To determine the coefficients of the Fourier series representation for a non-sinusoidal periodic function $f(t)$, i.e., to find the coefficients a_i, b_i, c_i, and ϕ_i, where

$$f(t) = a_0 + \sum_i (a_i \cos i\omega_0 t + b_i \sin i\omega_0 t) = a_0 + \sum_i c_i \cos (i\omega_0 t + \phi_i)$$

and where ω_0 is the fundamental harmonic frequency.

Additional Subprograms Required: The subroutine calls the trapezoidal integration subroutine ITRPZ, which in turn calls the function Y. This function must be provided in the form given in the notes. A function FCN(T) must also be provided to define the periodic function $f(t)$ over the range $0 \leq t \leq T$, where the period $T = 2\pi/\omega_0$.

Input Arguments:

- NF The number of the highest harmonic coefficient of the Fourier series that it is desired to compute.
- PERIOD The value of time required for one period of the non-sinusoidal periodic function $f(t)$.
- ITER The number of iterations to be used by the trapezoidal subroutine ITRPZ in determining each of the coefficients of the Fourier series.

Output Arguments:

- A0 The value of the coefficient a_0.
- A The one-dimensional array of variables A(I) in which are stored the values of the cosine coefficients a_i of the Fourier series.
- B The one-dimensional array of variables B(I) in which are stored the values of the sine coefficients b_i of the Fourier series.
- C The one-dimensional array of variables C(I) in which are stored the values of the magnitude coefficients c_i of the Fourier series.
- PH The one-dimensional array of variables PH(I) in which are stored the values of the phase coefficients ϕ_i (in radians) of the Fourier series.

Notes:

1. The variables of the subroutine are dimensioned as follows: A(10), B(10), C(10), PH(10).
2. The function Y(T) which must be used with the subroutine is given below.

```
      FUNCTION Y(T)
C     THIS FUNCTION IS DESIGNED TO BE USED WITH THE
C     SUBROUTINE FSER TO COMPUTE THE INTEGRAND OF THE
C     INTEGRALS WHICH DETERMINE THE FOURIER SERIES
C     COEFFICENTS. IT IS CALLED BY THE SUBROUTINE ITRPZ.
C     A FUNCTION FCN(T) MUST BE PROVIDED TO SPECIFY THE
C     NON-SINUSOIDAL PERIODIC FUNCTION
      COMMON OM, H, K
      IF (K. GT. 1) GO TO 4
      Y = COS (OM*H*T)*FCN(T)
      RETURN
    4 Y = SIN(OM*H*T)*FCN(T)
      RETURN
      END
```

TABLE 10-5.3 Characteristics of the Subroutine FPLT

Identifying Statement: SUBROUTINE FPLT (NT, PERIOD, A0, NC, C, PH, NP, P, SCALE)

Purpose: To compute and store the data required for plotting the waveform of one period of a non-sinusoidal periodic function $f(t)$ which is specified in the form

$$f(t) = a_0 + \sum_i c_i \cos(i\omega_0 t + \phi_i)$$

where ω_0 is the fundamental harmonic frequency.

Additional Subprograms Required: None.

Input Arguemnts:

- NT The number of points desired to be used for plotting one period of the non-sinusoidal periodic function $f(t)$.
- PERIOD The value of time required for one period of $f(t)$.
- A0 The value of the Fourier series coefficient a_0.
- NC The number of harmonic terms to be included in calculating the points used for plotting $f(t)$.
- C The one-dimensional array of variables C(I) in which are stored the values of the magnitude coefficients c_i of the Fourier series.
- PH The one-dimensional array of variables PH(I) in which are stored the phase coefficients ϕ_i (in radians) of the Fourier series.
- NP The position in the plotting array P(I, J) in which the computed data are to be stored, i.e., the value to be given to the index I.
- SCALE The scaling factor which is to be used when data are stored in the plotting array P(I, J). If no scaling is desired, this argument must be set to unity.

Output Argument:

- P The two-dimensional array of variables P(I, J) in which the values of the non-sinusoidal periodic function $f(t)$, as computed at different values of t, are stored for plotting. Specifically, these values are stored as P(NP, J) where J has a range from 1 to NT.

Note: The variables of the subroutine are dimensioned as follows: C(10), PH(10), P(5, 101).

10-5.10(a). A listing of the main program which implements this logic and the function subprogram FCN defining $v_1(t)$ is given in Fig. 10-5.10(b). A plot of the output waveform including terms through the fourth harmonic is shown in Fig. 10-5.11.

The subroutines FSER and FPLT introduced in this section provide a pair of computational tools which may readily be applied to solve a wide range of circuit analysis problems which require the use of the Fourier series to analyze the performance of a network which is excited by non-sinusoidal periodic inputs. Because of the tedious nature of the computations normally required for such applications, digital computational techniques are especially worthwhile for these types of problems.

10-6 THE FOURIER TRANSFORMATION

In this section we show how the Fourier series techniques developed in the preceding sections of this chapter may be extended so as to be applicable to non-periodic functions. Such an application is made possible by the use of the *Fourier transformation* (also referred to as the *Fourier integral*). In this section we will derive the Fourier transformation as a limiting case of the Fourier series. Before treating this topic, we will first consider a new way of writing the Fourier series which will be useful in the development of the transformation.

In Sec. 10–2 we developed two forms which could be used for expressing the Fourier series. These are

$$f(t) = a_0 + \sum_{k=1}^{\infty} a_k \cos k\omega_0 t + \sum_{k=1}^{\infty} b_k \sin k\omega_0 t \tag{10–98a}$$

and

$$f(t) = a_0 + \sum_{k=1}^{\infty} c_k \cos (k\omega_0 t + \phi_k) \tag{10–98b}$$

where the first series given above may be referred to as the *sine and cosine* form of the Fourier series and the second may be referred to as the *magnitude and phase* form. The relations between the coefficients of the two forms of the series in Eqs. (10–98a) and (10–98b) are given in Summary 10–2.2. Let us now consider a third form for the Fourier series. We begin by stating Euler's relations for expressing sine and cosine quantities. These are:

$$\cos x = \frac{e^{jx} + e^{-jx}}{2} \qquad \sin x = \frac{e^{jx} - e^{-jx}}{2j}$$
$$e^{\pm jx} = \cos x \pm j \sin x \tag{10–99}$$

Inserting the first two of these relations in Eq. (10–98a) and rearranging terms, we obtain

$$f(t) = a_0 + \sum_{k=1}^{\infty} \frac{a_k - jb_k}{2} e^{jk\omega_0 t} + \sum_{k=1}^{\infty} \frac{a_k + jb_k}{2} e^{-jk\omega_0 t} \tag{10–100}$$

Now let us define the constant C_k as follows:

$$C_k = \frac{a_k - jb_k}{2} \tag{10–101}$$

Using these constants, the Fourier series given in Eq. (10–100) may be written in the form

$$f(t) = a_0 + \sum_{k=1}^{\infty} C_k e^{j\omega_0 t} + \sum_{k=1}^{\infty} C_k^* e^{-j\omega_0 t} \tag{10–102}$$

where C_k^* is the complex conjugate of C_k. From the defining integral relations for the Fourier coefficients a_k and b_k given in (10–42) and (10–45), we may write the following integral expression[5] for the coefficients C_k:

$$C_k = \frac{1}{T}\int_{-T/2}^{T/2} f(t)\,(\cos k\omega_0 t - j\sin k\omega_0 t)\,dt$$

This equation may be simplified by using the last expression given in Eq. (10–99). Thus, we obtain

$$C_k = \frac{1}{T}\int_{-T/2}^{T/2} f(t)e^{-jk\omega_0 t}\,dt \tag{10–103}$$

We note that since the cosine function is even, i.e., $\cos(-x) = \cos x$, for a typical cosine term in the series in Eq. (10–98a), we may write

$$a_k \cos k\omega_0 t = a_{-k}\cos(-k\omega_0 t) = a_{-k}\cos k\omega_0 t \tag{10–104}$$

Comparing the first and last members of the above equation, we conclude that $a_k = a_{-k}$, i.e., in a Fourier series in which the integer constant k in the argument of the cosine terms is changed from k to $-k$, the coefficients a_k multiplying the cosine terms remain unchanged. Similarly, since the sine function is an odd function, i.e., $\sin(-x) = -\sin(x)$, for a typical sine term in the series of Eq. (10–98a) we may write

$$b_k \sin k\omega_0 t = b_{-k}\sin(-k\omega_0 t) = -b_{-k}\sin k\omega_0 t \tag{10–105}$$

Considering the first and last members of the above equation, we conclude that $b_k = -b_{-k}$. That is, if we change the integer constant k in the argument of the sine terms of a Fourier series from k to $-k$ we must also change the coefficients b_k multiplying these sine terms, to the negative of their original value to keep the value of the series the same. Substituting these results in (10–101), we find that

$$C_{-k} = \frac{a_k + jb_k}{2} \tag{10–106}$$

Comparing this expression with Eq. (10–101), we conclude that C_{-k} is the complex conjugate of C_k. We may now rewrite the exponential form of the Fourier series given in Eq. (10–102) using the constants C_k and C_{-k}. Thus, we obtain

$$f(t) = a_0 + \sum_{k=1}^{\infty} C_k e^{jk\omega_0 t} + \sum_{k=1}^{\infty} C_{-k} e^{-jk\omega_0 t} \tag{10–107}$$

[5]Since we may integrate over any full period, the upper and lower limits of integration have been changed from 0 and T to $-T/2$ and $T/2$, respectively, to facilitate future developments.

If we now substitute $-k$ for k in the second summation given in the right member of the above equation, we obtain

$$f(t) = a_0 + \sum_{k=1}^{\infty} C_k e^{jk\omega_0 t} + \sum_{k=-1}^{-\infty} C_k e^{jk\omega_0 t} \tag{10–108}$$

The only value of k not included in the two summations in the right member of the above equation is $k = 0$. To see what happens for this value, let us evaluate the integral relation given in (10–103) for the case where $k = 0$. Thus, we obtain

$$C_0 = \frac{1}{T} \int_{-T/2}^{T/2} f(t)\, dt \tag{10–109}$$

Comparing this relation with the one given in (10–37), we see that $a_0 = C_0$. We conclude that the entire right member of Eq. (10–108) may be consolidated into a single summation term which is run through limits of minus infinity to plus infinity. The expression for $f(t)$ may now be written

$$f(t) = \sum_{k=-\infty}^{\infty} C_k e^{jk\omega_0 t} \tag{10–110}$$

Thus, we have defined a new form of Fourier series in which the coefficients C_k are, in general, complex. Such a form is called the *exponential* form (or the *complex* form) of the Fourier series.

The relation given in Eq. (10–101) defines the manner in which the coefficients C_k of the exponential form of the Fourier series are related to the coefficients a_k and b_k of the sine and cosine form of such a series. The relations between the coefficients c_k and ϕ_k of the magnitude and phase form of the series given in Eq. (10–98b) and the coefficients C_k are also readily derived. To see this, let us first express the coefficients C_k defined in (10–101) in polar form. Thus, we obtain for positive values of k

$$C_k = \frac{1}{2}\sqrt{a_k^2 + b_k^2} \; \underline{\big/ -\tan^{-1}\left(\frac{b_k}{a_k}\right)} \tag{10–111a}$$

Similarly, for negative values of k

$$C_k = \frac{1}{2}\sqrt{a_{|k|}^2 + b_{|k|}^2} \; \underline{\big/ \tan^{-1}\left(\frac{b_{|k|}}{a_{|k|}}\right)} \tag{10–111b}$$

Comparing this with the relations defining the coefficients c_k and ϕ_k in terms of a_k and b_k (which are defined in Summary 10–2.2), we see that for positive values of k

$$|C_k| = \tfrac{1}{2}c_k \qquad \underline{/C_k} = \phi_k \tag{10–112a}$$

Similarly, for negative values of

$$|C_k| = \tfrac{1}{2}c_{|k|} \qquad \underline{/C_k} = -\phi_{|k|} \tag{10-112b}$$

In other words, the magnitude of C_k equals the quantity c_k divided by two and the argument of C_k is numerically equal to plus or minus ϕ_k. These relations hold for all values of k except $k = 0$, since we have already shown that the coefficient C_0 is equal to a_0 which, in turn, has been shown equal to c_0. The factor of $\frac{1}{2}$ that relates the magnitudes of the exponential and trigonometric forms of the Fourier series results from the fact that the summation index k for the coefficients C_k has a range from minus infinity to plus infinity, whereas the index for the coefficients c_k only has a range from 0 to infinity. Thus, in the exponential form of the series, there will always be two terms C_k which both contribute to the same harmonic frequency. The results given above may be summarized as follows:

SUMMARY 10-6.1

Exponential Form of the Fourier Series: The Fourier series for a non-sinusoidal function $f(t)$ has the form

$$f(t) \sum_{k=-\infty}^{\infty} C_k e^{jk\omega_0 t}$$

where the C_k are complex coefficients and $C_{-k} = C_k^*$. For positive values of k,

$$C_k = \frac{a_k - jb_k}{2} = \frac{c_k}{2}\underline{/\phi_k}$$

where the quantities a_k, b_k, c_k, and ϕ_k are defined in Summaries 10–2.1 and 10–2.2.

An example of the determination of the exponential form of a Fourier series follows.

EXAMPLE 10-6.1 It is desired to determine the exponential form of the Fourier series for the half-wave rectified sine wave shown in Fig. 10-6.1. Applying Eq. (10-103), and noting that $T/2 = \pi$ (thus $\omega_0 = 1$) and that the function is zero for $-\pi \leq t \leq 0$, we obtain

$$\begin{aligned} C_k &= \frac{1}{2\pi}\int_0^{\pi} \sin t \, e^{-jkt}\, dt \\ &= -\frac{1}{2\pi}\frac{1 + e^{-jk\pi}}{k^2 - 1} \end{aligned}$$

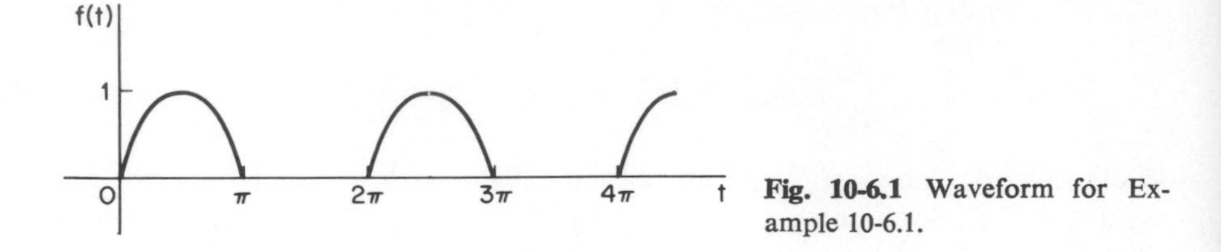

Fig. 10-6.1 Waveform for Example 10-6.1.

This relation is easily evaluated for the various values of k. Thus, for $k = 0$,

$$C_0 = -\frac{1}{2\pi}\frac{1+1}{-1} = \frac{1}{\pi}$$

For $k = 1$, the general expression for C_k is indeterminate, so this value must be determined directly from the defining integral. Thus, we obtain

$$C_1 = \frac{1}{2\pi}\int_0^\pi \sin t\, e^{-jt}\, dt$$

$$= \frac{1}{2\pi}\int_0^\pi \sin t\,(\cos t - j \sin t)\, dt = -j\frac{1}{4}$$

Continuing by applying the general relation for C_k we find $C_2 = -1/3\pi$, $C_3 = 0$, $C_4 = 1/15\pi$, etc. Thus the exponential form of the series for $f(t)$ is

$$f(t) = \cdots + \frac{1}{15\pi}e^{-j4t} - \frac{1}{3\pi}e^{-2jt} + j\frac{1}{4}e^{-jt} + \frac{1}{\pi} - j\frac{1}{4}e^{jt} - \frac{1}{3\pi}e^{j2t} + \frac{1}{15\pi}e^{j4t} + \cdots$$

The corresponding magnitude and phase form of the series is readily shown to be

$$f(t) = \frac{1}{\pi} + \frac{1}{2}\cos(t - 90^\circ) + \frac{2}{3\pi}\cos(2t - 180^\circ) + \frac{1}{15\pi}\cos(4t - 180^\circ) + \cdots$$

A line spectrum can be constructed for the magnitude and phase components of the exponential form of a Fourier series in the same manner as was done for the magnitude and phase form of the series. From the discussion given above and from (10–112a) and (10–112b), it is easily shown that there will be two main differences between the line spectra for the two series. These are the following:

1. All harmonic components (with the exception of the zero-frequency component) will have a magnitude one-half as large in the exponential form of the series as they do in the trigonometric form.
2. In the exponential form there will be a negative-frequency spectrum in addition to the positive-frequency spectrum. The negative-frequency spectrum will, of course, be symmetric about the origin with respect to the positive-frequency one.

As an example illustrating the two types of line spectra, consider the function shown in Fig. 10–6.2. The trigonometric form of the Fourier series for this function

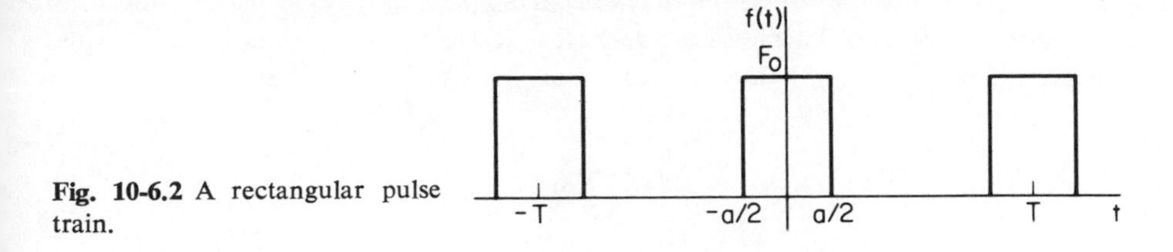

Fig. 10-6.2 A rectangular pulse train.

may be shown to be

$$f(t) = \frac{F_0 a}{T} + \sum_{k=1}^{\infty} \frac{2F_0 a}{T} \frac{\sin (k\omega_0 a/2)}{k\omega_0 a/2} \cos k\omega_0 t \tag{10-113}$$

The associated line spectrum for $T/a = 5$ is shown in Fig. 10–6.3(a). The exponental form of the Fourier series for this functon may be written as

$$f(t) = \sum_{k=-\infty}^{\infty} \frac{F_0 a}{T} \frac{\sin (k\omega_0 a/2)}{k\omega_0 a/2} e^{jk\omega_0 t} \tag{10-114}$$

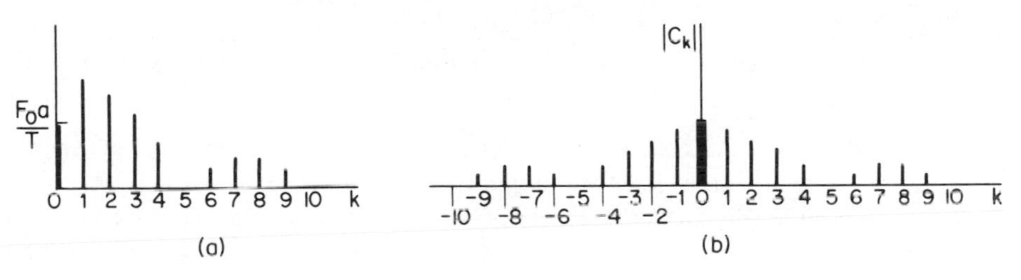

Fig. 10-6.3 Magnitude line spectra for the waveform shown in Fig. 10-6.2.

The line spectrum for this exponential form is shown in Fig. 10–6.3(b). A comparison of the two line spectra readily illustrates the differences between them for the given function.

Now let us see how the exponential form of the Fourier series presented above may be used to develop a Fourier transformation which may be applied to non-periodic functions. The key to our development here will be the concept of considering a non-periodic function as a limiting case of a periodic one, the limit being achieved by extending the period to infinity. Thus, over the entire range of t from minus infinity to plus infinity the non-periodic function may be considered as spanning only a single period. Now let us see how the exponential form of the Fourier series discussed above may be extended to apply to this situation. Specifically, we will be interested in examining the relations given in (10–103) and (10–110) in the limit as T approaches infinity. First let us consider the variable T itself. It is related to the fundamental harmonic frequency ω_0 by the expression

$$T = \frac{2\pi}{\omega_0} \tag{10-115}$$

The quantity ω_0 is not only the fundamental harmonic frequency, it is also the difference between adjacent harmonic frequencies. If we use the symbol $\Delta\omega$ for this difference, where $\Delta\omega = (k+1)\omega_0 - k\omega_0 = \omega_0$, we may define T as

$$T = \frac{2\pi}{\Delta\omega} \tag{10–116}$$

Now let us define the quantity $F(jk\omega_0)$ as the coefficient C_k multiplied by T. Thus, using (10–116) and the defining relation for C_k given in (10–103), we may write

$$F(jk\omega_0) = TC_k = \frac{2\pi C_k}{\Delta\omega} = \int_{-T/2}^{T/2} f(t)e^{-jk\omega_0 t}\, dt \tag{10–117}$$

As the period T is increased, we see from (10–115) that ω_0 becomes smaller. In the limit as T approaches infinity, the discrete set of quantities $k\omega_0$, where k is an integer, simply becomes a continuous variable which we will designate as ω. Thus, (10–117) may be written in the limit as

$$F(j\omega) = \int_{-\infty}^{\infty} f(t)e^{-j\omega t}\, dt \tag{10–118}$$

This relation is called the *Fourier transformation* or the *Fourier integral*. As such it provides the defining expression for the quantity $F(j\omega)$ which in turn is called the *Fourier transform* or the *Fourier spectrum* of the non-periodic time-domain function $f(t)$. Since $F(j\omega)$ has frequency, namely ω, as its variable, it is said to define a function in the *frequency domain*.

Now let us see how we may develop an expression specifying the inverse process, i.e., defining a function $f(t)$ in terms of a function $F(j\omega)$ in the frequency domain. Again we will start with the case where the period T is not infinite. From the first three members of Eq. (10–117) we may express C_k as a function of $F(jk\omega_0)$. Thus, we obtain

$$C_k = \frac{F(jk\omega_0)}{T} = \frac{\Delta\omega\, F(jk\omega_0)}{2\pi} \tag{10–119}$$

Now let us insert this result in the expression for $f(t)$ given in Eq. (10–110). We obtain

$$f(t) = \sum_{k=-\infty}^{\infty} \frac{\Delta\omega\, F(jk\omega_0)}{2\pi} e^{jk\omega_0 t} \tag{10–120}$$

Let us now consider what happens as the period T becomes larger. In such a case, from Eq. (10–116), we see that $\Delta\omega$ becomes smaller. Actually, in the limit as K approaches infinity, $\Delta\omega$ becomes the differential $d\omega$, and the infinite sum of terms given in Eq. (10–120), becomes a continuous integral. As before, the discrete set of variables $k\omega_0$ becomes a continuous variable ω. Inserting these results in (10–120), we obtain

$$f(t) = \frac{1}{2\pi}\int_{-\infty}^{\infty} F(j\omega)e^{j\omega t}\,d\omega \tag{10–121}$$

This relation defines the *inverse Fourier transformation*, i.e., it specifies how a function of the frequency variable $j\omega$ may be transformed back into a function of time. It is convenient to use the script letter $\mathscr{F}$ to symbolically indicate the Fourier transformation. Thus, the relations of Eqs. (10–118) and (10–121) may be indicated as

$$F(j\omega) = \mathscr{F}[f(t)] \qquad f(t) = \mathscr{F}^{-1}[F(j\omega)] \tag{10–122}$$

respectively. Taken collectively, these two relations may be said to comprise a *transform pair* for any function $f(t)$ for which the Fourier transformation exists. It may be shown that the Fourier transformation produces a unique result, i.e.,

$$f(t) = \mathscr{F}^{-1}[F(j\omega)] = \mathscr{F}^{-1}\{\mathscr{F}[f(t)]\} \tag{10–123}$$

The conditions for which the Fourier transformation of a given function may be found are similar to those enumerated for the Fourier series in Sec. 10–2. They are summarized in the following:

SUMMARY 10-6.2

The Fourier Transformation: Given a non-periodic function in the time domain $f(t)$ which has only a finite number of finite maxima, minima, and discontinuities and which satisfies the relation

$$\int_{-\infty}^{\infty} |f(t)|\,dt < \infty$$

we may determine a function $F(j\omega)$ by the Fourier transformation

$$F(j\omega) = \int_{-\infty}^{\infty} f(t)e^{-j\omega t}\,dt$$

where $F(j\omega)$ is a frequency-domain function called the Fourier transform of $f(t)$. Similarly, given a frequency-domain function $F(j\omega)$, we may apply the inverse Fourier transformation

$$f(t) = \frac{1}{2\pi}\int_{-\infty}^{\infty} F(j\omega)e^{j\omega t}\,d\omega$$

to find the time-domain function $f(t)$.

As an example of the determination of the Fourier transform of a non-periodic function, consider the single pulse shown in Fig. 10–6.4. Applying Eq. (10–118), we obtain

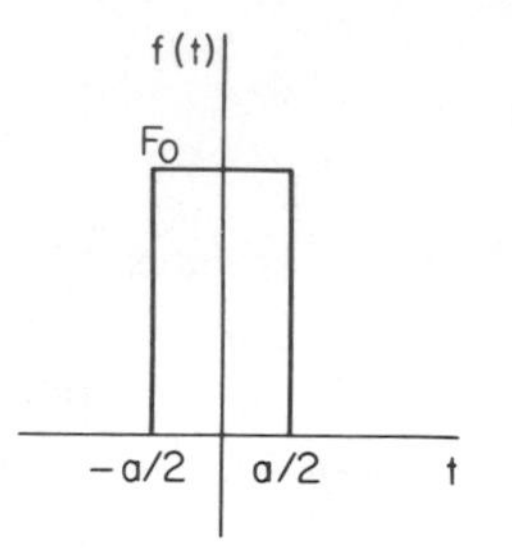

Fig. 10-6.4 A single pulse.

$$F(j\omega) = \int_{-a/2}^{a/2} F_0 e^{-j\omega t}\, dt = F_0 \frac{e^{-j\omega t}}{-j\omega}\bigg|_{-a/2}^{a/2}$$

$$= F_0 \frac{e^{j\omega a/2} - e^{-j\omega a/2}}{j\omega} \qquad (10\text{–}124)$$

If we apply Euler's identity of Eq. (10–99), we see that this result may be written in the form

$$F(j\omega) = F_0 a \frac{\sin \omega a/2}{\omega a/2} \qquad (10\text{–}125)$$

A plot of the magnitude of this function is given in Fig. 10–6.5. From this plot we note that the magnitude decreases as ω increases, thus, the higher-frequency signal content is less than the lower-frequency signal content. It should be noted, however,

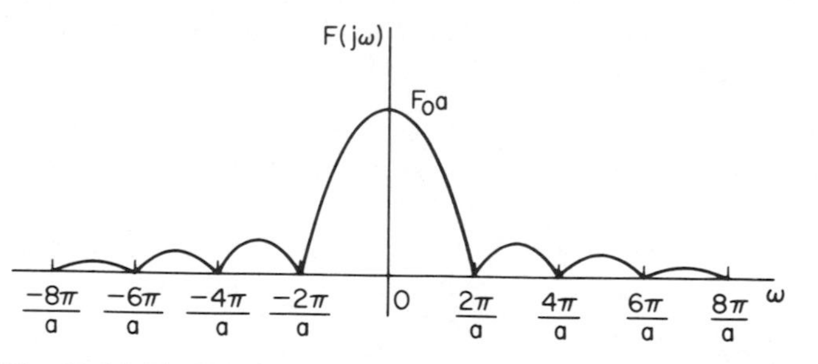

Fig. 10-6.5 The Fourier transform of the function shown in Fig. 10-6.4.

that if a is decreased, the magnitude of the plot stays higher for a greater frequency interval. Thus, we conclude that short duration pulses will contain more signal information at higher frequencies than longer duration pulses do. In the physical world we can find many examples of this phenomenon. For instance, a lightning stroke, which is of very short duration, produces observable signal information (static) over all the communication spectrum from the relatively low frequencies used in radio reception to the considerably higher ones used in television reception.

A spectrum or plot of the magnitude of some Fourier transform of the type shown in Fig. 10–6.5 may be considered as an extension of the discrete line spectrum discussed earlier in this section and in Sec. 10–2. It may be shown that the shape of such a continuous spectrum is the same as the *envelope* of the line spectrum that results when the identical waveform is repeated periodically. As an example of this, the periodic waveform shown in Fig. 10–6.2 is readily seen to consist of a periodic repetition of the same single pulse which was considered as a non-periodic waveform in Fig. 10–6.4. Thus the envelope of the discrete line spectrum shown in Fig. 10–6.3(b) for this periodic function has the same shape as the continuous spectrum shown in Fig. 10–6.5.

In the following section we shall develop some additional properties of the Fourier transformation, and illustrate the application of it to some basic problems in network analysis.

10-7 PROPERTIES OF THE FOURIER TRANSFORMATION

In this section we will present some properties of the Fourier transformation. The reader will note a considerable similarity between the properties developed here and those derived for the Laplace transformation in Chap. 8. These similarities are not merely coincidences. In the following section we shall show that the two transformations are directly related and we shall discuss the relation in detail. Just as we did for the Laplace transformation, in all the developments of this section we shall use a capital letter to indicate the transformation of a time-domain function defined by the same letter in lower case. Thus, we will use the notation $F(j\omega) = \mathscr{F}[f(t)]$, $G(j\omega) = \mathscr{F}[g(t)]$, etc. As a first property of the Fourier transformation let us investigate the transform of the function $af(t)$, where a is some real constant. Applying the defining relation for the Fourier transformation given in Eq. (10–118), we may write

$$\mathscr{F}[af(t)] = \int_{-\infty}^{\infty} af(t)e^{-j\omega t}\,dt = a\int_{-\infty}^{\infty} f(t)e^{-j\omega t}\,dt = aF(j\omega) \qquad (10\text{–}126)$$

The above result shows that the Fourier transformation has the property usually referred to as *homogeneity*. As a second property of the Fourier transformation, consider the transform of the sum of two functions $f_1(t)$ and $f_2(t)$. For this, we may write

$$\begin{aligned}\mathscr{F}[f_1(t) + f_2(t)] &= \int_{-\infty}^{\infty} [f_1(t) + f_2(t)]e^{-j\omega t}\,dt \\ &= \int_{-\infty}^{\infty} f_1(t)e^{-j\omega t}\,dt + \int_{-\infty}^{\infty} f_2(t)e^{-j\omega t}\,dt \\ &= F_1(j\omega) + F_2(j\omega) \qquad (10\text{–}127)\end{aligned}$$

where $F_1(j\omega) = \mathscr{F}[f_1(t)]$ and $F_2(j\omega) = \mathscr{F}[f_2(t)]$. This development shows that *superposition* is directly applicable to the Fourier transformation. Considering both of these results, namely, homogeneity and superposition, we may conclude that *the Fourier transformation is a linear transformation.*

Now let us derive the transformation of the derivative of a function $f(t)$. As a starting point, let us consider the defining relation for the inverse Fourier transformation. This is

$$f(t) = \frac{1}{2\pi}\int_{-\infty}^{\infty} F(j\omega)e^{j\omega t}\,d\omega \tag{10-128}$$

Differentiating both sides of this relation with respect to t, we obtain

$$\frac{df}{dt} = \frac{1}{2\pi}\int_{-\infty}^{\infty} j\omega F(j\omega)e^{j\omega t}\,d\omega \tag{10-129}$$

As indicated in the right member of the above equation, we see that the operation of differentiating a time-domain function $f(t)$ corresponds exactly with multiplication of a frequency-domain function $F(j\omega)$ by the factor $j\omega$. Thus, we may write

$$\mathscr{F}\left[\frac{df}{dt}\right] = j\omega F(j\omega) \tag{10-130}$$

where $F(j\omega)$ is the Fourier transform of $f(t)$.

A similar conclusion is reached if we investigate the Fourier transformation of the integral of a time-domain function. To do this, let us consider the function

$$g(t) = \int_{-\infty}^{t} f(\tau)\,d(\tau)$$

Since $dg/dt = f(t)$ and if $G(j\omega)$ is the Fourier transform of $g(t)$, then from the above result on differentiation we must have $j\omega\, G(j\omega) = F(j\omega)$, where $F(j\omega)$ is the Fourier transform of $f(t)$. However, in order for the function $g(t)$ to be transformable, $g(\infty)$ must equal 0. This requires that

$$g(\infty) = \int_{-\infty}^{\infty} f(t)\,dt = F(0) = 0 \tag{10-131}$$

In this case, we may write[6]

$$\mathscr{F}\left[\int_{-\infty}^{t} f(\tau)\,d\tau\right] = \frac{F(j\omega)}{j\omega} \tag{10-132}$$

Thus, we conclude that integration of a function in the time domain is transformed into division by the factor $j\omega$ for the corresponding frequency-domain function. The results given above are summarized as follows:

[6]If this condition is not satisfied, then the more general result is

$$F\left[\int_{-\infty}^{t} f(\tau)\,d\tau\right] = \frac{F(j\omega)}{j\omega} + \pi F(0)\,\delta(\omega)$$

An example of this result may be found in (10-141).

SUMMARY 10-7.1

Properties of the Fourier Transformation: The Fourier transformation is a linear transformation, i.e., the properties of homogeneity and superposition apply. In addition, the operations of differentiation and integration in the time domain are represented, respectively, by multiplication and division by the factor $j\omega$ in the frequency domain.

Now let us consider some other properties of the Fourier transformation. One of the most interesting of these is the effect of changing the time scale of a function. Basically, this consists of replacing the variable t by a new variable at, where a is some positive constant. If the original function is specified as $f(t)$, the time-scaled function thus becomes $f(at)$. Taking the Fourier transformation of such a function, we obtain

$$\mathscr{F}[f(at)] = \int_{-\infty}^{\infty} f(at)e^{-j\omega t}\,dt \tag{10–133}$$

To evaluate the integral in the right member of the above equation, let us make a substitution of variable of integration $x = at$. The differential dt now becomes dx/a. Substituting this in Eq. (10–133), we obtain

$$\mathscr{F}[f(at)] = \frac{1}{a}\int_{-\infty}^{\infty} f(x)e^{-j(\omega/a)x}\,dx = \frac{1}{a}F\left(j\frac{\omega}{a}\right) \tag{10–134}$$

From the above relation we see that the result of scaling the variable t of a time-domain function $f(t)$ is a reciprocal scaling of the variable ω in the transformed frequency-domain function $F(j\omega)$. In addition, there is a scaling of the magnitude of $F(j\omega)$ equal to $1/a$. This result provides a more rigorous justification for the conclusion reached in the preceding section with respect to shortening the duration of a pulse. Such a shortening can be considered as a scaling of the type indicated above in which $a > 1$. For example, a pulse $f(t)$ which occurs from 0 to 1 s may be scaled to a pulse of similar waveform but occurring from 0 to $\frac{1}{3}$ s if we express it as $f(3t)$. Correspondingly, from Eq. (10–134) the frequency spectrum of the transformed function $F(j\omega)$ will be tripled in width, and a scale factor of $\frac{1}{3}$ will be introduced.

Another interesting property of the Fourier transformation can be developed by finding the transform of the quantity $f(t) \cos \omega_0 t$. Such a time-domain quantity is frequently encountered in studies of the processes associated with radio transmission. Specifically, it is referred to as *amplitude modulation* in which the quantity $f(t)$ is a signal which is to be transmitted by modulating the amplitude of a radio-frequency carrier of ω_0 rad/s. Applying the Fourier transformation and using Euler's relation for the cosine function, we find

$$\mathscr{F}[f(t)\cos\omega_0 t] = \int_{-\infty}^{\infty} f(t)\frac{e^{j\omega_0 t} + e^{-j\omega_0 t}}{2} e^{-j\omega t}\, dt$$

$$= \frac{1}{2}\int_{-\infty}^{\infty} f(t)e^{-j(\omega-\omega_0)t}\, dt + \frac{1}{2}\int_{-\infty}^{\infty} f(t)e^{-j(\omega+\omega_0)t}\, dt$$

$$= \frac{1}{2}\Big\{F[j(\omega - \omega_0)] + F[j(\omega + \omega_0)]\Big\} \qquad (10\text{–}135)$$

where the quantities $F[j(\omega - \omega_0)]$ and $F[j(\omega + \omega_0)]$ are readily interpreted by recalling that $F(j\omega)$ is the Fourier transform of $f(t)$. Thus, we see that the Fourier spectrum of the resulting time function contains two distinct bands whose shapes differ only by a scale factor from the Fourier spectrum of $f(t)$ but whose positions are displaced along the frequency axis so that they are centered at $\pm j\omega_0$ rather than the origin.

As a final property of the Fourier transformation, let us consider the effect of shifting the time scale. That is, let us investigate the Fourier transformation of $f(t - a)$, namely, a function $f(t)$ which has been shifted in the direction of decreasing time by the quantity a. Applying the Fourier transformation, we obtain

$$\mathscr{F}[f(t - a)] = \int_{-\infty}^{\infty} f(t - a)\, e^{-j\omega t}\, dt \qquad (10\text{–}136)$$

The integral of the right member of the above equation may be evaluated by defining a new variable of integration $x = t - a$. Thus, we obtain

$$\mathscr{F}[f(t - a)] = \int_{-\infty}^{\infty} f(x)\, e^{-j\omega(x+a)}\, dx = e^{-j\omega a}\int_{-\infty}^{\infty} f(x)\, e^{-j\omega x}\, dx \qquad (10\text{–}137)$$

This may be written in the form

$$\mathscr{F}[f(t - a)] = e^{-j\omega a}\, F(j\omega) \qquad (10\text{–}138)$$

where $F(j\omega) = \mathscr{F}[f(t)]$. We conclude that the effect of shifting a function in the time domain is simply multiplication of the transformed function by an exponential. If s is substituted for $j\omega$, this result is identical with the result developed in connection with the Laplace transformation in Sec. 8–3. Other properties of the Fourier transformation are readily derived in a manner similar to that used in connection with the Laplace transformation. Table 10–7.1 (on page 829) summarizes the most important of the properties.

In addition to discussing the properties of the Fourier transformation, it is interesting to obtain the transforms of some of the more common time-domain functions. First let us find the transform of a unit impulse $\delta(t)$. Applying the Fourier transformation, we obtain

$$\mathscr{F}[\delta(t)] = \int_{-\infty}^{\infty} \delta(t)e^{-j\omega t}\, dt = e^{-j\omega t}\Big|_{t=0} = 1 \qquad (10\text{–}139)$$

where the evaluation of the integral follows the method outlined in Sec. 8–3. Thus, we see that the Fourier transformation of the unit impulse is unity. Note that this is the same result as was obtained for the Laplace transform of the unit impulse. Now let us consider the transformation of the unit step function $u(t)$. If we evaluate

$$\int_{-\infty}^{\infty} |u(t)|\, dt \tag{10–140}$$

we note that the integral is not finite; therefore, this function does not satisfy the criterion normally applied for Fourier transformability. Thus, we may conclude that the unit step does not have a Fourier transform. Note that this result is quite different than the result achieved for the Laplace transformation in which the integrating factor $e^{-\sigma t}$ permitted a transform to be found for the unit step. However, the application of a more involved branch of mathematics, called the theory of distributions, can be made to show that a Fourier transform for the unit step function may be defined as

$$\mathscr{F}[u(t)] = \pi\delta(\omega) + \frac{1}{j\omega} \tag{10–141}$$

The first term represents an impulse (in the frequency domain) of strength π occurring at ω equals zero. The second term is the same as the Laplace transformation of a unit step function in which s has been replaced by $j\omega$. There are many other functions which, like the unit step function, do not have a bounded integral and thus are not ordinarily considered to be transformable.

In considering general time-domain functions, it is important to note that the Fourier transformation is defined over the entire time spectrum and not just for positive values of time as the Laplace transformation is. As an example of this, we must distinguish between the functions $e^{-at}(a > 0)$ and $e^{-at}u(t)(a > 0)$. Only the latter function has a Fourier transformation. A list of some typical time-domain functions and their Fourier transforms is given in Table 10–7.2 (on page 829). The entries in this table represent transform pairs. Thus, they may be used as a means of finding the inverse transform of frequency domain functions, just as was done for the Laplace transformation.

Now let us apply Fourier transformation methods to some actual network problems. Suppose that such a problem is specified for a given network in terms of an excitation variable $g(t)$ and it is desired to find some response variable $f(t)$. From our study of time-domain networks in Chaps. 5 and 6 we know that these variables must be related by a differential equation having the form

$$a_0 f(t) + a_1\frac{df}{dt} + a_2\frac{d^2f}{dt^2} + \cdots = b_0 g(t) + b_1\frac{dg}{dt} + b_2\frac{d^2g}{dt^2} + \cdots \tag{10–142}$$

We may take the Fourier transformation of both members of this equation by first defining the quantities

$$\begin{aligned} F(j\omega) &= \mathscr{F}[f(t)] \\ G(j\omega) &= \mathscr{F}[g(t)] \end{aligned} \tag{10–143}$$

Applying the relations of Table 10–7.1, we obtain

$$\begin{aligned} F(j\omega)[a_0 + a_1(j\omega) + a_2(j\omega)^2 + \cdots] \\ = G(j\omega)[b_0 + b_1(j\omega) + b_2(j\omega)^2 + \cdots] \end{aligned} \tag{10–144}$$

This may be written in the form

$$\frac{F(j\omega)}{G(j\omega)} = N(j\omega) = \frac{b_0 + b_1(j\omega) + b_2(j\omega)^2 + \cdots}{a_0 + a_1(j\omega) + a_2(j\omega)^2 + \cdots} \tag{10–145}$$

Thus, we see that the relation between the transformed response variable $F(j\omega)$ and the transformed excitation variable $G(j\omega)$ is given by the quantity $N(j\omega)$ defined in Eq. (10–145). This is exactly the same result as was obtained in the application of phasor methods to the solution of differential equations in Sec. 7–3. In addition, the expression in the right member of Eq. (10–145) defining $N(j\omega)$ is identical with the quantity relating input and output phasors given in Summary 7–3.2. The reason for this is quite simple; namely, it is due to the fact that differentiation of a time-domain function is represented by multiplication by the factor $j\omega$ in both the phasor method and the Fourier transformation method. Since this is so, and since integration is similarly related to division by the factor $j\omega$, we may conclude that all the network functions defined in Chap. 7 may be directly applied to the Fourier transformation method of network analysis. The really great fact about this conclusion is that we can apply all the methods based on the use of impedances and admittances developed in Chap. 7 to the problem of finding the quantity $N(j\omega)$ which relates the Fourier transforms of response and excitation variables. The resulting function can then be used to determine the response due to arbitrary nonperiodic inputs by applying the Fourier transformation as well as being used in sinusoidal steady-state analysis. An example follows.

EXAMPLE 10-7.1 As an example of the application of Fourier transformation methods to network analysis consider the determination of the output voltage $v(t)$ when a pulse of current $i(t)$ with the waveform shown in Fig. 10-7.1 is applied to the parallel RC circuit

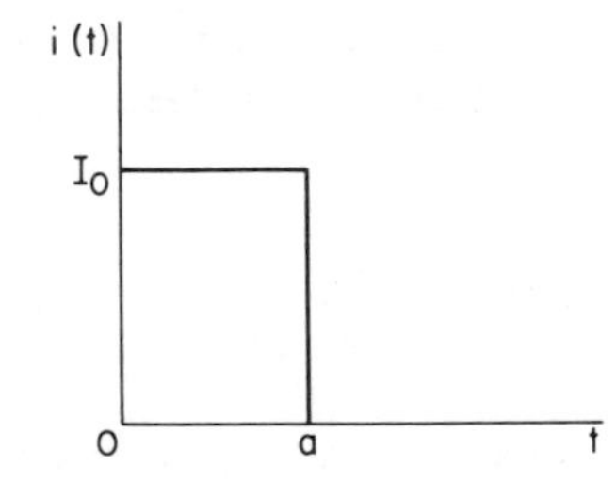

Fig. 10-7.1 Input waveform for Example 10-7.1.

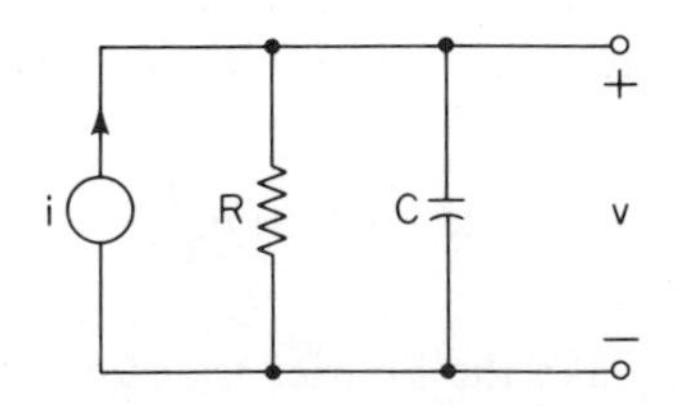

Fig. 10-7.2 Network for Example 10-7.1.

shown in Fig. 10-7.2. For this problem we may define the quantities $V(j\omega)$, $I(j\omega)$, and $Z(j\omega)$ as follows:

$$V(j\omega) = \mathscr{F}[v(t)] \qquad I(j\omega) = \mathscr{F}[i(t)] \qquad Z(j\omega) = \frac{V(j\omega)}{I(j\omega)}$$

The Fourier transform $I(j\omega)$ is found by applying the methods given in Sec. 10-6. The network function $Z(j\omega)$ is determined by applying the methods developed in Chap. 7. Thus, to find the Fourier transform $V(j\omega)$, we proceed as follows:

$$V(j\omega) = Z(j\omega)I(j\omega) = \frac{1}{1 + j\omega RC} \frac{I_0(1 - e^{-j\omega a})}{j\omega}$$

The time-domain function $v(t)$ is found by taking the inverse transformation of the expression for $V(j\omega)$. This may be done by applying the integral relation for the inverse transformation given in Eq. (10-128). Thus, we may write

$$v(t) = \frac{1}{2\pi} \int_{-\infty}^{\infty} \frac{I_0 R(1 - e^{-j\omega a})}{j\omega(1 + j\omega RC)} e^{j\omega t}\, d\omega$$

For simplicity, let us consider this expression as the sum of two integrals $v_1(t)$ and $v_2(t)$ as follows:

$$v(t) = v_1(t) + v_2(t) = \frac{I_0}{2\pi} \int_{-\infty}^{\infty} \frac{Re^{j\omega t}}{j\omega(1 + j\omega RC)}\, d\omega - \frac{I_0}{2\pi} \int_{-\infty}^{\infty} \frac{Re^{j\omega(t-a)}}{j\omega(1 + j\omega RC)}\, d\omega$$

Applying Euler's relation from (10-99) and rearranging terms in the integrand of the first integral, we may write

$$v_1(t) = \frac{I_0}{2\pi} \int_{-\infty}^{\infty} \frac{R(-j - \omega RC)}{\omega(1 - \omega^2 R^2 C^2)} (\cos \omega t + j \sin \omega t)\, d\omega$$

The imaginary component may be eliminated since the resulting integral must be real. Thus, we may write

$$v_1(t) = \frac{I_0}{2\pi} \int_{-\infty}^{\infty} \frac{-R^2 C}{1 - \omega^2 R^2 C^2} \cos \omega t\, d\omega + \frac{I_0}{2\pi} \int_{-\infty}^{\infty} \frac{R}{\omega(1 - \omega^2 R^2 C^2)} \sin \omega t\, d\omega$$

Integrating and applying the limits, we obtain

$$v_1(t) = I_0 R[u(t)\,(1 - e^{-t/RC}) - \tfrac{1}{2}]$$

wnere $u(t)$ is the unit step function which appears because of cancellation of positive and negative sign terms for $t < 0$. The second integral is evaluated by a similar process in which we replace the variable t by the quantity $t - a$. Thus, we obtain

$$v_2(t) = -I_0 R[u(t - a)\,(1 - e^{-(t-a)/RC}) - \tfrac{1}{2}]$$

Combining the above results, we obtain the following expression for $v(t)$:

$$v(t) = I_0 R[u(t)\,(1 - e^{-t/RC}) - u(t-a)\,(1 - e^{-(t-a)/RC})]$$

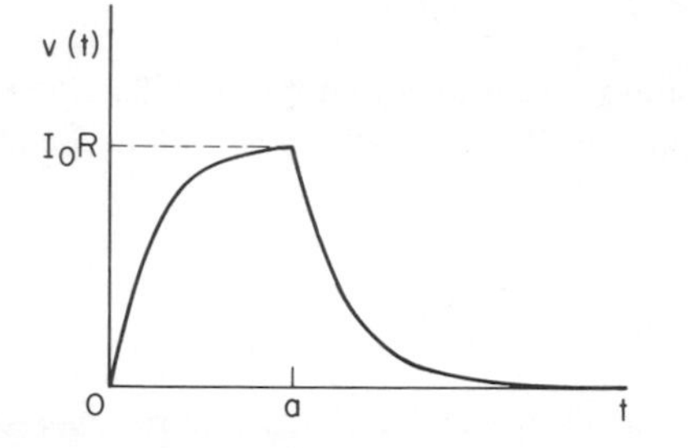

Fig. 10-7.3 Output waveform for Example 10-7.1.

A plot of this waveform is shown in Fig. 10-7.3.

In the above example, we have seen that a network function which relates the output and input phasors of a given network, and which is determined by the use of impedance and admittance concepts, is directly usable *without changes of any type* as the network function relating the corresponding output and input Fourier transformed variables. Nor need we stop here. As pointed out in Summary 8-6.2, the same network function (with s substituted for $j\omega$) gives the relation between the Laplace transformed output and input variables (in the absence of initial conditions). Thus, the network function is an invariant element common to sinusoidal steady-state methods, Fourier transformation methods, and Laplace transformation methods. Vive la network function!

SUMMARY 10-7.2

Fourier Transform Analysis of Networks: If a non-periodic function $g(t)$ with Fourier transformation $G(j\omega)$ is applied as an excitation to a network, the Fourier transform $F(j\omega)$ of the response $f(t)$ to this excitation is given as

$$F(j\omega) = N(j\omega)G(j\omega)$$

where $N(j\omega)$ is the same function as defined in Sec. 7-8 for use in phasor methods. If s is substituted for the argument $j\omega$, then $N(s)$ is the network function giving the ratio of response to excitation variables defined in Sec. 8-6 for use in Laplace transformation methods.

The results given above allow us to make another interesting observation about the network function. Assume that $N(j\omega)$ is such a function and that we apply a unit impulse $\delta(t)$ as an excitation. Since the Fourier transform of a unit impulse is 1, the resulting response is simply the inverse transform of $N(j\omega)$. If we define this unit impulse response in the time domain as $h(t)$, we may write

$$h(t) = \frac{1}{2\pi}\int_{-\infty}^{\infty} N(j\omega)e^{j\omega t}\,d\omega \tag{10-146}$$

Thus, we conclude that the network function is the Fourier transform of the unit impulse response of the network, or, conversely, taking the inverse Fourier transformation of the network function yields the unit impulse response for the network.

One of the most important practical problems in the analysis of networks and linear systems is the correlation between frequency-domain behavior and time-domain behavior. The reason for this is that the frequency-domain data are easier to obtain experimentally than almost any other type. As an example, consider the experiment shown in Fig 10–7.4 in which a sinusoidal signal generator with a variable output

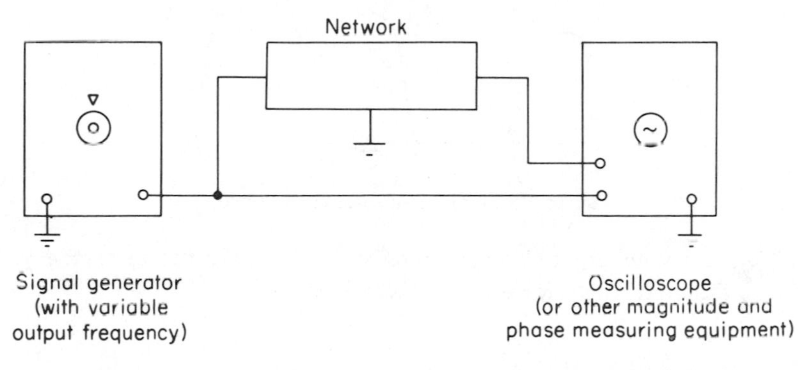

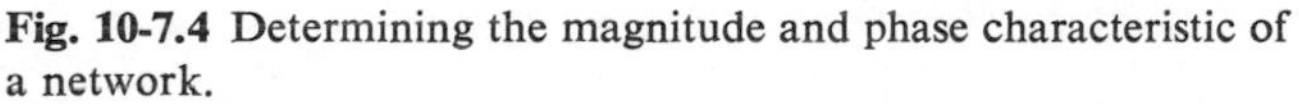

Fig. 10-7.4 Determining the magnitude and phase characteristic of a network.

frequency is used as an excitation source for a network and in which some suitable measuring equipment such as an oscillator or a set of magnitude and phase measuring meters are used to monitor the output. By varying the frequency of the signal generator over a wide range and recording the data characterizing the resulting output signal from the network, we may readily plot curves which show the magnitude and phase spectra of the network function. Such data, however, are not directly usable in applications which require that the network possess some specified *time-domain* behavior, such as a certain output rise time in response to a step input. In the following paragraphs, we shall show how the Fourier transformation method may be applied to establish a correlation between a network's performance in the frequency domain and in the time domain.

Let us begin our discussion by defining the concept of an *ideal transmission characteristic* for a network. Such a characteristic is assumed to reproduce the exact waveform of some input signal with two possible exceptions. Namely, the magnitude of the input signal waveform may be scaled and the entire waveform may be delayed, i.e., displaced in t. Thus, if an input signal $i(t)$ is applied to a network which has such an ideal transmission characteristic, the output signal $o(t)$ may be expressed as

$$o(t) = Ki(t - a) \qquad (10\text{–}147)$$

where k is a scaling factor and a is a delay factor. This result is illustrated in Fig.

10–7.5. Depending on whether k is greater than or less than unity, the network may be said to have "gain" or "loss," respectively. Applying the Fourier transformation to Eq. (10–147) and using the shifting property presented in Eq. (10–138), we obtain

$$O(j\omega) = KI(j\omega)e^{-j\omega a} \tag{10–148}$$

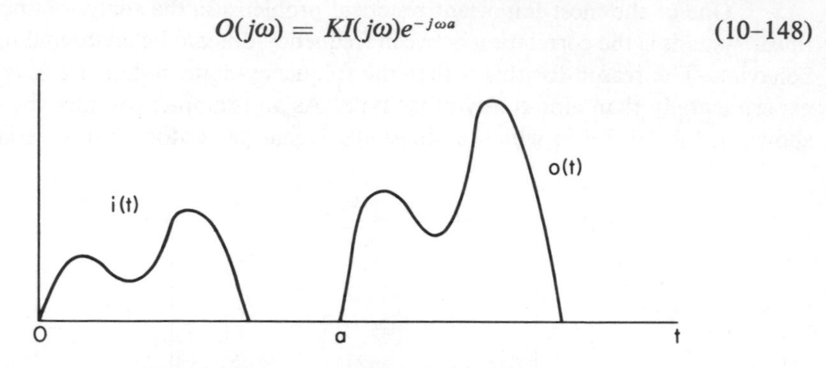

Fig. 10-7.5 Ideal transmission of a signal.

where $O(j\omega)$ and $I(j\omega)$ are the Fourier transforms of $o(t)$ and $i(t)$, respectively. Thus, the network function $N(j\omega)$ for such a system is defined as

$$N(j\omega) = \frac{O(j\omega)}{I(j\omega)} = Ke^{-j\omega a} \tag{10–149}$$

The magnitude and phase spectra for such an ideal transmission characteristic are shown in Fig. 10–7.6. These characteristics are physically unrealizable since they require that the phase be linear over an infinite bandwidth. Such a set of spectra can, however, be approximated over a limited frequency range. Ideally, such a *band-limited* characteristic will appear as shown in Fig. 10–7.7. Such a set of spectra is said to define an *ideal low-pass* characteristic.

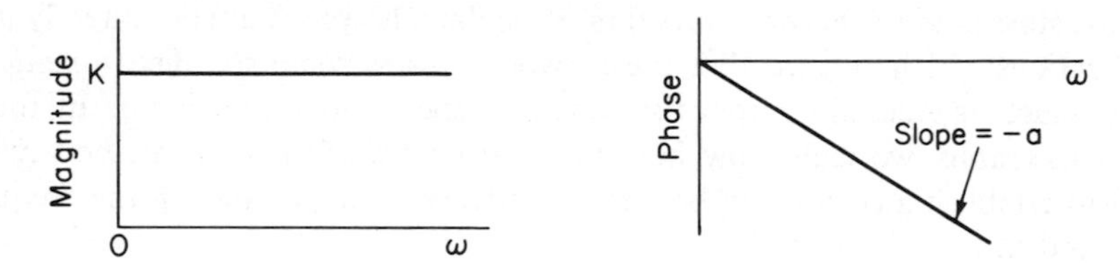

Fig. 10-7.6 An ideal transmission characteristic.

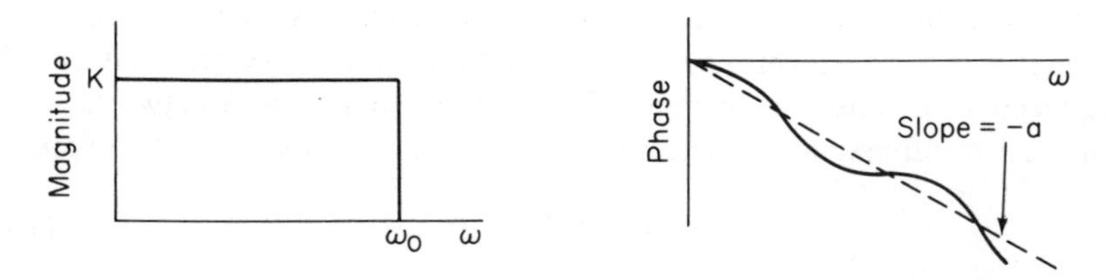

Fig. 10-7.7 A band-limited approximation to an ideal transmission characteristic.

TABLE 10-7.1 Properties of the Fourier Transformation

No.	$f(t)$	$F(j\omega)$	Description
1	$f_1(t) \pm f_2(t)$	$F_1(j\omega) \pm F_2(j\omega)$	superposition
2	$Kf(t)$	$KF(j\omega)$	homogeneity
3	$\frac{d^n f}{dt^n}$	$(j\omega)^n F(j\omega)$	differentiation
4	$\int_{-\infty}^{t} f(\tau)\,d\tau$	$\frac{1}{j\omega} F(j\omega)$	integration
5	$f(t - t_0)$	$F(j\omega)e^{-j\omega t_0}$	shifting of time axis
6	$f(t)e^{j\omega_0 t}$	$F[j(\omega - \omega_0)]$	shifting in frequency domain
7	$f(at)$	$\frac{1}{\lvert a\rvert}F\left(j\frac{\omega}{a}\right)$	time scaling
8	$f_1(t)*f_2(t)$	$F_1(j\omega)F_2(j\omega)$	convolution in time domain

TABLE 10-7.2 Basic Fourier Transform Pairs

No.	$f(t)$	$F(j\omega)$
1	$\delta(t)$	1
2	$u(t)$	$\pi\delta(\omega) + \frac{1}{j\omega}$
3	1	$2\pi\delta(\omega)$
4	$\lvert t\rvert$	$-\frac{2}{\omega^2}$
5	$e^{-at}u(t)$	$\frac{1}{a + j\omega}$
6	$e^{-a\lvert t\rvert}$	$\frac{2a}{a^2 + \omega^2}$
7	$te^{-at}u(t)$	$\frac{1}{(a + j\omega)^2}$
8	$\sin \omega_0 t\, u(t)$	$\frac{\pi}{2j}[\delta(\omega - \omega_0) + \delta(\omega + \omega_0)] + \frac{\omega_0}{\omega_0^2 - \omega^2}$
9	$\cos \omega_0 t\, u(t)$	$\frac{\pi}{2}[\delta(\omega - \omega_0) - \delta(\omega + \omega_0)] + \frac{j\omega}{\omega_0^2 - \omega^2}$
10	$\sin \omega_0 t$	$j\pi[\delta(\omega + \omega_0) - \delta(\omega - \omega_0)]$
11	$\cos \omega_0 t$	$\pi[\delta(\omega - \omega_0) + \delta(\omega + \omega_0)]$
12	$e^{-at} \sin \omega_0 t\, u(t)$	$\frac{\omega_0}{(a + j\omega)^2 + \omega_0^2}$

Now let us see what happens if we apply a simple pulse having the form shown in Fig. 10-6.4 as an input to such an ideal low-pass network. The amplitude spectrum of the pulse is shown in Fig. 10-6.5. The first lobe of the amplitude spectrum, which is also the part of the spectrum with the largest values, occurs around $\omega = \pm 2\pi/a$. If we let ω_0 be the cutoff frequency of the ideal low-pass filter, then the filter will pass most of the energy contained in the signal if $\omega_0 = 2\pi/a$. In this case, we see that

$$\omega_0 a = 2\pi \tag{10–150}$$

where a is the duration of the pulse and ω_0 is the cutoff frequency required in an ideal low-pass filter that transmits most of the energy continued in the pulse. Extending this result, it may be shown that, in general, for an arbitrarily shaped input pulse the following relation holds:

$$\begin{bmatrix}\text{Bandwidth required in an ideal}\\ \text{low-pass filter to transmit the}\\ \text{major portion of the energy}\\ \text{contained in an input pulse}\end{bmatrix} \times \begin{bmatrix}\text{Duration of}\\ \text{the input pulse}\end{bmatrix} = [\text{A constant}] \tag{10–151}$$

As another example of the use of the Fourier transformation to relate the frequency response characteristics of a network to its time-domain behavior, it may be shown that a relation similar to that given in (10–150) exists between the bandwidth of an ideal low-pass filter and the rise time of the output signal in response to a step input. Thus, we may write

$$\begin{bmatrix}\text{Bandwidth required in}\\ \text{an ideal low-pass filter}\end{bmatrix} \times \begin{bmatrix}\text{Rise time of output}\\ \text{waveform when a step}\\ \text{input is applied}\end{bmatrix} = [\text{A constant}]$$

Other examples of the use of the Fourier transformation to relate the frequency-domain characteristics and the time-domain performance of a network may be found in the references given in the Bibliography.

10-8 RELATIONS BETWEEN THE FOURIER TRANSFORMATION AND THE LAPLACE TRANSFORMATION

In the preceding section we derived many of the properties of the Fourier transformation. In several places we pointed out the similarity between these properties and those which were derived in Chap. 8 for the Laplace transformation. Now let us see if we can pinpoint the relations between the two transformations more specifically. To do this we will first show how the Laplace transformation may be derived from the Fourier transformation. Let us start with the expression for the Fourier transformation given in Eq. (10–118). Since the Laplace transformation applies only to functions of the type $f(t)u(t)$, where $f(t)$ is an arbitrary function and $u(t)$ is the unit step function, the value of the function to be transformed will always be zero for $t < 0$. In terms of such a function, we may revise the defining equation for the Fourier transform to the following form.

$$F(j\omega) = \int_0^\infty f(t)e^{-j\omega t}\,dt \tag{10–152}$$

Now let us consider a new function $f_E(t)$ which is equal to the original function $f(t)$ multiplied by an exponential term $e^{-\sigma t}$. Thus, we define

$$f_E(t) = f(t)e^{-\sigma t}u(t) \tag{10–153}$$

Taking the Fourier transformation of $f_E(t)$, we obtain $F_E(j\omega)$, where

$$F_E(j\omega) = \mathscr{F}[f_E(t)] = \int_0^\infty f(t)e^{-\sigma t}e^{-j\omega t}\,dt = \int_0^\infty f(t)e^{-(\sigma+j\omega)t}\,dt \tag{10–154}$$

Although we have derived this result as the Fourier transform of $f_E(t)$, if we examine the form of the right member, we see that it can also be considered as defining the Fourier transform of the original function $f(t)$ with the quantity $\sigma + j\omega$ used as the Fourier transformation variable. In this interpretation, we may define the function $F_L(\sigma + j\omega)$ as

$$F_L(\sigma + j\omega) = F_E(j\omega) = \int_0^\infty f(t)e^{-(\sigma+j\omega)t}\,dt \tag{10–155}$$

Considering the first and last members of the above, if we define the quantity $s = \sigma + j\omega$, we obtain

$$F_L(s) = \int_0^\infty f(t)e^{-st}\,dt \tag{10–156}$$

This result is identical with the definition of the Laplace transformation except for the L subscript in the transformed function. However, if we compare this result with Eq. (10–152), we see that the functions $F_L(s)$ and $F(j\omega)$ are identical, i.e.,

$$F(j\omega)|_{j\omega=s} = F_L(s) \qquad F_L(s)|_{s=j\omega} = F(j\omega) \tag{10–157}$$

Thus, we may write

$$F(s) = \int_0^\infty f(t)e^{-st}\,dt \tag{10–158}$$

which was shown in Eq. (8–12) to define the Laplace transformation (with the lower limit of the integral taken as 0). Thus, the Laplace transform and the Fourier transform are related by a simple substitution of variable. This result, of course, only holds true if the Fourier transformation is defined for the given function. We have already seen a case where this is not true, namely, the unit step function.

Now let us consider the inverse transformation process. Starting with $F_E(j\omega)$, the Fourier transform of $f_E(t)$ of Eq. (10–153) and applying (10–121), we obtain

$$f_E(t) = f(t)e^{-\sigma t}u(t) = \frac{1}{2\pi}\int_{-\infty}^{\infty} F_E(j\omega)e^{j\omega t}\,d\omega \tag{10–159}$$

Multiplying the last two members of the above expression by $e^{\sigma t}$, we obtain

$$f(t)u(t) = \frac{1}{2\pi}\int_{-\infty}^{\infty} F_E(j\omega)e^{(\sigma+j\omega)t}\,d\omega \tag{10–160}$$

Now let us substitute the relation $F_E(j\omega) = F_L(s)$ from Eq. (10–155). Thus, we obtain

$$f(t)u(t) = \frac{1}{2\pi}\int_{-\infty}^{\infty} F(s)e^{st}\,d\omega \tag{10–161}$$

Substituting $s = \sigma + j\omega$ as the variable of integration and noting that the differential $d\omega = ds/j$, we obtain

$$f(t)u(t) = \frac{1}{2\pi j}\int_{\sigma-j\infty}^{\sigma+j\infty} F(s)e^{st}\,ds \tag{10–162}$$

This is simply the expression for the inverse Laplace transformation as defined in Sec. 8–4. From the above development we see that, in a non-rigorous sense, we may derive the Laplace transformation from the Fourier transformation. The inverse process, i.e., determining the Fourier transformation from the Laplace transformation, is also readily demonstrated.

As an example of the application of the relation between the Laplace and Fourier transformations derived above, consider the function $f(t)$ defined as

$$f(t) = e^{-at}\cos bt\,u(t) \tag{10–163}$$

From Table 8–3.1, we find that $F(s)$, the Laplace transform of this function, is

$$F(s) = \frac{s+a}{(s+a)^2+b^2} \tag{10–164}$$

Since the envelope of the magnitude of $f(t)$ decreases exponentially with time, the integral of the magnitude is finite, thus $F(j\omega)$, the Fourier transform of $f(t)$, must exist. From the results given above, we see that we may write directly

$$F(j\omega) = \frac{j\omega+a}{(j\omega+a)^2+b^2} \tag{10–165}$$

This is the Fourier transform of the function defined in (10–163).

The results given above are summarized as follows:

SUMMARY 10-8.1

Relations Between the Fourier Transformation and the Laplace Transformation: If $f(t)u(t)$ is a time-domain function which is zero for $t < 0$ and which has a Laplace transform $F(s)$, the Fourier transform of $f(t)u(t)$, if it exists, is $F(j\omega)$. If $g(t)u(t)$ is a time-domain function which is zero for $t < 0$ and which has a Fourier transform $G(j\omega)$, the Laplace transform of $g(t)u(t)$ exists and is equal to $G(s)$.

Regarding the relations between the Laplace and Fourier transformations given above, it should be noted that in comparing the two transformations we must be careful to differentiate between network functions and the transforms of time-domain functions. Any network function $N(s)$ defined in the Laplace transformation variable s can be converted to a network function $N(j\omega)$ defined for the Fourier transformation variable $j\omega$ by substituting $j\omega$ for s. The converse is also always true. On the other hand, a Laplace transform $F(s)$ of some time-domain function $f(t)$ can be changed to a Fourier transform by substituting $j\omega$ for s only if $F(s)$ has no poles on the $j\omega$ axis or in the right half-plane. This is the same, of course, as requiring that the integral of the magnitude of the time-domain function represented by the inverse transform of $F(s)$ must be finite. Similarly, a Fourier transform $G(j\omega)$ of some time-domain function $g(t)$ can be changed to a Laplace transform by substituting s for $j\omega$ only if $g(t)$ is zero for $t < 0$.

Considering the above developments, we may conclude that time-domain functions may be divided into three classes. The smallest of these classes, and the one least frequently encountered, is that in which neither the Laplace nor the Fourier transformation exists. An example of such a function is e^{t^2}. The second class of time-domain functions is the one for which the Laplace transformation exists but the Fourier transformation does not (unless very special mathematical techniques such as the theory of distributions are employed). This is also a relatively small class of functions but it does include some which are frequently encountered, such as the unit step, the sine function, the cosine function, etc. The final class of time-domain functions, and the one which is of most general interest, is the one for which both the Laplace transformation and the Fourier transformation exist. We might also postulate a fourth class of time-domain functions, which includes those functions which are non-zero for $t < 0$ and for which the Fourier transformation exists, by claiming that the nonzero negative time behavior of such functions eliminates the possibility of obtaining the Laplace transforms of them. It may be shown, however, that a modification of the Laplace transformation called the *two-sided* Laplace transformation may be defined to cover such negative time behavior. Thus, in general, we will assume that any time-domain function which has a Fourier transform is also Laplace transformable.

It is interesting to note some of the other similarities and differences between the Laplace and Fourier transformations. First, let us consider some of the advantages of the Laplace transformation over the Fourier transformation. The Laplace transformation has the capability for introducing initial conditions explicitly in the network equations, while the Fourier transformation does not. This capability is readily verified by considering the expressions for the transformation of derivative functions given in Table 8-3.2 for the Laplace transformation and in Table 10-7.1 for the Fourier transformation. Another advantage of the Laplace transform method is that it is easier to apply to the analysis of networks than the Fourier transform method, since the algebraic manipulations necessary to find a transfer function are more readily made in terms of the variable s than in terms of the imaginary quantity $j\omega$. Finally, in the Laplace transform method, the time-domain behavior of network functions may be predicted from the pole and zero locations of the complex frequency function. In the Fourier transform method this, of course, is not possible. The Fourier transform method also has certain advantages over the Laplace transform one in that it provides more direct information on the frequency-domain behavior of a network. Thus, it is more readily applied to correlate experimentally obtained frequency-domain measurements than is the Laplace transform method.

10.9 CONCLUSION

In this chapter we have presented our final treatment concerning the frequency-domain analysis of networks by introducing the Fourier series and the Fourier transformation. The basic importance of the Fourier series is that it extends all the concepts of sinusoidal steady-state analysis to cover the situation where a non-sinusoidal (but still periodic) excitation is applied to a lumped, linear, time-invariant network. Just as was the case for the sinusoidal steady-state, the use of a Fourier series in analyzing a network only provides information as to the behavior of the network variables after any transient terms have disappeared. That is, it is only concerned with steady-state behavior.

The second technique discussed in this chapter is the Fourier transformation. This transformation is directly derivable from the Laplace transformation (or vice versa). It provides a means of determining the frequency spectrum, i.e., the frequency content of a time-domain waveform. Thus, it provides a readily applied link between the frequency-domain properties of a network and the effect that these properties will have on the energy contained in a specific time-domain waveform. The treatment given in this section can only be considered as an introduction to the subject, since many complete books have been devoted solely to the Fourier transformation.

Problems

Problem 10-1 (*Sec. 10-1*)

Decompose each of the periodic functions shown in Fig. P10-1 into the sum of an even and an odd function.

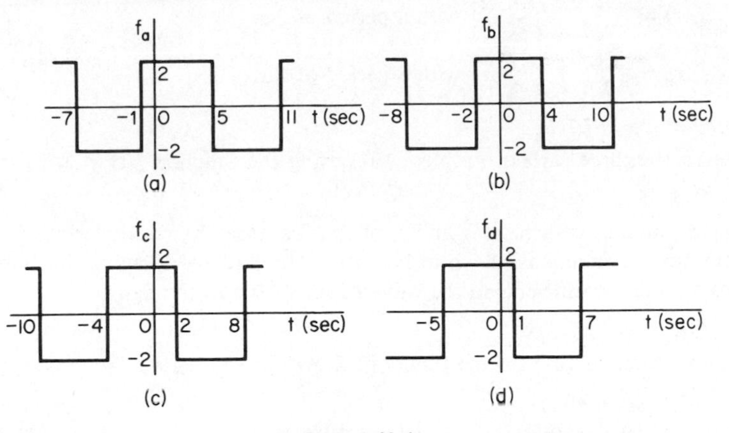

Figure P10-1

Problem 10-2 (Sec. 10-1)

Decompose each of the periodic functions shown in Fig. P10-2 into the sum of an even function and an odd one.

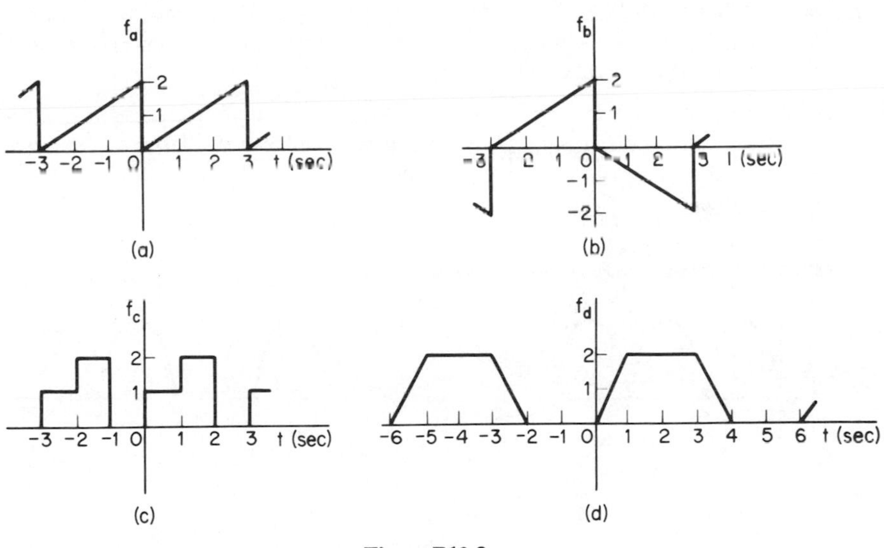

Figure P10-2

Problem 10-3 (Sec. 10-1)

Decompose each of the periodic functions defined below into the sum of an even function and an odd one:

(a) $f_a(t) = 3t$, $0 \leq t \leq 3$, with a period of 3 s.

(b) $f_b(t) = \begin{Bmatrix} 3t, & 0 \le t < 2 \\ 0, & 2 < t \le 3 \end{Bmatrix}$ with a period of 3 s.

(c) $f_c(t) = \begin{Bmatrix} 3t, & 0 \le t < 1 \\ 0, & 1 < t \le 3 \end{Bmatrix}$ with a period of 3 s.

Problem 10-4 (Sec. 10-1)

Repeat the three parts of Problem 10-3 using the function $f(t) = 3t^2$.

Problem 10-5 (Sec. 10-2)

Determine the coefficients a_i and b_i of the first three harmonic terms of the Fourier series for the functions shown in Fig. P10-1 by directly applying the integral formulas for these coefficients to the waveshapes shown in the figure.

Problem 10-6 (Sec. 10-2)

Repeat Problem 10-5 for the functions shown in Fig. P10-2.

Problem 10-7 (Sec. 10-2)

Determine the coefficients a_i and b_i of the first three harmonic terms of the Fourier series for the functions shown in Fig. P10-1. Do this by using the results found in Problem 10-1 to find the coefficients of the odd and even portions of the function separately, then add the results.

Problem 10-8 (Sec. 10-2)

Repeat Problem 10-7 for the functions shown in Fig. P10-2.

Problem 10-9 (Sec. 10-2)

Sketch the magnitude and phase spectra for the first three harmonics of each of the functions shown in Fig. P10-1.

Problem 10-10 (Sec. 10-2)

Determine the coefficients a_i and b_i of the first three harmonic terms of the functions shown in Fig. P10-10. Determine the magnitude and phase coefficients for these same terms.

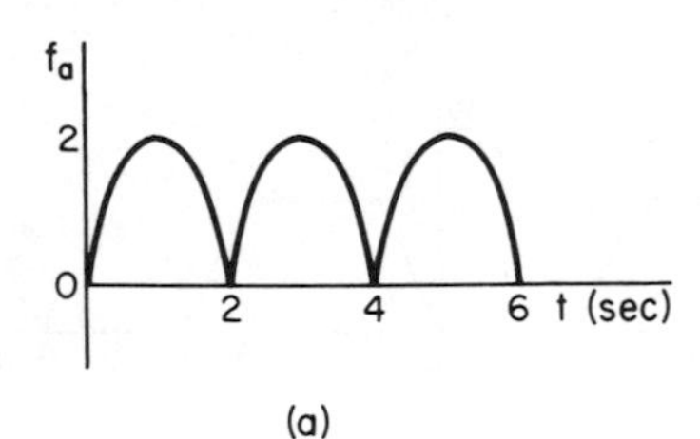

(a)

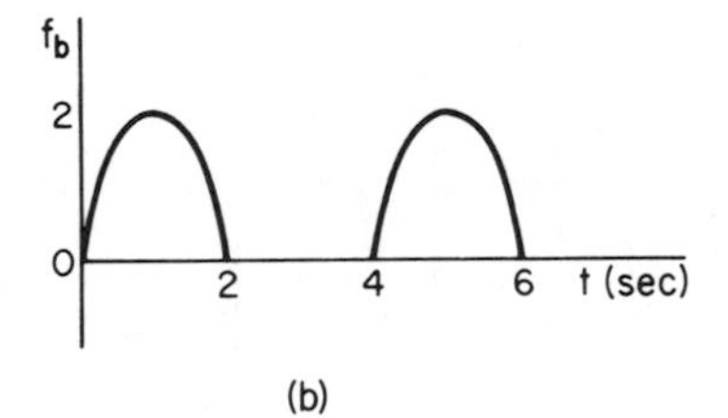

(b)

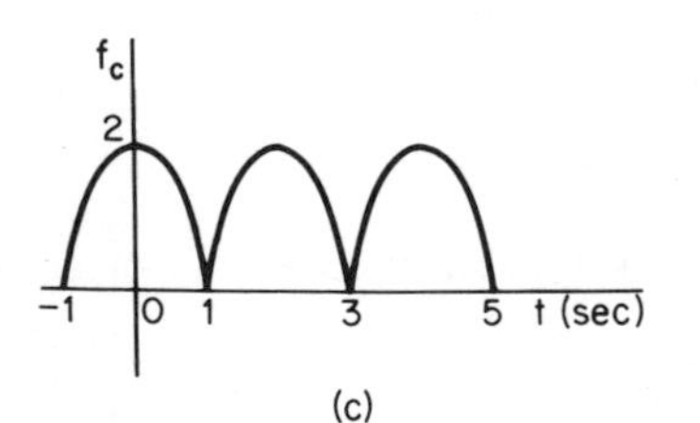

(c)

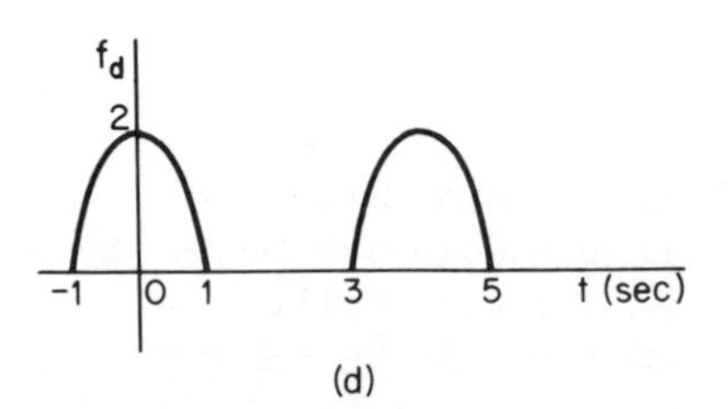

(d)

Figure P10-10

Problem 10-11 (*Sec. 10-2*)

A 1-Ω resistor has a square wave of current of amplitude ±1 A with a period of 1 s flowing through it. Find the average power dissipated in the resistor by (a) finding the rms value of the square wave, and (b) expressing the square wave as a Fourier series, and finding the power dissipated by each of the terms in the series through the fifth harmonic. Compare the answers obtained by the two methods.

Problem 10-12 (*Sec. 10-3*)

Predict upper bounds on the manner in which the magnitude coefficients of the Fourier series for the periodic functions defined below will diminish for higher-frequency harmonic terms, and determine enough coefficients for the series to verify your conclusion.

(a) $f_a(t) = (t + 1)(t - 1)$, $-1 < t < 1$, with a period of 2 s.
(b) $f_b(t) = (t + 1)^2(t - 1)^2$, $-1 < t < 1$, with a period of 2 s.

Problem 10-13 (*Sec. 10-3*)

Use as many of the following letters as are appropriate to describe the symmetry of each of the waveforms shown in Fig. P10-13.

a—even function c—half-wave symmetry
b—odd function d—only odd harmonics
e—only even harmonics

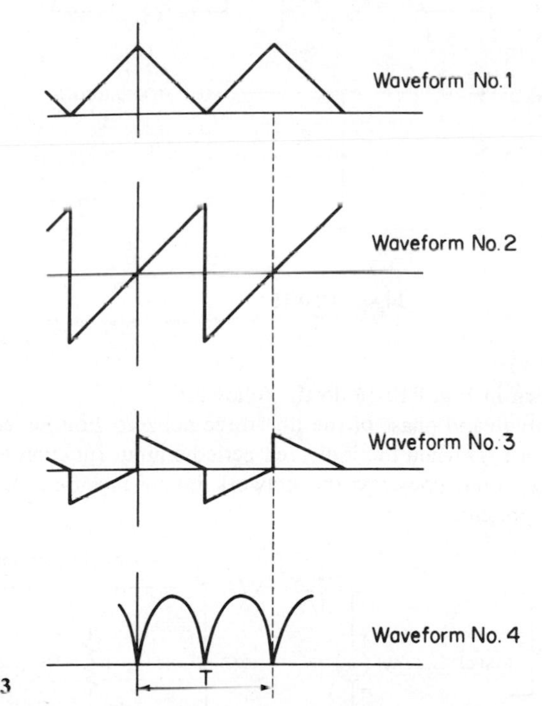

Figure P10-13

Problem 10-14 (*Sec. 10-3*)

For each of the cases listed below, sketch a non-trivial periodic function (which has *not* been discussed in the text) which meets the criteria.

(a) A function which has only even harmonics.
(b) A function which has only odd harmonics.
(c) A function which is an even function.
(d) A function which is an odd function.
(e) A function that has half-wave symmetry.

Problem 10-15 (Sec. 10-4)

(a) The Fourier series for one of the waveforms shown in Fig. P10-15(a) and (b) is

$$f(t) = \sin t + (\tfrac{1}{3}) \sin 3t + (\tfrac{1}{5}) \sin 5t + \cdots$$

Find the Fourier series of the other waveform.
(b) If the Fourier series given above in part (a) is applied as the input waveform $v_1(t)$ to the network shown in Fig. P10-15(c), find the first three terms of the Fourier series of the output waveform $v_2(t)$.

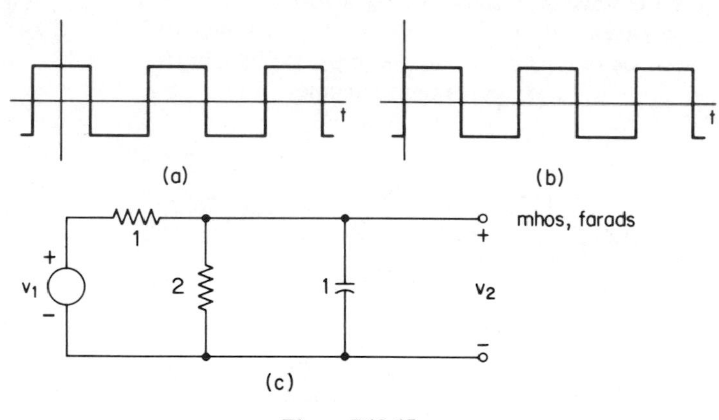

Figure P10-15

Problem 10-16 (Sec. 10-4)

For the circuit shown in Fig. P10-16 do the following.
(a) Find the magnitude and phase of the first three nonzero Fourier coefficients in the output waveform $v_2(t)$ when the indicated periodic input function is applied.
(b) Find the average input power to the network in the first three harmonic frequencies which are present.

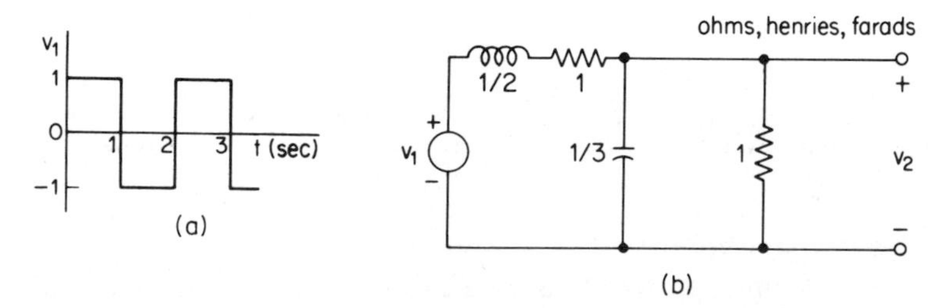

Figure P10-16

(c) Find the average output power in the resistor which is connected in parallel with the capacitor.

Problem 10-17 (*Sec. 10-4*)

Repeat Problem 10-16 using the input excitation shown in Fig. P10-17.

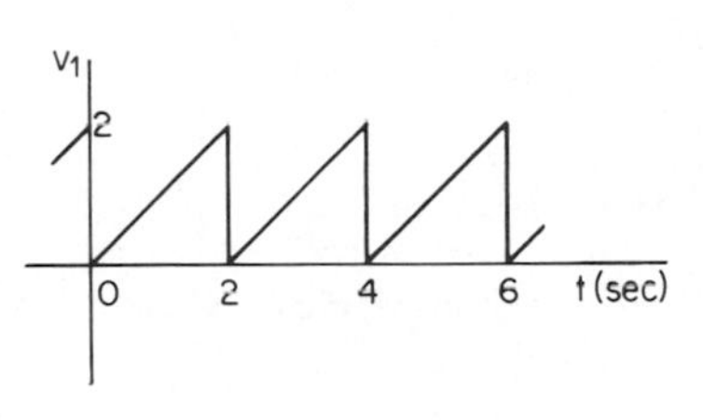

Figure P10-17

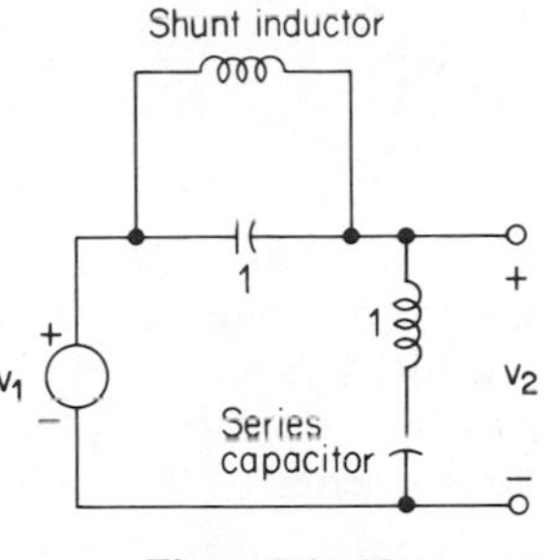

Figure P10-18

Problem 10-18 (*Sec. 10-4*)

The input waveform to the circuit shown in Fig. P10-18 has large unwanted harmonic components at ω equals 3 and 7 rad/s. Find values for the shunt inductance and the series capacitance so as to eliminate these frequency components from the output waveform $v_2(t)$.

Problem 10-19 (*Sec. 10-4*)

Find the values of G (in mhos) and C (in farads) for the network shown in Fig P10-19 such that the third harmonic component of $v_1(t)$ is reduced by 0.3 and the fifth harmonic component is reduced by 0.5. Assume $|\omega^2 C| \gg 1$, and $\omega_0 = 1$ rad/s.

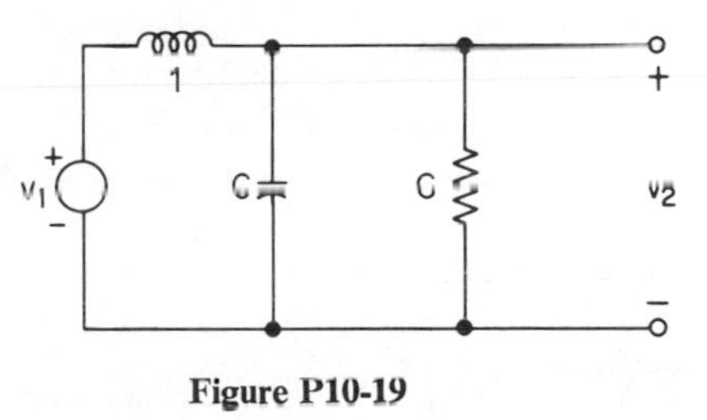

Figure P10-19

*Problem 10-20** (*Sec. 10-5*)[7]

Use the subroutine FSER to determine the coefficients of the first nine harmonic terms of the Fourier series for a periodic function with period of 100 s defined as $f(t) = -40 + 0.8t$ for $0 \leq t \leq 100$. Compare the results with those obtained by hand computation. Use the subroutine FPLT to plot (a) the function determined by the first three harmonic coefficients of the series, and (b) the function determined by the first nine coefficients of the series. Use 51 points on the plot to cover one period of $f(t)$.

*Problem 10-21** (*Sec. 10-5*)

Make plots of one cycle of $v_2(t)$ in the circuit shown in Fig. P10-21 for the following values of the parameter a: 0.01 and 0.1. Put both waveforms on the same plot. Use 61 values of time to cover the cycle and use seven harmonic coefficients in the Fourier series.

[7]Problems which require the use of digital computational techniques are identified by an asterisk.

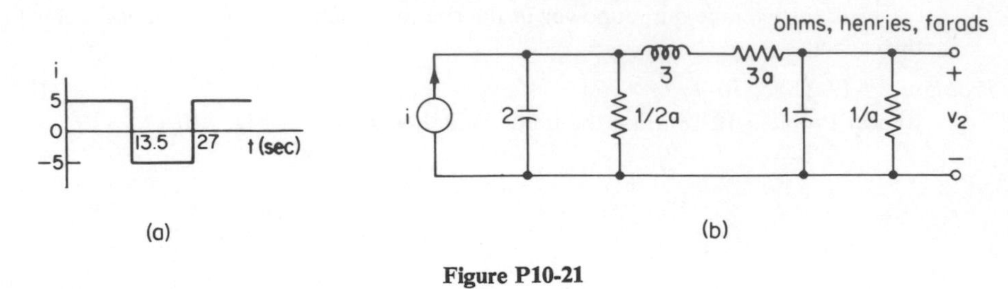

Figure P10-21

Problem 10-22(Sec. 10-5)*

Use the subroutine FSER to verify the results obtained in part (a) of Problem 10-16. Use FPLT to make a plot of one cycle of the output waveform. Scale the plot by 50. Use 61 points to cover one cycle and use seven harmonic coefficients in the Fourier series.

Problem 10-23(Sec. 10-5)*

Use the subroutine FSER to verify the results obtained in part (a) of Problem 10-17. Use FPLT to make a plot of one cycle of the output waveform. Scale the plot by 50. Use 61 points to cover one cycle and use seven harmonic coefficients in the Fourier series.

Problem 10-24(Sec. 10-5)*

Use the subroutine FSER and any other appropriate subroutines to determine the coefficients of the Fourier series of a periodic function, one cycle of which is defined by the tabulated data points given in Fig. P10-24. Determine the first five harmonic coefficients. Plot the original waveshape and, using FPLT, the waveshape reconstructed from the Fourier series. Compare the two plots. Use 51 points in the plot.

T	F	T	F	T	F
T=0.00	F= 23.284116	T= .34	F= 3.696341	T= .68	F= 5.937470
T= .02	F= 39.212407	T= .36	F= -1.838080	T= .70	F= 10.036618
T= .04	F= 52.947824	T= .38	F= -8.407568	T= .72	F= 10.409730
T= .06	F= 63.490688	T= .40	F=-16.038414	T= .74	F= 7.020851
T= .08	F= 70.225068	T= .42	F=-24.474329	T= .76	F= .312781
T= .10	F= 72.964336	T= .44	F=-33.190549	T= .78	F= -8.834970
T= .12	F= 71.939654	T= .46	F=-41.454828	T= .80	F=-19.204927
T= .14	F= 67.735247	T= .48	F=-48.427101	T= .82	F=-29.385127
T= .16	F= 61.181450	T= .50	F=-53.283585	T= .84	F=-37.939255
T= .18	F= 53.222101	T= .52	F=-55.347091	T= .86	F=-43.581737
T= .20	F= 44.775899	T= .54	F=-54.203927	T= .88	F=-45.335273
T= .22	F= 36.611670	T= .56	F=-49.789286	T= .90	F=-42.650604
T= .24	F= 29.254907	T= .58	F=-42.427067	T= .92	F=-35.473227
T= .26	F= 22.937992	T= .60	F=-32.816375	T= .94	F=-24.248824
T= .28	F= 17.599818	T= .62	F=-21.964365	T= .96	F= -9.867236
T= .30	F= 12.933108	T= .64	F=-11.072742	T= .98	F= 6.447103
T= .32	F= 8.470705	T= .66	F= -1.392026	T=1.00	F= 23.283415

Figure P10-24

Problem 10-25 (Sec. 10-6)

Sketch the magnitude of the Fourier transform of the function $f(t)$ shown in Fig. P10-25. Indicate the values of frequency at which the amplitude is zero.

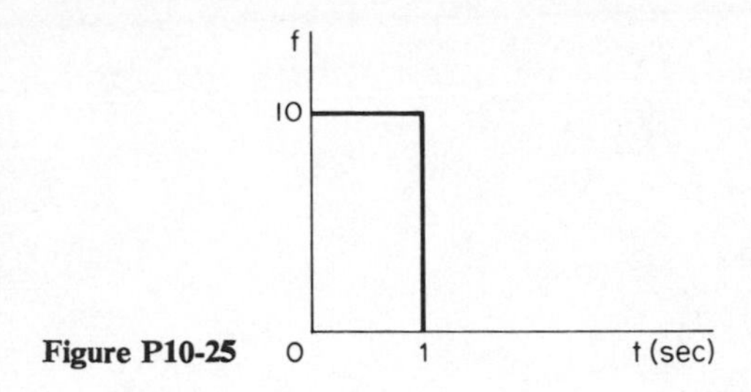

Figure P10-25

Problem 10-26 (Sec. 10-6)

Find the Fourier transform for the exponential pulse $f(t) = F_0 e^{-at} u(t)$ and sketch the magnitude and phase curves for the transform.

Problem 10-27 (Sec. 10-6)

Find the Fourier transform of the triangular pulse $f(t)$ shown in Fig. P10-27 and sketch the magnitude and phase curves for the transform.

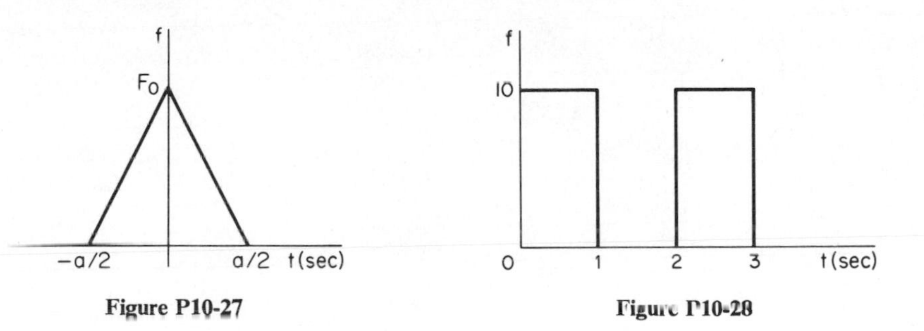

Figure P10-27

Figure P10-28

Problem 10-28 (Sec. 10-7)

Use the results of Problem 10-25 to sketch the magnitude of the Fourier transform of the function $f(t)$ shown in Fig. P10-28.

Problem 10-29 (Sec. 10-7)

Repeat the problem given in Example 10-7.1 for the case where the input is an exponential pulse similar to the one defined in Problem 10-26.

Bibliography

INTRODUCTORY CIRCUIT THEORY

The list of books which have been written on the subject of basic circuit theory is a very large one, and the following entries can only be considered as being representative samples. In most cases, however, the titles listed cover approximately the same topics as those covered in this text. Thus, they may be helpful to the student as sources of slightly different viewpoints on the material contained herein. They also include many examples which may be helpful in understanding the basic concepts.

Balabanian, N., *Fundamentals of Circuit Theory*, Allyn and Bacon, Inc., Boston, 1961.

Brenner, E., and M. Javid, *Analysis of Electric Circuits*, McGraw-Hill Book Company, New York, 1967.

Chirlian, P. M., *Basic Network Theory*, McGraw-Hill Book Company, New York, 1969.

Close, M., *The Analysis of Linear Circuits*, Harcourt, Brace & World, Inc., 1966.

Desoer, C. A., and E. S. Kuh, *Basic Circuit Theory*, McGraw-Hill Book Company, New York, 1969.

Friedland, B., O. Wing, and R. Ash, *Principles of Linear Networks*, McGraw-Hill Book Company, New York, 1961.

Hayt, W. H., Jr., and J. E. Kemmerly, *Engineering Circuit Analysis*, McGraw-Hill Book Company, New York, 1962.

Kuo, F. F., *Network Analysis and Synthesis*, John Wiley & Sons, Inc., New York, 1966.

Leon, B. J., and P. A. Wintz, *Basic Linear Networks for Electrical and Electronics Engineers*, Holt, Rinehart & Winston, Inc., New York, 1970.

Manning, L. A., *Electrical Circuits*, McGraw-Hill Book Company, New York, 1966.

Merriam, C. W., III, *Analysis of Lumped Electrical Systems*, John Wiley & Sons, Inc., New York, 1969.

Pearson, S. I., and G. J. Maler, *Introductory Circuit Analysis*, John Wiley & Sons, Inc., New York, 1965.

Skilling, H. H., *Electrical Engineering Circuits*, John Wiley & Sons, Inc., New York, 1965.

Van Valkenburg, M. E., *Network Analysis*, Prentice-Hall, Inc., Englewood Cliffs, N.J., 1964.

NUMERICAL TECHNIQUES AND USE OF COMPUTERS IN ENGINEERING

The following list gives examples of some of the more recent introductory books which have been written on numerical techniques and on the use of computers in solving engineering problems. These should provide useful references for the student who desires to learn more about the usage of computational digital computer methods of the type which have been introduced in this text.

Hamming, R. W., *Numerical Methods for Scientists and Engineers*, McGraw-Hill Book Company, New York, 1962.

James, M. L., G. M. Smith, and J. C. Wolford, *Applied Numerical Methods for Digital Computation with FORTRAN*, International Textbook Company, Scranton, Pa., 1967.

Kelly, L. G., *Handbook of Numerical Methods and Applications*, Addison-Wesley Publishing Company, Inc., Reading, Mass., 1967.

Kuo, S. S., *Numerical Methods and Computers*, Addison-Wesley Publishing Company, Inc., Reading, Mass., 1965.

Ley, B. J., *Computer Aided Analysis and Design for Electrical Engineers*, Holt, Rinehart & Winston, Inc., New York, 1970.

McCracken, D. D., and W. S. Dorn, *Numerical Methods and FORTRAN Programming*, John Wiley & Sons, Inc., New York, 1964.

Mischke, C. R., *An Introduction to Computer-Aided Design*, Prentice-Hall, Inc., Englewood Cliffs, N.J., 1968.

Ralston, A., *A First Course in Numerical Analysis*, McGraw-Hill Book Company, New York, 1965.

Ramey, R. L., and E. J. White, *Matrices and Computers in Electronic Circuit Analysis*, McGraw-Hill Book Company, New York, 1971.

Seely, S., N. H. Tarnoff, and D. Holstein, *Digital Computers in Engineering*, Holt, Rinehart & Winston, Inc., New York, 1970.

MATRIX ALGEBRA

The following books will serve as useful supplements to the material on matrix algebra which is included in Appendix A.

Dorf, R. C., *Matrix Algebra—A Programmed Introduction*, John Wiley & Sons, Inc., New York, 1969.

Hohn, F. E., *Elementary Matrix Algebra*, The Macmillan Company, New York, 1958.

Nering, E. D., *Linear Algebra and Matrix Theory*, John Wiley & Sons, Inc., New York, 1963.

Schwartz, J. T., *Introduction to Matrices and Vectors*, McGraw-Hill Book Company, New York, 1961.

Tropper, A. M., *Matrix Theory for Electrical Engineers*, Addison-Wesley Publishing Company, Inc., Reading, Mass., 1962.

ACTIVE RC CIRCUITS

Although the general field of active *RC* circuits is a relatively new one, there are already several books covering some of the important aspects of the subject. The most significant of these are listed below.

Chirlian, P. M., *Integrated and Active Network Analysis and Synthesis*, Prentice-Hall, Inc., Englewood Cliffs, N.J., 1967.

Huelsman, L. P., *Theory and Design of Active RC Circuits*, McGraw-Hill Book Company, New York, 1968.

Huelsman, L. P. (ed.), *Active Filters: Lumped, Distributed, Integrated, Digital, and Parametric*, McGraw-Hill Book Company, New York, 1970.

Mitra, S. K., *Analysis and Synthesis of Linear Active Networks*, John Wiley & Sons, Inc., New York, 1969.

Newcomb, R. W., *Active Integrated Circuit Synthesis*, Prentice-Hall, Inc., Englewood Cliffs, N.J., 1968.

Su, K. L., *Active Network Synthesis*, McGraw-Hill Book Company, New York, 1965.

APPENDIX A

Matrices and Determinants

Frequently in network analysis studies we have the need to write sets of simultaneous equations. Such a task can be facilitated by the use of matrices. There are two advantages to such a use:

1. The elements of the set of simultaneous equations are divided so as to emphasize the three different kinds of these elements, namely, the dependent variables, the independent variables, and the coefficients of the equations.
2. A symbolic shorthand notation is provided which may be conveniently used to indicate operations such as substitution of variables and solutions of a set of equations.

In this appendix we shall present a discussion of the general theory of matrices and the rules for manipulating them. The latter subject is frequently referred to as *matrix algebra.*

A-1 DEFINITIONS

A *matrix* is a rectangular array of elements arranged in horizontal and vertical lines. The array is usually enclosed in brackets. The horizontal lines are called *rows* and the vertical lines are called *columns*. The elements of the array may be literal symbols (usually with subscripts), real or complex numbers, or functions of some real or complex variable. Thus, typical elements may have the form

$$a_{ij} \qquad 17 \qquad 3 + j22 \qquad \frac{x^3}{x+1} \qquad \tanh\sqrt{s} \tag{A-1}$$

Typical arrays composed of such elements are

$$\begin{bmatrix} 4 & 5 \\ 6 & 7 \\ 8 & 9 \end{bmatrix} \qquad \begin{bmatrix} 1+2j \\ 3j \\ 4 \end{bmatrix} \qquad [x_1 \quad x_2] \qquad \begin{bmatrix} a_{11} & a_{12} \\ a_{21} & a_{22} \end{bmatrix} \tag{A-2}$$

In such arrays the columns are usually considered as being numbered from left to right and the rows from top to bottom. Thus, the element consisting of the number 4 in the first matrix in (A–2) may be referred to as the 11 element, implying it is located at the intersection of the first row and the first column. Similarly, the 5 is the 12 element, the 6 is the 21 element, etc. If we use a literal symbol to specify a matrix element, it is conventional to use a first subscript to indicate the row in which the element is located and a second subscript to indicate the column. Thus, the element a_{11} in the fourth matrix of (A–2) is the element located in the first row and the first column, a_{12} is in the first row and the second column, etc. More generally, a_{ij} is defined as the element located in the ith row and the jth column. Using this notation, we may write the literal form of the general matrix with m rows and n columns as

$$\begin{bmatrix} a_{11} & a_{12} & \cdots & a_{1n} \\ a_{21} & a_{22} & \cdots & a_{2n} \\ \vdots & \vdots & & \\ a_{m1} & a_{m2} & & a_{mn} \end{bmatrix} \tag{A-3}$$

A first means of characterizing matrices is by the number of rows and columns they contain. To indicate this information we use the term *order*. Thus, a matrix which has m rows and n columns is referred to as a matrix of order (m, n) or, more simply, as an $m \times n$ matrix (note that the number of rows is given first, then the number of columns). Thus, the first matrix given in (A–2) may be referred to as a 3×2 matrix, the second as a 3×1 matrix, the third as a 1×2 matrix, and the fourth as a 2×2 matrix. Some frequently encountered orders of matrices are usually given special names. For example, a matrix which has only a single column such as the second one shown in (A–2) is called a *column matrix* or a *vector*. This notation applies irrespective of how many rows are present. Frequently, for such a matrix only a single subscript is used for the elements of the array. Another type of matrix which is given a special name is one which contains only a single row. This is logically called a *row matrix*. An example is the third matrix given in (A–2). The last special classification given matrices covers the case of a matrix which has the same number of rows and columns. Such a matrix is called a *square matrix*. More specifically, if there are n rows and n columns, we say that such a matrix has order n. Finally, it should be noted that a 1×1 matrix which is comprised of a single element is simply referred to as a *scalar*.

There are several subclasses of square matrices which are frequently encountered. One of these is the *symmetric matrix*. If we use the symbol a_{ij} to refer to the elements of a symmetric matrix, then these elements satisfy the equality

$$a_{ij} = a_{ji} \tag{A-4}$$

for all values of i and j. Now let us define the *main diagonal* of a square matrix of order n as the n elements $a_{11}, a_{22}, a_{33}, \ldots, a_{nn}$. From (A-4) we see that a symmetric matrix is one which is symmetrical with respect to the main diagonal. A special case of a symmetric matrix is one in which all the elements except those on the main diagonal are zero. This is called a *diagonal matrix*. If, in addition, the diagonal elements are all unity, the matrix is referred to as an *identity matrix*. Some examples of symmetric, diagonal, and identity matrices follow.

$$\begin{bmatrix} a & d & e \\ d & b & f \\ e & f & c \end{bmatrix} \quad \begin{bmatrix} 1 & 0 & 0 & 0 \\ 0 & 2 & 0 & 0 \\ 0 & 0 & 3 & 0 \\ 0 & 0 & 0 & 4 \end{bmatrix} \quad \begin{bmatrix} 1 & 0 \\ 0 & 1 \end{bmatrix} \quad \begin{bmatrix} 1 & 0 & 0 \\ 0 & 1 & 0 \\ 0 & 0 & 1 \end{bmatrix}$$

Symmetric matrix — Diagonal matrix — Identity matrices

In order to simplify the notation when sets of simultaneous equations are to be considered, it is convenient to use a single symbol to represent a given matrix. In this text we use a boldface capital letter for such a represention. It is usually convenient to make this be the same letter as is used for the individual elements of the array. Thus, if the elements of the array are expressed in literal form as a_{ij}, the matrix is simply referred to as the matrix $\mathbf{A}$.

A-2 MATRIX EQUALITY

In order to start our discussion of matrix algebra, we begin by defining a matrix equality. Two matrices are equal if and only if each of the corresponding elements of the matrices is equal. Thus, if $\mathbf{A}$ and $\mathbf{B}$ are two matrices with elements a_{ij} and b_{ij}, respectively, $\mathbf{A} = \mathbf{B}$ if and only if $a_{ij} = b_{ij}$ for all values of i and j. As a consequence of this definition, we see that only matrices having the same number of rows and columns can be equated.

A-3 ADDITION AND SUBTRACTION OF MATRICES

The next operation of matrix algebra we shall consider is the summing of two matrices so as to form a third matrix. The sum of two matrices is equal to a third matrix if and only if the sum of the elements in corresponding rows and columns of the two matrices is equal to the value of the element in the same row and column of the third matrix. Thus, if **A**, **B**, and **C** are matrices with elements a_{ij}, b_{ij}, and c_{ij}, respectively, the relation $\mathbf{A} + \mathbf{B} = \mathbf{C}$ implies that $a_{ij} + b_{ij} = c_{ij}$ for all values of i and j. The operation of addition is clearly defined only for matrices which have the same number of rows and columns. As an example of matrix addition, consider the following.

$$\begin{bmatrix} 1 & 2 & 3 \\ 4 & 5 & 6 \end{bmatrix} + \begin{bmatrix} 2 & 3 & 4 \\ 5 & 6 & 7 \end{bmatrix} = \begin{bmatrix} 3 & 5 & 7 \\ 9 & 11 & 13 \end{bmatrix}$$

Subtraction is similarly defined, i. e., $\mathbf{A} - \mathbf{B} = \mathbf{C}$ implies $a_{ij} - b_{ij} = c_{ij}$ for all values of i and j.

A-4 MULTIPLICATION OF A MATRIX BY A SCALAR

Now let us define the operation of multiplication of an entire matrix by a scalar, i. e., a single number. Such an operation simply multiplies every element of the array by the scalar. Thus, if **A** is a matrix with elements a_{ij} and if k is a scalar, the elements of the matrix $k\mathbf{A}$ are ka_{ij}. The term scalar as we use it here may be interpreted as a real number, a complex number, or a function of some variable. As an example, consider the following.

$$\frac{s}{s^2+1}\begin{bmatrix} 1 & 2 \\ 3 & 4 \end{bmatrix} = \begin{bmatrix} \dfrac{s}{s^2+1} & \dfrac{2s}{s^2+1} \\ \dfrac{3s}{s^2+1} & \dfrac{4s}{s^2+1} \end{bmatrix}$$

It should be noted that this operation is considerably different than the operation of multiplying a determinant by a scalar constant, since the latter operation is equivalent to multiplying only the elements of a single row or a single column by such a constant.

A-5 MULTIPLICATION OF A SQUARE MATRIX BY A COLUMN MATRIX

One of our principal goals in the use of matrices is the representation of sets of simultaneous equations. As an example of this, let us consider the following such set:

$$\begin{aligned} y_1 &= a_{11}x_1 + a_{12}x_2 \\ y_2 &= a_{21}x_1 + a_{22}x_2 \end{aligned} \tag{A–5}$$

We will assume that we desire to use the matrix notation

$$\mathbf{Y} = \mathbf{AX} \tag{A–6}$$

to represent such a set of equations, and that the individual matrices **Y, A,** and **X** are defined as

$$\mathbf{Y} = \begin{bmatrix} y_1 \\ y_2 \end{bmatrix} \qquad \mathbf{A} = \begin{bmatrix} a_{11} & a_{12} \\ a_{21} & a_{22} \end{bmatrix} \qquad \mathbf{X} = \begin{bmatrix} x_1 \\ x_2 \end{bmatrix} \tag{A–7}$$

In order to use the notation of (A–6), the multiplication operation indicated as **AX** must be defined so as to produce the correct right members for the equations given in (A–5). Thus, we must be able to write

$$\mathbf{Y} = \begin{bmatrix} y_1 \\ y_2 \end{bmatrix} = \begin{bmatrix} a_{11}x_1 + a_{12}x_2 \\ a_{21}x_1 + a_{22}x_2 \end{bmatrix} = \begin{bmatrix} a_{11} & a_{12} \\ a_{21} & a_{22} \end{bmatrix} \begin{bmatrix} x_1 \\ x_2 \end{bmatrix} = \mathbf{AX} \tag{A–8}$$

Now consider the matrix equality relating the two column matrices given as the second and third members of the above equation. For the first element of these two column matrices (the elements in the first row) to be equal requires that

$$y_1 = a_{11}x_1 + a_{12}x_2 \tag{A–9}$$

Note that the right member of this equation may be interpreted as requiring that each element of the first row of the square matrix **A** be multiplied by a corresponding element of the column matrix **X** and the sum taken of the resulting products. Similarly, for the second element of the two column matrices to be equated requires that

$$y_2 = a_{21}x_1 + a_{22}x_2 \tag{A–10}$$

Thus, in the right member of this expression, we sum the products of the second row of the square matrix **A** with the corresponding elements of the column matrix **X**. This result defines the multiplication of a square matrix of order 2 by a column matrix with two elements. It is easily generalized to the case where **A** is an nth-order square matrix and **X** and **Y** are column matrices with n rows. In such a case, the matrices **Y, A,** and **X** are defined as

$$\mathbf{Y} = \begin{bmatrix} y_1 \\ y_2 \\ \vdots \\ y_n \end{bmatrix} \qquad \mathbf{A} = \begin{bmatrix} a_{11} & a_{12} & \cdots & a_{1n} \\ a_{21} & a_{22} & \cdots & a_{2n} \\ \vdots & \vdots & & \\ a_{n1} & a_{n2} & \cdots & a_{nn} \end{bmatrix} \qquad \mathbf{X} = \begin{bmatrix} x_1 \\ x_2 \\ \vdots \\ x_n \end{bmatrix} \tag{A–11}$$

The matrix equation $\mathbf{Y} = \mathbf{AX}$ is now simply defined by the relation

$$y_i = \sum_{k=1}^{n} a_{ik} x_k \qquad i = 1, 2, \ldots, n \tag{A-12}$$

As an example, applying this result to a third-order set of equations, we obtain

$$\mathbf{AX} = \begin{bmatrix} a_{11} & a_{12} & a_{13} \\ a_{21} & a_{22} & a_{23} \\ a_{31} & a_{32} & a_{33} \end{bmatrix} \begin{bmatrix} x_1 \\ x_2 \\ x_3 \end{bmatrix} = \begin{bmatrix} a_{11}x_1 + a_{12}x_2 + a_{13}x_3 \\ a_{21}x_1 + a_{22}x_2 + a_{23}x_3 \\ a_{31}x_1 + a_{32}x_2 + a_{33}x_3 \end{bmatrix} = \begin{bmatrix} y_1 \\ y_2 \\ y_3 \end{bmatrix} = \mathbf{Y} \tag{A-13}$$

Note that the number of elements in the two-column matrices must be the same as the order of the square matrix for the multiplication operation presented in this section to be defined. More specifically, the multiplication $\mathbf{Y} = \mathbf{AX}$, where $\mathbf{Y}$ is an $n \times 1$ matrix, is only defined for the case where $\mathbf{A}$ is an $n \times n$ matrix and $\mathbf{X}$ is an $n \times 1$ matrix. It should also be noted that the operation $\mathbf{XA}$ is not defined, i. e., $\mathbf{XA} \neq \mathbf{AX}$. Thus, matrix multiplication in general is not commutative.

A-6 MULTIPLICATION OF TWO RECTANGULAR MATRICES

In the preceding section we considered the operation of multiplying a square matrix and a column matrix. In this section we will consider the multiplication of two rectangular matrices. If we define this operation $\mathbf{C} = \mathbf{AB}$, then the elements c_{ij} of the matrix $\mathbf{C}$ are related to the elements a_{ij} and b_{ij} of the matrices $\mathbf{A}$ and $\mathbf{B}$ by the following equation.

$$c_{ij} = \sum_{k=1}^{n} a_{ik} b_{kj} \tag{A-14}$$

There are two properties of this relation that should be noted:

1. The summation index k that must be used for each term c_{ij} appears as the second subscript, i. e., the column subscript, of the a_{ik} terms, however, it appears as the first subscript, i. e., the row subscript, of the b_{kj} terms. Thus, it is necessary that the *number of columns of the first matrix that enters into such a multiplication must be the same as the number of rows of the second matrix* for the multiplication operation to be defined.
2. Every value of the row subscript i that the element a_{ik} can have, as determined by the number of rows in the $\mathbf{A}$ matrix, specifies a corresponding row subscript in the resulting element c_{ij} Thus, *the* $\mathbf{C}$ *matrix will have the same number of rows as the* $\mathbf{A}$ *matrix*. Similarly, every value of the column subscript j that the element b_{kj} can have determines the corresponding column subscript of the element c_{ij}. Thus, *the* $\mathbf{C}$ *matrix will have the same*

*number of columns as the **B** matrix,* i. e., the second matrix in the multiplication operation.

As an example of general matrix multiplication consider the following:

$$\begin{bmatrix} a & b & c \\ d & e & f \end{bmatrix} \begin{bmatrix} 1 & 2 \\ 3 & 4 \\ 5 & 6 \end{bmatrix} = \begin{bmatrix} a + 3b + 5c & 2a + 4b + 6c \\ d + 3e + 5f & 2d + 4e + 6f \end{bmatrix} \tag{A-15}$$

As an illustration of the non-commutativity of matrix multiplication, consider two matrices **A** and **B** defined as

$$\mathbf{A} = \begin{bmatrix} 1 \\ 2 \\ 3 \end{bmatrix} \qquad \mathbf{B} = [1 \quad 2 \quad 3] \tag{A-16}$$

The product **AB** gives the following square matrix.

$$\mathbf{AB} = \begin{bmatrix} 1 & 2 & 3 \\ 2 & 4 & 6 \\ 3 & 6 & 9 \end{bmatrix}$$

The product **BA**, however, yields the simple scalar 14. Clearly, the two results are not the same.

A-7 SOME BASIC RULES OF MATRIX ALGEBRA

In this section we present some general basic rules of matrix algebra. In specifying the indicated operations of equality, addition, subtraction, and multiplication, we assume that the order of the component matrices is such that the implied operations are defined as noted in the preceding sections. The proof of these rules is left to the reader as an exercise.

Rule 1 Matrix addition is associative. $\mathbf{A} + (\mathbf{B} + \mathbf{C}) = (\mathbf{A} + \mathbf{B}) + \mathbf{C}$.

Rule 2 Matrix addition is commutative. $\mathbf{A} + \mathbf{B} = \mathbf{B} + \mathbf{A}$.

Rule 3 Matrix multiplication is associative. $\mathbf{AB}(\mathbf{C}) = \mathbf{A}(\mathbf{BC})$.

Rule 4 Matrix multiplication is not, in general, commutative. Except in special cases, $\mathbf{AB} \neq \mathbf{BA}$.

Rule 5 Matrix multiplication is distributive with respect to addition. $\mathbf{A}(\mathbf{B} + \mathbf{C}) = \mathbf{AB} + \mathbf{AC}$.

Rule 6 Any matrix multiplied by an identity matrix equals itself. $\mathbf{1A} = \mathbf{A1} = \mathbf{A}$, where $\mathbf{1}$ is an identity matrix.

A-8 THE INVERSE OF MATRICES

The only major operation found in scalar algebra which has not been defined above for matrix algebra is division. The reason is that *division is not a permitted operation in the algebra of matrices.* Thus, we may *not* use a notation such as **A/B**. The operation of matrix algebra which most closely corresponds to the division process of scalar algebra is multiplication by the inverse of a matrix. A treatment of the determination of the inverse of a matrix may be found in Chap. 3.

A-9 DETERMINANTS AND COFACTORS

A *determinant* is a scalar quantity defined in terms of the elements of a square matrix. If the matrix has n rows and n columns, the determinant may be indicated as follows.

$$\det \mathbf{R} = \begin{vmatrix} r_{11} & r_{12} & \cdots & r_{1n} \\ r_{21} & r_{22} & \cdots & r_{2n} \\ \cdot & \cdot & \cdots & \cdot \\ \cdot & \cdot & \cdots & \cdot \\ \cdot & \cdot & \cdots & \cdot \\ r_{n1} & r_{n2} & \cdots & r_{nn} \end{vmatrix} \tag{A-17}$$

This is frequently referred to as a *determinant of order n*. Such a determinant has a definite (scalar) value which is determined by the elements r_{ij}. For low-order determinants, the value of the determinant may be found directly by simple rules. For example, for a second-order determinant, it is well-known that

$$\det \mathbf{R} = \begin{vmatrix} r_{11} & r_{12} \\ r_{21} & r_{22} \end{vmatrix} = r_{11}r_{22} - r_{12}r_{21} \tag{A-18}$$

The equation for evaluating the determinant of the second-order array given in Eq. (A–18) may be remembered by thinking of multiplying the elements on the *main diagonal* (the elements placed along a diagonal line from the upper left corner of the array to the lower right corner), and subtracting the product of the terms on the other diagonal as indicated below.

$$\begin{matrix} r_{11} & r_{12} \\ r_{21} & r_{22} \end{matrix} = r_{11}r_{22} - r_{12}r_{21} \tag{A-19}$$

A similar procedure may be applied to determine the determinant of a third-order array. This is shown below. For convenience, the first and second columns have been repeated at the right of the array.

$$\begin{matrix} r_{11} & r_{12} & r_{13} & r_{11} & r_{12} \\ r_{21} & r_{22} & r_{23} & r_{21} & r_{22} \\ r_{31} & r_{32} & r_{33} & r_{31} & r_{32} \end{matrix} \tag{A-20}$$

$$\det \mathbf{R} = r_{11}r_{22}r_{33} + r_{12}r_{23}r_{31} + r_{13}r_{21}r_{32} - r_{13}r_{22}r_{31} - r_{11}r_{23}r_{32} - r_{12}r_{21}r_{33}$$

An example follows.

EXAMPLE A-1. The rule given above is easily applied to find the determinant of the matrix for the three-mesh resistance network analyzed in Example 3-2.1. The resistance matrix **R** is repeated here for convenience.

$$\mathbf{R} = \begin{bmatrix} 3 & -2 & -1 \\ -2 & 5 & -3 \\ -1 & -3 & 8 \end{bmatrix}$$

Following the rule for the evaluation of the determinant given in Eq. (A-20), we obtain

$$\begin{aligned} \det \mathbf{R} &= (3 \times 5 \times 8) + (-2 \times -3 \times -1) + (-1 \times -2 \times -3) \\ &\quad -(-1 \times 5 \times -1) - (3 \times -3 \times -3) - (-2 \times -2 \times 8) \\ &= 44 \end{aligned}$$

A more general procedure for evaluating the third-order determinant is by *expanding* the determinant. Such an expansion is made in terms of the elements of any row or column and a set of second-order determinants known as cofactors. For example, expanding the determinant along the first row, we obtain

$$\det \mathbf{R} = r_{11}\begin{vmatrix} r_{22} & r_{23} \\ r_{32} & r_{33} \end{vmatrix} - r_{12}\begin{vmatrix} r_{21} & r_{23} \\ r_{31} & r_{33} \end{vmatrix} + r_{13}\begin{vmatrix} r_{21} & r_{22} \\ r_{31} & r_{32} \end{vmatrix} \tag{A-21}$$

Thus, we have expressed the determinant of a third-order array in terms of the determinants of second-order arrays. A general procedure for expressing the determinant of an nth order array in terms of the determinants of $(n - 1)$-order arrays is called *Laplace's expansion.* If such an expansion is made along the ith row of an array, it has the form

$$\det \mathbf{R} = \sum_{k=1}^{n} r_{ik} R_{ik} \tag{A-22}$$

where the r_{ik} are the elements of **R** and the R_{ik} are *cofactors.* These cofactors are formed by deleting the ith row and kth column of the array, taking the determinant of the remaining elements (a determinant of order $n - 1$), and prefixing the result by the positive or negative multiplier determined by the expression $(-1)^{i+k}$.[1] For example, for a third-order array with elements, r_{ij}, if we expand along the first row, then we see from Eq. (A–22) that the cofactors R_{11}, R_{12}, and R_{13} are required. These are determined as

$$
\begin{aligned}
R_{11} &= +\begin{vmatrix} r_{22} & r_{23} \\ r_{32} & r_{33} \end{vmatrix} \\
R_{12} &= -\begin{vmatrix} r_{21} & r_{23} \\ r_{31} & r_{33} \end{vmatrix} \\
R_{13} &= +\begin{vmatrix} r_{21} & r_{22} \\ r_{31} & r_{32} \end{vmatrix}
\end{aligned} \tag{A–23}
$$

It is readily apparent that these are the cofactors which were used in Eq. (A–21).

By the use of Eq. (A–22) we can express the determinant of an nth order array as a function of determinants of $(n - 1)$-order arrays. These, in turn, may be expressed as a function of the determinants of $(n - 2)$-order arrays, and the process may be continued until the value of the determinant is obtained. Such a process becomes quite tedious (as well as very liable to error) for arrays higher than third degree. In such cases, we are led to seek other methods of solving sets of simultaneous equations which do not require the use of determinants. A method which is readily implemented on the digital computer is described in Chap. 3.

A Laplace expansion may also be made along a *column* of a given array. For example, to evaluate the determinant in terms of the elements of the jth column, we may use the expression

$$\det \mathbf{R} = \sum_{k=1}^{n} r_{kj} R_{kj} \tag{A–24}$$

where the R_{kj} are cofactors as previously defined.

[1]The determinant of order $n - 1$ is called a first-order minor.

APPENDIX B

Complex Numbers

In network studies involving sinusoidal steady-state analysis, we must use numbers and variables which are complex rather than real. The rules for manipulating such complex quantities, however, are quite different than the corresponding rules for manipulating real quantities. In this appendix we will present a summary of such rules. Taken together, these rules define an *algebra for complex numbers.*

B-1 DEFINITIONS

A complex number a may be defined as an ordered pair of two real numbers (a_1, a_2). The two real numbers are referred to as the real part and the imaginary part, respectively, of the complex number a. Thus, we may write

$$\begin{aligned} a &= a_1 + ja_2 \\ a_1 &= \operatorname{Re}(a) \\ a_2 &= \operatorname{Im}(a) \end{aligned} \tag{B-1}$$

where Re stands for "the real part of," Im stands for "the imaginary part of," and j has the value $\sqrt{-1}$. A convenient graphical representation for such a complex number is obtained by defining a complex plane in which real quantities are plotted along the abscissa and imaginary quantities are plotted along the ordinate as shown in Fig. B-1.1(a). Such a representation is referred to as the rectangular form of a complex number. Using this rectangular form as a starting point, we may develop a

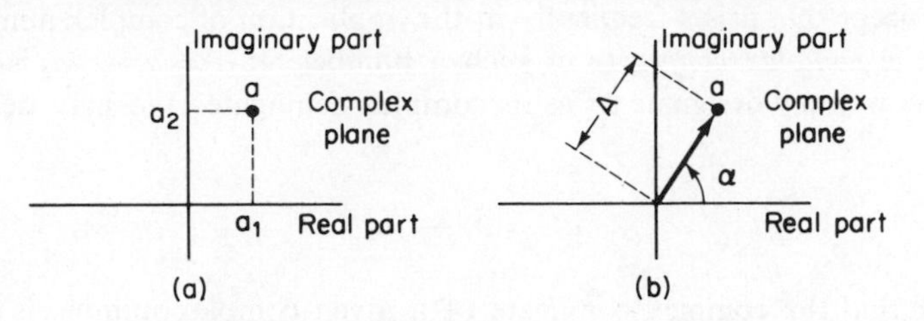

Fig. B-1.1 Representation of a complex number.

second representation for a complex number. This is obtained by drawing a vector from the origin of the complex plane to the point a as shown in Fig. B–1.1(b) and using this vector to represent the complex number. In terms of such a representation, we may define a complex number a as follows.

$$a = Ae^{j\alpha}$$
$$A = |a| \tag{B–2}$$
$$\alpha = \text{Arg}\,(a)$$

Thus, we may consider a complex number as a vector, and we may use the magnitude and the argument of the vector to define the complex number. Such a representation is referred to as the polar or exponential form of a complex number. Although the angle (or argument) α defined in Eq. (B–2) is given in radians, it is frequently convenient to use angle measure in degrees. In such a case, we may write the polar form of a complex number a as

$$a = Ae^{j\phi^\circ}$$

where φ° is the angle α given in degrees. The relations between the rectangular and polar forms are easily found by comparing the representations given in Fig. B–1.1. Thus, from (B–1) and (B–2) we see that

$$a_1 = A\cos\alpha \qquad A = \sqrt{a_1^2 + a_2^2}$$
$$a_2 = A\sin\alpha \qquad \alpha = \tan^{-1}\left(\frac{a_2}{a_1}\right) \tag{B–3}$$

Inserting the relations for a_1 and a_2 given in (B–1), we obtain the relation

$$e^{j\alpha} = \cos\alpha + j\sin\alpha \tag{B–4a}$$

Similarly, we may show that

$$e^{-j\alpha} = \cos\alpha + j\sin\alpha \tag{B–4b}$$

These are frequently referred to as *Euler's identity*.

A concept that arises frequently in the application of complex numbers is the concept of the *complex conjugate* of such a number. If $a = a_1 + ja_2$ is a complex number, then we may designate a^* as its complex conjugate. We may define a^* by the relation

$$a^* = a_1 - ja_2 \tag{B-5}$$

Thus, we see that the complex conjugate of a given complex number is formed by reversing the sign of its imaginary part. From (B–3) we see that this is equivalent to reversing the sign of the phase angle. Thus, if $a = Ae^{j\alpha}$, then a^* (the complex conjugate of a) is given as

$$a^* = Ae^{-j\alpha} \tag{B-6}$$

B-2 ADDITION AND SUBTRACTION

If a and b are complex numbers defined as $a = a_1 + ja_2$ and $b = b_1 + jb_2$, then the sum and difference of a and b are defined by the relations

$$\begin{aligned} a + b &= (a_1 + b_1) + j(a_2 + b_2) \\ a - b &= (a_1 - b_1) + j(a_2 - b_2) \end{aligned} \tag{B-7}$$

In other words, complex numbers are added (or subtracted) by separately adding (or subtracting) their real and imaginary parts. Now consider the determination of the sum of two complex numbers a and b if these numbers are expressed in polar form, i. e., $a = Ae^{j\alpha}$ and $b = Be^{j\beta}$. Substituting the relations of (B–3) in (B–7), we obtain

$$a + b = \sqrt{(A\cos\alpha + B\cos\beta)^2 + (A\sin\alpha + B\sin\beta)^2} \; e^{j\tan^{-1}[(A\sin\alpha + B\sin\beta)/(A\cos\alpha + B\cos\beta)]} \tag{B-8}$$

Obviously, it is much more convenient to express complex numbers using the rectangular form if addition is to be performed. As examples of adding and subtracting two complex numbers, consider the following.

$$\begin{aligned} (1 + j3) + (2 - j4) &= 3 - j1 \\ (1 + j3) - (2 - j4) &= -1 + j7 \end{aligned}$$

If is readily shown that the operations of addition and subtraction are commutative and associative. Thus, if a, b, and c are complex numbers, we may write

$$\begin{aligned} a + b &= b + a \\ a + (b + c) &= (a + b) + c \end{aligned} \tag{B-9}$$

B-3 MULTIPLICATION AND DIVISION

As a preliminary result which will be useful in studying the multiplication and division of complex numbers, let us consider the case where a complex number $a = a_1 + ja_2$ is multiplied by a real number k. This operation is defined as

$$ka = k(a_1 + ja_2) = ka_1 + jka_2 \tag{B-10}$$

Thus, we see that in such a case the real number separately multiplies the real and the imaginary part of the complex number. Similarly, division of a complex number $a = a_1 + ja_2$ by a real number k may be defined as

$$\frac{a}{k} = \frac{a_1 + ja_2}{k} = \frac{a_1}{k} + j\frac{a_2}{k} \tag{B-11}$$

It is readily shown that these operations are commutative and associative. Thus, if a is a complex number and k and m are real numbers, we may write

$$(km)a = k(ma) = m(ka) \tag{B-12}$$

Now let us consider the product of two complex numbers $a = a_1 + ja_2$ and $b = b_1 + jb_2$. If we separately multiply each of the terms of these two complex numbers, and note that $j^2 = -1$, we obtain

$$ab = (a_1b_1 - a_2b_2) + j(a_1b_2 + a_2b_1) \tag{B-13}$$

Let us convert this result to polar form. Substituting from (B-3) we obtain

$$\begin{aligned} ab &= AB[(\cos\alpha\cos\beta - \sin\alpha\sin\beta) + j(\cos\alpha\sin\beta + \sin\alpha\cos\beta)] \\ &= AB[\cos(\alpha+\beta) + j\sin(\alpha+\beta)] \\ &= AB\,e^{j(\alpha+\beta)} \end{aligned} \tag{B-14}$$

where the last result is obtained by applying Euler's identity as given in (B-4). Thus, we see that the product of two complex numbers which are expressed in polar form is found by multiplying the magnitudes and adding the arguments of the two numbers. Obviously, it is much more convenient to express complex numbers in polar form if it is desired to multiply these numbers. As an example of the multiplication of two complex numbers, consider the following.

$$(1 + j1)(-2 + j2) = \sqrt{2}\,e^{j45^\circ}\,2\sqrt{2}\,e^{j135^\circ} = 4e^{j180^\circ} = -4 + j0$$

As an application of the multiplication of two complex numbers, let us see what happens when we multiply a number by its complex conjugate. Let $a = Ae^{j\alpha}$ be the number. Applying the results given above, we obtain

$$aa^* = Ae^{j\alpha}Ae^{-j\alpha} = A^2e^{j0} = A^2 \tag{B-15}$$

We thus conclude that the product of a number and its complex conjugate yields a real number equal to the square of the magnitude of the complex number.

It is readily shown that multiplication of complex numbers is commutative and associative. Thus if a, b, and c are complex numbers, we may write

$$\begin{aligned} ab &= ba \\ a(bc) &= (ab)c \end{aligned} \tag{B-16}$$

Similarly, multiplication of complex numbers is distributive, i. e.,

$$a(b + c) = ab + ac \tag{B-17}$$

Now let us consider the division of two complex numbers $a = a_1 + ja_2$ and $b = b_1 + jb_2$. If the numerator and denominator of the quotient a/b are first multiplied by the complex conjugate of the denominator, the quotient may be expressed as a multiplication of two complex numbers and a division by a real number. Thus, we may write

$$\frac{a}{b} = \frac{ab^*}{bb^*} = \frac{(a_1 + ja_2)(b_1 - jb_2)}{(b_1 + jb_2)(b_1 - jb_2)} = \frac{(a_1b_1 + a_2b_2) + j(a_2b_1 - a_1b_2)}{b_1^2 + b_2^2} \tag{B-18}$$

This result may also be expressed in polar form as

$$\frac{a}{b} = \frac{Ae^{j\alpha}}{Be^{j\beta}} = \frac{A}{B}e^{j(\alpha-\beta)} \tag{B-19}$$

Thus, we see that the quotient of two complex numbers is simply equal to the quotient of their magnitudes and the difference of their arguments. Obviously, the division of two complex numbers is most conveniently performed if the numbers are first expressed in polar form. As an example of the division of two complex numbers, consider the following.

$$\frac{(1 - j1)}{(2 + j2)} = \frac{\sqrt{2}\, e^{-j45°}}{2\sqrt{2}\, e^{j45°}} = \frac{1}{2}e^{-j90°} = 0 - j\frac{1}{2}$$

B-4 POWERS AND ROOTS

To raise a complex number to a power, we need simply use the polar representation for a complex number and apply the usual rule of exponents. Thus, if $a = Ae^{j\alpha}$ is a complex number, raising it to the nth power ($n = 1, 2, 3, \ldots$) is done as follows:

$$a^n = (Ae^{j\alpha})^n = A^n e^{jn\alpha} \tag{B-20}$$

From the above equation, we see that raising a complex number to a power is done by raising the magnitude to the specified power and multiplying the argument by the value of the power. As an example of this, consider the following:

$$(1 + j1)^3 = (\sqrt{2}\, e^{j45^\circ})^3 = 2\sqrt{2}\, e^{j135^\circ} = -2 + j2$$

Now let us consider taking the root of a complex number. Again we will use a polar representation for the number. We first note that adding or subtracting any multiple of 2π radians to the argument of a complex number does not change its value. Thus, if $a = Ae^{j\alpha}$ is a complex number, we may write

$$a = Ae^{j\alpha} = Ae^{j(\alpha+2\pi)} = Ae^{j(\alpha+4\pi)} = \cdots = Ae^{j(\alpha+2k\pi)} \tag{B-21}$$

where, in the right member, k is a positive or negative integer. We may now apply the rule of exponents to find $a^{1/m}$. Using the expression given in the right member of (B-21), we obtain

$$a^{1/m} = A^{1/m}\, e^{j(\alpha/m + 2k\pi/m)} \tag{B-22}$$

If we investigate this result for values of k equal to $0, 1, 2, \ldots, m-1$, we find that there are exactly m unique values of the argument. Thus, there will be m roots whose locus is a circle of radius $A^{1/m}$ centered at the origin. As an example of this conclusion, consider the determination of the cube root of unity, which in complex form may be written $1 + j0$. Applying (B-21), we obtain three different polar representations for $1 + j0$. These are

$$(1 + j0) = 1e^{j0} = 1e^{j2\pi} = 1e^{j4\pi}$$

Now let $m = 3$ in (B-22). Inserting the values for $1 + j0$ given above, we obtain

$$(1 + j0)^{1/3} = (1e^{j0})^{1/3} = 1e^{j0} = (1 + j0)$$
$$(1 + j0)^{1/3} = (1e^{j2\pi})^{1/3} = 1e^{j2\pi/3} = -0.5 + j0.866$$
$$(1 + j0)^{1/3} = (1e^{j4\pi})^{1/3} = 1e^{j4\pi/3} = -0.5 - j0.866$$

The location of the three distinct cube roots of unity are shown in Fig. B-4.1. It is readily verified that cubing any of these complex numbers produces unity.

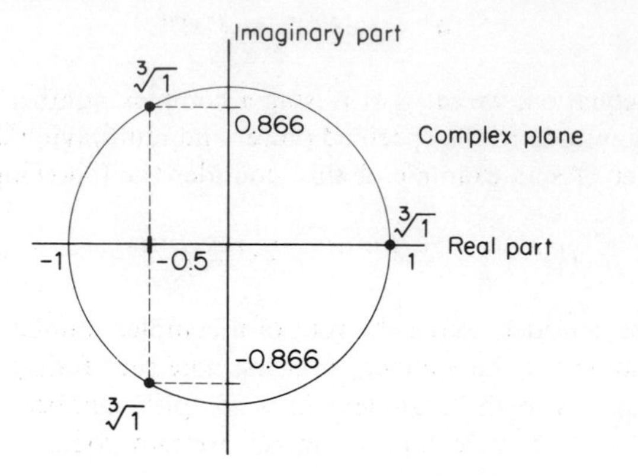

Fig. B-4.1 The cube roots of unity.

APPENDIX C

Descriptions of Subroutines

In this appendix, descriptions are given of four subroutines which form a part of the software package used with this text. In general, these subroutines have been designed with the minimum level of sophistication necessary to effectively perform their function. The obvious advantage of such an approach is the minimization of computer core requirements as well as the reduction of compilation and execution times. The student or the professor who wishes a higher level of performance will find that these programs form an excellent basis upon which embellishments may be readily added.

C-1 THE SUBROUTINE PLOT5

The subroutine PLOT5, which was originally introduced in Chap. 4, is used to provide a printer-constructed plot of one or more variables as defined by a sequence of values of data stored in a two-dimensional array. The plot is printed with the positive abscissa direction oriented vertically downward on the printed page, and with the positive ordinate direction going from left to right across the page. The identifying statement for this subroutine is

SUBROUTINE PLOT5 (Y, M, NF, MAX)

A summary of the characteristics of the subroutine is given in Table 4–5.1. Here we will concentrate on giving a detailed description of how the subroutine operates. A flow chart of the subroutine logic is given in Fig. C–1.1. A listing of the statements of the subroutine is given in Fig. C–1.2. The most significant variables encountered in this subroutine are

L A one-dimensional array of variables L(I) in which are stored the values of the ordinate scale.

LINE A one-dimensional array of 101 variables LINE(I) in which alpha-

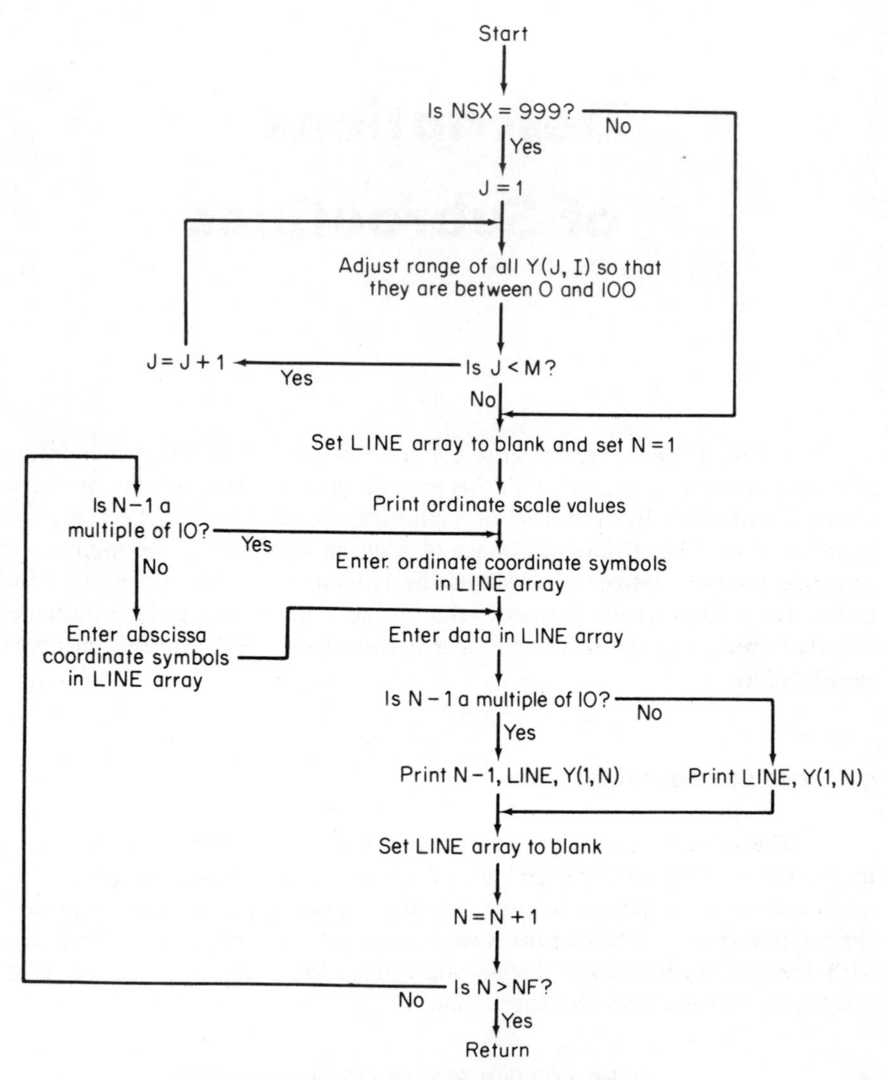

Fig. C-1.1 Flow chart for the subroutine PLOT5.

```
      SUBROUTINE PLOT5(Y,M,NF,MAX)
C     SUBROUTINE FOR PLOTTING AND SCALING A
C     5 X 101 INPUT ARRAY (FORTRAN 4)
      DIMENSION Y(5,101), LINE(101),L(11),JL(5)
      DATA (JL(I),I=1,5)/1HA,1HB,1HC,1HD,1HE/,JN,JP,JI,JBLANK,JZ/
     11H-,1H+,1HI,1H ,1H$/
      NS=MAX
      IF (NS.NE.999) GO TO 19
C     RESCALE DATA TO COVER ENTIRE ORDINATE RANGE
      NS=100
      DO 16 J=1,M
      YMAX=-1.E+50
      YMIN=1.E+50
      DO 9 I=1,NF
      IF (Y(J,I).GT.YMAX) YMAX=Y(J,I)
    9 IF (Y(J,I).LT.YMIN) YMIN=Y(J,I)
      RANGE=YMAX-YMIN
      TEMP=100./RANGE
      DO 13 I=1,NF
   13 Y(J,I)=(Y(J,I)-YMIN)*TEMP
      PRINT 15,JL(J),YMIN,YMAX,RANGE
   15 FORMAT (1X,5HPLOT ,A1,5H FROM,E10.3,3H TO,
     1E10.3,9H, RANGE =,E10.3)
   16 CONTINUE
      PRINT 18
   18 FORMAT ( )
   19 DO 20 I=1,101
   20 LINE(I)=JBLANK
      N=1
C     PRINT ORDINATE SCALE
      DO 23 I=1,11
   23 L(I)=10*I-110+NS
      PRINT 25,(L(I),I=1,11)
   25 FORMAT (3X,10(I4,6X),I4,2X,8HA-VALUES)
      GO TO 28
   27 IF ((N-1)/10-(N-2)/10) 37,37,28
C     CONSTRUCT ORDINATE GRAPH LINE
   28 ND=0
      DO 34 I=1,10
      ND=ND+1
      LINE(ND)=JP
      DO 34 J=1,9
      ND=ND+1
   34 LINE(ND)=JN
      LINE(101)=JP
      GO TO 39
C     CONSTRUCT 1 LINE OF ABSCISSA GRAPH LINES
   37 DO 38 I=1,101,10
   38 LINE(I)=JI
C     CHANGE NUMERICAL DATA TO LETTERS
   39 DO 49 I=1,M
      XNS=NS
      JA=Y(I,N)+101.49999-XNS
      IF (JA-101) 43,48,44
   43 IF (JA) 46,46,48
   44 LINE(101)=JZ
      GO TO 49
   46 LINE(1)=JZ
      GO TO 49
```

Fig. C-1.2 Listing of the subroutine PLOT5.

```
   48 LINE(JA)=JL(I)
   49 CONTINUE
C     PRINT LINE OF DATA
      IF (N.EQ.1) GO TO 51
      IF ((N-1)/10-(N-2)/10) 55,55,51
   51 N1=N-1
      PRINT 53,N1,LINE,Y(1,N)
   53 FORMAT (1X,I4,101A1,1X,  E12.5)
      GO TO 57
   55 PRINT 56,LINE,Y(1,N)
   56 FORMAT (5X,101A1,1X,E12.5)
C    SET LINE VARIABLES TO ZERO
   57 DO 58 I=1,101
   58 LINE(I)=JBLANK
   59 N=N+1
      IF (N-NF) 27,27,61
   61 RETURN
      END
```

Fig. C-1.2 Continued

numeric information is stored corresponding with the desired form of the line of the plot which is currently being printed.

M A variable specifying the number of quantities that are to be plotted.

N An index which is used internally in the program to indicate which line of the plot is currently being printed.

NF A variable specifying the number of values of each of the quantities to be plotted, i. e., the number of lines that the plot will have.

NS An internal program variable which has the same function as the variable MAX but whose value may be altered by the program during execution.

MAX An input variable specifying the maximum value desired for the ordinate scale. This variable is also used (MAX = 999) to select the automatic scaling option.

Y The two-dimensional array of variables Y(I, J) giving the jth value of the ith function which is to be plotted.

The first operation of the subroutine PLOT5 is to determine whether a scaled or an unscaled plot is required. To this end, the variable NS which has been set equal to the input variable, MAX, is tested. If this variable has the value 999, then the data representing the first function to be plotted, i.e., the data stored as Y(1, I) (I = 1, NF), are examined to find the maximum and minimum values. The data for this function are then scaled so that they cover a range of 0–100. The process is then repeated for the second function to be plotted, which consists of the data stored as Y(2, I) (I= 1, NF), etc. The maximum ordinate variable NS is then set to 100. Finally, data on the maximum and minimum values and the scale factor of each function to be plotted are printed. From this point the operation of the subroutine PLOT5 is the same regardless of whether the automatic scaling option is used or not.

At this point in the program a set of ordinate values are computed, stored in the array L, and printed. These values range from NS-100 to NS. The program now sets the variable N, which indicates the line of the plot which is currently being printed, to 1 and sets the variables in the LINE array to the alphanumeric values corresponding to the plus and minus signs which are used to form the coordinate grid lines parallel to the ordinate. The numerical data of the first point of the function to be plotted are now converted to an index variable JA whose value corresponds with the plotted position of the data. The alphanumeric value of the letter A is then stored as LINE (JA). This process is then repeated for the first data point for any other functions which are to be plotted, using the alphanumeric value of B for the second function, C for the third, etc. These alphanumeric characters A, B, C, and so forth representing the values of the functions replace the alphanumeric values of the plus and minus signs previously stored in the LINE array. A test is now made to determine whether N — 1 is an exact multiple of 10 by computing the value of the term

$$(N - 1)/10 - (N - 2)/10$$

Due to truncation of the integer variables, this term will be unity for the cases in which N — 1 is an exact multiple of 10; otherwise, it will be zero. In all those cases in which N — 1 is an exact multiple of 10, the value of N — 1 is printed as an abscissa scale value. The values stored in the LINE array and the numerical value of Y(1, N) are also printed. If N — 1 is not an exact multiple of 10, only the LINE array and Y(1, N) are printed. Following the printing of the LINE array, the variables of the array are set to the alphanumeric value for a blank and N is increased by 1. After testing to make certain that N is not greater than NF, a test is made to see if the new value of N — 1 is a multiple of 10. If it is, a new ordinate grid coordinate line consisting of plus and minus signs is stored in LINE. Otherwise, only the alphanumeric value of I, the character which is used to form the grid coordinate lines parallel to the abscissa, is stored in every tenth variable of the LINE array. The Nth values of data for the new value of N are then stored as letter symbols in the appropriate positions of the LINE array and the procedure described above is repeated. The process is terminated when the line with a value of N equal to NF is printed. At this point, control is returned to the main program.

C-2 THE SUBROUTINE ROOT

The subroutine ROOT introduced in Chap. 8 is used to find the roots of a given polynomial. It has the identifying statement

SUBROUTINE ROOT (B, N, P)

A summary of its characteristics is given in Table 8-5.2. Here we present a more detailed description of how the subroutine operates. A flow chart of the logic used in

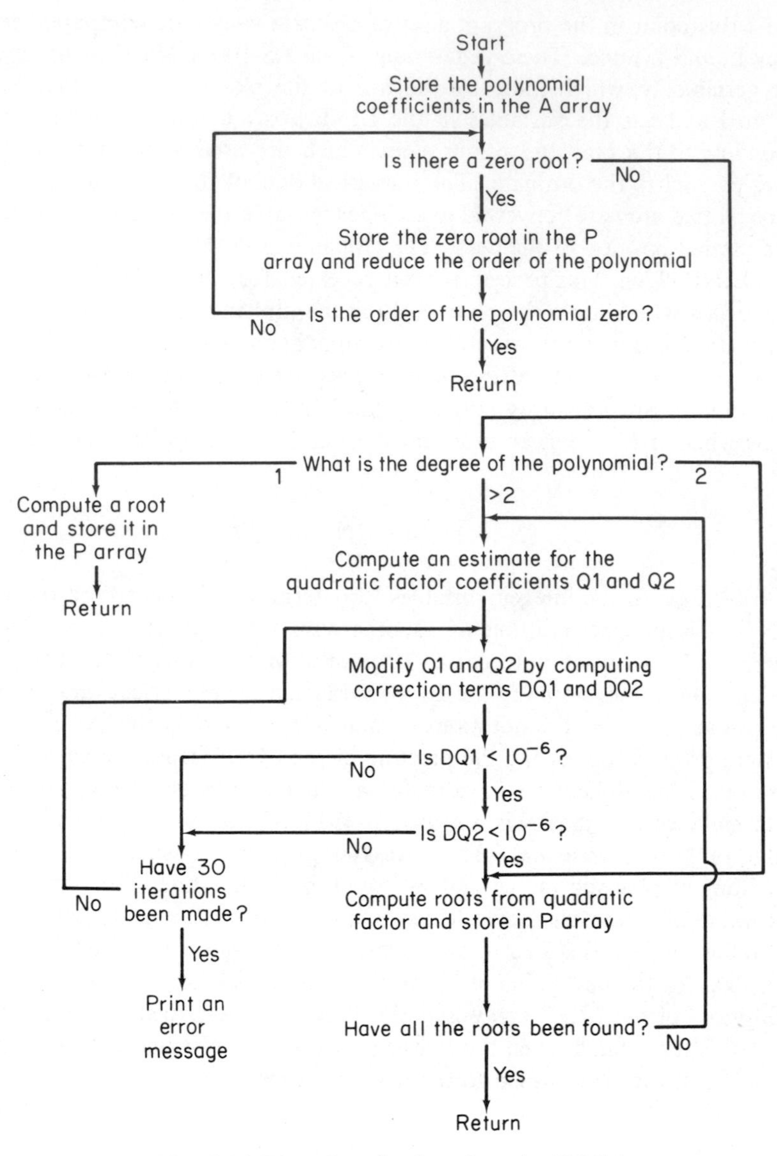

Fig. C-2.1 Flow chart for the subroutine ROOT.

this subroutine is given in Fig. C–2.1. A listing of the statements of the subroutine may be found in Fig. C–2.1. The subroutine ROOT uses a Lin-Bairstow method to determine a quadratic factor of the given polynomial. It then removes this factor to obtain a polynomial of reduced order. The process is repeated until all the roots have been found.

The first operation of the subroutine ROOT is to transfer the polynomial coefficients from the input array B to an internal program array A. This latter array is used for all subsequent root-finding operations on the polynomial. Thus, the values of the original polynomial coefficients are preserved in the input array B. The polynomial is now assumed to have the form

$$a_1 + a_2 s + a_3 s^2 + \cdots + a_{n+1} s^n$$

The subroutine now tests the polynomial to determine if there is a zero root, i. e., if the coefficient a_1 is zero. If there is such a root, it is stored in the P array, the coefficients in the A array are reordered, and the degree of the polynomial is lowered by 1. This process is continued until all the zero roots have been removed. The subroutine then tests the degree of the remaining polynomial. If this degree is less than or equal to 2, a quadratic or first-order explicit formula is applied to find the root or roots. These are then stored in the complex P array. However, if the degree of the polynomial remaining after any zero roots have been removed is greater than 2, an estimate of a quadratic factor

$$q_1 + q_2 s + s^2$$

as represented by the subroutine variables Q1 and Q2 is made using the relations

$$q_1 = \frac{a_1}{a_3} \qquad q_2 = \frac{a_2}{a_3}$$

An iterative process is then initiated to determine the quantities Δq_1 and Δq_2 (as represented by the variables DQ1 and DQ2) which are used to improve the initial estimate of the quantities q_1 and q_2 by the statements

$$\text{Q1} = \text{Q1} + \text{DQ1} \qquad \text{Q2} = \text{Q2} + \text{DQ2}$$

The terms Δq_1 and Δq_2 are computed by examining the first- and zero-order coefficients of the remainder obtained by dividing the original polynomial by the estimated quadratic factor, and making a Taylor expansion of these coefficients with respect to q_1 and q_2. This produces a set of two equations in which the quantities Δq_1 and Δq_2 are the unknowns. These equations are then solved to find the quantities Δq_1 and Δq_2. This iterative process is repeated until convergence is reached, as defined by the values of both Δq_1 and Δq_2 being less than 10^{-6}. A test of the number of iterations is also made, and the process is terminated if convergence is not reached within 30 iterations. In such a case, an error message is printed. When suitably accurate values for the quadratic factor coefficients q_1 and q_2 are found, the roots of the quadratic factor are determined and stored in the P array. The quadratic factor is then divided into the polynomial and the degree of the remaining polynomial is tested to see whether or not it is greater than 2. From this point on, the process described above is repeated until all the roots of the polynomial have been found.

```
      SUBROUTINE ROOT (N, B, P)
C     SUBROUTINE TO DETERMINE THE ROOTS OF A POLYNOMIAL OF THE FORM
C     B(1) + B(2)*S + B(3)*S**2 + ... + B(N+1)*S**N  USING THE
C     LIN-BAIRSTOW METHOD. THE POLYNOMIAL B(S) IS PRESERVED
C             N - DEGREE OF THE POLYNOMIAL B(S)
C             B - INPUT ARRAY OF COEFFICIENTS OF B(S)
C             P - COMPLEX OUTPUT ARRAY OF ROOTS OF B(S)
      DIMENSION A(10),B(10),C(10),D(10),E(10)
      COMPLEX P(10)
      M = N
      MP = M + 1
      DO  4   I = 1, MP
    4 A(I) = B(I)
    5 IF (ABS(A(1)).GT.1.E-06) GO TO 13
      DO 7    I = 1, M
    7 A(I) = A(I+1)
      P(M) = (0.,0.)
      M = M - 1
      IF (M.EQ.0) RETURN
      MP = M + 1
      GO TO 5
   13 IF (M - 2) 67, 64, 14
   14 ITER = 0
      Q1 = A(1) / A(3)
      Q2 = A(2) / A(3)
   17 ITER = ITER + 1
      C(MP) = A(MP)
      D(MP) = 0.
      E(MP) = 0.
      C(M) = A(M) - Q2 * C(MP)
      D(M) = -C(MP)
      E(M) = 0.
      IF (M.LT.4) GO TO 33
      MM = M - 3
      DO 32   I = 1, MM
      MI = M - I
      MI1 = MI + 1
      MI2 = MI + 2
      C(MI) = A(MI) - Q2*C(MI1) - Q1*C(MI2)
      D(MI) = -C(MI1) - Q2*D(MI1) - Q1*D(MI2)
   32 E(MI) = -C(MI2) - Q2*E(MI1) - Q1*E(MI2)
   33 Q1A = A(1) - Q1 * C(3)
      Q2A = A(2) - Q1 * C(4) - Q2 * C(3)
      G11 = -C(3) - Q1 * D(4) - Q2 * D(3)
      G12 = -C(4) - Q1 * E(4) - Q2 * E(3)
      G21 = -Q1 * D(3)
      G22 = -C(3) - Q1 * E(3)
      DET = G11 * G22 - G12 * G21
      DQ1 = (-G11 * Q1A + G21 * Q2A) / DET
      DQ2 = ( G12 * Q1A - G22 * Q2A) / DET
      Q1 = Q1 + DQ1
      Q2 = Q2 + DQ2
      IF (ABS(DQ2).LT.1.E-6) GO TO 47
   45 IF (ITER.GT.30) GO TO 69
      GO TO 17
   47 IF (ABS(DQ1).GT.1.E-06) GO TO 45
   48 DISC = Q2 * Q2 - 4. * Q1
      IF (DISC.LT.0.) GO TO 55
      DISCR = SQRT(DISC)
```

Fig. C-2.2 Listing of the subroutine ROOT.

```
      P(M) = CMPLX((-Q2 + DISCR) / 2., 0.)
      P(M-1) = CMPLX((-Q2 - DISCR) / 2., 0.)
      M = M - 2
      GO TO 59
   55 DISCR = SQRT(-DISC)
      P(M) = CMPLX(-Q2 / 2., DISCR / 2.)
      P(M-1) = CONJG(P(M))
      M = M - 2
   59 IF (M.EQ.0) RETURN
      MP = M + 1
      DO 62    I = 1, MP
   62 A(I) = C(I+2)
      IF (M-2) 67, 64, 14
   64 Q1 = A(M-1) / A(M+1)
      Q2 = A(M) / A(M+1)
      GO TO 48
   67 P(1) = CMPLX(-A(1) / A(2), 0.)
      RETURN
   69 PRINT 70
   70 FORMAT (//34H ROOT SUBROUTINE DOES NOT CONVERGE//)
      RETURN
      END
```

Fig. C-2.2 Continued

C-3 THE SUBROUTINE XYPLTS

The subroutine XYPLTS introduced in Chap. 7 is used to provide a printer-constructed $x - y$ plot of a set of data points (x_i, y_i) stored in two one-dimensional arrays. The identifying statement for this subroutine is:

SUBROUTINE XYPLTS (NDP, X, Y, NSCLX, NSCLY, NNPX)

The plot is printed with the positive abscissa direction oriented vertically downward on the page and the positive ordinate direction going across the page from left to right. A summary of the characteristics of the subroutine is given in Table 7–10.4. Here we present a detailed description of how the subroutine operates. The discussion is supplemented by the flow chart given in Fig. C–3.1 and the listing given in Fig. C–3.2.

The most significant variables used in the subroutine are:

K An internal program index which is used to determine which of the data points is currently being plotted.

L A one-dimensional array of variables L(I) in which are stored the values of the ordinate scale.

LINE A one-dimensional array of variables LINE(I) in which are stored the alphanumeric values of the symbols used to represent the coordinate grid lines and the points which are to be plotted.

N An internal program index which is used to indicate the line number of the plot which is currently being printed.

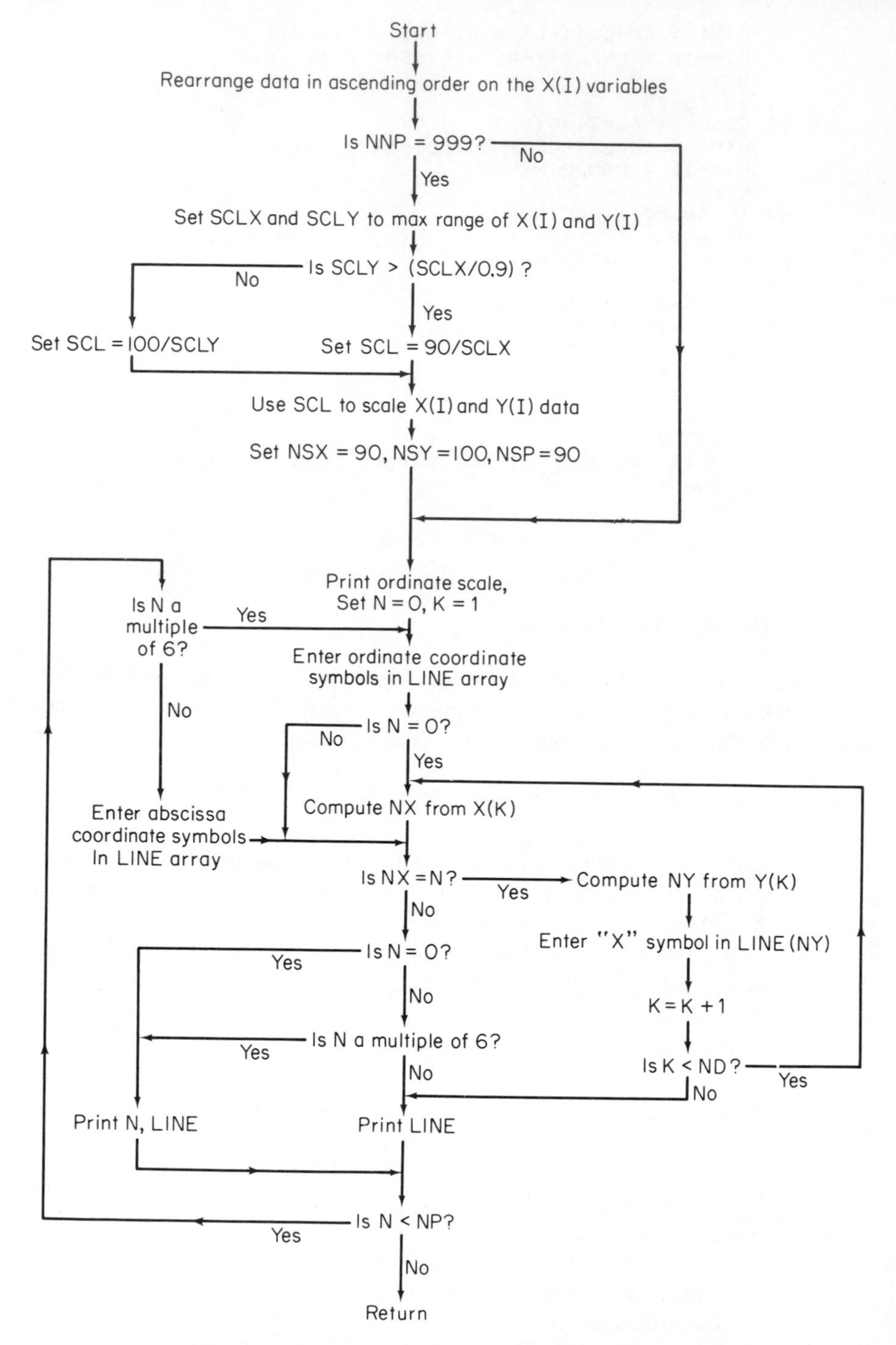

Fig. C-3.1 Flow chart for the subroutine XYPLTS.

```
      SUBROUTINE XYPLTS(NDP, X, Y, NSCLX, NSCLY, NNPX)
C     SUBROUTINE FOR X-Y PLOTTING (FORTRAN 4)
      DIMENSION X(200), Y(200), L(11), LINE(101)
      DATA JN,JP,JI,JBLANK,JZ,JX/1H-,1H+,1HI,1H ,1H$,1HX/
      ND  = NDP
      NSX = NSCLX
      NSY = NSCLY
      NNP = NNPX
      NDM=ND-1
C     ARRANGE DATA IN ASCENDING ORDER ON X
      DO 16 I=1,NDM
      IA=I+1
      DO 16 J=IA,ND
      IF(X(I)-X(J)) 16,16,10
   10 TEMP=X(I)
      X(I)=X(J)
      X(J)=TEMP
      TEMP=Y(I)
      Y(I)=Y(J)
      Y(J)=TEMP
   16 CONTINUE
      IF (NNP.NE.999) GO TO 43
      YMAX = -1.E+50
      YMIN = 1.E+50
      XMIN = X(1)
      DO 22  I = 1, ND
      IF(Y(I).GT.YMAX) YMAX = Y(I)
   22 IF(Y(I).LT.YMIN) YMIN = Y(I)
      SCLX = X(ND) - X(1)
      SCLY = YMAX - YMIN
      IF(SCLY.GT.SCLX/0.9) GO TO 28
      SCL = 90.0 / SCLX
      GO TO 29
   28 SCL = 100.0 / SCLY
   29 PRINT 30, X(1), X(ND), SCL
   30 FORMAT(1X,16HVALUES OF X FROM,E10.3,3H TO,
     1E10.3,11H, SCALED BY,E10.3)
      PRINT 32, YMIN, YMAX, SCL
   32 FORMAT(1X,16HVALUES OF Y FROM, E10.3,3H TO,
     1E10.3,11H, SCALED BY,E10.3/)
   34 DO 36  I = 1, ND
      Y(I) = (Y(I) - YMIN) * SCL
   36 X(I) = (X(I) - XMIN) * SCL
      NSX = 90
      NSY = 100
      NNP = 90
   43 NP=(NNP/10)*6
      XNP=NP
      XNS=(NSX/10)*6
      YNS=NSY
C     PRINT ORDINATE SCALE FIGURES
      DO 48 I=1,11
   48 L(I)=10*I-110+NSY
      PRINT 50,L
   50 FORMAT (3X,11(I4,6X))
      DO 53 I=1,101
      LINE(I)=JBLANK
   53 CONTINUE
      N=0
```

Fig. C-3.2 Listing of the subroutine XYPLTS.

```
      K=1
C     PREPARE DATA FOR ORDINATE GRAPH LINE
   57 NQ=0
      DO 63 I=1,10
      NQ=NQ+1
      LINE(NQ)=JP
      DO 63 J=1,9
      NQ=NQ+1
   63 LINE(NQ)=JN
      LINE(101)=JP
      IF (N) 72,66,72
C     SCALE ABSCISSA DATA
   66 NX=X(K)*.6-XNS+XNP+.499999
      IF (NX) 68,72,70
   68 NX=0
      GO TO 72
C     CHECK TO SEE IF DATA IS STORED FOR CURRENT ABSCISSA VALUE
   70 IF (NX-NP) 72,72,71
   71 NX=NP
   72 IF (NX-N) 83,73,83
C     SCALE ORDINATE DATA
   73 NY=Y(K)+101.499999-YNS
      IF (NY-1) 75,80,77
   75 LINE(1)=JZ
      GO TO 81
   77 IF (NY-101) 80,80,78
   78 LINE(101)=JZ
      GO TO 81
   80 LINE(NY)=JX
   81 K=K+1
      IF (K-ND) 66,66,84
   83 IF (N) 84,88,84
   84 IF (N/6-(N-1)/6) 85,85,88
C     PRINT LINE AND DATA WITHOUT ORDINATE GRAPH LINE
   85 PRINT 86, LINE
   86 FORMAT (5X,101A1)
      GO TO 91
   88 NN=(N*10)/6+NSX-NNP
C     PRINT LINE AND DATA WITH ORDINATE GRAPH LINE
      PRINT 90, NN,LINE
   90 FORMAT (1X,I4,101A1)
   91 IF (N-NP) 92,99,99
   92 N=N+1
      IF (N/6-(N-1)/6) 93,93,57
C     SET LINE VARIABLE TO BLANKS
   93 DO 95 I=1,101
      LINE(I)=JBLANK
   95 CONTINUE
C     SET UP ABSCISSA GRAPH LINES
      DO 97 I=1,101,10
   97 LINE(I)=JI
      GO TO 72
   99 RETURN
      END
```

Fig. C-3.2 Continued

- **NDP** An input argument specifying the number of points of data that it is desired to plot. This variable is redefined as ND in the subroutine.
- **NNPX** An input argument giving the total range of values of the abscissa scale desired for the plot. If this argument is set to the number 999, an automatic scaling option will be used. This variable is redefined in the subroutine as NNP.
- **NP** An internal program variable giving the number of lines which are to be printed. (This is approximately equal to $0.6 \times$ NNP.)
- **NSCLX** An input argument which specifies the maximum value of the abscissa scale which is to be used in the plot. This argument is redefined as NSX in the subroutine.
- **NSCLY** An input argument which specifies the maximum value of the ordinate scale which is to be used in the plot (the minimum value is NSCLY-100). This variable is redefined as NSY in the subroutine.
- **NX** An internal program variable indicating the line, i. e., the value of N corresponding with a given value of data stored in the X array.
- **NY** A variable indicating the position on a given line corresponding with a given value of data stored in the Y array.
- **SCL** A factor used for scaling the variables stored in the X and Y array when the automatic scaling option is used.
- **X** The one-dimensional array of variables X(I) in which are stored the abscissa coordinate values of each of the points it is desired to plot.
- **Y** The one-dimensional array of variables Y(I) in which are stored the ordinate coordinate values of each of the points it is desired to plot.

The first operation of the subroutine XYPLTS is to rearrange the order of the data points so that they are in descending order with respect to the values of the variables X(I). Next, a test is made on the variable NNP to determine whether the automatic scaling option is desired. If it is (NNP = 999), the values of the variables Y(I) are examined to determine the maximum and minimum values. Next, the range of the X(I) values and the Y(I) values are compared to determine which is the limiting factor in the automatic scaling operation. The process assumes a maximum range of 90 units for the abscissa and 100 units for the ordinate. Thus, the values of the points must be rescaled so that either the Y(I) values range from 0 to 100 or the X(I) values range from 0 to 90. The maximum and minimum unscaled values and the scale factor used are printed as output. Finally, the variables NSX, NSY, and NNP are set to 90, 100, and 90, respectively. From this point on, the operation of the program is the same whether the unscaled or the automatic scaling option has been selected.

At this point, the program computes NP, the number of lines required to encompass the range of the abscissa specified by the input argument NNP. Note that, to obtain a square grid, six printed lines (requiring a distance of 1 inch down the page) are equal to 10 units of the abscissa scale. This provides the same grid line spacing as is used for the ordinate in which 10 spaces (requiring 1 inch across the page) corresponds to 10 units of the ordinate scale. The program then computes the ordinate scale values, stores them in the L array, and prints them.

In a manner similar to that done in the subroutine PLOT5, the program now enters a set of alphanumeric values for + and − signs into the LINE array so as to construct an ordinate grid line. It then computes the x position of the variable X(K) and stores this as the index NX. The program now tests N to see if N is equal to NX. If not, it prints the nth line and increases N by one. The operation is repeated until the correct line for X(K) is found. The program now examines the value of Y(K) and computes an index NY indicating the position in the LINE array for this variable. The appropriate element of the array is changed to the alphanumeric value of the symbol X to plot the point. K is then increased by one and X(K) is tested to see if it should also be plotted on the current line. If not, the line array is then printed, the index N is increased, and the process continues. Logic is provided to test the value of N so that coordinate grid lines parallel to the ordinate and constructed of + and − symbols are drawn every six values of N. In between these lines, coordinate grid lines parallel to the abscissa constructed of the symbol I are drawn every 10 spaces across the page. The process continues until all NP lines have been printed, at which time control is returned to the main program.

C-4 THE SUBROUTINE PFEXSD

The subroutine PFEXSD which was originally introduced in Chap. 8 is used to find the constants associated with the first- and second-order terms in the partial-fraction expansion for a rational function. It has the identifying statement

SUBROUTINE PFEXSD (M, A, N, B, P, R1, R2)

A summary of its characteristics is given in Table 8–5.4. Here we present a more detailed description of the algorithm used by the subroutine in determining the residue for a second-order pole, i. e., the constant of the first-order term in the partial-fraction expansion for such a pole.

Let us assume that the proper rational function $F(s)$ has a second-order pole at $s = p_1$ and a series of other poles which are simple. The function may thus be expressed as

$$F(s) = \frac{N(s)}{B(s)} = \frac{K_{12}}{(s - p_1)^2} + \frac{K_{11}}{s - p_1} + \sum_{i=3}^{n} \frac{K_i}{s - p_i} = K\frac{\prod_{j=1}^{m}(s - z_j)}{\prod_{i=1}^{n}(s - p_i)} \tag{C-1}$$

where $N(s)$ and $B(s)$ are the numerator and denominator polynomials, respectively. If we multiply both members of the above equation by the term $(s - p_1)^2$, we obtain

$$\frac{(s-p_1)^2 N(s)}{B(s)} = K\frac{\prod_{j=1}^{m}(s-z_j)}{\prod_{i=3}^{n}(s-p_i)} = K_{12} + K_{11}(s-p_1) + \sum_{i=3}^{n}\frac{K_i(s-p_1)^2}{s-p_i} \qquad \text{(C-2)}$$

Evaluating this expression at $s = p_1$ we obtain the following expression for the constant K_{12}.

$$K_{12} = K\frac{\prod_{j=1}^{m}(s-z_j)}{\prod_{i=3}^{n}(s-p_i)}\Bigg|_{s=p_1} \qquad \text{(C-3)}$$

Now let us determine the constant K_{11}. We may begin this process by differentiating both members of (C-2). Thus, we obtain

$$\frac{d}{ds}\left[K\frac{\prod_{j=1}^{m}(s-z_j)}{\prod_{i=3}^{n}(s-p_i)}\right] = K_{11} + \frac{d}{ds}\sum_{i=3}^{n}\frac{K_i(s-p_1)^2}{s-p_i} \qquad \text{(C-4)}$$

Evaluating both members of this expression at $s = p_1$ we obtain

$$K_{11} = \frac{d}{ds}\left[K\frac{\prod_{j=1}^{m}(s-z_j)}{\prod_{i=3}^{n}(s-p_i)}\right] = \frac{d}{ds}K\frac{N(s)}{D(s)} \qquad \text{(C-5)}$$

where $D(s)$ is defined as $B(s)/(s-p_1)^2$. The right member of this expression may be rewritten as

$$\frac{d}{ds}\left[K\frac{N(s)}{D(s)}\right] = K\frac{D(s)\frac{d}{ds}N(s) - N(s)\frac{d}{ds}D(s)}{D(s)^2} \qquad \text{(C-6)}$$

Substituting from (C-5) in the above equation, we obtain

$$K_{11} = K\left\{\frac{\prod_{i=3}^{n}(s-p_i)\frac{d}{ds}\prod_{j=1}^{m}(s-z_j) - \prod_{j=1}^{m}(s-z_j)\frac{d}{ds}\prod_{i=3}^{n}(s-p_i)}{\left[\prod_{i=3}^{n}(s-p_i)\right]^2}\right\} \qquad \text{(C-7)}$$

Performing the indicated differentiation in the above equation, we obtain

$$K_{11} = K\frac{\prod_{i=3}^{n}(s-p_i)\left[\prod_{j=2}^{m}(s-z_j) + (s-z_1)\prod_{j=3}^{m}(s-z_j) + (s-z_1)(s-z_2)\prod_{j=4}^{m}(s-z_j) + \cdots\right] - \prod_{j=1}^{m}(s-z_j)\left[\prod_{i=4}^{n}(s-p_i) + (s-p_3)\prod_{i=5}^{n}(s-p_i) + (s-p_3)(s-p_4)\prod_{i=6}^{n}(s-p_i) + \cdots\right]}{\left[\prod_{i=3}^{n}(s-p_i)\right]^2}$$

Rearranging terms, we may write this in the form

$$K_{11} = K\frac{\prod_{j=1}^{m}(s-z_j)}{\prod_{i=3}^{n}(s-p_i)}\left[\sum_{j=1}^{m}\frac{1}{s-z_j} - \sum_{i=3}^{n}\frac{1}{s-p_i}\right] \tag{C–9}$$

The first term in this expression, however, is the constant K_{12} defined in (C–3). Thus, we may write

$$K_{11} = K_{12}\left[\sum_{j=1}^{m}\frac{1}{s-z_j} - \sum_{i=3}^{n}\frac{1}{s-p_i}\right] \tag{C–10}$$

This is the result given in Eq. (8–153).

Solutions to Selected Problems

CHAPTER 1

Problem 1–2: $q(t) - 1 - e^{-4t}$ C. Problem 1–3(c): $w(0.003) = 0.8595 \times 10^{-9}$ J. Problem 1–12: $i_1(t) = 3 + 2t, i_4(t) = 3 - \sin t, i_5(t) = -6 - 2t + \sin t$ A. Problem 1–16: $v_2(t) = 1 - 7t + e^{-t}$, $v_3(t) = \sin t - 7 - e^{-t}$ V. Problem 1–21: $i_2(t) = 1 + 6t - t^2 + 2 \sin t$ A. Problem 1–26: 4 nodes, 6 branches, 7 cutsets.

CHAPTER 2

Problem 2–1(b): $0.01 \sin^2 (2\pi t)$ W. Problem 2–5: Power in 5-V source is $\frac{25}{8}$ W, power in 10-V source is $-\frac{50}{8}$ W. Problem 2–10: $\frac{11}{8}$ Ω. Problem 2–19: $i_o(t) = \frac{9}{5} \cos 2t$ A.

CHAPTER 3

Problem 3–1: $i_1 = 2, i_2 = -1$ A. Problem 3–10: $i_1 = 6, i_2 = 7, i_3 = 8$ A. Problem 3–23: $a = 31$, $b = d = 19$, $c = e = 11$. Problem 3–26: $v_1 = 2$, $v_2 = -6$ V. Problem 3–34: $i_a = \frac{25}{13}$ A. Problem 3–44(a): $R_{eq} = \frac{8}{11}$ Ω, $V_{eq} = \frac{10}{11}$ V. Problem 3–48: $i_1 = 3$, $i_2 = -1$ A. Problem 3–53: $R_{eq} = \frac{7}{2}$ Ω. Problem 3–62: Gain $= (G_1/G_2) + 1$.

CHAPTER 4

Problem 4–7: $i(t) = (6 + 2t)\cos 2t + \sin 2t$ A. Problem 4–10(a): $f_a(t) = -3u(-t + 2)$. Problem 4–11(a): $f_a'(t) = 3\delta(-t + 2)$. Problem 4–12(b): $\int f_b(t)\,dt = 4(t + 1)u(t + 1)$. Problem 4–27: $q(10) = \frac{5}{2}$ C, $\phi(10) = \frac{1}{2}$ Weber-turn, $v(10) = \frac{1}{20}$ V. Problem 4–31: $i(0+) = -1$ A, $di(0+)/dt = 2$ A/s. Problem 4–36(a): $C = \frac{26}{33}$ F.

CHAPTER 5

Problem 5–4: $V_o = 44.5$ V. Problem 5–8: $v_C(t) = -6e^{-10t/21}$ V. Problem 5–16: $v(t) = -e^{-t} + \cos 2t + 2\sin 2t$ V. Problem 5–20: $i(0) = -\frac{2}{5}$ A. Problem 5–26: $i(t) = 3e^{-t}$ A. Problem 5–32: $v_C(t) = 2 - e^{-2t}$ V, $i(t) = e^{-2t}/3$ A.

CHAPTER 6

Problem 6–2: $i_L(t) = -5e^{-2t} + \frac{3}{2}e^{-4t}$ A. Problem 6–7(a): $v(t) = e^{-2t}(\cos t + 2\sin t)$ V. Problem 6–14: $v(t) = e^{-t}\sin t$ V. Problem 6–18: $v(t) = 3e^{-2t} - 3e^{-4t}$ V. Problem 6–23: $v(t) = -5e^{-t} + 15e^{-3t}$ V. Problem 6–34: $v_{SW}(0+) = 2$, $v_{SW}(\infty) = 20$ V. Problem 6–45: $8s^2 + 25s + 11 = 0$. Problem 6–50: $s^3 + 3s^2 + 3s + 1 = 0$. Problem 6–54: $v_2(t) = (4e^{-t}/5) + e^{-t/2}(-\frac{4}{5}\cos t + \frac{7}{5}\sin t)$ V.

CHAPTER 7

Problem 7–4: $K/\sqrt{3}$ for both waveshapes. Problem 7–10: $v(t) = \sqrt{5}\cos(2t - 63.5°)$ V. Problem 7–19: $v_1(t) = -4.4\cos(t - 63.4°)$ V. Problem 7–25: $i_1(t) = (\sqrt{2}/6)\cos(2t - 135°)$, $i_2(t) = \frac{2}{3}\cos(2t - 90°)$ A. Problem 7–32(a): $Y(j\omega) = (3 - 5\omega^2 + 7j\omega)/(6 + 5j\omega)$ mhos. Problem 7–46: $\mathscr{V} = 1/\sqrt{2}\,\angle 90°$, $Z = \frac{3}{2} + j\frac{1}{2}$. Problem 7–58: $\mathscr{V}_2/\mathscr{V}_1 = 1/(-j\omega^3 - 5\omega^2 + 6j\omega + 1)$. Problem 7–79: $C = 16\ \mu$F. Problem 7–87: $\mathscr{I}_a = 4.16\,\angle -31°$, $\mathscr{I}_b = 2.58\,\angle -165°$, $\mathscr{I}_c = 12.05\,\angle 69°$.

CHAPTER 8

Problem 8–4(a): $F_a(s) = e^{-2s}/s$. Problem 8–7(d): $f(t) = e^{-t}(\cos t - 2\sin t)$. Problem 8–8(d): $f(t) = (-t/2)\sin$ t. Problem 8–17: $v_2(t) = -e^{-t} + e^{-t}(\cos t + \sin t)$. Problem 8–31: $18v_2'''(t) + 21v_2''(t) + 12v_2'(t) + 4v_2(t) = 3v_1''(t) + 3v_1'(t) + 4v_1(t)$. Problem 8–39(b): $s = -1, -\frac{5}{6}$. Problem 8–42: $A(s) = F(s) = D(s)$, $B(s) = E(s) = H(s)$, $C(s) = G(s)$. Problem 8–48(c): Reciprocal. Problem 8–62(a): $V_{\text{Thevenin}} = 3/(2s^2 + 3s)$.

CHAPTER 9

Problem 9–1(c): $z_{11}(s) = (6s^2 + s)/(2s^2 + 6s + 1)$. Problem 9–5(c): $y_{11}(s) = (6s + 1)/(6s^2 + s + 3)$. Problem 9–11(b): $g_{11}(s) = s + 2$. Problem 9–16(d): $A(s) = (s^2 + 3s + 1)/(s^2 + 2s + 1)$. Problem 9–19(d): Z and $ABCD$ parameters.

CHAPTER 10

Problem 10–5(a): $b_1 = 4\sqrt{3}/\pi$, $a_1 = 4/\pi$, $a_2 = 8/3\pi$, others zero. Problem 10–13: Waveform 3 has symmetries c, d. Problem 10–18: $L = \frac{1}{9}$ H, $C = \frac{1}{49}$ F.

Index

J

K

L

M

S

U

V